Handbook of Research on Computational and Systems Biology:

Interdisciplinary Applications

Limin Angela Liu
Shanghai Jiao Tong University, China

Dong-Qing Wei
Shanghai Jiao Tong University, China

Yixue Li
Shanghai Jiao Tong University, China

Huimin Lei
Shanghai Jiao Tong University, China

Medical Information Science Reference

MEDICAL INFORMATION SCIENCE REFERENCE

Hershey · New York

Senior Editorial Director:	Kristin Klinger
Director of Book Publications:	Julia Mosemann
Editorial Director:	Lindsay Johnston
Acquisitions Editor:	Erika Carter
Development Editor:	Christina Bufton
Production Coordinator:	Jamie Snavely
Typesetters:	Deanna Zombro, Michael Brehm, & Milan Vracarich Jr.
Cover Design:	Nick Newcomer

Published in the United States of America by
Medical Information Science Reference (an imprint of IGI Global)
701 E. Chocolate Avenue
Hershey PA 17033
Tel: 717-533-8845
Fax: 717-533-8661
E-mail: cust@igi-global.com
Web site: http://www.igi-global.com/reference

Library of Congress Cataloging-in-Publication Data

Liu, Limin Angela, 1974-
Handbook of research on computational and systems biology : interdisciplinary applications / Limin Angela Liu, Dongqing Wei and Yixue Li, Editors.
p. cm.
Includes bibliographical references and index.
Summary: "This book offers information on the state-of-the-art development in the fields of computational biology and systems biology, presenting methods, tools, and applications of these fields by many leading experts around the globe"--Provided by publisher.
ISBN 978-1-60960-491-2 (hardcover) -- ISBN 978-1-60960-492-9 (ebook) 1. Computational biology--Research. 2. Systems biology--Research. I. Wei, Dongqing. II. Li, Yixue, 1955- III. Title.
QH324.2.L58 2011
570.285--dc22

2010035344

British Cataloguing in Publication Data
A Cataloguing in Publication record for this book is available from the British Library.

Editorial Advisory Board

List of Reviewers

List of Contributors

Table of Contents

Section 2
Method Development in Bioinformatics

Section 3
Biological Networks and Pathways

Detailed Table of Contents

Section 1
Drug Development and Medicine

This section contains four chapters summarizing the efforts and the challenges related to a wide range of research in drug discovery and medicine. Chapter 1 provides a review and discussion of the ethical and privacy issues related to predictive and personalized medicine in the post-genomic era, as increasing amount of personal medical data are now stored electronically and accessed over the Internet. Chapter 2 presents a thorough review of prevalent computational methods used for virtual screening in drug discovery, such as docking, QSAR, pharmacophore model, et cetera. The authors also offer their expert opinions on the advantages and limitations of each method and indicate important future research directions. Chapter 3 reviews the recent efforts in vaccine development using systems biology approaches. Many successful vaccine development examples are presented. Various resources for epitope predictions and immunoinformatics research are summarized. New directions for vaccine research, such as through synthetic whole organism vaccines, are discussed. Chapter 4 comprehensively reviews recent method development and applications in biomarker discovery in genomics, proteomics, transcriptomics, and metabolomics. The advantages and limitations of existing computational and experimental techniques are discussed.

Biomarkers play an increasingly important role in drug discovery and development and can be applied for many purposes, including disease mechanism study, diagnosis, prognosis, staging and treatment selection. Advances in high-throughput "omics" technologies, including genomics, transcriptomics, proteomics, and metabolomics significantly accelerate the pace of biomarker discovery. Comprehensive molecular profiling using these "omics" technology has become a field of intensive research aiming at identifying biomarkers relevant for improved diagnostics and therapeutics. Although each "omics" technology plays important roles in biomarker research, different "omics" platforms have different strengths and limitations. This chapter aims to give an overview of these "omics" technologies and their current application in the biomarker discovery.

Section 2
Method Development in Bioinformatics

This section contains five chapters that review the most recent advances in method development in sequence and high-throughput data analysis. Chapter 5 reviews the genome-wide association studies of human single nucleotide polymorphisms with their quantitative complex diseases and traits. Chapter 6 reviews several computational methods in genome-wide association studies and presents a novel approach to detecting epistatic interactions by employing expert knowledge, such as pathway and protein-protein interaction information. Chapter 7 reviews the theory, strengths and limitations of existing biclustering methods for the analysis of DNA microarray data. Several important applications to drug discovery and various problems in systems biology are also summarized. Chapter 8 reviews computational methods for the prediction of epigenetic target sites from DNA sequences based on nucleosome positioning, histone modification, and DNA methylation. Chapter 9 describes a novel method for protein sequence analysis. Evolutionary, structural, and functional information are taken into consideration to improve protein structure and function prediction. Application to the prestin protein is discussed.

Common diseases or traits in humans are often influenced by complex interactions among multiple genes as well as environmental and life style factors, rather than being attributable to a genetic variation within a single gene. Identification of genes that confer disease susceptibility can be facilitated by studying DNA markers such as single nucleotide polymorphism (SNP) associated with a disease trait. Genome-wide association approaches offer a systematic analysis of the association of hundreds of thousands of SNPs with a quantitative complex trait. This method has been successfully applied to a wide variety of common human diseases and traits, and has generated valuable findings that have improved our understanding of the genetic basis of many complex traits. This chapter outlines the general mapping process and methods, highlights the success stories, and describes some limitations and challenges that lie ahead.

Recent technological developments in the field of genetics have given rise to an abundance of research tools, such as genome-wide genotyping, that allow researchers to conduct genome-wide association studies (GWAS) for detecting genetic variants that confer increased or decreased susceptibility to disease. However, discovering epistatic, or gene-gene, interactions in high dimensional datasets is a problem due to the computational complexity that results from the analysis of all possible combinations of single-nucleotide polymorphisms (SNPs). A recently explored approach to this problem employs biological expert knowledge, such as pathway or protein-protein interaction information, to guide an analysis by the selection or weighting of SNPs based on this knowledge. Narrowing the evaluation to gene combinations that have been shown to interact experimentally provides a biologically concise reason why those two genes may be detected together statistically. Here we discuss the challenges of discovering epistatic interactions in GWAS and how biological expert knowledge can be used to facilitate genome-wide genetic studies.

In this chapter, different methods and applications of biclustering algorithms to DNA microarray data analysis that have been developed in recent years are discussed and compared. Identification of biologically significant clusters of genes from microarray experimental data is a very daunting task that emerged, especially with the development of high throughput technologies. Various computational and evaluation methods based on diverse principles were introduced to identify new similarities among genes. Mathematical aspects of the models are highlighted, and applications to solve biological problems are discussed.

Epigenetic regulation provides an extra layer of gene control in addition to the genomic sequence and is critical for the maintenance of cell-type specific gene expression programs. Significant changes of epigenetic patterns have been linked to developmental stages, environmental exposure, ageing, and diet. However, the regulatory mechanisms for epigenetic recruitment, maintenance, and switch are still poorly understood. Computational biology provides tools to uncover deeply hidden connections and these tools have played a major role in shaping our current understanding of gene regulation, but their application in epigenetics is still in its infancy. Here I review some recent developments of computational approaches to predict epigenetic target sites.

Understanding the structure-function relationship of proteins offers the key to biological processes, and can offer knowledge for better investigation of matters with widespread impact, such as pathological disease and drug intervention. This relationship is dictated at the simplest level by the primary protein sequence. Since useful structures and functions are conserved within biology, a sequence with known structure-function relationship can be compared to related sequences to aid in novel structure-function prediction. Sequence analysis provides a means for suggesting evolutionary relationships, and inferring structural or functional similarity. It is crucial to consider these parameters while comparing sequences as they influence both the algorithms used and the implications of the results. For example, proteins that are closely related on an evolutionary time scale may have very similar structure, but entirely different functions. In contrast, proteins which have undergone convergent evolution may have dissimilar primary structure, but perform similar functions. In this chapter, we detail how the aspects of evolution, structure, and function can be taken into account when performing sequence analysis, and propose an expansion on traditional approaches resulting in direct improvement of said analysis. We apply our model to a case study in the prestin protein and show that our proposed approach provides a better understanding of input and output and can improve the performance of sequence analysis by means of motif detection software.

Section 3
Biological Networks and Pathways

This section contains fourteen chapters, including four method reviews providing introductory material for this field, five specialized reviews, and five original research articles that focus on specific types of biological problems. Chapter 10 reviews the basic concepts of biological pathways and networks in detail, as well as available databases and tools for their storage and analysis. Integration of knowl-

edge and data is presented with several applications to target discovery and disease pathway analysis. Chapter 11 thoroughly reviews existing computational methods to identify modules in biological networks. Their applications in a variety of important biological problems, such as protein function and interaction predictions and disease studies, are also discussed in detail. Chapter 12 reviews methods for constructing a functional linkage network (FLN) that consists of genes that are functionally associated. Two important applications to disease study, including prediction of disease gene and disease-disease association, are discussed. Chapter 13 reviews network-driven analysis methods with a special focus on drug target identification. Chapter 14 reviews an important and helpful pathway analysis tool, the Rat Genome Database (RGD). The novel pathway ontology for gene annotation adopted by RGD is explained and examples of pathway visualization and analysis are demonstrated using their Web service. Chapter 15 reviews several novel methods that model cellular signaling networks, and where signaling network perturbation data are analyzed by integrating multivariate measurement data to gain much needed information and knowledge about these networks. Chapter 16 reviews recent advances in the experimental and computational analysis of MAPK (mitogen-activated protein kinase) cascades, providing original insights to these important signal transduction networks. Chapter 17 reviews computational methods for the classification of cancer subtypes and the identification of deregulated pathways in different cancer subtypes. Chapter 18 reviews novel computational methods based on structures and sequences of biosynthesis enzymes in the modeling of secondary metabolite biosynthetic pathways. Chapter 19 presents an original research paper describing the analysis and prediction of metastatic relapse in breast cancer by sub-network extraction. A novel interactome-transcriptome integration method for extracting sub-networks is presented by integrating protein-protein interaction and gene expression data. Chapter 20 presents an original research article describing a novel analysis method of time-course microarray data to predict transcription factors that temporally regulate differentially expressed genes under diverse stimuli. Chapter 21 presents an original research article on the dynamic modeling and parameter optimization of the DNA damage and repair network. Chapter 22 presents an original research paper in which a novel model for building causal biological networks based on high-throughput data is described. The model is built by unifying two complimentary methods (Granger Causality Model and Dynamic Causal Model). An application to the analysis of microarray data for gene circuit construction is presented. Chapter 23 presents an original research paper that describes the development of microfluidic cell arrays for high-throughput examination of host-pathogen interactions. A prototype is presented that enables the study of the infection of human cells by up to 16 different bacterial strains.

Target and biomarker selection in drug discovery relies extensively on the use of various genomics platforms. These technologies generate large amounts of data that can be used to gain novel insights in biology. There is a strong need to mine these information-rich datasets in an effective and efficient manner. Pathway and network based approaches have become an increasingly important methodology to mine bioinformatics datasets derived from 'omics' technologies. These approaches also find use in exploring the unknown biology of a disease or functional process. This chapter provides an overview of pathway databases and network tools, network architecture, text mining and existing methods used in knowledge-driven data analysis. We show examples of how these databases and tools can be integrated to apply existing knowledge and network-based approach in data analytics.

One of the most prominent properties of networks representing complex systems is modularity. Network-based module identification has captured the attention of a diverse group of scientists from various domains and a variety of methods have been developed. The ability to decompose complex biological systems into modules allows the use of modules rather than individual genes as units in biological studies. A modular view is shaping the way we do research in biology. Module-based approaches have found broad applications in protein complex identification, protein function prediction, protein expression prediction, as well as disease studies. Compared to single gene-level analyses, module-level analyses offer higher robustness and sensitivity. More importantly, module-level analyses can lead to a better understanding of the design and organization of complex biological systems.

A major challenge in the post-genomic era is to understand the specific cellular functions of individual genes, and how dysfunctions of these genes lead to different diseases. As an emerging area of systems biology, gene networks have been used to shed light on gene function and human disease. In this chapter, we first demonstrate the existence of functional association for genes working in a common biological process and implicated in a common disease. Next, we review approaches to construct the functional linkage gene network (FLN) to represent functional associations between genes based on genomic and proteomic data integration. Finally, two FLN-based applications related to diseases are reviewed: prediction of new disease genes and therapeutic targets, and identification of disease-disease associations at the molecular level. Both of these applications bring new insights into the molecular mechanisms of diseases, and provide new opportunities for drug discovery.

Drug discovery and development face tremendous challenges to find promising intervention points for important diseases. Any therapeutic agent targeting such an intervention point must prove its efficacy and safety in human patients. Success rates measured from first studies in human to registration average around 10% only. Over the last decade, massive knowledge on biological systems has been accumulated and genome-scale primary data are produced at an ever increasing rate. In parallel, methods to use that knowledge have matured. In this review, we will present some of the problems facing the pharmaceutical industry and elaborate on the current state of network-driven analysis methods. We will focus especially on semi-quantitative methods that are amenable to large-scale data analysis and point out their potential use in many relevant drug discovery challenges.

The set of interacting molecules representing a biological pathway or network is a central concept in biology. It is within the pathway context that the functioning of individual molecules acquires purpose and it is the integration of these molecular circuitries that underlies the functioning of biological systems. In order to provide the research community with a dynamic platform for accessing pathway information, the Rat Genome Database (RGD – http://rgd.mcw.edu) is using a multi-tiered approach. In this chapter, the pathway resources that RGD currently offers are presented. Issues covered include: the biological pathway (the concept and the ontology), pathway literature curation and annotation of genes, interactive pathway diagrams, and tools and resources to access and navigate between pathway data. A case study is presented, and future directions are discussed.

The inference of cellular architectures from detailed time-series measurements of intracellular variables is an active area of research. High throughput measurements of responses to cellular perturbations are usually analyzed using a variety of machine learning methods that typically only work within one type of measurement. Here, summaries of some recent research attempts are presented. These studies have expanded the scope of the problem by systematically integrating measurements across multiple layers of regulation including second messengers, protein phosphorylation markers, transcript levels, and functional phenotypes into signaling vectors or signatures of signal transduction. Data analyses through simple unsupervised methods provide rich insight into the biology of the underlying network, and in some cases reconstruction of key architectures of the underlying network from perturbation data. The methodological advantages provided by these efforts are examined using data from a publicly available database of responses to systematic perturbations of cellular signaling networks generated by the Alliance for Cellular Signaling (AfCS).

Signaling through mitogen-activated protein kinase (MAPK) cascades is a conserved and fundamental process in all eukaryotes. Here, we review recent progress made in the identification of components of MAPK signaling networks using novel large scale experimental methods. We also present recent landmarks in the computational modeling and simulation of the dynamics of MAPK signaling modules. The in vitro MAPK signaling network reconstructed from predicted phosphorylation events is dense, supporting the hypothesis of a combinatorial control of transcription through selective phosphorylation of sets of transcription factors. Despite the fact that additional co-factors and scaffold proteins may regulate the dynamics of signal transduction in vivo, the complexity of MAPK signaling networks supports a new model that departs significantly from that of the classical definition of a MAPK cascade.

Cancer is a complex disease that is associated with a variety of genetic aberrations. The diagnosis and treatment of cancer have been difficult because of poor understanding of cancer and lack of effective cancer therapies. Many studies have investigated cancer from different perspectives. It remains unclear what molecular mechanisms trigger and sustain the transition of normal cells to malignant tumor cells in cancer patients. This chapter gives an introduction to the genetic aberrations associated with cancer and a brief view of the topics key to decoding cancer, from identifying clinically relevant cancer subtypes to uncovering the pathways deregulated in particular subtypes of cancer.

Secondary metabolites belonging to polyketide and nonribosomal peptide families constitute a major class of natural products with diverse biological functions and a variety of pharmaceutically important properties. Experimental studies have shown that the biosynthetic machinery for polyketide and nonribosomal peptides involves multi-functional megasynthases like Polyketide Synthases (PKSs) and nonribosomal peptide synthetases (NRPSs) which utilize a thiotemplate mechanism similar to that for fatty acid biosynthesis. Availability of complete genome sequences for an increasing number of microbial organisms has provided opportunities for using in silico genome mining to decipher the secondary metabolite natural product repertoire encoded by these organisms. Therefore, in recent years, there have been major advances in development of computational methods which can analyze genome sequences to identify genes involved in secondary metabolite biosynthesis and help in deciphering the putative chemical structures of their biosynthetic products based on analysis of the sequence and structural features of the proteins encoded by these genes. These computational methods for deciphering the secondary metabolite biosynthetic code essentially involve identification of various catalytic domains present in this PKS/NRPS family of enzymes, prediction of various reactions these enzymatic domains would catalyze, and their substrate specificities, as well as precise identification of the order in which these domains would catalyze various biosynthetic steps. Structural bioinformatics analysis of known secondary metabolite biosynthetic clusters has helped in formulation of predictive rules for deciphering domain organization, substrate specificity, and order of substrate channeling. In this chapter, we describe the progress in development of various computational methods by different research groups and specifically discuss their utility in identification of novel metabolites by genome mining and rational design of natural product analogs by biosynthetic engineering studies.

Maxime Garcia, Inserm, Paoli Calmettes Institute, France
Olivier Stahl, Inserm, Paoli Calmettes Institute, France
Pascal Finetti, Inserm, Paoli Calmettes Institute, France
Daniel Birnbaum, Inserm, Paoli Calmettes Institute, France
François Bertucci, Inserm, Paoli Calmettes Institute, France
Ghislain Bidaut, Inserm, Paoli Calmettes Institute, France

The introduction of high-throughput gene expression profiling technologies (DNA microarrays) in molecular biology and their expected applications to the clinic have allowed the design of predictive signatures linked to a particular clinical condition or patient outcome in a given clinical setting. However, it has been shown that such signatures are prone to several problems: (i) they are heavily unstable and linked to the set of patients chosen for training; (ii) data topology is problematic with regard to the data dimensionality (too many variables for too few samples); (iii) diseases such as cancer are provoked by subtle misregulations which cannot be readily detected by current analysis methods. To find a predictive signature generalizable for multiple datasets, we devised a strategy of superimposition of a large scale of protein-protein interaction data (human interactome) over several gene expression datasets (a total of 2,464 breast cancer tumors were integrated), to find discriminative regions in the interactome (subnetworks) predicting metastatic relapse in breast cancer. This method, Interactome-Transcriptome Integration (ITI), was applied to several breast cancer DNA microarray datasets and allowed the extraction of a signature constituted by 119 subnetworks. All subnetworks have been stored in a relational database and linked to Gene Ontology and NCBI EntrezGene annotation databases for analysis. Exploration of annotations has shown that this set of subnetworks reflects several biological processes linked to cancer and is a good candidate for establishing a network-based signature for prediction of metastatic relapse in breast cancer.

Bin Li, Merrimack Pharmaceuticals, Inc., USA

Toll-like receptors (TLRs) are critical players in the innate immune response to pathogens. However, transcriptional regulatory mechanisms in the TLR activation pathways are still relatively poorly characterized. To address this issue, the author applied a systematic approach to predict transcription factors that temporally regulate differentially expressed genes under diverse TLR stimuli. Time-course microarray data were selected from mouse bone marrow-derived macrophages stimulated by six TLR agonists. Differentially regulated genes were clustered on the basis of their dynamic behavior. The author then developed a computational method to identify positional overlapping transcription factor (TF) binding sites in each cluster, so as to predict possible TFs that may regulate these genes. A second microarray dataset, on wild-type, Myd88-/- and Trif-/- macrophages stimulated by lipopolysaccharide (LPS), was used to provide supporting evidence on this combined approach. Overall, the author was able to identify known TLR TFs, as well as to predict new TFs that may be involved in TLR signaling.

DNA damage and repair processes are key cellular phenomena that are being intensely studied because of their implications in the onset and therapy of cancer. In this chapter, after introducing a general dynamic model of gene expression, we propose a genetic network modeling framework, based on the interconnection of a continuous-time model and a hybrid model. We apply this strategy to a network built around the p53 gene and protein, which detects DNA damage and activates the downstream nucleotide excision repair (NER) network, which carries out the actual repair tasks. We then present two different parameter identification techniques for the proposed models. One is based on a least squares procedure, which treats the signals provided by a high gain observer; the other one is based on a Mixed Extended Kalman Filter. Prior to the estimation phase, identifiability and sensitivity analyses are used to determine which parameters can be and/or should be estimated. The procedures are tested and compared by means of data obtained by in silico experiments.

As one of the most successful approaches to uncovering complex network structures from experimental data, Granger causality has been widely applied to various reverse engineering problems. In this chapter, we first review some current developments of Granger causality and then present our graphical user interface (GUI) to facilitate the application. To make Granger causality more computationally feasible and satisfy biophysical constraints for dealing with increasingly large dynamical datasets, two attempts are introduced including the combination of Granger causality and Basis Pursuit when faced with non-uniformly sampled data and the unification of Granger causality and the Dynamic Causal Model as a novel Unified Causal Model (UCM) to bring in the notion of stimuli and modifying coupling. Several examples, both from toy models and real experimental data, are included to demonstrate the efficacy and power of the Granger causality approach.

methods of reduced protein 3D structure representation such as the Cα trace method, not only lack essential molecular detail but also ignore the chemical properties of the component amino acid side chains. We describe a reduced protein 3D structure representation called "double-centroid reduced representation" and present a visualization tool called the "DCRR Web Server" that graphically displays a protein 3D structure in DCRR along with non-covalent intra- and intermolecular hydrogen bonding and van der Waals interactions. In the DCRR model, each amino acid residue is represented as two points: the centroid of the backbone atoms and that of the side chain atoms. In the visualization Web server, they and the non-bonded interactions are color-coded for easy identification. Our visualization tool is implemented in MATLAB and to our knowledge is the first for a reduced protein representation, as well as one that simultaneously displays non-covalent interactions in the molecule. The DCRR model reduces the atomicity of the protein structure by ~75% while capturing the essential chemical properties of the component amino acids. In the second half of this report, we describe the application of this reduced representation to the modeling and screening of ligand binding sites using a data model we term the "tetrahedral motif". This type of ligand binding site modeling and screening presents a novel type of pharmacophore modeling and screening dependant on a reduced protein representation.

The rolling of a single biological cell is analysed using modelling of the local kinetics of successive attachment and detachment of bonds occurring at the interface between a single cell and the wall of an ECM (extracellular matrix). Those kinetics correspond to a succession of creations and ruptures of ligand-receptor molecular connections, under the combined effects of mechanical, physical (both specific and non-specific) and chemical external interactions. A three-dimensional model of the interfacial molecular rupture and adhesion kinetic events is developed in the present contribution. From a mechanical point of view, we assume that the cell-wall interface is composed of two elastic shells, namely the wall and the cell membrane, linked by rheological elements, representing the molecular bonds. Both the time and space fluctuations of several parameters related to the mutual affinity of ligands and receptors are described by stochastic field theory; especially, the individual rupture limits of the bonds are modelled in Fourier space from the spectral distribution of power. The bonds are modelled as macromolecular chains undergoing a nonlinear elastic deformation according to the commonly used freely joined chains model, while the cell membrane facing the ECM wall is modelled as a linear elastic plate. The cell itself is represented by an equivalent constant rigidity. Numerical simulations predict the sequence of broken bonds, as well as the newly established connections on the 'adhesive part' of the interface. The interplay between adhesion and rupture entails a rolling phenomenon. In the last part of this contribution, a model of the deformation induced by the random fluctuation of the protrusion force resulting from the variation of affinity with chemiotactic sources is calculated, using stochastic finite element methods in combination with the theory of Gaussian random variables.

Preface

The biological sciences have been among the most exciting and intensely pursued fields of science for the past several decades. The advancement of high-throughput technologies that generate large scale biological data, as well as the development of related computational tools, has enabled global efforts at understanding complex biological systems and brought revolutionary changes to biological research. Increasingly, biologists work with scientists and engineers from a broad spectrum of disciplines to unravel how complex biological systems work. Biological phenomena are often studied quantitatively and on the scale of the whole organism. Such types of interdisciplinary research, broadly defined as computational biology and systems biology, are helping the transformation of biological science from a more descriptive and qualitative field to a more quantitative and precise science.

Here, we define computational biology broadly as an interdisciplinary field that applies computational methods developed in mathematics, statistics, computer science, etc., to the modeling and analysis of biological data. Structural modeling of biological molecules, sequence analysis and functional annotation, mathematical modeling of biological systems, et cetera, are a few examples of research work in this field. Computational biology has been instrumental in the interpretation of experimental findings and the elucidation of the mechanisms of many biological phenomena. In addition, it can provide predictions that help define research directions for both experimentalists and theoreticians.

We define systems biology as an interdisciplinary field that studies biological systems by examining the interactions of all relevant components of the living organisms simultaneously. A living organism is a complex and intricate system consisting of many inter-related components, such as nucleic acids, proteins, metabolites, and so on. The function of the living organism is realized by the proper interactions and organizations among these components and between these components and the environment. The success of systems biology relies on advances in both experimental technologies and computational models and tools. The former would include novel high-throughput experiments that allow all of the necessary molecular entities to be examined. The latter would include analytical tools for data interpretation and mathematical modeling that may encapsulate the biological system and allow explanations and predictions of biological behaviors.

One ultimate goal of computational biology and systems biology is to find cures for complex diseases such as cancer and to enable personalized medical diagnosis and treatment taking into account each individual's genetic makeup, metabolic level, and drug disposition. Despite rapid progress that we have made in biological and medical sciences, our understanding of biological systems remains limited, and we are still far from achieving this goal. Therefore, continued improvement in technology, theory, and computing tools in these interdisciplinary fields is much needed.

"*Handbook of Research on Computational and Systems Biology: Interdisciplinary Applications*" has been developed to summarize and present some of the most recent research carried out in these fields to encourage and guide future research. During the book development process, several hundred world-leading scientists and researchers in computational biology and systems biology were invited to contribute a chapter to the book. Each submitted manuscript was reviewed by at least three reviewers in a double-blind review process. The reviewers may be Editorial Advisory Board members, contributing authors, or external reviewers. From the forty-three submissions, twenty-eight were accepted to appear in the book.

The final book is a collection of eighteen thorough reviews and ten original research articles on the state-of-the-art development in the fields of computational biology and systems biology. Methods, tools, and applications of these fields are presented by many leading experts around the globe. The book chapters are written with the objective that novices in these fields will be able to learn the concepts and apply the techniques in their own studies and research. Active researchers in these fields may also appreciate the timely and in-depth review of existing literature and may be inspired to carry out innovative work to move biological sciences forward.

ORGANIZATION OF THE BOOK

A broad range of topics in systems biology and computational biology is covered by the book. The twenty-eight chapters of the book are divided into four sections:

- **Drug Development and Medicine (4 Chapters)**
- **Method Development in Bioinformatics (5 Chapters)**
- **Biological Networks and Pathways (14 Chapters)**
- **Structural and Mathematical Modeling (5 chapters)**

Section 1: Drug Development and Medicine contains four chapters summarizing the efforts and the challenges related to a wide range of research in drug discovery and medicine.

Chapter 1 provides a review and discussion of the ethical and privacy issues related to predictive and personalized medicine in the post-genomic era, as increasing amount of personal medical data are now stored electronically and accessed over the Internet.

Chapter 2 presents a thorough review of prevalent computational methods used for virtual screening in drug discovery, such as docking, QSAR, pharmacophore model, et cetera. The authors also offer their expert opinions on the advantages and limitations of each method and indicate important future research directions.

Chapter 3 reviews the recent efforts in vaccine development using systems biology approaches. Many successful vaccine development examples are presented. Various resources for epitope predictions and immunoinformatics research are summarized. New directions for vaccine research, such as through synthetic whole organism vaccines, are discussed.

Chapter 4 comprehensively reviews recent method development and applications in biomarker discovery in genomics, proteomics, transcriptomics, and metabolomics. The advantages and limitations of existing computational and experimental techniques are discussed.

Section 2: Method Development in Bioinformatics contains five chapters that review the most recent advances in method development in sequence and high-throughput data analysis.

Chapter 5 reviews the genome-wide association studies of human single nucleotide polymorphisms with their quantitative complex diseases and traits.

Chapter 6 reviews several computational methods in genome-wide association studies and presents a novel approach to detecting epistatic interactions by employing expert knowledge, such as pathway and protein-protein interaction information.

Chapter 7 reviews in depth the theory, strengths and limitations of existing biclustering methods for the analysis of DNA microarray data. Several important applications to drug discovery and various problems in systems biology are also summarized.

Chapter 8 reviews computational methods for the prediction of epigenetic target sites from DNA sequences based on nucleosome positioning, histone modification, and DNA methylation.

Chapter 9 presents an original research paper describing a novel method for protein sequence analysis. Evolutionary, structural, and functional information are taken into consideration to improve protein structure and function prediction. Application to the prestin protein is discussed.

Section 3: Biological Networks and Pathways contains fourteen chapters, including four method reviews providing introductory material for this field, five specialized reviews, and five original research articles that focus on specific types of biological problems.

Chapter 10 reviews the basic concepts of biological pathways and networks in detail, as well as available databases and tools for their storage and analysis. Integration of knowledge and data is presented with several applications to target discovery and disease pathway analysis.

Chapter 11 thoroughly reviews existing computational methods to identify modules in biological networks. Their applications in a variety of important biological problems, such as protein function and interaction predictions and disease studies, are also discussed in detail.

Chapter 12 reviews methods for constructing a functional linkage network (FLN) that consists of genes that are functionally associated. Two important applications to disease study, including prediction of disease gene and disease-disease association, are discussed.

Chapter 13 reviews network-driven analysis methods with a special focus on drug target identification.

Chapter 14 reviews an important and helpful pathway analysis tool, the Rat Genome Database (RGD). The novel pathway ontology for gene annotation adopted by RGD is explained and examples of pathway visualization and analysis are demonstrated using their Web service.

Chapter 15 reviews several novel methods that model cellular signaling networks, where signaling network perturbation data are analyzed by integrating multivariate measurement data to gain much needed information and knowledge about these networks.

Chapter 16 reviews recent advances in the experimental and computational analysis of MAPK (mitogen-activated protein kinase) cascades, providing original insights to these important signal transduction networks.

Chapter 17 reviews computational methods for the classification of cancer subtypes and the identification of deregulated pathways in different cancer subtypes.

Chapter 18 reviews novel computational methods based on structures and sequences of biosynthesis enzymes in the modeling of secondary metabolite biosynthetic pathways.

Chapter 19 presents an original research paper describing the analysis and prediction of metastatic relapse in breast cancer by sub-network extraction. A novel interactome-transcriptome integration method for extracting sub-networks is presented by integrating protein-protein interaction and gene expression data.

Chapter 20 presents an original research article describing a novel analysis method of time-course microarray data to predict transcription factors that temporally regulate differentially expressed genes under diverse stimuli.

Chapter 21 presents an original research article on the dynamic modeling and parameter optimization of the DNA damage and repair network.

Chapter 22 presents an original research paper in which a novel model for building causal biological networks based on high-throughput data is described. The model is built by unifying two complimentary methods (Granger Causality Model and Dynamic Causal Model). An application to the analysis of microarray data for gene circuit construction is presented.

Chapter 23 presents an original research paper that describes the development of microfluidic cell arrays for high-throughput examination of host-pathogen interactions. A prototype is presented that enables the study of the infection of human cells by up to 16 different bacterial strains.

Section 4: Structural and Mathematical Modeling contains five chapters, including three chapters on the structural modeling of biological molecules and two chapters on the mathematical modeling of specific biological phenomena.

Chapter 24 reviews state-of-the-art methods for non-coding RNA identification based on structural alignment of RNAs and with full consideration of pseudoknots.

Chapter 25 presents an original research paper on the computational modeling of RNA folding based on both folding kinetics and energetic considerations.

Chapter 26 presents an original research paper demonstrating a novel method for a reduced representation of protein structure in the application of ligand binding site modeling and screening.

Chapter 27 presents an original research paper on modeling the rolling of a cell on the surface of the extracellular matrix by simulating the successive attachment and detachment processes.

Chapter 28 presents an original research paper describing the modeling of chemotactic axon guidance, an important neurological process, at both microscopic and macroscopic scales.

These twenty-eight chapters represent only a small portion of the research work conducted in computational biology and systems biology. Nonetheless, we hope the readers may get a sense regarding the status of these fields from reading these chapters and become interested in carrying out additional research work to expand our understanding of biology.

August, 2010

Limin Angela Liu
Shanghai Jiao Tong University, China

Dong-Qing Wei
Shanghai Jiao Tong University, China

Yixue Li
Shanghai Jiao Tong University, China

Huimin Lei
Shanghai Jiao Tong University, China

Acknowledgment

We would like to thank all contributing authors for their strong submissions and timely preparations of book materials. Their dedication and contribution are essential to the success of the book.

We would like to thank all Editorial Advisory Board (EAB) members and reviewers for reviewing the book chapters. Their thorough evaluation and constructive suggestions are invaluable to the authors for preparing high quality chapters.

In addition, many EAB members also provided instrumental guidance and support during the development of the book. We would like to thank Russell Schwartz, Jason McDermott, Limsoon Wong, Zhaolei Zhang, Yu (Brandon) Xia, Haijun Gong, Esko Ukkonen, Jerzy Tiuryn, and Igor Jurisica, for their detailed suggestions and patient help regarding the design of chapter evaluation form as well as the preparation and organization of the book materials.

We would especially like to thank the following EAB members who have helped edit and proofread the book chapters shown in parentheses:

- **Tatsuya Akutsu** (Structural Alignment of RNAs with Pseudoknots)
- **Todor Dudev** (Finding Attractors on a Folding Energy Landscape; Structural Alignment of RNAs with Pseudoknots)
- **Thomas Manke** (Systems Biology-based Approaches Applied to Vaccine Development)
- **Jason McDermott** (Using Systems Biology Approaches to Predict New Players in the Innate Immune; System Granger causality: Its foundation and applications in systems biology)
- **Victoria Petri** (Modules in Biological Networks: Identification and Application)
- **Dennis R. Salahub** (Computational methods for identification of novel secondary metabolite biosynthetic pathways by genome analysis; A multiscale computational model of chemotactic axon guidance)
- **Alexander Schoenhuth** (Network-driven Analysis Methods and their Application to Drug Discovery; Connecting Microbial Population Genetics with Microbial Pathogenesis: Engineering Microfluidic Cell Arrays for High-throughput Interrogation of Host-Pathogen Interaction)
- **Russell Schwartz** (Mechanical models of cell adhesion incorporating nonlinear behavior and stochastic rupture of the bonds: concepts and preliminary results)
- **Fengzhu Sun** (Addressing the Challenges of Detecting Epistasis in Genome-Wide Association Studies of Common Human Diseases Using Biological Expert Knowledge)
- **Jerzy Tiuryn** (Biclustering of DNA Microarray Data: Theory, Evaluation, and Applications)
- **Jack Tuszynski** (Visualization of Protein 3D Structures in 'Double-Centroid' Reduced Representation: Application to Ligand Binding Site Modeling and Screening)

- **Esko Ukkonen (**Modules in Biological Networks: Identification and Application)
- **Lusheng Wang (**Biclustering of DNA Microarray Data: Theory, Evaluation, and Applications)
- **Yu (Brandon) Xia (**Current Omics Technologies in Biomarker Discovery; Linking interactome to disease: a network-based analysis of metastatic relapse in breast cancer; Cancer and Signaling Pathway Deregulation)
- **Zhaolei Zhang (**Prediction of epigenetic target sites by using genomic DNA sequence; Single nucleotide polymorphism and its application in mapping loci involved in developing human diseases and traits)
- **Xianghong Jasmine Zhou (**Unsupervised methods to identify cellular signaling networks from perturbation data)

We would like to thank the IGI Global for giving us the opportunity for developing this book. We would especially like to thank our book development coordinator, Christine Bufton, for her tireless and prompt assistance and guidance.

Last but not least, we would like to thank our family members for their patience, understanding and support throughout the book development process.

August, 2010

Limin Angela Liu
Shanghai Jiao Tong University, China

Dong-Qing Wei
Shanghai Jiao Tong University, China

Yixue Li
Shanghai Jiao Tong University, China

Huimin Lei
Shanghai Jiao Tong University, China

Section 1
Drug Development and Medicine

This section contains four chapters summarizing the efforts and the challenges related to a wide range of research in drug discovery and medicine. Chapter 1 provides a review and discussion of the ethical and privacy issues related to predictive and personalized medicine in the post-genomic era, as increasing amount of personal medical data are now stored electronically and accessed over the Internet. Chapter 2 presents a thorough review of prevalent computational methods used for virtual screening in drug discovery, such as docking, QSAR, pharmacophore model, et cetera. The authors also offer their expert opinions on the advantages and limitations of each method and indicate important future research directions. Chapter 3 reviews the recent efforts in vaccine development using systems biology approaches. Many successful vaccine development examples are presented. Various resources for epitope predictions and immuno-informatics research are summarized. New directions for vaccine research, such as through synthetic whole organism vaccines, are discussed. Chapter 4 comprehensively reviews recent method development and applications in biomarker discovery in genomics, proteomics, transcriptomics, and metabolomics. The advantages and limitations of existing computational and experimental techniques are discussed.

Chapter 1
Ethics and Privacy Considerations for Systems Biology Applications in Predictive and Personalized Medicine

Jake Y. Chen
Indiana Center for Systems Biology and Personalized Medicine, USA; Indiana University, USA & Purdue University, USA

Heng Xu
The Pennsylvania State University, USA

Pan Shi
The Pennsylvania State University, USA

Adam Culbertson
Indiana University, USA

Eric M. Meslin
Indiana University Center for Bioethics, USA & Indiana University, USA

ABSTRACT

Integrative analysis and modeling of the omics data using systems biology have led to growing interests in the development of predictive and personalized medicine. Personalized medicine enables future physicians to prescribe the right drug to the right patient at the right dosage, by helping them link each patient's genotype to their specific disease conditions. This chapter shares technological, ethical, and social perspectives on emerging personalized medicine applications. First, it examines the history and

DOI: 10.4018/978-1-60960-491-2.ch001

research trends of pharmacogenomics, systems biology, and personalized medicine. Next, it presents bioethical concerns that arise from dealing with the increasing accumulation of biological samples in many biobanking projects today. Lastly, the chapter describes growing concerns over patient privacy when large amount of individuals' genetic data and clinical data are managed electronically and accessible online.

INTRODUCTION

Predictive and personalized medicine based on the use of detailed genetic information is being regarded as a major development in modern medicine (Hood, Heath, Phelps, & Lin, 2004). Despite tremendous success in the past 50 years, conventional medicine which is based on observing and treating of patients' critical signs and symptoms at the physiological, organ, or pathobiological levels, has its limitations when treating complex, polygenic, and chronic disorders such as type-II diabetes, cancer, neurodegenerative diseases, and mental disorders. A primary reason is that, while treatment plans have been standardized, an individual's genetic background varies, the result of which is that some treatment work, some do not help at all and some actually harm. To cancer patients, for example, only by trial and error will the right regimen and drug dosage be eventually prescribed correctly, sometimes after long periods of receiving toxicological side effects or non-response. Genomics, pharmacogenomics, functional genomics, proteomics, and metabolomics—the so called "omics" technologies present biology and medicine with a new set of analytic tools to mine more deeply the genetic data contained in many types of biological molecules in biological samples (B. Palsson, 2002). Taken together, the potential for more integrative analysis and mining of the omics data with computer modeling techniques of multi-scale biological structure, or Systems Biology, have led to the creation of modern molecular medicine that is predictive and personalized (Aebersold et al., 2009). Many promising breakthrough drugs and biomarker molecules are being developed to improve targeted therapeutics of drugs, minimize drugs' side effect, and monitor clinical outcome of treatments. Given these development, it can be expected that more personalized "omic-based" medicine will be adopted as a standard clinical practice to improve health care quality and reduce health care cost (Naylor & Chen, 2010).

As we note below, systems biology aims to unify publicly accumulated knowledge of genes, proteins, molecular functional annotations, molecular interactions, and molecular measurements into integrated in silico models. Driven by advances in systems biology, personalized medicine is expected to gain wide-spread support for clinical adoption (Hood et al., 2004). It is expected that there will be concerns associated with the electronic storage, computerized processing, and online sharing of genomics and functional genomic information among biomedical researchers, health care providers, and patients. Ethical concerns about how this information may be misused or abused arise, as privacy concerns emerge about how to enforce patients' rights to protect their genetic information. In this paper, we provide a historical perspective on the emergence of predictive and personalized medicine, its relationship with systems biology, bioethics concerns when developing biobanks, and privacy considerations when dealing with large amount of electronic molecular medicine data.

PERSONALIZED MEDICINE: THE HISTORY

While medicine has always been, to some extent, "personalized" (for example, even ancient codes of medical ethics spoke of the ethical expectations of physicians to consider the well being of individual patients (Daikos, 2007), our current preoccupation with personalized medicine evolved from the new capacity to take "trial-and-error" out of the modern drug development process (Naylor & Chen, 2010). Unlike personalized medicine practiced for thousands of years, in which clinical symptoms derived at the physiological level (e.g., pulse rhythms, tongue color, temperature, pain points, and facial characteristics) were used to "personalize" diagnosis and treatment regimen (e.g., compound formulation of herbal medicine), modern personalized medicine seek to incorporate molecular information of the patient into modern medical decision makings. Optimal administration of drugs would be customized for each individual, after his/her unique genetic makeup is retrieved from molecular medical records to be created and taken into considerations.

The definition and scope of personalized medicine is still a popular topic up for debate within the scientific community. (Collins, Green, Guttmacher, & Guyer, 2003) defined personalized medicine as the answer to grand challenges of how to take advantage of human genetic variation to benefit human health and cure of diseases. The Personalized Medicine Coalition defined it as "the management of a patient's disease or disease predisposition, by using molecular analysis to achieve the optimal medical outcomes for that individual – thereby improving the quality of life and health, and potentially reducing overall health care costs". Leroy Hood's group defined it as being "predictive, preventive, participatory, and personalized" into what they called "P4 Medicine" (Aebersold et al., 2009). Hood's definition highlights the requirement for personalized medicine to have predictive power (e.g., predicting cancer metastasis), help people taking preventive measures by assessing disease onset risks, involve all stakeholders during treatment, and tailor treatment to individual's genetic variations. No matter how personalized medicine is defined, it is expected that personalized medicine will have the following characteristics:

- Use a molecular characterization approach to create a medical decision-making system for disease risk assessment, diagnostics, prognosis monitoring, and treatment outcome predictions.
- Its intervention will consider each individual's specific genetic/biochemical fingerprint.
- The power of cure will shift from symptom-relieving treatment to proactive measures that may reduce disease risks, prevent disease onset, and revert course of disease progression.
- The individualization will be done to minimize side effects.

In pharmacogenomics and functional pharmacogenomics studies, the need to maximize a drug's therapeutic benefit and minimize its toxicological side effect for every patient has led to the rebirth of "personalized medicine". Pharmacogenetics refers to the study of drug response in patients with diverse human genetic backgrounds. In classical pharmacogenetics studies, a drug's therapeutic effects are known to be influenced by genetic variations of critical genes, involving mechanisms such as single-nucleotide polymorphisms (SNPs), DNA copy number variations, and epigenetic modifications. The genetic differences of these genes (a notable example is the Cytochrome P450 subfamily of genes (Plant, 2007) may cause a drug to be metabolized differently from person to person, thus rendering a drug administered at an ideal dosage to one individual less effective to another. One example is the prescription of Warfarin, a blood-thinning drug that may cause internal bleeding when used at a heavy dosage over the "normal" range, which varies from one individual to another. Currently, genetic tests have been made available to classify patients who may benefit from Warfarin and to prescribe it "for the right patient at the right dose at the right time" (Klein et al., 2009). In clinical trials of certain cancer drugs with narrow effective dosage ranges, however, similar tests have been made available. This could spell great benefits for some patients receiving the cancer drug while for others, disastrous side effects. Therefore, considering each patient's genotype to customize a drug's dosage and to determine its utility to a subpopulation that may respond well is a primary research agenda for pharmacogenetics.

SNP analysis has served as the primary genetic basis for pharmacogenetic studies. In SNP analysis, researchers look for frequent single nucleotide variations between genes. These variations, often considered as a genetic code aberration, occur approximately every 300 base pairs in the human genome at an allele frequency level of at least 1 percent within the general population (Korkko, Milunsky, Prockop, & Ala-Kokko, 1998). Up to 30 million SNPs exist in the human genome and most of them fall outside of the gene coding regions (Sherry et al., 2001). Within the coding region, non-synonymous SNPs can cause amino acid differences and in the worst case, disruption of the entire protein. SNPs in promoter regions or other gene regulatory sites of the genome can also affect the gene expression and transcriptional regulation. SNPs may explain individual peculiarity and may cause one person to be more susceptible to developing certain diseases, or to be more responsive to certain medical therapies than others. Some SNPs can lead to higher susceptibility to diseases, e.g., in sickle cell anemia and cystic fibrosis. However, most SNPs have not been characterized and their association to human traits remains to be established. Experimental techniques to detect and analyze SNPs include: Denaturing High-Performance Liquid Chromatography (DHPLC) (Wolford, Blunt, Ballecer, & Prochazka, 2000), Single-strand Conformation Polymorphism (SSCP) analysis (Tahira, Suzuki, Kukita, & Hayashi, 2003), conformation sensitive gel electrophoresis (Korkko et al., 1998), solid-phase chemical cleavage (Bui, Babon, Lambrinakos, & Cotton, 2003), DNA sequencing (Kwok & Duan, 2003), allele-specific PCR (Gaudet, Fara, Beritognolo, & Sabatti, 2009), oligonucleotide ligation (Bruse, Moreau, Azaro, Zimmerman, & Brzustowicz, 2008), MALDI-TOF mass spectrometry (Pusch, Wurmbach, Thiele, & Kostrzewa, 2002), and multiplexed amplification coupled mini-sequencing (Gilbert et al., 2007; Shapero, Leuther, Nguyen, Scott, & Jones, 2001). For a comprehensive review of the SNP discovery, genotyping, and their significance in pharmacogenetic research, interested readers may refer to Giacomini et al. (2007), Suh and Vijg (2005), and Twyman (2004).

The completion of the human genome in 2001 was a critical milestone for modern personalized medicine leading to the emergence of pharmacogenomics (Baltimore, 2001). As an extension of pharmacogenetics in the post-genome context, pharmacogenomics is an advanced form of pharmacogenetic study that aims to consider the roles of multiple genes or genetic factors including SNPs at the genomic level. The genome-wide pharmacogenetic study is necessary, because a drug has to go through the pharmacokinetics of absorption, distribution, metabolism, and excretion (ADME) after it enters the human body and before it reaches its target site. Along the way, the drug must interact with many proteins—ion channels, transporters, enzymes, and receptors—that may demonstrate different degree of genetic variations. A holistic genomics-based approach could potentially unify the fragmented knowledge of pharmacology and toxicology of a particular drug and explain significant portion of a drug's adverse side effects.

Compared with pharmacogenetics, experimental techniques supporting pharmacogenomics can achieve massive parallelism and throughput in genotyping. Multiplexed PCR assays have been used to monitor more than one SNP at a time for genotyping analysis of many samples simultaneously (Gilbert et al., 2007). To achieve the maximal possible throughput, second-generation sequencing, or "pyrosequencing", has been used successfully at the genome scale (Ronaghi, 2003). In pyrosequencing, SNPs in their context of the surrounding sequence are discovered, therefore eliminating the need for other controls. However, the massive amount of information generated from this platform creates a daunting bottleneck for informatics data processing. Therefore, once the initial discovery of SNP variations is completed, simpler, faster, and more automated genotyping methods such as molecular beacons genotyping (Barreiro, Henriques, & Mhlanga, 2009) have been suggested for routine pharmacogenomics studies. However, due to the prohibitive cost of genotyping the entire human genome for individuals until recently and the huge challenge of determining individual trait-associated genetic variations for complex polygenic diseases, pharmacogenomics has yet a long way to deliver its original promise.

Functional pharmacogenomics studies that go beyond DNA-based platforms have also been established in the past two decades. By measuring the effects of individual genetic variations at the functional genomics or proteomics level, one can also identify within patient subpopulation those who respond well to particular drugs with a particular molecular expression profile. For example, DNA microarrays as a well-established molecular profiling platform have been applied by Genomic Health, Inc., USA, to produces a multiplexed expression panel—Oncotype DX. The 21-gene RT-PCR assay Oncotype DX stratifies and predicts the likelihood for breast cancer reoccurrence for women with HER2+ lymph-node negative metastatic breast cancer (Habel et al., 2006). A second major platform for functional pharmacogenomics studies is clinical proteomics, the study of the expression and function of all human proteins. While there are less than 25,000 human genes, there are as many as one million proteins in an individual's blood. Approximately 10,000 of these proteins can be detected with tandem mass spectrometry (Saha et al., 2008) and many of them have been associated with therapeutic drug targets or disease biomarkers (Saha, Harrison, & Chen, 2009). In many clinical trials, the serum proteomic profiles can help evaluate the pharmacological or toxicological effects of experimental drugs and tailor individualized therapies (Tchabo, Liel, & Kohn, 2005). Due to cost and detection sensitivity concerns, multiplexed immunoassays, e.g., protein microarrays, are often used to measure multiple protein analytes in clinical settings (Kricka, Master, Joos, & Fortina, 2006).

SYSTEMS BIOLOGY AND PERSONALIZED MEDICINE

Starting from the 1990's, advances in high-throughput biological data capturing instruments such as automated sequencers, DNA microarrays, tandem mass spectrometry, and NMR has made it possible to routinely perform functional genomic, proteomic, and metabolomic analyses of complex biological samples (B. Palsson, 2002). Separately, each of these platforms can produce massive datasets that need to be pre-processed, analyzed, and mined using the latest bioinformatics software data analysis tools. However, information accumulated from one "omics" platform often does not correlate well with information accumulated from a different platform, causing integrative data analysis to remain a challenging task. A primary reason is that different "omics" tend to measure different aspects of the biology, which is too complex to study without an integrative biological model or advanced informatics tools (Huan, Wu, & Chen, 2010).

Systems biology aims to unify publicly accumulated knowledge of genes, proteins, functional annotations, molecular interactions, and molecular measurements into integrated in silico models (Hood et al., 2004; Kitano, 2002; Bernhard Palsson, 2006). A systems view of physiology, biology, and disease provides fresh opportunities to unravel human biology complexity (Chen, Yan, Shen, Fitzpatrick, & Wang, 2007). In isolation, the study of genes, transcripts, proteins, and metabolites in the context of human physiology and disease has been challenging, because these molecular entities must assemble into molecular machinery and cooperate in molecular pathways to play their roles inside cells. By systematically offering new global insights into the structure, function, and dynamics of complex biological components, systems biology studies can help interpret molecular functions and bridge molecular understanding of human diseases.

Systems biology is essential to predictive and personalized medicine. Electronic databases and repositories play a central role to store and analyze clinical data, which has been useful for conventional evidence-based clinical decision-making. With the advent of molecular medicine, many molecular measurement data will be deposited into medical databases. The establishment of connections between clinical data (phenotype) and molecular data (genotype) requires integrative models to help enhance medical decision-making. Meanwhile, the complexity of systems biology has made the development and practice of predictive and personalized medicine ever illusive. Fast-forward 10 years: imagine that a technology-savvy physician wants to prescribe a drug regimen to her patient. She first prescribes a blood "proteomics profiling" test based on hundreds of proteins that are observed in the patient's blood, to determine whether the patient's disease is of a particular subtype and therefore deserves the use of drug X. Then, she prescribes a second "genome-wide genotyping" test of her patient's DNA sample. The test will scan the entire genome of the patient to look for possible genetic variability that help determine the right dosage of the drug X to be used. Unlike classical laboratory biochemical tests that only measure the level of a few dozen blood analytes, both the "proteomics profiling" test and the "genome-wide genotyping" test will generate mountains of raw data (at the Gigabyte to Terabyte level, orders of magnitude bigger than his entire clinical record coded in text)—a valuable gold mine of information that can be saved as part of the patient's molecular medical records. Luckily, insurance-approved standardized systems biology software tools will have been developed to aid the physician to come up with a necessary prediction to deliver customized treatment plan, without the patient worrying about any details.

This will be a great picture, for the clinical adoption of systems biology, and for the realization of computerized personalized medicine. The world is awaiting the arrival of the day personalized medicine comes true. It shall be an exciting day. Why won't it be?

PRIVACY AND ETHICAL DILEMMAS: A PROBLEM STATEMENT

The increasing complexity of molecular medicine information to be generated from future personalized medicine is sure to bring new challenges—technological, legal, social, and ethical—for the vast majority of physicians, basic biomedical researchers, and health care workers who are unaware of the incoming tidal wave of data soon to arrive. A notable prelude into the future can be seen in the recent development of Personal Genome Project (PGP) (Church, 2005) and the mushrooming of many commercial companies that offer consumers the opportunity to obtain whole-genome genotyping. will have their genome published on the world wide web for others to download. Although not initially required, these individuals will also be able to contribute additional biospecimens as additional research warrants, to help advance the understanding of genomics and health. This altruistic idea is based on the assumption that by sharing personal genomics data and all clinical information associated with it, the research community as a whole will be given the unprecedented opportunity to understand how personal genetic variation leads to disease predisposition and progression.

As with any new technology, the impact of sharing personalized genomics and personalized medicine data is usually not fully appreciated – at least not until unintended consequences occur. It is already known that the vast amount of genomics and functional genomics data from these genome-wide molecular tests may reveal information unique to each individual. In theory, to distinguish every two individuals in a world of six billion population, the detection of differences of merely 5 Tag SNPs (SNPs in a region of the genome with high linkage disequilibrium) at an allele frequency of 1% is needed ($0.01^5 < 1/(6*10^9)$). In reality, more Tag SNPs are needed to uniquely identify a single individual, since closely related human beings have highly similar SNPs and the allele frequency of many SNPs may exceed 1%; therefore, common SNPs may be necessary to uniquely identify an individual (Homer et al., 2008), although the exact number required is up for debate (Visscher & Hill, 2009). However, this amount of SNPs can be easily extracted from personalized genomics data, in which hundreds of thousands of SNPs (genotype) are recorded for each individual, along with disease trait-associated SNP information. Therefore, anonymizing individual patients based on personalized genomics data may no longer be technically feasible, and this remains a much-debated research topic (Fullerton, Anderson, Guzauskas, Freeman, & Fryer-Edwards, 2010). Nonetheless, as the progress of science and technology continue to unravel at an exciting pace, we ask, "Have we gone too far to bet the future of genomic health on disclosing and sharing genomic information in the name of science and personalized medicine?"

We have identified several challenges when deciding whether or not it is worth sharing emerging data derived from personalized medicine.

- Among biomedical researchers, a choice with ethical consequences exists between unfettered sharing of personalized genomics data and preventing unintended leaks of patient identity exists. We do not know for certain whether individual patients' identities may be compromised by advanced genomic analysis/deciphering techniques; meanwhile, the risk of not sharing this data set publicly may slow scientific progress of studying phenotype-genotype associations within the broad research community.
- As a policy matter, society must decide whether to seek "opt in" or "opt out" approaches to obtaining genomic information for scientific research. The current informed consent process for donating genomic samples and information is based on the right of individuals to make a positive decision regarding their participation in research (opt in). Switching to an opt-out system – where genomic information would be used unless patients specifically declined, challenges this historical principle of informed consent. However, immediate family members connected by blood are known to have highly similar genetic makeup; disclosing one may unwillingly disclose genetic makeup of the other with high degrees of inference powers. It would be unfair to either deny people the opportunity to consent or force them to consent to the use of personal genomics information, regardless of the patient's interest in sharing their personal data.
- The challenge faced by hospitals is, on the one hand to ensure that strong protections exist to limit the number of individuals who need to know a patient's personal information (and protect their privacy), and on the other, permitting greater access to more health care professionals whose access might provide the patient with better care. This is primarily because the large

size of the data makes it impossible for health care professionals, who are not always computer experts, to authorize the use of the exact portion of personal genomics information.

For the rest of the paper, we address in detail what level of ethical and privacy concerns exists in personalized medicine and what historical lessons may prepare us in proactively designing solutions that address these concerns.

ETHICS ISSUES IN THE ERA OF PERSONALIZED MEDICINE

For more than five decades, much of the focus of bioethical inquiry has been to examine the development and use of emerging technologies – with an eye to drawing cautionary lessons about their impact on individuals and society. Initial concerns about the potential for technology to harm and even 'dehumanize' -- especially reproductive, genetic, and end of life technology -- led to calls for a thorough-going regulatory regime and gave rise to a robust literature. Few still cling to this paradigm, but a principle of precaution can still be found particularly in areas involving research with human subjects. Now that the human genome has been mapped and sequenced (along with the genomes of other plants and animals), attention is turning to the myriad ways in which access to and use of genetic information can contribute to a better understanding of disease. The development of two technology platforms, biobanks and large electronic health records (EHR) systems are shifting the paradigm in bioethics thinking: from cautionary progress to an ethical imperative (Eric M. Meslin & Goodman, 2010).

There is a close connection between these platforms, the most obvious of which is that they are key components of biobanks – repositories of human biological materials linked to electronic health databases – which are now increasingly employed around the world to mine the relationship between genotype and phenotype (E M Meslin & Quaid, 2004). Biobanks are a key resource in pharmacogenomics in which individual samples may be studied to establish genotype-phenotype relationships between genetic variation, gene expression, or protein expressions and disease susceptibility or drug response. Pharmacogenomics may help improve targeting of therapies by developing tests to identify, in advance, individuals who are genetically disposed to respond favorably or unfavorably to particular medicines (Lesko et al., 2003). In addition to a "right-drug, right-dose, right-patient" model, pharmacogenomics also purports to reduce drug costs and stimulate research and development for the pharmaceutical and biotech industries (Evans, Flockhart, & Meslin, 2004). Therefore, the collection of patient biospecimens, patient clinical information, and molecular genotype information in biobanks is a prerequisite to personalized medicine.

Clinical Ethics Lessons

Biobanks implicate both *clinical* and *research* ethics issues in important and unique ways. For example, the long history of bioethics includes profound attention to the ethical issues that arise in the clinical encounter at the bedside between health care providers, patients, and families. The paradigmatic "western" encounter between a patient and their physician may be described as follows: the virtuous physician, respectful of the autonomous decision making rights of individual patients, seeks the patient's permission to act in the patient's best medical interest (usually through treatment, surgery, etc.) with the expectation that the patient will get better (or not worse) (Pellegrino & Thomasma, 1984; Ramsey, 1960; Veatch, 1981). Pellegrino and Thomasma give an account of the patient's expectation in the clinical encounter, which typifies the western paradigm:

"... the patient seeks not only to be protected from harm, but also to be healed and to have health restored or improved, pain and anxiety removed, disability lessened. .. [t]he patient desires these good ends within some definition of the good life that is uniquely and personally his or her own."

Each component of this somewhat caricatured relationship invokes ethical norms of behaviour and expectation: the virtuous physician is one who is properly disposed to act with a morally motivated character and without secondary interests that may conflict with those of the patient's welfare; the respect owed to decision making by a patient presumes an independence of thought and action sufficient to exercise full autonomy; seeking of permission requires a bi-directional conversation with sufficient information disclosed (and time to consider it) such that a voluntary, non-coerced, reasoned consent to treatment is provided (or declined); and throughout the encounter is an expectation of privacy – or at least the attention to limiting access to information about the person to those with a need to know. What lessons might we learn in a genomic world, where personalized medicine is the goal?

For starters, clinical medicine has always been personalized. Long before *Pervical's Medical Ethics* was published at the dawn of the 19th century (among the first modern statements of medical ethics for physicians), ancient oaths of medicine including the Indian *Charaka Samhita* (which preceded the Hippocratic Oath by six centuries) and the Oath of Maimonides (12th C. CE) included statements of ethical responsibility and obligation to the patient. The history of medicine has always focused on the patient, even if there was little that physicians could actually do for patients. In the classic text *An Introduction to the History of Medicine,* Garrison reminds readers that in 25 of the 42 cases in the Hippocratic corpus, the patients died and Hippocrates himself wrote: "I have written this down deliberately believing it is valuable to learn of unsuccessful experiments and to know the causes of their failure" (Garrison, 1917).

Reflecting on the relevance of these findings to the birth of modern medical records, Goodman has observed that Garrison's descriptions by Hippocrates were valuable particularly when comparing them to the records of the Roman physician Galen, which were boastful and limited to remarkable cures and the errors of other practitioners (Goodman, 1998; Goodman & Miller, 2006). More simply, Goodman notes, the practice of medicine and nursing has evolved to the point where we require the keeping of records not only to keep track of what is being done to and for patients, but also to assess whether what was done actually worked.

In this way, biobanks are very much like medical records insofar as they are repositories of information collected about patients. Where contemporary *research-focused* biobanks differ from common medical records, however, is that the latter (medical records) are designed principally for clinical use, whereas the former are now being purposely built and designed for research. That being said, the collection, storage, and mining of genotypic information for pharmacogenomic drug development invokes many of the same general ethical concerns that physicians have always faced in their encounter with patients. The prospect of personalized medicine translated from the "bench to the bedside" has been a dream for years. Yet for it to become a reality -- that is for personalized medicine to become the standard of practice – two key challenges will need to be overcome: first, there will need to be in place a significant cadre of genetically literate health care professionals – those sufficiently well trained in genetics and genomics to be able to teach it, train others, and to implement genetic findings in the clinic. Second, these clinicians will have to show that they are capable of undertaking conversations with patients and families using the complex language of risk, susceptibility, and probability – areas that are understandably problematic for both ethical and numerical reasons (Schwartz & Meslin, 2008).

Research Ethics Lessons

Many of the same ethical issues (and lessons) that arise in the clinical encounter are also present in the context of the relationship between researchers and research-subject participants. For example, clinicians and researchers must be mindful of the issues that arise whenever they seek informed consent: how much information to disclose? How well does the patient/subject understand? Similarly, patients and research subjects can be expected to wonder about who will have access to personal information about them and whether the institutional rules and procedures are adequate for protecting personal privacy and confidentiality. Where distinctions and differences arise, they do so because the goals differ: in basic research the primary goal is to seek knowledge and understand nature; in applied research, the goals include understanding the impact of science in individuals and populations.

But differences also arise from the perspective of the architecture used to assure ethical practice and conduct. In research, most notably in countries where ethical permission to carry out studies is a prerequisite for obtaining research funding, this architecture involves prior review by a peer committee of experts with the institutional authority to review, require modifications to or disapprove a research project. However, while improvements have been made in national and international governance documents in the past decade to accommodate genomic studies (Kaye & Stranger, 2009), certain challenges remain for assuring the protection of human subjects, including: persistent difficulties with harmonization across within and across countries (Chalmers, 2007; IOM, 2001; NBAC, 1999), a lack of professional guidance (Eric M. Meslin & Goodman, 2010; E M Meslin & Quaid, 2004), ethical, legal and social issues in genetic testing (Andrews, Fullarton, Holtman, & Motulsky, 1994; Holtzman & Watson, 1998), informed consent for genetic research (ASHG, 1996), and oversight of specialized areas such as research on population-based studies of low-penetrance gene variants (Beskow et al., 2001).

The clinical and research "paradigms" may be idealized descriptions, but there is no denying that genomic science challenges both in profound ways. The great Canadian physician illiam Osler wrote in 1892: "If it were not for the great variability among individuals, medicine might as well be a science and not an art" a statement hauntingly prescient of much of genomic medicine today even though Osler knew nothing of how genetic variations may explain why some people are vulnerable to certain diseases, might not respond to drugs, or have poor disease prognosis. Indeed, we can appreciate that greater certainty arising from biomarker research will give physicians more confidence to diagnose and treat patients, and raise public expectations at the same time. What we do not know is how much confidence physicians will be entitled to express given the status of the biomarker test development, i.e., sensitivity, specificity of tests. Moreover, the blurring of the line between science and the clinic through a global movement towards translational medicine will only further call to question the ways in which clinical and research ethics will inform personalized medicine.

Like all research, clinical and translational science will succeed only if it is scientifically valid, ethically sound, and clinically applicable. This means that, at a minimum, research must be designed and conducted by competent individuals within a system that protects the rights and welfare of human subjects. This system has long been based on the clinical drug trials paradigm where investigators address ethical issues chronologically: from the initial conception of a study to its completion and dissemination as new knowledge (Norton et al., 1994). This paradigm requires that investigators are trained sufficiently in their ethical responsibilities to submit high quality protocols to the IRB, recruit research subjects without coercion, obtain valid informed consent, conduct their studies without bias, and accurately report their results (Emanuel, Wendler, & Grady, 2000).

Preparing scientists to engage in the responsible conduct of research has its own challenges (Kalichman, 2007), but research also raises important ethical and policy issues for those other than investigators and IRBs, such as educators, institutions, private companies, and legislators (NBAC, 2001). For example, before the promise of genetic science is fully realized for clinical medicine, many ethical and policy issues must be addressed that extend beyond the design and conduct of studies, such as how state and federal laws apply to "ownership" of DNA or "invention" of molecular biomarkers (Charo, 2007); how the lack of harmonization in the federal system for the protection of human subjects may inhibit research in new areas such as pharmacogenomics (Evans & Meslin, 2006); how potential conflicts of interest between academic and private sector partners should be managed (Angell, 2000); and how privacy protections should apply in the face of an increased need for data-sharing (Wolf, Sieber, Steel, & Zarate, 2005). These and other issues are now part of a broader administrative rationale for education in the responsible conduct of research (Vasgird, 2007).

Paradigm Shifts When Science Meets Economics

Personalized medicine is now adapting to a new economic reality (Evans et al., 2004). It is becoming possible for predictive testing and drug targeting to become so accurate that physicians can virtually guarantee patients that the drug they were prescribed would help and not harm. Given that up to only ~ 60% of the prescriptions written produced desired benefits and 40% of prescribed medicines do not help (and up to 7% harmed patients) then the economic incentive for developed targeted medicine is obvious: in 2002 prescription drug spending was USD $162.4 billion. This means that as much as $65 billion (40% of $162.4 billion) was spent on prescription drugs that either did not help patients or actually harmed them (Evans et al., 2004) What if it were possible for personalized medicine to guarantee whether a patient would be in the 60% group of assured responders, or the 7% group of those who were harmed? Not only would this make sense medically, but economically. This presents an intriguing analogy from the world of consumer protection where guarantees come in the form of sales and service warranties. If refrigerators do not work as advertised, customers are entitled to repair or replacement, leading Evans to audaciously analogize that "if new refrigerators hurt 7% of customers and failed to work for another one third of them, customers would expect refunds". We don't offer such guarantees for medicine, but a personalized medicine paradigm shift may present this option for society.

Public Willingness and Governance

There is a growing body of data regarding the public's willingness to donate tissue or other biological material to science in general and to biobanks in particular (Axler, Irvine, Lipworth, Morrell, & Kerridge, 2008; Bates, Lynch, & Bevan, 2005; Bernhardt, Tambor, Fraser, Wissow, & Geller, 2002; R. E. Goldman et al., 2008; Helgesson & Swartling, 2008; Leiman, Lorenzi, Wyatt, Doney, & Rosenbloom, 2008; Murphy et al., 2009). A review of the empirical literature conducted on PubMed in early 2009 found more than 60 studies, with at least 20 surveys published between February 2008 and January 2009. These studies reveal a gradual *increase* in the public's support for genetic studies generally, and for the donation of specimens to biobanks, in particular (Haas, Renbarger, Meslin, Drabiak, & Flockhart, 2008; Helft, Champion, Eckles, Johnson, & Meslin, 2007; E M Meslin, 2010).

And yet, while evidence of public support for personalized medicine would appear to be growing, one must be cautious about drawing definitive conclusions. It is not enough to know *that* the public has concerns (as evidenced by the public opinion data above). Instead, it is critical to appreciate that the context for these concerns informs the type of tradeoffs that arise between protecting privacy and permitting access to information to advance research on human health. Altman's group have graphically illustrated this tradeoff by showing how the price paid for increased privacy protection is less access to SNP data and vice versa (Lin, Owen, & Altman, 2004). Not satisfied with such simple model of tradeoffs, there are growing studies that propose governance structures and instruments to assure the ethical construction and implementation of personalized medicine platforms (Kaye & Stranger, 2009). Indeed, there has been no shortage of guidance documents on these issues. A search of the authoritative *HumGen* database found more than 200 national guidance documents on the topic of biobanks alone (Humgen, 2010). In the United States, a set of federal regulations governs the oversight of research involving human subjects and this is the same regulatory structure for research on human biological materials (Eric M. Meslin & Goodman, 2010).

But research ethics governance is not limited to ethics review systems alone. With the domestic and international proliferation of biobanks and their associated connections to health information databases, scholarly attention has been turning from the ethical issues arising from the construction of biobanks to the ethical issues that emerge in their *operation and management* in society more broadly (E M Meslin, 2010). Accordingly, there have been demands for greater transparency in governance structures. Two different approaches have been adopted for addressing these types of demands: the first is a 'top-down' approach that focuses on developing policy, procedures, regulations, and guidelines to aid decision-makers. In contrast is a 'bottom-up' approach, which begins with those who are most affected by the issues and attempts to inductively develop consensus recommendations and policy. Although both approaches have merit, more work needs to be done on 'bottom-up' strategies if trust and transparency are to be more than mere slogans (E M Meslin, 2010).

One example of a successful system of public understanding and engagement can be found on the other side of the world in Western Australia. For more than three decades, Western Australia (WA) has been quietly collecting some of the most comprehensive administrative health datasets in the world (Hobbs & McCall, 1970; Holman, Bass, Rouse, & Hobbs, 1999; Fiona J. Stanley, Croft, Gibbins, & Read, 2008). When used in combination with medical record audits, the WA dataset provides a novel platform for comprehensive evaluation of health system performance and is now moving towards adoption for policy making in pharmacovigilance studies (F.J. Stanley & Meslin, 2007). Researchers and policy makers in "the WA" have now developed a data linkage system that both protects privacy and enables research (Kelman, Bass, & Holman, 2002). This "win-win" approach results from keeping any identifiable information from the researchers, who only need the linked data on exposures and outcomes for their analyses. Others note, since this program has been in place, general requests for access to identifiable data have declined markedly (Trutwein, Holman, & Rosman, 2006). The most convincing evidence of this program's success probably comes from the public itself:

"When people in the general community were asked if they approved of their information being used in this way, they were found to be not only supportive of it, but they questioned why it was not already being done." (F.J. Stanley & Meslin, 2007)

Enduring Bioethics Questions

Unlike the early years of bioethics inquiry -- when organ transplantation, *in vitro* fertilization, and end-of-life care caused near social upheaval -- the application of knowledge gained from the human genome to develop more tailored, personalized medical care is unlikely to wreak the same type of ethical havoc. At the same time, society's failure to learn the lessons from the history of clinical and research ethics would surely have undesirable results: building a biobank without public consultation will disincline them to support such efforts and result in an overall slowing of scientific progress (Campbell, 2007).

We also know that despite general public support for genetic studies, including the use of stored tissues to test and design new medicines, there remains a measurable distrust of efforts to commercialize the products of genomic research (Nicol & Critchley, 2009). These points suggest that researchers, organizations and institutions consider carefully how they construct consent forms to accurately disclose what will be done with specimens, how they will be shared, and what commercial interests are involved.

Similarly, there will be downstream ethical issues for physicians who will be expected to knowledgably discuss the latest biomarker study and its implications for disease management, diet, and other environmental factors affecting their patients. Given that such knowledge is ever-evolving, and that patient numeracy issues make discussions of risk and susceptibility difficult at the best of times, attention will need to be focused on how to close the 'risk-perception gap'.

Finally, it can be expected that personalized medicine will be proposed as a model for putting limits on rising health costs (Conti, Veenstra, Armstrong, Lesko, & Grosse, 2010). While we are still in the early days of the "genomic revolution" we know enough now to recognize that what is needed is a systematic effort to understand how economic strategies, including cost effectiveness and comparative effectiveness methods can be employed.

PRIVACY CONSIDERATIONS IN THE ERA OF PERSONALIZED MEDICINE

The Concept of Privacy

The definitions of privacy vary and depend on the field, ranging from a "right" or "entitlement" in law (Warren & Brandeis, 1890) to a "state of limited access or isolation" in philosophy (Schoeman, 1984) or "control" in social sciences and psychology (M. J. Culnan, 1993; Westin, 1967). The definitional approaches to privacy can be broadly classified as either "privacy as right" or "privacy as commodity". The right-based definition of privacy originated in the jurisdiction (Warren & Brandeis, 1890) – "right to be left alone" – was seen as integral part of the society's moral value system and should be cherished as such. However, when the "privacy as right" concept was applied to consumer behavior, the economists characterized the consumer behavior as paradoxical (Acquisti & Grossklags, 2005): Despite reported high privacy concerns, consumers still readily submit their personal information when purchasing or looking for deals and coupons. Thus, the notion of "privacy as commodity" was conceptualized (Bennett, 1995) where privacy is still an individual and societal value, but rather than being absolute, it can be assigned an economic value and enter a cost-benefit calculation as will be discussed below.

In the context of this research, privacy is concerned with the limits on access to personal information; anonymity, confidentiality and secrecy are dimensions of it (Lunshof, Chadwick, Vorhaus, & Church, 2008). *Anonymity* is the ability to conceal a person's identity (Marx, 1999; Zwick & Dholakia, 2004), which is central for the information collected for statistical purposes. Anonymity exists when someone is acting in a way that limits the availability of identifiers to others. The result is that information produced during transactions is not personally identifiable. Thus, the information cannot be correlated back to the individual, and this may enable privacy. *Confidentiality* concerns the externalization of restricted but accurate information to a specific entity (Zwick & Dholakia, 2004). Confidentiality implies that the data themselves and the information they represent must be protected and their use confined to authorized purposes by authorized people (Camp, 1999). *Secrecy* has been defined as intentional concealment of information (Bok, 1989) and implies having control over the disclosure of information (Lunshof et al., 2008). Secrecy enables individuals to manipulate and control environments by denying outsiders vital information about themselves (Tefft, 1980).

Health Information Privacy and Genetic Privacy

Early in the 1960s and 1970s, computerized methods were introduced to collect, transfer and store patients' health data and medical records in some large hospitals (Anonymous, 1966). Since then, medical records are stored in electronic databases by both government and private medical providers at an increasing pace (Hodge, Gostin, & Jacobson, 1999). Electronic medical records (EMRs) are designed to increase the accessibility and sharing of health records among authorized individuals (Barrows & Clayton, 1996). In February 2009, President Obama signed into law the Health Information Technology for Economic and Clinical Health Act (the "HITECH Act"), for the improvement of the nationwide healthcare infrastructure and the adoption of EMRs. Simultaneously, it has ignited unprecedented concerns regarding the protection of large quantities of personal information. In spite of the anticipated value potential of EMRs, the highly personal and sensitive nature of healthcare data and the associated privacy concerns may cause significant economic, psychological, and social harms to individuals in case of information disclosure or misuse (Davis, 1995). Therefore, the need for the protection of health information privacy has been highlighted in medical ethics and health related laws and regulations.

With recent advances in high-throughput genomic technologies and personalized medicine research, the concept of genetic information privacy has emerged over the last few decades. Databases containing large amounts of genotypic, phenotypic and demographic data about individuals are linked together to facilitate pharmacogenomics research to generate personalized medicine (Vaszar, Cho, & Raffin, 2003). Because such rich set of biological data contains an individual's stable and complete DNA sequences, some researchers posited that the individual's genome represents his or her probabilistic "future diary" so that they deserve special protection as "genetic exceptionalism" (Annas, Glantz, & Roche, 1995). Given this context, genetic privacy is defined as an individual's right to protection from non-voluntary disclosure of genetic information (Lunshof et al., 2008). Genetic privacy not only relates with an individual but also extended to families and communities due to its genetically identifiable characteristics.

Prior literature has identified sources of concerns for genetic privacy risks, including potential loss of confidentiality (Lunshof et al., 2008), linkability of identification data (Vaszar et al., 2003), revelation of family genetic disease information (Harman, 2001), decline of patients' autonomy (Marks & Steinberg, 2002) and the problem of "orphan patients" (Anderlik & Rothstein, 2001; Robertson, 2001; Rothstein & Epps, 2001). Moreover, individuals may not want to share their genetic profiles for the fear that they may be discriminated against by insurance companies, employers, and other entities (Brown, 2002).

Understanding Privacy in the Era of Personalized Medicine

Genetic privacy issues have become more complicated, with many new ethical, legal, and social implications and challenges. On the one hand, substantial risks to patients' or genomic donors' privacy do exist in transmission of medical information (Cassa, 2008). For example, personal or family identity, hereditary data, and medical conditions could be disclosed due to unauthorized access to genome database. On the other hand, the advances in personalized medicine research rely heavily on the open-access genome databases, which allows for accessing, sharing and linking genotype-phenotype data (Lunshof et al., 2008). To better understand genetic privacy issues in the era of personalized medicine, below we apply two theoretical lenses that incorporate the commodity versus fundamental rights view of privacy, respectively. Each of these theoretical lenses looks at information privacy differently and therefore provides a valid basis upon which the factors influencing judgments about the degree of privacy concerns could be reasonably proposed.

Privacy Calculus. One very important perspective views information privacy in terms of an exchange whereby personal information is given in return for certain benefits. This perspective is found in various works which viewed privacy as a *calculus* (Xu, Teo, Tan, & Agarwal, 2010). According to this perspective, Klopfer and Rubenstein (1977), for instance, found that the concept of privacy is not absolute but, rather, can be interpreted in "economic terms" (p.64). That is, individuals often make their decisions about the disclosure of information based on a "calculus of behavior" (Laufer & Wolfe, 1977, p.36). The decision of information disclosure will be made based on a cost-benefit assessment that their personal information will subsequently be used fairly and that they will not suffer negative consequences in the future.

Such privacy calculus perspective is evident in works of analyzing genetic privacy concerns. It has been noted that individuals' evaluations of potential benefits and risks of sharing their genetic data could influence their willingness to participate in medical research trials (B. R. Goldman, 4, 83., 2005). Individuals often consider the nature of the benefit being offered in exchange for information when deciding whether an activity violates their privacy (M. J. Culnan, 1993). The benefits of sharing genetic data could include the provision of more accurate treatment for diseases and illnesses (B. R. Goldman, 4, 83., 2005). The utilization of an individual's genetic profile in prescribing medicines is expected to prevent unwanted side-effects and make drugs work more efficiently (2002). As Goldman (2005) noted, "people should be willing to share their information because they and their family members will reap the future benefits of a database indicating which drugs will cause adverse reactions and which drugs will work most effectively for their conditions" (p. 92). Overall, such calculus perspective of privacy suggests the importance of educating the public with benefits of sharing their genetic information.

Social Contract and Trust. Because privacy has a strong normative component, it is not surprising that the ethical dimensions of information privacy have been discussed in the literature (Ashworth & Free, 2006; Caudill & Murphy, 2000; Foxman & Kilcoyne, 1993). Privacy has been examined in the literature from a number of ethical theoretical perspectives including social contract theory, duty-based theory, stakeholder theory, virtue ethics theory, and the power-responsibility equilibrium model (see Caudill & Murphy, 2000 for a review). Among these ethical theoretical perspectives, the Integrative Social Contract Theory (ISCT) (Donaldson & Dunfee, 1999) is the most widely used ethical theory in the context of privacy. Specifically, a social contract is held to occur when individuals provide personal information to certain organizations or research institutes. For the organization or research institution, one generally understood obligation accruing from this social contract is that it will undertake the responsibility to manage individuals' personal information properly (Caudill & Murphy, 2000).

This implied contract is considered breached if individuals are unaware that their information is being collected, if the organization shares individuals' personal information to a third party without permission, or if the organization uses individuals' personal information for other purposes without notification (M. J. Culnan, 1995). Thus, applying a social contract lens requires individuals' to place their trust on the organizational compliance and the assurance of privacy protection. The trust issue becomes even more salient in the context of genetic privacy, especially for the use of open genome databases. For a personal genome database that is maintained by researchers and physicians, confidentiality implies trust in private and in professional relationships between individuals, which is especially vital to the trust that the public places in physicians, lawyers and members of the research institutions (Lunshof et al., 2008). It has been noted in Robertson (2001) that "protecting human rights and dignity through an open, transparent system of DNA research is essential to engender public trust in the genomic enterprise". Similarly, Lin et al. (2004) highlighted that social concerns for genetic privacy are connected to both beliefs about benefits of pharmacogenomics research and trustworthiness of researchers and governmental agencies.

Privacy by Design (PbD)

The pioneering concept of Privacy by Design (Cavoukian, 2009) ensures the protection of privacy through the use of privacy enhancing features – embedding them into the design specifications of systems, organizational practices and procedures, and physical environments – making privacy the default. Such PbD notion could be employed in the context of protecting genetic privacy, which aims to assure anonymity, confidentiality and secrecy. In addition, informed consent often plays an important role in genomics research.

Technical Approaches in Protecting Genetic Data. Technical solutions are being designed and developed to protect genetic privacy in many ways. Against linking genetic data with other personal identifiable data, technical solutions which are based on the concept of bin size and data perturbation are developed (Cassa, 2008; Vaszar et al., 2003). The 'bin' is the subset of data which meets a set of combined characteristics. For example, the Social Security Administration (SSA) and other government agencies use the rule that any combination of characteristics in a public-use file yields a subset of not fewer than five individuals, whose minimum bin size is five. Thus, a lager minimum bin size is more appropriate given the sensitivity of the information (Vaszar et al., 2003). In some other cases, to eliminate non-essential data fields is a way to perturb data so as to decrease the possibility of identifying individuals.

Besides using bin size idea, there is a variation to aggregate records binning at the patient level instead of characteristic fields within individual records. The Human Genome Project (HGP) used participant pool selection as a privacy technique to assure anonymity (Cassa, 2008). In this approach, after a large number of samples from individuals were gathered, a very small subset was then anonymously selected to create a consensus hybrid of several participant genomes to prevent the identity of participants from being identified (Project, 2010). Other technical approaches to protect genetic privacy include adding noise to a genotypic sequence and synthesizing anonymized "individuals" using statistical data associations (Lasko & Vinterbo, 2010). Similarly, k-anonymization (Sweeney & Sweeney, 2002) and L-diversity (Machanavajjhala, Kifer, Gehrke, & Venkitasubramaniam, 2007) methods have been developed to provide robust protections against re-identification attacks such as dictionary attacks (Malin & Sweeney, 2004). In summary, technical methods to protect genetic privacy mainly lie in assuring data confidentiality and are used to prevent identity revelation. In reviewing these technical solutions, we could see that there is a tradeoff between data privacy and pharmacogenomics research needs. The achievement of data confidentiality such as data perturbation and adding noise will decrease the quality of data for genomics research (Cassa, 2008).

Approaches to Obtain Informed Consent. Several studies have shown that current implementation of the informed consent process appears inadequate for comprehensive genomic analysis (Vaszar et al., 2003). For instance, Weir and Horton (1995) conducted an analysis of 23 consent documents for long-term storage of DNA samples and the results showed that none of the documents is regarded to have adequate description in terms of how confidentiality and privacy would be maintained. In addition, although there are several guidelines for conducting informed consent procedures in genomics area (see Caplan, 2009), there is no standardized language in use for referring to either problems or solutions pertaining to biobanking (Elger & Caplan, 2006; B.M. Knoppers, 2005; B. M. Knoppers & Saginur, 2005). Therefore, novel ways to design informed consent are in need. Arguments and debates have occurred regarding this issue, whereby Caplan (2009) suggests abandoning informed consent and turning to trustworthy third-parties for protecting biological information. Based on the moral principle of veracity, Lunshof et al. (2008) introduces the open consent framework that was implemented in Personal Genome Project. Open consent means that "volunteers consent to unrestricted redisclosure of data originating from a confidential relationship, namely their health records, and to unrestricted disclosure of information that emerges from any future research on their genotype–phenotype data set, the information content of which cannot be predicted" (Lunshof et al., 2008, p.409). In this open consent framework, research volunteers should realize that they are potentially identifiable and that no promises of confidentiality, privacy, or anonymity are guaranteed (Lunshof et al., 2008).

Summary

The debate between fundamental rights versus a commodity view of privacy corresponds to the question on the relative effectiveness of regulation versus technology in ensuring genetic information privacy. Skepticism about the effectiveness of self-regulation in protecting privacy by organizations and research institutions often results in privacy advocates and consumers clamoring for strong and effective legislation to curtail ethics and privacy concerns. Regulatory or legislative action could mandate the implementation of certain design criteria for privacy protection. However, at the society level, although overarching government regulations can enhance consumer trust, regulation may not be socially optimal because of lower profit margins for organizations as well as reactive and outdated nature of law enactment (Tang, Hu, & Smith, 2008). From this perspective, the technological solution may play a significant role in protecting privacy, particularly because of its ability to cross international, political, regulatory, and industry boundaries (Turner & Dasgupta, 2003). In this chapter, we have discussed how ethics and privacy concerns could be addressed by policy (through regulation or legislation) or by design (through privacy enhancing technologies or the informed consent). Figure 1 summarizes the conceptual framework.

CONCLUSION

Predictive and personalized medicine has been promised to change the way future patients receive high-quality cost-effective health care. Driven primarily by advances first in pharmacogenomics and then in systems biology technologies, large amount of genome-wide genotype and clinical data will be collected of individuals. These data will

Figure 1. Conceptual framework: Privacy by policy vs. by design

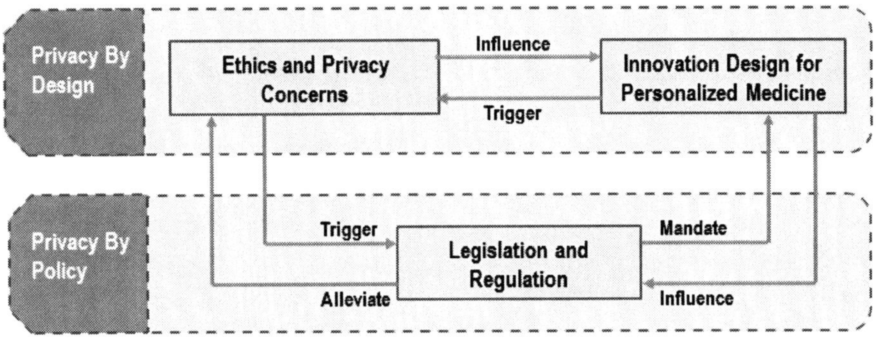

be stored in online electronic databases, often in association of biobanks, and analyzed by biomedical researchers and by health care professionals. In this process, significant biomedical knowledge discovery opportunities have been discussed within the scientific community, which has tend to overlook the serious ethical and privacy concerns over the potential abuse, misuse, and unauthorized use of this information, once it become part of an individual's personal health record. While a pessimistic viewpoint could be held what this could mean to jeopardize personal privacy and socioeconomic balances, we believe that there is chance for the community of researchers, including biomedical researchers, healthcare professionals, bioethics experts, and privacy experts, to come together and define the triangulated relationships between personalized medicine technological innovations, bioethics concerns, and best privacy practices.

ACKNOWLEDGMENT

The authors acknowledge the generous support of the following research centers in helping the writing of this manuscript: Indiana Center for Systems Biology and Personalized Medicine, Indianapolis, IN and Indiana University Center for Bioethics, Indianapolis, IN. Dr. Eric Meslin is supported by the following sources for this work: a grant from the Richard M. Fairbanks Foundation, Indianapolis to the IU Center for Bioethics Program in Predictive Health Ethics Research (PredictER); Grant #UL1RR025761-01, NCRR/NIH: Indiana Clinical and Translational Sciences Institute; the Institute for Advanced Studies, Professor-at-Large Program, University of Western Australia. Dr. Meslin is a consultant to Eli Lilly & Company.

REFERENCES

Acquisti, A., & Grossklags, J. (2005). Privacy and Rationality in Individual Decision Making. *IEEE Security & Privacy*, *3*(1), 26–33. doi:10.1109/MSP.2005.22

Aebersold, R., Auffray, C., Baney, E., Barillot, E., Brazma, A., & Brett, C. (2009). Report on EU-USA workshop: how systems biology can advance cancer research (27 October 2008). *Molecular Oncology*, *3*(1), 9–17. doi:10.1016/j.molonc.2008.11.003

Anderlik, M. R., & Rothstein, M. A. (2001). Privacy and confidentiality of genetic information: what rules for the new science? *Annual Review of Genomics and Human Genetics*, *2*(1), 401–433. doi:10.1146/annurev.genom.2.1.401

Andrews, L. B., Fullarton, J. E., Holtman, N. A., & Motulsky, A. G. (1994). *Assessing genetic risks: implications for health and social policy*. Washington, DC: National Academy Press.

Angell, M. (2000). Is academic medicine for sale? *The New England Journal of Medicine*, *342*(20), 1516–1518. doi:10.1056/NEJM200005183422009

Annas, G. J., Glantz, L. H., & Roche, P. A. (1995). Drafting the Genetic Privacy Act: Science, Policy, and Practical Considerations. *The Journal of Law, Medicine & Ethics*, *23*, 360. doi:10.1111/j.1748-720X.1995.tb01378.x

Anonymous,. (1966). Microfiche system saves time, cuts storage by 98 per cent. *Modern Hospital*, *107*, 66–67.

ASHG. (1996). American Society of Human Genetics ASHG report: statement on informed consent for genetic research. *American Journal of Human Genetics*, *59*, 471–474.

Ashworth, L., & Free, C. (2006). Marketing Dataveillance and Digital Privacy: Using Theories of Justice to Understand Consumers Online Privacy Concerns. *Journal of Business Ethics*, *67*(2), 107–123. doi:10.1007/s10551-006-9007-7

Axler, R. E., Irvine, R., Lipworth, W., Morrell, B., & Kerridge, I. H. (2008). Why might people donate tissue for cancer research? Insights from organ/tissue/blood donation and clinical research. *Pathobiology*, *75*(6), 323–329. doi:10.1159/000164216

Baltimore, D. (2001). Our genome unveiled. *Nature*, *409*(6822), 814–816. doi:10.1038/35057267

Barreiro, L. B., Henriques, R., & Mhlanga, M. M. (2009). High-throughput SNP genotyping: combining tag SNPs and molecular beacons. *Methods in Molecular Biology (Clifton, N.J.)*, *578*, 255–276. doi:10.1007/978-1-60327-411-1_17

Barrows, R. C., & Clayton, P. D. (1996). Privacy, Confidentiality, and Electronic Medical Records. *Journal of the American Medical Informatics Association*, *3*(2), 139–148.

Bates, B. R., Lynch, J. A., & Bevan, J. L. (2005). Condit CM: Warranted concerns, warranted outlooks: a focus group study of public understandings of genetic research. *Social Science & Medicine*, *60*, 331–344. doi:10.1016/j.socscimed.2004.05.012

Bennett, C. J. (1995). *The Political Economy of Privacy: A Review of the Literature*. Hackensack, NJ: Center for Social and Legal Research.

Bernhardt, B. A., Tambor, E. S., Fraser, G., Wissow, L. S., & Geller, G. (2002). Parents' and children's attitudes toward the enrollment of minors in genetic susceptibility research: Implications for informed consent. *American Journal of Medical Genetics*, *116A*(4), 315–323. doi:10.1002/ajmg.a.10040

Beskow, L. M., Burke, W., Merz, J. F., Barr, P. A., & Terry, S., V.B., P., et al. (2001). Informed consent for population-based research involving genetics. *Journal of the American Medical Association*, *286*, 2315–2321. doi:10.1001/jama.286.18.2315

Bok, S. (1989). *Secrets: On the Ethics of Concealment and Revelation*. New York: Vintage.

Brown, S. M. (2002). *Essentials of Medical Genomics*. Hoboken, NJ: John Wiley & Sons, Inc. doi:10.1002/0471483087

Bruse, S., Moreau, M., Azaro, M., Zimmerman, R., & Brzustowicz, L. (2008). Improvements to bead-based oligonucleotide ligation SNP genotyping assays. *BioTechniques*, *45*(5), 559–571. doi:10.2144/000112960

Bui, C. T., Babon, J. J., Lambrinakos, A., & Cotton, R. G. (2003). Detection of mutations in DNA by solid-phase chemical cleavage method. A simplified assay. *Methods in Molecular Biology (Clifton, N.J.)*, *212*, 59–70.

Camp, L. J. (1999). Web security and privacy: An American perspective. [Article]. *The Information Society*, *15*(4), 249–256. doi:10.1080/019722499128411

Campbell, A. V. (2007). The ethical challenges of genetic databases: safeguarding altruism and trust. *King's Law Journal*, *18*, 227–246.

Caplan, A. L. (2009). What No One Knows Cannot Hurt You: The Limits of Informed Consent in the Emerging World of Biobanking. In Solbakk, J. H., Holm, S., & Hofmann, B. (Eds.), *The Ethics of Research Biobanking* (pp. 25–32). Springer. doi:10.1007/978-0-387-93872-1_2

Cassa, C. A. (2008). *Privacy and identifiability in clinical research, personalized medicine, and public health surveillance*: PhD thesis, Massachusetts Institute of Technology.

Caudill, E. M., & Murphy, P. E. (2000). Consumer Online Privacy: Legal and Ethical Issues. *Journal of Public Policy & Marketing*, *19*(1), 7–19. doi:10.1509/jppm.19.1.7.16951

Cavoukian, A. (2009). *Privacy by Design: Take the Challenge*. Ontario, Canada: Information and Privacy Commissioner of Ontario.

Chalmers, D. (2007). International co-operation between biobanks: the case for harmonisation of guidelines and governance. In Stranger, M. (Ed.), *Human Biotechnology & Public Trust: Trends, Perceptions and Regulation* (pp. 237–246). Hobart: Centre for Law and Genetics.

Charo, R. A. (2007). Body of Research - Ownership and use of Human Tissue. *The New England Journal of Medicine*, *355*, 1517–1519. doi:10.1056/NEJMp068192

Chen, J. Y., Yan, Z., Shen, C., Fitzpatrick, D. P., & Wang, M. (2007). A systems biology approach to the study of cisplatin drug resistance in ovarian cancers. *Journal of Bioinformatics and Computational Biology*, *5*(2a), 383–405. doi:10.1142/S0219720007002606

Church, G. M. (2005). The personal genome project. *Mol Syst Biol*, *1*, 2005 0030.

Collins, F. S., Green, E. D., Guttmacher, A. E., & Guyer, M. S. (2003). A vision for the future of genomics research. *Nature*, *422*(6934), 835–847. doi:10.1038/nature01626

Conti, C., Veenstra, D. L., Armstrong, K., Lesko, J. L., & Grosse, S. D. (2010). (in press). Personalized medicine and genomics: challenges and opportunities in assessing effectiveness, cost effectiveness, and future research priorities. *Medical Decision Making*. doi:10.1177/0272989X09347014

Culnan, M. J. (1993). 'How Did They Get My Name'? An Exploratory Investigation of Consumer Attitudes toward Secondary Information Use. *Management Information Systems Quarterly*, *17*(3), 341–364. doi:10.2307/249775

Culnan, M. J. (1995). Consumer Awareness of Name Removal Procedures: Implication for Direct Marketing. *Journal of Interactive Marketing*, *9*, 10–19.

Daikos, G. K. (2007). History of medicine: our Hippocratic heritage. *International Journal of Antimicrobial Agents*, *29*(6), 617–620. doi:10.1016/j.ijantimicag.2007.01.008

Davis, R. (1995). Online medical records raise privacy fears. *USA Today March 22*.

Donaldson, T., & Dunfee, W. T. (1999). *Ties that Bind: A Social Contracts Approach to Business Ethics*. Cambridge, MA: Harvard Business School Press.

Elger, B. S., & Caplan, A. L. (2006). Consent and anonymization in research involving biobanks: Differing terms and norms present serious barriers to an international framework. *EMBO Reports*, *7*(7), 661–666. doi:10.1038/sj.embor.7400740

Emanuel, E., Wendler, D., & Grady, C. (2000). What makes clinical research ethical? *Journal of the American Medical Association*, *283*(20), 2701–2711. doi:10.1001/jama.283.20.2701

Evans, B. J., Flockhart, D. A., & Meslin, E. M. (2004). Creating incentives for genomics research to improve targeting of therapies. *Nature Medicine*, *10*, 1289–1291. doi:10.1038/nm1204-1289

Evans, B. J., & Meslin, E. M. (2006). Encouraging Translational Research Through Harmonization of FDA and Common-Rule Informed Consent Requirements for Research with Banked Specimens. *Journal of Legal Medicine*, *27*, 119–166. doi:10.1080/01947640600716366

Foxman, E. R., & Kilcoyne, P. (1993). Information Technology, Marketing Practice, and Consumer Privacy: Ethical Issues. *Journal of Public Policy & Marketing*, *12*(1), 106–119.

Fullerton, S. M., Anderson, N. R., Guzauskas, G., Freeman, D., & Fryer-Edwards, K. (2010). Meeting the Governance Challenges of Next-Generation Biorepository Research. *Science Translational Medicine*, *2*(15), cm3. doi:10.1126/scitranslmed.3000361

Garrison, F. H. (1917). *An Introduction to the History of Medicine* (2nd ed.). Philadelphia: W.B. Saunders Company.

Gaudet, M., Fara, A. G., Beritognolo, I., & Sabatti, M. (2009). Allele-specific PCR in SNP genotyping. *Methods in Molecular Biology (Clifton, N.J.)*, *578*, 415–424. doi:10.1007/978-1-60327-411-1_26

Giacomini, K. M., Brett, C. M., Altman, R. B., Benowitz, N. L., Dolan, M. E., & Flockhart, D. A. (2007). The pharmacogenetics research network: from SNP discovery to clinical drug response. *Clinical Pharmacology and Therapeutics*, *81*(3), 328–345. doi:10.1038/sj.clpt.6100087

Gilbert, M. T., Sanchez, J. J., Haselkorn, T., Jewell, L. D., Lucas, S. B., & Van Marck, E. (2007). Multiplex PCR with minisequencing as an effective high-throughput SNP typing method for formalin-fixed tissue. *Electrophoresis*, *28*(14), 2361–2367. doi:10.1002/elps.200600589

Goldman, B. R., 4, 83. (2005). Pharmacogenomics: Privacy in the Era of Personalized Medicine. *Northwestern Journal of Technology and Intellectual Property*, *4*(1), 140–143.

Goldman, R. E., Kingdon, C., Wasser, J., Clark, M. A., Goldberg, R., Papandonatos, G. D., et al. (2008). Rhode Islanders' attitudes towards the development of a statewide genetic biobank *Personalized Medicine, 5*(4), 339-359.

Goodman, K. W. (Ed.). (1998). *Ethics, Computing and Medicine: Informatics and the Transformation of Health Care*. New York: Cambridge University Press.

Goodman, K. W., & Miller, R. (2006). Ethics and Health Informatics: Users, Standards and Outcomes. In Shortliffe, E. H., Cimino, J., Garber, A. M., Owens, D. K., Singer, S. J., & Enthoven, A. C. (Eds.), *Medical Informatics: Computer Applications in Health Care and Biomedicine* (3rd ed., pp. 379–402). New York: Springer-Verlag.

Haas, D., Renbarger, J., Meslin, E. M., Drabiak, K., & Flockhart, D. (2008). Patient attitudes toward genotyping in an urban women's health clinic. *Obstetrics and Gynecology*, *112*, 1023–1028. doi:10.1097/AOG.0b013e318187e77f

Habel, L. A., Shak, S., Jacobs, M. K., Capra, A., Alexander, C., & Pho, M. (2006). A population-based study of tumor gene expression and risk of breast cancer death among lymph node-negative patients. *Breast Cancer Research*, *8*(3), R25. doi:10.1186/bcr1412

Harman, L. B. (2001). *Ethical challenges in the management of health information* (1 ed.): Aspen Pub.

Helft, P. R., Champion, V. L., Eckles, R., Johnson, C. S., & Meslin, E. M. (2007). Cancer patients' attitudes toward future research uses of stored human biological materials. *Journal of Empirical Research on Human Research Ethics; JERHRE*, *2*(3), 15–22. doi:10.1525/jer.2007.2.3.15

Helgesson, G., & Swartling, U. (2008). Views on data use, confidentiality and consent in a predictive screening involving children. *Journal of Medical Ethics*, *34*, 206–209. doi:10.1136/jme.2006.020016

Hobbs, M. S., & McCall, M. G. (1970). Health statistics and record linkage in Australia. *Journal of Chronic Diseases*, *23*(5), 375–381. doi:10.1016/0021-9681(70)90020-2

Hodge, J. G., Gostin, L. O., & Jacobson, P. D. (1999). Legal Issues Concerning Electronic Health Information: Privacy, Quality, and Liability. *Journal of the American Medical Association*, *282*(15), 1466–1471. doi:10.1001/jama.282.15.1466

Holman, C. D. J., Bass, A. J., Rouse, I. L., & Hobbs, M. S. T. (1999). Population-based linkage of health records in Western Australia: development of a health services research linked database. *Australian and New Zealand Journal of Public Health*, *23*, 453–459. doi:10.1111/j.1467-842X.1999.tb01297.x

Holtzman, N. A., & Watson, M. S. (Eds.). (1998). *Promoting safe and effective genetic testing in the United States: final report of the task force on genetic testing*. Baltimore: Johns Hopkins University Press.

Homer, N., Szelinger, S., Redman, M., Duggan, D., Tembe, W., & Muehling, J. (2008). Resolving individuals contributing trace amounts of DNA to highly complex mixtures using high-density SNP genotyping microarrays. *PLOS Genetics*, *4*(8), e1000167. doi:10.1371/journal.pgen.1000167

Hood, L., Heath, J. R., Phelps, M. E., & Lin, B. (2004). Systems biology and new technologies enable predictive and preventative medicine. *Science*, *306*(5696), 640–643. doi:10.1126/science.1104635

Huan, T., Wu, X., & Chen, J. Y. (2010). (in press). Systems Biology Visualization Tools for Drug Target Discovery. *Expert Opinion on Drug Discovery*. doi:10.1517/17460441003725102

Humgen. (2010). http://www.humgen.org/int/GB2_p.cfm?mod=1.

IOM. (2001). *Preserving public trust: accreditation and human research participant protection programs*. Washington, DC: Institute of Medicine.

Kalichman, M. (2007). Responding to challenges in educating for responsible conduct of research. *Academic Medicine*, *82*, 870–875. doi:10.1097/ACM.0b013e31812f77fe

Kaye, J., & Stranger, M. (Eds.). (2009). *Principles and Practice in Biobank Governance*. Surrey, UK: Ashgate.

Kelman, C. W., Bass, A. J., & Holman, C. D. (2002). Research use of linked health data--a best practice protocol. *Australian and New Zealand Journal of Public Health*, *26*, 251–255. doi:10.1111/j.1467-842X.2002.tb00682.x

Kitano, H. (2002). Systems biology: a brief overview. *Science*, *295*(5560), 1662–1664. doi:10.1126/science.1069492

Klein, T. E., Altman, R. B., Eriksson, N., Gage, B. F., Kimmel, S. E., & Lee, M. T. (2009). Estimation of the warfarin dose with clinical and pharmacogenetic data. *The New England Journal of Medicine*, *360*(8), 753–764. doi:10.1056/NEJMoa0809329

Klopfer, P. H., & Rubenstein, D. L. (1977). The concept privacy and its biological basis. *The Journal of Social Issues*, *33*, 52–65. doi:10.1111/j.1540-4560.1977.tb01882.x

Knoppers, B. M. (2005). Consent revisited: points to consider. *Health Law Review*, *13*(2/3), 33–38.

Knoppers, B. M., & Saginur, M. (2005). The Babel of genetic data terminology. *Nature Biotechnology*, *23*(8), 925–927. doi:10.1038/nbt0805-925

Korkko, J., Milunsky, J., Prockop, D. J., & Ala-Kokko, L. (1998). Use of conformation sensitive gel electrophoresis to detect single-base changes in the gene for COL10A1. *Human Mutation*, (Suppl 1), S201–S203.

Kricka, L. J., Master, S. R., Joos, T. O., & Fortina, P. (2006). Current perspectives in protein array technology. *Annals of Clinical Biochemistry*, *43*(Pt 6), 457–467. doi:10.1258/000456306778904731

Kwok, P. Y., & Duan, S. (2003). SNP discovery by direct DNA sequencing. *Methods in Molecular Biology (Clifton, N.J.)*, *212*, 71–84.

Lasko, T. A., & Vinterbo, S. A. (2010). Spectral Anonymization of Data. *IEEE Transactions on Knowledge and Data Engineering*, *22*(3), 437–446. doi:10.1109/TKDE.2009.88

Laufer, R. S., & Wolfe, M. (1977). Privacy as a Concept and a Social Issue - Multidimensional Developmental Theory. *The Journal of Social Issues*, *33*(3), 22–42. doi:10.1111/j.1540-4560.1977.tb01880.x

Leiman, D. A., Lorenzi, N. M., Wyatt, J. C., Doney, A. S. F., & Rosenbloom, S. T. (2008). US and Scottish Health Professionals' Attitudes toward DNA Biobanking. *Journal of the American Medical Informatics Association*, *15*(3), 357–362. doi:10.1197/jamia.M2571

Lesko, L. J., Salerno, R. A., Spear, B. B., Anderson, D. C., Anderson, T., & Brazell, C. (2003). Pharmacogenetics and pharmacogenomics in drug development and regulatory decision making: Report of the first FDA-PWG-PhRMA-DruSafe Workshop. *Journal of Clinical Pharmacology*, *43*, 342–358. doi:10.1177/0091270003252244

Lin, Z., Owen, A. B., & Altman, R. B. (2004). GENETICS: Genomic Research and Human Subject Privacy. *Science*, *305*(5681), 183. doi:10.1126/science.1095019

Lunshof, J. E., Chadwick, R., Vorhaus, D. B., & Church, G. M. (2008). From genetic privacy to open consent. *Nature Reviews. Genetics*, *9*(5), 406–411. doi:10.1038/nrg2360

Machanavajjhala, A., Kifer, D., Gehrke, J., & Venkitasubramaniam, M. (2007). L-diversity: Privacy beyond k-anonymity. *ACM Trans. Knowl. Discov. Data*, *1*(1), 3. doi:10.1145/1217299.1217302

Malin, B. A., & Sweeney, L. (2004). How (not) to protect genomic data privacy in a distributed network: using trail re-identification to evaluate and design anonymity protection systems. *Journal of Biomedical Informatics*, *37*(3), 179–192. doi:10.1016/j.jbi.2004.04.005

Marks, A. D., & Steinberg, K. K. (2002). The Ethics of Access to Online Genetic Databases: Private or Public? *American Journal of Pharmacogenomics*, *2*, 207–212. doi:10.2165/00129785-200202030-00006

Marx, G. T. (1999). What's in a name? Some reflections on the sociology of anonymity. [Article]. *The Information Society*, *15*(2), 99–112. doi:10.1080/019722499128565

Meslin, E. M. (2010). The Value of Using Top-Down and Bottom-Up Approaches for Building Trust and Transparency in Biobanking. *Public Health Genomics, 13*(4).

Meslin, E. M., & Goodman, K. W. (2010). An Ethics and Policy Agenda for Biobanks and Electronic Health. *Science Progress, February 25th*, http://www.scienceprogress.org/2010/2002/bank-on-it/.

Meslin, E. M., & Quaid, K. A. (2004). Ethical Issues in the Storage, Collection and Research Use of Human Biological Materials. *The Journal of Laboratory and Clinical Medicine*, *144*, 229–234. doi:10.1016/j.lab.2004.08.003

Murphy, J., Scott, J., Kaufman, D., Geller, G., LeRoy, L., & Hudson, K. (2009). Public Perspectives on Informed Consent for Biobanking. *American Journal of Public Health*, *99*(12), 2128–2134. doi:10.2105/AJPH.2008.157099

Naylor, S., & Chen, J. Y. (2010). Unraveling Human Complexity and Disease with Systems Biology and Personalized Medicine. *Personalized Medicine*, *7*(3), 275–289. doi:10.2217/pme.10.16

NBAC. (1999). *Research involving human biological materials: ethical issues and policy guidance, vol I: report and recommendations*. Bethesda, MD: National Bioethics Advisory Commission.

NBAC. (2001). *Ethical and Policy Issues in Research Involving Human Participants*. Bethesda, MD: National Bioethics Advisory Commission.

Nicol, D., & Critchley, C. (2009). What Benefit Sharing Arrangements Do People Want From Biobanks? A Survey of Public Opinion in Australia. In Kaye, J., & Stranger, M. (Eds.), *Principles and Practice in Biobank Governance* (pp. 17–31). Surrey, UK: Ashgate.

Norton, P. G., Bain, J., Birtwhistle, R., Davis, D., & Dunn, E. CP, H., et al. (1994). Guidelines for the Dissemination of New Information Discovered by Researchers. In M. A. Stewart, P. G. Norton, M. Bass, E. Dunn & F. Tudiver (Eds.), *Disseminating Research/Changing Practice. Research Methods for Primary Care* (Vol. 6, pp. 87-94). Thousand Oaks, CA: Sage Publications.

Palsson, B. (2002). In silico biology through "omics". *Nature Biotechnology*, *20*(7), 649–650. doi:10.1038/nbt0702-649

Palsson, B. (2006). *Systems biology: properties of reconstructed networks*. Cambridge, New York: Cambridge University Press. doi:10.1017/CBO9780511790515

Pellegrino, E. D., & Thomasma, D. C. (1984). *For the Patient's Good: The Restoration of Beneficence in Health Care*. New York: Oxford.

Plant, N. (2007). The human cytochrome P450 subfamily: transcriptional regulation, inter-individual variation and interaction networks. *Biochimica et Biophysica Acta*, *1770*(3), 478–488.

Project, H. G. (2010). *Human Genome Project*: www.ornl.gov/hgmis/home.shtml.

Pusch, W., Wurmbach, J. H., Thiele, H., & Kostrzewa, M. (2002). MALDI-TOF mass spectrometry-based SNP genotyping. *Pharmacogenomics*, *3*(4), 537–548. doi:10.1517/14622416.3.4.537

Ramsey, P. (1960). *For the Patient's Good*. Princeton University Press.

Robertson, J. A. (2001). Consent and privacy in pharmacogenetic testing. *Nature Genetics*, *28*(3), 207–209. doi:10.1038/90032

Ronaghi, M. (2003). Pyrosequencing for SNP genotyping. *Methods in Molecular Biology (Clifton, N.J.)*, *212*, 189–195.

Rothstein, M. A., & Epps, P. G. (2001). Ethical and legal implications of pharmacogenomics. *Nature Reviews. Genetics*, *2*(3), 228–231. doi:10.1038/35056075

Saha, S., Harrison, S. H., & Chen, J. Y. (2009). Dissecting the human plasma proteome and inflammatory response biomarkers. *Proteomics*, *9*(2), 470–484. doi:10.1002/pmic.200800507

Saha, S., Harrison, S. H., Shen, C., Tang, H., Radivojac, P., & Arnold, R. J. (2008). HIP2: An online database of human plasma proteins from healthy individuals. *BMC Medical Genomics*, *1*, 12. doi:10.1186/1755-8794-1-12

Schoeman, F. D. (1984). *Philosophical Dimensions of Privacy: an Anthology*. Cambridge, New York: Cambridge University Press. doi:10.1017/CBO9780511625138

Schwartz, P. H., & Meslin, E. M. (2008). The Ethics of Information: Absolute Risk Reduction and Patient Understanding of Screening. *Journal of General Internal Medicine*.

Shapero, M. H., Leuther, K. K., Nguyen, A., Scott, M., & Jones, K. W. (2001). SNP genotyping by multiplexed solid-phase amplification and fluorescent minisequencing. *Genome Research*, *11*(11), 1926–1934.

Sherry, S. T., Ward, M. H., Kholodov, M., Baker, J., Phan, L., & Smigielski, E. M. (2001). dbSNP: the NCBI database of genetic variation. *Nucleic Acids Research, 29*(1), 308–311. doi:10.1093/nar/29.1.308

Stanley, F. J., Croft, M. L., Gibbins, J., & Read, A. W. (2008). A population database for maternal and child health research in Western Australia using record linkage. *Paediatric and Perinatal Epidemiology, 8*(4), 433–447. doi:10.1111/j.1365-3016.1994.tb00482.x

Stanley, F. J., & Meslin, E., M. (2007). Australia Needs a Better System for Health Care Evaluation. *The Medical Journal of Australia, 186*, 220–221.

Suh, Y., & Vijg, J. (2005). SNP discovery in associating genetic variation with human disease phenotypes. *Mutation Research, 573*(1-2), 41–53. doi:10.1016/j.mrfmmm.2005.01.005

Sweeney, L., & Sweeney, L. (2002). Achieving K-Anonymity Privacy Protection Using Generalization and Suppression. *International Journal on Uncertainty, Fuzziness and Knowledge-based Systems, 10*, 2002.

Tahira, T., Suzuki, A., Kukita, Y., & Hayashi, K. (2003). SNP detection and allele frequency determination by SSCP. *Methods in Molecular Biology (Clifton, N.J.), 212*, 37–46.

Tang, Z., Hu, Y. J., & Smith, M. D. (2008). Gaining Trust Through Online Privacy Protection: Self-Regulation, Mandatory Standards, or Caveat Emptor. *Journal of Management Information Systems, 24*(4), 153–173. doi:10.2753/MIS0742-1222240406

Tchabo, N. E., Liel, M. S., & Kohn, E. C. (2005). Applying proteomics in clinical trials: assessing the potential and practical limitations in ovarian cancer. *American Journal of Pharmacogenomics, 5*(3), 141–148. doi:10.2165/00129785-200505030-00001

Tefft, S. K. (1980). *Secrecy, a Cross-Cultural Perspective*. New York, N.Y.: Human Sciences Press.

Trutwein, B., Holman, C. D., & Rosman, D. L. (2006). Health data linkage conserves privacy in a research-rich environment. *Annals of Epidemiology, 16*(4), 279–280. doi:10.1016/j.annepidem.2005.05.003

Turner, C. E., & Dasgupta, S. (2003). Privacy on the Web: An Examination of User Concerns, Technology, and Implications for Business Organizations and Individuals. *Information Systems Management*, (Winter): 8–18. doi:10.1201/1078/43203.20.1.20031201/40079.2

Twyman, R. M. (2004). SNP discovery and typing technologies for pharmacogenomics. *Current Topics in Medicinal Chemistry, 4*(13), 1423–1431. doi:10.2174/1568026043387656

Vasgird, D. (2007). Prevention Over Cure: The Administrative Rationale for Education in the Responsible Conduct of Research. *Academic Medicine, 82*, 835–837. doi:10.1097/ACM.0b013e31812f7e0b

Vaszar, L. T., Cho, M. K., & Raffin, T. A. (2003). Privacy issues in personalized medicine. *Pharmacogenomics, 4*(2), 107–112. doi:10.1517/phgs.4.2.107.22625

Veatch, R. M. (1981). *A Theory of Medical Ethics*. New York: Basic Books.

Visscher, P. M., & Hill, W. G. (2009). The limits of individual identification from sample allele frequencies: theory and statistical analysis. *PLOS Genetics, 5*(10), e1000628. doi:10.1371/journal.pgen.1000628

Warren, S. D., & Brandeis, D. L. (1890). The Right to Privacy. *Harvard Law Review, 4*(5), 193–220. doi:10.2307/1321160

Weir, R. F., & Horton, J. R. (1995). DNA banking and informed consent -- part 2. *IRB, 17*(5-6), 1–8.

Westin, A. F. (1967). *Privacy and Freedom*. New York: Atheneum.

Wolf, V. A. d., Sieber, J. E., Steel, P. M., & Zarate, A. O. (2005). Part I: What Is the Requirement for Data Sharing? *IRB: Ethics and Human Research*, *27*(6), 12–16. doi:10.2307/3563537

Wolford, J. K., Blunt, D., Ballecer, C., & Prochazka, M. (2000). High-throughput SNP detection by using DNA pooling and denaturing high performance liquid chromatography (DHPLC). *Human Genetics*, *107*(5), 483–487. doi:10.1007/s004390000396

Xu, H., Teo, H. H., Tan, B. C. Y., & Agarwal, R. (2010). The Role of Push-Pull Technology in Privacy Calculus: The Case of Location-Based Services. *Journal of Management Information Systems*, *26*(3), 137–176.

Zwick, D., & Dholakia, N. (2004). Whose Identity Is It Anyway? Consumer Representation in the Age of Database Marketing. *Journal of Macromarketing*, *24*(1), 31–43. doi:10.1177/0276146704263920

ADDITIONAL READING

Beauchamp, T. L., & Childress, J. F. (2008). *Principles of Biomedical Ethics*. Oxford University Press.

Bolouri, H. (2009). *Personal Genomics and Personalized Medicine*. Imperial College Press.

Etzioni, A. (1999). *The Limits of Privacy*. New York: Basic Books.

Gottweis, H., & Petersen, A. (2008). *Biobanks: Governance in Comparative Perspective*. Routledge.

Solove, D. J. (2008). *Understanding Privacy*. Cambridge, MA: Harvard University Press.

Waldo, J., Lin, H., & Millett, L. I. (2007). *Engaging Privacy and Information Technology in a Digital Age*. National Academies Press.

Weber, W. (2008). *Pharmacogenetics*. Oxford University Press.

Westin, A. F. (2003). Social and Political Dimensions of Privacy. *The Journal of Social Issues*, *59*(2), 431–453. doi:10.1111/1540-4560.00072

KEY TERMS AND DEFINITIONS

Biobank: Refers to a research resource or archive repository of biological samples taken from different individuals or species.

Bioethics: Is the philosophical study of the ethical controversies brought about by advances in biology and medicine. Bioethicists are concerned with the ethical questions that arise in the relationships among life sciences, biotechnology, medicine, politics, law, and philosophy.

Electronic Health Record (EHR): Is systematic collection of electronic health information about individual patients or populations. It is a digital record that may be access by different health care providers. EHRs may include patient individual or summarized information such as demographics, medical history, medication and allergies, immunization status, laboratory test results, radiology images, and billings.

Genetic Information Privacy: Is defined as an individual's right to protection from non-voluntary disclosure of genetic information.

Informed Consent: Is a legal phrase to indicate that a person has given consent based upon a clear appreciation and understanding of the facts, implications, and future consequences of an action.

Personalized Medicine: Is a medical model that emphasizes the systematic use of information about individual patients to select or optimize that patient's health care. Broadly, it can also be defined as products and services that leverage the science of omics and systems biology to empower consumer-centric health, wellness, and tailored medical care. It is often referred to as "the right treatment for the right person at the right time."

Pharmacogenetics: Is the branch of pharmacology which examines the influence of genetic variation on drug response, primiarly one gene at a time.

Pharmacogenomics: Is the whole genome application of pharmacogenetics, which examine the large-scale influence of genetic variation on drug response in patients by correlating genome-scale gene expression or single-nucleotide polymorphisms with a drug's efficacy or toxicity. By doing so, pharmacogenomics aims to develop rational means to optimize drug therapy to o ensure maximum efficacy with minimal adverse effects.

Privacy by Design (PbD): Ensures the protection of privacy through the use of privacy enhancing features – embedding them into the design specifications of systems, organizational practices and procedures, and physical environments – making privacy the default.

Single-Nucleotide Polymorphism (SNP): Refers to a DNA sequence variation that occurs when a single nucleotide — A, T, C, or G — in the genome (or other shared sequence) differs between members of a species or paired chromosomes in an individual.

Chapter 2
Virtual Screening:
An Overview on Methods and Applications

Khaled H. Barakat
University of Alberta, Canada

Jonathan Y. Mane
University of Alberta, Canada

Jack A. Tuszynski
University of Alberta, Canada

ABSTRACT

Virtual screening, or VS, is emerging as a valuable tool in discovering new candidate inhibitors for many biologically relevant targets including the many chemotherapeutic targets that play key roles in cell signaling pathways. However, despite the great advances made in the field thus far, VS is still in constant development with a relatively low success rate that needs to be improved by parallel experimental validation methods. This chapter reviews the recent advances in VS, focusing on the range and type of computational methods and their successful applications in drug discovery. The chapter also discusses both the advantages and limitations of the various techniques used in VS and outlines a number of future directions in which the field may progress.

INTRODUCTION

Once, a US General summarized his philosophy on warfare in just four concise statements, "The art of war is simple enough. Find out where your enemy is. Get at him as soon as you can. Strike him as hard as you can, and keep moving." Although these overarching statements formed the basic premise of modern war strategies, the same concepts have been applied in designing new drugs aimed at combating a broad range of diseases. In this context, rational drug design (Mandal et al., 2009) has been established as an exciting research approach aimed at developing safer and more efficacious drugs. The ultimate goal of this research is to design small organic non-peptidic compounds that bind to a specific molecular target, and result in the inhibition (or less frequently, activation) of a particular protein or enzyme involved in a given cellular pathway.

DOI: 10.4018/978-1-60960-491-2.ch002

The development of such drugs has been recognized early on by the pharmaceutical industry as a principal foundation that provides it with the necessary return on investment to fuel further research and development (Szymkowski, 2005) leading to a discovery and development cycle.

Our understanding of cell mechanisms and pathways becomes deeper and clearer every day. This is largely due to the great efforts and hard work of genomic and proteomic research groups who add novel targets for drug intervention on a regular basis (Drews, 2000; Fishman & Porter, 2005; Hopkins & Groom, 2002). Thus far, several hundred proteins have been synthetically reproduced and many of them are currently evaluated for their druggability (Hopkins & Groom, 2002). These targets involve several families comprised of G-protein coupled receptors (GPCRs), ligand-gated ion channels (LGICs), cytoskeleton proteins, phosphatases, kinases, nuclear receptors (NRs) and DNA repair proteins. The growing list of potential drug targets encourages a bold question if it is in principle possible to restore any diseased cell to a healthy state by uncovering a drug for every potential druggable target? Certainly, if this dream is ever realized, many diseases will be cured and relegated to the dustbin of history in a manner similar to the effect of the discovery of vaccines in the 19th and early 20th century.

Without a doubt, developing a new drug is a highly structured and expensive route that begins with the identification of the target and concludes with a phase III clinical trial followed by marketing (Fishman & Porter, 2005). A candidate drug may never materialize into a safe and efficacious medicine due to its failure to comply with stringent requirements at any stage of the drug discovery process. The further a potential drug progresses in the development process, the more costly its failure becomes. Accordingly, it is important to reduce the probability of late-stage expensive failures by identifying a diversity of lead compounds that are suitable for structural optimizations. Throughout the last two decades, experimental high throughput screening (HTS) and combinatorial chemistry formed the principal source for lead identification. However, as these approaches are particularly expensive and require considerable resources in terms of equipment and skills of the highly qualified personnel (Lahana, 1999), it is vital to search for an alternative or a complementary low-cost technique that aids in the discovery of new bioactive compounds while maintaining the high yield and rapidity of HTS. More recently, a new trend was born which has been named computational virtual screening (VS) or *in silico* screening (Stahura & Bajorath, 2004; Zoete, Grosdidier, & Michielin, 2009).

While the fundamentals of VS have been accumulated from a few studies in the early nineties (DesJarlais et al., 1990; Kuntz, 1992; Kuntz et al., 1982), the term "virtual screening" was first used by Horvath in 1997 in his study that led to the discovery of new Trypanothione Reductase inhibitors (Horvath, 1997). These pioneering efforts defined the overall concept of a typical VS computational protocol as "searching for bioactive molecules within large compound databases". These molecules are predicted to complement a specific binding site of a particular molecular target in terms of parameters such as shape, charge, the number of hydrogen-bond donors/acceptors, and several additional biochemical characteristics. Over the past decade, the method has undergone immense improvements and gained popularity as a result of an exponential increase in the performance of computer hardware, more efficient algorithms and methods as well as vastly enhanced human expertise. Driven by the combined efforts of many research groups these advances have been directed toward increasing the accuracy in selecting active compounds and amplifying hit (active) rates while keeping the computational cost as low as possible (Abagyan & Totrov, 2001; Schneider & Bohm, 2002; Shoichet et al., 1993). Currently, VS is considered to be a valuable prototype within the rational drug design tool box, helping in prioritizing compounds for experimental HTS as well as aiding in compound progression through lead optimization. Indeed, since its development, *in silico* methods have added at least 50 new compounds to the successful drug candidate arsenal that progressed through clinical trials, some of them are presently FDA-approved drugs (Jorgensen, 2004).

Once the structure of a target (typically a protein) is available, docking algorithms can be used to place each ligand (i.e. a molecule or a molecular fragment included in a typical library of compounds) and predict its most probable binding mode within the binding site of the target (Abagyan & Totrov, 2001; Schneider & Bohm, 2002). Moreover, most docking programs can rank the activity of each compound by analyzing the different ligand-target interactions and estimating the binding affinity of the complex. In addition to docking techniques, one can define the essential interactions between the ligand and the binding site of the receptor and translate this information into the formulation of binding-site pharmacophore models (Good at al., 2000). These models can be used to search the available chemical space for compounds that can complement the physico-chemical features of the receptor. As these two procedures require a comprehensive understanding of the structural arrangement of the target, they have been commonly termed as structure-based virtual screening (SBVS). On the other hand, and for most of the cases, the three-dimensional structure of the target, the binding site or even the target itself are not accurately known, although there may be a number of known active compounds that have been identified experimentally. In this case, data mining algorithms can be used to screen for compounds that are structurally similar to the known actives (similarity search) (Willett, 2006), comprise the chemical features of these compounds (pharmacophore search) (Krovat et al., 2005), or their structures are correlated with their bioactivity (quantitative structure–activity relationships) (Free et al., 1964), in what is called ligand-based virtual screening (LBVS). Thus, these two fundamental procedures, SBVS and LBVS, form the general layout of present-day VS protocols.

This review summarizes the recent advances in VS, focusing on the methods and their successful applications in drug discovery. Moreover, we will sum up the limitations of the various techniques used in VS pointing out to future directions of the field. First, we will describe SBVS which will be followed by an overview of LBVS.

STRUCTURE-BASED VIRTUAL SCREENING (SBVS)

SBVS requires the knowledge of the three-dimensional structure of the target protein (Lyne, 2002; Schneider & Bohm, 2002; Shoichet et al., 1993). Therefore, an essential step prior to the actual screening procedure is to generate or predict the structure of the target protein. The target structure can be obtained by experimental techniques such as NMR, X-ray crystallography, electron crystallography or it can be predicted computationally using homology modeling. In addition to the 3D-structure, it is also important to identify the relevant binding site(s) within the protein which is (are) deemed responsible for its biological activity. Generally, the binding site is a pocket, a groove or a protrusion having an assortment of apparent hydrogen bond donors and acceptors, hydrophobic features and it can be associated with molecular adherence surfaces. There may be a number of metal ions or water molecules as part of the active site that are essential for the activity of the protein and they must be considered during the screening procedure in order to produce a correct result. There are two basic approaches for SBVS namely, docking (Abagyan & Totrov, 2001) and receptor-based pharmacophore modeling (Good et al., 2000).

Docking

Molecular docking is a standard constituent of many SBVS studies described in the literature (Abagyan & Totrov, 2001; Halperin et. al, 2002). The idea of docking and scoring as a virtual screening tool has been proposed since the birth of docking methods (DesJarlais et al., 1990; Kuntz, 1992; Kuntz et al., 1982). The main problem which all docking algorithms try to solve can be stated as follows: given two interacting molecular structures, what is the most probable binding configuration to form a stable three-dimensional protein-ligand complex? In order to address this problem, docking procedures can be divided into two major steps. First, explore the conformational space of the ligand within the binding site of the target. At this stage, many conformations are generated for the ligand. Second, select the optimal target-ligand alignment by scoring their interactions and ranking the docking results (poses) according to their predicted binding affinity. Today, there are at least 30 docking programs commercially (or freely) available with different conformational sampling algorithms and a variety of scoring functions (Bissantz, Folkers, & Rognan, 2000). The most commonly used programs are AUTODOCK (Goodsell & Olson, 1990), GOLD (Jones et. al, 1995b), GLIDE (Friesner et al., 2004), DOCK (Kuntz et al., 1982), ICM (Totrov & Abagyan, 1997), IFREDA (McGann at. al, 2003) and FlexX (Rarey at. al, 1996). These programs differ mostly in the way they deal with protein-ligand flexibility or their scoring and ranking methods.

Regardless of the poor representation of the receptor flexibility during docking, most docking methods can handle the flexibility of ligands efficiently (Bissantz et al., 2000). In other words, for most of the cases, docking algorithms can reproduce the protein-ligand binding modes that have been observed experimentally using X-ray crystallography. As an example, Figure 1a shows the successful docking of nutlin, a well-known p53-MDM2 inhibitor, to the p53-binding site within MDM2 using AUTODOCK 4.0. In general, the degree of success for docking methods can be measured by comparing the predicted binding mode to the experimental conformation (the native binding mode) (Goodsell & Olson, 1990; Jones et al., 1995b; McGann et al., 2003). This assessment can be evaluated quantitatively by calculating the root-mean-square deviation (RMSD) between the two structures. However, in certain systems, where unexpected flexibility of the receptor is crucial for the binding reaction or the interaction of the ligand and protein is mediated by water molecules or metal ions, docking may fail to predict the correct binding conformation of the complex, leading to improper and unrealistic interactions.

Ligand Flexibility

The binding reaction between a ligand and a particular target involves numerous conformational changes in the two molecules as well as water molecules and ions located in their interface. Each entity in this reaction adapts its shape and distribution in order to maximize its interactions with the other entities, forcing the whole system to reach the global minimum. This binding interaction is somehow similar to a folding problem of a protein, comprising a huge number of degrees of freedom. Consequently, the majority of docking programs avoid this conformational flare-up problem by implementing almost full-flexibility for the ligands while keeping the target completely rigid with no flexibility allowed (Abagyan & Totrov, 2001; Bissantz et al., 2000). According to the nature of the searching method, the conformational sampling techniques that have been adopted by all docking programs to introduce ligand flexibility can be classified into three main categories: (1) systematic search routines, (2) stochastic exploration, and (3) simulation techniques.

Figure 1. Structural variations between MDM2 (yellow) and MDMX (red) and their effect on the binding modes of nutlin-3 (a) and two selected hits from the predicted MDM2/MDMX inhibitors (b and c). Tyr100 and Leu99 of MDM2 and the same residues in MDMX are shown in licorice representations with the same color as that of the two proteins. For each compound, the binding mode within MDM2 is shown in green and that within MDMX is shown in gray. Tyr99 and Leu98 prevent nutlin-3 from binding to MDMX with the same binding conformation adopted by nutlin-2 within the MDM2-pocket (blue). The conformation of nutlin-2 was extracted from the MDM2-nutlin crystal structure 1RV1. On the other hand, compounds Pub#11952782 (b) and ZINC04629876 (c) from the suggested MDM2/MDMX inhibitor list can tolerate the structural variations in the two binding sites in order to maximize their interactions with the proteins.

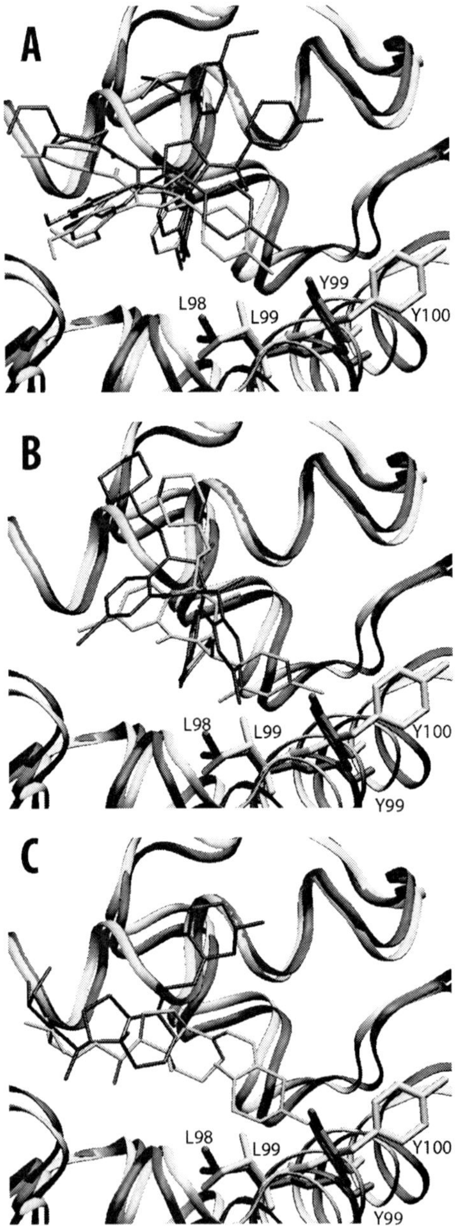

Systematic Search

In a typical systematic search, all rotatable bonds in the ligand are gradually rotated in order to cover all possible combinations among the dihedral angles. Evidently, the number of generated structures using this method increases dramatically with the number of rotatable bonds involved, leading to the problem of combinatorial explosion. In this way, applying a standard systematic search to explore the entire conformational space of a ligand requires massive calculations and considerable computational time. Although, a number of docking programs such as FLOG (Miller et. al, 1994) get around this hindrance by limiting the created structures to a pre-generated set of conformations recorded in structural databases, other docking algorithms adopt an incremental procedure to reconstruct the ligand within the binding site of the target. The main objective of these methods is to limit the number of degrees of freedom for the ligand, allowing for a less-expensive and rapid conformational search.

Essentially, there are two main approaches for the incremental reconstruction methods. First is the one that has been employed by LUDI (Bohm, 1992), FlexX (Rarey et al., 1996), DOCK (Kuntz et al., 1982), ADAM (Mizutani et al., 1994) and Hammerhead (Welch at al., 1996), where the ligand is split into a rigid core fragment that is docked first and a number of flexible regions that are subsequently and successfully added. This method is commonly referred to as the "incremental approach". The other method, known as "place and join", is to break the ligand into several fragments, dock them within the binding site of the target and finally connect them together in order to rebuild the final ligand conformation.

Stochastic Exploration

Stochastic exploration samples the conformational space of a ligand by generating random variations in the orientation of all rotatable bonds and in some cases random translations for the whole ligand within the binding site. This is done mainly to enable crossing the energy barriers and searching for local minima enclosed by the rugged energy surface of the ligand. This procedure can be applied to a single ligand or a population of conformations derived from the same molecular structure of the ligand. Each resultant conformation is then evaluated according to a probability distribution or by estimating its binding affinity with respect to the target. In fact, there are three methods that are derived from this technique, namely: Monte Carlo simulations, Genetic Algorithms and Tabu Search methods (Schneider & Bohm, 2002).

Monte Carlo (MC) simulations are one of the most powerful techniques ever developed to allow for overcoming potential energy barriers and sampling the conformational space of a typical system. Since this method has been successfully implemented to solve several molecular modeling problems, it was regarded as one of the natural options to explore ligand flexibility in docking. The method starts with a randomly generated conformation for the ligand by randomly changing one or more dihedral angles or even the whole orientation or position of the ligand with respect to the target. This new conformation (pose) is accepted or rejected according to a Metropolis algorithm that follows the Boltzmann probability distribution. Programs like ICM (Totrov & Abagyan, 1997), MCDOCK (Liu & Wang, 1999) and DockVision (Hart & Read, 1992) employ this approach in exploring the flexibility of ligands during docking.

Genetic algorithms (GA) exploit the biological concepts introduced by Darwin in order to explore all possible conformations of the ligand and predict its native structure. In contrast to MC-based algorithms, instead of manipulating a single ligand, GA generates a random population of the same molecular structure of the ligand (Goodsell & Olson, 1990; Jones et al., 1995b). Each member of this population is unique in terms of the internal orientation and the global placement and alignment within the binding site of the target. This random population forms the initial generation (seed) of a set of non-interacting ligand species. These poses are further subjected to a number of biological operators that add up more diversity to the generated structures. Among these operators are the mutation operator (generates new ligands from earlier ones by altering a rotatable bond or moving the whole ligand to a new position), and the crossover operator (merges two ligands in order to create a new structure comprising their common features). The fitness of each newly generated structure is evaluated by calculating its binding affinity to the target. The pose which retains the most predominant interactions with the binding site survives and becomes the parent of the new generation. This iterative procedure terminates after reaching a predefined number of generations or energy evaluations, or if no more improvement to the binding affinity has been observed (converged solution). Examples of programs that incorporate genetic algorithms in conformational sampling include AUTODOCK, GOLD and DARWIN (Taylor & Burnett, 2000).

As a memory-based stochastic exploration method, Tabu Search (TS) prevents the searching machinery from revisiting the same conformation more than once. PRO_LEADS is one of the most popular programs that employ this searching technique (Baxter et al., 1998). This is generally achieved by creating a list that records all previously visited solutions, which acts as a memory for the algorithm. A decision to accept or reject a new conformation is made after comparing its RMSD to the other recorded conformations in order to declare that no conformation is visited twice.

Simulation Techniques

Simulation techniques employ a deterministic approach that either: (1) passes through both time and space giving rise to an evolving trajectory describing the biological behavior of a typical system, or (2) reconfigures the system by rearranging its particle composition towards a more stable state. Molecular dynamics (MD) simulations and energy minimization methods are the most widely used techniques in a number of docking programs. Although the two approaches can handle the full-flexibility of both the ligand and target, their foremost disadvantage is that they can be readily trapped within a local energy minimum, which in turn precludes them from efficiently sampling the conformational space of the complex. Therefore, simulation techniques are usually used as complementary methods as a refining step subsequent to GA or MC simulations (Bissantz et al., 2000).

Protein Flexibility

Docking a ligand against a crystal or relaxed receptor structure is a commonly used approach in structure-based drug design (Abagyan & Totrov, 2001). However, in many cases, the degree of success that may be achieved in a typical docking experiment depends on the characteristics of the target and how important the protein flexibility is in the simulation. Most of the successful cases reported in the literature were either related to nearly rigid proteins or proteins having real binding mode of their respective ligands, or complexes whose structure was determined experimentally (Reddy et al., 2007). In spite of these studies, there are cases where the binding interaction has been shown to induce significant conformational changes to the target, ranging from local reorganization of sidechains to hinge movement of domains. Sampling these conformational changes during docking is impractical, as they involve too large a number of degrees of freedom. To address such problems, a number of docking packages like AUTODOCK, GOLD, FlexE and IFREDA, manage to include a modest amount of flexibility in the target during the docking simulations. These approaches include soft docking (Osterberg et al., 2002), sidechain flexibility (Schneider & Bohm, 2002), combined protein grid and united descriptors of the target (Knegtel et al., 1997).

Soft docking algorithms allow the ligand to penetrate through the surface of the protein in order to approximate and predict the dynamical changes that may take place within the active site as a result of ligand binding. This is generally achieved by attenuating Lennard-Jones repulsive parameters in the potential energy function that describes the system (Osterberg et al., 2002). Another commonly used technique to introduce active site dynamics in the context of docking is to allow key side chains that have been shown to mediate the interactions with the ligand to rotate freely and search for their preferred conformation. These side chain rotations are usually restricted to a number of pre-defined experimental conformations stored in rotamer libraries or predicted from a prior MD simulation. While this method reduces the risks associated with the lack of flexibility to some extent, it neglects backbone dynamics, which may affect the ultimate docking results (Reddy et al., 2007).

In order to account for a larger degree of receptor flexibility at a reasonable computational cost, a number of dominant protein conformations can be combined simultaneously to generate a comprehensive model that describes the essential dynamics of the binding site (Knegtel et al., 1997). This approach is generally termed as "combined protein grid" and is usually implemented in two steps. First, for each conformation, all possible protein-ligand atomic interactions are calculated and recorded in what is called a docking grid. Second, a combined grid is created by applying a weighted average for all the resulting grids representing various conformations. Alternatively, the averaging procedure may be applied to the atomic coordinates to generate an average structure for the protein.

Scoring Methods

As scoring of the poses is crucial in prioritizing and ranking of the compounds, it is important to use sensitive and accurate scoring functions that can replicate and predict experimental data. This is normally achieved using an objective scoring function that directs the conformational search algorithm in predicting the native conformation and ultimately estimates the binding affinity of the protein-ligand complex. Nevertheless, it has been broadly demonstrated that docking scoring functions are less successful at predicting the actual binding affinities and at discriminating true binders from inactive (decoy) compounds (Abagyan & Totrov, 2001; Schneider & Bohm, 2002; Shoichet et al., 1993). These puzzling results are direct outcomes of many factors that have been mistreated while analyzing the binding interactions of the resulting poses as a compromise to speed up the docking process. These factors mostly comprise the lack of proper solvation, the neglect of protein flexibility and the bias toward the training set of structures that have been used in optimizing the scoring process (Reddy et al., 2007). In fact, developing new scoring functions and innovative ranking schemes is a widely open area of research in the field of docking. Although a more precise scoring method can be practically implemented within docking, the large computational cost that is associated with such a function will be the actual barrier from using it. In this way, many assumptions have been proposed in the currently used docking scoring functions in order to reduce the complexity and computational time required to evaluate a particular pose. Overall, a typical scoring function includes at least among its ingredients, a descriptor for the hydrophobic effects, van de Waals dispersion interactions, hydrogen bonding, electrostatic interactions, and solvation effects. Based on their scoring functions, all docking programs that are in use today (Bissantz et al., 2000; Kitchen, Decornez et al., 2004; Reddy et al., 2007) can be divided into four major categories: force field-based, empirical, knowledge-based and consensus methods.

According to the energy landscape theory, the native conformation for a ligand within the binding site of its target is correlated with a profound deep minimum on the energy surface. Therefore, potential energy (force-field) functions have been used to describe protein-ligand interactions and assess their binding affinity by exploring the energy surface and locating these minima. Over the past 30 years, rigorous efforts have been devoted to build new force field models and make them available for a substantial number of applications ranging from molecular docking to molecular dynamics simulations (Guvench et al., 2008). One of the main problems of such models is the selection of a potential energy functional form and adjusting its various parameters to better represent experimental data or quantum mechanical predictions. These energy functions are commonly restricted to a number of assumptions and approximations for the sake of minimizing their computational time, reducing the efforts of refitting the parameters to more complex representations and, aligning them with many applications that are currently running with force fields of standard functional forms. An obvious example of such restrictions is the use of atom-centered charges in electrostatic calculations. Rationally, a more accurate representation of atomic charges should explicitly represent lone pairs on electronegative acceptors such as oxygen and take electronic polarization into account.

As docking algorithms usually deal with a single target conformation, the internal protein interactions are typically neglected. Accordingly, force field methods approximate the ligand-protein binding interactions by adding the interaction energy between the protein and the ligand to the ligand internal energy. These internal interactions are approximated by harmonic springs that describe the vibrations and rotations of the different bonds forming the ligands. The non-bonded interactions between the ligand and its target are estimated by van der Waals, hydrogen bonding, and electrostatic terms. For example, the potential energy function of the general AMBER force field (known as GAFF) (Wang et al., 2004) is shown below:

$$\begin{aligned} E_{pair} &= \sum_{bonds} k_r (r - r_{equ})^2 \\ &+ \sum_{angles} k_\theta (\theta - \theta_{equ})^2 \\ &+ \sum_{dihedrals} \frac{v_n}{2} \times \left[1 + \cos(n\varphi - \gamma)\right] \\ &+ \sum_{i<j} \left[\frac{A_{ij}}{R_{ij}^{12}} - \frac{B_{ij}}{R_{ij}^{6}} + \frac{q_i q_j}{\varepsilon R_{ij}} \right] \end{aligned} \quad (1)$$

where r_{equ} and θ_{wqu} are equilibrium structural parameters; k_r and k_θ, v_n are force constants; n is the multiplicity and γ is the phase angle for the torsional angular parameters. The A, B and q parameters represent the nonbonded potentials (charge-charge and van der Waals terms). Although nonbonded interactions can be obtained from liquid state calculations and available experimental data, parameters such as stretching, bending, and torsional terms are generally fit to quantum chemical calculations. Noticeably, the major drawback of standard force field scoring functions is the lack of solvation and entropy contributions to the binding energy. Moreover, such classical functions cannot be used to model complex interactions such as the formation or breaking of a covalent bond between a ligand and its target. Examples of force-field-based scoring functions include D-Score (B. Kramer et al., 1999) and GoldScore (Verdonk et al., 2003).

Another widely used scoring approach is to hypothesize an empirical scoring function that has been optimized to reproduce a collection of experimental data (Bohm, 1992). These data may include binding affinities or native conformations for known active compounds. Notable examples include F-Score (Rarey et al., 1996), ChemScore (Eldridge et al., 1997), SCORE (Tao & Lai, 2001) and Fresno (Rognan et al., 1999). The basic concept behind this type of scoring functions is that the binding energies can be approximated by a summation of unrelated contributions. Each element of this summation describes a certain binding interaction such as hydrophobic, hydrogen bonding, electrostatic or solvation effects. Some functions may comprise an approximation for the loss of entropy due to binding, which is proportional to the number of rotatable bonds included in the ligand. Overall, the terms that build up a typical empirical scoring scheme are simple enough to be rapidly evaluated in order to speed up the docking process. A fairly accurate estimate for the coefficients pre-multiplying these terms can be obtained by performing regression analysis and fit the whole function against the set of experimental data. An example of such functions is the AUTODOCK scoring function (Goodsell & Olson, 1990) (see below), which has an accuracy of ±2.177 kCal/mol:

$$\begin{aligned} \Delta G = {} & \Delta G_{vdW} \sum_{i,j} \left(\frac{A_{ij}}{r_{ij}^{12}} - \frac{B_{ij}}{r_{ij}^{6}} \right) \\ & + \Delta G_{hbond} \sum_{i,j} E(t) \left(\frac{C_{ij}}{r_{ij}^{12}} - \frac{D_{ij}}{r_{ij}^{10}} \right) \\ & + \Delta G_{elec} \sum_{i,j} \frac{q_i q_j}{\varepsilon(r_{ij}) r_{ij}} + \Delta G_{tor} N_{tor} \\ & + \Delta G_{sol} \sum_{i,j} \left(S_i V_j \right) e^{(-r_{ij}^{2}/2\sigma^2)} \end{aligned} \qquad (2)$$

here, the five ΔG terms on the right-hand side are coefficients empirically determined using linear regression analysis from the set of protein-ligand complexes with known binding constants. The function includes three in vacuo interaction terms, namely, a Lennard-Jones 12-6 dispersion/repulsion term; a directional 12-10 hydrogen bonding term, where E(t) is a directional weight based on the angle, t, between the probe and the target atom; and screened Columbic electrostatic potential. In addition, the unfavorable entropy contributions are estimated by a term that is proportional to the number of rotatable bonds in the ligand and solvation effects are represented by a pairwise volume-based term that is calculated by summing up, for all ligand atoms, the fragmental volumes of their surrounding protein atoms weighted by an exponential function and then multiplied by the atomic solvation parameter of the ligand atom (S_i).

It should be noted that, although several empirical functions like the above-mentioned AUTODOCK scoring function have been successfully used for many cases, they are generally biased toward the experimental data that have been used in their optimization and are not efficient at eliminating false binders from a set of tested compounds (Bissantz et al., 2000; Reddy et al., 2007).

Similar to empirical scoring functions, knowledge-based scoring methods attempt to reproduce experimentally determined structures using simpler atomic interaction-pair potentials. These potentials are based on the frequency of occurrence of all possible interactions between the ligand and its target. Using statistical analysis, knowledge-based models implicitly describe binding effects that are hard to represent explicitly during docking. Therefore, the accuracy of these methods depends on the extent of the manipulated protein-ligand data set and the diversity of atomic interactions included in these complexes. Example of such functions is DrugScore (Gohlke et al., 2000).

A more recent scoring technique, consensus scoring collects assessments from several scoring functions in order to evaluate a particular docking result (Charifson et al., 1999). Hence, the method is expected to reduce the errors that result from the individual scoring functions and improve the probability of selecting true binders. Nevertheless, it has been recommended to use different and uncorrelated scoring functions in constructing a successful consensus scheme. This is because correlated functions tend to produce similar results leading to error amplification and misleading results. Despite these constraints, several studies have pointed to the success of a number of consensus scoring functions when compared to using a single scoring method (Terp et al., 2001). These methods include X-CSORE (Wang et al., 2002) and FlexX (Rarey et al., 1996) scoring functions.

Structure-Based Pharmacophore Modeling (SBPM)

In spite of the great efforts of many research groups in both the pharmaceutical industry and academia, there still exist several barriers preventing docking methods from leading the evolution of drug discovery (Abagyan & Totrov, 2001; Zoete et al., 2009). These barriers include the following: (1) the inclusion of sufficient target flexibility and the simulation of the induced fit mechanism; (2) the druggability of the target; (3) the modeling of water-mediated interactions at the interface of the target-ligand complex; (4) the effect of metal ions located within the binding site of interest, and; (5) the fine-tuning of the protonation states of both the receptor and the ligand. Consequently, these difficulties motivated researchers to develop novel VS approaches that are far removed from standard docking methods. In this context, structure-based pharmacophore modeling (SBPM) originated as an alternative SBVS technique that can describe the essential interactions behind the binding of a ligand to its target (Z. Chen et al., 2009; Kim et al., 2006; Leach et al., 2009; Sun, 2008).

As a matter of fact, the concepts of pharmacophore modeling have been introduced by Ehrlich as early as in 1909 as "a molecular framework that carries (*phoros*) the essential features responsible for a drug's (*pharmacon's*) biological activity". These features are generally classified into two major categories, namely, chemical-based and shape-based features. The former include hydrogen bond acceptors or donors, charge centers, metal binding regions, aromatic rings and hydrophobic regions, while the latter mainly include volume-excluded regions and geometrical constraints like distances, angles and dihedral angles. By allocating the different features and including their three-dimensional distributions within the binding site, one can understand the essential properties required for bioactivity of known true binders (Sun, 2008).

Currently, there are two approaches to generate a pharmacophore model depending on the accessibility of the target structure. If the structure of the target is available, one can build a pharmacophore hypothesis that would complement the chemical features within the binding site of the target (Krovat et al., 2005). These pharmacophore models can be further improved if there is any active compound that has been co-crystallized with the target. This approach is called "binding sites-based pharmacophore models" or "structure-based pharmacophore models". The second approach, that is more widely used, relies only on the known active compounds and no information about the target is required. This technique is commonly referred to as "ligand-based pharmacophore modeling" and will be explained in detail later in this chapter. In this section, we will focus on the "structure-based pharmacophore models".

The most straightforward approach in designing a SBPM is to analyze the experimentally determined crystal structures of protein-ligand complexes. The program LigandScout (Wolber & Langer, 2005) introduced by Wolber and co-workers is very effective at manipulating these structures and automatically interpreting the various interactions between a particular macromolecule and its co-crystallized ligands into functional pharmacophore models. The program starts by cleaning up the structures of the ligands by assigning hybridization states and bond characteristics that are missing in the crystal structure. This is accomplished by using an extended heuristic approach combined with template-based numeric analysis. Following this step, pharmacophore models are created by analyzing the atomic interactions between the ligand and all residues located within the binding site of the target. These interactions are classified into complementing groups in terms of hydrogen bonding, electrostatic charges and hydrophobic contacts. Moreover, the flexibility of both the target and ligand can be incorporated by aligning several models in order to generate a common-feature pharmacophore model.

When only the structure of the active site is available, programs like structure-based focusing (SPF) aid in building up complementing pharmacophore hypotheses (Hoffren et al., 2001). The process starts by mapping favorable regions or "hot spots" for protein-ligand interactions within the binding site of the target. These regions are then clustered into hydrogen bond donating and hydrogen bond accepting vectors and hydrophobic interaction sites. The clustered groups are then used to build the pharmacophore model. Other algorithms that can construct SBPMs in addition to LBPMs are Unity (Tripos. http://www.tripos.com/) and MOE (Molecular Operating Environment; http://www.chemcomp.com/). Once a pharmacophore hypothesis is created, the model can be converted to a query that is used to screen chemical databases for molecules that satisfy these proposed hypotheses.

CASE STUDIES FOR SBVS

Condensing all the great discoveries and outstanding developments that took place in the last twenty years of SBVS studies is an unattainable task for the purpose of this review. However, here we will try to summarize the important studies that helped to advance the field and guide the drug discovery community towards the highest impact techniques. These studies were not only centered on uncovering new drug entities but also directed towards validating the methods developed, reducing their limitations and adding new approaches to the computational arsenal.

The initial evaluations of docking programs were focused on their ability to reproduce the native structures of co-crystallized ligands (Jones et al., 1997; Verdonk et al., 2003; Warren et al., 2006). It is only during the last 10 years that this focus has shifted towards determining the ability of docking techniques to filter chemical databases as a VS tool (Chen et al., 2006; Zhou et al., 2007). Nevertheless, even with the many studies published in the literature that ranked and benchmarked docking programs, there are still many practical problems preventing the robust and fair assessment of these programs (Chen et al., 2006; Cole et al., 2005). Warren et al. (2006) and Chen et al. (2006) carried out two independent studies to assess different docking programs considering a number of crucial factors. These factors included the variety of the screened compounds, the diversity of the targets and the effect of intrinsic docking parameters on the outcome of the VS protocol. The docking algorithms that were tested in these studies comprised Glide, GOLD, ICM, FlexX, DOCK, Dockit, Flo, Fred, LigFit, MOE and MVB. As a principal conclusion resulting from these two studies, there is no universal docking algorithm or scoring function that can work well for every target. In particular, most used scoring functions were inaccurate in predicating the actual binding affinities for a number of ligands. Unsurprisingly, recent studies suggested that consensus scoring might form a successful approach in reducing the imprecision of single scoring methods (Feher, 2006). However, this statement still requires more validation (Oda et al., 2006). On the other hand, it is worth mentioning that many improvements have been implemented for a number of well-established scoring functions. Relevant examples include potential of mean force (PMF) (Muegge, 2006), DrugScore (Velec et al., 2005) and Glide XP (Friesner et al., 2006).

Above and beyond scoring functions, protein flexibility and its effect on hit ranking is the other most important factor in SBVS (C, Subramanian, & Sharma, 2009; Mishra et al., 2009; Nettles et al., 2007). The importance of protein flexibility has been demonstrated in many studies and shown to have a remarkable influence on the final results. Bouzida et al. (1999) investigated the docking of sb203386 and skf107457 inhibitors to HIV-1 protease using MC simulations. These authors compared docking against a single protein conformation to using an ensemble of structures. Their recommendation was to use different protein structures in order to improve the docking results and enhance the final protein-ligand conformation. In another study by Murray et al. (1999), only 49% of the ligands were cross-docked correctly to another target that was co-crystallized with a different ligand. On the other hand, this showed that inducing small movements of the side chains in the binding site resulted in large variations in the predicted binding affinities. These early studies have drawn the attention of docking research groups to the importance of target flexibility and motivated them to implement various techniques to include this factor in the context of docking.

One way to accommodate receptor flexibility with more accurate scoring techniques is to implement a hybrid between docking and MD simulations. Originally, the use of MD simulations in VS studies was intended to create a set of receptor conformations (Broughton, 2000; Carlson et al., 2000). However, it was always debatable whether to use structures derived from MD simulations or NMR data. For example, Philippopoulos et al. suggested NMR structures as the most effective source for protein conformations (Philippopoulos & Lim, 1999). A set of 15 NMR conformations for ribonuclease HI was compared to a trajectory obtained from a 1.7 ns MD simulation. The NMR data explored the conformational space of the protein more efficiently than the conventional MD simulation. In spite of their findings, it should be noted that Philippopoulos et al. used a standard single trajectory MD simulations for a relatively short simulation time. In generating such ensembles, one should employ multiple trajectory MD simulations, commonly termed the replica exchange molecular dynamics (REMD), or run the simulation over longer times. For example, Figure 2 shows 28 dominant conformations representing the major backbone dynamics of the MDM2 target as extracted from ~100ns MD simulation. Another benefit of incorporating MD simulations with docking is to use more accurate scoring functions like molecular mechanics/Poisson-Boltzmann surface area (MMPBSA) (Thompson et al., 2008) or linear interaction (Aqvist et al., 1994) methods to re-rank the docking results.

In this context, a successful approach reported by McCammon et. al. (2000) is the relaxed complex scheme (RCS). MD simulations are applied to explore the conformational space of the protein receptor, while docking is subsequently used for the fast screening of drug libraries against an ensemble of receptor conformations. This methodology has been successfully applied to a number of cases (Lin et al., 2002). An excellent example is an HIV inhibitor, raltegravir (Markowitz et al., 2006), which became the first FDA approved drug targeting HIV integrase. MD simulations played a significant role in discovering a novel binding site, and compounds that can exchange between the two binding sites have formed a new generation of HIV integrase inhibitors.

In our recent work, we have applied an improved RCS to uncover dual-inhibitors for the MDM2/MDMX-p53 protein-protein interactions (Barakat, Mane, Friesen, & Tuszynski, 2009). In this study, we first filtered a set of ~6000 different ligands against 28 different MD-based conformations for the p53-binding site within MDM2 using AUTODOCK 4.0. Top 300 hits were redocked to the same binding site within MDMX. We finally rescored the top hits from both of the two screening experiments using MMPBSA. In this study, although the binding sites were fairly similar (see Figure 1), the MDMX pocket seemed to be more compact than that of MDM2. This is mainly due to the three residues Pro95, Ser96 and Pro97 in

Figure 2. Twenty-eight dominant conformations for MDM2. This ensemble comprised the crystal structure (red), five structures extracted from the holo (p53-bound)-trajectory (green) and twenty-two structures extracted from the apo (p53-free)-trajectory (blue).

MDMX that have been replaced by His96, Arg97 and Lys98 in MDM2. These substitutions shifted a helical domain in MDMX relative to MDM2 and caused Lys98 and Tyr99 to protrude into the p53-binding cleft within MDMX, making it shallower and less accessible to many of the MDM2 top hits we found. This observation is clear when comparing the binding modes of nutlin within the two pockets (see Figure 1a). While Tyr100 and Leu99 of MDM2 extend the binding site allowing nutlin to intimately bind to MDM2, the same residues in MDMX clash with the drug preventing it from taking the normal conformation that was adopted within MDM2. On the other hand, Figures 1b-c show how two compounds from the list of proposed MDM2/MDMX inhibitors were able to tolerate the structural variations between the two binding sites. This is apparent in Figure 1, where the compounds took on different conformations within the two binding pockets in order to maximize their interactions with the proteins.

In another interesting study, Doman *et al.* (2002) examined protein tyrosine phosphatase-1B (PTP1B) using both docking-based VS and HTS. In the course of the HTS run, approximately 400,000 compounds were screened against the target and only 85 compounds resulted in IC_{50} values less than 100 μM. Applying the docking-based protocol against the X-ray crystal structure and using 165,581 compounds, the top 350 hits were tested which resulted in 127 compounds with IC_{50} values less than 100 μM. While docking based VS yielded better results than HTS, it should be mentioned that the screened libraries and assay conditions were different in the two approaches. In a more difficult study, Vangrevelinghe et al. (2003) carried out a VS experiment using DOCK 4.0 in discovering the most potent inhibitor for human casein kinase II (CK2). This work started by building a homology model using a crystal structure template that shared 85% similarity with the human protein. Then, 400,000 different compounds were screened against the target followed by post-processing of the 12,428 top hits. The processing procedure followed two levels of iterations. Since the interaction with the hinge domain of the binding site was determined to be crucial for any potential activity, the top ranked compounds were first filtered for molecules that have at least two hydrogen bonds with this hinge structure. The second filter employed a more sophisticated scoring function that included an additional solvation term. This two-step filtering scheme revealed 1592 promising compounds. Only 12 compounds were finally tested experimentally and resulted in 4 hits that inhibited more than 50% of the enzyme activity. The most potent compound exhibited an IC_{50} of 80 nM and it turned out to be the best CK2 inhibitor discovered so far.

Using the induced-fit conformation for aspartic protease renin protein, Krovat et al. (2004) employed shape-based VS using the docking program LigandFit/Cerius to uncover novel renin-inhibitors and test a number of scoring schemes. Seven different scoring functions (LigScore1, LigScore2, PLP1, PLP2, JAIN, PMF, LUDI) were used to rank the compounds (Krovat & Langer, 2004). A collection of 990 diverse and drug-like compounds was used along with 10 known active renin inhibitors. Individually, all seven scoring functions that were used recovered at least 50% of the active compounds within the first 20% of the entire database. Using a consensus scoring that incorporated only four scoring functions (LigScore2, PLP1, PLP2, and JAIN), a hit rate of 90% in the top 1.4% was recorded. Finally, a more focused database was created and these compounds were docked to the binding site of the protein. A hit rate of 100% in the top 8.4% resulted from the use of the triple consensus scoring of PLP1, PLP2, and PMF.

Although the above-mentioned successful cases integrated docking as the key and principal ingredient of the VS protocol, other studies employed pharmacophore search techniques to reduce the size of the chemical database and extract a more focused library of compounds (Brenk et al., 2003; Cavasotto & Phatak, 2009; Rollinger et al., 2004; Rollinger et al., 2004; Sirois et al., 2004; Varady et al., 2003; Westerfors et al., 2003). Varady et al. (2003) introduced an exciting example of the method in their screening of 250,251 compounds for dopamine 3 (D_3) inhibitors. Similar to Vangrevelinghe's approach mentioned above, Varady's team started by building up a homology model for D_3 followed by applying a hybrid, stepwise computational screening approach. Then, a pharmacophore model was developed and used to identify promising compounds from a chemical database of approximately 250, 000 synthetic compounds and natural products. As a result, 6,727 molecules were identified which satisfied the pharmacophore hypotheses. These molecules were further screened through structure-based searching using multiple receptor models obtained from MD simulations via computational docking and scoring. Top-ranked potential D_3 inhibitors were further subjected to structural novelty screening by comparing them to the known D_3 ligands. Finally, the most promising 20 potential D_3 ligands were tested for their D_3 binding affinities. As a result, 11 of these 20 compounds showed IC_{50} values greater than 3 μM and 8 compounds showed nanomolar affinities.

In contrast to Varady's case, instead of initially filtering the database for potential hits, a pharmacophore model was used as a post-docking filter to validate the results and reduce the number of hits submitted to experimental verification (Betzi et al., 2007). Betzi used this technique to identify the first set of drug-like ligands that bind to the HIV-1 Nef SH3 surface. To accomplish this, the whole chemical database was filtered for drug-likeness properties which reduced the set of compounds to ~1,400 molecules. These structures were then docked to the binding interface between Nef and SH3 proteins. Top 335 hits were then filtered using a structure-based pharmacophore model derived from the binding site within the SH3 domain. This post-filtering procedure yielded 33 compounds that were visually inspected and reduced to 10 hits for experimental validation. Out of the final 10 tested compounds, one molecule resulted in a dissociation constant, K_D, of 1.8 μM. Following a similar procedure, our group constructed a dynamic pharmacophore model for an inhibitor that targets the ERCC1-XPA protein-protein interaction (K. H. Barakat et al., 2009). Figure 3 illustrates the procedure that was followed in constructing the ERCC1-XPA-inhibitor pharmacophore model.

Rollinger et al. (2004) used structure-based pharmacophore screening for the discovery of novel inhibitors for acetylcholinesterase (AChE). This study generated a pharmacophore model using the protein and one of the known inhibitors. Then, the in-house 3D database was screened which included about 110, 000 natural products within a molecular mass of 140–700 Da (Rollinger, Hornick et al., 2004). Finally, two compounds from the top VS hits were subjected to experimental validation and progressed to *in vitro* and *in vivo* testing. On another large scale SBVS study and using a pharmacophore searching strategy, Sirois et al. (2004) screened over 3.6 million compounds against SARS (severe acute respiratory syndrome). About 0.07% of the initial database of compounds satisfied at least five of the original six pharmacophoric points. Furthermore, a subsequent evaluation for the druggability of the top ranked compounds retrieved 17% that had a perfect score of 1.0.

LIGAND BASED VIRTUAL SCREENING (LBVS)

Despite the advances in macromolecular structure prediction methods, the number of protein structures that have been determined experimentally is still lagging compared to that of their sequenced counterparts. In this case, homology modeling may play a role in understanding and predicting the three-dimensional structure of the target. However, homology modeling has its own limitations and the degree of success of incorporating the method within the context of VS depends mainly on the quality of the predicted structure (Cavasotto & Phatak, 2009). Therefore, it is important to seek alternative routes that depend merely on known active compounds and in which no information about the target is required. These ligand-based filtering techniques have played a significant role in discovering potent inhibitors for many targets. In fact, ligand-based screening methods use known active and inactive compounds as templates and employ comparative algorithms to identify new compounds that are similar to the templates. Overall, one can classify the different LBVS methods into three approaches, namely, similarity search (Willett, 2006), pharmacophore search (Sun, 2008) and quantitative structure–activity relationships (QSAR) (Free et al., 1964).

Similarity Search

The fundamental theory behind this approach is Maggiora's "similar property principle" (Johnson & Maggiora, 1999), which states that similar molecules are more than likely to have similar properties. While not universally correct (Kubinyi, 1998), there are many cases where this simple idea showed great success and helped in the discovery of novel active molecules (Martin et al., 1993; Patterson, Cramer, Ferguson, Clark, & Weinberger, 1996). According to this concept, one can use known active compounds as reference structures and filter a given chemical library for ligands that are structurally similar to the active molecules. The filtered compounds are expected to display some activity that in some cases could be greater than the original reference structures. In fact, there are mainly two ways to assess the similarity between two molecular structures. These methods comprise of molecular alignment and molecular descriptors algorithms (Lemmen & Lengauer, 2000).

*Figure 3. ERCC1-XPA-inhibitor pharmacophore determination. (a) The equilibrated ERCC1 (grey surface) showing the excluded volume occupied by atoms from ligands obtained in the virtual screening experiments (green surface). Atoms included in this image were obtained by clustering the top ligands, from virtual screening experiments, and omitting those that were outside of a 90% RMSD cutoff. (b) Pharmacophores from each of the top 30 ligands were created with their interactions in the ERCC1 binding site. The type of pharmacophore interactions with each residue were scored and are represented schematically. Yellow patches indicate hydrophobic interactions with the pocket, red and blue patches represent hydrogen bond acceptor and donors respectively, while green patches indicate aromatic interactions. Orientation of the binding site is the same as in panel a. Tyr145 and His149 (indicated by an asterisk *) do not lie on the bottom of the pocket but are observed within a lip that overhangs the pocket (see panel a). (c) The averaged pharmacophore model obtained from the docked poses from virtual screening. Each sphere represents a specific chemical entity with the size being representative of the overall contribution at each position. Coloring is identical to that described for panel a. (d) The chemical structure of UCN-01, a weak ERCC1-XPA small molecule inhibitor.*

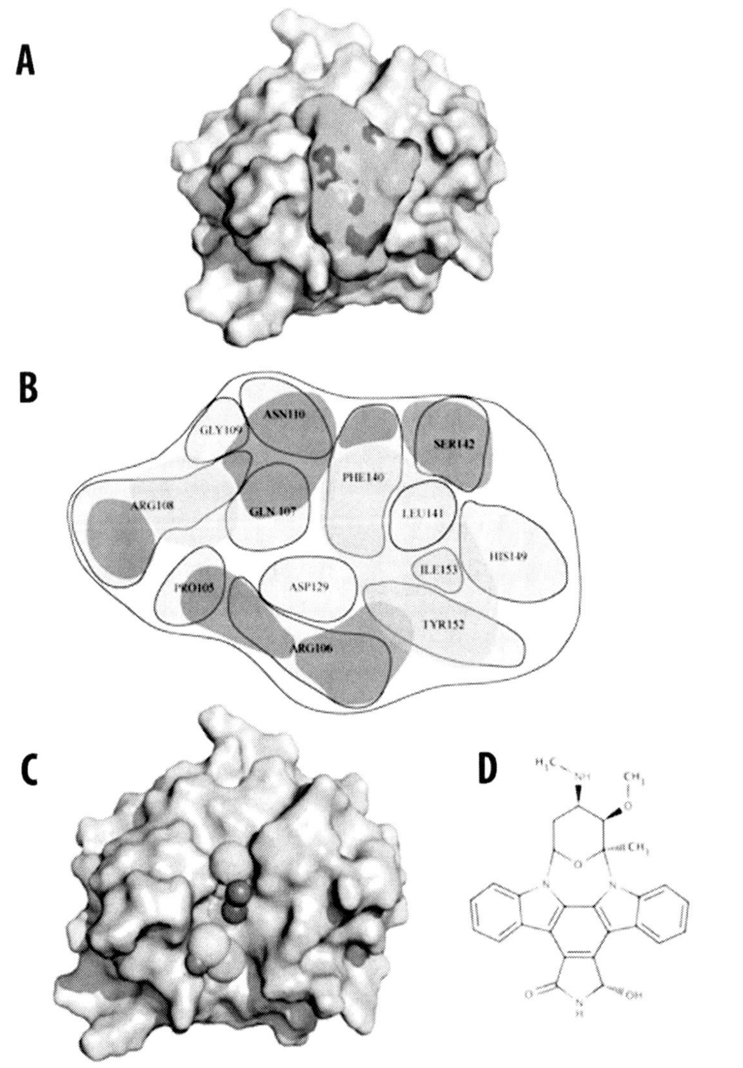

Molecular alignment algorithms such as FlexS (Lemmen et al., 1998) or GASP (Jones et al., 1995a) typically align the filtered compounds with the reference structure and rank them according to their degree of similarity. During the superimposition process, the two aligned molecules can be treated either as rigid or flexible. Similar to docking methods, flexibility can be introduced by employing an incremental construction approach (FlexS) or a genetic algorithm procedure (GASP). Other algorithms like Fflash (A. Kramer et al., 2003; Pitman et al., 2001) apply fragment-based techniques to incorporate ligand flexibility during the filtering process. Other algorithms adopted more complicated approaches to carry out more accurate molecular alignment. These methods include incorporation of Gaussian functions as in the program MIMIC (Mestres et al., 1997) or constructing interaction potential grids around molecules (Goodford, 1985).

A major drawback of molecular alignment techniques is that the time required for a single molecule comparison is long enough to discourage a user from employing the method in screening large databases (Willett, 2006). As a result, more efficient and accurate techniques have been developed to describe the information inherited in the molecular structure of a given ligand along with its physiochemical and topological properties. These molecular descriptors are generated on-the-fly and compared to the reference structure very rapidly. Based on the dimension of the information that is used, molecular descriptors can be classified into 1D-, or 2D-descriptors. Evidently, the higher the dimension of the descriptor approach, the longer its computational time will be and the higher the accuracy one can expect from the searching protocol. Generally speaking, bulk properties like molecular weight, molar refractivity or log P values are adequate to construct a 1D-molecular descriptor (Lipinski et al., 2001). However, since there is no information about the structural properties or chemical features of the ligand, it is impossible to only rely on such descriptors in filtering a typical chemical library for active molecules. Consequently, one should draw on a higher level of information and include structural properties as an additional descriptor in order to increase the accuracy of the method. This introduced molecular fingerprints as the most successful and widely used similarity search approach in LBVS.

Molecular fingerprint is a bit-string representation that reflects structural features and other properties of a molecule given its chemical structure (Willett, 2006). Key advantages of this approach over direct comparisons of molecules are that it is very simple to implement, remarkably fast to calculate and the final outcome is expressed as a single number that quantifies the degree of similarity. According to the complexity level and design scheme, one can recognize two basic approaches in generating a molecular fingerprint for a specified structure. The first approach is what is known as "keyed" representation (McGregor & Pallai, 1997). In this case, an individual bit within the string can be set as "on" or "off" reflecting the presence or absence of a pre-defined functional group (pattern) in the sub-structural space of the ligand. While the order of the bit-string map is the same for each molecule, the individual bits are turned on or off depending on if their representative substructure exists or not. A widely used VS algorithm that employs this procedure is MACCS (McGregor & Pallai, 1997) whose bit-strings may include up to 166 bits representing commonly known fragments. The second approach is known as the "hashed" representation (James & Weininger, 1995). This method resembles human fingerprints by not restricting the definition of bits to describe a pre-specified set of patterns. That is, like human fingerprints, which are very characteristic of individuals, a pattern's fingerprint characterizes the pattern, but the meaning of any particular bit is not well defined. To do so, a typical hashed representation algorithm starts with generating a pattern for each atom. Then it creates a pattern representing each atom and its nearest neighbors in addition to the bonds that join them. This hierarchal construction evolves to include higher order nearest neighbors until the complete structure is recovered.

In the heart of these similarity-based VS techniques lays a similarity measure, usually termed a similarity coefficient that is used to quantify the degree of resemblance between two molecules. In fact, the most commonly used parameter is Tanimoto (Jaccard) coefficient (described in Equation 1) (Willett, 2006) To understand the concept behind this parameter, let us consider the case of 2D fingerprints representing two molecules A and B that have *a* and *b* bits that are set as true, respectively. Now, if there are *c* common bits that are mutually set as true in the two molecules, where *c* is the intersection subset of *a* and *b*, one defines their Tanimoto coefficient as:

$$T = \frac{c}{a + b - c} \tag{3}$$

The Tanimoto coefficient gives values between zero (no similarity) and one (maximum similarity). Although this coefficient is the most popular choice for both in-house and commercial screening packages, a number of studies have indicated that it is dependent on the size of the fingerprint of the reference structure and on how many bits are set in its bit-string representation (Flower, 1988). However, other studies suggested that 85% of compounds that have a Tanimoto coefficient value of 0.85 or greater relative to an active compound are predicted to be active as well (Patterson et al., 1996).

Pharmacophore Search

While the basic concepts behind the pharmacophore search approach have been introduced in previous sections, in this part of the chapter we will focus on pharmacophore modeling techniques that have been broadly followed in the literature if no target structure is available (Barnum et al., 1996; Z. Chen et al., 2009; Sun, 2008). In this case, the only information that can be exploited is a set of known active compounds that are recognized experimentally and the general procedure can be summarized in two fundamental steps. First, this set of molecules is analyzed in order to identify all chemical features within their structures. Then, for each molecule, an ensemble of different conformations is generated and used to produce the best alignment between the different compounds to overlay their corresponding features. Although the main approach seems feasible and simple to implement, searching the conformational space is the most important and most difficult part of the method. This is because it is hard to predict the active conformer of a given ligand without understanding how it interacts with the target, with the solvent molecules and other elements of the binding environment. Nevertheless, there are several programs that have been successfully used in building ligand-based pharmacophore models for many targets. These programs differ mostly in the way they handle ligand flexibility and the method of searching a typical chemical database for promising hits. The most popular programs are Catalyst (Barnum et al., 1996), DiscoTech (Martin et al., 1993) and GASP (Jones et al., 1995a).

Catalyst introduces ligand flexibility very efficiently and, in the mean time, is extremely fast in searching 3D chemical databases (Barnum et al., 1996). In brief, the program extensively explores the conformational space of a ligand by using a random search algorithm along with a poling function which creates a large number of low-energy conformations. Catalyst follows two alternative algorithms in building up pharmacophore models. The first algorithm, HypoGen, is a quantitative approach in which each chemical feature allocated to the ligand structure is associated with a particular weighting factor that is related to its relative importance in describing the bioactivity of the molecule. Following this procedure, the algorithm builds up a number of pharmacophore hypotheses and ranks them based on their ability to explain available experimental data. In the other approach, Catalyst follows a qualitative procedure that is termed the HipHop algorithm. In this process, for each ligand, the algorithm checks for the surface accessibility for receptor interactions. Then, chemical features are defined based on their absolute coordinates in the different conformations of the molecule rather than by their inter-feature distances. This procedure usually starts with the most active compounds in the training set followed by highlighting their matching features from other less-important molecules. This results in a considerable number of proposed pharmacophore hypotheses that is significantly reduced by rejecting models that cannot explain the bioactivity of these molecules. Regardless of which approach is used, Catalyst can merge different models in order to generate a more comprehensive pharmacophore hypotheses.

Disco not only suggests pharmacophore models that demonstrate the important features in a ligand, but it also predicts their potential complementary regions that should be located within the binding site (Martin et al., 1993). This is accomplished by breaking up a pharmacophore to groups of ligand points and binding pocket interaction sites. Ligand points include atoms with hydrogen bonding properties, charge centers and hydrophobic characteristics. Binding pocket interaction sites are predicted to be complementary regions within the target and are calculated using the coordinates of the heavy atoms of the ligand. Similarly to Catalyst, the conformational flexibility of the ligands is explored using a set of pre-calculated conformations for each ligand in the training set. However, one pitfall of using Disco is that all chemical features that make up the final pharmacophore model must be identified in every molecule, which may result in the exclusion of talented models.

In contrast to both Catalyst and Disco, the program GASP (Jones et al., 1995a) handles the ligand conformational flexibility in a very sophisticated manner. Instead of using a pre-calculated set of ligand conformations, the program uses a genetic algorithm to explore the conformational space of the ligand during the pharmacophore generation process. GASP algorithm starts by detecting all possible chemical features in the structure of each ligand. The molecule with the least number of features is selected as a reference structure. Every structure in the training set is then fitted to the reference structure using a genetic algorithm that is similar to what is used in docking programs (see above). However, in this case, the fitness of a particular model is measured based on a combination of similarity, the number of overlaid features and the volume integral of the overlay. One more advantage for GASP over DISCO and Catalyst is that, models generated by GASP account for the steric clashes between the ligands in generating the final pharmacophore model. On the other hand, the other two programs propose their models by only matching the chemical features of the ligands without taking their overall shape into account.

No matter which approach is used to generate a pharmacophore model, which includes SBPM, there are two main ways to screen chemical libraries for compounds that satisfy the constraints of the pharmacophore hypotheses (Z. Chen et al., 2009; Sun, 2008). First, one can use a database file format that includes a set of pre-defined conformers for each compound in the database. Although this approach remarkably speeds up the search process, it requires massive storage of the different conformations. Alternatively, a single conformation can be used as a precursor for generating an ensemble of conformations followed by fitting these structures to the pharmacophore query during the screening process. While this procedure eliminates the need for substantial storage, it is much slower than the former method.

Quantitative Structure–Activity Relationships (QSAR)

Pioneered by the efforts of Fischer (Fischer, 1894), Hansch, Fujita, Free and Wilson (Hansch et al.,

1964; Free et al., 1964), quantitative structure–activity relationships (QSAR) are currently one of the widely used tools in the drug discovery process. In fact, QSAR has a praiseworthy history in discovering putative drugs that have an ideal balance of pharmacokinetics and safety, as well as potency and selectivity (Norinder, 2005). The basic idea behind this method is analogous to that of similarity search algorithms (described above), that is, compounds with similar physico-chemical properties elicit similar biological effects. In this way, QSAR are often used to correlate the structural arrangements of a set of potential drug candidates to their electronic properties and their binding affinities. This is commonly achieved through building virtual models that can predict quantities such as the binding affinity, adsorption, distribution, metabolism, elimination, toxicity (ADMET) properties or the oral bioavailability of existing or hypothetical molecules (Martin, 2005). Although the original efforts in developing and using QSAR models were based on a single physico-chemical property, such as the solubility or the pK_a value of a molecule (commonly known as 1D-QSAR), current methods can employ the connectivity of a compound by considering physico-chemical properties of single atoms and functional groups constituting the compound and their contribution to the biological activity (2D-QSAR). Moreover, other models can include 3D-structural descriptors such as the length or width of a substituent to build more accurate models (3D-QSAR) in which the binding affinity, pharmacodynamic and pharmacokinetic properties are predicted from the three-dimensional structure of the ligands.

In order to enhance the accuracy of VS in selecting active compounds with appealing pharmacodynamic and pharmacokinetic properties, recent studies integrated QSAR models within VS protocols leading to the design of new small-molecule drug candidates. This process reduces the probability of facing problems with the ADME characteristics of a given hit, which have been identified as a major cause of drug candidate failure in late stages of the drug discovery process. Examples of frequently used QSAR models are CoMFA (Comparative Molecular Field Analysis) (Cramer et al., 1989), VolSurf (Cruciani et al., 2000) and Hologram QSAR (Castilho et al., 2006). CoMFA allows the analysis of a large number of quantitative descriptors and uses chemometric methods such as partial least squares (PLS) to correlate changes in bioactivity with changes in chemical structure. VolSurf models are used to predict and optimize in silico pharmacokinetic properties for a given set of compounds. In this approach, 3D molecular interaction energy grid maps are transformed into 2D molecular descriptors, which are very simple to understand and to interpret. Hologram QSAR is an automated technique that utilizes weighted 2D fingerprints in conjunction with the PLS statistical methodology to build robust QSAR models.

Case Studies for LBVS

Similar to SBVS, a plethora of LBVS studies has been published in the literature in the last few years including methodology-development papers focused on enhancing the accuracy of current models or devising new methods (Bender et al., 2009; Hert et al., 2004). Others represent pure applications for the various techniques attempting to recognize bioactive molecules or identifying new scaffolds by screening chemical databases.

In an important assessment study, Bender et al. (2009) compared the performance of 37 different molecular descriptors. The judgment was based on the ability of the different descriptors to rank active compounds from a database using a number of randomly selected query molecular structures. The basic question in this study was which descriptors contain independent (orthogonal) information and which descriptors are redundant because they are correlated with each other. Consequently, Bender et al. identified four independent broad descriptor classes: (1) circular fingerprints; (2) circular fingerprints considering counts; (3) path-based and keyed fingerprints and; (4) pharmacophoric descriptors. In 2004, Hert et al. investigated the performance of a new similarity searching methodology, known as data fusion. This technique is similar to the consensus scoring approach that is widely used in current docking protocols. However, like consensus scoring, data fusion is mainly dependent on the combination(s) of similarity measures that are used to rank the compounds, which, if selected properly, will lead to a significant enhancement in searching performance. In another interesting study, Tresadern et al. (2009) compared a number of LBVS methods with a careful application to corticotropin releasing factor 1 receptor (CRF1). This list of assessment and development studies continues to grow and includes many entries that are committed to improve the LBVS techniques and increase their accuracy (Burton, Ijjaali, Petitet, Michel, & Vercauteren, 2009; Neves, Dinis, Colombo, & Sa e Melo, 2009).

In addition to the aforementioned evaluation studies, there have been many successful applications of the methods. For example, following a two-step ligand-based approach, Franke *et al.* filtered a database of natural product and natural-product-derived compounds for potential inhibitors of human 5-lipoxygenase (5-LO) activity (Franke et al., 2007). In the first step, a similarity searching was performed using a topological pharmacophore model constructed from 43 known inhibitors. Out of the 430 top hits, 18 compounds were selected that satisfied the pharmacophore features of the original reference inhibitors. Experimental investigation of these compounds retrieved two novel structures that showed significant activity. Using these two molecules, a second round of virtual screening was performed through different ligand-based virtual screening methods. The top ranked molecules from the final set of screening potently inhibited 5-LO activity in intact cells representing a novel class of 5-LO inhibitors. Using a similar approach, Carosati *et al.* used LBVS to identify novel activators for Kir6.2/SUR1 K_{ATP} channels (Carosati et al., 2007). Firstly, using pharmacokinetic filtering via the VolSurf program, the set of 65,208 commercially available compounds was reduced to 1,913 compounds. Subsequently, six molecules were selected to serve as reference structures in a similarity search procedure using different scoring methods. As a result, 32 hit candidates were identified and tested for their biological activity. This procedure resulted in the discovery of 3 novel compounds that were able to inhibit insulin release with micromolar potency.

In order to identify compounds that can block NAADP (nicotinic acid adenine dinucleotide phosphate) signaling, Naylor *et al.* used NAADP as the query ligand to screen the chemical library ZINC for compounds with its 3D-shape and electrostatic similarity (Naylor et al., 2009). By testing the top-ranking hits in a sea urchin egg bioassay, it was found that one hit, Ned-19, blocked NAADP signaling at nanomolar concentrations.

Sciabola *et al.* developed a new 3D pharmacophore fingerprinting approach (TOPP, Triplets Of Pharmacophore Points) to identify inhibitors for a specific P450 target, polymorphic CYP2D6 (Sciabola et al., 2007). Their new 3D pharmacophoric fingerprints were as accurate as other 3D descriptors and 2D features. Interestingly, it was found in the 3D models that the use of more realistic substrate conformations, and by an additional docking step using GOLD on a homology model structure, did not improve the statistical results significantly. Using a hybrid between LBPM and docking-based screening, Li *et al.* screened for inhibitors against Raf-1 kinase (Li et al., 2009). First, a 3D pharmacophore model was developed from a training set of structurally diverse ligand structures. Then a virtual database searching was performed with the pharmacophore model as a 3D query. Subsequently, molecular docking was carried out on the selected hits whose estimated IC_{50} was less than 1 μM which yielded 29 hits that were identified as potential leads against the Raf-1 kinase target. In another study, Neves *et al.* developed a new ligand-based strategy combining important pharmacophoric and structural features based on a known aromatase-inhibitor (Neves et al., 2009). The NCI database was screened and new potent aromatase inhibitors were found to be active in the low nanomolar range. Interestingly, all the molecules found exhibited a common binding mode.

Chen and his co-workers carried out a very exciting study to look for dual inhibitors for both neuraminidase (NA) type 1 (N1) and haemagglutinin type 1 (H1) (Chen et al., 2009). An N1- pharmacophore model was constructed and used to screen the NCI database for N1-inhibtors. Top hits resulted from the initial screening were docked into haemagglutinin type 1 (H1) in order to find out dual-inhibitors for both N1 and H1. It was suggested that the compound NCI0353858 could be effective in curing the worldwide disease N1H1 influenza. Finally, in an interesting study that integrated both molecular docking and QSAR analysis, Tintori and his co-workers built a robust model using a set of known c-Src tyrosine kinase inhibitors (Tintori et al., 2009). This study was conducted using a structure-based alignment and by applying the GRID/GOLPE approach. Their model proved to be capable of predicting the activities of an external test set of compounds. Using this model, they were able to identify two regions within the target whose occupation by hydrophobic portions of ligands favorably affected the activity. Moreover, they found that hydrogen bond interactions involving residues Met343, Asp406 and Ser347 are important in determining the affinity of the active inhibitors toward c-Src. Furthermore, the inhibitors bearing a basic nitrogen provided enhanced potency through protonation and salt bridge formation with Asp350. They also used Volsurf to predict the pharmacokinetic profile of the tested molecules.

CONCLUSION

Under a growing social pressure toward discovering new and improved drugs and lowering the cost of current brand names, the pharmaceutical industry is struggling to develop innovative, fast and economical drug discovery techniques. Although HTS is still the main approach driving the identification of new lead compounds, the overall methodology is still largely phenomenological and technology driven (Lahana, 1999). Moreover, the basic cost of a typical HTS laboratory includes collecting and maintaining screening libraries of thousands, if not millions, of compounds. In contrast, using VS methods for drug discovery, the compounds that are tested need not physically exist or be chemically synthesized. In addition, experimental difficulties that are associated with HTS such as limited solubility or aggregate formation are not relevant to VS and do not need to be considered (Stahura & Bajorath, 2004; Zoete et al., 2009).

Even when used perfectly, no computational method developed thus far can predict all possible solutions. It is important to use a combination of all possible approaches in order to substitute a weakness of one technique with strengths in the others (Muegge, 2006). Although SBVS is more computationally demanding than LBVS (Reddy et al., 2007), it is easier to rationalize and help in designing more specific drugs by targeting a precise domain or binding pocket within a particular target. Additionally, recent advances in crystallography allowed the automation of structure prediction for parallel crystallization of different protein targets. These new rapid techniques introduce protein crystallography as a powerful high throughput approach in reducing the gap between sequenced and structured proteins (Sharff & Jhoti, 2003). Moreover, new co-crystallized ligands, binding affinity assays and high performance computing facilities add more promise for improving and validating SBVS protocols. Nevertheless, as the majority of potential therapeutic targets, especially membrane proteins, continue to be hard to crystallize (Drews, 2000), LBVS will remain a dominant alternative means in identifying new lead compounds and designing new drugs. On the other hand, SBVS on its own cannot help to understand the functional activity of a typical ligand. Examples of such problems include agonism, antagonism (Mailman & Murthy, 2009) and side effects, which are hard to capture and are difficult to model. Hence, SBVS and LBVS together will continue being two optional and complementary methods in drug discovery and development. However, based on the discussion presented in this chapter, VS as a whole is still under development, and far from forming a mature field of science. This is apparent in the fact that the number of strategies followed in the field is nearly as considerable as the number of reported screening campaigns.

REFERENCES

Abagyan, R., & Totrov, M. (2001). High-throughput docking for lead generation. *Current Opinion in Chemical Biology*, *5*(4), 375–382. doi:10.1016/S1367-5931(00)00217-9

Aqvist, J., Medina, C., & Samuelsson, J. E. (1994). A new method for predicting binding affinity in computer-aided drug design. *Protein Engineering*, *7*(3), 385–391. doi:10.1093/protein/7.3.385

Barakat, K., Mane, J., Friesen, D.,& Tuszynski, J. (2009). Ensemble-based virtual screening reveals dual-inhibitors for the p53-MDM2/MDMX interactions. *Journal of Molecular Graphics Models*.

Barakat, K. H., Torin Huzil, J., Luchko, T., Jordheim, L., Dumontet, C., & Tuszynski, J. (2009). Characterization of an inhibitory dynamic pharmacophore for the ERCC1-XPA interaction using a combined molecular dynamics and virtual screening approach. *Journal of Molecular Graphics & Modelling*, *28*(2), 113–130. doi:10.1016/j.jmgm.2009.04.009

Barnum, D., Greene, J., Smellie, A., & Sprague, P. (1996). Identification of common functional configurations among molecules. *Journal of Chemical Information and Computer Sciences*, *36*(3), 563–571. doi:10.1021/ci950273r

Baxter, C. A., Murray, C. W., Clark, D. E., Westhead, D. R., & Eldridge, M. D. (1998). Flexible docking using Tabu search and an empirical estimate of binding affinity. *Proteins*, *33*(3), 367–382. doi:10.1002/(SICI)1097-0134(19981115)33:3<367::AID-PROT6>3.0.CO;2-W

Bender, A., Jenkins, J. L., Scheiber, J., Sukuru, S. C., Glick, M., & Davies, J. W. (2009). How similar are similarity searching methods? A principal component analysis of molecular descriptor space. *Journal of Chemical Information and Modeling*, *49*(1), 108–119. doi:10.1021/ci800249s

Betzi, S., Restouin, A., Opi, S., Arold, S. T., Parrot, I., & Guerlesquin, F. (2007). Protein protein interaction inhibition (2P2I) combining high throughput and virtual screening: Application to the HIV-1 Nef protein. *Proceedings of the National Academy of Sciences of the United States of America*, *104*(49), 19256–19261. doi:10.1073/pnas.0707130104

Bissantz, C., Folkers, G., & Rognan, D. (2000). Protein-based virtual screening of chemical databases. 1. Evaluation of different docking/scoring combinations. *Journal of Medicinal Chemistry*, *43*(25), 4759–4767. doi:10.1021/jm0010441

Bohm, H. J. (1992). The computer program LUDI: A new method for the de novo design of enzyme inhibitors. *Journal of Computer-Aided Molecular Design*, *6*(1), 61–78. doi:10.1007/BF00124387

Brenk, R., Naerum, L., Gradler, U., Gerber, H. D., Garcia, G. A., & Reuter, K. (2003). Virtual screening for submicromolar leads of tRNA-guanine transglycosylase based on a new unexpected binding mode detected by crystal structure analysis. *Journal of Medicinal Chemistry*, *46*(7), 1133–1143. doi:10.1021/jm0209937

Broughton, H.B. (2000). A method for including protein flexibility in protein-ligand docking: improving tools for database mining and virtual screening. *Journal of Molecular Graphics Models, 18*(3), 247-257, 302-244.

Burton, J., Ijjaali, I., Petitet, F., Michel, A., & Vercauteren, D. P. (2009). Virtual screening for cytochromes: Successes of machine learning filters. *Combinational Chemistry and High Throughput Screening, 12*(4), 369–382. doi:10.2174/138620709788167935

Carlson, H. A., Masukawa, K. M., Rubins, K., Bushman, F. D., Jorgensen, W. L., & Lins, R. D. (2000). Developing a dynamic pharmacophore model for HIV-1 integrase. *Journal of Medicinal Chemistry, 43*(11), 2100–2114. doi:10.1021/jm990322h

Carosati, E., Mannhold, R., Wahl, P., Hansen, J. B., Fremming, T., & Zamora, I. (2007). Virtual screening for novel openers of pancreatic K(ATP) channels. *Journal of Medicinal Chemistry, 50*(9), 2117–2126. doi:10.1021/jm061440p

Castilho, M. S., Postigo, M. P., de Paula, C. B., Montanari, C. A., Oliva, G., & Andricopulo, A. D. (2006). Two- and three-dimensional quantitative structure-activity relationships for a series of purine nucleoside phosphorylase inhibitors. *Bioorganic & Medicinal Chemistry, 14*(2), 516–527. doi:10.1016/j.bmc.2005.08.055

Cavasotto, C. N., & Phatak, S. S. (2009). Homology modeling in drug discovery: Current trends and applications. *Drug Discovery Today, 14*(13-14), 676–683. doi:10.1016/j.drudis.2009.04.006

C.B.R., Subramanian, J. & Sharma, S.D. (2009). Managing protein flexibility in docking and its applications. *Drug Discovery Today, 14*(7-8), 394–400. doi:10.1016/j.drudis.2009.01.003

Charifson, P. S., Corkery, J. J., Murcko, M. A., & Walters, W. P. (1999). Consensus scoring: A method for obtaining improved hit rates from docking databases of three-dimensional structures into proteins. *Journal of Medicinal Chemistry, 42*(25), 5100–5109. doi:10.1021/jm990352k

Chen, C. Y., Chang, Y. H., Bau, D. T., Huang, H. J., Tsai, F. J., & Tsai, C. H. (2009). Ligand-based dual target drug design for H1N1: Swine flu-a preliminary first study. *Journal of Biomolecular Structure & Dynamics, 27*(2), 171–178.

Chen, H., Lyne, P. D., Giordanetto, F., Lovell, T., & Li, J. (2006). On evaluating molecular-docking methods for pose prediction and enrichment factors. *Journal of Chemical Information and Modeling, 46*(1), 401–415. doi:10.1021/ci0503255

Chen, Z., Li, H. L., Zhang, Q. J., Bao, X. G., Yu, K. Q., & Luo, X. M. (2009). Pharmacophore-based virtual screening versus docking-based virtual screening: a benchmark comparison against eight targets. *Acta Pharmacologica Sinica, 30*(12), 1694–1708. doi:10.1038/aps.2009.159

Cole, J. C., Murray, C. W., Nissink, J. W., Taylor, R. D., & Taylor, R. (2005). Comparing protein-ligand docking programs is difficult. *Proteins, 60*(3), 325–332. doi:10.1002/prot.20497

Cramer, R. D. III, Patterson, D. E., & Bunce, J. D. (1989). Recent advances in comparative molecular field analysis (CoMFA). *Progress in Clinical and Biological Research, 291*, 161–165.

Cruciani, G., Pastor, M., & Guba, W. (2000). VolSurf: A new tool for the pharmacokinetic optimization of lead compounds. *European Journal of Pharmaceutical Sciences, 11*(Suppl 2), S29–S39. doi:10.1016/S0928-0987(00)00162-7

DesJarlais, R. L., Seibel, G. L., Kuntz, I. D., Furth, P. S., Alvarez, J. C., & Ortiz de Montellano, P. R. (1990). Structure-based design of nonpeptide inhibitors specific for the human immunodeficiency virus 1 protease. *Proceedings of the National Academy of Sciences of the United States of America*, *87*(17), 6644–6648. doi:10.1073/pnas.87.17.6644

Doman, T. N., McGovern, S. L., Witherbee, B. J., Kasten, T. P., Kurumbail, R., & Stallings, W. C. (2002). Molecular docking and high-throughput screening for novel inhibitors of protein tyrosine phosphatase-1B. *Journal of Medicinal Chemistry*, *45*(11), 2213–2221. doi:10.1021/jm010548w

Drews, J. (2000). Drug discovery: A historical perspective. *Science*, *287*(5460), 1960–1964. doi:10.1126/science.287.5460.1960

Eldridge, M. D., Murray, C. W., Auton, T. R., Paolini, G. V., & Mee, R. P. (1997). Empirical scoring functions: I. The development of a fast empirical scoring function to estimate the binding affinity of ligands in receptor complexes. *Journal of Computer-Aided Molecular Design*, *11*(5), 425–445. doi:10.1023/A:1007996124545

Feher, M. (2006). Consensus scoring for protein-ligand interactions. *Drug Discovery Today*, *11*(9-10), 421–428. doi:10.1016/j.drudis.2006.03.009

Fischer, E. (1894). Einfluss der Configuration auf die Wirkung der Enzyme. *Ber. Dtsch. Chem. Ges.*, *27*, 2984–2993.

Fishman, M. C., & Porter, J. A. (2005). Pharmaceuticals: A new grammar for drug discovery. *Nature*, *437*(7058), 491–493. doi:10.1038/437491a

Franke, L., Schwarz, O., Muller-Kuhrt, L., Hoernig, C., Fischer, L., & George, S. (2007). Identification of natural-product-derived inhibitors of 5-lipoxygenase activity by ligand-based virtual screening. *Journal of Medicinal Chemistry*, *50*(11), 2640–2646. doi:10.1021/jm060655w

Free, S. M. (1964). A mathematical contribution to structure–activity studies. *Journal of Medicinal Chemistry*, *7*, 395–399. doi:10.1021/jm00334a001

Friesner, R. A., Banks, J. L., Murphy, R. B., Halgren, T. A., Klicic, J. J., & Mainz, D. T. (2004). Glide: A new approach for rapid, accurate docking and scoring. 1. Method and assessment of docking accuracy. *Journal of Medicinal Chemistry*, *47*(7), 1739–1749. doi:10.1021/jm0306430

Friesner, R. A., Murphy, R. B., Repasky, M. P., Frye, L. L., Greenwood, J. R., & Halgren, T. A. (2006). Extra precision glide: Docking and scoring incorporating a model of hydrophobic enclosure for protein-ligand complexes. *Journal of Medicinal Chemistry*, *49*(21), 6177–6196. doi:10.1021/jm051256o

Gohlke, H., Hendlich, M., & Klebe, G. (2000). Knowledge-based scoring function to predict protein-ligand interactions. *Journal of Molecular Biology*, *295*(2), 337–356. doi:10.1006/jmbi.1999.3371

Good, A. C., Krystek, S. R., & Mason, J. S. (2000). High-throughput and virtual screening: Core lead discovery technologies move towards integration. *Drug Discovery Today*, *5*(12Suppl 1), 61–69. doi:10.1016/S1359-6446(00)00015-5

Goodford, P. J. (1985). A computational procedure for determining energetically favorable binding sites on biologically important macromolecules. *Journal of Medicinal Chemistry*, *28*(7), 849–857. doi:10.1021/jm00145a002

Goodsell, D. S., & Olson, A. J. (1990). Automated docking of substrates to proteins by simulated annealing. *Proteins*, *8*(3), 195–202. doi:10.1002/prot.340080302

Guvench, O., & MacKerell, A. D. Jr. (2008). Comparison of protein force fields for molecular dynamics simulations. *Methods in Molecular Biology (Clifton, N.J.)*, *443*, 63–88. doi:10.1007/978-1-59745-177-2_4

Halperin, I., Ma, B., Wolfson, H., & Nussinov, R. (2002). Principles of docking: An overview of search algorithms and a guide to scoring functions. *Proteins*, *47*(4), 409–443. doi:10.1002/prot.10115

Hansch, C. (1964). p–s–p Analysis. A method for the correlation of biological activity and chemical structure. *Journal of the American Chemical Society*, *86*, 1616–1626. doi:10.1021/ja01062a035

Hart, T. N., & Read, R. J. (1992). A multiple-start Monte Carlo docking method. *Proteins*, *13*(3), 206–222. doi:10.1002/prot.340130304

Hert, J., Willett, P., Wilton, D. J., Acklin, P., Azzaoui, K., & Jacoby, E. (2004). Comparison of topological descriptors for similarity-based virtual screening using multiple bioactive reference structures. *Organic & Biomolecular Chemistry*, *2*(22), 3256–3266. doi:10.1039/b409865j

Hoffren, A. M., Murray, C. M., & Hoffmann, R. D. (2001). Structure-based focusing using pharmacophores derived from the active site of 17beta-hydroxysteroid dehydrogenase. *Current Pharmaceutical Design*, *7*(7), 547–566. doi:10.2174/1381612013397870

Hopkins, A. L., & Groom, C. R. (2002). The druggable genome. *Nature Reviews. Drug Discovery*, *1*(9), 727–730. doi:10.1038/nrd892

Horvath, D. (1997). A virtual screening approach applied to the search for trypanothione reductase inhibitors. *Journal of Medicinal Chemistry*, *40*(15), 2412–2423. doi:10.1021/jm9603781

Jones, G., Willett, P., & Glen, R. C. (1995a). A genetic algorithm for flexible molecular overlay and pharmacophore elucidation. *Journal of Computer-Aided Molecular Design*, *9*(6), 532–549. doi:10.1007/BF00124324

Jones, G., Willett, P., & Glen, R. C. (1995b). Molecular recognition of receptor sites using a genetic algorithm with a description of desolvation. *Journal of Molecular Biology*, *245*(1), 43–53. doi:10.1016/S0022-2836(95)80037-9

Jones, G., Willett, P., Glen, R. C., Leach, A. R., & Taylor, R. (1997). Development and validation of a genetic algorithm for flexible docking. *Journal of Molecular Biology*, *267*(3), 727–748. doi:10.1006/jmbi.1996.0897

Jorgensen, W. L. (2004). The many roles of computation in drug discovery. *Science*, *303*(5665), 1813–1818. doi:10.1126/science.1096361

Kim, S. Y., Lee, Y. S., Kang, T., Kim, S., & Lee, J. (2006). Pharmacophore-based virtual screening: The discovery of novel methionyl-tRNA synthetase inhibitors. *Bioorganic & Medicinal Chemistry Letters*, *16*(18), 4898–4907. doi:10.1016/j.bmcl.2006.06.057

Kitchen, D. B., Decornez, H., Furr, J. R., & Bajorath, J. (2004). Docking and scoring in virtual screening for drug discovery: Methods and applications. *Nature Reviews. Drug Discovery*, *3*(11), 935–949. doi:10.1038/nrd1549

Knegtel, R. M., Kuntz, I. D., & Oshiro, C. M. (1997). Molecular docking to ensembles of protein structures. *Journal of Molecular Biology*, *266*(2), 424–440. doi:10.1006/jmbi.1996.0776

Kramer, A., Horn, H. W., & Rice, J. E. (2003). Fast 3D molecular superposition and similarity search in databases of flexible molecules. *Journal of Computer-Aided Molecular Design*, *17*(1), 13–38. doi:10.1023/A:1024503712135

Kramer, B., Rarey, M., & Lengauer, T. (1999). Evaluation of the FLEXX incremental construction algorithm for protein-ligand docking. *Proteins*, *37*(2), 228–241. doi:10.1002/(SICI)1097-0134(19991101)37:2<228::AID-PROT8>3.0.CO;2-8

Krovat, E. M., Fruhwirth, K. H., & Langer, T. (2005). Pharmacophore identification, in silico screening, and virtual library design for inhibitors of the human factor Xa. *Journal of Chemical Information and Modeling*, *45*(1), 146–159. doi:10.1021/ci049778k

Krovat, E. M., & Langer, T. (2004). Impact of scoring functions on enrichment in docking-based virtual screening: An application study on renin inhibitors. *Journal of Chemical Information and Computer Sciences*, *44*(3), 1123–1129. doi:10.1021/ci0342728

Kuntz, I. D. (1992). Structure-based strategies for drug design and discovery. *Science*, *257*(5073), 1078–1082. doi:10.1126/science.257.5073.1078

Kuntz, I. D., Blaney, J. M., Oatley, S. J., Langridge, R., & Ferrin, T. E. (1982). A geometric approach to macromolecule-ligand interactions. *Journal of Molecular Biology*, *161*(2), 269–288. doi:10.1016/0022-2836(82)90153-X

Lahana, R. (1999). How many leads from HTS? *Drug Discovery Today*, *4*(10), 447–448. doi:10.1016/S1359-6446(99)01393-8

Leach, A.R., Gillet, V.J., Lewis, R.A. & Taylor, R. (2009). Three-dimensional pharmacophore methods in drug discovery. *Journal of Medical Chemistry*.

Lemmen, C., & Lengauer, T. (2000). Computational methods for the structural alignment of molecules. *Journal of Computer-Aided Molecular Design*, *14*(3), 215–232. doi:10.1023/A:1008194019144

Lemmen, C., Lengauer, T., & Klebe, G. (1998). FLEXS: A method for fast flexible ligand superposition. *Journal of Medicinal Chemistry*, *41*(23), 4502–4520. doi:10.1021/jm981037l

Li, H. F., Lu, T., Zhu, T., Jiang, Y. J., Rao, S. S., & Hu, L. Y. (2009). Virtual screening for Raf-1 kinase inhibitors based on pharmacophore model of substituted ureas. *European Journal of Medicinal Chemistry*, *44*(3), 1240–1249. doi:10.1016/j.ejmech.2008.09.016

Lipinski, C. A., Lombardo, F., Dominy, B. W., & Feeney, P. J. (2001). Experimental and computational approaches to estimate solubility and permeability in drug discovery and development settings. *Advanced Drug Delivery Reviews*, *46*(1-3), 3–26. doi:10.1016/S0169-409X(00)00129-0

Liu, M., & Wang, S. (1999). MCDOCK: A Monte Carlo simulation approach to the molecular docking problem. *Journal of Computer-Aided Molecular Design*, *13*(5), 435–451. doi:10.1023/A:1008005918983

Lyne, P. D. (2002). Structure-based virtual screening: An overview. *Drug Discovery Today*, *7*(20), 1047–1055. doi:10.1016/S1359-6446(02)02483-2

Mailman, R.B. & Murthy, V. (2009). Third generation antipsychotic drugs: Partial agonism or receptor functional selectivity? *Current Pharmaceutical Design*.

Mandal, S., Moudgil, M., & Mandal, S. K. (2009). Rational drug design. *European Journal of Pharmacology*, *625*(1-3), 90–100. doi:10.1016/j.ejphar.2009.06.065

Martin, Y. C. (2005). A bioavailability score. *Journal of Medicinal Chemistry*, *48*, 3164–3170. doi:10.1021/jm0492002

Martin, Y. C., Bures, M. G., Danaher, E. A., DeLazzer, J., Lico, I., & Pavlik, P. A. (1993). A fast new approach to pharmacophore mapping and its application to dopaminergic and benzodiazepine agonists. *Journal of Computer-Aided Molecular Design*, *7*(1), 83–102. doi:10.1007/BF00141577

McGann, M. R., Almond, H. R., Nicholls, A., Grant, J. A., & Brown, F. K. (2003). Gaussian docking functions. *Biopolymers*, *68*(1), 76–90. doi:10.1002/bip.10207

Mestres, J., Rohrer, D.C. & Maggiora, G.M. (1997). A molecular field-based similarity approach to pharmacophoric pattern recognition. *Journal of Molecular Graphics and Modelling, 15*(2), 114-121, 103-116.

Miller, M. D., Kearsley, S. K., Underwood, D. J., & Sheridan, R. P. (1994). FLOG: A system to select 'quasi-flexible' ligands complementary to a receptor of known three-dimensional structure. *Journal of Computer-Aided Molecular Design*, *8*(2), 153–174. doi:10.1007/BF00119865

Mishra, N., Basu, A., Jayaprakash, V., Sharon, A., Basu, M., & Patnaik, K. K. (2009). Structure based virtual screening of GSK-3beta: importance of protein flexibility and induced fit. *Bioorganic & Medicinal Chemistry Letters*, *19*(19), 5582–5585. doi:10.1016/j.bmcl.2009.08.042

Mizutani, M. Y., Tomioka, N., & Itai, A. (1994). Rational automatic search method for stable docking models of protein and ligand. *Journal of Molecular Biology*, *243*(2), 310–326. doi:10.1006/jmbi.1994.1656

Muegge, I. (2006). PMF scoring revisited. *Journal of Medicinal Chemistry*, *49*(20), 5895–5902. doi:10.1021/jm050038s

Naylor, E., Arredouani, A., Vasudevan, S. R., Lewis, A. M., Parkesh, R., & Mizote, A. (2009). Identification of a chemical probe for NAADP by virtual screening. *Nature Chemical Biology*, *5*(4), 220–226. doi:10.1038/nchembio.150

Nettles, J. H., Jenkins, J. L., Williams, C., Clark, A. M., Bender, A., & Deng, Z. (2007). Flexible 3D pharmacophores as descriptors of dynamic biological space. *Journal of Molecular Graphics & Modelling*, *26*(3), 622–633. doi:10.1016/j.jmgm.2007.02.005

Neves, M. A., Dinis, T. C., Colombo, G., & Sa e Melo, M. L. (2009). Fast three dimensional pharmacophore virtual screening of new potent non-steroid aromatase inhibitors. *Journal of Medicinal Chemistry*, *52*(1), 143–150. doi:10.1021/jm800945c

Norinder, U. (2005). In silico modelling of ADMET—a mini review of work from 2000 to 2004. *SAR and QSAR in Environmental Research*, *16*, 1–11. doi:10.1080/10629360412331319835

Oda, A., Tsuchida, K., Takakura, T., Yamaotsu, N., & Hirono, S. (2006). Comparison of consensus scoring strategies for evaluating computational models of protein-ligand complexes. *Journal of Chemical Information and Modeling*, *46*(1), 380–391. doi:10.1021/ci050283k

Osterberg, F., Morris, G. M., Sanner, M. F., Olson, A. J., & Goodsell, D. S. (2002). Automated docking to multiple target structures: incorporation of protein mobility and structural water heterogeneity in AutoDock. *Proteins*, *46*(1), 34–40. doi:10.1002/prot.10028

Patterson, D. E., Cramer, R. D., Ferguson, A. M., Clark, R. D., & Weinberger, L. E. (1996). Neighborhood behavior: a useful concept for validation of "molecular diversity" descriptors. *Journal of Medicinal Chemistry*, *39*(16), 3049–3059. doi:10.1021/jm960290n

Philippopoulos, M., & Lim, C. (1999). Exploring the dynamic information content of a protein NMR structure: Comparison of a molecular dynamics simulation with the NMR and X-ray structures of Escherichia coli ribonuclease HI. *Proteins*, *36*(1), 87–110. doi:10.1002/(SICI)1097-0134(19990701)36:1<87::AID-PROT8>3.0.CO;2-R

Pitman, M. C., Huber, W. K., Horn, H., Kramer, A., Rice, J. E., & Swope, W. C. (2001). FLASH-FLOOD: A 3D field-based similarity search and alignment method for flexible molecules. *Journal of Computer-Aided Molecular Design*, *15*(7), 587–612. doi:10.1023/A:1011921423829

Rarey, M., Kramer, B., Lengauer, T., & Klebe, G. (1996). A fast flexible docking method using an incremental construction algorithm. *Journal of Molecular Biology*, *261*(3), 470–489. doi:10.1006/jmbi.1996.0477

Reddy, A. S., Pati, S. P., Kumar, P. P., Pradeep, H. N., & Sastry, G. N. (2007). Virtual screening in drug discovery-a computational perspective. *Current Protein & Peptide Science*, *8*(4), 329–351. doi:10.2174/138920307781369427

Rognan, D., Lauemoller, S. L., Holm, A., Buus, S., & Tschinke, V. (1999). Predicting binding affinities of protein ligands from three-dimensional models: application to peptide binding to class I major histocompatibility proteins. *Journal of Medicinal Chemistry*, *42*(22), 4650–4658. doi:10.1021/jm9910775

Rollinger, J. M., Haupt, S., Stuppner, H., & Langer, T. (2004). Combining ethnopharmacology and virtual screening for lead structure discovery: COX-inhibitors as application example. *Journal of Chemical Information and Computer Sciences*, *44*(2), 480–488. doi:10.1021/ci030031o

Rollinger, J. M., Hornick, A., Langer, T., Stuppner, H., & Prast, H. (2004). Acetylcholinesterase inhibitory activity of scopolin and scopoletin discovered by virtual screening of natural products. *Journal of Medicinal Chemistry*, *47*(25), 6248–6254. doi:10.1021/jm049655r

Schneider, G., & Bohm, H. J. (2002). Virtual screening and fast automated docking methods. *Drug Discovery Today*, *7*(1), 64–70. doi:10.1016/S1359-6446(01)02091-8

Sciabola, S., Morao, I., & de Groot, M. J. (2007). Pharmacophoric fingerprint method (TOPP) for 3D-QSAR modeling: Application to CYP2D6 metabolic stability. *Journal of Chemical Information and Modeling*, *47*(1), 76–84. doi:10.1021/ci060143q

Sharff, A., & Jhoti, H. (2003). High-throughput crystallography to enhance drug discovery. *Current Opinion in Chemical Biology*, *7*(3), 340–345. doi:10.1016/S1367-5931(03)00062-0

Shoichet, B. K., Stroud, R. M., Santi, D. V., Kuntz, I. D., & Perry, K. M. (1993). Structure-based discovery of inhibitors of thymidylate synthase. *Science*, *259*(5100), 1445–1450. doi:10.1126/science.8451640

Sirois, S., Wei, D. Q., Du, Q., & Chou, K. C. (2004). Virtual screening for SARS-CoV protease based on KZ7088 pharmacophore points. *Journal of Chemical Information and Computer Sciences*, *44*(3), 1111–1122. doi:10.1021/ci034270n

Stahura, F. L., & Bajorath, J. (2004). Virtual screening methods that complement HTS. *Combinatorial Chemistry & High Throughput Screening*, *7*(4), 259–269.

Sun, H. (2008). Pharmacophore-based virtual screening. *Current Medicinal Chemistry*, *15*(10), 1018–1024. doi:10.2174/092986708784049630

Szymkowski, D. E. (2005). Creating the next generation of protein therapeutics through rational drug design. *Current Opinions in Drug Discovery and Development*, *8*(5), 590–600.

Tao, P., & Lai, L. (2001). Protein ligand docking based on empirical method for binding affinity estimation. *Journal of Computer-Aided Molecular Design*, *15*(5), 429–446. doi:10.1023/A:1011188704521

Taylor, J. S., & Burnett, R. M. (2000). DARWIN: A program for docking flexible molecules. *Proteins*, *41*(2), 173–191. doi:10.1002/1097-0134(20001101)41:2<173::AID-PROT30>3.0.CO;2-3

Terp, G. E., Johansen, B. N., Christensen, I. T., & Jorgensen, F. S. (2001). A new concept for multidimensional selection of ligand conformations (MultiSelect) and multidimensional scoring (MultiScore) of protein-ligand binding affinities. *Journal of Medicinal Chemistry*, *44*(14), 2333–2343. doi:10.1021/jm0010901

Thompson, D. C., Humblet, C., & Joseph-McCarthy, D. (2008). Investigation of MM-PBSA rescoring of docking poses. *Journal of Chemical Information and Modeling*, *48*(5), 1081–1091. doi:10.1021/ci700470c

Tintori, C., Magnani, M., Schenone, S., & Botta, M. (2009). Docking, 3D-QSAR studies and in silico ADME prediction on c-Src tyrosine kinase inhibitors. *European Journal of Medicinal Chemistry*, *44*(3), 990–1000. doi:10.1016/j.ejmech.2008.07.002

Totrov, M., & Abagyan, R. (1997). Flexible protein-ligand docking by global energy optimization in internal coordinates. *Proteins*, (Supplement 1), 215–220. doi:10.1002/(SICI)1097-0134(1997)1+<215::AID-PROT29>3.0.CO;2-Q

Vangrevelinghe, E., Zimmermann, K., Schoepfer, J., Portmann, R., Fabbro, D., & Furet, P. (2003). Discovery of a potent and selective protein kinase CK2 inhibitor by high-throughput docking. *Journal of Medicinal Chemistry*, *46*(13), 2656–2662. doi:10.1021/jm030827e

Varady, J., Wu, X., Fang, X., Min, J., Hu, Z., & Levant, B. (2003). Molecular modeling of the three-dimensional structure of dopamine 3 (D3) subtype receptor: Discovery of novel and potent D3 ligands through a hybrid pharmacophore- and structure-based database searching approach. *Journal of Medicinal Chemistry*, *46*(21), 4377–4392. doi:10.1021/jm030085p

Velec, H. F., Gohlke, H., & Klebe, G. (2005). DrugScore(CSD)-knowledge-based scoring function derived from small molecule crystal data with superior recognition rate of near-native ligand poses and better affinity prediction. *Journal of Medicinal Chemistry*, *48*(20), 6296–6303. doi:10.1021/jm050436v

Verdonk, M. L., Cole, J. C., Hartshorn, M. J., Murray, C. W., & Taylor, R. D. (2003). Improved protein-ligand docking using GOLD. *Proteins*, *52*(4), 609–623. doi:10.1002/prot.10465

Wang, J., Wolf, R. M., Caldwell, J. W., Kollman, P. A., & Case, D. A. (2004). Development and testing of a general amber force field. *Journal of Computational Chemistry*, *25*(9), 1157–1174. doi:10.1002/jcc.20035

Wang, R., Lai, L., & Wang, S. (2002). Further development and validation of empirical scoring functions for structure-based binding affinity prediction. *Journal of Computer-Aided Molecular Design*, *16*(1), 11–26. doi:10.1023/A:1016357811882

Warren, G. L., Andrews, C. W., Capelli, A. M., Clarke, B., LaLonde, J., & Lambert, M. H. (2006). A critical assessment of docking programs and scoring functions. *Journal of Medicinal Chemistry*, *49*(20), 5912–5931. doi:10.1021/jm050362n

Welch, W., Ruppert, J., & Jain, A. N. (1996). Hammerhead: Fast, fully automated docking of flexible ligands to protein binding sites. *Chemistry & Biology*, *3*(6), 449–462. doi:10.1016/S1074-5521(96)90093-9

Westerfors, M., Tedebark, U., Andersson, H. O., Ohrman, S., Choudhury, D., & Ersoy, O. (2003). Structure-based discovery of a new affinity ligand to pancreatic alpha-amylase. *Journal of Molecular Recognition*, *16*(6), 396–405. doi:10.1002/jmr.626

Willett, P. (2006). Similarity-based virtual screening using 2D fingerprints. *Drug Discovery Today*, *11*(23-24), 1046–1053. doi:10.1016/j.drudis.2006.10.005

Wolber, G., & Langer, T. (2005). LigandScout: 3-D pharmacophores derived from protein-bound ligands and their use as virtual screening filters. *Journal of Chemical Information and Modeling*, *45*(1), 160–169. doi:10.1021/ci049885e

Zhou, Z., Felts, A. K., Friesner, R. A., & Levy, R. M. (2007). Comparative performance of several flexible docking programs and scoring functions: enrichment studies for a diverse set of pharmaceutically relevant targets. *Journal of Chemical Information and Modeling*, *47*(4), 1599–1608. doi:10.1021/ci7000346

Zoete, V., Grosdidier, A., & Michielin, O. (2009). Docking, virtual high throughput screening and in silico fragment-based drug design. *Journal of Cellular and Molecular Medicine*, *13*(2), 238–248. doi:10.1111/j.1582-4934.2008.00665.x

ADDITIONAL READING

Alvarez, J., & Shoichet, B. (2005). *Virtual Screening in Drug Discovery (Drug Discovery Series)*. New York: CRC Press.

Böhm, H. Schneider, G., Kubinyi, H., & Mannhold, R.(2000). Virtual Screening for Bioactive Molecules (Methods and Principles in Medicinal Chemistry). Weinbeim (Fedral Republic of Germeny): Wiley.

Dailey, M. M., Hait, C., Holt, P. A., Maguire, J. M., Meier, J. B., & Miller, M. C. (2009). Structure-based drug design: from nucleic acid to membrane protein targets. *Experimental and Molecular Pathology*, *86*(3), 141–150. doi:10.1016/j.yexmp.2009.01.011

Foloppe, N., & Chen, I. J. (2009). Conformational sampling and energetics of drug-like molecules. *Current Medicinal Chemistry*, *16*(26), 3381–3413. doi:10.2174/092986709789057680

Mahé, P., & Vert, J. P. (2009). Virtual screening with support vector machines and structure kernels. *Combinatorial Chemistry & High Throughput Screening*, *12*(4), 409–423. doi:10.2174/138620709788167926

Seifert, M. H. (2009). Targeted scoring functions for virtual screening. *Drug Discovery Today*, *14*(11-12), 562–569. doi:10.1016/j.drudis.2009.03.013

Song, C. M., Lim, S. J., & Tong, J. C. (2009). Recent advances in computer-aided drug design. *Briefings in Bioinformatics*, *10*(5), 579–591. doi:10.1093/bib/bbp023

Varnek, A., & Tropsha, A. (2008). *Chemoinformatics: An Approach to Virtual Screening. Cambridge CB4 0WF*. UK: The Royal Society of Chemistry. doi:10.1039/9781847558879

KEY TERMS AND DEFINITIONS

Binding Site: A pocket, a groove or a protrusion having an assortment of apparent hydrogen bond donors and acceptors, hydrophobic features and it can be associated with molecular adherence surfaces.

Docking: A method for predicting the preferred orientation of one molecule when bound to a target to form a stable complex.

Inhibitor: A small molecule that binds to a specific binding site on the surface of a target (usually a protein). This binding process prevents the target from interacting with other molecules inside the cell, which in turn, reduce the activity of the cellular pathway that is associated with this target.

Molecular Dynamics Simulations: Integration of Newtown's equations of motion for a particular system in order to predict its future dynamics.

Pharmacophore: "A molecular framework that carries (*phoros*) the essential features responsible for a drug's (*pharmacon's*) biological activity" (Ehrlich, 1909).

Scoring Function: An additive function that includes representations of various interactions between a ligand (small molecule) and a target. These representations describe the electrostatic, hydrophobic, solvation and hydrogen bonding interactions between the two molecules. Scoring functions are used by docking programs for estimating the binding affinity between the interacting molecules as well as for predicting the native structure of the protein-ligand complex.

Virtual Screening: Filtering chemical databases for bioactive molecules (usually inhibitors for a particular cellular pathway) using computational tools.

Chapter 3
Systems Biology-Based Approaches Applied to Vaccine Development

Patricio A. Manque
Universidad Mayor, Chile

Ute Woehlbier
University of Chile, Chile

ABSTRACT

Vaccines represent one of the most cost-effective ways to prevent and treat diseases. The use of vaccines in the control of viral diseases represents an important milestone in the history of medicine. The genomic revolution brought us the possibility to scan genomes in the search of new and more effective vaccine candidates and the advancement of bioinformatics provided the framework for the application of strategies that were focused not only on antigen discovery but also on comparative genomics, and pathogenic factor identification and data mining. In addition, the progress in post-genomic technologies including gene expression technologies such as microarray and proteomics gave us the opportunity to explore the host responses to vaccines leading to a better understanding of immune responses to pathogens and/or to vaccines, assisting in the development of new and better vaccines and adjuvants. This chapter will review how systems biology-based approaches including genomics, gene expression technologies, and bioinformatics have changed the way of thinking about antigen discovery and vaccine development. In addition, the chapter will discuss how the study of the host responses in combination with "in silico" approaches could help predict immunogenicity and improve the efficacy of vaccines.

DOI: 10.4018/978-1-60960-491-2.ch003

INTRODUCTION

Since the first historical experimental vaccination, the inoculation with the related cow-pox virus to induce immunity against the deadly scourge of smallpox, conducted by Edward Jenner more than 200 years ago, vaccines have represented one of the most successful approaches to control and cure disease in medical history. Indeed, at the end of the 20th century, the U.S. centers for disease control and prevention (CDC) cited vaccination as the number one public health achievement of the past century. As an example of the achievements obtained by vaccination, in 1980s, the World Health Organization declared the world free of endemic smallpox. Furthermore, diseases such as diphtheria, pertussis, tetanus, measles, mumps and rubella experienced, thanks to vaccination programs, a 95-100% reduction in case number (MMWR Morb Mortal Wkly Rep, 1999). Despite the success of vaccination programs implemented so far, pathogenic microorganisms are still the most important health threat worldwide, therefore the challenge of developing new, better and more efficient vaccines remains unanswered. For instance, for devastating diseases such as malaria, tuberculosis, Chagas disease or AIDS, no effective vaccine is available. Even if treatment is available, it is expensive, poorly effective or privative for poor and under developed countries. The appearance of newly emerging infectious agents like H1N1 swine flu virus or severe acute respiratory syndrome (SARS) coronavirus and re-emerging pathogens such as *Clostridium difficile*, *mumps virus*, *Streptococcus group A* and *Staphylococcus aureus*, reinforce the necessity to speed up the development of vaccines and immunotherapeutics, especially considering that the World Health Organization (WHO) expects at least one such new pathogen to appear every year (Dong, 2008; Yang, 2008). Thus, the technological breakthroughs in genomics, transcriptomics, proteomics, metabolomics and methodological advances in bioinformatics (Kandpal, 2009) set the perfect scenario for vaccine development. The omics era -driven promises accelerated antigen discovery and the framework to understand how organisms respond to infections. Fundamental questions are expected to be addressed, such as how the immune response is elicited by a particular organism and how this knowledge can be used to improve vaccine efficacy.

The development of vaccines has followed closely the history of biomedical research. Early vaccines were in general based on killed or attenuated microorganisms or in some cases chemically inactivated components. As we further advance our understanding of the molecular mechanisms associated with infection, novel pathogenic factors and determinants were characterized, including attachment and colonization factors, toxins, capsular polysaccharides, surface proteins, capsid coat proteins, internal core antigens, and proteases. The identification of several microbial antigens that can be targeted led to the identification of a potential single target for vaccine development (reviewed in Vivona, 2008). With the development of molecular biology techniques, a dramatic change in the field of vaccinology occurred. Previously identified single vaccine targets were now cloned, expressed in homologous or heterologous systems such as bacteria and more recently yeast, leading to the development of recombinant vaccines.

An example of a very successful recombinant vaccine is the Hepatitis B vaccine (Andre, 1990), which became first available in 1981. Hepatitis B is a contagious liver disease that results from infection with the hepatitis B virus (HBV). It can range in severity from a mild illness lasting a few weeks to a serious, lifelong illness occurring when HBV remains in a person's body (chronic hepatitis B), resulting in long-term health problems, and even death. The vaccine contains one of the viral envelope proteins, hepatitis B surface antigen (HBsAg). A course of three vaccinations is given to provide a long-term protection from HBV infection.

With the publication of the *Haemophilus influenzae* genome in 1995, the first free-living bacterial genome sequenced (Fleischmann, 1995) and a new era in biomedical research began. This milestone is known as the genomic revolution. Furthermore, this technological breakthrough provided a framework to approach biological problems at the systems level thus, complementing our reductionist methods, this new strategy is termed systems biology (Bruggeman & Westerhoff, 2007). This idea basically advocates the concept that biological systems display emergent properties which cannot be unraveled by breaking down the system into pieces as proposed by reductionism, but instead biological systems need to be studied as a whole. By understanding the properties of systems in the context of the whole, we thereby recognize the way that components interact with each other in a temporal fashion, and as a consequence, we observe the emergence of properties, ultimately giving us the ability to model and predict the behavior of biological systems. The application of the rapid evolving omics technologies to the dissection of biological systems using a systems biology-based approach often involved multidisciplinary teams that included biologists, biochemists, mathematicians, physicists and computer scientists. With this very fruitful marriage of disciplines and the massive amounts of data generated by the omics technologies, remarkable changes in the field of vaccinology in the areas of antigen discovery and vaccine efficacy were initiated. Furthermore, under the large umbrella of systems biology, the discipline of systems immunology has recently started to develop (Kleinstein, 2008). This new research field aims to study the immune system from a more integrated and holistic perspectives, attempting to define how the different components and the different systems levels observed in the immune systems interact in a temporal fashion. Thus, systems immunology represents a new and different approach to understanding the structure and function of the immune system based on systems based theory, omics techniques, as well as on mathematical and computational tools. The implications of this emergent discipline for vaccine development will be discussed below.

REVERSE VACCINOLOGY AND BEYOND, ACCELERATING VACCINE DISCOVERY

Since the beginning of the genomic revolution and with the more recent development of new sequencing technologies, commonly referred to as the next generation sequencers (i.e. 454 Titanium and Illumina GAIIX sequencing), hundreds of new completed genomes have been sequenced including those of *Plasmodium falciparum* responsible for malaria (Gardner, 2002), *Mycobacterium tuberculosis* causing tuberculosis (Cole, 1998) and *Trypanosoma cruzi* the protozoan parasite responsible for chagas disease (El-Sayed, 2005), as well as *Cryptosporidium hominis* an enteric pathogen (Xu, 2004) (for further information please visit www.genomesonline.org). Even more thrilling is the fact that the next generation sequencing technologies are capable to sequence whole bacterial genomes and small eukaryotic genomes in a matter of days for a very low amount of cost (see Box 1).

The number of large datasets comprising whole genome sequences has recently reached 1000 prokaryotic genomes including most if not all the genomes of organisms of medical interest. This achievement represented an amazing opportunity to identify new potential vaccinogens in an approach that is known as "reverse vaccinology" (Mora, 2003). The basic idea of this approach is to search in the genome of pathogens, using bioinformatics tools, for antigens that could potentially represent good vaccine candidates. Some of the criteria for selection of these potential vaccinogens include (i) molecular signatures that indicate that the protein is a surface protein, (e.g. signal peptide, GPI anchor, single transmembrane

Box 1. Performance of next generation sequencing technologies

The Roche 454 Sequencer generates ~500-600 Mb of sequence per run, with sequence reads between 400-450 bases. Assuming a genome size of 50 Mb, a reasonable estimate for a genome of protozoan parasites, it is possible to generate a >6 fold coverage with a single run of the instrument. Therefore, each genome will require ~2.5 runs on average to generate ~ 15 fold coverage. On the other hand, *Illumina GAIIX* sequencer provides 1-2 billion 'mappable' bases in short reads (50-75 bases). Because the short reads cannot cross even short segments of repeated sequence, the Illumina technology is of little use when used alone for *de novo* sequencing. However, it is possible to complement with the Roche sequencing technology because: 1) Illumina can provide high coverage with low cost and high accuracy to improve the quality of 454 based sequence assemblies; 2) The Illumina data can cross small gaps that frequently remain after 454 sequence assemblies. Thus, with this high sequencing capability, it is not only possible to sequence a single organism in a very short period of time, but also these technologies open the possibility to sequence several strains of the same organism even during the same run using tag technologies to identify polymorphism associated with the gene(s) of interest. The latter approach is known as pan reverse vaccinology which is further discussed in the main text.

domain), (ii) homology to known pathogenic factors, such as attachment or colonization factor, toxins, invasion associated proteases to name a few, and (iii) the presence of potential immunologically relevant epitopes identified using immunoinformatics (reviewed below). Moreover, this approach provides several interesting advantages over conventional vaccine development. For instance traditional antigen discovery is mainly focused on antigens recognized by the immune system, which are essentially those with high immunogenicity, whereas in reverse vaccinology all antigens expressed by a particular pathogen can be selected based on a set of criteria defined by the investigator. This has important implications for antigen discovery, e.g., antigens that display low immunogenicity or are expressed transiently during the life cycle of a pathogen or even antigens with attenuated expression in "*in vitro*" culture can be identified by reverse vaccinology. Another main advantage of reverse vaccinology is that non-cultivable microorganisms can be approached and classical molecular biology techniques can be used to clone and express these antigens in different systems. However, a main disadvantage of this approach is that it disregards all non-protein components such as polysaccharides, lipopolysaccharides and glycoproteins, which are in many cases important antigens. These non-protein antigens are the main constituents of several vaccines including the *Haemophilus* b polysaccharide vaccine (HbPV), developed for the prevention of invasive disease caused by *Haemophilus influenza* type b bacteria causing meningitis.

In a groundbreaking paper, the group of Dr. Rino Rappuoli applied for the first time this approach to identify new vaccine candidates against the deadly gram-negative bacteria *Neisseria meningitidis* group B (Pizza, 2000). This microorganism causes meningitides (infection of the lining and fluid surrounding the brain) and sepsis (infection of the bloodstream), two lethal diseases. The epidemiology of this pathogen shows that it is a major cause of morbidity and mortality in children in industrialized countries and is responsible for epidemics in Africa and in Asia. Using genome mining, Dr. Rappouli's group was able to identify 600 potential vaccine candidates for *N. meningitidis* in a very short period of time using bioinformatic tools, 350 of them were tested for immunogenicity and some of them are currently in clinical trials (Giuliani, 2006). This study not only highlighted the use of genomics to identify new vaccine targets but also demonstrated that this goal can be achieved in a very short time frame. Thus, one of the most important advantages of this approach is the dramatic acceleration of the process of antigen discovery. For instance, before this study was conducted there were only a handful of vaccine candidates against *Neisseria*. In fact, it is estimated that using traditional approaches for vaccine development a vaccine could reach the market after 12-15 years

Figure 1. Schematic representation of the various steps of conventional vaccine development in comparison to the reverse vaccinology approach or the pan-genome reverse vaccinology approach

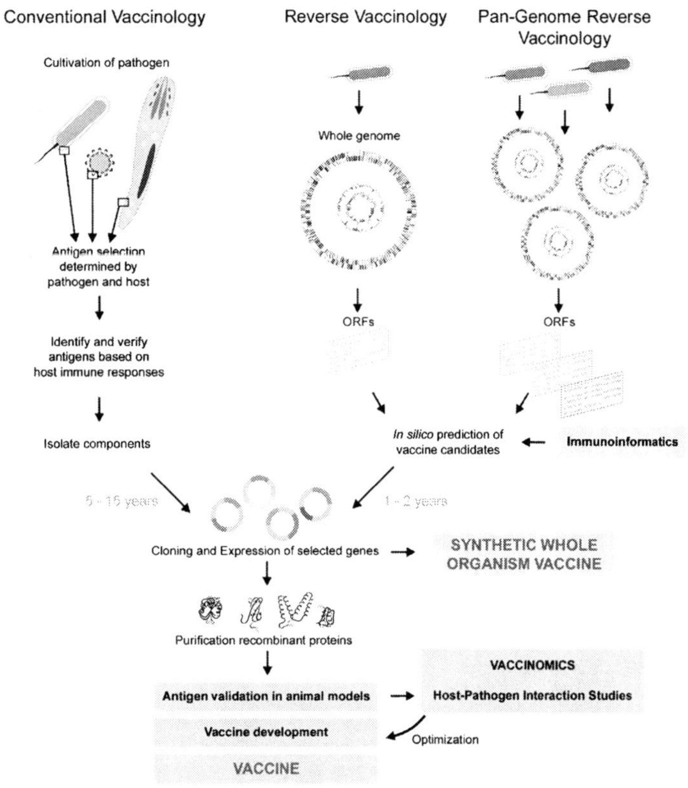

of research, during which about 2/3 of the time is used for antigen discovery and 1/3 for development, whereas by applying a reverse vaccinology approach the time dedicated to antigen discovery is reduced dramatically (Scarselli, 2005) (Figure 1). Another important challenge during the process of vaccine development is the possibility that the selected antigen displays high levels of variability among pathogen strains. This potential obstacle originates from the process of antigen discovery. In many cases, the selected potential vaccinogen has been isolated from the dominant strain of a pathogen circulating in a particular geographical location. However, often there are different pathogen strains circulating in different geographical regions complicating the identification of a vaccine target that could be effective worldwide. In the past, this represented a major setback leading to delays in the progress of vaccine development. Nowadays, the variability of the selected vaccine candidates can be easily studied by applying an approach known as pan-genome reverse vaccinology (Tettelin, 2009), whereby sequencing of several strains of a pathogen in a short period of time can be achieved using next generation sequencers. The basic idea of this strategy is to examine several whole genome sequences corresponding to different strains of the same pathogen to verify the polymorphisms associated with the selected antigen(s), overcoming two major problems: (i) gene variability caused by mutations such as insertions or deletions, and (ii) presence or absence of the selected vaccine target in the pathogen's population. By applying bioinformatics tools, the pan-genome can be defined as the global gene repertoire of a particular species. Further func-

Box 2. Reverse vaccinology applied to human pathogens

The reverse vaccinology approach has been applied to a large variety of pathogenic microorganisms including bacteria and parasites.
Example 1: Tackling large public health problems, using reverse vaccinology to develop a vaccine against malaria. Malaria is a mosquito-borne disease caused by the protozoan parasite *Plasmodium falciparum*, poses one of the world's biggest challenges for vaccine development (WHO, 2007). To date malaria vaccine development mostly relied on conventional approaches. Two main strategies were employed, the generation of (i) attenuated whole sporozoite vaccines, or (ii) recombinant subunit vaccines. There is an urgent need to accelerate vaccine discovery and development; reverse vaccinology and genome-wide scans pose promising strategies. In the past several years already such approaches have been employed. Extensive microarray and proteomics studies for *P. falciparum* have been reported (Daily, 2005; Hall, 2005), as well as genome-wide scans for polymorphisms (Mu, 2007), leading to the identification of potential vaccine targets. An example of a recent success is the identification of the essential *P. berghei* preerythrocytic stage-specific gene UIS3, which was identified by gene-profiling studies and now is the genetic target for a genetically attenuated sporozoite vaccine (Mueller, 2005).
Example 2: Whole genome approach to identifying vaccine candidates against *Streptococcus pneumonia*.*Streptococcus pneumoniae* is the leading cause of bacterial sepsis, pneumonia, meningitis, and otitis media in young children in the United States (Jacobs, 2004). The vaccines in current use are formulations of capsular carbohydrate from the 23 serotypes responsible for 85 to 90% of infections in the United States. These vaccines are poorly efficacious in infants and the elderly, the populations most at risk, suggesting the imperative need of new and more effective vaccines. Thus, Wizemann and collaborators (2001) screened the whole genome *S. pneumoniae* identifying 130 open reading frames encoding proteins with secretion motifs or similarity to predicted virulence factors. Challenge experiments using a mouse model were conducted with 108 of these proteins, six conferred protection against disseminated *S. pneumoniae* infection. Further characterization of these vaccine targets confirmed their surface localization. Moreover, these protective antigens showed broad strain distribution and essential requirement for an effective vaccine, making them ideal candidates for an improved vaccine against *S. pneumoniae*.
Example 3: Pan-genome reverse vaccinology approach to develop a vaccine against *S. agalactiae*, the group B *Streptococcus*.
As mentioned, a fundamental aspect of this approach is to utilize high-throughput technologies to characterize a large number of isolates and strains to verify the variability of the vaccine target in the pathogen's population. The goal is to obtain the most conserved vaccine candidate possible. A very illustrative example of this approach is the discovery of a universal vaccine against *Streptococcus agalactiae*, the group B *Streptococcus* (GBS). In newborns, GBS is the most common cause of sepsis and meningitis and a common cause of pneumonia (Nandyal, 2008). In adults, GBS usually causes no symptoms. However, in rare cases, it can lead to serious bloodstream, urinary tract, skin infections, and pneumonia, especially in people with immunosuppression and other chronic health problems, such as diabetes (Farley, 1995). Nine distinct capsular serotypes of GBS have been described; however, the major disease-causing isolates in Europe and US belong to only five serotypes: Ia, Ib, II, III and V (Johri, et al., 2006). A comparative genomic hybridization analysis revealed that there was significant variation in gene content among different clinical isolates of GBS (Tettelin, 2002). This observation led to the sequencing of additional strains belonging to the five serotypes. By applying bioinformatics algorithms to the pan-genome of GBS (Maione, 2005), it was possible to identify four antigens capable, to significantly increase the survival rate in animal models. Interestingly, only one of these antigens was part of the core-genome, the remaining three belonged to the dispensable genome. Thus, the final vaccine formulation comprises a combination of the four antigens, which provide broad strain coverage with levels of protection similar to those seen when using capsular carbohydrate-based vaccines.

tional characterizations divide the pan-genome in three main groups, (i) the core-genome, which includes the set of genes invariably present and conserved in all the isolates; (ii) the 'dispensable genome', comprised of genes present in some but not all the strains, and (iii) the strain-specific genes, which are present only in single isolates. Data mining of these three main groups allows the identification of those antigens belonging to the conserved group of the pathogen pan-genome and as a consequence representing promising vaccine targets. An alternative and complementary approach known as comparative reverse vaccinology (Serruto, 2004) takes advantage of data generated by large sequencing projects and aims to compare pathogenic strains with their non-pathogenic counterparts in order to identify potential genes associated with pathogenicity, which could also be good vaccine candidates. Examples of some success stories of reverse vaccinology are presented in Box 2 for human pathogens and in Box 3 for veterinary pathogens.

Hence, (pan-genome) reverse vaccinology represents a promising strategy to circumvent the usual challenges experienced in traditional vaccine development approaches. The continuous and vigorous development of high-throughput technologies during the 'omics' era not only impacted our ability to discover new antigens as mentioned above but also the speed of antigen

Box 3. Reverse vaccinology applied to veterinary pathogens with potential economical and public health implications

Example 1: Swine dysentery.
Swine dysentery (SD) is mucohaemorrhagic colitis of pigs resulting from infection of the large intestine with the anaerobic intestinal spirochaete *Brachyspira hyodysenteriae* (Moxley & Duhamel, 1999). The high cost of disease is associated with mortality (low), morbidity (high), depression of growth and feed conversion efficiency, and costs of continual in-feed medication. Although there is a vaccine available to control SD, the efficacy is highly variable demanding the search for new and more effective vaccines that help control this important veterinary pathogen. Recently, Song et al. (2009) applied a reverse vaccinology approach to identify *B. hyodysenteriae* proteins for use as recombinant vaccine components. An *in silico* analysis of partial genomic sequence data from this pathogen revealed 19 open reading frames (ORFs) predicted to encode potential vaccine candidates. Based on immunogenicity and distribution among strains of *B. hyodysenteriae*, a subset of candidates were cloned and used for challenging experiments. Thus, eight pigs were vaccinated twice intramuscularly with a combination of four of these proteins. The pigs developed antibodies to the proteins, and only one developed SD compared to five of nine non-vaccinated control pigs. These results demonstrated the possible use of these recombinant proteins as vaccinogens.
Example 2: Echinococciasis.
Another example is the attempt to develop a vaccine against the parasitic cestode *Echinococcus granulosus*. *E. granulosus* is the causative agent of a chronic, debilitating and widespread cystic echinococciasis, an important zoonosis with transmission between dog and domestic livestock with humans representing the intermediary host. Cystic hydatid disease as it is referred in humans remains a serious problem because it can cause extensive pathological damage due to the formation of cysts that can measure up to 35 cm in different tissues of the infected person (Siracusano, 2009). Recently, Gan and collaborators (Gan, 2010) identified *E. granulosus* tegumental membrane protein enolase as vaccine candidate using a reverse vaccinology strategy. This promising vaccine candidate displays several structural features that indicate that it carries B and T cells epitopes suggesting that this protein may be immunogenic and may provide protection against the disease.

discovery and vaccine development, as well as dramatically increased the number of diseases that can be addressed by vaccination.

BIOINFORMATICS AND THE EMERGENT ROLE OF IMMUNOINFORMATICS

The vertiginous development of genomic projects and other high-throughput analyses including transcriptomics, proteomics, and metabolomics have produced escalating volumes of 'omics' data. In this scenario, bioinformatics play a crucial role in the management and analysis of such data accelerating the progress in a wide variety of fields including understanding of omics data in the areas of antigen discovery, host cell response to pathogens and more recently immunoinformatics. Development of more refined algorithms for data mining, protein motif identification, and comparative genomics, among other applications in a user-friendly environment are providing more efficient, high confidence predictions that provide scientists with more resources to carry out "wet lab" experiments.

The advent of genomics has facilitated significantly among other achievements the identification of virulence factors, key targets for vaccine design. Thus, with rapid increase of genome projects that include hundreds of genomes of pathogenic organisms and their non-pathogenic counterparts currently in progress or finished, the identification of genes that are associated with pathogenesis provides key clues for rational vaccine design. This concept is based on the idea that interactions of pathogens with their host lead to changes in their transcriptional profile resulting in the expression of a variable set of genes, including those associated with virulence, known as virulence factors. Virulence factors represent ideal vaccine and therapeutic targets because an efficient interference with their function will impair the establishment of an infection. An example is the genomic analysis of the Group A *Streptococcus* (GAS), a gram-positive bacterium causing several diseases in humans, including pharyngitis and/or tonsillitis, skin infections (including impetigo, erysipelas), acute rheumatic fever, scarlet fever, poststreptococcal glomerulonephritis, a toxic shock–like syndrome, and necrotizing fasciitis (Olsen, 2009). By using traditional molecular and cellular biology techniques few pathogenic factors of this bacterium were identified but within a few years after the first GAS genome sequence became available, no fewer than 13 new proteins that contribute to pathogenesis were described (Musser & Shelburne, 2009). Even more importantly, given that numerous open reading frames in the GAS genome encode putative cell surface proteins with unknown function, it seems highly likely that many other virulence factors will continue to be discovered in the coming years.

As exemplified above, mining of large datasets obtained by genomics, expression technologies such as microarrays, proteomics, *in vivo* expression technology, and signature-tagged mutagenesis has proved to be very promising but several challenges remain to be addressed. For instance, in many cases these experiments identify genes that are associated with basic cellular functions, known as housekeeping genes, which cannot be directly involved in virulence. However, if these genes do not display homology with any human genes they can be used as vaccine or therapeutic targets. In addition, among the genes identified by these technologies a large number is categorized as hypothetical or unknown proteins potentially hiding some true virulence factors and/or new therapeutic targets. Recently, Hernandez et al. (2009) tackled this fundamental problem using a series of bioinformatics tools and refined algorithms designed to unravel hidden vaccine targets in omics data. The basic strategy taken by the authors was to re-annotate all hypothetical proteins from several respiratory pathogens by a careful and in-depth analysis of each one. Although some of the re-annotations match with functions that can be related to microbial virulence, the identification of virulence factors remains a very difficult task. Thus, the authors concluded that a careful re-annotation of the genomes from pathogens could be useful for finding additional virulence factors whose genes could be targeted. It is interesting to note that a high percentage of open reading frames and/or genes that are deposited in Genbank and other large databases are of unknown function, reinforcing the ideas mentioned above that many pathogenic factors and highly immunogenic antigens remain hidden in our databases.

Another important contribution of bioinformatics is to provide algorithms and data mining tools to study molecular epidemiology, population biology and ecology of pathogens and related organisms. Applications include pan-genomics approaches (described above) and the analysis of the genome plasticity and gene pools of pathogenic bacteria. An example of this is the study to characterize the genome plasticity of BCG and the impact on vaccine development (Brosch, 2007). BCG, the "Bacille de Calmette et Guerin" an attenuated derivative of *Mycobacterium bovis*, has been used in more than 3 billion individuals as vaccine against tuberculosis. Indeed BCG has been very efficient to prevent extrapulmonary tuberculosis in children, however, its efficacy in adults against pulmonary tuberculosis is variable (Colditz, 1995; Fine, 1995). Thus, to understand the evolution, attenuation and variable protective efficacy of BCG vaccines, the *Mycobacterium bovis* BCG1173P2 was subjected to comparative genome and transcriptome analysis revealing that gene amplification together with lesions in genes encoding transcriptional factors could affect gene expression levels, immunogenicity and possibly protection against tuberculosis. Besides genome plasticity, studies aimed to characterize antigenic diversity and antigenic variation would lead to more effective vaccines and vaccine implementation programs. Furthermore, molecular epidemiology plays a very important role in vaccine development, providing the ability to study pathogen microevolution and emergence of new clones or even new strains that could eventually impact vaccine effectiveness. A very important tool for these studies is multilocus sequence typing (MLST), a molecular biology technique used for genotyping and studies of population structure and genetic diversity of any given microorganism. Recently, a web-based database of MLST was launched (www.mlst.net). This database represents a nucleotide sequence-based approach for the characterization of isolates of bacteria and other organisms via the internet. As the website describes, "the aim of MLST is to provide a portable, accurate, and highly discriminating typing system that can be used for most bacteria and some other organisms".

A central element in the advancement of bioinformatic tools is the ability to develop databases that can store large datasets produced by OMICs technologies as well as be able to mining these datasets. For instance and as mentioned above, a malaria vaccine would represent a major milestone in vaccine development, medicine and public health. The sequencing of genomes of the *Plasmodium* species causing malaria, offers immense opportunities to aid in the identification of new therapeutic and vaccine targets through bioinformatics tools and resources. Thus, scientists developed MalVac (http://malvac.igib.res.in), a web-based database that contains a detailed analysis of proteins from various *Plasmodium* species (Table 1), especially those involved in the adhesion process, a critical step during the development of the infection, and are likely to play an important role as vaccine candidates (Chaudhuri, 2008). MalVac database is the collection of known vaccine candidates and a set of predicted vaccine candidates identified from the whole proteomes of *Plasmodium* species provided by PlasmoDb, a web-based integrated database for *Plasmodium* genome resources (Table 1). These predicted vaccine candidates were analyzed through several publicly available algorithms to obtain information regarding structure, putative function, immunological relevant sequences and identification of homologous, orthologs and paralogs for each gene. All of this information was collected and organized making it accessible and relevant from the view point of reverse vaccinology, facilitating decision making on the most probable choice for vaccine strategy.

The biodefense resource center (www.proteomicsresource.org): This is a NIAID (National Institute for Allergy and Infectious Diseases) initiative and another interesting example of a bioinformatics framework, employing a protein-centric approach to integrate and support mining and analysis of large and heterogeneous data aiming to identify targets for potential vaccines, therapeutics, and diagnostics for agents of concern in bioterrorism, including bacterial, parasitic, and viral pathogens. The program includes seven Proteomics Research Centers, generating diverse types of host-pathogen data, including mass spectrometry, microarray transcriptional profiles, protein interactions, protein structures and biological reagents. One of the most interesting features of this resource is the availability of host and pathogen omics data in an integrated framework, which allows global and integrated analysis of the data. Thus providing the opportunity to identify hidden relationships between host and pathogen proteins, unveiling for instance conserved strategies used by the host to control and eliminate pathogens, thereby offering a systems approach to understand pathogenicity and to facilitate target identification.

NERVE, a bioinformatics-driven immunology environment for vaccine development: To provide a bioinformatic framework to implement reverse vaccinology approaches, Vivona et al. (2006) developed NERVE (New Enhanced Reverse Vaccinology Environment), a user-friendly software environment for *in silico* identification of vaccine candidates from whole proteomes of bacterial pathogens. The software (http://www.bio.unipd.it/molbinfo) integrates multiple robust and well-known algorithms for protein analysis and comparison providing a rank of the best-selected candidates. Thus, NERVE represents a tool specifically designed to automatize *in silico* steps not only making reverse vaccinology easily

Table 1. Algorithms used in MalVac to predict molecular features of potential malaria vaccine candidates

Algorithm	Principle
MAAP	Predicts Malarial adhesins and adhesins-like proteins based on Support Vector Machines.
BLASTCLUST	Clusters protein or DNA sequences based on pair wise matches found using the BLAST algorithm in case of proteins or Mega BLAST algorithm for DNA.
TMHMM Server v. 2.0	Predicts the transmembrane helices in proteins based on Hidden Markov Model.
BetaWrap	Predicts the right-handed parallel beta-helix supersecondary structural motif in primary amino acid sequences by using beta-strand interactions learned from non-beta-helix structures.
TargetP1.1	Predicts the subcellular location of eukaryotic proteins based on the predicted presence of any of the N-terminal presequences: chloroplast transit peptide (**cTP**), mitochondrial targeting peptide (**mTP**) or secretory pathway signal peptide (**SP**).
SignalP 3.0	Predicts the presence and location of signal peptide cleavage sites in amino acid sequences from different organisms. The method incorporates a prediction of cleavage sites and a signal peptide/non-signal peptide prediction based on a combination of several artificial neural networks and hidden Markov models.
BlastP	It uses the BLAST algorithm to compare an amino acid query sequence against a protein sequence database.
Antigenic	Predicts potentially antigenic regions of a protein sequence, based on occurrence frequencies of amino acid residue types in known epitopes.
Conserved Domain Database and Search Service, v2.13	The Database is a collection of multiple sequence alignments for ancient domains and full-length proteins. It is used to identify the conserved domains present in a protein query sequence.
ABCPred	Predict *B cell epitope(s)* in an antigen sequence, using artificial neural network.
BcePred	Predicts linear B-cell epitopes, using physico-chemical properties.
Discotope 1.1	Predicts discontinuous B cell epitopes from protein three dimensional structures utilizing calculation of surface accessibility (estimated in terms of contact numbers) and a novel epitope propensity amino acid score.
CEP	The algorithm predicts epitopes of protein antigens with known structures. It uses accessibility of residues and spatial distance cut-off to predict antigenic determinants (ADs), conformational epitopes (CEs) and sequential epitopes (SEs).
NetMHC 2.2	Predicts binding of peptides to a number of different HLA alleles using artificial neural networks (ANNs) and weight matrices.
MHCPred 2.0	**MHCPred** uses the **additive method** to predict the binding affinity of major histocompatibility complex (**MHC**) class I and II molecules and also to the Transporter associated with Processing (TAP). Allele specific Quantitative Structure Activity Relationship (**QSAR**) models were generated using partial least squares (PLS).
Bimas	Ranks potential 8-mer, 9-mer, or 10-mer peptides based on a predicted half-time of dissociation to HLA class I molecules. The analysis is based on coefficient tables deduced from the published literature by Dr. Kenneth Parker, Children's Hospital Boston.
Propred	Predicts MHC Class-II binding regions in an antigen sequence, using quantitative matrices derived from published literature. It assists in locating promiscous binding regions that are useful in selecting vaccine candidates.
AlgPred	Predicts allergens in query protein based on similarity to known epitopes, searching MEME/MAST allergen motifs using MAST and assigning a protein allergen if it have any motif, searching based on SVM modules and searching with BLAST, searching against 2890 allergen-representative peptides obtained from Bjorklund et al. 2005 and assigning a protein allergen if it has a BLAST hit.
Allermatch	Predicts the potential allergenicity of proteins by bioinformatics approaches as recommended by the Codex alimentarius and FAO/WHO Expert consultation on allergenicity of foods derived through modern biotechnology.
Weballergen	Predicts the potential allergenicity of proteins. The query protein is compared against a set of pre-built allergenic motifs that have been obtained from 664 known allergen proteins.

available but also time and cost efficient. NERVE software pipeline can be divided in two parts: data production and storage and data selection. Six different scripts screen the entire proteome to mine and infer information that flows into a MySQL table. A seventh script uses four filters (LOC, localization; TOP topology; PAD, probability of being adhesin; SHP, similarity to human proteins) and analyzes values created by steps 1 through 5 to select and rank vaccine candidates that are then presented in a html table with links to relevant data.

Predicting epitopes, the holy grail of immunoinformatics: As described above, bioinformatics tools can assist in different ways to vaccine design and development. An important element that defines vaccine efficacy is the identification of antigens relevant to the immune response, capable to elicit protective immunity, thus predicting immunologically relevant regions or sequences in proteins represents a fundamental challenge for immunoinformatics.

An epitope, also known as antigenic determinant, is the region of a macromolecule, protein or carbohydrate which is recognized by the immune system, in particular by antibodies, B and T cells, leading to the activation of the immune system. Therefore, in immunological terms and for vaccine development epitopes are the key component of a molecule and their identification can help in the design of vaccine components and immunodiagnostic reagents. Most epitopes recognized are three-dimensional surface features of an antigen; exceptions are linear epitopes, which are determined by the amino acid sequence (the primary structure) rather than by the 3D shape (tertiary structure) of a protein. Although in minority in nature, most of the available epitope prediction methods focus on continuous epitopes. These prediction methods are based upon the amino acid physical-chemical properties such as hydrophilicity, solvent accessibility, secondary structure, flexibility, and antigenicity. In addition, based on linear epitope databases such as Bcipep and FIMM, there are also some methods that use machine learning algorithms. A series of such algorithms were designed to recognize complex patterns and make intelligent decisions based on data, to locate linear epitopes using proteins as input. Unlike linear epitope prediction, only few studies have been performed to predict discontinuous epitopes using structural information of a target protein. Although such studies are of great importance, the small number of available structures of antibody-antigen complexes limits these studies. Several databases (see Box 4), such as IEDB, SACS, and CED, collected all existing structures of antibody-antigen complexes from the PDB bank. Another important aspect of epitope identification is the assessment of epitope conservation. As discussed above, in vaccine design, a high level of conservation can provide broader

Box 4. Description of databases used in immunoinformatics

IEDB	The Immune Epitope Database (IEDB, www.iedb.org) provides a catalog of experimentally characterized B and T cell epitopes, as well as data on Major Histocompatibility Complex (MHC) binding and MHC ligand elution experiments.
SACS	Summary of Antibody Crystal Structures (SACS) database contains information on antibody structures present in the Protein Data Bank.
CED	The conformational Epitope database CED provides a collection of conformational epitopes and related information including the residue make up and location of the epitope, the immunological property of the epitope, the source antigen and corresponding antibody of the epitope.
FIMM	FIMM database (http://sdmc.krdl.org.sg:8080/fimm) contains fully referenced data on protein antigens, major histocompatibility complex (MHC) molecules, MHC associated peptides and relevant disease associations.
BCIPEP	BCIPEP is a database of experimentally determined linear B-cell epitopes of varying immunogenicity collected from literature and other publicly available databases.

protection across multiple strains of pathogens. Thus, tools for epitope conservation evaluation and epitope prediction are fundamental and represent a subject of intense research.

As example, Liang and colleagues (Liang, 2009) described a new antigen epitope prediction method, which uses ConsEnsus Scoring (EPCES) from six different scoring functions - residue epitope propensity, conservation score, side-chain energy score, contact number, surface planarity score, and secondary structure composition. Applied to unbounded antigen structures from an independent test set, EPCES was able to predict antigenic eptitopes with 47.8% sensitivity, 69.5% specificity and an AUC value of 0.632. The performance of the method is statistically similar to other published methods. The AUC value of EPCES is slightly higher compared to the best results of existing algorithms suggesting that the combination of criteria led to better performance.

VACCINOMICS: IMPROVING VACCINE DESIGN BY STUDYING HOST GENETIC VARIATION

Recently systems biology -based approaches are focusing on the role of human genetic variation in vaccine design. Thus, a new discipline in the omics world has been born, vaccinomics (Poland, 2008). This field refers to the examination of heterogeneity in host genetic markers at the individual or population level that may result in variations in humoral, cell-mediated, and/or innate immune responses to vaccines expecting to predict and optimize vaccine outcomes. The development of vaccinomics and personalized vaccinology was enabled by the completion of the first phase of the Human Genome Project (that provides a description of genetic similarities and differences between humans) and the first phase of the international HapMap, and accelerated by new molecular assay tools that allow high-throughput detection of gene variations, particularly single nucleotide polymorphism (SNP) and linkage disequilibrium maps. Among the main accomplishments of vaccinomics, we can mention the (i) demonstration that widespread polymorphism of immune response genes are critical to the development of protective immune responses, (ii) vaccinomics initially focused on individual gene polymorphism associations, broadened to haplotypes and extended haplotypes, and is now evolving towards the ultimate: the real-time ability to understand, at the whole genome level, the effects of whole genome gene/polymorphic activation, suppression and modification in the immune responses to antigens in a predictive manner, (iii) recognition that although gene polymorphisms throughout the pathway from infection through the development of immune responses are important, so far there seem to be few specific polymorphisms that are dominant determinants of the immune responses (i.e., few 'all or none' polymorphisms) and, (iv) immune response gene polymorphisms can have positive, negative or neutral effects on adaptive immune responses, and these polymorphisms explain individual variations in immune responses. In a seminal work, Querec et al. (Querec, 2009) used a systems biology approach to identify early gene 'signatures' that predicted immune responses in humans vaccinated with yellow fever vaccine YF-17D. Thus, the authors verified that vaccination induced genes that regulate virus innate sensing and type I interferon production. Computational analyses identified a gene signature, which correlates with the predicted YF-17D $CD8^+$ T cell responses with up to 90% accuracy in an independent, blinded trial. A distinct signature, including B cell growth factor TNFRS17, predicted the neutralizing antibody response with up to 100% accuracy. These findings suggest that it is possible to predict immunogenicity and/or protective efficacy of emerging vaccines by using systems biology-based approaches. In addition, this new method for measuring early vaccine efficacy provides testable hypotheses for the mechanisms that underlie immunogenicity thus offering the opportunity to intervene and improve vaccine efficacy.

A PROMISING TREND: SYNTHETIC WHOLE ORGANISM VACCINES

Genetic manipulation and DNA recombinant technologies have been driving forces in biomedical research since the early 1950s. In a groundbreaking work by the group of J. Craig Venter, published by Gibson et al. (2008), described a multistage process to construct the complete genome of *Mycoplasma genitalium*. Thus, the authors synthesized a 582,970-base pair genome named JCVI-1.0, containing all the genes of wild-type *M. genitalium* G37 except MG408, which was disrupted by an antibiotic marker to block pathogenicity and to allow for selection. To identify the genome as synthetic, they inserted "watermarks" at intergenic sites known to tolerate transposon insertions. The complete synthetic genome was assembled by transformation-associated recombination cloning in the yeast *Saccharomyces cerevisiae*, then isolated and sequenced.

This work marked the beginning of a new discipline, synthetic genomics. This new field advocates the generation of organisms artificially using genetic material. It involves the design and assembly of genes, gene pathways, chromosomes and even whole genomes by using a combination of methods for the chemical synthesis of DNA with computational techniques. The goal of synthetic genomics is to obtain new genomes able to code for new types of cells with desired properties. In vaccine development, the chemical synthesis of genetic material could involve the creation of new proteins, the reduction in cost for protein engineering and structural analysis or the possibility to generate recombinant vaccines against emergent microbial diseases. To further develop this idea we can envision a "synthetic whole organism vaccine". Recently the first steps in this direction were taken by the group of Craig Venter. In a groundbreaking paper Venter and coworkers created the first "artificial cell" by chemically synthesizing a full genome that was implanted in a recipient cytoplasm leading to the replication of this new cell (Gibson et al., 2010).

With all the knowledge generated by the omics technologies as part of the system-based approaches that is used to understand host-pathogen interaction and advance vaccine development, we could in the future design synthetic organisms that could efficiently (i) induce strong immune responses that could elicit protective immunity with no adverse reactions and/or (ii) deliver antigens with high efficiency leading to strong and protective immune responses. So far we have taken the first steps in this direction.

CONCLUSION

Classical approaches to vaccine development are time-consuming, biased towards the identification of abundant antigens which may not lead to immunity, and are dependent on the ability to cultivate the pathogen under laboratory conditions.

The advent of the 'omics' revolution has led us into a new era of vaccinology. With a continous flow and an enormous amount of data, the omics field has provided us with the opportunity to look at host-pathogen interactions in a holistic way, enabling us in the future to design vaccines using a comprehensive base of knowledge. Reverse vaccinology (i.e. genome-based approach to vaccine development) marked the first step towards this goal. With the sequencing of many genomes of bacteria, viruses and parasites already completed or that will be completed in the future, many vaccines previously thought to be impossible will become a reality. In addition, the comparison of the genomes of multiple strains of a single pathogen, known as 'pan-genomics', has further expanded our understanding of virulence mechanisms. Necessary advances in computational techniques in combination with integrative strategies provide us nowadays with the ability to make high-confidence predictions of complex biological processes, with further improvements possibly leading us in a future of computer-aided "synthetic whole organism vaccines".

The paradigm shift in thinking about biological systems led to dramatic changes in the world of vaccines. Highly efficient sophisticated techniques and an in-depth knowledge of biological processes of both host and pathogen will provide us with the tools to defeat infectious microorgansims. With the recent evolution in the field of vaccinology, we expect a tremendous acceleration of vaccine development. Our anticipations for the future are therefore marked by great optimism.

REFERENCES

Andre, F.E. (1990). Overview of a 5-year clinical experience with a yeast-derived hepatitis B vaccine. *Vaccine, 8 Suppl*, S74-78; discussion S79-80.

Brosch, R., Gordon, S. V., Garnier, T., Eiglmeier, K., Frigui, W., & Valenti, P. (2007). Genome plasticity of BCG and impact on vaccine efficacy. *Proceedings of the National Academy of Sciences of the United States of America, 104*(13), 5596–5601. doi:10.1073/pnas.0700869104

Bruggeman, F. J., & Westerhoff, H. V. (2007). The nature of systems biology. *Trends in Microbiology, 15*(1), 45–50. doi:10.1016/j.tim.2006.11.003

Chaudhuri, R., Ahmed, S., Ansari, F. A., Singh, H. V., & Ramachandran, S. (2008). MalVac: Database of malarial vaccine candidates. *Malaria Journal, 7*, 184. doi:10.1186/1475-2875-7-184

Colditz, G. A., Berkey, C. S., Mosteller, F., Brewer, T. F., Wilson, M. E., & Burdick, E. (1995). The efficacy of bacillus Calmette-Guerin vaccination of newborns and infants in the prevention of tuberculosis: Meta-analyses of the published literature. *Pediatrics, 96*(1 Pt 1), 29–35.

Cole, S. T., Brosch, R., Parkhill, J., Garnier, T., Churcher, C., & Harris, D. (1998). Deciphering the biology of Mycobacterium tuberculosis from the complete genome sequence. *Nature, 393*(6685), 537–544. doi:10.1038/31159

Daily, J. P., Le Roch, K. G., Sarr, O., Ndiaye, D., Lukens, A., & Zhou, Y. (2005). In vivo transcriptome of Plasmodium falciparum reveals overexpression of transcripts that encode surface proteins. *The Journal of Infectious Diseases, 191*(7), 1196–1203. doi:10.1086/428289

Dong, J., Olano, J. P., McBride, J. W., & Walker, D. H. (2008). Emerging pathogens: Challenges and successes of molecular diagnostics. *The Journal of Molecular Diagnostics, 10*(3), 185–197. doi:10.2353/jmoldx.2008.070063

El-Sayed, N. M., Myler, P. J., Bartholomeu, D. C., Nilsson, D., Aggarwal, G., & Tran, A. N. (2005). The genome sequence of Trypanosoma cruzi, etiologic agent of Chagas disease. *Science, 309*(5733), 409–415. doi:10.1126/science.1112631

Farley, M. M. (1995). Group B streptococcal infection in older patients. Spectrum of disease and management strategies. *Drugs & Aging, 6*(4), 293–300. doi:10.2165/00002512-199506040-00004

Fine, P. E. (1995). Variation in protection by BCG: Implications of and for heterologous immunity. *Lancet, 346*(8986), 1339–1345. doi:10.1016/S0140-6736(95)92348-9

Fleischmann, R. D., Adams, M. D., White, O., Clayton, R. A., Kirkness, E. F., & Kerlavage, A. R. (1995). Whole-genome random sequencing and assembly of Haemophilus influenzae Rd. *Science, 269*(5223), 496–512. doi:10.1126/science.7542800

Gan, W., Zhao, G., Xu, H., Wu, W., Du, W., & Huang, J. (2010). Reverse vaccinology approach identify an Echinococcus granulosus tegumental membrane protein enolase as vaccine candidate. *Parasitology Research*, *106*(4), 873–882. doi:10.1007/s00436-010-1729-x

Gardner, M. J., Hall, N., Fung, E., White, O., Berriman, M., & Hyman, R. W. (2002). Genome sequence of the human malaria parasite Plasmodium falciparum. *Nature*, *419*(6906), 498–511. doi:10.1038/nature01097

Gibson, D. G., Benders, G. A., Andrews-Pfannkoch, C., Denisova, E. A., Baden-Tillson, H., & Zaveri, J. (2008). Complete chemical synthesis, assembly, and cloning of a Mycoplasma genitalium genome. *Science*, *319*(5867), 1215–1220. doi:10.1126/science.1151721

Gibson, D. G., Glass, J. I., Lartigue, C., Noskov, V. N., Chuang, R. Y., & Algire, M. A. (2010). Creation of a bacterial cell controlled by a chemically synthesized genome. *Science*, *329*(5987), 52–56. doi:10.1126/science.1190719

Giuliani, M. M., Adu-Bobie, J., Comanducci, M., Arico, B., Savino, S., & Santini, L. (2006). A universal vaccine for serogroup B meningococcus. *Proceedings of the National Academy of Sciences of the United States of America*, *103*(29), 10834–10839. doi:10.1073/pnas.0603940103

Hall, N., Karras, M., Raine, J. D., Carlton, J. M., Kooij, T. W., & Berriman, M. (2005). A comprehensive survey of the Plasmodium life cycle by genomic, transcriptomic, and proteomic analyses. *Science*, *307*(5706), 82–86. doi:10.1126/science.1103717

Hernandez, S., Gomez, A., Cedano, J., & Querol, E. (2009). Bioinformatics annotation of the hypothetical proteins found by omics techniques can help to disclose additional virulence factors. *Current Microbiology*, *59*(4), 451–456. doi:10.1007/s00284-009-9459-y

Jacobs, M. R. (2004). Streptococcus pneumoniae: Epidemiology and patterns of resistance. *The American Journal of Medicine*, *117*(Suppl 3A), 3S–15S.

Johri, A. K., Paoletti, L. C., Glaser, P., Dua, M., Sharma, P. K., & Grandi, G. (2006). Group B Streptococcus: Global incidence and vaccine development. *Nature Reviews Microbiology*, *4*(12), 932–942. doi:10.1038/nrmicro1552

Kandpal, R., Saviola, B., & Felton, J. (2009). The era of 'omics unlimited. *BioTechniques*, *46*(5), 351–352, 354–355. doi:10.2144/000113137

Kleinstein, S. H. (2008). Getting started in computational immunology. *PLoS Computational Biology*, *4*(8), e1000128.

Liang, S., Zheng, D., Zhang, C., & Zacharias, M. (2009). Prediction of antigenic epitopes on protein surfaces by consensus scoring. *BMC Bioinformatics*, *10*, 302. doi:10.1186/1471-2105-10-302

Maione, D., Margarit, I., Rinaudo, C. D., Masignani, V., Mora, M., & Scarselli, M. (2005). Identification of a universal Group B streptococcus vaccine by multiple genome screen. *Science*, *309*(5731), 148–150. doi:10.1126/science.1109869

MMWR. (1999). Impact of vaccines universally recommended for children--United States, 1990-1998. *Morbidity and Mortality Weekly Report*, *48*(12), 243–248.

Mora, M., Veggi, D., Santini, L., Pizza, M., & Rappuoli, R. (2003). Reverse vaccinology. *Drug Discovery Today*, *8*(10), 459–464. doi:10.1016/S1359-6446(03)02689-8

Moxley, R. A., & Duhamel, G. E. (1999). Comparative pathology of bacterial enteric diseases of swine. *Advances in Experimental Medicine and Biology*, *473*, 83–101.

Mu, J., Awadalla, P., Duan, J., McGee, K. M., Keebler, J., & Seydel, K. (2007). Genome-wide variation and identification of vaccine targets in the Plasmodium falciparum genome. *Nature Genetics*, *39*(1), 126–130. doi:10.1038/ng1924

Mueller, A. K., Labaied, M., Kappe, S. H., & Matuschewski, K. (2005). Genetically modified Plasmodium parasites as a protective experimental malaria vaccine. *Nature*, *433*(7022), 164–167. doi:10.1038/nature03188

Musser, J. M., & Shelburne, S. A. III. (2009). A decade of molecular pathogenomic analysis of group A Streptococcus. *The Journal of Clinical Investigation*, *119*(9), 2455–2463. doi:10.1172/JCI38095

Nandyal, R. R. (2008). Update on group B streptococcal infections: Perinatal and neonatal periods. *The Journal of Perinatal & Neonatal Nursing*, *22*(3), 230–237.

Olsen, R. J., Shelburne, S. A., & Musser, J. M. (2009). Molecular mechanisms underlying group A streptococcal pathogenesis. *Cellular Microbiology*, *11*(1), 1–12. doi:10.1111/j.1462-5822.2008.01225.x

Pizza, M., Scarlato, V., Masignani, V., Giuliani, M. M., Arico, B., & Comanducci, M. (2000). Identification of vaccine candidates against serogroup B meningococcus by whole-genome sequencing. *Science*, *287*(5459), 1816–1820. doi:10.1126/science.287.5459.1816

Poland, G. A., Ovsyannikova, I. G., & Jacobson, R. M. (2008). Personalized vaccines: The emerging field of vaccinomics. *Expert Opinion on Biological Therapy*, *8*(11), 1659–1667. doi:10.1517/14712598.8.11.1659

Querec, T. D., Akondy, R. S., Lee, E. K., Cao, W., Nakaya, H. I., & Teuwen, D. (2009). Systems biology approach predicts immunogenicity of the yellow fever vaccine in humans. *Nature Immunology*, *10*(1), 116–125. doi:10.1038/ni.1688

Scarselli, M., Giuliani, M. M., Adu-Bobie, J., Pizza, M., & Rappuoli, R. (2005). The impact of genomics on vaccine design. *Trends in Biotechnology*, *23*(2), 84–91. doi:10.1016/j.tibtech.2004.12.008

Serruto, D., Adu-Bobie, J., Capecchi, B., Rappuoli, R., Pizza, M., & Masignani, V. (2004). Biotechnology and vaccines: Application of functional genomics to Neisseria meningitidis and other bacterial pathogens. *Journal of Biotechnology*, *113*(1-3), 15–32. doi:10.1016/j.jbiotec.2004.03.024

Siracusano, A., Teggi, A., & Ortona, E. (2009). Human cystic echinococcosis: Old problems and new perspectives. *Interdisciplinary Perspectives on Infectious Diseases*, *2009*, 474368. doi:10.1155/2009/474368

Song, Y., La, T., Phillips, N. D., Bellgard, M. I., & Hampson, D. J. (2009). A reverse vaccinology approach to swine dysentery vaccine development. *Veterinary Microbiology*, *137*(1-2), 111–119. doi:10.1016/j.vetmic.2008.12.018

Tettelin, H. (2009). The bacterial pan-genome and reverse vaccinology. *Genome Dynamics*, *6*, 35–47. doi:10.1159/000235761

Tettelin, H., Masignani, V., Cieslewicz, M. J., Eisen, J. A., Peterson, S., & Wessels, M. R. (2002). Complete genome sequence and comparative genomic analysis of an emerging human pathogen, serotype V Streptococcus agalactiae. *Proceedings of the National Academy of Sciences of the United States of America*, *99*(19), 12391–12396. doi:10.1073/pnas.182380799

Vivona, S., Bernante, F., & Filippini, F. (2006). NERVE: New enhanced reverse vaccinology environment. *BMC Biotechnology*, *6*, 35. doi:10.1186/1472-6750-6-35

Vivona, S., Gardy, J. L., Ramachandran, S., Brinkman, F. S., Raghava, G. P., & Flower, D. R. (2008). Computer-aided biotechnology: from immuno-informatics to reverse vaccinology. *Trends in Biotechnology*, *26*(4), 190–200. doi:10.1016/j.tibtech.2007.12.006

WHO. (2007). Report of the World Health Organization technical consultation on prevention and control of iron deficiency in infants and young children in malaria-endemic areas, Lyon, France, 12-14 June 2006. *Food and Nutrition Bulletin*, *28*(4Suppl), S489–S631.

Wizemann, T. M., Heinrichs, J. H., Adamou, J. E., Erwin, A. L., Kunsch, C., & Choi, G. H. (2001). Use of a whole genome approach to identify vaccine molecules affording protection against Streptococcus pneumoniae infection. *Infection and Immunity*, *69*(3), 1593–1598. doi:10.1128/IAI.69.3.1593-1598.2001

Xu, P., Widmer, G., Wang, Y., Ozaki, L. S., Alves, J. M., & Serrano, M. G. (2004). The genome of Cryptosporidium hominis. *Nature*, *431*(7012), 1107–1112. doi:10.1038/nature02977

Yang, X., Yang, H., Zhou, G., & Zhao, G. P. (2008). Infectious disease in the genomic era. *Annual Review of Genomics and Human Genetics*, *9*, 21–48. doi:10.1146/annurev.genom.9.081307.164428

ADDITIONAL READING

Ansari, H. R., Flower, D. R., & Raghava, G. P. (2010). AntigenDB: an immunoinformatics database of pathogen antigens. *Nucleic Acids Research*, *38*(Database issue), D847–D853. doi:10.1093/nar/gkp830

Bork, P., & Serrano, L. (2005). Towards cellular systems in 4D. *Cell*, *121*(4), 507–509. doi:10.1016/j.cell.2005.05.001

Evans, M. C. (2008). Recent advances in immunoinformatics: application of in silico tools to drug development. *Current Opinion in Drug Discovery & Development*, *11*(2), 233–241.

Friboulet, A., & Thomas, D. (2005). Systems Biology-an interdisciplinary approach. *Biosensors & Bioelectronics*, *20*(12), 2404–2407. doi:10.1016/j.bios.2004.11.014

Garfinkel, M. S., Endy, D., Epstein, G. L., & Friedman, R. M. (2007). Synthetic genomics | options for governance. *Biosecurity and Bioterrorism*, *5*(4), 359–362. doi:10.1089/bsp.2007.0923

Kanoi, B. N., & Egwang, T. G. (2007). New concepts in vaccine development in malaria. *Current Opinion in Infectious Diseases*, *20*(3), 311–316. doi:10.1097/QCO.0b013e32816b5cc2

Korber, B., LaBute, M., & Yusim, K. (2006). Immunoinformatics comes of age. *PLoS Computational Biology*, *2*(6), e71. doi:10.1371/journal.pcbi.0020071

Medini, D., Donati, C., Tettelin, H., Masignani, V., & Rappuoli, R. (2005). The microbial pan-genome. *Current Opinion in Genetics & Development*, *15*(6), 589–594. doi:10.1016/j.gde.2005.09.006

Medini, D., Serruto, D., Parkhill, J., Relman, D. A., Donati, C., & Moxon, R. (2008). Microbiology in the post-genomic era. *Nature Reviews Microbiology*, *6*(6), 419–430.

Mesarovic, M. D., Sreenath, S. N., & Keene, J. D. (2004). Search for organising principles: understanding in systems biology. *Systems Biology*, *1*(1), 19–27. doi:10.1049/sb:20045010

Mora, M., Donati, C., Medini, D., Covacci, A., & Rappuoli, R. (2006). Microbial genomes and vaccine design: refinements to the classical reverse vaccinology approach. *Current Opinion in Microbiology*, *9*(5), 532–536. doi:10.1016/j.mib.2006.07.003

Mora, M., Veggi, D., Santini, L., Pizza, M., & Rappuoli, R. (2003). Reverse vaccinology. *Drug Discovery Today*, *8*(10), 459–464. doi:10.1016/S1359-6446(03)02689-8

Muzzi, A., Masignani, V., & Rappuoli, R. (2007). The pan-genome: towards a knowledge-based discovery of novel targets for vaccines and antibacterials. *Drug Discovery Today*, *12*(11-12), 429–439. doi:10.1016/j.drudis.2007.04.008

Poland, G. A., Ovsyannikova, I. G., Jacobson, R. M., & Smith, D. I. (2007). Heterogeneity in vaccine immune response: the role of immunogenetics and the emerging field of vaccinomics. *Clinical Pharmacology and Therapeutics*, *82*(6), 653–664. doi:10.1038/sj.clpt.6100415

Rinaudo, C. D., Telford, J. L., Rappuoli, R., & Seib, K. L. (2009). Vaccinology in the genome era. *The Journal of Clinical Investigation*, *119*(9), 2515–2525. doi:10.1172/JCI38330

Seib, K. L., Dougan, G., & Rappuoli, R. (2009). The key role of genomics in modern vaccine and drug design for emerging infectious diseases. *PLOS Genetics*, *5*(10), e1000612. doi:10.1371/journal.pgen.1000612

Serruto, D., Adu-Bobie, J., Capecchi, B., Rappuoli, R., Pizza, M., & Masignani, V. (2004). Biotechnology and vaccines: application of functional genomics to Neisseria meningitidis and other bacterial pathogens. *Journal of Biotechnology*, *113*(1-3), 15–32. doi:10.1016/j.jbiotec.2004.03.024

Serruto, D., & Rappuoli, R. (2006). Post-genomic vaccine development. *FEBS Letters*, *580*(12), 2985–2992. doi:10.1016/j.febslet.2006.04.084

Serruto, D., Serino, L., Masignani, V., & Pizza, M. (2009). Genome-based approaches to develop vaccines against bacterial pathogens. *Vaccine*, *27*(25-26), 3245–3250. doi:10.1016/j.vaccine.2009.01.072

Tettelin, H. (2009). The Bacterial Pan-Genome and Reverse Vaccinology. *Genome Dynamics*, *6*, 35–47. doi:10.1159/000235761

Tettelin, H., & Feldblyum, T. (2009). Bacterial genome sequencing. *Methods in Molecular Biology (Clifton, N.J.)*, *551*, 231–247. doi:10.1007/978-1-60327-999-4_18

Tong, J. C., & Ren, E. C. (2009). Immunoinformatics: Current trends and future directions. *Drug Discovery Today*, *14*(13-14), 684–689. doi:10.1016/j.drudis.2009.04.001

van Vliet, A. H. (2009). Next generation sequencing of microbial transcriptomes: challenges and opportunities. *FEMS Microbiology Letters*, *302*(1), 1–7. doi:10.1111/j.1574-6968.2009.01767.x

Yang, X., & Yu, X. (2009). An introduction to epitope prediction methods and software. *Reviews in Medical Virology*, *19*(2), 77–96. doi:10.1002/rmv.602

KEY TERMS AND DEFENITIONS

Immunoinformatics: A set of bioinformatic tools designed to aid in vaccine discovery.

Pan-Genome Vaccinology: A genome comparative approach to identify vaccine candidates.

Reverse Vaccinology: A genome-based approach to identify vaccine candidates.

Vaccines: An immunogen designed to stimulate a protective immune response.

Vaccinomics: An emergent field aimed to personalized vaccines.

Chapter 4
Current Omics Technologies in Biomarker Discovery

Wei Ding
Merck & Co., Inc., USA

Ping Qiu
Merck & Co., Inc., USA

Yan-Hui Liu
Merck & Co., Inc., USA

Wenqing Feng
Accela Sciences, LLC, USA

ABSTRACT

Biomarkers are playing an increasingly important role in drug discovery and development and can be applied for many purposes, including disease mechanism study, diagnosis, prognosis, staging, and treatment selection. Advances in high-throughput "omics" technologies, including genomics, transcriptomics, proteomics and metabolomics, significantly accelerate the pace of biomarker discovery. Comprehensive molecular profiling using these "omics" technology has become a field of intensive research aiming at identifying biomarkers relevant for improved diagnostics and therapeutics. Although each "omics" technology plays important roles in biomarker research, different "omics" platforms have different strengths and limitations. This chapter aims to give an overview of these "omics" technologies and their current application in the biomarker discovery.

INTRODUCTION

Biomarker research is an interdisciplinary field that bridges basic scientific research and drug discovery with clinical development. Offering great potential to stratify patient populations, quantify drug benefits, improve risk assessment and evaluate the impact of regulatory actions, biomarker fits perfectly with the vision of personalized medicine. Biomarker research has been a central focus in many research labs across academia, government agencies, and the pharmaceutical industry and is evolving at a fast pace.

DOI: 10.4018/978-1-60960-491-2.ch004

There are many different ways to define biomarkers based on their applications and molecular properties. The U.S. Food and Drug Administration (FDA) defines a biomarker as "a characteristic that is objectively measured and evaluated as an indicator of normal biological or pathogenic processes or pharmacological responses to a therapeutic intervention". A biomarker can be genetic variations in DNA, gene expression profiles in tissue biopsies, proteins, metabolites or lipids in blood, etc. Hence, the discovery, analytical validation and qualification of biomarkers require multidisciplinary skills and a proper application of technologies available at present.

There are four types of biomarker application, *target engagement biomarkers, efficacy biomarkers, Safety biomarkers and Surrogate endpoint biomarker.*

- *Target engagement biomarkers*: occur early in the pathophysiological cascade and inform on physical or biological interactions with the drug target. used to establish pharmacological response in the pre-clinical animal model;
- *Efficacy biomarkers*: used to estimate the efficacy of drug;
- *Safety biomarkers*: used to assess predict or anticipate toxicity or adverse events;
- *Surrogate endpoint biomarkers:* used in therapeutic trials as a substitute for a clinically meaningful endpoint that is a direct measure of how a patient feels, functions, or survives and is expected to predict the effect of the therapy.

While many existing biological and analytical technologies have been directly applied to biomarker research, many novel high-throughput technologies, especially in the fields of genomics, transcriptomics, proteomics, and metabolomics, have been developed and made it easier to interrogate hundreds, or even thousands of potential biomarkers simultaneously, without prior knowledge of the underlying biology or pathophysiology of system being studied. These "omics" technologies have been used for identifying potential biomarkers at different molecular levels and revolutionized the methods of biomarker discovery. The goal of the chapter is to provide a review of these "omics" technologies and their application in the biomarker discovery. The issues on the analysis, validation, and qualification of biomarkers based on these technologies are also discussed.

GENETIC BIOMARKER

A genetic biomarker is a gene or DNA sequence with a known location on a chromosome and associated with a particular gene or trait. A genetic biomarker may be a short DNA sequence, such as a sequence surrounding a single base-pair change (single nucleotide polymorphism, SNP), or a long one, like microsatellites.

The history of human genetics has focused on mapping regions of the genome that can explain part or all of a disease or human trait. With the completion of the Human Genome Project in 2003, researchers began to pinpoint areas of the genome that varied between individuals. Shortly thereafter, they discovered that the most common type of DNA sequence variation found in the genome is the single nucleotide polymorphism (International HapMap Consortium, 2005; Sachidanandam et al., 2001; Chanock, 2001). A worldwide effort known as the HapMap Project seeks to identify and localize these and other genetic variants, and to learn how the variants are distributed within and among populations from different parts of the world. To date, the project has identified over 10 million SNPs across the human genome.

In the early days, linkage study and candidate gene approach are popular ways to identify genetic biomarkers.

Linkage Analysis

Linkage analysis is an approach that has used microsatellite markers across the genome to scan for markers that segregate within a family (NIH/CEPH Collaborative Mapping Group, 1992; Elston & Cordell, 2001). Disease genes are mapped by measuring recombination against a panel of different markers spread over the entire genome. In most cases, recombination will occur frequently, indicating that the disease gene and marker are far apart. However, some markers will tend not to recombine with the disease gene and these are said to be linked to the disease due to their proximity. Ideally, close markers are identified that flank the disease gene and define a candidate region of the genome between 1 and 5 million bp in length. The gene responsible for the disease lies somewhere in this region. Because of the wide spacing of markers across the genome, signals often pointed to regions over multiple megabases. To find the causative mutations requires deep sequencing of those megabase regions which is a daunting task.

Association Analysis

Allelic association refers to a correlation between a particular marker allele and a disease trait. There are two basic types of association studies: case-control studies and family-based studies. Case-control studies compare allele frequencies between a group of unrelated, affected individuals and an unrelated group of matched controls. The controls should be matched to the cases with respect to factors such as age, gender and ethnicity, so that they differ only in disease status. Controlling for these factors allows for the interpretation of differences in allele frequencies as evidence for association. The primary disadvantage of this approach is that false positive associations can result if there is underlying population substructure in the data of which the investigator is not aware of. Methods have been developed to test and correct for population stratification. To deal with population stratification at the study design level, family-based association was introduced. However they are in general less powerful per genotype than case-control methods and the necessary sample structure may be more difficult to collect, particularly for late onset diseases such as Alzheimer's disease etc. (Risch & Teng, 1998, 1999)

In an early genetic association study, the analysis consists of a comparison of the frequency of a handful of annotated marker allele between cases and controls, in search of a statistical difference that can be reflected in an estimated effect size. This is so called candidate gene association study. Most candidate gene case-control studies of complex traits to date have been disappointing. Many initially positive reports have not withstood replication in other cohorts (Helgadottir et al., 2005; Lohmussaar et al., 2005). There are many reasons for the overall failure of candidate gene approaches. One of them is the very low pretest probability that any given gene (out of the estimated 30,000 genes in the human genome) contributes to the susceptibility of a complex trait despite a priori hypotheses based on cell, tissue, or animal model experiments. Other reasons include the use of underpowered sample sizes, multiple testing, phenotypic heterogeneity, poor phenotype characterization, selection bias, population stratification, and incomplete knowledge of the complete set of allelic variants in the region of a candidate gene (Botstein & Risch, 2003;Tabor et al., 2002)

With the completion of the human genome sequencing project, comprehensive HapMap information and the advance of the genotyping technology, genome wide association study (GWAS) become feasible. In such a study, the distribution of SNPs is determined in hundreds or even thousands of people with and without a particular disease. By studying which SNPs co-occur with disease symptoms, researchers can make a statistical estimate regarding the level of increased risk associated with each SNP. For instance, in a 2007 study conducted in the United Kingdom, researchers identified people affected by seven common disorders, and they then genotyped 2,000 people in each disease category (for a total of 14,000 individuals studied). Next, these individuals were compared to 3,000 genotyped controls who did not have any of the seven disorders. As a result of these comparisons, the researchers were able to identify new genetic biomarkers that point to an increased risk for multifactorial disorders such as heart disease and diabetes (Wellcome Trust Case Control Consortium, 2007). It was announced in July 2008 that this study will be expanded to include an additional 36,000 individuals and will focus on examining the genetic contributions to a total of 14 common disorders, as well as to individuals' responses to certain drugs.

GWAS have several advantages over alternative candidate gene approach. In contrast to candidate gene studies, which select genes for study based on known or suspected disease mechanisms, GWAS permit a comprehensive scan of the genome in an unbiased fashion and thus have the potential to identify totally novel susceptibility factors. Genome wide association studies have two key advantages in comparison to family linkage-based approaches. First, they are able to capitalize on all meiotic recombination events in a population, rather than only those in the families studied. Because of this, association signals are localized to small regions of the chromosome containing only a single to a few genes which enable rapid detection of the actual disease susceptibility gene. Second, most common disorders are caused by multiple genes and mutations in one gene only cause modest increase in risk. In compare with linkage study, GWAS has more advantage in identifying this type of mutations.

Despite the advantages of GWAS, the power to detect association between genetic variation and disease is a function of several factors, including the frequency of the risk allele, the relative risk conferred by the disease-associated allele, the correlation between the genotyped marker and the risk allele, sample size, disease prevalence, and genetic heterogeneity of the sample population. Even though the first three factors are unknown prior to GWAS, a well designed study can influence their impact which can minimize the potential of false association signals and ensure high power to detect genes of modest risk. Power studies have shown that at least 2,000 to 5,000 samples for both cases and controls groups are required when using general populations. Matching of cases and controls with respect to geographic origin and ethnicity is critical for minimizing false positive signals due to population substructure. A second key success factor is having a comprehensive map of hundreds of thousands of carefully selected SNPs. Currently popular SNP arrays for genotyping, such as Affymetrix and Illumina both provide products containing more than 500,000 SNPs. Achieving high call rates and genotyping accuracy are also critically important, because small decreases in accuracy or increases in missing data can result in relatively large decreases in the power to detect disease genes. IT infrastructure and analytic tools/algorithms are also needed to properly store, manage, process and analyze the enormous data sets arising from GWAS for rapid data analysis. Since these analyses require considerable computing power to handle terabytes of data, genome-wide analyses are often limited to single SNPs with haplotype analyses performed once candidate regions are identified.

Like any other approaches, GWAS has its own limitations. Researchers must be cautious about giving too much weight to SNP profiles. Complex diseases are caused by the combination and interaction of environmental and genetic factors. Single genetic variant such as SNP only makes a small contribution to an individual's overall risk. Identifying a correlation between a genetic change and the incidence of a complex disease is limited to statistical estimation of increased risk for developing the disorder, rather than hard prediction. Therefore, findings from a GWAS normally cannot be directly applied to the prevention or treatment of disease. The full pathway of disease development and the involvement of all variables need to be understood before any medicinal interventions can be applied based on a SNP profile. For example, one situation that makes the link between SNPs and disease difficult to understand is the case in which a SNP is not located within an exon of a gene. In such instances, studies are required to investigate the possibility that the SNP lies in a promoter or enhancer region and somehow affects regulation of the causal gene. Occasionally, the results of a GWAS seem relatively straightforward; this is often the case when a variant is located in a causal gene of a multifactorial disorder (Steinthorsdottir et al., 2007). A well-known example of this is the link between certain alleles of the apolipoprotein E (ApoE) and the development of Alzheimer's disease. ApoE codes for a protein that helps carry cholesterol in the bloodstream, and it has three common alleles: e2, e3, and e4. Research has shown that having one or two copies of the ApoE e4 allele significantly increases a person's risk for developing Alzheimer's disease, but it does not guarantee development of this disorder (National Institute on Aging, 2008).

Over the past few years, scientific community experienced the explosion of data derived from GWAS studies. Savvy entrepreneurs are capitalizing on existing GWAS research. Consumer genomics companies such as 23andMe, deCODE genetics, Navigenics, and Knome now offer a range of personal genotyping and sequencing services to clients who are interested in learning their estimated genome-based risk for developing a number of common disorders. For about U.S. $1,000, you can have your entire genome scanned for markers that have been identified by GWAS and receive personalized risk calculations that are updated as new knowledge becomes available. Even though it's not clear whether these risk assessments will make a difference in patients' lives. However, it's certain that personal genetic profiles will continue to increase in their medical value as researchers cultivate more and more knowledge about the genetic and environmental factors that interact to contribute to the development of common disorders.

DNA Sequencing in Biomarker Discovery

DNA sequencing is routinely used in research and clinical for the detection of DNA sequence variants, single nucleotide changes, or small insertions or deletions, when the spectrum of DNA variation is unknown. Dideoxy sequencing (Sanger sequencing) is the most commonly used DNA sequencing methodology today. For example, somatic DNA sequence variation is often the cause of abnormal cell growth, regulation and ultimately tumorgenesis. This kind of mutation can be present in tumor tissue but not in normal tissue from the same individual. One successful example that a biomarker is successfully identified by sequencing is that of *BRAF*. Activating BRAF mutation were identified in 70% of malignant melanomas, 10% of colon cancer and a fraction of the other cancers in Cancer Genome Project by candidate gene sequencing approach (Davies et al., 2002). Inhibition of the activated version of BRAF could effectively treat these cancers, and BRAF inhibitors are currently in various stages of clinical development. A complete atlas of the molecular alterations present in different tumors of different cancer type is being carried out by the Cancer Genome Anatomy Project.

Since first introduced to the market in 2005, next-generation sequencing technologies start to partially supplant Sanger method due to their dramatic increases in cost-effective sequence throughput. The next-generation technologies commercially available today include the 454 GS20 pyrosequencing based instrument (Roche Applied Science) (Margulies et al., 2005), the Solexa 1G analyzer (Illumina, Inc.) (Bennett et al., 2005), the SOLiD instrument from Applied Biosystems (Shendure et al., 2005), and the Heliscope from Helicos, Inc. The next-generation technologies have been used for standard sequencing applications, such as genome sequencing, resequencing and for novel applications previously unexplored by Sanger sequencing. Compared to Sanger sequencing, advantages of the next-generation technologies mentioned above alleviate the need for in vivo cloning by clonal amplification of spatially separated single molecules using either emulsion PCR (454/Roche and ABI/SOLiD) or bridge amplification on solid surface (Illumina/Solexa). In addition to providing a means for cloning-free amplification, these methods use single-molecule templates allowing for the detection of heterogeneity in a DNA sample (e.g., identifying mutations present only in a subpopulation of cells), which is a significant advantage over Sanger sequencing (Bentley, 2006; Thomas et al., 2006; Morozova & Marra, 2008).

TRANSCRIPTOMIC BIOMARKER

Transcriptomic biomarkers are genes, or microRNAs (miRNAs) whose changes in expression are associated with a biological processes, pathogenic processes, or pharmacologic responses to a therapeutic intervention. The use of transcriptomic biomarkers in drug and diagnostic development and regulatory decision making is showing a lot of potential. There are two key application areas. First, at the drug discovery stage, transcriptomic biomarkers applied to compound profiling offer efficacy and toxicity data, allowing better decision making at earlier stages and reducing late-stage attrition. Second, in the clinical setting, transcriptomic biomarkers have the promise in disease diagnosis/prognosis; treatment, patient, or dose selection; and clinical safety and efficacy assessment.

Recent advances in the technology of massive parallel RNA expression profiling using microarrays are beginning to revolutionize biomedical research and pharmacological discovery. With these new and powerful tools, researchers can now examine the dynamics of a whole biological system by simultaneously interrogating the expression of tens of thousands of transcripts. In addition to providing important insights into biological systems, microarray technologies are being successfully integrated into the transcriptomic biomarker discovery.

Mechanistic Approach and Predictive Approach

There are two approaches for transcriptomic biomarker discovery, mechanistic approach and predictive approach. For mechanistic approach, the discovery process starts with identification of transcripts, and in most cases, genes, that involved in a specific cellular process or signaling pathway, followed by asking whether gene expression changes could predict the outcome of the biological process. However, very often the mechanistic approach could not yield the best predictive model since it's biased to our prior knowledge. Apparently we know too little about the complicated biological system. Most modern biomarkers are found using predictive approach. Mechanistic explanations can be found later (e.g. ABCB1 for drug resistance); sometimes they are unclear even after years of biomarker use (e.g. Ki-67 for cell proliferation).

On the other hand, the predictive approach follows a common theme. Typically, genome wide microarrays are used in a preliminary screen on a relatively small set of cell lines or tissue biopsies. Attempts to associate a specific biological effect with a single gene biomarker often fails due to the complexity of biological systems. Biomarker responses are often complex, resulting from changes in the expression of multiple genes simultaneously. The most powerful and successful approach is to use combinatorial biomarkers. A combinatorial biomarker is a finite set of genes whose joint expression profile can be statistically associated with a particular biological response. The identified candidate biomarkers need to be confirmed using alternative methods and larger number of independent sample to verify the predictive/classification power, before they can be considered validated and advanced for testing as useful biomarkers.

Statistical Analysis

Due to the nature and characteristics of microarray data, statistical evaluation is an integral part of data analysis and is essential to the biomarker validation process. Many statistical methods have found their applications in microarray studies, which include experimental design, quality assessment, data processing and normalization, biomarker selection, classification etc. Although here we are particularly discussing transcriptomic biomarker discovery using microarray technology, many of these methods can be further extended to other types of high-throughput technology such as proteomics and metabolomics.

Experimental Design

Microarray data analysis should start with experimental design. A carefully planned experiment will not only minimize experimental costs and yield the high quality data, but also be critical for the statistical analysis and data interpretation. In general, microarray experimental design should take several factors into consideration:

1. **Scientific purpose.** This is hypothesis-driven; a proper experimental design is instrumental to deriving the correct answer to the biological question. Many basic experimental protocols have been proposed and described in detail, e.g. randomization in sample selection and treatment assignment, direct or indirect design, single factor or multi-factorial design etc (Yang et al., 2002; Churchill, 2002). The objective is to make the analysis of the data and the interpretation of the results as simple and as powerful as possible. In addition to technical artifacts, physiological, pharmacological, and pathological features of human patients or animal models can also be the source of spurious biomarker results. Appropriate controls should also be included in the sample sets. It is seldom effective to simply compare data from a group of diseased individuals only to a group of healthy ones. Controls matched for such characteristics as age, gender, as well as samples from patients with other diseases with similar clinical profiles can improve the relevance of the data. Furthermore, other variables, such as location or date of sample collection, might also confound the distinction between the compared classes. Randomization of confounding variables related to experimental conditions under your control is important.
2. **Sample size calculation.** Sampling size should be determined by estimating the number of samples required to attain statistical relevance. The challenge is a lack of good estimates of variances of gene expression levels and test effects, especially at the design stage. So far, there are no comprehensive methods for planning sample size in microarray gene expression studies. But a few suggestions have appeared in the literature recently and some applications, e.g. BRB tools (http://brb.nci.nih.gov/) and Bioconductor packages sizepower and SSPA, are developed to help estimate sample size and power for the experiment planning (Dobbin et al., 2007, 2008; Hu et al., 2005; Jung, 2005; Qiu & Lee, 2006; Oura et al., 2009; Ruppert et al., 2007; van Iterson et al., 2009). A verification strategy with alternative technology, such as northern or western blot analyses, Taqman-based RT-PCR assays, or in situ hybridization, should also be planned as a follow up to the experimental results. The amount of verification that is carried out can influence the sample size calculation.

In addition, estimation of the availability of RNA sample and cost of reagents and chips is also an experimental design consideration.

3. **Sample and information collection.** Proper sample collection, handling, and storage are required to produce robust biomarker discovery results. Standardized protocols that maintain consistency in the timing of sample collection, equipment and reagents, and methods and timing of all processing steps are essential to minimize systematic bias. Information on potential confounders should be gathered as comprehensively as possible, and attention should be paid to whether the assay will be robust enough in the real world of sample procurement. This need has been addressed by FDA in a concept article on drug and diagnostic codevelopment and in a series of guidance documents (FDA, 2005). To ensure success, there must be access to systematically collected, well-annotated, well-preserved specimens with linked outcome data.

Data Processing and Analysis

For the completeness and to orient people who are new to the area of analysis of microarray data, we will provide an overview of microarray data analysis. We ignore the different microarray platforms, image analysis, and instead only focus on the biomarker discovery related statistics.

Quality Assessment

The assessment of data quality is an important issue in the analysis of gene expression microarray. Sample collection and handling can adversely affect microarray data quality, therefore, quality metrics are required to detect and assess the potential errors induced by these factors. The quality assessment is essential to ensure that the data to be subsequently analyzed are biologically relevant, with minimal contamination by technical artifacts. There are many quality assessment tools freely available. Some of them are array type-specific, such as Affymetrix (Parman & Halling, 2005), Illumina (Dunning et al., 2007), two-colour cDNA arrays (Buness et al., 2005). Others are particular problem oriented such as, for spot quality assessment (Li et al., 2005), for sample quality assessment (Gautier et al., 2004), for hybridization quality assessment (Petri et al., 2004), or for outlier array detection (Freue et al., 2007; Brettschneider et al., 2008). arrayQualityMetrics is a Bioconductor package (Gentleman et al., 2004) developed by Kauffmann et al. (2009) that provides access to a variety of quality metrics and comprehensive quality assessment reports. It works on most expression arrays and platforms.

Normalization

There are many sources of experimental variability that can affect the measurement of the gene expression level, so one has to normalize the raw data to compensate for differences caused by technical variation before making any between-chip comparisons. The challenge of normalization is to remove as much of the technical variation as possible while leaving the biological variation untouched.

There are many normalization methods proposed. The most popular one is linear normalization such as internal control normalization and total intensity normalization. This algorithm assumes that intensities between two or more arrays are linearly related with zero intercept. However, this approach does not deal particularly well with cases where there are non-linear relationships between arrays. Approaches using non-linear smooth curves have been proposed by Schadt et al. (2001, 2002) and Li and Wong (2001). The general approach of these methods is to select a set of approximately rank invariant probes (between the baseline and arrays to be normalized) and fit a non-linear relation like smoothing splines or a piecewise running median line. Another approach is to transform the data so that the distribution of probe intensities is the same across a set of arrays. Sidorov et al. (2002) propose both a parametric and a non-parametric method to achieve this. These approaches all depend on the choice of a baseline array. Bolstad et al. (2003) proposed three baseline array independent complete data methods, cyclic loess, contrast based method, quantile normalization. These methods combine information from all arrays to form the normalization relation. Be aware that all these data normalization methods rely on implicit assumptions on microarray data and one should choose suitable one where the assumptions underlying the approach are met.

Biomarker projects often include many batches of microarray experiments, where batch variations are commonly observed across different labs, array types, or platforms. Normalization procedures are often not sufficient to adjust data for batch effects. Benito et al. (2004) used distance weighted discrimination (DWD) for adjusting data for batch effects. However the method requires many samples (>25) in each batch for best performance. Johnson et al. (2007) proposed empirical Bayes frameworks for systematic bias adjustment that is robust to outliers in small sample sizes and performs comparable to DWD methods for large samples. Both approaches are shown to be very effective in removing systematic biases and can be used as powerful tools for combining microarray data sets.

GeneS

A key goal of gene selection via different expression patterns hidden behind microarray data is to remove "noise" features and identify the responsible genes as biomarkers for specific biological effects or responses, so that it is easy and accurate for suitable predictors to be designed. There are many potential benefits of gene selection: facilitating data visualization and data interpretation, gaining a deeper insight into the underlying process, avoiding overfitting and improving the prediction performance of models, and providing faster and more cost-effective models.

In the context of biomarker based classification, gene (feature) selection techniques can be organized into three categories, depending on how they combine the feature selection search with the construction of the classification model: filter methods, wrapper methods and embedded methods. Table 1 provides a general overview of feature selection methods, showing for each technique the description, pros and cons, as well as some examples of the most influential techniques.

Classification

The class prediction tries to predict the class membership of a set of subjects given their gene expression data. Some straightforward methods, including classic linear and quadratic discriminant analysis, nearest neighbor prediction, and more modern classification tree algorithms were described and compared by Dudoit et al. (2002). The comparison showed that the k-nearest neighbor and diagonal linear discriminant analysis performed the best in general while Fisher linear discriminant analysis performed the worst. The performance of classification tree–based approaches fell in between and can be improved with techniques known as "bagging" (Breiman, 1996) or "boosting" (Schapire, 1998). Combining Bagging and random feature selection methods to generate multiple classifier, Random Forest also shows excellent accuracy for microarray classification (Díaz-Uriarte et al., 2006).

More complicated machine learning algorithms that include collective and non-linear effects among samples have also been applied to the analysis of microarray data. Among them, support vector machine (SVM) is a most promising application (Brown, 2000, Furey et al., 2000, Gaasterland et al., 2000). SVM tries to map a hyper-plane to separate one class from the other in high-dimensional gene expression space with the maximal margin between training sets. Ar-

Table 1. The overview of feature selection techniques

Methods	Description	Advantages	Disadvantages	Examples
Filter Feature selection, then classification	Feature selection is defined as a pre-processing step and independent from the classifier. A filter method computes an informative score for each feature and then ranks and selects features according to the scores	• Sound statistical theory • Straight-forwarding biological understanding • Fast and scalable • Independent of the classifier	• Selection criteria may not match with classification goal • Redundancy in selected features • Risk of over-simplifying biological explanation	Univariate Approach: T test, ANOVA, Chi square, Bayesian (Baldi and Long 2001), Perason correlation, Information gain (Xing 2001) TNoM (Ben-Dor et al. 2000) Multivariate Approach: Markov blanket filter (MBF) (Xing (2001) Correlation-based feature selection(CFS) (Yu and Liu, 2004)
Wrapper Wrapping feature selection and classification	A search procedure in the space of feature subsets is "wrapped" around a specific classifier. Using cross validation, each feature subset is evaluated according to the performance of the trained classifier and their usefulness to the classifier	• Selection criterion consistent with classification criterion • Possible co-effect of genes considered	• require the implementation of efficient search mechanisms and various re-training and parameter tuning • Ranking represents the relative measure of whether a gene is informative for classification and is not comparable • Risk of overfitting: When cross-validation is not done properly (often 2-loop CV is needed), the result may be severely biased.	Deterministic Approach: Sequential forward selection (SFS) and backward elimination (SBE) (Kittler, 1978) Randomized approach Genetic algorithms (Jirapech-Umpai 2005) Simulated annealing
Embedded Performs feature selection within classification	Attempt simultaneous feature selection and classifier training. Often optimizes objective function that rewards classifier accuracy and penalizes the number of features used	• Selection criterion consistent with classification criterion • Possible co-effect of genes considered • Better computational complexity than wrapper methods	• Feature selection is depend on classifier	Nearest shrunken centroid (Tibshirani, 2002) Weighted naive Bayes (Duda et al., 2001) SVM-RFE (Guyon 2002) R-SVM (Zhang 2006) Decision Trees (Bagging and Boosting, Random Forest) (Breiman, 1984, 2001; Schapire et al.,1998; Díaz-Uriarte et al., 2006)

tificial intelligence (AI) models have also been proposed. Neural Networks (NN) (Ripley, 1996) are one of the most popular AI machine learning models. NN consists of multiple layers of perceptrons, which are calibrated by training sets to recognize the classes of interest. However the problem of "overtraining" is the major concern for the method. Table 2 presents key references and a simple comparison of the most popular classification techniques.

Another important issue in classifier design, besides feature selection, is the performance assessment. The prediction accuracy (or error rate) is usually estimated through cross validation (CV) or bootstrap methods. The performance assessment with CV or bootstrap is a critical step to avoid classifier overfitting problem and often is a required step in both feature and classifier selection.

Table 2. Comparison and key references for the most influential classification techniques

Method		Built-in Feature Selection	Accuracy	Interpretability	Computational Cost	Reference
Discriminant Analysis	Fisher Linear Discriminant Analysis (FLDA)	Yes	Poor	No	Low	Dudoit et al., 2002
	Diagonal Linear Discriminant Analysis (DLDA)	No	Good	No	Low	
	Diagonal Quadratic Discriminant Analysis (DQDA)	No	Fair	No	Low	
	Flexible Discriminant Analysis (FDA)	No	Fair	No	Low	Hastie et al.,1994
	Nearest Shrunken Centroid (NSC)	Yes	Good	Yes	Medium	Tibshirani et al., 2002
	Weight Voting	Yes	Fair	Yes	Low	Golub et al.,1999
Logistic Regression		Yes	Poor	No	Low	Liao et al., 2007
K-Nearest Neighbour		No	Good	No	Low	Dudoit et al., 2002
Support Vector Machines (SVM)		Yes	Good	No	High	Dudoit et al., 2002; Guyon et al., 2002; Zhang et al., 2006
Decision Tree	Classification and Regression Tree (CART)	Yes	Poor (Overfitting problem, pruning may help)	Yes	Low	Breiman, 1984, 2001; Liaw & Wiener, 2002; Dudoit et al., 2002; Bureau et al., 2003; Díaz-Uriarte et al.,2006
	Aggregating (Bagging) Tree	Yes	Fair (Overfitting problem, pruning may help)	No	Medium	
	Random Forest	Yes	Good	No	High	
Artificial Neural Network		Yes	Poor (Overfitting)	No	High	Ripley,1996; Ringnér et al., 2002; Rao et al., 2007;Khan et al., 2001

Biomarker Confirmation, Analytical Validation and Qualification

Before candidate biomarkers can be put into use, they must undergo several stages of confirmation, analytical validation, and for clinical biomarkers, qualification for clinical use. These verification studies remains a significant challenge in translating biomarker discovery to clinically useful test in a sensitive, reproducible, and time and cost-effective manner. Although this particular verification process addresses transcriptomic biomarkers, its application can be further extended to other types of biomarkers (e.g., DNA, protein or metabolite biomarkers).

This confirmation is a key initial step in the verification process for selecting genes for future study as potential biomarkers. The biomarker candidates are typically migrated to a different assay platform, that can better probe a small number of candidate genes for a large number of samples, to confirm and quantitatively measure their expression in independent, usually larger number of samples. Quantitative real-time polymerase chain reaction (qRT-PCR) technology, in particular Taqman-based RT-PCR gene expression arrays which combines the quantitative performance of RT-PCR with the multiple gene-profiling capabilities of microarrays, has shown great potential in bridging the gaps between biomarker discovery and clinical practice. The strengths of such a high-performance array platform include good reproducibility, specificity, high sensitivity and wide dynamic ranges. In particular, its flexibility and simplicity make it accessible for routine use in every laboratory, which is beneficial to large-scale analysis of biomarker validation across different laboratories.

Analytical validation is the process of assessing the assay or measurement performance characteristics, and the optimal conditions. The primary objectives of analytical validation are to evaluate accuracy and the precision of the assay, which include specificity, linearity, range, intermediate precision, reproducibility, detection limit, quantitative limit, and robustness. The analytical validation process begins with choosing the right assay followed by developing this assay into a validated method. Indeed, the choice of assay is pivotal to not only biomarker identification and validation but also biomarker qualification. In fact, the platform applied in biomarker discovery or validation can be further developed and used as a clinical analytic platform.

Biomarker qualification is the evidentiary process of linking biomarkers with the biology processes and clinical end points. The FDA has set up a pilot structure to start a qualification process for biomarkers in drug development and issued a guidance (FDA 2006) in classifying biomarkers as exploratory, probable valid, or known valid. A known valid biomarker is defined as "a biomarker that is measured in an analytical test system with well-established performance characteristics and for which there is widespread agreement in the medical or scientific community about the physiologic, toxicologic, pharmacologic, or clinical significance of the results." A probable valid biomarker is defined as "a biomarker that is measured in an analytical test system with well-established performance characteristics and for which there is a scientific framework or body of evidence that appears to elucidate the physiologic, toxicologic, pharmacologic, or clinical significance of the test results." The difference between these two classes of biomarkers is in the broad consensus indicated by classification as known valid. Exploratory biomarkers are potential precursors for probable or known valid biomarkers and can be used to fill in gaps of uncertainty about disease targets and variability in drug response, bridge the results of preclinical animal studies to human outcome, or select and prioritize new compounds and improve cost–benefit ratios and success rate for future drug development programs.

The qualification gap between exploratory and valid biomarkers is a gap not only between scientific proposals and consensus but also between the inefficient process through which biomarkers have been customarily introduced and accepted and a process through which these biomarkers could be seamlessly applied in drug development and regulatory review. A qualification process map has been proposed by the FDA that evaluates exploratory transcriptomic biomarkers to assess the potential of genomic technologies in mock submission (Leighton et al., 2006; Goodsaid et al., 2007) and identify key variables that can be used to determine the success of these biomarkers in voluntary genomic data submission (Salerno

& Lesko, 2004). The proposal transitions an exploratory biomarker to a known valid transcriptomic biomarker through a series of phases from discovery to method development to validation studies and cross-validation consortium (Goodsaid et al., 2006).

Public Resources

Microarray technology has led to an explosion of genomic expression profiling, generating an enormous volume of genomic data. The Gene Expression Ominbus (GEO, http://www.ncbi.nlm.nih.gov/geo) from the National Center for Biotechnology Information (NCBI) and ArrayExpress (http://www.ebi.ac.uk/microarray-as/ae/) from European Bioinformatics Institute (EBI) are the two major public repositories that archives and freely distributes microarray data submitted by the scientific community. The GEO and ArrayExpress may be freely browsed, queried, visualized, and downloaded to address specific biologic questions. With the rapid increase of microarray expression data in the public database over the past few years, it has become possible to monitor the general expression level of a gene in diverse biological samples under various conditions. It makes economic sense to fully take advantage of these public data to help experiment design, and for transcriptomic biomarker identification and validation.

MASS SPECTROMETRY-BASED QUANTITATIVE PROTEOMICS FOR BIOMARKER DISCOVERY

Proteomics is the large-scale study of proteins, particularly their structures and functions (Anderson & Anderson, 1998; Blackstock & Weir, 1999). Mass spectrometry-based proteomics, with recent advances in mass spectrometry (MS) (Hardman & Makarov, 2003; Schevchenko et al., 1997; Silva et al., 2005; Macek et al., 2006), gel-based (Gozal et al., 2009; Pietrogrande et al., 2006) and liquid phase protein separations (Fournier et al., 2007; Kay et al., 2007), and bioinformatics tools (Keller et al., 2002; Nesvizhskii et al., 2003; Carvalho et al., 2008), has made impact in biomarker discovery and development, through mapping of numerous tissue samples and biofluids, generation of quantitative protein profiles, and definition of post-translational modifications (PTMs) on a proteome-wide scale (Abersold & Mann, 2003). The ability of mass spectrometry to identify thousands of proteins from complex samples, and more importantly, to accurately quantify changes in protein expression in response to various stimuli and changes in protein expression over space and time can be expected to impact on finding highly specific biomarkers for disease diagnoses and treatment (Zhou et al., 2005; Kondo, 2008; Huettenhain et al., 2009).

An important aspect of proteomic applications and a key component toward discovery of novel biomarkers and validation of the biomarkers is quantitative proteomics - the ability to obtain a snapshot of concentrations of proteins associated with different states, hence allows deeper insights into how cells and organism function, and to detect or treat disease. There are a variety of techniques used to date in quantitative biology, including MS, 2-dimensional gel electrophoresis (DE), protein arrays, fluorescence microscopy, and ELISA, among others. Due to the scope of this article, here we focus only on MS-based techniques for quantitative proteomics. MS is a technology of choice for protein abundance measurement in cell lysates, tissue extracts and biofluids, through relative quantitative proteomics or absolute quantitative proteomics. Relative quantitation, also called protein profiling, is the detection of disease or treatment-related changes of protein expression within a large number of samples; the results are expressed as "fold" increases or decreases, whereas, absolute quantitation determines protein expression in terms of exact amount or concentration of proteins of interests, e.g. ng/mL of a protein in plasma or pmol per gram of tissue.

Top-Down Methods for Relative Quantitation

MS-based relative quantitation can be performed at protein level, e.g. top-down, or at peptide level through enzymatic digestion of proteins, e.g. bottom-up. Top-down quantitation methods study intact proteins for identification and quantitation. Two-dimensional gel electrophoresis combined with mass spectrometric identification has traditionally been applied as a top-down method for quantitative proteomics. Differential gel electrophoresis (DIGE), which uses fluorescent CyDyes to label proteins, was developed by GE Healthcare to quantify protein expression changes by fluorescence (Knowles et al., 2003). In a 2D-DIGE experiment, three fluorescent dyes, Cy2, Cy3, and Cy5, are used to label three different samples. The samples are then combined and separated on a 2D gel. The ratios of differential expression of proteins can be determined by DeCyder software (GE Healthcare) from the fluorescent images. Even though 2D-DIGE/MS is a relatively low throughput method compared with gel free MS-based techniques, the MS analysis time is relatively short as only the spots that showed differential expression are investigated. Two major challenges for 2D-DIGE/MS are (1) not all the differentially expressed proteins can be detected by MS due to the high sensitivity of fluorescence. In this case, a preparation gel with a higher sample loading has to be run for MS detection. (2) At times a 2D gel spot may contain more than one protein thus preventing accurate quantitation, unless the proteins can be resolved by a "zoomed in" IEF separation using a narrower range IPG strip (e.g. pI 4 – 7, instead of pI 3 – 10). Given that intact proteins are detected, the technique is more sensitive toward detection of protein degradation and better at distinguishing protein isoforms and various PTM forms of a protein than bottom-up techniques, where these possibly will be reported as a single protein. Several studies have demonstrated the utility of this technique for identification of disease biomarkers (Sheta et al., 2006; Pisitkun et al., 2006).

Quantitative proteomics combined with top-down mass spectrometry to capture differential expression of intact proteins also allows direct comparison of cells or biofluids at different states without trypsin digestion of proteins and subsequent peptide analysis. In this approach, the mass of an intact protein is first obtained by MS, followed by isolation in FTICR or Orbitrap for MS/MS to provide sequence information. Electron capture dissociation (ECD) (Zubarev et al., 1998) and electron transfer dissociation (ETD) (Syka et al., 2004), two fragmentation reactions alternative to collision induced dissociation (CID) commonly used for peptide analysis, are preferred fragmentation methods for MS detection of intact proteins to achieve higher backbone sequence coverage and retain labile PTMs (Siuti & Kelleher, 2007). Compared with bottom-up approaches, direct measurement of intact protein abundances instead of measuring abundances of peptides reduces the ambiguity of peptide-to-protein compilation and enables determination of specific form of a protein, locating PTMs and identifying alternative slicing variants and degradation products, hence improves reliability of protein quantification (Waanders et al., 2007; Du et al., 2006; Bunger et al., 2008). However, technological limitations, including difficulty of achieving high resolution LC separation of a complex mixture of intact proteins, restraint of mass spectrometers to high resolution FTMS and LTQ Orbitrap instruments in order to resolve isotopic envelopes of co-eluting proteins, the upper size limit of the protein that can be studied, underdeveloped database search algorithms, keeps this top-down approach to analysis of mostly single proteins and simple protein mixtures (McLafferty et al., 2007), till some of the recent studies extended it to complex mixture analysis (Ferguson et al., 2009; Collier et al.,

2008). Kelleher et al. demonstrated that by using top-down MS/MS, 15 proteins in *Methanosarcina acetivorans* were detected with mispredicted start sites with an additional 5 from small open reading frames (SORFs) (Ferguson et al., 2009).

Bottom-Up Methods for Relative and Absolute Quantitation

The bottom-up proteomics is a much more widely implemented strategy relative to top-down when one needs to track high-complexity samples for large-scale protein identification and quantitation. Bottom-up proteomics proteolytically digest proteins from cell lysates, tissue extracts, or biofluids into peptides, rendering the samples as a complex mixture of short peptides with defined C-termini. The peptides are then separated by liquid chromatography prior to electrospray mass spectrometry (ESI/MS) analysis (Fenn et al., 1989), that ensues peptide masses and sequences are used to identify and quantify corresponding proteins. At the peptide level, various MS approaches have been applied to both relative quantitation and absolute quantitation.

Relative Quantitation

Relative quantitative proteomics compares two or more samples using either isotope labeling, which is based on the introduction of a chemically equivalent differential mass tag that changes the mass of a peptide without affecting its analytical or biochemical properties, or label-free methods. In the labeling approaches, differential isotopic labels can be introduced metabolically in cell culture (e.g. SILAC) (Ong et al., 2002), chemically using isotopic tags (e.g. iTRAQ and TMT) (Lliuk et al., 2009), or enzymatically with ^{18}O (Reynolds et al., 2002).

Metabolic Labeling (SILAC)

Metabolic labeling incorporates stable isotopes during protein biosynthesis. SILAC (stable isotope labeling by amino acids in cell culture), a more widely used labeling method than using ^{15}N media (Oda et al., 1999), was mainly developed by Mann's group for quantitative proteomics by growing cells in media containing isotopically differentiated amino acid (Ong et al., 2002). The application of SILAC involves growing two populations of cells in isotopically distinct media, one with "light" amino acids (^{12}C-, ^{14}N-labeled) and the other one with "heavy" amino acids (^{13}C-, ^{15}N-labeled). Lysine and arginine are the two commonly used amino acids in cell culture media that are combined with trypsin digestion, which normally leaves a single isotopic label on each peptide, rendering easier MS identification and quantitation. After isotope incorporation, equal amount of proteins from both cell populations are combined, processed (i.e. protein purification, reduction/alkylation, trypsin digestion and HPLC separation) and analyzed by MS. The relative peak intensities of multiple isotopically distinct peptides from each protein are applied to determine the abundance changes of the protein in the samples. Due to the availability of several labels, SILAC minimizes the number of manipulations during the preparative procedures, hence allows for comprehensive comparisons with least amount of variation between samples. In addition, the differentially labeled peptides co-elute chromatographically, allowing for accurate quantitation. Despite the pitfalls associated with SILAC, such as difficulty in labeling tissues, biofluids and primary neuronal cells (Krueger et al., 2008), complete incorporation of isotopic amino acids is not the same for all cell lines (Harsha et al., 2008), and conversion of arginine to proline during cell division, it has been applied successfully to a variety of biological studies (Gronborg et al., 2006; Krueger et al., 2008; Bonenfant et al., 2007). The largest quantified proteome reported to date is using SILAC to identify and quantify 5,111 proteins in mouse embryonic stem cells (Graumann et al., 2008).

Chemical Labeling (ICAT and iTRAQ)

Chemical labeling of either proteins or peptides is perhaps the most frequently used method in current proteomics research in terms of quantitation. Among the numerous labeling molecules and large variety of chemical approaches, examples will be discussed herein for relative quantitative proteomics include (1) isotope-coded affinity tags (ICAT), used for labeling of free cysteines (Gygi et al., 1999) and (2) isobaric tags (iTRAQ), used for labeling of free amines (Ross et al., 2004). The iTRAQ approach can also be used for absolute quantitation.

ICAT was one of the first generation of chemical reagents introduced for quantitative proteomics to explore yeast proteome. The ICAT label has a thiol-specific iodoacetate group (to target cysteinyl residues), biotin (for affinity purification), and either a heavy ($^2H_8$) or light ($^1H_8$) isotope coded linker lies in between. To prevent the chromatographic retention time shift caused by deuterium in ICAT, a new generation of the reagent, cleavable ICAT (cICAT) was developed by using ^{13}C isotope tag, $^{13}C_9$ or $^{12}C_9$. The labeling is performed at the protein level, followed by proteolysis, affinity purification of the labeled peptides and MS analysis for quantitation and identification of the proteins. Due to the nature of cysteine labeling, ICAT greatly reduces the sample complexity therefore increases the potential for identifying low-abundant proteins. However, to the other end of the spectrum, it (1) reduces the reliability of quantitation as the experiment is based on a limited number of peptides per protein and (2) makes it impossible to detect changes in ~ 10-13% of proteins that do not contain cysteine residues (Vaughn et al., 2006). ICAT and cICAT have been applied to many areas in proteomic research, such as the study performed by Stewart et al. on the nuclear, cytosolic, and microsomal fractions from ovarian cancer cell lines, IGROV-1 (cisplatin-sensitive) and IGROV-1/CP (cisplatin-resistant) (Stewart et al., 2006). Out of the 1117 proteins identified and quantified by ICAT/MS/MS, 121 varied between the two cell lines. Their study revealed that the direction of changes in expression levels between proteins and mRNAs were not always in the same direction, with only a 44% similarity, which demonstrated the importance of profiling at the protein level.

Currently the most widely used chemical labeling, iTRAQ, is an amine-specific label targeted at the peptides. It is an isobaric compound containing a reporter group with variable mass of 114-117 Da (4-plex) or 113-121 Da (8-plex), a mass balance group, and an amino reactive group targets at lysine side chains and at peptide N-termini. In this approach, qantitation is carried out in the MS/MS mode to generate reporter ions of iTRAQ in the 114-121 Da mass range. The isobaric nature of iTRAQ-labeled peptides allows the signal from all peptides to be summed in both MS and MS/MS mode thus enhancing the sensitivity of detection and identification of the peptides. Instead of limited to two labels as ICAT, the multiplexing capability of iTRAQ increases throughput and reduces variation of quantitation; up to eight samples can be labeled, pooled, and run in a single experiment. In addition, the label is done at the peptide level, therefore multiple peptides can be detected for the same protein which increases the confidence of protein identification and multiple quantitation measurements can be obtained for each protein. The mass-balanced labels such as iTRAQ hold the most promise for quantitative biomarker discovery; iTARQ has been used to study cell systems and biofluids for biological systems (Zhang et al., 2008; Rajcevic et al., 2009); Zhou et al. (2007) used iTRAQ to measure quantitative changes in phosphorylation-dependent peptide-protein interaction; Tian et al. (2009) used iTRAQ to understand the molecular mechanism of paclitaxel resistance.

Label-Free Quantitation

Label-free quantitation uses global references where all measurements are related to a set of molecules that are chemically different from the quantified ions. Label-free quantitation methods are relatively high-throughput, requiring no time-consuming and expensive labeling step. Theoretically there is no limit to the number of samples that can be analyzed in a given experiment by label-free methods; therefore they are the most compatible methods for biological experiments. Spectral counting (Liu et al., 2004) is one of the label-free methods for protein quantitation which is based on the observation that the frequency of an ion trap MS fragment of a peptide is correlated to its quantity. An alternative method is to use the total ion intensity of the peptides for protein quantitation (America & Cordewener, 2008; Yates et al., 2007). Though the label-free principle sounds straightforward, in reality this approach imposes some practical constraints to instrumentation and processing software, especially choosing appropriate normalizations (Listgarten & Emili, 2005; Zimmer et al., 2006). Label-free method has been applied by Smith et al. to image the brain proteome in the adult and in development in order to understand the etiology of neurodegenerative and other brain disorders (Petyuk et al., 2010).

Absolute Quantitation

After the identity of a protein biomarker is known, it is essential to validate the novel protein biomarker. Multiple reaction monitoring (MRM), a highly sensitive and precise MS method routinely used to quantify small molecules for decades, is a widely used absolute quantitation method during validation and assay development phases of protein biomarker discovery (Kitteringham et al., 2009).

Traditionally, MRM is performed on a triple quadrupole mass spectrometer, where a selected peptide/metabolite is allowed to pass through the first quadrupole and enter the collision cell (the second quadrupole). In the third quadrupole, only a specific fragment ion generated in the collision cell of the peptide/metabolite is allowed to pass through and detected by the detector, rendering the method high specificity and sensitivity. Furthermore, HPLC retention time of the analyte adds an additional level of specificity to this approach. Hence MRM is ideal for proteomics where targeted detection of analytes from a complex background is required. Isotope-labeled internal standards have long been used in MRM for absolute quantitation of small molecules. Quantitation is achieved by spiking a known amount of the isotope labeled internal standard into the sample prior to MS, thus the relative levels of labeled and endogenous forms of the analyte can be calculated. Among many methods developed for use in proteomics, e.g. SISCAPA (stable isotope standards and capture by anti-peptide antibodies) (Anderson et al., 2004) and PC-IDMS (protein cleavable-isotope dilution mass spectrometry) (Barnidge et al., 2004), AQUA (absolute quantification of proteins) was introduced by Gygi et al. (2003) for absolute quantitation in proteomics in 2003. The proposed method showed that protein concentrations could be determined by quantifying their tryptic peptide components by adding a known amount of isotope labeled (^{2}H, ^{13}C, or ^{15}N) tryptic peptides to a tryptic digest of a complex mixture of proteins to determine changes in protein expression. This approach of using MRM with isotopically labeled internal standard has been demonstrated to be capable of absolute determination of peptide concentrations at low ng/mL level (Rifai et al., 2006; States et al., 2006) across a wide dynamic range of 10^3 – 10^4 (Addona et al., 2009; Kamiie et al., 2008) and with low coefficients of variation (CVs) of 5-10% (Kuzyk et al., 2009).

METABONOMICS IN BIOMARKER DISCOVERY

One of the technology platforms of biomarker discovery is metabonomics (often referred to as metabolomics), which is still under active development. Metabonomics is defined as the quantitative measurement of time-related multiparametric responses of multicellular systems to pathophysiological stimuli or genetic modifications (Nicholson et al., 1999). It is a comprehensive and simultaneous systematic profiling of metabolite levels and their systematic and temporal changes through such effects on diet, lifestyle, environment, genetics, and pharmaceuticals, both beneficial and adverse, in whole organisms (Lindon et al., 2006). Metabonomics is a non-targeted analysis of low molecular weight metabolites in the metabolome, which comprises the end products of changes at gene, RNA and protein levels. Metabonomics studies are reviewed extensively in recent years (Nicholson et al., 1999; Lindon et al., 2006; Lindon et al., 2003; Haleem et al., 2009; Dunn et al., 2008; Scalbert et al., 2009; Gowda et al., 2008), and the discussion here focuses on the metabonomics of biofluids. The major analytical techniques for metabonomics studies are NMR spectroscopy and Mass spectrometry. Other reported methods include CE, FTIR, and Coulometric detection. The large amount of experimental data generated for a metabonomics study requires data mining using intensive chemometric analysis including multivariate statistics and pattern analyses. Various in-house software and the metabolite databases are developed by individual labs and companies. How to extract useful biomarkers from the complexed data remains a big challenge. Integration of NMR and MS techniques or data analysis for metabonomics studies has been emerging (McKelvie et al., 2009; Crockford et al., 2006). The application of metabonomics for biomarker research ranges from disease diagnosis, drug efficacy, investigation of metabolic pathways, pharmaceutical development, and toxicology, to food science, nutrition, and environmental sciences.

The goal of a metabonomics study is usually to find biomarkers through comparison of a before and an after state upon changes in conditions. NMR can universally detect and quantify small molecular metabolites simultaneously, and therefore it can quickly provide a profile or fingerprint of the biofluids. It is the only direct detection technique which does not rely on separation in metabonomics technologies, whereas other methods all require hyphenated separation, such as GC-, CE-, HPLC-, or UPLC-MS-based metabonomics. Another advantage of NMR is that samples can be recovered for further analysis. The latest data analysis method, targeted profiling, can more accurately profile the metabolic makeup of the biofluids qualitatively and quantitatively (Weljie et al., 2006). With the cryoprobe technology, the sensitivity of NMR is dramatically improved. However, it remains a challenge to analyze NMR spectra of complex mixtures due to spectral overlap, and the sensitivity and resolution hinder the identification of biomarkers presence at a much lower level. MS has superior sensitivity for detecting low-level metabolites. GC-MS is widely used for microorganism and plant metabonomics (Fiehn, 2002). LC-MS is mostly used in studying biofluids, but the resulting intensity profile depends on the ionization of the small molecular metabolites. HPLC reproducibility and MS ionization may be a problem. The advent of UPLC improved the resolution, speed, and sensitivity of the chromatography (Wilson et al., 2005). In order to characterize as much metabolites as possible in the metabolome, the nature of MS makes it impossible to capture all in a single experiment. Multiple extractions with different solvents followed by multidimensional separation methods, and MS using positive and negative ionization mode are required. Still, the identification of all metabolites in a metabolome remains challenging. The technology of MS and MS/MS are best used for biomarker identification.

NMR-Based Metabonomics Methods

Study Design

Metabonomics relies on multidisplinary collaborations where samples are often collected at a different facility than sample analysis. A Study Protocol is written at the beginning of the study outlining the objective of the study, selection of objects, dosing regime of the objects, sample collection amount and time points, sample handling, shipping and storage conditions. It is important to include control samples and conditions in the protocol. Take into account that a number of factors, such as diet, stress, medications, age, exercise, fasting, and consumption of alcohol, will alter the metabolic composition of urine (Saude & Sykes, 2007) and blood. Keep in mind that signals dominating NMR spectra downstream should be avoid. For example, serum, instead of plasma, is collected from blood samples which are allowed to clot at room temperature for 2-3 hours without the addition of any anti-coagulate reagents.

NMR Sample Preparation

Metabonomics samples may come from a wide range of biofluids, however, the major sources are urine and plasma. Samples for small molecular analysis are usually urine and filtered serum (Lenz et al., 2003). Raw plasma can also be used, however, due to the presence of macromolecules, the NMR lines are usually broad and spectra are hard to analyze. The sample should be first centrifuged to remove debris. For urine, an NMR sample (600 μL) can be consisted of 360 μL clear urine, 60 mM sodium phosphate buffer (pH 7.2), 0.1% sodium azide for preservative, 0.1 mM internal standard TSP d4 or DSS, and 10% deuterium oxide for lock signal. For serum which are filtered using microcentrifuge filter tubes with 3K MWCO, the filter tubes need to be pre-washed by centrifuging 0.5 mL water at 13k rpm for 5 min a total of six times in order to remove contaminants on the filter. Serum (60 μL) will be then diluted with 40 μL water, and centrifuged at 13k rpm for 20 min at 4°C. The NMR sample (150 μL) can be consisted of 80 μL serum filtrate, 55 μL water and 15 μL 0.05% TSP d4 in deuterium oxide. Optimal conditions reported was the use of a 0.25 M phosphate buffer solution with urine to buffer ratio of 3:1 and added EDTA (2.5 μmol for 300 μl urine) (Asiago et al., 2008). Addition of EDTA avoids the apparent chemical shift variations of a large number of NMR signals, due to the high physiological variation of urine pH, metabolite concentration and ionic strength. Sample preparation or at least some steps can be done on an automated workstation. Serum samples need to be filtered to remove proteins prior to making into NMR samples.

NMR Experiments

Automated data collection is usually conducted for large number of samples either prepared in NMR tubes or 96-well plates. Automated NMR software conducts locking, shimming, temperature equilibration, NMR data collection, and saving of NMR spectra. The standard 1D NOESY experiment with presat water suppression is commonly used. Other reported experiments include selective TOCSY, spin-echo using CPMG, J-resolved, diffusion-edited and 1H-13C HMBC spectra (Lindon et al., 2003; Sandusky et al., 2005; Dumas et al., 2002). Large molecules in the samples such as proteins in serum can be removed using relaxation-based Carr-Purcell-Meiboom-Gill (CPMG) experiments to select signals from small molecules, or using diffusion-based gradient filter editing experiments such as stimulate echo (STE) and the gradient version bipolar pulse pair STE (BPPSTE). NMR spectral profiles of biofluids were often dominated by a few components at high concentration such as glucose. The change of the minor components was masked by the fluctuation of the large components, although changes in the minor components will be more interesting in determining the biochemical and physiological properties of the biofluids. The significant interest in developing improved methods in metabolic profiling stems from the high sensitivity of metabolite profiles to even subtle stimuli, which is potentially important for detecting the earliest onset of various adverse biological perturbations. A number of new NMR methodologies were reported including selective TOCSY (Sandusky et al., 2005) and chemical deriviterization of incorporating isotope tag such as 13C, 15N or 31P into the selected class of metabolites (Ye et al., 2009).

Data Processing and Preparation for Multivariate Data Analysis

The NMR FIDs were Fourier transformed, phased, referenced, baseline corrected, integrated, binned, and then output into a table for subsequent multivariate data analysis. Dominating resonances such as glucose, urea, and water can be excluded from analysis. Usually the processed spectra are imported into software such as AMIX-VIEWER for generating bucket tables which contained resonance frequencies and spectral integrals. For example, the spectral region of 0.5 to 9.5 ppm can be divided into 178 buckets of 0.04 ppm, excluding 4.5 – 6.4 ppm where the spectra have urea and distorted residual water. Each spectrum is normalized by its own sum of spectral integrals in the analyzed region of the 1D spectrum. The chemical shift buckets and the corresponding integrals are therefore output into a table for subsequent multivariate data analysis.

Chemometrics Methods

Commonly used multivariate data analysis methods include unsupervised data mining tools such as principal component analysis (PCA) (Eriksson et al., 2001) and hierarchical clustering, and supervised methods such as partial least squares discriminant analysis (PLS-DA). PCA is a starting point and provides fast overview of the data profiles. Additional pattern is extracted by using more advanced multivariate methods such as SIMCA, PLS by projections to latent structures, the orthogonal-PLS, batch modeling, and hierarchical PCA. Starting from PCA, models were refined using PLS-DA and OPLS-DA (Orthogonal PLS-DA) methods. PCA method yields maximum variance projection with outlier detection and overview of the data. PLS-DA and OPLS-DA are used to model two classes of data to increase the class separation, simplify data interpretation, and facilitate the identification of potential biomarkers (Bylesjo et al., 2007). PLS-DA maximizes separation of classes and therefore improves classification and identification of biomarkers. OPLS-DA further separates classes based on their differences, concentrates on the group variation, and makes the data interpretation much easier to visualize through the presentation of S-Plot. The S-Plot, in which p (corr) and w* are plotted, combines the information from loading plot and the column plot confidence limits. The variables centered in the middle are regarded as not related to the class separation, while the ones residing on the both ends of the "S"-shape are potential biomarkers.

Targeted profiling, developed by Chenomx, was also an excellent tool for quantitatively analyzing metabonomics data (Weljie et al., 2006). Individual NMR resonances of interest are mathematically modeled from pure compound spectra. This database is then interrogated to identify and quantify as much endogenous metabolites as possible in the NMR spectra obtained from biofluids. More recently, methods used in the analysis of DNA microarray dataset were evaluated for analyzing metabonomics data (Parsons et al., 2007).

MS-Based Metabonomics Methods

For GC-MS metabonomics, biofluid samples are first spiked with internal standards, deproteinized by acetonitrile, and followed by centrifugation and freeze drying (Michell et al., 2008). Urease is added to reduce urea in urine samples. The processed samples were then chemically derivatized and the final solution spiked with retention index solution. The response ratio, which is the metabolite peak area divided by the internal standard peak area, gives the relative amount of each metabolite in each sample. The data table containing metabolite peak vs. sample number is then used for chemometric modeling. Metabolite can be assigned based on comparison of its mass spectrum with a library database such as MPI/Golm which has over 80000 spectra.

For LC-MS metabonomics, MS methods with electronspray ionization are mostly used on biofluids. Urine can be analyzed by directed injection into the UPLC system, and the eluent was directly introduced to mass spectrometer without splitting (Wang et al., 2009). Better resolution is achieved from the recent introduction of UPLC. Each dataset contain the mass and retention time pair. The ion intensities are normalized by the sum of the total ion intensity in a LC-MS run. The MS run can be conducted in either or both positive or negative mode.

Both GC-MS and LC-MS data, when output as a data matrix, can be analyzed using multivariate data analysis methods described in Section 4.1.5. Potential biomarker ions can be identified and further analysis of the accurate mass and MS/MS fragmentation patterns may provide structure identification of the potential biomarkers. Identification of metabolites was achieved by first obtain the elemental composition of the unknown by accurate mass, database searching using the elemental composition, eliminating of certain candidates based on the MS/MS data, and comparison of the retention times. These biomarkers must be subsequently validated before put into use.

CONCLUSION

Biomarkers research is an exciting evolving field, with a variety of potential applications at the crossroads of scientific discovery, new drugs discovery and clinical research and has enormous potential for improving human health and welfare. The rapid advancement of the post-genomic technologies in the areas of genomics, transcriptomics, proteomics and metabolomics has led to the development of global strategies aimed at relating genotypic profile to phenotypic outcome in biological systems. The completion of a number of genome sequencing projects and recent advances in "omics" technologies, together with powerful bioinformatics tools, have made direct impact on the way the search for biomarkers is conducted. The selection and integration of different technologies prove pivotal to biomarker identification, characterization, and validation. Biomarkers can play diverse roles in clinical, facilitating the management of clinical conditions from diagnosis to prognosis and guidance of treatment. In the field of drug development, in addition to being surrogate end points in clinical trials, they inform important decisions in drug development and regulatory approvals. Also, aiming to exploit biomarkers to improve drug development productivity is becoming a strategically more viable goal than finding the next surrogate end point for regulatory approval of new drugs. Whilst post-genomic technologies hold great promise for human health research, substantial technical challenges remain. Along with the intrinsic problems involved in the global analysis of transcripts, proteins and metabolites, additional issues of feasibility, cost and practicality of using these technologies in a clinical environment should be considered. The evolution of biomarkers represents the coordinated and concerted effort of basic research scientists, clinicians, technology experts, epidemiologists, statisticians, academic and industrial sponsors, and regulatory agencies within a cooperative framework.

REFERENCES

Abersold, R., & Mann, M. (2003). Mass spectrometry-based proteomics. *Nature*, *422*(6928), 198–207. doi:10.1038/nature01511

Addona, T. A., Abbatiello, S. E., & Schilling, B. (2009). Multi-site assessment of the revision and reproducibility of multiple reaction monitoring-based measurements of proteins in plasma. *Nature Biotechnology*, *27*(7), 633–641. doi:10.1038/nbt.1546

America, A. H. P., & Cordewener, J. H. G. (2008). Comparative LC-MS: A landscape of peaks and valleys. *Proteomics*, *8*(4), 731–749. doi:10.1002/pmic.200700694

Anderson, N. L., & Anderson, N. G. (1998). Proteome and proteomics. New technologies, new concepts, and new words. *Electrophoresis*, *19*(11), 1853–1861. doi:10.1002/elps.1150191103

Anderson, N. L., Anderson, N. G., Haines, L. R., Hardie, D. B., Olafson, R. W., & Pearson, T. W. (2004). Mass spectrometric quantitation of peptides and proteins using stable isotope standards and capture by anti-peptide antibodies (SISCAPA). *Journal of Proteome Research*, *3*(2), 235–244. doi:10.1021/pr034086h

Asiago, V. M., Gowda, G. A. N., Zhang, S., Shanaiah, J. C., & Raftery, D. (2008). Use of EDTA to minimize ionic strength dependent frequency shifts in the H NMR spectra of urine. *Metabolomics*, *4*(4), 328–336. doi:10.1007/s11306-008-0121-7

Baldi, P., & Long, A. (2001). A Bayesian framework for the analysis of microarray expression data: Regularized t-test and statistical inferences of gene changes. *Bioinformatics (Oxford, England)*, *17*, 509–516. doi:10.1093/bioinformatics/17.6.509

Barnidge, D. R., Hall, G. D., Stocker, D. C., & Muddiman, D. C. (2004). Evaluation of a cleavable stable isotope labeled synthetic peptide for absolute protein quantification using LC-MS/MS. *Journal of Proteome Research*, *3*(3), 658–661. doi:10.1021/pr034124x

Ben-Dor, A. (2000). Tissue classification with gene expression profiles. *Journal of Computational Biology*, *7*, 559–584. doi:10.1089/106652700750050943

Benito, M. (2004). Adjustment of systematic microarray data biases. *Bioinformatics (Oxford, England)*, *20*(1), 105–114. doi:10.1093/bioinformatics/btg385

Bennett, S. T., Barnes, C., Cox, A., Davies, L., & Brown, C. (2005). Toward the 1,000 dollars human genome. *Pharmacogenomics*, *6*, 373–382. doi:10.1517/14622416.6.4.373

Bentley, D. R. (2006). Whole-genome re-sequencing. *Current Opinion in Genetics & Development*, *16*, 545–552. doi:10.1016/j.gde.2006.10.009

Blackstock, W. P., & Weir, M. P. (1999). Proteomics: Quantitative and physical mapping of cellular proteins. *Trends in Biotechnology*, *17*(3), 121–127. doi:10.1016/S0167-7799(98)01245-1

Bolstad, B. M., Irizarry, R. A., Astrand, M., & Speed, T. P. (2003). A comparison of normalization methods for high density oligonucleotide array data based on variance and bias. *Bioinformatics (Oxford, England)*, *19*(2), 185–193. doi:10.1093/bioinformatics/19.2.185

Bonenfant, D., Towbin, H., Coulot, M., Schindler, P., Mueller, D. R., & van Oostrum, J. (2007). Analysis of dynamic changes in post-translational modifications of human histones during cell cycle by mass spectrometry. *Molecular & Cellular Proteomics*, *6*(11), 1917–1932. doi:10.1074/mcp.M700070-MCP200

Botstein, D., & Risch, N. (2003). Discovering genotypes underlying human phenotypes: Past successes for Mendelian disease, future approaches for complex disease. *Nature Genetics*, *33*(Supplement), 228–237. doi:10.1038/ng1090

Breiman, L. (1996). Bagging predictors. *Machine Learning*, *24*, 123–140. doi:10.1007/BF00058655

Breiman, L. (2001). Random forests. *Machine Learning*, *45*, 5–32. doi:10.1023/A:1010933404324

Breiman, L., Friedman, J., Olshen, R., & Stone, C. (1984). *Classification and regression trees*. New York: Chapman & Hall.

Brettschneider, J., Collin, F., Bolstad, B. M., & Speed, T. P. (2008). Rejoinder for quality assessment for short oligonucleotide microarray data. *Technometrics*, *50*(3), 279–283. doi:10.1198/004017008000000389

Brown, M. P., Grundy, W. N., Lin, D., Cristianini, N., Sugnet, C. W., & Furey, T. S. (2000). Knowledge-based analysis of microarray gene expression data by using support vector machines. *Proceedings of the National Academy of Sciences of the United States of America*, *97*, 262–267. doi:10.1073/pnas.97.1.262

Buness, A., Huber, W., Steiner, K., Sltmann, H., & Poustka, A. (2005). arrayMagic: Two-colour cDNA microarray quality control and preprocessing. *Bioinformatics (Oxford, England)*, *21*, 554–556. doi:10.1093/bioinformatics/bti052

Bunger, M. K., Cargile, B. J., Ngunjiri, A., Bundy, J. L., & Stephenson, J. L. Jr. (2008). Automated proteomics of E. coli via top-down electron-transfer dissociation mass spectrometry. *Analytical Chemistry*, *80*(5), 1459–1467. doi:10.1021/ac7018409

Bureau, A., Dupuis, J., Hayward, B., Falls, K., & Van Eerdewegh, P. (2003). Mapping complex traits using random forests. *BMC Genetics*, *4*, S64. doi:10.1186/1471-2156-4-S1-S64

Bylesjo, M., Rantalainen, M., Cloarec, O., Nicholson, J. K., Holmes, E., & Trygg, J. (2007). OPLC discriminant analysis: Combining the strengths of PLS-DA and SIMCA classification. *Journal of Chemometrics*, *20*, 341–351. doi:10.1002/cem.1006

Carvalho, P. C. (2008). PatternLab for proteomics: A tool for differential shotgun proteomics. *BMC Bioinformatics*, *9*, 316–329. doi:10.1186/1471-2105-9-316

Chanock, S. (2001). Candidate genes and single nucleotide polymorphisms (SNPs) in the study of human disease. *Disease Markers*, *17*, 89–98.

Churchill, G. A. (2002). Fundamentals of experimental design for cDNA microarrays. *Nature Genetics*, *32*(Supplement), 490–495. doi:10.1038/ng1031

Collier, T. S., Hawkridge, A. M., Georgianna, D. R., Patne, G. A., & Muddiman, D. C. (2008). Top-down identification and quantification of stable isotope labeled proteins from Aspergillus flavus using online nano-flow reversed-phase liquid chromatography coupled to a LTQ-FTICR mass spectrometer. *Analytical Chemistry*, *80*(13), 4994–5001. doi:10.1021/ac800254z

Crockford, D. J., Holmes, E., Lindon, J. C., Plumb, R. S., Zirah, S., & Bruce, S. J. (2006). Statistical heterospectroscopy, (SHY), an approach to the integrated analysis of NMR and UPLC-MS data sets: Application in metabonomic toxicology studies. *Analytical Chemistry*, *78*(2), 363–371. doi:10.1021/ac051444m

Davies, H. (2002). Mutations of the BRAF gene in human cancer. *Nature*, *417*, 949–954. doi:10.1038/nature00766

Díaz-Uriarte, R., & Alvarez de Andrés, S. (2006). Gene selection and classification of microarray data using random forest. *BMC Bioinformatics*, *7*, 3. doi:10.1186/1471-2105-7-3

Dobbin, K., & Simon, R. (2007)... *Biostatistics (Oxford, England)*, *8*, 101–117. doi:10.1093/biostatistics/kxj036

Dobbin, K., Zhao, Y. D., & Simon, R. (2008)... *Clinical Cancer Research*, *14*, 108–114. doi:10.1158/1078-0432.CCR-07-0443

Du, Y., Parks, B. A., Sohn, S., Kwast, K. E., & Kelleher, N. L. (2006). Top-down approaches for measuring expression ratios of intact yeast proteins using Fourier transform mass spectrometry. *Analytical Chemistry*, *78*(3), 686–694. doi:10.1021/ac050993p

Duda, P. (2001). *Pattern classification*. New York: Wiley.

Dudoit, S., Fridlyand, J., & Speed, T. P. (2002). Comparison of discrimination methods for the classification of tumors using gene expression data. *Journal of the American Statistical Association*, *97*, 77–87. doi:10.1198/016214502753479248

Dumas, M. E., Canlet, C., Andre, F., Vercauteren, J., & Paris, A. (2002). Metabonomic assessment of physiological disruptions using ^{1}H-^{13}C HMBC-NMR spectroscopy combined with pattern recognition procedures performed on filtered variables. *Analytical Chemistry*, *74*(10), 2261–2273. doi:10.1021/ac0156870

Dunn, W. B. (2008). Current trends and future requirements for the mass spectrometric investigation of microbial, mammalian and plant metabolomes. *Physical Biology*, *5*, 1–24. doi:10.1088/1478-3975/5/1/011001

Dunning, M. J., Smith, M. L., Ritchie, M. E., & Tavare, S. (2007). beadarray: R classes and methods for Illumina bead-based data. *Bioinformatics (Oxford, England)*, *23*, 2183–2184. doi:10.1093/bioinformatics/btm311

Elston, R. C., & Cordell, H. J. (2001). Overview of model-free methods for linkage analysis. *Advances in Genetics*, *42*, 135–150. doi:10.1016/S0065-2660(01)42020-7

Eriksson, L., Johansson, E., Kettaneh-Wold, N., & Wold, S. (2001). *Multi- and megavariate data analysis*. Umetrics Academy.

Fenn, J. B., Mann, M., Meng, C. K., Wong, S. F., & Whitehouse, C. M. (1989). Electrospray ionization for mass spectrometry of large biomolecules. *Science*, *246*, 64. doi:10.1126/science.2675315

Ferguson, J. T., Wenger, C. D., Metcalf, W. W., & Kelleher, N. L. (2009). Top-down proteomics reveals novel protein forms expressed in Methanosarcina acetivorans. *JASMS*, *20*(9), 1743–1750.

Fiehn, O. (2002). Metabolomics–the link between genotypes and phenotypes. *Plant Molecular Biology*, *48*(1), 155–171. doi:10.1023/A:1013713905833

Fournier, M. L., Gilmore, J. M., Martin-Brown, S. A., & Washburn, M. P. (2007). Multidimensional separations-based shotgun proteomics. *Chemical Reviews*, *107*(8), 3654–3686. doi:10.1021/cr068279a

Freue, G. V. C., Hollander, Z., Shen, E., Zamar, R. H., Balshaw, R., & Scherer, A. (2007). MDQC: A new quality assessment method for microarrays based on quality control reports. *Bioinformatics (Oxford, England)*, *23*, 3162–3169. doi:10.1093/bioinformatics/btm487

Furey, T. S., Cristianini, N., Duffy, N., Bednarski, D. W., Schummer, M., & Haussler, D. (2000). Support vector machine classification and validation of cancer tissue samples using microarray expression data. *Bioinformatics (Oxford, England)*, *16*, 906–914. doi:10.1093/bioinformatics/16.10.906

Gaasterland, T., & Bekiranov, S. (2000). Making the most of microarray data. *Nature Genetics*, *24*, 204–206. doi:10.1038/73392

Gautier, L., Cope, L., Bolstad, B. M., & Irizarry, R. A. (2004). Affy–analysis of affymetrix genechip data at the probe level. *Bioinformatics (Oxford, England)*, *20*, 307–315. doi:10.1093/bioinformatics/btg405

Gentleman, R. C., Carey, V. J., Bates, D. J., Bolstad, B. M., Dettling, M., & Dudoit, S. (2004). Bioconductor: Open software development for computational biology and bioinformatics. *Genome Biology*, *5*, R80. doi:10.1186/gb-2004-5-10-r80

Gerber, S. A., Rush, J., Stemman, O., Kirschner, M. W., & Gygi, S. P. (2003). Absolute quantification of proteins and phosphoproteins from cell lysates by tandem MS. *Proceedings of the National Academy of Sciences of the United States of America*, *100*(12), 6940–6945. doi:10.1073/pnas.0832254100

Golub, T., Slonim, D., Tamayo, P., Huard, C., Gaasenbeek, M., & Mesirov, J. (1999). Molecular classification of cancer: Class discovery and class prediction by gene expression monitoring. *Science*, *286*, 531–537. doi:10.1126/science.286.5439.531

Goodsaid, F., & Frueh, F. (2006). Process map proposal for the validation of genomic biomarkers. *Pharmacogenomics*, *7*, 773–782. doi:10.2217/14622416.7.5.773

Goodsaid, F., & Frueh, F. (2007). Biomarker qualification pilot process at the US Food and Drug Administration. *The AAPS Journal*, *9*(1), E105–E108. doi:10.1208/aapsj0901010

Gowda, G. N., Zhang, S., Gu, H., Asiago, V., Shanaiah, N., & Raftery, D. (2008). Metabolomics-based methods for early disease diagnostics. *Expert Review of Molecular Diagnostics*, *8*, 617–633. doi:10.1586/14737159.8.5.617

Gozal, D. (2009). Two-dimensional differential in-gel electrophoresis proteomic approaches reveal urine candidate biomarkers in pediatric obstructive sleep apnea. *American Journal of Respiratory and Critical Care Medicine*, *180*(12), 1253–1261. doi:10.1164/rccm.200905-0765OC

Graumann, J., Hubner, N. C., Kim, J. B., Ko, K., Moser, M., & Kumar, C. (2008). Stable isotope labeling by amino acids in cell culture (SILAC) and proteome quantitation of mouse embroyonic stem cells to depth of 5,111 proteins. *Molecular & Cellular Proteomics*, *7*(4), 672–683. doi:10.1074/mcp.M700460-MCP200

Gronborg, M., Kristiansen, T. Z., & Iwahori, A. (2006). Biomarker discovery from pancreatic cancer secretome using a differential proteomic approach. *Molecular & Cellular Proteomics*, *5*(1), 157–171. doi:10.1074/mcp.M500178-MCP200

Guyon, I., Weston, J., Barnhill, S., & Vapnik, V. (2002). Gene selection for cancer classification using support vector machines. *Machine Learning*, *46*, 389–422. doi:10.1023/A:1012487302797

Gygi, S. P., Rist, B., Gerber, S. A., Turecek, F., Gelb, M. H., & Aebersold, R. (1999). Quantitative analysis of complex protein mixtures using isotope-coded affinity tags. *Nature Biotechnology*, *17*(10), 994–999. doi:10.1038/13690

Hall, M. (1999). Correlation-based feature selection for machine learning. Unpublished doctoral thesis, Department of Computer Science, Waikato University, New Zealand.

Hardman, M., & Makarov, A. A. (2003). Interfacing the orbitrap mass analyzer to an electrospray ion source. *Analytical Chemistry*, *75*(7), 1699–1075. doi:10.1021/ac0258047

Harsha, H. C., Molina, H., & Pandey, A. (2008). Quantitative proteomics using stable isotope labeling with amino acids in cell culture. *Nature Protocols*, *3*(3), 505–516. doi:10.1038/nprot.2008.2

Hastie, T., Tibshirani, R., & Buja, A. (1994). Flexible discriminant analysis. *Journal of the American Statistical Association*, *89*, 1255–1270. doi:10.2307/2290989

Helgadottir, A., Gretarsdottir, S., & St Clair, D. (2005). Association between the gene encoding 5-lipoxygenase-activating protein and stroke replicated in a Scottish population. *American Journal of Human Genetics*, *76*, 505–509. doi:10.1086/428066

Hu, J., Zou, F., & Wright, F. A. (2005). Practical FDR-based sample size calculations in microarray experiments. *Bioinformatics (Oxford, England)*, *21*(15), 3264–3272. doi:10.1093/bioinformatics/bti519

Huettenhain, R., Malmstroem, J., Picotti, P., & Aebersold, R. (2009). Perspectives of targeted mass spectrometry for protein biomarker verification. *Current Opinion in Chemical Biology*, *13*(5-6), 518–525. doi:10.1016/j.cbpa.2009.09.014

International HapMap Consortium. (2005). A haplotype map of the human genome. *Nature*, *437*(7063), 1299–1320. doi:10.1038/nature04226

Issaq, H. J., Van, Q. N., Waybright, T. J., Muschik, G. M., & Veenstra, T. D. (2009). Analytical and statistical approaches to metabonomics research. *Journal of Separation Science*, *32*, 2183–2199. doi:10.1002/jssc.200900152

Jirapech-Umpai, T., & Aitken, S. (2005). Feature selection and classification for microarray data analysis: Evolutionary methods for identifying predictive genes. *BMC Bioinformatics*, *6*, 148. doi:10.1186/1471-2105-6-148

Johnson, W. E., & Li, C. (2007). Adjusting batch effects in microarray expression data using empirical Bayes methods. *Biostatistics (Oxford, England)*, *8*(1), 118–127. doi:10.1093/biostatistics/kxj037

Jung, S. H. (2005). Sample size calculation for multiple testing in microarray data analysis. *Biostatistics (Oxford, England)*, *6*(1), 157–169. doi:10.1093/biostatistics/kxh026

Kamiie, J., Ohtsuki, S., & Iwase, R. (2008). Quantitative atlas of membrane transporter proteins: development and application of a highly sensitive simultaneous LC/MS/MS method combined with novel in-silico peptide selection criteria. *Pharmaceutical Research*, *25*(6), 1469–1483. doi:10.1007/s11095-008-9532-4

Kauffmann, A., Gentleman, R., & Huber, W. (2009). arrayQualityMetrics-a bioconductor package for quality assessment of microarray data. *Bioinformatics (Oxford, England)*, *25*(3), 415–416. doi:10.1093/bioinformatics/btn647

Kay, R. G., Gregory, B., Grace, P. B., & Pleasance, S. (2007). The application of ultra-performance liquid chromatography/tandem mass spectrometry to the detection and quantitation of apolipoproteins in human serum. *Rapid Communications in Mass Spectrometry*, *21*(16), 2585–2593. doi:10.1002/rcm.3130

Keller, A., Nesvizhskii, A. I., Kolker, E., & Aebersold, R. (2002). Empirical statistical model to estimate the accuracy of peptide identifications made by MS/MS and database search. *Analytical Chemistry*, *74*(20), 5383–5392. doi:10.1021/ac025747h

Khan, J., Wei, J. S., Ringner, M., Saal, L. H., Ladanyi, M., & Westermann, F. (2001). Classification and diagnostic prediction of cancers using gene expression profiling and artificial neural networks. *Nature Medicine*, *7*, 673–679. doi:10.1038/89044

Kitteringham, N. R., Jenkins, R. E., Lane, C. S., Elliott, V. L., & Park, B. K. (2009). Multiple reaction monitoring for quantitative biomarker analysis in proteomics and metabolomics. *Journal of Chromatography. B, Analytical Technologies in the Biomedical and Life Sciences*, *877*(13), 1229–1239. doi:10.1016/j.jchromb.2008.11.013

Kittler, J. (1978). Feature set search algorithms. *Pattern recognition and signal processing,* (pp. 41–60).

Knowles, M. R., Cervino, S., & Skynner, H. A. (2003). Multiplex proteomic analysis by two-dimensional differential in-gel electrophoresis. *Proteomics*, *3*(7), 1162–1171. doi:10.1002/pmic.200300437

Kondo, T. (2008). Cancer proteomics for biomarker development. *Journal of Proteomics and Bioinformatics*, *1*(9), 477–484. doi:10.4172/jpb.1000055

Krueger, M., Kratchmarova, I., Blagoev, B., Tseng, Y. H., Kahn, C. R., & Mann, M. (2008). Dissection of the insulin signaling pathway via quantitative phosphoproteomics. *Proceedings of the National Academy of Sciences of the United States of America*, *105*(7), 2451–2456. doi:10.1073/pnas.0711713105

Krueger, M., Moser, M., Ussar, S., Thievessen, I., & Luber, C. A. (2008). SILAC mouse for quantitative proteomics uncovers kindling-3 as an essential factor for red blood cell function. *Cell*, *134*(2), 353–364. doi:10.1016/j.cell.2008.05.033

Kuzyk, M., Smith, D., & Yang, J. (2009). Multiple reaction monitoring-based, multiplexed, absolute quantitation of 45 proteins in human plasma. *Molecular & Cellular Proteomics*, *8*(8), 1860–1877. doi:10.1074/mcp.M800540-MCP200

Lee, Y., & Lee, C. (2003). Classification of multiple cancer types by multicategory support vector machines using gene expression data. *Bioinformatics (Oxford, England)*, *19*(9), 1132–1139. doi:10.1093/bioinformatics/btg102

Leighton, J., Brown, P., & Ellis, A. (2006). Workgroup report: Review of genomics data based on experience with mock submissions-view of the CDER Pharmacology Toxicology Nonclinical Pharmacogenomics Subcommittee. *Environmental Health Perspectives*, *114*(4), 573–578. doi:10.1289/ehp.8318

Lenz, E. M., Bright, J., Wilson, I. D., Morgan, S. R., & Nash, A. F. P. (2003). A ^{1}H NMR-based metabonomic study of urine and plasma samples obtained from healthy human subjects. *Journal of Pharmaceutical and Biomedical Analysis*, *33*(5), 1103–1115. doi:10.1016/S0731-7085(03)00410-2

Li, C., & Wong, W. H. (2001). Model-based analysis of oligonucleotide arrays: Model validation, design issues and standard error applications. *Genome Biology*, *2*(8), 1–11.

Li, Q., Fraley, C., Bumgarner, R. E., Yeung, K. Y., & Raftery, A. E. (2005). Donuts, scratches and blanks: Robust model-based segmentation of microarray images. *Bioinformatics (Oxford, England)*, *21*, 2875–2882. doi:10.1093/bioinformatics/bti447

Liao, J. G., & Chin, K. V. (2007). Logistic regression for disease classification using microarray data: model selection in a large p and small n case. *Bioinformatics (Oxford, England)*, *23*(15), 1945–1951. doi:10.1093/bioinformatics/btm287

Liaw, A., & Wiener, M. (2003). Classification and regression by randomForest. *R News*, *2/3*, 18–22.

Lindon, J. C., Holmes, E., & Nicholson, J. K. (2006). Metabonomics techniques and applications to pharmaceutical research & development. *Pharmaceutical Research*, *23*, 1075–1088. doi:10.1007/s11095-006-0025-z

Lindon, J. C., Nicholson, J. K., Holmes, E., Antti, H., Bollard, M. E., & Keun, H. (2003). The role of metabonomics in toxicology and its evaluation by the COMET project. *Toxicology and Applied Pharmacology*, *187*, 137–146. doi:10.1016/S0041-008X(02)00079-0

Listgarten, J., & Emili, A. (2005). Statistical and computational methods for comparative proteomic profiling using liquid chromatography-tandem mass spectrometry. *Molecular & Cellular Proteomics*, *4*(4), 419–434. doi:10.1074/mcp.R500005-MCP200

Liu, H., Sadygov, R. G., & Yates, J. R. (2004). A model for random sampling and estimation of relative protein abundance in shotgun proteomics. *Analytical Chemistry*, *76*(14), 4193–4201. doi:10.1021/ac0498563

Lliuk, A., Galan, J., & Tao, W. A. (2009). Playing tag with quantative proteomics. *Analytical and Bioanalytical Chemistry*, *393*(2), 503–513. doi:10.1007/s00216-008-2386-0

Lohmussaar, E., Gschwendtner, A., & Mueller, J. C. (2005). ALOX5AP gene and the PDE4D gene in a central European population of stroke patients. *Stroke*, *36*, 731–736. doi:10.1161/01.STR.0000157587.59821.87

Macek, B., Waanders, L. F., Olsen, J. V., & Mann, M. (2006). Top-down protein sequencing and MS3 on a hybrid linear quadrupole ion trap-orbitrap mass spectrometer. *Molecular & Cellular Proteomics*, *5*(5), 949–958. doi:10.1074/mcp.T500042-MCP200

Margulies, M. (2005). Genome sequencing in microfabricated high-density picolitre reactors. *Nature*, *437*, 376–380.

McKelvie, J. R., Yuk, J., Xu, Y., Simpson, A. J., & Simpson, M. J. (2009). ^{1}H NMR and GC-MS metabonomics of earthworm responses to sub-lethal DDT and endosulfan exposure. *Metabolomics*, *5*(1), 84–94. doi:10.1007/s11306-008-0122-6

McLafferty, F. W., Breuker, K., & Jin, M. (2007). Top-down MS, a power complement to the high capabilities of proteolysis proteomics. *The FEBS Journal*, *274*(24), 6256–6268.

Michell, A. W., Mosedale, D., Grainger, D. J., & Barker, R. A. (2008). Metabolomic analysis of urine and serum in Parkinson's disease. *Metabolomics*, *4*(3), 191–201. doi:10.1007/s11306-008-0111-9

Morozova, O., & Marra, M. A. (2008). Applications of next-generation sequencing technologies in functional genomics. *Genomics*, *92*(5), 255–264. doi:10.1016/j.ygeno.2008.07.001

National Institute on Aging. (2009). Alzheimer's disease genetics facts sheet. Retrieved from http://www.nia.nih.gov/Alzheimers/Publications/geneticsfs.htm

Nesvizhskii, A. I., Keller, A., Kolker, E., & Aebersold, R. (2003). A statistical model for identifying proteins by tandem mass spectrometry. *Analytical Chemistry*, *75*(17), 4646–4658. doi:10.1021/ac0341261

Nicholson, J. K., Lindon, J. C., & Holmes, E. (1999). Metabonomics: Understanding the metabolic responses of living systems to pathophysiological stimuli via multivariate statistical analysis of biological NMR spectroscopic data. *Xenobiotica*, *29*, 1181–1189. doi:10.1080/004982599238047

NIH/CEPH Collaborative Mapping Group. (1992). A comprehensive genetic linkage map of the human genome. *Science*, *258*, 67–86. doi:10.1126/science.1439770

Oda, Y., Huang, K., Cross, F. R., Cowburn, D., & Chait, B. T. (1999). Accurate quantitation of protein expression and site specific phosphorylation. *Proceedings of the National Academy of Sciences of the United States of America*, *96*(12), 6591–6596. doi:10.1073/pnas.96.12.6591

Ong, S.-E., Blagoev, B., Kratchmarova, I., Kristensen, D. B., Steen, H., & Pandey, A. (2002). Stable isotope labelling by amino acids in cell culture, SILAC, as a simple and accurate approach to expression proteomics. *Molecular & Cellular Proteomics*, *1*(5), 376–386. doi:10.1074/mcp.M200025-MCP200

Oura, T., Matsui, S., & Kawakami, K. (2009). Sample size calculations for controlling the distribution of false discovery proportion in microarray experiments. *Biostatistics (Oxford, England)*, *10*(4), 694–705. doi:10.1093/biostatistics/kxp024

Parman, C. & Halling, C. (2005). affyQCReport: QC report generation for affyBatch objects. *R package,* version 1.17.0.

Parsons, H. M., Ludwig, C., Gunther, U. L., & Viant, M. R. (2007). Improved classification accuracy in 1- and 2-dimensional NMR metabolomics data using the variance stabilizing generalized logarithm transformation. *BMC Bioinformatics*, *8*, 234. doi:10.1186/1471-2105-8-234

Petri, A., Fleckner, J., & Matthiessen, M. W. (2004). Array-a-lizer: A serial DNA microarray quality analyzer. *BMC Bioinformatics*, *5*, 12. doi:10.1186/1471-2105-5-12

Petyuk, V. A., Qian, W.-J., Smith, R. D., & Smith, D. J. (2010). Mapping protein abundance patterns in the brain using voxelation combined with liquid chromatography and mass spectrometry. *Methods (San Diego, Calif.)*, *50*(2), 77–84. doi:10.1016/j.ymeth.2009.07.009

Pietrogrande, M. C., Marchetti, N., Dondi, F., & Righetti, P. G. (2006). Decoding 2D-PAGE complex maps: Relevance to proteomics. *Journal of Chromatography B – Analytical. Technological, and Biomedical Life Sciences*, *833*(1), 51–62. doi:10.1016/j.jchromb.2005.12.051

Pisitkun, T., Johnstone, R., & Knepper, M. A. (2006). Discovery of unrinary biomarkers. *Molecular & Cellular Proteomics*, *5*(10), 1760–1771. doi:10.1074/mcp.R600004-MCP200

Qiu, W., & Lee, M. T. (2006). SPCalc: A Web-based calculator for sample size and power calculations in micro-array studies. *Bioinformation*, *1*(7), 251–252.

Rajcevic, U., Petersen, K., & Knol, J. C. (2009). iTRAQ-based protoemics profiling reveals increased metabolic activity and cellular cross-talk in angiogenic compared with invasive glioblastoma phenotype. *Molecular & Cellular Proteomics*, *8*(11), 2595–2612. doi:10.1074/mcp.M900124-MCP200

Rao, K. V. G., Chand, P. P., & Murthy, M. V. R. (2007). A neural network approach in medical decision systems. *Journal of Theoretical and Applied Information Technology*, *3*(4).

Reynolds, K. J., Yao, X., & Fenselau, C. (2002). Proteolytic 18O labeling for comparative proteomics: Evaluation of endoprotease glu-C as the catalytic agent. *Journal of Proteome Research*, *1*(1), 27–33. doi:10.1021/pr0100016

Rifai, N., Gillette, M. A., & Carr, S. A. (2006). Protein biomarker discovery and validation: The long and uncertain path to clinical utility. *Nature Biotechnology*, *24*(8), 971–983. doi:10.1038/nbt1235

Ringnér, M., & Peterson, C. (2003). Microarray-based cancer diagnosis with artificial neural networks. *BioTechniques*, *34*, S30–S35.

Ripley, B. (1996). *Pattern recognition and neural networks*. Cambridge, UK: Cambridge University Press.

Risch, N., & Teng, J. (1998). The relative power of family-based and case-control designs for linkage disequilibrium studies of complex human diseases I. DNA pooling. *Genome Research*, *8*(12), 1273–1288.

Ross, P. L., Huang, Y. N., Marchese, J. N., Williamson, B., & Parker, K. (2004). Multiplexed protein quantitation in Saccharomyces cerevisiae using amine-reactive isobaric tagging reagents. *Molecular & Cellular Proteomics*, *3*(12), 1154–1169. doi:10.1074/mcp.M400129-MCP200

Ruppert, D., Nettleton, D., & Hwang, J. T. G. (2007). Exploring the information in p-values for the analysis and planning of multiple-test experiments. *Biometrics*, *63*(2), 483–495. doi:10.1111/j.1541-0420.2006.00704.x

Sachidanandam, R. (2001). A map of human genome sequence variation containing 1.42 million single nucleotide polymorphisms. *Nature*, *409*, 928–933. doi:10.1038/35057149

Salerno, R. A., & Lesko, L. J. (2004). Pharmacogenomic data: FDA voluntary and required submission guidance. *Pharmacogenomics*, *5*, 503. doi:10.1517/14622416.5.5.503

Sandusky, P., & Raftery, D. (2005). Use of semiselective TOCSY and the Pearson correlation for the metabonomic analysis of biofluid mixtures: Application to urine. *Analytical Chemistry*, *77*, 7717–7723. doi:10.1021/ac0510890

Saude, E. J., & Sykes, B. D. (2007). Urine stability for metabolomic studies: Effects of preparation and storage. *Metabolomics*, *3*(1), 19–27. doi:10.1007/s11306-006-0042-2

Scalbert, A., Brennan, L., Fiehn, O., Hankemeier, T., Kristal, B. S., & Ommen, B. V. (2009). Mass-spectrometry-based metabolomics: limitations and recommendations for future progress with particular focus on nutrition research. *Metabolomics*, *5*, 435–458. doi:10.1007/s11306-009-0168-0

Schadt, E., Li, C., Eliss, B., & Wong, W. H. (2002). Feature extraction and normalization algorithms for high-density oligonucleotide gene expression array data. *Journal of Cellular Biochemistry*, *84*(S37), 120–125. doi:10.1002/jcb.10073

Schadt, E., Li, C., Su, C., & Wong, W. H. (2001). Analyzing highdensity oligonucleotide gene expression array data. *Journal of Cellular Biochemistry*, *80*, 192–202. doi:10.1002/1097-4644(20010201)80:2<192::AID-JCB50>3.0.CO;2-W

Schapire, R., Freund, Y., Bartlett, P., & Lee, W. S. (1998). Boosting the margin: A new explanation for the effectiveness of voting methods. *Annals of Statistics*, *26*, 1651–1686. doi:10.1214/aos/1024691352

Schevchenko, A., Chernushevich, I., & Ens, W. (1997). Rapid de novo peptide sequencing by a combination of nanoelectrospray, isotopic labeling and a quadrupole/time-of-flight mass spectrometer. *Rapid Communications in Mass Spectrometry*, *11*(9), 1015–1024. doi:10.1002/(SICI)1097-0231(19970615)11:9<1015::AID-RCM958>3.0.CO;2-H

Shendure, J. (2005). Accurate multiplex polony sequencing of an evolved bacterial genome. *Science*, *309*, 1728–1732. doi:10.1126/science.1117389

Sheta, E. A., Appel, S. H., & Goldknopf, I. L. (2006). 2D gel blood serum biomarkers reveal differential clinical proteomics of the neurodegenerative diseases. *Expert Review of Proteomics*, *3*(1), 45–62. doi:10.1586/14789450.3.1.45

Sidorov, I. A., Hosack, D. A., Gee, D., Yang, J., Cam, M. C., & Lempicki, R. A. (2002). Oligonucleotide microarray data distribution and normalization. *Information Sciences*, *146*, 65–71. doi:10.1016/S0020-0255(02)00215-3

Silva, J. C., Denny, R., & Dorschel, C. A. (2005). Quantitative proteomic analysis by accurate mass retention time pairs. *Analytical Chemistry*, *77*(7), 2187–2200. doi:10.1021/ac048455k

Siuti, N., & Kelleher, N. L. (2007). Decoding protein modifications using top-down mass spectrometry. *Nature Methods*, *4*(10), 817–821. doi:10.1038/nmeth1097

States, D.J., Omenn, G.S., Blackwell, T.W., & Fermin, D., Eng., J., Speicher, D.W., et al. (2006). Challenges in deriving high-confidence protein identifications from data gathered by a HUPO plasma proteome collaborative study. *Nature Biotechnology*, *24*(3), 333–338. doi:10.1038/nbt1183

Steinthorsdottir, V. (2007). A variant in CDKAL1 influences insulin response and risk of type 2 diabetes. *Nature Genetics*, *39*, 770–775. doi:10.1038/ng2043

Stewart, J. J., White, J. T., Yan, X., Collins, S., Drescher, C. W., & Urban, N. D. (2006). Proteins associated with cisplatin resistance in ovarian cancer cells identified by quantitative proteomic technology and integrated with mRNA expression levels. *Molecular & Cellular Proteomics*, *5*(3), 433–443. doi:10.1074/mcp.M500140-MCP200

Syka, J. E., Coon, J. J., Schroeder, M. J., Shabanowitz, J., & Hunt, D. F. (2004). Peptide and protein sequence analysis by electron transfer dissociation mass spectrometry. *Proceedings of the National Academy of Sciences of the United States of America*, *101*(26), 9528–9533. doi:10.1073/pnas.0402700101

Tabor, H. K., Risch, N. J., & Myers, R. M. (2002). Opinion: Candidate-gene approaches for studying complex genetic traits: Practical considerations. *Nature Reviews. Genetics*, *3*, 391–397. doi:10.1038/nrg796

Teng, J., & Risch, N. (1999). The relative power of family-based and case-control designs for linkage disequilibrium studies of complex human diseases. II. Individual genotyping. *Genome Research*, *9*(3), 234–241.

Thomas, R. K. (2006). Sensitive mutation detection in heterogeneous cancer specimens by massively parallel picoliter reactor sequencing. *Nature Medicine*, *12*, 852–855. doi:10.1038/nm1437

Tian, Y., Tan, A., Sun, X., & Olson, M. T. (2009). Quantitative proteomic analysis of ovarian cancer cells identified mitochondrial proteins associated with paclitaxel resistance. *Proteomics: Clinical Applications*, *3*(11), 1288–1295. doi:10.1002/prca.200900005

Tibshirani, R., Hastie, T., Narasimhan, B., & Chu, G. (2002). Diagnosis of multiple cancer types by shrunken centroids of gene expression. *Proceedings of the National Academy of Sciences of the United States of America*, *99*, 6567–6572. doi:10.1073/pnas.082099299

US Food and Drug Administration. (FDA). (2005). *Drug-diagnostic co-development concept paper.* Retrieved January 3, 2010, from http://www.fda.gov/downloads/Drugs/ScienceResearch/ResearchAreas/Pharmacogenetics/UCM116689.pdf

US Food and Drug Administration. (FDA). (2006). *Guidance for industry—pharmacogenomic data submissions*. Retrieved January 3, 2010, from http://www.fda.gov/downloads/RegulatoryInformation/Guidances/ucm126957.pdf

van Iterson, M., 't Hoen, P. A., Pedotti, P., Hooiveld, G. J., den Dunnen, J. T., & van Ommen, G. J. (2009). Relative power and sample size analysis on gene expression profiling data. Find Similar. *BMC Genomics*, *10*(1), 439. doi:10.1186/1471-2164-10-439

Vaughn, C. P., Crockett, D. K., Lim, M. S., & Elenitoba-Johnson, K. S. J. (2006). Analytical characteristics of cleavable isotope-coded affinity tag-LC-tandem mass spectrometry for quantitative proteomic studies. *The Journal of Molecular Diagnostics*, *8*(4), 513–520. doi:10.2353/jmoldx.2006.060036

Waanders, L. F., Hanke, S., & Mann, M. (2007). Top-down quantitation and characterization of SILAC-labeled proteins. *Journal of the American Society for Mass Spectrometry*, *18*(11), 2058–2064. doi:10.1016/j.jasms.2007.09.001

Wang, J., Reijmers, T., Chen, L., Heijden, R. V. D., Wang, M., & Peng, S. (2009). System toxicology study of doxorubicin on rats using ultra performance liquid chromatography coupled with mass spectrometry based metabolomics. *Metabolomics*, *5*, 407–418. doi:10.1007/s11306-009-0165-3

Weljie, A. M., Newton, J., Mercier, P., Carlson, E., & Slupsky, C. M. (2006). Targeted profiling: Quantitative analysis of ^{1}H NMR metabolomics data. *Analytical Chemistry*, *78*(13), 4430–4442. doi:10.1021/ac060209g

Wellcome Trust Case Control Consortium. (2007). Genome-wide association study of 14,000 cases of seven common diseases and 3,000 shared controls. *Nature*, *447*, 661–682. doi:10.1038/nature05911

Wilson, I. D., Nicholson, J. K., Castro-Perez, J., Granger, J. H., Johnson, K. A., & Smith, B. W. (2005). High resolution ultra performance liquid chromatography coupled to as-TOF mass spectrometry as a tool for differential metabolic pathway profiling in functional genomic studies. *Journal of Proteome Research*, *4*, 591–598. doi:10.1021/pr049769r

Xing, E. P., Jordan, M. I., & Karp, R. M. (2001). Feature selection for high-dimensional genomic microarray data. *In Proceedings of the Eighteenth International Conference on Machine Learning*, 601–608.

Yang, Y. H., & Speed, T. P. (2003). *Design and analysis of comparative microarray experiments. Statistical analysis of gene expression microarray data*. Chapman & Hall.

Yates, N. A., Deyanova, E. G., Geissler, W., & Wiener, M. C. (2007). Identification of peptidase substrates in human plasma by FTMS based differential mass spectrometry. *International Journal of Mass Spectrometry*, *259*(1-3), 174–183. doi:10.1016/j.ijms.2006.09.020

Ye, T., Mo, H., Shanaiah, N., Gowda, G. A. N., Zhang, S., & Raftery, D. (2009). Chemoselective ^{15}N Tag for sensitive and high-resolution nuclear magnetic resonance profiling of the carboxy-containing metabolome. *Analytical Chemistry*, *81*(12), 4882–4888. doi:10.1021/ac900539y

Yu, L., & Liu, H. (2004). Redundancy based feature selection for microarray data. *Proceedings of the Tenth ACM SIGKDD international conference on Knowledge discovery and data mining*, Seattle, WA, USA.

Zhang, J., Sui, J., Ching, C. B., & Chen, W. N. (2008). Protein profile in neuroblastoma cells incubated with S- and R-enantiomers of ibuprofen by iTRAQ-coupled 2-D LC-MS/MS analysis: Possible action of induced proteins on Alzheimer's disease. *Proteomics*, *8*(8), 1595–1607. doi:10.1002/pmic.200700556

Zhang, X. G., Lu, X., Xu, X. Q., Leung, H. E., Wong, W. H., & Liu, J. S. (2006). RSVM: A SVM based strategy for recursive feature selection and sample classification with proteomics mass-spectrometry data. *BMC Bioinformatics*, *7*, 197. doi:10.1186/1471-2105-7-197

Zhou, F., Galan, J., Geahlen, R. L., & Tao, W. A. (2007). A novel quantitative proteomics strategy to study phosphorylation-dependent peptide-protein interactions. *Journal of Proteome Research*, *6*(1), 133–140. doi:10.1021/pr0602904

Zhou, M., Conrads, T. P., & Veenstra, T. D. (2005). Proteomics approaches to biomarker detection. *Briefings in Functional Genomics & Proteomics*, *4*(1), 69–75. doi:10.1093/bfgp/4.1.69

Zimmer, J. S., Monroe, M. E., Qian, W. J., & Smith, R. D. (2006). Advances in proteomics data analysis and display using an accurate mass and time tag approach. *Mass Spectrometry Reviews*, *25*(3), 450–482. doi:10.1002/mas.20071

Zubarev, R. A., Kelleher, N. L., & McLafferty, F. W. (1998). Electron capture dissociation of multiply charged protein cations. A nonergodic process. *Journal of the American Chemical Society*, *120*(13), 3265–3266. doi:10.1021/ja973478k

KEY TERMS AND DEFINITIONS

Biomarker: A characteristic that is objectively measured and evaluated as an indicator of a biological state.

Genomics: In general it is the study of the genomes of organisms. The field includes intensive efforts to determine the entire DNA sequence of organisms and fine-scale genetic mapping efforts.

Metabolomics: It is the systematic study of the unique chemical fingerprints that specific cellular processes leave behind - specifically, the study of their small-molecule metabolite profiles

Proteomics: It is the large-scale study of proteins, particularly their structures and functions.

Transcriptomics: It's the study of transcriptome of organisms. Transcriptome is the set of all RNA molecules, including mRNA, rRNA, tRNA, and non-coding RNA produced in one or a population of cells. Transcriptomics is also referred to as expression profiling, examines the expression level of RNAs in a given cell population.

Section 2
Method Development in Bioinformatics

This section contains five chapters that review the most recent advances in method development in sequence and high-throughput data analysis. Chapter 5 reviews the genome-wide association studies of human single nucleotide polymorphisms with their quantitative complex diseases and traits. Chapter 6 reviews several computational methods in genome-wide association studies and presents a novel approach to detecting epistatic interactions by employing expert knowledge, such as pathway and protein-protein interaction information. Chapter 7 reviews the theory, strengths and limitations of existing biclustering methods for the analysis of DNA microarray data. Several important applications to drug discovery and various problems in systems biology are also summarized. Chapter 8 reviews computational methods for the prediction of epigenetic target sites from DNA sequences based on nucleosome positioning, histone modification, and DNA methylation. Chapter 9 describes a novel method for protein sequence analysis. Evolutionary, structural, and functional information are taken into consideration to improve protein structure and function prediction. Application to the prestin protein is discussed.

Chapter 5
Single Nucleotide Polymorphism and its Application in Mapping Loci Involved in Developing Human Diseases and Traits

Rui-Ru Ji
Bristol-Myers Squibb, USA

ABSTRACT

Common diseases or traits in humans are often influenced by complex interactions among multiple genes as well as environmental and lifestyle factors rather than being attributable to a genetic variation within a single gene. Identification of genes that confer disease susceptibility can be facilitated by studying DNA markers such as single nucleotide polymorphism (SNP) associated with a disease trait. Genome-wide association approaches offers a systematic analysis of the association of hundreds of thousands of SNPs with a quantitative complex trait. This method has been successfully applied to a wide variety of common human diseases and traits, and has generated valuable findings that have improved the understanding of the genetic basis of many complex traits. This chapter outlines the general mapping process and methods, highlights the success stories, and describes some limitations and challenges that lie ahead.

INTRODUCTION

SNP, or single nucleotide polymorphism, is a genetic variation in a person's DNA sequence that occurs when a single nucleotide is replaced by one of the other three nucleotides. SNPs are very common in the human population, occurring in the genome more than one percent of the time (http://www.ncbi.nlm.nih.gov/About/primer/snps.html). Since only three to five percent of the human genome encodes protein sequences, the majority of SNPs are outside of the so called

DOI: 10.4018/978-1-60960-491-2.ch005

coding regions. SNPs within a coding region are of particular interest to researchers because they are more likely to alter the biological function of a protein. Because of their high prevalence in the human genome, SNPs can be used as genetic markers to pinpoint the susceptibility loci for a disease in an association study.

Botstein and colleagues (Botstein et al., 1980) first proposed the genome-wide association approach. They suggested that the naturally occurring DNA sequence polymorphisms could be used as genetic markers to examine systematically the transmission of phenotypes in families. When a polymorphism shows significant linkage to a disease, additional markers can be genotyped in the region, termed fine mapping, to identify the responsible gene or variant.

In Mendelian traits and diseases, the underlying gene can usually be mapped unambiguously and precisely to a small chromosomal interval because of the strong correlation between genotype and phenotype. Subsequently, discovery of coding sequence variants in one of a small number of candidate genes in affected individuals usually provides sufficient evidence to establish the identity of the causal gene (Botstein & Risch, 2003). The same certainties do not apply to complex disease and traits, which are polygenic in nature. It is much more difficult to identify genes that contribute to complex traits because of low penetrance, epistasis, locus heterogeneity, and variable expressivity. The candidate gene approach was shown to be woefully inadequate as many disease genes were completely unsuspected based on prior knowledge of biological pathways.

A possible path forward was suggested by advances in population genetics and genomics: instead of mapping disease genes by tracing phenotype in families, one could identify them by association studies by comparing the frequencies of genetic variants among case and control individuals (Altshuler et al., 2008).

The completion of the human genome sequence (International Human Genome Sequencing Consortium, 2004) and the provision of an initial catalog of human genetic variations and a haplotype map (known as the International Human HapMap Project; International HapMap Consortium, 2005) have made it possible to perform genome-wide association study (GWAS) utilizing SNPs as genetic markers. The rapid technology development in high throughput genotyping platforms and data analysis approaches have now permitted GWAS to be undertaken in a large number of samples (McCarthy et al., 2008).

The underlying rationale for GWAS is the so called "common disease, common variant" hypothesis, which predicts that common variants (classically defined as having a minor allele frequency larger than 1%) in the human populations manifest a common disease (Risch & Merikangas, 1996; Collins et al., 1997). In the past few years we have witnessed the success of GWA studies in identifying hundreds of common genetic variants associated with common diseases and traits (Goldstein, 2009; Hirschhorn, 2009; Kraft & Hunter, 2009). An updated list of published GWA studies can be found at the National Human Genome Research Institute's catalog of published genome-wide association studies (http://www.genome.gov/GWAStudies/).

The findings from GWAS have provided valuable novel insights into the complex allelic architecture of common diseases and traits. However, for most conditions studied to date, the implicated genetic variants only explain a fraction of the familial aggregation, limiting the early application potential for predicting individual risk. Much work remains to obtain a complete catalog of susceptibility loci and to elucidate the molecular mechanisms through which these variants operate. As such, it remains a distant objective to translate these findings into clinical practice.

The purpose of this chapter is to review the current status of the application of SNPs as genetic markers in mapping quantitative complex traits in humans. We will outline the general mapping process and the analytical methods, highlight the success stories, and describe the major limitations and challenges that lie ahead.

THE POWER OF GENOME-WIDE ASSOCAITION STUDIES

Conventionally, the search for a disease gene begins with linkage analysis, a simple form of genetic mapping conceived by Sturtevant for the fruit fly in 1913 (Sturtevant, 1913). In this approach, the aim is to find out the rough location of the disease gene relative to another DNA sequence called a genetic marker, whose position is already known. Despite its success in the studies of Mendelian disease genes (Botstein & Risch, 2003), the utility of genetic linkage mapping to human common diseases is very limited. In common forms of common disease, risk in relatives is lower than in Mendelian cases, and linkage analysis to detect a single causal gene yielded equivocal results, suggesting a polygenic model for common diseases.

The "common disease, common variant" hypothesis predicts that common diseases are attributable to common polymorphisms present in more than 1-5% of the population (Risch & Merikangas, 1996; Collins et al., 1997). In order to perform a GWA study to identify susceptible loci for common diseases and traits, a dense set of common variants across the whole genome and techniques to genotype these variants in a large number of study objects are required. By 2006, these tools were well in place: the completion of the human genome project has facilitated the discovery of millions of common variants such as SNPs in the human population (International HapMap Consortium, 2005); in addition, several technologies have been developed to simultaneously genotype hundreds of thousands of SNPs with high accuracy (Altshuler et al., 2008). In the meantime, an analytical framework has been developed to distinguish true association from noise or artifacts (Balding, 2006).

The GWAS has a clear advantage over family-based linkage analysis to map complex traits as the latter has lower power and resolution for variants of modest effects. GWAS also represent an important advance compared to the so called "candidate gene" approach, which is woefully inadequate as most of the disease genes were completely unsuspected based on prior knowledge of biological pathways. In addition, the "candidate gene" studies often involved small sample sizes and the genetic markers assayed were limited to a selected few. Consequently, association results were susceptible to false positives and few could be replicated in subsequent studies (Hirschhorn et al., 2002; Todd, 2006).

Since 2006, scores of GWA studies have identified hundreds of common SNPs associated with a wide range of common diseases and clinical conditions (breast cancer, prostate cancer, colorectal cancer, asthma, restless leg syndrome, gallstones, glaucoma, coronary disease, atrial fibrillation, multiple sclerosis, celiac disease, systemic lupus erythematosus, rheumatoid arthritis, inflammatory bowel disease, obesity, type 1 and type 2 diabetes, age-related macular degeneration, etc.), as well as individual traits (height, hair color, freckles, eye color, etc.) (http://www.genome.gov/GWAStudies/). This explosion represents one of the most prolific periods of discovery in human genetics, and has provided valuable insights into the complexities of the genetic architectures of common human diseases and traits.

GWA studies have re-discovered many genes previously known to be important to the disease of interest. For example, of 19 loci meeting statistical significance in recent GWA studies of LDL, HDL, or triglyceride levels in blood, 12 involve genes encoding apolipoproteins, lipases, and other key molecules in lipid metabolism (Kathiresan et al., 2008; Willer et al., 2008). Studies in other diseases and traits also highlight equally relevant genes (Hirschhorn, 2009). The number of overlapping loci is overwhelmingly greater than what would be expected by chance.

In other cases, GWA studies have highlighted biological pathways that were not suspected to be involved in a particular disease or trait. For example, components of the complement system are strongly associated with age-related macular degeneration (Edwards et al., 2005; Klein et al., 2005; Yates et al., 2007). Similarly, loci associated with Crohn's disease point unambiguously to autophagy and IL-23 pathways (Lettre & Rioux, 2008). Again, this clustering into biological pathways is highly nonrandom (Raychaudhuri et al., 2009). Studies are already under way to translate the knowledge into new therapeutic leads.

STAGES OF A GENOME-WIDE ASSOCIATION STUDY

A typical genome-wide association study involves multiple stages that require manipulation of large volumes of data and scientists from different areas of expertise (Hardy & Singleton, 2009). As in all epidemiological studies, the first step is to define the phenotypic trait and to decide whether to use a quantitative or qualitative trait. Diseases by definition are qualitative but can be measured quantitatively, for instance, the level of low-density lipoprotein cholesterol (LDL-C) is often measured to predict risk of developing heart disease. Similarly, quantitative traits can be analyzed qualitatively by applying a threshold value. Analyzing quantitative traits usually increases power because of larger information content, but may also decrease the statistical power if the measurement is not accurate. Importantly, the phenotype should be defined as precisely as possible with regard to the fundamental mechanism so as to improve the power by reducing heterogeneity, yet retaining the simplicity to facilitate subsequent studies and not reducing sample size excessively (Newton-Cheh & Hirschhorn, 2005).

Another important step to design a GWA study is the selection of population samples. There are four commonly used sampling schemes: case-control, cohort-based, family-based, and population isolates-based. The choice of sample structure is generally based on the prevalence and familial segregation of the phenotype of interest (Smith & Newton-Cheh, 2009). Case-control sampling is the most widely used approach as it is relatively easy and inexpensive to collect. However, this design is also highly susceptible to various forms of bias, most notably confounding by ancestry. Cohort studies are difficult and costly to collect as they involve a large number of individuals and lengthy follow-up to detect incident events. They have the advantage of being more representative of the population and thus less prone to selection bias. The family-based approach is merited if the phenotype is qualitative, relatively rare, and exhibits significant familial segregation. Population isolates have the advantage of low genetic and environmental heterogeneity. In addition, they may be enriched in rare alleles that are causal to a disease. However, the power of the studies may be limited by small sample sizes, and excessive relatedness of the samples. Estimate of the sample size needed to achieve certain levels of power can be obtained using statistical tools such as the Genetic Power Calculator (http://pngu.mgh.harvard.edu/~purcell/gpc/).

Several high throughput genotyping platforms assaying marker sets of different density and selection criterion are available. Marker density currently ranges from 100k to 1 million per array, selected: (a) randomly; (b) using a tagging approach to maximize coverage based on haplotype patterns; (c) by focusing on SNPs known or likely to be functional. The choice of the genotyping method should take into account the sample size, ancestral origin, and marker selection criteria. To minimize batch effects due to DNA sources, extraction protocols, genotyping procedures, and plate effects, sample collection and handling should be as uniform as possible. Data quality control is necessary since genotyping errors are a potential cause of spurious associations. For example, the genotyped sex needs to be matched to the recorded sex for individual samples. Samples within a group need to be assessed and outliers may be removed based on general pattern of genetic variability (Chanock et al., 2007). Systematic biases should also be checked in this step. For instance, ethnically distant subjects may be removed or adjusted for any systematic differences between or within cohorts (Devlin & Roeder, 1999; Pritchard & Rosenberg, 1999; Price et al., 2006).

In the next step, each SNP that survives quality control is tested for its association with a disease or phenotypic trait. For quantitative traits, this is typically done by a t test or analysis of variance (ANOVA). Because of the number of statistical tests performed, there is a high rate of false positives. To alleviate this problem, a stringent statistical cutoff is often employed. The result of the association analysis is often visualized in a so called Manhattan plot, where the position of the SNP is on the horizontal axis and the negative logarithm of the P value of the association is on the vertical axis (Hardy & Singleton, 2009). The highly significant SNPs are selected based on statistical significance or a combination of statistical significance and biological plausibility. These SNPs are usually retested in an independent sample sets, ideally of the same or larger size than the original samples. In addition, bioinformatic and data mining approaches are utilized to identify transcripts and other genetic variation next to the unequivocally associated loci. Fine mapping may be performed to discover new variants and genotyping of untyped variants to determine which are most significantly associated with the disease trait. Further analysis of the region is performed to determine the most critical variants, the pathologically relevant gene, and the likely biologic effect.

UNEXPLAINED HERITABILITY

To this date, several hundreds of genetic loci can be reproducibly associated with complex disease and traits in humans. In a few cases, common variants with an effect size of >= 2 fold have been found: for example, complement factor H (CFH) in age-related macular degeneration (Edwards et al., 2005; Klein et al., 2005). However, in vast majority of cases, common variants individually or in combination confer small increments in risk (1.1 to 1.5 fold) and explain only a small proportion of heritability, the proportion of observed variation in a particular trait that can be attributed to inherited genetic factors in contrast to environmental ones. For example, three studies identified 54 variants significantly associated with human height, a classic complex trait with an estimated 80% heritability, yet these associations only explain about 5% of the phenotypic variance despite a sample size of about 63000 individuals (Visscher, 2008).

The question arises as to why so much of the heritability is not explained by the GWAS studies. Although it is not expected to find all or most of the variants associated with disease in the initial studies, one would expect to find at least the ones with the strongest effects. Several explanations for this missing heritability have been suggested and are summarized below (Altschuler et al., 2008; Maher, 2008; Manolio et al., 2009).

The first possible cause is poor experimental design of GWAS studies, especially the earlier ones. For example, the control and case groups may be of questionable comparability, leading to inaccurate estimate of effect sizes. Technical artifacts are also problematic if cases and controls are not genotyped in parallel. Methods have now been developed to detect and adjust for such biases (Devlin & Roeder, 1999; Pritchard & Rosenberg, 1999; Price et al., 2006).

Secondly, most common variants identified so far have small effect sizes, suggesting that much larger sample sizes are required to uncover additional variants. Indeed, it has been clearly demonstrated that the number of detected variants increases with increasing sample sizes (Zeggini et al., 2008; Ahmed et al., 2009; Kathiresan et al.,

2009). Meta-analysis of matching studies can also increase the power of the association.

Third, the majority of the GWAS studies published so far are based on populations of European ancestry. Studies of non-European populations are likely to yield new variants and provide insight into the allelic architecture of complex traits (Yasuda et al., 2008; Zheng et al., 2009). By contrast, study of isolated populations may be of value as they may be enriched in unique variants (Sabatti et al., 2009).

Fourth, the current heritability estimate may not be accurate. It can be inflated due to non-additive genetic effects such as epistasis or gene-gene interaction, shared environments, and correlation or interaction among genotypes and environment (Visscher et al., 2008). Therefore, experimentally identified variants could never account for an erroneously inflated heritability estimate. Teasing apart the contribution from environmental factors shared among relatives is now possible utilizing the identity-by-descent (IBD) index that can be empirically estimated using genome-wide markers. On average full siblings share half of these markers, thus by relating phenotypic differences to the observed IBD among siblings pairs it is possible to estimate the heritability due to environmental factors (Visscher et al., 2006).

Lastly, it has been proposed that variants not captured by the available genotyping technologies may contribute to the missing heritability (Manolio et al., 2009). These variants include variants of low minor allele frequency (MAF), defined as roughly 0.5% < MAF < 5%, rare alleles (MAF < 0.5%), and structural variants such as copy number variants (CNVs). Further development in detection technologies as well as analytical methods is needed to study the association of these genetic variants with common diseases.

RARE ALLELES AND COMMON DISEASES AND TRAITS

In his review of GWA studies, Goldstein (2009) used an exponential model to fit the observed data of effect sizes and concluded that approximately 93000 SNPs would be needed to account for the heritability of human height. In doing so, he employed two assumptions: first, all SNPs yet to be identified have weaker effect sizes than the weakest identified so far; second, the effects of these SNPs are additive. Only a dramatic departure from these assumptions would result in a manageable number of common variants to account for a sizable fraction of the heritability. Similarly, one can draw the same conclusion for type 2 diabetes and other common diseases. Therefore Goldstein concludes that it is more likely that rare alleles could contribute substantially to the missing heritability. For example, 20 variants with MAF of 1% and allelic odd ratio (OR, defined as the probability of an event occurring divided by the probability of it not occurring, compared in individuals with versus without the allele) of three would account for most familial aggregation of type 2 diabetes (Manolio et al., 2009).

One problem in studying rare alleles is that they are not captured by the current genotyping arrays. These alleles are also difficult to be detected in a classical linkage analysis in family studies unless they carry very large effect sizes as in monogenic conditions (McCarthy et al., 2008). The launch of the 1000 genome project in 2008 (http://www.1000genomes.org/page.php) will facilitate the discovery of the association of low frequency and rare alleles with common diseases and traits. This multinational effort is designed to provide comprehensive catalog of human genetic variations with MAF as low as 1%. So far the pilot study has already identified more than 11 million new SNPs from 172 individuals (Manolio et al., 2009).

Currently the primary technology to detect rare alleles is sequencing and the sample size required for detecting sequence variants increases linearly with 1/MAF. For association studies, sample size also scales approximately quadratically with 1/|(OR-1)| (Manolio et al., 2009). As the odds ratio (OR) decreases, sample size needed to detect an association increases sharply, presenting a burden in both cost and capacity to the traditional capillary sequencing technology. The development of "next generation" sequencing technologies, which can generate millions of sequence reads in parallel, make it possible to quickly survey genetic variations in hundreds of samples (Mardis, 2008). Presently two deep sequencing strategies are employed to identify rare alleles associated with common diseases: targeted sequencing of genomic regions that have strong and replicated associations with common traits; whole genome sequencing of individuals with extreme phenotypes.

There is no guarantee that associations with rare alleles will provide immediate biological insight to disease pathology. However, given the limited role of common variants in many highly heritable diseases and traits, it is reasonable to hope that some of them will provide novel therapeutic targets, predict individual risk, and help in patient care. Recently, re-sequencing of 10 candidate genes has identified four rare alleles in IFIH1 (interferon induced with helicase C domain 1), a gene located in a region previously associated with type 1 diabetes, that lower the risk to the disease. This finding firmly establishes the role of IFIH1 in T1D and demonstrates that re-sequencing studies can pinpoint disease-causing genes in genomic regions initially identified by GWA studies (Nejentsev et al., 2009).

STRUCTURAL VARIANTS AND COMMON DISEASES AND TRAITS

Structural variations include insertions and deletions, which affect DNA copy number, and inversions and translocations, which preserve copy number. Structural variations are shown to be widespread in the human genome, and may contribute more to phenotypic variations than SNPs (Redon et al., 2006; Stranger et al., 2007). To this date structural variations have also been implicated in gene expression variations, female fertility, systematic autoimmunity, and other clinical conditions (Stefansson et al., 2005; Fanciulli et al., 2007; Stranger et al., 2007). It is likely that structural variations may account for some of the unexplained heritability of common human diseases.

Innovation in the genotyping array design makes it possible to integrate analysis of copy number variations (CNVs) into GWAS. Using the hybrid genotyping array (Affymetrix SNP 6.0), McCarroll et al (2008) have shown that approximately 80% of observed copy number difference between any two individuals arise from common copy number polymorphisms (CNPs) with an allele frequency > 5%, and more than 99% derived from inheritance. In addition, most common, diallelic CNPs are in strong linkage disequilibrium with SNPs.

As CNV detection algorithms evolve and comprehensive, high-resolution map of CNPs can be cataloged and measured in large reference panels, it is possible to evaluate the impact of CNV in common human diseases in the coming years. Clearly understanding the full extent of structural variations is important for elucidating the molecular basis for human phenotypic variations and genetic diseases (Conrad et al., 2010).

THE UTILITY OF EXPRESSION QTLS

The functional effects of genetic variations on complex diseases can be mediated through several mechanisms. Variant alleles affecting protein coding sequence and consequently protein function can have drastic effect. For example, a frameshift variant and two missense variants of NOD2, a member of the Apaf-1/Ced-4 superfamily of apoptosis regulators that is expressed in monocytes, confers susceptibility to Crohn's disease (Hugot et al., 2001). However, the majority of variants identified through GWAS do not map to recognizable protein coding regions. Understanding of the contribution of these variants to the disease etiology represents a substantial gap in a GWAS study (Hardy & Singleton, 2009). It has been proposed that these variants might influence disease phenotypes through regulating transcript abundance.

Indeed gene expression can be affected by DNA polymorphisms in the regulatory elements (Cookson et al., 2009). Consequently, transcript levels can be used as quantitative traits that can be mapped and the genomic loci that regulate gene expression are termed expression quantitative trait loci (eQTLs). Systematic identification of eQTLs can be achieved by assaying gene expression and genetic variation simultaneously on the genomic scale in a large number of samples. The resulting eQTL map can be used to categorize both *cis* and *trans* effect of genetic variants on gene expression. In addition, this information also helps interpret the GWAS results. For example, genetic markers associated with disease trait can be examined to see if they are also associated with the gene expression variations of one or several genes. Causality analysis can then be employed to examine which variation, DNA polymorphism or gene expression, is the more immediate cause of the disease phenotype (Schadt et al., 2005; Kulp & Jagalur, 2006; Aten et al., 2008; Charlesworth et al., 2009; Millstein et al., 2009).

eQTL Mapping

The microarray technology that allows the measurement of tens of thousands of genes simultaneously is the driving force for systematic mapping of eQTLs. In principle, the methodology to map quantitative phenotypic traits such as body weight can be directly applied to eQTL mapping. Interpretation of the eQTL results can be further assisted by sophisticated methodologies developed for gene expression analysis, for example, the analysis of regulatory networks (Fuller et al., 2007; Chen et al., 2008; Emilsson et al., 2008).

Many human eQTLs have been shown to be highly heritable (heritability > 0.8) through family studies (Dixon et al., 2007; Visscher et al., 2008). These genetic factors can influence gene expression either in *cis* or in *trans*. The definition of a *cis*-effect is somewhat arbitrary, typically any eQTL within 150kb upstream and downstream of the effected gene are considered to be *cis*-acting. Detailed analysis of the position of *cis*-acting eQTLs has revealed that they are usually around transcription start and end sites, and are rarely more than 20kb away from the effected gene (Veyrieras et al., 2008). By contrast, *trans*-acting eQTLs are more numerous and usually have weaker effects than the *cis*-acting ones (Schadt et al., 2003; Morley et al., 2004). It is not clear if the *trans* effect is mediated by transcription factors or other mechanisms. However, master regulators, the trans-acting factors that affect gene expression of many genes, are not enriched in transcription factors, rather they are a group of genes of very diverse molecular functions (Yvert et al., 2003).

eQTL and Disease Gene Mapping

The analysis of eQTL provides the link between genetic markers of a disease and gene expression of a gene or genes. In particular, association analysis can reveal important differences in gene expression within the available sample pool that cannot be readily detected by a simple comparison of gene expression in cases and controls (Cookson et al., 2009). The value of incorporating eQTL analysis in GWA studies has been demonstrated in several recent studies. For instance, a study of post-mortem, neuropathologically normal human brain samples has identified eQTLs affecting genes such as MAPT (microtubule-associated protein tau) and APOE (apolipoprotein E), which are known to play important role in the Alzheimer's disease (Saunders et al., 1993; Myers et al.,

2005). These genetic-expression effects will help understand the underlying molecular mechanism of the common disease and guide the analysis of genomic regions involved in the control of normal gene expression.

eQTL can also facilitate interpretation of GWAS results. The recent studies of Crohn's disease (CD) illustrate this approach (Barrett et al, 2008; Libioulle et al., 2007). The initial GWA study identified multiple markers on chromosome 5 with strong disease association. However, biological interpretation of these associations was elusive since these markers are contained within a 1.25Mb gene desert. However, subsequent examination of eQTL database revealed that the disease-associated alleles correlate with quantitative expression levels of the prostaglandin receptor EP4, PTGER4, the gene that resides closest to the associated region. The homologue of this gene has been implicated in phenotypes similar to CD in mouse (Kabashima et al., 2002).

Since phenotypic traits are often the outcome of the interplay of multiple genes in a network, network analysis incorporating eQTL data has recently provided important novel insight into the molecular mechanism underlying human common diseases. Systematic identification of gene networks involved in disease processes is now possible using sophisticated algorithms developed for analyzing global gene expression, protein, and metabolite data (Gardner et al., 2003; Sontag et al., 2004). Application of co-expression network analysis to liver and adipose gene expression data generated from a segregating mouse population resulted in the identification of a macrophage-enriched network that were significantly enriched in expression traits supported as having a causal relationship with metabolic syndrome. Three genes in this network, lipoprotein lipase (Lpl), lactamase beta (Lactb) and protein phosphatase 1-like (Ppm1l), were validated by gene knockouts as novel obesity genes, strengthening the association between this network and metabolic disease traits (Chen et al., 2008). Moreover, network analysis allowed the identification of a core network module that is enriched for genes involved in the inflammatory and immune response and has been found to be causally associated to obesity-related traits in parallel studies in humans and mice (Emilsson et al., 2008).

Several eQTL databases have been created and they serve as important reference source for GWA studies (Cookson et al., 2009). Indeed, the comprehensive catalog of human eQTLs will facilitate the identification of expression traits that are affected by genetic variations, and will provide a valuable basis for studying the mechanism of gene regulation. As such, the National Institute of Health has recently announced the Genotype-Tissue Expression (GTEx) project that aims to create a whole body map of eQTLs so that any risk allele for a disease can be readily checked for its effect on global gene expression across all tissues. The pilot project will collect multiple tissues from approximately 160 donors. If the pilot phase proves successful, the project will be scaled up to involve approximately 1000 donors (http://nihroadmap.nih.gov/GTEx/).

Limitations of eQTL

Although it is well established that gene expression may be considered as quantitative traits and mapped with considerable power, there are many limitations to current methodologies and potentials for further improvement. As with conventional QTL, currently mapped eQTLs account for only a portion of estimated heritability. Similarly, the causes for unexplained heritability for conventional QTL also apply to eQTL. Importantly, not all factors that affect transcript abundance are accounted for in a SNP-based association study. For example, epigenetic modification can lead to monoallelic gene expression depending on parent of origin (Jaenisch & Bird, 2003). In addition, transcript abundance is a function of transcript stability as well as transcript production. Many factors can affect transcript stability, either through protein-RNA complex, or by small interfering RNA (siRNA) (Cookson et al., 2009).

IDENTIFICATION OF CAUSAL ALLELES

Advances in technologies and analytical methods have led to unprecedented catalog of genetic variants associated with a broad range of complex diseases and traits in humans. However, the molecular mechanisms underlying the overwhelming majority of these phenotypes remain elusive. It is reasonable to predict that our understanding of complex diseases and traits will be enhanced by increased sample size, higher density genotyping, expanded ethnic diversity, and deep sequencing of risk regions. However, none of these approaches will unequivocally enable the transition from association to causality (McCarthy et al., 2008; Katsanis, 2009).

At best, most arguments put forth to date to link a specific gene with disease susceptibility are correlative. It is likely that that most of the variants identified to date are markers, not causal variants (Goldstein, 2009). As such, the mammalian quantitative genetics community (Members of the Complex Trait Consortium, 2003) has proposed several lines of experimentation that are necessary to identify genetic loci that govern quantitative traits:

1. Polymorphisms in coding or regulatory regions: the sequence differences are associated with different phenotypic effects.
2. Gene function: there exists a mechanistic link between the gene function and the trait of interest, for example, gene expression in the appropriate tissue or cell, or gene involvement in the appropriate pathway.
3. *In vitro* functional study: one allele can be shown to have, for example, different biochemical properties reflective of the *in vivo* phenotype.
4. Transgenesis: disease phenotype can be rescued by transgenic complementation.
5. Knock-ins: allele replacement by homologous recombination should alter the quantitative trait as expected.
6. Deficiency-complementation test: a variant allele has a different phenotypic effect when in trans to a knock-out of the candidate QTL.
7. Mutational analysis: induced or spontaneous mutations in the candidate QTL should change the phenotype in a predictable fashion.
8. Homology searches: natural genetic variant at an orthologous locus affects the same trait in another species.

Clearly some of the conditions provide stronger evidence than the others. In addition, it is rare and not necessary that all these criteria need to be fulfilled. However, several of the conditions listed above need to be met in order to declare a causal relationship between a genetic variant and a phenotypic trait.

CONCLUSION

Since 2006 GWA studies have reproducibly identified hundreds of common SNP variants associated with many dozens of traits, considerably surpassing early expectations. In spite of these advances, our understanding of the genetic architecture of common diseases and traits is still very limited. For many traits only a small proportion of estimated heritability can be explained by these variants (Maher, 2008), casting doubt over the validity of the "common disease, common variant" hypothesis. New technologies and analytical methods are currently being developed to evaluate the contribution from other genetic variants such as rare variants and CNVs in future GWA studies. The integrated analysis of SNP, haplotypes, CNV, and epigenetic elements will help elucidate the molecular etiology of common human diseases, and advance our understanding of how these different genetic variations act in concert to influence human phenotypes.

ACKNOWLEDGMENT

The author is grateful to Karl-Heinz Ott, Nathan Siemers, and the three unknown reviewers for their critical review of this chapter.

REFERENCES

Ahmed, S., Thomas, G., Ghoussaini, M., Healey, C. S., Humphreys, M. K., & Platte, R. (2009). Newly discovered breast cancer susceptibility loci on 3p24 and 17q23.2. *Nature Genetics*, *41*(5), 585–590. doi:10.1038/ng.354

Altshuler, D., Daly, M. J., & Lander, E. S. (2008). Genetic mapping in human disease. *Science*, *322*(5903), 881–888. doi:10.1126/science.1156409

Aten, J. E., Fuller, T. F., Lusis, A. J., & Horvath, S. (2008). Using genetic markers to orient the edges in quantitative trait networks: the NEO software. *BMC Systems Biology*, *2*, 34. doi:10.1186/1752-0509-2-34

Balding, D. J. (2006). A tutorial on statistical methods for population association studies. *Nature Reviews. Genetics*, *7*(10), 781–791. doi:10.1038/nrg1916

Barrett, J. C., Hansoul, S., Nicolae, D. L., Cho, J. H., Duerr, R. H., & Rioux, J. D. (2008). Genome-wide association defines more than 30 distinct susceptibility loci for Crohn's disease. *Nature Genetics*, *40*(8), 955–962. doi:10.1038/ng.175

Botstein, D., & Risch, N. (2003). Discovering genotypes underlying human phenotypes: Past successes for Mendelian disease, future approaches for complex disease. *Nature Genetics*, *33*(Supplement), 228–237. doi:10.1038/ng1090

Botstein, D., White, R. L., Skolnick, M., & Davis, R. W. (1980). Construction of a genetic linkage map in man using restriction fragment length polymorphisms. *American Journal of Human Genetics*, *32*(3), 314–331.

Chanock, S. J., Manolio, T., Boehnke, M., Boerwinkle, E., Hunter, D. J., & Thomas, G. (2007). Replicating genotype-phenotype associations. *Nature*, *447*(7145), 655–660. doi:10.1038/447655a

Charlesworth, J. C., Peralta, J. M., Drigalenko, E., Göring, H. H., Almasy, L., & Dyer, T. D. (2009). Toward the identification of causal genes in complex diseases: A gene-centric joint test of significance combining genomic and transcriptomic data. *BMC Proceedings*, *3*(Supplement 7), S92. doi:10.1186/1753-6561-3-s7-s92

Chen, Y., Zhu, J., Lum, P. Y., Yang, X., Pinto, S., & MacNeil, D. J. (2008). Variations in DNA elucidate molecular networks that cause disease. *Nature*, *452*(7186), 429–435. doi:10.1038/nature06757

Collins, F. S., Guyer, M. S., & Charkravarti, A. (1997). Variations on a theme: Cataloging human DNA sequence variation. *Science*, *278*(5343), 1580–1581. doi:10.1126/science.278.5343.1580

Conrad, D. F., Pinto, D., Redon, R., Feuk, L., Gokcumen, O., & Zhang, Y. (2010). Origins and functional impact of copy number variation in the human genome. *Nature*, *464*(7289), 704–712. doi:10.1038/nature08516

Cookson, W., Liang, L., Abecasis, G., Moffatt, M., & Lathrop, M. (2009). Mapping complex disease traits with global gene expression. *Nature Reviews. Genetics*, *10*(3), 184–194. doi:10.1038/nrg2537

Devlin, B., & Roeder, K. (1999). Genomic control for association studies. *Biometrics*, *55*(4), 997–1004. doi:10.1111/j.0006-341X.1999.00997.x

Dixon, A. L., Liang, L., Moffatt, M. F., Chen, W., Heath, S., & Wong, K. C. (2007). A genome-wide association study of global gene expression. *Nature Genetics*, *39*(10), 1202–1207. doi:10.1038/ng2109

Edwards, A. O., Ritter, R. III, Abel, K. J., Manning, A., Panhuysen, C., & Farrer, L. A. (2005). Complement factor H polymorphism and age-related macular degeneration. *Science*, *308*(5720), 421–424. doi:10.1126/science.1110189

Emilsson, V., Thorleifsson, G., Zhang, B., Leonardson, A. S., Zink, F., & Zhu, J. (2008). Genetics of gene expression and its effect on disease. *Nature*, *452*(7186), 423–428. doi:10.1038/nature06758

Fanciulli, M., Norsworthy, P. J., Petretto, E., Dong, R., Harper, L., & Kamesh, L. (2007). FCGR3B copy number variation is associated with susceptibility to systemic, but not organ-specific, autoimmunity. *Nature Genetics*, *39*(6), 721–723. doi:10.1038/ng2046

Fuller, T. F., Ghazalpour, A., Aten, J. E., Drake, T. A., Lusis, A. J., & Horvath, S. (2007). Weighted gene coexpression network analysis strategies applied to mouse weight. *Mammalian Genome*, *18*(6-7), 463–472. doi:10.1007/s00335-007-9043-3

Gardner, T. S., di Bernardo, D., Lorenz, D., & Collins, J. J. (2003). Inferring genetic networks and identifying compound mode of action via expression profiling. *Science*, *301*(5629), 102–105. doi:10.1126/science.1081900

Goldstein, D. B. (2009). Common genetic variation and human traits. *The New England Journal of Medicine*, *360*(17), 1696–1698. doi:10.1056/NEJMp0806284

Hardy, J., & Singleton, A. (2009). Genomewide association studies and human disease. *The New England Journal of Medicine*, *360*, 1759–1768. doi:10.1056/NEJMra0808700

Hirschhorn, J. N. (2009). Genomewide association studies – illustrating biologic pathways. *The New England Journal of Medicine*, *360*, 1699–1701. doi:10.1056/NEJMp0808934

Hirschhorn, J. N., Lohmueller, K., Byrne, E., & Hirschhorn, K. (2002). A comprehensive review of genetic association studies. *Genetics in Medicine*, *4*(2), 45–61. doi:10.1097/00125817-200203000-00002

Hugot, J. P., Chamaillard, M., Zouali, H., Lesage, S., Cézard, J. P., & Belaiche, J. (2001). Association of NOD2 leucine-rich repeat variants with susceptibility to Crohn's disease. *Nature*, *411*(6837), 599–603. doi:10.1038/35079107

International HapMap Consortium. (2005). A haplotype map of the human genome. *Nature*, *437*, 1229–1320.

International Human Genome Sequencing Consortium. (2004). Finishing the euchromatic sequence of the human genome. *Nature*, *431*(7011), 931–945. doi:10.1038/nature03001

Jaenisch, R., & Bird, A. (2003). Epigenetic regulation of gene expression: how the genome integrates intrinsic and environmental signals. *Nature Genetics*, *33*(Supplement), 245–254. doi:10.1038/ng1089

Kabashima, K., Saji, T., Murata, T., Nagamachi, M., Matsuoka, T., & Segi, E. (2002). The prostaglandin receptor EP4 suppresses colitis, mucosal damage and CD4 cell activation in the gut. *The Journal of Clinical Investigation*, *109*(7), 883–893.

Kathiresan, S., Melander, O., Guiducci, C., Surti, A., Burtt, N. P., & Rieder, M. J. (2008). Six new loci associated with blood low-density lipoprotein cholesterol, high-density lipoprotein cholesterol or triglycerides in humans. *Nature Genetics*, *40*(2), 189–197. doi:10.1038/ng.75

Kathiresan, S., Willer, C. J., Peloso, G. M., Demissie, S., Musunuru, K., & Schadt, E. E. (2009). Common variants at 30 loci contribute to polygenic dyslipidemia. *Nature Genetics*, *41*(1), 56–65. doi:10.1038/ng.291

Katsanis, N. (2009). From association to causality: The new frontier for complex traits. *Genome Medicine*, *1*(2), 23. doi:10.1186/gm23

Klein, R. J., Zeiss, C., Chew, E. Y., Tsai, J. Y., Sackler, R. S., & Haynes, C. (2005). Complement factor H polymorphism in age-related macular degeneration. *Science*, *308*(5720), 385–389. doi:10.1126/science.1109557

Kraft, P., & Hunter, D. J. (2009). Genetic risk prediction-are we there yet? *The New England Journal of Medicine*, *360*(17), 1701–1703. doi:10.1056/NEJMp0810107

Kulp, D. C., & Jagalur, M. (2006). Causal inference of regulator-target pairs by gene mapping of expression phenotypes. *BMC Genomics*, *7*, 125. doi:10.1186/1471-2164-7-125

Lettre, G., & Rioux, J. D. (2008). Autoimmune diseases: Insights from genome-wide association studies. *Human Molecular Genetics*, *17*(R2), R116–R121. doi:10.1093/hmg/ddn246

Libioulle, C., Louis, E., Hansoul, S., Sandor, C., Farnir, F., & Franchimont, D. (2007). Novel Crohn disease locus identified by genome-wide association maps to a gene desert on 5p13.1 and modulates expression of PTGER4. *PLOS Genetics*, *3*(4), e58. doi:10.1371/journal.pgen.0030058

Maher, B. (2008). Personal genomes: The case of the missing heritability. *Nature*, *456*(7218), 18–21. doi:10.1038/456018a

Manolio, T. A., Collins, F. S., Cox, N. J., Goldstein, D. B., Hindorff, L. A., & Hunter, D. J. (2009). Finding the missing heritability of complex diseases. *Nature*, *461*(7265), 747–753. doi:10.1038/nature08494

Mardis, E. R. (2008). The impact of next-generation sequencing technology on genetics. *Trends in Genetics*, *24*(3), 133–141.

McCarroll, S. A., Kuruvilla, F. G., Korn, J. M., Cawley, S., Nemesh, J., & Wysoker, A. (2008). Integrated detection and population-genetic analysis of SNPs and copy number variation. *Nature Genetics*, *40*(10), 1166–1174. doi:10.1038/ng.238

McCarthy, M. I., Abecasis, G. R., Cardon, L. R., Goldstein, D. B., Little, J., & Ioannidis, J. P. (2008). Genome-wide association studies for complex traits: Consensus, uncertainty and challenges. *Nature Reviews. Genetics*, *9*(5), 356–369. doi:10.1038/nrg2344

Millstein, J., Zhang, B., Zhu, J., & Schadt, E. E. (2009). Disentangling molecular relationships with a causal inference test. *BMC Genetics*, *10*, 23. doi:10.1186/1471-2156-10-23

Morley, M., Molony, C. M., Weber, T. M., Devlin, J. L., Ewens, K. G., & Spielman, R. S. (2004). Genetic analysis of genome-wide variation in human gene expression. *Nature*, *430*(7001), 743–747. doi:10.1038/nature02797

Myers, A. J., Kaleem, M., Marlowe, L., Pittman, A. M., Lees, A. J., & Fung, H. C. (2005). The H1c haplotype at the MAPT locus is associated with Alzheimer's disease. *Human Molecular Genetics*, *14*(16), 2399–2404. doi:10.1093/hmg/ddi241

Nejentsev, S., Walker, N., Riches, D., Egholm, M., & Todd, J. A. (2009). Rare variants of IFIH1, a gene implicated in antiviral responses, protect against type 1 diabetes. *Science*, *324*(5925), 387–389. doi:10.1126/science.1167728

Newton-Cheh, C., & Hirschhorn, J. N. (2005). Genetic association studies of complex traits: Design and analysis issues. *Mutation Research*, *573*(1-2), 54–69. doi:10.1016/j.mrfmmm.2005.01.006

Pricc, A. L., Patterson, N. J., Plenge, R. M., Weinblatt, M. E., Shadick, N. A., & Reich, D. (2006). Principal components analysis corrects for stratification in genome-wide association studies. *Nature Genetics*, *38*, 904–909. doi:10.1038/ng1847

Pritchard, J. K., & Rosenberg, N. A. (1999). Use of unlinked genetic markers to detect population stratification in association studies. *American Journal of Human Genetics*, *65*(1), 220–228. doi:10.1086/302449

Raychaudhuri, S., Plenge, R. M., Rossin, E. J., Ng, A. C., Purcell, S. M., & Sklar, P. (2009). Identifying relationships among genomic disease regions: Predicting genes at pathogenic SNP associations and rare deletions. International Schizophrenia Consortium. *PLOS Genetics*, *5*(6), e1000534. doi:10.1371/journal.pgen.1000534

Redon, R., Ishikawa, S., Fitch, K. R., Feuk, L., Perry, G. H., & Andrews, T. D. (2006). Global variation in copy number in the human genome. *Nature*, *444*(7118), 444–454. doi:10.1038/nature05329

Risch, N., & Merikangas, K. (1996). The future of genetic studies of complex human diseases. *Science*, *273*(5281), 1516–1517. doi:10.1126/science.273.5281.1516

Sabatti, C., Service, S. K., Hartikainen, A. L., Pouta, A., Ripatti, S., & Brodsky, J. (2009). Genome-wide association analysis of metabolic traits in a birth cohort from a founder population. *Nature Genetics*, *41*(1), 35–46. doi:10.1038/ng.271

Saunders, A. M., Strittmatter, W. J., Schmechel, D., George-Hyslop, P. H., Pericak-Vance, M. A., & Joo, S. H. (1993). Association of apolipoprotein E allele epsilon 4 with late-onset familial and sporadic Alzheimer's disease. *Neurology*, *43*(8), 1467–1472.

Schadt, E. E., Lamb, J., Yang, X., Zhu, J., Edwards, S., & Guhathakurta, D. (2005). An integrative genomics approach to infer causal associations between gene expression and disease. *Nature Genetics*, *37*(7), 710–717. doi:10.1038/ng1589

Schadt, E. E., Monks, S. A., Drake, T. A., Lusis, A. J., Che, N., & Colinayo, V. (2003). Genetics of gene expression surveyed in maize, mouse and man. *Nature*, *422*(6929), 297–302. doi:10.1038/nature01434

Smith, J. G., & Newton-Cheh, C. (2009). Genome-wide association study in humans. *Methods in Molecular Biology (Clifton, N.J.)*, *573*, 231–258. doi:10.1007/978-1-60761-247-6_14

Sontag, E., Kiyatkin, A., & Kholodenko, B. N. (2004). Inferring dynamic architecture of cellular networks using time series of gene expression, protein and metabolite data. *Bioinformatics (Oxford, England)*, *20*(12), 1877–1886. doi:10.1093/bioinformatics/bth173

Stefansson, H., Helgason, A., Thorleifsson, G., Steinthorsdottir, V., Masson, G., & Barnard, J. (2005). A common inversion under selection in Europeans. *Nature Genetics*, *37*(2), 129–137. doi:10.1038/ng1508

Stranger, B. E., Forrest, M. S., Dunning, M., Ingle, C. E., Beazley, C., & Thorne, N. (2007). Relative impact of nucleotide and copy number variation on gene expression phenotypes. *Science*, *315*(5813), 848–853. doi:10.1126/science.1136678

Sturtevant, A. H. (1913). The linear arrangement of six sex-linked factors in Drosophila, as shown by their mode of association. *The Journal of Experimental Zoology*, *14*, 43–59. doi:10.1002/jez.1400140104

Todd, J. A. (2006). Statistical false positive or true disease pathway? *Nature Genetics*, *38*(7), 731–733. doi:10.1038/ng0706-731

Veyrieras, J. B., Kudaravalli, S., Kim, S. Y., Dermitzakis, E. T., Gilad, Y., & Stephens, M. (2008). High-resolution mapping of expression-QTLs yields insight into human gene regulation. *PLOS Genetics*, *4*(10), e1000214. doi:10.1371/journal.pgen.1000214

Visscher, P. M. (2008). Sizing up human height variation. *Nature Genetics*, *40*(5), 489–490. doi:10.1038/ng0508-489

Visscher, P. M., Hill, W. G., & Wray, N. R. (2008). Heritability in the genomics era-concepts and misconceptions. *Nature Reviews. Genetics*, *9*(4), 255–266. doi:10.1038/nrg2322

Visscher, P. M., Medland, S. E., Ferreira, M. A., Morley, K. I., Zhu, G., & Cornes, B. K. (2006). Assumption-free estimation of heritability from genome-wide identity-by-descent sharing between full siblings. *PLOS Genetics*, *2*(3), e41. doi:10.1371/journal.pgen.0020041

Willer, C. J., Sanna, S., Jackson, A. U., Scuteri, A., Bonnycastle, L. L., & Clarke, R. (2008). Newly identified loci that influence lipid concentrations and risk of coronary artery disease. *Nature Genetics*, *40*(2), 161–169. doi:10.1038/ng.76

Yasuda, K., Miyake, K., Horikawa, Y., Hara, K., Osawa, H., & Furuta, H. (2008). Variants in KCNQ1 are associated with susceptibility to type 2 diabetes mellitus. *Nature Genetics*, *40*(9), 1092–1097. doi:10.1038/ng.207

Yates, J. R., Sepp, T., Matharu, B. K., Khan, J. C., Thurlby, D. A., & Shahid, H. (2007). Complement C3 variant and the risk of age-related macular degeneration. *The New England Journal of Medicine*, *357*(6), 553–561. doi:10.1056/NEJMoa072618

Yvert, G., Brem, R. B., Whittle, J., Akey, J. M., Foss, E., & Smith, E. N. (2003). Trans-acting regulatory variation in Saccharomyces cerevisiae and the role of transcription factors. *Nature Genetics*, *35*(1), 57–64. doi:10.1038/ng1222

Zeggini, E., Scott, L. J., Saxena, R., Voight, B. F., Marchini, J. L., & Hu, T. (2008). Meta-analysis of genome-wide association data and large-scale replication identifies additional susceptibility loci for type 2 diabetes. *Nature Genetics*, *40*(5), 638–645. doi:10.1038/ng.120

Zheng, W., Long, J., Gao, Y. T., Li, C., Zheng, Y., & Xiang, Y. B. (2009). Genome-wide association study identifies a new breast cancer susceptibility locus at 6q25.1. *Nature Genetics*, *41*(3), 324–328. doi:10.1038/ng.318

KEY TERMS AND DEFINITIONS

Complex Diseases and Traits: Diseases and traits that are influenced by more than one factor, which can be a gene or an environmental factor.

Genome-Wide Association Study (GWAS): An examination of genetic variation across a given genome, designed to identify genetic associations with observable traits.

Haplotype: In genetics, a haplotype is a combination of alleles at multiple loci that are transmitted together on the same chromosome. Haplotype also refers to a set of single-nucleotide polymorphisms (SNPs) on a single chromatid that are statistically associated.

Heritability: The proportion of phenotypic variation in a population that is attributable to genetic variation among individuals. It is estimated by the ratio of genetic variance to total trait variance, so that 0 indicates no genetic effect on trait variance and 1 indicates that all variance are under genetic control.

Linkage Disequilibrium (LD): Non-random association of alleles at two or more loci, not necessarily on the same chromosome. It describes a situation where some combinations of alleles or genetic markers occur more or less frequently in a population than would be expected from a random formation of haplotypes based on allelic frequencies.

Quantitative Trait (QT): Trait that has measurable phenotypic variation.

Quantitative Trait Locus (QTL): A genetic locus that affects a quantitative trait.

Single Nucleotide Polymorphism (SNP): A variation in a person's DNA, molecules inside cells that carry genetic information. It occurs when a single nucleotide is replaced with another. These changes may cause disease, and may affect how a person reacts to bacteria, viruses, drugs, and other substances.

Chapter 6
Addressing the Challenges of Detecting Epistasis in Genome-Wide Association Studies of Common Human Diseases Using Biological Expert Knowledge

Kristine A. Pattin
Dartmouth Medical School, USA

Jason H. Moore
Dartmouth Medical School, USA

ABSTRACT

Recent technological developments in the field of genetics have given rise to an abundance of research tools, such as genome-wide genotyping, that allow researchers to conduct genome-wide association studies (GWAS) for detecting genetic variants that confer increased or decreased susceptibility to disease. However, discovering epistatic, or gene-gene, interactions in high dimensional datasets is a problem due to the computational complexity that results from the analysis of all possible combinations of single-nucleotide polymorphisms (SNPs). A recently explored approach to this problem employs biological expert knowledge, such as pathway or protein-protein interaction information, to guide an analysis by the selection or weighting of SNPs based on this knowledge. Narrowing the evaluation to gene combinations that have been shown to interact experimentally provides a biologically concise reason why those two genes may be detected together statistically. This chapter discusses the challenges of discovering epistatic interactions in GWAS and how biological expert knowledge can be used to facilitate genome-wide genetic studies.

DOI: 10.4018/978-1-60960-491-2.ch006

INTRODUCTION

The fields of human genetics and genetic epidemiology have benefited greatly from the completion of the Human Genome Project in 2003 and the HapMap Project in 2005. The availability of dense maps of single-nucleotide polymorphisms (SNPs) along with high-throughput genotyping technologies has set the stage for routine genome-wide association studies (GWAS) that are expected to significantly improve our ability to identify susceptibility loci across the human genome. To be able to identify genetic variants that are associated with susceptibility to common-complex diseases is an important goal of the aforementioned fields, and the end goal of this endeavor is to utilize these genetic association results to develop better strategies for disease diagnosis, prevention, and treatment.

The GWAS is the current strategy for identifying and characterizing genetic predictors of disease and provides the capability to assess the role of one million or more SNPs in determining disease susceptibility (Hirschhorn & Daly, 2005; Wang et al., 2005). While great strides have been taken to optimize and establish the technical details of measuring a large representative set of SNPs in an accurate and efficient manner (Spencer et al., 2009), the analytical methods for determining which SNPs are important are in their infancy. These methods are based on assumptions such as each SNP having a large and independent effect on disease risk (Clark et al., 2005). It is recognized that most SNPs discovered have small effects on disease susceptibility making them less than ideal targets for medical research or genetic testing. One potential reason for this is that the current analytical framework follows the assumption that each associated SNP will have a detectible effect on disease risk that is independent of all the other variations in the genome as well as independent of the ecological context of each sampled human subject. While the one SNP at a time analytical approach is logical in the sense that it is time efficient and the results are easy to interpret, it is not comprehensive because it fails to acknowledge the complexity of the diseases at hand. If we assume a disease to have a complex genetic architecture, single SNP analyses may only reveal a small portion of the total genetic effects. It is evident that there needs to be an analytical retooling to address the complexity of common diseases (Thorton-Wells et al., 2004).

Common-complex diseases have a much more complex etiology that is due to phenomena such as epistasis (gene-gene interaction), plastic reaction norms (gene-environment interaction), and locus heterogeneity. Therefore, epistasis is a critical genetic component in determining disease susceptibility, where numerous points of genetic variation interact to influence risk. To be able to detect and characterize these interactions is pertinent to our understanding of the biological mechanisms underlying these diseases. However, there are many important challenges that need to be addressed if we wish to completely explore epistasis in a GWAS in order to gain a more coherent understanding of the genetic architecture of a complex trait and its interacting elements. The complexity of the genotype-to-phenotype mapping relationship for common diseases suggests that we are unlikely to identify important genetic variants until we acknowledge and address the many phenomena that create nonlinear patterns in genetic association data (Templeton, 2000; Moore, 2003; Sing et al., 2003; Thornton-Wells et al., 2004; Rea et al., 2006; Moore & Williams, 2009). Not only is there a need for statistical methods powerful enough to model the relationship between SNP interactions and disease susceptibility, but there is a technical challenge that needs to be addressed as well.

In order to detect and characterize epistasis in GWAS, there needs to be an effective way to statistically explore all possible combinations of SNPs, and while many methods have been developed to do so in smaller data sets, analysis of all SNP combinations in GWAS remains computationally daunting with all current existing methods. Recently, there has been a major emphasis placed on incorporating biological expert knowledge from biochemical pathways, gene networks, and protein-protein interactions into the GWAS analysis as a solution to this analytical problem (Moore, 2009; Pattin & Moore, 2009). Ideker and Sharan (2008) highlight the importance of protein interaction networks as tools that will help to unravel the underlying mechanisms of human disease. They suggest that the application of such networks for identifying new disease genes, disease-related sub-networks, as well as conducting network –based disease classification has revolutionized the single-gene analysis approach to common disease by demonstrating the importance of such inter-relationships (Ideker & Sharan, 2008). Emily et al. (2008) similarly demonstrate how biological networks, specifically protein-protein interactions, can be used to guide a GWAS analysis. Since one of the strongest demonstrations of the functional interactions between genes is protein-protein interactions, it is plausible that the expert knowledge extracted from protein interaction databases will allow for a more directed analysis of genome-wide studies as well as facilitate the biological interpretation of the data (Pattin & Moore, 2009). While current work has fostered promise in the idea of combining many types of existing biological data as an analysis strategy, such as demonstrated by (Lage et al., 2007; Mani et al., 2008), to narrow the evaluation to pathways or gene combinations that have been shown to interact experimentally, provides a biologically concise reason why interactions may be detected together statistically.

While ultimately we anticipate that the GWAS may be a tool that is capable of being applied in the clinical setting, there are many challenges that need to be addressed before that can become a reality. In this chapter we not only discuss the technical challenges of conducting a GWAS (i.e. the potential for false positives and attaining adequate patient sample size) but also the computational challenges of detecting epistasis in the GWAS. As mentioned, it has been observed that for many association studies, there exists a failure to replicate a genetic association in a second independent sample, indicating that a SNP may contribute to disease susceptibility through nonlinear, epistatic, interactions with one or more other SNPs. However, to detect epistasis, all SNP combinations must be searched so that we may attain a more complete understanding of the etiology of these common complex diseases. We elaborate on this computational problem and propose a potential solution to this by suggesting that biological expert knowledge both from protein-protein interactions or pathway information be employed to guide GWAS analyses.

GENOME-WIDE ASSOCIATION STUDIES (GWAS)

When the GWAS became a tool available to the scientific community, it was revolutionary in that it allowed researchers to investigate the entirety of human genome in thousands of unrelated individuals without prior hypothesis about disease association. Currently, technological advances in high-throughput genotyping allow for the GWAS to assess the role of one million or more single-nucleotide polymorphisms (SNPs) and look for associations between DNA sequence variants and phenotypes of interest (Hirschhorn & Daly, 2005; Wang et al., 2005).

Background/Study Design

The most common study design of the GWAS often represents a group of individuals who are either affected or not affected by a disease (i.e. a case-control study). This approach entails that each individual is genotyped at the positions of thousands or more SNPs. Those SNPs for which one variant is statistically more common in one group individuals, either affected or non-affected, are reported as being associated with the phenotype. The case-control design has its advantages over other formats because it tends to be easier, more time efficient, and less expensive to conduct, however it does have its limitations which we will further discuss in the following section.

Other known study designs for the GWAS are the trio and cohort design. The trio design assesses the affected individual and both of the parents of this individual. Unlike the case-control design, only affected offspring are genotyped. The goal of this study design is to measure the frequency at which an allele is transmitted from the parents to offspring. Under the null-hypothesis of no association, transmission should be 50%. However, this frequency will be much higher for alleles associated with disease. While this design is not susceptible to population stratification, genotyping error can have a more prominent effect on results, so particular care should be taken to ensure the quality of materials and procedure (Pearson & Manolio, 2008.) A third design, the cohort design, aims to collect baseline information pertaining to a larger group of individuals, who are then assessed for the incidence of disease in subgroups that are determined by the genetic variants. This design has a disadvantage in that it requires a large sample size, and can be more lengthy and expensive. However, individuals are often more representative of the population from which they were drawn, and cases develop during the duration of experimental observation, making the study free of survival bias (Pearson & Manolio, 2008).

Regardless of the study design used, a major challenge confronting GWAS analysis is determining the many false-positive associations that can arise. It has become common practice, often even prior to publication, to confirm promising results in subsequent experiments where the same findings are replicated in a new group of individuals. The first GWAS was aimed to determine genetic variants of type II diabetes and was published in *Nature* by Sladek et al. (2007). They tested 392,935 SNPs in a French case-control group and were able to determine four loci that contained variants that confer type II diabetes risk. These included the known association with the *TCF7L2* gene, a non-synonymous polymorphism in the zinc transporter *SLC30A8,* and two linkage disequilibrium blocks that contain genes potentially involved in beta-cell development or function, *IDE–KIF*, *11–HHEX*, and *EXT2–ALX4*. To date, a total of more than 400 GWAS have been replicated spanning more than 80 diseases and traits (www.genome.gov). The National Human Genome Research Institute supplies a list of all currently published GWAS studies that fall under the criteria of having at least 100,000 SNPs in the initial stage, before quality control filters are applied, and that demonstrate a statistical significance (SNP-trait p-value $<1.0 \times 10^{-5}$) in the overall (initial GWAS and replication) population with few certain exceptions (Donnelly, 2008; Hindorff et al., 2009).

Among the multiple GWAS that have been reported, there are several successful publications that have emerged. However, it can be noted that in many of these publications, only the strongest associations were detected using the traditional analyses methods and it's possible that many more associations may remain undetected (Couzin & Kaiser, 2007; Williams et al., 2007; Ritchie, 2009). Certainly this is not the only limitation of the GWAS, and while there is a great push for this "revolutionary" tool to be of use in the clinical setting for diagnostics, prognostics, treatment, and prevention, for this to become a reality, these limitations need to be acknowledged and appropriately addressed.

GWAS Limitations

As previously noted, one problematic feature of the GWAS, especially in the case-control design, can be the potential for false positives due to the large number of statistical tests performed. This leads to the requirement of a more stringent statistical significance threshold. However, defining a common p-value threshold for all studies would serve no purpose given that this value is determined by several factors, including the number of genetic variants as well as the types of variants (Spencer et al., 2009). It's also important to consider carefully the number of individuals to be genotyped. Acquiring a reasonable sample size and the appropriate selection of individuals is also a challenge. This can be attributed to a lack of well characterized clinical samples and the ability to posses the funding needed to attain larger sample sizes that increase the power of the study. Therefore, a balance must be attained in order for researchers to have adequate power to support their results and continue on to analyze these results within the means of their budget (Wang et al., 2005; Spencer et al., 2009). While it may not be possible to obtain larger sample sizes in order to detect smaller genetic effects, it is possible to conduct a meta-analysis by combining results from previous studies (Mufano & Flint, 2005). As mentioned, in practice, the research community feels strongly that the results of association studies should not be relied on too heavily without being replicated in independent samples or in a meta-analysis, partially on account of the potential for false positives. Many journals, along with the National Human Genome Research Institute, list replication as a criterion required for publication. However, although this idea is highly supported by the research community, as discussed below, replication may not be the gold standard it was once thought to be.

The replication criteria of Chanock et al. (2007) and Hunter & Kraft (2007) assume that SNPs have independent effects, and Greene & Moore (2009) have demonstrated that failure to replicate a genetic association in a second independent sample can be an indication that the SNP contributes to disease susceptibility through nonlinear interactions with one or more other SNPs. Greene et al. (2009) showed that the power to replicate a SNP with a significant main effect can drop from > 80% to < 20% with a change in allele frequency at a second interacting SNP of < 0.1. Such small changes in allele frequency are very often observed even when the replication sample is taken from the same population. This study recommended that SNPs that fail to replicate be followed up with epistasis analysis to check for interaction. As Greene et al. (2009) discuss, the validity of a result relies more on the biological interpretation and experimental evidence than it does on the actual statistical finding. Wilke et al. (2008) have suggested that we should not even begin to analyze a GWAS until we have exhaustively studied each candidate gene in each pathway represented, encompassing biological evidence of interaction as well. Only then will we have the appropriate knowledge base to make sense of GWAS results. As Moore (2009) noted, there is major shift in the field of genetic epidemiology away from the purely statistical approaches to these problems to a more bioinformatics approach that considers knowledge about gene function, gene networks, and biochemical pathways. These current years perhaps mark the turning point towards more of a systems approach that addresses the challenges of detecting epistasis and other complexities in the genetic architecture of common diseases.

CHALLENGES OF DETECTING EPISTASIS IN GWAS

As Clark et al. (2005) note, the success with GWAS is entirely dependent on the assumption that is made about genetic architecture. Until recently, most GWAS have assumed that there is a single SNP with a large effect on disease risk that is independent of the genome and ecology. As the frequency of the less common variants decrease, so does the ability to detect an association between that SNP and the disease at hand, meaning that studies thus far have been better suited for discovering associations with more common variants. There have been but few reports of associations with rare variants. Not only this, but variants detected in GWAS have been shown to exhibit little effect on diseases risk. Future success with GWAS will depend critically on our ability to address complexity in the genetic architecture of common diseases due to epistasis (gene–gene interaction), plastic reaction norms (gene–environment interaction) and locus heterogeneity.

Modeling Epistasis: A Statistical Challenge

As mentioned, common-complex diseases have a much more complex etiology that results from phenomena such as epistasis, locus heterogeneity, epigenetics, phenocopy, and gene–environment interaction. However, detecting and characterizing epistasis in order to gain an understanding of the genetic susceptibility to these diseases is not a simple task. Because of their accepted and strong theoretical foundation, traditional methods of analysis such as linear and logistic regression have been a vital component of modern genetic epidemiology. Also, such methods are easy to implement and interpret using a wide-range of different software packages such as SAS and R. However, such methods have their limitations when used for detecting non-linear patterns of interaction, a pattern exemplified by epistatic interactions (Moore &Williams 2002; Moore et al., 2010).

For example, when interactions among multiple SNPs are considered, there are many multilocus genotype combinations that may be represented by very few or no individuals being genotyped in the GWAS. Parametric linear models are generally implemented such that interaction effects are only modeled using factors that exhibit independent marginal effects. Even though this may be an easier approach to fitting a genetic model, it assumes the genetic architecture of common complex diseases to be simple and that important predictors will have detectable marginal effects (Moore et al., 2010). This can lead to an increase in type I and type II errors due to parameter estimates with very large standard errors. Given this, these linear models are not likely to explain a large part of the variance of any given trait (Moore, 2003) and may be a plausible explanation for some of the missing heritability that has not been accounted for by GWAS (Manolio et al., 2009). In order to begin to explore epistasis in the GWAS, a statistical or computational modeling of nonlinear interactions that requires looking at multiple-way combinations of SNPs is needed.

The limitations of the linear model and other parametric statistical approaches have motivated the development of computational approaches such as those from machine learning and data mining to analyze higher order interactions that also make fewer assumptions about the functional form of the model and the effects being modeled (Mitchell, 2009; Hastie et al., 2009; McKinney et al., 2007). Several review papers highlight the need for new methods (Thorton-Wells et al., 2004) and discuss and compare different strategies for detecting epistasis (Motsinger et al., 2006; Cordell, 2009). Here we highlight a few methods: multifactor dimensionality reduction (MDR) (Ritchie et al., 2003), combinatorial partitioning method (Nelson et al., 2001), symbolic discriminant analysis (SDA) (Reif et al., 2004), Monte Carlo logistic regression (Kooperberg & Ruczinski, 2005), recursive partitioning method (RPM) (Young & Ge, 2005), focused interaction testing framework (FITF) (Millstein et al., 2006), backward genotype-trait association (BGTA) (Zheng et al., 2006), Bayesian epistasis association mapping (BEAM) (Zhang & Liu, 2007), a forest- based approach (Chen et al., 2007), penalized logistic regression (Park & Hastie, 2008), grammatical evolution neural network (GENN) (Motsinger-Reif et al., 2008), and MegaSNPHunter (Xiang et al., 2009). Each method has its advantages and disadvantages, yet is capable of analyzing higher order interactions. However, the computational burden of examining multi-way interactions in datasets as large as a GWAS is immense and infeasible beyond two-way interactions for any method.

Detecting Epistasis: A Computational Conundrum

It was nearly a decade ago that Risch and Merikangas first suggested the testing of all known SNPs in the human genome for disease association either directly or by linkage disequilibrium with other SNPs (Risch & Merikangas, 1996), and today this "genome-wide" approach is expected to revolutionize the genetic analysis of common human diseases (Hirschhorn & Daly, 2005; Wang et al., 2005; Kingsmore et al., 2007; Greene & Moore, 2008; Moore, 2009). Currently, it is possible to measure more than a million SNPs with a genome-wide human SNP array available from Affymetrix, and Illumina has released the Human1 M DNA Analysis BeadChip that is capable of profiling 1,000,000 SNPs on a single array across the human genome. With these technologies now available to the scientific community at more affordable costs, researchers are capable of producing large amounts of data efficiently and rapidly. Unfortunately, as mentioned there is a present computational obstacle when attempting to detect higher order interactions in a GWAS.

To illustrate the scope of an epistasis analysis using computational methods to test all interaction combinations, consider a recent report from the International HapMap Consortium (Altshuler et al., 2005) that suggests that approximately 300,000 carefully selected SNPs may be sufficient to represent all of the relevant genetic variation across the human Caucasian genome. If this is to be regarded as the lower limit of a GWAS, then approximately 4.5 x 10^{10} pairwise combinations (300,000 choose 2) and 4.5 x 10^{15} three-way combinations (300,000 choose 3) would need to be exhaustively analyzed to detect low-order epistasis. To put this into computing time, assuming one had one-million PCs that were each processing one model per second, this would result in an 11 hour analysis for all 2-way combinations, 127 years for all 3-way combinations, and 9,513 millennia and greater for 4-way and more combinations.

Our ability to measure genetic information, and biological information in general, is far outpacing our ability to interpret it. It is evident that GWAS harbor a wealth of information about susceptibility genes that can be used to improve the prevention, diagnosis, and treatment of common diseases. However, to access this information to our full advantage, we need to address the specific technical challenges that confront researchers in the analysis process, such as these computational limitations. Yet, we also need to be capable of interpreting these results such that they retain their biological relevance. This issue has been recognized by many groups that have proposed utilizing sources of prior biological expert knowledge, such as information from protein-protein interaction databases or biological pathways, as a method to aid in the GWAS analysis process while maintaining biological integrity of the results.

BIOLOGICAL EXPERT KNOWLEDGE

Expert knowledge can be defined as existing biological or statistical information about the problem at hand that can be employed to guide an analysis process or an algorithm in a more directed fashion. For example, when considering SNPs or genes in a GWAS, biological expert knowledge may be derived from many sources that describe what is known about the function of the SNP or the gene. Some of this information may involve biochemical pathways, Gene Ontology (GO), expression, or interaction information for that gene or SNP.

Protein-Protein Interaction Expert Knowledge

Currently, there exist numerous publicly available protein-protein interaction (PPI) databases that contain information about human specific interactions (Table 1). The majority of PPIs in these databases are from curation of the literature by biologists. However, some are incorporated by

Table 1. PPI databases. A table listing of publicly available protein interaction databases and additional features of each

PPI Database	Citation	Website	# of Interactions	Additional features
HPRD	Prasad TSK *et al. Nucleic Acids Res* 37, D767-D762 (2009).	http://www.hprd.org	>38,000	Additional information on phosphorylation motifs, signaling pathways, domain architecture, protein functions, enzyme–substrate relationships, subcellular localization, tissue expression, and disease association of genes
BioGRID	Breitkreutz BJ *et al. Nucleic Acids Res* 36, D637-D640 (2008).	http://www.thebiogrid.org	>200,000	Interaction directionality, phenotype, post-translational modification, domains and motifs are being added
BIND	Alfarano C *et al. Nucleic Acids Res* 33, D418-D424 (2005).	http://bind.ca	> 100,000	Includes genetic, biopolymer-biopolymer and protein-small molecule interactions, DNA, RNA, and protein sequence information
MINT	Chatr-aryamontri A *et al. Nucleic Acids Res* 35, D572-D574 (2007).	http://mint.bio.uniroma2.it	>100,000	Represents complexes, biomolecules, detailed experimental descriptions, protein structure information
DIP	Salwinski L *et al. Nucleic Acids Res* 32, D449-D451(2004).	http://dip.doe-mbi.ucla.edu	>57,000	Provides experimental quality assessment to identify most reliable interactions, represents complexes
Reactome	Matthews L *et al. Nucleic Acids Res.* 37, D619-D622 (2009).	http://www.reactome.org	>2,900 Reactions	Extensive cover of human pathways in 46 domains of human biology, hypergeometric testing is used to display statistically over-represented events in the event hierarchy
STRING	Jensen LJ *et al. Nucleic Acids Res.* 37, D412-D416 (2009).	http://string.embl.de	Incorporates a number of databases >2,500,000 proteins	Imports pathway information, provides confidence score based on evidence from conserved genomic neighborhoods, gene fusion events, co-occurrence events, co-expression data, experimental data, database information, text mining, and homology
UniHI	Chaurisa G *et al. Nucleic Acids Res* 37, D657-D660 (2009).	http://www.mdc-berlin.de/unihi	Incorporates a number of interaction databases > 250,000	Allows for the construction of tissue specific networks, statistical interaction validation by gene co-expression data and validation by shared path length according to GO co-annotation hierarchy

direct deposit prior to publication by the investigator. When querying proteins, these databases will usually return a list of protein interactors, information pertaining to the experimental evidence for that interaction, as well as information about the protein itself.

One of the largest publicly available databases is the Human Protein Reference Database (HPRD), which to date has over 38,000 PPIs, over 270,000 pub-med links, access to curated pathways, as well as information about post-translational modifications (PTMs), domain architecture, protein functions, enzyme–substrate relationships, subcellular localization, tissue expression, and disease association of genes (Prasat et al., 2009). Another large and growing database that has similar components is BioGrid, which currently houses approximately 42,800 human PPI's, but altogether contains > 200,000 interactions from *Saccharomyces cerevisiae*, *Schizosaccharomyces pombe*, *Caenorhabditis elegans*, *Mus musculus*, and *Drosophila melanogaster* in addition to *Homo sapiens*. Other available databases that are smaller than the

HPRD and BioGrid yet offer additional unique features are The Biomolecular Interaction Network Database (BIND), which is a component of the Biomolecular Object Network Database (BOND), the Molecular Interaction database (MINT), the Database of Interacting Proteins (DIP), and Reactome.

Resources such as the Search Tool for the Retrieval of Interacting Genes/Proteins (STRING) and Unified Human Interactome (UniHI), access a number of the reviewed databases to integrate protein interaction information. The newest version of STRING, 8.0, covers approximately 2.5 million proteins from 630 different organisms, and incorporates PPI information from a number of interaction databases such as HPRD, BioGrid, MINT, BIND, DIP, and also imports known reactions from Reactome and KEGG pathways. Recent additions to this database incorporate interactions from IntACT, EcoCyc, NCI-Nature Pathway Interaction Database, and Gene Ontology (GO). Automated text-mining of PubMed abstracts, OMIM, and information from other databases such as the *Saccharomyces* Genome Database, Wormbase, and the Interactive Fly supplement this. Each interaction is given a numerical confidence score based on the experimental evidence and orthologous evidence behind that interaction (Jensen et al., 2009). UniHI integrates protein-protein interactions not only from large Y2-H screens and curated databases such as HPRD, DIP, BIND, Reactome, but also predicts interactions based on orthology and computational text-mining approaches. This database also provides detailed information about each interaction including statistical interaction validation by gene co-expression data and validation by shared path length according to GO co-annotation hierarchy (Chaurasia et al., 2009).

In the previously mentioned study, Emily et al. (2009) exemplify how interaction information, specifically from protein-protein interactions, can be utilized. They use experimental knowledge about biological networks to narrow the search for two-locus epistasis that confer susceptibility to Crohn's disease, bipolar disorder, hypertension, and rheumatoid arthritis. In this study, the protein interaction database STRING was queried, and expert knowledge derived from the confidence of protein-protein interactions was extracted to guide their search for epistasis. They were able to identify 71,000 high confidence potential protein-protein interactions in the database. From the pool of all interactions, they then identified all of the SNPs that corresponded to the genes for the relevant protein involved in these interactions. Subsequently this information was used to extract these prioritized SNPs in Wellcome Trust Case-Control Consortium data. At the completion of the study, they were able to identify four significant cases of epistasis between unlinked loci in all four diseases. This otherwise may have been a task insurmountable without the guidance of prior expert knowledge about these datasets. Not only were they able to identify these epistatic interactions, but they were able to place them into a working biological context.

Similarly, using prior interaction and disease-gene information, Bush et al. (2009) show that using biological knowledge to guide genetic association studies may provide more meaningful results. They present a tool, BioFilter, that integrates multiple public databases comprised of gene groupings and sets of disease-related genes to identify multi-SNP models of disease susceptibility. By doing so, these models possess an established biological foundation. Expert knowledge pertaining to disease related genes is derived from the Genetic Association Database (GAD). To indentify genes with some prior evidence of epistasis, databases that link two or more genes together are used such as GO, the Database of Interacting Proteins (DIP), the Protein Families Database (PFAM), Kyoto Encyclopedia of Genes and Genomes (KEGG), Reactome, and Netpath. From these sources, BioFilter identifies gene combinations and quantifies the degree of knowledge-based support for a model with an implication index derived from the number of data sources that provide evidence of gene-gene relationships. Applying BioFilter to two large genotyping platforms, Affymetrix 1M and the Illumina 1M, search space for each was reduced to 0.241% and 0.40% of the original platforms, respectively.

Pathway-Based Expert Knowledge

Another approach that utilizes expert knowledge is a pathway-based approach or pathway based analyses (PBA's), which employ computational methods to define sets of genes based on common biological attributes such as gene ontology or biological pathways. Over the past few years, many research groups have published bioinformatic methods that implement gene annotations such as expression profile, gene length, or patterns of duplication as a way to narrow a list of candidate genes (Shriner et al., 2008; Tiffin et al., 2006). This information is used to define a measure of enrichment of each gene set among disease associated markers. For example, Askland et al. (2009) recently showed that patterns of SNPs in biological pathways are more likely to replicate than individual SNPs in GWAS.

Shriner et al. (2008) demonstrate how this concept of PBA can be employed in a similar way by applying it to linkage studies of complex traits. Much like the challenge described for detecting epistasis in GWAS, in linkage studies of complex traits, testing each candidate gene from every region is a computational challenge. They describe their commonality of functional annotation method (CFA), which operates by testing individual Gene Ontology terms for enrichment in candidate gene pools and ultimately ranks genes based on the number of quantitative trait loci (QTL) regions where genes with such annotations are found. The reasoning behind this method is based on the idea that if multiple genes show a correlation to the same trait, then it is valid to hypothesize that those genes have a higher probability of sharing one or more annotations as compared to those genes that show no correlation with the trait (Badano & Katsanis, 2002). Attempting to reduce the list of potential positional candidates by means of wet-lab experimentation is not only a laborious endeavor, but expensive, and methods such as CFA would greatly reduce this burden. When this method is applied to published linkage studies that examine age-of onset of Alzheimer's and body mass index, new candidate genes as well as previously published candidate genes are identified. Due to CFA's ability to generate a set of prioritized candidate genes that may be linked to complex traits, for this study, defining these new genes, if confirmed, may offer new targets for diagnosis and treatment.

Saccone et al. (2008), integrate additional sources of biological and statistical expert knowledge to compliment pathway information, but take a different approach to prioritizing SNPs in a GWAS. This prioritization method was successfully applied to a nicotine dependence study. Greater weight is given to biologically relevant SNPs via a systematically defined algorithm. These weights are based on sources of evidence that incorporate statistical expert knowledge of genotype–phenotype correlation with known pathways involved in the pathologic development of disease, SNP and gene functional properties, comparative genomics, evidence of genetic linkage, and linkage disequilibrium (LD). More specifically, they considered SNPs within genes to be an order of magnitude more relevant than those that are not. They also defined 3 tiers of biologically relevant gene systems and categories for nicotine dependence through an expert committee within the NIDA Genetics Consortium. Tier 2 genes, represented basic neurotransmitter systems, and were considered to be an order of magnitude more relevant than arbitrary genes. Tiers 1 and 3 were scored a half order higher and lower than Tier 2. Human/mouse standard evolutionary conserved regions (ECRs) from the ECRbase were also used to prioritize genes on a more conservative level along with evidence of linkage. For linkage, if the link is weak, such as a

SNP being in LD with a gene rather than actually in the gene, their priority is diminished. If the link is strong, such as a SNP being a non-synonymous coding change in a gene, it is increased.

Additional Sources of Expert Knowledge

The studies discussed that employ both interaction and pathway-based expert knowledge merely highlight two of the many proposed methods by which biological expert knowledge can be integrated in order facilitate GWAS analyses. Certainly, other sources of biological expert knowledge can be implemented in a similar fashion. For example, Tu et al. (2006), note that inferring causal genes underlying expression variation that is associated with disease phenotypes is a challenge. This is due to the presence of many genes in regions of the chromosome where the expression variation is linked. They develop a method that integrates genotype information, gene expression, protein-protein interactions, protein phosphorylation, and transcription

factor (TF)–DNA binding information in a network based stochastic algorithm aimed to infer causal genes and identify the underlying regulatory pathways. The number of possible pathways needed to be considered is significantly reduced while pathway identification ultimately helps to answer which gene is the causal gene (Tu et al., 2006).

In addition to biological information, it has also been shown that using prior statistical knowledge to guide an epistasis analysis has been useful. For example, we have mentioned that LOD scores from a prior linkage analysis could be used to weight SNPs in certain chromosomal regions higher during a combinatorial epistasis analysis. That is, SNPs from a certain pathway or chromosomal region would be statistically evaluated for interactions with a higher probability than others in the dataset (Pattin & Moore, 2009). Statistical knowledge could also come from filter algorithms that explicitly assess the quality of SNPs based on their relationship with the clinical endpoint. The Tuned ReliefF (TuRF) algorithm is an example of an algorithm that can assign high quality scores to SNPs involved in complex interactions (Moore & White, 2007). The TuRF algorithm uses a nearest neighbor approach to assess SNP quality and thus doesn't suffer from the computational limitations of an algorithm that explicitly considers combinations of SNPs. As such, it is very useful for preprocessing the data prior to analysis. Once computed, the TuRF scores can be used to select some reduced number of SNPs for combinatorial analysis or can be used to help guide a computational search algorithm (Greene et al., 2008). There is certain potential for both biological and statistical expert knowledge to be incorporated simultaneously as a method to weight or prioritize SNPs in a GWAS. While it has been demonstrated that both sources may be useful for facilitating GWAS, for this chapter, we turn our focus to how biological expert has potential for addressing the analytical challenges confronting GWAS studies. Detecting epistasis at the statistical level in the GWAS may be indicative of biological epistasis and vice-versa, making this a valid application as well as a potential way to gain an understanding of the relationship between the two and fuel the interpretation of the results.

Limitations of Expert Knowledge

While it is apparent that many groups are addressing the analytical challenges of the GWAS using biological expert knowledge, it is important to acknowledge the potential limitations of using the various sources mentioned for this purpose. Keep in mind that just because an interaction is not detected on the biological level, it does not mean that this interaction does not exist or will not be seen at the statistical level, and alternatively, what is detected statistically, may not have any biological relevance (Moore & Williams, 2005). Therefore we must be aware of and concerned about the amount of important information we may potentially be missing due to bias and lack of annotation as well as recognize the types of studies this expert knowledge is appropriate for.

It's important to understand that bias may exist across all databases used as sources of expert knowledge and that genes and SNPs represented in the GWAS may be unannotated or anonymous. How does one deal with anonymous SNPs or ones that are not in coding regions? It may be that a researcher wishes to consider this SNP as part of the gene that is within a certain number of kilobases of a nearby gene, or perhaps consider these variations to be defined as those that it is in linkage disequilibrium (LD) with. Also, there are many proteins, genes, and pathways that have not been studied thoroughly or even studied at all rendering them to be underrepresented or non-existent in these databases. To add to this, there exists a bias in experimental methods for capturing certain information represented in these databases. For example, in terms of protein-protein interaction information, yeast two-hybrid (Y2-H) experiments are not entirely adequate to detect interactions with integral membrane proteins (Mathivanan et al., 2006).

Additionally, concerning protein-protein interaction databases, one must also acknowledge the dynamics of PPIs and the fact that just because a database claims an existing interaction, for that interaction to have meaning in the context of a disease, these two proteins must be both spatially and temporally coordinated (Shrabanek et al., 2007). PPIs are largely context dependent and require the appropriate cellular conditions in order for certain structural modifications to occur that enable the interaction to physically take place. While mammalian two-hybrid systems that allow for assayed proteins to undergo these modifications in the appropriate cellular context have been developed to complement yeast-two hybrid systems, these tools are still under development. Currently, a large number or PPI networks available are representative of a static and not a dynamic network (Lievens et al., 2009). The pure fact that there exists a wide variety of databases to explore is an issue as well, and as Mathivanan et al. (2006) discovered, that while there may be good overlap at the protein level between these databases, the level of overlap of curated PPIs is not as great. They also discover that for PPIs that do overlap between databases, there exists a variation in protein annotation partly on account of differences that arise according to how biologists interpret the experimental results. This presents an obstacle when attempting to integrate interaction information as expert knowledge that is extracted from multiple databases. Such a situation may lead to the exclusion of important interactions or the inclusion of non-influential interactions in a GWAS analysis.

Concerning pathway-based approaches, there are limitations that have been noted, as Shriner et al. (2008) describe, concerning the use of GO terms and ontologies as a source of expert knowledge. The common method for evaluating if a certain set of genes displays enrichment in GO terms is the Fisher's exact test based on hypergeometric distribution for sampling without replacement (Khatri et al, 2005; Curtis et al., 2005; Rivais et al., 2007; Shriner et al., 2008). Using this approach, terms are tested one at a time and relationships between terms are ignored, thus multiple hypothesis testing corrections are required to be stringent. Even though the three main sub-ontologies of GO: biological process, molecular function, and cellular component, are structurally independent, there may still exist terms within and between these that are highly correlated because genes can have many annotations across the subontologies.

Aside from the issues that arise concerning databases, some may argue that using expert knowledge is bias in and of itself. The ability to conduct a genome wide association study has been said to 'relax' the need for a strong prior hypothesis because the whole genome can be analyzed at once (Chanock et al., 2007). Those in support of 'genomic agnosticism' believe that when conducting a genome-wide analysis, they will assume every SNP in the genome to be equally functional (Carlson et al., 2006). This brings us back to the issue of what information we may be missing by applying expert knowledge. While the benefits of conducting an unbiased GWAS study are valid, such as having no prior hypothesis, elimination of bias, and inclusion of all information, we emphasize that we are still at a loss for computational power to conduct these studies and fully explore all potential epistatic interactions. We argue that biological expert knowledge offers a logical solution to this problem.

OUTLOOK/ CONCLUDING REMARKS

The goals projected when designing the GWAS had been to create a platform that would be capable of indicating genetic markers influential on disease and trait mechanisms, aid in disease prediction and prevention, and represent the basis for disease and trait variability. While it is recognized that this was a "revolutionary" technical advancement in the field of genetics and there have been successful GWAS publications, it cannot be said that the GWAS has fully met these goals. Goldstein (2009) expresses concern about the fact that most SNPs in the GWAS are responsible for contributing only a small fraction to disease risk. As Moore et al. (2010) note, a possible explanation for these mixed results of GWAS is that the current biostatistical analysis paradigm takes agnostic or unbiased approach that ignores all prior knowledge about disease pathobiology. To add to this, the linear modeling framework that is commonly employed in GWAS often considers only one SNP at a time thus ignoring their genomic and environmental context.

As Moore & Ritchie (2004) summarize, there are at least three important challenges to detecting epistasis in GWAS. First, modeling interactions inherently requires looking at combinations of variables. We have summarized the issues concerning the linear modeling framework and a variety of different methods that have been designed to statistically evaluate higher order interactions. The second challenge to detecting epistasis in GWAS is computational in nature. The detection of epistasis in the absence of significant main effects requires combinations of SNPs to be systematically evaluated. The combinatorial assessment of SNPs in a GWAS is a computationally daunting task beyond exploring two-way and three-way combinations. Third, perhaps the most important challenge that is faced in detecting and characterizing epistasis is interpretation. Going from a population-level statistical summary of epistasis to inferences about the biological interactions occurring at the cellular level is a significant and difficult leap. Conversely, translating knowledge of gene networks and cellular function at the individual level to predictions about public health is equally difficult. As discussed by Moore & Williams (2005), systems biology holds the promise of helping us to traverse this conceptual and practical divide. The integration of GWAS with systems biology will be necessary to advance beyond the "one SNP at a time" approach to the genetics of common diseases. Not only will we be able to apply this biological knowledge in order to logically begin to analyze these large amounts of data, but we will also be providing a biological foundation by which to base results and interpretation upon.

The goal of this chapter was to place emphasis on the need to address the second and third of these challenges and propose how biological expert knowledge may reduce the computational burden of the GWAS as well as facilitate the biological interpretation of the data. Complexity at the statistical level may be indicative of biological complexity, further supporting why we observe "missing heritability" when testing single genetic variants. We foresee these methods that have been developed and other similar approaches to not only be applicable to SNP studies, but also to studies involving other forms of genetic variation (i.e. as copy number variation, sequence repeats, and epigenetic modifications) that may encounter the same computational challenges as the GWAS. While there is anticipation that the power to examine a GWAS will improve, it is not expected that within the next decade there will exist a sufficient level of computational power available to conduct epistasis analyses at higher levels. However, being able to embark on an analyses that contains the entire genomic spectrum of SNPs and genes, allows greater room for exploration and novel discovery. Researchers have the option to explore many pathways represented in the GWAS and not simply focus on one particular pathway when designing an analysis. Until we can fully analyze a GWAS, the exploitation of expert knowledge will eventually not only ease our computational burdens, but also aid in understanding the relationship between biological and statistical epistasis.

While acknowledging the limitations and concerns when employing expert knowledge, in order to fully exploit these sources, there needs to be a logical method to evaluate this information in order to incorporate it into the GWAS analyses such that relevant interaction information is retained within valid biological context. Certainly, this is no simple task, yet we have exhibited in this chapter success in similar endeavors. The challenge is worthwhile and needs to be explored if the research community wishes to make the GWAS applicable in the clinical setting.

Currently, there is a lack of publications concerning the clinical application of the GWAS, yet it is not unlikely that the coming years will exhibit a more collaborative effort between basic research and clinical medicine to bring this tool from the bench to the bedside. We are just beginning to understand the genetic basis of these diseases at a finer resolution, and forming a collaboration between individual research programs involved in the more translational aspects of these projects will be beneficial to accelerating this process in a time where sequencing each individual's entire genome is still not financially feasible (Motsinger et al., 2006). The results of these studies, will eventually lead to important preventive measures for those predisposed to particular diseases, and the foundations for new therapies will be born from newly identified genomic targets. It would be a foreseeable goal to develop pharmaceutical options that are capable of targeting the cause of many complex diseases and not just the symptoms themselves. Future success in these important clinical objectives will of course depend on our ability to address the complexities both of complex diseases and interpreting the results of the GWAS appropriately. Once we have been able to successfully develop these methods, not only will we improve the ease at which we will be able to identify important epistatic interactions in GWAS, but we will gain an understanding of the physical biology that underlies these interactions and perhaps their role in a given disease.

REFERENCES

Altshuler, D., Brooks, L. D., Chakravarti, A., Collins, F. S., Daly, M. J., & Donnelly, P. (2005). A haplotype map of the human genome. *Nature*, *437*, 1299–1320. doi:10.1038/nature04226

Askland, K., Read, C., & Moore, J. H. (2009). Pathway-based analyses of whole-genome association study data in bipolar disorder reveal genes mediating ion channel activity and synaptic neurotransmission. *Human Genetics*, *125*, 63–79. doi:10.1007/s00439-008-0600-y

Badano, J. L., & Katsanis, N. (2002). Beyond Mendel: An evolving view of human genetic disease transmission. *Nature Reviews. Genetics*, *3*, 779–789. doi:10.1038/nrg910

Bush, W. S., Dudek, S. M., & Ritchie, M. D. (2009). BioFilter: A knowledge-integration system for the multi-locus analysis of genome-wide association studies. *Pacific Symposium on Biocomputing*, 368-379.

Carlson, C. S. (2006). Agnosticism and equity in genome-wide association studies. *Nature Genetics, 38*(6), 605–606. doi:10.1038/ng0606-605

Chanock, S. J., Manolio, T., Boehnke, M., Boerwinkle, E., Hunter, D. J., & Thomas, G. (2007). Replicating genotype-phenotype associations. *Nature, 447*, 655–660. doi:10.1038/447655a

Chaurasia, G., Malhotra, S., Russ, J., Schnoegl, S., Hänig, C., & Wanker, E. E. (2009). UniHI 4: New tools for query, analysis and visualization of the human protein–protein interactome. *Nucleic Acids Research, 37*, D657–D660. doi:10.1093/nar/gkn841

Chen, X., Lie, C. T., Zhang, M., & Zhang, H. (2007). A forest based approach to identifying gene and gene-gene interactions. *Proceedings of the National Academy of Sciences of the United States of America, 104*, 19199–19203. doi:10.1073/pnas.0709868104

Clark, A. G., Boerwinkle, E., Hixson, J., & Sing, C. F. (2005). Determinants of the success of whole-genome association testing. *Genome Research, 15*, 1463–1467. doi:10.1101/gr.4244005

Cordell, H. J. (2009). Genome-wide association studies: Detecting gene-gene interactions that underlie human diseases. *Nature Reviews. Genetics, 10*(6), 392–404. doi:10.1038/nrg2579

Couzin, J., & Kaiser, J. (2007). Genome-wide association: Closing the net on common disease genes. *Science, 316*, 820–822. doi:10.1126/science.316.5826.820

Curtis, R. K., Oresic, M., & Vidal-Puig, A. (2005). Pathways to the analysis of microarray data. *Trends in Biotechnology, 23*, 429–435. doi:10.1016/j.tibtech.2005.05.011

Donnelly, P. (2008). Progress and challenges in genome-wide association studies in human. *Nature, 456*(7223), 728–731. doi:10.1038/nature07631

Emily, M., Mailund, T., Hain, J., Schauser, L., & Schierup, M. H. (2009). Using biological networks to search for interacting loci in genome-wide association studies. *European Journal of Human Genetics, 17*(10), 1231–1240. doi:10.1038/ejhg.2009.15

Goldstein, D. B. (2009). Common genetic variation and human traits. *The New England Journal of Medicine, 360*, 1696–1698. doi:10.1056/NEJMp0806284

Greene, C. S., Gilmore, J. M., Kiralis, J., Andrews, P. C., & Moore, J. H. (2009). *Optimal use of expert knowledge in ant colony optimization for the analysis of epistasis in human disease*. (LNCS 5483), (pp. 92-103).

Greene, C. S., & Moore, J. H. (2008). *Ant colony optimization for genome-wide genetic analysis*. (LNCS 5217), (pp. 27-47).

Greene, C. S., & Moore, J. H. (2009). Solving complex problems in human genetics using nature-inspired algorithms requires strategies which exploit domain-specific knowledge. *Nature Inspired Informatics, 7*, 166-180. Hershey, PA: IGI Global.

Greene, C. S., Penrod, N. M., Williams, S. M., & Moore, J. H. (2009). Failure to replicate a genetic association may provide important clues about genetic architecture. *Public Library of Science ONE, 4*, e5639.

Hastie, T., Tibshirani, R., & Friedman, J. (2009). *The elements of statistical learning: Data mining, inference, and prediction*. New York: Springer-Verlag.

Hindorff, L. A., Sethupathy, P., Junkins, H. A., Ramos, E. M., Mehta, J. P., & Collins, F. S. (2009). Potential etiologic and functional implications of genome-wide association loci for human diseases and traits. *Proceedings of the National Academy of Sciences of the United States of America, 106*(23), 9362–9367. doi:10.1073/pnas.0903103106

Hirschhorn, J. N., & Daly, M. J. (2005). Genome-wide association studies for common diseases and complex traits. *Nature Reviews. Genetics*, *6*, 95–108. doi:10.1038/nrg1521

Hunter, D. J., & Kraft, P. (2007). Drinking from the fire hose–statistical issues in genome-wide association studies. *The New England Journal of Medicine*, *357*(5), 436–439. doi:10.1056/NEJMp078120

Ideker, T., & Sharan, R. (2008). Protein networks in disease. *Genome Research*, *18*, 644–652. doi:10.1101/gr.071852.107

Jensen, L. J., Kuhn, M., Stark, M., Chaffron, S., Creevey, C., & Muller, J. (2009). STRING 8-a global view on proteins and their functional interactions in 630 organisms. *Nucleic Acids Research*, *37*, D412–D416. doi:10.1093/nar/gkn760

Khatri, P., & Draghici, S. (2005). Ontological analysis of gene expression data: Current tools, limitations, and open problems. *Bioinformatics (Oxford, England)*, *21*, 3587–3595. doi:10.1093/bioinformatics/bti565

Kingsmore, S. F., Lindquist, I. E., Mudge, J., & Beavis, W. D. (2007). Genome-wide association studies: Progress in identifying genetic biomarkers in common, complex diseases. *Biomarker Insights*, *2*, 283–292.

Kooperberg, C., & Ruczinski, I. (2005). Indentifying interaction SNPs using Monte Carlo logic regression. *Genetic Epidemiology*, *28*, 157–170. doi:10.1002/gepi.20042

Lage, K., Karlberg, O. E., Størling, Z. M., Ólason, P. Í., Pedersen, A. G., & Rigina, O. (2007). A human phenome-interactome network of protein complexes implicated in genetic disorders. *Nature*, *25*, 309–316. doi:10.1038/nbt1295

Lievens, S., Lemmens, I., & Tavernier, J. (2009). Mammalian two-hybrids come of age. *Trends in Biochemical Sciences*, *34*(11), 579–588. doi:10.1016/j.tibs.2009.06.009

Mani, K. M., Lefebvr, C., Wang, K., Lim, W. K., Basso, K., & Dalla-Favera, R. (2008). A systems biology approach to prediction of oncogenes and molecular perturbation targets in B-cell lymphomas. *Molecular Systems Biology*, *4*, 169. doi:10.1038/msb.2008.2

Mathivanan, S., Periaswamy, B., Gandi, T., Kandasamy, K., Suresh, S., Mohmood, R., et al. (2006). An evaluation of human protein-protein interaction data in the public domain. *BioMed Central Bioinformatics, 7*, Suppl 5S19.

McKinney, B. A., Reif, D. M., White, B. C., Crowe, J. E. Jr, & Moore, J. H. (2007). Evaporative cooling feature selection for genotypic data involving interactions. *Bioinformatics (Oxford, England)*, *23*, 2113–2120. doi:10.1093/bioinformatics/btm317

Millstein, J., Conti, D. V., Gilliland, F. D., & Gauderman, J. W. (2006). A testing framework for identifying susceptibility genes in the presence of epistasis. *American Journal of Human Genetics*, *78*, 15–27. doi:10.1086/498850

Mitchell, T. (2009). *Machine learning*. New York: McGraw-Hill.

Moore, J. H. (2003). The ubiquitous nature of epistasis in determining susceptibility to common human diseases. *Human Heredity*, *56*, 73–82. doi:10.1159/000073735

Moore, J. H. (2009). From genotype to genometype: Putting the genome back in genome-wide association studies. *European Journal of Human Genetics*, *17*(10), 1231–1240. doi:10.1038/ejhg.2009.39

Moore, J. H., Asselbergs, F. W., & Williams, S. M. (2010). Bioinformatics challenges for genome-wide association studies. *Bioinformatics (Oxford, England)*, *26*(4), 445–455. doi:10.1093/bioinformatics/btp713

Moore, J. H., & White, B. C. (2007). Tuning ReliefF for genome-wide genetic analysis. *Lecture Notes in Computer Science*, *4447*, 166–175. doi:10.1007/978-3-540-71783-6_16

Moore, J. H., & Williams, S. M. (2002). New strategies for identifying gene-gene interactions in hypertension. *Annals of Medicine*, *34*, 88–95. doi:10.1080/07853890252953473

Moore, J. H., & Williams, S. M. (2005). Traversing the conceptual divide between biological and statistical epistasis: Systems biology and a more modern synthesis. *BioEssays*, *27*, 637–646. doi:10.1002/bies.20236

Moore, J. H., & Williams, S. M. (2009). Epistasis and its implications for personal genetics. *American Journal of Human Genetics*, *85*(3), 309–320. doi:10.1016/j.ajhg.2009.08.006

Motsinger, A. A., Ritchie, M. D., & Dobrin, S. E. (2006). Clinical applications of whole-genome association studies: Future applications at the bedside. *Expert Review of Molecular Diagnostics*, *6*(4), 551–565. doi:10.1586/14737159.6.4.551

Motsinger-Reif, A. A., Dudek, S. M., Hahn, L. W., & Ritchie, M. D. (2008). Comparison of approaches for machine-learning optimization of neural networks for detecting gene-gene interactions in genetic epidemiology. *Genetic Epidemiology*, *32*, 325–340. doi:10.1002/gepi.20307

Mufano, M. R., & Flint, J. (2005). Meta-analysis of genetic association studies. *Trends in Genetics*, *21*(5), 268–269.

Nelson, M. R., Kardia, S. L., Ferrell, R. E., & Sing, C. F. (2001). A combinatorial partitioning method to identify multilocus genotypic partitions that predict quantitative trait variation. *Genome Research*, *11*, 458–470. doi:10.1101/gr.172901

Park, M. Y., & Hastie, T. (2008). Penalized logistic regression for detecting gene interactions. *Biostatistics (Oxford, England)*, *9*(1), 30–50. doi:10.1093/biostatistics/kxm010

Pattin, K. A., & Moore, J. H. (2009). Role for protein-protein interaction databases in human genetics. *Expert Review of Proteomics*, *6*, 647–659. doi:10.1586/epr.09.86

Pearson, T. A., & Manolio, T. A. (2008). How to interpret a genome-wide association study. *Journal of the American Medical Association*, *299*(11), 1335–1344. doi:10.1001/jama.299.11.1335

Prasad, T. S. K., Goel, R., Kandasamy, K., Keerthikumar, S., Kumar, S., & Mathivanan, S. (2009). Human protein reference database-2009 update. *Nucleic Acids Research*, *37*, D411–D414.

Rea, T. J., Brown, C. M., & Sing, C. F. (2006). Complex adaptive system models and the genetic analysis of plasma HDL-cholesterol concentration. *Perspectives in Biology and Medicine*, *49*(4), 490–503. doi:10.1353/pbm.2006.0063

Reif, D. M., White, B. C., & Moore, J. H. (2004). Integrated analysis of genetic, genomic and proteomic data. *Expert Review of Proteomics*, *1*, 1095–1104. doi:10.1586/14789450.1.1.67

Risch, N. J., & Merikangas, K. R. (1996). The future of genetic studies of complex human disease. *Science*, *273*, 1516–1517. doi:10.1126/science.273.5281.1516

Ritchie, M. D. (2009). Using prior knowledge and genome-wide association to identify pathways involved in multiple sclerosis. *Genome Medicine*, *1*(6), 65. doi:10.1186/gm65

Ritchie, M. D., Hahn, L. W., & Moore, J. H. (2003). Power of multifactor dimensionality reduction for detecting gene-gene interactions in the presence of genotyping error, missing data, phenocopy, and genetic heterogeneity. *Genetic Epidemiology*, *24*, 150–157. doi:10.1002/gepi.10218

Rivals, I., Personnaz, L., Taing, L., & Potier, M. C. (2007). Enrichment or depletion of a GO category within a class of genes: Which test? *Bioinformatics (Oxford, England)*, *23*, 401–407. doi:10.1093/bioinformatics/btl633

Saccone, S. F., Saccone, N. L., Swan, G. E., Madden, P. A., Goate, A. M., & Rice, J. P. (2008). Systematic biological prioritization after a genome-wide association study: An application to nicotine dependence. *Bioinformatics (Oxford, England)*, *24*, 1805–1811. doi:10.1093/bioinformatics/btn315

Shrabanek, L., Saini, H. K., Bader, G. D., & Enright, A. J. (2007). Computational prediction of protein-protein interactions. *Molecular Biotechnology*, *38*, 1–17. doi:10.1007/s12033-007-0069-2

Sing, C. F., Standard, J. H., & Kardia, S. L. (2003). Genes, environment, and cardiovascular disease. *Arteriosclerosis, Thrombosis, and Vascular Biology*, *23*, 1190–1196. doi:10.1161/01.ATV.0000075081.51227.86

Spencer, C. C., Su, Z., Donnelly, P., & Marchini, J. (2009). Designing genome-wide association studies: Sample size, power, imputation, and the choice of genotyping chip. *Public Library of Science Genetics*, *5*(5), e1000477.

Templeton, A. R. (2000). Epistasis and complex traits. In Wolf, J., Wade, M., & Brodie, B. III, (Eds.), *Epistasis and evolutionary process*. New York: Oxford University Press.

Thornton-Wells, T. A., Moore, J. H., & Haines, J. L. (2004). Genetics, statistics and human disease: Analytical retooling for complexity. *Trends in Genetics*, *20*, 640–647. doi:10.1016/j.tig.2004.09.007

Tiffin, N., Adie, E., Turner, F., Brunner, H. G., van Driel, M. A., & Oti, M. (2006). Computational disease gene identification: A concert of methods prioritizes type 2 diabetes and obesity candidate genes. *Nucleic Acids Research*, *34*, 3067–3081. doi:10.1093/nar/gkl381

Tu, Z. D., Wang, L., Arbeitman, M., Chen, T., & Sun, F. Z. (2006). An integrative approach for causal gene identification and gene regulatory pathway inference. *Bioinformatics (Oxford, England)*, *22*, e489–e496. doi:10.1093/bioinformatics/btl234

Wang, W. Y., Barratt, B. J., Clayton, D. G., & Todd, J. A. (2005). Genome-wide association studies: Theoretical and practical concerns. *Nature Reviews. Genetics*, *6*, 109–118. doi:10.1038/nrg1522

Wilke, R. A., Mareedu, R. K., & Moore, J. H. (2008). The pathway less traveled: Moving from candidate genes to candidate pathways in the analysis of genome-wide data from large scale pharmacogenetic association studies. *Current Pharmacogenomics and Personalized Medicine*, *6*, 150–159.

Williams, S. M., Canter, J. A., Crawford, D. C., Moore, J. H., Ritchie, M. D., & Haines, J. L. (2007). Problems with genome-wide association studies. *Science*, *316*, 1840–1842. doi:10.1126/science.316.5833.1840c

Xiang, W., Yang, C., Yang, Q., Xue, H., Tang, N. L., & Yu, W. (2009). MegaSNPHunter: A learning approach to detect disease predisposition SNPs and high level interactions in genome wide association studies. *BioMed Central Bioinformatics*, *10*, 13.

Young, S. S., & Ge, N. (2005). Recursive partitioning analysis of complex disease pharmicogenetic studies I. motivation and overview. *Pharmicogenetics*, *6*, 65–75. doi:10.1517/14622416.6.1.65

Zhang, Y., & Liu, J. S. (2007). Bayesian inference of epistatic interactions in case-control studies. *Nature Genetics*, *39*(9), 1167–1173. doi:10.1038/ng2110

Zheng, T., Wang, H., & Lo, S. H. (2006). Backward genotype-trait association (BGTA) - based dissection of complex traits in case-control design. *Human Heredity*, *62*, 196–212. doi:10.1159/000096995

ADDITIONAL READING

Abou Jamra, R., Fuerst, R., Kaneva, R., Orozco Diaz, G., Rivas, F., & Mayoral, F. (2007). The first genome-wide interaction and locus-heterogeneity linkage scan in bipolar affective disorder: strong evidence of epistatic effects between loci on chromosomes 2q and 6q. *American Journal of Human Genetics*, *81*, 974–986. doi:10.1086/521690

Bauer-Mehern, A., Furlong, L. I., Raustschka, M., & Sanz, F. (2009). From SNPs to pathways: integration of functional effect of sequence variations on models of cell signaling pathways. *BioMed Central Bioinformatics*, *10*(suppl 8), S6.

Chautard, E., Thierry-Meig, N., & Ricard-Blum, S. (2009). Interaction networks: from protein function to drug discovery. *Pathologie Biologie*, *57*(4), 324–333. doi:10.1016/j.patbio.2008.10.004

Cheverud, J. M., & Routman, E. J. (1995). Epistasis and its contribution to genetic variance components. *Genetics*, *139*, 1455–1461.

de Bakker, P. I., Neale, B. M., & Daly, M. J. (2010). Meta-analysis of genome-wide association studies. *.Cold Spring Harbor Protocol, 6*, pdb.top81.

Easton, D. F., & Eeles, R. A. (2008). Genome-wide association studies in cancer. *Human Molecular Genetics*, *17*(R2), R109–R115. doi:10.1093/hmg/ddn287

Eichler, E. E., Flint, J., Gibson, G., Kong, A., Leal, S. M., & Moore, J. H. (2010). Missing heritability and strategies for finding the underlying causes of complex disease. *Nature Reviews. Genetics*, *11*(6), 446–450. doi:10.1038/nrg2809

Eppstein, M. J., Payne, J. L., White, B. C., & Moore, J. H. (2007). Genomic mining for complex disease traits with 'random chemistry'. *Genetic Programming and Evolvable Machines*, *8*(4), 395–411. doi:10.1007/s10710-007-9039-5

Feldman, I., Rzhetsky, A., & Vitkup, D. (2008). Network properties of genes harboring inherited disease mutations. *Proceedings of the National Academy of Sciences of the United States of America*, *105*, 4323–4328. doi:10.1073/pnas.0701722105

Goh, K., Cusick, M. E., Valle, D., Childs, B., Vidal, M., & Barabási, A. (2007). The human disease network. *Proceedings of the National Academy of Sciences of the United States of America*, *104*(21), 8685–8690. doi:10.1073/pnas.0701361104

Greene, C. S., Himmelstein, D. S., Kelsey, K. T., Williams, S. M., Andrew, A. S., Karagas, M. R., et al. (2010). Enabling personal genomics with an explicit test of epistasis. *Pacific Symposium on Biocomputing*, 327-336.

Greene, C. S., White, B. C., & Moore, J. H. (2009). An expert knowledge-guided mutation operator for genome-wide genetic analysis using genetic programming. *Lecture Notes in Bioinformatics*, *4774*, 30–40.

Hartman, M., Loy, E. Y., Ku, C. S., & Chia, K. S. (2010). Molecular epidemiology and its current clinical use in cancer management. *The Lancet Oncology*, *11*(4), 383–390. doi:10.1016/S1470-2045(10)70005-X

Lehne, B., & Schlitt, T. (2009). Protein-protein interaction databases: keeping up with growing interactomes. *Human Genomics*, *3*(3), 291–297.

Liu, Y. J., Guo, Y. F., Zhang, L. S., Pei, Y. F., Yu, N., & Yu, P. (2010). (in press). Biological pathway-based genome-wide association analysis identified the vasoactive intestinal peptide (VIP) pathway important for obesity. *Obesity (Silver Spring, Md.)*. doi:10.1038/oby.2010.83

Makino, T., & Gojobori, T. (2007). Evolution of protein-protein interaction network. *Genome Dynamics*, *3*, 13–29. doi:10.1159/000107601

Moore, J. H., Boczko, E. M., & Summar, M. L. (2005). Connecting the dots between genes, biochemistry, and disease susceptibility: Systems biology modeling in human genetics. *Molecular Genetics and Metabolism*, *84*, 104–111. doi:10.1016/j.ymgme.2004.10.006

Moore, J. H., & White, B. C. (2006). Exploiting expert knowledge in genetic programming for genome-wide genetic analysis. *Lecture Notes in Computer Science*, *4193*, 969–977. doi:10.1007/11844297_98

Moore, J. H., & White, B. C. (2007). Genome-wide genetic analysis using genetic programming: The critical need for expert knowledge. In Riolo, R. L., Soule, T., & Worzel, B. (Eds.), *Genetic Programming Theory and Practice IV, Genetic and Evolutionary Computation*. New York: Springer. doi:10.1007/978-0-387-49650-4_2

Moore, J. H., & Williams, S. M. (2009). Epistasis and its implications for personal genetics. *American Journal of Human Genetics*, *85*, 309–320. doi:10.1016/j.ajhg.2009.08.006

Need, A. C., & Goldstein, D. B. (2010). Whole genome association studies in complex diseases: where do we stand*? Dialogues in Clinical Neuroscience*, *12*(1), 37–46.

Pellegrini, M., Haynor, D., & Johnson, J. M. (2004). Protein Interaction Networks. *Expert Review of Proteomics*, *1*, 239–249. doi:10.1586/14789450.1.2.239

Rosenberg, N. A., Huang, L., Jewett, E. M., Szpiech, Z. A., Jankovic, I., & Boehnke, M. (2010). Genome-wide association studies in diverse populations. *Nature Reviews. Genetics*, *11*(5), 356–366. doi:10.1038/nrg2760

Suthram, S., Beyer, A., Karp, R. M., Eldar, Y., & Ideker, T. (2008). eQED: an efficient method for interpreting eQTL associations using protein networks. *Molecular Systems Biology*, *4*, 162. doi:10.1038/msb.2008.4

Velez, D. R., White, B. C., Motsinger, A. A., Bush, W. S., Ritchie, M. D., & Moore, J. H. (2007). A balanced accuracy metric for epistasis modeling in imbalanced datasets using multifactor dimensionality reduction. *Genetic Epidemiology*, *31*, 306–315. doi:10.1002/gepi.20211

KEY TERMS AND DEFINITIONS

Common-Complex Disease: A disease that does not follow a Mendelian pattern of inheritance and has a complex etiology that is due to phenomena such as epistasis (gene-gene interaction), plastic reaction norms (gene-environment interaction) and locus heterogeneity.

Epistasis: Gene-gene interaction.

Expert Knowledge: existing biological or statistical information about the problem at hand that can be employed to guide an analysis process or an algorithm in a more directed fashion.

Genome-Wide Association Study (GWAS): Used to assess the role thousands of single-nucleotide polymorphisms SNPs and their association with phenotypes of interest in genotyped individuals.

Pathway Based Analysis (PBA): Employs computational methods to define sets of genes based on common biological attributes such as gene ontology or biological pathways.

Protein-Protein Interaction (PPI): Can be a form of expert knowledge.

SNP: A form of genetic variation represented as a single nucleotide polymorphism.

Chapter 7
Biclustering of DNA Microarray Data:
Theory, Evaluation, and Applications

Alain B. Tchagang
National Research Council, Canada

Youlian Pan
National Research Council, Canada

Fazel Famili
National Research Council, Canada

Ahmed H. Tewfik
University of Minnesota, USA

Panayiotis V. Benos
University of Pittsburgh, USA

ABSTRACT

In this chapter, different methods and applications of biclustering algorithms to DNA microarray data analysis that have been developed in recent years are discussed and compared. Identification of biological significant clusters of genes from microarray experimental data is a very daunting task that emerged, especially with the development of high throughput technologies. Various computational and evaluation methods based on diverse principles were introduced to identify new similarities among genes. Mathematical aspects of the models are highlighted, and applications to solve biological problems are discussed.

INTRODUCTION

Recent developments in genomics and highthroughput technology have shown that biclustering is an emerging and powerful methodology for gene expression data analysis. This is driven by the fact that biclustering performs simultaneous row-column clustering and is able to identify local behaviors of the dataset. When dealing with

DOI: 10.4018/978-1-60960-491-2.ch007

DNA microarray data, biclustering is capable to find subgroups of genes that are intimately related across subgroups of attributes, *e.g.* experimental conditions, time points, or tissue samples. In other words, by simultaneously clustering the rows and columns of the gene expression matrix, one can identify candidate subsets of attributes that are associated with specific biological functions, in which only a subset of genes potentially plays a role. Biological analysis and experimentation could then confirm the significance of the candidate subsets.

Since the introduction of biclustering algorithms in DNA microarray data analysis in 2000 by Cheng and Church, biclustering has received a great deal of attention. Thousands of research papers have been published, presenting new algorithms or improvements to solve this biological data mining problem more efficiently. In this chapter, we explain the biclustering problem, some of its variations, and the main techniques to solve them. Obviously, given the huge amount of work on this topic, it is impossible to explain or even mention all proposed algorithms. Instead, in this chapter, we attempt to give a comprehensive survey of the most influential algorithms and results. It begins with a description of the biological problem motivating the underlying methodology. At each step, an attempt is made to describe both the relevant biological and relevant statistical assumptions so that it is accessible to biologists, statisticians, and computer scientists, and can be of use to those starting to do research on biclustering of microarray data as well as users experienced with this technique. Furthermore, we give more insights regarding the methodologies available for statistical and biological evaluations of the biclusters, and demonstrate the applicability of biclustering algorithms to solve specific problems in computational biology and gene expression data analysis in particular.

This chapter is divided into five sections with several examples at the end. The section on biclustering of DNA microarray data introduces the application of biclustering to microarray data, illustrating the practical aspects of these techniques. The section on bicluster models interpretations and validations discusses the available procedures to measure the validity of the resulting biclusters. The section on algorithms for biclusters identification presents the implementation of popular algorithms. The section on biological applications shows several examples of biclustering applied to microarray data to answer specific biological questions. Lastly, in the final section, we conclude and provide some insights on future research directions.

BICLUSTERING OF DNA MICROARRAY DATA

Quantitative gene expression measurements using microarrays were first performed by Schena et al. (1995) on 45 *Arabidopsis thaliana* genes and shortly after, on thousands of genes or even a whole genome (DeRisi et al., 1996; DeRisi et al., 1997). Since that time, various methods for the analysis of such data have been developed. This includes the biclustering techniques.

DNA Microarray

Microarrays are solid substrates hosting hundreds of single stranded DNAs with a specific sequence, which are found on localized features arranged in grids. These molecules, called probes, hybridize with single stranded cDNA molecules, named targets, which have been labeled during a reverse transcription procedure. The targets reflect the amount of mRNA isolated from a sample obtained under a particular condition. Thus, the amount of fluorescence emitted by each spot is proportional to the amount of mRNA transcribed from corresponding DNA sequence. The microarray is scanned and the resulting image is analyzed using signal and image processing techniques so that the signal from each probe can be quantified into numerical values. Such values represent the expression level of the gene in the given condition (Simon et al., 2003).

Microarrays can be fabricated by depositing cDNAs or previously synthesized oligonucleotides; this approach is usually referred to as printed microarrays. In contrast, *in situ* manufacturing encompasses technologies that synthesize the probes directly on the solid support. Slightly different oligonucleotides array platforms are manufactured by companies such as Affymetrix, Agilent, and NimbleGen. Each technology has its advantages and disadvantages and serves a particular research goal. A good review on DNA microarray technology can be found in (Irizarry et al., 2005).

Microarray data acquired during time-course experiments allow the temporal variations in gene expression to be monitored. Therefore, this kind of data has been widely used to study the dynamic behavior of the cells in a variety of biological processes, including cell proliferation (Spellman et al., 1998), development (Arbeitman et al., 2002), and response to extracellular stimuli (Gash et al., 2000; Guillemin et al., 2002). On the other hand, steady state microarray data are acquired under different experimental conditions when the cell reaches the steady state, such as homeostasis (Gardner & Faith, 2005). When dealing with DNA microarray experiments, a significant question is whether to use a time-series design or a steady-state design. The steady-state design may miss dynamic events that are critical for correctly inferring the control mechanism of a transcription relational network, but enables one to observe more diverse experimental conditions. On the other hand, time-series experiments can capture dynamics, but many of the data points may contain redundant information leading to inefficient use of experimental resources.

Gene Expression Matrix

A gene expression matrix can be defined using either an $N \times M$ matrix: Equation 1A, or a set Equation 1B.

$$A = \begin{bmatrix} g(1) \\ g(2) \\ \vdots \\ g(n) \\ \vdots \\ g(N) \end{bmatrix} = \begin{bmatrix} a(1,1) & a(1,2) & \cdots & a(1,m) & \cdots & a(1,M) \\ a(2,1) & a(2,2) & \cdots & a(2,m) & \cdots & a(2,M) \\ \vdots & \vdots & \vdots & \vdots & \vdots & \vdots \\ a(n,1) & a(n,2) & \cdots & a(n,m) & \cdots & a(n,M) \\ \vdots & \vdots & \vdots & \vdots & \vdots & \vdots \\ a(N,1) & a(N,2) & \cdots & a(N,m) & \cdots & a(N,M) \end{bmatrix} \tag{1A}$$

$$R = \{G, C\} \tag{1B}$$

where $G = \{g(1), g(2), \ldots, g(n), \ldots, g(N)\}$ represents the set of genes. Each element of G corresponds to a row of the gene expression matrix. $C = \{c(1), c(2), \ldots, c(m), \ldots, c(M)\}$ represents the set of experimental conditions, time points, or tissue samples. Each element of C corresponds to the columns of the gene expression matrix. The entry $a(n,m)$ or simply a_{nm} of the gene expression matrix corresponds to a real value representing the relationship between row n and column m, which is the expression level of the nth gene: $g(n)$, under the mth experimental condition: $c(m)$. The nth row $g(n) = a(n,:) = [a(n,1)\ a(n,2)\ \ldots\ a(n,m)\ \ldots\ a(n,M)]$ is a $1 \times M$ vector. It corresponds to the expression level of the nth gene under the M experimental conditions. The mth column $c(m) = a(:,m) = [a(1,m)\ a(2,m)\ \ldots\ a(n,m)\ \ldots\ a(N,m)]^T$ is an $N \times 1$ vector. It corresponds to the expression level of the N genes under the mth experimental condition.

As we mentioned above, the microarray technology has a very high throughput interrogating thousands of genes at the same time. However, the process includes numerous sources of variability (Kerr & Churchill, 2001; Yang & Speed, 2002). Several tools such as statistical experimental

Figure 1. Schematic illustration of conventional clustering vs. biclustering algorithms: (A) gene-based clustering, (B) sample-based clustering, and (C) biclustering or gene-sample-based clustering.

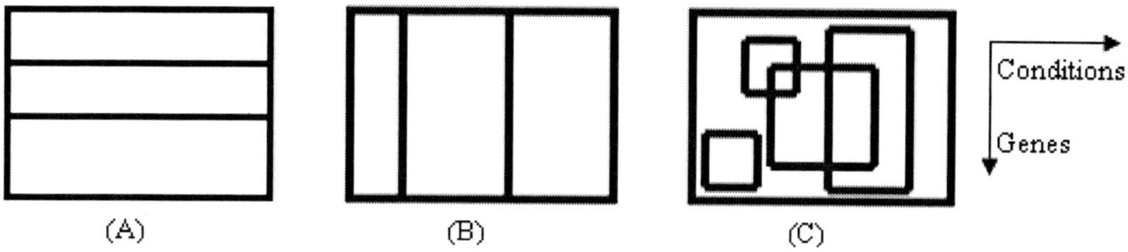

design and data normalization can help to obtain high quality results from microarray experiments

DNA Microarray Data Preprocessing and Normalization

Preprocessing of the data is an important step prior to biclustering. The aim of normalization is to account for systematic differences across different datasets and eliminate artifacts. The challenge of normalization is to remove as much technical variations as possible while leaving the biological variation untouched. Few general preprocessing techniques such as the logarithmic transformation are useful for all microarray platforms, whereas many others are specific to a given technology. A good review of normalization of microarray data can be found in (Quackenbush, 2002).

Depending on the type of biclusters (see Section Bicluster Models and Statistical Evaluations below) one seeks to identify; a second step is the removal of all genes that show low variation across samples or experimental time points, which may affect negatively the subsequent biclustering process. Researchers may be tempted to apply principal component analysis (PCA) to reduce the dimensionality of the data prior to clustering or biclustering, but it is proved to not improve the results especially in the classical clustering context. It has been suggested that the quality of the clusters is not necessarily higher with PCA than without it; in most cases the first principal components does not necessarily yield the best clusters (Yeung & Ruzzo, 2001).

Clustering vs. Biclustering Algorithms

The objective of any clustering approach is to identify groups of co-expressed genes that are potentially co-regulated. Genes exhibiting similar responses to certain treatments are likely to be controlled by similar regulatory mechanisms. This is often referred to as the guilt by association principle (Androulakis et al., 2007). Therefore, identifying coherent expression profiles is important in order to identify co-regulation and to understand the underlying machinery driving the co-expression. From a computational perspective, this is a clustering problem. Clustering of co-expressed gene has been active biological data mining topic and advances in parallel with the development of DNA microarray technology.

Clustering is the process of classifying data objects into a set of disjoint classes, called clusters, so that objects within a class are highly similar to each other, while objects in separate classes are dissimilar (Xu & Wunch, 2005). One of the characteristics of gene expression data is its coherent structure with regard to subspaces in either or both dimensions (genes and/or samples). Therefore, co-expressed genes can be clustered based on their expression patterns along either or both of these two dimensions. In gene-based clustering (Figure 1A) the genes are treated as the objects, while the samples are the features. Sample-based clustering (Figure 1B), on the other hand, regards the samples as the objects and the genes as the features.

The distinction of gene-based clustering and sample-based clustering is that the clustering tasks are based on different characteristics in gene expression data. Some clustering algorithms, such as K-means (see Xu & Wunch, 2005) and hierarchical (Eisen et al., 1998) approaches can be applied to classify either genes or samples based on expression profiles along the respective dimension. Both the gene-based and sample-based clustering approaches perform exclusive and exhaustive partitions of objects that share the same feature space.

The main difference between clustering and biclustering is that clustering can be applied to either the rows or the columns of the data matrix, separately. Biclustering, on the other hand, performs clustering in these two dimensions simultaneously (Figure 1C). That is, unlike clustering algorithms, biclustering algorithms identify subgroups of genes that show similar activity patterns under a specific subset of the experimental conditions. Note that biclustering algorithms are also known as co-clustering, two-dimensional clustering, or two-way clustering in other research fields. Biclustering is an unsupervised learning technique. That is, no prior knowledge about the data one seeks to cluster is available, as opposed to a supervised learning approach.

Problem Formulation

Mathematical Definition of Biclusters

Given the representation of the gene expression matrix as in Equations 1A or 1B, we will define a bicluster using Equations 2A or 2B:

$$B = [b(i,j)] = [b_{ij}]; \qquad (2A)$$

$$S = \{I, J\}. \qquad (2B)$$

Mathematically, a bicluster can be viewed as a submatrix B of A (Equation 2A), or as a subset S of R (Equation 2B) whose elements are I and J, with $I \subseteq G$, and $J \subseteq C$, furthermore, $i \in I$ and $j \in J$. In a set notation, the entire set of biclusters is defined as: $\Omega = \{2\text{-}tuples\ (I_k, J_k), I_k \subseteq G \text{ and } J_k \subseteq C\}$, $k = 1$ to K, and K is the total number of biclusters.

In the following, the cardinality (defined as $card(.)$ or $|.|$), of a set corresponds to the number of its elements. For example, $card(G) = |G| = N$ is the number of genes in a gene expression matrix, and $card(C) = |C| = M$ is the number of experimental conditions, time points, or tissue samples, $card(\Omega) = |\Omega| = K$ is number of biclusters. Below, when dealing with biclusters: $S = \{I, J\}$, we use $|I| = I$ and $|J| = J$ unless specified.

Problem Formulation

Biclustering algorithms should be able to address the following three problems.

Problem 1: Given the gene expression matrix as define above, the specific problem addressed by biclustering algorithms is to identify the set of biclusters $\Omega = \{2\text{-}tuples\ (I_k, J_k), I_k \subseteq G \text{ and } J_k \subseteq C\}$ such that the entries of each bicluster $B_k = [b(i,j)]$ (with $i \in I_k$ and $j \in J_k$) satisfies some specific characteristics of homogeneity (see Section on Bicluster Models and Statistical Evaluations below).

Problem 2: Because of the noisy nature of DNA microarray experiments, not only biclustering algorithm should be capable to identify the sets of biclusters that are statistically significant, but also, they should be robust to noise.

Problem 3: Since the end goal of any DNA microarray experimental study is to answer specific biological questions, biclustering algorithms should be able to identify sets of biclusters that are biologically meaningful.

BICLUSTER MODELS, INTERPRETATIONS, AND EVALUATIONS

Several mathematical models have been defined in the literature not only to model biclusters, but also to aid in biclustering algorithm designed and to statistically and biologically evaluate their significance.

Definitions

Recall that a bicluster is defined as $B = [b(i,j)] = [b_{ij}]$. Let's define the following statistical parameters: b_{iJ} is the mean of the ith row in the bicluster and it is defined using Equation 3A, b_{Ij} is the mean of the jth column in the bicluster, and it is defined using Equation 3B, and b_{IJ} is the mean of all the elements of the bicluster and it is defined using Equation 3C.

$$b_{iJ} = \frac{1}{|J|}\sum_{j \in J} b_{ij} \tag{3A}$$

$$b_{Ij} = \frac{1}{|I|}\sum_{i \in I} b_{ij} \tag{3B}$$

$$b_{IJ} = \frac{1}{|I|}\sum_{j \in J} b_{iJ} = \frac{1}{|J|}\sum_{i \in I} b_{Ij} = \frac{1}{|I||J|}\sum_{i \in I}\sum_{j \in J} b_{ij} \tag{3C}$$

The variance and the mean square residue of a bicluster or a matrix can be defined using Equation 4 and Equation 5 respectively.

$$V(B) = \sum_{i \in I}\sum_{j \in J}(b_{ij} - b_{IJ})^2 \tag{4}$$

$$H(B) = \frac{1}{|I||J|}\sum_{i \in I}\sum_{j \in J}(b_{ij} - b_{iJ} - b_{Ij} + b_{IJ})^2 \tag{5}$$

Bicluster Models and Statistical Evaluations

Biclustering algorithms are usually designed to identify one of the following five models: biclusters with constant values, biclusters with constant values on rows, biclusters with constant values on columns, biclusters with coherent values, and biclusters with coherent evolutions.

Biclusters with Constant Values

Biclusters with constant values in a gene expression matrix describe subsets of genes with equal expression values within a subset of experimental conditions. Mathematically they can be modeled using Equation 6 assuming that the data is noise free.

$$B = [b(i,j)] = [b_{ij}] = [\mu] \tag{6}$$

Equation 7 shows an example of a noise free bicluster with constant values. This synthetic bicluster has 4 genes (rows) and 4 experimental conditions (columns).

$$\begin{bmatrix} \mu & \mu & \mu & \mu \\ \mu & \mu & \mu & \mu \\ \mu & \mu & \mu & \mu \\ \mu & \mu & \mu & \mu \end{bmatrix} \tag{7}$$

One can easily show that the variance of such noise free bicluster is zero.

Proof.

Recall that the variance of a matrix or a bicluster is defined as:

$$V(B) = \sum_{i \in I}\sum_{j \in J}(b_{ij} - b_{IJ})^2 \tag{8}$$

Since $B = [b(i,j)] = [b_{ij}] = [\mu]$ and $S = \{I, J\}$, we have:

$$V(B) = \sum_{i\in I}\sum_{j\in J}(b_{ij} - b_{IJ})^2 = \sum_{i\in I}\sum_{j\in J}(b_{ij} - \frac{1}{IJ}\sum_{i\in I}\sum_{j\in J}b_{ij})^2 = \sum_{i\in I}\sum_{j\in J}(\mu - \frac{1}{IJ}\sum_{i\in I}\sum_{j\in J}\mu)^2 \quad (9)$$

$$V(B) = \sum_{i\in I}\sum_{j\in J}(\mu - \frac{1}{IJ}IJ\mu) = \sum_{i\in I}\sum_{j\in J}(\mu - \mu)^2 = 0 \quad (10)$$

In real applications, because of the noisy nature of the data, a bicluster with constant values becomes $B = [b_{ij}] = [\mu+\varepsilon_{ij}]$, where ε_{ij} represent the expected level of noise generated during microarray experiments. Hence of, the variance of biclusters with constant values becomes:

$$V(B) = \sum_{i\in I}\sum_{j\in J}(b_{ij} - b_{IJ})^2 < \delta \quad (11)$$

where δ is a small positive number, which is usually defined by the user. This merit function (Equation 11) is used to evaluate the statistical significance of biclusters with constant values.

Biclusters with Constant Values on Rows

Biclusters with constant values along rows indicate a subset of genes with expression levels that do not change across a subset of conditions, irrespective of the actual expression levels of the individual genes.

Mathematically, a noise free bicluster with constant values on rows can be modeled using the following equations:

$B = [b(i,j)] = [b_{ij}] = [\mu+\alpha_i] \rightarrow$ additive model. (12A)

$B = [b(i,j)] = [b_{ij}] = [\mu\alpha_i] \rightarrow$ multiplicative model. (12B)

Equation 13 shows an example of a noise free bicluster with constant values on rows. This synthetic bicluster has 4 genes (rows) and 4 experimental conditions (columns).

$$\begin{bmatrix} \alpha_1 & \alpha_1 & \alpha_1 & \alpha_1 \\ \alpha_2 & \alpha_2 & \alpha_2 & \alpha_2 \\ \alpha_3 & \alpha_3 & \alpha_3 & \alpha_3 \\ \alpha_4 & \alpha_4 & \alpha_4 & \alpha_4 \end{bmatrix} \quad (13)$$

Using either the additive or the multiplicative model, we can show that $max(b(i,:))-min(b(i,:)) = 0$. In noisy dataset, we have $max(b(i,:))-min(b(i,:)) = \delta_i$, and the statistical evaluation of this type of bicluster is done using Equation 14.

$max(b(i,:)) - min(b(i,:) < \delta$, for all $i \in I_k$. (14)

Biclusters with Constant Values on Columns

Biclusters with constant columns isolate a subset of conditions for which a subset of genes have constant expression values that may differ from condition to condition. Mathematically, a noise free bicluster with constant values on columns can be modeled using the following equations:

$B = [b(i,j)] = [b_{ij}] = [\mu+\beta_j] \rightarrow$ additive model. (15A)

$B = [b(i,j)] = [b_{ij}] = [\mu\beta_j] \rightarrow$ multiplicative model. (15B)

Equation 16 shows an example of a noise free bicluster with constant values along columns.

This synthetic bicluster has 4 genes (rows) and 4 experimental conditions (columns).

$$\begin{bmatrix} \beta_1 & \beta_2 & \beta_3 & \beta_4 \\ \beta_1 & \beta_2 & \beta_3 & \beta_4 \\ \beta_1 & \beta_2 & \beta_3 & \beta_4 \\ \beta_1 & \beta_2 & \beta_3 & \beta_4 \end{bmatrix} \quad (16)$$

Using either the additive or the multiplicative model, we can show that $max(b(:,j))-min(b(:,j)) = 0$. In noisy dataset, we have $max(b(:,j))-min(b(:,j)) = \delta_j$, and the statistical evaluation of this type of bicluster is done using Equation 17.

$$max(b(:,j)) - min(b(:,j)) < \delta, \text{ for all } j \in J_k. \quad (17)$$

Biclusters with Coherent Values

Biclusters with coherent values identify subset of genes that are up-regulated and down-regulated coherently across subsets of conditions, *i.e.* same magnitude and same direction across experimental conditions. Mathematically, a noise free bicluster with coherent values can be modeled using the following equations:

$$B = [b(i,j)] = [b_{ij}] = [\mu+\alpha_i+\beta_j] \rightarrow \text{additive model.} \quad (18A)$$

$$B = [b(i,j)] = [b_{ij}] = [\mu\alpha_i\beta_j] \rightarrow \text{multiplicative model.} \quad (18B)$$

Equation 19 shows an example of a noise free bicluster with coherent values. This synthetic bicluster has 4 genes (rows) and 4 experimental conditions (columns).

$$\begin{bmatrix} \alpha_1+\beta_1 & \alpha_1+\beta_2 & \alpha_1+\beta_3 & \alpha_1+\beta_4 \\ \alpha_2+\beta_1 & \alpha_2+\beta_2 & \alpha_2+\beta_3 & \alpha_2+\beta_4 \\ \alpha_3+\beta_1 & \alpha_3+\beta_2 & \alpha_3+\beta_3 & \alpha_3+\beta_4 \\ \alpha_4+\beta_1 & \alpha_4+\beta_2 & \alpha_4+\beta_3 & \alpha_4+\beta_4 \end{bmatrix} \text{ or } \begin{bmatrix} \alpha_1\beta_1 & \alpha_1\beta_2 & \alpha_1\beta_3 & \alpha_1\beta_4 \\ \alpha_2\beta_1 & \alpha_2\beta_2 & \alpha_2\beta_3 & \alpha_2\beta_4 \\ \alpha_3\beta_1 & \alpha_3\beta_2 & \alpha_3\beta_3 & \alpha_3\beta_4 \\ \alpha_4\beta_1 & \alpha_4\beta_2 & \alpha_4\beta_3 & \alpha_4\beta_4 \end{bmatrix} \quad (19)$$

The mean square residue (MSR) of such biclusters is shown to be zero.

Proof.

Recall that the MSR of a matrix or a bicluster is defined as:

$$H(B) = \frac{1}{|I||J|}\sum_{i\in I}\sum_{j\in J}(b_{ij} - b_{iJ} - b_{Ij} + b_{IJ})^2. \quad (20)$$

Given that $B = [b(i,j)] = [b_{ij}] = [\mu+\alpha_i+\beta_j]$ and $S = \{I, J\}$, we have shown in Box 1.

Same *proof* can be done using the multiplicative model. In real applications, because of the noisy nature of the data, a bicluster with coherent values becomes $B = [b_{ij}] = [\mu+\alpha_i+\beta_j+\varepsilon_{ij}]$, where

Box 1.

$$H(B) = \frac{1}{IJ}\sum_{i\in I}\sum_{j\in J}((\mu+\alpha_i+\beta_j) - \frac{1}{I}\sum_{i\in I}(\mu+\alpha_i+\beta_j) - \frac{1}{J}\sum_{j\in J}(\mu+\alpha_i+\beta_j) + \frac{1}{IJ}\sum_{j\in J}\sum_{i\in I}(\mu+\alpha_i+\beta_j))^2$$

(21)

$$H(B) = \frac{1}{IJ}\sum_{i\in I}\sum_{j\in J}(\mu+\alpha_i+\beta_j-\mu-\frac{1}{I}\sum_{i\in I}\alpha_i-\beta_j-\mu-\alpha_i-\frac{1}{J}\sum_{j\in J}\beta_j+\mu+\frac{1}{I}\sum_{i\in I}\alpha_i+\frac{1}{J}\sum_{j\in J}\beta_j)^2 = 0$$

(22)

ε_{ij} represent the expected level of noise generated during microarray experiments. Hence of, real biclusters with coherent values are evaluated using the following merit function:

$$H(B) = \frac{1}{|I||J|}\sum_{i\in I}\sum_{j\in J}(b_{ij} - b_{iJ} - b_{Ij} + b_{IJ})^2 < \delta \tag{23}$$

where δ is a very small positive number, which is usually defined by the user.

In Equations 6, 12, 15, 18, μ is usually referred to as the background effect, α_i the row or gene effect, and β_j the column or condition effect. Hence of, constant biclusters: μ correspond to background effect, constant values on rows biclusters: $\mu+\alpha_i$ to the background effect + rows or genes effect, constant values on columns biclusters: $\mu+\beta_j$ to the background effect + column or condition effect, whereas coherent values biclusters: $\mu+\alpha_i+\beta_j$ correspond to the summation of the background effect + gene effect + condition effect.

Biclusters with Coherent Evolutions

Unlike biclusters with coherent values, biclusters with coherent evolutions identify subsets of genes that are up-regulated or down-regulated coherently across subsets of conditions irrespective of their actual values, *i.e.* same directions but varying magnitude. Unlike other types of biclusters, coherent evolution biclusters are difficult to model using a mathematical equation. But, depending on how coherent evolution is defined, several merit function can be defined for their statistical validation. For example, Ben-Dor et al. (2003) define a coherent evolution bicluster using the order preserving submatrix (OPSM) framework, in which the expression levels of all genes induce the same linear ordering of the experiments, and used the upper bound on the probability (Equation 24) of having a bicluster with J experimental conditions and I or greater number of genes, to estimate their statistical significance.

$$Z(J,I) = M...(M-J+1)\sum_{n=I}^{N}\binom{N}{n}\left(\frac{1}{J!}\right)^n\left(1-\frac{1}{J!}\right)^{N-n} \tag{24}$$

N and M are the number of genes and experimental conditions respectively. Equation 24 is derived as follows. $1/J!$ represents the probability that a random row (gene) belongs to an order preserving model that has J columns. Since the rows are assumed to be independent, the probability of having at least I rows in the order preserving model is the I-tail of the (N, ($1/J!$)) binomial distribution *i.e.* the summation term of Equation 24. Finally, $M...(M-J+1)$ represents the number of ways to choose a complete order preserving model. Equation 25 shows an example of a bicluster with coherent evolution. This synthetic bicluster has 4 genes (rows) and 4 experimental conditions (columns) and they all have the same induced permutation [2 3 1 4].

$$\begin{bmatrix} 1 & 2 & 0 & 4 \\ 2 & 3 & 1 & 5 \\ 4 & 6 & 2 & 7 \\ 5 & 7 & 3 & 8 \end{bmatrix} \tag{25}$$

Such coherent evolution patterns might arise, for example, if the set of experimental conditions J represents distinct stages in the progress of a disease or in a cellular process and the expression levels of all genes in I vary across the stages in the same way. Similar approach has also been used by several other authors: (Tewfik et al., 2006 and references therein).

Remark 1. Mathematically, one can show that biclusters with constant values are special cases of biclusters with constant values along rows (max([μ]) – min([μ]) = 0, for all $i\in I_k$) or along columns (max([μ]) – min([μ]) = 0, for all $j\in J_k$).

Remark 2. It can also be shown that biclusters with constant values, with constant values on rows or columns are special cases of biclusters with coherent values: $H([\mu]) = 0$; $H([\mu+\alpha_i]) = 0$; $H([\mu\alpha_i]) = 0$; $H([\mu+\beta_j]) = 0$; $H([\mu\beta_j]) = 0$.

Remark 3. Likewise, it can also be shown that each of above four types of biclusters is a special case of biclusters with coherent evolutions.

Nevertheless, the biological interpretation of each type of bicluster should be unique and specific to the problem that the DNA microarray experimental design sought to solve.

There are several properties that could be inferred from the above definitions and that could be used in biclusters identification designed. Recall that an $N \times M$ gene expression matrix is defined as: $A = [a_{nm}] = [a(n,m)]$, with set of genes G and set of experimental conditions C. Furthermore, let us define L as the number of distinct entries of the gene expression matrix A.

Property 1. In a gene expression matrix the total number of biclusters with constant values on M experimental conditions is $\leq L$.

Property 2. From property 1, we can infer that the total number of biclusters with constant values in a gene expression matrix will always be $\leq L \times 2^{M-1}$.

Property 3. In a gene expression matrix the total number of biclusters with constant values on rows on M experimental conditions is either 0 or 1.

Property 4. From property 3, one can also infer that the total number of biclusters with constant values on rows in a gene expression matrix is $\leq 2^{M-1}$.

Property 5. If a subset of I_1 genes are co-expressed across a subset of J_1 experimental conditions, they will always be co-expressed across a subset of J_2 experimental conditions, with $J_2 \subseteq J_1$.

Property 1 can easily be shown as follows. Since the gene expression matrix has L distinct entries, it can be written as in Equation 26.

$$A = \sum_{l=1}^{L} \lambda_l A_l \tag{26}$$

λ_ls are the distinct values and A_ls the binary matrices associated to each λ_l ($l = 1$ to L). Thus in each A_l, we can only identify one bicluster with constant value on the entire M experimental conditions. Since we have L A_l, we can conclude that the total number of biclusters with constant values on M conditions in a gene expression matrix is $\leq L$. The other properties can be shown similarly.

Biological Evaluation of Biclusters

As we described above, a good biclustering algorithm identifies genes with similar expression patterns. These co-expressed genes are probably controlled by the same transcription factors. In addition, genes that are co-expressed frequently participate in the same biological pathways. In both cases, if the identified biclusters are biologically meaningful, then their corresponding sets of genes should be enriched with and annotated under the same gene ontology terms, involved in the same biological pathways, or regulated by the same transcription factors. The biological role of the genes in the biclusters can be assessed using several external biological datasets and knowledge bases such as Gene Ontology, transcription factor gene interactions data, biological pathways databases, protein-protein interactions data, microRNA, and epigenetic modifications data.

Existing Biological Knowledge for Bicluster Evaluations

Biological evaluation of the significance of biclusters can be tested using biological knowledge from publicly available knowledge and databases.

Gene Ontology. The Gene Ontology provides controlled vocabularies for the description of the molecular function, biological process, and cellular component of gene products (The Gene

Ontology Consortium, 2000; DayRichter et al., 2007; Carbon et al., 2009). Biclustering algorithms should be able to identify sets of genes that are annotated under the same or related Gene Ontology terms (Prelic et al., 2006; Tchagang et al., 2008).

Transcription factor gene interactions and transcription factor binding sites databases. Transcription factor (TF) gene associations data describe with a given probability whether or not a given gene is regulated by a specific transcription factor under a given experimental condition. Ideally, TF-gene associations data are derived from Chromatin Immunoprecipitation (ChIP) (Buck et al., 2004; Lee et al., 2002) experiments. ChIP is a well-established methodology used to investigate interactions between TFs and their genomic DNA targets *in vivo*. They can also be compiled from literature search and/or TF databases of well known and well characterized biological interactions such as TRANSFAC (Matys et al., 2006), JASPAR (Sandelin et al., 2004), PRODORIC (Munch et al., 2003), RegulonDB (Salgado et al., 2006), YEASTRACT (Teixeira et al., 2006), and SCPD7 (Zhu et al., 1999). For example, TRANSFAC is a database on eukaryotic transcriptional regulation. The database contains data on TFs, their target genes and their experimentally-proven binding motifs. Good biclustering algorithms identify sets of genes that are co-regulated by the same TFs (Tchagang et al., 2008; Tchagang et al., 2009).

Biological pathways and protein-protein interactions databases. In addition to GO annotations and TF-gene interactions data, other types of biological knowledge such as metabolic and protein-protein interaction networks, that have been derived from other types of data than gene expression can be used. Although each type of data reveals other aspects of the underlying biological system, one can expect to a certain degree that genes that participate in the same pathway respectively form a protein complex also show similar expression patterns (Zien et al., 2000; Ideker et al., 2002; Ihmels et al., 2002). The question here is whether the computed biclusters reflect this correspondence. Hence of, biclustering algorithms should be able to identify sets of genes that are associated with the same pathways (Tchagang et al., 2010). In this regard, there are several pathways and protein-protein interactions databases such as KEGG pathways (Shujiro et al., 2008) that can be used to verify this hypothesis. The KEGG database for example provides a very rich resource for pathways. Its aim is to link individual level information such as genes, proteins, enzymes, with system level information such as interactions, enzymatic reactions, and pathways.

There are several other biological data such as microRNA and epigenetic modifications data that could also be used for biclusters biological evaluation. In higher organisms it is well documented that response to environments occurs at the transcriptional level. TFs, microRNAs, and epigenetic modifications can combine to form a complex regulatory network (Huttenhower et al., 2009 and references therein).

Example of Biological Evaluation Tools

There are several packages available for functional characterization of gene groups (Table 1), among many others that are referenced in the microarray tools' section at the GO website, and new ones are being released (Coulibaly & Page, 2008).

Several of these testing tools can be easily implemented using computational packages. One common strategy is to create a custom data analysis pipeline using statistical analysis software packages such as R and MATLAB. Both allow great flexibility, customized analysis, and access to many specialized packages designed for analyzing gene expression data. Not only is R freely available, but also allows the use of BioConductor (Gentleman et al., 2005), a collection of R tools including many powerful current gene expression analysis methods written and tested by experts from the growing microarray community.

Table 1. Example of biological evaluation packages

Package	P-value computation	Multiple testing	Type of array	GO level	Other annotations	Platforms	Reference
BiNGO	Hypergeometric test, binomial	FDR, Bonferroni	Commercial arrays	Available, GOSlim	Not available	Cytoscape plug-in	Maere et al. (2005)
CLENCH2	Hypergeometric test, binomial, χ^2	N/A	User-provided	Static global	Not available	Windows	Shah & Fedoroff (2004)
DAVID	Hypergeometric, Fisher's exact	Bonferroni, Benjamini, FDR	User-provided	Available	KEGGpathways TRANSFAC	Web-based, standalone	Dennis et al. 2003; Huang et al. (2009)
Fatigo	Fisher's exact	FDR	User-provided	Available	KEGG pathways, SwissPROT keywords	Any	Al-Shahrour et al. (2003)
FuncAssociate	Fisher's exact	Monte Carlo simulation	User-provided	Not available	Not available	Web-based	Berriz et al. (2003)
GOAL	Hypergeometric, Fisher's exact	Bonferroni, Benjamini, FDR	User-provided	Available	TF-gene data, KEGG pathways	Any, standalone, plug-in, web server	Tchagang et al. (2010)
GOSt	Hypergeometric, Fisher's exact	Bonferroni, Benjamini,	User-provided	Available	KEGG pathways	Linux	Jüri Reimand et al. 2006
GOstat	Hypergeometric, Fisher's exact	Holm Benjamini, Yekutieli	User-provided	Available	Not available	Web-based	Beißbarth, & Speed (2004)
GoSurfer	χ^2	FDR	Affymetrix only	Lowest level	Not available	Windows	Zhong et al. (2004)
GoToolBox	Hypergeometric test, Fisher's exact test, binomial	Bonferroni	User-provided	Available	Not available	Any	Martin et al. (2004)
Onto-Express	Hypergeometric test, Fisher's exact test, binomial, χ^2	Bonferroni, Holm, Sidak, FDR	commercial array	Available	Chromosomal position	Any	Draghici et al. (2003)

ALGORITHMS FOR BICLUSTERS IDENTIFICATION

Given the huge amount of work on biclustering algorithms, it is impossible to cover all proposed algorithms. Instead, we give a comprehensive survey of some of the most influential biclustering algorithms and their variations. Thus providing a platform for researchers from different background to develop novel biclustering algorithms, improve or use the existing ones to solve various biological problems. Some of the algorithms discussed below aim at finding only one bicluster. This is always viewed as the best bicluster based on some statistical criteria. Other algorithms aim at finding the *K* best biclusters. Furthermore, some algorithms iteratively find one bicluster or a group of biclusters during each iteration, whereas others find the entire set of biclusters simultaneously. Most of the pioneering biclustering algorithms only make use of the gene expression data, whereas a few of the most recent ones take an integrative data approach (Huttenhower et al., 2009; Halperin et al., 2009; Reiss et al., 2006), *i.e.*, they consume data from several different platforms to tune their search into more parsimonious results. Below, biclustering algorithms are classified into two categories: non-integrative and integrative approach.

Cheng and Church (CC-Algorithm) 2000

```
Input
-     A → Gene expression matrix
-     G → Set of genes
-     C → Set of experimental conditions
-     δ → Maximum mean square residue
Output
-     Ω = {2-tuples (I_k,J_k), I_k⊆G and J_k ⊆C}, k = 1 to K.
Begin
(i)     Initialize bicluster:
a.     B_k = A
b.     I_k = G
c.     J_k = C
d.     Compute H.
 (ii)     Deletion phase
a.     While H > δ
i.     Remove rows and columns that maximize H.
ii.     Compute H.
 (iii)     Addition phase
a.     While H < δ
i.     Add previously deleted rows and columns.
ii.     Compute H.
(iv)     Output
a.     I_k = Set of co-expressed genes;
b.     J_k = Set of conditions under which they (I_k) are co-expressed;
c.     B_k = Gene expression level in bicluster;
 (v)     Mask identified bicluster B_k with random number in A and go to (i).
End Begin
```

Non-Integrative Approach

In the non-integrative approach, the biclustering algorithm takes the gene expression matrix and outputs one or more biclusters.

Cheng and Church Algorithm (CC-Algorithm) and its Variants

Cheng and Church (2000) define a bicluster to be a submatrix of the gene expression matrix, for which the mean squared residue score (Equation 20) is below a user defined threshold δ (Equation 23). Each entry b_{ij} in the bicluster is the superposition of: the background level, the row (gene) effect, the column (condition) effect, and the noise: $B = [b_{ij}] = [\mu+\alpha_i+\beta_j+\varepsilon_{ij}]$ (Equation 18A). They hypothesized that a dataset can contain a number of K biclusters, which are not necessarily disjoint, and showed that the problem of finding the largest δ–biclusters is *NP*-complete.

Cheng and Church (2000) proposed a two-phase greedy algorithm to identify δ–biclusters. (i) ***Deletion phase***: rows and columns are removed from the original expression matrix until the above constraint ($H < \delta$) is fulfilled. (ii) ***Addition phase***:

previously deleted rows and columns are added to the resulting submatrix as long as the bicluster score does not exceed δ. This procedure is composed of several iterations and each iteration is restricted to the identification of only one bicluster while previously defined biclusters are masked with random values. The masking of previously discovered biclusters with random numbers and the subsequent discovery of new ones as in the original CC-algorithm may result in the phenomenon of random interference which in turn impacts the discovery of high quality biclusters. Also, the CC-algorithm is unlikely to identify overlapping biclusters. To address some of these issues and to further accelerate the biclustering process, Yang et al. (2002 & 2003) proposed the flexible overlapping biclustering (FLOC) algorithm, which is basically a generalized model of δ–bicluster that incorporates null values to handle missing values in a bicluster up to a threshold. Then, they developed the FLOC algorithm to discover a set of K possibly overlapping biclusters simultaneously.

The CC and the FLOC algorithms use MSR to evaluate the score of a bicluster during the bicluster identification procedure. Unfortunately, when this formula is used, it is common to find submatrices of δ-biclusters that are not δ-biclusters, thus contradicting the basic definition of a bicluster (Wang et al., 2002). Furthermore, many δ-biclusters found in a given dataset may differ only in one or two outliers they contain. Some of these issues are due to the fact that when the MSR is used as scoring function, it measures the square deviation from the sum of the mean value of expression levels in the entire bicluster, and the mean values of expression levels along each row and column in the bicluster. In conclusion, even if the MSR can be used to statistically evaluate the biclusters already identified, its use as scoring function during the bicluster identification procedure inhibits the design of accurate biclustering algorithms.

Wang et al. (2002) introduced the pCluster to solve the problems related to MSR. A submatrix $B = [b_{ij}]$ or $S = \{I,J\}$, with, $i \in I$ and $j \in J$ is considered a δ-pCluster if the absolute difference of the differences of the attribute (condition) values of two objects (genes) is less than a threshold δ for every pair of objects and attributes. That is

$$\forall x, y \in I, \forall u, v \in J, | (b_{xu} - b_{xv}) - (b_{yu} - b_{yv}) | \leq \delta \quad (27)$$

The first advantage of their pCluster model is that any submatrix of a δ-pCluster is also a δ-pCluster. This property is fundamental to their developed pCluster algorithm, which locates δ-pClusters by first identifying two-object and two-attribute pClusters and incrementally builds larger sets. They then developed a depth-first algorithm to efficiently and effectively discover all the pClusters with a size larger than a user-specified threshold. The pCluster algorithm is deterministic can mine multiple clusters simultaneously, can detect overlapping clusters, and is resilient to outliers, and will not miss any qualified clusters, unlike CC and FLOC, which only provide approximations of the full bicluster set.

The key concept behind pCluster is the maximum dimension set (MDS), which are maximum length contiguous subsequences of the sorted values of the difference of two objects across all attributes bounded by the threshold δ. Although pCluster can effectively address the problem related to MSR, it only works well on small datasets. Tewfik et al. (2006) addressed the complexity issue by developing a deterministic parallel biclustering algorithm for coherent biclusters identification. Their algorithm relies on an exhaustive enumeration and an early pruning strategy to search for biclusters of genes with similar patterns. Note that the pClusters of Wang et al. (2002) are the equivalent or subclasses of coherent evolution biclusters.

There have been several other versions of the initial CC-algorithms that uses more elaborate algorithmic approaches such as genetic algorithms, fuzzy algorithms, evolutionary algorithms, multi-objective evolutionary algorithms to handle the interference problem and the quality of biclusters. Mitra and Banka (2006) employ a combination of a multi-objective evolutionary biclustering framework and local search strategies for biclustering gene expression data, in order to generate and iteratively refine an optimal set of biclusters. In a multi-objective evolutionary biclustering framework, a non-dominated sorting genetic algorithm is used to converge to the global Pareto front and to simultaneously maintain the diversity of the population. Local search strategies are used to speed up convergence by refining the chromosomes. A vector of decision variables $x_p \in F$ is Pareto optimal if there does not exist another $x \in F$ such that $f_i(x) \leq f_i(x_p)$ for all $i = 1,\ldots, k$ and $f_j(x) < f_j(x_p)$ for at least one j. F denotes the feasible region of the problem (*i.e.*, where the constraints are satisfied). Divina & Aguilar-Ruiz (2006) proposed an algorithm based on evolutionary computation, named Sequential Evolutionary Biclustering (SEBI). The algorithm employs a sequential covering strategy and an evolutionary algorithm, in order to find biclusters of maximum volume, with a mean squared residue lower than a given threshold, with a relatively high row variance, and minimizing the effect of overlapping among biclusters.

Plaid Model and its Variants

The plaid model is a biclustering algorithm developed by Lazzeroni and Owen (2002) for the analysis of gene expression data. It combines a statistical modeling approach and linear algebra techniques for biclusters identification. In plaid model, the gene expression matrix is modeled as a superposition of layers, corresponding to biclusters.

Given the gene expression matrix $A = [a_{nm}] = [a(n,m)]$, the plaid model consists of a bias plus a sum of K layers, where each layer is a bicluster. The expression value in a bicluster corresponds to the sum of the background effect, the gene effect, and the condition effect, *i.e.*:

$$a_{nm} = \mu_0 + \sum_{k=1}^{K} (\mu_k + \alpha_{nk} + \beta_{mk})\rho_{nk}\theta_{mk}. \quad (28)$$

ρ and θ are matrices of size $N \times K$ and $M \times K$, respectively, containing binary membership variables, *i.e.* $\rho_{nk} \in \{0,1\}$ and $\theta_{mk} \in \{0.1\}$. For example, $\rho_{nk} = 1$ if and only if gene n belongs to bicluster k, and $\theta_{mk} = 1$ if and only if condition m belongs to bicluster k.

The authors then derive an iterative heuristic algorithm that attempts to find the model parameters ($K, \mu_0, \mu_k, \alpha_{nk}, \beta_{mk}, \rho_{nk}$, and θ_{mk}) by minimizing the quadratic error (Equation 29) between the original gene expression matrix and the model given in (Equation 28).

$$\sum_{n=1}^{N}\sum_{m=1}^{M}[a_{nm} - \mu_0 + \sum_{k=1}^{K}(\mu_k + \alpha_{nk} + \beta_{mk})\rho_{nk}\theta_{mk}]^2. \quad (29)$$

Subject to:

$$\sum_{n=1}^{N}\alpha_{nk}\rho_{nk} = 0 \quad \text{and} \quad \sum_{m=1}^{M}\beta_{mk}\theta_{mk} = 0 \quad (30)$$

The constraints (Equation 30) are added to reduce the number of parameters. Furthermore, the following approaches can be used to determine the initial memberships of ρ and θ. That is: (i) all parameters set to *0.5*; (ii) all parameters set to random values near *0.5*; (iii) more complicated heuristics such as fixing all $\omega_{nmk} = \mu_k + \alpha_{nk} + \beta_{mk}$ to 1, perform several iterations that update ρ and θ only, and scale ρ and θ so that they sum to *N/2* and *M/2* respectively (Lazzeroni & Owen, 2002).

Lagrange multipliers can be used to obtain θ_k from ρ_{k-1} and κ_{k-1}, *i.e.* the best fit of the models, subject to the condition that every row and column has a zero mean.

$$\mu_k = \frac{\sum_{n=1}^{N}\sum_{m=1}^{M} \rho_{nm}\theta_{mk} z_{nm}^{K-1}}{\left(\sum_{n=1}^{N}\rho_{nk}^2\right)\left(\sum_{m=1}^{M}\theta_{mk}^2\right)},$$

$$\alpha_{nk} = \frac{\sum_{m=1}^{M}(z_{nm}^{K-1} - \mu_k\rho_{nk}\theta_{mk})\theta_{mk}}{\rho_{nk}\sum_{m=1}^{M}\theta_{mk}^2},$$

$$\beta_{mk} = \frac{\sum_{n=1}^{N}(z_{nm}^{K-1} - \mu_k\rho_{nk}\theta_{mk})\rho_{nk}}{\kappa_{nk}\sum_{n=1}^{N}\theta_{mk}^2} \qquad (31)$$

The residual z_{nm}^{K-1} from the first *K-1* layers is defined as:

$$z_{nm}^{K-1} = a_{nm} - \omega_{nm0} - \sum_{k=1}^{K-1}\omega_{nmk}\rho_{nk}\theta_{mk} \qquad (32)$$

Similarly, the membership parameters can be determined by:

$$\rho_{nk} = \frac{\sum_{m=1}^{M}\omega_{nmk}\theta_{mk}z_{nm}^{K-1}}{\sum_{m=1}^{M}\omega_{nmk}^2\theta_{mk}^2},\ \theta_{mk} = \frac{\sum_{n=1}^{N}\omega_{nmk}\rho_{nk}z_{nm}^{K-1}}{\sum_{n=1}^{N}\omega_{nmk}^2\rho_{nk}^2} \qquad (33)$$

During the iterative procedure, new layers are added to the model one at a time, and it stops when a layer has a smaller size than expected (found by random permutation of data, *i.e.,* after finding a layer, permute (residual) elements of a_{nm} (*i.e.*, the z_{nm}) by row and by column. Compute importance of permuted layers, and see if larger or smaller than importance of layer *K*). The algorithm can also stop if K_{max} (a user defined parameter) layers have been found (Lazzeroni & Owen, 2002).

The features of the plaid model make it an attractive method for biclustering of gene expression data. It can effectively address several problems posed by the CC-algorithm and some of its variants (see above). However, the efficiency of the original plaid model algorithm maybe compromised by the relaxation of binary constraints on the bicluster membership parameters in the model. Furthermore, in the original plaid model, the authors implicitly assumed that the noise in the gene expression matrix is Gaussian. To address the first problem, Turner et al. (2005) introduced an extended version of the plaid model algorithm that takes advantage of the binary constraints on the cluster membership parameters, providing a simpler and more direct method of optimization. Their model-based method partially supervises the plaid model algorithm to favor biclusters corresponding to the external grouping information, in order to determine whether the biclusters are related to one or more a priori groups. In addition, their model-based method extends the original plaid model to biclustering whole time series of expression levels.

Other difficulties that exist with the plaid model are due to the fact that there are mixed binary and continuous variables in its modeling framework, for which the traditionally used optimization algorithms suitable for continuous variables cannot be employed in the realization of the biclustering process. Zhang et al. (2008) developed a neural-network approach to tackle such a mixed binary and continuous optimization problem. Note that, their method is proposed only to tackle the mixed continuous and binary optimization problem that is intrinsic in the plaid model. Two mutually interactive parts in the neural network, with one corresponding to the binary value of θ and the other to the binary value of ρ, are introduced. The two variables of θ and ρ are firstly relaxed to two continuous variables which are constrained in the range of [0, 1] by an appropriate definition of the activation function. Upon convergence of the network, the variables are forced automatically to the binary values of 0 or 1.

Caldas and Kaski (2008) reformulated the original plaid model in a Bayesian framework (Equation 34), and developed a collapsed Gibbs

Plaid Models; Lazzeroni & Owen 2002

```
Input
-      A → Gene expression matrix
-      G → Set of genes
-      C → Set of experimental conditions
-      S → maximum cycle per iteration
Output
-      Ω = {2-tuples (I_k,J_k), I_k⊆G and J_k ⊆C}, k = 1 to K
Begin
(i)      set K = 0
 (ii)      Compute initial value: ρ_nk and θ_mk
 (iii)      Adding a new layer
a.      K = K+1; s = 1;
b.      While s < S, do
i.      Compute: μ_k, α_nk, and β_mk (Equation 30)
ii.      Compute: ρ_nk and θ_mk (Equation 32)
iii.      If the importance of layer k is not random
1.      Record the layer and go to (iii).
iv.      End
 (iv)      Output
a.      I_k = Set of co-expressed genes;
b.      J_k = Set of conditions under which they (I_k) are co-expressed;
c.      B_k = Expression level in biclusters;
End Begin
```

sampler for inferring the posterior distribution of bicluster memberships, more specifically the binary membership variables that indicate which genes and which conditions belong to each bicluster.

$$a_{nm} \sim N\left(\mu_0 + \sum_{k=1}^{K}(\alpha_{nk} + \beta_{mk})\rho_{nk}\theta_{mk}, \sigma^2\right) \tag{34}$$

In Equation 34, each entry a_{nm} is conditioned on the parameters μ_0, α_n, β_m, ρ_n, θ_m, and σ^2, is assumed to follow a Gaussian distribution. Like Caldas & Kaski, Gu and Liu (2008) also reformulated the original plaid model in the Bayesian framework and implemented a Gibbs sampler for biclusters inference. For multiple biclusters identification, they constrain the overlapping of biclusters to only one direction (*i.e.*, either gene or condition direction), and use a more flexible error model, which allows the error term of each bicluster to have a different variance.

The Order Preserving Submatrix (OPSM) Algorithm and its Variants

Most clustering models, including those used in subspace clustering described above, define similarity among different objects by distances over either all or only a subset of the dimensions. Some well-known distance functions include Euclidean distance, Manhattan distance, and cosine distance. However, distance functions are not always adequate in capturing correlations among

OPSM; Ben-Dor et al. 2003

```
Input:
-      A → gene expression matrix
-      G → Set of genes
-      C → Set of experimental conditions
Output
-      Ω = {2-tuples (Ik,Jk), Ik⊆G and Jk ⊆C}, k = 1 to K
Begin
(i)      Evaluate all (1, 1) partial models and keep the best l of them.
 (ii)      Expand them to (2, 1) models and keep the best l of them.
 (iii)      Expand them to (2, 2) models and keep the best l of them.
 (iv)      Expand them to (3, 2) models and keep the best l of them.
 (v)      -------------------------
 (vi)      Until getting l ([s/2], [s/2]) models, which are complete models.
 (vii)      Output the best one.
a.      Ik = Set of co-expressed genes;
b.      Jk = Set of conditions under which they (Ik) are co-expressed;
a.      Bk = Expression level in biclusters;
End Begin
```

the objects. In fact, strong correlations may still exist among a set of objects even if they are far apart from each other as measured by the distance functions.

In cell biology, investigations have shown that more often than not, several genes contribute to a disease, which motivates researchers to identify a subset of genes whose expression levels rise and fall coherently under a subset of conditions, *i.e.*, they exhibit fluctuation of a similar shape when conditions change. Discovery of such clusters of genes is essential in revealing the significant connections in gene regulatory networks. As in the case of pCluster (Wang et al., 2002) described above, algorithms discussed here focus on finding cluster of objects that have coherent evolutions (*i.e.* same directions but varying magnitudes) rather than objects that are physically close to each other (*i.e.* same directions and same magnitudes).

Ben-Dor et al. (2003) tackled coherent evolution patterns identification as an order preserving submatrix (OPSM) problem. They defined a bicluster as a submatrix that preserves the order of the selected columns for all of the selected rows. In other words, the expression values of the genes within a bicluster induce an identical linear ordering across the selected samples. Based on a stochastic model (Equation 24), the authors developed a deterministic heuristic algorithm to find the largest and statistically significant bicluster. To motivate their heuristic approach to the OPSM problem, they showed that the OPSM problem is NP-Hard.

Given the gene expression matrix as defined above, Ben-Dor et al. (2003) defines a complete model as (C, π), where, C is a set of conditions (columns), π is an ordering of the conditions in C. Furthermore, they define a partial model as $(<c_1, c_2, \ldots, c_a>, <c_{s-b+1}, \ldots, c_s>, s)$, where the first a and last b conditions are specified, but not the remaining s-a-b conditions, s is the size of the model. A row "supports" a model if by applying the permutation to the row; it yields a set of monotonically increasing values. Having these definitions in mind, the basic idea behind the algorithm is to grow partial models until they become complete models.

OPSM focuses on the uniformity of the relative order of the conditions rather than on the uniformity of the actual expression levels as in the plaid model or the MSR models. This approach is potentially more robust to the stochastic nature of the expression levels, and to the variation caused by the measurements process. OPSMs have been accepted as a biologically meaningful subspace cluster model, capturing the general tendency of gene expressions across a subset of conditions. As we mentioned earlier, in an OPSM, the expression levels of all genes induce the same linear ordering of the conditions. The OPSM cluster model focuses on the relative order of columns rather than the uniformity of actual values in data matrices. By sorting the row vectors and replacing the entries with their corresponding column labels, the data matrix can be transformed into a sequence database, and OPSM mining is reduced to a special case of the sequential pattern mining problem with some unique properties. In particular, the sequence database is extremely dense since each column label appears exactly once (assuming no missing values) in each sequence. A sequential pattern uniquely specifies an OPSM cluster, with all the supporting sequences as the cluster contents. The number of supporting sequences is the support for the pattern. The original OPSM algorithm of Ben-Dor et al. (2003) can be used to identify OPSMs. The authors mentioned without *proof* that it can also be adapted to handle relaxations and extensions of the OPSM conditions. For example, allowing the different rows of biclusters ($I \times J$) to induce similar but not identical orderings of the columns, or allowing the set of conditions (J) to include more than one representation of each condition of a biological process.

A recent extension to OPSM finds multiple, overlapping coherent biclusters in noisy datasets, but like the original OPSM, it is very expensive in the number of features and requires excessive computational resources if applied to large gene expression matrices. To tackle this issue, Liu et al. (2004) also defined a bicluster as an Order-Preserving Cluster (OP-Cluster) and used OPC-tree, an exhaustive bicluster enumeration algorithm to perform simultaneous biclusters identification. Essentially, the biclustering problem is transformed into a problem of finding longest common subsequences. Each row in the input data matrix is arranged as a non-decreasing sequence of columns. These rows are then stored in a tree and simultaneously, common subsequences and number of rows supporting these subsequences are determined by doing operations on the tree. The OP-Cluster approach uses a tree data structure, which is exhaustive in nature.

After converting each gene vector into an ordered label sequence, Teng and Chan (2007) transformed the OPSM problem into finding frequent orders appearing in the sequence set. They then developed an algorithm for finding the frequent orders by iteratively combining the most frequent prefixes and suffixes in a statistical way. Griffith et al. (2009) introduced the KiWi mining framework for massive datasets, that exploits two parameters k and w to provide a biased testing on a bounded number of candidates, substantially reducing the search space and problem scale, targeting on highly promising seeds that lead to significant clusters and twig clusters.

The Iterative Signature Algorithm (ISA) and its Variants

The Iterative Signature Algorithm (ISA) considers a bicluster $S = \{I, J\}$ to be a transcription module, *i.e.*, a set of co-regulated genes together with the associated set of regulating conditions (Bergmann et al., 2003), and it is required to verify the following criteria:

$$S = \{I, J\} = \begin{cases} I = \{n \in G / \mid a_{nJ}^{G} \mid > t_{g}\sigma_{g}\} \\ J = \{m \in C / \mid a_{\mathrm{Im}}^{C} \mid > t_{c}\sigma_{c}\} \end{cases} \tag{35}$$

where t_g and t_c are the gene and condition z-score threshold respectively. a_{Im}^{C} represents the mean expression of genes from I in the sample m, a_{nJ}^{G} the mean expression of the gene n in samples from J, and σ_c and σ_g their respective standard deviation.

Given the gene expression matrix as defined above, the ISA normalizes the expression matrix A to obtain the two matrices $A^G = [a_{nm}^{G}]$ and $A^C = [a_{nm}^{C}]$, to eliminate any experimental errors, such that:

$$\sum_{n=1}^{N} a_{nm}^{G} = 0, \sum_{n=1}^{N} (a_{nm}^{G})^2 = 1, \text{ for } m \in C \tag{36}$$

$$\sum_{m=1}^{M} a_{nm}^{C} = 0, \sum_{m=1}^{M} (a_{nm}^{C})^2 = 1, \text{ for } n \in G \tag{37}$$

Next, starting with a random input seed gene I, it computes: a_{Im}^{C} for each condition m and retains the conditions for which the average $\mid a_{\mathrm{Im}}^{C} \mid > t_g\sigma_g$. It then computes a_{nJ}^{G} for each gene n and retains the new set of genes for which $\mid a_{nJ}^{G} \mid > t_c\sigma_c$. The algorithm then iterates until the following stopping criteria is met:

$$\frac{I_{n+1} - I_n}{I_{n+1} + I_n} < \varepsilon \tag{38}$$

I_n is the cardinality obtained in the nth iteration. Multiple biclusters can be identified by running the iterative signature algorithm on several initial gene sets. Note that, this approach requires identification of a reference gene set which needs to be carefully selected for good quality results. In the absence of pre- specified reference gene set, random set of genes is selected at the cost of quality of overall biclustering solution.

Although the ISA algorithm has been successfully applied to identify transcription module in the yeast, its drawback lies in its predilection for strong signals, which are found hundreds of times before weaker signals are, if at all, detected. In cases where genes with a strong signal have been selected into the initial sample, they dominate the average, driving the module towards their signal. In the plaid model of Lazzeroni and Owen (2003) described above, similar problems do exist, and it is addressed by subtracting signals which are contained in the already detected bicluster. To address the above mentioned issue in the ISA context, Kloster et al. (2005) extended the ISA, stipulating that the condition vector of each new module be orthogonal to the condition vectors of the previously identified modules.

Gupta & Aggarwal (2007) implemented the seeded iterative signature algorithm (SISA), an improved and fast version of the original ISA that works in three phases. The main idea of SISA algorithm is to create a set of k medoids randomly. Medoids are chosen far apart from each other. For each medoid, the algorithm computes a set of genes which are similar to it in some sense. These are then taken as the input gene seeds to compute the set of transcription modules simultaneously. Supper et al. (2007) developed EDISA (Extended Dimension Iterative Signature Algorithm), a probabilistic clustering approach for gene-condition-time datasets. Based on the mathematical definitions of gene expression modules, the EDISA samples initial modules from the dataset which are then refined by removing genes and conditions until they comply with the module definition.

The Robust Biclustering Algorithm (RoBA) and its Variants

The Robust Biclustering Algorithm (RoBA) is a linear algebra inspired algorithmic approach de-

ISA: Bergmann et al. (2003)

```
Input:
-     A → gene expression matrix
-     G → Set of genes
-     C → Set of experimental conditions
-     t_g → gene z-score threshold
-     t_c → condition z-score threshold
-     ε → stopping criteria
Output
-     Ω = {2-tuples (I_k,J_k), I_k⊆G and J_k ⊆C}, k = 1 to K
Begin
(i)     Normalize the gene expression matrix;
c.     Compute A^C;
d.     Compute A^G;
 (ii)     Start with an initial set of genes (I);

(iii)     Score all samples with respect to the selected gene set;
e.     Retain samples for which the score exceeds a predefined threshold;

(iv)     Score all genes with respect to the selected samples;
f.     Select a new set of genes for which the score exceeds a predefined
threshold;

(v)     Repeat (iii) and (iV) until the set of genes and the set of samples
converge to the stopping criteria;

(vi)     Output bicluster
g.     I_k = Set of co-expressed genes;
h.     J_k = Set of conditions under which they (I_k) are co-expressed;
i.     B_k = Expression level in biclusters;
End Begin
```

veloped by Tchagang and Tewfik (2006) to tackle the biclustering problem. RoBA first quantizes the gene expression matrix then expresses it as the sum of the product of its distinct entries with their corresponding elementary matrices whose entries are binary (Equation 39). Using these binary matrices, RoBA is able to identify any type of biclusters from the given gene expression matrix.

Recall that an $N \times M$ gene expression matrix is defined as: $A = [a_{nm}] = [a(n,m)]$, with set of genes G and set of experimental conditions C. Let us define L as the number of distinct entries of the gene expression matrix A obtained after data quantization. The first step of RoBA consists of expressing the gene expression matrix A as the sum of the products of each of its distinct entries λ_l with their corresponding binary matrix A_l:

$$A = \sum_{l=1}^{l=L} \lambda_l A_l = \lambda_1 A_1 + \lambda_2 A_2 + \ldots + \lambda_l A_l + \ldots + \lambda_L A_L \quad (39)$$

After the gene expression matrix is written as in Equation 39, obtaining any type of biclusters is straightforward.

For example, one will notice that biclusters with constant values will correspond to the submatrices of each A_l whose entries are all 1. Biclusters with constant values on rows will correspond to the submatrices of the $NL \times M$ matrix Z_1 (Equation 40) whose entries are all 1.

$$Z_1 = \begin{bmatrix} A_1 \\ A_2 \\ \vdots \\ A_{L-1} \\ A_L \end{bmatrix} \quad (40)$$

Biclusters with constant values on columns will correspond to submatrices of the $N \times ML$ matrix Z_2 (Equation 41) whose entries are all 1.

$$Z_2 = \begin{bmatrix} A_1 & A_2 & \cdots & A_{L-1} & A_L \end{bmatrix} \quad (41)$$

Biclusters with coherent values and biclusters with coherent evolutions will correspond to submatrices of the $NL \times ML$ matrix Z_3 (Equation 42) whose entries are all 1 (Tchagang &Tewfik, 2006).

$$Z_3 = \begin{bmatrix} A_1 & A_2 & \cdots & A_{L-1} & A_L \\ A_2 & A_3 & \cdots & A_L & A_1 \\ \vdots & \vdots & \vdots & \vdots & \vdots \\ A_{L-1} & A_L & \cdots & A_{L-3} & A_{L-2} \\ A_L & A_1 & \cdots & A_{L-2} & A_{L-1} \end{bmatrix} \quad (42)$$

The critical steps in RoBA are the data quantization or discretization step which should be done with care and the identification of all 1 submatrices which should be done in a timely manner. There are several data quantization techniques defined in the literature and that can be efficiently used to achieve this step (Tchagang & Tewfik, 2006 and references therein). On the other hand, there are several algorithms in the literature known as frequent itemset algorithms that can be used efficiently for the identification of all 1 submatrices (Goethals & Zaki, 2004; Agrawal & Srikant, 1994; Tchagang & Tewfik, 2006).

The BiMax algorithm proposed by Prelic et al. (2006) also rely on a discretization scheme for bicluster identification. Their approach only focuses on finding constant biclusters. BiMax first discretizes the input gene expression matrix into a binary matrix (two levels: up or down). Because of the two levels discretization scheme used in BiMax, it is harder for it to determine coherent biclusters. Okada et al. (2007) proposed an exhaustive enumeration biclustering algorithm, (BIMODULE), based on a closed itemset enumeration algorithm. As in RoBA, their algorithm starts by normalizing and discretizing the data matrix into L levels and the discretized data are given as input to a closed itemset miner (Goethals & Zaki 2004; Agrawal & Srikant 1994) in form of transaction items. Liu et al. (2007) proposed an algorithm based on closed itemset. Unlike Okada et al. (2007), they introduced a distance-based subspace clustering model that uses a more flexible method to partition the dimensions to preserve meaningful and significant clusters that may not be discovered by a grid based approach used by Okada et al. (2007). Their algorithm considers only those biclusters containing a nontrivial number of objects and attributes, and do not mine for coherent bicluster.

Mahfouz and Ismail (2009) presented an iterative algorithm termed BIDENS, which can approximate a number of K possibly overlapping biclusters. Their algorithm start with K initial biclusters and iteratively move rows and columns from or to biclusters such that the resulting biclustering is accepted and the average space of biclusters is minimized from one iteration to the next until it terminates. Their model is based on the real values of the input matrix. But to reduce the complexity of the proposed algorithm they

RoBA; Tchagang & Tewfik (2006)

```
Input
-     A → Gene expression matrix
-     G → Set of genes
-     C → Set of experimental conditions
-     Discretization parameters
Output
-     Ω = {2-tuples (I_k,J_k), I_k⊆G and J_k ⊆C}, k = 1 to K
Begin
(i)     Gene expression matrix quantization;
 (ii)     Gene expression matrix decomposition ; (Equations 26, 39)
 (iii)     Biclusters Identification
a.     Biclusters with constant values
i.     All 1 submatrices of each A_l
b.     Biclusters with constant values on rows
i.     All 1 submatrices of Z_1 (Z_2 is defined below Equation 40);;
c.     Biclusters with constant values on columns
i.     All 1 submatrices of Z_2 (Z_2 is defined below Equation 41);
d.     Biclusters with coherent values
i.     All 1 submatrices of Z_3 (Z_3 is defined below Equation 42);
e.     Biclusters with coherent evolutions
i.     All 1 submatrices of Z_3 (Z_3 is defined below Equation 42);
 (iv)     Output
a.     I_k = Set of co-expressed genes;
b.     J_k = Set of conditions under which they (I_k) are co-expressed;
c.     B_k = Expression level in biclusters;
End Begin
```

discretize the matrix such that the threshold is an integer value corresponding to the range of bins in which the values of each column lie. This is similar to BIMODULE of Okada et al. (2007), but unlike BIMODULE, their discretization is done on the whole values in the input matrix using histogram and a level corresponding to several contiguous bins. More recently, Ibrahim et al. (2009) presented a time and space efficient implementation of RoBA.

Other Biclustering Algorithms and Their Variants

There are several other non-integrative based biclustering algorithms that we did not cover in this chapter among which some are worth mentioning. Murali and Kasif (2003) have proposed a representation for gene-expression data called conserved gene expression motifs or xMOTIFs (biclusters). A gene's expression level is conserved across a set of samples if the gene is expressed with the same abundance in all the samples. A conserved gene-expression motif is a subset of genes that is simultaneously conserved across a subset of samples. Murali and Kasif assumed that a gene could be in a fixed number of states. These states can simply be up-regulated and down-regulated when only two states are considered. They also assumed that data may contain several xMOTIFs, and the goal is to find the largest xMOTIF, *i.e.*, the one that contains the maximum number of conserved rows by evaluating a merit function.

The SAMBA algorithm (Statistical-Algorithmic Method for Bicluster Analysis) of Tanay et al. (2004) uses probabilistic modeling of the data and graph theoretic techniques to identify subsets of genes that jointly respond across a subset of conditions, where a gene is termed responding in some condition if its expression level changes significantly at that condition with respect to its normal level. Within the SAMBA framework, the expression data are modeled as a bipartite graph whose two parts correspond to conditions and genes, respectively, with edges for significant expression changes. The vertex pairs in the graph are assigned weights according to a probabilistic model, so that heavy subgraphs correspond to biclusters with high likelihood. Under this weighting scheme, the discovery of the most significant biclusters in the data is reduced to finding the heaviest subgraphs in the model bipartite graph. SAMBA employs a practical heuristic to search for heavy subgraphs. The search algorithm is motivated by a combinatorial algorithm for finding heavy bicliques that is exponential in the maximum gene degree in the graph.

Spectral biclustering approaches use techniques from linear algebra to identify bicluster structures in the input data (Kluger et al., 2003). In this model, it is assumed that the expression matrix has a hidden checkerboard-like structure that the algorithm tries to identify using eigenvector computations. The structure assumption is argued to hold for clinical data, where tissues cluster to cancer types and genes cluster to groups, each distinguishing a particular tissue type from the other types.

Coupled two-way clustering (CTWC), introduced by Getz et al. (2000) defines a generic scheme for transforming a one-dimensional clustering algorithm into a biclustering algorithm. The algorithm relies on having a one-dimensional (standard) clustering algorithm that can discover significant clusters. Given such an algorithm, the coupled two-way clustering procedure will recursively apply the one-dimensional algorithm to submatrices, aiming to find subsets of genes giving rise to significant clusters of conditions and subsets of conditions giving rise to significant gene clusters.

Liu & Wang (2007) recently proposed and developed a polynomial time biclustering algorithm, referred to as RMSBE, which requires reference genes for biclusters identification. The algorithm MSB is shown to find optimal biclusters with the maximum similarity score. Unlike other biclustering algorithms, which only find biclusters with rectangular shapes, RMSBE is able to find biclusters of any shape.

Data Integration Approach

In the biclustering approaches discussed above, one usually assumes that co-expression may imply co-regulation, *i.e.* may share the same regulatory controls, thereby implying biological relevance for such a pre-clustering step. However, gene transcript levels can be correlated either by chance (due to experimental noise or systematic error) or because of indirect effects, and therefore they might not actually be directly co-regulated. The integration of additional biological data or biologically- relevant evidence into a clustering procedure may be used to provide constraints on the identification of groups of co-regulated genes.

Similarly, most existing approaches to regulatory module discovery such as the ones described above break the biclustering and motif discovery tasks into separate stages: first, expression data is biclustered and afterwards, each bicluster is analyzed for enrichment of sequence motifs. To *cMonkey: Reiss et al. (2006)*

```
Input:
-      A → Gene expression matrix
-      G → Set of genes
-      C → Set of experimental conditions
-      Seq → Sequence data
Output
-      Ω = {2-tuples (I_k,J_k), I_k⊆G and J_k ⊆C}, k = 1 to K
-      de novo motifs
Begin
 (i)      Seed a new bicluster
(ii)      Search for motifs in bicluster
(iii)      Compute the conditional probability that each gene/condition is a
member of the bicluster
(iv)      Perform moves sampled from conditional probability
(v)      Did the cluster change?
a.      If YES go back to (ii)
b.      If NO
i.      Output
1.      I_k = Coregulated genes
2.      J_k = Conditions where they are co-regulated
3.      Putative regulating motifs
ii.      Go back to (i)
End Begin
```

discover regulatory modules most effectively, though, it would be natural to perform both tasks at the same time, discovering clusters of genes that are both co-expressed and enriched for regulatory motifs. Recent work (Huttenhower et al., 2009; Halperin et al., 2009; Reiss et al., 2006) has indeed confirmed the intuition that regulatory module discovery by simultaneous analysis of expression and sequence data for example can be extremely effective.

Data integration biclustering algorithms such as cMonkey (Reiss et al., 2006), Allegro (Harlepin et al., 2009), and COALESCE (Huttenhower et al., 2009) integrate de novo motif detection into the biclustering scheme to guide the process towards more biologically parsimonious solutions. Unlike the biclustering algorithms described above, this group detects putative co-regulated gene groupings by integrating the biclustering of gene expression data and various functional associations with the de novo detection of sequence motifs.

The cMonkey Algorithm

The cMonkey algorithm (Reiss et al., 2006) is based on the reformulation of the FLOC algorithm (Yang el al., 2003) using a basic probabilistic model for expression data. This enables a more rigorous and intuitive integration of the model of expression data with models for the additional data types, as well as with prior distributions for constraining bicluster sizes and redundancy.

In cMonkey, each bicluster is modeled via a Markov chain process, in which the bicluster is iteratively optimized, and its state is updated based upon conditional probability distributions computed using the cluster's previous state. This enables to define the probabilities that each gene or condition belongs in the bicluster, *conditioned upon* the current state of the bicluster, as opposed to requiring to build a complete (joint) model for the bicluster, *a priori*. The components of this conditional probability are modeled independently (one for each of the different types of information which is integrated) as *p*-values based upon individual data likelihoods, which are then combined into a regression model to derive the full conditional probability.

In cMonkey, three major distinct data types are used (gene expression, upstream sequences, and association networks), and accordingly *p*-values for three such model components are computed: the *expression* component, the *sequence* component, and the *network* component. Each bicluster begins as a *seed*, or starting cluster, that is iteratively optimized by adding/removing genes and conditions to/from the cluster by sampling from the conditional probability distribution using a Monte Carlo procedure, to prevent premature convergence. Additional clusters are seeded and optimized until a given number *K* of clusters have been generated, or significant optimization is no longer possible.

The Allegro algorithm of Harlepin et al. (2009) is another example of such data integration biclustering algorithm. Like cMonkey, Allegro has not been developed to handle heterogeneous data integration, nor has it been scaled for application to complex metazoan genomes. COALESCE was recently proposed by Huttenhower et al. (2009) to fill some of these needs.

COALESCE Algorithm

The COALESCE algorithm provides an efficient, iterative framework for predicting regulatory modules (co-regulated genes, the conditions under which they are co-regulated, and putative regulatory motifs) from very large collections of gene expression data. Additional data types such as nucleosome positioning or evolutionary conservation can be integrated in a Bayesian framework, and the algorithm scales sufficiently to handle very large genomes (>25 000 genes) and gene expression collections compendia (>15 000 conditions).

The basic COALESCE algorithm consumes gene expression and DNA sequence data as input to produce putative co-regulated modules as output. Each resulting module consists of a set of co-regulated genes, one or more expression conditions under which they are co-expressed and zero or more motifs predicted to drive the co-regulation. The algorithm finds modules in a serial manner by seeding each new module with a set of co-expressed genes and iteratively refining the module to convergence.

Each iteration begins with a process of feature extraction, in which expression conditions and sequence motifs showing differential expression/enrichment are associated with the developing module. This is followed by a Bayesian integration step, in which each gene's values for the selected features are combined probabilistically to determine whether the gene should be included in the module, with priors proportional to the fraction of features actually selected. After these two stages are alternated to convergence, the module's centroid is subtracted from the selected genes and features and the process begins again for the next cluster (Huttenhower et al., 2009).

Table 2 lists example of biclustering packages that have been implemented in the literature and that are freely available.

Validation of Biclustering Algorithms and Robustness to Noise

As we mentioned earlier, the main drawback of gene expression data is their high noise levels. DNA chips provide only rough approximation of expression levels, and are subject to errors. Any

COALESCE: Huttenhower et al. (2009)

```
Input:
-     Primary data
o   A → Gene expression matrix
o   G → Set of genes
o   C → Set of experimental conditions
o   Seq → Sequence data
-     Supporting data
o   Evolutionary conservation
o   Nucleosome positioning
Output
-     Ω = {2-tuples (I_k,J_k), I_k⊆G and J_k ⊆C}, k = 1 to K
-     de novo motifs
Begin
 (i)     Initialize a new regulatory module
a.     Select a small set of correlated genes
 (ii)     Iterate until convergence
a.     Feature selection
i.     Identify conditions where genes are co-expressed
ii.     Identify motif enriched in genes' sequences
b.     Bayesian integration
iii.     Identified genes based on selected expression motifs and supporting
data
 (iii)     Output regulatory modules
a.     I_k = Coregulated genes
b.     J_k = Conditions where they are co-regulated
c.     Putative regulating motifs
 (vi)     Subtract module mean from all data and go to (i)
End Begin
```

analysis method such as biclustering algorithms should therefore be robust enough to cope with significant levels of noise. Thus a very crucial point of a biclustering framework is the validation of the inference algorithm. This has to be done by means of artificially generated or real biological data. The validation should identify the strengths and weakness of the algorithm and indicate under which condition it shows reliable results. Each method has to cope with noisy, high dimensional, and incomplete data, hence the performance with these kind of data should be validated.

The use of artificially generated data has several advantages. The underlying biclusters are known (domain knowledge) as well as their properties. One can easily add different level of noise or missing values and analyze the impact on the results obtained. Furthermore, an arbitrary number of datasets with diverse sizes can be generated. Hence of, the scalability pertaining to large datasets can be addressed. Nevertheless the significance of these statistical evaluations is strongly connected with the artificial data generator model.

Table 2. Example of implemented and available biclustering algorithms

Algorithms	Availability	Reference of implementation
δ-bicluster	Java (BiCat) R package (Biclust)	Barkow et al. (2006) Kaiser& Leisch (2008)
FLOC	R package (biclust)	Kaiser& Leisch (2008)
Plaid model	C	Lazzeroni & Owen (2002)
Plaid_t	R package (Biclust)	Kaiser& Leisch (2008)
Spectral biclustering	R package (Biclust)	Kaiser& Leisch (2008)
cHawk	Java	Ahmad & Khokhar (2007)
BiMax	Java (BiCat) R package (Biclust)	Barkow et al. (2006) Kaiser& Leisch (2008)
RoBA	MATLAB	Tchagang et al. (2006)
ISA	Java (BiCat) R package	Barkow et al. (2006) CSardi et al. (2010)
PBA	MATLAB	Tewfik et al. (2006)
OPSM	Java (BiCat)	Barkow et al. (2006)
SAMBA	EXPANDER	Shamir et al. (2005)
xMotifs	Java (BiCat) R package (Biclust)	Barkow et al. (2006) Kaiser& Leisch (2008)
cMonkey	R package	Reiss et al. (2006)
Allegro		Halperin et al. (2009)
COALESCE	C++	Huttenhower et al. (2009)
FABIA/ FABIAS	R package, MATLAB	Hochreiter et al. (2010)

Adjusted Rand Index (ARI)

The robustness to noise of biclustering algorithms can be tested using the adjusted rand index (ARI). It is computed as follows:

$$ARI(Y,Z) = \frac{2(ux - vw)}{(u+v)(v+x)+(u+w)(w+x)} \quad (43)$$

In Equation 43, Y represents the domain knowledge, Z the clustering results, and u, v, w and x correspond respectively to the number of object pairs that belong to the same cluster in both Y and Z, belong to the same cluster in Y but not Z, belong to the same cluster in Z but not Y, and belong to different clusters in both Y and Z. ARI's values lies between 0 and 1, and larger value means higher similarity between the domain knowledge and the clustering results. If the experimental result is perfectly consistent to the domain knowledge, the index value will be 1. If a clustering is no more than a random choice, the index will be zero.

Other Validation Approaches

There are several other validation approaches that have been developed in the literature especially in the classical clustering context, and that could also be adjusted to test the robustness of biclustering algorithms. Such approaches include: Rand statistic, Jaccard coefficient, Folkes and Mallow (FM) index, Hubert Γ statistic, and normalized Γ statistic (Halkidi et al., 2001; Lubovac et al., 2001; Salazar et al., 2002).

BIOLOGICAL APPLICATIONS OF BICLUSTERING ALGORITHMS

Biological application of biclustering algorithms in the DNA microarray studies ranges from gene sample classification, genetic pathways identification, gene co-regulation study, transcriptional regulatory modules identification, biomarkers discovery, drug designed, and genetic interactions identification.

Gene Functional Annotation and Genetic Pathways

The basic idea behind biclustering of DNA microarray data is to identify subsets of genes that are co-expressed across subsets of experimental conditions. In functional genomics applications, the goal is to understand the functions of each of the genes operating in a biological system in a set of conditions or a time domain. The rationale is that genes with similar expression patterns are likely to be regulated by the same factors and therefore may share function. By collecting expression profiles from many different biological conditions and identifying joint patterns of gene expression among them, researchers have characterized transcriptional programs and assigned putative function to thousands of genes (Lazzeroni & Owen 2000; Tanay et al., 2004; Tchagang & Tewfik, 2006; Tewfik et al., 2006; Ihmels et al., 2002).

A gene may have multiple functions. Transcriptional programs are often based on combinatorial regulation. Biclustering is highly applicable in this domain. The function of genes could be inferred through "guilt by association" or appearance in the same bicluster. Subgroups of genes clustered together under the subset of experimental conditions could have related functions or be co-regulated (as demonstrated by other evidence such as common promoter regulatory sequences and experimental verification). More precisely, clustering of co-expressed genes, into biologically meaningful groups, helps to infer the biological role of an unknown gene that is co-expressed with a set of known genes. Furthermore, identification of such functionally related genes allow the association of the genes and their interconnections with known biological pathways.

Sample Classification, Biomarkers Discovery, and Drug Designed

Another common use of clustering analysis is the grouping of samples (arrays) by relatedness in expression profiles. The expression profile is effectively a complex phenotype and clustering analysis is used to identify samples with similar phenotypes. In medical research, this approach allows the discrimination between pathologies based on differential patterns of gene expression, rather than relying on traditional histological methods. In clinical applications, gene expression analysis is done on tissues taken from patients with a medical condition as compared with normal condition. Using such assays, biologists and medical doctors have identified molecular fingerprints that can help in the classification and diagnosis of the patient status and guide treatment protocols (Hochreiter et al. 2010; Caldas& Kaski 2010; Tchagang et al., 2008; Kluger et al., 2003; Murali & Kasif 2003; Getz et al., 2000).

In these studies, the focus is primarily on identifying profiles of expression over a subset of the genes that can be associated with clinical conditions and treatment outcomes, where ideally, the set of samples is equivalent only in some subtypes or the stages of the disease. However, a patient may be a part of more than one clinical group, *e.g.*, may suffer from syndrome *X*, have a genetic background *Y* and be exposed to environment *Z*. Biclustering analysis is thus highly appropriate for identifying and distinguishing the biological factors affecting the patients along with the corresponding gene subsets.

Gene Co-Regulation and Transcriptional Regulatory Modules

Transcriptional regulatory modules are sets of co-regulated genes, the conditions under which they are co-regulated, and their sequence-level regulatory motifs. It has been observed that much of a cell's regulatory response to changing environments occurs at the transcriptional level. In higher organisms for example, transcription factors, microRNAs, and epigenetic modifications can combine to form a complex regulatory network (Huttenhower et al., 2009). Furthermore, biological systems are inherently modular and grouping genes into modules dramatically reduces the effective complexity of any given dataset (Reiss et al., 2006; Ihmels et al., 2002).

However, if the grouping is done incorrectly, very little of the downstream analysis is likely to be correct. The problem is complicated by the fact that many genes are active in only subsets of cell states and environmental conditions, the noise in available data, and the complexity of the underlying regulatory system. Biclustering algorithms appear to be an important computational tool to tackle this problem (Huttenhower et al., 2009; Halperin et al., 2009; Reiss et al., 2006; Ihmels et al., 2002). Although taking advantage of modularity is key to success in learning biological networks from data, it is still a tough problem and an active area of research.

Genetic Interactions

Sometimes mutations in two genes produce a phenotype that is surprising in light of each mutation's individual effects. This phenomenon, which defines genetic interaction, can reveal functional relationships between genes and pathways. For example, double mutants with surprisingly slow growth define synergistic interactions that can identify compensatory pathways or protein complexes (Boone et al., 2007). Studying genetic interactions can reveal gene function, the nature of the mutations, functional redundancy, and protein interactions. Biclustering algorithms appear to be a powerful tool to tackle this problem.

CONCLUSION, DISCUSSION AND FUTURE RESEARCH DIRECTIONS

We have presented a comprehensive survey of the models, methods, evaluations, and applications developed in the field of biclustering algorithms. From the list of models and approaches it is our opinion that the scientific community has a significant number of models and algorithms to choose from. The different types of biclusters described above are rich enough to appropriately model very complex interactive processes. Many issues in biclustering algorithm design also remain open and should be addressed by the scientific community. From these open issues, we have selected the algorithmic approach and the analysis of the statistical significance of biclusters, the biological evaluation of biclusters and biological data integration, and the development of visualization tools that can be used by biologists to explore the biclusters, their biological significance, and their effective integration and utilization to solve the underlying biological questions, as some of the most important ones.

Algorithmic Approach and Statistical Evaluation

Regardless of the concept behind which the biclustering problem is formulated and given the complexity of the biclustering problem, NP-hard, a number of different heuristic approaches have been used to find biclusters from a given gene expression data (Madeira & Oliveira, 2004). These heuristic approaches include: iterative row and column, divide and conquer, greedy iterative search, exhaustive bicluster enumeration, and distribution parameter identification. Each one of these approaches has its own advantages and disadvantages.

The iterative row and column approach apply clustering algorithms to rows and columns of the data matrix separately (*e.g.* couple two way clustering algorithm of Getz et al. (2000)). Then using an iterative procedure the results of the two cluster arrangements are combined to obtained biclusters. A divide and conquer algorithm works by recursively breaking down a problem into two or more sub-problems of the same (or related) type until these become simple enough to be solved directly (*e.g.* block clustering algorithm, Hartigan (1972)). The solutions to the sub-problems are then combined to give a solution to the original problem. The significant advantage of divide-and-conquer algorithms is that they are potentially very fast. A very significant drawback of this approach is that it is very likely to miss good biclusters that may be split before they can be identified. Greedy iterative search methods choose a solution which seems to be optimal locally (*e.g.* δ-bicluster of Cheng & Church (2000)). It is then considered that this gives a globally optimal solution. As far as the weakness of this approach is concerned it may make wrong decisions to consider due to its greedy nature and may lose good biclusters. Exhaustive bicluster enumeration works by assuming restrictions on the size of the biclusters that should be listed, a number of methods have been used to speed up exhaustive search (*e.g.* RoBA of Tchagang & Tewfik (2006)). These algorithms certainly find the best biclusters, if they exist, but have a very serious drawback. They can only be executed by assuming restrictions on the size of the biclusters due to their high complexity. Finally, in the distribution parameter identification approach, biclusters are generated using a given statistical model and the distribution parameters that fit, in the best way, the available data, by minimizing certain criteria (*e.g.* plaid model of Lazzeroni & Owen (2000)).

Even though existing biclustering approaches have weaknesses, they still produce acceptable results. As research continues, new ideas come into light, and new methodologies and strategies will be developed, and the existing ones can be improved. The basic idea behind these newly proposed methods is either to propose a totally novel way for solving a particular problem or based on some previous method, proposing a better way instead of the existing approach. Many new ideas can be combined and adapted to develop new algorithms that are potentially more effective in particular applications. The possibility to use different biclustering algorithms inside a single graphical tool such as BicAT (Barkow et al., 2006) and biclust (Kaiser & Leisch, 2008) allows the user to compare biclustering results and choose the algorithm that best fits to specific biological scenario. The extraction of a large number of biclusters in real data may lead to results that are difficult to interpret. Hence of, the development and/or integration of statistical tools and biological means into biclustering algorithm to tackle this issue will certainly be a direction to follow.

Biological Evaluation of Biclusters and Biological Data Integration

Biological evaluation and interpretation of gene groups, and biological data integration into biclustering algorithms will certainly be a direction to follow. It has been observed that co-regulated genes are often functionally (physically, spatially, genetically, and/or evolutionarily) linked (Reiss et al., 2006 and references therein). For example, genes whose products form a protein complex are likely to be co-regulated (Huttenhower et al., 2009 and references therein). Other types of associations among genes, or their protein products, that (can) imply functional associations include: common cis-regulatory motifs; same metabolic pathway (s); cis-binding by common regulator(s); physical interaction; common ontology; evolutionary conservation; common synthetic phenotypes; subcellular co-location; and proximity in the genome. These associations can be inferred experimentally or computationally, and used as priors in subsequent analysis. Indeed it is common practice as we saw above (Table 1) to use one or more of these associations as a measure of the biological quality of a gene group.

However, each of these tools (Table 1) has its limitations (Khatri & Draghici, 2005). Among these limitations has been the lack of appropriate corrections for multiple hypothesis testing and the failure to consider which genes were actually assayed in the experiment that produced the list of interesting genes (Berriz et al., 2003). Several of these pioneering functional evaluation tools have mainly been implemented using only the GO annotations (Berriz et al., 2003; Beißbarth, & Speed, 2004). Some of these tools, such as BINGO (Maere et al., 2005) and CLENCH (Shah & Fedoroff, 2004), are available as a part of some broader software packages or dedicated to specific applications, and do not offer flexibility in terms of deployment and biological applications. BINGO only offers GO evaluation and it is designed as a plug-in to Cytoscape (Shannon et al., 2003). CLENCH is specifically designed to allow Arabidopsis thaliana researchers to perform automated retrieval of GO annotations from TAIR database (Swarbreck et al., 2008). DAVID (Dennis et al., 2003; Huang et al., 2009), FATIGO (Al-Shahrour et al., 2003), and GOAL (Tchagang et al., 2010) offer the use of more advanced functions, such as protein-protein interactions and/or KEGG pathways, TF-gene interactions, than the others.

Although the use of the above described experimentally and/or computationally derived biological associations among genes have been successful in validating several groups of coherent behavior genes, it is important to note that some of these associations, to varying degrees, can contain a high rate of false positives, or may imply relationships that have no implication for co-regulation. Therefore in their consideration as evidence for co-regulation, these different sources of evidence should be treated as priors, with appropriately different weights, based upon prior knowledge (or assumptions) of their quality and/or relevance.

Biclusters Visualization

Biclustering algorithms are widely used to analyze gene expression data because of their advantage of extracting local behaviors and biological processes from the dataset analyzed. Unfortunately, biclustering algorithm usually output many biclusters of genes making their post-processing step a very daunting task. Because of the metadata curse, most researchers usually restrict themselves to the analysis of few selected biclusters. These selections are most of the time very subjective and may miss important information. Thus, another interesting research area is to develop visualization tools to explore the biclusters, their biological significance, and their effective integration and utilization to solve the underlying biological problems.

There have already been some pioneering works in this direction: BiVisu of Cheng et al. (2007) and BicOverlapper of Santamaría et al., (2008a, 2008b). BiVisu uses a parallel coordinates approach for biclusters visualization whereas BicOverlapper represents biclusters using a bubble. More precisely, BicOverlapper represents biclusters in two dimensions, the position of biclusters in the graph is a two dimensional representation of the row and column combination in the cluster, the size of the bubble corresponds to the size of the bicluster. Hence, a bicluster containing another bicluster is drawn as a big bubble around a smaller one. Although these pioneering works have been useful, more still has to be done.

ADDITIONAL READING

Biclustering algorithms and applications presented in this chapter represent only a small fraction (*i.e.* DNA microarray data analysis only) of the potential applications of this technique. Several other domain of application in data mining, database research, text mining, market research, and collaborative filtering are worth exploring by the interested readers. For example, biclustering algorithms have been used for: analysis of nutritional data and currency exchange (Lazzeroni et al., 2000), text mining (Dhillon, 2001), market research (Gaul & Schader, 1996), dimensionality reduction and association rules in databases (Agrawal et al., 1998), electoral data analysis (Hartigan, 1972), and analysis of EEG data from epileptic patients treated with vagus nerve stimulation (Busygin et al., 2007). Good reviews on frequent itemsets mining and sequential pattern mining can be found in (Goethals *et al.*, 2003; Agrawal & Srikant 1994).

REFERENCES

Agrawal, R., Imielinski, T., & Swami, A. (1993). *Mining association rules between sets of items in large databases*. SIGMOD.

Agrawal, R., & Srikant, R. (1994). Fast algorithms for mining association rules. In. *Proceedings of VLDB*, *94*, 487–499.

Al-Shahrour, F., Minguez, P., Tárraga, J., Montaner, D., Alloza, E., & Vaquerizas, J. M. M. (2003). BABELOMICS: A systems biology perspective in the functional annotation of genome-scale experiments. *Nucleic Acids Research*, *34*, W472–W476. doi:10.1093/nar/gkl172

Androulakis, I. P., Yang, E., & Almon, R. R. (2007). Analysis of time-series gene expression data: Methods, challenges, and opportunities. *Annual Review of Biomedical Engineering*, *9*, 205–228. doi:10.1146/annurev.bioeng.9.060906.151904

Arbeitman, M. N., Furlong, E. E., Imam, F., Johnson, E., Null, B. H., & Baker, B. S. (2002). Gene expression during the life cycle of Drosophila melanogaster. *Science*, *297*(5590), 2270–2275. doi:10.1126/science.1072152

Barkow, S., Bleuer, S., Prelic, A., Zimmermann, P., & Zitzler, E. (2006). BicAT: A biclustering analysis toolbox. *Bioinformatics (Oxford, England)*, *22*(10), 1282–1283. doi:10.1093/bioinformatics/btl099

Beißbarth, T., & Speed, T. P. (2004). GOstat: Find statistically overrepresented gene ontologies within a group of genes. *Bioinformatics Applications Note*, *20*(9), 1464–1465.

Ben-Dor, A., Chor, B., Karp, R., & Yakhini, Z. (2003). Discovering local structure in gene expression data: The order-preserving submatrix problem. *Journal of Computational Biology*, *10*(3-4), 373–384. doi:10.1089/10665270360688075

Bergmann, S., Ihmels, J., & Barkai, N. (2003). Iterative signature algorithm for the analysis of largescale gene expression data. *Physical Review E: Statistical, Nonlinear, and Soft Matter Physics*, *67*(3 Pt 1), 03190201–03190218.

Berriz, G. F., King, O. D., Bryant, B., Sander, C., & Roth, F. P. (2003). Characterizing gene sets with FuncAssociate. *Bioinformatics Applications Note*, *19*(18), 2502–2504.

Boone, C., Howard Bussey, H., & Andrews, B. J. (2007). Exploring genetic interactions and networks with yeast. *Nature Reviews. Genetics*, *8*, 437–449. doi:10.1038/nrg2085

Buck, M. J., & Lieb, J. D. (2004). ChIP-chip: Considerations for the design, analysis, and application of genome-wide chromatin immunoprecipitation experiments. *Genomics*, *83*, 349–360. doi:10.1016/j.ygeno.2003.11.004

Busygin, S., Boyko, N., Pardalos, P. M., Bewernitz, M., & Ghacibeh, G. (2007). Biclustering EEG data from epileptic patients treated with vagus nerve stimulation. *AIP Conference Proceedings*, *953*, 220. doi:10.1063/1.2817345

Caldas, J., & Kaski, S. (2008). Bayesian biclustering with the plaid model. *In Proceedings of the IEEE International Workshop on Machine Learning for Signal Processing*, 291-296.

Caldas, J., & Kaski, S. (2010). Generative tree biclustering for information retrieval and microRNA biomarker discovery. *In Proceedings of RECOMB 2010*, April 25-28, Lisbon, Portugal.

Carbon, S., Ireland, A., Mungall, C. J., Shu, S., Marshall, B., & Lewis, S. (2009). AmiGO hub, Web presence working group. AmiGO: Online access to ontology and annotation data. *Bioinformatics (Oxford, England)*, *25*(2), 288–289. doi:10.1093/bioinformatics/btn615

Cheng, K. O., Law, N. F., Siu, W. C., & Lau, T. H. (2007). Bivisu: Software tool for bicluster detection and visualization. *Bioinformatics (Oxford, England)*, *23*, 2342–2344. doi:10.1093/bioinformatics/btm338

Cheng, Y., & Church, G. M. (2000). Biclustering of expression data. *Proceedings of the International Conference on Intelligent Systems for Molecular Biology*, *8*, 93–103.

Christinat, Y., Wachmann, B., & Zhang, L. (2008). Gene expression data analysis using a novel approach to biclustering combining discrete and continuous data. *IEEE/ACM Transactions on Computational Biology and Bioinformatics*, *5*(4), 583–593. doi:10.1109/TCBB.2007.70251

Coulibaly, I., & Page, G. P. (2008). Bioinformatics tools for inferring functional information from plant microarray data II: Analysis beyond single gene. *International Journal of Plant Genomics*, (893941): 13.

Creighton, C., & Hanash, S. (2003). Mining gene expression databases for association rules. *Bioinformatics (Oxford, England)*, *19*(1), 79–86. doi:10.1093/bioinformatics/19.1.79

Csardi, G., Kutalik, Z., & Bergmann, S. (2010). Modular analysis of gene expression data with R. *Bioinformatics Applications Note*, *26*(10), 1376–1377.

Day-Richter, J., Harris, M. A., & Haendel, M. (2007). Gene ontology OBO-edit working group, OBO-edit-an ontology editor for biologists. *Bioinformatics (Oxford, England)*, *23*(16), 2198–2200. doi:10.1093/bioinformatics/btm112

Dennis, G. Jr, Sherman, B. T., Hosack, D. A., Yang, J., Gao, W., & Lane, H. C. (2003). DAVID: Database for Annotation, Visualization, and Integrated Discovery. *Genome Biology*, *4*(5), 3. doi:10.1186/gb-2003-4-5-p3

DeRisi, J., Penland, L., Brown, P. O., Bittner, M. L., Meltzer, P. S., & Ray, M. (1996). Use of a cdna microarray to analyse gene expression patterns in human cancer. *Nature Genetics*, *14*(4), 457–460. doi:10.1038/ng1296-457

DeRisi, J. L., Iyer, V. R., & Brown, P. O. (1997). Exploring the metabolic and genetic control of gene expression on a genomic scale. *Science*, *278*(5338), 680–686. doi:10.1126/science.278.5338.680

Dhillon, I. S. (2001). Co-clustering documents and words using bipartite spectral graph partitioning. *In Proceedings of the 7th ACM SIGKDD International Conference on Knowledge Discovery and Data Mining*, 269–274.

Divina, F., & Aguilar-Ruiz, J. S. (2006). Biclustering of expression data with evolutionary computation. *IEEE Transactions on Knowledge and Data Engineering*, *18*, 590–602. doi:10.1109/TKDE.2006.74

Draghici, S., Khatri, P., Bhavsar, P., Shah, A., Krawetz, S. A., & Tainsky, M. A. (2003, July). Onto-Tools, the toolkit of the modern biologist: Onto-Express, Onto-Compare, Onto-Design and Onto-Translate. *Nucleic Acids Research*, *31*(13), 3775–3781. doi:10.1093/nar/gkg624

Eisen, M. B., Spellman, P. T., Brown, P. O., & Botstein, D. (1998). Cluster analysis and display of genome-wide expression patterns. *Proceedings of the National Academy of Sciences of the United States of America*, *95*(25), 14863–14868. doi:10.1073/pnas.95.25.14863

Gardner, T. S., & Faith, J. J. (2005). Reverse-engineering transcription control networks. *Physics of Life Reviews*, *2*, 65–88. doi:10.1016/j.plrev.2005.01.001

Gasch, A. P., Spellman, P. T., Kao, C. M., Carmel-Harel, O., Eisen, M. B., & Storz, G. (2000). Genomic expression programs in the response of yeast cells to environmental changes. *Molecular Biology of the Cell*, *11*(12), 4241–4257.

Gaul, W., & Schader, M. (1996). A new algorithm for two-mode clustering. In Bock, H.-H., & Polasek, W. (Eds.), *Data analysis and Information Systems* (pp. 15–23). Heidelberg: Springer.

Gene Ontology. (2010). *Home page*. Retrieved from http://www.geneontology.org

Gentleman, R. C., Garey, V. J., Huber, W., Irizarry, R., & Dudoit, S. (2005). *Bioinformatics and computational biology solutions using R and Bioconductor*. New York: Springer. doi:10.1007/0-387-29362-0

Getz, G., Levine, E., & Domany, E. (2000). Coupled two-way clustering analysis of gene microarray data. *Proceedings of the National Academy of Sciences of the United States of America*, *97*(22), 12079–12084. doi:10.1073/pnas.210134797

Goethals, B., & Zaki, M. (2003). Advances in frequent itemset mining implementations: Report on FIMI'03. *SIGKDD Explorations*, *6*(1), 109–117. doi:10.1145/1007730.1007744

GOSt, [http://biit.cs.ut.ee/gprofiler/]

Griffith, O. L., Gao, B., Bilenky, M., Prychyna, Y., Ester, M., & Jones, S. (2009). KiWi: A scalable subspace clustering algorithm for gene expression analysis. In *Proceedings of the 3rd International Conference on Bioinformatics and Biomedical Engineering,* June 11–13, Beijing, China.

Gu, J., & Liu, J. S. (2008). Bayesian biclustering of gene expression data. *BMC Genomics*, *9*(Suppl 1), S4. doi:10.1186/1471-2164-9-S1-S4

Guillemin, K., Salama, N. R., Tompkins, L. S., & Falkow, S. (2002). Cag pathogenicity island-specific responses of gastric epithelial cells to Helicobacter pylori infection. *Proceedings of the National Academy of Sciences of the United States of America*, *99*(23), 15136–15141. doi:10.1073/pnas.182558799

Gupta, N., & Aggarwal, S. (2008). SISA: Seeded Iterative Signature Algorithm for biclustering gene expression data. *IADIS, European Conference on Data Mining.*

Halkidi, M., Batistakis, Y., & Vazirgiannis, M. (2001). On clustering validation techniques. *Intelligent Information Systems Journal*, *17*(2-3), 107–145. doi:10.1023/A:1012801612483

Halperin, Y., Linhart, C., Ulitsky, I., & Ron Shamir, R. (2009). Allegro: Analyzing expression and sequence in concert to discover regulatory programs. *Nucleic Acids Research*, *37*(5), 1566–1579. doi:10.1093/nar/gkn1064

Hartigan, J. A. (1972). Direct clustering of a data matrix. [JASA]. *Journal of the American Statistical Association*, *67*(337), 123–129. doi:10.2307/2284710

Hochreiter, S., Bodenhofer, U., Heusel, M., & Mayr, A. (2010). *FABIA: Factor Analysis for Bicluster Acquisition*. Bioinformatics Advance Access.

Huang, D. W., Sherman, B. T., & Lempicki, R. A. (2009). Systematic and integrative analysis of large gene lists using DAVID Bioinformatics Resources. *Nature Protocols*, *4*(1), 44–57. doi:10.1038/nprot.2008.211

Hussain, A., & Abdullah, A. (2006). A new biclustering technique based on crossing minimization. *Neurocomputing Journal*, *69*, 1882–1896. doi:10.1016/j.neucom.2006.02.018

Huttenhower, C., Mutungu, T., Indik, N., Yang, W., Schroeder, M., & Forman, J. J. (2009). Detailing regulatory networks through large scale data integration. *Bioinformtics*, *25*(24), 3267–3274. doi:10.1093/bioinformatics/btp588

Ibrahim, M., Noman, N., & Iba, H. (2009). *Genome Informatics*, December 14-16, Yokohama Pacifico, Japan.

Ideker, T. (2002). Discovering regulatory and signaling circuits in molecular interaction networks. *Bioinformatics (Oxford, England)*, *18*(Suppl. 1.), S233–S240.

Ihmels, J., Friedlander, G., Bergmann, S., Sarig, O., Ziv, Y., & Barkai, N. (2002). Revealing modular organization in the yeast transcriptional network. *Nature Genetics*, *31*, 370–377.

Irizarry, R. A., Warren, D., Spencer, F., Kim, I. F., & Biswal, S. (2005). Multiple-laboratory comparison of microarray platforms. *Nature Methods*, *2*, 345–350. doi:10.1038/nmeth756

Kaiser, S., & Leisch, F. (2008). Biclust-a toolbox for bicluster analysis in R. *In Proceedings of Computational Statistics*.

Kerr, M. K., & Churchill, G. A. (2001). Experimental design for gene expression microarrays. *Biostatistics (Oxford, England)*, *2*, 183–201. doi:10.1093/biostatistics/2.2.183

Khatri, P., & Draghici, S. (2005). Ontology analysis of gene expression data: Current tools, limitations, and open problems. *Bioinformatics (Oxford, England)*, *21*, 3587–3595. doi:10.1093/bioinformatics/bti565

Kloster, M., Tang, C., & Wingreen, N. S. (2005). Finding regulatory modules through large-scale gene-expression data analysis. *Bioinformatics (Oxford, England)*, *21*, 1172–1179. doi:10.1093/bioinformatics/bti096

Kluger, Y., Barsi, R., Cheng, J. T., & Gerstein, M. (2003). Spectral biclustering of microarray data: Coclustering genes and conditions. *Genome Research*, *13*(4), 703–716. doi:10.1101/gr.648603

Lazzeroni, L., & Owen, A. (2000). Plaid models for gene expression data. *Statistica Sinica*, *12*, 61–86.

Lee, T. I., Rinaldi, N. J., Robert, F., Odom, D. T., Bar-Joseph, Z., & Gerber, G. K. (2002). Transcriptional regulatory networks in Saccharomyces cerevisiae. *Science*, *298*, 799–804. doi:10.1126/science.1075090

Liu, J., Yang, J., & Wang, W. (2004). Biclustering in gene expression data by tendency. *IEEE Computational Systems Bioinformatics Conference Proceedings, 182*(193), 16-19.

Liu, X., & Wang, L. (2007). Computing the maximum similarity bi-clusters of gene expression data. *Bioinformatics (Oxford, England)*, *23*(1), 50–56. doi:10.1093/bioinformatics/btl560

Lubovac, Z., Olsson, B., Jonsson, P., Laurio, K., & Anderson, M. L. (2001). Biological and statistical evaluation of clusterings of gene expression profiles. In C.E. D'Attellis, V.V. Kluev & N.E. Mastorakis, (Eds.), *Proc. Mathematics and Computers in Biology and Chemistry (MCBC '01)*, (pp. 149–155). Skiathos Island, Greece, September.

Madeira, S. C., & Oliveira, A. L. (2004). Biclustering algorithms for biological data analysis: A survey. *IEEE/ACM Transactions on Computational Biology and Bioinformatics*, *1*(1), 24–45. doi:10.1109/TCBB.2004.2

Maere, S., Heymans, K., & Kuiper, M. (2005). BiNGO: A cytoscape plugin to assess overrepresentation of gene ontology categories in biological networks. *Bioinformatics (Oxford, England)*, *21*, 3448–3449. doi:10.1093/bioinformatics/bti551

Mahfouz, M.A. & Ismail, M.A. (2009). BIDENS: Iterative density based biclustering algorithm with application to gene expression analysis. *Proceedings of World Academy of Science, Engineering and Technology, 37*(2070-3740), 342–348.

Martin, D., Brun, C., Remy, E., Mouren, P., Thieffry, D., & Bernard Jacq, B. (2004). GOToolBox: Functional analysis of gene datasets based on gene ontology. *Genome Biology, 5*(12). doi:10.1186/gb-2004-5-12-r101

Matys, V. (2006). TRANSFAC® and its module TRANSCompel®: Transcriptional gene regulation in eukaryotes. *Nucleic Acids Research, 34*, D108–D110. doi:10.1093/nar/gkj143

Mitra, S., & Banka, H. (2006). Multi-objective evolutionary biclustering of gene expression data. *Pattern Recognition, 39*(12), 2464–2477. doi:10.1016/j.patcog.2006.03.003

Munch, R. (2003). PRODORIC: Prokaryotic database of gene regulation. *Nucleic Acids Research, 31*, 266–269. doi:10.1093/nar/gkg037

Murali, T. M., & Kasif, S. (2003). Extracting conserved gene expression motifs from gene expression data. *In Proceedings of the 8th Pacific Symposium on Biocomputing, 8*, 77-88.

Okada, Y., Fujibuchi, W., & Horton, P. (2007). Module discovery in gene expression data using closed itemset mining algorithm. *IPSG Transactions in Bioinformatics, 48*, 39–48.

Prelic, A., Bleuler, S., Zimmermann, P., Wille, A., Buhlmann, P., & Gruissem, W. (2006). A systematic comparison and evaluation of biclustering methods for gene expression data. *Bioinformatics (Oxford, England), 22*, 1122–1129. doi:10.1093/bioinformatics/btl060

Quackenbush, J. (2002). Microarray data normalization and transformation. *Nature Genetics, 32*, 496–501. doi:10.1038/ng1032

Reiss, D. J., Baliga, S. N., & Bonneau, R. (2006). Integrated biclustering of heterogeneous genome-wide datasets. *BMC Bioinformatics, 7*(1), 280. doi:10.1186/1471-2105-7-280

Salazar, E. J., Veléz, A. C., Parra, C. M., & Ortega, O. (2002). A cluster validity index for comparing non-hierarchical clustering methods. In *Memorias Encuentro de Investigaci 'on sobre Tecnologias de Informacion Aplicadas a la Soluci 'on de Problemas* (EITI2002), Medell'ın, Colombia, 2002.

Salgado, H. (2006). RegulonDB (version 5.0): Escherichia coli K-12 transcriptional regulatory network, operon organization, and growth conditions. *Nucleic Acids Research, 34*, D394–D397. doi:10.1093/nar/gkj156

Sandelin, A., Alkema Engström, W. P., Wasserman, W. W., & Lenhard, B. (2004). JASPAR: An open-access database for eukaryotic transcription factor binding profiles. *Nucleic Acids Research, 32*, D91–D94. doi:10.1093/nar/gkh012

Santamaría, R., Therón, R., & Quintales, L. (2008a). A visual analytics approach for understanding biclustering results from microarray data. *BMC Bioinformatics, 9*, 247. doi:10.1186/1471-2105-9-247

Santamaría, R., Therón, R., & Quintales, L. (2008b). A tool for bicluster visualization. *Bioinformatics (Oxford, England), 24*, 1212–1213. doi:10.1093/bioinformatics/btn076

Schena, M., Shalon, D., Davis, R. W., & Brown, P. O. (1995). Quantitative monitoring of gene expression patterns with a complementary DNA microarray. *Science, 270*(5235), 467–740. doi:10.1126/science.270.5235.467

Shah, N. H., & Fedoroff, N. V. (2004). CLENCH: A program for calculating Cluster ENriCHment using the gene ontology. *Bioinformatics (Oxford, England), 20*(7), 1196–1197. doi:10.1093/bioinformatics/bth056

Shannon, P., Markiel, A., & Ozier, O. (2003). Cytoscape: A software environment for integrated models of biomolecular interaction networks. *Genome Research*, *13*, 2498–2504. doi:10.1101/gr.1239303

Shujiro, O., Takuji, Y., Masami, H., Masumi, I., Toshiaki, K., & Peer, B. (2008). KEGG atlas mapping for global analysis of metabolic pathways. *Nucleic Acids Research*, *36*(2), W423.

Simon, R. M., McShane, L. M., Korn, E. L., & Radmacher, M. D. (2003). *Design and analysis of DNA microarray investigations*. New York: Springer.

Spellman, P. T., Sherlock, G., Zhang, M. Q., Iyer, V. R., Anders, K., & Eisenm, M. B. (1998). Comprehensive identification of cell cycle-regulated genes of the yeast Saccharomyces cerevisiae by microarray hybridization. *Molecular Biology of the Cell*, *9*(12), 3273–3297.

Supper, J., Strauch, M., Wanke, D., Harter, K., & Zell, A. (2007). EDISA: Extracting biclusters from multiple time-series of gene expression profiles. *BMC Bioinformatics*, *8*, 334. doi:10.1186/1471-2105-8-334

Swarbreck, D., Wilks, C., & Lamesch, P. (2008). The Arabidopsis Information Resource (TAIR): Gene structure and function annotation. *Nucleic Acids Research*, *36*, D1009–D1014. doi:10.1093/nar/gkm965

Tanay, A., Sharan, R., & Shamir, R. (2002). Discovering statistically significant biclusters in gene expression data. *Bioinformatics (Oxford, England)*, *18*, S136–S144.

Tanay, A., Sharan, R., & Shamir, R. (2006). Biclustering algorithms: A survey. In Aluru, S. (Ed.), *Handbook of computational molecular biology* (pp. 26-1–26-17). Chapman and Hall/CRC Press.

Tchagang, A. B., Bui, K. V., McGinnis, T., & Benos, P. V. (2009). Extracting biologically significant patterns from short time series gene expression data. *BMC Bioinformatics*, *10*, 255. doi:10.1186/1471-2105-10-255

Tchagang, A. B., Gawronski, A., Bérubé, H., Phan, S., Famili, F., & Pan, Y. (2010). GOAL: A software tool for assessing biological significance of genes group. *BMC Bioinformatics*, *11*, 229. doi:10.1186/1471-2105-11-229

Tchagang, A. B., & Tewfik, A. H. (2006). DNA microarray data analysis: A novel biclustering algorithm approach. *EURASIP Journal on Applied Signal Processing*, *59809*, 12.

Tchagang, A. B., Tewfik, A. H., & Benos, P. V. (2008). Biological evaluation of biclustering algorithms using gene ontology and ChIP-chip data. *In Proceedings of IEEE, International Conference on Acoustics, Speech and Signal Processing*, Las Vegas, Nevada.

Tchagang, A. B., Tewfik, A. H., Skubitz, K. M., DeRycke, M. S., & Skubitz, A. P. N. (2008). Early detection of ovarian cancer using group biomarkers. *Molecular Cancer Therapeutics*, *7*(1), 27–37. doi:10.1158/1535-7163.MCT-07-0565

Teixeira, M. C. (2006). The YEASTRACT database: A tool for the analysis of transcription regulatory associations in Saccharomyces cerevisiae. *Nucleic Acids Research*, *34*, D446–D451. doi:10.1093/nar/gkj013

Teng, L. & Chan, L. (2007). *Order preserving clustering by finding frequent orders in gene expression data*. (LNCS 4774).

Tewfik, A. H., Tchagang, A. B., & Vertatschitsch, L. (2006). Parallel identification of gene biclusters with coherent evolutions. *IEEE Transactions on Signal Processing*, *54*, 2408–2417. doi:10.1109/TSP.2006.873720

The Gene Ontology Consortium. (2000). Gene ontology: Tool for the unification of biology. *Nature Genetics*, *25*(1), 25–29. doi:10.1038/75556

Turner, H. L., Bailey, T. C., Krzanowski, W. J., & Hemingway, C. A. (2005). Biclustering models for structured microarray data. *IEEE/ACM Transactions on Computational Biology and Bioinformatics*, *2*(4), 316–329. doi:10.1109/TCBB.2005.49

Wang, H., Wang, W., Yang, J., & Yu, P. S. (2002). Clustering by pattern similarity in large data sets. *Proceedings of 2002 ACM SIGMOD International Conference on the Management of Data*, (pp. 394-405).

Xu, R., & Wunsch, D. (2005). Survey of clustering algorithms. *IEEE Transactions on Neural Networks*, *16*(3), 645–678. doi:10.1109/TNN.2005.845141

Yang, J., Wang, W., Wang, H., & Yu, P. S. (2002). δ-clusters: Capturing subspace correlation in a large data set. *In ICDE*, 517-528.

Yang, J., Wang, W., Wang, H., & Yu, P. S. (2003). Enhanced biclustering on expression data. *Proceedings of the Third IEEE Conference on Bioinformatics and Bioengineering*, 321-327.

Yang, Y. H., & Speed, T. (2002). Design issues for cDNA microarray experiments. *Nature Reviews. Genetics*, *3*, 579–588.

Yeung, K. Y., & Ruzzo, L. W. (2001). Principal component analysis for clustering gene expression data. *Bioinformatics (Oxford, England)*, *17*(9), 763–774. doi:10.1093/bioinformatics/17.9.763

Zhang, J., Wang, J. J., & Yan, H. (2008). A neural-network approach for biclustering of gene expression data based on the plaid model. *International Conference on Machine Learning and Cybernetics, 2*(2008), 1082-1087.

Zhong, S., Storch, F., Lipan, O., Kao, M. J., Weitz, C., & Wong, W. H. (2004). GoSurfer: A graphical interactive tool for comparative analysis of large gene sets in gene ontology space. *Applied Bioinformatics*, *3*(4), 1–5.

Zhu, J., & Zhang, M. Q. (1999). SCPD: A promoter database of the yeast Saccharomyces cerevisiae. *Bioinformatics (Oxford, England)*, 607–611. doi:10.1093/bioinformatics/15.7.607

Zien, A. (2000). Analysis of gene expression data with pathway scores. *Proceedings of the International Conference on Intelligent Systems for Molecular Biology*, *8*, 407–417.

KEY TERMS AND DEFINITIONS

Biclustering: Class of clustering algorithms that perform clustering on the rows and columns of a data matrix at the same time. It is also known as co-clustering, two-dimensional clustering, or two-way clustering.

Bicluster: A subset of rows in a data matrix that exhibit similar behavior across a subset of columns.

DNA Microarray: A small solid support, usually a membrane or glass slide, on which sequences of DNA are fixed in an orderly manner. DNA microarrays are used to measure the expression of many genes simultaneously. It is also called *DNA chip*.

Gene Expression: It is the process by which information from a gene is used in the synthesis of a functional gene product. It can also be viewed as the level of activity of a gene or group of genes.

Tissue Samples: A part of an organism used for biological, medical or research testing, for example, a small piece of a tumor.

Chapter 8
Prediction of Epigenetic Target Sites by Using Genomic DNA Sequence

Guo-Cheng Yuan
Harvard School of Public Health, USA & Dana-Farber Cancer Institute, USA

ABSTRACT

Epigenetic regulation provides an extra layer of gene control in addition to the genomic sequence and is critical for the maintenance of cell-type specific gene expression programs. Significant changes of epigenetic patterns have been linked to developmental stages, environmental exposure, ageing, and diet. However, the regulatory mechanisms for epigenetic recruitment, maintenance, and switch are still poorly understood. Computational biology provides tools to deeply uncover hidden connections and these tools have played a major role in shaping the current understanding of gene regulation, but its application in epigenetics is still in the infancy. This chapter reviews some recent developments of computational approaches to predict epigenetic target sites.

INTRODUCTION

Epigenetics refers to heritable changes of gene expression or genotypes without change of the DNA sequence (Waddington, 1942). In a multicellular organism, the DNA sequence is constant in all cell lineages, but the gene activities in different cell-types are highly variable. Such cell-type specific gene expression patterns are controlled by epigenetic mechanisms, including nucleosome positioning, histone modifications, and DNA methylation. Together these mechanisms control the accessibility of the genomic DNA to regulatory proteins. Only a small part of the genomic blueprint is used in any cell type.

DOI: 10.4018/978-1-60960-491-2.ch008

The rapid advance of microarray and DNA sequencing technologies (Barski et al., 2007; Mikkelsen et al., 2007; Ren et al., 2000) has allowed researchers to identify genome-wide epigenetic patterns in various species. Recent epigenomic studies have identified dramatic epigenetic differences between different cell-types (Barski et al., 2007; Heintzman et al., 2009; Meissner et al., 2008; Mikkelsen et al., 2007; Mohn et al., 2008), between normal and disease tissues (Schlesinger et al., 2007; Seligson et al., 2005; TCGA, 2008), and between stimulated and resting cells (Saccani & Natoli, 2002; Wei et al., 2009). These differences are also highly correlated with gene expression level changes. Importantly, the epigenetic changes are not permanent but can be reversed. Strikingly, the entire epigenetic state in an adult cell can be reprogrammed to a pluripotent cell state (called iPS cell) that is highly similar to an embryonic stem (ES) cell (Okita et al., 2007; Takahashi & Yamanaka, 2006; Wernig et al., 2007; Yu et al., 2007). This reversibility makes epigenetic marks the ideal targets for therapeutic treatment (Sharma et al., 2010). Indeed, a number of drugs have been developed and currently used to treat a number of diseases including cancer (Yoo & Jones, 2006). However, a major challenge is to avoid off-target interactions, since our current understanding of the targeting mechanisms of epigenetic factors is still limited.

The epigenetic pattern is not randomly distributed across the genome (Bernstein et al., 2007). A fundamental question is how target specificity is achieved. The targeting mechanism is complex and involves many factors such as the genomic DNA sequence, chromatin modifiers, transcription factors (TFs), and non-coding RNAs. How these factors work together to regulate epigenetic patterns is still poorly understood. Among these factors, the association with DNA sequence has been most studied. Perhaps the most commonly studied factor is the DNA sequence. Here I review some of computational studies aimed at prediction of epigenetic patterns based on the DNA sequence. Additional information on certain specific aspects can be found in some excellent reviews (Bock & Lengauer, 2008; Kaplan et al. 2010). In addition, the readers are referred to some excellent reviews for biological background (Jiang & Pugh, 2009; Kouzarides, 2007; Rando & Chang, 2009; Hawkins et al. 2010; Zhou et al. 2011).

METHODS TO PREDICT EPIGENETIC TARGETS

Nucleosome Positioning

The eukaryotic DNA is packaged into chromatin. The fundamental unit of chromatin is the nucleosome, consisting of two copies each of four core histone proteins: H2A, H2B, H3, and H4 (Kornberg & Lorch, 1999). Each nucleosome wraps around 146 bp DNA in about 2 turns. Ever since the initial discovery of nucleosomes in the 70's (Kornberg, 1974), the regulatory mechanisms underlying nucleosome positioning have been intensely investigated. A potentially important role of DNA sequences was noticed decades ago. The investigators recognized that the structural properties of DNA are dependent upon the base pair composition; therefore specific DNA sequences might be favored for nucleosome binding. By extracting nucleosome core particles from chicken red blood cells and then analyzing the DNA sequences attached to these nucleosome particles, Satchwell et al. (1986) observed an approximately 10 bp periodic pattern of the frequency of the dinucleotide pair AA/TT. Similar results were found by several other groups (Ioshikhes et al., 1996; Widom, 2001). The 10 bp periodicity pattern agrees well with the high-resolution nucleosome structure (Luger et al., 1997; Richmond & Davey, 2003), where the histones interact with the DNA sequence approximately once every 10 base pair.

However, early studies were limited by the fact that only a handful of sequences were known to be nucleosome bound. During the past decade, rapid progress has been made thanks to the development of high throughput technologies such as microarrays and DNA sequencing. As a result, high-resolution, genome-scale maps of nucleosome positions have been identified in various organisms including human (Chodavarapu et al. 2010; Johnson et al., 2006; Lanterman et al. 2010; Lee et al., 2007; Mavrich, Ioshikhes et al., 2008; Mavrich, Jiang et al., 2008; Ozsolak et al., 2007; Schones et al., 2008; Yuan et al., 2005). Despite the wide-range of species, the overall nucleosome positioning pattern is strikingly conserved. A common signature at the promoter regions is characterized by a nucleosome-free region (NFR) adjacent to the transcription start sites (TSS) flanked by well-positioned nucleosomes.

Segal et al. (2006) were the first to develop computational methods to predict genome-wide nucleosome positions (Figure 1A). Based on a set of 199 high-resolution nucleosome DNA sequences obtained by direct sequencing, the authors constructed a position specific weight matrix to model the nucleosome-DNA binding affinity. The approach is similar to standard methods for TF motif detection, but an important difference is that the basic units are dinucleotides (AA, CpG, etc.) instead of single nucleotides (A, C, G, and T). Consistent with previous studies, the average AA/TT/TA pattern is approximately periodic with about 10 bp periodicity. In order to predict genome-wide nucleosome positions, Segal et al. (2006) further developed a thermodynamic model, which took into account the steric hindrance effect between neighboring nucleosomes. Impressively, their model is able to predict 54% of nucleosome positions within 35 bp accuracy. However, the significance of this result is compromised by the observation that even random guesses could provide 38% prediction accuracy when evaluated in the same way. The underlying problem with this accuracy measure is that it increases with the total number of predicted sites, making it difficult to interpret the results. Around the same time, Ioshikhes et al. (2006) used a different approach to predict genome-wide nucleosome positions. They computed the correlation between an input sequence and a pre-determined nucleosome positioning sequence (NPS) pattern and then detected regions of high correlation score (Figure 1B). They observed that the two approaches shared similar prediction accuracy.

The above studies considered only information from the nucleosome sequences. On the other hand, it has been recognized that certain DNA sequences such as poly dA:dT runs are inhibitive for nucleosome binding (Bernstein et al., 2004; Sekinger et al., 2005; Yuan et al., 2005). Such sequences impose an important constraint on the nucleosome positions. Recognizing the role of nucleosome-free (linker) sequences, several groups have developed computational models to incorporate both nucleosome and linker DNA sequences for prediction (Lee et al., 2007; Peckham et al., 2007; Yuan & Liu, 2008), but the classification strategies are quite different. Specifically, Peckman et al. (2007) represented a sequence pattern by the number of word counts corresponding to a set of short *k*-mers (*k* up to 6), and the differences between nucleosome and linker sequences were detected by using a support vector machine (SVM) (Figure 1C). Lee et al. (2007) characterized sequence signatures based on TF motif scores and DNA structural parameters and used a Lasso regression method as a classifier. A third approach converted the dinucleotide frequencies to wavelet coefficients in order to detect discriminative periodic patterns, and a stepwise logistic regression model was used to distinguish nucleosome bound sequences from those located in the NFRs (Yuan & Liu, 2008) (Figure 1D). In the latter study, it was also found that a more objective measure for model performance was the false positive error rate rather than the false negative error rate which were used in previous studies. Despite the variation of the predictor selection and classification schemes, these models all significantly improve the model performance. More recently, Segal and colleagues

Figure 1. Schematic diagrams to illustrate the concepts behind various nucleosome positioning prediction methods. (A). Segal et al. (2006) modeled the nucleosome sequence pattern by using a position specific weight matrix. Each entry represents the probability of observing a specific dinucleotide at a specific position. (B) Ioshikhes et al. (2006) defined a NPS pattern based on the bias of AA/TT distribution. (C) Peckham et al. (2007) extracted a large number of sequence features by counting the occurrence of various short words. (D) Yuan and Liu (2008) studied the role of periodicity by decomposing a multiscale signal to various wavelet components, each varying at a specific length scale. (E) Miele et al. (2008) and Morozov et al. (2010) calculated the free energy required for any DNA sequence to wrap around a nucleosome. A sequence associated with lower energy is more favored for nucleosome binding.

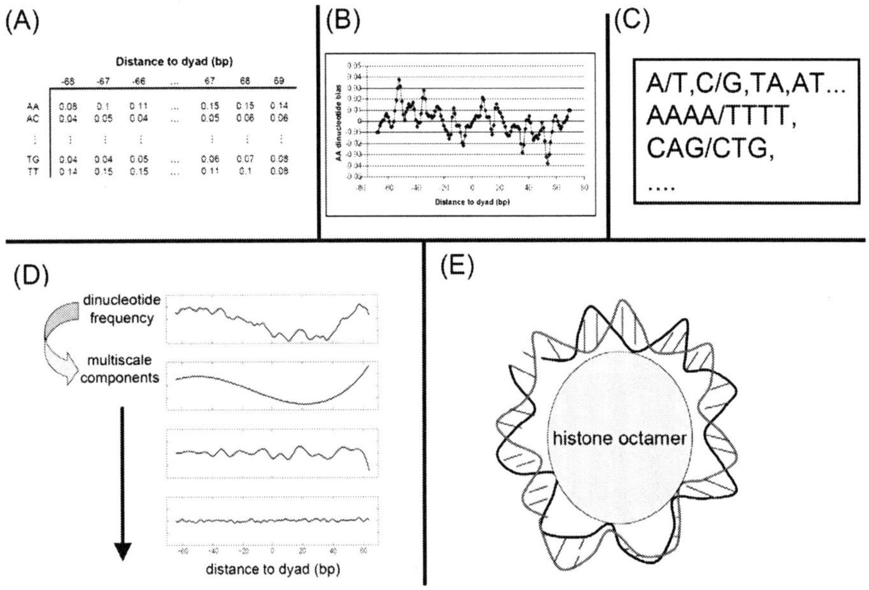

have also incorporated negative controls in their model framework and obtained much improved performance (Field et al. 2008).

The methods mentioned above are all based on empirical data. While these models can provide good model accuracy, they do not necessarily offer mechanistic insights. In the meantime, a different class of models has been recently developed based on calculation of the biophysical properties (Miele et al., 2008; Morozov et al., 2009) (Figure 1E). Both studies showed that a biophysically-based model may offer competitive performance for genome-wide predictions.

It is also important to note that, just because the DNA sequence and nucleosome positions are correlated, it does not mean that it is deterministic. In fact, Kornberg and Stryer (1988) pointed out that positioned nucleosomes can also be predicted based on a statistical model. In this model, only the nucleosome boundaries are determined by the DNA sequences, whereas the nucleosomes themselves are randomly packed. This mechanism results in highly aligned nucleosomes near the boundary whereas fuzzier configuration elsewhere, which is in fact consistent with experimental data (Mavrich, Ioshikhes et al., 2008). In addition, the *in vivo* nucleosome positions are inevitably affected by additional factors, such as the perturbation by chromatin remodeling complexes, competition with TF binding, and influence of transcriptional events, it is unclear to what extent the resulting positions are intrinsically coded in DNA sequences. To overcome this challenge, two groups (Kaplan et al., 2009; Zhang et al., 2009) have recently used next generation DNA sequencing methods to map the *in vitro* nucleosome positioning for sequences extracted from the yeast genome. Interestingly, these studies draw different conclusions although their data are similar. Kaplan et al. (2009) concluded that the intrinsic DNA sequence preference of nucleosomes have a central role in determining nucleosome positioning *in vivo*, noting that the nucleosome occupancy level is similar between *in vivo* and *in vitro* environments. On the other hand, Zhang et al. (2009) concluded against a genomic code for nucleosome positioning, pointing out the observed similarity can be simply explained by a statistical positioning model (Kornberg & Stryer, 1988).

Histone Modification

The N-terminal ends of the core histone proteins are unstructured and referred to as the histone tails, which can be post-translationally modified in multiple ways at multiple sites. Early studies were focused on histone acetylation, the modification that an acetyl group is added to a lysine residue. While promoter histone acetylation is generally associated with gene activation (Dion et al., 2005; Roh et al., 2005), histone acetylation in coding regions may lead to gene repression (Wang et al., 2002). Another well-characterized mark is histone methylation. The function of histone methylation is more complex and still not well understood. For example, H3K9 methylation is highly correlated with gene repression but H3K4 methylation is correlated with gene activation (Barski et al., 2007; Pokholok et al., 2005). An additional complexity for histone methylation is that it can happen in three flavors: mono-, di-, and tri-methylation, each may have its own role. For example, H3K4me3 is highly correlated with active promoters (Barski et al., 2007; Guenther et al., 2007), whereas H3K4me1 tends to be depleted at promoters but enriched at tissue-specific enhancers (Heintzman et al., 2007). In addition to acetylation and methylation, a large number of histone modifications have been identified including phosphorylation, ubiquitylation, ADP ribsylation, deimination, and praline isomerization (Kouzarides, 2007). The functionality for most of these modifications is still poorly understood and their function may be also context dependent. The task for identifying the function of the combinations of different histone modification marks is generally referred to as the "histone code" hypothesis, originally proposed by Allis and colleagues (Jenuwein & Allis, 2001; Strahl & Allis, 2000).

Although histone modifying enzymes do not interact with the DNA sequence, they may be recruited to specific loci by interacting with TFs, non-coding RNAs, or other DNA interacting regulators. There is abundant evidence that the distribution of many histone modification marks is associated with the DNA sequence each in its own way (Bernstein et al., 2007). For example, the H3K4me3 mark is mainly associated with high density and can be well-predicted by a CpG density alone (Bernstein et al., 2006), while the H3K9me3 mark is weakly associated with a number of repetitive sequences (Martens et al., 2005).

Computational studies for histone modifications have been mainly limited to H3K27me3 through investigation of Polycomb group (PcG) proteins targeting. PcG was first discovered in *Drosophila* for controlling Homeotic (Hox) genes (Lewis, 1978) but later found to also play an important role in early development in vertebrates (Schuettengruber et al., 2007; Sparmann & van Lohuizen, 2006). A major function of PcG proteins is to repress the transcriptional activities of their target genes through tri-methylation of the histone H3 on lysine 27 residue (H3K27me3) (Francis & Kingston, 2001). Much effort has been taken to identify the DNA elements that are responsible for PcG recruitment, called the Polycomb response elements (PRE). In *Drosophila*, Ringrose et al. showed that the PREs are well-characterized by the motif sequences of distinct TFs among which

PHO is the most important one (Ringrose et al., 2003). On the other hand, the mammalian PREs are still poorly characterized (Schuettengruber et al., 2007; Simon & Kingston, 2009). Experimentally validated PREs have only been recently identified (Sing et al., 2009; Woo et al., 2010).

Computational predictions of PREs are mainly centered around TF motifs. For *Drosophila*, Ringrose et al. found that individual TF motifs are insufficient for discriminating PREs from non-PREs (Ringrose et al., 2003). However, by pairing different motifs together, the discriminative power was much improved. These authors then derived a score based on a linear combination of the motif-pair scores. To test their prediction accuracy, they experimentally validated 43 regions randomly selected from a total of 167 predicted sites. 29 out of the tested regions were verified. More recently, additional TF motifs were incorporated in the same modeling framework, which led to improved model performance (Hauenschild et al., 2008).

In comparison, mammalian PREs are less characterized. Ku et al. (2008) investigated the association between TF motifs and genome-wide PcG targets in mouse ES cells. They identified several distinct TF motif patterns and used these patterns to predict PcG targets. Among the top 2836 predicted targets, about 60% are correct predictions. Similar results were obtained by using a more sophisticated model (Liu et al., 2010). In this study, Liu et al. (2010) found that the highly scoring genes tend to be marked by PcG in multiple cell-types, suggesting the DNA sequence is strongly related to target plasticity.

Currently, general methods for histone modification target predictions are still limited. A major challenge is that there are a large number of possible combinations, each has its own distribution profile. Some are focal (e.g. H3K4me3), others are broader (e.g. H3K9me3), and yet others are mixed (e.g. H3K27me3) (Barski et al., 2007). Another complexity is that the factors that regulate histone modifications are more complex. A histone modification mark can be either added or removed by specific enzymes. There are a large number of such enzymes, many of which share overlapping roles (Kurdistani & Grunstein, 2003; Lan et al., 2008). Since each factor functions differently, it is likely each only contributes to a small subset of targets.

A modified version of the wavelet model mentioned above has been applied to predict histone modification patterns in human (Yuan, 2009). The model performance is highly variable among different histone modification marks. For a few well-studied histone modification marks, such as H3K4me3 and H3K4me1, the model indeed performs well and the performance cannot simply be explained the local enrichment of CpG. On the other hand, the model predicts H3K9me3 rather poorly. The performance of the model is correlated with the overall spread of a histone modification mark. Interestingly, the H3K4me2 and H3K27me3 marks do not overlap in adult cells, yet their target sequences are highly similar. A possible explanation is that the two sets of marks both target same regions but only one mark can be established. Experimental evidence supporting this possibility is that the H3K4me2 pattern at the HoxA cluster in one tissue (lung) is similar to the H3K27me3 pattern in a different tissue (foot) (Rinn et al., 2007), suggesting that the targeted competition between the two marks may be responsible for epigenetic switching.

DNA Methylation

The genomic DNA itself can also be covalently modified and the modification has important implication on gene regulation (Bird, 2002). In this case, the cytosine nucleotide can be methylated. With the exception of a few special cases (Cokus et al., 2008; Lister et al., 2009), DNA methylation almost always occurs in the context of a CpG dinucleotide. Since CpG dinucleotide is self-complementary, the DNA methylation pattern on one DNA strand can be faithfully reproduced on the other strand, a property that is important for epigenetic inheritance. While promoter DNA methylation is often correlated with gene repression, recent epigenomic studies have shown that DNA methylation can also occur at coding region and its functional consequence is still poorly understood. In cancer, it has been found that the genome-wide DNA is widely hypomethylated, whereas specific loci, such as certain tumor repressor genes, are associated with hypermethylation (Esteller, 2007; Jones & Baylin, 2007). The overall methylation level can be influenced by food intake such as folic acid (Jirtle & Skinner, 2007).

Not surprisingly, the DNA methylation status is highly correlated with the local CpG density. The majority of CpG is located in low CpG density regions and tends to be methylated. On the other hand, CpG can also form clusters called the CpG islands, which tend to be unmethylated. However, some CpG islands are methylated in certain cell-types but not others. The sequence characteristics of such differentially methylated regions (DMR) are still incompletely understood, although it has been shown that DMRs are typically associated with intermediate CpG density (Bock et al., 2006). In cancers, it was found that a number of tumor repressor genes are silenced by DNA methylation (Keshet et al., 2006; TCGA, 2008). A challenge is to understand which set of CpG islands can be methylated.

A number of computational methods have been developed to predict the CpG island methylation from the underlying DNA sequence (Bock et al., 2006; Das et al., 2006; Fang et al., 2006; Feltus et al., 2006; Keshet et al., 2006). The overall strategies in these studies are similar, although there is a variation of the predicting sequence features that are used in these studies. For example, Das et al. (2006) used 102 sequence features as predictors, including GC content, word counts, and repetitive sequences. The classification was done by using support vector machine. Feltus et al. (2006) obtained discriminative sequence pattern by de novo motif searching followed by a decision tree as the classification model. In addition, Bock et al. (2006) incorporated the DNA structural parameters as predictors. In this study, the authors also used a similar approach to predict several other epigenetic marks including histone modifications and DNA hypersensitivity and then combine the information together to predict the overall strength of a CpG island. The biological interpretation of the strength of a CpG island is the likelihood of being kept in an open chromatin state and targeted by TFs. Recent analysis has shown that integrating the histone modification pattern information can improve model performance (Fan et al.).

DISCUSSION

The Role of DNA Sequence in Defining Epigenetic Patterns

The targeting mechanism for epigenetic factors is complex and involves a large number of factors and this complexity is only beginning to be investigated systematically. Here we discuss an important first step, which is the role of DNA sequence in shaping the global epigenetic landscape. The results reviewed in this paper strongly indicate that the DNA sequence plays an important role in the targeting of many epigenetic marks. There are some important exceptions. For example, the H3K36me3 pattern is mainly determined by transcription rather than coded in the DNA sequence.

Although the detailed mechanism is still unclear, we can think of the DNA sequence as defining the intrinsic stability of an epigenetic mark. In the case of nucleosome positioning, the predicted stability has been directly validated by genetic experiments and it is found that the nucleosome occupancy indeed change as predicted (Segal et al., 2006; Sekinger et al., 2005). These results suggest that the DNA sequences are indeed required for establishment of the proper nucleosome positions. Interestingly, the DNA sequence at the 5' end of a coding region is typically coded for high nucleosome occupancy, which may act as important barrier for passage of transcriptional machineries. Recent studies have found that PolII occupies many inactive genes but only is paused near TSS (Core & Lis, 2008; Guenther et al., 2007), whereas can be paused at the 5'end and used only to generate short incomplete transcripts (Core & Lis, 2008; Guenther et al., 2007). The requirement to overcome this barrier to finish a full transcript suggests an important transcription control mechanism (Mavrich, Ioshikhes et al., 2008).

Although less understood, the DNA sequence is also related to the overall plasticity of other epigenetic marks such as DNA methylation and histone modification. For DNA methylation, the regions with high CpG density tend be unmethylated, whereas those associated with low CpG tend to be methylated. Interestingly, the most variable regions seem to be related to intermediate CpG content (Bock et al., 2008; Das et al., 2006; Feinberg & Irizarry, 2010). A similar but more intricate pattern has been found for histone medications as well.

Recent studies have found that cancer is not only characterized with high genetic changes but also high epigenetic changes (TCGA, 2008). Interestingly, the aberrant DNA methylation pattern is correlated with genetic mutations. For example, in treated samples which display DNA methylation at the MGMT promoter, 81% of all mutations are of the G:C to A:T type in non-CpG dinucleotides, compared to a mere 4% within CpG dinucleotide. In comparison, in samples without MGMT methylation, the frequencies of the two types of mutations are roughly equal (29% vs 23%, respectively). It is still unclear whether other epigenetic changes are also correlated with genetic mutations in cancer.

Finally, we recognize that a lot of work is still needed to gain mechanistic insights. For example, despite the success of DNA sequence in prediction of nucleosome occupancy, the *in vivo* nucleosome positioning pattern can be simply explained by a statistical positioning model (Kornberg & Stryer, 1988; Mavrich, Ioshikhes et al., 2008; Zhang et al., 2009), suggesting that the DNA sequence may only be important for delineating the boundaries of nucleosome occupied regions.

Beyond the Sequence

While the DNA sequence is constant across cell-types, the actual epigenetic pattern is tissue-specific and cannot be determined by the DNA sequence alone. There are a large number of potential regulators, including chromatin modifying enzymes, TFs, and non-coding RNAs. For example, the ATP dependent chromatin remodelers can remove nucleosomes from their favored positions. A classical example is the regulation of PHO5 (Svaren & Horz, 1997). At normal conditions, the PHO5 promoter is occupied by well-positioned nucleosomes, one of which is centered at -275 bp relative to the ATG codon, occluding Pho4 from binding to its target site at -247 bp (Almer et al., 1986). This and three other nucleosomes are depleted upon phosphate starvation, making the Pho4 binding site accessible. The eviction of nucleosome is caused by the activity of SWI/SNF, an ATP-dependent chromatin remodeler. Similarly, the tissue-specific patterns of histone modification and DNA methylation are also highly dependent on the activities of various histone and DNA modification enzymes.

Since chromatin modifiers can be recruited to specific target sites by interacting with sequence-specific TFs. The activity of these TFs can also significantly affect the overall epigenetic pattern. For example, the histone deacetylase Hst1 is recruited by a single TF Sum1 in yeast (Robert et al., 2004). The genome-wide targets of Hst1 and Sum1 are nearly identical. Genetic deletion of SUM1 completely abolishes the binding of Hst1 and causes increased H3 and H4 acetylation level at their target sites. Similarly, in *Drosophila*, the TF PHO plays an important role in PcG recruitment, and deletion of PHO results in derepression of Hox genes, indicating disrupted PcG binding (Wang et al., 2004).

Another class of regulators that have been recently described is the non-coding RNAs (Guttman et al., 2009; Rinn et al., 2007; Zhao et al., 2008). For example, in mammals one of the X chromosomes in females is completely silenced for dosage compensation. This whole chromosome silencing is mediated by the DNA methylation. The large Xist RNA is produced at the inactive X-chromosome and thought to initiate the establishment of DNA methylation and X-inactivation (Lee, 2009). In addition, small RNAs can also interact with chromatin modifiers thereby regulating the local histone modification and DNA methylation patterns (Moazed, 2009).

These examples demonstrate that there are a large number of potential regulators for epigenetic patterns. A fundamental task is to understand their respective roles in establishing the global epigenetic patterns. Computational methods suitable for this task have yet been developed.

Epigenetics and Evolution

Modern evolutionary theory is firmly based on genetic variation and natural selection. The role of epigenetics in evolution remains unclear. Feinberg and Irizarry have recently proposed that the stochastic epigenetic variation originated from genetic variation may play an important role in evolutionary adaptation (Feinberg & Irizarry, 2010). Such variation does not change the mean phenotype, but stochastic variation is advantageous for adaptation to environmental changes. By using numerical simulation, the authors demonstrated that the increased variation can indeed increase fitness in a varying environment. They also found experimental evidence supporting that the locations of variably methylated regions across different samples are correlated with the local CpG density. Interestingly, the variability changes between human and mouse accompanied by CpG density changes. A core component of this hypothesis is that genetic variation is closely related to epigenetic variation, which is supported by the numerous studies reviewed in this paper.

Several studies have taken a comparative genomic approach to investigate whether genomic variations may be associated with expression changes via difference in epigenetic patterns. TF binding sites are found to be typically associated rigid DNA (Tirosh et al., 2007), consistent with nucleosome depletion at these sites. Interestingly, these authors also found that the locations of rigid DNA elements are conserved in TATA-less promoters but vary substantially at the promoters containing the TATA element. These differences are thought to be related to be developed during evolution to initiate species-specific responses to environmental changes. Along the same line, Field et al. predicted the promoter nucleosome occupancy level for various yeast species based on the genomic sequences (Field et al., 2009). Interestingly, the predicted nucleosome occupancy level is substantially different at genes which have different expression patterns between different yeast species. In particular, the respiratory genes are active in aerobic yeast species but inactive in anaerobic ones. In accordance to this difference, the predicted nucleosome level is low at the respiratory promoters for the aerobic species but much higher in other ones.

ACKNOWLEDGMENT

This research was supported by a Claudia Adams Barr Award.

REFERENCES

Almer, A., Rudolph, H., Hinnen, A., & Horz, W. (1986). Removal of positioned nucleosomes from the yeast PHO5 promoter upon PHO5 induction releases additional upstream activating DNA elements. *The EMBO Journal*, *5*(10), 2689–2696.

Barski, A., Cuddapah, S., Cui, K., Roh, T. Y., Schones, D. E., & Wang, Z. (2007). High-resolution profiling of histone methylations in the human genome. *Cell, 129*(4), 823–837.

Bernstein, B. E., Liu, C. L., Humphrey, E. L., Perlstein, E. O., & Schreiber, S. L. (2004). Global nucleosome occupancy in yeast. *Genome Biology, 5*(9), R62.

Bernstein, B. E., Meissner, A., & Lander, E. S. (2007). The mammalian epigenome. *Cell, 128*(4), 669–681.

Bernstein, B. E., Mikkelsen, T. S., Xie, X., Kamal, M., Huebert, D. J., & Cuff, J. (2006). A bivalent chromatin structure marks key developmental genes in embryonic stem cells. *Cell, 125*(2), 315–326.

Bird, A. (2002). DNA methylation patterns and epigenetic memory. *Genes & Development, 16*(1), 6–21.

Bock, C., & Lengauer, T. (2008). Computational epigenetics. *Bioinformatics (Oxford, England), 24*(1), 1–10.

Bock, C., Paulsen, M., Tierling, S., Mikeska, T., Lengauer, T., & Walter, J. (2006). CpG island methylation in human lymphocytes is highly correlated with DNA sequence, repeats, and predicted DNA structure. *PLOS Genetics, 2*(3), e26.

Bock, C., Walter, J., Paulsen, M., & Lengauer, T. (2008). Inter-individual variation of DNA methylation and its implications for large-scale epigenome mapping. *Nucleic Acids Research, 36*(10), e55.

Chodavarapu, R. K., Feng, S., Bernatavichute, Y. V., Chen, P. Y., Stroud, H., & Yu, Y. (2010). Relationship between nucleosome positioning and DNA methylation. *Nature, 466*(7304), 388–392.

Cokus, S. J., Feng, S., Zhang, X., Chen, Z., Merriman, B., & Haudenschild, C. D. (2008). Shotgun bisulphite sequencing of the Arabidopsis genome reveals DNA methylation patterning. *Nature, 452*(7184), 215–219.

Core, L. J., & Lis, J. T. (2008). Transcription regulation through promoter-proximal pausing of RNA polymerase II. *Science, 319*(5871), 1791–1792.

Das, R., Dimitrova, N., Xuan, Z., Rollins, R. A., Haghighi, F., & Edwards, J. R. (2006). Computational prediction of methylation status in human genomic sequences. *Proceedings of the National Academy of Sciences of the United States of America, 103*(28), 10713–10716.

Dion, M. F., Altschuler, S. J., Wu, L. F., & Rando, O. J. (2005). Genomic characterization reveals a simple histone H4 acetylation code. *Proceedings of the National Academy of Sciences of the United States of America, 102*(15), 5501–5506.

Esteller, M. (2007). Cancer epigenomics: DNA methylomes and histone-modification maps. *Nature Reviews. Genetics, 8*(4), 286–298.

Fan, S., Zhang, M. Q., & Zhang, X. (2008). Histone methylation marks play important roles in predicting the methylation status of CpG islands. *Biochemical and Biophysical Research Communications, 374*(3), 559–564.

Fang, F., Fan, S., Zhang, X., & Zhang, M. Q. (2006). Predicting methylation status of CpG islands in the human brain. *Bioinformatics (Oxford, England), 22*(18), 2204–2209.

Feinberg, A., & Irizarry, R. (2010). Stochastic epigenetic variation as a driving force of development, evolutionary adaptation, and disease. *Proceedings of the National Academy of Sciences, Early Edition.*

Feltus, F. A., Lee, E. K., Costello, J. F., Plass, C., & Vertino, P. M. (2006). DNA motifs associated with aberrant CpG island methylation. *Genomics*, *87*(5), 572–579.

Field, Y., Fondufe-Mittendorf, Y., Moore, I. K., Mieczkowski, P., Kaplan, N., & Lubling, Y. (2009). Gene expression divergence in yeast is coupled to evolution of DNA-encoded nucleosome organization. *Nature Genetics*, *41*(4), 438–445.

Field, Y., Kaplan, N., Fondufe-Mittendorf, Y., Moore, I. K., Sharon, E., & Lubling, Y. (2008). Distinct modes of regulation by chromatin encoded through nucleosome positioning signals. *PLoS Computational Biology*, *4*(11), e1000216.

Francis, N. J., & Kingston, R. E. (2001). Mechanisms of transcriptional memory. *Nature Reviews. Molecular Cell Biology*, *2*(6), 409–421.

Guenther, M. G., Levine, S. S., Boyer, L. A., Jaenisch, R., & Young, R. A. (2007). A chromatin landmark and transcription initiation at most promoters in human cells. *Cell*, *130*(1), 77–88.

Guttman, M., Amit, I., Garber, M., French, C., Lin, M. F., & Feldser, D. (2009). Chromatin signature reveals over a thousand highly conserved large non-coding RNAs in mammals. *Nature*, *458*(7235), 223–227.

Hauenschild, A., Ringrose, L., Altmutter, C., Paro, R., & Rehmsmeier, M. (2008). Evolutionary plasticity of polycomb/trithorax response elements in Drosophila species. *PLoS Biology*, *6*(10), e261.

Hawkins, R. D., Hon, G. C., & Ren, B. (2010). Next-generation genomics: an integrative approach. *Nature Reviews. Genetics*, *11*(7), 476–486.

Heintzman, N. D., Hon, G. C., Hawkins, R. D., Kheradpour, P., Stark, A., & Harp, L. F. (2009). Histone modifications at human enhancers reflect global cell-type-specific gene expression. *Nature*, *459*(7243), 108–112.

Heintzman, N. D., Stuart, R. K., Hon, G., Fu, Y., Ching, C. W., & Hawkins, R. D. (2007). Distinct and predictive chromatin signatures of transcriptional promoters and enhancers in the human genome. *Nature Genetics*, *39*(3), 311–318.

Ioshikhes, I., Bolshoy, A., Derenshteyn, K., Borodovsky, M., & Trifonov, E. N. (1996). Nucleosome DNA sequence pattern revealed by multiple alignment of experimentally mapped sequences. *Journal of Molecular Biology*, *262*(2), 129–139.

Ioshikhes, I. P., Albert, I., Zanton, S. J., & Pugh, B. F. (2006). Nucleosome positions predicted through comparative genomics. *Nature Genetics*, *38*(10), 1210–1215.

Jenuwein, T., & Allis, C. D. (2001). Translating the histone code. *Science*, *293*(5532), 1074–1080.

Jiang, C., & Pugh, B. F. (2009). Nucleosome positioning and gene regulation: Advances through genomics. *Nature Reviews. Genetics*, *10*(3), 161–172.

Jirtle, R. L., & Skinner, M. K. (2007). Environmental epigenomics and disease susceptibility. *Nature Reviews. Genetics*, *8*(4), 253–262.

Johnson, S. M., Tan, F. J., McCullough, H. L., Riordan, D. P., & Fire, A. Z. (2006). Flexibility and constraint in the nucleosome core landscape of Caenorhabditis elegans chromatin. *Genome Research*, *16*(12), 1505–1516.

Jones, P. A., & Baylin, S. B. (2002). The fundamental role of epigenetic events in cancer. *Nature Reviews. Genetics*, *3*(6), 415–428.

Kaplan, N., Hughes, T. R., Lieb, J. D., Widom, J., & Segal, E. (2010, Nov 30). Contribution of histone sequence preferences to nucleosome organization: proposed definitions and methodology. *Genome Biology*, *11*(11), 140.

Kaplan, N., Moore, I. K., Fondufe-Mittendorf, Y., Gossett, A. J., Tillo, D., & Field, Y. (2009). The DNA-encoded nucleosome organization of a eukaryotic genome. *Nature*, *458*(7236), 362–366.

Keshet, I., Schlesinger, Y., Farkash, S., Rand, E., Hecht, M., & Segal, E. (2006). Evidence for an instructive mechanism of de novo methylation in cancer cells. *Nature Genetics*, *38*(2), 149–153.

Kornberg, R. D. (1974). Chromatin structure: A repeating unit of histones and DNA. *Science*, *184*(139), 868–871.

Kornberg, R. D., & Lorch, Y. (1999). Twenty-five years of the nucleosome, fundamental particle of the eukaryote chromosome. *Cell*, *98*(3), 285–294.

Kornberg, R. D., & Stryer, L. (1988). Statistical distributions of nucleosomes: Nonrandom locations by a stochastic mechanism. *Nucleic Acids Research*, *16*(14A), 6677–6690.

Kouzarides, T. (2007). Chromatin modifications and their function. *Cell*, *128*(4), 693–705.

Ku, M., Koche, R. P., Rheinbay, E., Mendenhall, E. M., Endoh, M., & Mikkelsen, T. S. (2008). Genomewide analysis of PRC1 and PRC2 occupancy identifies two classes of bivalent domains. *PLOS Genetics*, *4*(10), e1000242.

Kurdistani, S. K., & Grunstein, M. (2003). Histone acetylation and deacetylation in yeast. *Nature Reviews. Molecular Cell Biology*, *4*(4), 276–284.

Lan, F., Nottke, A. C., & Shi, Y. (2008). Mechanisms involved in the regulation of histone lysine demethylases. *Current Opinion in Cell Biology*, *20*(3), 316–325.

Lantermann, A. B., Straub, T., Strålfors, A., Yuan, G. C., Ekwall, K., & Korber, P. (2010). Schizosaccharomyces pombe genome-wide nucleosome mapping reveals positioning mechanisms distinct from those of Saccharomyces cerevisiae. *Nature Structural & Molecular Biology*, *17*(2), 251–257.

Lee, J. T. (2009). Lessons from X-chromosome inactivation: Long ncRNA as guides and tethers to the epigenome. *Genes & Development*, *23*(16), 1831–1842.

Lee, W., Tillo, D., Bray, N., Morse, R. H., Davis, R. W., & Hughes, T. R. (2007). A high-resolution atlas of nucleosome occupancy in yeast. *Nature Genetics*, *39*(10), 1235–1244.

Lewis, E. B. (1978). A gene complex controlling segmentation in Drosophila. *Nature*, *276*(5688), 565–570.

Lister, R., Pelizzola, M., Dowen, R. H., Hawkins, R. D., Hon, G., & Tonti-Filippini, J. (2009). Human DNA methylomes at base resolution show widespread epigenomic differences. *Nature*, *462*(7271), 315–322.

Liu, Y., Shao, Z., & Yuan, G. C. (2010). Prediction of polycomb target genes in mouse embryonic stem cells. *Genomics*, *96*(1), 17–26.

Luger, K., Mader, A. W., Richmond, R. K., Sargent, D. F., & Richmond, T. J. (1997). Crystal structure of the nucleosome core particle at 2.8. A resolution. *Nature*, *389*(6648), 251–260.

Martens, J. H., O'Sullivan, R. J., Braunschweig, U., Opravil, S., Radolf, M., & Steinlein, P. (2005). The profile of repeat-associated histone lysine methylation states in the mouse epigenome. *The EMBO Journal*, *24*(4), 800–812.

Mavrich, T. N., Ioshikhes, I. P., Venters, B. J., Jiang, C., Tomsho, L. P., & Qi, J. (2008). A barrier nucleosome model for statistical positioning of nucleosomes throughout the yeast genome. *Genome Research*, *18*(7), 1073–1083.

Mavrich, T. N., Jiang, C., Ioshikhes, I. P., Li, X., Venters, B. J., & Zanton, S. J. (2008). Nucleosome organization in the Drosophila genome. *Nature*, *453*(7193), 358–362.

Meissner, A., Mikkelsen, T.S., Gu, H., Wernig, M., Hanna, J., Sivachenko, A., et al. (2008). Genome-scale DNA methylation maps of pluripotent and differentiated cells. *Nature*.

Miele, V., Vaillant, C., d'Aubenton-Carafa, Y., Thermes, C., & Grange, T. (2008). DNA physical properties determine nucleosome occupancy from yeast to fly. *Nucleic Acids Research*, *36*(11), 3746–3756.

Mikkelsen, T. S., Ku, M., Jaffe, D. B., Issac, B., Lieberman, E., & Giannoukos, G. (2007). Genome-wide maps of chromatin state in pluripotent and lineage-committed cells. *Nature*, *448*(7153), 553–560.

Moazed, D. (2009). Small RNAs in transcriptional gene silencing and genome defence. *Nature*, *457*(7228), 413–420.

Mohn, F., Weber, M., Rebhan, M., Roloff, T. C., Richter, J., & Stadler, M. B. (2008). Lineage-specific polycomb targets and de novo DNA methylation define restriction and potential of neuronal progenitors. *Molecular Cell*, *30*(6), 755–766.

Morozov, A. V., Fortney, K., Gaykalova, D. A., Studitsky, V. M., Widom, J., & Siggia, E. D. (2009). Using DNA mechanics to predict in vitro nucleosome positions and formation energies. *Nucleic Acids Research*, *37*(14), 4707–4722.

Okita, K., Ichisaka, T., & Yamanaka, S. (2007). Generation of germline-competent induced pluripotent stem cells. *Nature*, *448*(7151), 313–317.

Ozsolak, F., Song, J. S., Liu, X. S., & Fisher, D. E. (2007). High-throughput mapping of the chromatin structure of human promoters. *Nature Biotechnology*, *25*(2), 244–248.

Peckham, H. E., Thurman, R. E., Fu, Y., Stamatoyannopoulos, J. A., Noble, W. S., & Struhl, K. (2007). Nucleosome positioning signals in genomic DNA. *Genome Research*, *17*(8), 1170–1177.

Pokholok, D. K., Harbison, C. T., Levine, S., Cole, M., Hannett, N. M., & Lee, T. I. (2005). Genome-wide map of nucleosome acetylation and methylation in yeast. *Cell*, *122*(4), 517–527.

Rando, O. J., & Chang, H. Y. (2009). Genome-wide views of chromatin structure. *Annual Review of Biochemistry*, *78*, 245–271.

Ren, B., Robert, F., Wyrick, J. J., Aparicio, O., Jennings, E. G., & Simon, I. (2000). Genome-wide location and function of DNA binding proteins. *Science*, *290*(5500), 2306–2309.

Richmond, T. J., & Davey, C. A. (2003). The structure of DNA in the nucleosome core. *Nature*, *423*(6936), 145–150.

Ringrose, L., Rehmsmeier, M., Dura, J. M., & Paro, R. (2003). Genome-wide prediction of Polycomb/Trithorax response elements in Drosophila melanogaster. *Developmental Cell*, *5*(5), 759–771.

Rinn, J. L., Kertesz, M., Wang, J. K., Squazzo, S. L., Xu, X., & Brugmann, S. A. (2007). Functional demarcation of active and silent chromatin domains in human HOX loci by noncoding RNAs. *Cell*, *129*(7), 1311–1323.

Robert, F., Pokholok, D. K., Hannett, N. M., Rinaldi, N. J., Chandy, M., & Rolfe, A. (2004). Global position and recruitment of HATs and HDACs in the yeast genome. *Molecular Cell*, *16*(2), 199–209.

Roh, T. Y., Cuddapah, S., & Zhao, K. (2005). Active chromatin domains are defined by acetylation islands revealed by genome-wide mapping. *Genes & Development*, *19*(5), 542–552.

Saccani, S., & Natoli, G. (2002). Dynamic changes in histone H3 Lys 9 methylation occurring at tightly regulated inducible inflammatory genes. *Genes & Development*, *16*(17), 2219–2224.

Satchwell, S. C., Drew, H. R., & Travers, A. A. (1986). Sequence periodicities in chicken nucleosome core DNA. *Journal of Molecular Biology*, *191*(4), 659–675.

Schlesinger, Y., Straussman, R., Keshet, I., Farkash, S., Hecht, M., & Zimmerman, J. (2007). Polycomb-mediated methylation on Lys27 of histone H3 pre-marks genes for de novo methylation in cancer. *Nature Genetics*, *39*(2), 232–236.

Schones, D. E., Cui, K., Cuddapah, S., Roh, T. Y., Barski, A., & Wang, Z. (2008). Dynamic regulation of nucleosome positioning in the human genome. *Cell*, *132*(5), 887–898.

Schuettengruber, B., Chourrout, D., Vervoort, M., Leblanc, B., & Cavalli, G. (2007). Genome regulation by polycomb and trithorax proteins. *Cell*, *128*(4), 735–745.

Segal, E., Fondufe-Mittendorf, Y., Chen, L., Thastrom, A., Field, Y., & Moore, I. K. (2006). A genomic code for nucleosome positioning. *Nature*, *442*(7104), 772–778.

Segal, E., & Widom, J. (2009). What controls nucleosome positions? *Trends in Genetics*, *25*(8), 335–343.

Sekinger, E. A., Moqtaderi, Z., & Struhl, K. (2005). Intrinsic histone-DNA interactions and low nucleosome density are important for preferential accessibility of promoter regions in yeast. *Molecular Cell*, *18*(6), 735–748.

Seligson, D. B., Horvath, S., Shi, T., Yu, H., Tze, S., & Grunstein, M. (2005). Global histone modification patterns predict risk of prostate cancer recurrence. *Nature*, *435*(7046), 1262–1266.

Sharma, S., Kelly, T. K., & Jones, P. A. (2010). Epigenetics in cancer. *Carcinogenesis*, *31*(1), 27–36.

Simon, J. A., & Kingston, R. E. (2009). Mechanisms of polycomb gene silencing: Knowns and unknowns. *Nature Reviews. Molecular Cell Biology*, *10*(10), 697–708.

Sing, A., Pannell, D., Karaiskakis, A., Sturgeon, K., Djabali, M., & Ellis, J. (2009). A vertebrate Polycomb response element governs segmentation of the posterior hindbrain. *Cell*, *138*(5), 885–897.

Sparmann, A., & van Lohuizen, M. (2006). Polycomb silencers control cell fate, development and cancer. *Nature Reviews. Cancer*, *6*(11), 846–856.

Strahl, B. D., & Allis, C. D. (2000). The language of covalent histone modifications. *Nature*, *403*(6765), 41–45.

Svaren, J., & Horz, W. (1997). Transcription factors vs nucleosomes: Regulation of the PHO5 promoter in yeast. *Trends in Biochemical Sciences*, *22*(3), 93–97.

Takahashi, K., & Yamanaka, S. (2006). Induction of pluripotent stem cells from mouse embryonic and adult fibroblast cultures by defined factors. *Cell*, *126*(4), 663–676.

TCGA. (2008). Comprehensive genomic characterization defines human glioblastoma genes and core pathways. *Nature*, *455*(7216), 1061–1068.

Tirosh, I., Berman, J., & Barkai, N. (2007). The pattern and evolution of yeast promoter bendability. *Trends in Genetics*, *23*(7), 318–321.

Waddington, C. (1942). The epigenotype. *Endeavour*, *1*, 18–20.

Wang, A., Kurdistani, S. K., & Grunstein, M. (2002). Requirement of Hos2 histone deacetylase for gene activity in yeast. *Science*, *298*(5597), 1412–1414.

Wang, L., Brown, J. L., Cao, R., Zhang, Y., Kassis, J. A., & Jones, R. S. (2004). Hierarchical recruitment of polycomb group silencing complexes. *Molecular Cell*, *14*(5), 637–646.

Wei, G., Wei, L., Zhu, J., Zang, C., Hu-Li, J., & Yao, Z. (2009). Global mapping of H3K4me3 and H3K27me3 reveals specificity and plasticity in lineage fate determination of differentiating CD4+ T cells. *Immunity*, *30*(1), 155–167.

Wernig, M., Meissner, A., Foreman, R., Brambrink, T., Ku, M., & Hochedlinger, K. (2007). In vitro reprogramming of fibroblasts into a pluripotent ES-cell-like state. *Nature*, *448*(7151), 318–324.

Widom, J. (2001). Role of DNA sequence in nucleosome stability and dynamics. *Quarterly Reviews of Biophysics*, *34*(3), 269–324.

Woo, C. J., Kharchenko, P. V., Daheron, L., Park, P. J., & Kingston, R. E. (2010). A region of the human HOXD cluster that confers polycomb-group responsiveness. *Cell*, *140*(1), 99–110.

Yoo, C. B., & Jones, P. A. (2006). Epigenetic therapy of cancer: Past, present and future. *Nature Reviews. Drug Discovery*, *5*(1), 37–50.

Yu, J., Vodyanik, M. A., Smuga-Otto, K., Antosiewicz-Bourget, J., Frane, J. L., & Tian, S. (2007). Induced pluripotent stem cell lines derived from human somatic cells. *Science*, *318*(5858), 1917–1920.

Yuan, G. C. (2009). Targeted recruitment of histone modifications in humans predicted by genomic sequences. *Journal of Computational Biology*, *16*(2), 341–355.

Yuan, G. C., & Liu, J. S. (2008). Genomic sequence is highly predictive of local nucleosome depletion. *PLoS Computational Biology*, *4*(1), e13.

Yuan, G. C., Liu, Y. J., Dion, M. F., Slack, M. D., Wu, L. F., & Altschuler, S. J. (2005). Genome-scale identification of nucleosome positions in S. cerevisiae. *Science*, *309*(5734), 626–630.

Zhang, Y., Moqtaderi, Z., Rattner, B. P., Euskirchen, G., Snyder, M., & Kadonaga, J. T. (2009). Intrinsic histone-DNA interactions are not the major determinant of nucleosome positions in vivo. *Nature Structural & Molecular Biology*, *16*(8), 847–852.

Zhao, J., Sun, B. K., Erwin, J. A., Song, J. J., & Lee, J. T. (2008). Polycomb proteins targeted by a short repeat RNA to the mouse X chromosome. *Science*, *322*(5902), 750–756.

Zhou, V. W., Goren, A., & Bernstein, B. E. (2011). Charting histone modifications and the functional organization of mammalian genomes. *Nature Reviews. Genetics*, *12*(1), 7–18.

KEY TERMS AND DEFINITONS

Epigenetics: The study of inherited changes in phenotype or gene expression caused by mechanisms other than changes in the underlying DNA sequence.

Nucleosome: The basic unit of DNA packaging in eukaryotes, consisting of a segment of DNA wound around a histone protein core.

Chapter 9
A New Approach for Sequence Analysis:
Illustrating an Expanded Bioinformatics View through Exploring Properties of the Prestin Protein

Kathryn Dempsey
University of Nebraska at Omaha, USA & University of Nebraska Medical Center, USA

Benjamin Currall
Creighton University, USA

Richard Hallworth
Creighton University, USA

Hesham Ali
University of Nebraska at Omaha, USA & University of Nebraska Medical Center, USA

ABSTRACT

Understanding the structure-function relationship of proteins offers the key to biological processes, and can offer knowledge for better investigation of matters with widespread impact, such as pathological disease and drug intervention. This relationship is dictated at the simplest level by the primary protein sequence. Since useful structures and functions are conserved within biology, a sequence with known structure-function relationship can be compared to related sequences to aid in novel structure-function prediction. Sequence analysis provides a means for suggesting evolutionary relationships, and inferring structural or functional similarity. It is crucial to consider these parameters while comparing sequences as they influence both the algorithms used and the implications of the results. For example, proteins that are closely related on an evolutionary time scale may have very similar structure, but entirely different functions. In contrast, proteins which have undergone convergent evolution may have dissimilar primary structure, but perform similar functions. This chapter details how the aspects of evolution,

DOI: 10.4018/978-1-60960-491-2.ch009

structure, and function can be taken into account when performing sequence analysis, and proposes an expansion on traditional approaches resulting in direct improvement of said analysis. This model is applied to a case study in the prestin protein and shows that the proposed approach provides a better understanding of input and output and can improve the performance of sequence analysis by means of motif detection software.

INTRODUCTION

Computational methods have simplified the analysis of the massive genetic code. Sequence analysis, as it is known, takes advantage of the inherent conservation of genomes by comparing set of nucleotide or amino acid sequences to infer relationships. One of the first described methods for sequence comparison was published in 1970 by Gibbs and McIntyre *et al.* in defining a means for comparing two biopolymer sequences with the allowance of gaps to discern homology using the dot method. In 1990, Altschul *et al.* described a method for the approximation of sequence alignments that would later become known as NCBI's BLAST. Since then, various methods for sequence comparison have been proposed reducing computational cost, runtimes, and improving accuracy. Currently, a variety of reputable methods exist for creating a multiple sequence alignment and one may choose the tool appropriate for their domain with the best speed, the least computational burden, or the best accuracy. This has only highlighted that sequence comparison, by means of alignment or pattern search, remains a consistent genre of tools for discerning basic characteristics about a set of sequences.

Even with the many advances in sequence comparison since 1970, one may beg the question of improvement, especially with the recent explosion of DNA and protein sequence data gathered from high-throughput methods. Sequence comparison methods have come under fire due to their over-simplifying the analysis process. The call for 'intelligent' sequence analysis has been proposed by Wagner *et al.*(2008) by describing alignment methods that incorporate domain expertise with sophisticated methods to achieve the best alignment possible. The idea of intelligent sequence analysis implies that the user can apply their expertise to appropriately prepare the input data and thus better understand and interpret the output, and even adjust software parameters to best suit their focus. This intelligent sequence analysis requires that users have knowledge in both biological and informatics fields. By breaking down sequence analysis, these ideas can be better applied.

In our analysis of nucleotide and amino acid sequences, we first divide methods into experimental analysis or computational analysis. Experimental analysis includes all experimental methods (e.g. sequencing and mass spectrometry) used to determine the sequence of nucleotides and amino acid residues pertaining to genomes as well as experimental methods that determine the evolution, structure, and/or function of those nucleotide and amino acid sequences. Computational analysis includes methods for preparing, accessing, and analyzing sequence data, with capabilities for examination of massive volumes of data using the parallel architectures of supercomputing.

Traditional Approach to Sequence Analysis

Computational sequence analysis can be divided into three stages: input, method, and output. The input of the traditional approach may contain one to several nucleotide or amino acid sequences and are chosen based upon user discretion. Several methods exist, but typically corresponding nucleotides or amino acid residues are evaluated to determine if they match (*i.e.* have identical or similar properties) or mismatch (*i.e.* gaps in sequence or dissimilar properties). Various outputs also exist, but all somehow provide a measure of comparison depending on scoring technique, providing an estimate based on the cellular level of conservation.

When computational analysis is coupled with experimental analysis, a canonical pathway for genomic analysis is formed. The standard flow of information is that some biological observation or query based on experimental analysis becomes a means for providing data with the intention of identifying common and unique sequence characteristics computationally. The relationship inferred by software outputs can then be used to further describe the sequence data as shown in Figure 1.

This traditional pathway is often supplemented by various non-canonical pathways. For example, ClustalW takes DNA or protein sequences as an input and ultimately produces a multiple sequence alignment. First, however, it iterates through a non-canonical pathway by performing several pair-wise alignments to produce a pre-

Figure 1. The canonical approach to sequence analysis. Experimental analysis produces the DNA or protein sequence of biopolymers through various methods. These sequences then become inputs for bioinformatics algorithms, which infer knowledge from these data and output results that reflect the implied sequence conservation.

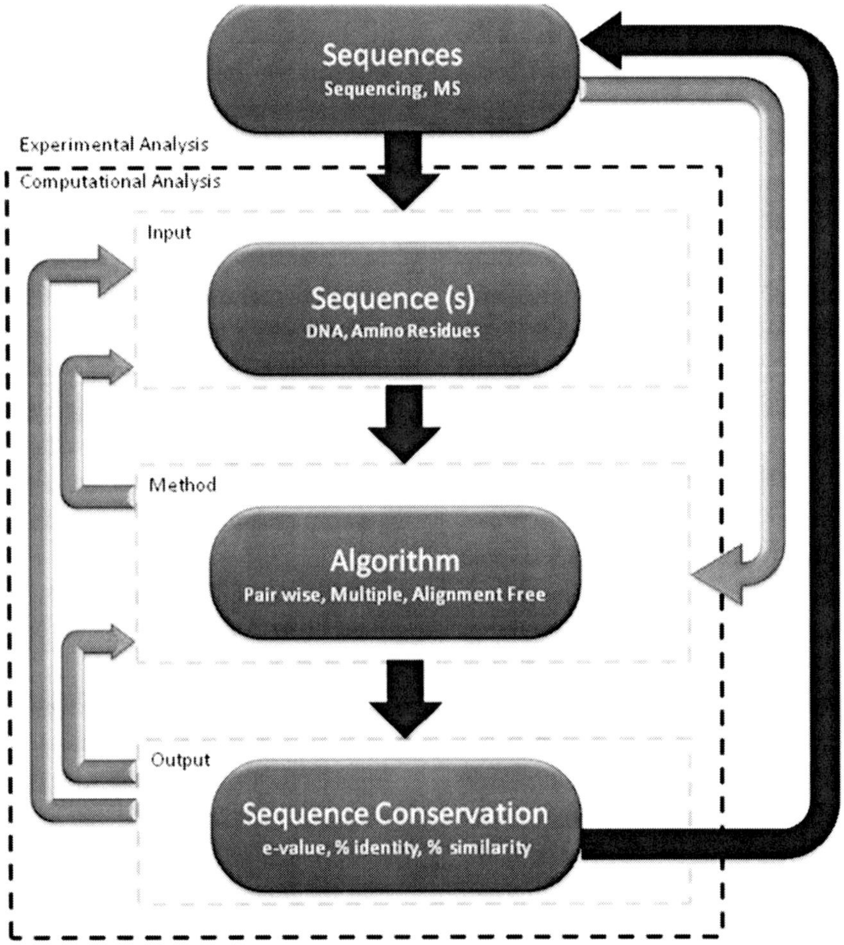

liminary phylogenetic output or "guide tree", before using this tree to weight the initial input. The weighted input then becomes a critical factor in producing the heuristic multiple sequence alignment output (Markel & Leon, 2003). The Roundup tool, also utilizes a non-canonical pathway by performing a reciprocal BLAST based on the initial input sequences to retrieve additional input sequences before producing an output (Deluca *et al.*, 2006). Most often, the non-canonical pathways use multiple iterations of single or multiple algorithms. These algorithms, however, often rely on inferred relations from the original inputs rather than empirical or auxiliary relationships (Benson *et al.,* 2004; Hedges *et al.,* 2004; Kumar *et al.,* 2005).

The continued improvement of both these canonical and non-canonical pathways will decrease the computational requirements and increase the precision of the analysis.

Even with this improvement however, they often disregard error in both input and the interpretation of the output which may do little to improve accuracy at best and decrease the accuracy of the analysis at worst. (One of the most basic, and surprisingly common, mistakes is to assume, for example, that a sequence with many mammalian paralogs whose input may be dominated by mammalian homologous sequences is better conserved than a sequence with few mammalian paralogs, whose input may be dominated by bacterial homologous sequences).

Expanded Approach to Sequence Analysis

In the commonly used traditional approach to sequence analysis, input is left up to the discretion of the individual users. Often, the input is gathered through unfiltered database searches such as BLAST (Altschul *et al.* 1990). Depending on the database contents and the users' preferences employed for the input, the output of sequence may vary widely from user to user. Both precision and accuracy of sequence analysis could be improved by further consideration of input. Particularly, that nucleotide and amino acid residue conservation are dependent upon at least three variables: evolution, structure, and function.

Evolutionary distances between species' sequences dictate the amount of basal mutation. For example, if sequence analysis is performed comparing orthologous human and mouse sequences, then a high number of matched residues are expected due to short evolutionary distance. If sequence analysis is performed comparing orthologous human and *D. melanogaster* sequences, then far fewer conserved residues are expected due to a greater evolutionary distance. Structural relationships between proteins can also affect conservation. For example, though no evolutionary or functional relationship between different types of membrane proteins appear to exist, many have similar α-helical structures containing hydrophobic residues. Functional relationships may also affect conservation. Two evolutionarily close human paralogs would be expected to have a high number of conserved residues. If they have different functions, however, it is likely that a non-conserved residue may be more important for that protein's discrete function and thus, the organism's well-being. As genomic analysis becomes more refined, systematic errors become apparent in sequence analysis such as the under-prediction of evolutionary distance between phylogenies in sequence analysis (Hedges *et al.,* 2004; Kumar *et al.* 2005).

The proposed expanded approach, shown in Figure 2, acknowledges and builds the evolutionary, structural, and functional variables on top of the traditional approach. In the expanded approach, both the canonical and non-canonical pathways remain intact. The experimental analysis, however, is expanded to recognize the additional sequence information provided by the biological disciplines that examine evolution, structure, and function. These correspond to the division of the computational analysis input into evolution, structure, and

Figure 2. The suggested expanded approach to sequence analysis. We propose that input and output be further modified with little focus on change of the algorithm can improve results and thus further the experimental analyses that drive evolutionary biology. Here we propose the use of structural genomics, molecular physiology, and evolutionary biology to tailor data creating an expanded input.

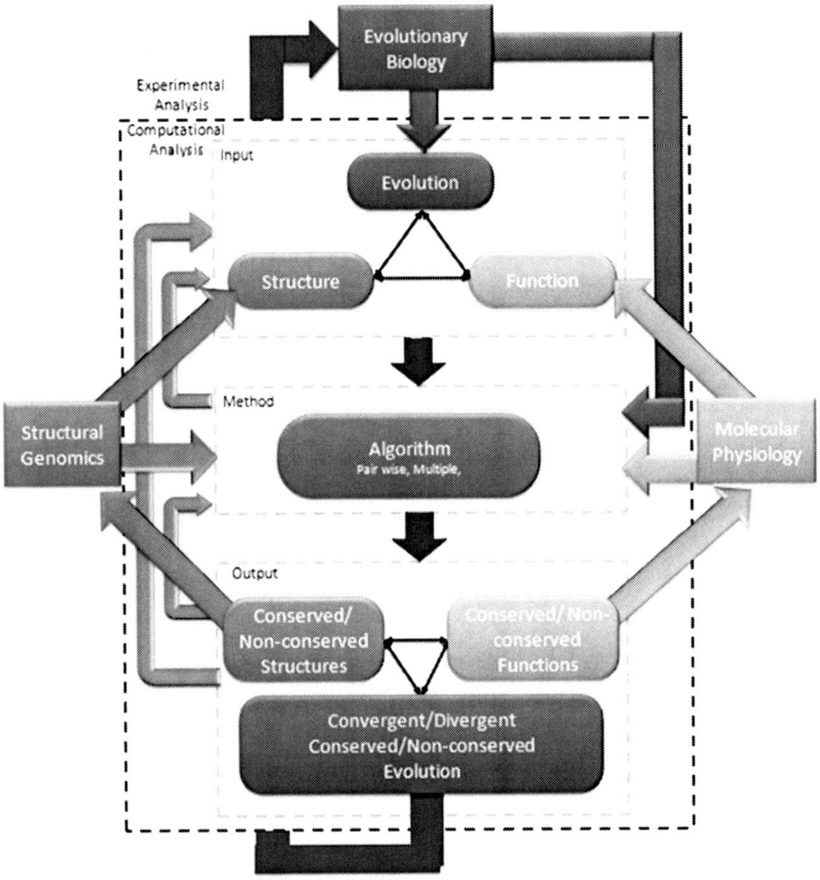

function, implying that all three variables will be taken into account before submitting to a method. While similar computation analysis methods, and their associated algorithms, are used in both approaches, the output must also be analyzed in context of evolution, structure, and function.

Experimental Analysis

The first step in our expanded approach sequence analysis is the experimental analysis of genetic or protein material to determine the nucleotide or amino acid sequences.

Experimental Analysis: Theory

The dominant method to obtain nucleotide sequences, polymerase chain reaction (PCR) sequencing of DNA has become routine laboratory technique, and advancing technologies have improved both the speed and accuracy. Using shotgun sequencing, whole genomes can be sequenced in a relatively short period. Coding sequences can also be obtained by sequencing cDNA (complimentary DNA synthesized by reverse transcribing mRNA), which helps distinguish between the exons and introns found in eukaryotic genomic DNA. Protein sequences can also be obtained directly using mass spectrometry, but isolating sufficient amounts of protein for accurate sequencing can be difficult.

Experimental analysis can also supply evidence about the evolution, structure, and function between sequences. Evolutionary distances can be calculated using paleontological data such as radiocarbon dating of common ancestors. Structure of proteins can be solved using either X-ray crystallography or NMR spectroscopy. Function of proteins can be analyzed using techniques such as electrophysiology and enzyme kinetics, all of which can inform sequence analysis.

Experimental Analysis: Case Study

Prestin (Slc26a5) is a member of the Slc26 protein family of ion transporters. Slc26 proteins can be found in a variety of organs ranging from the brain to the kidney. All family members contain both a xanathine uracil permease (XUP) and a sulphate transporter anti-sigma factor antagonist (STAS) superfamily domain as well as a sulphate transporter motif (Mount & Romero, 2004). While the STAS superfamily domain shares homology with the well described bacterial sporulation factor (SpoIIaa), little is known about the structure-function relationship of the XUP superfamily domain (Dorwart *et al.* 2008). There are currently 11 identified human SLC26 paralogs (a1-a11), however, one gene, SLC26A10, is believed to be a pseudogene. Several diseases have been correlated to mutations within this family of proteins including Pendred syndrome (SLC26A4), congenital chloride losing diarrhea (SLC26A3), and chrondodysplasia (SLC26A2) (Dorwart *et al.* 2008). While most Slc26s act as ion transporters, mammalian prestin acts as a motor protein (Schaechinger & Oliver, 2007). Mammalian prestin is found in the membrane of the inner ear outer hair cell (OHC) (Zhang *et al.*, 2000). When the OHC is depolarized, the cell shrinks; and when hyperpolarized, the cell elongates, in a process known as somatic motility. Somatic motility has been shown to operate at rates of greater than 70 kHz and play a role in the amplification of sound input within the inner ear (Ashmore, 2008). This theory is strengthened by the demonstration that knocking out Slc26a5 results in loss of hearing in mice (Cheatham *et al.*, 2005). Even more intriguing, this motor function of the prestin proteins occurs only in mammals, while the non-mammalian prestin functions as an anion transporter (Boekhoff-Falk, 2005; Schaechinger *et al.*, 2007). This unique function of mammalian prestin presents an opportunity to analyze both shared and unique structural components of the Slc26 family to help deduce the structure-function relationship.

COMPUTATIONAL ANALYSIS: INPUT

Input: Theory

Input into any analysis method should be considered in light of the questions being asked and the expected output. In many modern sequence analysis methods, nucleotide or amino acid sequences can be inserted given a standardized format (e.g. FASTA format). These standardized formats typically only take the composition and order of the sequence into account with little consideration of additional information related to the sequence (e.g. species, functions, domains). The type of input is based upon number and the type of method to be used is described in Box 1.

Some methods will combine several of these input types to simplify user interface or to increase method precision. This allows even novice users to produce data from sequence analysis methods.

Input: Traditional Approach

In the traditional approach, input varies widely from user to user. In many circumstances, however, users will input their particular sequence of interest or a list of sequences based on availability.

Box 1. General input types in sequence comparison

1. **Query**: One sequence searched against a database of sequences or profiles.
2. **Pairwise**: Comparison of exactly two sequences at a time, hence the name.
3. **Multiple**: Greater than two sequences being compared, although it should be noted that multiple sequence alignment can also be used for pairwise alignment when appropriate.

Although quite convenient, this approach does not take into consideration information that may influence the results. In addition, most current sequence databases tend to have greater amount of bacterial (due to their small size and ease of sequencing) or mammalian (due to their short evolutionary distance from human) sequences than sequences from other clades. This creates an inherent bias of databases that is translated to sequence analysis.

Input: Expanded Approach

The proposed expanded approach to sequence alignment tries to capture some of the information lost in the traditional approach by curating sequences based upon variables that affect sequence conservation (i.e. evolution, structure, and function).

Evolution: Estimating phylogenetic relationships can be circular: A multiple sequence alignment (MSA) is needed to create a phylogenetic tree, but first an *ad hoc* "guide tree" is generally created to create an MSA (Edgar, 2004; Larkin *et al.* 2007). This circular nature can be addressed by incorporating true phylogenetic relationships; unfortunately discerning true phylogeny is an extremely difficult task. However, there are methods to estimate to our best ability: the concept of the "molecular clock" attempts to use methods of paleontological and molecular dating to estimate the true phylogenetic relationships between organisms.

Molecular clock. The concept of the molecular clock brings the bias of evolution to the forefront by looking at changes between genomic and protein data versus paleontological or fossil dating. Variable rates of evolution and lack of sequence and fossil data are just a few of the issues compounding the ability to accurately determine true phylogenetic relationships (Kumar *et al.,* 2005). When examining the tree of life, there are two general methods for determining a clade split, and both are estimates at best: molecular dating and fossil or paleontological dating (Benson *et al.,* 2004; Hedges *et al.,* 2004; Kumar *et al.* 2005). Molecular dating involves looking at current rates of evolution and differences in genome and proteome between those organisms closely and distantly related, to determine when the evolutionary "split" occurred to result in two new species. However, with differing rates of reproductive turnover and speciation events, the matter of determining true evolutionary distance becomes impossible and age is often overestimated (Benson *et al.,* 2004; Lin & Moret, 2008). Paleontological dating is the other method used to estimate the true molecular clock by determining fossil age, species type, and relation, but tends to *under*estimate actual age, as fossils cannot be assumed to be the first of their species (Benson *et al.,* 2004,; Hedges *et al.* 2004).

Structure: Protein structure is often subdivided into primary, secondary, tertiary, and quaternary structure. The primary structure is the sequence of amino acids produced during translation from mRNA. Protein secondary structure generally refers to the most basic folds of a protein before it reaches its native state, the most common of these structures being α-helices or β-sheets. The tertiary structure of proteins involves the protein being folded into its final structure as a whole protein or as a component of a larger complex. The quaternary structure of a protein is achieved when many protein components come together to form an active complex that acts toward one discrete function.

There are numerous methods for obtaining tertiary or quaternary structural information such as X-ray crystallography or NMR technology. However, structures can only be predicted at best for certain groups, such as membrane proteins which are hard to crystallize (Brenner, 2001). For these proteins, *ad hoc* or homology prediction algorithms can suggest structure but caution should be exercised until the structure can be experimentally verified.

It has been observed that two sequences with a high structural similarity on the tertiary level can have very low primary sequence similarity based on the 20-letter amino acid alphabet (Benson *et al.*, 2004). Using reduced amino acid alphabet is one way to discern sequence similarity from seemingly different proteins; perhaps most important to note is that two sequences do not need to observe any level of conservation to share structure (Peterson *et al.*, 2009). However, in protein super-families, common tertiary folding patterns combined with similar or same function has been repeatedly observed compared to non-family sequences (Pei & Grishin, 2001), suggesting that some level of conservation is observed to maintain structure and function separately. It has been proposed that residues that exhibit high sequence conservation tend to be clustered together or 'tightly packed' in structure (Atchley *et al.*, 2000, Gouveia-Oliveira *et al.*, 2009; Poirot *et al.*, 2004), whereas those residues that are more diverse but possibly co-evolving are responsible for maintenance of function (or restoration of function lost) (Madej *et al.*, 2007).

Function: As research continues, functions associated with primary sequences are being revealed. While some of this information is being captured in publicly available databases, any known function associated with primary sequence is not usually taken into account when performing sequence comparison which can misconstrue the results. For example, in more complex species, several paralogs of a gene may exist. These paralogs often have different function, different interactions, and different expression patterns, although they share a common ancestor and conserved sequence homology. When aligning such paralogs, much of the structural elements will be similar, however, the motifs for function, binding, or localization may be completely different from one paralog to the next. If the user's intention was to identify these motifs, then aligning such paralogs may prove fruitless. In the expanded approach, awareness of common and unique function is encouraged by eliminating any homolog that has a function different from that of the sequence of interest and thus unexpected results may be better understood.

Input: Case Study

Input: Traditional Approach

In the traditional approach, prestin was analyzed for motifs. The prediction software of choice (see Computational Analysis: Methods) required multiple sequence inputs. These sequences were obtained by performing a BLAST search with the human prestin protein as a query (*H.* sapiens SLC26a5 isoform a, RefSeq ID: NP_945350.1) and retrieving the top ten blastp results from the non-redundant sequence database (all organisms included). This resulted in a total of 11 sequences in our dataset for the traditional approach, which included prestin homologs from the following species: *C. familiaris, R. leschenaultii, F. catus, E. spelaea, C. sphinx, B. Taurus, B. physalus, A. melanoleuca,* and *M. novaeangliae.*

Input: Expanded Approach

In the expanded approach, sequence analysis was performed to determine common transport motifs across the entire Slc26 family. To accomplish this, a broad range of sequences was obtained through multiple searches using human prestin as our original query. Sequences were then curated based upon their evolutionary, structural, and functional properties as described below:

Sequence retrieval: Sequences were obtained through BLAST, Conserved Domain Architectural Retrieval Tool (CDART), or naming conventions through NCBI, Ensembl, or Transport Classification Database (TCDB). Only sequences containing both a XUP and STAS domain, in the order of N—XUP—STAS—C, were kept for further analysis.

Evolution: Over a hundred homologous sequences to prestin were obtained including sequences from nearly every kingdom. Figure 3 reflects some of the orthologs found from this search, highlighting the known divisions of the sequences according to classification. Since the goal is to identify motifs common to all Slc26 family, sequences from each kingdom were considered. In order to remove bias towards any particular kingdom only one representative of each kingdom (Animalia, Plantae, Fungi, Protista, and Bacteria) was chosen for use in the dataset for our Expanded Approach. In addition, we created two additional datasets, one containing human prestin and the orthologs listed in Figure 3 (17 sequences in total), and one containing human prestin and one sequence per each of prestin's human paralogs (11 sequences in total). By grouping sequences in this manner, we expect to observe some struggle on part of the algorithm to detect motifs in the paralog group because the high level of conservation creates a background that makes it difficult to discern patterns conserved for functional reasons versus residues conserved due to low levels of evolutionary distance. In contrast, we expect that signals will be easier to discern in the orthologous group because more time has passed between evolutionary breaks and thus, there exists a wider evolutionary distance between organism homologs.

Structure: The structure for prestin, or any of its homologs, has yet to be solved by either NMR

Figure 3. The classification of some prestin orthologs. Naming conventions (SLC26a5) are maintained throughout chordate orthologs (and one outside this distinction, S. purpuratus.). Outside this group, orthologs are still acknowledged as SLC26 superfamily members.

				Kingdom	Phylum	Class	InfraClass	Species	Gene
SLC26	Slc26a5	Motor Protein Slc26a5	Eutheria Slc26a5	Animalia	Chordata	Mammalia	Eutheria	*Homo sapiens*	SLC26A5
								Cynopterus sphinx	Slc26a5
								Sus scrofa	Slc26a5
								Mus musculus	Slc26a5
								Equus callabus	Slc26a5
							Metatheria	*Monodelphis domestica*	Slc26a5
							Protheria	*Ornithorhynchus anatinus*	Slc26a5
						Aves		*Gallus gallus*	Slc26a5
						Amphibia		*Xenopus laevis*	Slc26a5
						Actinopterygii		*Danio rerio*	Slc26a5
					Arthropoda			*Drosophila melanogaster*	CG5485
					Nematoda			*Caenorhabditis elegans*	Sulp-6
					Echinodermata			*Strongylocentrotus purpuratus*	Slc26a5
				Plantae				*Arabidopsis thaliana*	Sultr1;2
				Fungi				*Saccharomyces cerevisiae*	Sulp1
				Protista				*Dictyostelium discoideum*	Sultr
				Bacteria				*Mycobacterium tuberculosis*	Sultr

spectroscopy or X-ray crystallography. The Slc26 family, however, does contain both the XUP and STAS domain. While there is no solved structure for the XUP domain, the STAS domain is homologous to the SpoIIaa protein homologs which do have solved structures (PDB #1AUZ). Therefore, the full length of the sequence and both domains were analyzed separately.

Function: While the components of transport are slightly different between prestin homologs in each of the different kingdom, they all act as transporters. Mammalian prestin, however, acts as a motor protein rather than a transporter. Therefore, it is important to analyze the mammalian prestin separately from the rest of the Slc26 family. In this study, an ortholog of prestin with transport function, chicken Slc26a5, was used instead of a mammalian prestin ortholog.

COMPUTATIONAL ANALYSIS: METHOD

Method: Theory

Many methods for sequence analysis *in silico* exist with varied applicability. Highlighted here are the general concepts that drive sequence analysis. The advent of public databases containing genomic and protein sequence information has reinforced the immediate need for deciphering differences and similarities between sequence data. As a means for clarification of terms and as a brief review, here we provide details on a select set of common approaches with a corresponding table including references and I/O formats (Figure 4). It should be noted that this list is not intended to be comprehensive or complete.

Method: Traditional Approach

Alignment-free methods: Alignment-free comparison is a non-linear approach to sequence analysis, recognizing that genomic recombination events often change not only DNA (and therefore protein) content, but also the linear continuity of a genome is not necessarily maintained. A gene translocation event, for example, can result in loss of the continuity of a sequence exploited by traditional sequence alignment and in turn a failure to identify conservation where conservation has not been lost, but misplaced. Enter alignment free methods (Figure 5), which break a sequence down into smaller sequences or "tuples" (Vinga & Almeida, 2003). One can then determine the frequency of tuples between two or more sequences,

Figure 4. Some of the popular methods for sequence comparison, highlighting the method type, and input and output types. In addition to the traditional method, three main genera of input parameters that should be addressed based on the proposed approach: evolution, structure, and function are indicated.

		Method	Input	Output	Source
Alignment	*Global*	CLUSTAL	PW, MULTI	ALIGNMENT, PHYLOGENY	http://www.clustal.org/
		T-Coffee	PW, MULTI	ALIGNMENT, 2S	http://www.tcoffee.org/
		MUSCLE	PW, MULTI	ALIGNMENT	http://www.drive5.com/muscle/
	Local	BLAST	QUERY	E-VAL	http://blast.ncbi.nlm.nih.gov/Blast.cgi
		PSI-BLAST	QUERY	E-VAL	http://blast.ncbi.nlm.nih.gov/Blast.cgi
		Gibbs Motif Sampler	MULTI	MOTIF	http://bayesweb.wadsworth.org/gibbs/gibbs.html
		MEME	MULTI	MOTIF	http://meme.sdsc.edu/meme4_3_0/intro.html
		Dialign	MULTI	ALIGNMENT, PHYLOGENY	http://bibiserv.techfak.uni-bielefeld.de/dialign/
Alignment-Free	*	DOT-plot	*	DOT	[32],[33]
		Compression based	*	*	*

Figure 5. Example of how alignment-free methods work. Assuming three hypothetical proteins with domains a-e, alignment-free methods examine content without assuming linear continuity to determine which sequences are more homologous.

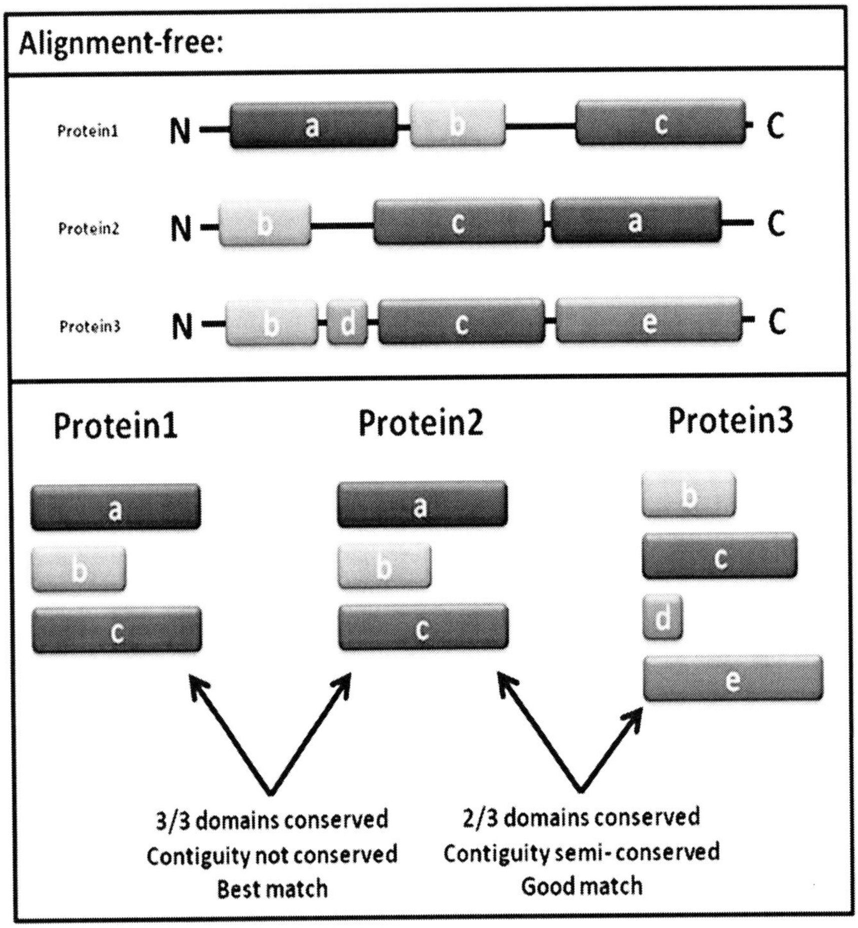

and define a metric of similarity by comparing tuple composition (Vinga & Almeida, 2003). First introduced in the late 1980's, alignment-free methods have recently been gaining speed as a non-linear method for sequence analysis.

***Alignment based methods**:* Alignment based methods (Figure 6) have long been established in bioinformatics as the main method for sequence comparison among DNA or protein sequences. The aim is to 'align' two or more linear sequences by allowing the insertion of gaps and the mutation of bases or residues, resulting in a scored visual alignment that highlights positional similarities and differences between inputs (Clote & Backhofen, 2000; Markel & Leon, 2003). There are many flavors of sequence alignment that each serves a certain purpose.

Global versus local sequence alignment. Global sequence alignment looks at the entire span of two or more sequences and creates a scored output based on the entire length of all sequences compared. Local sequence alignment does not analyze entire datasets but instead aligns smaller regions with highest relative conservation.

Pairwise versus multiple sequence alignment. Pairwise sequence alignment, as suggested by

Figure 6. The goal of sequence alignment. Given two hypothetical proteins, sequence alignment will align regions with highest homology while maintaining continuity of the linear protein sequences.

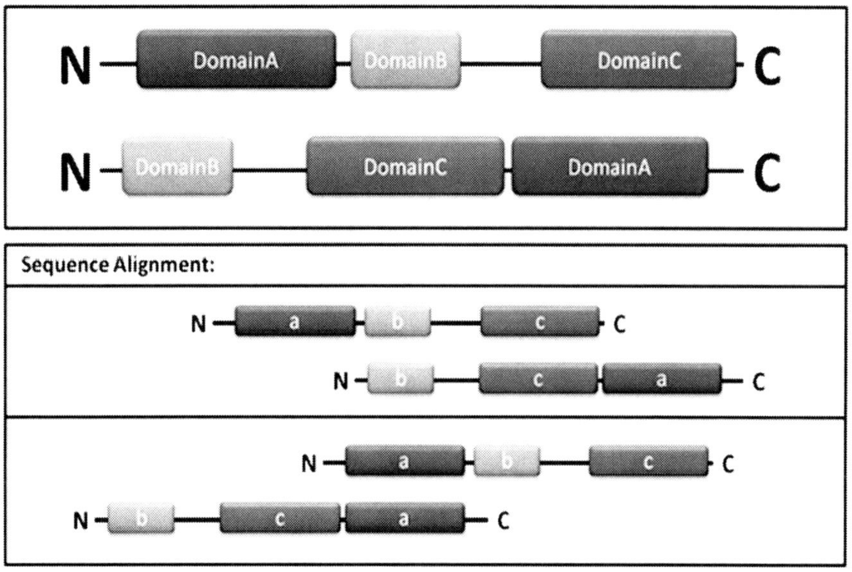

its name, is a comparison of two sequences only. *Multiple sequence alignment (MSA)* is designed to handle many inputs at once (for an example, see Figure 7). Generally speaking, a tool designed for multiple sequence alignment can also be used for pairwise sequence alignment, although the domain for which the method was designed should be taken into account. Most methods for MSA first create a rudimentary "guide tree" to determine an initial relation of input sequences. This guide tree can be constructed through the Neighbor Joining (NJ) method, which produces an unrooted tree, or

Figure 7. An example multiple sequence alignment based on homologs of human prestin (SLC26A5) using ClustalW.

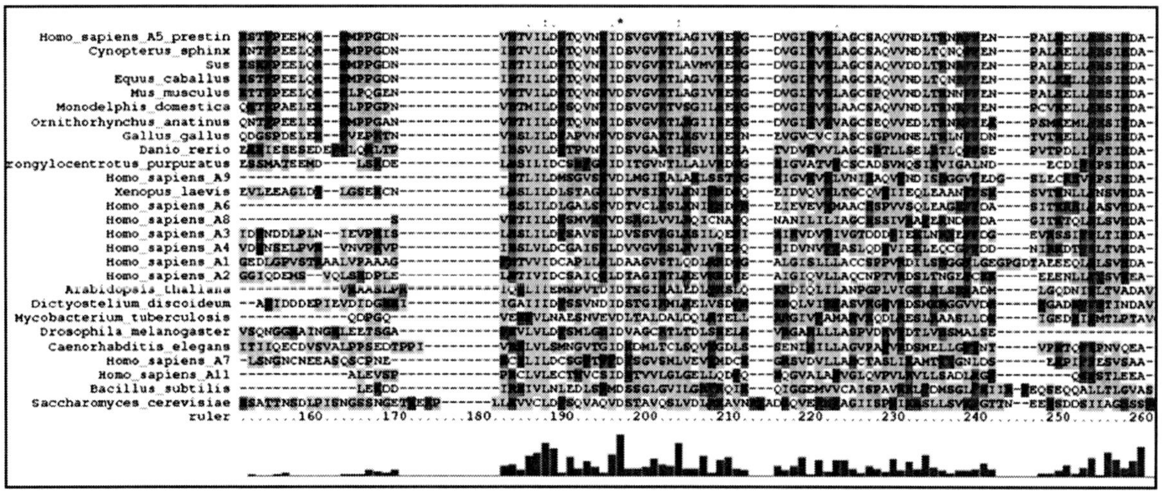

Box 2. Variables in motif detection software

1. Use of auxiliary data: It is possible that additional data are needed for accurate analysis, such as incorporating histone positional information when looking at DNA.
2. Training data: A set of background sequences is often available as an optional parameter that provides an expected level of "noise" for comparing to input sequences.
3. Relation of input sequences: The user should take care to compile the input data set with the algorithm in mind, and avoid biasing of the input data.
4. Output formatting: There are pros and cons to each output format of motif detection programs. Users should familiarize themselves with the output format for each program used.
5. Algorithm and scoring method (Tompa *et al.*, 2005): With the volume of motif detection programs always increasing, it is important to choose the best algorithm(s) for the user's domain of interest. Some programs will perform better on DNA versus protein, or prokaryotes versus eukaryotes. A general knowledge of domain focus is advised.

UPGMA method, which assumes a constant evolution rate to produce a rooted tree (Edgar, 2004; Vinga & Almeida, 2003; Wagner *et al.*, 2008).

Motif detection. Motif detection is a sub-class of local multiple sequence alignment designed for comparing sequences that exploits the presence of patterns in DNA and protein sequences. Generally intended for sequences related by evolution or co-regulation, these algorithms use a variety of methods to search for shorter regions of conservation where sequence is conserved but positional conservation is rarer, such as transcription factor binding sites in DNA. The working concept behind detecting transcription factor binding sites (TFBSs) is that binding sites tend to be more highly conserved than other regions of sequence as they are the part of a higher regulatory mechanism needed for the regulation of transcription and thus survival of the cell or organism. Naturally, detection of TFBSs is but one application of motif detection methods, as these same concepts can be applied to a variety of patterns such as protein dimerization sites, post-translational modification motifs, and other patterns in sequence data. As such, a variety of motif detection software are able to detect patterns without the specific goal of finding TFBS; indeed, many of these software are created to search for any type of short conserved pattern within DNA or protein.

Over 150 algorithms exist for detecting those short patterns in biological sequence data. As with sequence alignment, motif detection comes in many flavors. Software can be focused to certain domains such as binding sites in mainly prokaryotes. The method for motif finding can be general, i.e. looking for patterns only, or can be specific, looking specifically for transcription factor binding sites. Some of the variables in motif detection software to note are highlighted in Box 2.

As the number and variety of motif detection programs grows, there rises a need for assessment of performance when possible. Studies on performance generally look at methods that are popular in literature, publicly available, command-line operable and version maintained. In Quest *et al.* (2008), four indicators of algorithm performance in comparable methods were identified; these are identified in Box 3.

Method: Expanded Approach

The expanded approach is specifically designed to avoid further modification of algorithms and focus on better understanding of input and output. Therefore, the algorithm part remains the same for the Expanded Approach.

Method: Case Study

Method: Traditional Approach

As previously stated, it is up to the discretion of the user to choose which method is best suited for their analysis based on expertise and working

Box 3. Indicators of algorithm performance

1. Preprocessing: General sequence characteristics are identified, for example GC content. Massive regions of repeats are known to exist in genomic data as well, such as AT repeat regions and CpG islands. Algorithms will easily pick up on these areas for their frequency and strong conservation; performance can be enhanced by "masking" these regions.
2. Input: The importance of intelligent input is again re-iterated for a motif detection program. Sequences that are unrelated by evolution, regulation mechanism, or some other metric input to a pattern finding algorithm can result in motif discovery, but the biological impact of patterns found, if any, will be harder to deduce.
3. Parameter tuning: Required inputs for any motif detection software will be a set of input sequences, but there will undoubtedly be a set of optional, "tunable" parameters such as motif length ranges, expected motifs per sequence, etc. It is generally accepted that running motif detection software on default settings versus modified parameter settings will yield similar results (Quest *et al.* 2008); however, it should be noted that these options are available to optimize performance if possible.
4. Advancement: General availability and executability of a program are important keys to determining which detection suite is the best for each user. It is encouraged to follow-up on software versioning and maintenance for major and minor program upgrades. Interaction between user and developer though communication such as bug-reporting continues to be a valuable tool for improvement of motif detection softwares as well.

knowledge of the biological processes at work. For our case study, we chose the Gibbs motif sampler (Gibbs) for our sequence analysis. Gibbs can be used as a general pattern finder and in performance studies has shown comparable sensitivity and specificity on a position and site-specific level (*nSn, nSp, sSp*), compared to other tools (Quest *et al.*, 2008; Tompa *et al.*, 2005). The criteria for choosing Gibbs are highlighted in Box 4.

Proteins were selected with basis in multiple stages of literature review and working knowledge of the prestin protein.

Gibbs Algorithm: The Gibbs algorithm was originally defined in 1993 by Lawrence *et al.* for local multiple alignment. The goal of the method is to find "subtle" patterns assumed to be present in a set of varied protein or DNA sequences. It is heralded as a fast method for finding ungapped common patterns in linear time with a relatively high sensitivity. The method requires no prior information and works solely off the input set of data, highlighting two ideas: one, that no expertise or knowledge of secondary or tertiary structures is required, and two, the fact that again it is of utmost importance to provide the program with intelligent input. Gaps are not used in Gibbs pattern finding; it is maintained that the need for gaps generally comes from variable loop size in secondary structures (such as in RNA loops) but the active site is conserved, resulting in short conserved patterns. Also allowed for is differential positioning of patterns among sequences, due to genomic rearrangement events. Thus the Gibbs method of pattern detection allows for the discovery of short conserved sequences in DNA or protein sequences that are varied among patterns.

Method: Expanded Approach

The expanded approach uses the same algorithms as the traditional approach.

Box 4. Criteria for choosing Gibbs motif sampler

1. No auxiliary data or phylogenetic disposition required (able to perform a general pattern search)
2. Training/background data is randomized based upon the suggested defaults
3. Available for protein sequences input
4. Popularity in literature
5. Performance in multiple assessments
6. Ability to run on command-line
7. Establishment of method in community
8. Maintained and versioned; most stable and current versions used

Box 5. General output types in sequence comparison

1. **Alignment:** A visual representation (accompanied by flat-file format available for parsing) using color, score, and gaps ("-") to represent positional residue conservation of all sequences.
2. **Motif:** A "local" alignment of shorter conserved patterns found within all or some input sequences. Continuity of sequences is not expected to be maintained.
3. **DOT:** Used only for pairwise evaluation, the DOT-plot method was originally intended to increase homology between two sequences by introducing gaps (Altschul *et* al., 1990; Fitch, 1969).
4. **E-value:** The e-value is the metric used by BLAST and BLAST variations to express the reliability that the found result is truly similar to the query (and is usually accompanied by a similarity score). A smaller e-value tends to indicate a better match.
5. **Secondary Structure (2S):** The continued improvement of alignment tools and incorporation of auxiliary data such as structural information allows for the output of predicted secondary and in some cases tertiary structural information.

COMPUTATIONAL ANALYSIS: OUTPUT

Output: Theory

Output of sequence comparison must be evaluated for the answer to initial hypothesis and for correctness. The variety of outputs available are addressed in Box 5.

The frequency measure used in the results is defined per result. If the program takes 10 sequences as input and finds a motif in 7 out of the 10 results, the frequency will be 7/10, or 0.70. This is a very basic reflection of the motif strength without looking at content. As the number of sequences increases, a motif strength of 1.00 becomes a much more significant result than a frequency of 0.20. For the sake of equal comparison, this is the method used to evaluate both the traditional and expanded approaches. A more in-depth explanation of this method is described in Dempsey *et al.*(in press) which uniformly evaluates not only frequency but also motif content. Motifs that were duplicates were omitted from our results, with the exception of motifs with similar content but different length or starting/ending positions. An example of a kept pattern of similar content with varying length is the motifs DSVG and DSVGVA. These were considered separate results and were kept if found.

Experimental Validation of Results. Experimental validation of *in silico* results is of the utmost importance to confirm or deny the functional implications of motifs found and to further suggest improvements on algorithm. One approach for doing this is by searching for results in databases that contain common patterns with known functional correspondence, such as ExPASy's PROSITE pattern search. The results of these searches will inform the user whether the sequence entered matches a known pattern that has been experimentally verified and is in published literature. As such, any known associations of motifs found were included in our results.

Output: Case Study

Output: Traditional Approach

Results. We examined the results from the input given as the Traditional Approach and found that at least 5 motifs were found with one instance per motif in every sequence (Table 1). Only one of these motifs, #3, had known significance. In addition, the algorithm only found one motif per length, which suggests that the program struggled to find multiple strong patterns at varying lengths.

Output: Expanded Approach

Controlling for evolution, structure, and function: In the expanded approach, only transport prestin homologs from each kingdom were chosen for analysis. The results must be interpreted in this context. Mammalian prestin is the only Slc26 homolog acting as a motor protein. By excluding

Table 1. Results from Gibbs run on lengths 6, 8, 10, 12, 14, and 16 (default params) on traditional input

	Traditional Approach		
	Motif	**Score**	**Association**
1	VIMYCF	1.00	-
2	QFCLGVCR	1.00	-
3	FFWRTSKIEL	1.00	PKC Phosphorylation, Casein kinase II phosphorylation
4	RPIFSHPVLQER	1.00	-
5	KGMFMQFSDLPFFW	1.00	-

it from the analysis, we are examining what is conserved in its transport homologs rather than the prestion motor function. By examining the homologs from different kingdoms, we are looking at roughly equidistant evolution of the entire transport family and have removed bias towards any particular branch of the family. The analysis of this output, therefore, should suggest motifs that are conserved throughout the *entire* Slc26 family that are important for *transport* function. Since mammalian prestin does not have a transport function, changes or loss of these motifs could correlate to the transport loss-of-function seen in mammalian prestin. In addition, motifs found in mammalian prestin, but not found in these results, may include motifs involved in the motor gain-of-function seen in mammalian prestin homologs.

Results: Table 2 describes the top 5 results of a Gibbs search using default parameters over lengths 6, 8, 10, 12, 14, and 16 on the Expanded Approach for orthologs and paralogs. The frequency describes the number of sequences the motif was

Table 2. Results of the semi -expanded approach, by paralogs and orthologs. The paralog dataset contained all 10 human orthologs of prestin without the actual prestin protein itself (because of the functional distinction).

	Expanded Approach: Paralogs & Orthologs			
	Motif	**Score**	**Empirical Association**	**Comment**
Para-logs	PQALAYA	0.80	-	*Slc26a11 (pseudogene) found in none of these sequences
	KSVLGVITIVNL	0.60	-	
	IDANQELLAIGLTN	1.00	-	
	SFGRTAVNAQSGVCTP	1.00	-	
	KVWDLPRLWRMSPADALVW	0.90	Casein kinase II phosphorylation	
	Motif	**Score**	**Empirical Association**	**Comment**
Or-tho-logs	RTSKIEL	0.56	PKC Phosphorylation, Casein kinase II phosphorylation	Motif in chordates only
	LDFTQVNFIDSV	0.69	none	Motif in chordates, yeast, and protists
	GDLVSGISTGVLQLP	0.75	N-myristoylation	
	TCTPKKIRNIIYMFL	0.56	PKC Phosphorylation, Casein kinase II phosphorylation	
	KTKRYSGIFSVVYSTVA	0.75	PKC Phosphorylation, cAMP dependent protein kinase phosphorylation	

found versus the total number of sequences given as input. The parameter settings allowed for only one motif to be found per sequence so maximum frequency is 1.00.

The results of the paralog run are similar to those of the results from the traditional run. This result was not unexpected, and based on the model for the expanded approach we can speculate that these sequences are too conserved to allow the distinction of strong signals for structural or functional regions to be observed. One peculiar finding is that the known pseudogene paralog was not included in 3 of the 5 results, indicating that there may be a link between the loss of function and loss of conservation.

The orthologous results return a more varied output. The top two motifs returned were found in specific classes only, and almost all motifs had some known empirical association. According to our model, this confirms that the orthologs are (1) less conserved than the paralogs and (2) some loss of function and conservation occurs between the break of paralogs and orthologs, especially based on the frequencies of the motifs found. No motif in this set was found in 100% of the sequences.

Table 3 highlights the motifs from the Expanded Approach according to our model. Evolutionary restrictions are reflected in the choice of one sequence per kingdom, structural restrictions are reflected in the separation based on known XUP and STAS domains, and functional bias is removed by removing human prestin from the original input (which again has a novel function compared to the rest of its SLC26 family members). Three motifs in the STAS domain had known associations and every motif found had exactly one instance per

Table 3. Results of the expanded approach, by evolution, structure, and function

	Expanded Approach: XUP and STAS Domains			
	Motif	**Score**	**Empirical Association**	**Comment**
Entire	GFLRFGFVAIYL	1.00	-	*All motifs found had one instance per kingdom
	SAVNFMAGCQTAV	0.60	N-myristoylation	
	PVFGLYSSFYPVFLYTFF	1.00	N-myristoylation	
	GDIISGISTGVMQLPQ	1.00	N-myristoylation	
	FALKHGYTIDGNQELIAL-GICN	1.00	-	
	Motif	**Score**	**Empirical Association**	**Comment**
STAS	PDTDIYC	1.00	-	*All motifs found had one instance per kingdom
	RPQYRILGQI	1.00	-	
	DAGVQDGSPDELEHF	1.00	N-myristoylation, casein kinase II phosphorylation	
	FFDNTVTRELLFHSIHDA	1.00	Casein kinase II phosphorylation	
	ILDFAPVNFVDSVGAK-TLKSV	1.00	STAS domain profile, PKC phosphorylation, DSVG phosphorylation site	
	Motif	Score	**Empirical Association**	**Comment**
XUP	SMSRSLV	1.00	-	*All motifs Found had One instance Per kingdom
	NQELIALGICN	1.00	Glutamine amidotransferase type 1 domain profile, N-myristoylation	
	NSVGSFFQSFPITCSMSR	1.00	-	
	RFGFVAIYLTEPLVRGFTTA	1.00	-	
	GFVAIYLTEPLVRGFTTA-AAVHV	1.00	Sulfate transporter family profile	

sequence, showing that these signals are strong. Those motifs without known associations, then, become prime targets for further experimental investigation. The XUP domain motifs only had 2 known associations out of the top 5 motifs but retained high frequency and each motif again had exactly one instance per input sequence. Thus, we have removed signals that are weak from the results and kept those frequently occurring motifs that have the potential to correspond to known functions. These results and the use of other motif databases can help to further narrow the search space for definition of important patterns and residues within our data.

Figure 8 is the 3D structure (PDB entry ID: 1AUZ) of the hypothetical STAS domain structure based on bacterial SpoIIAA homolog. Specifically, the DSVG sub-motif found from the expanded approach STAS domain results sits in the α3 region and is a known phosphorylation site for bacterial SpoIIAA. Other STAS domain motifs found spatially fall under the following areas: Motifs 1 and 2, with no known associations, fall in the β1 and N-terminus regions, respectively. Motifs 3 and 4, which are potential sites for casein kinase II phosphorylation, fall in the β3 (back purple) and C-terminus regions, respectively. The location of potential phosphorylation sites fall on regions that would hypothetically be exposed and thus able to be accessed by the proper enzymes.

CONCLUSION

In this report, we compared the traditional approach to an expanded approach to sequence analysis. Results suggest that our model for expanded approach can lead to more informative results by lending explanation for results that are less than optimal, or by improving the focus by which results are obtained and thus narrowing the search.

The expanded approach establishes that there are three biological relationships that should be addressed when performing any sequence analysis: evolution, structure, and function. These relationships affect both the preparation of the input and interpretation of the output. Currently, these relationships must be addressed by the user of methods, because automation is currently infeasible due to issues such as lack of diverse species' sequences, an established molecular clock, and known crystal structure. However, the expanded approach represents a movement

Figure 8. Proposed structure of the STAS domain based upon the bacterial homolog SpoIIAA.

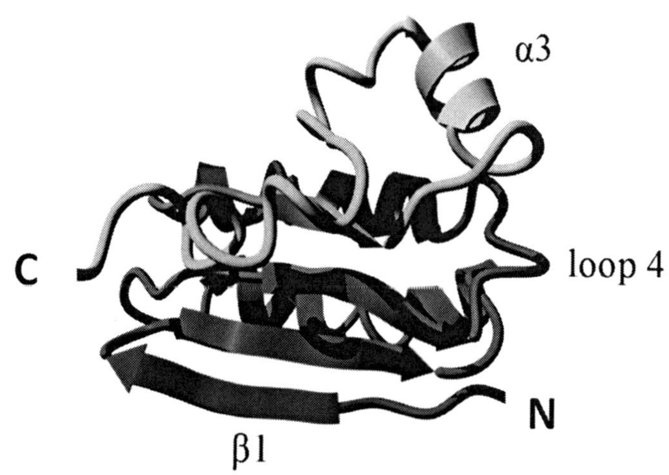

towards a more 'intelligent' sequence analysis which takes into account more of the available information within sequence databases. It further highlights the burden of both the users, typically biologists, and designers, typically the informaticians, to understand the limits and expectations of sequence analysis. Proper use of this intelligent sequence analysis has the potential to be a more focused pathway for experimental validation, saving time, resources and labor on behalf of the researcher and their laboratory.

The typical sequence analysis is simply a means for revealing information from data, for discerning knowledge from noise. When viewed this way, it becomes apparent that careful preparation of data is critical to reveal common and unique structure and function. By including discriminatory criteria for evolutionary relationships, we can infer common structure and/or function within a set of sequences.

ACKNOWLEDGMENT

The authors would like to thank Dr. Dhundy Bastola for his help in concepts regarding this work. The project was partially funded by the NIH grant number P20 RR016469 from the INBRE Program of the National Center of Research Resources.

REFERENCES

Altschul, S. F., Gish, W., Miller, W., Myers, E. W., & Lipman, D. J. (1990). Basic local alignment search tool. *Journal of Molecular Biology*, *215*(3), 403–410.

Ashmore, J. (2008). Cochlear outer hair cell motility. *Physiological Reviews*, *88*(1), 173–210. doi:10.1152/physrev.00044.2006

Atchley, W. R., Wollenberg, K. R., Fitch, W. M., Terhalle, W., & Dress, A. W. (2000). Correlations among amino acid sites in bHLH protein domains: An information theoretic analysis. *Molecular Biology and Evolution*, *17*(1), 164–178.

Benson, S. D., Bamford, J. K., Bamford, D. H., & Burnett, R. M. (2004). Does common architecture reveal a viral lineage spanning all three domains of life? *Molecular Cell*, *16*(5), 673–685. doi:10.1016/j.molcel.2004.11.016

Boekhoff-Falk, G. (2005). Hearing in drosophila: Development of Johnston's organ and emerging parallels to vertebrate ear development. *Developmental Dynamics: An Official Publication of the American Association of Anatomists*, *232*(3), 550–558.

Brenner, S. E. (2001). A tour of structural genomics. *Nature Reviews. Genetics*, *2*(10), 801–809. doi:10.1038/35093574

Cheatham, M. A., Zheng, J., Huynh, K. H., Du, G. G., Gao, J., & Zuo, J. (2005). Cochlear function in mice with only one copy of the prestin gene. *The Journal of Physiology*, *569*(Pt 1), 229–241. doi:10.1113/jphysiol.2005.093518

Clote, P., & Backhofen, R. (2000). *Computational molecular biology: An introduction*. Hoboken, NJ: Wiley.

Deluca, T. F., Wu, I. H., Pu, J., Monaghan, T., Peshkin, L., & Singh, S. (2006). Roundup: A multi-genome repository of orthologs and evolutionary distances. *Bioinformatics (Oxford, England)*, *22*(16), 2044–2046. doi:10.1093/bioinformatics/btl286

Dempsey, K., Currall, B., Hallworth, R., & Ali, H. (In press). An intelligent data-centric approach toward identification of conserved motifs in protein sequences. In *ACM International Conference on Bioinformatics and Computational Biology 2010*.

Dorwart, M. R., Shcheynikov, N., Yang, D., & Muallem, S. (2008). The solute carrier 26 family of proteins in epithelial ion transport. *Physiology (Bethesda, MD)*, *23*, 104–114. doi:10.1152/physiol.00037.2007

Edgar, R. C. (2004). MUSCLE: A multiple sequence alignment method with reduced time and space complexity. *BMC Bioinformatics*, *5*, 113. doi:10.1186/1471-2105-5-113

Fitch, W. M. (1969). Locating gaps in amino acid sequences to optimize the homology between two proteins. *Biochemical Genetics*, *3*(2), 99–108. doi:10.1007/BF00520346

Gibbs, A.J. & McIntyre, G.A. (1970). The diagram, a method for comparing sequences. Its use with amino acid and nucleotide sequences. *European Journal of Biochemistry / FEBS, 16*(1), 1-11.

Gouveia-Oliveira, R., Roque, F. S., Wernersson, R., Sicheritz-Ponten, T., Sackett, P. W., & Molgaard, A. (2009). InterMap3D: Predicting and visualizing co-evolving protein residues. *Bioinformatics (Oxford, England)*, *25*(15), 1963–1965. doi:10.1093/bioinformatics/btp335

Hedges, S. B., Blair, J. E., Venturi, M. L., & Shoe, J. L. (2004). A molecular timescale of eukaryote evolution and the rise of complex multicellular life. *BMC Evolutionary Biology*, *4*, 2. doi:10.1186/1471-2148-4-2

Kumar, S., Filipski, A., Swarna, V., Walker, A., & Hedges, S. B. (2005). Placing confidence limits on the molecular age of the human-chimpanzee divergence. *Proceedings of the National Academy of Sciences of the United States of America*, *102*(52), 18842–18847. doi:10.1073/pnas.0509585102

Larkin, M. A., Blackshields, G., Brown, N. P., Chenna, R., McGettigan, P. A., & McWilliam, H. (2007). Clustal W and clustal X version 2.0. *Bioinformatics (Oxford, England)*, *23*(21), 2947–2948. doi:10.1093/bioinformatics/btm404

Lawrence, C. E., Altschul, S. F., Boguski, M. S., Liu, J. S., Neuwald, A. F., & Wootton, J. C. (1993). Detecting subtle sequence signals: A Gibbs sampling strategy for multiple alignment. *Science*, *262*(5131), 208–214. doi:10.1126/science.8211139

Lin, Y., & Moret, B. M. (2008). Estimating true evolutionary distances under the DCJ model. *Bioinformatics (Oxford, England)*, *24*(13), i114–i122. doi:10.1093/bioinformatics/btn148

Madej, T., Panchenko, A. R., Chen, J., & Bryant, S. H. (2007). Protein homologous cores and loops: Important clues to evolutionary relationships between structurally similar proteins. *BMC Structural Biology*, *7*, 23. doi:10.1186/1472-6807-7-23

Markel, S., & Leon, D. (2003). *Sequence analysis in a nutshell*. Sebastopol, CA: O'Reilly.

Mount, D. B., & Romero, M. F. (2004). The SLC26 gene family of multifunctional anion exchangers. *Pflugers Archive: European Journal of Physiology*, *447*(5), 710–721. doi:10.1007/s00424-003-1090-3

Pei, J., & Grishin, N. V. (2001). AL2CO: Calculation of positional conservation in a protein sequence alignment. *Bioinformatics (Oxford, England)*, *17*(8), 700–712. doi:10.1093/bioinformatics/17.8.700

Peterson, E. L., Kondev, J., Theriot, J. A., & Phillips, R. (2009). Reduced amino acid alphabets exhibit an improved sensitivity and selectivity in fold assignment. *Bioinformatics (Oxford, England)*, *25*(11), 1356–1362. doi:10.1093/bioinformatics/btp164

Poirot, O., Suhre, K., Abergel, C., O'Toole, E., & Notredame, C. (2004). 3DCoffee@igs: A Web server for combining sequences and structures into a multiple sequence alignment. *Nucleic Acids Research*, *32*, W37-40. doi:10.1093/nar/gkh382

Quest, D., Dempsey, K., Shafiullah, M., Bastola, D., & Ali, H. (2008). MTAP: The motif tool assessment platform. *BMC Bioinformatics*, *9*(9), S6. doi:10.1186/1471-2105-9-S9-S6

Schaechinger, T. J., & Oliver, D. (2007). Nonmammalian orthologs of prestin (SLC26A5) are electrogenic divalent/chloride anion exchangers. *Proceedings of the National Academy of Sciences of the United States of America*, *104*(18), 7693–7698. doi:10.1073/pnas.0608583104

Tompa, M., Li, N., Bailey, T. L., Church, G. M., De Moor, B., & Eskin, E. (2005). Assessing computational tools for the discovery of transcription factor binding sites. *Nature Biotechnology*, *23*(1), 137–144. doi:10.1038/nbt1053

Vinga, S., & Almeida, J. (2003). Alignment-free sequence comparison-a review. *Bioinformatics (Oxford, England)*, *19*(4), 513–523. doi:10.1093/bioinformatics/btg005

Wagner, H., Morgenstern, B., & Dress, A. (2008). Stability of multiple alignments and phylogenetic trees: An analysis of ABC-transporter proteins family. *Algorithms for Molecular Biology; AMB*, *3*, 15. doi:10.1186/1748-7188-3-15

Zheng, J., Shen, W., He, D. Z., Long, K. B., Madison, L. D., & Dallos, P. (2000). Prestin is the motor protein of cochlear outer hair cells. *Nature*, *405*(6783), 149–155. doi:10.1038/35012009

ADDITIONAL READING

Frank, G., Hemmert, W., & Gummer, A. W. (1999). Limiting dynamics of high-frequency electromechanical transduction of outer hair cells. *Proceedings of the National Academy of Sciences of the United States of America*, *96*(8), 4420–4425. doi:10.1073/pnas.96.8.4420

Ito, K., Ikebe, M., Kashiyama, T., Mogami, T., Kon, T., & Yamamoto, K. (2007). Kinetic mechanism of the fastest motor protein, chara myosin. *The Journal of Biological Chemistry*, *282*(27), 19534–19545. doi:10.1074/jbc.M611802200

Jankun-Kelly, T. J., Lindeman, A. D., & Bridges, S. M. (2009). Exploratory visual analysis of conserved domains on multiple sequence alignments. *BMC Bioinformatics*, *10*(Suppl 11), S7. doi:10.1186/1471-2105-10-S11-S7

Koehl, P. (2001). Protein structure similarities. *Current Opinion in Structural Biology*, *11*(3), 348–353. doi:10.1016/S0959-440X(00)00214-1

Lichtarge, O., Bourne, H. R., & Cohen, F. E. (1996). An evolutionary trace method defines binding surfaces common to protein families. *Journal of Molecular Biology*, *257*(2), 342–358. doi:10.1006/jmbi.1996.0167

Marschall, T., & Rahmann, S. (2009). Efficient exact motif discovery. *Bioinformatics (Oxford, England)*, *25*(12), i356–i364. doi:10.1093/bioinformatics/btp188

Martin, L. C., Gloor, G. B., Dunn, S. D., & Wahl, L. M. (2005). Using information theory to search for co-evolving residues in proteins. *Bioinformatics (Oxford, England)*, *21*(22), 4116–4124. doi:10.1093/bioinformatics/bti671

Mio, K., Kubo, Y., Ogura, T., Yamamoto, T., Arisaka, F., & Sato, C. (2008). The motor protein prestin is a bullet-shaped molecule with inner cavities. *The Journal of Biological Chemistry*, *283*(2), 1137–1145. doi:10.1074/jbc.M702681200

Mount, D. B. (2001). *Bioinformatics: Sequence and Genome Analysis*. Cold Spring Harbor, NY: Cold Spring Harbor Press.

Rouached, H., Berthomieu, P., El Kassis, E., Cathala, N., Catherinot, V., & Labesse, G. (2005). Structural and functional analysis of the C-terminal STAS (sulfate transporter and anti-sigma antagonist) domain of the arabidopsis thaliana sulfate transporter SULTR1.2. *The Journal of Biological Chemistry*, *280*(16), 15976–15983. doi:10.1074/jbc.M501635200

Sadreyev, R. I., & Grishin, N. V. (2004). Estimates of statistical significance for comparison of individual positions in multiple sequence alignments. *BMC Bioinformatics*, *5*, 106. doi:10.1186/1471-2105-5-106

Thompson, W., Rouchka, E. C., & Lawrence, C. E. (2003). Gibbs recursive sampler: Finding transcription factor binding sites. *Nucleic Acids Research*, *31*(13), 3580–3585. doi:10.1093/nar/gkg608

Yang, A. S., & Honig, B. (2000). An integrated approach to the analysis and modeling of protein sequences and structures. III. A comparative study of sequence conservation in protein structural families using multiple structural alignments. *Journal of Molecular Biology*, *301*(3), 691–711. doi:10.1006/jmbi.2000.3975

KEY TERMS AND DEFINITIONS

Homologs: A term for a set of two or more sequences that share some level of sequence conservation with another.

Motif Detection: The search for patterns in biopolymer sequences (e.g. DNA, RNA, protein). This is one type of sequence analysis.

Orthologs: A set of homologs that are from different organisms.

Paralogs: A set of homologs that are all from one organism.

Sequence Analysis: A term that describes any method used for comparing biological sequences or examining characteristics, typically investigated computationally.

Section 3
Biological Networks and Pathways

This section contains fourteen chapters, including four method reviews providing introductory material for this field, five specialized reviews, and five original research articles that focus on specific types of biological problems. Chapter 10 reviews the basic concepts of biological pathways and networks in detail, as well as available databases and tools for their storage and analysis. Integration of knowledge and data is presented with several applications to target discovery and disease pathway analysis. Chapter 11 thoroughly reviews existing computational methods to identify modules in biological networks. Their applications in a variety of important biological problems, such as protein function and interaction predictions and disease studies, are also discussed in detail. Chapter 12 reviews methods for constructing a functional linkage network (FLN) that consists of genes that are functionally associated. Two important applications to disease study, including prediction of disease gene and disease-disease association, are discussed. Chapter 13 reviews network-driven analysis methods with a special focus on drug target identification. Chapter 14 reviews an important and helpful pathway analysis tool, the Rat Genome Database (RGD). The novel pathway ontology for gene annotation adopted by RGD is explained and examples of pathway visualization and analysis are demonstrated using their Web service. Chapter 15 reviews several novel methods that model cellular signaling networks, and where signaling network perturbation data are analyzed by integrating multivariate measurement data to gain much needed information and knowledge about these networks. Chapter 16 reviews recent advances in the experimental and computational analysis of MAPK (mitogen-activated protein kinase) cascades, providing original insights to these important signal transduction networks. Chapter 17 reviews computational methods for the classification of cancer subtypes and the identification of deregulated pathways in different cancer subtypes. Chapter 18 reviews novel computational methods based on structures and sequences of biosynthesis enzymes in the modeling of secondary metabolite biosynthetic pathways. Chapter 19 presents an original research paper describing the analysis and prediction of metastatic relapse in breast cancer by sub-network extraction. A novel interactome-transcriptome integration method for extracting sub-networks is presented by integrating protein-protein interaction and gene expression data. Chapter 20 presents an original research article describing a novel analysis method of time-course microarray data to predict transcription factors that temporally regulate differentially expressed genes under diverse stimuli. Chapter 21 presents an original research article on the dynamic modeling and parameter optimization of the DNA damage and repair network. Chapter 22 presents an original research paper in which a novel model for building causal biological networks based on high-throughput data is described. The model is built by unifying two complimentary methods (Granger Causality Model and Dynamic Causal Model). An application to the analysis of microarray data for gene circuit construction is presented. Chapter 23 presents an original research paper that describes the development of microfluidic cell arrays for high-throughput examination of host-pathogen interactions. A prototype is presented that enables the study of the infection of human cells by up to 16 different bacterial strains.

Chapter 10
Knowledge-Driven, Data-Assisted Integrative Pathway Analytics

Padmalatha S. Reddy
Pfizer, USA

Stuart Murray
Agios Pharmaceuticals Inc, USA

Wei Liu
Agios Pharmaceuticals Inc, USA

ABSTRACT

Target and biomarker selection in drug discovery relies extensively on the use of various genomics platforms. These technologies generate large amounts of data that can be used to gain novel insights in biology. There is a strong need to mine these information-rich datasets in an effective and efficient manner. Pathway and network based approaches have become an increasingly important methodology to mine bioinformatics datasets derived from 'omics' technologies. These approaches also find use in exploring the unknown biology of a disease or functional process. This chapter provides an overview of pathway databases and network tools, network architecture, text mining and existing methods used in knowledge-driven data analysis. It shows examples of how these databases and tools can be used integratively to apply existing knowledge and network-based approach in data analytics.

INTRODUCTION

Target and Biomarker Selection in Drug discovery

A critical step in the drug discovery process is the effective selection of candidate molecular targets. Target identification and selection requires a thorough understanding of the cellular role of the target, the signaling and metabolic pathways it is involved in, and the network of interactions that are involved in the functional role of the target. Perturbations in one or more of these may be responsible for a disease state or an off-target effect during drug treatment. Companies must deploy effective methods to select the targets since

DOI: 10.4018/978-1-60960-491-2.ch010

the drug discovery and development process is expensive and time-consuming. Furthermore, it is essential to fully understand the target and disease pathways to minimize expensive late-stage failures and to successfully translate animal models into the clinical development of therapies. With the advent of high-throughput 'omics' technologies and the rise of informatics technologies, it has become possible to routinely and systematically explore targets and disease-related cellular pathways, as well as cross-talk between pathways and interaction networks. Thus, a rational pathway and network based approach for target and biomarker identification has begun to be adopted by pharmaceutical companies in the recent years.

Biological Networks and Their Descriptors

Cellular functions are carried out through a complex network of interactions between biomolecules (genes, transcripts, proteins, metabolites, miRNAs, etc.). The various interactions can be biochemical or physical, and the interconnected assembly of 'cellular machinery' can be effectively presented as an "interactome" or "network" that enables visualization of molecular relationships and the logic of their function. The topology and dynamics of these complex networks can be readily studied by graph theory. The terms "interactome", "graph" and "network" have been used interchangeably. However, "interactome" and "network" describe the physical or biological system, whereas "graph" denotes the mathematical object representing the topology of the system. Topological analysis of the networks provides information about the networks, and these are described by the following parameters (Zhu, Gerstein, & Snyder, 2007) (i) **Degree:** The number of edges connected to one node is defined as its degree. In directed networks these can be further subdivided into incoming degree, outgoing degree and total degree. A node with high degree is well connected (also called "hubs") and may play a role in maintaining network structure. Thus the number of interactions a node has positively correlates with its importance in the network. Hub nodes that represent essential genes/proteins are generally conserved in evolution (Barabasi & Oltvai, 2004). (ii) **Clustering coefficient:** The ratio of the actual number of links between a node's neighbors and the maximum possible number of links between them. A high clustering coefficient for a network indicates a small world network. (iii) **Shortest path:** For any pair of nodes, the minimum number of network edges that need to be traversed to travel from one node to another. (iv) **Characteristic path length:** The average length of "shortest paths" for all pairs of nodes. (v) **Diameter:** The maximum distance between any two nodes. The average shortest path length and diameter of a network measure the approximate distance between nodes in a network. A network with a small diameter is termed "small world", in which any two nodes can be connected with relatively short paths. (vi) **Betweenness:** The fraction of the shortest paths between all pairs of nodes that pass through one node or edge, and provides an estimate of the information flow through one node or edge. It is a better indicator for the essentiality of a gene than degree centrality (Han, 2008). (vii) **Eigenvector centrality:** A measure of the contribution of degree centrality by its neighbors. (viii) **Closeness centrality:** A measure of the centrality of a node based on how close it is to other nodes in a network.

It is generally hypothesized that perturbation of one or more nodes (gene or protein) in a network disrupts cellular pathways, functions and cellular processes, giving rise to various disease conditions. There are over 6000 human diseases caused by a defect in a single gene (McKusick, 2007). In these disorders, single gene defects are sufficient to perturb the network, resulting in the disruption of normal cellular, tissue and organ functions. A recent study demonstrated that disease causing alleles that result in truncated proteins lead to the removal of nodes from networks. Disease alleles that result in near full-length proteins lead to the removal of edges in networks. The latter group of disease alleles are more likely to be autosomal dominant and associated with structural proteins, implying that the perturbation of protein-protein interactions (edges) has a dominant negative effect (Zhong, et al., 2009). Furthermore, it has been proposed that while disease genes encode hubs in networks, they are positioned either peripherally or at topologically neutral places in the networks, so disruption of the hub (mutation) would have minimal influence on the overall network (Yildirim, Goh, Cusick, Barabasi, & Vidal, 2007).

Network Topology

The nature of biological networks is closely associated with the topology and the biological functions of the nodes and edges. Biological networks are not randomly organized; these networks have a heterogeneous edge distribution, few well connected nodes (defined as hubs) and a large number of nodes with fewer connections. This topology is defined as "scale-free", meaning that the node connectivity obeys a power law (Nikolsky, Nikolskaya, & Bugrim, 2005; Zhu, et al., 2007). A scale free system is more robust than random networks as random loss of individual nodes (through mutations, etc.) is less disruptive in scale free networks. Robustness is the ability of a network to function despite perturbations in the system (such as loss of individual nodes and edges) owing to node and edge redundancies. Robustness is a feature of cellular networks that enables the network to function in the face of cellular insult (Bauer-Mehren, Furlong, & Sanz, 2009). Signaling pathways represent examples of robust networks, with their "bow-tie toplogy with multiple inputs and outputs constrained via a conserved middle tier of signaling components" (Bauer-Mehren, et al., 2009; Hellerstein, 2008). A caveat when building an experimental network is that the observed scale-free topology may not represent the topology of the complete *in vivo* network.

Network perturbations have minimal effects on such systems as the redundancy of node-edge connections compensates for the effects of perturbation. However, perturbations of key nodes and edges in the system can lead to significant functional disruptions. It is likely that these points of fragility represent the nodes and processes that are dysfunctional in diseases. In parallel, drug treatments can be partial or ineffective due to redundancies in the target network, or have off-target effects owing to the propagation of their perturbations across fragile points in the network. Therefore, understanding the target's biological context and network of interactions at the molecular and cellular levels is crucial.

Network Modules and Motifs

Biological networks are modular in nature, meaning that various cellular functions can be represented by modules described as small, well connected networks of biomolecules, with each module representing a distinct function. These can be stable or transient modules (Spirin & Mirny, 2003). Different approaches have been proposed to identify modules from complex networks. These include: (i) Monte Carlo method for tightly connected clusters, (ii) clustering based on shortest-paths-length distribution, (iii) unsupervised graph clustering, (iv) subnetwork enrichment based on structural and functional features, one example being feedback and feed-forward loops (Nikolsky, et al., 2005). Functional modules can also be identified using causal reasoning techniques after overlaying high-throughput 'omics' data (e.g., gene expression) onto the network. Other tools and methods that are used to build and analyze subnetworks include: CFinder (Palla, Derenyi, Farkas, & Vicsek, 2005), CellCircuits (http://cellcircuits.org/search/index.html), DICS (Dietmann, Georgii, Antonov, Tsuda, & Mewes, 2009), MCL (http://www.micans.org/mcl/), NetworkBlast (Sharan, et al., 2005), DPClus (http://kanaya.naist.jp/DPClus/) and others.

Meta-Analysis and Integrated Analysis

Meta-analysis is the analysis of previously analyzed data relating to the same or similar biological phenomena or treatment studied across the same or similar technology platforms. It is a means to gain a deeper and more comprehensive understanding of biology. Expression profiling is one of the most widely used platforms in research and a dominant theme in meta-analysis publications (Kong, Mas, & Archer, 2008; Zhou, et al., 2005). Many of the gene expression datasets come from NCBI'S GEO (Barrett, et al., 2007) and EBI's Array Express databases (Parkinson, et al., 2009). Many companies often couple these with their own internal expression datasets as a standard workflow process.

It has become clear that meta analysis of many studies of the same data type is not sufficient to understand the "cellular machinery" of an experimental system. Biological complexity requires a system-wide approach to data analysis, typified by the integration of bioinformatics data using computational methods. In recent years, we have seen the emergence of broader types of meta-analysis, in particular, integrative analysis. Integrative analysis distinguishes itself from the more traditional meta-analysis as it relates to the analysis of heterogeneous types of data from inter-platform technologies rather than the more homogenous data-types typically seen in meta-analysis. Integrative analysis thus incorporates meta-analysis of multiple 'omics' data-types (transcriptional profiling, HTS, proteomics, etc.), and includes analyses of text mined and curated literature, and the use of conventional databases such as KEGG, Biocarta, OMIM, PANTHER (see Table 1), etc. This type of an approach is less prone to the influence of the limitations of a particular technology platform and attempts to provide a broader picture of the biological changes to build sounder hypotheses. Thus the gap between data generated from various 'omics' platforms and biology, which has been a traditional bottleneck (Pitluk & Khalil, 2007), can be bridged by the assembly of these various components in a manner that attempts to recapitulate the cellular milieu to enable a better comprehension of biological processes and the consequences of their deregulation. In this chapter, we provide an overview of the processes and workflows in pathway and network analytics that can help bridge this gap thereby translating data and information into knowledge.

PATHWAY AND NETWORK ANALYSIS TOOLS

Pathway building is the process of identifying and integrating entities, interactions, and associated annotations into a reusable knowledgebase format; often visually represented in the form of a network. A range of pathway and network tools have emerged, and some matured somewhat, in the last few years to help address the needs of pathway and network building. A recent survey identified and reviewed the most commonly used databases and analytical tools for biological network analysis (Pavlopoulos, Wegener, & Schneider, 2008). One of the prominent tools that is widely used in the academic community is Cytoscape (Shannon, et al., 2003). Cytoscape is an open source software, with additional plugins for expression profiling analysis, network annotation using ontologies, network topological analysis, subnetwork enrichment, alternative layout algorithms, and many more functions (http://www.cytoscape.org/index.php).

In general, there are two main objectives behind pathway and network building; "data-driven objective (DDO)" and a "knowledge-driven objective (KDO)". DDO is used to generate relationships between entities of interest based on data, perhaps to explain findings from a high-throughput

Table 1. List of all databases and tools described in this chapter

Pathway, interactome databases and network analysis tools	
BIND	www.bind.ca
Biobase	http://www.biobase-international.com
BIOCARTA	www.biocarta.com/
BioGrid	thebiogrid.org
GeneGo	http://www.genego.com
GVKBio	http://www.gvkbio.com/informatics.html
HPRD	www.hprd.org
Ingenuity	http://www.ingenuity.com
INTACT	www.ebi.ac.uk/intact/
Jubilant	http://www.jubilantbiosys.com/pathart.html
KEGG	http:/www.genome.jp/kegg
MINT	mint.bio.uniroma2.it/mint/
MIPS	http://mips.helmholtz-muenchen.de/proj/ppi/
OMIM	www.ncbi.nlm.nih.gov/omim
PANTHER	www.pantherdb.org
Reactome	www.reactome.org
Xtractor	http://www.xtractor.in
Network building and analysis tools	
Cfinder	http://cfinder.org/
DPClus	http://kanaya.naist.jp/DPClus/
DICS	http://mips.helmholtz-muenchen.de/cgi-bin/dics/dics.pl?content=about
MCL	http://www.micans.org/mcl/
Cytoscape	http://www.cytoscape.org/index.php
Gene expression databases	
Geo	http://www.ncbi.nlm.nih.gov/geo/
Array Express	http://www.ebi.ac.uk/microarray-as/ae/
BioExpress	http://www.genelogic.com/knowledge-suites/bioexpress-system
ONCOMINE	www.oncomine.org
Gene Atlas	http://biogps.gnf.org
Text Mining	
GoPubMed	http://www.gopubmed.org
Textpresso	http://www.textpresso.org/
PubGene	http://www.pubgene.com/
iHOP	http://www.ihop-net.org/UniPub/iHOP/
I2E	www.linguamatics.com
Pathway Studio	www.ariadne.com

continued on following page

Table 1. continued

Pathway, interactome databases and network analysis tools	
Other	
mirBASE	www.mirbase.org
Pathway Express	http://vortex.cs.wayne.edu/projects.htm#Pathway-Express
GSEA	www.broadinstitute.org/gsea

omics experiment, while KDO involves gathering detailed information around an area of interest such as a specific disease or a pathway or a biological process (Viswanathan, Seto, Patil, Nudelman, & Sealfon, 2008). Both approaches seek to elucidate biology and/or disease mechanisms, mechanism of action, toxicity and off-target effects of drugs, as well as biomarker discovery.

Many pathway and network tools use an underlying content database from which they derive their networks. Content falls into one of three general categories: curated extracts and findings from biomedical literature, statements from biomedical literature automatically extracted via text-mining, and genomics and large-scale study results dominated by protein-protein interactions from yeast two-hybrid studies. The biological context of extracted interactions is important in pathway and network analysis. Context can be derived by extracting information relevant to species, cells, tissues, organs, etc. The value of species-specific context lies in exploring similarities and differences between animal models and human disease since preclinical models do not always translate well into human clinical trials. This can have profound consequences for patients or clinical volunteers, as was observed in the anti-CD28 clinical trial (Suntharalingam, et al., 2006). The cellular, tissue or organ-based context of extracted interactions is equally important as protein-protein interactions may be constrained temporally or spatially by the local environment, a limitation not often considered during *in silico* modeling. It is possible to apply context-constraints during text-mining by careful selection of document corpuses and application of additional context-defining search criteria, however, this type of context-based annotation is best managed by manually curated databases.

Curated Content for Pathway and Network Analytics

Curated content from literature includes information on entities (nodes) such as endogenous and non-endogenous metabolites, chemicals, drugs, miRNA, genes, proteins and protein complexes, etc. Interactions (edges) between nodes are often represented directionally (where available) and categorized as binding, phosphorylation, dephosphorylation, activation, expression, regulation, inhibition, etc. Curated information is represented in a structured manner that simplifies and speeds accumulation and retrieval. It typically involves reading and redrafting of information in a structured format and undergoes quality checks to ensure accurate representation of the findings. Manual curation is a slow process and does not scale to the exponentially growing information in literature. Typical examples of providers of curated information include: GeneGo (http://www.genego.com), Xtractor (http://www.xtractor.in), Ingenuity (http://www.ingenuity.com), Biobase (http://www.biobase-international.com), GVK (http://www.gvkbio.com/informatics.html), and Jubilant (http://www.jubilantbiosys.com/pathart.html).

Text-Mined Content for Pathway and Network Analytics

With the exponential rise in biomedical literature, text-mining is becoming an integral part of data mining and pathway/network analytics. Text mining generally refers to the automated extraction of information from unstructured text (literature); more strictly, it can be defined as "the discovery by computer of new, previously unknown information, by automatically extracting information from different written resources" (Jensen, Saric, & Bork, 2006). The typical sources of unstructured text are biomedical abstracts or journal articles. Text-mining biomedical literature uncovers associations and patterns between related facts, often revealing hitherto unknown or overlooked knowledge. Biomedical literature mining can be divided into several disciplines related to text mining (Erhardt, Schneider, & Blaschke, 2006): (i) **Information retrieval:** finding documents about a specific biological problem or disease. (ii) **Document clustering and classification:** organizing groups of documents based on their contents. (iii) **Entity recognition:** classifies individual elements of text into predefined categories. (iv) **Information extraction:** automatic detection and retrieval of entities and assertions. (v) **Natural language processing:** 'understanding' statements in biomedical text.

There are numerous academic software tools and web-based applications for biomedical text mining, many of which have been designed to query PubMed. Typical examples of retrieval and analysis tools include: (i) **GoPubMed:** Clustering and categorization of abstracts using the Gene Ontology (GO) (Doms & Schroeder, 2005) (http://www.gopubmed.org/). (ii) **Textpresso:** Querying a sub-set of PubMed using a custom ontology for information on specific biological concept and their relations (Muller, Kenny, & Sternberg, 2004). (http://www.textpresso.org/). (iii) **PubGene:** Presents biomedical terms as graphical networks based on their co-occurrence in PubMed abstracts (Jenssen, Laegreid, Komorowski, & Hovig, 2001) (http://www.pubgene.com/). (iv) **iHOP:** NLP-based analysis of abstracts to identify and visualize interactions between genes (Hoffmann, et al., 2005) (http://www.ihop-net.org/UniPub/iHOP/). Typical examples of commercial providers of text-mining-based tools include: I2E (Linguamatics; www.linguamatics.com), Pathway Studio (Ariadne Genomics; www.ariadne.com) and Pathway Architect (Stratagene); both Pathway Studio and Pathway Architect build underlying knowledge-bases using text-mining to provide extracted facts which are then semi-curated. Table1 lists all databases and tools and the web sites described in this chapter.

Content from Genomic and Large-Scale Studies

Many interaction-based databases incorporate protein-protein interactions derived from academic and commercial genomic-scale yeast two-hybrid screens. Examples include protein-protein interactions from BIND (Gilbert, 2005), BioGrid (Breitkreutz, et al., 2008), DIP (Salwinski, et al., 2004), HPRD (Peri, et al., 2003), INTACT (Kerrien, et al., 2007), MINT (Ceol, et al.), MIPS (Pagel, et al., 2005) and others (Lehne & Schlitt, 2009).

High confidence data derived from organized, systematic and biologically relevant experimental datasets serve as an excellent source to explore and understand complex biology. The Connectivity Map, a compendium of gene signatures derived from cultured human cells treated with bioactive small molecules, is widely used to explore gene-drug-disease relationships and possible drug-repositioning (Lamb, 2007). Commercial expression profiling databases of rat tissues treated with a number of drugs are useful in exploring toxicity mechanisms. Examples include ICONIX 's Drug Matrix (Ganter, Snyder, Halbert, & Lee, 2006) and Gene Logic's BioExpress (http://www.genelogic.com/knowledge-suites/bioexpress-system) (Katz,

Irizarry, Lin, Tripputi, & Porter, 2006). Oncology-specific databases such as ONCOMINE contains gene expression profiles of normal tissues vs. tumors from humans and mice (Rhodes, et al., 2004). Investigators can also generate their own specific and relevant compendia for analysis, using repositories such as the NCBI's Gene Expression Omnibus (GEO) (Barrett, et al., 2007) and Array Express (Parkinson, et al., 2009). When creating a specific gene signature set from public databases, statistically significant changes are identified based on *p*-value (or FDR, false discovery rate) and/or fold changes, the choice of which lacks a rationale (Zhang & Cao, 2009). Imposing stringent cutoffs ensures that changes observed are reliable, and represent key perturbations in experimentally relevant pathways and or biological processes; however, subtle and significant differences may also need to be investigated to understand the disease mechanisms as they too could represent a state of disequilibrium.

Other Content

In addition, these databases include canonical pathways, "collections of reference pathways that reflect the understanding of the experts in the field" (Bauer-Mehren, et al., 2009). These metabolic and signaling pathways are often sourced from public resources such as KEGG, Biocarta, Reactome and others. Commercial providers often supplement these with their own curated pathways associated with specific cells, functions and/or diseases. In addition to canonical pathways, it is usual for data-providers to incorporate public resources such as the Gene Ontology (GO) (Ashburner, et al., 2000), OMIM (McKusick, 2007), miRNA target predictions from miRBase (Griffiths-Jones, Grocock, van Dongen, Bateman, & Enright, 2006), etc.

Recently, an important and growing source is the wealth of results from published large-scale studies, such as disease-protein complexes-tissue matrix (Lage, et al., 2008), gene-disease networks (Goh, et al., 2007), disease-drug networks (Yildirim, et al., 2007), gene-drug networks (Yildirim, et al., 2007), and gene-gene networks via functional processes (Washington, et al., 2009). Inclusion of these types of information into an analysis can help address specific biological questions.

Hypothesis Generation Using Pathway Tools

The amount of data in biomedical publications is increasing at such a pace that it is difficult for an individual to completely read all publications from a specific field of biology. A single researcher can only read a subset of the available biomedical literature, therefore, it is increasingly likely for one to be unaware of all the facts required to make logical inference or hypothesis, especially if facts are published within disconnected research areas. Hypothesis generation by literature-mining relies on the emergence of meaningful connections or associations between disconnected entities or facts. Inferring hypotheses from literature uses facts that have been extracted from different publications, which enables the assembly of new, indirect relationships around known entities. For example, one paper publishes that kinase A phosphorylates protein B, while another, unrelated paper publishes that phosphorylated protein B activates receptor C, thereby leading to the inferred relationship: kinase A -> protein B -> receptor C (A leads to C). This approach was pioneered by Swanson (Swanson, 1986) who used a simple semi-automated method to infer the new relationships to help patients suffering from Raynaud disease (Arrowsmith; http://arrowsmith.psych.uic.edu/arrowsmith_uic/index.html). The speed and scale of current literature-mining tools has enabled the development of knowledge-driven inductive methods of scientific discovery to complement the traditional hypothesis-driven deductive science. This can be characterized by the rapid "mining" of candidate hypotheses from literature, which can then be tested or validated against published or proprietary experimental data.

Pathway analysis usually incorporate common pathway and network building algorithms such as preferentially connecting nodes (grow-out/build-out), shortest path walking between nodes, finding/adding neighbors of nodes, in addition to building feature-driven subnetworks around regulators, targets or specific node types: receptors, transcription factors, kinases, or cytokines, etc. These can be presented using visualization algorithms such as spring-embedded, hierarchical or cell-location-based layouts, etc. The goal of such analysis is to uncover patterns and structures in the networks that would, otherwise, be hidden in the unstructured data. However, what is generally lacking in pathway tools is a capability to predict hypothesis. This has been somewhat addressed by a small number of tools: Pathway Express (Draghici, et al., 2007) (http://vortex.cs.wayne.edu/projects.htm#Pathway-Express) incorporates gene expression changes, molecular function and topological location plus interactions to create a "causal" network; GeneGo recently introduced a topological scoring tool driven by experimental datasets that creates a ranked list of testable hypotheses (Dezso, et al., 2009); NetworKin is an algorithm developed by Linding *et al.*, which builds phosphorylation networks using experimentally identified phosphorylation sites (Linding, et al., 2008), and Genstruct has their own proprietary algorithm that builds gene-regulatory networks using raw experimental data (Kightley, Chandra, & Elliston, 2004). Causal reasoning has emerged from perturbation analysis as a novel means to predict regulatory networks that are likely to be responsible for the changes seen in experimental data. To date, there are no commercial providers of causal reasoning software.

Frequently several hypotheses are generated from this type of analyses and they can be prioritized in a data-assisted fashion. For instance, activation of pathway A may occur by repression of pathway B or pathway C, and if the data support the repression of pathway B, the hypothesis that pathway A represses B should be scored with a higher ranking. In another scenario, pathways X and Y are significantly dysregulated, but because of database limitations, links between pathways X and Y may not exist. In these situations text-mining a broader literature range can add information about the relationships between pathways X and Y, and a relevant context can further improve the ranking of the hypothesis around the pathways X and Y.

Datasets for Network Analysis

The typical type of datasets applied to pathway and network analysis represent biomolecules that exhibit an altered state, either in a disease or through a natural biological process such as ageing, or by the administration of drugs, chemicals, or other biomolecules such as genes, proteins, or microRNAs, etc. The challenge in the past has been the translation of the gene (entity) lists into biological processes. Researchers tend to focus on a small number of genes in the list that are significantly dysregulated, the biology of which they were familiar with and represent druggable targets. A routine methodology is to uncover enrichment or over-representation of canonical pathways, molecular and/or cellular functions and disease pathways among a set of differentially regulated genes. This has limitations, one being the reliance on imposing arbitrary fold change and *p*-value cutoffs to determine the degree of enrichment (Zhang & Cao, 2009). An alternate approach is to use GSEA (gene set enrichment analysis) to score a gene list for enrichment ranked by their correlation to the study under investigation (Lamb, 2007; Mootha, et al., 2003; Subramanian, et al., 2005). GSEA approaches use a nonparametric, rank-based pattern-matching strategy to overcome having to impose arbitrary cutoffs.

Datasets used in network analysis can be sparse or non-sparse. Non-sparse datasets are well represented across the underlying corpus of pathways, functions and/or diseases, from which a meaningful understanding of the dataset can be derived. Non-sparse gene sets are readily amenable for generating enriched data-driven subnetworks. These subnetworks, coupled with over-represented canonical pathways and enrichment of ontological terms (molecular functions and/or cellular processes, and disease states) can be used to shed light on the underlying experimental biology. In contrast, sparse datasets are challenging, as their data-points are poorly represented across the underlying analysis corpus due to fewer experimental data points. Additional workflows have to be incorporated to understand sparse data. Sparse data can be transformed into non-sparse data by a number of means: (i) Supplement the filtered differential set (that defines a focus set) with a reasonably sized non-focus set, genes that still constitute as significantly changing, but with lower filter stringency (i.e. differential set with less stringency). (ii) Add in immediate neighbors using genome-scale protein-protein interaction data. (iii) Use subnetwork enrichment to add in upstream regulators and downstream targets of the sparse dataset. (iv) Incorporate known, biologically relevant genes that are present in the experimental condition but do not show differential changes in the experiment. The adopted approach to convert a sparse dataset to a non-sparse dataset depends greatly upon the hypothesis and the biology under investigation. In some cases it may not be appropriate to expand a sparse dataset and an investigator may choose to work with the sparse dataset. The decision to expand a sparse set relies upon the experience of the experimentalist and the available relevant knowledge or content. In general, data added into the sparse data must be biologically relevant as deviation from the biology of interest can profoundly distort the analysis.

INTEGRATIVE ANALYSIS AND EXAMPLES OF ITS APPLICATION

Integrating curated information, text-mined statements and experimentally derived datasets is key in pathway and network analysis as use of single data-types can prove to be overly limiting to the analysis. Integration of multiple data-types compensates for the weaknesses and incompleteness of individual data-types, thereby enabling the construction of comprehensive interactomes leading to the generation of more reliable hypotheses.

There are many examples of data integration published in literature; our goal is to provide some specific illustrative examples that can be used by researchers as a guide for their own studies. In our examples, we consider a key theme in the pathway/network workflow analytics paradigm discussed in this chapter to be the use of *a priori* knowledge to drive data analysis.

Example 1: Data Integration to Reconcile Technological Differences

The first example presented here was a study to assess pathway changes upon drug treatment by gene expression profiling and phosphoproteomics studies (unpublished, see Figure 1). Upon selecting genes whose mRNA change satisfied the desired fold change and statistical significance, and proteins whose phosphorylation status changed upon drug treatment, it was apparent that the drug was effective in inhibiting the desired pathway based on decreased phosphorylation and mRNA levels detected for proteins in this pathway. It was also noted that, for some genes whose mRNA changed upon drug treatment, there were no detectable phosphorylation status changes. It was predicted that GSK3B, a key regulatory molecule in the pathway of interest, would be inhibited upon drug treatment. Somewhat unexpectedly, its mRNA level was increased. Just by examining GSK3B gene expression data, one might mistakenly assume this represents activation of the pathway. However, while GSK3B mRNA was increased, its phosphorylation was decreased, which would lead to inactivation of GSK3B and repression of the pathway by the drug treatment. It was hypothesized that the overexpression of GSK3B

Figure 1. Example showing the integration of different data types (gene chip, phosphproteomics, text-mined relationships, and protein-protein interactions). Up-regulated genes based on gene expression profiling results include EDN1, APP, PLAB, NMT1, PRKCL1, CAPN1, PPP1R9B, protein phospahatse 1, GSK3B, H11, PPP1R5A, BCKHDA, SP1, ATF3, MAP2K7, AHR, PXN, TSC2, DDIT3, PSEN1 and down-regulated genes include CALM1, HNPRD, EIF4E, PRO1489 and LRP8 (indicated as "U" & "D" respectively). All other genes that are called "present" genes whose mRNA levels do not change upon drug treatment. These genes are used to expand the sparse dataset. A haze around the nodes reflects decreased phosphorylation status upon drug treatment, which is in agreement with the expected inhibitory function of the drug. Dark lines are protein-protein interaction evidence and all other lines are gene-gene relationships text-mined by MedScan (Ariadne Genomics). Arrow indicates the gene (GSK3B) whose mRNA and phosphorylation status changes were in opposite directions.

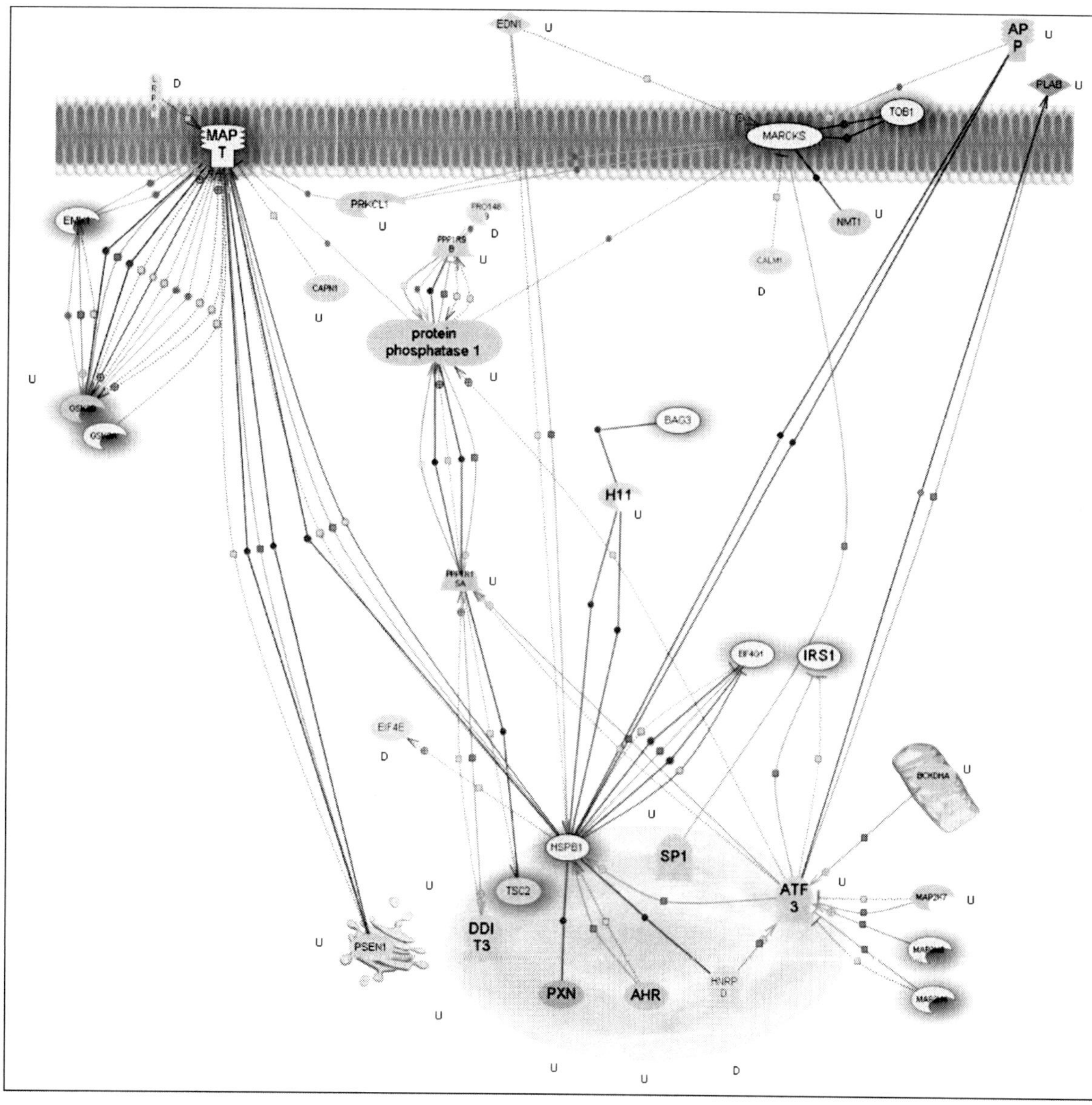

is probably due to a compensatory mechanism as a consequence of reduced pathway activity. Furthermore, by incorporating both text-mined gene-gene relationships and publically available genome-scale protein-protein interaction datasets, a model interactome could be built to account for both the observed phosphorylation and gene expression changes.

Data integration as seen in this example not only helped resolve the seemingly conflicting changes in data from two different technology platforms, but also helped uncover novel relationships among genes in the pathway, subsequently revealing novel players within the pathway.

Example 2: Data Integration for Target Discovery for Non-Small Cell Lung Cancer (NSCLC)

Lung cancer is the second most common cancer type, of which 85% are non-small cell lung cancer (NSCLC) (Weinberg, 2007). NSCLC is a complex disease that can be categorized into three subtypes: squamous cell carcinoma (25-30% of NSCLC) found in the middle of the lung near bronchi, adenocarcinoma (40% of NSCLC) usually found in the outer part of the lung, and large-cell carcinoma (10-15% of NSCLC) that can form in any part of the lung.

Due to the complexity of this disease, it is desirable to increase the "information content" for potential drug-program targets by integrating gene expression data, potential mutation sources/causes of over-expression, essentiality of genes for NSCLC cell and patient survival, and patient selection criteria for clinical trials using drugs for these targets. Targets selected from this combined information will thus carry more information for scientists to make rational decisions.

For gene expression changes, one can use public data sets such as those found in the publications of Beer et al. (Beer, et al., 2002; Bhattacharjee, et al., 2001) and Bhattacharjee et al. (Beer, et al., 2002; Bhattacharjee, et al., 2001), and the Director's challenge Consortium (Shedden, et al., 2008). These three studies represent expression profiles for 86 primary lung adenocarcinomas profiled on the Affymetrix Hu6800 chipset, 203 tumors (125 of which are adenocarcinomas with clinical information) profiled on the Affymetrix HG-U95Av2 chipset, and 442 adenocarcinoma samples profiled on the Affymetrix HG-U133A chipset, respectively, with a total of 731 different NSCLC tumors profiles on various Affymetrix microarrays. As these three data sets are compiled using different Affymetrix platforms, meta-analysis is first applied to these data sets to derive 371 genes associated with patient survival using a Cox proportional hazards model (logrank $p < 0.002$ in either direction) Beer et al. (Beer, et al., 2002; Bhattacharjee, et al., 2001). Of these 371 genes, 9 are identified in at least two data sets, these 9 include ERBB2, which is known to be involved in NSCLC progression and is a target of many developmental and clinical programs focused on NSCLC (Bianco, 2004; von Minckwitz, et al., 2005). Thus one can propose that the remaining 8 genes are of similar key importance as ERBB2 to NSCLC. The essentiality of the 371 genes to cancer cell survival can be deduced using the published studies of review on cancer genes (Futreal, et al., 2004) and oncoantigens (Cavallo, Calogero, & Forni, 2007), essential genes in cancers (Luo, et al., 2008; Silva, et al., 2008), and copy number changes in lung cancer (Weir, et al., 2007). For the 9 overlapping genes mentioned above, 5 have been shown to be essential for cancer cell survival in these studies. Therefore, inhibition of these genes is likely to affect NSCLC tumor survival.

By meta-analysis one can also identify a subset of the 371 genes that are associated with patient survival and are also over-expressed in tumors. Therefore, it would be of interest to identify the cause of their over-expression. One possible cause is increased copy number variations (CNV) of genomic loci harboring these genes. Such focal regions with CNV changes scan be estimated from gene expression profiling data using the

ACE algorithm (Hu, et al., 2009), or from array Comparative Genomic Hybridization (aCGH) studies (Tonon, et al., 2005), which used aCGH on 33 cell lines and 44 primary adenocarcinoma and squamous cell carcinoma samples, or from dense SNP chips (Weir, et al., 2007), which has 242 adenocarcinoma samples with matched normals, or that by Zhao et al.(Zhao, et al., 2005), which has 70 primary lung tumors (51 NSCLC and 19 SCLC) and 31 cell lines. Meta-analysis is again applied to these data sets from different technology platforms to derive commonly observed copy number variations that correspond to our 371-gene set. Subsequently, one can focus on focal regions with CNVs that harbor genes known to be over-expressed in tumors, associated with patient survival, and potentially could be "driver genes" for tumor progression (Bauer-Mehren, et al., 2009; Hu, et al., 2009). The gene list that satisfies all of these criteria is of particular interest as potential therapeutic targets. One could extend this further by following antibody-based therapeutic strategy to select only cell surface proteins coded by these genes as our particular interest.

In addition to harboring potential therapeutic targets these focal regions with CNV changes can also serve as biomarkers to stratify patients that could potentially benefit from therapeutics that target these genes. Text-mining was applied to more than 32,000 NSCLC-related PubMed abstracts to further strengthen the findings from data-mining. The incidence frequency (% of patient population) of genomic aberrations in the above target gene list can thus be identified from different NSCLC subtypes from both smokers and non-smokers. In addition, since data-mining points to focal genomic region changes, others genes in these focal regions that have been described in literature are also of interest as potential risk, treatment, and prognostic biomarkers. From a commercial perspective, incidence frequency provides a good estimate of the market size for a potential therapeutic target.

Example 3: Knowledge and Data Integration to Explore the Role of mTOR Pathway in Mouse Models of Lupus Nephritis and Human Lupus (Reddy, Et Al., 2008)

Our recently published study using a mouse lupus nephritis model identified a set of genes associated with lupus nephritis in an expression profiling study of kidneys from the NZBxNZW F1 strain. Treatment with Sirolimus, a mammalian target of rapamycin (mTOR) inhibitor, was efficacious in this and other mouse models of lupus nephritis, indicating a critical role for the mTOR pathway. We carried out network analysis to understand the basis of Sirolimus efficacy using 3 datasets: (i) Lupus-nephritis genes, (ii) Sirolimus-modulated lupus-nephritis genes, and (iii) Sirolimus-resistant lupus nephritis genes. The goal of this analysis was to explain the efficacy of Sirolimus in lupus nephritis, and explore the role of mTOR in human lupus. Since Sirolimus inhibits mTOR, a knowledge-based approach was taken to build the rapalog-mTOR pathway. This pathway consisted of the mTORC1 complex (mTOR, GBL, Raptor), the mTORC2 complex (mTOR, GBL, AVO3), the immediate downstream targets of mTOR (RPS6KB1 and RPS6KB2) and the upstream effectors of mTOR – (AKT1, AKT2, TSC1, TSC2). Sirolimus and other Rapamycin analogs (Rapalogs) that all target mTOR, were included in the pathway to account for Rapalog-mediated non-mTOR effects reported in literature. A shortest path network was built from the rapalog-mTOR pathway to a set of lupus nephritis genes. 15% of the lupus nephritis genes were either immediately downstream or one-step downstream from the rapalog-mTOR pathway. Interestingly, the lupus nephritis genes that are one-step downstream of the rapalog-mTOR pathway were identified as genes associated with lupus in literature. Expanding the network analysis to include all genes associated with lupus in literature places many of these known lupus disease genes either immediately downstream or one-step downstream of the rapalog-mTOR pathway, implicating mTOR as a key pathway modulator in lupus. From a commercial perspective, the lupus nephritis-related genes immediately downstream of the rapalog-mTOR pathway constitute an important group of modulators of the disease.

In a data-independent fashion, we explored the role of mTOR in human lupus through an interactome-based analysis. We built an mTOR-pathway interactome defined as a network of proteins that interacted with the rapalog-mTOR pathway. All human disease networks significantly associated with the mTOR-pathway interactome were identified using MetaCore (GeneGo). Of the 87 human diseases represented in Metacore, human systemic lupus erythmatosus (SLE) was identified as highly significant. Cancer- and non-cancer diseases were also identified through this process, including Alzheimer's disease and other autoimmune diseases such as multiple sclerosis, diabetes and arthritis.

Having investigated the Metacore-defined human disease-mTOR pathway connectivity, we explored the validity of these connections by mining literature and clinical databases for data showing the effects of rapalogs on these human diseases. Nine SLE patients who had been unsuccessfully treated with immunosuppressive drugs had significantly improved disease scores after Sirolimus treatment (BILAG $p = 0.0218$, SELDAI $p = 0.00002$). There was additional evidence in literature to support the relationship between the mTOR pathway and multiple sclerosis, diabetes, arthritis and cancers. A search of clinical trial databases revealed ongoing clinical studies with rapalogs in a number of these diseases.

From these results, the experimental design and microarray data proved to be extremely useful to identify the key pathway(s) involved in the animal model of lupus nephritis. Pathway and disease related information coupled with information specific for human diseases were key drivers in the analysis of the role of mTOR pathway in human SLE. Text mining and information from clinical trials offered additional lines of evidence to support the reported findings (Figure 2). This approach highlights the means to identifying the role of investigational pathways in other diseases, and is commercially important when considering drug positioning strategies (Reddy, et al., 2008).

Example 4: Knowledge and Data Driven Approach to Identify Subnetworks in the ErbB-MAPK Signaling Pathway that are Dysregulated Across Subsets of Breast Cancers (Heiser, et al., 2009)

Breast cancer is a heterogeneous disease, with its heterogeneity being contributed by many factors. One of the pathways dysregulated in many breast cancers is the ErbB-MAPK pathway and dysregulation in this pathway is often heterogeneous across various breast cancer subtypes. To identify subtype-specific networks in the ErbB-MAPK signaling, Heiser et al. (Heiser, et al., 2009) used a knowledge-driven approach to identify differences in the breast cancer subsets by exploring the gene expression patterns and signaling pathways, in particular the ErbB-MAPK pathway. Since there is extensive cross talk between the ErbB-MAPK pathway and other signaling pathways, the authors built an ErbB-MAPK network using Pathway Logic. Pathway Logic is a system used to model and analyze signal transduction and metabolic networks. The two components used in this system are the rules and an initial state, i.e., a representation of all proteins present in the system where components are defined as either present or absent in the experimental datasets. Rules are defined as interactions between various nodes derived from curation of literature. The ErbB-MAPK network thus comprises the 4 ErbB receptors, 11 known ligands for these receptors, direct association of

Figure 2. The analytical workflow described in (Reddy, et al., 2008)

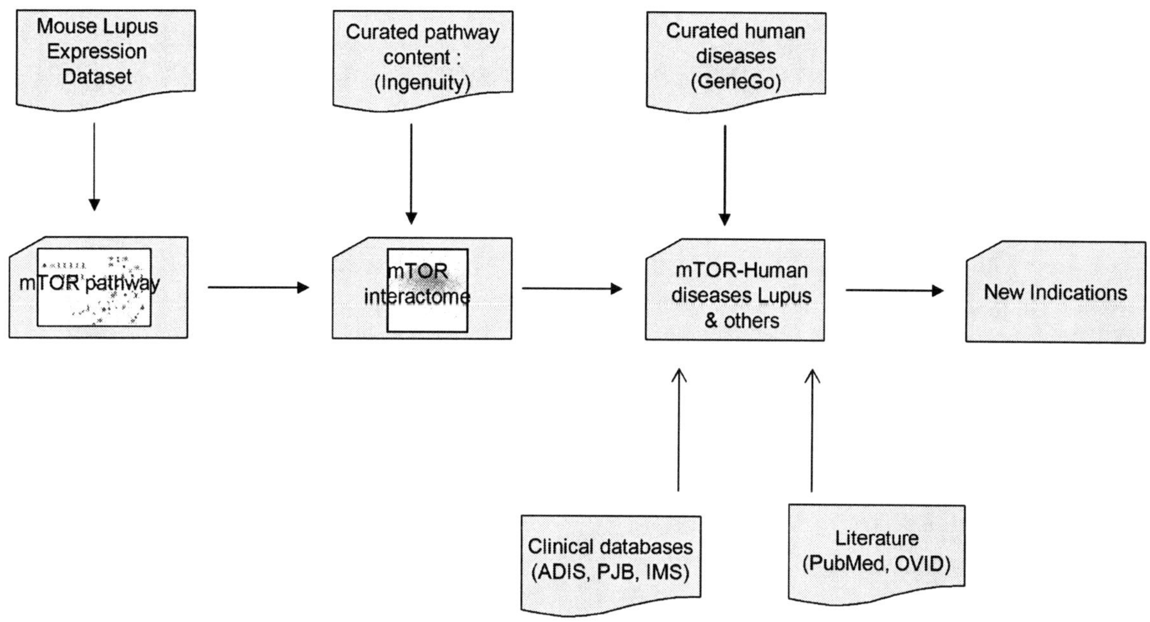

intracellular signaling proteins with the phosphorylated receptors, components of the canonical Raf-Mek-Erk signaling pathway and the cross-talk with members of the Pi3K and Jak/Stat pathways, activation of immediate early transcription factors (e.g. Jun and Fos), and upstream receptors whose activation influences MAPK signaling such as EphA2 and integrins. Expression data from a panel of 51 breast cancer cell-lines were used to define the state. The panel of cell-lines represents 2 clusters — basal and luminal subtypes, which differ in their morphology, invasive potential and other characteristics. An unsupervised hierarchical clustering of network features that comprise of the initial states, and rules and states that change in the various cell lines resulted in 3 subgroups — basal, luminal and mixed, suggesting that the grouping of cell lines is based on site of origin, and is also influenced by signaling pathways. 30 unique subnetworks were identified that were differentially present across cell lines. One example is the RhoB subnetwork or module that distinguishes basal and luminal cell lines by its presence in the latter group. A correlation between lower levels of Rho B and cancer progression supports the more invasive nature of basal cancer cell lines. Another example is the presence of three Src modules in the mixed group. PAK1-mediated regulation of MAPK signaling module was another difference noted in some cell-lines and was supported by the increased sensitivity of the PAK1 overexpressing cells to Mek inhibitors. Thus over-expression of PAK1 resulting in activation of the MAPK cascade can be used as a biomarker to identify patients that may respond to drugs that target Mek. The knowledge-based, network approach reported here led to the discovery of various modules that are activated in the ErbB-MAPK signaling network in various cell-lines, and has implications in the clinical setting as it may point the way towards individual therapeutic interventions for various types of breast cancers.

Example 5: Bridging High-Throughput Genetic and Transcriptional Data using Known Molecular Interactions to Highlight Major Response Pathways

Yeger-Lotem et al. (Yeger-Lotem, et al., 2009) recently published an elegant and novel approach to integrating high-throughput data with known molecular interaction data. Their work demonstrated the power of integrative approaches to illuminate underexplored cellular processes.

Yeger-Lotem et al. applied their integrated analytical framework to a study of alpha-synuclein toxicity implicated in Parkinson's Disease. Their screen identified genes whose over-expression altered alpha-synuclein toxicity. This set encompassed genes involved in vesicle trafficking, protein degradation, cell cycle regulation, nitrosative stress, osmolyte biosynthesis and manganese transport. Furthermore, their gene set established novel links between alpha-synuclein and cellular and environmental factors previously linked to neuropathology and Parkinson's disease. Underpinning the approach to Yeger-Lotem et al.'s study are the biases that typical high-throughput 'omic screens introduce into experimental data and the limitations associated with relying upon a single data-type. High-throughput screens are routinely used to determine genome- and proteome-wide molecular changes associated with perturbations such as disease or drug treatment. Two commonly used high-throughput approaches are genetic screens and mRNA profiling. Genetic screens include mutation, deletion, over-expression and RNAi library screens. These identify mutations/alterations in genes that are capable of influencing the phenotype of treated cells. In contrast, mRNA profiling experiments identifies genes that are differentially expressed following treatments. Yeger-Lotem et al. illustrated that these techniques limit the identification of the full nature of cellular responses, as genetic screens do not identify the same genes as mRNA assays do in the same conditions. They uncovered a marked bias in each technique: genetic assays identify the regulation of cellular responses, while mRNA profiling assays identify metabolic aspects of cellular responses. Yeger-Lotem et al. bridged this informational gap by using an algorithm that exploits molecular interaction data to reveal the functional context of genetic hits by introducing proteins that participate in the response but were not detected by either the genetic or the mRNA profiling assays.

Yeger-Lotem et al. assembled a model of the yeast interactome containing previously published protein–protein interactions, metabolic relations and protein–DNA interactions. This interactome represents 5,622 proteins and 5,510 regulated genes (nodes) via 57,955 molecular interactions (edges). As is frequently seen, connecting genetic hits to differentially expressed genes resulted in difficult-to-interpret "hairball" networks. Yeger-Lotem et al. overcame this by applying a "flow algorithm" to interpreting high-throughput 'omics' data (ResponseNet). A Flow algorithm is a computational method used previously to analyze known signaling or metabolic pathways. Flow goes from a source node to a sink node through graph edges; edges have capacity that associates the flow with a cost. By applying a minimum-cost flow optimization, the ResponseNet Algorithm gives preference to high-probability paths. The ResponseNet creates a sparse network connecting genetic hits to many differentially expressed genes through known interactions and intermediary proteins. While predicted to be involved in relevant pathways, high-throughput genetic analysis or mRNA profiling did not detect the intermediary proteins. Notably, the ResponseNet algorithm/network is consistent with general observations of biological networks: proteins are ranked by the amount of flow they carry (Betweenness) and the more flow that passes through a protein, the more important it is in connecting the input sets.

The discord between genetic screens and differentially expressed genes has significant implications for the search for therapeutic strategies: many regulatory proteins are not detected by transcriptional assays because (1) they are regulated post-transcriptionally; (2) they have a low transcript concentration; or (3) their differential expression is transient and hard to detect. Conversely, genes that are differentially transcribed are often involved in metabolic processes or redundant functions that tend to be robust to single mutations. Yeger-Lotem et al. illustrated clearly that genetic, physical and transcriptional data complement each other in the context of cellular response to biological perturbations for revealing intervention points that may provide new therapeutic opportunities.

FUTURE DIRECTIONS

In addition to the integrative analysis of the typical 'omics' data types at the knowledge, pathway and network level, there are other 'omics' data types that can benefit from the same knowledge-driven analytics approach. The first GWAS study was published in 2007 (Sladek, et al., 2007) and many GWAS studies have been undertaken since. Some limitations of GWAS studies are the high false positive rate and hidden low-frequency variants. Network based approaches are finding an increased use in prioritizing hits from these studies to explain the association with a disease or phenotype (Peng, et al.2010).

Microarray technology now allows for a large-scale evaluation of changes in microRNA (miRNA) levels. miRNAs regulate genes at the transcriptional and post-transcriptional level. There are many available tools to predict targets of miRNAs, however for a given miRNA the prediction can result in hundreds or thousands of potential targets. Interpreting relevant biology from the predicted targets of miRNAs poses a huge challenge. There is an exponential growth of information relating to miRNAs and their targets in literature, and text-mining can be used to extract miRNA-targets information. This information, coupled with curated information from pathway databases, serves as an excellent starting point for exploring miRNA biology.

Alternative splicing is prevalent in biological systems and exon arrays provide the platform to interrogate these events in disease and normal states (Okoniewski & Miller, 2008). Further analysis of hits from exon arrays poses a challenge due to the sparse information in literature on the biology of splice forms. In the absence of information on splice forms, one has to rely on what is known for the full-length transcripts. Structural genomics examining the effect of splicing on the protein structure, presence/absence of domains in the protein that result in gain/loss of interactions and functions would be the approach in such studies. Using high-confidence exon-array signature sets in a GSEA based manner will prove useful for future studies.

Knowledge gaps have an impact on the ability to explore novel biology, as can be seen in the above exon array example. There are still many hypothetical, predicted, and poorly studied genes; for these genes, experimentalists often revert to structural and evolutionary genomics. Non-functional pseudogenes with sequences similar to their normal counterparts fall into the same category. There is some evidence in literature for pseudogenes to play functional or regulatory roles (Lai, et al., 2008; Tam, et al., 2008; Zou, et al., 2009). As such, databases ought to include pseudogenes as entities. In the absence of information in literature for these genes, data-driven approaches such as co-expression patterns and other methods need to be incorporated into any relevant analysis.

Automation of the analyses described above is a challenge, since there is a strong element of judgment and skill required in the process. It is difficult to apply a set of rules as parameters and the steps needed for analysis change from one set of data to another. Nonetheless, network and pathway analytics have paved the way to target pathways rather than genes for drug-discovery. Network analytics thus represents a universal platform for data integration and analysis and has applications in drug discovery for target and biomarker discovery, mechanism of drug action, toxicology, and drug repurposing.

REFERENCES

Ashburner, M., Ball, C. A., Blake, J. A., Botstein, D., Butler, H., & Cherry, J. M. (2000). Gene ontology: Tool for the unification of biology. The Gene Ontology Consortium. *Nature Genetics*, *25*(1), 25–29. doi:10.1038/75556

Barabasi, A. L., & Oltvai, Z. N. (2004). Network biology: Understanding the cell's functional organization. *Nature Reviews. Genetics*, *5*(2), 101–113. doi:10.1038/nrg1272

Barrett, T., Troup, D. B., Wilhite, S. E., Ledoux, P., Rudnev, D., & Evangelista, C. (2007). NCBI GEO: Mining tens of millions of expression profiles-database and tools update. *Nucleic Acids Research*, *35*(Database issue), D760–D765. doi:10.1093/nar/gkl887

Bauer-Mehren, A., Furlong, L. I., & Sanz, F. (2009). Pathway databases and tools for their exploitation: Benefits, current limitations and challenges. *Molecular Systems Biology*, *5*, 290. doi:10.1038/msb.2009.47

Beer, D. G., Kardia, S. L., Huang, C. C., Giordano, T. J., Levin, A. M., & Misek, D. E. (2002). Gene-expression profiles predict survival of patients with lung adenocarcinoma. *Nature Medicine*, *8*(8), 816–824.

Bhattacharjee, A., Richards, W. G., Staunton, J., Li, C., Monti, S., & Vasa, P. (2001). Classification of human lung carcinomas by mRNA expression profiling reveals distinct adenocarcinoma subclasses. *Proceedings of the National Academy of Sciences of the United States of America*, *98*(24), 13790–13795. doi:10.1073/pnas.191502998

Bianco, A. R. (2004). Targeting c-erbB2 and other receptors of the c-erbB family: Rationale and clinical applications. *Journal of Chemotherapy (Florence, Italy)*, *16*(4), 52–54.

Breitkreutz, B. J., Stark, C., Reguly, T., Boucher, L., Breitkreutz, A., & Livstone, M. (2008). The BioGRID interaction database: 2008 update. *Nucleic Acids Research*, *36*(Database issue), D637–D640. doi:10.1093/nar/gkm1001

Cavallo, F., Calogero, R. A., & Forni, G. (2007). Are oncoantigens suitable targets for anti-tumour therapy? *Nature Reviews. Cancer*, *7*(9), 707–713. doi:10.1038/nrc2208

Ceol, A., Chatr Aryamontri, A., Licata, L., Peluso, D., Briganti, L., & Perfetto, L. (2009). MINT, the molecular interaction database: 2009 update. *Nucleic Acids Research*, *38*(Database issue), D532–D539. doi:10.1093/nar/gkp983

Dezso, Z., Nikolsky, Y., Nikolskaya, T., Miller, J., Cherba, D., & Webb, C. (2009). Identifying disease-specific genes based on their topological significance in protein networks. *BMC Systems Biology*, *3*, 36. doi:10.1186/1752-0509-3-36

Dietmann, S., Georgii, E., Antonov, A., Tsuda, K., & Mewes, H. W. (2009). The DICS repository: Module-assisted analysis of disease-related gene lists. *Bioinformatics (Oxford, England)*, *25*(6), 830–831. doi:10.1093/bioinformatics/btp055

Doms, A. & Schroeder, M. (2005). GoPubMed: Exploring PubMed with the gene ontology. *Nucleic Acids Research, 33*(Web Server issue), W783-786.

Draghici, S., Khatri, P., Tarca, A. L., Amin, K., Done, A., & Voichita, C. (2007). A systems biology approach for pathway level analysis. *Genome Research*, *17*(10), 1537–1545. doi:10.1101/gr.6202607

Erhardt, R. A., Schneider, R., & Blaschke, C. (2006). Status of text-mining techniques applied to biomedical text. *Drug Discovery Today*, *11*(7-8), 315–325. doi:10.1016/j.drudis.2006.02.011

Futreal, P. A., Coin, L., Marshall, M., Down, T., Hubbard, T., & Wooster, R. (2004). A census of human cancer genes. *Nature Reviews. Cancer*, *4*(3), 177–183. doi:10.1038/nrc1299

Ganter, B., Snyder, R. D., Halbert, D. N., & Lee, M. D. (2006). Toxicogenomics in drug discovery and development: Mechanistic analysis of compound/class-dependent effects using the DrugMatrix database. *Pharmacogenomics*, *7*(7), 1025–1044. doi:10.2217/14622416.7.7.1025

Gilbert, D. (2005). Biomolecular interaction network database. *Briefings in Bioinformatics*, *6*(2), 194–198. doi:10.1093/bib/6.2.194

Goh, K. I., Cusick, M. E., Valle, D., Childs, B., Vidal, M., & Barabasi, A. L. (2007). The human disease network. *Proceedings of the National Academy of Sciences of the United States of America*, *104*(21), 8685–8690. doi:10.1073/pnas.0701361104

Griffiths-Jones, S., Grocock, R. J., van Dongen, S., Bateman, A., & Enright, A. J. (2006). MiRBase: MicroRNA sequences, targets and gene nomenclature. *Nucleic Acids Research*, *34*(Database issue), D140–D144. doi:10.1093/nar/gkj112

Han, J. D. (2008). Understanding biological functions through molecular networks. *Cell Research*, *18*(2), 224–237. doi:10.1038/cr.2008.16

Heiser, L. M., Wang, N. J., Talcott, C. L., Laderoute, K. R., Knapp, M., & Guan, Y. (2009). Integrated analysis of breast cancer cell lines reveals unique signaling pathways. *Genome Biology*, *10*(3), R31. doi:10.1186/gb-2009-10-3-r31

Hellerstein, M. K. (2008). Exploiting complexity and the robustness of network architecture for drug discovery. *The Journal of Pharmacology and Experimental Therapeutics*, *325*(1), 1–9. doi:10.1124/jpet.107.131276

Hoffmann, R., Krallinger, M., Andres, E., Tamames, J., Blaschke, C., & Valencia, A. (2005). Text mining for metabolic pathways, signaling cascades, and protein networks. *Science's STKE*, (283): e21.

Hu, G., Chong, R. A., Yang, Q., Wei, Y., Blanco, M. A., & Li, F. (2009). MTDH activation by 8q22 genomic gain promotes chemoresistance and metastasis of poor-prognosis breast cancer. *Cancer Cell*, *15*(1), 9–20. doi:10.1016/j.ccr.2008.11.013

Jensen, L. J., Saric, J., & Bork, P. (2006). Literature mining for the biologist: From information retrieval to biological discovery. *Nature Reviews. Genetics*, *7*(2), 119–129. doi:10.1038/nrg1768

Jenssen, T. K., Laegreid, A., Komorowski, J., & Hovig, E. (2001). A literature network of human genes for high-throughput analysis of gene expression. *Nature Genetics*, *28*(1), 21–28. doi:10.1038/88213

Katz, S., Irizarry, R. A., Lin, X., Tripputi, M., & Porter, M. W. (2006). A summarization approach for Affymetrix GeneChip data using a reference training set from a large, biologically diverse database. *BMC Bioinformatics*, *7*, 464. doi:10.1186/1471-2105-7-464

Kerrien, S., Alam-Faruque, Y., Aranda, B., Bancarz, I., Bridge, A., & Derow, C. (2007). IntAct-open source resource for molecular interaction data. *Nucleic Acids Research*, *35*(Database issue), D561–D565. doi:10.1093/nar/gkl958

Kightley, D. A., Chandra, N., & Elliston, K. (2004). Inferring gene regulatory networks from raw data-a molecular epistemics approach. *Pacific Symposium of Biocomputing*, 510-520.

Kong, X., Mas, V., & Archer, K. (2008). A non-parametric meta-analysis approach for combining independent microarray datasets: Application using two microarray datasets pertaining to chronic allograft nephropathy. *BMC Bioinformatics*, 9.

Lage, K., Hansen, N. T., Karlberg, E. O., Eklund, A. C., Roque, F. S., & Donahoe, P. K. (2008). A large-scale analysis of tissue-specific pathology and gene expression of human disease genes and complexes. *Proceedings of the National Academy of Sciences of the United States of America*, *105*(52), 20870–20875. doi:10.1073/pnas.0810772105

Lai, P. C., Bahl, G., Gremigni, M., Matarazzo, V., Clot-Faybesse, O., & Ronin, C. (2008). An olfactory receptor pseudogene whose function emerged in humans: A case study in the evolution of structure-function in GPCRs. *Journal of Structural and Functional Genomics*, *9*(1-4), 29–40. doi:10.1007/s10969-008-9043-x

Lamb, J. (2007). The connectivity map: A new tool for biomedical research. *Nature Reviews. Cancer*, *7*(1), 54–60. doi:10.1038/nrc2044

Lehne, B., & Schlitt, T. (2009). Protein-protein interaction databases: Keeping up with growing interactomes. *Human Genomics*, *3*(3), 291–297.

Linding, R., Jensen, L. J., Pasculescu, A., Olhovsky, M., Colwill, K., & Bork, P. (2008). NetworKIN: A resource for exploring cellular phosphorylation networks. *Nucleic Acids Research*, *36*(Database issue), D695–D699. doi:10.1093/nar/gkm902

Luo, B., Cheung, H. W., Subramanian, A., Sharifnia, T., Okamoto, M., & Yang, X. (2008). Highly parallel identification of essential genes in cancer cells. *Proceedings of the National Academy of Sciences of the United States of America*, *105*(51), 20380–20385. doi:10.1073/pnas.0810485105

McKusick, V. A. (2007). Mendelian inheritance in man and its online version, OMIM. *American Journal of Human Genetics*, *80*(4), 588–604. doi:10.1086/514346

Mootha, V. K., Lindgren, C. M., Eriksson, K. F., Subramanian, A., Sihag, S., & Lehar, J. (2003). PGC-1alpha-responsive genes involved in oxidative phosphorylation are coordinately downregulated in human diabetes. *Nature Genetics*, *34*(3), 267–273. doi:10.1038/ng1180

Muller, H. M., Kenny, E. E., & Sternberg, P. W. (2004). Textpresso: An ontology-based information retrieval and extraction system for biological literature. *PLoS Biology*, *2*(11), e309. doi:10.1371/journal.pbio.0020309

Nikolsky, Y., Nikolskaya, T., & Bugrim, A. (2005). Biological networks and analysis of experimental data in drug discovery. *Drug Discovery Today*, *10*(9), 653–662. doi:10.1016/S1359-6446(05)03420-3

Okoniewski, M., & Miller, C. (2008). Comprehensive analysis of affymetrix exon arrays using BioConductor. *PLoS Computational Biology*, *4*(2). doi:10.1371/journal.pcbi.0040006

Pagel, P., Kovac, S., Oesterheld, M., Brauner, B., Dunger-Kaltenbach, I., & Frishman, G. (2005). The MIPS mammalian protein-protein interaction database. *Bioinformatics (Oxford, England)*, *21*(6), 832–834. doi:10.1093/bioinformatics/bti115

Palla, G., Derenyi, I., Farkas, I., & Vicsek, T. (2005). Uncovering the overlapping community structure of complex networks in nature and society. *Nature*, *435*(7043), 814–818. doi:10.1038/nature03607

Parkinson, H., Kapushesky, M., Kolesnikov, N., Rustici, G., Shojatalab, M., & Abeygunawardena, N. (2009). ArrayExpress update-from an archive of functional genomics experiments to the atlas of gene expression. *Nucleic Acids Research*, *37*(Database issue), D868–D872. doi:10.1093/nar/gkn889

Pavlopoulos, G. A. G., Wegener, A. L. A., & Schneider, R. R. (2008). A survey of visualization tools for biological network analysis. *BioData Mining*, *1*(1), 12. doi:10.1186/1756-0381-1-12

Peng, G., Luo, L., Siu, H., Zhu, Y., Hu, P., & Hong, S. (2010). Gene and pathway-based second-wave analysis of genome-wide association studies. *European Journal of Human Genetics*, *18*(1), 111–117. doi:10.1038/ejhg.2009.115

Peri, S., Navarro, J. D., Amanchy, R., Kristiansen, T. Z., Jonnalagadda, C. K., & Surendranath, V. (2003). Development of human protein reference database as an initial platform for approaching systems biology in humans. *Genome Research*, *13*(10), 2363–2371. doi:10.1101/gr.1680803

Pitluk, Z., & Khalil, I. (2007). Achieving confidence in mechanism for drug discovery and development. *Drug Discovery Today*, *12*(21-22), 924–930. doi:10.1016/j.drudis.2007.10.001

Reddy, P. S., Legault, H. M., Sypek, J. P., Collins, M. J., Goad, E., & Goldman, S. J. (2008). Mapping similarities in mTOR pathway perturbations in mouse lupus nephritis models and human lupus nephritis. *Arthritis Research & Therapy*, *10*(6), R127. doi:10.1186/ar2541

Rhodes, D. R., Yu, J., Shanker, K., Deshpande, N., Varambally, R., & Ghosh, D. (2004). ONCOMINE: A cancer microarray database and integrated data-mining platform. *Neoplasia (New York, N.Y.)*, *6*(1), 1–6.

Salwinski, L., Miller, C. S., Smith, A. J., Pettit, F. K., Bowie, J. U., & Eisenberg, D. (2004). The database of interacting proteins: 2004 update. *Nucleic Acids Research*, *32*(Database issue), D449–D451. doi:10.1093/nar/gkh086

Shannon, P., Markiel, A., Ozier, O., Baliga, N. S., Wang, J. T., & Ramage, D. (2003). Cytoscape: A software environment for integrated models of biomolecular interaction networks. *Genome Research*, *13*(11), 2498–2504. doi:10.1101/gr.1239303

Sharan, R., Suthram, S., Kelley, R. M., Kuhn, T., McCuine, S., & Uetz, P. (2005). Conserved patterns of protein interaction in multiple species. *Proceedigns of the National Academy of Sciences USA*, *102*(6), 1974–1979. doi:10.1073/pnas.0409522102

Shedden, K., Taylor, J. M., Enkemann, S. A., Tsao, M. S., Yeatman, T. J., & Gerald, W. L. (2008). Gene expression-based survival prediction in lung adenocarcinoma: A multi-site, blinded validation study. *Nature Medicine*, *14*(8), 822–827. doi:10.1038/nm.1790

Silva, J. M., Marran, K., Parker, J. S., Silva, J., Golding, M., & Schlabach, M. R. (2008). Profiling essential genes in human mammary cells by multiplex RNAi screening. *Science*, *319*(5863), 617–620. doi:10.1126/science.1149185

Sladek, R., Rocheleau, G., Rung, J., Dina, C., Shen, L., & Serre, D. (2007). A genome-wide association study identifies novel risk loci for type 2 diabetes. *Nature*, *445*(7130), 881–885. doi:10.1038/nature05616

Spirin, V., & Mirny, L. A. (2003). Protein complexes and functional modules in molecular networks. *Proceedings of the National Academy of Sciences of the United States of America*, *100*(21), 12123–12128. doi:10.1073/pnas.2032324100

Subramanian, A., Tamayo, P., Mootha, V. K., Mukherjee, S., Ebert, B. L., & Gillette, M. A. (2005). Gene set enrichment analysis: A knowledge-based approach for interpreting genome-wide expression profiles. *Proceedings of the National Academy of Sciences of the United States of America*, *102*(43), 15545–15550. doi:10.1073/pnas.0506580102

Suntharalingam, G., Perry, M. R., Ward, S., Brett, S. J., Castello-Cortes, A., & Brunner, M. D. (2006). Cytokine storm in a phase 1 trial of the anti-CD28 monoclonal antibody TGN1412. *The New England Journal of Medicine*, *355*(10), 1018–1028. doi:10.1056/NEJMoa063842

Swanson, D. R. (1986). Fish oil, Raynaud's syndrome, and undiscovered public knowledge. *Perspectives in Biology and Medicine, 30*(1), 7–18.

Tam, O. H., Aravin, A. A., Stein, P., Girard, A., Murchison, E. P., & Cheloufi, S. (2008). Pseudogene-derived small interfering RNAs regulate gene expression in mouse oocytes. *Nature, 453*(7194), 534–538. doi:10.1038/nature06904

Tonon, G., Wong, K. K., Maulik, G., Brennan, C., Feng, B., & Zhang, Y. (2005). High-resolution genomic profiles of human lung cancer. *Proceedings of the National Academy of Sciences of the United States of America, 102*(27), 9625–9630. doi:10.1073/pnas.0504126102

Viswanathan, G. A., Seto, J., Patil, S., Nudelman, G., & Sealfon, S. C. (2008). Getting started in biological pathway construction and analysis. *PLoS Computational Biology, 4*(2), e16. doi:10.1371/journal.pcbi.0040016

von Minckwitz, G., Harder, S., Hovelmann, S., Jager, E., Al-Batran, S. E., & Loibl, S. (2005). Phase I clinical study of the recombinant antibody toxin scFv(FRP5)-ETA specific for the ErbB2/HER2 receptor in patients with advanced solid malignomas. *Breast Cancer Research, 7*(5), R617–R626. doi:10.1186/bcr1264

Washington, N. L., Haendel, M. A., Mungall, C. J., Ashburner, M., Westerfield, M., & Lewis, S. E. (2009). Linking human diseases to animal models using ontology-based phenotype annotation. *PLoS Biology, 7*(11), e1000247. doi:10.1371/journal.pbio.1000247

Weinberg, R. A. (2007). *The bology of cancer.* Garland Science.

Weir, B. A., Woo, M. S., Getz, G., Perner, S., Ding, L., & Beroukhim, R. (2007). Characterizing the cancer genome in lung adenocarcinoma. *Nature, 450*(7171), 893–898. doi:10.1038/nature06358

Yeger-Lotem, E., Riva, L., Su, L. J., Gitler, A. D., Cashikar, A. G., & King, O. D. (2009). Bridging high-throughput genetic and transcriptional data reveals cellular responses to alpha-synuclein toxicity. *Nature Genetics, 41*(3), 316–323. doi:10.1038/ng.337

Yildirim, M. A., Goh, K. I., Cusick, M. E., Barabasi, A. L., & Vidal, M. (2007). Drug-target network. *Nature Biotechnology, 25*(10), 1119–1126. doi:10.1038/nbt1338

Zhang, S., & Cao, J. (2009). A close examination of double filtering with fold change and t test in microarray analysis. *BMC Bioinformatics, 10*, 402. doi:10.1186/1471-2105-10-402

Zhao, X., Weir, B. A., LaFramboise, T., Lin, M., Beroukhim, R., & Garraway, L. (2005). Homozygous deletions and chromosome amplifications in human lung carcinomas revealed by single nucleotide polymorphism array analysis. *Cancer Research, 65*(13), 5561–5570. doi:10.1158/0008-5472.CAN-04-4603

Zhong, Q., Simonis, N., Li, Q. R., Charloteaux, B., Heuze, F., & Klitgord, N. (2009). Edgetic perturbation models of human inherited disorders. *Molecular Systems Biology, 5*, 321. doi:10.1038/msb.2009.80

Zhou, X., Kao, M.-C., Huang, H., Wong, A., Nunez-Iglesias, J., & Primig, M. (2005). Functional annotation and network reconstruction through cross-platform integration of microarray data. *Nature Biotechnology, 23*(2). doi:10.1038/nbt1058

Zhu, X., Gerstein, M., & Snyder, M. (2007). Getting connected: Analysis and principles of biological networks. *Genes & Development, 21*(9), 1010–1024. doi:10.1101/gad.1528707

Zou, M., Baitei, E. Y., Alzahrani, A. S., Al-Mohanna, F., Farid, N. R., & Meyer, B. (2009). Oncogenic activation of MAP kinase by BRAF pseudogene in thyroid tumors. *Neoplasia (New York, N.Y.), 11*(1), 57–65.

KEY TERMS AND DEFINITIONS

Canonical Pathways: Collections of reference pathways that reflect the understanding of the experts in the field.

Hubs: Define a well connected node or a node with high degree.

Integrative Analysis: Analysis of heterogeneous types of data from inter-platform technologies.

Interactome or Network: Describe the interactions in a system.

Meta-Analysis: Analysis of previously analyzed data relating to the same or similar biological phenomena or treatment studied across the same or similar technology platforms.

Chapter 11
Modules in Biological Networks:
Identification and Application

Bing Zhang
Vanderbilt University School of Medicine, USA

Zhiao Shi
Vanderbilt University, USA

ABSTRACT

One of the most prominent properties of networks representing complex systems is modularity. Network-based module identification has captured the attention of a diverse group of scientists from various domains and a variety of methods have been developed. The ability to decompose complex biological systems into modules allows the use of modules rather than individual genes as units in biological studies. A modular view is shaping research methods in biology. Module-based approaches have found broad applications in protein complex identification, protein function prediction, protein expression prediction, as well as disease studies. Compared to single gene-level analyses, module-level analyses offer higher robustness and sensitivity. More importantly, module-level analyses can lead to a better understanding of the design and organization of complex biological systems.

INTRODUCTION

The twentieth-century biology has been focused on individual cellular components and their functions. Despite the huge success of this approach, a discrete biological function can rarely be attributed to an individual molecule (Hartwell et al., 1999). It is increasingly clear that the cell can be understood as a complex network of interacting components (Barabasi & Oltvai, 2004). Unraveling the interactions between the components of a cell constitutes a major goal of the post-genomic era.

With recent advances in high-throughput experimental technologies, genomic data are now available for the reconstruction of large-scale biological networks, in which nodes are biological molecules (e.g. proteins, genes, metabolites, microRNAs, etc.) and edges are functional re-

DOI: 10.4018/978-1-60960-491-2.ch011

lationships among the molecules (e.g. protein interactions, genetic interactions, transcriptional regulations, protein modifications, metabolic reactions, etc.). Biological networks that have been studied include protein interaction networks constructed from protein-protein interaction data (Gavin et al., 2006; Ito et al., 2001; Krogan et al., 2006; Rual et al., 2005; Stelzl et al., 2005; Uetz et al., 2000), gene co-expression networks constructed from gene expression profiling data (Oldham et al., 2006; Stuart et al., 2003; van Noort et al., 2004), transcriptional regulation networks constructed from protein-DNA interaction data (Harbison et al., 2004; Lee et al., 2002), and metabolic networks constructed from bioreaction data (Duarte et al., 2007; Jeong et al., 2000). At a more abstract level, functional association networks have been used to represent integrated information from various types of functional association data (Franke et al., 2006; Jensen et al., 2009).

One of the most important properties of networks representing complex systems is *modularity*, i.e., the organization of nodes in clusters, with many edges connecting nodes of the same cluster and comparatively few edges connecting nodes of different clusters (Girvan & Newman, 2002). Indeed, modularity has been observed in protein interaction (Gavin et al., 2006), transcriptional regulation (Ihmels et al., 2002), and metabolic networks (Ravasz et al., 2002). Network-based module identification has captured the attention of a diverse group of scientists from various domains such as statistical physics, computer science, discrete mathematics, sociology, and computational biology (Fortunato, 2009). Although an ideal solution remains to be reached, the enormous effort of a large interdisciplinary community of scientists has generated a variety of approaches to this problem. In the biology community, besides network-based inference, modules can also be derived from existing knowledge on pathways and biological processes (Wang et al., 2008).

Modules, by definition, are sub-groups of elements (e.g. nodes and edges in the context of networks) that function in a semi-autonomous fashion and serve as building blocks of complex systems. The ability to decompose complex biological systems into modules allows the use of modules rather than individual genes as units in biological studies. Module-based analyses have several advantages over gene-based methods, including improved robustness against the inherent noise that exists in the sample population and increased sensitivity in identifying patterns that are too subtle to discern when considering individual genes (Chuang et al., 2007; Mootha et al., 2003). More importantly, module-based analyses can achieve a higher-level understanding of the design and organization of biological systems (Gavin et al., 2006; Segal et al., 2004).

In this review, we first summarize module identification algorithms that have been developed in different research communities. Next, we provide examples of module-based applications in biological research, including protein complex identification, protein function prediction, protein expression prediction, and disease studies.

COMPUTATIONAL METHODS FOR MODULE IDENTIFICATION

Module identification in networks is the process of grouping network elements that are similar in some way into different sets, i.e. modules. Depending on different research communities, modules are also called *communities* or *clusters*. Loosely stated, a module is a cohesive set of nodes that are connected "more tightly" to each other than to other nodes in the graph. Despite of various attempts, there is no universally accepted quantitative definition of a module. In many cases, it is induced by the algorithms without a precise *a priori* definition. Not surprisingly, many module identification algorithms have been proposed based on their needs in different scientific domains (Fortunato, 2009). In this section, we present representative module

Table 1. A summary of representative module identification methods

Methods	References	Time complexity
Cluster analysis		
Hierarchical clustering	(Hastie et al., 2009)	$O(n^2 \log n)$ (complete and average linkage) $O(n^2)$ (single linkage)
k-means clustering	(MacQueen, 1967)	$O(kn)$
Kernighan-Lin algorithm	(Kernighan & Lin, 1970)	$O(n^2 \log n)$
Betweenness-based methods		
Shortest-path betweenness	(Newman & Girvan, 2004)	$O(mn)$
Random walk betweenness	(Newman, 2005)	$O((m+n)n^2)$
Current-flow betweenness	(Newman, 2005)	$O((m+n)n^2)$
Modularity optimization		
Greedy agglomerative	(Newman, 2004)	$O((m+n)n)$
Using max-heap data structure	(Clauset et al., 2004)	$O(n\log^2 n)$
Simulated annealing	(Guimerà & Amaral, 2005)	Parameter dependent
Modularity optimization in the neighborhood of each node	(Blondel et al., 2008)	$O(m)$, much more storage requirement
Extremal optimization	(Duch & Arenas, 2005)	$O(n\log^2 n)$
Spectral partitioning		
Bipartition, minimizing a normalized cut	(Shi & Malik, 2000)	$O(n^3)$
k-means clustering on *k* smallest eigenvectors	(Ng et al., 2001).	$O(n^3)$
Eigenvectors of the Laplacian matrix	(Donetti & Muñoz, 2004)	$O(n^3)$
Hierarchical clustering	(Alves, 2007)	$O(n^3)$
Clique-based methods		
CFinder	(Palla et al., 2005)	$O(e^n)$
Applying CFinder on bipartite graphs	(Lehmann et al., 2008)	$O(e^n)$
Dynamic methods		
Biased random walker	(Zhou & Lipowsky, 2004)	$O(n^3)$
Minimizing of the Hamiltonian	(Reichardt & Bornholdt, 2004)	Parameter dependent

identification methods categorized according to the approaches they take (Table 1).

Notations and Definitions

A network can be formally represented as a graph. Let $G=(V,E)$ be a simple undirected graph, where V is the set of nodes (or vertices) and E contains the edges of the graph. We use n and m to denote the number of nodes and edges in G.

Clustering

In the field of data mining and machine learning, *cluster analysis* (or *clustering*) (Hartigan, 1975) aims to organize collection of patterns into clusters based on similarity.

Clustering methods can be broadly divided into two basic types: hierarchical and partitional clustering. In hierarchical clustering, higher-level clusters are composed of subclusters. A hierarchical clustering algorithm produces a tree-structured graph called a *dendrogram*. It represents the nested grouping of patterns and the similarity levels where groupings change. The root of the dendrogram is the cluster that includes all patterns whereas the bottom level contains all the singleton clusters. An *agglomerative methods* starts with individual nodes and proceeds successively by merging smaller clusters into larger ones. Following the opposite direction, *divisive algorithms* continuously split coarse clusters into finer ones starting with the largest cluster that contains all nodes. In order to decide which cluster should be combined for agglomerative algorithms or which cluster should be split for divisive algorithms, a measure of similarity between pairs of existing clusters is required. After the measure is chosen, one computes the similarity matrix D where element d_{ij} is the similarity between node i and node j. In practice, however, one usually computes the "distance" between a pair of nodes. This is actually a measure of dissimilarity. If the graph nodes can be embedded in a q-dimensional Euclidean space, one could use measures such as Euclidean distance (L_2-norm), Manhattan distance (L_1-norm) and L_∞-norm. Given the two data points $X = (x_1, x_2, \ldots, x_q)$ and $Y = (y_1, y_2, \ldots y_q)$, these measures can be computed as follow:

$$L_2\text{-norm: } d^E = \sqrt{\sum_{i=1}^{q} (x_i - y_i)^2}$$

$$L_1\text{-norm: } d^M = \sum_{i=1}^{q} | x_i - y_i |$$

$$L_\infty\text{-norm: } d^\infty = \max_{i \in [1,q]} | x_i - y_i |$$

Another popular set of distance measures are related to correlation coefficient. For example, the Pearson distance is defined as $d^P = 1 - r_P$, where r_P is the Pearson correlation coefficient. Spearman rank correlation is more robust against outliers than the Pearson correlation. To compute the Spearman rank correlation, one replaces each data value with their ranks after ordering the data in each vector by their value. Pearson correlation is then calculated between the two rank vectors instead of the data vectors. As in the case of Pearson distance, we can define Spearman distance as $d^S = 1 - r_S$, where r_S is the Spearman correlation coefficient.

Because clusters are combined or divided based on their similarity, it is necessary to characterize how similar two clusters are based on node similarity. In *single-linkage* clustering, the distance between two clusters is the *minimum* of the distances between all pairs of nodes with one node drawn from each cluster. In *complete-linkage* clustering, the distance between two clusters is the *maximum* of the distances between all pairs of nodes from different clusters. In *average-linkage* clustering, the *average* of the distances between all pairs is used.

Unlike the hierarchical approach, partitional clustering methods produce a single flat partition of the data. They are typically robust in the sense that they are suitable for analyzing large data sets. However the number of clusters (k) needs to be predetermined. An appropriate value of k can be decided based on prior knowledge of the properties of the data set. If such knowledge does not exist, it can be decided based on the Bayesian information criterion (BIC) or an information theoretic approach (Sugar & James, 2003). Partitional clustering approaches usually start with a randomly assigned or user-defined clustering and subsequently optimize the initial clustering according to some criterion function. Perhaps the simplest and most popular partitional clustering technique is *k-means clustering* (MacQueen, 1967). The algorithm starts with randomly assigning each node to one of the k clusters. It then iterates until some criterion is met. During each iteration, two steps are performed. In the assignment step, each data point x is assigned to the cluster whose center is the closest to x. In the update step, the cluster means are adjusted to reflect the node reassignment during the previous assignment step. The iteration process stops when either there is no node reassignment in one step or the error measurement (the squared distance between each node and its cluster center) stops decreasing.

Other popular partitional clustering methods include the graph partitioning algorithms such as the Kernighan-Lin algorithm (Kernighan & Lin, 1970) and the neural network clustering methods such as the self-organizing map (SOM) (Jain et al., 1999). For a detailed discussion of these techniques, we refer to (Gan et al., 2007; Jain et al., 1999).

Betweenness-Based Methods

One way to detect modules in a graph is to identify the edges that connect the modules and remove them. However, the difficulties lie in finding a property of such edges to be used for their identification. In social network analysis, in order to understand networks and their participants, it is necessary to evaluate the "importance" of nodes or edges. The exact definition of "importance" usually depends on the applications (Freeman, 1979). A *centrality index* of a node or an edge is a numerical value that indicates such "importance". Many centrality indices have been introduced since 1950s when the first indices appeared (Brandes et al., 2007). Recently, the *betweenness centrality* has received great attention since its introduction to network community research in the field of statistical physics (Girvan & Newman, 2002). Generally speaking, betweenness centrality is an index that measures the importance of edges according to some features of the graph or certain processes conducted on the graph. There are three important betweenness centrality indices found in the literature: *shortest-path betweenness*, *random walk betweenness* and *current-flow betweenness*.

Shortest-path betweenness was first proposed to estimate how much "work" is done by each node in a communication network. It has since been adapted to measure the edge betweenness. Let $\delta_{st}(e)$ denote the fraction of shortest paths between node s and node t that pass edge e and σ_{st} the number of shortest paths between s and t. Then,

$$\delta_{st}(e) = \frac{\sigma_{st}(e)}{\sigma_{st}}$$

The shortest-path betweenness index is defined as:

$$c_B^{SP}(e) = \sum_{s \in V} \sum_{t \in V} \delta_{st}(e)$$

It sums up the relative number of shortest paths for each pair of end nodes and can be interpreted as the extent to which an edge e controls the communication between such pairs. The shortest path betweenness of all edges of a graph can be calculated in time $O(mn)$ based on breadth-first-search techniques (Newman & Girvan, 2004).

Although conceptually appealing, in most networks, however, information spreading does not necessarily follow shortest paths. It may be more reasonable to assume that information flows randomly. A random walk betweenness index of an edge thus can be similarly defined as the relative frequency of a random work traversal passes across the edge (Newman, 2005). This so-called random walk simply means that the path follows each adjacent edge with equal probability. It takes $O((m+n)n^2)$ to compute the random walk betweenness.

The idea of current-flow betweenness originates from resistor networks and is based on the flow of electric current in such networks. Each pair of nodes acts as the unit voltage source and sink and edges have unit resistance. The current flows from source to sink following different paths with lower resistance path carrying more current. By solving a system of Kirchoff's equations for each node pair, one can calculate the amount of current carried on each edge with respect to a unit supply on each node pair. The current-flow betweenness of an edge is defined as the average value of the current carried by the edge. Newman shows that current-flow betweenness is equivalent to random walk betweenness numerically (Newman, 2005). The time complexity for the calculation is also $O((m+n)n^2)$.

The betweenness based module identification methods first calculate the betweenness indices for all the edges. They then remove the edge with the largest betweenness and randomly pick one in case of tie. The indices need to be updated once the edge is removed. The algorithms repeat the previous two steps until desired number of modules is formed.

Modularity-Based Approaches

Modularity-based approaches use a quantitative metric to evaluate how good a partition is. A *quality function* is such a metric that assigns a numeric value to each partition of a graph. Using the value given by the function, one can rank the partitions. Partition with the largest value has the best quality by definition. One of the most used and best known quality functions is the so called *modularity* Q (Newman & Girvan, 2004). It is based on the idea of comparing the edge density of considered partition to that of a null model, i.e. a graph with the same similar structural properties but without any modular structure. The function is formally defined as:

$$Q \equiv \sum_{s=1}^{n_m} (e_{ss} - a_s^{\,2})$$

where e_{ss} the fraction of edges that both originate from and end in module s,

$$a_s \equiv \sum_j e_{sj}$$

is the fraction of edges that originate from module s and end in modules other than s, and n_m the number of modules. Alternatively, the function can also be written as:

$$Q \equiv \sum_{s=1}^{n_m} \left[\frac{l_s}{m} - \left(\frac{d_s}{2m} \right)^2 \right]$$

where l_s is the total number of edges joining nodes of module s and d_s the sum of the degrees of the nodes in module s. Modularity Q is always less than 1.

Since it was first proposed, many algorithms have been developed to optimize modularity based on the assumption that large values of modularity indicate good partition. However, exact optimization of Q requires enumeration of exponential number of possible partition configuration which is not computationally feasible (Brandes et al., 2007). Several heuristics have been demonstrated to obtain good approximation of the optimal solutions in reasonable time.

Newman proposed a greedy agglomerative module identification method where modules are merged only when the merging increases the modularity (Newman, 2004). Initially all modules is a singleton cluster with no edges. By adding a new edge, the number of modules decreases by one. The edge that gives the maximum increase (minimum decrease if no increase is possible) of modularity compared to the previous partition configuration is added. The process continues until a single module is formed. The partition configuration that attains the largest modularity is selected as the final solution. The time complexity for this algorithm is $O((m+n)n)$. Clauset et al. improved the algorithm by using efficient data structures such as *max-heap* (Clauset et al., 2004). It has a complexity of $O(n\log^2 n)$ on sparse graphs.

Guimerà and Amaral employed simulated annealing technique to find the global optimum of the modularity (Guimerà & Amaral, 2005). Simulated annealing is a widely used procedure for global optimization in different fields. In their implementation, a node can move from one module to another randomly. Modules are merged and split to reduce the risk of getting trapped in local minima. The method is shown to obtain results very close to the true maximum at the expense of higher computing cost (Guimerà et al., 2004).

Blondel et al. used a multistep technique based on the modularity optimization in the neighborhood of each node in a weighted graph (Blondel et al., 2008). After a partition is established, modules are replaced by supernodes. Two supernodes are connected if there is at least an edge between nodes of the corresponding modules. The weight of the edge between two supernodes is the sum of the weights of the edges between the lower level modules. The procedure is repeated until the modularity stops increasing. Note that the modularity is computed from the original graph. This method costs $O(m)$ in time but needs much more storage than the previous algorithms.

Other modularity-based approaches for identifying modules include spectral partitioning (Newman, 2006a, 2006b) which will be discussed below, extremal optimization (Duch & Arenas, 2005) and others (Fortunato, 2009). As pointed out by (Reichardt & Bornholdt, 2006), the modularity-based methods can detect a significant module structure only if the maximum modularity is considerably larger than the maximum of random graphs of the same size and expected degree sequence. Furthermore, Fortunato and Barthélemy showed that modularity optimization has a resolution limit that can possibly prevent it from finding modules that are relatively small with respect to the size of the graph even when they are well defined modules like cliques (Fortunato & Barthelemy, 2007).

Spectral Partitioning

Spectral methods that use the eigenvalues and eigenvectors of a matrix representation of a graph are widely employed to partition graphs. Typically the Laplacian matrix is used to represent a graph. The Laplacian matrix L of a graph G is the $n \times n$ matrix with the degrees of the nodes of G on the diagonal and $l_{ij} = -1$ if G has the edge (v_i, v_j) and 0 otherwise. Let $\lambda_1 \leq \lambda_2 \leq \ldots \leq \lambda_n$ be the eigenvalues and $\mu_1, \mu_2, \ldots, \mu_n$ be the corresponding eigenvectors of the Laplacian matrix. Depending on how many partitions are produced, spectral methods can be categorized into two classes. The first class generates a bi-partition of the graph with its leading eigenvector of the Laplacian (Gasch et al., 2000; Shi & Malik, 2000; Watts & Strogatz, 1998). Algorithms in the second class use multiple eigenvectors to generate several partitions of a graph.

Shi and Malik proposed the SM algorithm to create a bipartition of a graph. It first computes the eigenvector μ_2 that corresponds to the second smallest eigenvalue (Shi & Malik, 2000). A linear search on μ_2 is conducted to find a partition of the graph that minimizes a normalized cut criterion. The authors showed that when certain constraints are met, the algorithm can find the optimum of normalized cut. It is also possible to extend the algorithm to find more than two modules by applying it recursively.

The NJW algorithm finds a k-way partition of a graph using the k eigenvectors that correspond to the k smallest eigenvalues (Ng et al., 2001). The k eigenvectors are combined to form a matrix $Y = [\mu_1\ \mu_2 \ldots \mu_k]$. Then rows are normalized and embedded to a k-dimensional space. A standard k-means clustering algorithm is applied to group the rows into k groups.

Donetti and Muñoz developed an algorithm based on the idea that the values of the eigenvector components should be closer for nodes in the same module than in different modules (Donetti & Muñoz, 2004). Similarly, they used the Euclidian distance in an n-dimensional space as a metric to measure the closeness between two eigenvectors. The algorithm works as follow. First, D eigenvalues and eigenvectors of the graph Laplacian matrix are computed with the relatively fast Lanczos method (Golub & Van Loan, 1996). The number D is not given *a priori* and can be determined heuristically. It then performs either a single-linkage or complete-linkage hierarchical clustering procedure. The similarity measure is based on the Euclidian distance of eigenvectors in high dimensional space. During every step of the clustering process, modularity is computed. Once the clustering process is completed and the dendrogram is generated, the splitting that attains the maximum modularity is chosen as the output for the corresponding D.

Recently, Alves used random walk process to obtain a similarity matrix between node pairs in an electric network with edges of unit resistance (Alves, 2007). Hierarchical clustering is performed to put nodes into groups. The algorithm can easily be extended to weighted graphs.

Other types of matrices such as the *transfer matrix*, *adjacency matrix* and *right stochastic matrix* can also be used instead of the Laplacian (Capocci et al., 2005; Eriksen et al., 2003; Simonsen et al., 2004; Slanina & Zhang, 2005).

Clique-Based Methods

In graph theory, *cliques* represent perfectly dense groups (Luce & Perry, 1949). In a clique, each node has a connection with every other node. Therefore it is also perfectly compact and connected. A clique C is a maximal clique if and only if there is no clique C' in the graph with $C \subset C'$. In real world networks, large cliques are rare due to the strong structural requirement. Finding maximal cliques is a computationally difficult problem. In fact, it is one of the original problems that have been confirmed as NP-complete (Garey & Johnson, 1979). A number of algorithms since have been developed to enumerate all maximal cliques in a graph (Bron & Kerbosch, 1973; Harary & Ross, 1957; Pardalos & Xue, 1994; Tsukiyama et al., 1977). In the field of social networks, several types of relaxations of the cliques have been proposed to study the interaction between individuals, organizations and other entities (Seidman, 1983a, 1983b; Wasserman & Faust, 1994). An *n-clique* L of a graph G is a maximal subgraph of G such that for all pairs of nodes u, v of L, the distance in G $d_G(u,v) \leq n$. The distance between two nodes in L is based on their shortest path in G and may involve nodes not belonging to L. Consequently, the diameter of L may be larger than n and L might not be a connected graph. Therefore, the concept *n-clan* was introduced. An *n-clan* M of a graph G is an *n-clique* of G such that for all pairs of nodes u, v of M, the distance in M $d_M(u,v) \leq n$. A related term is *n-club* that means a maximal subgraph of diameter n. Another set of definitions is based on the degree of the module members. A *k-plex* is a maximal subgraph with the following property: each node of the induced subgraph is connected to at least n-k other nodes, where n is the number of nodes in the induced subgraph (Seidman, 1980; Seidman & Foster, 1978). A *k-core* is a maximal subgraph such that each node has at least k connections to other members of the k-core.

Palla et al. developed an algorithm called Clique Percolation Method (CPM) that uses *k-cliques* to uncover the overlapping communities for large networks (Palla et al., 2005). It is based on the idea that a larger module can be interpreted as a union of smaller maximal cliques that share subset of nodes. The smaller maximal cliques are called *k-cliques* where k refers to the number of nodes in that clique. The authors define a *k-clique-community* as the union of all k-cliques that can be reached from each other through a series of adjacent k-cliques, where two k-cliques are said to be adjacent if they share k-1 nodes. The algorithm first finds all maximal cliques for the graph. Then a clique-clique overlap matrix (Everett & Borgatti, 1998) is constructed to encode all information necessary to obtain the communities for any value of k. The k-clique-communities for a given value of k are equivalent to connected clique components in which the neighboring cliques are linked to each other by at least k-1 common nodes. The algorithm is implemented as a freely available software package called *CFinder* (Adamcsek et al., 2006) .

Lehmann et al. extended the idea of *k*-clique-community to bipartite graphs (Lehmann et al., 2008). A bipartite network is a network with two non-overlapping sets of nodes *A* and *B*, where all links must have one end node belonging to each set. Define $K_{a,b}$ clique as a complete subgraph with *a* nodes in node set *A* and *b* nodes in the node set *B*. A $K_{a,b}$ clique can be identical to a maximal complete subgraph or it can exist on a subset of the nodes of a maximal complete subgraph. Similar to *k*-clique-community, a $K_{a,b}$ clique community is defined as a union of all $K_{a,b}$ cliques that can be reached from each other through a series of adjacent $K_{a,b}$ cliques. Two $K_{a,b}$ cliques are considered as adjacent if their overlap is at least a $K_{a-1,b-1}$ biclique. The algorithm is similar to the original CPM algorithm by finding the connected components of the biclique-biclique overlap matrix.

Dynamic Methods

The methods in this category use dynamic processes, such as spin-spin interactions and random walk (Pons & Latapy, 2005; Zhou & Lipowsky, 2004), running on the graph to detect the modules. We mainly focus on the method that is based on the minimization of the Hamiltonian of a Potts-like spin model in this section.

In statistical mechanics, the Potts model (Wu, 1982) is a model of interacting spins on a crystalline lattice. It is a generalization of the Ising model to more than two components. Consider a system of spins in a plane, with each spin pointing to one of the *q* equally spaced directions. Assume that the nearest neighbor interaction depends only on the relative angle between the two vectors. The Hamiltonian of such a system is:

$$H = -\sum_{ij} J_{ij}\delta(\sigma_i, \sigma_j)$$

where $-J_{ij}$ is the interaction energy between spins *i* and *j* if they are in the same state (otherwise it is 0). σ_k represents the state of spin *k*. $\delta(\sigma_{i,}\ \sigma_j)$ is the Kronecker delta function. In a *q*-state Potts model with *N* spins, each spin can have one of the *q* states. Therefore, the system has a total of q^N configurations.

The problem of network module detection can be mapped to the problem of minimizing the Hamiltonian of a spin system (Reichardt & Bornholdt, 2004). The original Hamiltonian is modified to include a second term in favoring the diversification of spin distribution:

$$H = -J\sum_{ij}\delta(\sigma_i, \sigma_j) + \gamma\sum_{s=1}^{q}\frac{n_s(n_s - 1)}{2}$$

where n_s is the number of spins in state *s* such that

$$\sum_{s=1}^{q} n_s = N \,.$$

J denotes the ferromagnetic interaction strength. γ is a positive parameter that determines how strongly the minimum of the combined Hamiltonian depends on the topology of the graph. The minimization of *H* can be carried out with optimization methods such as simulated annealing and is relatively fast. The number of modules *q* is not a critical parameter and needs only to be large enough to accommodate for all possible modules. The algorithm can also be extended to the analysis of weighted graphs.

MODULE-BASED APPLICATIONS IN BIOLOGICAL STUDIES

Module-level analyses can help reveal the higher-level order of biological networks and have practical applications in areas such as protein complex identification, protein function prediction, protein expression prediction, disease prognosis and therapeutics, etc. In this section, we will review representative applications categorized according to the biological problems they address (Table 2).

Protein Complex Identification

Detecting protein complexes from protein interaction networks has been an active research area during the last decade. Module identification methods described in the previous section can be customized for this task. Some representative algorithms specifically applied on protein complex identification include the Molecular Complex Detection (MCODE) (Bader & Hogue, 2003), Super Paramagnetic Clustering (SPC) (Spirin & Mirny, 2003), Restricted Neighborhood Search Clustering (RNSC) (King et al., 2004), and Markov Clustering (MCL) (Enright et al., 2002).

The MCODE algorithm is based on a local search for densely connected regions in a network. It operates in three stages: vertex weighting, complex prediction and post-processing. During the first stage of vertex weighting, it weights all vertices based on their local network density using the highest *k*-core of the vertex neighborhood. Next, it takes as input the vertex weighted graph, seeds a complex with the highest weighted vertex and recursively moves outward from the seed vertex, including vertices in the complex whose weight is above a threshold, typically set as a given percentage away from the weight of the seed vertex. Finally, it filters or adds proteins in the resulting complexes by predefined connectivity criteria. MCODE is available as a plug-in in Cytoscape (Shannon et al., 2003). Because of its user friendliness and intuitiveness, it has become a popular tool for protein complex identification.

The SPC algorithm is based on spin-spin interactions. It assigns a spin to each node in the graph. Each spin can be in several (more than two) states. Spins belonging to connected nodes interact and have the lowest energy when they are in the same state. The system is subject to equilibration at nonzero temperature, making spins fluctuate. The concept behind this method is that spins belonging to a highly connected cluster fluctuate in a correlated fashion. By detecting correlated spins, the algorithm identifies nodes belonging to a highly connected area of the graph.

The RNSC algorithm is a clustering-based algorithm. It searches for a low-cost clustering by first composing an initial random clustering, then iteratively moving one node from one cluster to another in a randomized fashion to improve the clustering's cost. In general, a move that reduces the clustering cost by a near-optimal amount is selected. The program ends up when a pre-specified number of moves have been reached without decreasing the cost function.

The MCL algorithm is based on simulation of stochastic flow in graphs. The assumption is that random walks on a graph will infrequently go from one natural cluster to another. The algorithm simulates random walks within a graph by alternation of two operators called expansion and inflation. Expansion coincides with taking the power of a stochastic matrix using the normal matrix product (i.e. matrix squaring). Inflation corresponds with taking the Hadamard power of a matrix (taking powers entry-wise), followed by a scaling step, such that the resulting matrix is stochastic again, i.e. the matrix elements (on each column) correspond to probability values. The MCL process causes flow to spread out within natural clusters and evaporate between different clusters and eventually results in the separation of the graph into different clusters.

Because proteins frequently have multiple functions and may be involved in multiple complexes, allowing overlapping modules is a desired feature for protein complex identification algorithms. Among the above four algorithms, only MCODE allows overlapping modules. Other methods that allow overlapping modules include the Iterative Hierarchical Clustering (IHC) (Gavin

Table 2. A summary of representative module-based biological applications

Method	Reference	Comments
	Protein complex identification	
Molecular Complex Detection (MCODE)	(Bader & Hogue, 2003)	• Overlapping modules • Available in Cytoscape
Super Paramagnetic Clustering (SPC)	(Spirin & Mirny, 2003)	• Disjoint modules
Restricted Neighborhood Search Clustering (RNSC)	(King et al., 2004)	• Disjoint modules • Robust to parameter choice
Markov Clustering (MCL)	(Enright et al., 2002)	• Disjoint modules • Robust to false and missing edges
Iterative Hierarchical Clustering (IHC)	(Gavin et al., 2006)	• Overlapping modules • Core and attachments
Clique merging (CM)	(Zhang et al., 2008)	• Overlapping modules • Core and attachments
Heavy Subgraph Search	(Sharan et al., 2005)	• Integrate orthology information
Semantic Weights for MODule Elucidation (SWEMODE)	(Lubovac et al., 2006)	• Integrate functional information
	Protein function/expression prediction	
Neighborhood counting	(Chua et al., 2006; Hishigaki et al., 2001; Schwikowski et al., 2000)	• Easy to implement
Graph theoretic approaches	(Karaoz et al., 2004; Nabieva et al., 2005; Vazquez et al., 2003)	• Global optimization
Probabilistic approaches	(Deng et al., 2003; Letovsky & Kasif, 2003)	• Global optimization
Module-assisted methods	(Li et al., 2009; Ramakrishnan et al., 2009)	• Consider the dynamic modular organization networks
	Disease gene prediction and prioritization	
Gaussian kernel scoring	(Franke et al., 2006)	• One disease at a time
Random walk scoring	(Kohler et al., 2008)	• One disease at a time
Integrated genome-phenome network	(Lage et al., 2007; Wu et al., 2008)	• Capture the complex relationships between phenotypes and genotypes
	Elucidating disease mechanisms at a modular level	
Gene Set Enrichment Analysis (GSEA)	(Mootha et al., 2003)	• Rely on predefined modules (gene sets)
Gene module map	(Segal et al., 2004)	• Based on but not limited by predefined modules
Iterative Clique Enumeration (ICE)	(Shi et al., 2010)	• Data driven, not rely on existing knowledge
	Module-based disease prediction	
Protein interaction network-based approaches	(Chuang et al., 2007)	• Greedy search for subnetwork biomarkers
Gene module map-based approaches	(van Vliet et al., 2007)	• Extension of the unsupervised gene module map framework

et al., 2006) and clique merging (CM) (Zhang et al., 2008).

The IHC algorithm starts with a weighted graph in which the edge weight represents interaction affinity between two proteins connected by the edge. Hierarchical clustering is applied on the graph to generate an initial list of complexes. Next, the algorithm subtracts a penalty from the initial interaction affinities and repeats clustering. Tight associations are not drastically affected by the penalty, while loose ones are gradually eroded, and can be replaced by others not present initially. This procedure not only identifies overlapping complexes, but also partitions proteins in complexes into two types: *core components* that are present in most protein complex variations (or isoforms), and *attachments* present in only some of them. This important feature helps reveal the dynamic organization property of the complexes.

The CM algorithm first enumerates all maximal cliques from a protein interaction network. Next, highly overlapping cliques are merged in an iterative fashion to identify protein complexes. Similar to the IHC algorithm, this procedure can infer dynamic organization of protein complexes by separating proteins in the complexes into core proteins and attachment proteins. Besides the CM algorithm, the clique percolation method described in the previous section can also uncover overlapping structure of protein complexes (Adamcsek et al., 2006). Compared to other complex identification algorithms, the clique-based approaches have an obvious advantage of intuitiveness at the cost of higher computation complexity (Table 1).

Protein complex identification algorithms described above are based on very different principles and techniques. It is critical to evaluate the performance of the algorithms and help biologists decide which algorithm(s) to use to meet their specific needs. To this end, Brohee and Van Helden have performed a systematic quantitative evaluation of the capability of four clustering methods, including MCODE, SPC, RNSC, and MCL for identifying protein complexes from protein interaction networks (Brohee & van Helden, 2006). A reference network was built based on 220 complexes annotated in the MIPS database. Each complex was represented as a clique in the reference network. Through randomly removing edges from or adding edges to the reference network, 41 altered networks were further generated. Each clustering algorithm was applied to these networks with various parameter settings, and the clusters were compared with the annotated complexes. First, they analyzed the sensitivity of the algorithms to the parameters and determined their optimal parameter values. Results showed that under most conditions, RNSC and MCL outperformed MCODE and SPC. In general, RNSC is remarkably robust to variations in the choice of parameters, whereas the other algorithms require appropriate tuning in order to yield relevant results. Next, they evaluated the robustness of the algorithms to alterations of the reference network, using fixed parameters. Results clearly showed differences between the algorithms, highlighting the robustness of MCL, and to a lesser extent RNSC, to network alterations. They also applied the same four algorithms to interaction networks obtained from six high-throughput studies and confirmed the general superiority of MCL over the three other algorithms tested.

In a manuscript describing the CM algorithm for protein complex identification, Zhang et al. applied their proposed algorithm on a pull-down data set in yeast and assessed the qualities of the identified complexes by calculating recall and precision based on manually annotated protein complexes in the MIPS database (Zhang et al., 2008). They compared the CM results to those generated from MCODE and MCL and showed that CM performed better than both MCL and MCODE. They also evaluated the functional relevance of the protein complexes using a hypergeometric enrichment test against all Gene Ontology (GO) categories and most complexes showed high functional homogeneity according to the test.

Protein complex identification inevitably depends on the quality of protein interaction networks. Unfortunately, current protein interaction data are still sparse and unreliable. Additional information can be incorporated to improve protein complex identification.

Because many protein complexes are conserved in evolution, Sharan et al. used conservation to find complexes that are common to the yeast *Saccharomyces cerevisiae* and the bacteria *Helicobacter pylori* (Sharan et al., 2005). Their analysis combined protein interaction data that were available for each of the two species with orthology information based on protein sequence comparison. A probabilistic model for protein complexes in a single species and another for the conservation of complexes between two species were developed. Using these models, they formulated the question of finding conserved complexes as a problem of searching for heavy subgraphs in an edge- and node-weighted graph, whose nodes are orthologous protein pairs. The results demonstrated that incorporating conservation information helped achieve a much higher specificity for protein complex identification.

Lubovac et al. combined GO annotation and network connectivity for protein complex identification (Lubovac et al., 2006). Specifically, they described two alternative network measures that combine functional information with topological properties of the networks for the analysis of protein interaction networks. These measures, called *weighted clustering coefficient* and *weighted average nearest-neighbors degree*, use weights representing the strengths of interactions between the proteins. Weights are calculated according to the semantic similarity based on the GO annotations of the proteins. An algorithm named SWEMODE (Semantic WEights for MODule Elucidation) was developed to identify dense sub-graphs containing functionally similar proteins as functional modules. Using a yeast two-hybrid data set of experimentally determined protein-protein interactions, they demonstrated that SWEMODE was able to identify dense clusters containing proteins that are functionally similar. Furthermore, many of the identified modules correspond to known complexes or subunits of these complexes.

Protein Function Prediction

A major challenge of the post-genomic era is to understand the function of the proteins. One of the most successful projects in this area is the GO project (Ashburner et al., 2000). GO uses a structured, precisely defined, common, controlled vocabulary (i.e. GO terms) for describing the roles of genes and gene products in different species. The three organizing principles of GO are *cellular component*, *biological process* and *molecular function*. Genes are associated with the GO terms through manual curation as well as computational inference. Although great progresses have been made during the last decade, a significant proportion of proteins remain uncharacterized even in human and well-studied model organisms. For example, according to the Ensembl database (Version 56) (Hubbard et al., 2009), 31% of human proteins do not have a biological process annotation, 27% do not have a molecular function annotation, and 26% do not have a cellular component annotation.

Traditional approaches for the computational prediction of protein function transfer functional annotations from characterized proteins to uncharacterized ones on the basis of sequence and structure similarity (Whisstock & Lesk, 2003). Owing to the newly available large-scale protein interaction networks, network-based methods are becoming important complementary approaches for protein function prediction (Sharan et al., 2007). Unlike the traditional methods that focus on the pairwise relationship between proteins, these approaches study protein function in the context of a network.

Network-based protein function prediction can be broadly divided into direct methods and module-assisted methods and have been described in an excellent review (Sharan et al., 2007). Direct methods share the same assumption that proteins that lie closer to one another in a network are more likely to have similar function. Based on this assumption, various methods have been proposed, including neighborhood counting (Chua et al.,

2006; Hishigaki et al., 2001; Schwikowski et al., 2000), graph theoretic approaches (Karaoz et al., 2004; Nabieva et al., 2005; Vazquez et al., 2003), and probabilistic approaches (Deng et al., 2003; Letovsky & Kasif, 2003).

Instead of predicting functions for individual proteins, the module-based approaches use modules as units for function prediction. First, modules or coherent groups of proteins are identified from a protein interaction network. Methods described in the previous section can be applied directly for this purpose. Next, functions can be assigned to unannotated proteins in a given module based on the function of annotated proteins in the same module. Usually, for each module, the hypergeometric test is used to evaluate the enrichment for different functions, and significantly enriched functions are assigned to the unannotated proteins in the module. This workflow is conceptually similar to module-based protein expression prediction, which will be described in detail in the next section.

An important aspect of protein function prediction is to carefully benchmark the predictions. Although information on the performance of individual methods is usually available in corresponding studies, comprehensive comparisons of the available methods, including both direct and module-assisted methods, are in great need.

Effective prediction relies on the quality of the protein interaction network. False predictions arise when edges are made between functionally unrelated genes or edges are missing between functionally related ones. It has been estimated that our knowledge on human protein interaction is currently 10–30% complete (Hart et al., 2006). Besides protein interaction networks, functional association networks that incorporate protein interaction, gene co-expression, and other types of functional associations could also be used for protein function prediction. Functional association network should have better coverage and this type of data can be obtained from online databases such as the STRING (Jensen et al., 2009).

Prediction of Protein Expression

Recently, network-assisted approaches have been employed to improve protein identification in shotgun proteomics (Li et al., 2009; Ramakrishnan et al., 2009). Shotgun proteomics is a powerful technology for protein identification in complex samples with remarkable applications in elucidating cellular and subcellular proteomes (Foster et al., 2006; Kislinger et al., 2006), mapping protein interaction networks (Gavin et al., 2006; Krogan et al., 2006), and discovering disease biomarkers (Decramer et al., 2006; Whiteaker et al., 2007). In a typical shotgun proteomics experiment, proteins in a complex mixture are digested by sequence-specific enzymes and the resulting peptides are analyzed by tandem mass spectrometry (MS/MS). Next, MS/MS data acquired from the analyses are processed to identify peptides that gave rise to observed spectra. Finally, proteins are inferred based on peptide identifications and reported.

In classical protein assembly pipelines (Nesvizhskii et al., 2003; Yang et al., 2004; Zhang et al., 2007), proteins are considered as independent entities. To ensure the reliability of protein identification, a large number of possible but non-confident proteins are eliminated, including those supported by single peptide and those without distinct peptide evidence (Nesvizhskii & Aebersold, 2005; Zhang et al., 2007). Such conservative assembly may eliminate more than half of all possible proteins, including some truly expressed proteins that could contribute to the systematic understanding of the biological systems.

We have developed a protein interaction network-assisted *clique-enrichment approach* (CEA) to improve protein identification by taking into consideration the functional relationship among proteins as embedded in protein interaction networks (Li et al., 2009). The underlying hypothesis of the approach is that an eliminated protein is more likely to be expressed in the original sample if it is a member of a complex for which other members have been confidently identified in the same sample. The workflow is illustrated in Figure 1. First, peptide identification and protein assembly are processed using standard methods. All possible proteins are grouped into confident proteins and non-confident proteins after protein assembly and then mapped to the protein interaction network. In this network, vertices representing confident proteins are labeled as positive (red), vertices representing proteins with no experimental evidence are labeled as negative (blue), and vertices representing non-confident proteins are unlabeled (grey). Next, all maximal cliques in the protein interaction network are identified. For each identified clique, an enrichment score derived from the Fisher's exact test is used to evaluate the enrichment of confident proteins in the clique. All non-confident proteins that coexist in a clique enriched with confident proteins are thus rescued and added to the final list, whereas others are discarded. The enrichment threshold is set to achieve desired sensitivity and specificity using cross-validation.

In several data sets tested, CEA increased protein identification by 8-23% with an estimated accuracy of 85%. Rescued proteins were supported by existing literature or transcriptome profiling studies at similar levels as confident proteins and at a significantly higher level than abandoned ones. Applying CEA on a breast cancer data set rescued proteins coded by well-known breast cancer genes.

It has also been demonstrated in the study that the module-based CEA approach compares favorably to direct methods including the neighborhood counting method and the Hopfield method (Karaoz et al., 2004). Better performance of the CEA was attributed to its ability to capture the modular architecture of protein interaction networks. Although all three methods are based on the evaluation of neighborhood enrichment, neighborhood counting and Hopfield do not investigate proteins in a modular context. Instead, all interacting proteins are considered equally and simultaneously. A protein interaction network only represents a

Figure 1. Workflow of the clique-enrichment approach (CEA) for protein identification in shotgun proteomics. Adapted from (Li et al., 2009).

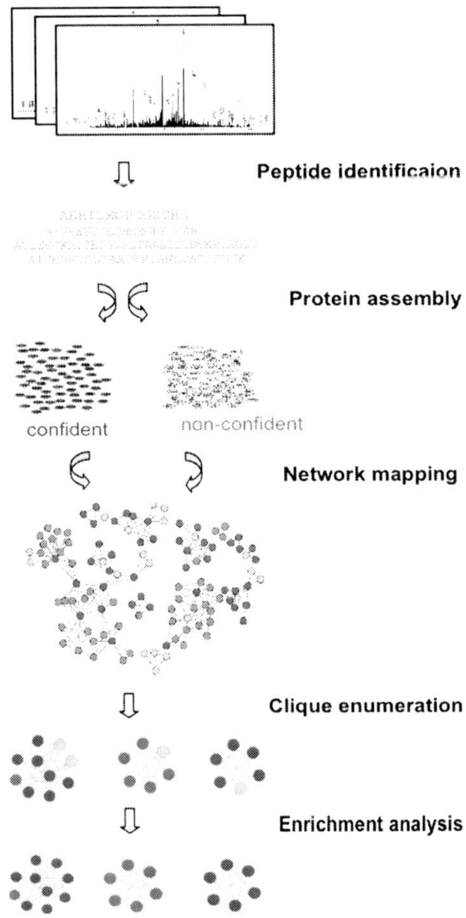

collection of possible interactions under many different conditions. Although a protein may be involved in many modules, not all of them are required for a given condition. The evidence of expression for one of the modules is enough to infer the expression of a protein within the module. Considering the dynamic modular organization of the network, CEA focuses on the most enriched clique and gains sensitivity. On the other hand, even under specific conditions, one protein can be involved in multiple modules to perform different functions. False negative identifications in one module will not necessarily affect other modules. Given the multifunctional nature of proteins, CEA gained robustness by evaluating all possible cliques separately. This result highlights the potential advantage of module-based prediction method as compared to direct methods. However, whether this holds true in the context of network-based protein function prediction will require a separate evaluation.

Module-Based Approaches in Disease Studies

Genes associated with similar disorders show both higher likelihood of physical interactions between their products and higher expression profiling similarity for their transcripts, supporting the existence of distinct disease-specific functional modules (Goh et al., 2007). In recent years, module-based approaches have been used to predict and prioritize disease genes, elucidate disease mechanisms, and build prediction models for disease prognosis.

Disease Gene Prediction and Prioritization

Linkage analysis and genome-wide association studies typically generate large sets of potential candidate disease genes. Network-based approaches have been developed to prioritize the candidates for future validation. The common hypothesis underlying these approaches is that disease genes tend to cluster into a few functional modules in a protein interaction network or functional association network.

Franke et al. developed a network-based approach for prioritizing positional candidate genes identified in linkage studies (Franke et al., 2006). A functional human gene network was constructed through the integration of information on protein interaction, pathways, gene ontology and gene co-expression. Using the functional association network, positional candidate genes that resided in different loci but that are close to each other in the network are assigned higher scores than those that are far apart from each other. The distance between two genes is defined as the shortest path length in the network. They showed that the method could significantly reduce the cost and effort of pinpointing true disease genes when numerous loci have been reported for a disease.

Kohler et al. presented another method based on the same principle (Kohler et al., 2008). However, they used a global network distance measure based on random walk analysis for the definition of gene similarities in the network. Results showed that the global network-similarity measure is better suited to capture relationships between disease genes than algorithms based on direct interactions or shortest paths between disease genes. In the comparative evaluation, the proposed method outperformed the non-network based method ENDEAVOUR (Aerts et al., 2006).

The above methods focus on one disease at a time. Because diseases with overlapping clinical manifestations are usually caused by mutations in different genes that are part of the same functional module, recent works attempt to improve disease gene prediction by using information on related diseases (Lage et al., 2007; Wu et al., 2008). The framework introduced by Lage et al. integrates human protein interaction network with computationally derived phenotype similarity score to prioritize positional candidate genes for a disease (Lage et al., 2007). First, a virtual pull-down of each candidate identifies putative protein complexes containing the candidate. Next, for each complex, proteins known to be involved in disorders similar to the disease of interest are identified, where similarity between diseases is measured by text-mining. Finally, a Bayesian predictor is used to score the complexes, and all candidates in the interval are ranked based on corresponding complex scores.

Similarly, Wu et al. proposed a computational framework that integrates human protein interactions, disease phenotype similarities and known gene-phenotype associations to capture the complex relationships between phenotypes and genotypes (Wu et al., 2008). They showed that the global concordance between the human protein network and the phenotype network could reliably predict disease genes. The success of the framework was attributed to its ability to exploit the modularity of genetic diseases more comprehensively. The disease gene prediction and prioritization framework is available through a tool named CIPHER.

Besides candidate gene prioritization, module-based methods can also be used to expand disease gene lists. Starting with four known genes encoding tumor suppressors of breast cancer, Pujana et al. used a network modeling strategy to identify genes potentially associated with high risk of breast cancer (Pujana et al., 2007). This interesting study not only identified novel breast cancer-associated genes, but also linked breast cancer susceptibility and centrosome dysfunction.

Elucidating Disease Mechanisms at a Modular Level

Microarray and other high throughput technologies generate quantitative measurements at a genome scale and hold great potential to improve our understanding of disease mechanisms at a systems level. Single gene level analyses usually generate a long list of genes that correlate with disease phenotypes, but these lengthy lists are often difficult to interpret. Modular approaches aiming at a higher-order of transcriptional changes have been developed to address this problem. These approaches focus on the coherent expression changes within groups of functionally related genes, or modules (Mootha et al., 2003; Segal et al., 2004; Wang et al., 2008). In these studies, modules are usually derived from existing knowledge on pathways, biological processes, and protein interaction networks. They are also called *gene sets* or *metagenes* in literatures.

Mootha et al. introduced the *Gene Set Enrichment Analysis* (GSEA), which was designed to detect modest but coordinate changes in the expression of groups of functionally related genes (Mootha et al., 2003). GSEA is based on the non-parametric Kolmogorov-Smirnov test. Using this approach, they identified a set of genes involved in oxidative phosphorylation whose expression was coordinately decreased in human diabetic muscle. The results associate this gene set with clinically important variation in human metabolism and illustrate the value of modular level analysis of genomic profiling data in decoding disease mechanisms.

This pioneering work has been followed up with many extensions attempting to improve the statistical test. The Kolmogorov-Smirnov test used by GSEA, which detects any changes in the distribution, is often not optimally powerful for detecting specific location changes. Methods that directly test for location changes include PAGE (Kim & Volsky, 2005) and Functional Class Scoring (Pavlidis et al., 2004). PAGE uses normal distribution to approximate test statistics based on differences in means for gene-set genes and other genes; Functional Class Scoring method computes mean (-log(*p*-value)) from *p*-values for all genes in a gene set, and compares this score to an empirically derived distribution based on randomly selected gene sets of the same size using a statistical resampling approach. Other examples of permutation- and bootstrap-based methods include SAFE (Barry et al., 2005), iGA (Breitling et al., 2004) and GSA (Efron & Tibshirani, 2007). Recently, a mixed model-based approach for gene set enrichment analysis has been developed (Wang et al., 2009; Wang et al., 2008). The method can readily accommodate complex designs under standard parametric assumptions. Besides improvements in statistical test, it has been demonstrated that better results can be achieved by including information on the topology or position of the differentially expressed genes in the pathway (Draghici et al., 2007).

Gene sets reflect biological modules only approximately. Under a specific condition, only a subset of genes in a set may be relevant to the condition (i.e. expression signature), and different gene sets may be associated with the same set of conditions, owing to either an overlap between the gene sets or coregulation of nonoverlapping gene sets. Segal et al. have developed a gene module map method, which is based on but not limited by predefined gene sets (Segal et al., 2004). Starting from predefined gene sets, the method identifies experimental conditions in which each gene set has a prominent expression signature by testing whether the expression of a statistically significant fraction of the genes in the set changes coordinately under the condition. Gene sets sharing similar signatures are integrated to derive a module, which both refines the gene composition of each gene set and combines several related gene sets. Applying this method on a cancer compendium of 1975 microarray data sets identified 456 modules. Some of them are shared across a diverse set of clinical condition, suggesting possible common tumor progression mechanisms. Others are specific to particular types of tumor. Recently, it has been shown that the gene module map method can identify specific functional pathways associated with disease subtypes that might be susceptible to targeted therapies (Wong et al., 2008). Therefore, this approach may enable rapid translation of complex genomic signatures in human disease to targeted therapeutic strategies.

Both GSEA and the module map method start with predefined gene sets. In contrast, unsupervised clustering methods are frequently applied on gene expression data to identify modules of co-expressed genes in a data-driven fashion (Kerr et al., 2008). The motivation behind clustering is the assumption that co-expressed genes are coordinately regulated and functionally similar. Based on this assumption, co-expression modules can be further analyzed to identify shared *cis*-regulatory elements, through which specific transcriptional regulators (e.g. transcription factors or microRNAs) operate. This type of analysis does not rely on existing functional annotation and holds great potential in discovering novel regulatory programs in human disease and identifying potential therapeutic targets (Gargalovic et al., 2006; Goodarzi et al., 2009; Mobini et al., 2009; Presson et al., 2008). Recently, we have developed an Iterative Clique Enumeration (ICE) algorithm for identifying relatively independent maximal cliques as co-expression modules and a module-based approach to the analysis of gene expression data. Applying this approach on public breast cancer datasets demonstrated its ability to provide a robust, interpretable, and mechanistic characterization of transcriptional changes (Shi et al., 2010).

Module-Based Disease Prediction Models

A major concern with the development of a predictive model using high-throughput data is the small ratio of samples relative to the number of genes/proteins, a problem known as "curse of dimensionality", i.e. the difficulty of obtaining accurate estimates when there are many parameters to be estimated simultaneously. Because modules are relatively independent functional units in complex systems, one solution to this problem is to use modules, instead of individual genes, as features for module construction.

Chuang et al. integrated gene expression data and protein interaction network to identify subnetworks as biomarkers for breast cancer metastasis (Chuang et al., 2007). Specifically, the algorithm overlays the expression values of each gene on its corresponding protein in the network and searches for subnetworks whose activities across the patients were highly discriminative of metastasis. It has been demonstrated that the subnetwork markers are more reproducible than individual marker genes, and prediction models built upon subnetwork markers increased the classification accuracy of metastasis. Moreover, the resulting subnetwork biomarkers provide novel hypotheses for pathways involved in tumor progression.

Van Vliet et al. extended the unsupervised gene module map framework described above to the supervised classification domain (van Vliet et al., 2007). This extension allows the identification of module-based prognostic markers, rather than gene-based markers. Results from their study suggest that modules-based models can achieve better performance on the validation data compared to gene-based models. Additionally, the module-based models provide a much richer insight into the underlying biology.

CONCLUSION

Modularity is one of the most important properties of the networks representing complex systems, including biological systems. Identifying modules in networks has attracted attention from various research communities and significant progress has been made in this area. Nevertheless, only a small number of the methods have found real applications in biological network analysis, suggesting a weak interaction between biologists and scientists in other domains. A good understanding of biology is needed to develop or customize computational methods for biological studies. As an example, despite of their high computational cost, clique-based methods have found their way in biological network analysis partially because of their ability to identify overlapping modules, an important property of biological networks. On the other hand, user-friendly software is essential for biologists to put the algorithms into practice, as exemplified by MCODE. Better communication between biologists and scientists in other domains will certainly allow fast and appropriate adaptation of newly developed methods to biological network analysis.

Besides module identification methods, the performance of module-based analysis is also determined by the quality of networks. As new methods for measuring transcript, protein, metabolite, and protein-modification levels are becoming more affordable, one obvious solution to improve network quality is through the integration of multiple types of -omics data, and to transfer information from model organisms to human. Owing to the difference in species, scales, and technologies, computational methods and frameworks are needed for better data preprocessing, integration, and network representation. Moreover, most studies described in this review treat network as undirected, unweighted, and static. In practice, the structure of networks can vary over time and space. Indeed, dynamic modularity is one of the important features of molecular networks. Although dynamic modeling poses significant challenge for both data acquisition and model development, the next wave of the biological network analysis will likely be driven by dynamic modularity (Bonneau, 2008; Han, 2008; X. Wang et al., 2008), which will ultimately lead to a more accurate understanding of biological systems.

ACKNOWLEDGMENT

This work was supported by the National Institutes of Health (NIH)/ National Institute of General Medical Sciences (NIGMS) through grant R01GM088822, the NIH/ National Institute of Mental Health (NIMH) through grant P50MH078028, and NIH/ National Cancer Institute (NCI) through grant R01CA126218.

REFERENCES

Adamcsek, B. (2006). CFinder: Locating cliques and overlapping modules in biological networks. *Bioinformatics (Oxford, England)*, *22*(8), 1021–1023. doi:10.1093/bioinformatics/btl039

Aerts, S. (2006). Gene prioritization through genomic data fusion. *Nature Biotechnology*, *24*(5), 537–544. doi:10.1038/nbt1203

Alves, N. (2007). Unveiling community structures in weighted networks. *Physical Review E: Statistical, Nonlinear, and Soft Matter Physics*, *76*(3), 036101. doi:10.1103/PhysRevE.76.036101

Ashburner, M. (2000). Gene ontology: Tool for the unification of biology. The Gene Ontology Consortium. *Nature Genetics*, *25*(1), 25–29. doi:10.1038/75556

Bader, G. D., & Hogue, C. W. (2003). An automated method for finding molecular complexes in large protein interaction networks. *BMC Bioinformatics*, *4*, 2. doi:10.1186/1471-2105-4-2

Barabasi, A. L., & Oltvai, Z. N. (2004). Network biology: Understanding the cell's functional organization. *Nature Reviews. Genetics*, *5*(2), 101–113. doi:10.1038/nrg1272

Barry, W. T. (2005). Significance analysis of functional categories in gene expression studies: A structured permutation approach. *Bioinformatics (Oxford, England)*, *21*(9), 1943–1949. doi:10.1093/bioinformatics/bti260

Blondel, V. (2008). Fast unfolding of communities in large networks. *Journal of Statistical Mechanics*, (10): P10008. doi:10.1088/1742-5468/2008/10/P10008

Bonneau, R. (2008). Learning biological networks: From modules to dynamics. *Nature Chemical Biology*, *4*(11), 658–664. doi:10.1038/nchembio.122

Brandes, U. (2007). On modularity clustering. *IEEE Transactions on Knowledge and Data Engineering*, *20*(2), 172–188. doi:10.1109/TKDE.2007.190689

Breitling, R. (2004). Iterative Group Analysis (iGA): A simple tool to enhance sensitivity and facilitate interpretation of microarray experiments. *BMC Bioinformatics*, *5*, 34. doi:10.1186/1471-2105-5-34

Brohee, S., & van Helden, J. (2006). Evaluation of clustering algorithms for protein-protein interaction networks. *BMC Bioinformatics*, *7*, 488. doi:10.1186/1471-2105-7-488

Bron, C., & Kerbosch, J. (1973). Algorithm 457: Finding all cliques of an undirected graph. *Communications of the ACM*, *16*(9), 575–577. doi:10.1145/362342.362367

Capocci, A. (2005). Detecting communities in large networks. *Physica A. Statistical and Theoretical Physics*, *352*(2-4), 669–676. doi:10.1016/j.physa.2004.12.050

Chua, H. N. (2006). Exploiting indirect neighbours and topological weight to predict protein function from protein-protein interactions. *Bioinformatics (Oxford, England)*, *22*(13), 1623–1630. doi:10.1093/bioinformatics/btl145

Chuang, H. Y. (2007). Network-based classification of breast cancer metastasis. *Molecular Systems Biology*, *3*, 140. doi:10.1038/msb4100180

Clauset, A. (2004). Finding community structure in very large networks. *Physical Review E: Statistical, Nonlinear, and Soft Matter Physics*, (6): 066111. doi:10.1103/PhysRevE.70.066111

Decramer, S. (2006). Predicting the clinical outcome of congenital unilateral ureteropelvic junction obstruction in newborn by urinary proteome analysis. *Nature Medicine*, *12*(4), 398–400. doi:10.1038/nm1384

Deng, M. (2003). Prediction of protein function using protein-protein interaction data. *Journal of Computational Biology*, *10*(6), 947–960. doi:10.1089/106652703322756168

Donetti, L., & Muñoz, M. (2004). Detecting network communities: A new systematic and efficient algorithm. *Journal of Statistical Mechanics*, *2004*, P10012. doi:10.1088/1742-5468/2004/10/P10012

Draghici, S. (2007). A systems biology approach for pathway level analysis. *Genome Research, 17*(10), 1537–1545. doi:10.1101/gr.6202607

Duarte, N. C. (2007). Global reconstruction of the human metabolic network based on genomic and bibliomic data. *Proceedings of the National Academy of Sciences of the United States of America, 104*(6), 1777–1782. doi:10.1073/pnas.0610772104

Duch, J., & Arenas, A. (2005). Community detection in complex networks using extremal optimization. *Physical Review E: Statistical, Nonlinear, and Soft Matter Physics, 72*(2), 027104. doi:10.1103/PhysRevE.72.027104

Efron, B., & Tibshirani, R. (2007). On testing the significance of sets of genes. *Annual Applied Statistics, 1*, 107–129. doi:10.1214/07-AOAS101

Enright, A. J. (2002). An efficient algorithm for large-scale detection of protein families. *Nucleic Acids Research, 30*(7), 1575–1584. doi:10.1093/nar/30.7.1575

Eriksen, K. (2003). Modularity and extreme edges of the Internet. *Physical Review Letters, 90*(14). doi:10.1103/PhysRevLett.90.148701

Everett, M. G., & Borgatti, S. P. (1998). Analyzing clique overlap. *Connections, 21*, 49–61.

Fortunato, S. (2009). *Community detection in graphs*. arXiv:0906.0612.

Fortunato, S., & Barthelemy, M. (2007). Resolution limit in community detection. *Proceedings of the National Academy of Sciences of the United States of America, 104*(1), 36–41. doi:10.1073/pnas.0605965104

Foster, L. J. (2006). A mammalian organelle map by protein correlation profiling. *Cell, 125*(1), 187–199. doi:10.1016/j.cell.2006.03.022

Franke, L. (2006). Reconstruction of a functional human gene network, with an application for prioritizing positional candidate genes. *American Journal of Human Genetics, 78*(6), 1011–1025. doi:10.1086/504300

Freeman, L. (1979). Centrality in social networks conceptual clarification. *Social Networks, 1*(3), 215–239. doi:10.1016/0378-8733(78)90021-7

Gan, G. (2007). *Data clustering: Theory, algorithms, and applications*. Society for Industrial and Applied Mathematics. doi:10.1137/1.9780898718348

Garey, M. R., & Johnson, D. S. (1979). *Computers and intractability: A guide to the theory of NP-completeness*. W. H. Freeman.

Gargalovic, P. S. (2006). Identification of inflammatory gene modules based on variations of human endothelial cell responses to oxidized lipids. *Proceedings of the National Academy of Sciences of the United States of America, 103*(34), 12741–12746. doi:10.1073/pnas.0605457103

Gasch, A. P. (2000). Genomic expression programs in the response of yeast cells to environmental changes. *Molecular Biology of the Cell, 11*(12), 4241–4257.

Gavin, A. C. (2006). Proteome survey reveals modularity of the yeast cell machinery. *Nature, 440*(7084), 631–636. doi:10.1038/nature04532

Girvan, M., & Newman, M. E. (2002). Community structure in social and biological networks. *Proceedings of the National Academy of Sciences of the United States of America, 99*(12), 7821–7826. doi:10.1073/pnas.122653799

Goh, K. I. (2007). The human disease network. *Proceedings of the National Academy of Sciences of the United States of America, 104*(21), 8685–8690. doi:10.1073/pnas.0701361104

Golub, G., & Van Loan, C. (1996). *Matrix computations (Johns Hopkins studies in mathematical sciences, 3rd ed.)*. The Johns Hopkins University Press.

Goodarzi, H. (2009). Revealing global regulatory perturbations across human cancers. *Molecular Cell*, *36*(5), 900–911. doi:10.1016/j.molcel.2009.11.016

Guimerà, R. (2004). Modularity from fluctuations in random graphs and complex networks. *Physical Review E: Statistical, Nonlinear, and Soft Matter Physics*, *70*(2), 025101. doi:10.1103/PhysRevE.70.025101

Guimerà, R., & Amaral, L. A. N. (2005). Functional cartography of complex metabolic networks. *Nature*, *433*(7028), 895–900. doi:10.1038/nature03288

Han, J. D. (2008). Understanding biological functions through molecular networks. *Cell Research*, *18*(2), 224–237. doi:10.1038/cr.2008.16

Harary, F., & Ross, I. (1957). A procedure for clique detection using the group matrix. *Sociometry*, *20*, 205–215. doi:10.2307/2785673

Harbison, C. T. (2004). Transcriptional regulatory code of a eukaryotic genome. *Nature*, *431*(7004), 99–104. doi:10.1038/nature02800

Hart, G. T. (2006). How complete are current yeast and human protein-interaction networks? *Genome Biology*, *7*(11), 120. doi:10.1186/gb-2006-7-11-120

Hartigan, J. A. (1975). *Clustering algorithms (Probability & mathematical statistics)*. John Wiley & Sons Inc.

Hartwell, L. H. (1999). From molecular to modular cell biology. *Nature*, *402*(6761 Suppl), C47–C52. doi:10.1038/35011540

Hastie, T. (2009). *The elements of statistical learning: Data mining, inference, and prediction* (2nd ed.). New York: Springer.

Hishigaki, H. (2001). Assessment of prediction accuracy of protein function from protein-protein interaction data. *Yeast (Chichester, England)*, *18*(6), 523–531. doi:10.1002/yea.706

Hubbard, T. J. (2009). Ensembl 2009. *Nucleic Acids Research*, *37*(Database issue), D690–D697. doi:10.1093/nar/gkn828

Ihmels, J. (2002). Revealing modular organization in the yeast transcriptional network. *Nature Genetics*, *31*(4), 370–377.

Ito, T. (2001). A comprehensive two-hybrid analysis to explore the yeast protein interactome. *Proceedings of the National Academy of Sciences of the United States of America*, *98*(8), 4569–4574. doi:10.1073/pnas.061034498

Jain, A. K. (1999). Data clustering: A review. *ACM Computing Surveys*, *31*(3), 264–323. doi:10.1145/331499.331504

Jensen, L. J. (2009). STRING 8-a global view on proteins and their functional interactions in 630 organisms. *Nucleic Acids Research*, *37*(Database issue), D412–D416. doi:10.1093/nar/gkn760

Jeong, H. (2000). The large-scale organization of metabolic networks. *Nature*, *407*(6804), 651–654. doi:10.1038/35036627

Karaoz, U. (2004). Whole-genome annotation by using evidence integration in functional-linkage networks. *Proceedings of the National Academy of Sciences of the United States of America*, *101*(9), 2888–2893. doi:10.1073/pnas.0307326101

Kernighan, B. W., & Lin, S. (1970). An efficient heuristic procedure for partitioning graphs. *The Bell System Technical Journal*, *49*, 291–307.

Kerr, G. (2008). Techniques for clustering gene expression data. *Computers in Biology and Medicine*, *38*(3), 283–293. doi:10.1016/j.compbiomed.2007.11.001

Kim, S. Y., & Volsky, D. J. (2005). PAGE: Parametric analysis of gene set enrichment. *BMC Bioinformatics*, *6*, 144. doi:10.1186/1471-2105-6-144

King, A. D. (2004). Protein complex prediction via cost-based clustering. *Bioinformatics (Oxford, England)*, *20*(17), 3013–3020. doi:10.1093/bioinformatics/bth351

Kislinger, T. (2006). Global survey of organ and organelle protein expression in mouse: combined proteomic and transcriptomic profiling. *Cell*, *125*(1), 173–186. doi:10.1016/j.cell.2006.01.044

Kohler, S. (2008). Walking the interactome for prioritization of candidate disease genes. *American Journal of Human Genetics*, *82*(4), 949–958. doi:10.1016/j.ajhg.2008.02.013

Krogan, N. J. (2006). Global landscape of protein complexes in the yeast Saccharomyces cerevisiae. *Nature*, *440*(7084), 637–643. doi:10.1038/nature04670

Lage, K. (2007). A human phenome-interactome network of protein complexes implicated in genetic disorders. *Nature Biotechnology*, *25*(3), 309–316. doi:10.1038/nbt1295

Lee, T. I. (2002). Transcriptional regulatory networks in Saccharomyces cerevisiae. *Science*, *298*(5594), 799–804. doi:10.1126/science.1075090

Lehmann, S. (2008). Biclique communities. *Physical Review E: Statistical, Nonlinear, and Soft Matter Physics*, *78*(1), 016108. doi:10.1103/PhysRevE.78.016108

Letovsky, S., & Kasif, S. (2003). Predicting protein function from protein/protein interaction data: A probabilistic approach. *Bioinformatics (Oxford, England)*, *19*(1), i197–i204. doi:10.1093/bioinformatics/btg1026

Li, J. (2009). Network-assisted protein identification and data interpretation in shotgun proteomics. *Molecular Systems Biology*, *5*, 303. doi:10.1038/msb.2009.54

Lubovac, Z. (2006). Combining functional and topological properties to identify core modules in protein interaction networks. *Proteins*, *64*(4), 948–959. doi:10.1002/prot.21071

Luce, R., & Perry, A. (1949). A method of matrix analysis of group structure. *Psychometrika*, *14*(2), 95–116. doi:10.1007/BF02289146

MacQueen, J. (1967). *Some methods for classification and analysis of multivariate observations.* Paper presented at the Proceedings of the Fifth Berkeley Symposium on Mathematical Statistics and Probability. Volume I: Statistics.

Mobini, R. (2009). A module-based analytical strategy to identify novel disease-associated genes shows an inhibitory role for interleukin 7 receptor in allergic inflammation. *BMC Systems Biology*, *3*, 19. doi:10.1186/1752-0509-3-19

Mootha, V. K. (2003). PGC-1alpha-responsive genes involved in oxidative phosphorylation are coordinately downregulated in human diabetes. *Nature Genetics*, *34*(3), 267–273. doi:10.1038/ng1180

Nabieva, E. (2005). Whole-proteome prediction of protein function via graph-theoretic analysis of interaction maps. *Bioinformatics (Oxford, England)*, *21*(1), i302–i310. doi:10.1093/bioinformatics/bti1054

Nesvizhskii, A. I. (2003). A statistical model for identifying proteins by tandem mass spectrometry. *Analytical Chemistry*, *75*(17), 4646–4658. doi:10.1021/ac0341261

Nesvizhskii, A. I., & Aebersold, R. (2005). Interpretation of shotgun proteomic data: The protein inference problem. *Molecular & Cellular Proteomics*, *4*(10), 1419–1440. doi:10.1074/mcp.R500012-MCP200

Newman, M. E. J. (2004). Fast algorithm for detecting community structure in networks. *Physical Review E: Statistical, Nonlinear, and Soft Matter Physics*, *69*(6), 066133. doi:10.1103/PhysRevE.69.066133

Newman, M. E. J. (2005). A measure of betweenness centrality based on random walks. *Social Networks*, *27*(1), 39–54. doi:10.1016/j.socnet.2004.11.009

Newman, M. E. J. (2006a). Finding community structure in networks using the eigenvectors of matrices. *Physical Review E: Statistical, Nonlinear, and Soft Matter Physics*, *74*(3), 036104. doi:10.1103/PhysRevE.74.036104

Newman, M. E. J. (2006b). Modularity and community structure in networks. *Proceedings of the National Academy of Sciences of the United States of America*, *103*(23), 8577–8582. doi:10.1073/pnas.0601602103

Newman, M. E. J., & Girvan, M. (2004). Finding and evaluating community structure in networks. *Physical Review E: Statistical, Nonlinear, and Soft Matter Physics*, *69*(2), 026113. doi:10.1103/PhysRevE.69.026113

Ng, A., et al. (2001). *On spectral clustering: Analysis and an algorithm.* Paper presented at the Advances in Neural Information Processing Systems 14.

Oldham, M. C. (2006). Conservation and evolution of gene coexpression networks in human and chimpanzee brains. *Proceedings of the National Academy of Sciences of the United States of America*, *103*(47), 17973–17978. doi:10.1073/pnas.0605938103

Palla, G. (2005). Uncovering the overlapping community structure of complex networks in nature and society. *Nature*, *435*(7043), 814–818. doi:10.1038/nature03607

Pardalos, P., & Xue, J. (1994). The maximum clique problem. *Journal of Global Optimization*, *4*(3), 301–328. doi:10.1007/BF01098364

Pavlidis, P. (2004). Using the gene ontology for microarray data mining: A comparison of methods and application to age effects in human prefrontal cortex. *Neurochemical Research*, *29*(6), 1213–1222. doi:10.1023/B:NERE.0000023608.29741.45

Pons, P., & Latapy, M. (2005). *Computing communities in large networks using random walks.* In (LNCS 3733). (pp. 284-293).

Presson, A. P. (2008). Integrated weighted gene co-expression network analysis with an application to chronic fatigue syndrome. *BMC Systems Biology*, *2*, 95. doi:10.1186/1752-0509-2-95

Pujana, M. A. (2007). Network modeling links breast cancer susceptibility and centrosome dysfunction. *Nature Genetics*, *39*(11), 1338–1349. doi:10.1038/ng.2007.2

Ramakrishnan, S. R. (2009). Mining gene functional networks to improve mass-spectrometry-based protein identification. *Bioinformatics (Oxford, England)*, *25*(22), 2955–2961. doi:10.1093/bioinformatics/btp461

Ravasz, E. (2002). Hierarchical organization of modularity in metabolic networks. *Science*, *297*(5586), 1551–1555. doi:10.1126/science.1073374

Reichardt, J., & Bornholdt, S. (2004). Detecting fuzzy community structures in complex networks with a Potts model. *Physical Review Letters*, *93*(21), 218701. doi:10.1103/PhysRevLett.93.218701

Reichardt, J., & Bornholdt, S. (2006). When are networks truly modular? *Physica D. Nonlinear Phenomena*, *224*(1-2), 20–26. doi:10.1016/j.physd.2006.09.009

Rual, J. F. (2005). Towards a proteome-scale map of the human protein-protein interaction network. *Nature*, *437*(7062), 1173–1178. doi:10.1038/nature04209

Schwikowski, B. (2000). A network of protein-protein interactions in yeast. *Nature Biotechnology*, *18*(12), 1257–1261. doi:10.1038/82360

Segal, E. (2004). A module map showing conditional activity of expression modules in cancer. *Nature Genetics*, *36*(10), 1090–1098. doi:10.1038/ng1434

Seidman, S. (1980). Clique-like structures in directed networks. *Journal of Social and Biological Structures*, *3*, 43–54. doi:10.1016/0140-1750(80)90019-6

Seidman, S. (1983a). Internal cohesion of LS sets in graphs. *Social Networks*, *5*(2), 97–107. doi:10.1016/0378-8733(83)90020-5

Seidman, S. (1983b). Network structure and minimum degree. *Social Networks*, *5*, 269–287. doi:10.1016/0378-8733(83)90028-X

Seidman, S., & Foster, B. (1978). A graph-theoretic generalization of the clique concept. *The Journal of Mathematical Sociology*, *6*, 139–154. doi:10.1080/0022250X.1978.9989883

Shannon, P. (2003). Cytoscape: A software environment for integrated models of biomolecular interaction networks. *Genome Research*, *13*(11), 2498–2504. doi:10.1101/gr.1239303

Sharan, R. (2005). Identification of protein complexes by comparative analysis of yeast and bacterial protein interaction data. *Journal of Computational Biology*, *12*(6), 835–846. doi:10.1089/cmb.2005.12.835

Sharan, R. (2007). Network-based prediction of protein function. *Molecular Systems Biology*, *3*, 88. doi:10.1038/msb4100129

Shi, J., & Malik, J. (2000). Normalized cuts and image segmentation. *IEEE Transactions on Pattern Analysis and Machine Intelligence*, *22*(8), 888–905. doi:10.1109/34.868688

Shi, Z. (2010). Co-expression module analysis reveals biological processes, genomic gain, and regulatory mechanisms associated with breast cancer progression. *BMC Systems Biology*, *4*, 74. doi:10.1186/1752-0509-4-74

Simonsen, I. (2004). Diffusion on complex networks: A way to probe their large-scale topological structures. *Physica A. Statistical and Theoretical Physics*, *336*(1-2), 163–173. doi:10.1016/j.physa.2004.01.021

Slanina, F., & Zhang, C. (2005). Referee networks and their spectral properties. *Acta Physica Polonica B*, *36*(9), 2797.

Spirin, V., & Mirny, L. A. (2003). Protein complexes and functional modules in molecular networks. *Proceedings of the National Academy of Sciences of the United States of America*, *100*(21), 12123–12128. doi:10.1073/pnas.2032324100

Stelzl, U. (2005). A human protein-protein interaction network: A resource for annotating the proteome. *Cell*, *122*(6), 957–968. doi:10.1016/j.cell.2005.08.029

Stuart, J. M. (2003). A gene-coexpression network for global discovery of conserved genetic modules. *Science*, *302*(5643), 249–255. doi:10.1126/science.1087447

Sugar, C. A., & James, G. M. (2003). Finding the number of clusters in a data set: An information theoretic approach. *Journal of the American Statistical Association*, *98*, 750–763. doi:10.1198/016214503000000666

Tsukiyama, S. (1977). A new algorithm for generating all the maximal independent sets. *SIAM Journal on Computing*, *6*(3), 505–517. doi:10.1137/0206036

Uetz, P. (2000). A comprehensive analysis of protein-protein interactions in Saccharomyces cerevisiae. *Nature*, *403*(6770), 623–627. doi:10.1038/35001009

van Noort, V. (2004). The yeast coexpression network has a small-world, scale-free architecture and can be explained by a simple model. *EMBO Reports*, *5*(3), 280–284. doi:10.1038/sj.embor.7400090

van Vliet, M. H. (2007). Module-based outcome prediction using breast cancer compendia. *PLoS ONE*, *2*(10), e1047. doi:10.1371/journal.pone.0001047

Vazquez, A. (2003). Global protein function prediction from protein-protein interaction networks. *Nature Biotechnology*, *21*(6), 697–700. doi:10.1038/nbt825

Wang, L. (2008). An integrated approach for the analysis of biological pathways using mixed models. *PLOS Genetics*, *4*(7), e1000115. doi:10.1371/journal.pgen.1000115

Wang, L. (2009). A unified mixed effects model for gene set analysis of time course microarray experiments. *Statistical Applications in Genetics and Molecular Biology*, *8*(1), 47. doi:10.2202/1544-6115.1484

Wang, X. (2008). Gene module level analysis: Identification to networks and dynamics. *Current Opinion in Biotechnology*, *19*(5), 482–491. doi:10.1016/j.copbio.2008.07.011

Wasserman, S., & Faust, K. (1994). *Social network analysis: Methods and applications*. Cambridge University Press.

Watts, D. J., & Strogatz, S. H. (1998). Collective dynamics of small-world networks. *Nature*, *393*(6684), 440–442. doi:10.1038/30918

Whisstock, J. C., & Lesk, A. M. (2003). Prediction of protein function from protein sequence and structure. *Quarterly Reviews of Biophysics*, *36*(3), 307–340. doi:10.1017/S0033583503003901

Whiteaker, J. R. (2007). Integrated pipeline for mass spectrometry-based discovery and confirmation of biomarkers demonstrated in a mouse model of breast cancer. *Journal of Proteome Research*, *6*(10), 3962–3975. doi:10.1021/pr070202v

Wong, D. J. (2008). Revealing targeted therapy for human cancer by gene module maps. *Cancer Research*, *68*(2), 369–378. doi:10.1158/0008-5472.CAN-07-0382

Wu, F. Y. (1982). The Potts model. *Reviews of Modern Physics*, *54*(1), 235. doi:10.1103/RevModPhys.54.235

Wu, X. (2008). Network-based global inference of human disease genes. *Molecular Systems Biology*, *4*, 189. doi:10.1038/msb.2008.27

Yang, X. (2004). DBParser: Web-based software for shotgun proteomic data analyses. *Journal of Proteome Research*, *3*(5), 1002–1008. doi:10.1021/pr049920x

Zhang, B. (2007). Proteomic parsimony through bipartite graph analysis improves accuracy and transparency. *Journal of Proteome Research*, *6*(9), 3549–3557. doi:10.1021/pr070230d

Zhang, B. (2008). From pull-down data to protein interaction networks and complexes with biological relevance. *Bioinformatics (Oxford, England)*, *24*(7), 979–986. doi:10.1093/bioinformatics/btn036

Zhou, H., & Lipowsky, R. (2004). *Network Brownian motion: A new method to measure vertex-vertex proximity and to identify communities and subcommunities*. In (LNCS 3038). (pp. 1062-1069).

KEY TERMS AND DEFINITIONS

Betweenness: A centrality measure of a vertex or an edge within a graph. A simple calculation is based on the involvement of the vertex/edge in shortest paths between other vertices, i.e. vertices/

edges that occur on many shortest paths between other vertices have higher betweenness than those that do not.

Centrality: An index that measures the relative importance of nodes or edges in a graph.

Clique: A subgraph in which every two vertices are connected by an edge.

Clustering: The assignment of a set of observations into subsets (called *clusters*) so that observations in the same cluster are similar in some sense.

Module: Sub-groups of elements (e.g. nodes and edges in the context of networks) that function in a semi-autonomous fashion and can serve as building blocks of complex systems.

Network Modularity: The organization of nodes in clusters, with many edges connecting nodes of the same cluster and comparatively few edges connecting nodes of different clusters.

Chapter 12
Using Functional Linkage Gene Networks to Study Human Diseases

Bolan Linghu
Novartis Institutes for BioMedical Research, USA

Guohui Liu
Millennium Pharmaceuticals Inc, USA

Yu Xia
Boston University, USA

ABSTRACT

A major challenge in the post-genomic era is to understand the specific cellular functions of individual genes and how dysfunctions of these genes lead to different diseases. As an emerging area of systems biology, gene networks have been used to shed light on gene function and human disease. In this chapter, first the existence of functional association for genes working in a common biological process or implicated in a common disease is demonstrated. Next, approaches to construct the functional linkage gene network (FLN) based on genomic and proteomic data integration are reviewed. Finally, two FLN-based applications related to diseases are reviewed: prediction of new disease genes and therapeutic targets, and identification of disease-disease associations at the molecular level. Both of these applications bring new insights into the molecular mechanisms of diseases, and provide new opportunities for drug discovery.

INTRODUCTION AND BACKGROUND

With the development of sequencing technologies, whole genome sequencing has been achieved for diverse species (Flicek et al., 2008). For a fully sequenced organism, most protein-encoding genes can be readily identified by available bioinformatics approaches (Flicek et al., 2008). By contrast, it remains a challenging task to understand the

DOI: 10.4018/978-1-60960-491-2.ch012

specific biological functions of these genes, and how dysfunctions of these genes lead to diverse human disease phenotypes.

A particular cellular function usually requires the collaboration between a specific group of genes or proteins. Rather than acting alone, these genes interact and communicate with each other in diverse ways, but all for the common purpose of maintaining the normal status of a specific biological process (Kanehisa et al., 2004). On the other hand, when one or more genes involved in a particular biological process are dysfunctional, the normal status of the biological process might be perturbed, which might further cause the organism to show abnormal physiological phenotypes referred to as a disease (Goh et al., 2007). Correspondingly, therapeutic drugs aim to target the genes or proteins involved in these perturbed biological processes such that the normal status of the biological processes can be reestablished (Janga & Tzakos, 2009). Therefore, for gene function and human disease research, it is very important to consider individual genes as functional related components within a coherent biological system.

Recent network-based approaches have demonstrated great success in representing functional relationships among genes with applications to understand gene function and human disease (Ahmed & Xing, 2009; Franke *et al.*, 2006; Huttenhower *et al.*, 2009; Kohler *et al.*, 2008; Lage *et al.*, 2007; I. Lee *et al.*, 2008b; Linghu *et al.*, 2008; Linghu *et al.*, 2009; McGary *et al.*, 2007; Oti & Brunner, 2007; Oti *et al.*, 2006; Schadt, 2009). In these networks, nodes represent genes, and edges represent functional associations between linked genes. These networks are referred to as functional linkage gene networks (FLN). In this chapter, we first review the molecular basis for genes working as a functional group, and demonstrate the existence of functional associations between genes implicated in a common disease. Next, we review ways to construct different types of FLNs, as well as two important FLN-based applications related to human diseases: prediction of new disease genes and therapeutic targets, and identification of disease-disease associations at the molecular level (Figure 1).

MAIN FOCUS OF THE CHAPTER

Functional Associations between Genes Underlying the Same or Related Diseases

Genes Work in Groups to Carry Out Particular Cellular Functions

Most genes or proteins do not work alone. Instead, a group of genes or proteins collaborate with each other efficiently as a specific functional module to carry out a particular cellular task (Hartwell et al., 1999; Ravasz et al., 2002). Such functional modules can be represented as specific biological processes or pathways. Genes or proteins within the same biological process or pathway can have multiple types of functional associations. For instance, certain proteins physically interact with each other to form a protein complex and function as a whole unit (Stelzl et al., 2005); certain transcription factors regulate a group of target genes in order to coordinate cellular activities related to a particular biological process such as cell cycle (Chen et al., 2000); certain genes with similar sequences encode members of a protein family such that different members can be used under different cellular conditions (Iwabe et al., 1996). These functional associations can be inferred from various data sources generated by different experimental and computational approaches. For instance, yeast two-hybrid and mass spectrometry experiments can detect protein-protein binary physical and co-complex interactions (Ewing et al., 2007; Rual et al., 2005); microarray experiments can detect co-expression relationships among genes at the transcriptional level (Griffith

Figure 1. Use functional linkage gene network to identify new drug targets and identify disease-disease associations. From multiple biological data sources, functional associations between genes are retrieved and integrated into a functional linkage gene network, where nodes represent individual genes and edges represent the degree of functional association upon data integration. Disease information can be incorporated into the FLN by labeling genes known to be associated with specific diseases. Two disease related applications can be obtained from the FLN. (i) Given known disease genes for a particular disease, other genes that are closely associated with known disease genes in the network can be predicted as new candidate disease genes. These new candidate disease genes can be further prioritized for therapeutic targets by evaluating their disease relevancy, safety issues, druggability, etc. (ii) Given gene-disease associations for multiple diseases, disease-disease associations at the molecular level can be predicted by determining how closely associated the two corresponding disease gene sets are in the network. For two closely related diseases, known disease genes for one disease may serve as new candidate disease genes for the other disease. Furthermore, if some of these genes have been launched for drugs, these genes as well as their therapeutic drugs may be directly tested for the other disease as new targets and new therapies.

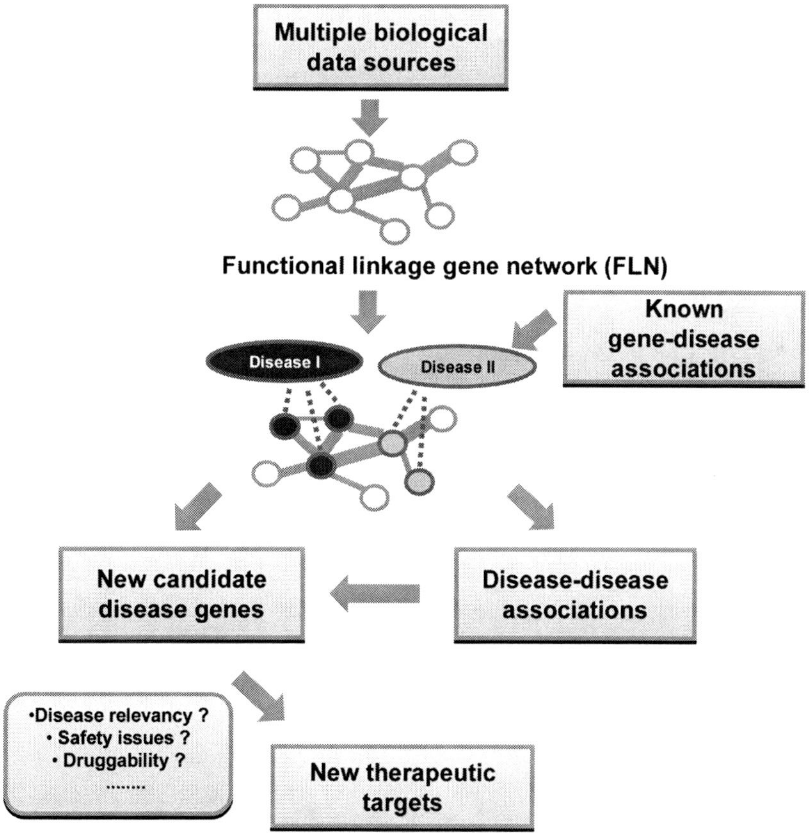

et al., 2005; Jensen et al., 2004; H. K. Lee et al., 2004; von Mering et al., 2007); using sequence data, computational approaches such as the phylogenetic profile method can predict gene pairs with similar function based on their correlated occurrence patterns across a set of species (von Mering et al., 2007).

Disease Phenotypes are Caused by Perturbations of Specific Functional Gene Groups at the Molecular Level

Genes collaborate in groups to perform specific cellular tasks to maintain the normal functions of cells, which will in turn support the normal activities of tissues, organs, and the whole organism. When one or more genes within a functional gene group are dysfunctional, the coherence and the effectiveness of the whole gene group can be affected, resulting in abnormal cellular states (Goh et al., 2007; Park et al., 2009). In human, these abnormal states can be clinically detected as diseases. From the point of view of disease phenotypes, human diseases are extremely diverse with vastly different phenotypic features (Oti et al., 2008). However, from the point of view of molecular mechanisms, the same or related diseases tend to be caused by perturbations of the same or related gene functional groups (Goh et al., 2007; Oti et al., 2008). In other words, genes underlying the same or related diseases tend to be functionally associated with each other. This concept is supported by functional analyses of genes associated with diverse diseases (Goh et al., 2007; Oti & Brunner, 2007; Oti et al., 2008; van Driel et al., 2006). It is found that genes implicated in the same or related diseases tend to show at least one type of functional associations, such as co-expression in specific tissues, correlated expression levels, binary physical interaction or co-complex membership, co-localization, sharing similar molecular function, or participation in similar biological processes (Goh et al., 2007; Oti et al., 2008).

Functional associations among genes implicated in the same or related diseases are further illustrated by detailed analyses of the molecular mechanisms of diseases. For example, inherited ataxia, a neurodegenerative disorder, can be caused by at least 23 known genes. Moreover, proteins encoded by 18 of these 23 genes are known to interact directly or indirectly (Lim et al., 2006). Walker-Warburg syndrome, a rare form of autosomal recessive congenital muscular dystrophy, is caused by dysfunction in genes involved in the O-linked glycosylation of the glycoprotein alpha-dystroglycan (van Reeuwijk et al., 2005). Ceroid lipofuscinosis, a metabolic disease which affects the nerve cells of the body, and mucopolysaccharidosis, another metabolic disease which affects physical abilities and mental development, are both caused by dysfunction in genes involved in lysosomal functions (Sedel et al., 2007).

In summary, genes or proteins tend to form specific functional groups for specific cellular functions. A particular disease represents the perturbation of a specific functional group as a result of the dysfunction of one or more gene members of the group. Thus, an important goal of disease therapy is to find the appropriate drugs which can perturb appropriate gene members of this functional group again in the right direction, so that the overall activity of the functional gene group is restored to be close to normal.

Functional Linkage Gene Network and Its Construction

Functional Linkage Gene Network

Since genes underlying the same or related diseases tend to be functionally associated, to study the molecular mechanisms of diseases, we need to consider the functional associations between genes. Recently, network-based approaches have been employed to represent the functional associations between human genes with further applications to disease studies (Feldman et al., 2008; Franke et al., 2006; Ideker & Sharan, 2008; Jiang et al., 2008; Kohler et al., 2008; Lage et al., 2007; Linghu et al., 2009; Pan, 2008; Pujana et al., 2007; Wu et al., 2008). These methods typically construct a functional linkage gene network (FLN) where nodes represent genes, and edges (links) represent functional associations between the linked genes (Figure 1). The evidence supporting functional associations can be derived from one or more types of genomic and proteomic data sources. The links in the FLN can be weighted, reflecting the degree of functional associations between genes (Jensen et al., 2009; Linghu et al., 2009).

Constructing Functional Linkage Gene Network

Some studies used one single type of evidence to construct the functional-association links in the FLN. For instance, Wu et al. (2008) constructed a network with functional links representing protein-protein physical interactions only. Similarly, Ala et al. (2008) constructed a network with functional links representing co-expression relationships only. Although these networks are successful for disease-related studies, the restriction to only one type of functional-association evidence potentially limits their applications for the following reasons: (i) functional associations between genes have diverse dimensions, and (ii) one single type of data source has limited coverage. In order to represent gene functional associations comprehensively with high coverage, multiple types of evidence based on genomic and proteomic data sources should be integrated. Recently, several research groups have successfully constructed integrated human functional linkage gene networks. Franke et al. (2006) constructed such a network by integrating protein-protein interaction, co-expression, and functional associations based on shared Gene Ontology (GO) annotations between genes. Jansen et al. (2009) constructed a weighed human FLN by integrating sequence-based genomic context data, high-throughput experiments, co-expression data, and literature-based text mining. Finally, in our previous study, we constructed a genome-scale weighted human functional linkage gene network with over 21,000 genes and over 22,000,000 links, by integrating 16 diverse functional genomic and proteomic data sources (Linghu et al., 2009). This integrated human FLN is the most comprehensive to date. In addition to the data sources mentioned above, we also integrated functional associations mapped from other model organisms via orthology. Because current available human data are far from complete, this step serves to maximize the coverage of the resulting integrated human FLN.

An FLN can be constructed by integrating diverse data sources. These different data sources usually vary in reliability and coverage. As a result, data integration should not be a simple union or intersection of individual data sources. Instead, data integration should take into account the differences among individual sources in terms of false positive and false negative error rates, as well as predictive power for functional linkage. A rigorous solution to this problem is methods based on (supervised) machine learning, such as Bayesian approaches, support vector machines, and logistic regression (Calvo *et al.*, 2006; Franzosa *et al.*, 2009; Linghu et al., 2009; Zhong & Sternberg, 2006). For instance, in our previous study, a naïve Bayes method was used to combine different data sources for the FLN construction (Linghu et al., 2009). By defining functional association in the network as biological-process sharing relationship based on Gene Ontology (GO) (Berardini et al., 2010), we constructed a benchmark dataset composed of both gold-standard positives (gene pairs sharing the same biological-process terms in GO) and gold-standard negatives (gene pairs with both members annotated in GO but not sharing any GO biological-process term). Using this benchmark dataset as training set, each individual data source was calibrated by its ability to predict the biological-process sharing relationship between genes, and each data source was weighted correspondingly prior to data integration such that stronger evidence is given higher weight, and weaker evidence is given lower weight. The final FLN is weighted and the linkage weight denotes the probability that the linked gene pair participate in the same biological process based on the summary of all available evidence from diverse sources. Such data integration approaches based on machine learning may achieve optimal performance in terms of accuracy and coverage of the functional links, because links supported by high-confidence pieces of evidence are given higher weights.

The FLNs can have different sizes. Some FLNs incorporate all genes in the genome and map the functional relationships between genes in a global manner (Franke et al., 2006; Jensen et al., 2009; Kohler et al., 2008; Linghu et al., 2009). To study a particular disease reflecting the perturbations of a functionally coherent gene group, a subset of the network where this specific gene group occupies can be examined. Since genes underlying the particular disease are functionally related, the corresponding gene subsets also tend to be clustered together in the global network. The benefit of a global FLN is that the same network can be used to study multiple diseases by focusing on different subnetworks corresponding to different diseases. Alternatively, for the purpose of studying a specific disease, a local network composed of only those genes related to this particular disease can be constructed and examined instead of the global network (Emilsson et al., 2008; Jiang et al., 2008; Lim et al., 2006). Such disease-specific networks have smaller gene coverage than the global network, and are therefore computationally less demanding in terms of data integration and analysis. These local, specific disease networks also allow the convenient integration of disease-specific data sources. However, to study multiple diseases systematically, the global FLN is still preferred.

FLN Visualization

The FLNs can be conveniently visualized by software tools designed for exploring and analyzing biological networks. One such software tool is VisANT, a web-based open-source platform for the visualization and analysis of different types of biomolecular networks (Hu et al., 2009; Hu et al., 2007; Hu et al., 2008). Users can upload and explore their own FLNs or the FLNs stored in the VisANT database, which includes integrated FLNs and FLNs derived from single types of data sources. VisANT enables users to interactively query genes of interest in an FLN, explore their network neighborhood, and perform topological analyses or calculate network degrees of selected nodes. Users can also filter the FLN with different linkage weight thresholds, and visualize the weights with edge color or edge thickness. As described earlier, genes work in functional groups for specific cellular tasks, and genes underlying the same diseases tend to be functionally associated and belong to the same functional groups. VisANT introduces meta-nodes, a special type of node that contains associated sub-nodes, to represent these functional groups, such as protein complexes, molecular pathways, or gene sets underlying the same diseases. As a result, the hierarchical structures of the FLN can be visualized: users can visualize not only the functional associations between individual genes in a low-level map, but also functional modules composed of gene groups in a high-level map. Besides VisANT, another popular FLN visualization tool is Cytoscape, an open-source software which can also be used for FLN visualization and analysis (Cline et al., 2007).

Use Functional Linkage Gene Network to Predict Drug Targets

Use FLN for New Disease Gene Prediction

The key assumption behind the utility of FLN in disease research, which is supported by diverse empirical evidence, is that genes underlying the same or related diseases tend to be functionally related. Because an FLN represents functional associations among genes, these genes associated with the same or related diseases are expected to be located close to each other in the same neighborhood of the network (Janga & Tzakos, 2009; Kohler et al., 2008; Lage et al., 2007; Linghu et al., 2009; Wu et al., 2008). Based on this assumption,

FLNs have been successfully used to predict new disease genes in recent studies. Given a particular disease of interest, these FLN-based approaches typically start with the identification of known disease genes in the network as "seeds", followed by exploring the network neighborhoods of these seeds, and prioritizing new candidate disease genes based on how closely connected they are to the seeds (Figure 1). For certain diseases, only the clinically defined phenotypes are known, but few genes are known to be related to these diseases. In these cases, additional seed disease genes can be borrowed from other diseases with similar phenotypes, with proper adjustments of seed strength (Lage et al., 2007; Wu et al., 2008). The adjustment of seed strength can be based on how similar other diseases are to the disease under study in terms of phenotypic descriptions (Lage et al., 2007; van Driel et al., 2006; Wu et al., 2008). Known gene-disease associations can be obtained from the Online Mendelian Inheritance in Man (OMIM) database, which is the most comprehensive compendium of human disease genes and phenotypes (Hamosh et al., 2005).

As described earlier, different FLNs can be constructed from single data sources such as protein-protein physical interaction (PPI) data and co-expression data, or from genomic data integration. Recent studies have demonstrated that these FLNs can all be used to predict new disease genes (Ala et al., 2008; Lage et al., 2007; Wu et al., 2008). Although FLNs based on single data sources are successful in disease gene predictions, the restriction to only one type of functional association potentially limits their predictive ability. For instance, both our previous study and Kohler et al.'s study demonstrated that FLNs based on genomic data integration significantly outperform those based on PPI data alone (Kohler et al., 2008; Linghu et al., 2009).

Most FLN-based disease gene prediction approaches can be summarized as the following two-step scheme: (i) build a functional linkage gene network as input network based on one or more types of data sources, and (ii) apply an appropriate decision rule to the input gene network to rank candidate genes based on the strength of their connection to the seed disease genes (Figure 1). The choice of the optimal decision rule is determined by the nature of the FLN. For instance, in our previous study, we constructed a weighted and integrated FLN from extensive data integration (Linghu et al., 2009). This particular FLN is very dense with high coverage (over 21,000 genes and over 22,000,000 edges), and each gene has over 2,000 neighbors on average. Since links in this FLN are weighted, the high density of the network allows one to directly measure the strength of functional associations between one gene and thousands of other genes. Taking these features into account, we used a simple and yet effective neighborhood weighting rule to rank candidate genes for a particular disease by focusing on the immediate neighborhood of the seed genes (Linghu et al., 2009). In particular, candidate genes are rank ordered by the sum of the weights of their functional links to the seed genes in the network. This approach can successfully rank ~12,000 genes for each disease. Neighborhood weighting rule is a simple local decision rule, and it uses only immediate network neighbors for new disease gene prediction. Yet it is very effective because in this particular case the FLN based on genomic data integration is weighted and very dense. Other non-local decision rules have been proposed for disease gene predictions, such as the shortest path rule and the random walk rule (Kohler et al., 2008; Wu et al., 2008). These rules make use of indirect connections between non-adjacent nodes for disease gene predictions. These rules are usually applied to networks that are much sparser. For example, in Wu et al.'s study where the shortest path rule is used, the input network is an unweighted protein-protein physical interaction network (14,433 nodes, 72,431 edges, and ~10 edges per node) (Wu et al., 2008). In Kohler et al.' study where a random walk rule is used, the input network is an unweighted functional linkage network (13,726 nodes, 258,314 edges, and ~38 edges per node) (Kohler et al., 2008). If the same seed gene sets from OMIM are used as in our previous study (~10 seeds per disease), a local rule using only the direct neighbors of seeds will only be able to rank hundreds of neighboring genes in both of these networks. Therefore, non-local rules such as the shortest path rule and the random walk rule are more desirable in these sparse networks: they allow more candidate disease genes to be prioritized based on their indirect connections to the seed genes.

From Candidate Disease Genes to Therapeutic Targets

Once the putative disease genes are predicted base on the FLN, an important yet very challenging task is to further identify therapeutic drug targets from the predicted gene list. The predicted disease genes are functionally related with the known disease genes at the molecular level, which means that these genes might be new disease causal genes or etiologically related factors. To further identify feasible drug targets from these predicted genes, additional analyses need to be performed (Figure 1).

First, for a given candidate target gene from the list, its molecular relevancy to the disease needs to be further confirmed. This is because links in the FLNs still contains false positive and false negative errors. For instance, many data sources used for FLN construction are based on high-throughput experiments such as yeast two-hybrid experiments and microarray experiments, which are subject to significant false positive errors (Hart et al., 2006; Yamada & Ueda, 2009). In addition, these FLNs miss a lot of functional links due to the incompleteness of data sources in human for inferring functional associations between genes. To validate disease relevancy of a candidate gene, many types of biological evidence need to be considered. For example, is there evidence based on genetic association study indicating that mutations in the gene are significantly associated with the disease phenotype in human? Does the expression of the gene change in the disease tissue relative to the normal tissue? In an animal model mimicking the disease phenotype, does the administration of specific therapeutic compound or antibody targeting the protein encoded by the candidate gene have therapeutic effects? Are there specific experiments supporting the role of the candidate gene as a key component regulating a known disease gene or being regulated by a known disease gene? To answer these questions, collaboration with biomedical scientists with specific disease domain knowledge is required, and additional biological validation experiments will need to be performed.

Second, the safety issue of the candidate target gene also needs to be considered. For instance, some candidate genes may have essential functions, and targeting these genes may put the patients into life-threatening risks (Berger & Iyengar, 2009). One way to predict whether a gene is essential or not is to check the phenotypes of the corresponding mouse ortholog mutant (Liao & Zhang, 2007; Zhang & Lin, 2009). If the corresponding mouse phenotype is lethal, it is very likely that the candidate gene is also essential in human. Additionally, computational approaches can be used to select candidate target genes which are not key regulators of important pathways, and thus targeting these genes are unlikely to affect the cellular networks globally with severe side effects (Berger & Iyengar, 2009; Hwang et al., 2008).

Third, for a candidate gene to be an attractive target, it needs to be druggable (Hopkins & Groom, 2002). Currently, the orally bioavailable small molecule compounds are still the most favorable therapeutic drugs in pharmaceutical industries. However, not all proteins encoded in the human genome are able to bind to these compounds with enough potency. Analysis of protein families for the current on-market drugs revealed that druggable protein families are limited to a subset of protein families such as G-protein coupled receptors, serine/threonine and tyrosine kinases, serine proteases, nuclear hormone receptors, and peptidases, which all have known binding sites for small molecules (Hopkins & Groom, 2002). For proteins lacking binding pockets, it will be very challenging to identify potent small molecule drugs. In addition, in order for a drug to be delivered to the target, the target needs to be easily accessible. For instance, analysis of protein targets with known approved drugs revealed that the most popular targets are membrane proteins located on the cell surfaces which are easy to access for small molecules (Yildirim et al., 2007). On the contrary, transcription factors located in the nucleus are very hard to access for small molecules, and are therefore very challenging targets for drug development.

Fourth, many diseases, especially complex diseases, are manifested not by the dysfunctions of single genes, but by the dysfunctions of multiple genes and their interactions (Berger & Iyengar, 2009; Hopkins, 2008). Therefore, targeting multiple genes rather than single genes is more desirable for these diseases. Target selection should take this into account, and focus on combinations of targets rather than single targets. Indeed, many successful drugs in oncology, psychiatry, and infectious diseases act on multiple targets rather than single targets (Hopkins, 2007). Recently, network-based approaches were used to predict combinations of targets for disease therapies such that the undesired processes can be blocked by disrupting a specific set of genes in the network but the desired biological processes remain unaffected (Dasika et al., 2006; Ruths et al., 2006).

Finally, due to the genetic heterogeneity of many diseases, different patients with the same disease may have the same symptoms which are caused by dysfunctions of different genes within the same or related biological processes. Therapeutic targets here can be either the dysfunctional genes themselves or other genes etiologically related to these genes (Yildirim et al., 2007). In either case, for a heterogeneous disease, targeting a single gene or a single set of genes may not have the same universal therapeutic effects on different patients. Instead, personalized targets, which tailor the variability of different patients, should be selected to match the genetic specificity of different patients (Conti et al., 2010). With the rapid development of new technologies such as next generation sequencing, whole genome and transcriptome analysis for individual patients can be achieved at relatively low costs in a high-throughput fashion. This will greatly facilitates the development of personalized medicine.

Use FLN to Identify Disease-Disease Associations at the Molecular Level

Disease-Disease Associations at the Molecular Level

Clinically, different diseases can affect different physiological systems in the human body with diverse phenotypic symptoms. However, recent studies indicate that human diseases are not just a catalog of isolated diseases. Instead, human diseases form an interrelated landscape, where different diseases can be linked together due to perturbation of the same or related biological processes at the molecular level (Goh et al., 2007; Oti & Brunner, 2007; Oti et al., 2008; van Driel et al., 2006). Not surprisingly, it is found that diseases with similar phenotypes tend to be caused by dysfunctions of the same genes. Less anticipated is the finding that diseases with dissimilar phenotypes can also be related at the molecular level (Oti & Brunner, 2007; Oti et al., 2008). For instance, anemia, a disease with decreased level of red blood cells, and porphyria, which can affect skin and nervous systems, can both be caused by genes involved in heme biosynthesis (Ajioka et al., 2006).

Why are we Interested in Disease-Disease Associations at the Molecular Level?

There are several advantages for studying disease-disease associations at the molecular level. First, not all human diseases are well studied and explored. For certain well studied diseases, detailed molecular and pathological mechanisms have been established through years of efforts by disease experts. On the other hand, there are still many diseases for which detailed molecular mechanisms are largely unknown. If one can establish molecular associations between a well studied disease and a poorly studied disease, one can then use the known molecular and pathological mechanisms of the well studied disease to guide the research of the other disease. Second, if two diseases are known to perturb the same or related biological processes, and if their known disease genes do not overlap completely, the non-overlapping part of the disease genes may serve as putative disease genes for the other disease (Figure 1). Furthermore, if one or more of the non-overlapping genes have been launched as drug targets in the market, then these genes and their drugs may be directly tested for therapy of the other disease. This will save a lot of resources in comparison with developing a completely new drug. Finally, diseases are traditionally classified clinically based on the specificity of affected physiological systems (Goh et al., 2007). However, diseases affecting the same physiological system may involve completely different pathological mechanisms at the molecular level. Likewise, diseases affecting different physiological systems may share similar molecular mechanisms (Linghu et al., 2009; Park et al., 2009). Identifying disease-disease associations at the molecular level will help refine the current disease classification system to take into account molecular causal factors, which is the key for accurate diagnosis and effective therapeutics.

Approaches to Identify Disease-Disease Associations at the Molecular Level

The first large-scale identification of disease-disease associations at the molecular level is the work of Goh et al. (Goh et al., 2007). This work included over 1,200 genetic disorders, and built a global disease-disease association network where nodes represent diseases and edges represent the sharing of at least one common disease genes between the two linked diseases. Goh et al.'s work demonstrated for the first time the existence of molecular associations between diverse diseases in a systematic manner. At the same time, their approach is limited by the restriction of disease-disease associations to the sharing of common disease genes. This is because in addition to disease gene sharing, molecular associations between diseases can involve dysfunctions of distinct genes which are functionally related (Linghu et al., 2009; Park et al., 2009). Because functional linkage gene networks provide information on functional relatedness between disease genes, they can also be further used to uncover disease-disease associations based on functional relatedness of the corresponding sets of disease genes.

Lee et al. explored this concept using one specific type of functional links, the metabolic links, to identify associations between diseases (D. S. Lee et al., 2008a). In their approach, two diseases are connected in a disease-disease association network if their associated genes encode enzymes which catalyze adjacent reactions in a metabolic pathway. According to this definition, two linked diseases share the perturbation of the same metabolic processes at the molecular level. As the disease progresses, these shared molecular-level perturbations are expected to propagate to higher levels such as cell level, organ level, and even organism level. To further validate the identified disease-disease associations, Lee et al. showed that connected diseases occur more frequently in the same patient than unconnected diseases, i.e., connected diseases have higher comorbidity. Additionally, Lee et al. compared their approach with Goh et al.'s disease gene sharing method, and showed that metabolic links are stronger predictors for disease comorbidity than shared disease genes. Lee et al.'s work focused on using one specific type of functional relatedness between genes (i.e., metabolic links) to uncover disease-disease associations. More recently, the same research group used two additional types of functional links, protein-protein interaction (PPI) and co-expression, to predict disease-disease associations. Again, they demonstrated that disease pairs whose corresponding disease genes interact with each other or share correlated gene expressions tend to have higher comorbidity (Park et al., 2009). All these approaches are successful in identifying disease-disease associations at the molecular level. However, they are all restricted by limiting functional associations between disease genes to only one specific type rather than integrating diverse types of functional associations. In fact, genes can be functionally related in multiple ways, such as transcriptional regulation, protein-protein interaction, co-expression, common protein domain sharing, subcellular co-localization, etc. To identify disease-disease associations comprehensively at the molecular level, diverse types of evidence for functional relatedness should be included instead of considering only one single type. Intuitively, the integrated FLN derived from diverse types of evidence for functional relatedness should further improve disease-disease association mapping.

To demonstrate this, in our previously study, we applied the integrated FLN for the first time to identify disease-disease associations (Linghu & Delisi, 2010; Linghu et al., 2009). As described earlier, we constructed a comprehensive human FLN by integrating 16 diverse functional genomic and proteomic data. Links in the network are weighted by the degree of functional association between the linked genes, as predicted by genomic data integration. To evaluate the degrees of association between diseases, known disease genes for each disease are first identified in the network. The strength of association between two diseases is quantified based on network connectivity of their corresponding disease gene sets (Figure 1). A disease pair is predicted to be closely related if there are dense and strong functional links between the corresponding disease gene sets, while a disease pair is predicted to be unrelated if there are no or only weak and sparse functional links between the corresponding disease gene sets. A disease-disease association network was subsequently constructed, where nodes represent individual diseases and edges represent the presence of molecular-level associations for the linked diseases. The application of the integrated FLN to disease-disease association mapping uncovered many novel disease-disease associations, where the associated disease pair shares no known disease genes and exhibits dissimilar phenotypes. One such example is the association between Alzheimer's disease, a neurological disorder, and hypercholesterolemia, a metabolic disorder. These two diseases share no known disease genes but have strong and dense functional links between their disease gene sets. These disease-disease associations provide immediate insight into the molecular mechanisms underlying different diseases, and generate novel hypotheses for therapeutic strategies. For instance, the association of hypercholesterolemia and Alzheimer's diseases suggests that high cholesterol may play an important role in the development of Alzheimer's disease, and that modulation of cholesterol levels may help to reduce or delay the risk of Alzheimer's disease. These hypotheses are indeed supported by recent literature (Anstey et al., 2008; Hooijmans & Kiliaan, 2008; Xiong et al., 2008). In addition, topological analysis of the disease-disease association network revealed high-level modular organization of the network. Diseases within the same network module tend to have relatively high connectivity to each other and share related molecular mechanisms, while diseases in different modules tend to be much less connected and share no related mechanisms.

FUTURE RESEARCH DIRECTIONS

Current human FLN research is still in its infancy, as real cellular networks in human are much more complex and dynamic. As a multi-cellular organism, humans have various types of cells, tissues, and organs. Different types of cells and tissues make use of specific sets of expressed genes and gene-gene interactions. Even within the same cell, cellular gene networks are highly dynamic and can change rapidly in response to constant changes in internal and external environments. Human diseases usually occur in one or more specific tissues and organs, and they reflect the specific states of cellular networks in certain abnormal physiological conditions. As the disease progresses, cellular networks may also change correspondingly. Additionally, the molecular interactions for constructing the FLN can be transient and depend on the context of the cellular compartments. However, most current genome-scale data based on functional genomics and proteomics do not capture the entire dynamics as well as the various aspects of context-dependency of the cellular systems, but rather take snapshots at single time points in specific experimental settings, and from samples covering limited cell and tissue types (Schadt, 2009). As a result, current FLNs derived from these data sources remain static with few tissue and cell specificities. Recently, Bossi & Lehner (2009) have made some progress by incorporating tissue specificity in a global protein-protein interaction (PPI) network. With the rapid development of new technologies such as next generation sequencing, analysis of whole genomes and transcriptomes can be achieved at relatively low costs in a high-throughput fashion. This will facilitate data generation from diverse samples collected under diverse biological conditions and time points. Together with further developments in bioinformatics approaches for analyzing high-throughput and multi-dimensional data, it is possible to construct FLNs which can capture the dynamic nature of cellular networks, and take into account the specificity of biological systems at various levels. With these improved FLNs, it is expected that disease mechanisms can be studied in more realistic biological contexts.

Many diseases can be triggered by various environmental factors, such as viral and bacterial infection, unhealthy life styles and diets, psychological stress, etc. (Van Heyningen & Yeyati, 2004). One hypothesis is that these environmental factors trigger a specific disease by affecting the cellular biological processes that are relevant to the disease, i.e., through affecting the underlying molecular mechanisms. By including the interactions between genes and environmental factors as another dimension of the FLN, it is expected that our understanding of the molecular mechanisms of human diseases and their relationships to environments will increase as well.

Besides FLNs, networks based on therapeutic aspects such as target and drug interactions can also be used to bring insights to drug discovery (Keiser et al., 2009; Yildirim et al., 2007). For instance, based on the knowledge from existing targets and drugs, a network of therapeutic targets can be constructed by connecting target pairs sharing the same drugs. Similarly, a network of drugs can be constructed by connecting drug pairs sharing the same targets (Yildirim et al., 2007). Systematic analysis of these target-drug networks can uncover rules and principles characterizing known targets and drugs, which can then be used to guide the selections of new targets and drugs (Yildirim et al., 2007). However, most current network-based approaches for disease studies either focus solely on using the FLN to identify molecular mechanisms of diseases and predict new disease genes, or focus solely on the therapeutic aspects in terms of interactions of targets and drugs. A major challenge is to integrate the functional linkage gene networks and target-drug interaction networks together to construct a comprehensive drug discovery pipeline. Such a pipeline will start from identifying the molecular basis of a disease to target identification and further to drug selection, or reversely to assess or predict wanted and unwanted drug effects by tracing back to the molecular processes affected by the drug. By integrating and analyzing normal cellular networks as well as disease and therapeutic perturbations within a unified framework, it is possible to gain more insights into disease mechanisms and corresponding therapies. This will require close collaborations of experts with different domain knowledge such as bioinformaticians, biologists, chemists, and clinicians.

CONCLUSION

Current bioinformatics approaches have successfully integrated diverse types of functional genomic and proteomic data to construct functional linkage gene networks. In this way, functional associations between genes can be mapped in the context of a functionally coherent biological system. More importantly, new disease genes and therapeutic targets for diverse diseases can be successfully predicted from these FLNs. In addition, FLNs can also reliably uncover associations between disease at the molecular level, even for diseases exhibiting dissimilar phenotypes or sharing no disease genes. Such associations can drive novel hypotheses on molecular mechanisms of diseases and therapies. With the rapid development of new high-throughput biotechnologies and the availability of data sources covering more biological conditions, FLNs can be further improved to capture the dynamic nature and specificities under different conditions of real cellular systems. Moreover, with the integration of FLNs and drug-target interaction based networks, it will be possible to study genes, diseases, and therapeutic drugs within a unified framework. This unified approach has the potential to bring more insights into the drug discovery pipeline.

REFERENCES

Ahmed, A., & Xing, E. P. (2009). Recovering time-varying networks of dependencies in social and biological studies. *Proceedings of the National Academy of Sciences of the United States of America*, *106*(29), 11878–11883. doi:10.1073/pnas.0901910106

Ajioka, R. S., Phillips, J. D., & Kushner, J. P. (2006). Biosynthesis of heme in mammals. *Biochimica et Biophysica Acta*, *1763*(7), 723–736. doi:10.1016/j.bbamcr.2006.05.005

Ala, U., Piro, R. M., Grassi, E., Damasco, C., Silengo, L., & Oti, M. (2008). Prediction of human disease genes by human-mouse conserved coexpression analysis. *PLoS Computational Biology*, *4*(3), e1000043. doi:10.1371/journal.pcbi.1000043

Anstey, K. J., Lipnicki, D. M., & Low, L. F. (2008). Cholesterol as a risk factor for dementia and cognitive decline: A systematic review of prospective studies with meta-analysis. *The American Journal of Geriatric Psychiatry*, *16*(5), 343–354.

Berardini, T. Z., Li, D., Huala, E., Bridges, S., Burgess, S., & McCarthy, F. (2010). The gene ontology in 2010: Extensions and refinements. *Nucleic Acids Research*, *38*(Database issue), D331–D335. doi:10.1093/nar/gkp1018

Berger, S. I., & Iyengar, R. (2009). Network analyses in systems pharmacology. *Bioinformatics (Oxford, England)*, *25*(19), 2466–2472. doi:10.1093/bioinformatics/btp465

Bossi, A., & Lehner, B. (2009). Tissue specificity and the human protein interaction network. *Molecular Systems Biology*, *5*, 260. doi:10.1038/msb.2009.17

Calvo, S., Jain, M., Xie, X., Sheth, S. A., Chang, B., & Goldberger, O. A. (2006). Systematic identification of human mitochondrial disease genes through integrative genomics. *Nature Genetics*, *38*(5), 576–582. doi:10.1038/ng1776

Chen, K. C., Csikasz-Nagy, A., Gyorffy, B., Val, J., Novak, B., & Tyson, J. J. (2000). Kinetic analysis of a molecular model of the budding yeast cell cycle. *Molecular Biology of the Cell*, *11*(1), 369–391.

Cline, M. S., Smoot, M., Cerami, E., Kuchinsky, A., Landys, N., & Workman, C. (2007). Integration of biological networks and gene expression data using cytoscape. *Nature Protocols*, *2*(10), 2366–2382. doi:10.1038/nprot.2007.324

Conti, R., Veenstra, D.L., Armstrong, K., Lesko, L.J. & Grosse, S.D. (2010). Personalized medicine and genomics: Challenges and opportunities in assessing effectiveness, cost-effectiveness, and future research priorities. *Medical Decision Making*.

Dasika, M. S., Burgard, A., & Maranas, C. D. (2006). A computational framework for the topological analysis and targeted disruption of signal transduction networks. *Biophysical Journal*, *91*(1), 382–398. doi:10.1529/biophysj.105.069724

Emilsson, V., Thorleifsson, G., Zhang, B., Leonardson, A. S., Zink, F., & Zhu, J. (2008). Genetics of gene expression and its effect on disease. *Nature*, *452*(7186), 423–428. doi:10.1038/nature06758

Ewing, R. M., Chu, P., Elisma, F., Li, H., Taylor, P., & Climie, S. (2007). Large-scale mapping of human protein-protein interactions by mass spectrometry. *Molecular Systems Biology*, *3*, 89. doi:10.1038/msb4100134

Feldman, I., Rzhetsky, A., & Vitkup, D. (2008). Network properties of genes harboring inherited disease mutations. *Proceedings of the National Academy of Sciences of the United States of America*, *105*(11), 4323–4328. doi:10.1073/pnas.0701722105

Flicek, P., Aken, B. L., Beal, K., Ballester, B., Caccamo, M., & Chen, Y. (2008). Ensembl 2008. *Nucleic Acids Research*, *36*(Database issue), D707–D714. doi:10.1093/nar/gkm988

Franke, L., Bakel, H., Fokkens, L., de Jong, E. D., Egmont-Petersen, M., & Wijmenga, C. (2006). Reconstruction of a functional human gene network, with an application for prioritizing positional candidate genes. *American Journal of Human Genetics*, *78*(6), 1011–1025. doi:10.1086/504300

Franzosa, E., Linghu, B., & Xia, Y. (2009). Computational reconstruction of protein-protein interaction networks: Algorithms and issues. *Methods in Molecular Biology (Clifton, N.J.)*, *541*, 89–100. doi:10.1007/978-1-59745-243-4_5

Goh, K. I., Cusick, M. E., Valle, D., Childs, B., Vidal, M., & Barabasi, A. L. (2007). The human disease network. *Proceedings of the National Academy of Sciences of the United States of America*, *104*(21), 8685–8690. doi:10.1073/pnas.0701361104

Griffith, O. L., Pleasance, E. D., Fulton, D. L., Oveisi, M., Ester, M., & Siddiqui, A. S. (2005). Assessment and integration of publicly available sage, cdna microarray, and oligonucleotide microarray expression data for global coexpression analyses. *Genomics*, *86*(4), 476–488. doi:10.1016/j.ygeno.2005.06.009

Hamosh, A., Scott, A. F., Amberger, J. S., Bocchini, C. A., & McKusick, V. A. (2005). Online Mendelian inheritance in man (OMIM), a knowledgebase of human genes and genetic disorders. *Nucleic Acids Research*, *33*(Database issue), D514–D517. doi:10.1093/nar/gki033

Hart, G. T., Ramani, A. K., & Marcotte, E. M. (2006). How complete are current yeast and human protein-interaction networks? *Genome Biology*, *7*(11), 120. doi:10.1186/gb-2006-7-11-120

Hartwell, L. H., Hopfield, J. J., Leibler, S., & Murray, A. W. (1999). From molecular to modular cell biology. *Nature, 402*(6761Suppl), C47–C52. doi:10.1038/35011540

Hooijmans, C. R., & Kiliaan, A. J. (2008). Fatty acids, lipid metabolism and Alzheimer pathology. *European Journal of Pharmacology, 585*(1), 176–196. doi:10.1016/j.ejphar.2007.11.081

Hopkins, A. L. (2007). Nctwork pharmacology. *Nature Biotechnology, 25*(10), 1110–1111. doi:10.1038/nbt1007-1110

Hopkins, A. L. (2008). Network pharmacology: The next paradigm in drug discovery. *Nature Chemical Biology, 4*(11), 682–690. doi:10.1038/nchembio.118

Hopkins, A. L., & Groom, C. R. (2002). The druggable genome. *Nature Reviews. Drug Discovery, 1*(9), 727–730. doi:10.1038/nrd892

Hu, Z., Hung, J.H., Wang, Y., Chang, Y.C., Huang, C.L., Huyck, M., et al. (2009). Visant 3.5: Multiscale network visualization, analysis and inference based on the gene ontology. *Nucleic Acids Research, 37*(Web Server issue), W115-121.

Hu, Z., Mellor, J., Wu, J., Kanehisa, M., Stuart, J. M., & DeLisi, C. (2007). Towards zoomable multidimensional maps of the cell. *Nature Biotechnology, 25*(5), 547–554. doi:10.1038/nbt1304

Hu, Z., Snitkin, E. S., & DeLisi, C. (2008). Visant: An integrative framework for networks in systems biology. *Briefings in Bioinformatics, 9*(4), 317–325. doi:10.1093/bib/bbn020

Huttenhower, C., Haley, E. M., Hibbs, M. A., Dumeaux, V., Barrett, D. R., & Coller, H. A. (2009). Exploring the human genome with functional maps. *Genome Research, 19*(6), 1093–1106. doi:10.1101/gr.082214.108

Hwang, W. C., Zhang, A., & Ramanathan, M. (2008). Identification of information flow-modulating drug targets: A novel bridging paradigm for drug discovery. *Clinical Pharmacology and Therapeutics, 84*(5), 563–572. doi:10.1038/clpt.2008.129

Ideker, T., & Sharan, R. (2008). Protein networks in disease. *Genome Research, 18*(4), 644–652. doi:10.1101/gr.071852.107

Iwabe, N., Kuma, K., & Miyata, T. (1996). Evolution of gene families and relationship with organismal evolution: Rapid divergence of tissue-specific genes in the early evolution of chordates. *Molecular Biology and Evolution, 13*(3), 483–493.

Janga, S.C. & Tzakos, A. (2009). Structure and organization of drug-target networks: Insights from genomic approaches for drug discovery. *Molecular Biosystems.*

Jensen, L. J., Kuhn, M., Stark, M., Chaffron, S., Creevey, C., & Muller, J. (2009). String 8-a global view on proteins and their functional interactions in 630 organisms. *Nucleic Acids Research, 37*(Database issue), D412–D416. doi:10.1093/nar/gkn760

Jensen, L.J., Lagarde, J., von Mering, C. & Bork, P. (2004). Arrayprospector: A Web resource of functional associations inferred from microarray expression data. *Nucleic Acids Research, 32*(Web server issue), W445-448.

Jiang, W., Li, X., Rao, S., Wang, L., Du, L., & Li, C. (2008). Constructing disease-specific gene networks using pair-wise relevance metric: Application to colon cancer identifies interleukin 8, desmin and enolase 1 as the central elements. *BMC Systems Biology, 2*(1), 72. doi:10.1186/1752-0509-2-72

Kanehisa, M., Goto, S., Kawashima, S., Okuno, Y., & Hattori, M. (2004). Thc kegg resource for deciphering the genome. *Nucleic Acids Research, 32*(Database issue), D277–D280. doi:10.1093/nar/gkh063

Keiser, M. J., Setola, V., Irwin, J. J., Laggner, C., Abbas, A. I., & Hufeisen, S. J. (2009). Predicting new molecular targets for known drugs. *Nature*, *462*(7270), 175–181. doi:10.1038/nature08506

Kohler, S., Bauer, S., Horn, D., & Robinson, P. N. (2008). Walking the interactome for prioritization of candidate disease genes. *American Journal of Human Genetics*, *82*(4), 949–958. doi:10.1016/j.ajhg.2008.02.013

Lage, K., Karlberg, E. O., Storling, Z. M., Olason, P. I., Pedersen, A. G., & Rigina, O. (2007). A human phenome-interactome network of protein complexes implicated in genetic disorders. *Nature Biotechnology*, *25*(3), 309–316. doi:10.1038/nbt1295

Lee, D. S., Park, J., Kay, K. A., Christakis, N. A., Oltvai, Z. N., & Barabasi, A. L. (2008a). The implications of human metabolic network topology for disease comorbidity. *Proceedings of the National Academy of Sciences of the United States of America*, *105*(29), 9880–9885. doi:10.1073/pnas.0802208105

Lee, H. K., Hsu, A. K., Sajdak, J., Qin, J., & Pavlidis, P. (2004). Coexpression analysis of human genes across many microarray data sets. *Genome Research*, *14*(6), 1085–1094. doi:10.1101/gr.1910904

Lee, I., Lehner, B., Crombie, C., Wong, W., Fraser, A. G., & Marcotte, E. M. (2008b). A single gene network accurately predicts phenotypic effects of gene perturbation in caenorhabditis elegans. *Nature Genetics*, *40*(2), 181–188. doi:10.1038/ng.2007.70

Liao, B. Y., & Zhang, J. (2007). Mouse duplicate genes are as essential as singletons. *Trends in Genetics*, *23*(8), 378–381. doi:10.1016/j.tig.2007.05.006

Lim, J., Hao, T., Shaw, C., Patel, A. J., Szabo, G., & Rual, J. F. (2006). A protein-protein interaction network for human inherited ataxias and disorders of purkinje cell degeneration. *Cell*, *125*(4), 801–814. doi:10.1016/j.cell.2006.03.032

Linghu, B., & Delisi, C. (2010). Phenotypic connections in surprising places. *Genome Biology*, *11*(4), 116. doi:10.1186/gb-2010-11-4-116

Linghu, B., Snitkin, E. S., Holloway, D. T., Gustafson, A. M., Xia, Y., & DeLisi, C. (2008). High-precision high-coverage functional inference from integrated data sources. *BMC Bioinformatics*, *9*, 119. doi:10.1186/1471-2105-9-119

Linghu, B., Snitkin, E. S., Hu, Z., Xia, Y., & Delisi, C. (2009). Genome-wide prioritization of disease genes and identification of disease-disease associations from an integrated human functional linkage network. *Genome Biology*, *10*(9), R91. doi:10.1186/gb-2009-10-9-r91

McGary, K. L., Lee, I., & Marcotte, E. M. (2007). Broad network-based predictability of saccharomyces cerevisiae gene loss-of-function phenotypes. *Genome Biology*, *8*(12), R258. doi:10.1186/gb-2007-8-12-r258

Oti, M., & Brunner, H. G. (2007). The modular nature of genetic diseases. *Clinical Genetics*, *71*(1), 1–11. doi:10.1111/j.1399-0004.2006.00708.x

Oti, M., Huynen, M. A., & Brunner, H. G. (2008). Phenome connections. *Trends in Genetics*, *24*(3), 103–106. doi:10.1016/j.tig.2007.12.005

Oti, M., Snel, B., Huynen, M. A., & Brunner, H. G. (2006). Predicting disease genes using protein-protein interactions. *Journal of Medical Genetics*, *43*(8), 691–698. doi:10.1136/jmg.2006.041376

Pan, W. (2008). Network-based model weighting to detect multiple loci influencing complex diseases. *Human Genetics*.

Park, J., Lee, D. S., Christakis, N. A., & Barabasi, A. L. (2009). The impact of cellular networks on disease comorbidity. *Molecular Systems Biology*, *5*, 262. doi:10.1038/msb.2009.16

Pujana, M. A., Han, J. D., Starita, L. M., Stevens, K. N., Tewari, M., & Ahn, J. S. (2007). Network modeling links breast cancer susceptibility and centrosome dysfunction. *Nature Genetics*, *39*(11), 1338–1349. doi:10.1038/ng.2007.2

Ravasz, E., Somera, A. L., Mongru, D. A., Oltvai, Z. N., & Barabasi, A. L. (2002). Hierarchical organization of modularity in metabolic networks. *Science*, *297*(5586), 1551–1555. doi:10.1126/science.1073374

Rual, J. F., Venkatesan, K., Hao, T., Hirozane-Kishikawa, T., Dricot, A., & Li, N. (2005). Towards a proteome-scale map of the human protein-protein interaction network. *Nature*, *437*(7062), 1173–1178. doi:10.1038/nature04209

Ruths, D. A., Nakhleh, L., Iyengar, M. S., Reddy, S. A., & Ram, P. T. (2006). Hypothesis generation in signaling networks. *Journal of Computational Biology*, *13*(9), 1546–1557. doi:10.1089/cmb.2006.13.1546

Schadt, E. E. (2009). Molecular networks as sensors and drivers of common human diseases. *Nature*, *461*(7261), 218–223. doi:10.1038/nature08454

Sedel, F., Turpin, J. C., & Baumann, N. (2007). Neurological presentations of lysosomal diseases in adult patients. *Revista de Neurologia*, *163*(10), 919–929. doi:10.1016/S0035-3787(07)92635-1

Stelzl, U., Worm, U., Lalowski, M., Haenig, C., Brembeck, F. H., & Goehler, H. (2005). A human protein-protein interaction network: A resource for annotating the proteome. *Cell*, *122*(6), 957–968. doi:10.1016/j.cell.2005.08.029

van Driel, M. A., Bruggeman, J., Vriend, G., Brunner, H. G., & Leunissen, J. A. (2006). A text-mining analysis of the human phenome. *European Journal of Human Genetics*, *14*(5), 535–542. doi:10.1038/sj.ejhg.5201585

Van Heyningen, V., & Yeyati, P. L. (2004). Mechanisms of non-Mendelian inheritance in genetic disease. *Human Molecular Genetics*, *13*(2), R225–R233. doi:10.1093/hmg/ddh254

van Reeuwijk, J., Brunner, H. G., & van Bokhoven, H. (2005). Glyc-o-genetics of Walker-Warburg syndrome. *Clinical Genetics*, *67*(4), 281–289. doi:10.1111/j.1399-0004.2004.00368.x

von Mering, C., Jensen, L. J., Kuhn, M., Chaffron, S., Doerks, T., & Kruger, B. (2007). String 7-recent developments in the integration and prediction of protein interactions. *Nucleic Acids Research*, *35*(Database issue), D358–D362. doi:10.1093/nar/gkl825

Wu, X., Jiang, R., Zhang, M. Q., & Li, S. (2008). Network-based global inference of human disease genes. *Molecular Systems Biology*, *4*, 189. doi:10.1038/msb.2008.27

Xiong, H., Callaghan, D., Jones, A., Walker, D. G., Lue, L. F., & Beach, T. G. (2008). Cholesterol retention in Alzheimer's brain is responsible for high beta- and gamma-secretase activities and abeta production. *Neurobiological Discoveries*, *29*(3), 422–437. doi:10.1016/j.nbd.2007.10.005

Yamada, R., & Ueda, H. (2009). Problems in analysis of large-scale data: Gene expression microarray analysis. *Tanpakushitsu Kakusan Koso*, *54*(10), 1307–1315.

Yildirim, M. A., Goh, K. I., Cusick, M. E., Barabasi, A. L., & Vidal, M. (2007). Drug-target network. *Nature Biotechnology*, *25*(10), 1119–1126. doi:10.1038/nbt1338

Zhang, R., & Lin, Y. (2009). Deg 5.0, a database of essential genes in both prokaryotes and eukaryotes. *Nucleic Acids Research*, *37*(Database issue), D455–D458. doi:10.1093/nar/gkn858

Zhong, W., & Sternberg, P. W. (2006). Genome-wide prediction of c. Elegans genetic interactions. *Science*, *311*(5766), 1481–1484. doi:10.1126/science.1123287

ADDITIONAL READING

Chautard, E., Thierry-Mieg, N., & Ricard-Blum, S. (2009). Interaction networks: From protein functions to drug discovery. A review. *Pathologie Biologie*, *57*(4), 324–333. doi:10.1016/j.patbio.2008.10.004

Costello, J. C., Dalkilic, M. M., Beason, S. M., Gehlhausen, J. R., Patwardhan, R., & Middha, S. (2009). Gene networks in drosophila melanogaster: Integrating experimental data to predict gene function. *Genome Biology*, *10*(9), R97. doi:10.1186/gb-2009-10-9-r97

Goutsias, J., & Lee, N. H. (2007). Computational and experimental approaches for modeling gene regulatory networks. *Current Pharmaceutical Design*, *13*(14), 1415–1436. doi:10.2174/138161207780765945

Guan, Y., Myers, C. L., Lu, R., Lemischka, I. R., Bult, C. J., & Troyanskaya, O. G. (2008). A genomewide functional network for the laboratory mouse. *PLoS Computational Biology*, *4*(9), e1000165. doi:10.1371/journal.pcbi.1000165

Hase, T., Tanaka, H., Suzuki, Y., Nakagawa, S., & Kitano, H. (2009). Structure of protein interaction networks and their implications on drug design. *PLoS Computational Biology*, *5*(10), e1000550. doi:10.1371/journal.pcbi.1000550

Hecker, M., Lambeck, S., Toepfer, S., van Someren, E., & Guthke, R. (2009). Gene regulatory network inference: Data integration in dynamic models-a review. *Bio Systems*, *96*(1), 86–103. doi:10.1016/j.biosystems.2008.12.004

Hidalgo, C. A., Blumm, N., Barabasi, A. L., & Christakis, N. A. (2009). A dynamic network approach for the study of human phenotypes. *PLoS Computational Biology*, *5*(4), e1000353. doi:10.1371/journal.pcbi.1000353

Hu, G., & Agarwal, P. (2009). Human disease-drug network based on genomic expression profiles. *PLoS ONE*, *4*(8), e6536. doi:10.1371/journal.pone.0006536

Ideker, T. E. (2007). Network genomics. *Ernst Schering Research Foundation Workshop*, (61): 89–115. doi:10.1007/978-3-540-31339-7_5

Kell, D. B. (2006). Systems biology, metabolic modelling and metabolomics in drug discovery and development. *Drug Discovery Today*, *11*(23-24), 1085–1092. doi:10.1016/j.drudis.2006.10.004

Larranaga, P., Calvo, B., Santana, R., Bielza, C., Galdiano, J., & Inza, I. (2006). Machine learning in bioinformatics. *Briefings in Bioinformatics*, *7*(1), 86–112. doi:10.1093/bib/bbk007

Lee, I., & Marcotte, E. M. (2009). Effects of functional bias on supervised learning of a gene network model. *Methods in Molecular Biology (Clifton, N.J.)*, *541*, 463–475. doi:10.1007/978-1-59745-243-4_20

Li, H., Xuan, J., Wang, Y., & Zhan, M. (2008). Inferring regulatory networks. *Frontiers in Bioscience*, *13*, 263–275. doi:10.2741/2677

Li, Z., Wang, R. S., Zhang, X. S., & Chen, L. (2009). Detecting drug targets with minimum side effects in metabolic networks. *IET Systems Biology*, *3*(6), 523–533. doi:10.1049/iet-syb.2008.0166

Ma'ayan, A. (2008). Network integration and graph analysis in mammalian molecular systems biology. *IET Systems Biology*, *2*(5), 206–221. doi:10.1049/iet-syb:20070075

Marcotte, E. M., & Tsechansky, M. (2009). Disorder, promiscuity, and toxic partnerships. *Cell*, *138*(1), 16–18. doi:10.1016/j.cell.2009.06.024

McGary, K. L., Park, T. J., Woods, J. O., Cha, H. J., Wallingford, J. B., & Marcotte, E. M. (2010). Systematic discovery of nonobvious human disease models through orthologous phenotypes. *Proceedings of the National Academy of Sciences of the United States of America*, *107*(14), 6544–6549. doi:10.1073/pnas.0910200107

Navlakha, S., & Kingsford, C. (2010). The power of protein interaction networks for associating genes with diseases. *Bioinformatics (Oxford, England)*, *26*(8), 1057–1063. doi:10.1093/bioinformatics/btq076

Ozgur, A., Vu, T., Erkan, G., & Radev, D. R. (2008). Identifying gene-disease associations using centrality on a literature mined gene-interaction network. *Bioinformatics (Oxford, England)*, *24*(13), i277–i285. doi:10.1093/bioinformatics/btn182

Paolini, G. V., Shapland, R. H., van Hoorn, W. P., Mason, J. S., & Hopkins, A. L. (2006). Global mapping of pharmacological space. *Nature Biotechnology*, *24*(7), 805–815. doi:10.1038/nbt1228

Schrattenholz, A., & Soskic, V. (2008). What does systems biology mean for drug development? *Current Medicinal Chemistry*, *15*(15), 1520–1528. doi:10.2174/092986708784638843

Spiro, Z., Kovacs, I. A., & Csermely, P. (2008). Drug-therapy networks and the prediction of novel drug targets. *Journal of Biology*, *7*(6), 20. doi:10.1186/jbiol81

Suthram, S., Dudley, J. T., Chiang, A. P., Chen, R., Hastie, T. J., & Butte, A. J. (2010). Network-based elucidation of human disease similarities reveals common functional modules enriched for pluripotent drug targets. *PLoS Computational Biology*, *6*(2), e1000662. doi:10.1371/journal.pcbi.1000662

Weber, G. W., Ozogur-Akyuz, S., & Kropat, E. (2009). A review on data mining and continuous optimization applications in computational biology and medicine. *Birth Defects Research. Part C, Embryo Today*, *87*(2), 165–181. doi:10.1002/bdrc.20151

Wilkinson, D. J. (2007). Bayesian methods in bioinformatics and computational systems biology. *Briefings in Bioinformatics*, *8*(2), 109–116. doi:10.1093/bib/bbm007

KEY TERMS AND DEFINITIONS

Data Integration: Integration of diverse types of functional genomics and proteomics data to gain a comprehensive view of gene function.

Disease Gene Prediction: Predict new genes related to the molecular mechanisms of a disease.

Disease-Disease Associations: Existence of overlap for the underlying molecular mechanisms between diseases.

Diseases: Clinical phenotypes when one or more normal biological processes are perturbed in human.

Drug Discovery: Identify new compounds to target one or more proteins related to a disease with therapeutic effects.

Gene Networks: A graph representation of gene-gene associations with nodes representing genes and edges representing functional associations between genes.

Human Gene Networks: Gene networks composed of human genes.

Chapter 13
Network-Driven Analysis Methods and their Application to Drug Discovery

Daniel Ziemek
Pfizer Inc., USA

Christoph Brockel
Pfizer Inc., USA

ABSTRACT

Drug discovery and development face tremendous challenges to find promising intervention points for important diseases. Any therapeutic agent targeting such an intervention point must prove its efficacy and safety in patients. Success rates measured from first studies in human to registration average around 10% only. Over the last decade, massive knowledge on biological systems has been accumulated and genome-scale primary data are produced at an ever increasing rate. In parallel, methods to use that knowledge have matured. This chapter will present some of the problems facing the pharmaceutical industry and elaborate on the current state of network-driven analysis methods. It will focus especially on semi-quantitative methods that are applicable to large-scale data analysis and point out their potential use in many relevant drug discovery challenges.

INTRODUCTION

Drug discovery and development continues to be a high-risk endeavor. Success rates for small molecule therapeutics average to around 10% from entry into Phase 1 ('First in Human') to successful registration across therapeutic indications and companies (Kola & Landis, 2004).

While historically much of the attrition was based on safety issues identified in Phase 1, more recent data indicate that the average success rate in Phase 1 has climbed to 60% and the success rates in Phase 2 and Phase 3 have reached 30-40% and 60%, respectively. Failure of a compound in a late

DOI: 10.4018/978-1-60960-491-2.ch013

stage of development is particularly undesirable, since it has then already incurred most of its total cost. With the total cost of drug discovery and development estimated to approach, and frequently exceed, $1B per registration, reduction in late stage attrition will be critical to affordable therapeutics.

In the pharmaceutical industry, the path to a therapy starts in many cases with the selection of a target, the intervention point that promises to cure or at least treat the disease of interest. Finding the right target is both crucial and non-trivial and there are several different approaches in use. The association of a clinical phenotype with a genetic mutation is very promising, since observations are made directly in human. Unfortunately, even in the case of monogenic diseases where the cause for the clinical phenotype can be linked to a single mutation, this information does not always lead directly to selection of a target. Reasons can be that the activity of the mutated protein cannot be modulated by a small molecule compound or that the mutation doesn't result in any direct change of a protein at all. Good examples for this problem are the known MODY ('Maturity Onset Diabetes of the Young') gene mutations causing early-onset diabetes by rendering the corresponding proteins non-functional. No compounds have been found to counteract this loss-of-function directly. The 'phenotype-mutation-target' approach fails completely for multigenic, complex diseases. Methodologies are needed to mechanistically connect the genetic layer to the clinical phenotype to then enable selection of optimal intervention points.

Once the target is identified, the search for compounds or biologicals that demonstrate selective activity against the target, in in-vitro assays and in-vivo models of the disease begins. These programs are typically lengthy and produce optimized molecules that still fail about half of the time in safety studies in human - often caused by off-target effects. The ones that pass that hurdle then frequently fail to demonstrate efficacy.

The development of technologies that enable the collection of comprehensive 'genome wide', data in biological systems – frequently referred to as 'OMICs' technologies has created the promise to understand human biology and disease on a molecular and mechanistic level.

Dramatic improvements in speed, throughput and cost to sequence DNA and RNA are adding to this promise and introduce the possibility to understand genetic drivers for complex diseases and possibly differences in susceptibility to treatment in different sub-populations. Personalized medicine, the matching of the right treatment for the individual patient seems within reach.

While system wide generation of experimental data in disease models and to some degree in human is feasible today, the interpretation of these data for specific, testable hypotheses is still a significant bottleneck.

As illustrated in Figure 1, the approach to understand disease and drug action involves several stages from experiment to data generation and analysis. Integrating experimental results – possibly recorded at different levels (transcript, protein abundance, protein modification, etc) - with existing biological knowledge is a key step towards interpretation and hypothesis generation. Pathway or network analysis promises to provide the missing link between the lists of statistically significant findings from experiments and the understanding of the underlying biological mechanisms.

In the following, we describe network-based approaches that can help understand disease at a molecular level, select alternate and hopefully better targets for cases where genetic drivers of the disease themselves are not suitable for small molecule approaches, understand the mechanism of action and toxicity of candidate compounds and select mechanistic markers that can be used to confirm exposure of the target to the compound in human. We present some example applications of network analysis, provide a review of the currently available methods/algorithms and close with a discussion of gaps and future perspectives.

Figure 1. Mechanistic understanding of disease and drug action involves several stages: (a) multiple model systems need to be perturbed and interrogated; (b) data from these experiments are generated on multiple molecular levels; (c) primary statistical analysis of these data leads to lists of observed molecular changes that need to be integrated with existing knowledge and finally (d) interpretation of the results to create hypotheses for target selection and compound progression.

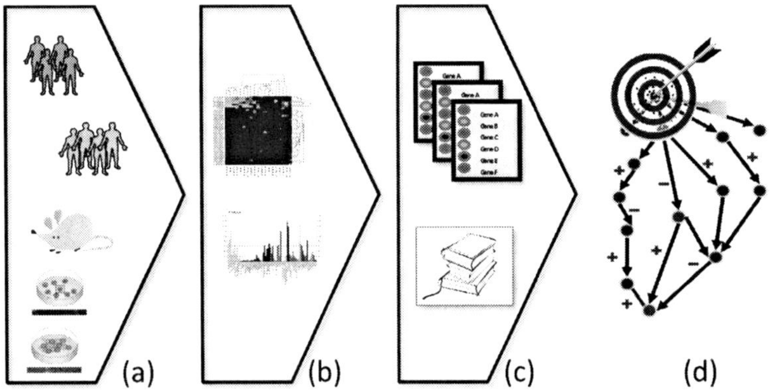

APPLICATIONS IN DRUG DISCOVERY

Fundamentally, the applications of network driven analysis in the context of drug discovery are in the areas of

a. understanding the mechanism of disease for the identification of targets and biomarkers and
b. elucidating or predicting the full effect of a therapeutic agent on the biological system to confirm or discover mechanism of action, understand the origin of adverse effects and for the identification of new indications for already approved drugs.

Understanding Disease

While there are a plethora of possible approaches to finding a therapeutically relevant agent, pharmaceutical drug discovery processes have mostly been optimized to work on the premise that a 'target' molecule for the therapeutic intervention has been identified. This target will then be used to discover chemical matter through screening for activity in biochemical or binding assays or to develop antibodies or other selective bio-molecules that modulate its activity or abundance.

The mechanistic origins or drivers of disease are key to the identification of such targets as they present possible intervention points to cure or at least treat the disease. Comparison of disease state with non-disease either in population studies or experimental model systems typically leads to long lists of statistically significant observations. These can be hard to reduce to understandable biological phenomena in practice.

Over the last decade a number of methods have been developed to address this problem. A high level understanding of the biological processes represented in the molecular observations can be obtained through comparison of the list of observations with pre-defined lists of molecules that are assigned to biological processes or pathways. This gene set enrichment approach has been broadly applied with gene expression data (e.g. Mootha et al., 2003; Ebert et al., 2008; Alles et al., 2009) and more recently genetic variation data, e.g. Bonifaci et al. (2008), and is helpful to abstract from the molecular observations to larger biological processes. The utility of the approach is limited to homogeneous data-types. Analyzing datasets that go across e.g. gene expression, protein modification and metabolite data from the same sample using gene set enrichment is not meaningful. Moreover, the result from gene set enrichment analysis is a list of processes/pathways that does not directly lead to more detailed hypotheses about the mechanism behind the experimentally observed data. Perhaps most importantly, the approach is limited by the availability of reference lists as well as the static nature of the lists.

Complementary to these approaches to link molecular observations to high level biological concepts is the attempt to identify densely connected 'modules' in the network of molecular interactions that can be linked to specific disease phenotypes. Over the past decades, evidence has emerged that similar phenotypes are often caused by genes and proteins acting in concert in complexes, common pathways or other coherent processes (Oti & Brunner, 2007). 'Active subnetworks' described by Ideker et al. (2002) are a specific example of such functional modules that can be discovered by combining gene expression data with protein-protein interaction networks.

Identification of modules can be particularly useful to suggest alternate intervention points for indications where a target is known, but is hard to modulate or has safety liabilities. Beyond the specific use for identification of targets, such methods will also help the detailed understanding of disease at a level of granularity between the high level process (e.g. apoptosis) and the single molecular observation.

The ultimate goal, however, remains the mechanistic understanding of disease that takes into account phenotypic observations, genetic variations and molecular data from disease models and patients. Methods for the integration of these different data types into one cohesive model and subsequent creation of testable hypothesis are still in their infancy, but are beginning to show promising results. For example, Zhu et al. (2008) reported a Bayesian approach to causal inference that demonstrated the feasibility of combining genetic variation and genome wide expression data to create mechanistic insights. In this example, the authors connected the association of Insig2 mutations with cholesterol levels previously identified by Cervino et al. (2005) mechanistically to molecular observations at the transcript level. Thereby, they not only create possible explanations as to *how* Insig2 might be involved in the regulation of cholesterol levels but also expose alternate targets for therapeutic intervention.

Understanding Drug Action

A molecular understanding of the effects of therapeutic agents on the biological system can improve survival in several areas of drug discovery, the most important one being the understanding of adverse effects driven by interactions of the agent with unanticipated targets ('off-target effects'). 'Toxicogenomics', an assessment of adverse effects on the molecular level with the promise to be able to predict adverse effects from gene expression data has had limited success so far. However, network-based analyses of system wide molecular responses to compounds could enable a mechanistic understanding of the drivers behind adverse effects and lead to identification of markers that help identify patients with higher (or lower) risk for such effects.

The ability to understand whether an observed adverse effect phenotype is caused by an off-target interaction of the agent or a direct result of the modulation of the activity of the target molecule can help direct the design of molecules to be more selective or possibly drive a decision to abandon the target. As an example, Xie et al. (2009) describe the creation of an 'off-target network' for Cholesteryl Ester Transfer Proteins (CETPs) in an attempt to explain clinically observed side effects like hypertension for a number of CETP inhibitors.

Complementary to the desire to understand adverse effects is the need to demonstrate efficacy on a mechanistic level. All too often, compounds with strong effect in *in-vitro* assays show very little effect in cellular or in vivo models or between different in vivo models. Even with demonstrated efficacy in model organisms, the therapeutic benefit in humans is often not achieved – leading to Phase 2 failure. A mechanistic understanding of the compound effect can lead to better understanding of the relevance of the model systems for the disease and better decisions for progression of the compound.

While the 'promiscuity' of therapeutic agents can lead to undesired adverse effects, it can also have a rather beneficial outcome. For instance, this is the case if the off-target activity can be exploited to use an already established, safe compound for indications it was not originally developed for. Identification of such opportunities can result in dramatically shortened timelines and reduced cost which can lead to treatments for diseases that would otherwise remain 'orphans'. As an example, Suthram et al. (2010) describe a network-based approach to elucidate disease similarities and identify common functional modules. Performing an integrated analysis of mRNA data for 54 diseases combined with a protein interaction network, they identified interaction modules shared between diseases. These are then used to explore new indications for drugs known to modulate a given module.

METHODS AND RESOURCES

In the following, we will give an overview of the most relevant algorithmic approaches utilizing biological networks. In many cases, direct applications to problems in drug discovery have been reported. In others, we see the potential for the method to be of use in the near future. Methods that rely on prior knowledge in the form of networks and pathways can only be as good as the underlying data sources they use. A number of public domain and commercial resources for pathways and networks have been developed ranging from experimentally determined protein-protein interaction networks to tested mathematical models of signaling cascades to manually drawn depictions of canonical biological processes. We will point out some of the most relevant sources.

We distinguish three types of methods based on the degree to which they incorporate or try to infer biological relationships between molecular entities (Figure 2). On the one end of the spectrum are methods relying exclusively on a defined collection of gene sets reflecting some form of coherent biology. While these are not strictly network-based methods, we are including them here because of their popularity and to contrast them to the other method types. Then, topology-based methods aim to explore neighborhood relationships in biological networks of different types. Finally, directed methods try to utilize or infer semantically richer relationships between the entities such as dependency, causality, or directionality. In this review, we focus on methods applicable to large scale data analysis. The field of in-depth kinetic models based on mathematical formalisms such as ordinary differential equations is not covered in detail. Rather, we give some pointers to excellent reviews of this also very important field.

Static Pathway / Gene Set Enrichment Methods

Probably the most commonly employed analysis approaches utilizing pre-defined pathways and biological processes are gene set enrichment methods. All such methods take two inputs:(1) a pre-defined collection of gene sets, e.g. derived from metabolic pathways in the KEGG database (Kanehisa et al., 2010) and (2) the measured outcome of a biological experiment, e.g. the dif-

Figure 2. Methods can be distinguished by how much prior information they try to infer or utilize. (a) Gene set methods label a list of biologically coherent entities with a common tag. Sets can be defined based on underlying pathways and will overlap. (b) Topology-driven methods make use of explicit neighborhood relationships. (c) Directed methods model dependencies, causality, or strength of relationships explicitly.

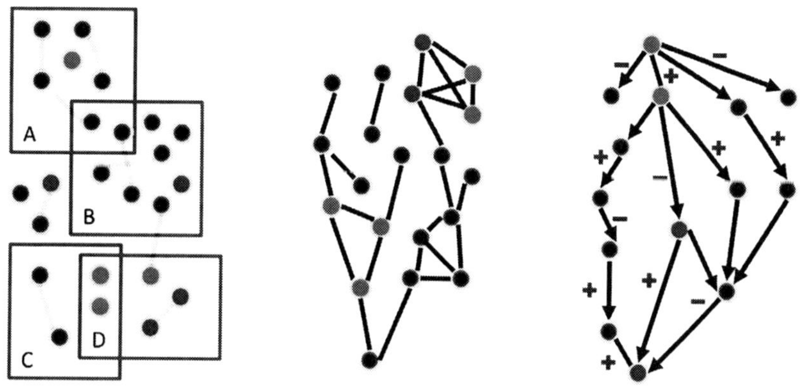

ferentially expressed genes in a microarray study. The output is usually a set of *p*-values quantifying the degree of association of each gene set with the experimentally derived data. This type of analysis is routinely used to get a first understanding of the readout of a genomic assay. For instance, Mootha, et al. (2003) analyzed expression data from diabetic patients and established a phenotype correlation with a gene set involved in oxidative phosphorylation. Ebert et al. (2008) assessed the effect of an RNAi-based assay to detect the disease gene for 5q⁻ syndrome, a specific cancer type using gene set enrichment. Another example is the work by Alles et al. (2009) in which they used such methods to give additional evidence for MYC as a potential target gene in estrogen receptor negative (ER−) breast cancer subtypes. While most applications of this general framework are in the context of gene expression data, more recently a number of studies have applied similar concepts to the interpretation of genome-wide association studies (GWAS). For instance, Bonifaci et al. (2008) analyzed the properties of a set of low-penetrance breast cancer-susceptibility genes from a GWAS. They compared the candidate genes to a set of biological processes from the Gene Ontology (Ashburner et al., 2000) and established links to cell communication and cell death processes.

Contingency Tables

By far the most common approach to determine whether a defined gene set is differentially expressed is based on 2x2 contingency tables. The table is compiled by assessing each measured gene with respect to two variables: (1) a gene is a member of the gene set and (2) a gene is differentially expressed. Then, a test for independence of the two variables is conducted, leading to a *p*-value quantifying the randomness of the association. Popular tests include Fisher's exact test and the Chi-Square test for independence. Many authors have suggested minor variations of this idea (e.g. Draghici et al., 2003; Hosack et al., 2003).

Gene Set Enrichment (GSEA)

A shortcoming of the contingency table approach is the need to define a hard threshold to label genes significantly differentially expressed. If not chosen appropriately, a large number of member genes might fall just below the threshold, leading to a failure to detect a true association with a gene set. The GSEA procedure of Mootha et al. (2003) popularized the idea of using the entire vector of expression measurements to identify differential expression in a gene set. GSEA starts from a ranked list of genes in which each gene is scored by its correlation with the phenotype of interest. The procedure then implements a signed and weighted variant of a Kolmogorov-Smirnov statistic and assesses significance through permutation of phenotype labels. GSEA has become a widely used method, in part because a free implementation and a comprehensive collection of gene sets, MSigDB (Subramanian et al., 2005), are readily available for download.

Other Methods

Recently, Ackermann & Strimmer (2009) gave an excellent overview of the zoo of different gene set enrichment methods and include simulation results as well as tests on real world data sets. Their conclusions are surprising in that they recommend simple statistics, such as the mean or median of the *t*-statistics combined with a gene or phenotype re-sampling procedure. In their simulations, these methods as e.g. proposed by Tian et al. (2005) outperform well-established methods like GSEA.

Gene Set Resources

In principle, any set of annotations for genes or proteins can be used for gene set enrichment. The Gene Ontology (Ashburner et al., 2000) is among the most popular annotation schemes used, but introduces an additional layer of complexity as the represented gene sets are organized in a membership hierarchy. In the context of reflecting biological networks and pathways as gene sets, a few highly popular sources stand out. The KEGG database (Kanehisa et al., 2010) is a comprehensive collection of initially primary metabolic pathways in several species. Over time, some signaling pathways as well as disease pathways were added. KEGG makes the relationships within each pathway explicit and is therefore also suitable for the topology-based methods described in a later section. BioCarta (http://www.biocarta.com), a company selling reagents and assays, started a community effort around the curation of pathways using a simple, but visually appealing template. The resulting database of pathway pictures including pointers to referenced genes has become a popular source of gene sets, even though the curation effort does not seem to continue. PantherDB (Thomas et al., 2003) and the collection of pathways maintained by the high-profile journals Nature and Science, called Pathway Interaction Database (Schaefer et al., 2009) and STKE Connection Maps (http://stke.sciencemag.org/cm/) respectively, are good sources as well. The easiest way to get started is probably an already compiled resource such as MSigDB (Subramanian et al., 2005). This database bundles several of the resources mentioned above among many others in one convenient package.

Pros and Cons

Substantial criticism against some gene set enrichment methods has been voiced. For instance, Goemann & Buhlmann (2007) pointed out that gene set enrichment methods are fundamentally of either a *competitive* or a *self-contained* nature. In a competitive test, the differential expression within one gene set is weighed against differential expression outside of that set. In this case, the genes itself become the sampling unit. Consequently, permutation of gene labels is usually used to assess significance. In contrast, a self-contained test takes into account only the association of the genes in a gene set with the phenotype. Genes outside of the set do not play a role. Here, significance assessment is accomplished through permutation of phenotype labels. Goeman and

Buhlmann (2007) argued that competitive tests are fundamentally of limited utility as they do not make any statements about the differential expression of gene sets when the experiment is re-run with new biological samples from the same phenotypes. Rather, they propose to use genes as the sampling unit and make a statement about the association of a new set of genes in the same set of samples. Also, the competitive test assumes independence of the genes which is problematic in many cases. Most proposed procedures based on the 2x2 contingency table are of the competitive type. GSEA embodies a hybrid approach in that its null model is motivated from a gene sampling standpoint, but the re-sampling scheme is based on phenotype labels. This combination makes the *p*-values hard to interpret statistically (Tian et al., 2005; Goemann & Buhlmann, 2007).

Still, gene set enrichment methods are very useful in practice and are ubiquitously applied to analyze expression data – at least as a first step to determine relevant biological processes or pathways. Software tools on the web and as standalone applications are readily available, usually accompanied by large collection of gene sets. DAVID (Huang et al., 2009) is an example of a user-friendly, well-documented web application for accomplishing this type of analysis. For the statistical programming language R (http://www.R-project.org), implementations of many of these methods are available as packages and allow rapid experimentation. This includes all methods tested in the review by Ackermann and Strimmer (2009).

Topology-Driven Methods

The methods described so far have fundamental limitations in that they rely on a predefined collection of gene sets and cannot combine information across several gene sets. With an ever increasing number of sets being defined - often with high degrees of overlap, the interpretation of the outcome of these methods also becomes more and more difficult. While defined gene sets often correspond to biological processes or canonical pathways, no information on the topology of the pathways is used in the analysis. Topology-driven methods attempt to address these shortcomings for tasks ranging from interpretation of expression data to prediction of protein function and determination of protein complexes. In general, the methods described in this section treat the entities and relationships as an undirected graph and make use of its topological properties only.

Insights Based on Network Descriptors

The application of graph theory and related descriptors of graph topology can lead to interesting insights into biological phenomena. Early on, Jeong et al. (2001) observed that highly connected proteins in a yeast protein-protein interaction network are more likely to be essential for yeast survival than proteins that are not as highly connected. More recently, Goh et al. (2007) constructed a gene-disease network to analyze the properties of disease genes with respect to network topology. Interestingly, they found that the majority of known disease genes are not in a central position of the network, are less correlated with other genes than expected by chance and have a tendency to be only expressed in a few tissues. This is in contrast to findings from other groups that detect a tendency towards higher degree nodes, e.g. Jonsson & Bates (2006). Possible explanations for these discrepancies include a focus on cancer genes in the latter study and a potential bias of literature-curated networks as disease-relevant genes tend to be well-studied. Similarly, Yildirim et al. (2007) construct a drug-target network to analyze the characteristics of current drugs and their positions in a PPI network. A relevant result is that only the targets of few drugs are topologically close to known disease genes in their specific indication. In general, the distance distribution of drug targets to known disease genes is the same as for random genes to those targets. They also detect that the distance between target and known disease gene appears to shrink in recent years, which might be an indication for tangible impact of rational drug design.

Detection of Active Pathways and Sub-Networks

Zien et al. (2000) proposed one of the earliest methods to exploit biological networks for gene expression analysis. In their pathway scoring method, they define rules to compile plausible pathways from elementary metabolic reactions and subsequently score the resulting pathways based on expression data. In principle, this method is applicable on any network, but in their paper the authors demonstrated its use specifically in the context of the yeast glycolysis pathway. The idea of using network information to score pre-defined pathways is taken further by Rahnenfuehrer et al. (2004). One contribution of that paper is the introduction of a topological factor to score the statistical association of a pathway with the phenotypes of interest. Basically, this factor weights the similarity of two gene expression profiles by the distance of the corresponding enzymes in the metabolic network. Very recently, Hung et al. (2010) proposed the PSEA method to score association of pathways with phenotypes by incorporating topological information as well. They explicitly compared their approach to GSEA and showed that higher sensitivity can be obtained with their method, if topological information is available.

Ideker et al. (2002) pioneered a class of methods to support the analysis of gene expression data in the context of protein-protein interaction networks. They started from a set of expression experiments and a corresponding network of protein-protein interactions and proposed to find 'active sub-networks', i.e., connected set of genes with significantly high differential gene expression. In contrast to the gene set enrichment methods described above, active sub-networks can span the boundaries of several canonical pathways to better reflect underlying biology. Ideker and coworkers proposed a scoring scheme to integrate *p*-values of differential expression and detected high-scoring sub-networks using a greedy approach as well as a heuristic method based on simulated annealing. This method is readily available as a plug-in for the popular biological network visualization software Cytoscape (Shannon et al., 2003). Several groups proposed variations of this method suggesting alternative scoring schemes or statistical significance estimations, e.g. Cabusora et al. (2005); Nacu et al. (2007). Recently, a significant improvement has been achieved by Dittrich et al. (2008) who mapped the original problem into an integer linear-programming framework and solved it to optimality in reasonable time on relevant instances. In their work, they demonstrated significant differences between the greedy and optimal solutions. Applications of these methods include the derivation of better predictors for the risk of breast cancer metastasis (Chuang et al., 2007) and the definition of novel modules regulated in response to bacterial endotoxins (Calvano et al., 2005). Liu et al. (2007) applied a hybrid method called GNEA that consists of first detecting active subnetworks and subsequently applying a gene set enrichment method. In a diabetes context, they identified gene sets related to insulin signaling and nuclear hormone receptors as differentially expressed and show that this would not have been possible using gene set enrichment alone.

For the interpretation of correlational data in the context of biological networks, Hanisch et al. (2002) suggested a co-clustering method that is based on a distance function combining network topology and expression correlation. Their method was tested on metabolic networks in yeast. Later, Ulitsky & Shamir (2007) proposed two methods with similar inputs. They define a probabilistic framework in order to extract functional modules from PPI networks and correlational expression data in yeast. Their second method, called CEZANNE, can incorporate confidence scores on the edges of the input network into the analysis (Ulitsky & Shamir, 2009). This is relevant as PPI networks generally are of varying quality and several authors proposed confidence measures to deal with this problem, e.g. von Mering et al. (2007). An interesting application of this type of approach was given by Mueller et al (2008) in the context of stem cell biology. They uncovered a network of protein-protein interactions that seems to be shared by all pluripotent stem cells and used expression states of this network to classify stem cell subtypes. Given the right data, this approach could also be applied to identify shared protein-protein interactions for a variety of cancer cells and possibly lead to intervention points and treatments that work for multiple cancer types.

Dense Sub-Graph Detection

Another topic of interest is the detection of dense sub-graphs in protein-protein interaction networks. It has been shown that such sub-graphs tend to represent protein complexes with coherent function (Sharan et al., 2007) and can therefore be used to assign biological function to previously uncharacterized genes. This capability can be of significant value where e.g. a genetic driver for a disease is identified, but the corresponding normal biology is not known. The literature abounds with methods to extract dense sub-graphs from networks and the review of Sharan et al. (2007) on the general topic of gene function prediction gives a very useful overview on many of these methods. Two examples not mentioned there are the methods of Georgii et al. (2009) and Pradines et al. (2005). Georgii et al. (2009) took an edge-weighted network as an input and created as output sub-graphs with maximal average pair-wise edge-weight. The edge-weight in the proposed application corresponds to a confidence in the specific protein-protein interaction. An interesting factor of this combinatorial algorithm is that it solves the problem to optimality and can take additional discrete constraints into account. For instance, the authors demonstrated how to use the method to extract tissue specific modules from a human PPI network. The software is freely available from the authors. A very different approach to the same problem is presented by Pradines et al. (2005). They suggested a novel framework to estimate probability distributions of edge counts for a set of genes in a protein-protein interaction network. Based on a plausible random-graph null model, they gave analytic approximations to estimate the probability for a set of n genes to have m or more connections in that graph. Based on this expression, they mine a protein-protein interaction graph for densely connected sub-graphs to suggest new biological functions for proteins. Recently, a somewhat surprising application of this framework was presented by Kaplow et al. (2009). They applied the edge-count statistic to determine plausible cut-offs in RNAi screens using the hypothesis that the n top scoring RNAi hits should form a dense sub-graph in a PPI network. Repeatedly applying the statistics of Pradines et al. (2005) to the top ranked hits, they define a fully automated procedure to determine a plausible threshold to separate true hits from noise. Since RNAi screens can lead to both the identification of targets and possible therapeutics directly, this approach has particular promise for drug discovery.

User-Defined Graph Extraction

Finally, a number of articles addressed the problem of extracting sub-graphs from large biological networks according to user-defined constraints. The underlying subgraph isomorphism problem is computationally hard and all approaches either simplify the problem or try to use heuristics to achieve reasonable running times on relevant instances. The early work of Kelley et al. (2003) was motivated by the problem of detecting conserved pathways across species boundaries. The suggested PathBLAST method can find linear pathways with gaps in a given protein-protein network and is freely available. More recently, Dost et al. (2008) extended this work to more complex query graphs with similar goals. A general framework for finding user-constrained sub-graphs in biological networks was suggested by Sohler & Zimmer (2005). Their method transforms the matching problem to a clique search in an appropriately defined graph, thereby enabling arbitrary complex query graphs at the possible expense of performance. Their use case is motivated similarly to the pioneering pathway scoring work of Zien et al. (2000) in that they extract small biologically plausible pathways snippets and score them using expression data. Finally, Banks et al. (2008) proposed a heuristic approach based on hashing techniques. Their freely available NetGrep application can be used to find complex user-defined subgraphs in large interaction networks and yields acceptable running times on many relevant queries. The NetGrep application seems appealing for experimentation as its focus is the ease of use for computational biologists using a graphical user interface as well as its performance on most queries.

Network Resources

Over recent years, a number of large scale networks of biological networks have become available. For instance, large scale protein-protein interaction (PPI) networks can be determined experimentally by a number of methods now. Chautard et al. (2009) give a recent review. IntAct (Aranda et al., 2010), KEGG (Kanehisa et al., 2010), Reactome (Matthews et al., 2009), and the Science's Connection Maps (http://stke.sciencemag.org/cm/) are examples of resources that provide complex, directed interactions hand-curated from the literature and are suited for constructing gene sets as well as larger graphs for topology-driven methods. Several commercial companies have created proprietary databases of relationships curated from the literature. Ingenuity (http://www.ingenuity.com/), GeneGO (http://www.genego.com), BioBase (http://biobase-international.com), and Ariadne (http://ariadnegenomics.com) are among the most relevant vendors. These data sets are not freely available and must be licensed separately or can only be used with proprietary tools and are therefore rarely used to assess method quality or to drive methods developed in academia.

Pros and Cons

Topology-driven methods promise to exploit the inherent dependencies between biological entities in a functional context. In addition to the above-mentioned assignment of biological function to proteins, this can lead to more robust predictions of phenotypes, e.g. risk of breast cancer metastasis (Chuang et al, 2007) and characterization of stem cell lines as demonstrated by Mueller et al. (2008). We expect that a more widespread adoption of these techniques, especially with the availability of more methods in the public domain, will lead to even more examples directly relevant to drug discovery.

A disadvantage of purely topology-driven methods is that the direction of modulation of a putative target is hard to predict if no information on the direction of regulation of the used biological relationships is available. Also, the methods outlined in this section neglect potentially available information on the dynamics of interactions in time and do not try to impose any form of causal flow in the network. Clearly, many processes of interest are dynamic in time and should be analyzed accordingly.

Directed Network Models

The dimension of time and the related concept of causality are highly relevant to model the etiology of disease or the effect of a compound in a cell line. In a drug development context, the most pressing question are 'what causes the disease' and 'are there ways to counteract this initial perturbation'. On a molecular level, signaling cascades represent dynamic processes that can ultimately lead to transcript changes and high level phenotypic observations.

Two classes of problems are often addressed using more sophisticated network models, (1) specification of a model based on literature data and subsequent interpretation or prediction of experimental findings and (2) the inference of such a model purely based on primary experimental data. In general, the models and algorithms that try to capture dynamic effects are more restricted in scope than the ones in the previous sections. This is mostly due to the high manual curation effort (cost) or large experimental data requirements to define dynamic or causal networks. Methodically, many approaches refer to the seminal work of Judea Pearl on causality in which he laid many of the foundations for inferring and reasoning with causal networks from data. This work is summarized in Pearl (2000).

Boolean Networks

Boolean networks represent biological processes similar to electric circuit diagrams and are fully deterministic in their simplest form. Each entity in the diagram corresponds to a biological entity. The entities are associated with one of two Boolean states, either ON (1) or OFF (0). Entities are then connected to other entities via directed edges. To determine the state of any entity, the states of all input entities are combined via an arbitrary Boolean function, e.g. AND, NOT, XOR. Such a model can describe complex interactions in a qualitative way and has been introduced by Kauffman (1969) more than 40 years ago. Boolean networks have so far mostly been used to model small systems. An early example is given by Mendoza et al. (1999) who analyzed genetic control in a plant model using a network of 10 genes. More recently, larger scale models have been proposed. For instance, Saez-Rodriguez et al (2007) used a model containing 94 nodes and 123 interactions to analyze T Cell Receptor Signaling. In that work, predictions in response to perturbation of specific proteins in the network were derived and successfully validated experimentally. Members from the same group further extended this approach in an interesting direction (Saez-Rodriguez et al., 2009). After compiling a Boolean network of 82 nodes and 116 edges to model the immediate-early responses of human cells to seven cytokines, they used experimental data in Human liver to focus the network on relevant edges in that context. They reported a good experimental fit to the data on a reduced network.

Further evolution of such methods might pave the way towards building larger models based on a collection of known relationships, only some of which might be relevant in a given cellular context. Complementary to the approaches in these examples, research has also been devoted to infer Boolean networks directly from experimental data. Especially, the extensions called Probabilistic Boolean Networks (Shmulevich et al., 2002) are well suitable for this task. Hickman & Hodgman (2009) gave a recent review of this field.

Causal Networks

An interesting approach to the large-scale analysis of expression data is given by Pollard et al. (2005). Their approach is related to Boolean networks in that they compiled a directed graph of biological relationships called a causal model. Each node in that graph can have one of three states: up-regulated, down-regulated, and unchanged. Relationships then describe transfer functions between those states. However, in contrast to Boolean networks they do not seem to support arbitrary logic combinations of the inputs. As there are no details on the approach in their article, it is hard to assess it in-depth. Interestingly though, their compiled network is large compared to many other discrete models: it contains 24.000 relationships that were curated from the literature. Based on this model, they presented an analysis of most likely regulators in expression data derived from type 2 diabetes patients, recovered known key players in diabetes and proposed new regulators.

Bayesian Networks

Bayesian network and related graphical models try to capture relationships among a set of biological entities probabilistically. In many applications, Bayesian networks are used to reflect statistical dependencies between biological entities in steady state, but the model can include the explicit notion of causality (Pearl, 2000). Also, variants exist to model a time dimension in so-called Dynamic Bayesian Networks (e.g. Dojer et al., 2006). Common to all Bayesian networks is the notion to describe the joint distribution of a set of potentially time-dependent random variables. The crucial idea is that usually many variables become mutually statistically independent, if conditioned on a small set of other variables. The set of independence assumptions is encoded in a directed, acyclic graph containing biological entities as nodes. A node in that graph is then assumed to be statistically independent of its non-descendants given its parents. In terms of biological network, this assumption seems sensible. For instance, for a set of regulatees of a common regulator, the states of the regulatees will be correlated. However, once the state of the common regulator is factored in, there is no additional information in knowing the state of any regulatee. Friedman et al. (2000) published an influential paper demonstrating the use of Bayesian networks to model the yeast cell cycle based on expression data experiments. A large body of work refining methods and applying it to various biological settings followed. Friedman (2004) gave a very good introduction to Bayesian networks, their underlying concepts and pointers to additional reading.

The early article of Hartemink (2001) demonstrates how to use the Bayesian network framework to distinguish between two alternative versions of the galactose pathway. In this work, the Bayesian network structure was manually derived, similar to most of the work in the Boolean network space. Overall, by far more work has been devoted to inferring the network structure itself from the primary data. Given a lack of standard data sets to compare performance on, it is very hard to judge which methods are best suited in a specific case. The recent DREAM (Dialogue for Reverse Engineering Assessments and Methods) initiative is a laudable step towards defining test challenges that are posed in a blinded fashion, so that methods can be compared in an unbiased way. The analysis of the DREAM2 challenge run in 2007 (Stolovitzky et al., 2009) shows that many methods do not fare better than random in some DREAM challenges. If the quality and quantity of the experimental data are high enough, however, some methods are able to recover partial networks.

However, there have been some interesting practical successes of Bayesian inference for recovering molecular relationships. For instance, Sachs et al. (2005) demonstrated how to recover a map of signaling in Human primary T cells. They used flow-cytometry data on 11 phosphorylated proteins and phospholipids in response to various perturbations as an input. An important characteristic of this data set is the large number of samples generated through flow-cytometry and the limited number of variables in the inferred network: a situation well suited for statistical inference. One of the most successful examples of using Bayesian networks for causal inference in a drug discovery setting is given by Schadt et al., 2005. A crucial characteristic of their approach is that they harness naturally occurring genetic variation as perturbations in a population of cross-bred mice. They conduct corresponding expression measurements in each individual for all target tissues of interest. Each detected genetic variation in the population is then tested for its association with a change in a transcript level of a target tissue. Furthermore, the individuals are assessed for a number of complex phenotypes, e.g. obesity, lipid levels. Given these integrated data, it is possible to infer which transcript level changes are causally linked to variations in phenotype. If the phenotype corresponds to a relevant disease, this procedure directly leads to new potential targets. The authors demonstrate this in reporting the experimental validation of three novel genes predicted to be causally linked to susceptibility to obesity. Zhu et al. (2008) and Schadt (2009) gave updates on recent refinements of this very interesting approach. The non-profit organization SAGE (http://www.sagebase.org) has recently been founded to build a research community around the wealth of genetic and transcriptomic data generated with this approach and to foster new methods and applications in this field.

Other Models

A wide range of more complex models based on kinetic parameters and detailed knowledge of the underlying system have been proposed. For instance, models based on ordinary differential equations have been used to model the Wnt pathway (Lee et al., 2008) and in principle such approaches can be used for drug target finding. For all such models, highly specific data on all aspects of the process to be modeled must be available for the simulation to be reasonable. Therefore, many of these models are not directly applicable to large-scale data analysis. For a general introduction into the field of complex models, the reviews of Kitano (2002) and Fisher and Henziger (2007) are good starting points. Materi & Wishart (2007) gave a review of recent development in the unfolding field of highly detailed dynamic models as it pertains to drug development.

Network and Data Resources

The above mentioned databases (e.g. KEGG, PID, STKE Connection Maps) can provide good starting points for manually constructing smaller scale detailed models of biological systems. Another resource is the growing set of models in the BioModels (Le Novere et al., 2006) database. Unfortunately, starting points for large scale models are limited as there is no public large-scale repository of high-quality biological relationship. To obtain such data in a structured way, commercial vendors such as Ingenuity (http://www.ingenuity.com/), GeneGO (http://www.genego.com), BioBase (http://biobase-international.com), Genstruct (http://genstruct.com) or Ariadne (http://ariadnegenomics.com) seem the only choice at this point. Public efforts to create such a large scale repository could have substantial impact on the improvement of the methods described in this section.

In terms of inferring networks from expression data, the quickly growing public expression repositories GEO (Barrett et al., 2008) and ArrayExpress (Parkinson et al., 2009) are excellent choices to identify suitable data sets.

Pros and Cons

The advantages of successfully incorporating time and causality into predictive models in the form of Boolean networks, causal graphs, Bayesian networks or other methods are obvious: the potential to precisely model biological processes and their response to stimuli and arrive at better testable hypotheses. Especially, the high-impact work of Schadt et al. (2005) demonstrates the power of more complex models in a target discovery context. Conversely, more complex models require better control of model parameters to prevent overfitting and also need more data to correctly infer them. These demands can be prohibitive for system-wide modeling and necessitate the use of simpler models in specific situations.

CURRENT GAPS

With technologies to record experimental data on a molecular and frequently system-wide level continuing to improve, one of the biggest gaps for the understanding of disease biology and drug action is the relative low availability of 'computable' biological information and structured data. This gap exists on two levels: (1) the existing biomedical knowledge is largely contained in scientific publications in free text and not accessible for computational approaches and (2) we simply do not understand biology completely and therefore the information just does not exist yet. While there are several providers of structured information extracted from a subset of the biomedical literature (e.g. Genego, Ingenuity, Ariadne, Genstruct), the curation and/or extraction processes are not harmonized, the cost for generation of high quality structured information is high, and the coverage incomplete. Computational approaches to infer biological relationships from experimental data against the 'backdrop' of existing knowledge are just starting to emerge and while network inference methods are showing some promise, the amount of 'cohesive' experimental data that is needed to power them is currently both cost prohibitive and in the context of the human system not feasible.

On the algorithmic or methodological side, one of the clear gaps is in the area of efficient dynamic analysis for time courses and/or dose escalation studies. While for some of the described approaches an extension to longitudinal or dose-escalation studies is conceivable, realization of such extensions have not been reported so far. Highly parameterized ODE-based approaches have been reported, but are so far limited to reasonably small networks and are additionally hampered by the proportionally large amount of mechanistic information that is required for their creation.

Lastly, with the increasing number of network analysis methods applied in biomedical space, the specific advantages of a new method over an already existing one are frequently non-obvious and improvements often not quantified. Reference data sets ('gold standards'), clear performance metrics and test protocols could help to assess the value of new approaches and potentially help focus development of methods and algorithms to solve fundamental problems, rather than marginal improvements.

PERSPECTIVES AND CONCLUSION

'Grand Challenges' recently posed in the context of US-President Barack Obama's 'Strategy for American Innovation' illustrate the desire to improve healthcare with demands for 'personalized medicine that enables the prescription of the right dose of the right drug for the right person' and 'Complete DNA sequencing of every case of cancer' amongst others ('Grand challenges for the 21st century', 2010). Network-based analysis holds great promise for the ability to efficiently interpret biological data towards better treatment options across a wide range of diseases. Crucially, we see several trends unfolding that could help to realize this potential.

Little more than a decade ago, the generation of system wide experimental data was extremely costly and therefore mostly generated by or on behalf of large pharmaceutical companies. Data were assets and computational scientists had limited access outside of the boundaries of these organizations. Today, such data is increasingly created in the context of academic and non-profit organizations and widely available. The 1000 genomes project (http://www.1000genomes.org/), the continued growth of deposition of data in GEO (Barrett et al., 2008), activities of the BROAD institute and the National Cancer Institute are just a few examples highlighting this trend.

Complementary to more (human) data, better coverage and broader availability of computable biomedical information will be needed. Technical feasibility to deliver the relevant information in a structured format has been demonstrated – perhaps not at a level where the structured information can completely capture the nuances of language, but at a level that is sufficient to explore computational approaches to hypothesis generation. A concerted effort to structure existing information and to publish new information in both human readable and computationally accessible format should greatly improve network quality and increase our ability to efficiently interpret experimental data and create testable hypotheses.

To turn these data into meaningful outcomes for healthcare – safer treatments at lower cost, improved diagnostics to enable prevention rather than treatment and possibly regenerative approaches where disease has progressed beyond treatment options, better computational approaches will be required to turn the data into insights. Adoption will be made easier, if a new method either clearly addresses a recognized gap, or its superiority over existing methods is easily recognizable. Very clearly formulated challenges posed by e.g. the pharmaceutical industry could help computational scientists to 'home in' on specific problems. Rigorous validation of new methods against existing approaches with mutually accepted 'gold standard' test sets and clear performance metrics will help to recognize approaches that deliver dramatic opportunities.

Initiatives like DREAM (Stolovitzky et al., 2009) are examples for good practice that should be adopted more broadly. The trend to make implementations of new algorithms freely available should become standard practice and will enable efficient direct comparisons by the community.

Of course, new approaches and methods will only be able to demonstrate their value if the pharmaceutical industry adopts the established methods and shares feedback on its applicability, success rate, and encountered deficiencies to enable a constructive dialogue.

To summarize, we see four main requirements ahead, namely (1) continued structured accumulation of primary genetics and genomics data in repositories such as GEO, (2) a similar effort to capture structured biological information, (3) common relevant gold standards to measure innovative computational methods against, and (4) a strong move by the pharmaceutical and biotech industry as well as academia and non-profits to pose clear challenges and embrace emerging methods. At this point, concrete examples for the impact of network methods on drug discovery exist, but are still limited in scope. Given that current trends continue and some appropriate steps will be taken in the near future, we expect a tremendous boost in our abilities for target discovery and the successful translation of knowledge of disease biology into drugs that improve the lives of patients based on network methods.

REFERENCES

Ackermann, M., & Strimmer, K. (2009). A general modular framework for gene set enrichment analysis. *BMC Bioinformatics*, *10*, 47. doi:10.1186/1471-2105-10-47

Alles, M., Gardiner-Garden, M., Nott, D., Wang, Y., Foekens, J., & Sutherland, R. (2009). Meta-analysis and gene set enrichment relative to er status reveal elevated activity of MYC and E2F in the basal breast cancer subgroup. *PLoS ONE*, *4*(3). doi:10.1371/journal.pone.0004710

Aranda, B., Achuthan, P., Alam-Faruque, Y., Armean, I., Bridge, A., & Derow, C. (2010). The IntAct molecular interaction database in 2010. *Nucleic Acids Research*, *38*(Database issue), D525–D531. doi:10.1093/nar/gkp878

Ashburner, M., Ball, C., Blake, J., Botstein, D., Butler, H., & Cherry, J. (2000). Gene ontology: Tool for the unification of biology. The Gene Ontology Consortium. *Nature Genetics*, *25*(1), 25–29. doi:10.1038/75556

Banks, E., Nabieva, E., Peterson, R., & Singh, M. (2008). NetGrep: Fast network schema searches in interactomes. *Genome Biology*, *9*, R138. doi:10.1186/gb-2008-9-9-r138

Barrett, T., Troup, D., Wilhite, S., Ledoux, P., Rudnev, D., & Evangelista, C. (2009). NCBI GEO: Archive for high-throughput functional genomic data. *Nucleic Acids Research*, *37*(Database issue), D885–D890. doi:10.1093/nar/gkn764

Bonifaci, N., Berenguer, A., Díez, J., Reina, O., Medina, I., & Dopazo, J. (2008). Biological processes, properties and molecular wiring diagrams of candidate low-penetrance breast cancer susceptibility genes. *BMC Medical Genomics*, *1*, 62. doi:10.1186/1755-8794-1-62

Cabusora, L., Sutton, E., Fulmer, A., & Forst, C. (2005). Differential network expression during drug and stress response. *Bioinformatics (Oxford, England)*, *21*(12), 2898–2905. doi:10.1093/bioinformatics/bti440

Calvano, S., Xiao, W., Richards, D., Felciano, R., Baker, H., & Cho, R. (2005). A network-based analysis of systemic inflammation in humans. *Nature*, *437*(7061), 1032–1037. doi:10.1038/nature03985

Cervino, A., Li, G., Edwards, S., Zhu, J., Laurie, C., & Tokiwa, G. (2005). Integrating QTL and high-density SNP analyses in mice to identify Insig2 as a susceptibility gene for plasma cholesterol levels. *Genomics*, *86*(5), 505–517. doi:10.1016/j.ygeno.2005.07.010

Chautard, E., Thierry-Mieg, N., & Ricard-Blum, S. (2009). Interaction networks: From protein functions to drug discovery. A review. *Pathologie Biologie*, *57*(4), 324–333. doi:10.1016/j.patbio.2008.10.004

Chuang, H.-Y., Lee, E., Liu, Y.-T., Lee, D., & Ideker, T. (2007). Network-based classification of breast cancer metastasis. *Molecular Systems Biology*, *3*, 140. doi:10.1038/msb4100180

Dittrich, M., Klau, G., Rosenwald, A., Dandekar, T., & Müller, T. (2008). Identifying functional modules in protein-protein interaction networks: An integrated exact approach. *Bioinformatics (Oxford, England)*, *24*(13), i223–i231. doi:10.1093/bioinformatics/btn161

Dojer, N., Gambin, A., Mizera, A., Wilczyński, B., & Tiuryn, J. (2006). Applying dynamic Bayesian networks to perturbed gene expression data. *BMC Bioinformatics*, *7*, 249. doi:10.1186/1471-2105-7-249

Dost, B., Shlomi, T., Gupta, N., Ruppin, E., Bafna, V., & Sharan, R. (2008). Qnet: A tool for querying protein interaction networks. *Journal of Computational Biology*, *15*, 1–15. doi:10.1089/cmb.2007.0172

Draghici, S., Khatri, P., Tarca, A., Amin, K., Done, A., & Voichita, C. (2007). A systems biology approach for pathway level analysis. *Genome Research*, *17*(10), 1537–1545. doi:10.1101/gr.6202607

Ebert, B., Pretz, J., Bosco, J., Chang, C., Tamayo, P., & Galili, N. (2008). Identification of RPS14 as a 5q- syndrome gene by RNA interference screen. *Nature*, *451*(7176), 335–339. doi:10.1038/nature06494

EOP. (2010). *Grand challenges for the 21st century*. Retrieved February 18, 2010, from http://www.whitehouse.gov/administration/eop/ostp/grand-challenges-request-information.

Fisher, J., & Henzinger, T. (2007). Executable cell biology. *Nature Biotechnology*, *25*(11), 1239–1249. doi:10.1038/nbt1356

Friedman, N. (2004). Inferring cellular networks using probabilistic graphical models. *Science*, *303*(5659), 799–805. doi:10.1126/science.1094068

Friedman, N., Linial, M., Nachman, I. & Pe'er, D. (2000). Using Bayesian networks to analyze expression data. *Journal of Computational Biology: A Journal of Computational Molecular Cell Biology, 7*(3-4), 601-20.

Georgii, E., Dietmann, S., Uno, T., Pagel, P., & Tsuda, K. (2009). Enumeration of condition-dependent dense modules in protein interaction networks. *Bioinformatics (Oxford, England)*, *25*(7), 933–940. doi:10.1093/bioinformatics/btp080

Goeman, J., & Buhlmann, P. (2007). Analyzing gene expression data in terms of gene sets: Methodological issues. *Bioinformatics (Oxford, England)*, *23*(8), 980. doi:10.1093/bioinformatics/btm051

Goh, K.-I., Cusick, M., Valle, D., Childs, B., Vidal, M., & Barabási, A.-L. (2007). The human disease network. *Proceedings of the National Academy of Sciences of the United States of America*, *104*(21), 8685–8690. doi:10.1073/pnas.0701361104

Hanisch, D., Zien, A., Zimmer, R., & Lengauer, T. (2002). Co-clustering of biological networks and gene expression data. *Bioinformatics (Oxford, England)*, *18*(Suppl 1), S145–S154.

Hartemink, A., Gifford, D., Jaakkola, T., & Young, R. (2001). Using graphical models and genomic expression data to statistically validate models of genetic regulatory networks. *Pacific Symposium on Biocomputing Pacific Symposium on Biocomputing*, 422-33.

Hickman, G., & Hodgman, T. (2009). Inference of gene regulatory networks using boolean-network inference methods. *Journal of Bioinformatics and Computational Biology*, *7*(6), 1013–1029. doi:10.1142/S0219720009004448

Hosack, D., Dennis, G., Sherman, B., Lane, H., & Lempicki, R. (2003). Identifying biological themes within lists of genes with EASE. *Genome Biology*, *4*(10), R70. doi:10.1186/gb-2003-4-10-r70

Huang, D., Sherman, B., & Lempicki, R. (2009). Systematic and integrative analysis of large gene lists using DAVID bioinformatics resources. *Nature Protocols*, *4*(1), 44–57. doi:10.1038/nprot.2008.211

Hung, J.-H., Whitfield, T., Yang, T.-H., Hu, Z., Weng, Z., & Delisi, C. (2010). Identification of functional modules that correlate with phenotypic difference: The influence of network topology. *Genome Biology*, *11*(2), R23. doi:10.1186/gb-2010-11-2-r23

Ideker, T., Ozier, O., Schwikowski, B., & Siegel, A. (2002). Discovering regulatory and signalling circuits in molecular interaction networks. *Bioinformatics (Oxford, England)*, *18*(Suppl 1), S233–S240.

Jeong, H., Mason, S., Barabási, A., & Oltvai, Z. (2001). Lethality and centrality in protein networks. *Nature*, *411*(6833), 41–42. doi:10.1038/35075138

Jonsson, P., & Bates, P. (2006). Global topological features of cancer proteins in the human interactome. *Bioinformatics (Oxford, England)*, *22*(18), 2291–2297. doi:10.1093/bioinformatics/btl390

Kanehisa, M., Goto, S., Furumichi, M., Tanabe, M., & Hirakawa, M. (2010). KEGG for representation and analysis of molecular networks involving diseases and drugs. *Nucleic Acids Research*, *38*(Database issue), D355–D360. doi:10.1093/nar/gkp896

Kaplow, I., Singh, R., Friedman, A., Bakal, C., Perrimon, N., & Berger, B. (2009). RNAiCut: Automated detection of significant genes from functional genomic screens. *Nature Methods*, *6*(7), 476–477. doi:10.1038/nmeth0709-476

Kauffman, S. (1969). Metabolic stability and epigenesis in randomly constructed genetic nets. *Journal of Theoretical Biology*, *22*(3), 437–467. doi:10.1016/0022-5193(69)90015-0

Keller, A., Backes, C., Gerasch, A., Kaufmann, M., Kohlbacher, O., & Meese, E. (2009). A novel algorithm for detecting differentially regulated paths based on gene set enrichment analysis. *Bioinformatics (Oxford, England)*, *25*(21), 2787–2794. doi:10.1093/bioinformatics/btp510

Kelley, B., Sharan, R., Karp, R., Sittler, T., Root, D., & Stockwell, B. (2003). Conserved pathways within bacteria and yeast as revealed by global protein network alignment. *Proceedings of the National Academy of Sciences of the United States of America*, *100*(20), 11394–11399. doi:10.1073/pnas.1534710100

Kitano, H. (2002). Systems biology: A brief overview. *Science*, *295*(5560), 1662–1664. doi:10.1126/science.1069492

Kola, I., & Landis, J. (2004). Can the pharmaceutical industry reduce attrition rates? *Nature Reviews. Drug Discovery*, *3*(8), 711–715. doi:10.1038/nrd1470

Le Novère, N., Bornstein, B., Broicher, A., Courtot, M., Donizelli, M., & Dharuri, H. (2006). BioModels Database: A free, centralized database of curated, published, quantitative kinetic models of biochemical and cellular systems. *Nucleic Acids Research*, *34*(Database issue), D689–D691. doi:10.1093/nar/gkj092

Lee, E., Salic, A., Krüger, R., Heinrich, R., & Kirschner, M. (2003). The roles of APC and Axin derived from experimental and theoretical analysis of the Wnt pathway. *PLoS Biology*, *1*(1), E10. doi:10.1371/journal.pbio.0000010

Liu, M., Liberzon, A., Kong, S., Lai, W., Park, P., & Kohane, I. (2007). Network-based analysis of affected biological processes in type 2 diabetes models. *PLOS Genetics*, *3*(6), e96. doi:10.1371/journal.pgen.0030096

Matthews, L., Gopinath, G., Gillespie, M., Caudy, M., Croft, D., & de Bono, B. (2009). Reactome knowledgebase of human biological pathways and processes. *Nucleic Acids Research*, *37*(Database issue), D619–D622. doi:10.1093/nar/gkn863

Mendoza, L., Thieffry, D., & Alvarez-Buylla, E. (1999). Genetic control of flower morphogenesis in Arabidopsis thaliana: A logical analysis. *Bioinformatics (Oxford, England)*, *15*(7-8), 593–606. doi:10.1093/bioinformatics/15.7.593

Mootha, V., Lindgren, C., Eriksson, K.-F., Subramanian, A., Sihag, S., & Lehar, J. (2003). PGC-1alpha-responsive genes involved in oxidative phosphorylation are coordinately downregulated in human diabetes. *Nature Genetics*, *34*(3), 267–273. doi:10.1038/ng1180

Müller, F.-J., Laurent, L., Kostka, D., Ulitsky, I., Williams, R., & Lu, C. (2008). Regulatory networks define phenotypic classes of human stem cell lines. *Nature*, *455*(7211), 401–405. doi:10.1038/nature07213

Nacu, S., Critchley-Thorne, R., Lee, P., & Holmes, S. (2007). Gene expression network analysis and applications to immunology. *Bioinformatics (Oxford, England)*, *23*(7), 850–858. doi:10.1093/bioinformatics/btm019

Oti, M., & Brunner, H. (2007). The modular nature of genetic diseases. *Clinical Genetics*, *71*(1), 1–11. doi:10.1111/j.1399-0004.2006.00708.x

Parkinson, H., Kapushesky, M., Kolesnikov, N., Rustici, G., Shojatalab, M., & Abeygunawardena, N. (2009). ArrayExpress update-from an archive of functional genomics experiments to the atlas of gene expression. *Nucleic Acids Research*, *37*(Database issue), D868–D872. doi:10.1093/nar/gkn889

Pearl, J. (2000). *Causality: Models, reasoning, and inference*. Cambridge, UK: Cambridge University Press.

Pollard, J., Butte, A., Hoberman, S., Joshi, M., Levy, J., & Pappo, J. (2005). A computational model to define the molecular causes of type 2 diabetes mellitus. *Diabetes Technology & Therapeutics*, *7*(2), 323–336. doi:10.1089/dia.2005.7.323

Pradines, J., Farutin, V., Rowley, S. & Dancík, V. (2005). Analyzing protein lists with large networks: Edge-count probabilities in random graphs with given expected degrees. *Journal of Computational Biology: A Journal of Computational Molecular Cell Biology, 12*(2), 113-28.

Rahnenführer, J., Domingues, F., Maydt, J., & Lengauer, T. (2004). Calculating the statistical significance of changes in pathway activity from gene expression data. *Statistical Applications in Genetics and Molecular Biology*, 3.

Sachs, K., Perez, O., Pe'er, D., Lauffenburger, D., & Nolan, G. (2005). Causal protein-signaling networks derived from multiparameter single-cell data. *Science*, *308*(5721), 523–529. doi:10.1126/science.1105809

Saez-Rodriguez, J., Alexopoulos, L., Epperlein, J., Samaga, R., Lauffenburger, D., & Klamt, S. (2009). Discrete logic modelling as a means to link protein signalling networks with functional analysis of mammalian signal transduction. *Molecular Systems Biology*, *5*, 331. doi:10.1038/msb.2009.87

Saez-Rodriguez, J., Simeoni, L., Lindquist, J., Hemenway, R., Bommhardt, U., & Arndt, B. (2007). A logical model provides insights into T cell receptor signaling. *PLoS Computational Biology*, *3*(8), e163. doi:10.1371/journal.pcbi.0030163

Schadt, E., Lamb, J., Yang, X., Zhu, J., Edwards, S., & Guhathakurta, D. (2005). An integrative genomics approach to infer causal associations between gene expression and disease. *Nature Genetics*, *37*(7), 710–717. doi:10.1038/ng1589

Schaefer, C., Anthony, K., Krupa, S., Buchoff, J., Day, M., & Hannay, T. (2009). PID: The pathway interaction database. *Nucleic Acids Research*, *37*(Database issue), D674–D679. doi:10.1093/nar/gkn653

Shannon, P., Markiel, A., Ozier, O., Baliga, N., Wang, J., & Ramage, D. (2003). Cytoscape: A software environment for integrated models of biomolecular interaction networks. *Genome Research*, *13*(11), 2498–2504. doi:10.1101/gr.1239303

Shmulevich, I., Dougherty, E., Kim, S., & Zhang, W. (2002). Probabilistic Boolean networks: A rule-based uncertainty model for gene regulatory networks. *Bioinformatics (Oxford, England)*, *18*(2), 261–274. doi:10.1093/bioinformatics/18.2.261

Sohler, F., & Zimmer, R. (2005). Identifying active transcription factors and kinases from expression data using pathway queries. *Bioinformatics (Oxford, England)*, *21*(Suppl 2), ii115–ii122. doi:10.1093/bioinformatics/bti1120

Stolovitzky, G., Prill, R., & Califano, A. (2009). Lessons from the DREAM2 challenges. *Annals of the New York Academy of Sciences*, *1158*, 159–195. doi:10.1111/j.1749-6632.2009.04497.x

Subramanian, A., Tamayo, P., Mootha, V., Mukherjee, S., Ebert, B., & Gillette, M. (2005). Gene set enrichment analysis: A knowledge-based approach for interpreting genome-wide expression profiles. *Proceedings of the National Academy of Sciences of the United States of America*, *102*(43), 15545–15550. doi:10.1073/pnas.0506580102

Suthram, S., Dudley, J., Chiang, A., Chen, R., Hastie, T., & Butte, A. (2010). Network-based elucidation of human disease similarities reveals common functional modules enriched for pluripotent drug targets. *PLoS Computational Biology*, *6*(2). doi:10.1371/journal.pcbi.1000662

Thomas, P., Campbell, M., Kejariwal, A., Mi, H., Karlak, B., & Daverman, R. (2003). PANTHER: A library of protein families and subfamilies indexed by function. *Genome Research*, *13*(9), 2129–2141. doi:10.1101/gr.772403

Tian, L., Greenberg, S., Kong, S., Altschuler, J., Kohane, I., & Park, P. (2005). Discovering statistically significant pathways in expression profiling studies. *Proceedings of the National Academy of Sciences of the United States of America*, *102*(38), 13544–13549. doi:10.1073/pnas.0506577102

Ulitsky, I., & Shamir, R. (2007). Identification of functional modules using network topology and high-throughput data. *BMC Systems Biology*, *1*, 8. doi:10.1186/1752-0509-1-8

Ulitsky, I., & Shamir, R. (2009). Identifying functional modules using expression profiles and confidence-scored protein interactions. *Bioinformatics (Oxford, England)*, *25*(9), 1158–1164. doi:10.1093/bioinformatics/btp118

von Mering, C., Jensen, L., Kuhn, M., Chaffron, S., Doerks, T., & Krüger, B. (2007). STRING 7-recent developments in the integration and prediction of protein interactions. *Nucleic Acids Research*, *35*(Database issue), D358–D362. doi:10.1093/nar/gkl825

Xie, L., Li, J., Xie, L., & Bourne, P. (2009). Drug discovery using chemical systems biology: Identification of the protein-ligand binding network to explain the side effects of CETP inhibitors. *PLoS Computational Biology*, *5*(5). doi:10.1371/journal.pcbi.1000387

Yildirim, M., Goh, K.-I., Cusick, M., Barabási, A.-L., & Vidal, M. (2007). Drug-target network. *Nature Biotechnology*, *25*(10), 1119–1126. doi:10.1038/nbt1338

Zhu, J., Zhang, B., & Schadt, E. (2008). A systems biology approach to drug discovery. *Advances in Genetics*, *60*, 603–635. doi:10.1016/S0065-2660(07)00421-X

Zien, A., Kuffner, R., Zimmer, R., & Lengauer, T. (2000). Analysis of gene expression data with pathway scores. *Proceedings of the International Conference on Intelligent Systems for Molecular Biology, ISMB International Conference on Intelligent Systems for Molecular Biology*.

ADDITIONAL READING

Barabási, A.-L., & Oltvai, Z. (2004). Network biology: understanding the cell's functional organization. *Nature Reviews. Genetics*, *5*(2), 101–113. doi:10.1038/nrg1272

Berger, S. I., & Iyengar, R. (2009). Network analyses in systems pharmacology. *Bioinformatics (Oxford, England)*, *25*(19), 466–472. doi:10.1093/bioinformatics/btp465

Bornholdt, S. (2008). Boolean network models of cellular regulation: prospects and limitations. *Journal of the Royal Society, Interface / the Royal Society, 5 Suppl 1*, S85-94.

Bosl, W. J. (2007). Systems biology by the rules: hybrid intelligent systems for pathway modeling and discovery. *BMC Systems Biology*, *1*, 13. doi:10.1186/1752-0509-1-13

Brennan, R. J., Nikolskya, T., & Bureeva, S. (2009). Network and pathway analysis of compound-protein interactions. *Methods in Molecular Biology (Clifton, N.J.)*, *575*, 225–247. doi:10.1007/978-1-60761-274-2_10

Duarte, N. C., Becker, S. A., Jamshidi, N., Thiele, I., Mo, M. L., & Vo, T. D. (2007). Global reconstruction of the human metabolic network based on genomic and bibliomic data. *Proceedings of the National Academy of Sciences of the United States of America*, *104*(6), 1777–1782. doi:10.1073/pnas.0610772104

Ganter, B., & Giroux, C. N. (2008). Emerging applications of network and pathway analysis in drug discovery and development. *Current Opinion in Drug Discovery & Development*, *11*(1), 86–94.

Huang, J., Zhu, H., Haggarty, S. J., Spring, D. R., Hwang, H., & Jin, F. (2004). Finding new components of the target of rapamycin (TOR) signaling network through chemical genetics and proteome chips. *Proceedings of the National Academy of Sciences of the United States of America*, *101*(47), 16594–16599. doi:10.1073/pnas.0407117101

Karp, P. D., Krummenacker, M., Paley, S., & Wagg, J. (1999). Integrated pathway-genome databases and their role in drug discovery. *Trends in Biotechnology*, *17*(7), 275–281. doi:10.1016/S0167-7799(99)01316-5

Ma, H., & Goryanin, I. (2008). Human metabolic network reconstruction and its impact on drug discovery and development. *Drug Discovery Today*, *13*(9-10), 402–408. doi:10.1016/j.drudis.2008.02.002

Mestres, J., Gregori-Puigjané, E., Valverde, S., & Solé, R. V. (2009). The topology of drug-target interaction networks: implicit dependence on drug properties and target families. *Molecular BioSystems*, *5*(9), 1051–1057. doi:10.1039/b905821b

Nikolsky, Y., Nikolskaya, T., & Bugrim, A. (2005). Biological networks and analysis of experimental data in drug discovery. *Drug Discovery Today*, *10*(9), 653–662. doi:10.1016/S1359-6446(05)03420-3

Rahman, S. A., & Schomburg, D. (2006). Observing local and global properties of metabolic pathways: 'load points' and 'choke points' in the metabolic networks. *Bioinformatics (Oxford, England)*, *22*(14), 1767–1774. doi:10.1093/bioinformatics/btl181

Rang, P. H. (Ed.). (2006). *Drug discovery and development: technology in transition*. Edinburgh: Churchill Livingstone, Elsevier.

Sharom, J. R., Bellows, D. S., & Tyers, M. (2004). From large networks to small molecules. *Current Opinion in Chemical Biology*, *8*(1), 81–90. doi:10.1016/j.cbpa.2003.12.007

KEY TERMS AND DEFINITIONS

Biological Network: A set of biological objects (molecules, concepts) and relationships. In this article, network is used synonymously with 'graph'.

Biomarker: Simple or complex measurable indicator of disease state, therapeutic efficacy or adverse effect. In the context of this article, the term is specifically used to describe molecular readouts with mechanistic relationship to the intervention.

Gene Set: List of biological objects associated with a particular context, e.g., result of an experiment or biological process. The term 'gene' is used loosely to include references to genes, transcripts and proteins.

Pathway: Defined set of biological molecules and interactions that are associated with a biological concept (e.g. apoptosis, EGF signaling, etc.). In this context, a pathway is typically associated with a defined graphical layout.

Target: Key molecular entity that is causally involved in the pathogenesis or manifestation of a disease and can be modulated by therapeutic agents (e.g., small molecule drugs or antibodies).

Transcriptomic Data: Results of transcript abundance measurements (typically large scale), used interchangeably with 'expression' or 'gene expression' data.

Chapter 14
Pathway Resources at the Rat Genome Database:
A Dynamic Platform for Integrating Gene, Pathway and Disease Information

Victoria Petri
Medical College of Wisconsin, USA

ABSTRACT

The set of interacting molecules representing a biological pathway or network is a central concept in biology. It is within the pathway context that the functioning of individual molecules acquires purpose and it is the integration of these molecular circuitries that underlies the functioning of biological systems. In order to provide the research community with a dynamic platform for accessing pathway information, the Rat Genome Database (RGD – http://rgd.mcw.edu) is using a multi-tiered approach. In this chapter, the pathway resources that RGD currently offers are presented. Issues covered include: the biological pathway, the concept and the ontology, pathway literature curation and annotation of genes, interactive pathway diagrams, and tools and resources to access and navigate between pathway data. A case study is presented; future directions are discussed.

INTRODUCTION

Pathways represent a central biological concept. The reactions biological macromolecules carry out and the interactions they establish with one another form small circuitries referred to as pathways or networks. Their cross-talk, synergy and co-regulation underlie the functioning of biological systems. When the molecular functioning falters such that the network gets perturbed, the malfunctioning can propagate to the point where

DOI: 10.4018/978-1-60960-491-2.ch014

the system as a whole is affected as manifested in the diseased state. In order to comprehensively capture the pathway universe as well as its alterations, the Rat Genome Database (RGD) has delineated several goals. Objectives include the ability to associate individual genes with the pathway or pathways in which they participate as well as to retrieve this information in an easy and efficient manner, the ability to visualize individual pathways in a manner that is both dynamic and integrated and finally, the ability to easily navigate between these components.

To achieve these goals, RGD is employing a multi-tiered approach, as delineated below:

1. It is developing an ontology solely dedicated to pathways, the Pathway Ontology (PW) in order to integrate the various types of biological pathways – metabolic, regulatory, signaling, drug, disease as well as altered pathways – and the relationships between them, within a hierarchical structure.
2. It is using the published scientific review literature to identify the individual components of particular pathways and the identified set of genes is annotated to the PW term for the rat, human and mouse complement.
3. It is building and publishing interactive pathway diagrams that can be accessed and linked via their unique PW term identifier. Elements in the diagram link to entries in RGD as well as other databases, whenever applicable.
4. It is using and building tools to provide easy access to and navigation between the objects stored in the database, analyses and downloads, and links to various outside resources.
5. It is actively seeking to add new dimensions to the current provision of pathway information.

BACKGROUND

The Pathway Concept - Reactions, Interactions and Regulation

A pathway is represented by a set of interacting molecules. The set may have 'boundaries' i.e., a beginning and an end as in the conversion of a particular compound into another in a metabolic pathway. For instance, in the glycolysis pathway of carbohydrate metabolism, glucose (the beginning) is broken down to pyruvate (the end) via 10 enzymatic reactions (Voet, Voet & Pratt, 2008). In signaling and regulatory pathways, the boundaries represent a consensus established by researchers for ease of communication as these circuitries fluidly transition from, to and between one another, a delicate balance choreographed by modulatory loops and molecules (Marks, Klingmüler & Müller-Decker, 2009; Kholodenko, 2007; Santos, Verveer & Bastiaens, 2007; Hunter, 2000; Schlessinger, 2000). For instance, insulin – a hormone of central relevance to normal physiological homeostasis – upon binding to its receptor, 'activates' the insulin signaling pathway which in turn triggers downstream intracellular pathways to modulate cellular and nuclear gene expression events. The conformational changes in the insulin receptor resulting from binding to its ligand lead to autophosphorylation of distinct tyrosine residues, followed by additional conformational changes, activation of the tyrosine kinase function of the receptor and modification of downstream substrates. What follows is a cascade of protein-protein recognitions and interactions that trigger the two main conduits of the insulin signal – the phosphatidylinositol 3-kinase-Akt and the extracellular signal-regulated Raf/Mek/Erk signaling pathways. Several phosphatases and regulatory proteins act to tightly control the activity of the insulin receptor; substrates are also targeted for regulation (Taniguchi, Emanuelli & Kahn, 2006).

The reactions are chemical modifications carried out by enzymes. The modification of the substrate (S) into the product (P) may be carried out in both directions by the same enzyme as in the case of reversible reactions where S ↔ P indicates that S → P can also go P → S. In the case of exergonic, or energetically highly favorable reactions, the S → P is basically irreversible and the reverse P → S modification needs to be carried out by a different enzyme. The substrate can be a chemical compound and the modifications are used to provide or to store nutrients and fuels as they are needed by various cells and tissues at particular points in time. This is exemplified by the various reactions found in metabolic pathways. The substrate can also be a biological macromolecule; in this case, the modification is used to modulate its function and for that matter, the direction and outcome of the pathway. Such regulatory modifications occur uni-directionally and the enzymes exist in 'pairs' in that for a modifying enzyme there exists a demodifying one. For instance, phosphorylation of proteins, which can have either an inhibitory or an activating effect, is carried out by protein kinases. The effect is reversed, i.e., the protein is dephosphorylated, by protein phosphatases. Several other tandem modifications are being used to modulate the function of macromolecules and the topology of pathways; phosphorylation/de-phosphorylation is one of the most widely used (Kholodenko, 2009). Note that nature has 'engineered' enzymes, probably originating from the fusion of two separate ancestral genes, which carry out both activities via two distinct domains. A notable example is the bifunctional enzyme of fructose 2,6-bisphophate synthesis and hydrolysis – 6-phosphofructo-2-kinase/fructose-2,6-bisphosphatase. The compound is a key allosteric regulator of carbohydrate metabolism and glucose homeostasis – it activates glycolysis and inhibits gluconeogenesis by regulating the levels of two essential enzyme activities in the two individual pathways (Michels and Rigden, 2006; Rider et al., 2004). Whether the substrate is a compound or a biological macromolecule, an enzyme is capable of greatly speeding up – or catalyzing, the reaction. Enzymes have evolved over time to catalyze selected reaction types and they are the most efficient catalysts. The enzyme-catalyzed reactions are several orders of magnitude faster than the un-catalyzed ones. As catalysts, they lower the activation energy of the reaction – the difference in free energy between the reactant(s) and the transition state of the reaction, by stabilizing the transition state. Some enzymes also employ aides, or co-factors, to carry out the reaction. Enzymes are named and classified based on the type of reaction they catalyze (Bugg, 2004).

However, the vast majority of encounters between biological macromolecules do not involve a chemical modification. This is the case of protein-protein and protein-nucleic acid non-covalent interactions. The non-covalent interactions and the covalent modifications that modulate them underlie and shape the tapestry of signaling and regulatory pathways. Any given pathway is defined by a particular combination of reactions and interactions representing the steps within the pathway and the regulatory actions exerted upon them. Together they finely chisel the topology and the possible outcomes of the pathway. Characteristic of these reactions and interactions is their proficiency, specificity and selectivity; attributes determined by the sequence dependent conformation and conformational propensities of the individual molecules. Biological macromolecules, proteins in particular, are not rigid bodies but dynamic entities fluctuating between or sampling a range of conformational states whose distribution can be shifted by environmental or molecular signals and which, upon an encounter can be distorted and can in turn, distort. These features are exploited to modulate their function, the outcome of their interactions and the behavior of the network or networks within which they occur. Functionality is generally conferred by small portions of a protein usually represented by domains. Domains are structural units that can independently fold; they are used to specify protein-protein and protein-other molecule recognition and define function. Compared to the number of proteins, the repertoire of domains is rather limited and variation is achieved via combination of these modules. Due to the dynamic nature of proteins, the formation of protein complexes leads to conformational changes that are used as signals for downstream events. The unstructured regions that are often found in proteins can also be used as recognition cues for interacting partners. In such cases, the ensuing disorder to order transition, akin to the conformational changes resulting from domain-domain interactions, can become a signal for the downstream occurring events. Covalent modifications, in addition to the regulatory effect they have on the function of proteins, can be used to give rise to recognition motifs. For instance, phosphorylated tyrosine residues within short peptide sequences of a protein, are recognized by binding partners with SH2 (Src homology 2) or PTB (phospho-tyrosine binding) domains. Finally, binding of small molecules can lead to conformational changes that in turn shape the course of downstream events. Signaling pathways are largely a tale of protein-protein interactions and accompanying conformational changes that together modulate in manifold ways the possible venues any given pathway may follow (Marks, Klingmüler & Müller-Decker, 2009; Smock & Gierasch, 2009; Gibson, 2009; Pawson & Nash, 2003; Moore et al., 2008; Liu et al., 2006)

Biological Ontologies

Ontologies, or controlled vocabularies, organize the concepts of a particular domain of knowledge in a hierarchical manner. The advantage is that concepts as terms in the vocabulary, unlike those same concepts as they may be expressed in free-text, are uniquely specified – a feature that renders the vocabulary usable by both people and computers. It is the latter aspect and the associated ability to informatically store, retrieve and manipulate data, particularly at a time when the scale of available biological data is ever-growing, that sparked the development and use of biological ontologies. Probably the best known and widely used ontology, and likely the one that paved the way for the development of many others, is the Gene Ontology (GO). GO provides three structured vocabularies for the annotations of genes to molecular function, biological process and cellular component terms in a species-independent manner (Ashburner et al., 2000). The GO vocabularies are DAG (Directed Acyclic Graph) tree type. A particular feature of the tree is that a more specialized, or child term, can have several more general or parent terms. In other words, there can be more than one path to a given term (Day-Richter et al., 2007). Many biological ontologies, covering many domains of the biological world, are being developed. While in some cases there may be a certain degree of overlap and similarity of terms between different bio-ontologies, the perspectives they offer and for that matter the relationships between terms and their positions within the tree, are distinct. A complete list of all biological ontologies is provided by the BioPortal at the National Center for Biomedical Ontology, funded by the National Institute of Health Roadmap - http://bioportal.bioontology.org/. In order to organize the various types of biological pathways, including disease and altered pathways, and the relationships between them within a hierarchical structure, RGD is developing the Pathway Ontology (PW). The Pathway Ontology and applications are presented in the next section.

MAIN FOCUS OF THE CHAPTER

The Pathway Ontology (PW)

The Pathway Ontology (PW) is being developed in order to organize the various types of biological pathways, including disease and altered pathways, and the relationships between them in a hierarchical manner. The ontology has five major nodes: classic metabolic, regulatory, signaling, drug and disease pathways. Resources such as the Kyoto Encyclopedia of Genes and Genomes (KEGG) - http://www.genome.jp/kegg/, Reactome - http://www.reactome.org/, PharmGKB - http://www.pharmgkb.org/index.jsp and others along with extensive use of the available review literature have been used to populate the nodes (Ogata et al, 1999; Kanehisa et al, 2010; Joshi-Tope et al., 2005; Matthews et al., 2009; Hewett et al., 2002; Thorn, Klein & Altman, 2010). The Pathway Interaction Database (PID) - http://pid.nci.nih.gov/ - a collaborative project between the US National Cancer Institute (NCI) and Nature Publishing Group (NPG) (Schaefer et al., 2009) – is using PW to map the manually curated human signaling and regulatory pathways to PW terms and allows users to browse these pathways by categories - http://pid.nci.nih.gov/browse_categories.shtml. As needed, PID requests the provision of new ontology terms and/or changes or re-arrangements for current terms.

The nodes are disjoint, meaning that a particular pathway term can be a child of parents belonging to different nodes. To give an example, the insulin signaling pathway exerts a plethora of actions and effects that can be viewed from various perspectives. The action of insulin is essential for the maintenance of proper glucose and energy homeostasis. 'Glucose homeostasis pathway' and 'energy homeostasis pathway' are higher order terms within the regulatory node. The 'insulin signaling pathway' along with other pathway terms, is in a 'part_of' relationship to the two homeostasis terms. The insulin molecule is a peptide hormone; as such the 'insulin signaling pathway' is a child of 'peptide and protein hormone signaling' term in the signaling node and in an 'is_a' relationship to it. The 'part_of' and 'is_a' relationship types are the two used in the ontology. Other terms in 'part_of' relationships to 'glucose homeostasis pathway' include the insulin secretion pathway, the glucagon secretion and signaling pathways and the glucose biosynthesis, utilization and transport pathways. Like insulin, the 'glucagon signaling pathway' is also a child of the 'peptide and protein hormone signaling pathway' parent term. Note that the latter term, along with several siblings, is a child of 'hormone signaling pathway' – a higher order term in the signaling node. The glucose biosynthesis, utilization and transport pathway terms are also higher order parent terms for several children terms which in turn, are parents of more specialized terms. For instance, one of the child terms of 'glucose utilization pathway' is 'glucose oxidation pathway' which has several child terms for individual pathways pertinent to glucose oxidation. The latter are also children of the 'carbohydrate metabolic pathway' term in the metabolic node. See Figure 1A for the major nodes of PW and 1B for entries associated with the 'glucose homeostasis pathway' term. See Figure 2A and 2B for the paths leading to 'insulin signaling pathway' and 'glycolysis pathway' terms, respectively.

The ontology has a node dedicated to disease pathway and also provides terms for the altered version of a pathway; the latter is a feature unique to the Pathway Ontology. An altered pathway is viewed as one that deviates from the normal course that it would otherwise have, due to changes in the normal functioning of one or several molecules within the network. A disease pathway is viewed as the sum of alterations in one or several pathways, the combination of which overcomes the ability of the system to respond and adjust to adverse effects and culminates in the diseased state (Hanahan & Weinberg, 2000). Genes within a pathway that are known to harbor mutations which interfere with its normal proceeding and are associated to various conditions can be annotated to the altered and the disease pathway terms, in addition to being annotated to the regular pathway term (gene annotations are discussed in the sub-section Applications: Part I). The interactive pathway diagrams that RGD publishes present the disease pathway from the standpoint of altered pathways that have been associated with the condition (interactive pathway diagrams

Figure 1. The major nodes and the hierarchical structure of the ontology. Panel A shows the nodes of the Pathway ontology. Panel B shows a partial view of the hierarchical organization of entries associated with 'glucose homeostasis pathway' in the regulatory node; the 'glucose homeostasis pathway' term with immediate children and the children of the 'glucose utilization pathway' child term are shown (a plus sign in front of a term denotes the presence of children).

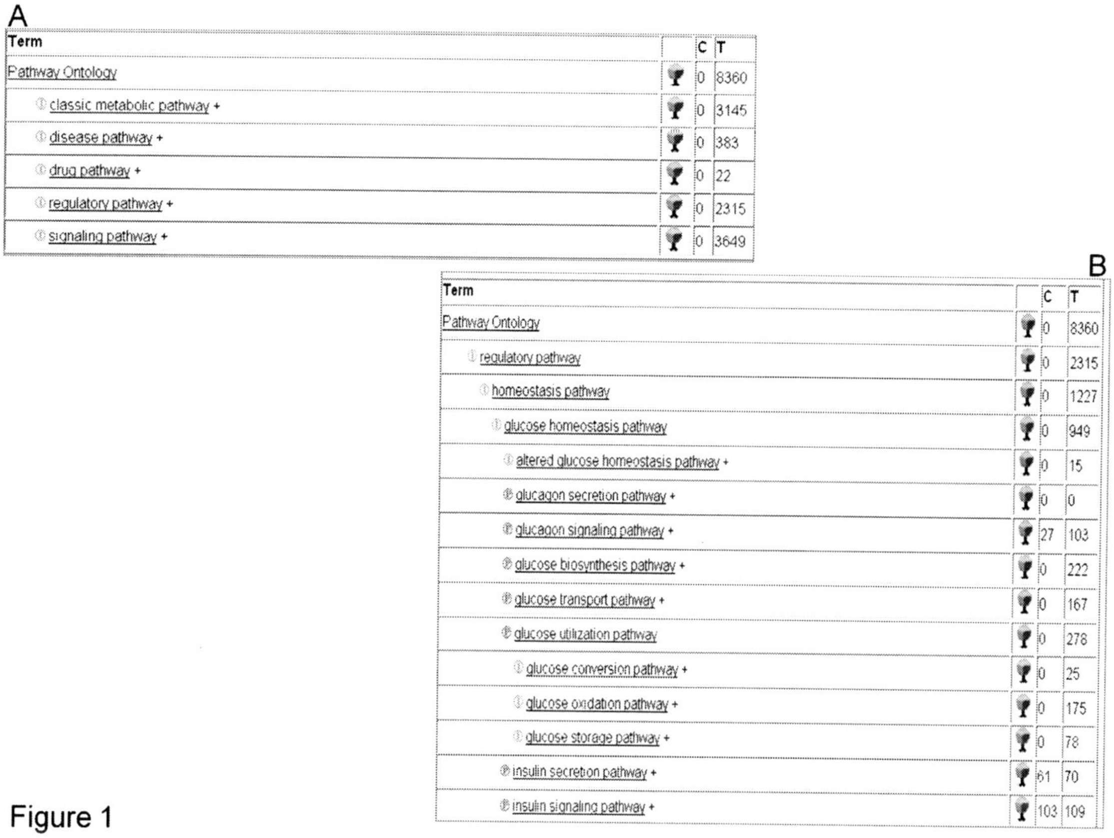

Figure 1

are discussed in the sub-section Applications: Part II). One has to mention here that KEGG provides maps for human diseases across several disease categories - http://www.genome.jp/kegg/pathway.html#disease. More recently, Reactome has provided a few entries for human diseases. Terms for the altered version(s) of a pathway are added when necessary in that the vocabulary is not populated with altered terms for any 'regular' pathway it contains, but only when there is evidence and a need for adding them.

While at times a 'pathway' and a 'process' concept may be used interchangeably, there is nonetheless a clear distinction between them. A process can be simple such as 'protein phosphorylation' or more complex such as 'angiogenesis'. In either instance, it is a plan of action and, depending on the complexity, many pathways may contribute to its unfolding or participate in its regulation. At the same time, any given pathway may contribute to and/or regulate many processes. To take the example of angiogenesis, signaling by vascular endothelial growth factor (VEGF), Notch and ephrins

Figure 2. The disjoint nature of nodes and the hierarchical organization of terms are demonstrated by showing the paths leading to two selected terms. Panel A shows the paths leading to the term 'insulin signaling pathway'. Panel B shows the paths leading to the term 'glycolysis pathway'.

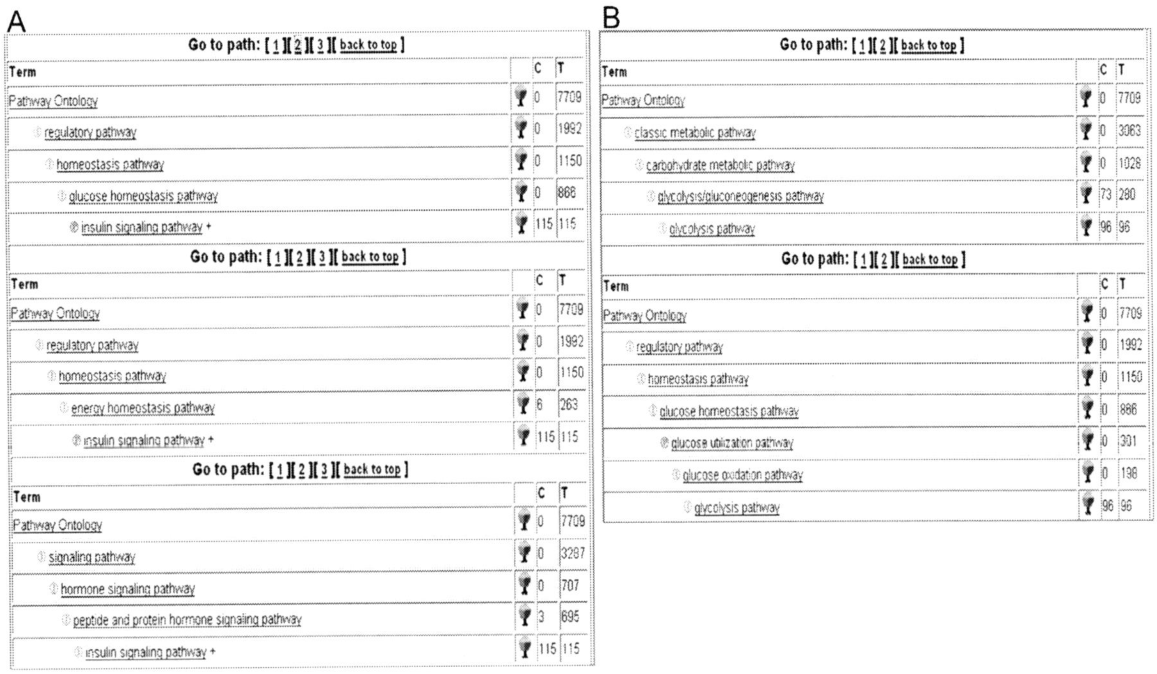

among others, is involved in the process (Bicknell & Harris, 2004). The Notch signaling pathway is also involved in embryonic development and cell fate determination, in addition to angiogenesis and other processes (Harper et al., 2003; Miele, 2006). In certain cases however, the distinction may be more blurred. For instance, the intrinsic apoptotic pathway (mostly referred to as such in the literature) results from the balance between the on and off states of several interacting proteins along with other reactions and interactions that shape its outcome. Intrinsic apoptosis can also be viewed as a process – the plan of action whose goal is the removal of unnecessary or damaged cells. Neither view is wrong; they are simply different ways of representing apoptosis and as such, they conjure up different perspectives and require different parent-child relationships. As an example, there are terms in PW that have associated or related GO terms in the process ontology. For instance, GO has mapped the KEGG metabolic pathways to GO process terms and thus concepts pertinent to metabolism can be found in both the GO process and PW ontologies. However, as in KEGG, the metabolic pathway is viewed as a network or a set of reactions, rather than a process. To take glycolysis again as an example, the pathway is represented by the set of ten enzyme-catalyzed reactions that, through a number of intermediates, convert one molecule of glucose into two molecules of pyruvate in the first round of glucose oxidation. Glycolysis can nonetheless be viewed as the process whose goal is to catabolize or break down the hexose monosaccharide – one of the many paths to 'glycolysis' in the GO process ontology. The GO process vocabulary has several terms for signaling via various receptors and for signal transduction. Both instances are 'derived' from 'regulation of [biological or cellular] process' terms, from response-associated terms or more recently, as GO is trying to accommodate the signaling concept within the process ontology, from the 'signaling' term. Yet, the perspective is dutifully the one of a process rather than the one of a network. Thus, the pathway concept conjures up the set of interacting molecules, the reactions and interactions that underlie the functioning network; the process concept conjures up the end result that the combined actions, the reactions and interactions engaging these molecules, produce.

In conclusion, the PW ontology is solely dedicated to pathways; it seeks to integrate all pathway types and the relationships between them within the hierarchical structure of the DAG tree and within the category or categories they can be associated with. The nodes are not disjoint in that, as shown, a particular term can be a child of parents belonging to different nodes. The ontology has terms for disease and for the altered version of pathways – the latter representing a unique feature of PW. While overlap between certain process and pathway terms can exist, the distinction between the two concepts is exemplified by the unique perspectives to and relationships between terms. The ontology is being developed using the OBO-Edit software (former DAG-Edit), developed by the Gene Ontology Consortium (Day-Richter et al., 2007). Every term in the ontology has a unique identifier in the form of PW:XXXXXXX, where X is a number from 0 to 9. The IDs are created sequentially by the software as terms are added. The actual value of the ID is irrelevant to the concept the term represents but is essential for the manipulation of data, especially if this is to be done by computers. The obo file is available for download from the RGD ftp site - ftp://rgd.mcw.edu/pub/, under the data_release directory - ftp://rgd.mcw.edu/pub/data_release/. The ontology can be browsed using the Ontology Browser available at RGD (presented in more detail in the section Navigational Capabilities) - http://rgd.mcw.edu/tools/ontology/ont_search.cgi, the one at the BioPortal site mentioned above - http://bioportal.bioontology.org/ontologies, or the one provided by the Ontology Lookup Service (OLS) - http://www.ebi.ac.uk/ontology-lookup/) at the European Bioinformatics Institute.

Applications: Part I

Annotations to Pathway Ontology Terms

RGD is using PW to annotate rat, human and mouse genes to pathway terms. In order to provide its users with the ability to make comparisons between the three mammalian species with sequenced genomes, RGD provides information on human and mouse genes as well as human and mouse QTL (Quantitative Trait Loci). The rat, human and mouse gene data come from the National Center for Biotechnology Information (NCBI) Entrez Gene - http://www.ncbi.nlm.nih.gov/gene/. RGD is providing ontology annotations for rat genes, QTL and strains, human genes and QTL and mouse genes and QTL via four ontologies. In this section, only annotations generated by RGD and in particular PW annotations are being discussed. As a note, annotations generated by other groups are brought in the database electronically via pipelines. To briefly describe, RGD is using the Gene Ontology (GO) for the annotation of rat genes to function, component and process terms; the Mammalian Phenotype Ontology (MP) for the annotation of rat genes and as applicable, human or mouse genes, of rat and human QTL and of rat strains; Disease Ontology (DO) for the annotation of rat, human and mouse genes, rat and human QTL and rat strains; PW for the annotations of rat, human and mouse genes. MP is developed by MGI (Mouse Genome Informatics – the Mouse Genome Database; http://www.informatics.jax.org/) and RGD has contributed many terms; DO is based on the C branch of the Medical Subject Headings (MeSH) - http://www.nlm.nih.gov/mesh/ from the National Library of Medicine (NLM) - http://www.nlm.nih.gov/.

Pathways are targeted for annotations based on the role they play in the broader context of the system's networks and physiological pathways, or the role they may have in the context of particular diseases that constitute topics of a Disease Portal. (The Disease Portals are an important project at RGD and will be referenced in the sub-section Navigational Capabilities). Once a pathway has been selected for annotation, the review literature is extensively used in order to identify all of its components as well as its boundaries as they are or are agreed upon. When the set of genes in a pathway has been identified, the complement of rat, human and mouse genes is annotated to that pathway term. The set of rat genes is also functionally annotated using the Gene Ontology and the available experimental rat literature. Disease and associated altered pathway are selected based on the various topics of the Disease Portals as well as the prevalence and impact of particular conditions. Genes whose mutations have been associated with particular conditions are annotated to the altered version of the pathway, in addition to the regular pathway term, the disease pathway term and the disease ontology term. When an ontology search is being made, the returning Ontology Report page shows all the genes annotated to the selected pathway. Individual gene report pages which can be accessed by simply clicking on a gene in the list the ontology report generates, or by searching for a gene in the general search, will show all the ontology annotations associated with that gene, including pathway or pathways. Thus, the user can find information regarding all the genes associated with a pathway; he/she can also find information regarding all the pathways, including altered and disease pathways a gene may be associated with (ontology search, ontology and gene reports are presented in more details in the sub-section Navigational Capabilities). The PW annotation files for the three species as well as the annotations files for other ontologies can be downloaded from the RGD ftp site under the subdirectory annotated_rgd_objects_by_ontology in the data_release directory, ftp://rgd.mcw.edu/pub/data_release/annotated_rgd_objects_by_ontology/. See Figure 3 for a view of the Ontology Report and Gene Report pages.

Applications: Part II

Interactive Pathway Diagrams

The graphical depiction of biological networks is of great interest as visualization of these networks offers an intuitive means to grasp and follow the role individual reactions and interactions play within the network and the ways various regulators may shape its outcome. The interactive pathway diagrams, one of the newest additions to the pathway data RGD provides (Dwinell et al., 2009), are built using Pathway Studio software package from Ariadne Genomics – http://www.ariadnegenomics.com; the diagram page is put together within a Content Management System (CMS). The diagrams can be accessed directly from the Pathway entry point of RGD's home page or via the Ontology Report page. Every diagram page contains a legend for entities, relationships and shapes, a brief description of the pathway and a variety of links. Every component in the diagram links to gene report pages in RGD or as necessary, to dynamic lists of genes. Small molecules link to their entries in PubChem (http://www.ncbi.nlm.nih.gov/pccompound/), domains mentioned in the description link to their entries in the Pfam database (http://pfam.sanger.ac.uk/), PubMed IDs of references provide links to their abstracts in PubMed (http://www.ncbi.nlm.nih.gov/pubmed/). Every page provides a link to the Ontology Report for that pathway term (The ontology reports are presented in the sub-section Navigational Capabilities). The diagram page also provides links to the Ontology Report of related GO terms, to KEGG maps or entries at Reactome, when applicable. A downloadable diagram link is provided for users of the Pathway Studio tool. Finally, pathways, which may be triggered, regulated or in some

Figure 3. Partial views of an Ontology Report page and an associated Gene Report page are presented. Panel A shows the top portion of the Ontology Report page for the term 'insulin secretion pathway'. Panel B shows the top portion of the gene report page for Abcc8 along with a partial list of ontology annotations including PW.

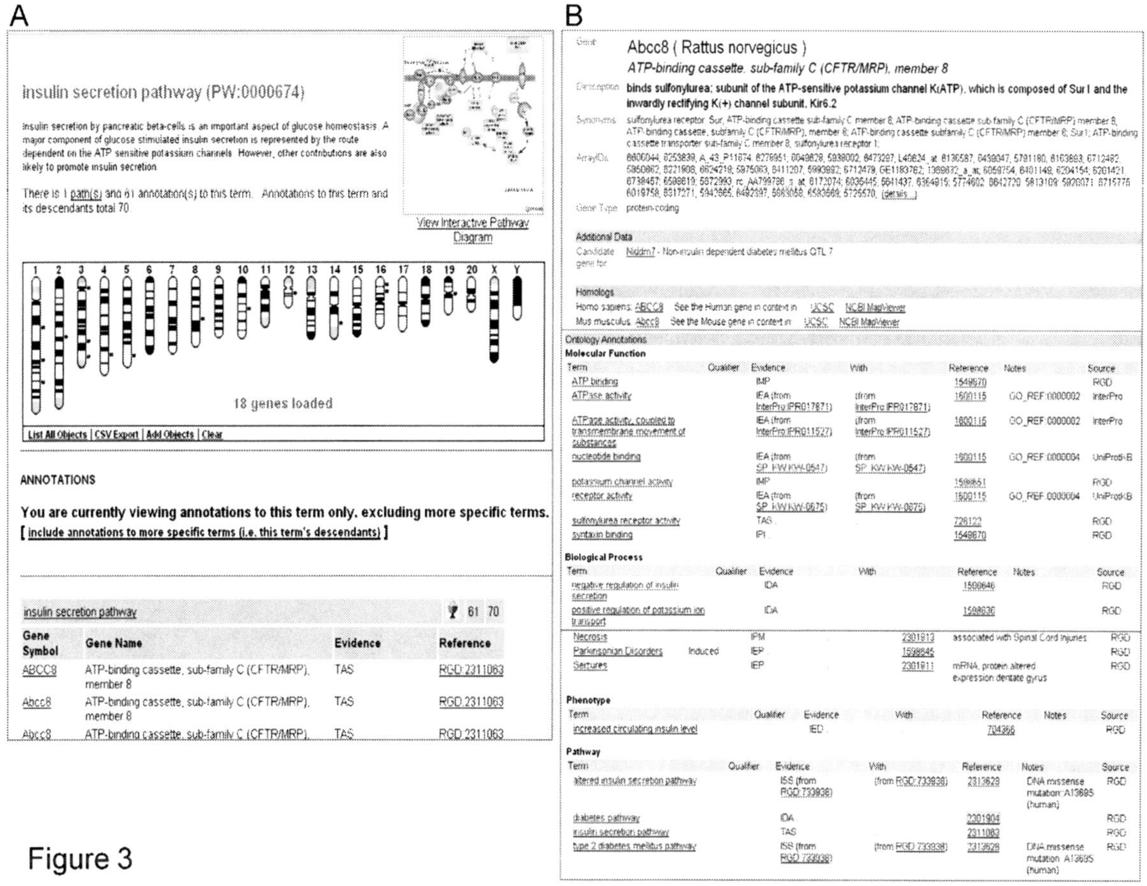

fashion associated and are present in the diagram, link to their Ontology Report pages from where the users can access the accompanying diagram, if one exists. This is a very important feature of the interactive pathway diagrams as it allows the user to 'walk' from one pathway to another and explore the pathway landscape. What makes this feature possible is the unique ID of individual pathways in the Pathway Ontology and the ability to associate the ID with the pathway object in the diagram in the Pathway Studio software tool. See Figure 4 for the diagram of 'insulin signaling pathway' showing the components of insulin signaling and the downstream cascades it triggers.

The Pathway Studio tool includes a large database, the mammalian ResNet database, that contains various type of information for rat, mouse and human genes, including in the case of rat genes, their RGD:ID. In addition, for any object in the diagram, new properties can be added with desired values assigned to them. In the case of a pathway present in the diagram, the property is PW.ID and the value is the PW:XXXXXXX. This feature has also been exploited to make links to PubChem for small molecules present in the dia-

Figure 4. The diagram part of the interactive diagram page for the 'insulin signaling pathway'. The arrows point to the two downstream signaling pathways triggered by insulin. The ontology report and accompanying diagram pages can be accessed by clicking on either of the two pathways.

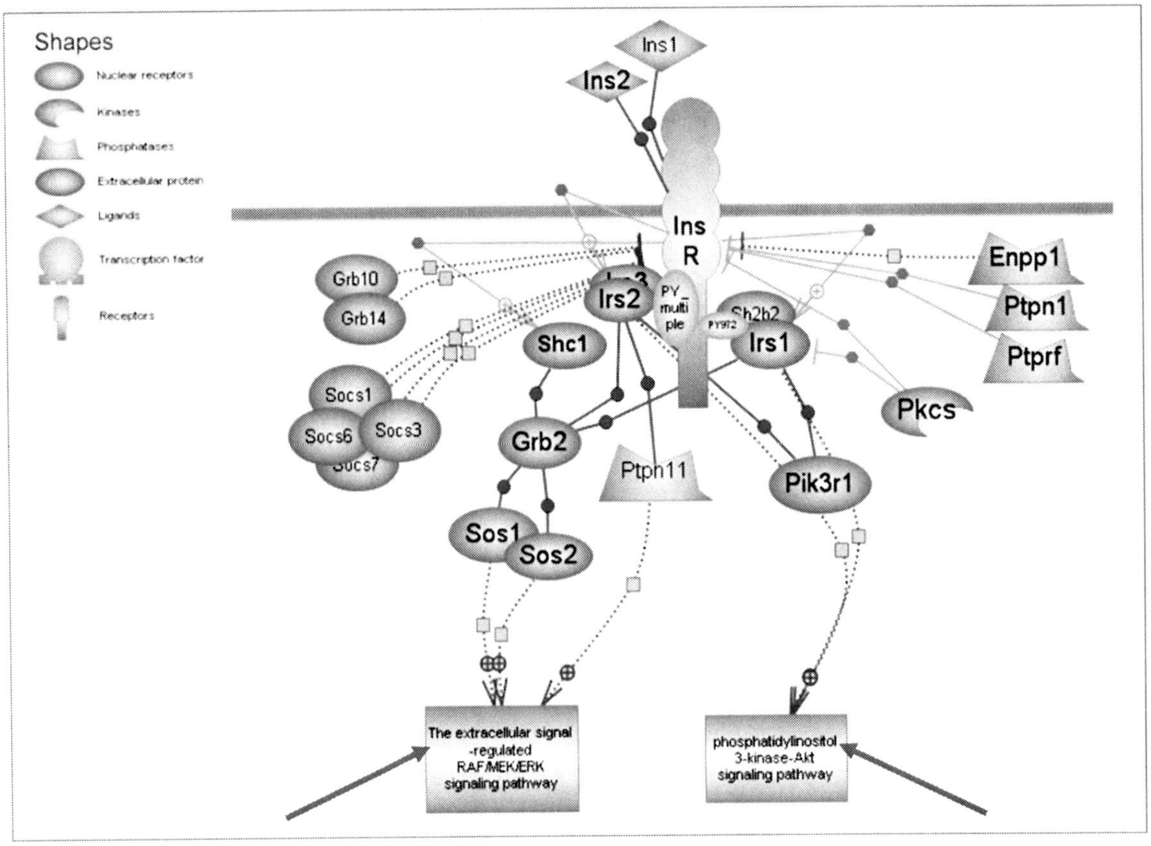

gram or to dynamic lists of genes. The lists are made when there are several members of a class in the pathway. In order to keep the diagram easy to follow, instead of placing objects for every member of the class, a generic object is made and linked to the webpage created for it, as will be explained. They are also made for target genes. Target genes, unless they in turn have targets and therefore mediate the signal, do not get annotated to the pathway term. A dynamic list created for the generic 'target name' allows the user to find information about them. The lists, for both the members of a class or for the target genes, are created within the CMS in the context of the pathway with which they are associated. The url of the page thus created is added as the value of a property called 'ListLink' that has been added in the ResNet database. In every list, the genes are listed with their symbol, aliases, full names and links to their gene report pages. As an example, in the integrin mediated signaling pathway, the 24 known pairs of receptor heterodimers are formed through the various combination of 18 alpha and 8 beta subunits. In the diagram page, generic alpha and generic beta subunit objects are added. In the CMS, within the page put together for 'integrin mediated signaling pathway' two new pages are created: the 'integrin-alpha-subunits' and the 'integrin-beta-subunits'. In the two pages, the genes are listed as mentioned and the url of the gene report page for each subunit is then added in the 'link' feature of the 'edit' mode. The url of the list page itself is added as the value of 'ListLink' in the Resnet database for the generic alpha and the generic beta objects of the 'integrin mediated signaling pathway' diagram. When the diagram is saved in Pathway Studio as html, a folder is created with files for every object and relationship in the diagram containing the information they have in the ResNet database. The folder is ftp-ed to the pathway website of RGD; a script parses every file for RGD:IDs, PW:IDs, ListLink and LinkUrl thus allowing for objects in the diagram to link to appropriate pages (LinkUrl has been added to link to PubChem entries for the small molecules present in a diagram page). See Figure 5 for the diagram of 'integrin mediated signaling pathway' showing the list of genes for the integrin beta subunits.

Links from the diagram page, such as the links to PubMed abstracts, ontology report pages, or protein domains mentioned in the description are made within CMS using, as already mentioned in the context of dynamic lists, the 'link' feature of the 'edit' mode. Thus, a combination of features present in the Ariadne Genomics software tool, available in the CMS along with bioinformatics approaches are being used to render the diagram

Figure 5. The diagram part of the interactive diagram page for the 'integrin mediated signaling pathway'. The inset shows the list of genes for the integrin beta subunits. The list can be accessed by clicking on the Itgb object in the diagram (star). Every entry gene in the list links to the respective gene report page.

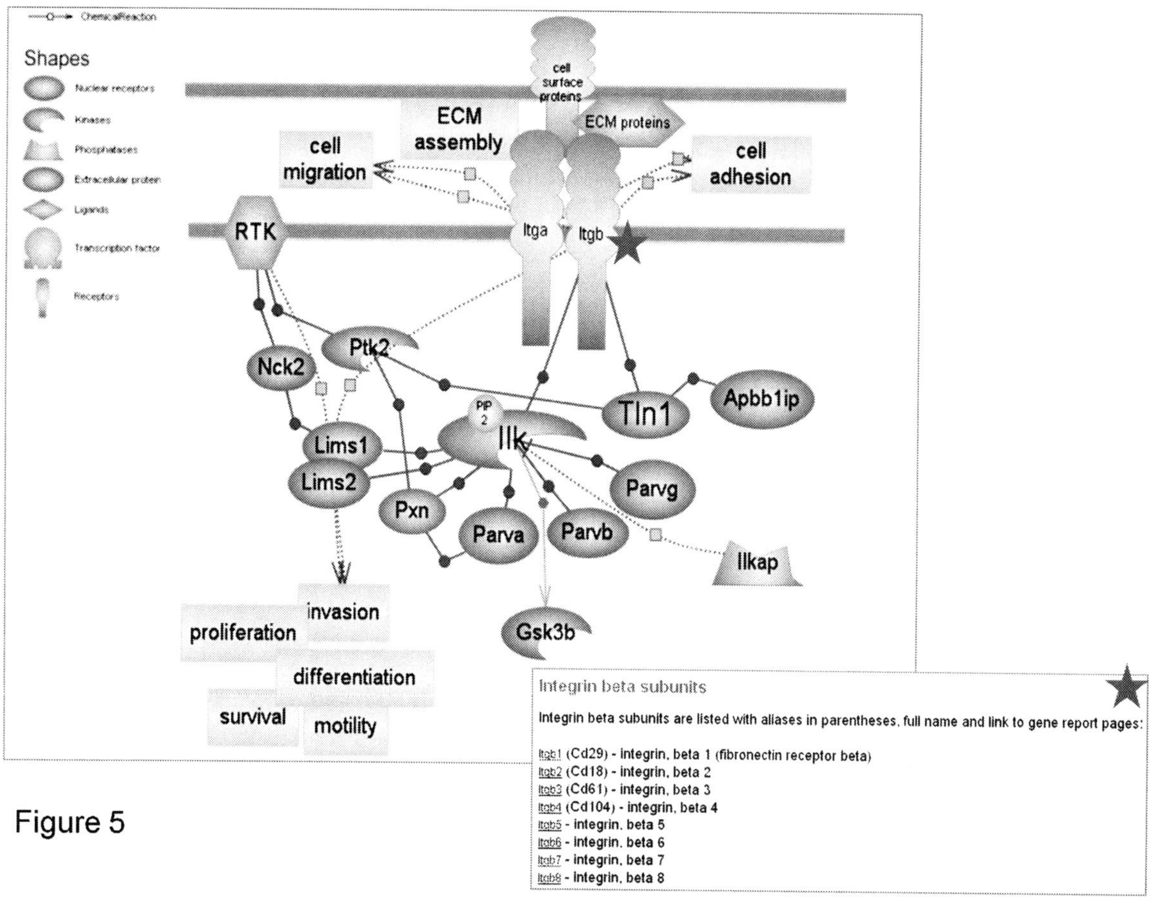

Figure 5

pages dynamic, integrated with other resources and open-ended. New features are envisioned to be added.

Navigational Capabilities

In this sub-section several tools and resources, particularly those allowing for accessing and navigating between pathway-related data are presented. For a complete list of tools and resources that RGD has to offer, the user is invited to RGD's home page to check the main entries listed within the body as well as on the top of the page, for instance Data and Genome Tools. – http://rgd.mcw.edu.

Ontology Search, Ontology Browser and GViewer, Ontology and Gene Report pages

RGD is offering a suite of tools and resources to make the stored data easily available and to allow the user to navigate between various types of objects and the information associated with them. The Ontology Browser and GViewer are tools that allow users to search for and retrieve ontology terms. The ensuing ontology reports allow the users to see the annotations made to selected terms, download them if desired and in the case of the Pathway Ontology, to link to interactive pathway diagrams. As mentioned, all available diagrams can be accessed directly from the Pathway entry point of RGD's home page. The browser can be accessed from the Data entry on the top of the RGD's home page; the GViewer can be accessed from the main Genome Tools entry point in the body or the Genome Tools entry on the top of the home page. In the Ontology Browser the user can select any or all the ontologies RGD is using to perform a search. In addition, in the GViewer the user can select any combination of ontologies. In either case, typing a word or a phrase will retrieve all the terms containing that word or phrase in the selected ontology. For instance, selecting the Pathway Ontology, and typing in 'insulin' will retrieve all the terms the ontology has that contain the word 'insulin' such as the signaling and the altered insulin signaling pathway, the insulin secretion and the altered insulin secretion pathway, among others. The results will show the icon of a tree associated with any of the terms retrieved. Clicking on the icon will bring up the ontology itself showing the position of the term within the tree and the path(s) to it. For instance, clicking on the tree for 'insulin signaling pathway' will bring up the Pathway Ontology showing the position of this pathway term within the ontology – the three path(s) already mentioned along with the term's siblings. The user can select to go up and down the ontology tree and explore it by clicking on the tree icon displayed on the right of every term. A plus sign in front of the term indicates that the term has children (see Figures 1 or 2). Clicking on the term itself, either in the list of terms brought up by a search or from within the ontology will bring up the Ontology Report for that term. The Ontology Report contains a brief description of the term, the GViewer, all the genes annotated to the term and the path(s) to the term. If there is an interactive pathway diagram, a clickable icon is displayed on the top of the report, to the right of the description. Clicking on any of the genes brings up the gene report page of that gene (see Figure 3). The GViewer tool provides a genome-wide view of the rat chromosomes with the position of annotated genes indicated. The tool allows the user to slide along the chromosomes and view the position of annotated genes on a bottom panel. Mousing over the gene shows the intron-exon positions from the GBrowse tool; clicking on it brings up the gene report page. In GViewer, users have the option to download the annotated/displayed objects with information regarding chromosome, start and end positions of genes. Users can also choose to add objects. For instance, the user may choose 'insulin signaling pathway' and then decide that she/he would like to see 'side-by-side' the genes that are also involved in the 'insulin secretion pathway'. One would first look for insulin signaling, and then use the 'add objects' feature to add the genes annotated to the insulin secretion term and select for instance, green as a display color (the default is brown). If a gene happens to be shared, the gene is colored yellow. See Figure 6 for an example of GViewer use and results.

As already mentioned, clicking on a gene from the pathway diagram (object or list), the ontology report or from within the GViewer, brings up the gene report page. A detailed description of the gene report page is beyond the scope of this chapter. It should be mentioned, however, that besides chromosome and mapping information and links to many external databases, the page provides all the annotations that have been associated with the gene across all the ontologies RGD is using, along with associated references and evidence codes (see Figure 3B). Clicking on any of the ontology terms in the gene report page will bring up the ontology report of the selected term. Thus, from interactive pathway diagrams to gene and/or ontology report pages, from ontology searches to navigating along the ontology tree or to ontology report pages, from ontology report pages to interactive diagrams, to navigating the ontology tree or looking up the available information for a particular gene, the user can tailor the 'trip' according to one's particular interests. With respect to pathways, the navigational features allow users not only to go from diagrams to ontology and gene report pages, external databases and back, but also, and very importantly, to navigate between related pathways, between disease and

Figure 6. A GViewer result is presented with the genes for insulin signaling in brown (default) and the 'added' genes for the insulin secretion pathway (green). The locked segment on Chromosome 5 displays, in a bottom panel, the region containing the Ptprf gene (involved in insulin signaling pathway) and the Cdc42 gene (involved in insulin secretion pathway).

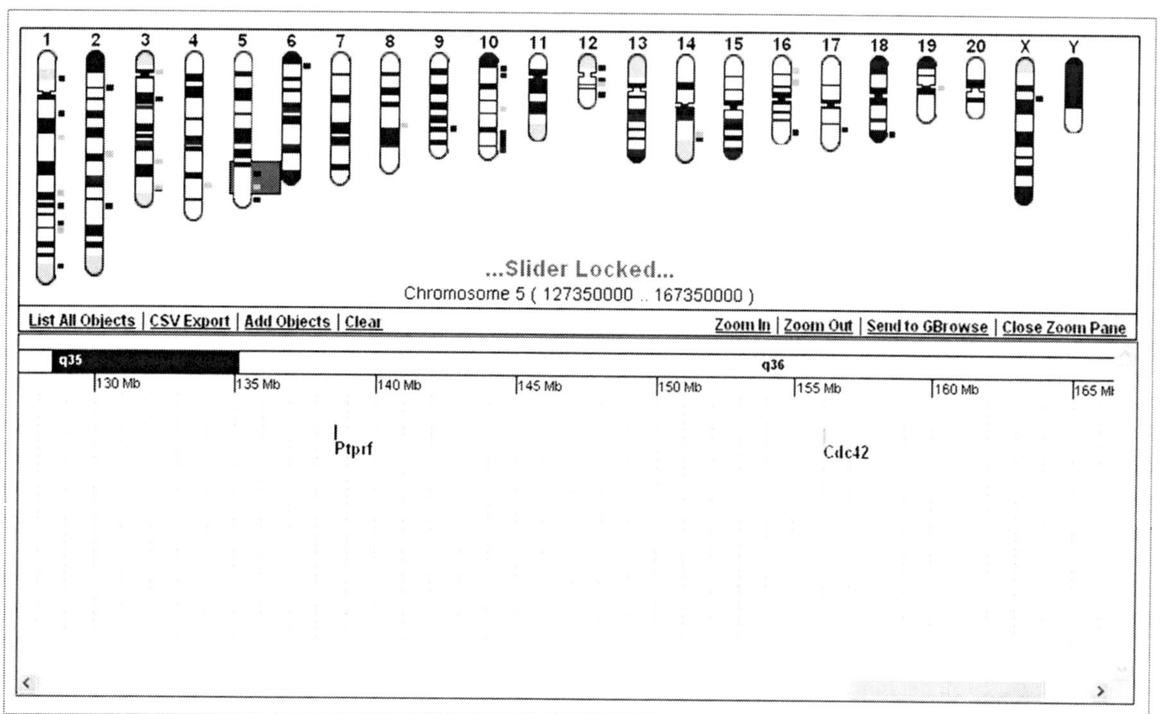

altered pathways, between the altered and the normal version of a given pathway.

Disease Portals

Another entry point to pathway information is via the disease portals. The portals have been designed to allow users from many fields of investigation access to various types of data in their areas of interest. Several portals have been launched since the inception of the project targeting a number of general areas such as cardiovascular, neurological, cancer (breast and urogenital), obesity/metabolic syndrome and the more recent diabetes portal; many more are planned for the future. Several options are offered by each of the portals and they include searches by diseases (the default), phenotypes, biological processes and pathways as well as tools, related links and rat strain models. By selecting pathways as entry point, the users can get all the genes that have been annotated to pathways associated with that particular portal, genes annotated within a broader category term or genes annotated to selected pathways within that category. For instance, glucose homeostasis pathway which plays such a central role in normal physiology and which if altered can have serious consequences, such as the development of diabetes, is one such broader category in the Diabetes Portal. Several individual pathways are available within this category, including insulin-related ones such as insulin secretion and signaling or pathways of glucose metabolism and transport. Any selection will show the GViewer, which in the context of portals allows for viewing the synteny between rat and human or rat and mouse, the total number of genes as well as the list of individual genes for rat, human and mouse species and an overview of Gene Ontology annotations. Every individual gene links to its report page and from there to many other places, as already mentioned.

A Case Study

As a case-study example, the 'type 2 diabetes mellitus pathway' is presented - http://rgd.mcw.edu/wg/pathway/type_2_diabetes_mellitus_pathway.

It is well known that type 2 diabetes mellitus (T2D) is reaching epidemic proportions across the world. Environmental factors such as diet and lifestyle contribute to the development of the condition and heighten the importance of identifying the genetic components which continue to be elusive. Glucose homeostasis whose maintenance relies on the proper and balanced functioning of a number of metabolic, regulatory and signaling pathways, appears to be impaired. Insulin resistance and elevated glucose levels are a hallmark of the condition and pathways underlying insulin secretion and action are known to be affected. The 'insulin secretion pathway' and the 'insulin signaling pathway' are amongst the most important candidate 'altered pathways'. It is of note that the plethora of effects exerted by the hormone insulin relies on triggering two very important intracellular signaling cascades: the phosphatidylinositol 3-kinase-Akt (PI3K-Akt) and the extracellular signal-regulated Raf/Mek/Erk signaling pathways (see Figure 4). The two, in turn, impact on both gene expression and the outcomes of other pathways. If insulin secretion is impaired, that can obviously affect the insulin signal. If the insulin signaling is impaired as well, the effect will be compounded by both the initial impact of diminished insulin secretion and the subsequent impact on the pathways downstream of insulin and the events associated with them. Such a scenario is worsened if any of the pathways downstream of insulin signaling happen to be altered. For instance, the Akt2 member of the PI3K-Akt pathway plays an important role in the insulin responsive translocation of Glut4 glucose transporter to the plasma membrane and therefore in the insulin dependent uptake of glucose in the heart, skeletal muscle and adipose tissue. While knockdown and gene silencing experiments confirm that loss of Akt2 diminishes insulin mediated transporter activity, few mutations have been identified in humans. Insulin secretion on the other hand is known to be severely affected in T2D patients. When plasma glucose gets elevated, the pancreatic beta cells of the islets of Langerhans secrete insulin. The exocytosis of insulin containing granules is biphasic: a first, rapid release (referred to as the 'triggering' phase) is followed by a second, slower but sustained phase (referred to as the 'amplifying' phase). The first phase is almost completely lost while the second is severely affected in T2D patients. At the molecular level, the first phase is the better understood one: the changes in ATP/ADP ratio resulting from glucose metabolism, lead to closure of the ATP-sensitive potassium channels (KATP), membrane depolarization, opening of voltage-gated calcium channels and the rise in calcium concentration which triggers fusion of insulin granules with the plasma membrane. The second phase is largely independent of KATP channels and is rather poorly understood. It may engage metabolic coupling factors resulting from glucose metabolism and their possible impact on voltage-dependent potassium channels. These channels which also get activated upon depolarization of the membrane can lead to membrane re-polarization. Channel inactivation (which the coupling factors might induce) will prevent this from happening, thus allowing for the events leading to insulin exocytosis to occur. Mutations in both the pore-forming and the regulatory subunits of KATP channels have been reported. KATP inhibitors, such as repaglinide and nateglinide have been developed in order to promote channel blockage, stimulate insulin secretion and lower blood glucose levels in patients with T2D. Mutations in insulin and its receptor have also been reported. While mutations in the hormone gene reduce its affinity for the receptor, mutations in the receptor are of two types – those that reduce the affinity of the receptor for insulin and those that reduce its tyrosine kinase activity. The phosphorylation of tyrosine residues on both the receptor and its downstream effectors is of the essence for signaling to proceed. Yet, these mutations are either rare or the role they play in the condition is controversial. It is worth mentioning here that linkage analyses and Genome Wide Association Studies (GWAS) have identified a number of candidate genes; some of the SNPs identified are in non-coding regions. In addition, microRNAs (miRNAs) have been identified that may play a role in beta-cell development, insulin gene expression, secretion and action as well as in diabetic and other complications.

The picture that emerges is one where accumulated effects, arising from alterations in pathways important in glucose homeostasis and possibly, from non-coding elements that regulate a range of underlying processes, may contribute to the onset and the development of T2D. Each of these aspects can be affected by environmental factors. The interactive pathway diagram for the 'type 2 diabetes mellitus pathway' shows several altered pathways such as insulin secretion and signaling, insulin responsive glucose transport and PI3K-AKT signaling along with the general anti-diabetic drug pathway. The altered version of a pathway displays a diagram with the same settings as the normal counterpart except that culprit genes are color-coded and their relationships to other elements deleted. For instance, in the 'altered insulin signaling pathway', the insulin and its receptor have lost their binding relationship; the receptor has also lost its modifying relationships to the effector molecules that it would otherwise phosphorylate. An 'altered' version of a diagram provides direct links to the 'normal' pathway. The anti-diabetic drug pathway diagram shows the pharmacokinetics of repaglinide and nateglinide and the pharmacodynamics pathway of the two drugs along with the related insulin secretion and T2D pathways. Information on candidate genes and miRNAs is provided via two linked lists; the processes which they impact are shown. Thus, the user can go from the T2D pathway to an altered pathway and from there instantly compare the altered to the normal version, go back to the disease pathway and select another altered pathway or choose to further explore triggered or associated pathways. Alternatively, the user can select the anti-diabetic drug pathway and go on to explore how the drugs are metabolized or how they affect their targets. The user can also look up the identity of candidate genes as well as check for their ontology-annotation-based characterization including pathway and disease participation, or check the miRNAs thought to have potential roles and explore their features and predicted targets. Finally, the user can choose to read the literature on the basis of which the diagrams and their descriptions have been built. But whatever the contours of the journey (or journeys) the user wishes to take, the experience will be rich and informative, and very importantly, open-ended. See Figure 7 for the diagram of 'type 2 diabetes mellitus pathway'.

Figure 7. A partial view of the interactive pathway page for the 'type 2 diabetes mellitus pathway' with the associated altered pathways and culprit genes for the altered pathway they are associated with. Also shown are the lists for the candidate genes and miRNAs (the generic 'potential T2D miRNAs' and 'other T2D candidate genes') along with their associated processes and the anti-diabetic drug pathway that targets KATP. The insets show a partial view of the diagram of the 'altered insulin signaling pathway' (a), the dynamic list for the 'other T2D candidate genes' (b), both of which can be accessed by clicking on the appropriate object.

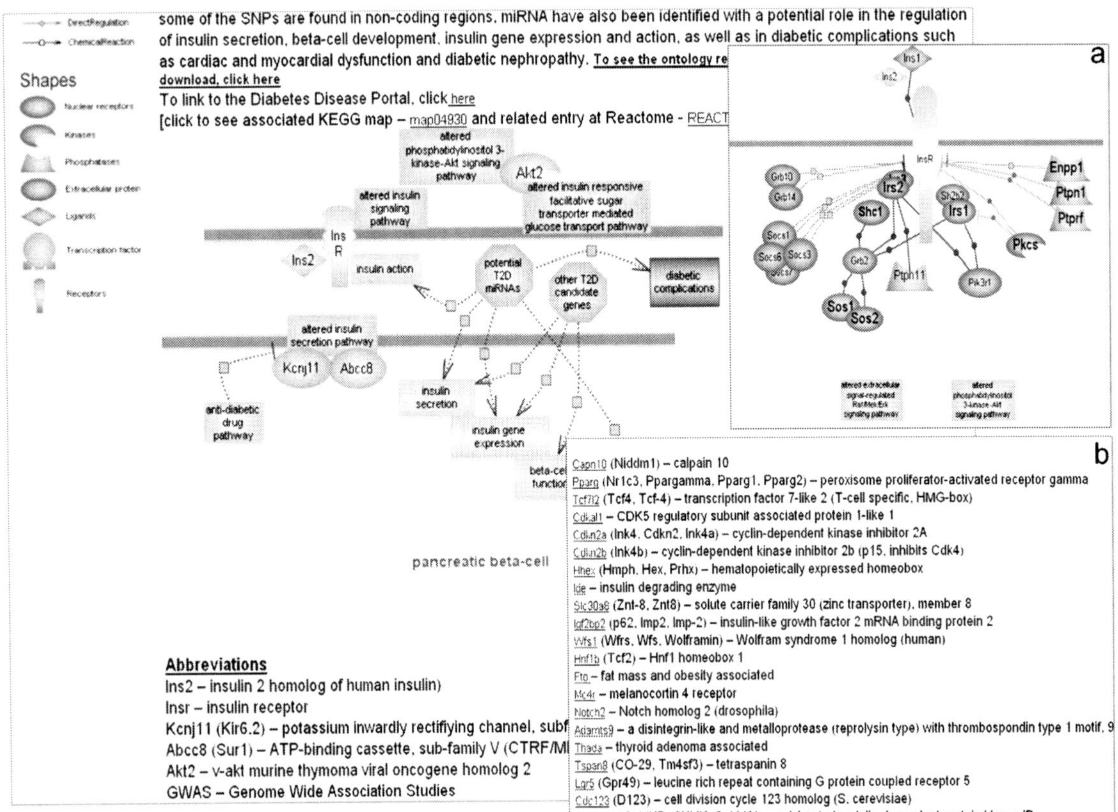

Figure 7

NEW DEVELOPMENTS

Drug pathway information is the latest dimension RGD has added to the pathway resources it provides (at the time of the chapter's writing). As mentioned, the PW ontology now has a node for drug pathways (see Figure 1A); terms within the node are added on a regular basis and drug pathway diagram pages are now provided (see the case-study and accompanying Figure 7; also check the Pathway entry point of RGD's home page). The terms and the diagrams cover and address the issues of both drug pharmacokinetics and drug pharmacodynamics. Pharmacokinetics refers to the uptake, metabolism and efflux of the drug; pharmacodynamics refers to the drug-target interaction. Of note is the fact that genetic variation can affect the way a particular organism processes a drug and makes it available; it can also affect how it responds to a given drug. Like with the other pathway topics, various resources and the available literature are being reviewed for the development of drug-related entries. Resources include but are not limited to the previously mentioned PharmGKB, the Comparative Toxicogenomics Database (CTD) - http://ctd.mdibl.org/ and Small Molecule Pathway Database (SMPDB) - http://www.smpdb.ca/, among others (Davis et al., 2009; Frolkis et al., 2010). The selection of drug pathways for building interactive diagrams is based on the disease the drug targets, which is represented in one of the Disease Portals. The addition of this section points to the very dynamic nature of the pathway and pathway-related projects at RGD; thus, users are encouraged to visit the site on a regular basis for updates and new features.

CONCLUSION AND FUTURE DIRECTIONS

RGD is using a multi-faceted approach to provide information on pathways and associated genes and diseases. The Pathway Ontology (PW) integrates the various types of biological pathways, including disease and altered pathways, as well as the relationships between them within a hierarchical structure. While the ontology provides a means to annotate rat, mouse and human genes to pathway terms, it is also a tool that allows the users to retrieve, visualize and navigate across the pathway landscape. The interactive pathway diagram pages offer a wide array of links and associated information – from ontology and gene report pages to references in PubMed, from compound information in PubChem to domains in Pfam. Very importantly, links to triggered or associated pathways allows the user to travel along the unfolding journey of a signal or the fate of a compound. While disease maps are provided by KEGG, the representation of disease pathways as the combined effect of several altered pathways is a feature unique to the diagrams provided by RGD. It allows the user to 'dissect' a disease pathway, explore each of the 'faulty' pathways, and compare an altered pathway to its normal counterpart. The information not captured directly in a diagram, such as the multiple members of a class, target genes, candidate genes and/or miRNAs of potential interest, is provided via dynamic lists. Each entry in the list has links to appropriate report pages. Whether directly from the diagram or indirectly via the linked lists, the reports that can be accessed provide a range of additional information, from ontology based characterization of genes, including disease and pathway participation along with accompanying literature to a variety of external database links. The ability to 'travel' from one diagram to another, a feature made possible via the unique ontology identifiers of individual pathway terms, will be further exploited. The continuous addition of metabolic, signaling and regulatory pathway diagrams is intended to provide the user with a comprehensive 'view' of the pathway landscape. The integration of these circuitries within the broader context of physiological pathways will further provide users with a more systems biology approach and view of the topic(s) of interest.

The continuing provision of drug pathways along with disease and altered pathways will allow users to investigate how particular conditions are being viewed and addressed from a medical and clinical standpoint. Together, they may prompt the exploration of new venues to further our understanding of how the interaction between genes and the environment is affected by genetic variation, how small changes combine and propagate to alter larger circuitries and eventually the system as a whole, and to develop novel methodologies and therapeutics. Along with these developments, RGD is committed to enhance the features of existing tools as well as to adopt and/or develop new ones to further improve and facilitate access to and navigation between data types. The Rat Genome Database is freely accessible at: http://rgd.mcw.edu.

AUTHOR'S NOTE

The RGD's Pathway Resources page - http://rgd.mcw.edu/wg/pathway?100, has recently added the first physiological pathway and several pathway suites. Each suite offers an instant snapshot of the bigger picture that brings together various types of pathways and as such, it provides a roadmap of the network that connects them. Every suite page contains a brief description of the suite; small icons of individual diagrams are grouped by categories and displayed with the pathway term as title and a short caption. From the titles, icons or captions, the users can navigate to the individual diagram pages.

ACKNOWLEDGMENT

I would like to thank Tom Hayman for proofreading of this chapter, Weisong Liu and Jennifer Smith for help with figures and document formatting. RGD is funded by grant HL64541 from the National Heart, Lung, and Blood Institute on behalf of the NIH.

REFERENCES

Ashburner, M. (2000). Gene ontology: Tool for the unification of biology. The Gene Ontology Consortium. *Nature Genetics*, *25*(1), 25–29. doi:10.1038/75556

Bicknell, R., & Harris, A. L. (2004). Novel angiogenic signaling pathways and vascular targets. *Annual Review of Pharmacology and Toxicology*, *44*, 219–238. doi:10.1146/annurev.pharmtox.44.101802.121650

Bugg, T. (2004). *An introduction to enzyme and coenzyme chemistry*. Oxford, Cambridge, MA: Blackwell Science. doi:10.1002/9781444305364

Davis, A.P., et al. (2009). Comparative toxicogenomics database: A knowledgebase and discovery tool for chemical-gene-disease networks. *Nucleic Acids Research, 37*(database issue), D786-792.

Day-Richter, J. (2007). OBO-Edit an ontology editor for biologists. *Bioinformatics (Oxford, England)*, *23*(16), 2198–2200. doi:10.1093/bioinformatics/btm112

Dwinell, M. R., et al. (2009). The Rat Genome Database 2009: Variation, ontologies and pathways. *Nucleic Acids Research, 37*(database issue), D744-749.

Frolkis, A., et al. (2010). SMPDB: The Small Molecule Pathway Database. *Nucleic Acids Research, 38*(database issue), D480-487.

Gibson, T. J. (2009). Cell regulation: Determined to signal discrete cooperation. *Trends in Biochemical Sciences*, *34*(10), 471–482. doi:10.1016/j.tibs.2009.06.007

Hanahan, D., & Weinberg, R. A. (2000). The hallmark of cancer. *Cell, 100*, 57–70. doi:10.1016/S0092-8674(00)81683-9

Harper, J. A. (2003). Notch signaling in development and disease. *Clinical Genetics*, *64*(6), 461–472. doi:10.1046/j.1399-0004.2003.00194.x

Hewett, M. (2002). PharmGKB: The Pharmacogenetics Knowledge Base. *Nucleic Acids Research*, *30*(1), 163–165. doi:10.1093/nar/30.1.163

Hunter, P. (2000). Signaling–2000 and beyond. *Cell*, *100*(1), 113–127. doi:10.1016/S0092-8674(00)81688-8

Joshi-Tope, G. (2005). Reactome: A knowledgebase of biological pathways. *Nucleic Acids Research*, *33*(Database issue), D428–D423. doi:10.1093/nar/gki072

Kanehisa, M. (2010). KEGG for representation and analysis of molecular networks involving diseases and drugs. *Nucleic Acids Research*, *38*(Database issue), D355–D360. doi:10.1093/nar/gkp896

Kholodenko, B. N. (2007). Untangling the signaling wires. *Nature Cell Biology*, *9*(3), 247–249. doi:10.1038/ncb0307-247

Kholodenko, B. N. (2009). Spatially distributed cell signaling. *FEBS Letters*, *583*(24), 4006–4012. doi:10.1016/j.febslet.2009.09.045

Liu, B. A. (2006). The human and mouse complement of SH2 domain proteins–establishing the boundaries of phosphotyrosine signaling. *Molecular Cell*, *22*(6), 851–868. doi:10.1016/j.molcel.2006.06.001

Marks, F., Klingműler, U., & Műller-Decker, K. (Eds.). (2009). *Cellular signal processing: An introduction to the molecular mechanisms of signal transduction*. New York: Garland Science.

Matthews, L. (2009). Reactome knowledgebase of human biological pathways and processes. *Nucleic Acids Research*, *37*(Database issue), D619–D622. doi:10.1093/nar/gkn863

Mickels, P. A., & Rigden, D. J. (2006). Evolutionary analysis of fructose 2,6-bisphohate metabolism. *International Union of Biochemistry and Molecular Biology (IUBMB). Life (Chicago, Ill.)*, *58*(3), 133–141.

Miele, L. (2006). Notch signaling. *Clinical Cancer Research*, *12*(4), 1074–1078. doi:10.1158/1078-0432.CCR-05-2570

Moore, A. D. (2008). Arrangements in the modular evolution of proteins. *Trends in Biochemical Sciences*, *33*(9), 444–451. doi:10.1016/j.tibs.2008.05.008

Ogata, H. (1999). KEGG: Kyoto Encyclopedia of Genes and Genomes. *Nucleic Acids Research*, *27*(1), 29–34. doi:10.1093/nar/27.1.29

Pawson, T., & Nash, P. (2003). Assembly of cell regulatory systems through protein interaction domains. *Science*, *300*, 445–452. doi:10.1126/science.1083653

Rider, M. H. (2004). 6-Phosphofructo-2-kinase/fructose-2,6-bisphosphatase: Head-to-head with a bifunctional enzyme that controls glycolysis. *The Biochemical Journal*, *381*, 561–579. doi:10.1042/BJ20040752

Santos, S. D., Verveer, P. J., & Bastiaens, P. I. (2007). Growth factor-induced MAPK network topology shapes Erk response determining PC-12 cell fate. *Nature Cell Biology*, *9*(3), 324–330. doi:10.1038/ncb1543

Schaefer, C. F. (2009). PID: The Pathway Interaction Database. *Nucleic Acids Research*, *37*(Database issue), D674–D679. doi:10.1093/nar/gkn653

Schlessinger, J. (2000). Cell signaling by receptor tyrosine kinases. *Cell*, *103*(2), 277–280. doi:10.1016/S0092-8674(00)00114-8

Smock, R. G., & Gierasch, L. M. (2009). Sending signals dynamically. *Science*, *324*, 198–203. doi:10.1126/science.1169377

Taniguchi, C. M., Emanuelli, B., & Kahn, C. R. (2006). Critical nodes in signaling pathways: Insights into insulin action. *Nature Reviews. Molecular Cell Biology*, *7*(2), 85–96. doi:10.1038/nrm1837

Thorn, C. F., Klein, T. E., & Altman, R. B. (2010). Pharmacogenomics and bioinformatics: PharmGKB. *Pharmacogenomics*, *11*(4), 501–505. doi:10.2217/pgs.10.15

Voet, D., Voet, J. C., & Pratt, C. W. (2008). *Fundamentals of biochemistry: Life at the molecular level*. Hoboken, NJ: Wiley.

ADDITIONAL READING

Bader, G. D., Cary, M. P., & Sander, C. (2006). Pathguide: a Pathway Resource List. [http://www.pathguide.org/]. *Nucleic Acids Research*, *34*(Database issue), D504–D506. doi:10.1093/nar/gkj126

Fenton, A. W. (2008). Allostery: an illustrated definition for the 'second secret of life'. *Trends in Biochemical Sciences*, *33*(9), 420–425. doi:10.1016/j.tibs.2008.05.009

Gomperts, B. D., Kramer, I. M., & Tatham, P. E. R. (Eds.). (2009). *Signal Transduction*. Amsterdam, Boston, London: Elsevier/Academic Press.

Tsai, C.-J., Ma, B., & Nussinov, R. (2009). Protein-protein interaction networks: how can a hub protein bind so many different partners? *Trends in Biochemical Sciences*, *34*(12), 594–600. doi:10.1016/j.tibs.2009.07.007

Yamada, T., & Bork, P. (2009). Evolution of biomolecular networks: lessons from metabolic and protein interactions. *Nature Reviews. Molecular Cell Biology*, *10*(11), 1–14. doi:10.1038/nrm2787

KEY TERMS AND DEFINITIONS

Content Management System (CMS): Is a collection of procedures that can be either manual or computer-based and that are used to manage work flow in a shared environment. In computing, it is a software system for organizing and facilitating document and other content creation for loading to a website.

File Transfer Protocol (FTP): Is a standard protocol to exchange and manipulate files over the Internet.

Non-Covalent Interactions: Weak energetic contributions arising from hydrogen bonds, hydrophobic interactions, ionic bonds and van der Waals forces that work in combinations. They are characteristic of macromolecular interactions such as protein-protein and protein-nucleic acids. Intramolecular non-covalent interactions are responsible for the secondary and tertiary structure of proteins and for holding together the two strands of double helical DNA.

Uniform Resource Locator (URL): Is the address of a web page.

Chapter 15
Unsupervised Methods to Identify Cellular Signaling Networks from Perturbation Data

Madhusudan Natarajan
Pfizer, USA

ABSTRACT

The inference of cellular architectures from detailed time-series measurements of intracellular variables is an active area of research. High throughput measurements of responses to cellular perturbations are usually analyzed using a variety of machine learning methods that typically only work within one type of measurement. Here, summaries of some recent research attempts are presented–these studies have expanded the scope of the problem by systematically integrating measurements across multiple layers of regulation including second messengers, protein phosphorylation markers, transcript levels, and functional phenotypes into signaling vectors or signatures of signal transduction. Data analyses through simple unsupervised methods provide rich insight into the biology of the underlying network, and in some cases reconstruction of key architectures of the underlying network from perturbation data. The methodological advantages provided by these efforts are examined using data from a publicly available database of responses to systematic perturbations of cellular signaling networks generated by the Alliance for Cellular Signaling (AfCS).

INTRODUCTION

How does cellular machinery function? What is the network of molecular entities that governs equally well resting basal homeostasis, as well as specific and precise response to cellular stimuli that drive a diverse range of cellular function? The dissection of cellular architectures and accompanying function has followed those precise routes – to first identify molecular entities, fol-

DOI: 10.4018/978-1-60960-491-2.ch015

lowed by a reductionist approach to place it in a simplified physiological context. With the advent of high-throughput experimentation and rapid technological advances, focused biochemical experiments are being replaced with high content experiments. These experiments permit the sampling of large numbers of molecular entities with sufficient resolution and accuracy to develop a comprehensive parts list. However, quantitative descriptions of both (relative) amounts and activity levels of a large number of molecular entities within cells pose novel problems. First, identification of multiple causal relationships between vast numbers of measured molecular entities is needed to define the architecture of the interactions between molecules. Second, the relative contribution of a molecule in the context of a network of interactions is needed to delineate functional consequences.

The inferencing of cellular architectures and functions from biological measurements has rich precedents in engineering - especially in systems identification in the fields of systems engineering and control systems, but mostly in non-biological areas (Ljung, 1999). The goal in systems identification is to be able to build dynamical models from measured data. Therefore, the characterization of cellular networks using mathematical methods is an extension of systems identification theories in that it is simply an attempt to formalize and summarize the system's resting behavior and its response to perturbation. In an ideal scenario, the cellular architecture inferred from the data is isomorphic (i.e., structurally identical) and is identifiable – given sufficient observations from the system, it is possible to uniquely infer the parameters of the model producing the data. In reality, these are almost impossible to achieve – first, cellular architectures are vastly more complex than engineered systems such that even if resource limitations are not considered, sufficient observations may not be available to satisfy the identifiability criterion. Second, while we have made remarkable progress in the inference of biological networks and functions from data, the algorithmic methods are still far from perfect. The availability of additional data is currently not a bottleneck. Thus, state of the art is still focused on developing better methods to deal with existing data sets.

One classical method of exposition of biological function is through perturbation analysis. Classical biological perturbation studies have typically attempted to make incremental change to biological systems through tools such as pharmacological agents, environmental stressors, or more recently anti-sense technologies and measure functional performance after each change. With high-throughput data available to monitor responses to each such perturbation, studies have shown that responses to perturbation are directly amenable to mapping network topology and can yield significant insight into network architecture (Tegner, Yeung, Hasty, & Collins, 2003; Yeung, Tegner, & Collins, 2002).

The problem of inferring cellular networks from measured data is certainly not novel, and has been an active area of interest for several years now (Friedman, 2004; Kalir & Alon, 2004; Nachman, Regev, & Friedman, 2004; Ronen, Rosenberg, Shraiman, & Alon, 2002). In the past, multiple studies have simplified the problem by typically focusing on one type of measurement that is amenable to an analysis method. For example, investigators have reconstructed gene expression regulatory networks or protein-protein interaction networks using statistical and machine learning methods (Friedman, Linial, Nachman, & Pe'er, 2000; Kim, Imoto, & Miyano, 2003; Nariai, Kim, Imoto, & Miyano, 2004; Patil & Nakamura, 2005; Perrin et al., 2003). The use of deterministic modeling of enzyme kinetics and metabolic networks has provided significant insight into signaling machineries (Hendriks et al., 2006; Janes & Lauffenburger, 2006; Palsson, Price, & Papin, 2003). Restricting the focus to only one such measurement allows easy comparison within datasets and avoids the problems faced with integration of disparate data-types. For example, in measurements of transcriptional regulation, investigators can identify en-bloc patterns such as correlated or anti-correlated gene expression and attempt to describe functional linkages between genetic elements (Niehrs & Pollet, 1999). While such descriptions are excellent first pass estimations towards reconstruction of the system, they are still inherently simplistic as such descriptive functional linkages are typically agnostic to mechanism. For example, a putative linkage between two genetic elements bundles or lumps all the inherent mechanistic complexity implicit in such a linkage such as transcription and translation. Clearly, a linkage between any two genetic elements in reality has to involve translational and post-translational downstream elements that are necessary to implement these functions – however, most analysis methods often completely ignore these latter mechanisms to reduce complexity. Thus, further expansion of the analysis methods is needed to encompass multiple sources of dynamic data to accurately recapitulate cellular architecture and function.

Attempts to bridge this divide, while not manifold, have systematically integrated measurements across multiple layers of regulation including second messengers, protein phosphorylation markers, transcript levels and functional phenotypes into signaling vectors or signatures of signal transduction (Cosgrove et al., 2009; Janes et al., 2005; Janes et al., 2004; Natarajan et al., 2006; Perlman et al., 2004). It is important to note that the primary focus around these efforts is to create comprehensive snapshots of signaling that can eventually be reflective of all information transduction that arises as a result of a stimulus or perturbation. By attempting to understand how different variables can be related, these efforts permit initial forays into reconstruction of a unified signaling response. Thus, primary emphasis is hugely dependent upon how to relate different data types together and subsequently on the nature of the analysis performed that permits biological insight. In the research efforts described here, the analysis techniques employed are often simple unsupervised methods but as shown below these methods are still successful in identifying known biological mechanisms implicit in these data sets. Furthermore, the analysis also identified novel connectivity leading to experimentally testable hypotheses.

COMBINING MULTI-VARIATE MEASUREMENT DATA INTO QUANTITATIVE ESTIMATES OF SIGNAL TRANSDUCTION

We now consider a typical response to a specific cellular perturbation such as ligand-induced activation of cell surface receptors. While the nature of the response to the perturbation depends upon the signaling cascade activated by the receptor, here we would consider what some typical responses could entail. Rapid activation of second messengers such as Ca^{2+} or cAMP can occur within seconds and can last for minutes; phosphorylation of proteins can happen within minutes and signaling can last for tens of minutes; transcriptional regimes can be engaged within minutes but are capable of staying activated for hours resulting in post-transcriptional and post-translational effects such as cytokine production and secretion. Figure 1 shows a schematic cell depicting receptor-mediated activation of downstream signaling elements; specifically, it shows sample measurements of assays of four such signaling elements in response to different ligands acting on RAW264.7 macrophages. Platelet Activating Factor (PAF), a GPCR agonist activating Gαq increases intracellular Ca^{2+} (top left), isoproterenol (ISO) eliciting increases in cAMP levels through Gαs (bottom left), bacterial endotoxin (lipopolysaccaride, LPS) stimulating release of Granulocyte colony-stimulating factor (GCSF, bottom right), and interferon-γ causing phosphorylation of STAT3 (top right). Detailed descriptions of experimental protocols for generation of these data and cell culture preparation and stimulation of RAW 264.7 macrophages as well as the data are available publicly at the website for the Alliance for Cellular Signaling [AfCS, http://www.afcs.org/, (Gilman et al., 2002)]. Further descriptions of experimental details are available on the website and beyond the scope of this chapter - this chapter liberally draws upon the data in this database to help illustrate and understand aspects of analyses applied by different research groups.

In the examples shown in Figure 1, the observed responses differ in the magnitude of response, in the units of measurement (calculated concentrations, fold changes, and absolute values), the number of sample measurements (ranging from hundreds for Ca^{2+} to three for cytokine measurements), the periodicity of samples (regular samples for Ca^{2+} every few seconds to irregular samples for protein phosphorylation), etc. If a ligand elicits responses in more than one assay, an accurate representation of all signaling in response to the ligand must integrate all these responses into a common framework. Thus, some fundamental issues need to be resolved – specifically, how we merge different kinds of data, and subsequently, to reduce over-parameterized data such that, when merged with other data types, oversampled (and non-independent) data do not outweigh all other data in analysis.

Data Integration: Combining Multivariate Output Data

Different research groups have applied different numerical methodologies to data integration, but almost all these approaches follow a similar theme in that they result in construction of a unified experiment space of all data, regardless of sources or differences in intrinsic dynamic range and signal to noise ratios. Two such approaches are described here and they describe parametric

Figure 1. Measurements of spatially and temporally distinct signaling elements in response to a perturbation of a cellular network. Examples of four such assayed responses are shown – in counter-clockwise direction Ca^{2+}, cAMP, cytokine secretion and protein phosphorylation in response to multiple applications of PAF, ISO, IFG and LPS, respectively. Measurements of protein phosphorylation and cytokine secretion subsequent to transcriptional regulation are illustrated using pSTAT3 and GCSF. These measurements can easily be multiplexed to assay multiple dozens of phosphorylated species and secreted cytokines. Measurements of transcriptional regulation can include many thousands of species.

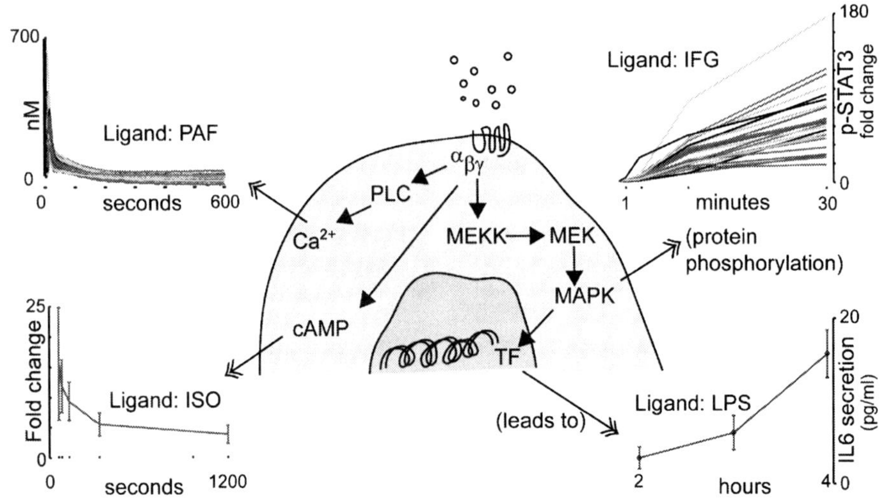

and non-parametric methods of combining multivariate data, respectively. In the first instance, we (Natarajan et al., 2006) assembled a distribution for basal or unperturbed measurements of each variable. This provided an estimate of the central tendency and noise associated with the unstimulated condition. For every subsequent condition where a ligand was applied, it was thus possible to estimate the significance of each measurement using a Z-score like metric (Figure 2A).

In contrast, Perlman and colleagues adopted a ks-statistic method to estimate the significance of difference between treated and untreated distributions (Perlman et al., 2004). Subsequent analysis resembled the earlier parametric analysis where basal distributions of ks-statistics for the untreated condition were used to derive a Z-score like parameter for the treated condition. In either instance, every data measurement regardless of type, time scale, and method of collection was rescaled into a dimensionless quantity that allowed a common basis for comparison and subsequent analysis such as clustering, regression, principal components analysis (PCA), etc.

There are many important considerations that are implicitly folded into such an analysis which is worthwhile to explicitly delineate. First, note that in either case the central theme adopted is to estimate an error-model that determines the propensity of a change given a baseline or unstimulated case. Thus the estimation of the error model is the parameterization of the distribution of the unstimulated case – described here as μ_{basal} and σ_{basal} (either as a Gaussian distribution or via a non-parametric estimate that converts it to a similar formalism). Subsequently, the significance of each stimulated response (Xi) is estimated with respect to this parameterization as a dimensionless Z-score like metric. The formula to reconstruct the dimensionless measure of signaling is a very simple and straightforward calculation as shown in Figure 2. The two methods however diverge on some basic assumptions associated with the datasets. In one instance, the unstimulated case

Figure 2. Assembling a unified experimental matrix from disparate multivariate data. A. Multiple measurements of the unstimulated or basal condition can be parameterized as a distribution with mean μ and standard deviation σ. The significance of any response to stimulation with ligand can be estimated with a Z-score like parameter, or as the number of standard deviations removed from the unstimulated mean value. B. In case we cannot assume a normal (or any parametric) distribution for the treated and untreated conditions, the ks-statistic that compares the cdfs of the two conditions allows estimation of significance. The distribution of ks-statistics is then amenable to a parametric interpretation such as in A.

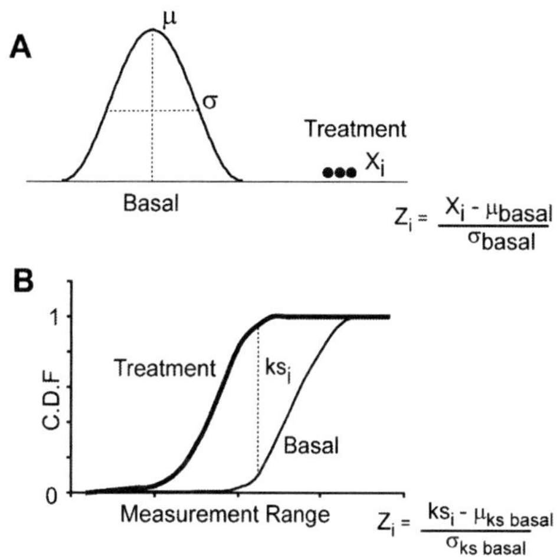

is assumed to be normally distributed, whereas the second follows a non-parametric distribution. The differences between the two approaches result from the intrinsic nature of the measurements that need to be integrated. Specifically, all measurements within the AfCS datasets follow relative or absolute amounts of actual cellular molecules (e.g., levels of second messengers, protein phosphorylation, amounts of cytokines secreted, etc.) where the measurement of the basal unstimulated state follows a normal distribution. The analysis by Perlman et al. measures many higher order combinations of cellular parameters such as nuclear area, nuclear volume, cell size, etc. In contrast to the AfCS datasets, clearly the relationship between the relative amounts of the most ubiquitous nuclear protein and nuclear area and nuclear volume are non-linear; furthermore, by definition, the tendencies of area and volume will not be normally distributed. To generalize, the dependencies between amounts of molecular entities and such higher order processes are not always clear and need not follow any known distributions. In such cases, the pragmatic solution is to follow a non-parametric method which at its limit will follow a distribution that can be parameterized.

Second, any approach that is critically dependent upon characterization of an error model is potentially very sensitive to noise; in the unstimulated case this is a natural cause for concern since unstimulated conditions typically tend to be non-actuated and close to zero or a nominal non-zero value. It is therefore crucial to eliminate any artificial source of noise or bias in the system. It follows that pre-processing steps such as normalization and scaling are critical, and that there needs to be careful consideration of experimental design that will accommodate such processing (Gardiner & Gettinby, 1998). The methods of normalization and scaling are critically dependent upon the nature of the experiment, and cannot intrinsically be compared across research endeavors. Within our analysis of the AfCS datasets, we followed a threefold approach to eliminate potential sources of systematic bias. First, to eliminate batch effects, we normalized across data-sets using co-processed controls within each experimental dataset. Furthermore, as the responses to ligands follow deterministic trajectories, we normalized all datasets to the area under the curve of a mean response to stimulation. Finally, for high-resolution datasets such as the calcium traces where extrinsic factors such as temperature, humidity etc., may play a role in influencing the short term temporal kinetics, we applied both peak amplitude and time-to-peak normalization to the calcium datasets (Flyvbjerg, Jobs, & Leibler, 1996). The goal with these successive normalization steps was to eliminate batch effects and preserve the characteristic response shapes within assays while retaining sources of variability that were due to inherent biological stochasticity. Ideally, any error model developed after normalization would be reflective only of relevant and intrinsic biological variability.

Eliminating Data Redundancy: Collapsing Non-Independent Outputs

The development of a dimensionless metric as described earlier allows subsequent analysis ranging from classification of signatures through unsupervised methods (e.g., hierarchical clustering) to complex analysis schemes. However, even the simplest form of analysis may be dramatically skewed by accidental bias introduced by the measurement of non-independent outputs. In an ideal scenario, the result of any perturbation can be categorized as a unique set of measured responses. Since we do not possess *a priori* knowledge on which outputs will be relevant, a typical experiment will attempt to cover the maximal number of feasible measurements that are pragmatically possible. Coverage can include increases in the number of unique molecules measured as well as multiple temporal measurements to track the dynamics of response. This sampling can introduce redundancy in two ways. First, estimation of biological redundancy in that the measurements can sample multiple downstream responses to the perturbation all of which are real responses but are not necessarily unique. Transcriptional profiling would be one such case - we can estimate thousands of measurements in response to a perturbation. Most expression changes are not likely to be significant. When considering only the significant changes, the activity of several genes is likely to change in concert introducing redundancy in categorizing the unique responses to the perturbation. Second, estimation of temporal redundancy – certain assays are more amenable to higher temporal resolution. Estimation of intracellular small second messengers such as Ca^{2+} would be one such case.

The calcium response to a perturbation is typically a time-series measured at a resolution of seconds. For example, measurement of Ca^{2+} in response to application of PAF in Figure 1 is a time-series measured every three seconds for 600 seconds. Note how the response is characterized by a large, rapid and transient increase followed by a plateau phase of reduced magnitude. Depending upon the nature of the cell system and the ligand, this can result in phenomena such as extended elevated maintenance of intracellular Ca^{2+} (a sustained plateau phase), oscillation, etc. If a sustained elevated phase is maintained or an oscillatory behavior is observed, should the time-series be measured until the response to the perturbation disappears? If so, should the Ca^{2+} responses to other ligands also be measured to the same temporal resolution for consistency? Consider what the latter would mean for analysis such as hierarchical clustering – the results of similarity of ligand signatures will be skewed such that all ligands that do not evoke a long sustained Ca^{2+} tail will appear to be similar by virtue of not producing a response. Thus the sampling interval artificially injects biological interpretation into the analysis – an occurrence that should be eschewed. Historical analysis of calcium transients have attempted to circumvent this problem by objective parameterization of calcium traces. Figure 3A shows some examples – typical parameters estimated include the magnitude of peak response, plateau amplitude, time to peak, decay time-estimation, etc. While these have worked remarkably well for the specific field of Ca^{2+} transient analysis, they do not translate well to other assays. Furthermore, in cases where ligands introduce novel responses, parameterization results in the loss of information. For systematic analysis, it is preferable to eliminate redundancy through unsupervised methods.

We parameterized the matrix of Ca^{2+} transients by identifying patterns of self-similarity along the time axis using unsupervised methods (Natarajan et al., 2006). In brief, the entire calcium matrix (mean Z-scores for all 22 ligands for 600 seconds) was hierarchically clustered (distance metric: city-block, linkage: complete) by column (time-axis) to identify regions of the calcium time series that are naturally correlated across all the data. Each data vector at every time-point was treated as an independent sample and compared across other time-points for self-similarity. Figure 3B shows the matrix of Z-scores for calcium responses to all the ligands in the AfCS dataset. The responses are shown on a colorimetric scale ranging from -10σ (blue) to +10σ (red) with insignificant values within $\pm 1\sigma$ colored white. The row of Z-scores corresponding to the responses to PAF is labeled accordingly and can be compared to Figure 1A. Other ligands also elicited significant responses, but the discussion of the biology associated with these responses are outside the scope of this paper. The results of the clustering of the entire calcium matrix identify regions of the calcium time series that are naturally correlated across all the data and therefore might be reasonably averaged to provide a reduced parameter set. Four prominent clusters were identified (Figure 3C, clusters 1-4) by this method using a threshold for an inconsistency coefficient, which provides an optimal cluster distribution.

If the dendrogram in Figure 3C were to be segmented further, this analysis would only decompose the smaller clusters (clusters 1,2) further rather than identifying sub-regions in the larger clusters (clusters 3,4). Analysis of the clustered time-series shows that the four clusters mostly correspond to contiguous sections of the calcium profile and identified the initial rise to peak of calcium (Figure 3D, filled squares), the rapid decay of calcium transient (Figure 3D, open squares), the slow decay to steady state (Figure 3D, filled inverted triangles), and the steady state (plateau) region (Figure 3D, filled inverted triangles). For

Figure 3. Reduction of the matrix of calcium transients into blocks of self-similarity. A. Canonical supervised methods of parameterization of calcium transients. Typical characterization includes amplitude parameters like averaged basal and peak amplitude, amplitude of the plateau phase, and temporal parameters like time to peak and decay time. B. The matrix of all calcium transients in the AfCS RAW 264.7 dataset. Mean Z-scores for calcium responses to all ligand stimuli are depicted colorimetrically using the color scale shown (red: +10σ, blue:-10σ). The trace corresponding to PAF is indicated for comparison with Figure 1. C. Hierarchical clustering of the matrix in B along the time axis is shown. The dendrogram of distances is superimposed above the clustered matrix. The inconsistency coefficient providing maximum separation (dotted line) results in the identification of four clusters (labeled 1-4). D. The time-points in each cluster were shown superimposed on an average response to PAF for comparison. Contiguous sections of the time-series were identified in each cluster. E. Averaged responses for the time-series in each cluster are shown colorimetrically. The entire calcium matrix of responses to all ligands across 600 seconds is parameterized into these four columns.

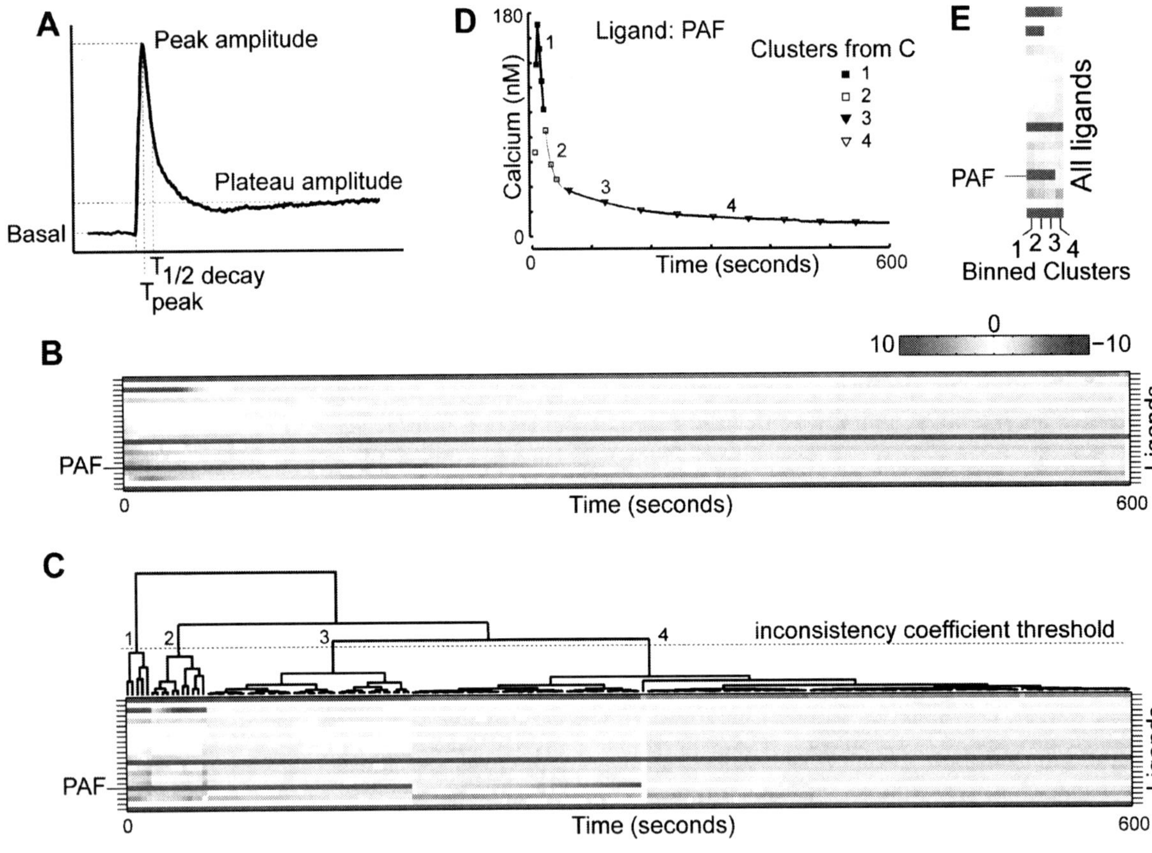

illustrative purposes and for comparison, the time points associated with each cluster in Figure 3C are projected onto a single average response to PAF in Figure 3D. Note that the time-points within each cluster form contiguous blocks along the PAF trace. The information in each block of data could then be averaged into a mean value and associated standard deviation resulting in the representation in Figure 3E.

Note that this method is unsupervised and agnostic to details of the calcium transient. If the response to another ligand is added to this data set such that the evoked calcium transients are sufficiently different from the other traces, the clustering pattern would segment differently and result in a different number of clusters. The correspondence to known mechanistic methods of calcium rise and fall such as rapid rise corresponding to the fast influx of extracellular calcium, and slower calcium-induced calcium release mechanisms corresponding to the plateau region provide confidence that this data-reduction adequately captures essential features of the biology involved in these processes. The methods involved here can clearly be extrapolated to other assays such as measurement of transcription factors. Moreover, note that this method is independent on the nature of temporal sampling – since each data vector at each time-point is treated as an independent variable, there are no temporal dependencies between data vectors. Thus, this method is equally applicable to instances where the samples are measured at aperiodic intervals, e.g., there are multiple early samples to capture transient dynamics, and fewer data-points at later stages which focus more on steady-state levels.

Alternative methods to identify redundant components can just as easily be applied. For instance, Janes and colleagues worked with 7980 molecular signals and 1440 response outputs while examining cytokine-induced apoptosis (Janes et al., 2005). To eliminate non-independent signals within their dataset, they utilized partial least-squares (PLS) regression to simplify the dimensions on the basis of their covariance with a specified dependent variable. While this method is not completely unsupervised, this still allows for a semi-independent decomposition of the dataset into its principal components.

A Unified Experimental Space Representing Signaling Signatures

The ultimate goal in devising methods to integrate multivariate data is to allow analysis and interpretation to derive biological insight. The easiest analysis is one of identification of self-similarity. Hierarchical clustering identifies and groups signatures that mostly resemble each other into clusters. The utility of such approaches is readily apparent – for example, Perlman and colleagues used multi-dimensional profiling of signaling responses to various drugs to create a compendium of signatures. Using the measured signaling responses to blinded drugs, they were able to identify mechanism of action of these drugs (Perlman et al., 2004). It is easy to extrapolate how this may be applicable to drugs of unknown mechanism of action, and in principle consistent with the chemical genetics approaches used to identify drug action through transcriptional profiling (Lamb et al., 2006). In the analysis of the publicly available AfCS datasets we have used so far, the compilation of the responses to individual applications of ligand are shown in Figure 4. The matrix shows responses to ligands (rows) for multiple assays (columns). Detailed descriptions of the biology underlying the ligands and the responses they evoke are beyond the scope of this chapter and are discussed elsewhere (Gilman et al., 2002; Natarajan et al., 2006). For purposes of understanding of methodological analysis, the individual responses to four specific ligands across four different assays shown in Figure 1 are highlighted in the unified experimental matrix.

Connectivity: Correlations vs. Causality

In both research studies described, each ligand signature is essentially a snapshot of the cellular response to the specific perturbation of the cell. The advantage with developing an error model of the unstimulated condition is that any significant response observed in the ligand signature has a direct dependence upon ligand stimulation in a causal fashion. At the level of the simple unified experimental matrix, the only possible interpretation between the inputs (ligand stimulation) and outputs (all measured variables) is one of causal-

Figure 4. A unified experimental matrix incorporates multivariate signal transduction data into ligand-evoked signatures. The consolidated matrix of Z-scores of responses to each ligand (rows) along the four types of assays (columns: calcium, cAMP, protein phosphorylation, and cytokine secretion) are shown colorimetrically (color scale as in Figure 3). For purposes of comparison with Figure 1, the ligands PAF, ISO, IFG and LPS are indicated. The corresponding responses for each of those ligands shown in Figure 1 are highlighted using dotted lines. Note that the nature of the matrix allows comparison across rows (ligands) or columns (responses).

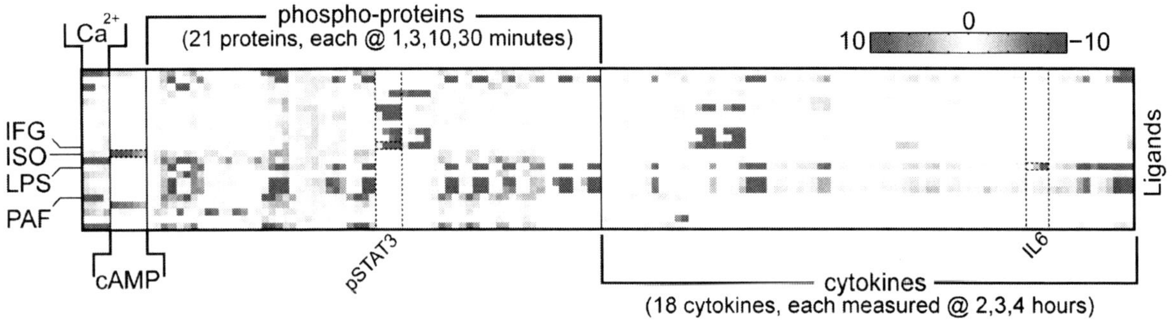

ity. However, the ligand signature includes both fast-acting and long-term events, it includes events proximal to the receptor being stimulated and close to the cell surface, and it includes events further downstream such as within the cellular nucleus and subsequently in the cytoplasm after any translational processing. Thus ligand stimulation may (and likely does) activate multiple downstream signaling pathways – a scenario where there is a clear dependency of evoked responses upon ligand stimulation, but within the set of evoked responses all dependencies are causal at best, uncorrelated at worst. With increased data availability, there is an ever growing need to discriminate between correlated relationships within measurements and causality of relationships across measurements (Pearl, 2000). It is therefore logical to try and determine whether any relationships exist within the ligand signatures – do any components of the early signaling events actuate later signaling events in a causal manner? Since the ligand signatures are parsimonious samples of the activity of the network, early identification of relationships between measurements can greatly help in understanding the underlying network.

To trace the path of signal initiation, transduction and propagation through the network of intermediaries, additional experimentation is necessary. Perturbations using multiple individual and combinatorial stimuli can establish whether the correlations are reproducibly carried over to other stimulation paradigms, and can eventually be verified to be indeed causal. One such initial set of experiments to explore the correlations of signaling intermediaries is the comparison of two individual ligand responses with the combined perturbation of the two ligands. This simple experiment has been used in biochemistry to identify non-additivity in biochemical interactions for decades. Here we simply use this to ask the question whether two ligands are independent of each other. If they are, the responses evoked in combination must be additive (simple addition), while non-additive responses imply interactions between the signaling networks mediating the responses. The publicly available AFCS double ligand screen dataset allows one to systematically test for non-additivity of interactions as shown in Figure 5. This examination of relationships can be carried out at two levels: first, this identification of synergy or anergy of ligand combinations can be examined as input-output relationships; and secondly, the examination can be extended to include intermediate signaling elements. Identification of non-linear responses that are transduced through the signaling network (i.e., observed both in "early" intermediary nodes, as well as in "later" output nodes) form the first line of evidence when examining whether observed correlations between signaling elements are indeed causal. Each correlation observed forms a testable hypothesis of connectivity within the signaling network and can be corroborated from prior studies in literature or through experimental verification.

Non-additive input-output relationships: An interesting finding is the commonality of interactions between activation of similar but not identical receptor classes. For example in cluster 1, activation of one of a class of TLR receptors by ligands for TLRs [LPS, PAM2CSK4 (P2C), PAM3CSK4 (P3C) and resiquimod (848)] in conjunction with interferons [interferon-β (IFB) and interferon-γ (IFG)] yields remarkably similar patterns of cytokine expression. This similarity of signaling patterns allows us to map these input-output relationships (indicated by the diagram to the right of cluster 1 in Figure 5) indicating how pathways downstream of TLRs and interferons can interact. Note that these pathways are devoid of mechanistic detail and merely provide a representation of cause and effect relationships and suggest interactions between signaling pathways downstream of TLR-receptors and interferon-mediated activation of signal transduction. Interestingly, this corresponds with a known role for TLR-induced autocrine adaptors implicated in effective innate responses, especially in viral signaling (Hertzog, O'Neill, & Hamilton, 2003).

Extending non-additive relationships to include intermediate elements: Now consider the relationships seen in cluster 2. Like the previous example, cluster 2 indicates a commonality of interactions between activation of similar but not identical receptor classes. In this case, similar patterns of cytokine production are observed for all combinations of ligands where one ligand of the pair activates TLRs and the second ligand is either ISO or prostaglandin E2 (PGE). Note that both ISO and PGE are known to activate Gαs and elicit increases in intracellular levels of the second-messenger cAMP. The cAMP response is triggered within seconds to minutes of ligand stimulation (Figures 1 - left bottom panel, and Figure 4), while the non-additivity is observed in secretion of cytokines that are measured hours after the initial stimulus. Thus, we can now introduce a test whether the early immediate cAMP signaling response evoked by ISO and PGE exert an influence on TLR-mediated secretion of cytokines hours later. This is a direct attempt to relate measurements between intermediary nodes within the experimental matrix, and serves as a test of causality. We thus hypothesized that cytokine secretion induced by activation of TLRs is altered by ISO and PGE via intracellular increases in cAMP. This inference from two independent data observations was experimentally verified. Replacing ISO and PGE with 8-Bromo-cAMP, a non-hydrolyzable analog of cAMP in combination with stimulation of TLRs elicited similar responses (Hsueh et al., 2009; Natarajan et al., 2006).

By extension, data from clusters 3-5 provide more instances of how disparate signaling architectures underlying different receptors may interact in altering the secretion of these cytokines. Every significant observation here indicates an interaction between signal transduction networks which either corroborates an observed phenomenon in literature [examples in (Hsueh et al., 2009; Natarajan et al., 2006)], or represents a novel signal transduction interaction which can be used to frame a hypothesis around mechanism of action.

FUTURE RESEARCH DIRECTIONS

Here we have shown how integration of multivariate data into composite experimental matrices generates biological insight into the cellular

Figure 5. The extent of non-additivity in the production of six cytokines (G-CSF, IL10, IL6, MIP1α, RANTES, TNFa) for all possible two-way combinations of ligands expressed colorimetrically (red: significantly synergistic and non-additive, blue: significantly anergic and non-additive, white: additive). The matrix of responses is clustered along rows (ligands) to identify which combinations of ligands evoked similar non-additive responses. Select clusters of ligand combinations are highlighted (1-5) along the length of the dendrogram. Schematic representations of the input-output relationships inferred from the data within each cluster are drawn to the right of each cluster. Ligand abbreviations that are not described previously are elaborated in the Abbreviations section.

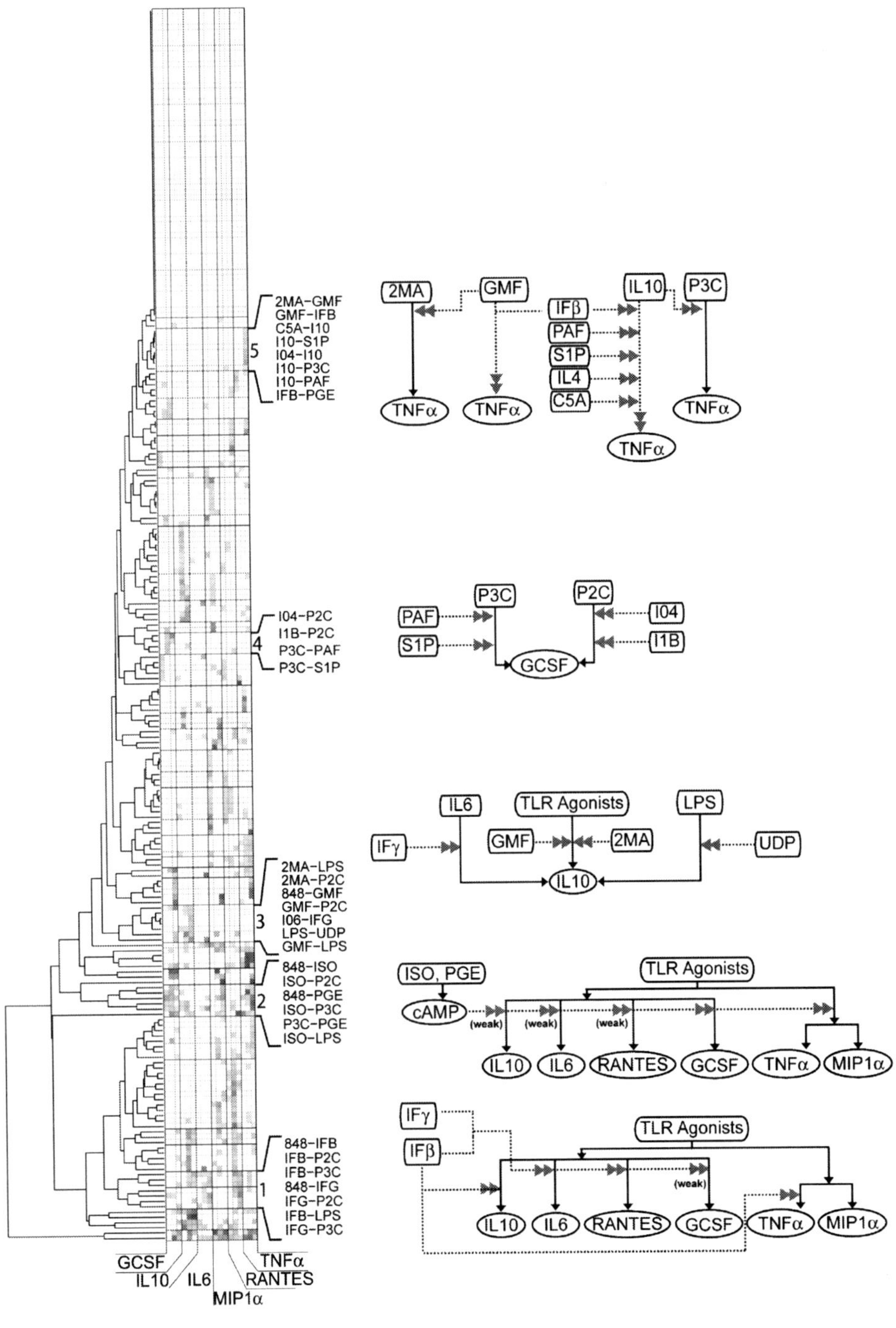

architectures generating these responses. The generation of experimentally testable hypotheses greatly facilitates expansion of the network. It is clear that these studies described here, while novel and exciting, are still early forays and do not provide comprehensive dissections of cellular signaling architectures. That said, simply expanding the numbers of perturbations to identify more cellular interactions is not a trivial task. Each perturbation should be factored in the context of all possible combinations of ligands – a task that gets harder exponentially and will impose pragmatic considerations such as decreases in the numbers of assays (Hsueh et al., 2009). Similarly, the logical expansion of more measurements using high-throughput methodology increases the problem of signal to noise ratio in identifying relationships between measurements. Clearly, the initial analysis methods, while rudimentary, do show that we can successfully map cellular interactions with these datasets. However, for revolutionary advancement in this field, novel analysis methods must attempt to simultaneously analyze multiple perturbations. Investigators have successfully adopted game theory methods to perform rigorous and quantitative multi-perturbation analysis of gene knockout and neuronal ablation experiments (Kaufman et al., 2005). It remains to be seen how well these methods can be extrapolated to the cell signaling context.

CONCLUSION

In summary, this chapter describes research efforts by different groups in examining multivariate signaling responses to specific cellular perturbations. While the nature of the question addressed is similar, differences in experimental systems and computational methodologies allows detailed examination of the advantages of the respective methods employed. Illustrative analysis using data from such a publicly available dataset permits detailed exposition of examples of the methodology adopted.

ACKNOWLEDGMENT

The author thanks Drs. Rama Ranganathan, Paul Sternweis, Elliot Ross, Alfred Gilman, Ronald Taussig (University of Texas Southwestern Medical Center, Dallas, TX), Dr. Mclvin Simon (California Institute of Technology, Pasadena, CA) and Dr. Henry Bourne (University of California, San Francisco, CA) for their invaluable guidance, mentorship and collaboration during the analysis of this dataset. The principal investigators, scientists and technicians of the AfCS deserve all the credit for compiling such a high quality dataset.

REFERENCES

Cosgrove, B. D., King, B. M., Hasan, M. A., Alexopoulos, L. G., Farazi, P. A., & Hendriks, B. S. (2009). Synergistic drug-cytokine induction of hepatocellular death as an in vitro approach for the study of inflammation-associated idiosyncratic drug hepatotoxicity. *Toxicology and Applied Pharmacology*, *237*(3), 317–330. doi:10.1016/j.taap.2009.04.002

Flyvbjerg, H., Jobs, E., & Leibler, S. (1996). Kinetics of self-assembling microtubules: An inverse problem in biochemistry. *Proceedings of the National Academy of Sciences of the United States of America*, *93*(12), 5975–5979. doi:10.1073/pnas.93.12.5975

Friedman, N. (2004). Inferring cellular networks using probabilistic graphical models. *Science*, *303*(5659), 799–805. doi:10.1126/science.1094068

Friedman, N., Linial, M., Nachman, I., & Pe'er, D. (2000). Using Bayesian networks to analyze expression data. *Journal of Computational Biology*, *7*(3-4), 601–620. doi:10.1089/106652700750050961

Gardiner, W. P., & Gettinby, G. (1998). *Experimental design techniques in statistical practice: A practical software-based approach*. Chichester, W. Sussex, UK: Horwood Pub.

Gilman, A. G., Simon, M. I., Bourne, H. R., Harris, B. A., Long, R., & Ross, E. M. (2002). Overview of the Alliance for Cellular Signaling. *Nature*, *420*(6916), 703–706. doi:10.1038/nature01304

Hendriks, B. S., Cook, J., Burke, J. M., Beusmans, J. M., Lauffenburger, D. A., & de Graaf, D. (2006). Computational modelling of ErbB family phosphorylation dynamics in response to transforming growth factor alpha and heregulin indicates spatial compartmentation of phosphatase activity. *Systems Biology*, *153*(1), 22–33. doi:10.1049/ip-syb:20050057

Hertzog, P. J., O'Neill, L. A., & Hamilton, J. A. (2003). The interferon in TLR signaling: More than just antiviral. *Trends in Immunology*, *24*(10), 534–539. doi:10.1016/j.it.2003.08.006

Hsueh, R. C., Natarajan, M., Fraser, I., Pond, B., Liu, J., & Mumby, S. (2009). Deciphering signaling outcomes from a system of complex networks. *Science Signaling*, *2*(71), ra22. doi:10.1126/scisignal.2000054

Janes, K. A., Albeck, J. G., Gaudet, S., Sorger, P. K., Lauffenburger, D. A., & Yaffe, M. B. (2005). A systems model of signaling identifies a molecular basis set for cytokine-induced apoptosis. *Science*, *310*(5754), 1646–1653. doi:10.1126/science.1116598

Janes, K. A., Kelly, J. R., Gaudet, S., Albeck, J. G., Sorger, P. K., & Lauffenburger, D. A. (2004). Cue-signal-response analysis of TNF-induced apoptosis by partial least squares regression of dynamic multivariate data. *Journal of Computational Biology*, *11*(4), 544–561. doi:10.1089/cmb.2004.11.544

Janes, K. A., & Lauffenburger, D. A. (2006). A biological approach to computational models of proteomic networks. *Current Opinion in Chemical Biology*, *10*(1), 73–80. doi:10.1016/j.cbpa.2005.12.016

Kalir, S., & Alon, U. (2004). Using a quantitative blueprint to reprogram the dynamics of the flagella gene network. *Cell*, *117*(6), 713–720. doi:10.1016/j.cell.2004.05.010

Kaufman, A., Keinan, A., Meilijson, I., Kupiec, M., & Ruppin, E. (2005). Quantitative analysis of genetic and neuronal multi-perturbation experiments. *PLoS Computational Biology*, *1*(6), e64. doi:10.1371/journal.pcbi.0010064

Kim, S. Y., Imoto, S., & Miyano, S. (2003). Inferring gene networks from time series microarray data using dynamic Bayesian networks. *Briefings in Bioinformatics*, *4*(3), 228–235. doi:10.1093/bib/4.3.228

Lamb, J., Crawford, E. D., Peck, D., Modell, J. W., Blat, I. C., & Wrobel, M. J. (2006). The connectivity map: Using gene-expression signatures to connect small molecules, genes, and disease. *Science*, *313*(5795), 1929–1935. doi:10.1126/science.1132939

Ljung, L. (1999). *System identification: Theory for the user* (2nd ed.). Paramus, NJ: Prentice Hall.

Nachman, I., Regev, A., & Friedman, N. (2004). Inferring quantitative models of regulatory networks from expression data. *Bioinformatics (Oxford, England)*, *20*(Suppl 1), i248–i256. doi:10.1093/bioinformatics/bth941

Nariai, N., Kim, S., Imoto, S., & Miyano, S. (2004). Using protein-protein interactions for refining gene networks estimated from microarray data by Bayesian networks. *Pacific Symposium on Biocomputing*, 336-347.

Natarajan, M., Lin, K. M., Hsueh, R. C., Sternweis, P. C., & Ranganathan, R. (2006). A global analysis of cross-talk in a mammalian cellular signalling network. *Nature Cell Biology*, *8*(6), 571–580. doi:10.1038/ncb1418

Niehrs, C., & Pollet, N. (1999). Synexpression groups in eukaryotes. *Nature*, *402*(6761), 483–487. doi:10.1038/990025

Palsson, B. O., Price, N. D., & Papin, J. A. (2003). Development of network-based pathway definitions: The need to analyze real metabolic networks. *Trends in Biotechnology*, *21*(5), 195–198. doi:10.1016/S0167-7799(03)00080-5

Patil, A., & Nakamura, H. (2005). Filtering high-throughput protein-protein interaction data using a combination of genomic features. *BMC Bioinformatics*, *6*, 100. doi:10.1186/1471-2105-6-100

Pearl, J. (2000). *Causality: Models, reasoning, and inference*. Cambridge, UK: Cambridge University Press.

Perlman, Z. E., Slack, M. D., Feng, Y., Mitchison, T. J., Wu, L. F., & Altschuler, S. J. (2004). Multidimensional drug profiling by automated microscopy. *Science*, *306*(5699), 1194–1198. doi:10.1126/science.1100709

Perrin, B. E., Ralaivola, L., Mazurie, A., Bottani, S., Mallet, J., & d'Alche-Buc, F. (2003). Gene networks inference using dynamic Bayesian networks. *Bioinformatics (Oxford, England)*, *19*(Suppl 2), ii138–ii148. doi:10.1093/bioinformatics/btg1071

Ronen, M., Rosenberg, R., Shraiman, B. I., & Alon, U. (2002). Assigning numbers to the arrows: Parameterizing a gene regulation network by using accurate expression kinetics. *Proceedings of the National Academy of Sciences of the United States of America*, *99*(16), 10555–10560. doi:10.1073/pnas.152046799

Tegner, J., Yeung, M. K., Hasty, J., & Collins, J. J. (2003). Reverse engineering gene networks: Integrating genetic perturbations with dynamical modeling. *Proceedings of the National Academy of Sciences of the United States of America*, *100*(10), 5944–5949. doi:10.1073/pnas.0933416100

Yeung, M. K., Tegner, J., & Collins, J. J. (2002). Reverse engineering gene networks using singular value decomposition and robust regression. *Proceedings of the National Academy of Sciences of the United States of America*, *99*(9), 6163–6168. doi:10.1073/pnas.092576199

ADDITIONAL READING

Alliance for Cellular Signaling (AfCS)

Analysis of cellular states, signal transduction, and response to perturbation using automated microscopy methods.

Gilman, A. G., Simon, M. I., Bourne, H. R., Harris, B. A., Long, R., & Ross, E. M. (2002). Overview of the Alliance for Cellular Signaling. *Nature*, *420*(6916), 703–706. doi:10.1038/nature01304

Hsueh, R. C., Natarajan, M., Fraser, I., Pond, B., Liu, J., & Mumby, S. (2009). Deciphering signaling outcomes from a system of complex networks. *Science Signaling*, *2*(71), ra22. doi:10.1126/scisignal.2000054

Loo, L. H., Lin, H. J., Singh, D. K., Lyons, K. M., Altschuler, S. J., & Wu, L. F. (2009). Heterogeneity in the physiological states and pharmacological responses of differentiating 3T3-L1 preadipocytes. *The Journal of Cell Biology*, *187*(3), 375–384. doi:10.1083/jcb.200904140

Loo, L. H., Lin, H. J., Steininger, R. J. III, Wang, Y., Wu, L. F., & Altschuler, S. J. (2009). An approach for extensibly profiling the molecular states of cellular subpopulations. *Nature Methods*, *6*(10), 759–765. doi:10.1038/nmeth.1375

Loo, L. H., Wu, L. F., & Altschuler, S. J. (2007). Image-based multivariate profiling of drug responses from single cells. *Nature Methods*, *4*(5), 445–453.

Natarajan, M., Lin, K. M., Hsueh, R. C., Sternweis, P. C., & Ranganathan, R. (2006). A global analysis of cross-talk in a mammalian cellular signalling network. *Nature Cell Biology*, *8*(6), 571–580. doi:10.1038/ncb1418

Perlman, Z. E., Slack, M. D., Feng, Y., Mitchison, T. J., Wu, L. F., & Altschuler, S. J. (2004). Multidimensional drug profiling by automated microscopy. *Science*, *306*(5699), 1194–1198. doi:10.1126/science.1100709

Slack, M. D., Martinez, E. D., Wu, L. F., & Altschuler, S. J. (2008). Characterizing heterogeneous cellular responses to perturbations. *Proceedings of the National Academy of Sciences of the United States of America*, *105*(49), 19306–19311. doi:10.1073/pnas.0807038105

Website for the Alliance for Cellular Signaling (AfCS). From http://afcs.org

Analysis of Protein-Phosphorylation and Cytokine Signaling Networks

Albeck, J. G., MacBeath, G., White, F. M., Sorger, P. K., Lauffenburger, D. A., & Gaudet, S. (2006). Collecting and organizing systematic sets of protein data. *Nature Reviews. Molecular Cell Biology*, *7*(11), 803–812. doi:10.1038/nrm2042

Aldridge, B. B., Haller, G., Sorger, P. K., & Lauffenburger, D. A. (2006). Direct Lyapunov exponent analysis enables parametric study of transient signalling governing cell behaviour. *Systems Biology*, *153*(6), 425–432. doi:10.1049/ip-syb:20050065

Alexopoulos, L. G., Saez Rodriguez, J., Cosgrove, B. D., Lauffenburger, D. A., & Sorger, P. K. (2010). Networks inferred from biochemical data reveal profound differences in TLR and inflammatory signaling between normal and transformed hepatocytes. *Molecular & Cellular Proteomics*, *9*, 1849–1865. doi:10.1074/mcp.M110.000406

Cosgrove, B. D., Alexopoulos, L. G., Hang, T. C., Hendriks, B. S., Sorger, P. K., & Griffith, L. G. (2010). Cytokine-associated drug toxicity in human hepatocytes is associated with signaling network dysregulation. *Molecular BioSystems*, *6*, 1195–1206. doi:10.1039/b926287c

Cosgrove, B. D., King, B. M., Hasan, M. A., Alexopoulos, L. G., Farazi, P. A., & Hendriks, B. S. (2009). Synergistic drug-cytokine induction of hepatocellular death as an in vitro approach for the study of inflammation-associated idiosyncratic drug hepatotoxicity. *Toxicology and Applied Pharmacology*, *237*(3), 317–330. doi:10.1016/j.taap.2009.04.002

Gaudet, S., Janes, K. A., Albeck, J. G., Pace, E. A., Lauffenburger, D. A., & Sorger, P. K. (2005). A compendium of signals and responses triggered by prodeath and prosurvival cytokines. *Molecular & Cellular Proteomics*, *4*(10), 1569–1590. doi:10.1074/mcp.M500158-MCP200

Janes, K. A., Albeck, J. G., Gaudet, S., Sorger, P. K., Lauffenburger, D. A., & Yaffe, M. B. (2005). A systems model of signaling identifies a molecular basis set for cytokine-induced apoptosis. *Science*, *310*(5754), 1646–1653. doi:10.1126/science.1116598

Janes, K. A., Albeck, J. G., Peng, L. X., Sorger, P. K., Lauffenburger, D. A., & Yaffe, M. B. (2003). A high-throughput quantitative multiplex kinase assay for monitoring information flow in signaling networks: application to sepsis-apoptosis. *Molecular & Cellular Proteomics*, *2*(7), 463–473.

Janes, K. A., Kelly, J. R., Gaudet, S., Albeck, J. G., Sorger, P. K., & Lauffenburger, D. A. (2004). Cue-signal-response analysis of TNF-induced apoptosis by partial least squares regression of dynamic multivariate data. *Journal of Computational Biology*, *11*(4), 544–561. doi:10.1089/cmb.2004.11.544

Janes, K. A., Reinhardt, H. C., & Yaffe, M. B. (2008). Cytokine-induced signaling networks prioritize dynamic range over signal strength. *Cell*, *135*(2), 343–354. doi:10.1016/j.cell.2008.08.034

Saez-Rodriguez, J., Goldsipe, A., Muhlich, J., Alexopoulos, L. G., Millard, B., & Lauffenburger, D. A. (2008). Flexible informatics for linking experimental data to mathematical models via DataRail. *Bioinformatics (Oxford, England)*, *24*(6), 840–847. doi:10.1093/bioinformatics/btn018

KEY TERMS AND DEFINITIONS

Cytokine: A cytokine is a class of small peptides or proteins secreted by an immune cell that serves to regulate the functioning of the same or other immune cells. They function by acting as ligands to receptors expressed on the same cell (autocrine signaling) or other cells (paracrine signaling). Here, cytokines are secreted from the immune cells upon stimulation by ligand and are thus considered an "end-point" for purposes of this analysis.

Ligand: A ligand is a molecule that binds specifically to a receptor on a cell to activate a signaling pathway. The response triggered is specific and concentration dependent. Here, ligands are used as reagents to perturb cells, and thus permitting subsequent responses to be measured.

Phospho-Protein: A method for transduction of information within proteins by the reversible addition or removal of a phosphate group onto specific amino acids within a protein. Post-translational modifications such as protein phosphorylation permit implementation of a large number of signal transduction and manipulation processes – for example, biological instances of manipulation of gain functions, positive and negative feedback signals, lag-time, analog-to-binary signal translation etc can be observed using simple protein phosphorylation and de-phosphorylation.

Receptor: A receptor is a protein on the cellular membrane that allows the transmission of a signal from outside the cell to inside the cell. Receptors are typically activated by their own cognate ligands.

Second-Messenger: A molecule that transduces signals from receptors on the cell surface to other intra-cellular molecules. Typically, second-messengers can amplify signals, thus acting analogous to a gain function in electronic circuits. Proximity to the receptors typically implies that these signals are activated early in signal transduction and are thus likely to influence the behavior of other intra-cellular molecules that are subsequently activated.

APPENDIX

Abbreviations for Ligands Used In the AfCS Datasets (With Alternate Names)

2MA	2-Methylthioadenosine 5'-triphosphate tetrasodium; 2-methyl-thio-ATP
848	Resiquimod (R-848)
C5A	Complement C5a, recombinant human
GMF	Granulocyte-macrophage colony-stimulating factor, recombinant mouse, G-MCSF.
IL-4	Interleukin-4, I04, IL4, recombinant mouse.
IL-6	Interleukin-6, I06, IL6, recombinant mouse.
IL-10	Interleukin 10, I10, IL10, recombinant human.
IL1β	Interleukin-1b, I1B, IL1b, recombinant mouse.
IFα	Interferon alpha, IFA, IFNa, recombinant mouse.
IFβ	Interferon beta, IFB, IFNb, recombinant mouse.
IFγ	Interferon gamma, IFG, IFNg, recombinant mouse.
ISO	Isoproterenol; isoprenaline, isopropylnoradrenaline, isoproptcrenol hydrochloride.
LPA	Lysophosphatidic acid, 1-oleoyl-2-hydroxy-sn-glycero-3-phosphate.
LPS	Lipopolysaccharide; with added LPS-binding protein (LBP).
MCF	Macrophage colony-stimulating factor, recombinant mouse; M-CSF.
P2C	Pam2Cys-SKKKK x 3 TFA; S-[2,3-bis(palmitoyloxy)-(2RS)-propyl]-[R]-cysteinyl-[S]-seryl-[S]-lysyl-[S]-lysyl-[S]-lysyl-[S]-lysine x 3 CF3COOH, PAM 2.
P3C	Pam3Cys-SKKKK x 3 HCl; (N-Palmitoyl-S-[2,3-bis(palmitoyloxy)-(2RS)-propyl]-[R]-cysteinyl-[S]-seryl-[S]-lysyl-[S]-lysyl-[S]-lysyl-[S]-lysine x 3 HCl), PAM 3.
PAF	Platelet activating factor, L-alpha-phosphatidylcholine, beta-acetyl-gamma-O-alkyl.
PGE	Prostaglandin E2, (5Z,11alpha,13E,15S)-11,15-Dihydroxy-9-oxoprosta-5,13-dienoic acid.
S1P	Sphingosine-1-phosphate.
TGFβ	Transforming growth factor-beta 1, recombinant human, Chinese hamster ovary cells; TGF- β1; TGF; TGFb.
UDP	Uridine 5'-diphosphate trisodium salt dehydrate.

Chapter 16
Complexity and Modularity of MAPK Signaling Networks

George V. Popescu
University Politehnica Bucharest, Romania

Sorina C. Popescu
Boyce Thompson Institute for Plant Research, USA

ABSTRACT

Signaling through mitogen-activated protein kinase (MAPK) cascades is a conserved and fundamental process in all eukaryotes. This chapter reviews recent progress made in the identification of components of MAPK signaling networks using novel large scale experimental methods. It also presents recent landmarks in the computational modeling and simulation of the dynamics of MAPK signaling modules. The in vitro MAPK signaling network reconstructed from predicted phosphorylation events is dense, supporting the hypothesis of a combinatorial control of transcription through selective phosphorylation of sets of transcription factors. Despite the fact that additional co-factors and scaffold proteins may regulate the dynamics of signal transduction in vivo, the complexity of MAPK signaling networks supports a new model that departs significantly from that of the classical definition of a MAPK cascade.

INTRODUCTION

Mitogen-activated protein kinase (MAPK) cascades are components of intracellular signaling activated in response to a wide array of external and internal signals, contributing towards development of diverse cellular responses such as growth, differentiation, response to pathogens, and cell death. A MAPK cascade contains several key components: a MAPK kinase kinase (MAP3K), a MAPK kinase (MAP2K) and a MAPK. MAP3Ks, activated by upstream kinases or receptor-associated molecules, activate in turn the MAP2Ks. Activated MAP2Ks phosphorylate the MAPK components which, in turn, phosphorylate diverse substrates such as transcription and translation factors, protein kinases and phosphatases, thus regulating many cellular processes in response

DOI: 10.4018/978-1-60960-491-2.ch016

to the initial stimulus (Chen & Thorner, 2007; Zhang et al., 2006).

In plants, the diversity of processes regulated through MAPK signaling cascades combined with the large number of predicted members within MAPK families implies a robust and well synchronized control of kinase signaling. In support of this hypothesis, recent large-scale screens to identify signaling proteins in eukaryotic organisms challenged the classical view of signaling cascades as simple, linear conduits composed of a handful of elements. Current models reflect the apparent complexity of signaling pathways, the cross-talk between parallel pathways, and the dynamic nature of the protein interactions (Friedman & Perrimon, 2006, 2007; Mackay, 2004; Popescu et al., 2007). A recent study (Popescu, Popescu, & Bachan et al., 2009) re-constituted a MAPK phosphorylation network based on experimental data generated using high-density protein microarrays. The study found that MAPKs are able to phosphorylate *in vitro* a large diversity of transcription factors with known or predicted roles in disease resistance, flower development, cellular differentiation and auxin signaling. The analysis of the re-constituted signaling network supported the hypothesis of a combinatorial control of cellular processes through MAP2Ks and MAPKs cascades.

In this chapter, we present a systems view of the molecular interactions of the MAPK proteins and review current results on the architecture and dynamics of MAPK signaling networks.

BACKGROUND

Signaling Components Represent a Major Part of Higher Eukaryotic Genomes

Identification of signaling networks is a central research topic in the systematic analysis of cellular organization. The employed methods range from direct, low throughput screening of kinase phosphorylation targets to indirect, large scale predictions of protein-protein interactions from protein microarrays data, protein similarity search, analysis of protein domains and gene coexpression. Signaling through MAPK cascades is a fundamental and conserved process in eukaryotes. The MAPK signaling network has a hierarchical structure (Figure 1) composed of at least three levels of nodes: (1) the MAP3K proteins which, when activated directly or indirectly by receptor molecules, phosphorylate MAP2K proteins; (2) the activated MAP2Ks which activate and phosphorylate MAPK proteins; (3) a multitude of cytoplasmic and nuclear substrates acted upon by activated MAPK; in addition, a MAP4K may activate the MAP3K component.

A prototypical multicellular model organism useful for the study of the architecture of MAPK signaling networks is *Arabidopsis thaliana*. Direct functional assays and analysis of sequence conservation with other eukaryotic organisms identified 128 *Arabidopsis* genes, putative members of four MAPKs groups (MAP4Ks, MAP3Ks, MAP2Ks, and MAPKs) (Champion et al., 2004; Ichimura et al., 2002). The large size of the *Arabidopsis* MAPK gene complement is comparable only with the mammalian one (Uhlik et al., 2004). Out of this large group of genes, only a handful have been verified functionally and proven to have roles in signal transduction within MAPK cascades or as adaptors (scaffolds) that bring together members of a MAPK module. The 10 *Arabidopsis* predicted MAP4K genes were classified in several groups such as PAK-like (p21 Ras-activated protein kinase), MST-like (mammalian sterile 20-like) and SOC-like (STE20 oxidant stress kinase), based on similarity with the yeast and mammalian counterparts (Champion et al., 2004). It is possible that some MAP4Ks function as adaptor proteins rather than enzymatic components of the MAPK signaling modules. The MAP3Ks constitute a heterogeneous family including 20 plant MEKKs ("true" MAP3Ks) and 58 plant MEKK-related genes (48

Figure 1. The MAPK signaling cascade. Plant MAPK cascade showing the key elements of a pathway: MAP4K, MAP3K, MAP2K, MAPK and the substrates. Scaffolds and phosphatases act as regulators. External or intracellular signals activate the MAPK pathway. MAPK activation induces a cascade of cellular events that changes gene transcription patterns and generates an appropriate biological response. The scaffold proteins may bind two or more elements of the MAPK module to confer signaling specificity and to localize it to specific cellular compartments. Phosphatases are key regulators of kinase activities at various levels within the MAPK cascade. Substrates may also act as specificity determinants. Tissue-specific substrate expression patterns regulated by development and/or interactions with environmental factors are instrumental in generating specific cellular outputs following MAPK activation. The putative substrates shown are taken from protein microarray data analysis. Arrows represent the direction of the phosphorylation. Dotted lines indicate reactions that are not part of the classical MAPK cascade model (such as MAP3K activation by GTPases, MAP2K trans-phosphorylation). The P symbols represent phosphorylation events.

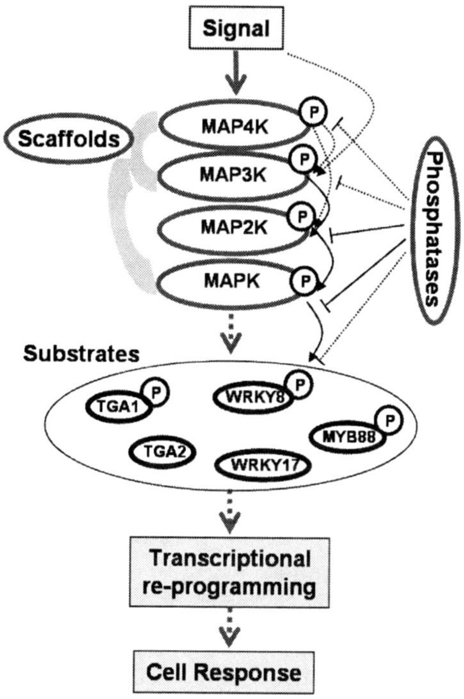

Raf-like and 10 ZIKs) (Champion et al., 2004). Some MAP3Ks may function as adaptor proteins, as well. For example, an alfalfa MAP3K, OMTK1, was suggested to function as a scaffold (Nakagami et al., 2004). A similar number of kinases, a total of 21, were shown to function as MEKKs in humans (Symons et al., 2006). Although the roles of several plant MEKKs in MAPK cascades are known (Ichimura et al., 1998; Krysan et al., 2002; Mizoguchi et al., 1996; Mizoguchi et al., 1998; Suarez-Rodriguez et al., 2007), the functions of the majority of the predicted MEKKs and MEKK-related proteins remain to be determined (Champion et al., 2004).

Specificity and Control of MAPK Pathways

MAPK signaling is carried out by interconnected networks with overlapping roles in controlling various cellular processes such as growth, development, hormones, and response to stress (Brader et al., 2007; Ichimura et al., 2000; Meszaros et al., 2006; Popescu, Popescu, Bachan et al., 2009; Teige et al., 2004). The diversity of processes regulated through MAPK cascades implies a robust and synchronized control of signaling pathways. The set of MAPK cascades activated by specific stimuli may be restricted by developmental stage, cellular compartmentalization, scaffolds and the action of other regulatory elements. It has been hypothesized that MAPK regulators participate not only in pathway isolation/specificity but also in enhancing the efficiency of signal transduction through signal amplification. Based on work in yeast and animal systems, MAPK signaling is regulated by proteins which fall into two groups: (1) activators (e.g. MAP2Ks) and negative regulators of MAPK phosphorylation (e.g. phosphatases), with enzymatic functions, and (2) scaffolds, proteins without a catalytic function towards MAPKs. Scaffolds provide spatial and temporal regulation of MAPK pathways by tethering signaling components to specific cellular compartments, organizing MAPKs in functional modules or, possibly, providing intracellular transport (Morrison and Davis, 2003; Yoshioka, 2004).

A fundamental challenge is to understand the rules that govern MAPK complex assembly. Signaling mechanisms are regulated and operate due to mainly post-transcriptional and post-translational processes (Menges et al., 2008). In consequence, inferring signaling networks exclusively from large-scale gene expression data is inherently unreliable. Moreover, it was found that gene expression, with the exception of genes encoding permanent protein complexes (such as the ribosome and the proteasome), does not correlate well with known protein-protein interactions and thus, cannot be used to reliably predict protein complexes (Jansen et al., 2002; Soong et al., 2008). System-level approaches that combine high-throughput experimental data with mathematical and computational modeling are emerging as a comprehensive way to study signaling regulation (Bader et al., 2008; Devos & Russell, 2007; Kolch et al., 2005; Morsy, 2008).

MAPK Networks Identification using Phosphorylation Motifs

Identification of protein phosphorylation motifs is a key aspect of current computational work on MAPK signaling. Several public databases with information on linear protein motifs are available. The *PROSITE* database (Hiscock & Allen, 2008) is a comprehensive resource of motifs classified by their functionality. Other, more specialized databases catalogue motifs with role in protein signaling (Mishra, 2009). A comprehensive resource for eukaryotic protein motifs is the *ELM* database (Puntervoll et al., 2003). A review of eukaryotic linear motifs is presented in (Diella et al., 2008).

Experimental methods are instrumental in protein motif discovery. For example, Sheridan et al. (2008) characterized the DEF-type docking motifs in MAPK proteins using a peptide library. Experiments using positional scanning peptide array confirmed the presence of DEF motifs in ERK2 and also in the p38α protein sequence. A significant finding of this study was that the presence of docking motifs in MAPK substrates dramatically increased the efficiency of the phosphorylation reactions catalyzed by p38α. In a similar study, a kinome profiling of subcellular fractions of Arabidopsis cells was performed using microarrays of peptides corresponding to *in vivo* phosphorylation sites. The study identified distinct kinase activities in the cytosolic and nuclear compartments with specific and distinct requirements for substrate recognition (de la Fuente van Bentem et al., 2008). Another study (Linding et al., 2007) identified *in vivo* phosphorylation networks using a motif-based prediction method. New phosphorylation sites were predicted computationally by mining multiple experimental phosphorylation data sets. The authors reported a large (2.5 fold) improvement in the accuracy of the re-constructed phosphorylation network. This demonstrates that, although defining MAPK signaling networks by exclusively analyzing experimental data may lead to ambiguities due to the low specificity of phosphorylation motifs, some of these ambiguities may be resolved using computational motif analysis.

Dynamics of MAPK Signaling Modules

Understanding MAPK signaling dynamics is a challenging endeavor. Pathway specificity in MAPK-mediated responses has been attributed to an extensive regulatory network. This control consists of positive and negative feedback regulatory loops assembled into complex molecular switches. We review here several key studies on MAPK signaling dynamics in eukaryotes.

Analysis of biochemical networks using stochastic methods has roots in statistical simulation methods (master equations) proposed by Gillespie (1976) and the theoretical work of McQuarrie (1967) on stochastic methods for chemical kinetics. At equilibrium, simulations of a MAPK cascade of reactions under both non-competitive (Michaelis-Menten) and competitive (Hill) binding are used to understand the dynamics of biochemical pathways. The first dynamics models of MAPK signaling cascades that explained the observed bistable responses were proposed by Kholodenko (2000) and Levchenko et al. (2000).

A prototypical model for studying the dynamics of MAPK signaling module is shown in Figure 2. The classical Huang-Ferrell cascade model (Huang and Ferrell, 1996) has been extended recently to consider the role of scaffolds in isolating pathways and ensuring signal specificity.

Scaffold proteins act as platforms on which signaling modules are assembled. They were also shown to localize signaling molecules at specific sites in the cell and coordinate positive and negative feedback through the signaling pathways by enhancing the specificity of kinases or limiting their ability to phosphorylate more than one downstream target (Shaw, 2009). Levchenko et al. (2000) performed simulations of kinase cascades in the presence of scaffold proteins and demonstrated the importance of an optimal scaffold concentration for efficient amplification of the signal. A biphasic model for scaffold-mediated kinase signaling was demonstrated computationally in the same paper. Several recent simulation studies analyzed the regulatory role of scaffold proteins. Locasale et al. (2007) showed, using Monte-Carlo MAPK cascade simulation, that scaffold proteins may play a role in amplifying some signals while suppressing others, thus conferring specificity to the kinase signaling cascades. They hypothesized that a possible

Figure 2. Biochemical model of MAPK cascades; P: phosphorylation; P'PASE: Phosphatase; SCAFFOLD: scaffold proteins

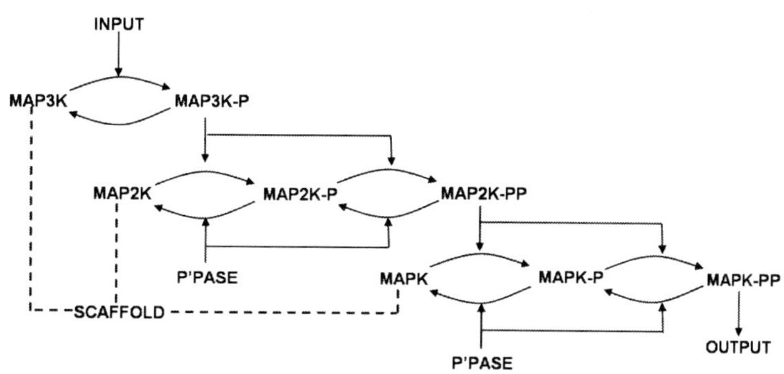

mechanism for regulating the specificity of MAPK signaling pathways consists of controlling the timing and place of scaffold gene expression. A different mechanism, called the kinetic insulation, was proposed by Behar et al. (2007) to achieve pathway specificity in kinase signaling networks. Using Michaelis-Menten kinetics, they showed that signaling cascades are able to separate signals that are transient from signals that increase in time, thus effectively achieving specificity.

In conclusion, the progress resulted from recent biochemical simulations has contributed to a better understanding of the complex dynamics of MAPK signaling modules and has identified new challenges for future research.

MAIN FOCUS OF THE CHAPTER

The Inherent Complexity of MAPK Signaling Networks

Current research sheds light on the complex structure and dynamics of MAPK signaling networks, in which diverse signals activate the same cascade, and multiple, parallel cascades are activated by the same stimulus. Signaling modules, identified through genetic and biochemical methods, have overlapping roles in controlling cell division, development, hormone signaling and synthesis, response to abiotic stress, pathogens and pathogen elicitors. This complexity of signaling pathways is reflected in the current views that describe MAPK signaling as a dense network rather than a set of compartmentalized linear circuits (Friedman and Perrimon, 2007; Kholodenko, 2006; Pedley and Martin, 2005).

Recently the MAP2K/MAPK/Substrates signaling pathways were studied using Arabidopsis protein microarrays (Popescu, Popescu, Bachan et al., 2009). A MAP2K/MAPK/Substrate phosphorylation network including previously known and novel signaling pathways was re-constructed using a microarray containing over 2000 protein preparations (Figure 3). The phosphorylation events in the network were identified by building Bayesian models for the *a priori* information on the hierarchical structure of the MAPK network. The decision method included two dimensions, one addressing the biological process (modeling phosphorylation events) and the other addressing the network structure (modeling the MAPK hierarchy as a causal network). The probability of a MAPK->Substrate phosphorylation event was conditioned on the probability that the upstream kinases were phosphorylated in each of the *in planta* identified MAP2K/MAPK combinations. The causal network used to compute the weights of MAPK->Substrate phosphorylation network was built from multiple observations of upstream MAP2K/MAPK modules. The MAPK-> Substrate phosphorylation score computed as follows:

$$w(i,j) = -\log(P(H(i,j)|\ KK_1(i))) -\log(P(H(i,j)|\ KK_2(i))),\ i=1.10, j=1.2158,$$

$KK_1(i)$, $KK_2(i)$ refer to two *MAP2Ks* used for in planta activation of *MAPK(i)*. The scores were used to test for significantly enhanced pathways observed from protein microarrays data. This method identified 570 MAPK substrates. The predicted substrates spanned a variety of protein families including transcription and translation factors, protein kinases, metabolic enzymes, and protein with unknown function.

A subset of the predicted MAPK substrates was subsequently verified in the plant by monitoring the predicted substrate phosphorylation state in the presence and in the absence of the upstream activating MAP2K/MAPK modules. Table 1 shows representative Gene Ontology (GO) categories enriched among MAPK predicted substrates relative to the proteins present on the microarray, along with calculated p-values. Significantly overexpressed GO categories include transcription factor activity, response to hormones and flower development. This analysis suggested that plant MAPK cascades are important regulators of gene

Figure 3. The MAP2K-> MAPK->Substrates phosphorylation network. K1 to K16 represent the 10 MAPK components studied (MPK1 to MPK8, MPK10 and MPK16), while KK1 to KK10 represent the 10 MAP2K components.

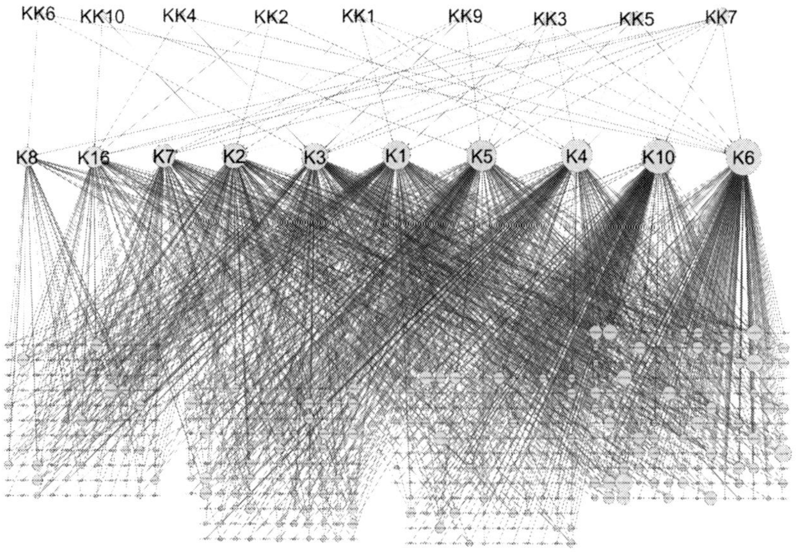

Table 1. Gene Ontology (GO) term enrichment in the MAPK target dataset relative to all proteins printed on the protein microarray. The GO term enrichment was determined using AmiGO (Boyle et al., 2004). The p-values were calculated using the hypergeometric distribution in the GO-TermFinder module; the Bonferroni correction was applied within the AmiGO computation method. The p-value cutoff was 0.01.

GO Term	Aspect	P-value
GO:0005634 nucleus	Cellular Component	4.68E-20
GO:0003700 transcription factor activity	Molecular Function	2.52E-83
GO:0009751 response to salicylic acid stimulus	Biological Process	2.79E-09
GO:0009753 response to jasmonic acid stimulus	Biological Process	5.73E-08
GO:0009723 response to ethylene stimulus	Biological Process	4.33E-06
GO:0009733 response to auxin stimulus	Biological Process	0.0000175
GO:0009908 flower development	Biological Process	0.00186

transcription through the modulation of transcription factor activity levels by phosphorylation.

Cluster analysis (using hierarchical clustering with average linking method from Cluster 2.11 (Eisen et al., 1998)) identified a relation between MAPKs structural similarity and the commonality of their substrates. The analysis distinguished four groups of MAPK proteins according to their substrates. Several MAPK proteins, MPK2 and MPK7 from the structural class C, MPK6 and MPK10 from class A2, and MPK8 and MPK16 from class D1 clustered together, having similar sets of substrates (Figure 4). These results suggest a common mechanism of target recognition for MAPKs in the same structural class, and potentially common recognition and/or phosphorylation motifs in their targets.

Figure 4. Clustering of MAPK phosphorylation targets

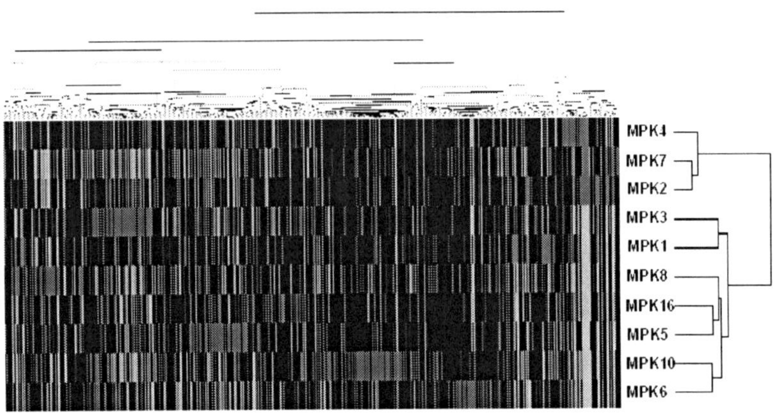

Integrated Analysis of Phosphorylation and Gene Expression Reveals Functional MAPK Pathways

Further analysis of the putative MAPK network identified in (Popescu, Popescu, Snyder et al., 2009) was performed by integrating phosphorylation and gene expression information. The objective of the analysis was to increase the confidence in predicted MAPK substrates by using gene co-expression data from the Bio-Array Resource for Arabidopsis Functional Genomics (BAR) database (Toufighi et al., 2005). Gene co-expression was measured using the Pearson correlation coefficient. A high coefficient is associated with functional interaction, influence, or dependence. Using gene co-expression data extracted from the *Arabidopsis* Interactions Viewer at BAR, the data set from (Popescu, Popescu, Snyder et al., 2009) was filtered for phosphorylation events with correlation coefficients higher than a computed threshold. The threshold was selected as a function of significance level α in a test of correlation between probability of interaction in protein microarray experiments and gene co-expression. Figure 5 shows the analysis of expression co-variation/dependencies in the MAPK network for all phosphorylation events with $r \geq 0.7$. The analysis found 108 phosphorylation events (network edges) that occur among 96 annotations (network nodes) containing gene pairs with highly similar expression profiles.

An interesting finding was the strong co-expression of the putative substrates of MPK8 and MPK6. MPK8 shared almost all targets with MPK6, indicating a possible co-participation in signaling pathways. On the other hand, MPK4, MPK6 and MPK10 network clusters were well separated in terms of target co-expression, indicating that they may function in independent cascades and/or within specific time frames. MPK1, MPK2 and MPK16 clustered close to MPK4, and shared several co-expressed transcription factors and genes with role in development. MPK6, MPK4 and MPK10 shared a strongly co-expressed R2R3 transcription factor. The high confidence MAPK->Substrates phosphorylation events selected in the phosphorylation/co-expression network are the starting point for *in vivo* validation of the identified MAPK network.

Understanding MAPK Networks Dynamics

The dynamics of a large network as shown in Figure 3 is complex. A possible approach to understand its dynamics can be based on the theory of decomposition of complex cellular circuits into monotone systems with input/output (MIOS)

Figure 5. MAPK network with integrated gene co-expression and phosphorylation information. Nodes represent proteins and are shown as circles with size proportional to the number of links. MAPKs are represented by dark grey nodes; MAP2Ks are shown as light grey nodes. Edges represent protein microarray-identified phosphorylation events with r≥0.7; Edge thickness is proportional with r-values.

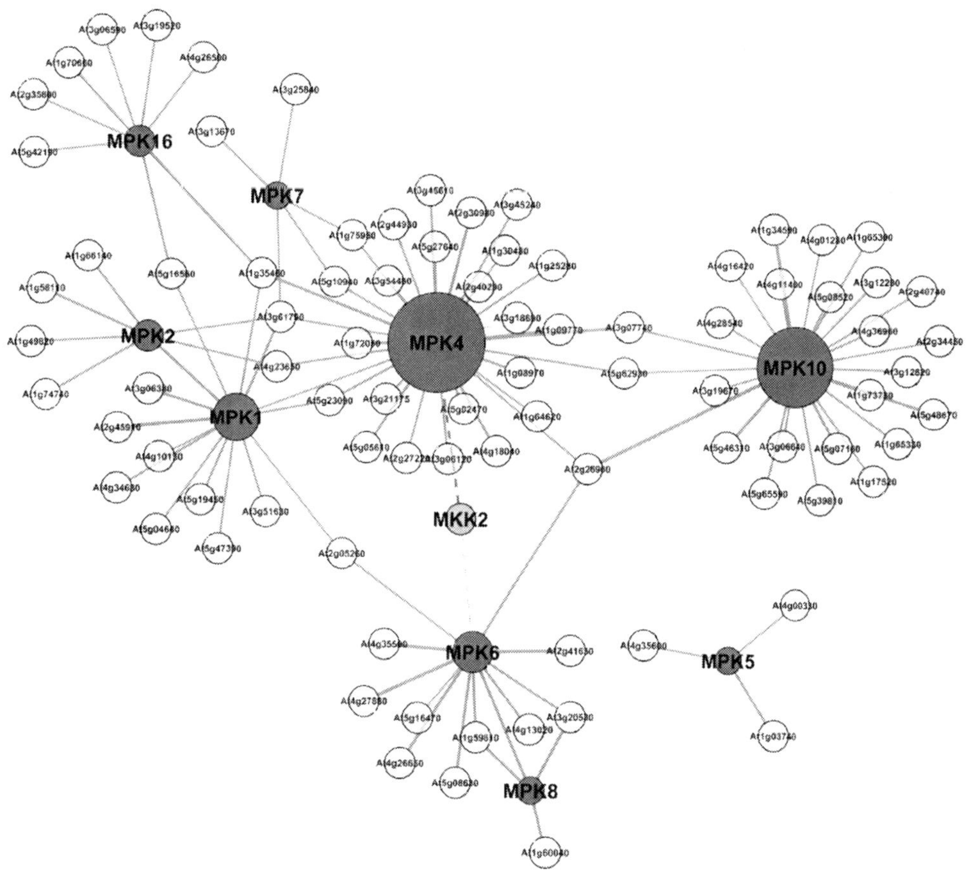

(Sontag, 2007). MIOS are mathematical models with monotone dynamics that can be combined to approximate the dynamics of nonlinear systems. In addition to monotonicity, MIOS have consistency requirements which relate their dynamics to the topology of the network described as a signed graph. Starting with the observation that kinetics of enzymatic reactions at equilibrium can be described with hyperbolic and sigmoidal responses, (Angeli & Sontag, 2003) showed that MAPK modules with nonlinear dynamics can be decomposed into MIOS elements. Combining sigmoidal MIOS, they obtained the bistable dynamics observed for MAPK modules. Applying this decomposition approach to the large scale MAPK network presented above has the added combinatorial complexity of identifying the best MIOS decomposition. In addition to modeling nonlinear network dynamics, the MIOS decomposition can be used for a more robust identification of MAPK network, to identify missing network elements, and to disentangle complex networks into functional modules. In addition to protein microarray observations and *apriori* knowledge of network structure, this will add dynamic behavior constraints on the structure of the MAPK network inferred from experimental data (in order to reject structures with chaotic dynamics).

DISCUSSION

Recent results are changing the view of MAPK signaling networks structure and dynamics. The architecture of the plant MAPK signaling network reconstructed in (Popescu, Popescu, Bachan et al., 2009) is significantly different from the traditional signaling cascade model and supports recent studies in other model organisms. The MAPKs phosphorylated network is denser than expected for a superposition of classic signaling cascades, supporting the hypothesis of a combinatorial control of transcription through selective phosphorylation of a large number of transcription factors. The number of phosphorylation events per kinase predicted in (Popescu, Popescu, Bachan et al., 2009) is consistent with those reported for other eukaryotic organisms. The Arabidopsis signaling network has an average of 157 phosphorylation substrates per MAPK. By comparison, the human ERK1/2 phosphorylates 160 well-characterized substrates, while earlier work estimated that mammalian MAPKs may regulate up to a thousand cellular targets (Avruch, 2007). In yeast, a whole proteome approach identified an average of 47 in vitro substrates per kinase; an exception, Tpk1, recognized a maximum of 256 targets (Ptacek, et al., 2005). It should be noted that *in vivo* specificity determinants, such as protein adaptors, tissue-specific and developmental patterns of gene expression, may significantly reduce the number of the true cellular phosphorylation events among the ones identified using *in vitro* methods.

The models proposed for MAPK cascades are evolving as well. Scaffolds proteins are now being considered key factors to enhance specificity of MAPK modules and insulate them in space (by restricting the number of downstream components they can interact with) and time (by separating transient signals from those that increase in time). Still, the complexity of MAPK networks is not yet addressed in the current signaling dynamics analyses. A possible approach is using graph and system analysis theory as illustrated by the decomposition method proposed by Sontag (2007). Systematic study of MAPK network dynamics is at present a significant challenge for understanding cellular signaling.

CONCLUSION

We have reviewed here recent progress on the identification of components of a plant MAPK signaling network based on high-throughput protein microarray data. A categorical (Gene Ontology) analysis of MAPK substrates provided evidence for the role of MAPK signaling cascades in many fundamental cellular processes: biotic and abiotic stress response, development, and cell cycle regulation. We have also reviewed results on the analysis of MAPK modules dynamics, with the goal of providing a basis for integrating experimental and computational methodologies to study MAPK signaling. Novel high-throughput approaches such as functional protein microarrays give new perspectives on MAPK signaling pathways and provide a framework for future experimental studies. However, the dynamics of these large networks is challenging to study. Under the hypothesis of "near-monotonicity" of biological systems, MAPK networks may be decomposed into modules with monotone input/output system dynamics. A system-level approach that combines high-throughput experimental data, mathematical and computational modeling is currently emerging as a comprehensive way to study MAPK signaling networks.

REFERENCES

Angeli, D., & Sontag, E. D. (2003). Monotone control systems. *IEEE Transactions on Automatic Control*, *48*(10), 1684–1698. doi:10.1109/TAC.2003.817920

Avruch, J. (2007). MAP kinase pathways: The first twenty years. *Biochimica et Biophysica Acta*, *1773*(8), 1150–1160. doi:10.1016/j.bbamcr.2006.11.006

Bader, S., Kühner, S., & Gavin, A.-C. (2008). Interaction networks for systems biology. *FEBS Letters*, *582*(8), 1220–1224. doi:10.1016/j.febslet.2008.02.015

Behar, M., Dohlman, H. G., & Elston, T. C. (2007). Kinetic insulation as an effective mechanism for achieving pathway specificity in intracellular signaling networks. *Proceedings of the National Academy of Sciences of the United States of America*, *104*(41), 16146–16151. doi:10.1073/pnas.0703894104

Boyle, E. I., Shuai, W., Jeremy, G., Heng, J., David, B., & Michael, C. J. (2004). GO: TermFinder-open source software for accessing Gene Ontology information and finding significantly enriched Gene Ontology terms associated with a list of genes. *Bioinformatics (Oxford, England)*, *20*(18), 3710–3715. doi:10.1093/bioinformatics/bth456

Brader, G., Djamei, A., Teige, M., Palva, E. T., & Hirt, H. (2007). The MAP Kinase Kinase MKK2 affects disease resistance in Arabidopsis. *Molecular Plant-Microbe Interactions*, *20*(5), 589–596. doi:10.1094/MPMI-20-5-0589

Champion, A., Picaud, A., & Henry, Y. (2004). Reassessing the MAP3K and MAP4K relationships. *Trends in Plant Science*, *9*(3), 123–129. doi:10.1016/j.tplants.2004.01.005

Chen, R. E., & Thorner, J. (2007). Function and regulation in MAPK signaling pathways: Lessons learned from the yeast Saccharomyces cerevisiae. *Biochimica et Biophysica Acta (BBA)-. Molecular Cell Research*, *1773*(8), 1311–1340.

de la Fuente van Bentem, S., Anrather, D., Dohnal, I., Roitinger, E., Csaszar, E., & Joore, J. (2008). Site-specific phosphorylation profiling of Arabidopsis proteins by mass spectrometry and peptide chip analysis. *Journal of Proteome Research*, *7*(6), 2458–2470. doi:10.1021/pr8000173

Devos, D., & Russell, R. B. (2007). A more complete, complex, and structured interactome. *Current Opinion in Structural Biology*, *17*(3), 370–377. doi:10.1016/j.sbi.2007.05.011

Diella, F., Niall, H., Chica, C., Budd, A., Michael, S., & Brown, N. P. (2008). Understanding eukaryotic linear motifs and their role in cell signaling and regulation. *Frontiers in Bioscience*, *13*, 6580–6603. doi:10.2741/3175

Eisen, M. B., Spellman, P. T., Brown, P. O., & Botstein, D. (1998). Cluster analysis and display of genome-wide expression patterns. *Proceedings of the National Academy of Sciences of the United States of America*, *95*(25), 14863–14868. doi:10.1073/pnas.95.25.14863

Friedman, A., & Perrimon, N. (2006). High-throughput approaches to dissecting MAPK signaling pathways. *Nature*, *40*(3), 262–271.

Friedman, A., & Perrimon, N. (2007). Genetic screening for signal transduction in the era of network biology. *Cell*, *128*(2), 225–231. doi:10.1016/j.cell.2007.01.007

Gillespie, D. (1976). A general method for numerically simulating the stochastic time evolution of coupled chemical reactions. *Journal of Computational Physics*, *22*(4), 403–434. doi:10.1016/0021-9991(76)90041-3

Hiscock, S. J., & Allen, A. M. (2008). Diverse cell signalling pathways regulate pollen-stigma interactions: The search for consensus. *The New Phytologist*, *179*(2), 286–317. doi:10.1111/j.1469-8137.2008.02457.x

Huang, C. Y., & Ferrell, J. E. Jr. (1996). Ultrasensitivity in the mitogen-activated protein kinase cascade. *Proceedings of the National Academy of Sciences of the United States of America*, *93*(19), 10078–10083. doi:10.1073/pnas.93.19.10078

Ichimura, K., Kazuo, S., Guillaume, T., Jen, S., Champion, H. Y., & Martin, A. K. (2002). Mitogen-activated protein kinase cascades in plants: A new nomenclature. *Trends in Plant Science*, *7*(7), 301–308. doi:10.1016/S1360-1385(02)02302-6

Ichimura, K., Mizoguchi, T., Irie, K., Morris, P., Giraudat, J., & Matsumoto, K. (1998). Isolation of ATMEKK1 (a MAP Kinase Kinase Kinase)-interacting proteins and analysis of a MAP Kinase cascade in Arabidopsis. *Biochemical and Biophysical Research Communications*, *253*(2), 532–543. doi:10.1006/bbrc.1998.9796

Ichimura, K., Mizoguchi, T., Yoshida, R., Yuasa, T., & Shinozaki, K. (2000). Various abiotic stresses rapidly activate Arabidopsis MAP kinases ATMPK4 and ATMPK6. *The Plant Journal*, *24*(5), 655–665. doi:10.1046/j.1365-313x.2000.00913.x

Jansen, R., Greenbaum, D., & Gerstein, M. (2002). Relating whole-genome expression data with protein-protein interactions. *Genome Research*, *12*(1), 37–46. doi:10.1101/gr.205602

Kholodenko, B. N. (2000). Negative feedback and ultrasensitivity can bring about oscillations in the mitogen-activated protein kinase cascades. *European Journal of Biochemistry*, *267*(6), 1583–1588. doi:10.1046/j.1432-1327.2000.01197.x

Kholodenko, B. N. (2006). Cell-signalling dynamics in time and space. *Nature Reviews. Molecular Cell Biology*, *7*(3), 165–176. doi:10.1038/nrm1838

Kolch, W., Calder, M., & David, G. (2005). When kinases meet mathematics: The systems biology of MAPK signalling. *FEBS Letters-Systems Biology*, *579*(8), 1891–1895. doi:10.1016/j.febslet.2005.02.002

Krysan, P. J., Jester, P. J., Gottwald, J. R., & Sussman, M. R. (2002). An Arabidopsis mitogen-activated protein kinase gene family encodes essential positive regulators of cytokinesis. *The Plant Cell*, *14*(5), 1109–1120. doi:10.1105/tpc.001164

Levchenko, A., Bruck, J., & Sternberg, P. W. (2000). Scaffold proteins may biphasically affect the levels of mitogen-activated protein kinase signaling and reduce its threshold properties. *Proceedings of the National Academy of Sciences of the United States of America*, *97*(11), 5818–5823. doi:10.1073/pnas.97.11.5818

Linding, R., Jensen, L. J., Ostheimer, G. J., van Vugt, M. A., Jorgensen, C., & Miron, I. M. (2007). Systematic discovery of in vivo phosphorylation networks. *Cell*, *129*(7), 1415–1426. doi:10.1016/j.cell.2007.05.052

Locasale, J. W., Shaw, A. S., & Chakraborty, A. K. (2007). Scaffold proteins confer diverse regulatory properties to protein kinase cascades. *Proceedings of the National Academy of Sciences of the United States of America*, *104*(33), 13307–13312. doi:10.1073/pnas.0706311104

Mackay, T. F. C. (2004). The genetic architecture of quantitative traits: Lessons from Drosophila. *Current Opinion in Genetics & Development*, *14*(3), 253–257. doi:10.1016/j.gde.2004.04.003

McQuarrie, D. A. (1967). Stochastic approach to chemical kinetics. *Journal of Applied Probability*, *4*(3), 413–478. doi:10.2307/3212214

Menges, M., Dóczi, R., Ökrész, L., Morandini, P., Mizzi, P., & Soloviev, M. (2008). Comprehensive gene expression atlas for the Arabidopsis MAP kinase signalling pathways. *The New Phytologist*, *179*(3), 643–662. doi:10.1111/j.1469-8137.2008.02552.x

Meszaros, T., Helfer, A., Hatzimasoura, E., Magyar, Z., Serazetdinova, L., & Rios, G. (2006). The Arabidopsis MAP kinase kinase MKK1 participates in defence responses to the bacterial elicitor flagellin. *The Plant Journal*, *48*(4), 485–498. doi:10.1111/j.1365-313X.2006.02888.x

Mishra, N.S., Tuteja, Renu, Tuteja & Narendra. (2009). Signaling through MAP kinase networks in plants. *Archives of Biochemistry and Biophysics*, *452*(1).

Mizoguchi, T., Hirayama, I. K., Hayashida, T., Yamaguchi-Shinozaki, N., Matsumoto, K., & Shinozaki, K. (1996). A gene encoding a mitogen-activated protein kinase kinase kinase is induced simultaneously with genes for a mitogen-activated protein kinase and an S6 ribosomal protein kinase by touch, cold, and water stress in Arabidopsis thaliana. *Proceedings of the National Academy of Sciences of the United States of America*, *93*(2), 765–769. doi:10.1073/pnas.93.2.765

Mizoguchi, T., Ichimura, K., Irie, K., Morris, P., Giraudat, J., & Matsumoto, K. (1998). Identification of a possible MAP kinase cascade in Arabidopsis thaliana based on pairwise yeast two-hybrid analysis and functional complementation tests of yeast mutants. *FEBS Letters*, *437*(1-2), 56–60. doi:10.1016/S0014-5793(98)01197-1

Morrison, D. K., & Davis, R. J. (2003). Regulation of MAP kinase signaling modules by scaffold proteins in mammals. *Annual Review of Cell and Developmental Biology*, *19*(1), 91–118. doi:10.1146/annurev.cellbio.19.111401.091942

Morsy, M., Gouthu, S., Orchard, S., Thorneycroft, D., Harper, J. F., & Mittler, R. (2008). Charting plant interactomes: Possibilities and challenges. *Trends in Plant Science*, *13*(4), 183–191. doi:10.1016/j.tplants.2008.01.006

Nakagami, H., Kiegerl, S., & Hirt, H. (2004). OMTK1, a novel MAPKKK, channels oxidative stress signaling through direct MAPK interaction. *The Journal of Biological Chemistry*, *279*(26), 26959–26966. doi:10.1074/jbc.M312662200

Pedley, K. F., & Martin, G. B. (2005). Role of mitogen-activated protein kinases in plant immunity. *Current Opinion in Plant Biology*, *8*(5), 541–547. doi:10.1016/j.pbi.2005.07.006

Popescu, S. C., Popescu, G. V., Bachan, S., Zhang, Z., Gerstein, M., & Snyder, M. (2009). MAPK target networks in Arabidopsis thaliana revealed using functional protein microarrays. *Genes & Development*, *23*(1), 80–92. doi:10.1101/gad.1740009

Popescu, S. C., Popescu, G. V., Bachan, S., Zhang, Z., Seay, M., & Gerstein, M. (2007). Differential binding of calmodulin-related proteins to their targets revealed through high-density Arabidopsis protein microarrays. *Proceedings of the National Academy of Sciences of the United States of America*, *104*(11), 4730–4735. doi:10.1073/pnas.0611615104

Popescu, S. C., Popescu, G. V., Snyder, M., & Dinesh-Kumar, S. P. (2009). Integrated analysis of co-expressed MAP kinase substrates in Arabidopsis thaliana. *Plant Signaling & Behavior*, *4*(6), 524–527. doi:10.4161/psb.4.6.8576

Puntervoll, P., Linding, R., Gemund, C., Chabanis-Davidson, S., Mattingsdal, M., & Cameron, S. (2003). ELM server: A new resource for investigating short functional sites in modular eukaryotic proteins. *Nucleic Acids Research*, *31*(13), 3625–3630. doi:10.1093/nar/gkg545

Shaw, A. S., & Filbert, E. L. (2009). Scaffold proteins and immune-cell signalling. *Nature Reviews. Immunology*, *9*(1), 47–56. doi:10.1038/nri2473

Sheridan, D. L., Kong, Y., Parker, S. A., Dalby, K. N., & Turk, B. E. (2008). Substrate discrimination among mitogen-activated protein kinases through distinct docking sequence motifs. *The Journal of Biological Chemistry*, *283*(28), 19511–19520. doi:10.1074/jbc.M801074200

Sontag, E. D. (2007). Monotone and near-monotone biochemical networks. *Systems and Synthetic Biology*, *1*(2), 59–87. doi:10.1007/s11693-007-9005-9

Soong, T.-t., Wrzeszczynski, K. O., & Rost, B. (2008). Physical protein-protein interactions predicted from microarrays. *Bioinformatics (Oxford, England)*, *24*(22), 2608–2614. doi:10.1093/bioinformatics/btn498

Suarez-Rodriguez, M. C., Adams-Phillips, L., Liu, Y., Wang, H., Su, S.-H., & Jester, P. J. (2007). MEKK1 is required for flg22-Induced MPK4 activation in Arabidopsis plants. *Plant Physiology*, *143*(2), 661–669. doi:10.1104/pp.106.091389

Symons, A., Beinke, S., & Ley, S. C. (2006). MAP kinase kinase kinases and innate immunity. *Trends in Immunology*, *27*(1), 40–48. doi:10.1016/j.it.2005.11.007

Teige, M., Scheikl, E., Eulgem, T., Doczi, R., Ichimura, K., & Shinozaki, K. (2004). The MKK2 pathway mediates cold and salt stress signaling in Arabidopsis. *Molecular Cell*, *15*(1), 141–152. doi:10.1016/j.molcel.2004.06.023

Toufighi, K., Siobhan, M., Brady, R. A., Ly, E., & Provart, N. J. (2005). The botany array resource: e-Northerns, expression angling, and promoter analyses. *The Plant Journal*, *43*(1), 153–163. doi:10.1111/j.1365-313X.2005.02437.x

Uhlik, M. T., Abell, A. N., Cuevas, B. D., Nakamura, K., & Johnson, G. L. (2004). Wiring diagrams of MAPK regulation by MEKK1, 2, and 3. *Biochemistry and Cell Biology*, *82*, 658–663. doi:10.1139/o04-114

Yoshioka, K. (2004). Scaffold proteins in mammalian MAP Kinase cascades. *Journal of Biochemistry*, *135*(6), 657–661. doi:10.1093/jb/mvh079

Zhang, T., Liu, Y., Yang, T., Zhang, L., Xu, S., & Xue, L. (2006). Diverse signals converge at MAPK cascades in plant. *Plant Physiology and Biochemistry*, *44*(5-6), 274–283. doi:10.1016/j.plaphy.2006.06.004

KEY TERMS AND DEFINITIONS

Kinome: The set of protein kinase genes within the genome of an organism.

MAPK Signaling Network: A network composed of MAPK signaling pathways, including MAP3Ks, MAP2Ks, MAPKs and their phosphorylation targets.

Michaelis-Menten Kinetics: An equation describing steady state dynamics of enzymatic reactions.

Mitogen Activated Protein Kinases (MAPK): A class of kinase proteins that are phosphorylated upon stimulation by an extracellular stimulus (signal) and participate in the transduction of this signal.

Protein Interaction Network: A network representation of protein-protein interactions in a cell.

Protein Phosphorylation: An enzymatic reaction consisting in addition of a phosphate group to specific residues of a protein.

Chapter 17
Cancer and Signaling Pathway Deregulation

Yingchun Liu
Dana-Farber Cancer Institute, USA & Harvard Medical School, USA

ABSTRACT

Cancer is a complex disease that is associated with a variety of genetic aberrations. The diagnosis and treatment of cancer have been difficult because of poor understanding of cancer and lack of effective cancer therapies. Many studies have investigated cancer from different perspectives. It remains unclear what molecular mechanisms have triggered and sustained the transition of normal cells to malignant tumor cells in cancer patients. This chapter gives an introduction to the genetic aberrations associated with cancer and a brief view of the topics key to decode cancer, from identifying clinically relevant cancer subtypes to uncovering the pathways deregulated in particular subtypes of cancer.

INTRODUCTION

With the development of high-throughput biotechnologies, various large-scale genomic data are available for screening genetic aberrations in cancer genome. Transforming such genome-wide data into meaningful biological interpretation of cancer is challenging. As much as information is provided by the data, confusion arises about which aberrations are the driver of cancer among the diverse genetic aberrations identified in cancer genome. Recent studies have found that cancer is a pathway disease. The genetic aberrations accumulated in all human cancers ultimately deregulate several biological pathways that control key cell functions. It is therefore important to look at cancer from pathway perspective rather than one gene at a time.

DOI: 10.4018/978-1-60960-491-2.ch017

BACKGROUND

Cancer is a genetic disease that involves various aberrant changes in the genome. These changes may be induced by external factors, such as radiation, chemicals, or viral infection. Other changes may be inherited from previous generations or randomly occur in DNA replication. The genetic aberrations found in cancer exhibit high diversity: from gain or loss of entire chromosomes to a single mutation in a gene, structural changes, or epigenetic alterations. These aberrations consequently affect the activity of cancer-promoting oncogenes and tumor suppressor genes. Oncogenes are typically activated in cancer cells, promoting cell growth and evading programmed cell death, while tumor suppressors are inactivated in cancer cells, resulting in the loss of control over accurate DNA replication, normal cellular signaling, and immune protection. Through multiple processes, eventually, the normal cells are transformed into highly malignant derivatives.

RB, a tumor suppressor gene, is absent or mutated in at least one-third of all human tumors (Berman et al., 2009). Inactivation of RB happens when RB is mutated, which in turn activates the E2F proteins that control the activity of genes required for S-phase progression, rendering cells insensitive to antigrowth factors. Ras oncogene proteins are structurally altered in about 25% of human tumors due to mutations in the gene that encodes them (Tsatsanis & Spandidos, 2000), which enables RAS proteins to release proliferation signal into cells without stimulation by their normal upstream regulators (Medema & Bos, 1993). As a result, cells undergo uncontrolled self-sufficient proliferation. High proliferation and growth rates are necessary but not sufficient for the development of cancer. Tumor cell populations must expand so that the progressive errors can accumulate. Normal cells have programmed systems to correct mistakes or initialize cell death when sensing errors. Cancer cells must evade such processes for a tumor to grow. The P53 tumor suppressor protein, which can elicit apoptosis by activating proapoptotic Bax in response to DNA damage, is seen inactivated in greater than 50% of human tumors (Harris, 1996). Additionally, inappropriate activation of other oncogenes like EGFR, MYC, and PTEN is seen in many human tumors as well (Nicholson et al., 2001; Rodriguez-Pinilla et al, 2007; Freeman et al., 2003).

The observed genetic aberrations in cancer genome can either be the cause or be the consequence of cancer. Among them, few may be the drivers that trigger the transformation from normal cells to malignant cells. Identifying such drivers is critical for the diagnosis and treatment of cancer. Fortunately, a number of high-throughput technologies have been developed to screen genetic aberrations in human tumors. In particular, mutations in tumors can be detected by using SNP arrays (LaFramboise, 2009); DNA copy number of genes can be measured by CGH arrays (Vissers et al., 2005); DNA methylation can be studied by ChIP-chip experiments (Wu et al., 2006). In all these arrays, tumor genome is compared with normal or population genome. Recently, with the advances in the development of New Generation of Sequencing (NGS) technology, we expect that each of the array-based technologies will be replaced with its sequencing alternative (Korbel et al., 2007). The NGS technology can sequence the exact DNA sequences in cancer genome, so higher resolution can be achieved compared with arrays.

Regardless of the high diversity in the genetic aberrations observed in cancer, all human tumor cells share a common set of characteristics: self-sufficiency in growth signals, insensitivity to antigrowth signals, evasion of programmed cell death, limitless replicative potential, sustained angiogenesis, and tissue invasion and metastasis (Hanahan & Weinberg, 2000). Each of these processes is governed by specific signaling pathways. A signaling pathway is a regulatory system that involves a cascade of biochemical reactions in response to extracellular stimulations. See Figure 1. Cell proliferation and death, alterations in metabolism, and activation of genes are cellular responses. In living cells, many genes work in concert to support cellular functions. Activation of genes in one pathway can, in turn, activate or inactivate genes in other pathways. Ultimately, an initial stimulus can lead to the activation or inactivation of a number of complex physiological events. Cancer is an outcome of the perturbed network interactions.

MAIN FOCUS OF THE CHAPTER

Classification of Cancer

There are many types of human cancer in terms of its host organ, for example, lung cancer, breast cancer, colon cancer, etc. Different technologies and treatments have been developed to diagnose and treat them specifically. Our ability to diagnose and effectively treat cancers is still at its early stage though. Up to date, classification of cancer is primarily based on the histopathological appearance of the tumor. Tumors with similar histologic appearance can, however, have developed from different genetic aberrations and have different responses to therapy. For example, different breast cancer patients have different responses to chemotherapy. A major challenge of cancer treatment has been to find specific therapies to pathogenetically distinct tumor types.

Figure 1. A sketch of signaling pathway. In signal transduction, signaling molecules, usually from outside the cell, interact with receptors on the surface of the cell membrane or with nuclear receptors. These interactions trigger a cascade of biochemical reactions. Proteins called transcription factors (TFs) are eventually transported to, or activated in, the nucleus of the cell where they turn transcription of target genes on or off.

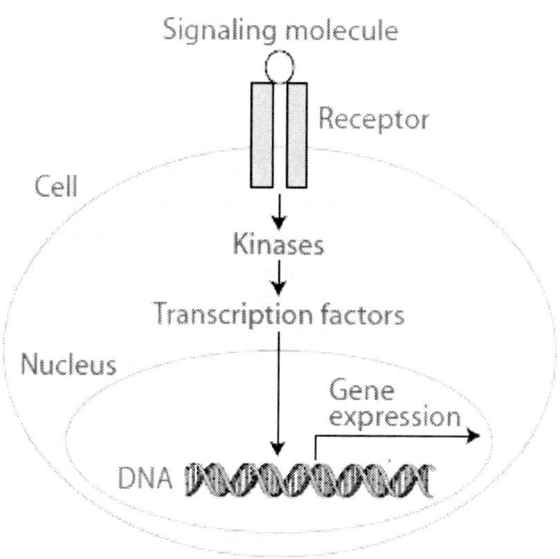

Recent studies have found that some morphologically similar tumors can be molecularly divided into subtypes with distinct pathogeneses. For example, microarray-based gene expression studies have identified several subtypes of cutaneous melanoma (Bittner et al., 2000), two subtypes of leukemia (Gelsi-Boyer et al., 2007), and cell-of-origin breast cancer subtypes (Van Laere et al., 2006). Figure 2 illustrates two subtypes of ovarian cancer identified by clustering gene expression data.

A possible explanation for these observations is that the genetic aberrations accumulated during cancer development ultimately affect the expression of a variety of genes. Different genetic aberrations are likely to affect the expression of different sets of genes. Gene expression pattern is, therefore, reflective of specific genetic aberrations and is a potential indicator of cancer subtypes. The ability to identify cancer subtypes using DNA microarray-based gene expression patterns has been demonstrated in multiple studies (Ramaswamy & Golub, 2002). Alternatively, unknown cancer subtypes could be uncovered based on mutation, DNA copy number or methylation patterns as well (Figueroa et al., 2010). Typically, identification of unknown cancer subtypes in tumor samples is performed by partitioning the samples into subgroups with distinct expression or genetic variation patterns. Many clustering methods have been developed for this purpose, where hierarchical clustering, *k*-means, and principle component analysis (PCA) are most widely used (Liu et al., 2009; D'haeseleer, 2005; Yeung, & Ruzzo, 2001). Each of these methods has its advantages or disadvantages. The limitations of these three methods are: hierarchical clustering

Figure 2. Ovarian cancer subtypes identified by clustering gene expression data. Two subtypes were identifieded, genes as rows and samples as columns. Red and blue colors indicate high and low expression levels, respectively.

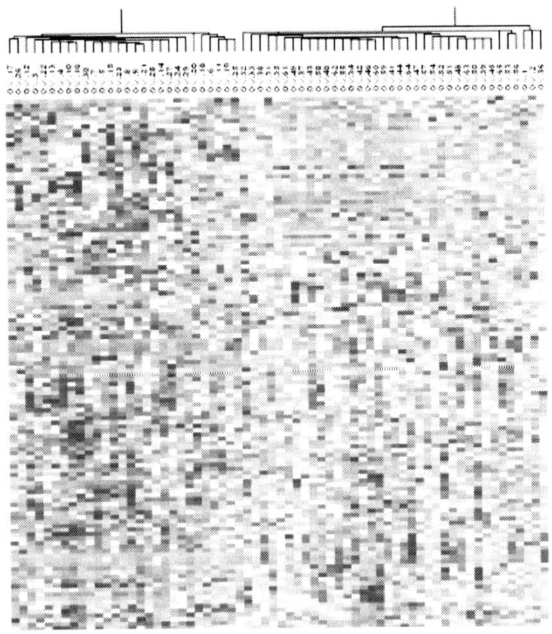

gives no clear clusters, so the user has to decide where to split the tree into groups; for *k*-means, the number of clusters, *k*, must be predefined; interpretation of genes contributed to the separation of clusters identified by PCA is hard.

The above mentioned methods to identify cancer subtypes are gene-by-gene approaches. Functional interpretation of the genes with their expression or genetic variation patterns associated with cancer subtypes provides valuable information on drug candidates. However, in cells, many genes function cooperatively for complex functions. For example, P53 is a tumor suppressor responsible for DNA repair. It inhibits cell growth in response to DNA damage. But p53 function is controlled by the Mdm2 protein interacting with it. Mdm2 enhances degradation of P53 (Piette et al., 1997). If a cancer patient had P53 mutations as well as an abnormal overabundance of MDM2 protein in the tumor cells, this patient could not be cured by drugs that simply increase P53 transcription. It is important to have statistical methods that classify human tumors into subgroups capturing such network interactions.

Bild et al. (2006) have developed an approach to uncover cancer subtypes in nonsmall cell lung carcinoma (NSCLC), breast cancer, and ovarian cancer samples based on gene expression signatures of several signaling pathways. A signaling pathway involves many genes working in concert. Different pathways can share a common set of genes. The set of all known pathways form a dynamic network of genes. Therefore, signaling pathways provide insight into the functional relationships of many genes reflected in the observed gene expression patterns. Particular subtypes of cancer usually have different pathways abnormally activated or inactivated. In addition, pathways can serve as the basis for identifying clinically relevant cancer subtypes. Bild et al. (2006) also reported that there was a close correlation between the activity of Ras and Src in breast cancer cell lines and the extent of cell proliferation inhibition by inhibitors of the Ras pathway. More recently, Chang et al. (2009) developed a strategy to deconstruct pathways into modules represented by gene expression signatures, and studied the relations between these modules and drug response. They assessed clinical response to cetuximab, an EGFR-specific therapy, in 68 patients with advanced colorectal cancer (CRC), and showed that patients with response and no response can be significantly distinguished by one of the EGFR modules. Therefore, pathways not only provide information about cancer subtypes but also guide anticancer drug response. Pathway patterns associated with cancer subtypes provide a path to identify potential biomarkers and drug candidates for the diagnosis and treatment of cancer. Often, biomarkers and anticancer drugs are proteins in pathways (Adjei & Hidalgo, 2005).

Identification of Pathways Associated with Particular Cancer Subtype

A consequence of the genetic aberrations accumulated in cancer is the deregulation of pathways that control cell proliferation and death. All human tumors have one or several signaling pathways inappropriately activated or inactivated. Identifying such pathways is essential for cancer treatment. Typically, a pathway is represented by a set of genes involved in the pathway, based on known studies or pathway databases, such as Kyoto Encyclopedia of Genes and Genomes (KEGG), Pathway Interaction Database (PID), TRANSPATH, BioCarta, Gene Octology (GO), etc. The activation of a pathway is measured by the phosphorylation status of one or a few enzymes in the pathway. However, a pathway can be activated at different points. If the enzyme at the top of a signaling cascade was not phosphorylated, the pathway might still have been activated at a lower point of the signaling cascade. Therefore, the phosphorylation status of single enzymes may not be a reliable indicator of pathway activation. Recently with genome-wide gene expression data available, pathway activation can be measured by the expression level of genes in the pathway. More importantly, the association of all known pathways with specific tumors can be examined simultaneously.

Gene Set Enrichment Analysis (GSEA) (Subramanian et al., 2005) provides an approach to search for gene sets, including pathways, associated with biological phenotypes using expression profiles. GSEA takes the expression profiles of samples belonging to different classes or time series, and ranks the genes according to their correlations with the phenotype labels. It then calculates an enrichment score (ES) for each gene set (pathway). A positive or negative ES indicates an overrepresentation of the genes in this gene set (pathway) at the top or bottom of the ranked list of genes. The statistical significance of the ES for each gene set (pathway) is evaluated by the ESs obtained when randomly assigning the labels to samples. By this means, pathways that have significantly different activity in the two classes can be identified. See Figure 3.

A limitation of GSEA for pathway analysis is that it is unclear whether the identified pathways were activated or inactivated in the two classes. High expression of a majority of genes in a pathway does not necessarily mean that this pathway is activated, as some genes might actually inhibit the pathway, and vice versa. This limitation exists for other applications for pathway analysis as well, including Ingenuity Pathways Analysis (IPA), EASE (Hosack et al., 2003), and ArrayX-Path (Chung et al., 2005).

The aforementioned methods are developed with the assumption that the expression level of genes in a pathway is indicative of pathway activity. However, when a pathway is activated, the expression level of the genes in the pathway is not necessarily affected. Mutation of a transcription factor (TF) or the phosphorylation status of an enzyme in the pathway can activate a pathway without affecting the expression level of the TF itself or other genes in the pathway. Such affected pathways would have been missed if pathway identification is merely based on the expression level of the genes in the pathway. As a matter of fact, expression of the genes regulated by the signaling pathways will definitely be affected, so is a better indicator of pathway activation.

Based on this understanding, our recent study (Liu & Ringnér, 2007) has introduced an approach to identify signaling pathways that are deregulated in caner by using gene expression and regulatory motif analysis. In this method, each signaling pathway is defined by a set of TFs that mediate it. Whether a pathway is associated with a particular type of cancer is determined by the enrichment of the TF motifs in the promoter of genes that distinguish cancer samples from normal samples. A score is calculated for each pathway, based on the enrichment of the TF motifs for

Figure 3. An example of GSEA for pathway analysis. GSEA identified the RAS pathway that was associated with Glioblastoma (GBM) tumors with NF1 mutations, data from http://tcga-data.nci.nih.gov/tcga/homepage.htm. The RAS pathway genes were significantly enriched in highly expressed genes in the GBM samples with NF1 mutations.

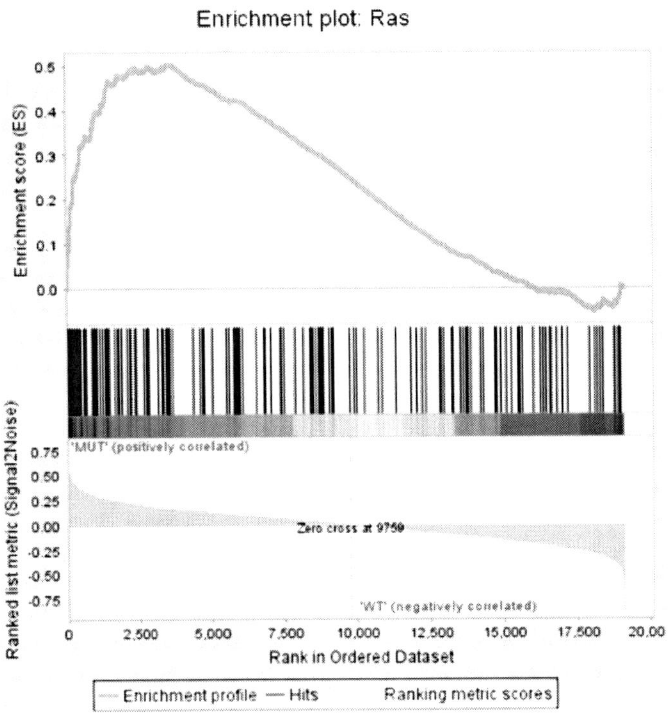

this pathway, and is compared with that of genes randomly selected from the genome. This method rediscovered the oncogenic pathways in the gene expression signature of three oncogenic pathways, and identified pathways relevant to breast cancer matastasis. More interestingly, it identified TGF-β and the same set of other pathways in two TGF-β gene signatures established in two different studies and having very low overlap (Liu & Ringnér, 2007). See Figure 4 for the signaling pathways differently affected in two subtypes of ovarian cancer identified by using this method. We also compared this method with GSEA and EASE. Both methods identified some of the downstream metabolic pathways rather than the oncogenic and TGF-β pathways in their corresponding gene expression signatures. Thus, this method is complimentary to GSEA and other similar applications for pathway analysis.

All the introduced methods for pathway analysis are limited to the prior understanding of pathways, either from previous studies or from available pathway databases. Statistical methods that can reveal unknown pathways or new molecular components of known pathways associated with particular types of cancer remain a distant prospect. In addition, integration of different sources of genomic data for pathway analysis is necessary.

Pathway and Cancer Therapy

Cancer treatment has been extremely challenging. One of the major reasons is that there are few effective therapies available. Most cancer

Figure 4. Pathways identified in ovarian cancer. It identified five signaling pathways that had significantly different activity in two subtypes of ovarian cancer samples. For each identified pathway, the TFs with their motifs enriched in the genes differentially expressed between these two subtypes were also listed.

Pathway	TF	Significant TF	P-value
IFN-map2	IRF-9, STAT1, STAT5	IRF1, IRF7, STAT1, STAT5	0.0026
TPO-map	STAT1, STAT3, STAT5	STAT1, STAT5	0.0026
Epo	Elk-1, FOS, STAT1, STAT3, STAT5A, STAT5B, STAT5	STAT1, STAT5A, STAT5B	0.0094
Fas-pathway	SRF	SRF	0.0146
Beta-catenin	MADH4, SOX17	SOX17	0.0374

Immune   Grow & Differentiation
Apoptosis   Cell adhension & mobility

therapies are designed to target a specific genetic aberration. Gefitinib is such an example for treating non-small-cell lung cancer, and has shown efficacy only in a small subset of patients carrying mutations in the epidermal growth factor receptor (EGFR). Treating cancer by targeting single aberration at a time is thus hard. Notably, regardless of the diverse genetic aberrations associated with cancer, all human tumors have only a few of about a dozen key signaling pathways abnormally activated or inactivated (Jones, 2008). This scenario provides an opportunity to develop new therapeutic options for patients targeting deregulated pathways. For example, the RAS and BRAF oncogenes are mutated in 40% of lung cancer and 60% of melanomas, respectively (Downward, 2003). Both mutations lead to inappropriate activation of the RAS pathway. Instead of designing different therapies that target these two mutations separately, cancer therapies that inhibit the RAS pathway have the potential to treat the lung cancer and melanoma with RAS or BRAF mutations.

Increasingly, cancer drugs are designed to target signaling pathways. The Hh pathway is abnormally activated in several human cancers. Hedgehog pathway inhibitor IPI-926 binds to and inhibits the receptor SMO, which suppresses the Hh pathway, potentially decreasing tumor cell proliferation and survival. In addition, DNA repair pathways and apoptosis pathways are linked to many human tumors and are therapeutic targets for particular cancers (Helleday et al., 2008; Ghobrial et al., 2005). Many more pathway-specific drug candidates are in the early stages of clinical trials. Novartis, an international pharmaceutical company, is developing innovative anticancer agents that target multiple pathways (Ma & Adjei, 2009). The combined use of pathway inhibitors with standard anticancer treatments enables us to attack human cancers on many fronts.

IMPLICATIONS AND FUNTURE DIRECTIONS

For decades, identifying driver genetic aberrations underlying cancer has been the focus of cancer research. The sequencing of human genome and other high-throughput technologies have offered scientists the opportunities to identify genetic aberrations in cancer genome and to describe biological states quantitatively. In fact, few human cancers can be precisely defined by individual genetic aberrations, and anticancer drugs that target single mutations have shown poor efficacy (Jones,

2008). Rather, a comprehensive understanding of the consequence of mutations, DNA copy number gain or loss, structural changes, and epigenetic alterations in terms of their network interactions will be essential for the diagnosis and improved treatment of cancer.

The accumulated genetic aberrations in cancer ultimately deregulate pathways that control cell proliferation and death. Although tumors in cancer patients are different, they all have a common group of such pathways abnormally activated or inactivated. Studies have shown that cancer is a pathway disease (Hanahan & Weinberg, 2000; Jones, 2008). We expect that much effort will be put toward identifying cancer subtypes and signature pathways associated with particular types of cancer. Perry Nisen, Senior Vice President of Cancer Research at GlaxoSmithKline, emphasized the need to understand the key pathways leading to caners. "One cannot separate genetic association data from the fundamental understanding of the biology and pathways", says Nisen (Jones, 2008). The search for cancer drugs will accordingly shift dramatically from targeting individual genetic aberrations to targeting deregulated pathways. Bert Vogelstein of the Johns Hopkins School of Medicine in Baltimore, Maryland, USA, and Emmanuel Petricoin, Co-Director of the Centre for Applied Proteomics and Molecular Medicine at George Mason University, Virginia, USA, envisages a future where drugs targeting entire signaling pathways rather than just one aberration at a time will be the key to drug development (Jones, 2008).

At present, our ability to describe human cancer at the molecular level remains poor. The gap between genetic aberrations and cancer will be bridged in part by new technologies and effective bioinformatics tools that aim to identify key biological mechanisms underlying cancer. The Cancer Genome Atlas (TCGA) project, funded by the National Institute of Health, is a coordinate effort to accelerate our understanding of the molecular basis of human cancer through integrated analysis of different sources of large-scale genetic and epigenetic data. The TCGA project has identified four molecular subtypes of Glioblastoma (GBM) as well as their associated pathways (McLendon, 2008). In the next five years, TCGA will provide genomic fingerprint of 20-25 different cancers. The Sanger Institute in UK also launched a Cancer Genome Project to identify DNA mutations critical in the development of cancer. With joint efforts from different research centers, we anticipate far deeper understanding of the network interaction of genes and pathways in human tumors in the next decade.

ACKNOWLEDGMENT

Yingchun Liu is supported by Dr. Lynda Chin and Belfer Institute of Applied Cancer Science in Dana-Farber Cancer Institute / Harvard Medical School, USA.

REFERENCES

Adjei, A. A., & Hidalgo, M. (2005). Intracellular signal transduction pathway proteins as targets for cancer therapy. *Journal of Clinical Oncology*, *23*(23), 5386–5403. doi:10.1200/JCO.2005.23.648

Berman, S. D., West, J. C., Danielian, P. S., Caron, A. M., Stone, J. R., & Lees, J. A. (2009). Mutation of p107 exacerbates the consequences of Rb loss in embryonic tissues and causes cardiac and blood vessel defects. *Proceedings of the National Academy of Sciences of the United States of America*, *106*(35), 14932–14936. doi:10.1073/pnas.0902408106

Bild, A. H., Yao, G., Chang, J. T., Wang, Q., Potti, A., & Chasse, D. (2006). Oncogenic pathway signatures in human cancers as a guide to targeted therapies. *Nature*, *439*(7074), 353–357. doi:10.1038/nature04296

Bittner, M., Meltzer, P., Chen, Y., Jiang, Y., Seftor, E., & Hendrix, M. (2000). Molecular classification of cutaneous malignant melanoma by gene expression profiling. *Nature, 406*(6795), 536–540. doi:10.1038/35020115

Chang, J. T., Carvalho, C., Mori, S., Bild, A. H., Gatza, M. L., & Wang, Q. (2009). A genomic strategy to elucidate modules of oncogenic pathway signaling networks. *Molecular Cell, 34*(1), 104–114. doi:10.1016/j.molcel.2009.02.030

Chung, H.J., Park, C.H., Han, M.R., Lee, S., Ohn, J.H., Kim, J., et al. (2005). ArrayXPath II: Mapping and visualizing microarray gene-expression data with biomedical ontologies and integrated biological pathway resources using Scalable Vector Graphics. *Nucleic Acids Research, 33*(Web server issue), W621-6.

D'haeseleer, P. (2005). How does gene expression clustering work? *Nature Biotechnology, 23*(12), 1499–1501. doi:10.1038/nbt1205-1499

Downward, J. (2003). Targeting RAS signalling pathways in cancer therapy. *Nature Reviews. Cancer, 3*(1), 11–22. doi:10.1038/nrc969

Figueroa, M. E., Lugthart, S., Li, Y., Erpelinck-Verschueren, C., Deng, X., & Christos, P. J. (2010). DNA methylation signatures identify biologically distinct subtypes in Acute Myeloid Leukemia. *Cancer Cell, 17*(1), 13–27. doi:10.1016/j.ccr.2009.11.020

Freeman, D. J., Li, A. G., Wei, G., Li, H. H., Kertesz, N., & Lesche, R. (2003). PTEN tumor suppressor regulates p53 protein levels and activity through phosphatase-dependent and -independent mechanisms. *Cancer Cell, 3*(2), 117–130. doi:10.1016/S1535-6108(03)00021-7

Gelsi-Boyer, V., Cervera, N., Bertucci, F., Trouplin, V., Remy, V., & Olschwang, S. (2007). Gene expression profiling separates chronic myelomonocytic leukemia in two molecular subtypes. *Leukemia, 21*(11), 2359–2362. doi:10.1038/sj.leu.2404805

Ghobrial, I. M., Witzig, T. E., & Adjei, A. A. (2005). Targeting apoptosis pathways in cancer therapy. *CA: a Cancer Journal for Clinicians, 55*(3), 178–194. doi:10.3322/canjclin.55.3.178

Hanahan, D., & Weinberg, R. A. (2000). The hallmarks of cancer. *Cell, 100*(1), 57–70. doi:10.1016/S0092-8674(00)81683-9

Harris, C. C. (1996). p53 tumor suppressor gene: From the basic research laboratory to the clinic—an abridged historical perspective. *Carcinogenesis, 17*, 1187–1198. doi:10.1093/carcin/17.6.1187

Helleday, T., Petermann, E., Lundin, C., Hodgson, B., & Sharma, R. A. (2008). DNA repair pathways as targets for cancer therapy. *Nature Reviews. Cancer, 8*(3), 193–204. doi:10.1038/nrc2342

Hosack, D. A., Dennis, G., Sherman, B. T., Lane, H. C., & Lempicki, R. A. (2003). Identifying biological themes within lists of genes with EASE. *Genome Biology, 4*(10), R70. doi:10.1186/gb-2003-4-10-r70

Jones, D. (2008). Pathways to cancer therapy. *Nature Reviews. Drug Discovery, 7*(11), 875–876. doi:10.1038/nrd2748

Jones, S., Zhang, X., Parsons, D. W., Lin, J. C., Leary, R. J., & Angenendt, P. (2008). Core signaling pathways in human pancreatic cancers revealed by global genomic analyses. *Science, 321*(5897), 1801–1806. doi:10.1126/science.1164368

Korbel, J. O., Urban, A. E., Affourtit, J. P., Godwin, B., Grubert, F., & Simons, J. F. (2007). Paired-end mapping reveals extensive structural variation in the human genome. *Science, 318*(5849), 420–426. doi:10.1126/science.1149504

LaFramboise, T. (2009). Single nucleotide polymorphism arrays: A decade of biological, computational and technological advances. *Nucleic Acids Research, 37*(13), 4181–4193. doi:10.1093/nar/gkp552

Liu, M., Matsumura, N., Mandai, M., Li, K., Yagi, H., & Baba, T. (2009). Classification using hierarchical clustering of tumor-infiltrating immune cells identifies poor prognostic ovarian cancers with high levels of COX expression. *Modern Pathology*, *22*(3), 373–384. doi:10.1038/modpathol.2008.187

Liu, Y., & Ringnér, M. (2007). Revealing signaling pathway deregulation by using gene expression signatures and regulatory motif analysis. *Genome Biology*, *8*(5), R77. doi:10.1186/gb-2007-8-5-r77

Ma, W. W., & Adjei, A. A. (2009). Novel agents on the horizon for cancer therapy. *CA: a Cancer Journal for Clinicians*, *59*(2), 111–137. doi:10.3322/caac.20003

McLendon, (2008). Comprehensive genomic characterization defines human glioblastoma genes and core pathways. *Nature*, *455*(7216), 1061–1068. doi:10.1038/nature07385

Medema, R. H., & Bos, J. L. (1993). The role of p21-ras in receptor tyrosine kinase signaling. *Critical Reviews in Oncogenesis*, *4*, 615–661.

Nicholson, R. I., Gee, J. M., & Harper, M. E. (2001). EGFR and cancer prognosis. *European Journal of Cancer*, *37*(Suppl 4), S9–S15. doi:10.1016/S0959-8049(01)00231-3

Parsons, D. W., Jones, S., Zhang, X., Lin, J. C., Leary, R. J., & Angenendt, P. (2008). An integrated genomic analysis of human glioblastoma multiforme. *Science*, *321*(5897), 1807–1812. doi:10.1126/science.1164382

Piette, J., Neel, H., & Marechal, V. (1997). Mdm2: Keeping p53 under control. *Oncogene*, *15*(9), 1001–1010. doi:10.1038/sj.onc.1201432

Ramaswamy, S., & Golub, T. R. (2002). DNA microarrays in clinical oncology. *Journal of Clinical Oncology*, *20*(7), 1932–1941.

Rodriguez-Pinilla, S. M., Jones, R. L., Lambros, M. B., Arriola, E., Savage, K., & James, M. (2007). MYC amplification in breast cancer: A chromogenic in situ hybridisation study. *Journal of Clinical Pathology*, *60*(9), 1017–1023. doi:10.1136/jcp.2006.043869

Subramanian, A., Tamayo, P., Mootha, V. K., Mukherjee, S., Ebert, B. L., & Gillette, M. A. (2005). Gene set enrichment analysis: A knowledge-based approach for interpreting genome-wide expression profiles. *Proceedings of the National Academy of Sciences of the United States of America*, *102*(43), 15545–15550. doi:10.1073/pnas.0506580102

Tsatsanis, C., & Spandidos, D. A. (2000). The role of oncogenic kinases in human cancer [Review]. *International Journal of Molecular Medicine*, *5*(6), 583–590.

Van Laere, S. J., Van den Eynden, G. G., Van der Auwera, I., Vandenberghe, M., van Dam, P., & Van Marck, E. A. (2006). Identification of cell-of-origin breast tumor subtypes in inflammatory breast cancer by gene expression profiling. *Breast Cancer Research and Treatment*, *95*(3), 243–255. doi:10.1007/s10549-005-9015-9

Vissers, L.E., Veltman, J.A., van Kessel, A.G. & Brunner, H.G. (2005). Identification of disease genes by whole genome CGH arrays. *Human Molecular Genetics, 14*(Spec No. 2), R215-23.

Wu, J., Smith, L. T., Plass, C., & Huang, T. H. (2006). ChIP-chip comes of age for genome-wide functional analysis. *Cancer Research*, *66*(14), 6899–6902. doi:10.1158/0008-5472.CAN-06-0276

Yeung, K. Y., & Ruzzo, W. L. (2001). Principal component analysis for clustering gene expression data. *Bioinformatics (Oxford, England)*, *17*(9), 763–774. doi:10.1093/bioinformatics/17.9.763

ADDITIONAL READING

BioCarta Pathway Database. http://www.biocarta.com/genes/index.asp

Chen, J., Odenike, O., & Rowley, J. D. (2010). Leukaemogenesis: more than mutant genes. *Nature Reviews. Cancer*, *10*(1), 23–36. doi:10.1038/nrc2765

Das, P. M., & Singal, R. (2004). DNA methylation and cancer. *Journal of Clinical Oncology*, *22*(22), 4632–4642. doi:10.1200/JCO.2004.07.151

Efroni, S., Schaefer, C. F., & Buetow, K. H. (2007). Identification of key processes underlying cancer phenotypes using biologic pathway analysis. *PLoS ONE, 2*, 5:e425.

Gene Ontology (GO). http://amigo.geneontology.org/cgi-bin/amigo/go.cgi

Ingenuity Pathways Analysis (IPA). http://www.ingenuity.com/products/pathways_analysis.html

Korbel, J. O., Urban, A. E., Affourtit, J. P., Godwin, B., Grubert, F., & Simons, J. F. (2007). Paired-end mapping reveals extensive structural variation in the human genome. *Science*, *318*(5849), 420–426. doi:10.1126/science.1149504

LaFramboise, T. (2009). Single nucleotide polymorphism arrays: a decade of biological, computational and technological advances. *Nucleic Acids Research*, *37*(13), 4181–4193. doi:10.1093/nar/gkp552

Lopez-Bergami, P., Lau, E., & Ronai, Z. (2010). Emerging roles of ATF2 and the dynamic AP1 network in cancer. *Nature Reviews. Cancer*, *10*(1), 65–76. doi:10.1038/nrc2681

Ma, S., & Kosorok, M. R. (2009). Identification of differential gene pathways with principal component analysis. *Bioinformatics (Oxford, England)*, *25*(7), 882–889. doi:10.1093/bioinformatics/btp085

Pathway DatabaseK. E. G. G.http://www.genome.jp/kegg/pathway.html

Pathway Interaction Database (PID). http://pid.nci.nih.gov/

Todd, J. A. (2006). Statistical false positive or true disease pathway? *Nature Genetics*, *38*(7), 731–733. doi:10.1038/ng0706-731

Wagner, E. F., & Nebreda, A. R. (2009). Signal integration by JNK and p38 MAPK pathways in cancer development. *Nature Reviews. Cancer*, *9*(8), 537–549. doi:10.1038/nrc2694

KEY TERMS AND DEFINITIONS

Epigenetics: Any biological events that affect the structure or functions of the genome but are not encoded in the DNA, for example, biochemical modifications in the histone proteins entangled in the DNA and DNA methylations.

Gene Expression: The abundance of mRNAs transcribed from the DNA of a gene.

Chapter 18
Computational Methods for Identification of Novel Secondary Metabolite Biosynthetic Pathways by Genome Analysis

Swadha Anand
National Institute of Immunology, India

Debasisa Mohanty
National Institute of Immunology, India

ABSTRACT

Secondary metabolites belonging to polyketide and nonribosomal peptide families constitute a major class of natural products with diverse biological functions and a variety of pharmaceutically important properties. Experimental studies have shown that the biosynthetic machinery for polyketide and nonribosomal peptides involves multi-functional megasynthases like Polyketide Synthases (PKSs) and nonribosomal peptide synthetases (NRPSs) which utilize a thiotemplate mechanism similar to that for fatty acid biosynthesis. Availability of complete genome sequences for an increasing number of microbial organisms has provided opportunities for using in silico genome mining to decipher the secondary metabolite natural product repertoire encoded by these organisms. Therefore, in recent years there have been major advances in development of computational methods which can analyze genome sequences to identify genes involved in secondary metabolite biosynthesis and help in deciphering the putative chemical structures of their biosynthetic products based on analysis of the sequence and structural features of the proteins encoded by these genes. These computational methods for deciphering the secondary metabolite biosynthetic code essentially involve identification of various catalytic domains present in

DOI: 10.4018/978-1-60960-491-2.ch018

this PKS/NRPS family of enzymes; a prediction of various reactions in these enzymatic domains and their substrate specificities and also precise identification of the order in which these domains would catalyze various biosynthetic steps. Structural bioinformatics analysis of known secondary metabolite biosynthetic clusters has helped in formulation of predictive rules for deciphering domain organization, substrate specificity, and order of substrate channeling. In this chapter, the progress in development of various computational methods is discussed by different research groups, and specifically, the utility in identification of novel metabolites by genome mining and rational design of natural product analogs by biosynthetic engineering studies.

INTRODUCTION

Polyketides and non-ribosomal peptides constitute the largest family of small molecule natural products biosynthesized by microbes, fungi and plants as secondary metabolites (Linne *et al.*, 2003; Schwarzer *et al.*, 2003; Shen, 2003). These small molecule natural products not only show enormous diversity in their chemical structure, they also have a variety of biomedical and pharmaceutical applications in view of their therapeutic potentials. The elucidation of polyketide and nonribosomal peptide biosynthetic machinery by pioneering genetic and biochemical studies have revealed that these secondary metabolites are biosynthesized by multi functional megasynthases like Polyketide Synthases (PKSs) and nonribosomal peptide synthetases (NRPSs), using an assembly line mechanism which resembles fatty acid biosynthesis. The availability of complete genome sequences in an increasing number of organisms has opened up the possibility of discovering novel secondary metabolite natural products by genome mining (Van Lanen & Shen, 2006). Major advances in biosynthetic engineering during the last decade have also demonstrated the feasibility of obtaining novel engineered natural products by rational manipulation of known secondary metabolite biosynthetic pathways using biosynthetic engineering approaches (Baltz, 2006; Zhang & Wilkinson, 2007). Hence, during the last decade, the research on PKS and NRPS biosynthetic pathways in various organisms has been pursued with two major goals, namely, identification and experimental characterization of new secondary metabolites in various microbial and fungal species and production of novel rationally designed natural products by manipulation of known PKS/NRPS biosynthetic machinery using a biosynthetic engineering approach.

The remarkable conservation of secondary metabolite gene clusters across organisms has offered abundant scope for obtaining novel insights into the secondary metabolite biosynthetic code by computational analysis. Hence, development of computational methods for relating the chemical structure of the complex secondary metabolites to the amino acid sequence of their corresponding biosynthetic proteins has been an area of active research. Such computational methods (Minowa *et al.*, 2007; Yadav *et al.*, 2003a) have played a major role in guiding various experimental approaches involving genetics, biochemistry, proteomics and metabolomics for discovery of new secondary metabolites by genome mining and reprogramming of known biosynthetic pathways for producing novel natural products by a rational design approach.

In this article, we attempt to give a brief overview of various computational methods which have facilitated easy correlation of the chemical structure of the secondary metabolites to the amino acid sequence of the various PKS and NRPS megasynthases present in the corresponding biosynthetic gene clusters. We first provide background information on various different polyketide and non-ribosomal peptide biosynthetic paradigms. This is followed by a description of different types of computational studies that have been carried out on secondary metabolite biosynthetic clusters and the rationale behind it. The sections that follow, give a detailed account of the computational techniques, which can help in predicting the organization of various PKS/NRPS catalytic domains in a secondary metabolite gene cluster as well as their substrate specificities. These sections also mention the various software or web servers available for such analysis. In the subsequent section, we describe the computational methods for predicting the order of substrate channeling in a secondary metabolite biosynthetic cluster based on analysis of inter subunit interactions. The last section describes a few examples where bioinformatics analyses have guided experimental studies to discover new metabolites, as well as studies to generate novel metabolites by reprogramming the known biosynthetic machinery.

BACKGROUND INFORMATION ON VARIOUS DIFFERENT PARADIGMS FOR BIOSYNTHESIS OF POLYKETIDES AND NONRIBOSOMAL PEPTIDES

The biosynthesis of polyketides resembles the assembly line mechanism involved in fatty acid biosynthesis and the major steps consist of recognition of various carboxyl acid starter/extender substrates by the acyltransferase (AT) domain, chain elongation by decarboxylative condensation of carboxylic acid substrates by the ketosynthase (KS) domain, and a series of optional modification to the resulting ketide group by ketoreductase (KR), dehydratase (DH) and enoylreductase (ER) domains (Cox, 2007). During the biosynthesis process, the growing polyketide chain remains tethered to the acyl carrier protein (ACP) domain and is transferred to different catalytic centers by the phosphopantetheine group of ACP. The polyketide synthases (PKSs) have been classified into different groups depending on the type of organization of these catalytic domains. The conventional type I, type II and type III biosynthetic paradigms (Shen, 2003) for PKS and known deviations (Moss *et al.*, 2004) from these text book biosynthetic logics have been discussed in detail in several recent reviews. The type I PKSs are multifunctional enzymes consisting of single or multiple sets of catalytic domains on a single polypeptide chain (Figure 1). In contrast, type II PKSs are multi-enzyme complexes, having each domain as a stand-alone protein, which acts iteratively to add functional units and form the final product. The domains, which carry out one round of polyketide chain extension and associated modifications, constitute a module in type I PKSs. The KS, AT and ACP domains form the minimal module while the optional reductive domains may or may not be present. The type I PKSs are further categorized into two groups. In type I modular PKSs separate modules are present for each round of chain extension while in Type I iterative PKSs, a single set of catalytic domains is repeatedly used to carry out the polyketide extension (Figure 1).

Type I modular PKSs often occur as clusters consisting of multiple ORFs or polypeptide chains and during biosynthesis the growing polyketide chain is transferred from the last module of the preceding ORF to the first module of the succeeding ORF. In both modular as well as iterative PKSs, the release of the final polyketide chain from ACP and its cyclization is carried out by thioesterase (TE) or Cyclization (CY) domains which are present in the last module or as stand-alone proteins in the PKS cluster. In contrast to type I and type II PKSs, type III PKSs are homodimeric condensing enzymes where a single catalytic site is iteratively used to form the final polyketide product. Type III PKSs directly utilize the acyl CoA substrates unlike type I and II PKSs which act on ACP bound substrates (Shen, 2003).

Similar to the PKSs, Nonribosomal peptide synthetases (NRPSs) perform sequential condensations of proteinogenic as well as nonpro-

Figure 1. Typical examples of domain organization in type I PKS and NRPS biosynthetic clusters. The top panel depicts presence of six modules arranged on three ORFs and each module adding a functional group as in the case of erythromycin. In contrast, the second panel depicts an iterative Type I PKS which has one module which is utilized six times to yield 1,3,6,8-tetrahydroxynapthalene. The third panel depicts the tyrocidine biosynthetic gene cluster showing modular organization where each module adds an amino acid extender group. The cluster contains 3 ORFs with 10 modules

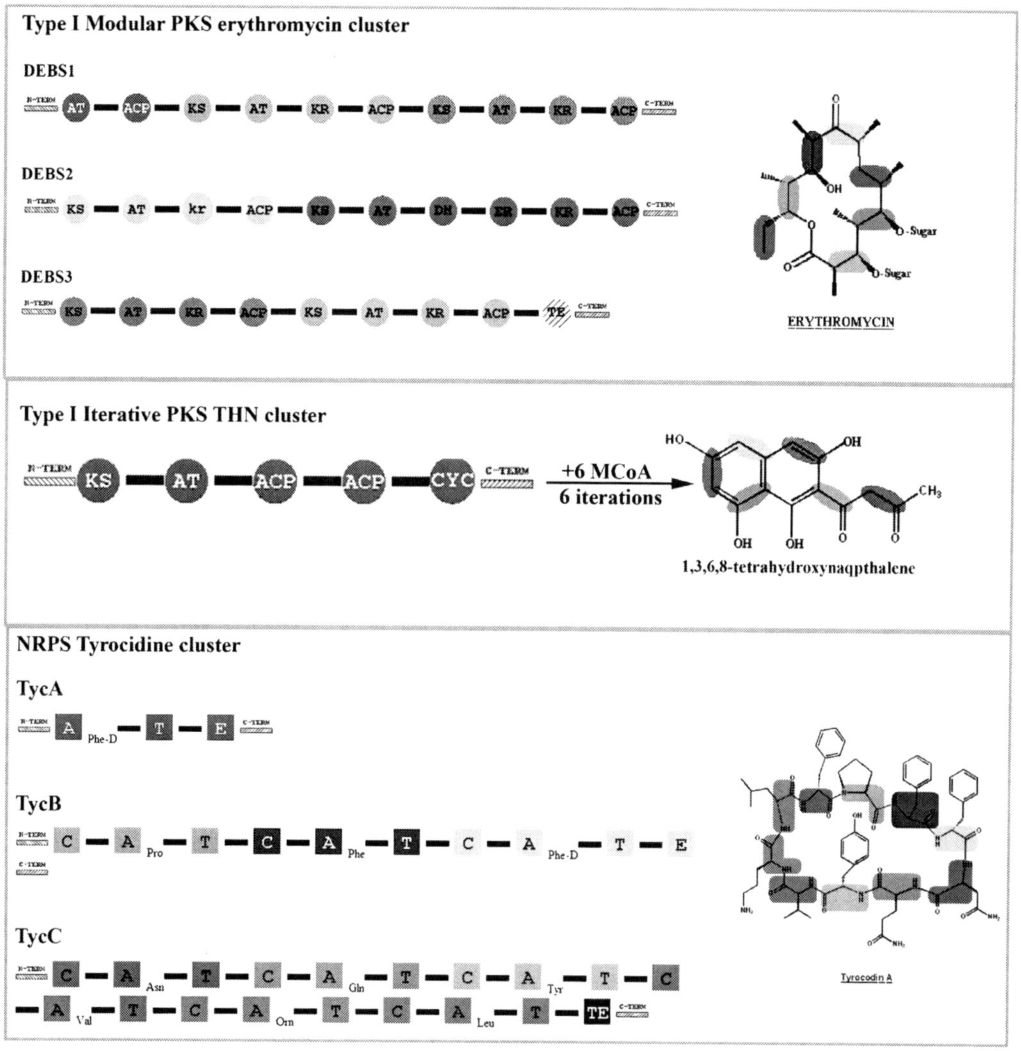

teinogenic amino acid monomers to produce diverse peptide products. The catalytic domains of NRPSs have been defined on the basis of their function. The core domains are the adenylation domain (*A*), for selection of starter and extender units, the peptidyl carrier protein (*PCP*) domain with a phosphopantetheine swinging arm for transferring intermediates from one active site to another and the condensation (*C*) domain responsible for peptide bond formation (Figure 1). Additional optional domains have also been identified, which are responsible for the modification of the peptide backbone. These optional domains include cyclization (*Cy*), epimerization

(*E*), methyltransferase (*MT*) etc. Upon reaching its full length, the peptide product is released from the protein in linear or cyclized form by a thioesterase (*TE*) domain which is generally the last functional domain for synthesis of the peptide product and is present at the distal C-terminus of these enzymes. Generally, NRPSs follow collinear biosynthetic logic and each module catalyzes addition of one monomeric unit to the growing nonribosomal peptide chain and, as seen in modular PKSs, the number of monomeric units in the final metabolic product correlates with the number of modules in the corresponding NRPS protein or NRPS cluster consisting of multiple Open Reading Frames (ORFs) (Fischbach & Walsh, 2006). However, there are several examples of nonlinear NRPS clusters where the number of modules does not correlate with the number of peptide units in the final product (Mootz *et al.*, 2002).

Structural diversity of secondary metabolites is often increased by cross talk between PKS and NRPS biosynthetic pathways, where a polyketide product is further elongated by NRPSs or *vice versa* to produce hybrid natural products. Different tailoring enzymes like glycosyltransferases, halogenases, methyltransferases and oxidoreductases (Rix *et al.*, 2002) also further alter the chemical structures of secondary metabolites by addition of various types of functional groups.

DECIPHERING THE SECONDARY METABOLITE BIOSYNTHETIC CODE BY *IN SILICO* ANALYSIS

Analysis of various PKS and NRPS biosynthetic paradigms indicate that the PKS/NRPS biosynthetic pathways generate enormous diversity in chemical structures of secondary metabolites by combinatorial use of a limited number of catalytic domains and subtle variations in their substrate specificities. For example, in type I modular PKS clusters and most NRPS clusters, the number of modules dictate the number of ketide or peptide units in the secondary metabolite product, while in iterative PKS clusters, the number of monomers is governed by the number of iterative condensations catalyzed by the module (Bachmann & Ravel, 2009; Hill, 2006). Similarly, the chemical structure of each monomer depends on the substrate specificity of the acyltransferase (AT) or adenylation (A) domains and types of optional chain modification domains present in the corresponding module. Figure 2 shows the various types of modules in PKS and NRPS proteins and the chemical moieties added by them. Therefore, the "secondary metabolite biosynthetic code" i.e. rules for relating chemical structures of the secondary metabolites to the sequences of the corresponding biosynthetic proteins, can in principle be deciphered by computational analysis of the domain organization and substrate specificities of PKS and NRPS proteins. Figure 3 shows a flow chart depicting various different computational steps involved in relating sequences of type I PKS or NRPS enzymes to the chemical structure of secondary metabolites encoded by them.

In the case of most type I modular PKS and NRPS proteins, the chemical structure of the secondary metabolite product can be predicted by the so called "collinearity rule" when all modules are on a single polypeptide chain. However, in case of modular PKS or NRPS clusters consisting of multiple ORFs, it is often observed that there are deviations from the collinearity rule, because the order of substrate channeling during biosynthesis might differ from the order of organization of the ORFs on the genome (Du *et al.*, 2000; Schwecke *et al.*, 1995). In such cases, the order of substrate channeling can be deciphered by analysis of inter subunit interactions between the PKS or NRPS proteins present in the biosynthetic cluster (Thattai *et al.*, 2007; Yadav *et al.*, 2009). The correct prediction of the organization of the catalytic domains, their substrate specificities and inter-subunit interactions in the case of modular PKS/NRPS and the number of iterations in the case of iterative PKSs, can only help

Figure 2. The chemical moieties being added by various domains in NRPS and PKS gene clusters

	Module	Chemical structure
PKS	KS-AT-ACP	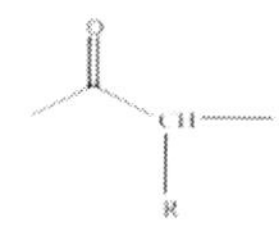
	KS-AT-KR-ACP	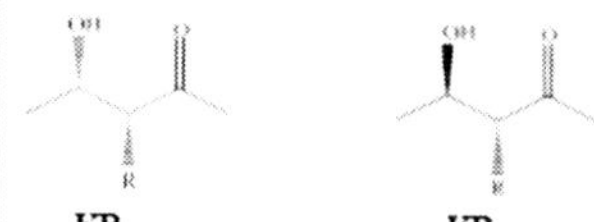
	KS-AT-DH-KR-ACP	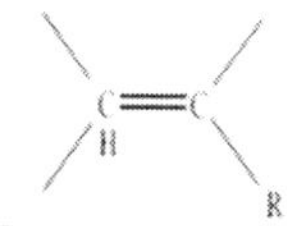
	KS-AT-DH-ER-KR-ACP	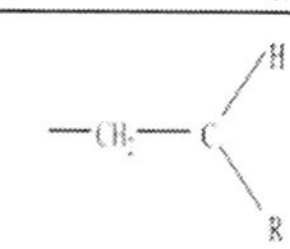
NRPS	C-A-T	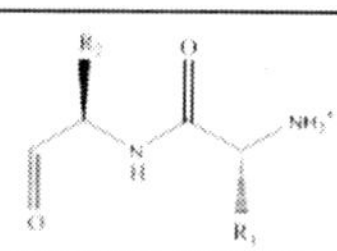
	C-A-T-E	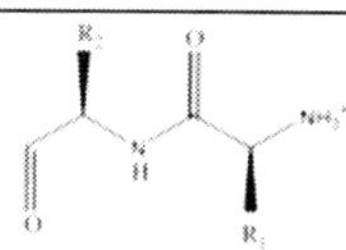
	Cy-A-T	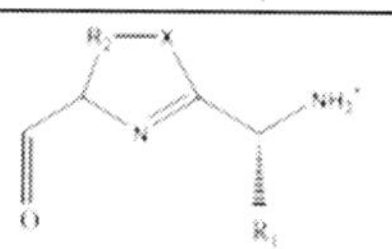 X=nucleophile of R_2 func. gp

in identifying the chemical structure of the linear polyketide or nonribosomal peptide. However, for deciphering the cyclization pattern and post-PKS/NRPS modifications, it will be necessary to understand the substrate specificities of thioesterase domains and other tailoring enzymes. Even though, as of today, no perfect computational tools exist for reliable prediction of chemical structure of secondary metabolites from genome sequences, during the last decade significant advances have been made in the development of knowledge-based computational approaches for each of the above mentioned tasks and such tools have provided valuable guidelines for discovery of new secondary metabolite biosynthetic pathways.

In recent years, there has been a tremendous increase in the number of PKS/NRPS gene clusters with experimentally characterized biosynthetic products. Evolutionary analysis of PKS/NRPS genes has not only provided novel insights into their phylogeny, but also helped in formulating predictive rules for deciphering their function as well as substrate specificity etc. For example, phylogenetic analysis of KS domains not only shows distinct clustering as per the modular or iterative condensations they catalyze, iterative

Figure 3. The steps involved in deciphering the biosynthetic product of a type I PKS or NRPS gene clusters by computational approach

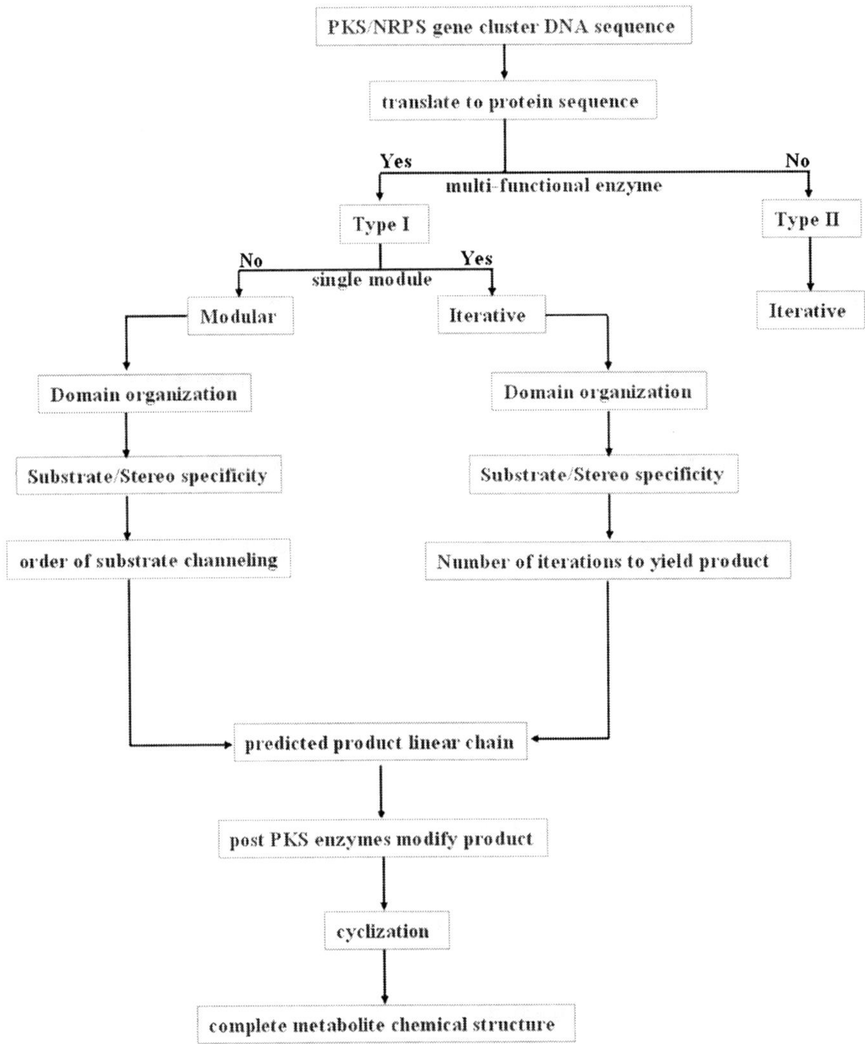

KS sequences also show further sub-groups correlating with the number of iterations (Moffitt & Neilan, 2003; Yadav *et al*., 2009). In contrast to most KS domains in modular PKS clusters, KS domains that accept the peptidyl substrates in PKS/NRPS hybrids, KSq domains in PKS loading modules, KS domains in trans-AT systems also cluster according to the substrates they bind (Ginolhac *et al*., 2005; Moffitt & Neilan, 2003; Nguyen *et al*., 2008). Similarly, condensation (C) domains of NRPS show clustering as per their stereo-selectivity rather than substrate specificity or species phylogeny (Belshaw *et al*., 1999; Clugston *et al*., 2003; Rausch *et al*., 2007). Similar clustering based on function or reaction types is also seen for most other catalytic domains as well as inter-polypeptide linkers or docking domains in secondary metabolite gene clusters. For example, the phylogenetic analysis of adenylation (A) domains of NRPS and acyltransferase (AT) domains of PKS indicate that they cluster as per their substrate specificity rather than their species of origin (Challis *et al.*, 2000; Stachelhaus *et al.*, 1999; Yadav *et al.*, 2003a). Hence, based on conserved patterns in groups of sequences which cluster together in the phylogenetic tree, profiles/motifs which are determinants of reaction types or substrate specificity have been derived. It has also been found that the incorporation of binding pocket information from homologous crystal structures along with evolutionary analysis further improves the predictive power of the derived specificity determining profiles/motifs (Ansari *et al.*, 2004; Challis *et al.*, 2000; Stachelhaus *et al.*, 1999). Therefore, most of the computational methods for relating sequences of PKS/NRPS enzymes to the chemical structures of their metabolic products have been developed based on phylogenetic as well as structure based analysis of various catalytic domains in experimentally characterized PKS/NRPS clusters.

We describe below, the recent advances in development of such computational methods, list various available bioinformatics software for carrying out similar analysis and discuss their utility in the discovery of new biosynthetic pathways by genome mining.

PREDICTION OF DOMAIN ORGANIZATION IN PKS/NRPS GENE CLUSTERS

The domain organization of any multifunctional protein can usually be predicted by aligning the protein sequence to profiles of various functional domains stored in standard databases like CDD (Conserved Domain Database) (Marchler-Bauer *et al.*, 2005), InterPro (Mulder *et al.*, 2003) etc. However, these generalized domain identification tools often fail to detect the presence of certain catalytic domains or do not depict the correct domain boundaries. Therefore, based on a comprehensive analysis of 20 PKS and 22 NRPS gene clusters, Yadav *et al* and Ansari *et al* developed a knowledge based method for automated identification of PKS and NRPS catalytic domains (Ansari *et al.*, 2004; Yadav *et al.*, 2003b). The approach essentially involved BLAST alignment of the query sequence with templates of various catalytic domains and in case of domains showing high sequence divergence, multiple templates were used. Domain boundaries were identified based on alignment with crystal structures of homologous standalone proteins. This computational protocol for automated identification of PKS and NRPS domains was implemented in the NRPS-PKS web server and prediction accuracy was benchmarked on an independent set of 32 experimentally characterized PKS and NRPS gene clusters based on correlation of the organization of catalytic domains to the chemical structures of the secondary metabolites (Ansari *et al.*, 2004). The utility of a specialized domain identification tool like NRPS-PKS has been demonstrated by the recent discovery of a missing trans enoylreductase (ER) domain in the PDIM gene cluster (Simeone *et al.*, 2007). Yadav *et al* had hypothesized the possible involvement of trans ER based on the lack of correlation between predicted domain organization of the PDIM cluster and known chemical structure of PDIM. Apart from NRPS-PKS, other tools like ASMPKS (Tae *et al.*, 2007), CLUSTSCAN (Starcevic *et al.*, 2008) and NP.searcher (Li *et al.*, 2009) have been developed recently. Even though they use similar knowledge-based approaches for domain depiction, many of them have additional features like HMM-based domain search and scanning of complete genomes for PKS/NRPS gene clusters. Table 1 shows a comparison of various features available in each of these tools.

The domain depiction by NRPS-PKS as well as these recently available tools is based on the boundaries of various catalytic domains as defined by early studies of Donadio and Katz (Donadio & Katz, 1992). Thus, long amino acid stretches intervening between the catalytic domains are defined as linker regions. However, recently available crystal structures of modular PKS fragments and mammalian FAS have indicated new domain boundaries for DH and KR domains. Very recently, the SBSPKS (Anand *et al.*, 2010) web server has been developed for the depiction of PKS structural and catalytic domains based on comparisons with recently available crystal structures (Table 1).

PREDICTION OF SUBSTRATE SPECIFICITY OF VARIOUS CATALYTIC DOMAINS

Substrate Specificity of AT and A Domains

The starter and extender units are recognized by AT domains of PKS and A domains of NRPS during the biosynthesis of polyketides and nonribosomal peptides. Therefore, apart from the organization of catalytic domains, the substrate specificity of AT and A domains are major determinants of structural diversity of secondary metabolites. Hence, prediction of the substrate specificity of AT and A domains has been major focus of several bioinformatics and phylogenetic studies. While AT domains are known to select 10 to 15 mono- or dicarboxylic acid substrates, the number of possible A domains of NRPS can be as high as 200 including many non-proteinogenic amino acids.

Even though the idea of deciphering the NRPS biosynthetic code from signature sequences dates back to the 1970s, only in 1995, De Crecy-Lagard *et al* (De Crecy-Lagard *et al.*, 1995) attempted to predict the substrate specificity of A domains based on phylogenetic clustering of a set of 55 A domain sequences. However, this approach resulted in a prediction accuracy of less than 50% because of the small data set and lack of substrate binding pocket information in the absence of any three-dimensional structures of A domains of NRPS. Secondly, organism-specific clustering dominated over the substrate specific clustering of A domains. After the availability of the substrate bound crystal structure (Conti *et al.*, 1997) of the N-terminal A domain of PheA from the gramicidin S synthetase NRPS cluster, the putative substrate

Table 1. Comparison of available software for analysis of PKS and NRPS biosynthetic pathways

Program name	NRPS/ PKS	Domain prediction	BLAST/ HMM/ SVM	AT /A domain specificity	KR stereo-specificity	Identification of inactive catalytic domains	Interface for scanning complete genome sequences	Prediction of product chemical structure	Order of substrate channeling	Structure modeling
NRPS-PKS	Both	Yes	BLAST	Yes	No	No	No	No	No	No
ASMPKS/MAPSI	PKS	Yes	BLAST	Yes	No	No	Yes	Yes	No	No
CLUSTSCAN	PKS	Yes	HMM	Yes	Yes	Yes	Yes	Yes	No	No
NP.searcher	Both	Yes	BLAST	Yes	No	No	Yes	Yes	No	No
NRPSpredictor	NRPS	No	TSVM	Yes	NA	-	-	-	No	No
PKS/NRPS analysis website Univ. of Maryland (Bachmann & Ravel, 2009)	Both	Yes	HMM	No	No	No	No	No	No	No
SBSPKS	Both	Yes	BLAST	Yes	Yes	No	No	Yes	Yes	Yes (PKS only)

binding pocket residues could be identified for various A domains based on their alignment to the crystal structure. Analysis by Stachelhaus *et al* as well as Challis *et al* demonstrated that, if A domains are phylogenetically clustered based on their putative substrate binding pocket residues, various clusters showed a much better correlation with substrate specificity than was possible using complete sequences of A domains (Challis *et al.*, 2000; Stachelhaus *et al.*, 1999). These studies facilitated the prediction of the substrate specificities of A domains based on a limited number of specificity-determining residues (SDR). Alteration of the substrate specificities of A domains by experimental studies from Stachelhaus *et al* (Stachelhaus *et al.*, 1999) involving site directed mutagenesis of SDRs further established the validity of the NRPS recognition code elucidated by these two pioneering studies. Based on the specificity code proposed by Stachelhaus *et al* and Challis *et al* (Challis *et al.*, 2000; Stachelhaus *et al.*, 1999), Ansari *et al* (Ansari *et al.*, 2004) developed the first automated tool for the prediction of A domain specificity in the NRPS-PKS web server and by systematic benchmarking on a data set of 90 A domains, demonstrated that, using this tool, substrate specificity of A domains can be predicted with an accuracy of 85% (Ansari *et al.*, 2004). The predictions by Ansari *et al* (Ansari *et al.*, 2004) were based on the simple comparison of putative binding pocket residues in the query sequence of an adenylation domain to the corresponding binding pocket residues in a data set of A domains with known specificity. However, subsequently, Rausch *et al* developed the NRPSpredictor (Rausch *et al.*, 2005) tool using a machine-learning method like the transductive support vector machine (TSVM) and a feature vector consisting of 12 different physico-chemical characteristics of the binding pocket residues. Currently, apart from NRPS-PKS and NRPSpredictor (Rausch *et al.*, 2005), NP.searcher (Li *et al.*, 2009) software can also predict A domain specificity, but the substrate specificity prediction tool for A domains in NP.searcher has been implemented using NRPSpredictor.

In order to predict the substrate specificity of AT domains in PKS, Yadav *et al* (Yadav *et al.*, 2003a) identified the specificity-determining residues in 187 PKS AT domains using the crystal structure (PDB ID 1MLA) of acyltransferase from the E.coli FAS structure as a template, as no structure of an AT domain from PKS was available at the time of the study. The phylogenetic analysis of these active site residues showed distinct clusters for malonate and methylmalonate specific AT domains, revealing conserved binding pocket motifs QQGHS[QMI]GRSHT[NS]V for methylmalonate-specific AT domains, and QQGHS[LVIFAM]GR[FP]H[ANTGEDS][NHQ]V for malonate-specific AT domains. Starter AT domains specific for monocarboxyl acid substrates also formed a separate cluster and lacked the conserved arginine present in GRSH or GRFH motifs. Earlier studies (Rangan & Smith, 1997) had proposed the involvement of arginine in the GRSH or GRFH motif in the recognition of dicarboxylic acid substrates. Molecular modeling studies (Yadav *et al.*, 2003a) also provided an elegant rationale for Ser to Phe mutation at position 200 (numbering as in 1MLA) in the GRSH motif resulting in the alteration of substrate specificity of AT domains. The results of this study were in agreement with an experimental mutagenesis study by Reeves *et al* (Reeves *et al.*, 2001) and were subsequently confirmed by Trivedi *et al* (Trivedi *et al.*, 2005). Thus the studies by Yadav *et al* (Yadav *et al.*, 2003a) demonstrated the feasibility of predicting substrate specificity of AT domains based on 13 putative binding pocket residues and this approach was implemented in the NRPS-PKS software for automated prediction of AT specificity. Other software like ASMPKS (Tae *et al.*, 2007), Clustscan (Starcevic *et al.*, 2008) and NP.searcher (Li *et al.*, 2009) have also implemented a similar approach for the prediction of AT specificity.

In contrast to the above mentioned structure-based methods used for predicting specificities of AT and A domains, Minowa *et al* have used a quantitative evolutionary tracing method for identifying the substrate-specific residues in AT, A and CoA ligase domains which are present in the starter modules of PKS and NRPS (Minowa *et al.*, 2007). The tracing involved the partitioning of available sequences by phylogenetic analysis into various clusters comprising homologous sequences. The multiple sequence alignment of all these sequences was done to obtain the variability score at each position as described by Landgraf *et al.* (1999). The variability of a residue in the overall alignment but its conservation in a particular cluster would indicate its criticality in substrate determination. They created HMM profiles from the residues obtained by evolutionary tracing and used the HMM profiles for predicting substrate specificities of AT and A domains. It is interesting to note that, Minowa *et al* were able to find specificity-determining residues for distinction between methylmalonyl-CoA, ethylmalonyl-CoA and methoxymalonyl-CoA which were not effectively distinguished by previous studies (Minowa *et al.*, 2007).

Substrate Specificity of KS and C Domains

The KS and C domains are responsible for condensation of bound substrates in PKSs and NRPSs, respectively. Therefore, several attempts have been made to investigate whether KS and C domains show stringent specificity towards the substrates they condense. In contrast to AT domain phylogeny, which shows evidence of both gene duplication and horizontal gene transfer (Jenke-Kodama *et al.*, 2005), analysis of KS domains from modular PKSs shows a monophylectic relationship indicating evolution by gene duplication. Therefore, KS domains of modular PKSs do not cluster as per their substrates, except for KS domains that accept the peptidyl substrates in PKS/NRPS hybrids and those involved in PKS loading modules like the KSq domains (Ginolhac *et al.*, 2005; Moffitt & Neilan, 2003). However, a recent study (Nguyen *et al.*, 2008) has indicated that phylogenetic clustering of KS domains in trans-AT systems shows a very high degree of correlation with the substrate specificity. In fact, this study demonstrated the feasibility of predicting the chemical structure of polyketide products of trans-AT systems based on analysis of KS domain sequences and the prediction was verified experimentally. Similarly to the KS domains of modular PKSs, the C domains of NRPSs do not show clustering as per the types of substrates they condense. However, C domains show phylogenetic clustering as per the chirality of the amino acids they condense (Rausch *et al.*, 2007) and based on this clustering, motifs responsible for stereo selectivity of condensation domains have been identified. It may be noted that, even though various bioinformatics studies have analyzed the substrate specificity of KS and C domains, none of the automated tools available, as of today, permit prediction of substrate specificity of KS and C domains.

Prediction of Stereo Specificity of KR Domains

KR domains catalyze the reduction of a carbonyl group to a hydroxyl moiety. The orientation of the hydroxyl group is dictated by the stereo specificity of the corresponding KR domain. The stereo specificity determining residues for KR domains in PKSs have been predicted using sequence and structure based studies (Caffrey, 2003; Keatinge-Clay, 2007) and these predictions have been verified using site-directed mutagenesis studies (Baerga-Ortiz *et al.*, 2006) and biochemical experiments (Valenzano *et al.*, 2009). Analysis of a data set consisting of 68 KRs indicated that the motif LDD from 93 to 95 (numbering as per alignment with DEBS KR1) was conserved in KRs catalyzing B-type alcohol stereo-chemistry. In this group, D95 is invariant and D94 is sometimes replaced by E, whereas L93 is less conserved. In contrast to B-type KRs, none of the A-type KRs in this dataset had the LDD motif. The region from residue 141 to 148 also differs between the two groups. In the B group, residues 144 and 148 are typically P and N. It has also been shown that some KRs control the chirality of the substituent at the α position (Caffrey, 2003). A recent study indicated that KRs can be divided into 6 categories on the basis of α and β substituent chirality and elucidated the mechanistic aspects governing the stereo-control (Keatinge-Clay, 2007). Based on these bioinformatics analysis, rules for automated prediction of KR stereo specificity have been incorporated into newly available software like Clustscan (Starcevic *et al.*, 2008) and SBSPKS (Anand *et al.*, 2010).

PREDICTION OF NUMBER OF ITERATIONS CATALYZED BY TYPE I ITERATIVE PKSS

In case of type I iterative PKS proteins a single PKS module can catalyze multiple condensation reactions in an iterative manner. Therefore, for *in silico* identification of novel polyketides, apart from deciphering the organization of various catalytic domains and predicting their substrate specificity, it is also necessary to distinguish type I iterative PKS proteins from their modular counterparts and then predict the number of chain condensations they are likely to catalyze. In a very recent study, Yadav *et al* have addressed these questions (Yadav *et al.*, 2009). In view of the distinct phylogenetic origin of iterative PKS proteins, based on bioinformatics analysis of KS domains in experimentally characterized modular and iterative PKSs, Yadav *et al* built HMM profiles which could successfully classify type I iterative, modular, hybrid PKS/NRPS and enedyne type KS domains into different groups. Phylogenetic analysis of iterative KS domain sequences by Yadav *et al* as well as by Moffitt and Neilan (Moffitt & Neilan, 2003), also showed that they show distinct clustering as per the number of iterations they catalyze and the type of reductive chain modifications associated with each chain condensation step. The crystal structure of KS-CLF (Zhu *et al.*, 2007) and other biochemical studies (Morita *et al.*, 2007; Tang *et al.*, 2003) also provided valuable clues regarding the relationship between polyketide products of iterative PKSs and cavity volume as well as the chemical environment of the KS domain active site pocket. Based on this information, Yadav *et al* systematically analyzed the structural models of several iterative KS domains and demonstrated that size and hydrophobicity of the KS active site pocket shows interesting correlations with the number of iterations and the degree of saturation of the corresponding polyketide products. This study not only demonstrated the feasibility of predicting the number of iterations based on analysis of residues lining the KS active site pocket, but also identified crucial amino acids which can be mutated to alter the number of iterations catalyzed by a given KS domain (Yadav *et al.*, 2009).

PREDICTION OF THE ORDER OF SUBSTRATE CHANNELING IN MODULAR PKS CLUSTERS

In type I modular PKS or NRPS clusters, within a single ORF, the various modules typically add different chemical moieties to a growing polyketide or nonribosomal peptide chain in the same order as they occur in the ORF. Thus, the polyketide product of a single ORF can usually be predicted by the so called collinearity rule, provided the organization of catalytic domains and their substrate specificities can be deciphered correctly. However, in the case of modular PKS or NRPS clusters consisting of multiple ORFs the order of substrate channeling between the multiple ORFs often deviates from the order of their occurrence on the genome. As can be seen from the example in Figure 4, the ORFs in the rapamycin cluster are used in the order RapA, RapB, RapP followed by RapC for metabolite biosynthesis while their arrangement in the genome is RapB, RapA, RapP followed by RapC.

Therefore, the final secondary metabolite product of the complete cluster cannot be predicted by simplistic application of the collinearity rule. For example, in a secondary metabolite gene cluster consisting of N ORFs, in the absence of knowledge about the cognate order of substrate channeling, the chemical moieties added by the N ORFs can be joined in N! ways resulting in a combinatorial explosion of theoretically possible chemical structures. For a gene cluster consisting of 6 ORFs, the number of possible chemical structures would be 720. Even though the identity of the first and the last ORFs can be deciphered based on the presence of typical loading domains (Moffitt & Neilan, 2003) in the first module and

Figure 4. The importance of the determination of the order of substrate channeling using rapamycin gene cluster as an example. The order of the genes comprising the cluster, on the genome and the way they are translated and utilized in the biosynthesis of rapamycin is different. In such cases, to determine the product structure from the genome sequence, it would be important to know the order in which the translated ORFs channel the substrate.

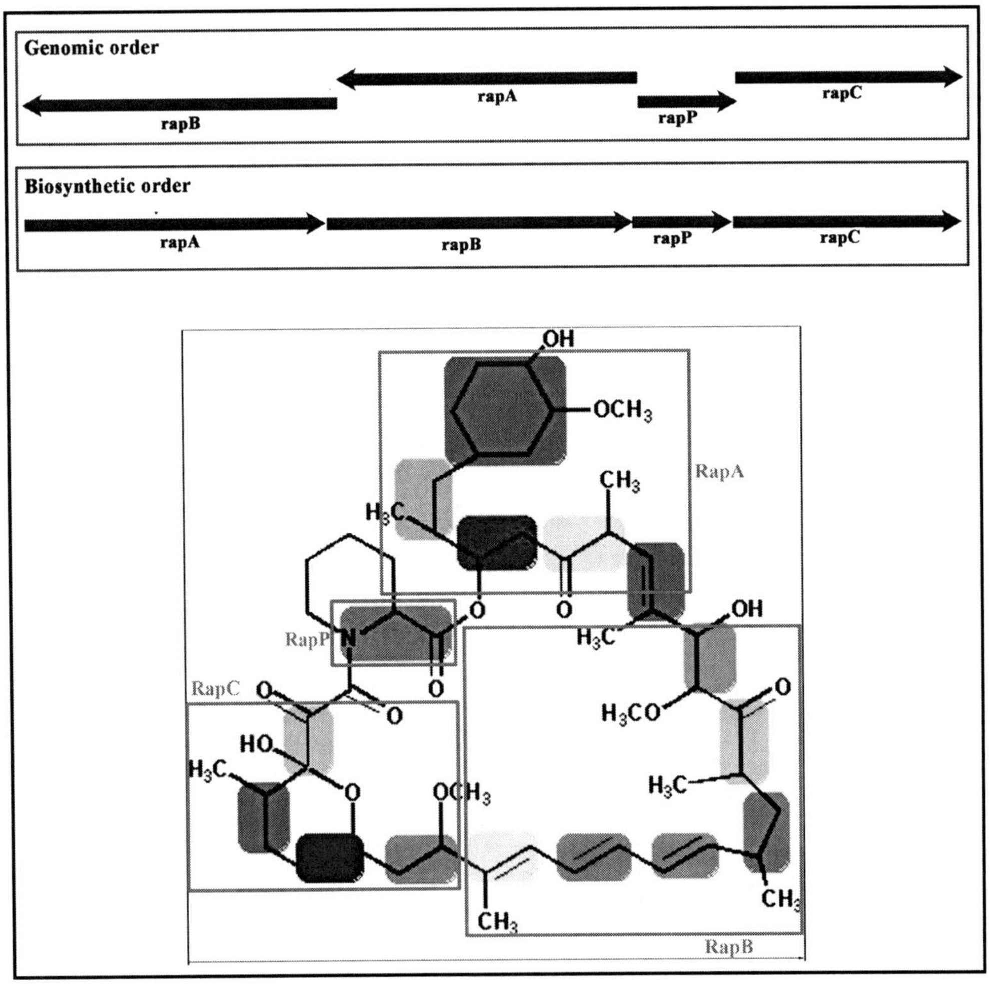

thioesterase (TE) domains for cyclization and chain release in the last module, the number of combinatorial possibilities of chain transfer for the remaining ORFs still remains very large. In view of this several theoretical studies have attempted to develop computational methods for predicting the cognate order of substrate channeling.

The various biochemical studies (Wu *et al.*, 2002) have suggested that the specificity of recognition between various ORFs during biosynthesis arises primarily from two different sources. One source of specificity is the recognition between the last ACP domain of the preceding ORF and the first KS domain of the succeeding ORF. On the other hand, other experimental studies suggest interactions between the C-terminus linkers of the preceding ORF and N-terminus linker of the succeeding ORF to be the primary determinant of intersubunit interactions in modular PKS clusters. Therefore, various theoretical studies have analyzed cognate interactions between ACP and KS domains as well as interactions mediated by inter polypeptide linkers or the so called 'docking domains' for the prediction of the cognate order of substrate channeling.

Minowa *et al.* have analyzed interdomain interactions involving KS and ACP in PKS and C and PCP in NRPS to distinguish cognate pairs from non-cognate pairs and have used this information to predict the order of substrate channeling (Minowa *et al.*, 2007). Since cognate pairs of domains are likely to interact physically, they co-evolve to maintain complementarities of interacting residue pairs. Therefore, Minowa *et al* assigned a log-likelihood score for physical interaction to each pair of ACP-KS and PCP-C domains based on their co-evolution rates.

In contrast to the work of Minowa *et al*, which attempts to predict inter subunit interactions in modular PKSs using sequences of ACP and KS domains, other groups have developed methods for predicting inter subunit interactions using sequences of inter polypeptide linkers (Figure 5).

Both sequence- as well as structure-based approaches have been used for deciphering intersubunit interactions in modular PKS clusters. Thattai *et al* used a sequence-based approach and analyzed N- and C-terminus linkers of type I modular PKS proteins from clusters with known subunit order for substrate channeling (Thattai *et al.*, 2007). They found the 'head regions' corresponding to the C-terminus linkers and 'tail regions' corresponding to the N-terminus linkers clustered into three distinct groups each. Interestingly, all the cognate tail regions corresponding to head regions from one cluster also formed a single cluster and thus each cluster of head regions paired with only a single cluster of tail regions. This study not only revealed the origin of specificity in intersubunit recognition, but also demonstrated the feasibility of predicting intersubunit interactions based on clustering of linker sequences. In this study, the authors also attempted to identify specificity-determining residues in the head and tail regions using the CRoSS algorithm which analyzes sites having statistically significant correlated mutations. Subsequently, Burger and van Nimwegen have also attempted to predict interacting pairs of inter polypeptide linkers using a Bayesian network algorithm on the same data set as used by Thattai *et al.* Burger and van Nimwegen have reported better prediction accuracy compared to the CRoSS algorithm (Burger & van Nimwegen, 2008) .

Novel prediction tools for identifying intersubunit interactions have also been developed based on the three-dimensional structure of inter-polypeptide linkers, which is also called the 'docking domain'. The structure of a docking domain from the erythromycin PKS cluster has been elucidated both by NMR (Broadhurst *et al.*, 2003) as well as crystallographic studies (Buchholz *et al.*, 2009). The three-dimensional structure of the docking domain indicates that intersubunit interactions involve a single helical stretch from the C-terminus linker of the preceding ORF and three helical stretches from the N-terminus linker

Figure 5. The interaction between C-terminal helix of preceding ORF with the three N-terminal helices of the succeeding one. It highlights two different methods used to predict these interactions. The sequence based method (left) is based on the co-evolution of interacting residues within N and C terminal domains. The structure based method predicts the inter subunit interactions on the based on alignment of the inter polypeptide linker sequences to the three dimensional structure of docking domain.

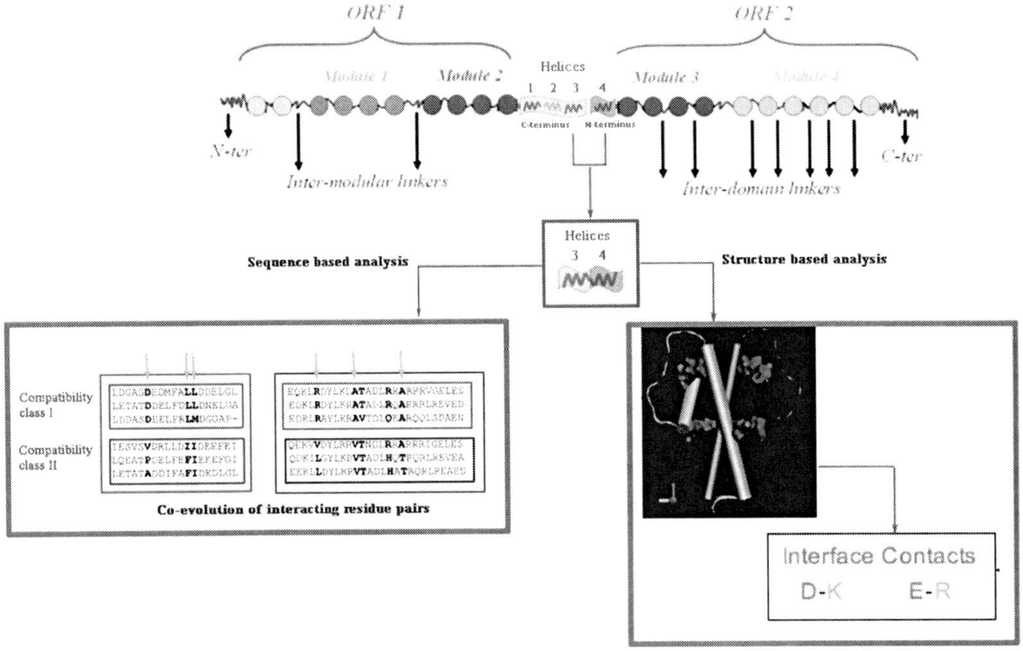

of the succeeding ORF, which together form a four-helix bundle structure (Broadhurst *et al.*, 2003; Weissman & Muller, 2008). It has been proposed that two crucial electrostatic residue pairs in the docking-domain structure mediate intersubunit association during substrate channeling between multiple ORFs in a modular PKS cluster, while unfavorable contacts at equivalent positions in the docking domain are believed to discriminate non-cognate intersubunit associations. This has been referred to as the 'docking code' in the literature (Broadhurst *et al.*, 2003; Weissman, 2006b; Weissman & Muller, 2008). Site directed mutagenesis experiments (Weissman, 2006a) as well as evolutionary analysis (Thattai *et al.*, 2007) of cognate and non-cognate residue pairs in experimentally characterized modular PKS clusters have provided evidence in support of the docking code. Based on this information on docking domains of modular PKS, Yadav *et al* have recently developed a novel structure-based approach to distinguish the cognate combination of ORFs in a modular PKS cluster from all possible non-cognate combinations (Yadav *et al.*, 2009). This automated approach of Yadav *et al* essentially involves identification of these two pairs of crucial specificity-determining residues for each of the intersubunit interfaces based on alignment of inter polypeptide linker sequences to the NMR structure of the docking domain and ranking the interacting residue pairs as favorable, unfavorable and neutral. Using this approach the total number of favorable, unfavorable and neutral interacting residue pairs at all the intersubunit interfaces are computed for all possible combinatorial orders of substrate channeling and the combination with the highest score is ranked as the preferred order of substrate channeling for the given PKS cluster. Benchmarking studies by Yadav *et al* on a data set of 17 modular PKS clusters indicated that, in 14 out of 17 cases the true cognate combination can be ranked within the top 20% in terms of total score. The recent version of SBSPKS (Anand *et al.*, 2010) software provides a web interface for automated extraction of linker sequences and prediction of the order of substrate channeling in case of modular PKS clusters.

Similar to the docking domains in modular PKSs, for NRPS clusters it has been proposed that, COM domains present in the inter polypeptide linker regions mediate intersubunit interactions (Hahn & Stachelhaus, 2004, 2006; Richter *et al.*, 2008). However, no automated bioinformatics tools are yet available for analysis of COM domains and prediction of the order of substrate channeling in case of NRPS clusters.

PREDICTION OF SPECIFICITY OF THIOESTERASE DOMAINS AND TAILORING ENZYMES

The various computational methods discussed earlier only permit the prediction of the chemical structure of the linear polyketide or nonribosomal peptide chain. However, the chemical structure of the final secondary metabolite is also governed by the types of cyclization reaction catalyzed by the thioesterase (TE) domains and also the substrate specificity of other tailoring enzymes like methyltransferases (MT) and glycosyltransferases (GTrs) etc. Even though crystal structures have been available for TE domains of both PKS and NRPS, no computational methods are available, as of today, for predicting the cyclization pattern. On the other hand, bioinformatics based analysis of MT and GTr enzymes of secondary metabolite biosynthetic pathways has led to the development of knowledge-based methods for predicting substrate specificities of these enzymes.

The SearchGTr web server has been developed for the prediction of donor/acceptor specificity of glycosyltransferases present in NRPS/PKS clusters (Kamra *et al.*, 2005). Ansari *et al* (Ansari *et al.*, 2008) have developed profile HMMs for the identification of MTs and their classification as C-methyl, N-methyl and O-methyl transferases. The NP.Searcher (Li *et al.*, 2009) software also predicts different post-assembly chain modification reactions like halogenation, hydroxylation, glycosylation etc catalyzed by different tailoring enzymes.

APPLICATION OF *IN SILICO* PREDICTIONS IN EXPERIMENTAL STUDIES INVOLVING DISCOVERY OF NEW SECONDARY METABOLITES & REPROGRAMMING OF KNOWN BIOSYNTHETIC PATHWAYS

The various knowledge-based computational approaches discussed earlier have played a major role in the elucidation of novel secondary metabolite biosynthetic pathways and the generation of novel secondary metabolites by reprogramming of known biosynthetic pathways. Most of the computational methods for analysis of PKS and NRPS proteins have also been made available as web servers and their utility in secondary metabolite biosynthesis research has been discussed in recent reviews (Bachmann & Ravel, 2009; Jenke-Kodama & Dittmann, 2009) on bioinformatics methods for analysis of secondary metabolite biosynthetic pathways.

Bioinformatics analysis at Thallion Pharmaceuticals (formerly Ecopia BioSciences) has led to the genomics-driven discovery of cryptic biosynthetic pathways (Zazopoulos *et al.*, 2003) and the identification of novel secondary metabolites (McAlpine *et al.*, 2005). Similar bioinformatics analysis by Lautru *et al* indicated the presence of a novel NRPS cluster which could biosynthesize a peptide with hydroxamic groups and hence having a siderophore type of function. This knowledge helped to purify the coelichelin siderophore, whereby an iron deficient media was provided and the ferric hydroxamate complex could be detected by UV-vis spectroscopy (Lautru *et al.*, 2005). However, the experimental results showed the presence of a tetrapeptide instead of the expected tripeptide due to the iterative use of the first module. A similar study led to the discovery of lipopeptide orfamide A from Pseudomonas fluorescens Pf-5 genome (Gross *et al.*, 2007). The prediction of A domain substrate specificity indicated the synthesis of lipopeptides with 4 Leu residues. This helped to devise a strategy which involved feeding ^{15}N-labelled Leu to cultures and ^{1}H-^{15}N HMBC NMR experiments were then used to identify the labeled metabolites. The domain organization and substrate specificity of AT domains in PKSs have also been used to determine and characterize various natural products using their physicochemical properties (Tohyama *et al.*, 2006). Similar strategy has also helped in the identification of new products in actinobacteria (Banskota, McAlpine, Sorensen, Aouidate *et al.*, 2006; Banskota, McAlpine, Sorensen, Ibrahim *et al.*, 2006; McAlpine *et al.*, 2005) and Saccharopolyspora (Zirkle *et al.*, 2004) species. Recently, two new gene clusters were found in the Aspergillus nidulans genome. Examination of flanking genes indicated the presence of a regulatory gene CtnR, a citrinin biosynthesis regulator. The new product found to be synthesized by these PKS clusters was named asperfuranone (Chiang *et al.*, 2009). Apart from the discovery of new biosynthetic pathways and novel secondary metabolites, bioinformatics analysis has also helped in elucidation of missing links in known biosynthetic pathways, some notable examples being, the discovery of the so called missing trans acting enoyl reductase (Simeone *et al.*, 2007) in the PDIM biosynthetic pathway and identification of a novel N-acyltransferase (NAT) gene involved in N-acylation of Lys in the mycobactin biosynthetic pathway (Krithika *et al.*, 2006) of *Mycobacterium tuberculosis*.

Bioinformatics analysis has also helped in the generation of novel secondary metabolites by a rational design approach. For example, reprogramming (Trivedi *et al.*, 2005) of the PDIM biosynthetic pathway by site directed mutagenesis of the acyltransferase domain to produce PDIM analogues lacking methyl branches. In this study, the S200F mutation was carried out in the mas gene of the PDIM cluster based on the bioinformatics analysis of Yadav *et al* (Yadav *et al.*, 2003a). Similar bioinformatics guided site-directed mutagenesis studies have also succeeded in altering the substrate specificity of A domains in NRPSs (Stachelhaus *et al.*, 1999) and have demonstrated the feasibility of altering the primary structure of important pharmaceutical compounds by a rational design approach. Comparison of predicted A domain specificities for a NRPS cluster with that of the amino acids in the identified product has led to identification of modules that have been skipped during biosynthesis. The pentapeptides, Myxochromides S1-S3 were shown to be synthesized by a hexamodular NRPS encoding gene cluster. The prediction of substrate specificity of it's A domains indicated skipping of module 4 in the assembly process (Wenzel *et al.*, 2005). Further studies showed that the absence of conserved Serine in the PCP domain of this module was responsible for the skipping of this module (Wenzel *et al.*, 2006).

FUTURE DIRECTIONS

The various computational methods discussed earlier essentially use a knowledge-based approach and the prediction rules have been derived from our current knowledge of various different paradigms of secondary metabolite biosynthesis. On the other hand, experimental characterization of PKS clusters in an increasing number of organisms has highlighted major deviations from these canonical biosynthetic paradigms like module skipping (Oh *et al.*, 2008), programmed iterations in modular PKS (Tatsuno *et al.*, 2007), intermolecular iterations (Chopra *et al.*, 2008), presence of core catalytic domains in trans (Lopanik *et al.*, 2008; Moldenhauer *et al.*) and hybrids of type I and type III PKS (Austin *et al.*, 2006; Sankaranarayanan, 2006). The secondary metabolite products of gene clusters having such deviations from the standard biosynthetic paradigm cannot be predicted by the currently available computational tools. This necessitates a comprehensive analysis of the evolution of PKS/NRPS biosynthetic pathways to derive new predictive rules for correlating genes to metabolites.

It might be appropriate to use a chemical systems biology approach and describe the secondary metabolite biosynthetic pathways as networks of catalytic domains, which evolve to generate metabolic diversity not only by sequence changes in individual catalytic domains or nodes of the network, but also by addition or deletion of nodes, addition or deletion of complete sub-networks or constituent network modules. Since the phenotype of a given network topology is a small molecule with a well defined chemical structure, it might be possible to establish easy correlations between substructures in chemical compound space with network modules in genomic space. Such a systems biology description of secondary metabolite biosynthesis must also have a framework for including factors governing regulation of secondary metabolism. Secondary metabolite biosynthetic gene clusters are under regulatory control of different transcriptional regulators, which are in turn controlled by different environmental signals. Several recent studies have unraveled the regulatory mechanisms controlling biosynthesis of diverse groups of secondary metabolites (Bate *et al.*, 2006; Bunet *et al.*, 2008). These regulatory genes may be present in the flanking regions of the gene cluster. The identification of these genes may be useful to harness the complete biosynthetic potential of diverse microorganisms as well as to find techniques to induce these cryptic genes in experimental studies. Experimental characterization (Chiang *et al.*, 2009; Krithika *et al.*, 2006) of regulatory genes for several secondary metabolite biosynthetic pathways has opened up the possibility of developing computational methods for identifying such regulatory networks of secondary metabolism by genome analysis.

CONCLUSION

During the last six to seven years, enormous progress has been made in the development of novel computational methods for the identification of new secondary metabolites by analysis of genome sequences. These computational methods are playing an increasingly important role in the discovery of new biosynthetic pathways by genome mining and the generation of novel secondary metabolites by rational design. Several studies have demonstrated the feasibility of predicting chemical structures of secondary metabolites by genome mining. Most of these computational methods use a knowledge-based approach involving the bioinformatics analysis of experimentally characterized secondary metabolite gene clusters with known natural products. Apart from information derived from known natural product biosynthetic pathways, these computational methods also use information from the structural modeling of various enzymatic domains. Thus, the availability of structural information on various catalytic domains and interdomain interactions will help in further improving the predictive ability of these methods. In this review we have given a brief overview of these methods, while the details on each of these methods can be found from the original published literatures cited in this work. We have also discussed a few examples where such computational methods have successfully complemented experimental efforts towards discovery of new secondary metabolites and generation of rationally altered metabolites with desired chemical structures. Since many of these computational methods are also available as web servers, new genomic sequences can be easily analyzed by these tools for searching novel natural product biosynthetic pathways and rational design of novel secondary metabolites by biosynthetic engineering approach.

ACKNOWLEDGMENT

SA thanks CSIR, India for award of senior research fellowship. The work has been supported by grants to National Institute of Immunology from Department of Biotechnology (DBT), India and grants to DM under BTIS project of DBT, India.

REFERENCES

Anand, S., Prasad, M.V., Yadav, G., Kumar, N., Shehara, J., Ansari, M.Z., et al. (2010). SBSPKS: Structure based sequence analysis of polyketide synthases. *Nucleic Acids Research, 38*(Web server issue), W487-496.

Ansari, M. Z., Sharma, J., Gokhale, R. S., & Mohanty, D. (2008). In silico analysis of methyltransferase domains involved in biosynthesis of secondary metabolites. *BMC Bioinformatics*, *9*, 454. doi:10.1186/1471-2105-9-454

Ansari, M.Z., Yadav, G., Gokhale, R.S. & Mohanty, D. (2004). NRPS-PKS: A knowledge-based resource for analysis of NRPS/PKS megasynthases. *Nucleic Acids Research, 32*(Web Server issue), W405-413.

Austin, M. B., Saito, T., Bowman, M. E., Haydock, S., Kato, A., & Moore, B. S. (2006). Biosynthesis of Dictyostelium discoideum differentiation-inducing factor by a hybrid type I fatty acid-type III polyketide synthase. *Nature Chemical Biology*, *2*(9), 494–502. doi:10.1038/nchembio811

Bachmann, B. O., & Ravel, J. (2009). Chapter 8. Methods for in silico prediction of microbial polyketide and nonribosomal peptide biosynthetic pathways from DNA sequence data. *Methods in Enzymology*, *458*, 181–217. doi:10.1016/S0076-6879(09)04808-3

Baerga-Ortiz, A., Popovic, B., Siskos, A. P., O'Hare, H. M., Spiteller, D., & Williams, M. G. (2006). Directed mutagenesis alters the stereochemistry of catalysis by isolated ketoreductase domains from the erythromycin polyketide synthase. *Chemistry & Biology*, *13*(3), 277–285. doi:10.1016/j.chembiol.2006.01.004

Baltz, R. H. (2006). Molecular engineering approaches to peptide, polyketide and other antibiotics. *Nature Biotechnology*, *24*(12), 1533–1540. doi:10.1038/nbt1265

Banskota, A. H., McAlpine, J. B., Sorensen, D., Aouidate, M., Piraee, M., & Alarco, A. M. (2006). Isolation and identification of three new 5-alkenyl-3,3(2H)-furanones from two streptomyces species using a genomic screening approach. *The Journal of Antibiotics*, *59*(3), 168–176. doi:10.1038/ja.2006.24

Banskota, A. H., McAlpine, J. B., Sorensen, D., Ibrahim, A., Aouidate, M., & Piraee, M. (2006). Genomic analyses lead to novel secondary metabolites. Part 3. ECO-0501, a novel antibacterial of a new class. *The Journal of Antibiotics*, *59*(9), 533–542. doi:10.1038/ja.2006.74

Bate, N., Bignell, D. R., & Cundliffe, E. (2006). Regulation of tylosin biosynthesis involving SARP-helper activity. *Molecular Microbiology*, *62*(1), 148–156. doi:10.1111/j.1365-2958.2006.05338.x

Belshaw, P. J., Walsh, C. T., & Stachelhaus, T. (1999). Aminoacyl-CoAs as probes of condensation domain selectivity in nonribosomal peptide synthesis. *Science*, *284*(5413), 486–489. doi:10.1126/science.284.5413.486

Broadhurst, R. W., Nietlispach, D., Wheatcroft, M. P., Leadlay, P. F., & Weissman, K. J. (2003). The structure of docking domains in modular polyketide synthases. *Chemistry & Biology*, *10*(8), 723–731. doi:10.1016/S1074-5521(03)00156-X

Buchholz, T. J., Geders, T. W., Bartley, F. E. III, Reynolds, K. A., Smith, J. L., & Sherman, D. H. (2009). Structural basis for binding specificity between subclasses of modular polyketide synthase docking domains. *ACS Chemical Biology*, *4*(1), 41–52. doi:10.1021/cb8002607

Bunet, R., Mendes, M. V., Rouhier, N., Pang, X., Hotel, L., & Leblond, P. (2008). Regulation of the synthesis of the angucyclinone antibiotic alpomycin in Streptomyces ambofaciens by the autoregulator receptor AlpZ and its specific ligand. *Journal of Bacteriology*, *190*(9), 3293–3305. doi:10.1128/JB.01989-07

Burger, L., & van Nimwegen, E. (2008). Accurate prediction of protein-protein interactions from sequence alignments using a Bayesian method. *Molecular Systems Biology*, *4*, 165. doi:10.1038/msb4100203

Caffrey, P. (2003). Conserved amino acid residues correlating with ketoreductase stereospecificity in modular polyketide synthases. *ChemBioChem*, *4*(7), 654–657. doi:10.1002/cbic.200300581

Challis, G. L., Ravel, J., & Townsend, C. A. (2000). Predictive, structure-based model of amino acid recognition by nonribosomal peptide synthetase adenylation domains. *Chemistry & Biology*, *7*(3), 211–224. doi:10.1016/S1074-5521(00)00091-0

Chiang, Y. M., Szewczyk, E., Davidson, A. D., Keller, N., Oakley, B. R., & Wang, C. C. (2009). A gene cluster containing two fungal polyketide synthases encodes the biosynthetic pathway for a polyketide, asperfuranone, in Aspergillus nidulans. *Journal of the American Chemical Society*, *131*(8), 2965–2970. doi:10.1021/ja8088185

Chopra, T., Banerjee, S., Gupta, S., Yadav, G., Anand, S., & Surolia, A. (2008). Novel intermolecular iterative mechanism for biosynthesis of mycoketide catalyzed by a bimodular polyketide synthase. *PLoS Biology*, *6*(7), e163. doi:10.1371/journal.pbio.0060163

Clugston, S. L., Sieber, S. A., Marahiel, M. A., & Walsh, C. T. (2003). Chirality of peptide bond-forming condensation domains in nonribosomal peptide synthetases: The C5 domain of tyrocidine synthetase is a (D)C(L) catalyst. *Biochemistry*, *42*(41), 12095–12104. doi:10.1021/bi035090+

Conti, E., Stachelhaus, T., Marahiel, M. A., & Brick, P. (1997). Structural basis for the activation of phenylalanine in the non-ribosomal biosynthesis of gramicidin S. *The EMBO Journal*, *16*(14), 4174–4183. doi:10.1093/emboj/16.14.4174

Cox, R. J. (2007). Polyketides, proteins and genes in fungi: Programmed nano-machines begin to reveal their secrets. *Organic & Biomolecular Chemistry*, *5*(13), 2010–2026. doi:10.1039/b704420h

De Crecy-Lagard, V., Marliere, P., & Saurin, W. (1995). Multienzymatic non ribosomal peptide biosynthesis: Identification of the functional domains catalysing peptide elongation and epimerisation. *Comptes Rondus de l'Academie des Sciences III*, *318*(9), 927–936.

Donadio, S., & Katz, L. (1992). Organization of the enzymatic domains in the multifunctional polyketide synthase involved in erythromycin formation in Saccharopolyspora erythraea. *Gene*, *111*(1), 51–60. doi:10.1016/0378-1119(92)90602-L

Du, L., Sanchez, C., Chen, M., Edwards, D. J., & Shen, B. (2000). The biosynthetic gene cluster for the antitumor drug bleomycin from Streptomyces verticillus ATCC15003 supporting functional interactions between nonribosomal peptide synthetases and a polyketide synthase. *Chemistry & Biology*, *7*(8), 623–642. doi:10.1016/S1074-5521(00)00011-9

Fischbach, M. A., & Walsh, C. T. (2006). Assembly-line enzymology for polyketide and nonribosomal Peptide antibiotics: Logic, machinery, and mechanisms. *Chemical Reviews*, *106*(8), 3468–3496. doi:10.1021/cr0503097

Ginolhac, A., Jarrin, C., Robe, P., Perriere, G., Vogel, T. M., & Simonet, P. (2005). Type I polyketide synthases may have evolved through horizontal gene transfer. *Journal of Molecular Evolution*, *60*(6), 716–725. doi:10.1007/s00239-004-0161-1

Gross, H., Stockwell, V. O., Henkels, M. D., Nowak-Thompson, B., Loper, J. E., & Gerwick, W. H. (2007). The genomisotopic approach: A systematic method to isolate products of orphan biosynthetic gene clusters. *Chemistry & Biology*, *14*(1), 53–63. doi:10.1016/j.chembiol.2006.11.007

Hahn, M., & Stachelhaus, T. (2004). Selective interaction between nonribosomal peptide synthetases is facilitated by short communication-mediating domains. *Proceedings of the National Academy of Sciences of the United States of America*, *101*(44), 15585–15590. doi:10.1073/pnas.0404932101

Hahn, M., & Stachelhaus, T. (2006). Harnessing the potential of communication-mediating domains for the biocombinatorial synthesis of nonribosomal peptides. *Proceedings of the National Academy of Sciences of the United States of America*, *103*(2), 275–280. doi:10.1073/pnas.0508409103

Hill, A. M. (2006). The biosynthesis, molecular genetics, and enzymology of the polyketide-derived metabolites. *Natural Product Reports*, *23*(2), 256–320. doi:10.1039/b301028g

Jenke-Kodama, H., & Dittmann, E. (2009). Bioinformatic perspectives on NRPS/PKS megasynthases: Advances and challenges. *Natural Product Reports*, *26*(7), 874–883. doi:10.1039/b810283j

Jenke-Kodama, H., Sandmann, A., Muller, R., & Dittmann, E. (2005). Evolutionary implications of bacterial polyketide synthases. *Molecular Biology and Evolution*, *22*(10), 2027–2039. doi:10.1093/molbev/msi193

Kamra, P., Gokhale, R.S. & Mohanty, D. (2005). SEARCHGTr: A program for analysis of glycosyltransferases involved in glycosylation of secondary metabolites. *Nucleic Acids Research, 33*(Web Server issue), W220-225.

Keatinge-Clay, A. T. (2007). A tylosin ketoreductase reveals how chirality is determined in polyketides. *Chemistry & Biology*, *14*(8), 898–908. doi:10.1016/j.chembiol.2007.07.009

Krithika, R., Marathe, U., Saxena, P., Ansari, M. Z., Mohanty, D., & Gokhale, R. S. (2006). A genetic locus required for iron acquisition in Mycobacterium tuberculosis. *Proceedings of the National Academy of Sciences of the United States of America*, *103*(7), 2069–2074. doi:10.1073/pnas.0507924103

Landgraf, R., Fischer, D., & Eisenberg, D. (1999). Analysis of heregulin symmetry by weighted evolutionary tracing. *Protein Engineering*, *12*(11), 943–951. doi:10.1093/protein/12.11.943

Lautru, S., Deeth, R. J., Bailey, L. M., & Challis, G. L. (2005). Discovery of a new peptide natural product by Streptomyces coelicolor genome mining. *Nature Chemical Biology*, *1*(5), 265–269. doi:10.1038/nchembio731

Li, M. H., Ung, P. M., Zajkowski, J., Garneau-Tsodikova, S., & Sherman, D. H. (2009). Automated genome mining for natural products. *BMC Bioinformatics*, *10*, 185. doi:10.1186/1471-2105-10-185

Linne, U., Stein, D. B., Mootz, H. D., & Marahiel, M. A. (2003). Systematic and quantitative analysis of protein-protein recognition between nonribosomal peptide synthetases investigated in the tyrocidine biosynthetic template. *Biochemistry*, *42*(17), 5114–5124. doi:10.1021/bi034223o

Lopanik, N. B., Shields, J. A., Buchholz, T. J., Rath, C. M., Hothersall, J., & Haygood, M. G. (2008). In vivo and in vitro trans-acylation by BryP, the putative bryostatin pathway acyltransferase derived from an uncultured marine symbiont. *Chemistry & Biology*, *15*(11), 1175–1186. doi:10.1016/j.chembiol.2008.09.013

Marchler-Bauer, A., Anderson, J. B., Cherukuri, P. F., DcWccsc-Scott, C., Geer, L. Y., & Gwadz, M. (2005). CDD: A Conserved Domain Database for protein classification. *Nucleic Acids Research*, *33*(Database issue), D192–D196. doi:10.1093/nar/gki069

McAlpine, J. B., Bachmann, B. O., Piraee, M., Tremblay, S., Alarco, A. M., & Zazopoulos, E. (2005). Microbial genomics as a guide to drug discovery and structural elucidation: ECO-02301, a novel antifungal agent, as an example. *Journal of Natural Products*, *68*(4), 493–496. doi:10.1021/np0401664

Minowa, Y., Araki, M., & Kanehisa, M. (2007). Comprehensive analysis of distinctive polyketide and nonribosomal peptide structural motifs encoded in microbial genomes. *Journal of Molecular Biology*, *368*(5), 1500–1517. doi:10.1016/j.jmb.2007.02.099

Moffitt, M. C., & Neilan, B. A. (2003). Evolutionary affiliations within the superfamily of ketosynthases reflect complex pathway associations. *Journal of Molecular Evolution*, *56*(4), 446–457. doi:10.1007/s00239-002-2415-0

Moldenhauer, J., Gotz, D. C., Albert, C. R., Bischof, S. K., Schneider, K., & Sussmuth, R. D. (2010). The final steps of bacillaene biosynthesis in Bacillus amyloliquefaciens FZB42: Direct evidence for beta, gamma dehydration by a trans-acyltransferase polyketide synthase. *Angewandte Chemie International Edition*, *49*(8), 1465–1467.

Mootz, H. D., Schwarzer, D., & Marahiel, M. A. (2002). Ways of assembling complex natural products on modular nonribosomal peptide synthetases. *ChemBioChem*, *3*(6), 490–504. doi:10.1002/1439-7633(20020603)3:6<490::AID-CBIC490>3.0.CO;2-N

Morita, H., Kondo, S., Oguro, S., Noguchi, H., Sugio, S., & Abe, I. (2007). Structural insight into chain-length control and product specificity of pentaketide chromone synthase from Aloe arborescens. *Chemistry & Biology*, *14*(4), 359–369. doi:10.1016/j.chembiol.2007.02.003

Moss, S. J., Martin, C. J., & Wilkinson, B. (2004). Loss of co-linearity by modular polyketide synthases: A mechanism for the evolution of chemical diversity. *Natural Product Reports*, *21*(5), 575–593. doi:10.1039/b315020h

Mulder, N. J., Apweiler, R., Attwood, T. K., Bairoch, A., Barrell, D., & Bateman, A. (2003). The InterPro Database, 2003 brings increased coverage and new features. *Nucleic Acids Research*, *31*(1), 315–318. doi:10.1093/nar/gkg046

Nguyen, T., Ishida, K., Jenke-Kodama, H., Dittmann, E., Gurgui, C., & Hochmuth, T. (2008). Exploiting the mosaic structure of trans-acyltransferase polyketide synthases for natural product discovery and pathway dissection. *Nature Biotechnology*, *26*(2), 225–233. doi:10.1038/nbt1379

Oh, D. C., Gontang, E. A., Kauffman, C. A., Jensen, P. R., & Fenical, W. (2008). Salinipyrones and pacificanones, mixed-precursor polyketides from the marine actinomycete Salinispora pacifica. *Journal of Natural Products*, *71*(4), 570–575. doi:10.1021/np0705155

Rangan, V. S., & Smith, S. (1997). Alteration of the substrate specificity of the malonyl-CoA/acetyl-CoA: Acyl carrier protein S-acyltransferase domain of the multifunctional fatty acid synthase by mutation of a single arginine residue. *The Journal of Biological Chemistry*, *272*(18), 11975–11978. doi:10.1074/jbc.272.18.11975

Rausch, C., Hoof, I., Weber, T., Wohlleben, W., & Huson, D. H. (2007). Phylogenetic analysis of condensation domains in NRPS sheds light on their functional evolution. *BMC Evolutionary Biology*, *7*, 78. doi:10.1186/1471-2148-7-78

Rausch, C., Weber, T., Kohlbacher, O., Wohlleben, W., & Huson, D. H. (2005). Specificity prediction of adenylation domains in nonribosomal peptide synthetases (NRPS) using transductive support vector machines (TSVMs). *Nucleic Acids Research*, *33*(18), 5799–5808. doi:10.1093/nar/gki885

Reeves, C. D., Murli, S., Ashley, G. W., Piagentini, M., Hutchinson, C. R., & McDaniel, R. (2001). Alteration of the substrate specificity of a modular polyketide synthase acyltransferase domain through site-specific mutations. *Biochemistry*, *40*(51), 15464–15470. doi:10.1021/bi015864r

Richter, C. D., Nietlispach, D., Broadhurst, R. W., & Weissman, K. J. (2008). Multienzyme docking in hybrid megasynthetases. *Nature Chemical Biology*, *4*(1), 75–81. doi:10.1038/nchembio.2007.61

Rix, U., Fischer, C., Remsing, L. L., & Rohr, J. (2002). Modification of post-PKS tailoring steps through combinatorial biosynthesis. *Natural Product Reports*, *19*(5), 542–580. doi:10.1039/b103920m

Sankaranarayanan, R. (2006). A type III PKS makes the difference. *Nature Chemical Biology*, *2*(9), 451–452. doi:10.1038/nchembio0906-451

Schwarzer, D., Finking, R., & Marahiel, M. A. (2003). Nonribosomal peptides: From genes to products. *Natural Product Reports*, *20*(3), 275–287. doi:10.1039/b111145k

Schwecke, T., Aparicio, J. F., Molnar, I., Konig, A., Khaw, L. E., & Haydock, S. F. (1995). The biosynthetic gene cluster for the polyketide immunosuppressant rapamycin. *Proceedings of the National Academy of Sciences of the United States of America*, *92*(17), 7839–7843. doi:10.1073/pnas.92.17.7839

Shen, B. (2003). Polyketide biosynthesis beyond the type I, II and III polyketide synthase paradigms. *Current Opinion in Chemical Biology*, *7*(2), 285–295. doi:10.1016/S1367-5931(03)00020-6

Simeone, R., Constant, P., Guilhot, C., Daffe, M., & Chalut, C. (2007). Identification of the missing trans-acting enoyl reductase required for phthiocerol dimycocerosate and phenolglycolipid biosynthesis in Mycobacterium tuberculosis. *Journal of Bacteriology*, *189*(13), 4597–4602. doi:10.1128/JB.00169-07

Stachelhaus, T., Mootz, H. D., & Marahiel, M. A. (1999). The specificity-conferring code of adenylation domains in nonribosomal peptide synthetases. *Chemistry & Biology*, *6*(8), 493–505. doi:10.1016/S1074-5521(99)80082-9

Starcevic, A., Zucko, J., Simunkovic, J., Long, P. F., Cullum, J., & Hranueli, D. (2008). ClustScan: An integrated program package for the semi-automatic annotation of modular biosynthetic gene clusters and in silico prediction of novel chemical structures. *Nucleic Acids Research*, *36*(21), 6882–6892. doi:10.1093/nar/gkn685

Tae, H., Kong, E. B., & Park, K. (2007). ASMPKS: An analysis system for modular polyketide synthases. *BMC Bioinformatics*, *8*, 327. doi:10.1186/1471-2105-8-327

Tang, Y., Tsai, S. C., & Khosla, C. (2003). Polyketide chain length control by chain length factor. *Journal of the American Chemical Society*, *125*(42), 12708–12709. doi:10.1021/ja0378759

Tatsuno, S., Arakawa, K., & Kinashi, H. (2007). Analysis of modular-iterative mixed biosynthesis of lankacidin by heterologous expression and gene fusion. *The Journal of Antibiotics*, *60*(11), 700–708. doi:10.1038/ja.2007.90

Thattai, M., Burak, Y., & Shraiman, B. I. (2007). The origins of specificity in polyketide synthase protein interactions. *PLoS Computational Biology*, *3*(9), 1827–1835. doi:10.1371/journal.pcbi.0030186

Tohyama, S., Kakinuma, K., & Eguchi, T. (2006). The complete biosynthetic gene cluster of the 28-membered polyketide macrolactones, halstoctacosanolides, from Streptomyces halstedii HC34. *The Journal of Antibiotics*, *59*(1), 44–52. doi:10.1038/ja.2006.7

Trivedi, O. A., Arora, P., Vats, A., Ansari, M. Z., Tickoo, R., & Sridharan, V. (2005). Dissecting the mechanism and assembly of a complex virulence mycobacterial lipid. *Molecular Cell*, *17*(5), 631–643. doi:10.1016/j.molcel.2005.02.009

Valenzano, C. R., Lawson, R. J., Chen, A. Y., Khosla, C., & Cane, D. E. (2009). The biochemical basis for stereochemical control in polyketide biosynthesis. *Journal of the American Chemical Society*, *131*(51), 18501–18511. doi:10.1021/ja908296m

Van Lanen, S. G., & Shen, B. (2006). Microbial genomics for the improvement of natural product discovery. *Current Opinion in Microbiology*, *9*(3), 252–260. doi:10.1016/j.mib.2006.04.002

Weissman, K. J. (2006a). Single amino acid substitutions alter the efficiency of docking in modular polyketide biosynthesis. *ChemBioChem*, *7*(9), 1334–1342. doi:10.1002/cbic.200600185

Weissman, K. J. (2006b). The structural basis for docking in modular polyketide biosynthesis. *ChemBioChem*, *7*(3), 485–494. doi:10.1002/cbic.200500435

Weissman, K. J., & Muller, R. (2008). Protein-protein interactions in multienzyme megasynthetases. *ChemBioChem*, *9*(6), 826–848. doi:10.1002/cbic.200700751

Wenzel, S. C., Kunze, B., Hofle, G., Silakowski, B., Scharfe, M., & Blocker, H. (2005). Structure and biosynthesis of myxochromides S1-3 in Stigmatella aurantiaca: Evidence for an iterative bacterial type I polyketide synthase and for module skipping in nonribosomal peptide biosynthesis. *ChemBioChem*, *6*(2), 375–385. doi:10.1002/cbic.200400282

Wenzel, S. C., Meiser, P., Binz, T. M., Mahmud, T., & Muller, R. (2006). Nonribosomal peptide biosynthesis: Point mutations and module skipping lead to chemical diversity. *Angewandte Chemie International Edition*, *45*(14), 2296–2301. doi:10.1002/anie.200503737

Wu, N., Cane, D. E., & Khosla, C. (2002). Quantitative analysis of the relative contributions of donor acyl carrier proteins, acceptor ketosynthases, and linker regions to intermodular transfer of intermediates in hybrid polyketide synthases. *Biochemistry*, *41*(15), 5056–5066. doi:10.1021/bi012086u

Yadav, G., Gokhale, R. S., & Mohanty, D. (2003a). Computational approach for prediction of domain organization and substrate specificity of modular polyketide synthases. *Journal of Molecular Biology*, *328*(2), 335–363. doi:10.1016/S0022-2836(03)00232-8

Yadav, G., Gokhale, R. S., & Mohanty, D. (2003b). SEARCHPKS: A program for detection and analysis of polyketide synthase domains. *Nucleic Acids Research*, *31*(13), 3654–3658. doi:10.1093/nar/gkg607

Yadav, G., Gokhale, R. S., & Mohanty, D. (2009). Towards prediction of metabolic products of polyketide synthases: An in silico analysis. *PLoS Computational Biology*, *5*(4), e1000351. doi:10.1371/journal.pcbi.1000351

Zazopoulos, E., Huang, K., Staffa, A., Liu, W., Bachmann, B. O., & Nonaka, K. (2003). A genomics-guided approach for discovering and expressing cryptic metabolic pathways. *Nature Biotechnology*, *21*(2), 187–190. doi:10.1038/nbt784

Zhang, M. Q., & Wilkinson, B. (2007). Drug discovery beyond the rule-of-five. *Current Opinion in Biotechnology*, *18*, 1–11. doi:10.1016/j.copbio.2007.10.005

Zhu, X., Yu, F., Li, X. C., & Du, L. (2007). Production of dihydroisocoumarins in Fusarium verticillioides by swapping ketosynthase domain of the fungal iterative polyketide synthase Fum1p with that of lovastatin diketide synthase. *Journal of the American Chemical Society*, *129*(1), 36–37. doi:10.1021/ja0672122

Zirkle, R., Black, T. A., Gorlach, J., Ligon, J. M., & Molnar, I. (2004). Analysis of a 108-kb region of the Saccharopolyspora spinosa genome covering the obscurin polyketide synthase locus. *DNA Sequencing*, *15*(2), 123–134.

ADDITIONAL READING

Bachmann, B. O., & Ravel, J. (2009). Chapter 8. Methods for in silico prediction of microbial polyketide and nonribosomal peptide biosynthetic pathways from DNA sequence data. *Methods in Enzymology*, *458*, 181–217. doi:10.1016/S0076-6879(09)04808-3

Jenke-Kodama, H., & Dittmann, E. (2009). Bioinformatic perspectives on NRPS/PKS megasynthases: advances and challenges. *Natural Product Reports*, *26*(7), 874–883. doi:10.1039/b810283j

Li, M. H., Ung, P. M., Zajkowski, J., Garneau-Tsodikova, S., & Sherman, D. H. (2009). Automated genome mining for natural products. *BMC Bioinformatics*, *10*, 185. doi:10.1186/1471-2105-10-185

Starcevic, A., Zucko, J., Simunkovic, J., Long, P. F., Cullum, J., & Hranueli, D. (2008). ClustScan: an integrated program package for the semi-automatic annotation of modular biosynthetic gene clusters and in silico prediction of novel chemical structures. *Nucleic Acids Research*, *36*(21), 6882–6892. doi:10.1093/nar/gkn685

Tae, H., Kong, E. B., & Park, K. (2007). ASMPKS: an analysis system for modular polyketide synthases. *BMC Bioinformatics*, *8*, 327. doi:10.1186/1471-2105-8-327

Yadav, G., Gokhale, R. S., & Mohanty, D. (2003a). Computational approach for prediction of domain organization and substrate specificity of modular polyketide synthases. *Journal of Molecular Biology*, *328*(2), 335–363. doi:10.1016/S0022-2836(03)00232-8

Yadav, G., Gokhale, R. S., & Mohanty, D. (2009). Towards prediction of metabolic products of polyketide synthases: an in silico analysis. *PLoS Computational Biology*, *5*(4), e1000351. doi:10.1371/journal.pcbi.1000351

KEY TERMS AND DEFINITIONS

Docking Domain: The term used for the structure formed by terminal linkers of interacting subunits in a gene cluster. The structure constitutes two four-alpha-helix bundles that constitute interacting residues which bring about recognition specificity.

Genome Mining: Genome mining refers to deriving various information about the organism based on genome analysis.

Non-Ribosomal Peptides: A class of peptide secondary metabolites synthesized from proteinogenic or non-proteinogenic amino acid monomers, by large multifunctional proteins called nonribosomal peptide synthetases (NRPS). Unlike ribosomal synthesis, NRPSs do not require messenger RNA.

Polyketides: Polyketides are a diverse class of natural products with various biological activities and pharmacological properties. They are usually biosynthesized through the decarboxylative condensation of malonyl-CoA derived extender units in a process similar to fatty acid biosynthesis by the action of multifunctional megasynthases

Secondary Metabolite: Products of metabolism, which do not influence the growth, development and reproduction in an organism.

Substrate Channeling: The passage of a substrate across multiple ORFs constituting a PKS cluster, often determined by inter subunit interactions between various ORFs.

Type I Polyketide Synthases: They are multifunctional polypeptides, which can be modular (constitute multiple modules) or iterative (one module may act multiple times). Each module constitutes a set of domains, each with a specific catalytic function.

Type II Polyketide Synthases: They are multienzyme complexes containing a single set of domains where each catalytic domain is present on a separate polypeptide chain.

Chapter 19
Linking Interactome to Disease:
A Network-Based Analysis of Metastatic Relapse in Breast Cancer

Maxime Garcia
Inserm, Paoli Calmettes Institute, France

Olivier Stahl
Inserm, Paoli Calmettes Institute, France

Pascal Finetti
Inserm, Paoli Calmettes Institute, France

Daniel Birnbaum
Inserm, Paoli Calmettes Institute, France

François Bertucci
Inserm, Paoli Calmettes Institute, France

Ghislain Bidaut
Inserm, Paoli Calmettes Institute, France

ABSTRACT

The introduction of high-throughput gene expression profiling technologies (DNA microarrays) in molecular biology and their expected applications to the clinic have allowed the design of predictive signatures linked to a particular clinical condition or patient outcome in a given clinical setting. However, it has been shown that such signatures are prone to several problems: (i) they are heavily unstable and linked to the set of patients chosen for training; (ii) data topology is problematic with regard to the data dimensionality (too many variables for too few samples); (iii) diseases such as cancer are provoked by subtle misregulations which cannot be readily detected by current analysis methods. To find a predictive signature generalizable for multiple datasets, a strategy of superimposition of a large scale of protein-protein interaction data (human interactome) was devised over several gene expression datasets (a total of 2,464 breast cancer tumors were integrated), to find discriminative regions in the interactome (subnetworks) predicting metastatic relapse in breast cancer. This method, Interactome-Transcriptome

DOI: 10.4018/978-1-60960-491-2.ch019

Integration (ITI), was applied to several breast cancer DNA microarray datasets and allowed the extraction of a signature constituted by 119 subnetworks. All subnetworks have been stored in a relational database and linked to Gene Ontology and NCBI EntrezGene annotation databases for analysis. Exploration of annotations has shown that this set of subnetworks reflects several biological processes linked to cancer and is a good candidate for establishing a network-based signature for prediction of metastatic relapse in breast cancer.

INTRODUCTION

Since introduction of high-throughput technologies in molecular biology in the late nineties, a number of technologies for deciphering the genomic origin of several diseases has flourished. Among these, cDNA microarrays (Schena et al., 1995) have allowed measuring Gene Expression Profiles (GEP) at the genome scale and have shed light on large scale gene regulation/misregulation under varied conditions. Many diseases, including several forms of cancer [leukemia (Golub et al., 1999), colon cancer (Li et al., 2001), breast cancer (Wang et al., 2005)], diabetes (Kaestner et al., 2003), and others (Munro & Perreau. 2009) have been studied that way. Of particular interest in the context of cancer, a particularly heterogeneous disease, is the use of GEPs to either predict drug resistance (de Lavallade et al, 2010), or the metastatic recurrence, for instance in breast cancer (van de Vijver et al., 2002). Tumor microenvironment studies have allowed understanding the influence of immune system on patient outcome (Pagès et al., 2009).

There is an increasing number of controversial cases for the use of systemic adjuvant therapy due to the clinical and pathological heterogeneity of the disease to treat. In node-negative early breast cancer, most patients undergo adjuvant chemotherapy even though 70-80% of them would have survived without it (Bertucci & Birnbaum, 2008). The refinement of current prognostic histopathological methods using molecular diagnostics can also lead to increase of detection of disease subtypes that necessitate specific treatments, such as T1 breast cancer (Mook et al., 2010). In all cases, the goal is to refine and individualize treatment and lead the way to personalized medicine for a growing number of pathologies.

In cancer, the understanding of molecular basis of metastasis is of primary importance. Several studies have attempted to obtain a molecular portrait for a large number of patients using DNA microarray analysis, performed supervised analysis and published list of genes predicting patient outcome. Two of these signatures are currently under clinical trials in breast cancer: the MINDACT trial, based on the 70-genes Mammaprint signature [van't Veer et al. (2002) van de Vijver et al. (2002), Bueno-de-Mesquita et al. (2007)]), and the TAYLORx trial, an RT-PCR-based 21-genes OncotypeDX signature (Paik et al., 2006).

However, most prognostic signatures reported for breast cancer show very little or no overlap, and do not appear generalizable from one study to another, and this un-reproducibility was widely criticized (Chuang et al., 2007, Bertucci et al., 2008). Two studies in particular are often cited for their lack of agreement, although they addressed similar questions, which were the two breast cancer studies performed by van de Vijver et al. (2002) and Wang et al. (2005), who reported two prognosis signatures for metastatic relapse in breast cancer. Two different signatures comprising respectively 70 and 76 genes predictive of breast cancer patient outcome were reported but presented only three genes in common. Even more concerning was the study by Michiels et al. (2005) which showed that hundreds of 70-genes signatures with equal classification power can be drawn by shuffling the training and test sets in the van't Veer et al. (2002) dataset, showing the instability and dependency of the resulting gene lists on the training data.

Microarray technology itself was blamed at first for these inconsistencies, and DNA microarrays were suspected to be extremely noisy and leading to non reproducible results. Once the stability and inherent reproducibility were demonstrated by comparing several platforms in the Microarray Quality Control project (Shi et al., 2006), the reasons for the lack of uniqueness in gene signatures had to be found elsewhere.

This chapter deals with addressing the problem of signature instability and proposes a new computational model, the Interactome-Transcriptome Integration (ITI), which simultaneously integrates multiple datasets, to compensate for data dimensionality, and uses the human interactome to include genes with weaker signal in the signature.

BACKGROUND

Instability of Signatures

The aim of any classification study is to provide a good prediction model. From that viewpoint (for pure classification and prognosis prediction), the gene set/classifier does not have to be unique (Dobbin et al., 2008), especially since it has been shown that established signatures have a high rate of concordance in regard to other datasets (Fan et al., 2006). However, from a biological point of view, the fact that signatures are not repeatable among studies is not acceptable. This shows a lack of robustness in detection methods that could prevent widespread acceptance of DNA microarray profiling methods for routine clinical use (Ein-Dor et al., 2006). The case holds for Next Generation Sequencing profiling, for which similar issues are likely to appear.

The trivial and most simplistic reasons are frequently cited and are insufficient to explain this situation. Are the discrepancies due to the heterogeneity of platforms and analytic tools used by microarray core among different institutions, or the genetic background inherent to each patient, or to the variations among statistical methods and classifiers?

Besides these, reasons behind the lack of generalization are twofold: (i) DNA microarray data structure and (ii) the biology of cancer. The data topology - too few patients profiled on too many variables - prevents any classifier to be trained according to proper statistical standards. Analysis is therefore suffering from a double curse (Fishel et al., 2007): *the curse of dimensionality* (too many variables), and the *curse of sparsity* (too few samples).

In addition, Chuang et al. (2007) showed that microarray data are highly sensitive to subtle misregulation of a few genes. Highly differentiated genes from an experiment are therefore resulting from subtle misregulation (or mutation) of a smaller set of genes that are at the origin of the disease. These genes are the true perpetrator of the clinical condition studied we wish to predict, but are not detected. For instance, mutation in RAS family of oncogenes, which are very common in multiple types of tumors, have a disastrous effect on regulation of a mitogen-activated protein kinases which in turn phosphorylates several receptors which can be potentially very different from one patient to another (Goodsell 1999). Availability of large scale protein-protein interaction data gives the opportunity of retrieving a large number of these hubs, critical for breast cancer relapse, whose activity is measured on several datasets.

Meta-Analysis Methods

Several solutions have been envisioned to tackle the curse of dimensionality issue. The most practical methods are based on a meta-analysis of several datasets, to increase sample size computationally. Meta-analysis consists of comparing gene statistics inferred from several datasets and combining them to obtain an integrated gene list. Methods of meta-normalization could also be envisioned, where datasets could be combined before analysis, but data heterogeneity prevents the adoption of these methods (Iziarry et al., 2005). Application to real data showed that meta-analysis in general achieves higher reproducibility of results than independent studies. Hong & Breitling (2008) have reported a comparison of three previously published methods: the T-based hierarchical clustering, rank products, and Fisher's Inverse chi-2 test. Fisher's Inverse chi-2 test [Fisher (1925), Bioconductor package GeneMeta, Gentleman et al.,(2004)] is a straightforward method combining *p*-values measured on different datasets to obtain a combined score for each gene. The T-based hierarchical clustering relies on the measurement of individual *t*-statistics on each dataset and the assessment of intra and inter-study variation by hierarchical modeling, also implemented in GeneMeta (Gentleman et al., 2004). Rank products method (Breitling et al., 2004) is also available as a Bioconductor package [RankProd, Hong et al. (2006)].

Other methods have been proposed. Conlon et al. (2006) proposed a Bayesian model to pool multiple independent studies and provided a Bayesian model of False Discovery Rates. Top-scoring pairs (TSP) method was used for cancer data integration (Xu et al., 2005). Van Vliet et al. (2007) proposed the use of meta-features (for instance, modules related to functional grouping of genes), which, by reducing the number of input variables, helps alleviate the curse of dimensionality. This is an extension of the model proposed by Segal et al. (2004) with training on a cancer compendia and inclusion of a classification system validated on independent data. Recently, more advanced techniques have been proposed, such as a neural network-based multi conditional classifier applied to developmental biology: In our previous studies, (Bidaut & Stoeckert, 2009a; Bidaut & Stoeckert, 2009b) we combined several stem cell profiles using vector projection technique to discover a multi-stage reproducible signature. Fishel et al. (2005) proposed a predictor-based approach (repeatability-based gene list, RGL) to find a stable lung cancer differentiation signature.

In addition, several reports have developed methods to answer the critical question of estimating the necessary sample size. Ein-Dor et al. (2006) suggested that several thousand patients are needed to obtain a signature that is robust among several studies, i.e., independent on the training set. Dobbin et al. (2008) tempered that conclusion and stated that such a high number of samples is not necessary when expression measurement information such as the largest standardized fold change, and the proportion of samples in each class are taken into account. They proposed a formula to calculate sample size based on a minimal set of information including the largest standardized fold change, the number of features and the data structure, i.e., the proportion of cases and control in the data.

The use of prior biological information helps reduce data dimensionality (Bidaut el al., 2006). In the past five years, several alternative approaches using network analysis have been proposed. For instance, Chuang et al. (2007) superimposed GEP over human interactome to generalize a signature for breast cancer metastasis relapse but using only a single dataset for training, as opposed to our approach. Wachi et al. (2005) showed that genes expressed in lung cancer tissues have a higher connectivity and are centrally located in the protein network. For a review of responsive functional modules identification in PPI networks, see the review by Wu et al. (2009).

INTERACTOME-TRANSCRIPTOME INTEGRATION

To discover a stable and robust signature predicting breast cancer metastatic potential and to infer genes subtly misregulated but crucial for such prediction, we proceeded by linking the human interactome on the largest body of available breast tumors profiled on DNA microarray, using a framework named as Interactome-Transcriptome Integration (ITI). Basically, we created a breast cancer compendium from several DNA microarray datasets and superimposed it over the human interactome. The compendium was built by selecting individual datasets on the basis of clinical information availability and large overlap with existing protein-protein interaction data. Several DNA microarray platforms are represented in the compendium (7 distinct platforms in total, see Table 1) in order to avoid platform biases. This gives the ability to recover common subtly differentiated genes correlated with distant metastatic relapse. A signature is searched on all data simultaneously by parsing the interactome and aggregating subsets of nodes correlated with distant metastatic relapse in a number of datasets (See ITI Algorithm section for details). No extra normalization step was necessary to integrate individual datasets, as gene expression is not used directly. Correlation of gene expression profiles with clinical situation was rather superimposed on the interaction data.

After superimposition of expression information over the interactome, the list of interactions was searched for with consistent agreement of discrimination power over multiple datasets, leading to a database of subnetworks linked to metastasis in breast cancer. This database is available from the ITI web site main page. Several expression datasets combinations were tested to assess platform bias, as shown in Table 2, to compare discriminative subnetworks assessed from all datasets with subnetworks derived only with datasets profiled on Affytmetrix platforms (data not shown).

Human Interactome: Combining Several Sources of Large Scale Interaction Data

To build our set of interaction data, we integrated two existing human interactomes. The first is a recent version of the Human Protein Reference Database [HPRD version 8, released on June 7th, 2009, (Prasad et al. (2009)]. We used the flat file version available from the HPRD web site (http://www.hprd.org) after registration for non-commercial use. This file includes 35880 binary interactions between 8769 proteins after removal of unidentified interactors. The second set of interactions is the *in silico* predicted interactome described in Ramani et al. (2005). The Ramani interactome is available as a flat file and including 31609 interactions between 7500 proteins. We chose to omit self interactions present in HPRD (already filtered out in the Ramani interactome for benchmarking reasons) as they are not quantified in the subnetwork search process. Both interactomes were integrated by uniqueness of NCBI EntrezGene identifiers, leading to a final set of 57991 interactions among 10943 proteins (with an overlap of 7165 interactions between them).

Building a Breast Cancer Compendium

We integrated 12 distinct DNA microarray datasets of breast tumor profiles in our compendium, after examination of about two tens of datasets for clinical information availability. Most datasets are accessible from NCBI Gene Expression Omnnibus (Barrett et al., 2009) repository with exception of the van de Vijver data, available from the original publication supplementary web site. Each dataset was downloaded either as raw data from GEO and normalized within Bioconductor using Affy and GCRMA packages, or directly loaded from GEO as a GSE file when raw data were unavailable, or from the author's web site. Correspondence tables between gene IDs and probe IDs were constructed with the method described in Reynal et al. (2005). Briefly, one probe was kept per gene by filtering out all probes carrying "x_at" extension and keeping the probe with highest expression profile median.

Table 1 summarizes the breast cancer compendium datasets, links to GEO, platforms, publications and sample size. Thereafter, we name

Table 1. List of datasets included in the Breast Cancer Compendium for training. Datasets in light grey were considered but not included because of lack of clinical data or lack of platform information, but are potentially includable if such information could be gathered. Some tumors were filtered out if they were already present in other datasets (for instance van't Veer dataset is filtered out since it has been included in van de Vijver). The compendium result from integration of white datasets, resulting in a total set of 2464 untreated tumors annotated with e.DFS or e.DMFS. Datasets spanning over multiple platforms (133A and B) were integrated into one (see Methods).

Dataset	NCBI Accession number (if available)	Platform	Number of samples before filtering	Number of samples after filtering	Presence of clinical information (e.DFS or e.DMFS)
Anders	GSE7849	U95v2	78	78	No
Bild	GSE3143	U95v2	158	158	No
Campone	GSE7017	UMGC-IRCNA 9k A	150	150	No
Chang	GSE3945	cDNA array	50	50	No
Chang-Kyu	GSE2845	Merck GEL Breast Tumor Profiles	311	311	No
Chanrion	GSE9893	MLRG Human 21K V12.0	155	155	No
Desmedt	GSE7390	U133A	198	198	Yes
Finetti		U133 Plus 2.0	129	129	Yes
Ivshina	GSE4922	U133 Plus 2.0	289	249	Yes
Jezequel	GSE11264	UMGC-IRCNA 9k A	252	252	No
Kreike	GSE4913	NKI-AVL 18K cDNA	59	59	Yes
Loi	GSE6532	U133A + U133B	327	293	Yes
		U133 Plus 2.0	87	87	
Miller	GSE3494	U133A + U133B	251	251	Yes
Parker	GSE10886	Agilent-011521 1A G4110A	2	2	Yes
		Agilent-012097 1A G4110B	27	22	
		Agilent 1A Oligo UNC Custom	196	177	
Pawitan	GSE1456	U133A + U133B	159	159	Yes
Perou	GSE61	SCV	84	84	No
Schmidt	GSE11121	U133A	200	200	Yes
Sorlie	GSE3193		85	85	No
Sotiriou	GSE2990	U133A	189	179	Yes
van de Vijver		Agilent whole human genome	295	295	Yes
van't Veer		Agilent whole human genome	117	117	Yes
Wang	GSE2034	U133A	286	286	Yes
Wong	GSE7930	U133A	6	6	No
Yu	GSE5364	U133A	341	341	No
Zhang	GSE12093	U133A	136	136	Yes
Zhou	GSE7378	U133Av2	54	54	Yes
Total: 12		7 distinct	2572	2464	

Table 2. Cross validation training organization. Two series were trained on all data from the breast cancer compendium but one (A1, B1), whereas two other series were trained only on datasets profiles on Affymetrix platforms (A2, B2) to assess inter-platform subnetwork stability.

Run	Training datasets
A1	All but van de Vijver
B1	All but Wang
A2	All Affymetrix platforms
B2	All Affymetrix Platforms but Wang

datasets after the corresponding paper's first author name. Since we are building a prognostic classifier for metastatic relapse, we gathered the clinical information related to e.DMFS (Distant Metastasis-Free Survival) or e.DFS (Disease-Free survival) for every dataset when available. Availability of this information was required to include a given dataset within our analysis. For the Parker, Pawitan and Wang studies, distant metastatic relapse information was not mentioned, and disease relapse information was used instead (variable e.DFS). The initial dataset from Ivshina contained 58 samples from the Pawitan study that were removed to avoid duplicates, and datasets profiled on Affymetrix HG-U133A and B platforms were merged (Ivshina, Loi and Pawitan) by creating a virtual platform annotation file of 44692 probes using the methodology previously employed for probe to gene expression conversion. Tumors without distant metastatic relapse information were further removed, leading to a final compendium of 2464 tumors. Clinical information was binarized in order to compute Pearson correlations with GEPs. Annotations were gathered for each platform from the Resourcerer database (Tsai et al. (2001), Data downloaded on Nov. 1st 2008). The Gene_info file was downloaded from NCBI the On Sept. 1st 2009 and was used as a table of correspondence between NCBI geneID accession numbers and NCBI gene Symbols (Sayers et al., 2010).

ITI Algorithm

To superimpose physical interaction data (human interactome) to several transcriptome datasets, we constructed an algorithm named ITI, Interactome-Transcriptome Integration, derived from the one described in Chuang et al. (2007) but extended to perform the analysis on several datasets. This algorithm allows one to superimpose GEP from several datasets to a map of physical interactions and to extract subnetworks that consistently discriminate two opposite clinical conditions across a number of datasets. To do so, a heuristic method examines all nodes present in interaction data and tries to construct a subnetwork by recursive aggregation of neighboring nodes. Aggregation is done on the basis of consistency of gene expression across subnetwork and high correlation of the whole subnetwork with the clinical condition. This is quantified by a subnetwork score computed as the absolute value of average correlation of gene expression profiles of genes included in the subnetwork with a numerical vector representing clinical situation for each dataset.

Several variables are set before starting the algorithm: *th* is the minimal score threshold that a subnetwork must meet to be accepted, *mi* is the minimal score increase when adding a new node to an existing subnetwork, and *c* (consensus) is the minimal number of datasets on which a gene must meet conditions on *th* and *mi* to be added on the subnetwork. The following formula *Sc(S, d)* details score calculation of a subnetwork *S* over a single dataset *d*. A global subnetwork score is computed for each subnetwork by averaging scores obtained over all datasets for information (see supporting web site at http://bioinformatics.marseille.inserm.fr/iti).

$$Sc(S,d) = Pearson\_corr(\frac{1}{p}\sum_{g \in S} GEP(g)), Cc(d))$$

S being the current subnetwork, *Sc(S)* the score, *d* the current dataset index, *p* the number of genes contained in the subnetwork *S*, *g* the current gene, *GEP(g)* the gene *g* expression profile of gene *g*, and *Cc*, the numerical vector representing clinical condition (1 = relapse, 0 = healthy).

To construct subnetworks, the following recursive algorithm is used. The subnetwork is first constructed from a candidate seed, and a recursive method aggregates neighbors' nodes if the score *Sc* stays above threshold *th* over at least *c* datasets and does not vary below the minimal value *mi* (minimal increase). The following pseudo code details the method, also represented in Figure 1. *Nc(S)* is the current consensus value for subnetwork *S*.

```
testSubnetwork = Subnetwork = empty;
RoutineconstructNodeFor Each node In
interactome
testSubnetwork = concatenate(node,
Subnetwork)
Nc=0;
For Each dataset
            Sc(testSubnetwork)= Com-
pute-subnetwork_score(testSubnetwork,
dataset)
            IfSc(Subnetwork,
dataset)>th and (Sc(testSubnetwork)
- Sc(Subnetwork))
>miThenNc(Subnetwork) ++
            End
End
IfNc(Subnetwork)>=cThen Subnetwork =
testSubnetwork
            For Each node In
neighbor(node) constructNode(node)End
Else break
End
End
```

Choice of variables has an impact on detected subnetworks' number and size. Obviously, lowering *th* and *c* will increase the number of detected subnetwork, and lowering *mi* increases subnetwork size. However, lowered scored subnetwork will be filtered out by the statistical validation step. Parameters have been set to the following values: *mi* = 0.01, *th* = 0.05 and *c* = 6 to obtain a reasonably sized subnetwork set. Subnetworks overlapping by more than 80% were removed. To tackle the computational cost of this algorithm, we parallelized it on a 96 cores [12 nodes] Beowulf cluster. Parallelization was done by partitioning the interactome over the nodes, leading to a 45 minutes approximate execution time, including random distribution drawing for statistical validation (see following section). As a point of comparison, execution time on a single CPU is about 10 hours.

Subnetworks Statistical Validation

Subnetworks are validated over three *p*-values, related to (i) the node aggregation decision (type 1 *p*-values), (ii) the link between expression data and interaction data (type 2 *p*-values), and (iii) the network topology (type 3 *p*-values). Each *p*-value is computed by drawing a random distribution of scores, and setting up a score threshold. In this framework, three random distributions are computed for each dataset. The first is computed by randomizing aggregation decision (nodes are added randomly until subnetwork size reaches a normally randomly distributed value having the same distribution as the regularly detected subnetworks). The second is computed by shuffling experimental conditions over datasets. The third is computed by shuffling all interactome interactions while keeping the original connectivity ±1 for each node. This allows drawing a *p*-value distribution evaluating only the link between protein-protein interactions and co-expression while conserving

Figure 1. Scheme represents the ITI Algorithm of DNA microarray datasets integration, their superimposition over the human interactome, and the construction of discriminating subnetwork. Briefly, all nodes are considered as a seed, and neighboring nodes are aggregated if discriminative over clinical condition over a number of datasets.

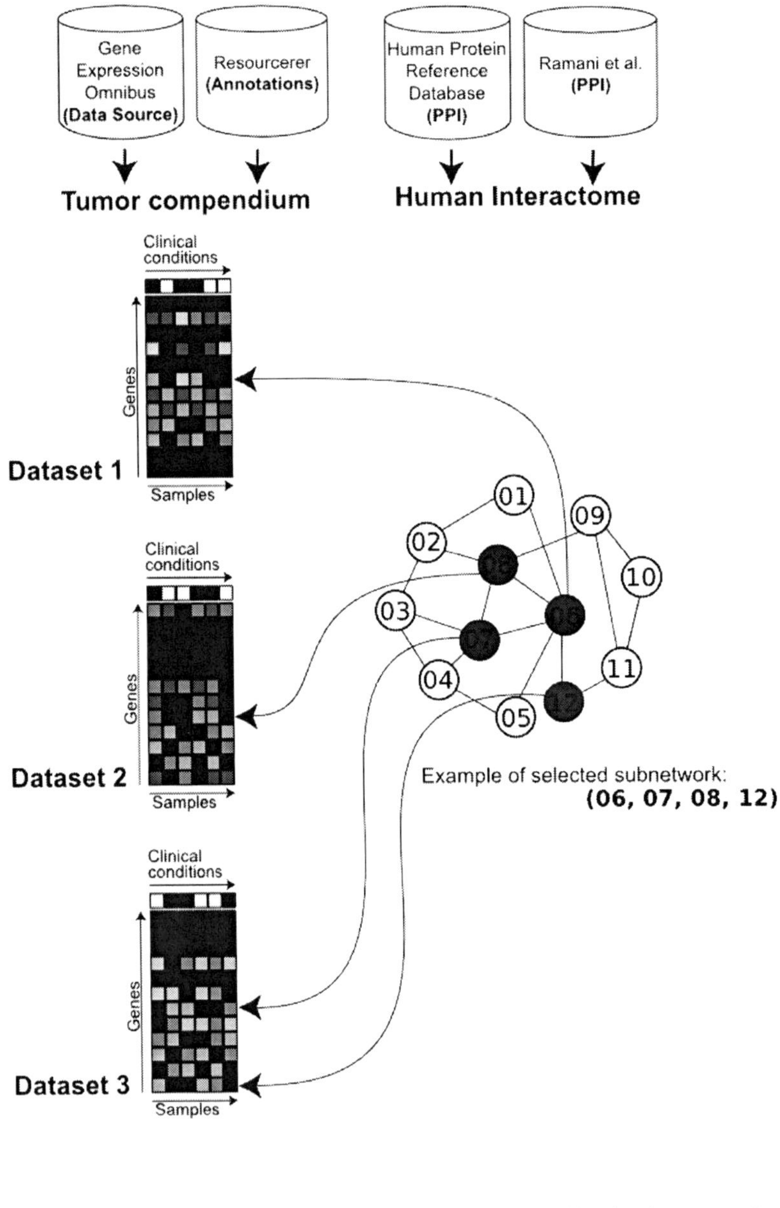

the human interactome power law distribution. *P*-values are then computed from these distributions by generating 3000 random subnetworks within these three settings. As an argument to the program, distribution model type can be given to properly model the random distributions (Gamma distribution, Normal distribution, and bimodal normal distribution). Examples of random distributions are presented in Figure 2. In the case of bimodal distribution, distributions were separated (Matlab © *gmdistribution* object, Statistical Toolbox, Matlab ©, The Mathworks), and the upper score distribution could be used. In this report, a Gamma distribution model was used.

Figure 2. Type 1 Random Distributions for three datasets for A1 configuration. The light grey histogram represents random distribution of scores, black histogram is the actual subnetworks score distribution, and black cure is the gamma model.

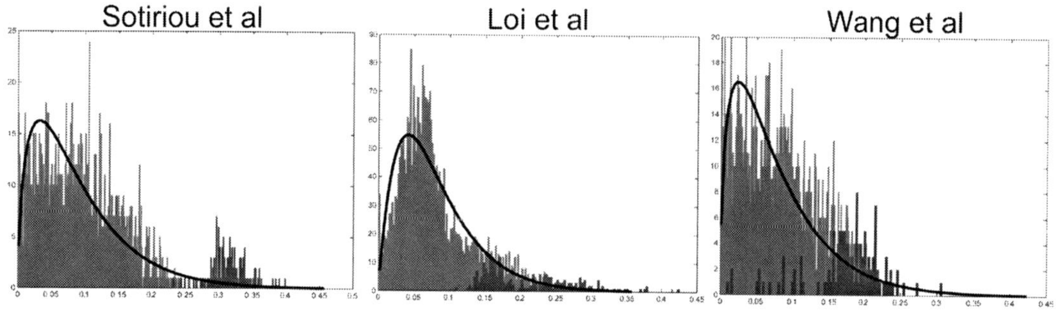

Additionally, functional biology is inferred by measurement of enriched Gene Ontology terms. For each subnetwork over each dataset, Benjamini-Hotchberg corrected *p*-values were calculated for Gene Ontology terms (GO, The Gene Ontology Consortium, 2009) using an hypergeometric distribution. Version 2.1.18 of ErmineJ (Lee et al., 2005) was used (GO data downloaded on Sept 1st, 2009).

Constructing a Gene Signature

To construct a gene signature reflecting the subnetwork information, a list of discriminative genes must be extracted from the subnetworks set. Two metrics are available: co-occurrence (number of times a gene appears in the subnetwork set) and correlation (Pearson correlation of its expression profile with clinical condition vector *Cc*). However, some genes appear in several subnetworks and have a high occurrence rate, but a low discriminative power, whereas some genes have a lower occurrence rate but a high correlation with the studied clinical situation. Therefore, a ranked metric is needed to reflect both the number of gene occurrences in subnetworks as well as the relative discriminative power of each gene in each subnetwork. To equitably rank genes from these two situations, a so-called 'general rank' is computed as the average rank obtained with co-occurrence and correlation ranking. Co-occurrence ranking is computed by counting gene occurrence on a subnetwork set, and correlation ranking is produced by ranking genes according to GEP-*Cc* Pearson correlation (see section ITI Algorithm). For instance, *LUC7L3* is ranked only 38th in our signature according to its occurrence in the subnetworks, but ranked 5th by general ranking, as it belongs to the subnetwork ranked 1st. Genes ordered by these different metrics are reported in the database.

Several Training Sets were used to Test Subnetwork Stability

To understand the impact of each dataset on results and generalization of subnetworks, discriminating subnetworks for different combinations of input datasets and validation datasets from the breast cancer compendium were generated. Four combinations were used, named A1, A2, B1, and B2 (See Table 2). The combination run A1 is using all but van de Vijver datasets for input. Combination B1 is using all but Wang datasets for input, combination A2 is using datasets profiled on Affymetrix platforms, and B2 is using all Affymetrix datasets but Wang. For each run, subnetworks were validated using *p*-values computed section Subnetwork statistical validation. Subnetwork set constructed with A1 datasets was validated by

Table 3. P-value thresholds and consensus chosen for the three random distribution types for each input dataset configuration

Run	Type 1 *p*-values	Type 2 *p*-values	Type 3 *p*-values
A1	1.10^{-2} on at least 2 datasets	1.10^{-1} on at least 11 datasets	1.10^{-1} on at least 1 datasets
B1	1.10^{-1} on at least 8 datasets	1.10^{-2} on at least 2 datasets	1.10^{-1} on at least 1 datasets
A2	1.10^{-1} on at least 11 datasets	1.10^{-2} on at least 2 datasets	1.10^{-1} on at least 2 datasets
B2	1.10^{-1} on at least 6 datasets	1.10^{-2} on at least 2 datasets	1.10^{-1} on at least 1 datasets

keeping subnetwork meeting type 1 *p*-values of 1.10^{-2} over at least 8 datasets, type 2 *p*-values of 1.0^{-2} over at least 9 datasets, and type 3 *p*-values of 1.10^{-1} over at least 2 datasets, yielding a final set of 119 discriminative subnetworks containing 406 genes. Lower *p*-values yielded by type 3 random subnetworks are discussed in results section. Subnetworks with more significant *p*-values on individual datasets were present but not consistent over all datasets and thus not retained. *P*-values thresholds for A2, B1, and B2 configurations were summarized in Table 3.

Examination of subnetworks found for each run (see ITI subnetwork Database in following section) shows little discrepancies among subnetwork sets found.

ITI Subnetwork Database

We stored the ITI algorithm results – discriminative subnetworks for a given clinical condition into a relational database, the ITI database. This database, publicly available on the web (http://bioinformatique.marseille.inserm.fr/iti), is the first one to explicitly link gene interaction models to disease, as opposed to many others, which store either plain gene lists, such as the Candidate List of yoUr Biomarker (CLUB, Lee et al., 2008), or subnetworks of interests but with no explicit link to the disease (CellCircuits database, Ideker Lab). CLUB database allows for sharing and comparison of putative candidate biomarkers, including lists of genes along with the protocol that obtained them. This database is cross-platforms to allow repository of studies made in heterogeneous platforms, such as proteomics, DNA microarrays, etc. Other gene list repositories exist, such as List Of Lists Annotated (LOLA) which allows inter-studies comparisons. CellCircuits contains network models extracted from different studies with little possibility of data mining in its current version. Its main interest is to have a platform for data sharing that allows searching for gene products and related Gene Ontology terms.

In our ITI implementation, the code produces web pages which can be placed on a plain web server and navigated through with a web browser. Web pages were generated for each runs, which allow qualitative comparison of subnetwork sets. Figure 3 shows information displayed by the database for each subnetwork. Briefly, each subnetwork is identified by a unique number (e.g. 387-4) and is described on a unique web page containing score and *p*-values for each dataset, heatmaps of expression data, subnetwork layout with superimposed correlation with the biological question asked, correlation and annotation for each gene contained in the subnetwork, and GO enrichment for each dataset. In parallel, one can find complete gene lists and annotations links to NCBI EntrezGene. A complete list of enriched Gene Ontology terms (Hypergeometric distribution) is also provided for the whole set of subnetworks. Figure 3 details the report page for subnetwork 387-4.

Figure 3. ITI Database Web Interface. (A) is the subnetwork ID (here 387-4), B is a table representing scores and p-values for each dataset, (C) is a series of links to gene expression heatmaps for each dataset, (D) shows the network topology as well as correlation (light grey= correlation, dark grey = anticorrelation) with clinical condition for each node. (E) is the individual genes score table, and (F) shows the top 10 enriched GO terms for each dataset, with Benjamini-Hotchberg-corrected p-values.

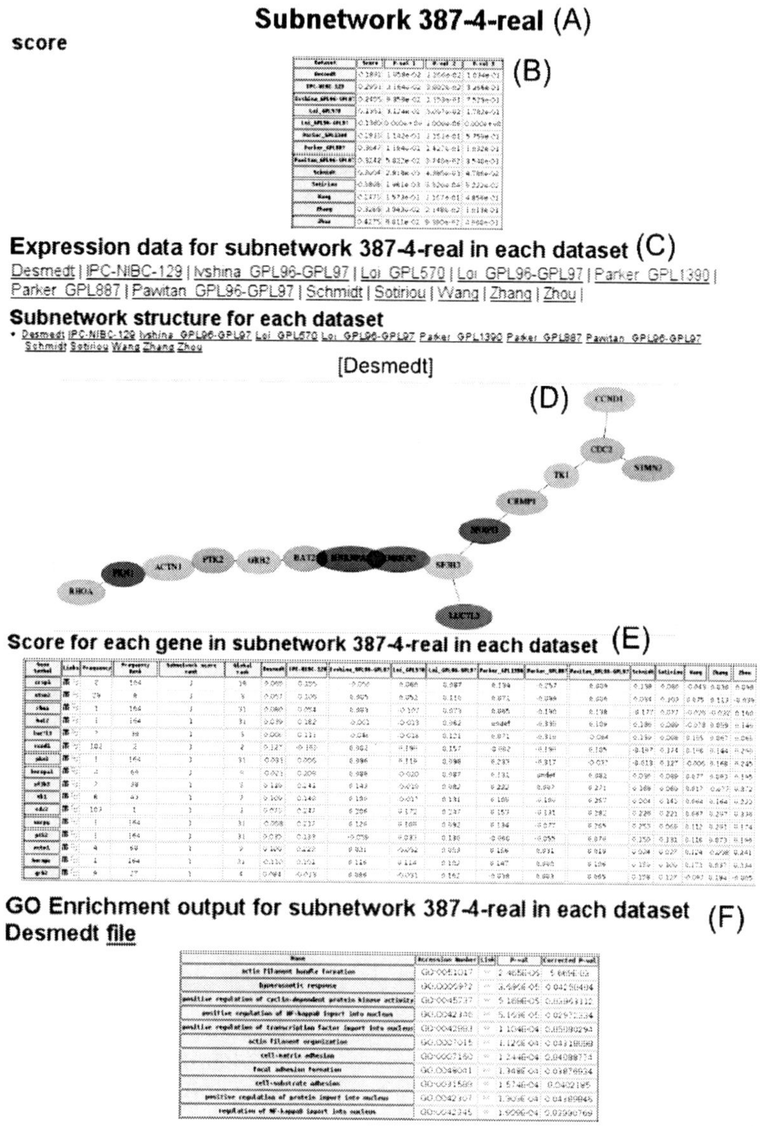

Implementation and Code Availability

A Matlab © license and a Matlab Statistics Toolbox © license are necessary to compute *p*-values.

Code availability: the Perl code is available from the ITI database web site ([1]http://bioinformatique.marseille.inserm.fr/iti/iti-1.0.tar.gz), and has been licensed under the CeCILL public license (French GPL extension developed by a consortium of French Research Organisms).

RESULTS

We build a database containing links between interactome and metastatic relapse in breast cancer using results from the ITI algorithm applied on our breast cancer compendium. The ITI database will be extended over time, as we refine the algorithm and process data from other datasets from public repositories. In the meantime, we are using the database to understand the biology of the signature found under the form of a subnetwork set.

Biology of Extracted Subnetwork is Meaningful

Intrinsic biology of the 119 extracted subnetworks (containing 406 genes) from A1 combination was examined using annotation information from NCBI EntrezGene database and Gene Ontology Consortium database linked directly to the ITI database. First, the biology of each gene group included in a subnetwork was assessed by statistical enrichment of Gene Ontology terms (see section Methods and ITI web site). We found that subnetworks formed complexes functionally supporting the studied disease for metabolism, cell cycle control, proliferation, cell-cell adhesion and immunological response, which are known mechanisms of the hallmarks of cancer and metastatic process. The first ranked subnetwork (387-4, score S=0.283) shows significant enrichment for 'actin filament bundle formation' (GO: 0051017), which is a biological process linked to cell development and polarity. The second ranked subnetwork shows enrichment for 'activation of caspase activity by cytochrome C' (GO 0008635) which is linked to apoptosis. It also shows enrichment for 'B cell lineage commitment' (GO:0002326), which reveals immune response to metastasis. The third ranked subnetwork (2810-3) has a similar function (score S=0.278). Lower in the list, subnetwork 58-7 (Score S=0.271, ranked 6th) shows enrichment for functions related to microtubule formation: 'microtubule organizing center organization' (GO:0031023) term is significantly enriched. Functional 'regulation of centrosome cycle' is also significantly enriched (GO:0046605). Subnetwork 29959-4 (ranked 7th, score S=0.270) is functionally linked to metabolism, as seen with the terms 'glucose catabolic process' (GO:0006007), 'fructose metabolic process' (GO:0006000) and 'alditol metabolic process' (GO:0019400). This subnetwork is also functionally involved in cell migration through the formation of cell surface protrusions, such as lamellipodium or filopodium, at the leading edge of a migrating cell (Gene Ontology term 'substrate-bound cell migration, cell extension', GO:0006930). Subnetwork 581-7 (Score S=0.267) is involved in cell adhesion: 'focal adhesion formation' GO term is significantly enriched (GO:0048041). Cellular differentiation is functionally represented by subnetwork 1452-7 through enrichment of genes involved in the Wnt pathway signaling (GO term 'regulation of Wnt receptor signaling pathway' - GO:0030111). Subnetwork involved in cell proliferation is 5155-5 (S=0.254) having the GO terms 'positive regulation of endothelial cell proliferation' (GO:0001938) and 'establishment or maintenance of epithelial cell apical/basal polarity' (GO: 0045197) significantly enriched. Other subnetworks are of course of potential interest. Global list of enriched ontology terms is also stored in the database.

At the gene level, several markers show obvious links to cancer and involvement in cell cycle, proliferation, cell adhesion and other biological mechanisms involved in the disease. We examined the gene list ordered by 'mixed ranks' (see methods).

CDK1 (cyclin-dependent kinase 1) is the highest ranked gene. The protein encoded by this gene is a catalytic subunit of the highly conserved protein kinase complex known as M-phase promoting factor (MPF), which is essential for G1/S and G2/M phase transitions of the eukaryotic cell cycle (EntrezGene). Other genes from this process are also found, such as CCND1 (Cyclin D1, ranked second). GRB2 (ranked 4th) is a growth factor associated with several cancer types and may have a role in metastasis (Yu et al., 2008). TK1 (thymidine kinase 1, soluble) is known as a proliferation marker in breast cancer, and its overexpression has been linked to thyroid carcinogenesis. TSC1 (tuberous sclerosis 1, ranked 8th) is known to play a central role in regulating cell survival and proliferation signaling pathways. Other genes of interest are present, including LAMA4 (laminin alpha-4), which has roles in *in vitro* migration and *in vivo* tumorigenicity of prostate cancer cells and others, and PGK1 (phosphoglycerate kinase 1), with proven involvement in prostate cancer, and many others.

Extracted Subnetworks Shows Identical Gene Expression Trend over Several Datasets

Figure 4 represents the top scoring subnetwork for A1 configuration: the subnetwork 387-4. It shows that gene expression of its components is consistent for several datasets (superimposition of 387-4 over Sotiriou, Finetti, Desmedt and Wang is shown) with high significance for most, showing a high power of discrimination and a high confidence of the correlation of gene expression of this subnetwork with the clinical condition. *P*-values obtained over several datasets were examined and are represented Figure 4. This subnetwork has significant scores for Finetti (*p*-values $<5.10^{-2}$), Sotiriou. (*p*-values $<=1.10^{-3}$) and Loi datasets (*p*-values $<=1.10^{-6}$) (see all obtained *p*-values Table 4). This subnetwork constitutes an interacting complex of proteins including the oncogene cyclin D1 (CCND1), and the Ras homolog gene family member (this protein may regulate the invasion and metastasis of breast cancer cells as an upstream signaling of Ezrin). The subnetwork also contains the Protein Tyrosine Kinase 2 (PTK2). It has been shown that increased PTK2 levels due to mutations of p53 are associated with breast and colon

Figure 4. Subnetwork 387-4 with correlation measured for each node and represented as light grey (correlated) or dark grey (anticorrelated). Global score for each dataset is shown as well as p-values, illustrating the expression consensus obtained for most genes within the network across different datasets.

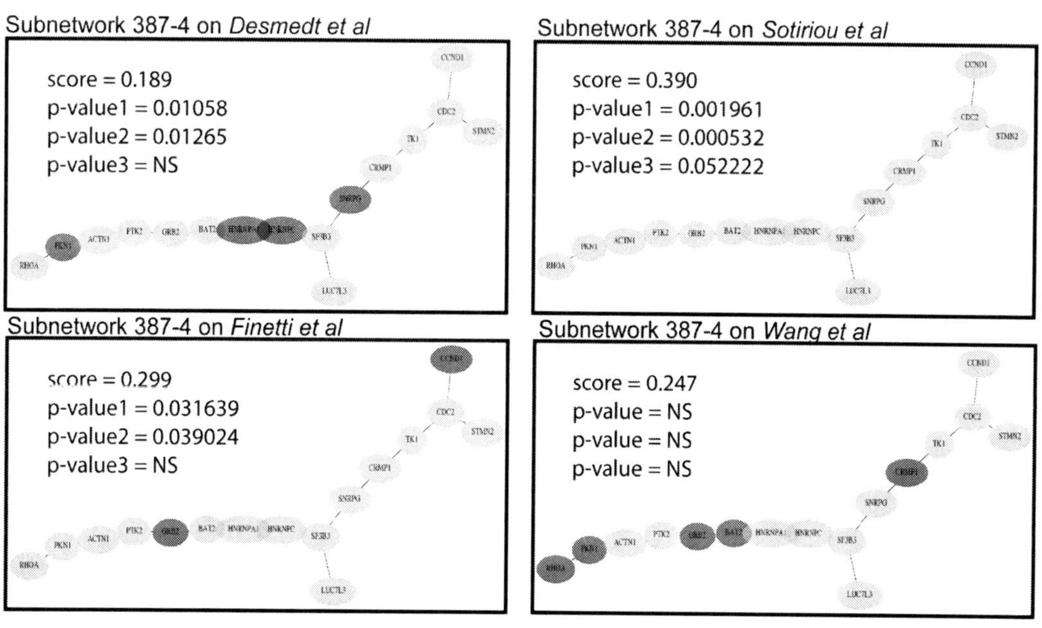

Table 4. Types 1, 2 and 3 p-values measured for subnetwork 387-4 for each dataset. P-values show high significance over several datasets, demonstrating the consensus of gene expression data over different datasets.

Dataset	Type 1 *p*-value	Type 2 *p*-value	Type 3 *p*-value
Desmedt	0.010584	0.012657	0.103381
IPC-NIBC-129	0.031639	0.039024	0.326597
Ivshina GPL96-GPL97	0.093584	0.115279	0.752917
Loi GPL570	0.031238	0.030974	0.178192
Loi GPL96-GPL97	$<1.10^{-6}$	1.10^{-6}	$<1.10^{-6}$
Parker GPL1390	0.114192	0.115112	0.575853
Parker GPL887	0.119422	0.142705	0.103176
Pawitan GPL96-GPL97	0.056216	0.037460	0.353977
Schmidt	0.002918	0.004385	0.047856
Sotiriou	0.001961	0.000532	0.052222
Wang	0.157285	0.116712	0.485642
Zhang	0.039429	0.021484	0.101313
Zhou	0.086106	0.093900	0.496010

cancers. Other subnetworks can be examined in the same way from the ITI database web interface.

CONCLUSION AND FUTURE RESEARCH DIRECTIONS

We present an Interactome-Transcriptome Integration framework (ITI) to isolate prognostic signatures generalizable over multiple datasets of breast cancer. We performed large scale integration of 15 DNA microarray datasets to create a breast cancer compendium. We also constructed a large coverage human interactome by integrating two existing human protein-protein interaction datasets (HPRD and Ramani dataset). These data, used conjointly with a discriminative subnetwork detection algorithm and significance scoring, allowed extraction of subnetworks linked with metastatic potential in breast cancers. Isolated subnetworks functionally cover biological functions related to metastasis and breast cancer, such as cell differentiation, cell cycle signaling, cell adhesion and proliferation, as well as functional links to immune response.

This database is the first of its kind to allow linking a human interactome to diseases or clinical situations. This resource can be mined for isolating potential drug targets as well as prognostic signatures for metastasis of breast cancer as well as other diseases. It has the potential of becoming the starting point to establish finer disease models by systems biology techniques. Improvement of public resources, such as extension of data repositories, refinement of platform annotations, and increased coverage of interaction data will help improve the resource. For instance, interaction data could be extended by inclusion of canonical pathways from the Kyoto Encyclopedia of Genes and Genomes (Kanehisa & Goto, 2000).

The ITI algorithm has also the potential of aiding in mining other large scale repositories such as ArrayExpress (Parkinson et al., 2008), and Stanford Microarray Database, (Hubble et al., 2009), and could be used conjointly with other technologies as well, such as proteomics (PRIDE database, Vizcaíno et al., 2009). Other diseases, especially pathologies with limited number of available samples (prostate cancer for instance) are planned for further analysis with ITI.

Future developments include algorithm improvements, such as a better network parsing heuristics to find regions of interest. For instance, interactions with high reported confidence should have a higher probability to be included within a subnetwork than *in silico* predicted ones. Error rates in interaction data must also be taken into account for future developments, especially as coverage of interaction database grows. We are also considering integration with promoter specific methylation events, and genomic alteration data (Comparative Genomic Hybridization arrays).

An obvious extension of the presented framework will be the inclusion of a classification system (like Support-Vector Machine) to predict clinical outcome on independent data and to compare signature robustness with previous studies.

ACKNOWLEDGMENT

Research is funded by the Institut National du Cancer and the Institut National de la Santé et de la Recherche Médicale. Code development and calculation were performed on a Beowulf cluster funded by a grant from Fondation pour la Recherche Médicale. Maxime Garcia is funded by a Région Provence-Alpes-Côte d'Azur Fellowship. We thank Françoise Birg and Wahiba Gherraby for their suggestions for improving the manuscript.

REFERENCES

Anders, C. K., Acharya, C. R., Hsu, D. S., Broadwater, G., Garman, K., & Foekens, J. A. (2008). Age-specific differences in oncogenic pathway deregulation seen in human breast tumors. *PLoS ONE*, *3*(1), e1373. doi:10.1371/journal.pone.0001373

Barrett, T., Troup, D. B., Wilhite, S. E., Ledoux, P., Rudnev, D., & Evangelista, C. (2009). NCBI GEO: Archive for high-throughput functional genomic data. *Nucleic Acids Research*, *37*(Database issue), D885–D890. doi:10.1093/nar/gkn764

Bertucci, F., Finetti, P., Cervera, N., & Birnbaum, D. (2008). Prognostic classification of breast cancer and gene expression profiling. *Medecine Sciences*, *24*(6-7), 599–606.

Bidaut, G., & Stoeckert, C. J., Jr. (2009). Characterization of unknown adult stem cell samples by large scale data integration and artificial neural networks. *Pacific Symposium on Biocomputing*, 356-367.

Bidaut, G., & Stoeckert, C. J. Jr. (2009). Large scale transcriptome data integration across multiple tissues to decipher stem cell signatures. *Methods in Enzymology*, *467*, 229–245. doi:10.1016/S0076-6879(09)67009-9

Bidaut, G., Suhre, K., Claverie, J. M., & Ochs, M. F. (2006). Determination of strongly overlapping signaling activity from microarray data. *BMC Bioinformatics*, *7*, 99. doi:10.1186/1471-2105-7-99

Bild, A. H., Yao, G., Chang, J. T., Wang, Q., Potti, A., & Chasse, D. (2006). Oncogenic pathway signatures in human cancers as a guide to targeted therapies. *Nature*, *439*(7074), 353–357. doi:10.1038/nature04296

Breitling, R., Armengaud, P., Amtmann, A., & Herzyk, P. (2004). Rank products: A simple, yet powerful, new method to detect differentially regulated genes in replicated microarray experiments. *FEBS Letters*, *573*(1-3), 83–92. doi:10.1016/j.febslet.2004.07.055

Bueno-de-Mesquita, J. M., van Harten, W. H., Retel, V. P., van't Veer, L. J., van Dam, F. S., & Karsenberg, K. (2007). Use of 70-gene signature to predict prognosis of patients with node-negative breast cancer: A prospective community-based feasibility study (RASTER). *The Lancet Oncology*, *8*(12), 1079–1087. doi:10.1016/S1470-2045(07)70346-7

Campone, M., Campion, L., Roche, H., Gouraud, W., Charbonnel, C., & Magrangeas, F. (2008). Prediction of metastatic relapse in node-positive breast cancer: Establishment of a clinicogenomic model after FEC100 adjuvant regimen. *Breast Cancer Research and Treatment*, *109*(3), 491–501. doi:10.1007/s10549-007-9673-x

Chang, H. Y., Sneddon, J. B., Alizadeh, A. A., Sood, R., West, R. B., & Montgomery, K. (2004). Gene expression signature of fibroblast serum response predicts human cancer progression: Similarities between tumors and wounds. *PLoS Biology*, *2*(2), E7. doi:10.1371/journal.pbio.0020007

Chanrion, M., Negre, V., Fontaine, H., Salvetat, N., Bibeau, F., & MacGrogan, G. (2008). A gene expression signature that can predict the recurrence of tamoxifen-treated primary breast cancer. *Clinical Cancer Research*, *14*(6), 1744–1752. doi:10.1158/1078-0432.CCR-07-1833

Chuang, H. Y., Lee, E., Liu, Y. T., Lee, D., & Ideker, T. (2007). Network-based classification of breast cancer metastasis. *Molecular Systems Biology*, *3*, 140. doi:10.1038/msb4100180

Conlon, E. M., Song, J. J., & Liu, J. S. (2006). Bayesian models for pooling microarray studies with multiple sources of replications. *BMC Bioinformatics*, *7*, 247. doi:10.1186/1471-2105-7-247

de Lavallade, H., Finetti, P., Carbuccia, N., Khorashad, J. S., Charbonnier, A., & Foroni, L. (2010). A gene expression signature of primary resistance to imatinib in chronic myeloid leukemia. *Leukemia Research*, *34*(2), 254–257. doi:10.1016/j.leukres.2009.09.026

Desmedt, C., Piette, F., Loi, S., Wang, Y., Lallemand, F., & Haibe-Kains, B. (2007). Strong time dependence of the 76-gene prognostic signature for node-negative breast cancer patients in the TRANSBIG multicenter independent validation series. *Clinical Cancer Research*, *13*(11), 3207–3214. doi:10.1158/1078-0432.CCR-06-2765

Dobbin, K. K., Beer, D. G., Meyerson, M., Yeatman, T. J., Gerald, W. L., & Jacobson, J. W. (2005). Interlaboratory comparability study of cancer gene expression analysis using oligonucleotide microarrays. *Clinical Cancer Research*, *11*(2 Pt 1), 565–572.

Ein-Dor, L., Zuk, O., & Domany, E. (2006). Thousands of samples are needed to generate a robust gene list for predicting outcome in cancer. *Proceedings of the National Academy of Sciences of the United States of America*, *103*(15), 5923–5928. doi:10.1073/pnas.0601231103

Fan, C., Oh, D. S., Wessels, L., Weigelt, B., Nuyten, D. S., & Nobel, A. B. (2006). Concordance among gene-expression-based predictors for breast cancer. *The New England Journal of Medicine*, *355*(6), 560–569. doi:10.1056/NEJMoa052933

Fishel, I., Kaufman, A., & Ruppin, E. (2007). Meta-analysis of gene expression data: A predictor-based approach. *Bioinformatics (Oxford, England)*, *23*(13), 1599–1606. doi:10.1093/bioinformatics/btm149

Fisher, R. A. (1925). *Statistical methods for research workers*. London: Edinburg.

Gentleman, R. C., Carey, V. J., Bates, D. M., Bolstad, B., Dettling, M., & Dudoit, S. (2004). Bioconductor: Open software development for computational biology and bioinformatics. *Genome Biology*, *5*(10), R80. doi:10.1186/gb-2004-5-10-r80

Golub, T. R., Slonim, D. K., Tamayo, P., Huard, C., Gaasenbeek, M., & Mesirov, J. P. (1999). Molecular classification of cancer: Class discovery and class prediction by gene expression monitoring. *Science*, *286*(5439), 531–537. doi:10.1126/science.286.5439.531

Goodsell, D. S. (1999). The molecular perspective: The ras oncogene. *The Oncologist*, *4*(3), 263–264.

Hong, F., & Breitling, R. (2008). A comparison of meta-analysis methods for detecting differentially expressed genes in microarray experiments. *Bioinformatics (Oxford, England)*, *24*(3), 374–382. doi:10.1093/bioinformatics/btm620

Hong, F., Breitling, R., McEntee, C. W., Wittner, B. S., Nemhauser, J. L., & Chory, J. (2006). RankProd: A bioconductor package for detecting differentially expressed genes in meta-analysis. *Bioinformatics (Oxford, England)*, *22*(22), 2825–2827. doi:10.1093/bioinformatics/btl476

Hubble, J., Demeter, J., Jin, H., Mao, M., Nitzberg, M., & Reddy, T. B. (2009). Implementation of GenePattern within the Stanford Microarray Database. *Nucleic Acids Research*, *37*(Database issue), D898–D901. doi:10.1093/nar/gkn786

Irizarry, R. A., Warren, D., Spencer, F., Kim, I. F., Biswal, S., & Frank, B. C. (2005). Multiple-laboratory comparison of microarray platforms. *Nature Methods*, *2*(5), 345–350. doi:10.1038/nmeth756

Ivshina, A. V., George, J., Senko, O., Mow, B., Putti, T. C., & Smeds, J. (2006). Genetic reclassification of histologic grade delineates new clinical subtypes of breast cancer. *Cancer Research*, *66*(21), 10292–10301. doi:10.1158/0008-5472.CAN-05-4414

Jezequel, P., Campone, M., Roche, H., Gouraud, W., Charbonnel, C., & Ricolleau, G. (2009). A 38-gene expression signature to predict metastasis risk in node-positive breast cancer after systemic adjuvant chemotherapy: A genomic substudy of PACS01 clinical trial. *Breast Cancer Research and Treatment*, *116*(3), 509–520. doi:10.1007/s10549-008-0250-8

Kaestner, K. H., Lee, C. S., Scearce, L. M., Brestelli, J. E., Arsenlis, A., & Le, P. P. (2003). Transcriptional program of the endocrine pancreas in mice and humans. *Diabetes*, *52*(7), 1604–1610. doi:10.2337/diabetes.52.7.1604

Keshava Prasad, T. S., Goel, R., Kandasamy, K., Keerthikumar, S., Kumar, S., & Mathivanan, S. (2009). Human Protein Reference Database-2009 update. *Nucleic Acids Research*, *37*(Database issue), D767–D772. doi:10.1093/nar/gkn892

Kreike, B., Halfwerk, H., Kristel, P., Glas, A., Peterse, H., & Bartelink, H. (2006). Gene expression profiles of primary breast carcinomas from patients at high risk for local recurrence after breast-conserving therapy. *Clinical Cancer Research*, *12*(19), 5705–5712. doi:10.1158/1078-0432.CCR-06-0805

Lee, B. T., Liew, L., Lim, J., Tan, J. K., Lee, T. C., & Veladandi, P. S. (2008). Candidate List of yoUr Biomarker (CLUB): A Web-based platform to aid cancer biomarker research. *Biomarker Insights*, *3*, 65–71.

Lee, H. K., Braynen, W., Keshav, K., & Pavlidis, P. (2005). ErmineJ: Tool for functional analysis of gene expression data sets. *BMC Bioinformatics*, *6*, 269. doi:10.1186/1471-2105-6-269

Li, L., Weinberg, C. R., Darden, T. A., & Pedersen, L. G. (2001). Gene selection for sample classification based on gene expression data: Study of sensitivity to choice of parameters of the GA/KNN method. *Bioinformatics (Oxford, England)*, *17*(12), 1131–1142. doi:10.1093/bioinformatics/17.12.1131

Loi, S., Haibe-Kains, B., Desmedt, C., Wirapati, P., Lallemand, F., & Tutt, A. M. (2008). Predicting prognosis using molecular profiling in estrogen receptor-positive breast cancer treated with tamoxifen. *BMC Genomics*, *9*, 239. doi:10.1186/1471-2164-9-239

Michiels, S., Koscielny, S., & Hill, C. (2005). Prediction of cancer outcome with microarrays: A multiple random validation strategy. *Lancet*, *365*(9458), 488–492. doi:10.1016/S0140-6736(05)17866-0

Miller, L. D., Smeds, J., George, J., Vega, V. B., Vergara, L., & Ploner, A. (2005). An expression signature for p53 status in human breast cancer predicts mutation status, transcriptional effects, and patient survival. *Proceedings of the National Academy of Sciences of the United States of America*, *102*(38), 13550–13555. doi:10.1073/pnas.0506230102

Mook, S., Knauer, M., Bueno-de-Mesquita, J. M., Retel, V. P., Wesseling, J., & Linn, S. C. (2010). Metastatic potential of T1 breast cancer can be predicted by the 70-gene MammaPrint signature. *Annals of Surgical Oncology*, *17*(5), 1406–1413. doi:10.1245/s10434-009-0902-x

Munro, K. M., & Perreau, V. M. (2009). Current and future applications of transcriptomics for discovery in CNS disease and injury. *Neuro-Signals*, *17*(4), 311–327. doi:10.1159/000231897

Pages, F., Galon, J., Dieu-Nosjean, M. C., Tartour, E., Sautes-Fridman, C., & Fridman, W. H. (2009). Immune infiltration in human tumors: A prognostic factor that should not be ignored. *Oncogene*, *29*(8), 1093–1102. doi:10.1038/onc.2009.416

Paik, S., Tang, G., Shak, S., Kim, C., Baker, J., & Kim, W. (2006). Gene expression and benefit of chemotherapy in women with node-negative, estrogen receptor-positive breast cancer. *Journal of Clinical Oncology*, *24*(23), 3726–3734. doi:10.1200/JCO.2005.04.7985

Parker, J. S., Mullins, M., Cheang, M. C., Leung, S., Voduc, D., & Vickery, T. (2009). Supervised risk predictor of breast cancer based on intrinsic subtypes. *Journal of Clinical Oncology*, *27*(8), 1160–1167. doi:10.1200/JCO.2008.18.1370

Parkinson, H., Kapushesky, M., Kolesnikov, N., Rustici, G., Shojatalab, M., & Abeygunawardena, N. (2009). ArrayExpress update-from an archive of functional genomics experiments to the atlas of gene expression. *Nucleic Acids Research*, *37*(Database issue), D868–D872. doi:10.1093/nar/gkn889

Pawitan, Y., Bjohle, J., Amler, L., Borg, A. L., Egyhazi, S., & Hall, P. (2005). Gene expression profiling spares early breast cancer patients from adjuvant therapy: Derived and validated in two population-based cohorts. *Breast Cancer Research*, *7*(6), R953–R964. doi:10.1186/bcr1325

Perou, C. M., Sorlie, T., Eisen, M. B., van de Rijn, M., Jeffrey, S. S., & Rees, C. A. (2000). Molecular portraits of human breast tumours. *Nature*, *406*(6797), 747–752. doi:10.1038/35021093

Ramani, A. K., Bunescu, R. C., Mooney, R. J., & Marcotte, E. M. (2005). Consolidating the set of known human protein-protein interactions in preparation for large-scale mapping of the human interactome. *Genome Biology*, *6*(5), R40. doi:10.1186/gb-2005-6-5-r40

Reyal, F., Stransky, N., Bernard-Pierrot, I., Vincent-Salomon, A., de Rycke, Y., & Elvin, P. (2005). Visualizing chromosomes as transcriptome correlation maps: Evidence of chromosomal domains containing co-expressed genes-a study of 130 invasive ductal breast carcinomas. *Cancer Research*, *65*(4), 1376–1383. doi:10.1158/0008-5472.CAN-04-2706

Sayers, E. W., Barrett, T., Benson, D. A., Bolton, E., Bryant, S. H., & Canese, K. (2010). Database resources of the National Center for Biotechnology Information. *Nucleic Acids Research*, *38*(Database issue), D5–D16. doi:10.1093/nar/gkp967

Schena, M., Shalon, D., Davis, R. W., & Brown, P. O. (1995). Quantitative monitoring of gene expression patterns with a complementary DNA microarray. *Science*, *270*(5235), 467–470. doi:10.1126/science.270.5235.467

Schmidt, M., Bohm, D., von Torne, C., Steiner, E., Puhl, A., & Pilch, H. (2008). The humoral immune system has a key prognostic impact in node-negative breast cancer. *Cancer Research*, *68*(13), 5405–5413. doi:10.1158/0008-5472.CAN-07-5206

Segal, E., Friedman, N., Koller, D., & Regev, A. (2004). A module map showing conditional activity of expression modules in cancer. *Nature Genetics*, *36*(10), 1090–1098. doi:10.1038/ng1434

Shi, L., Reid, L. H., Jones, W. D., Shippy, R., Warrington, J. A., & Baker, S. C. (2006). The MicroArray Quality Control (MAQC) project shows inter- and intra-platform reproducibility of gene expression measurements. *Nature Biotechnology*, *24*(9), 1151–1161. doi:10.1038/nbt1239

Sorlie, T., Perou, C. M., Tibshirani, R., Aas, T., Geisler, S., & Johnsen, H. (2001). Gene expression patterns of breast carcinomas distinguish tumor subclasses with clinical implications. *Proceedings of the National Academy of Sciences of the United States of America*, *98*(19), 10869–10874. doi:10.1073/pnas.191367098

Sotiriou, C., Wirapati, P., Loi, S., Harris, A., Fox, S., & Smeds, J. (2006). Gene expression profiling in breast cancer: Understanding the molecular basis of histologic grade to improve prognosis. *Journal of the National Cancer Institute*, *98*(4), 262–272. doi:10.1093/jnci/djj052

The Gene Ontology Consortium. (2009). The gene ontology's reference genome project: A unified framework for functional annotation across species. *PLoS Computational Biology*, *5*(7), e1000431. doi:10.1371/journal.pcbi.1000431

van de Vijver, M. J., He, Y. D., van't Veer, L. J., Dai, H., Hart, A. A., & Voskuil, D. W. (2002). A gene-expression signature as a predictor of survival in breast cancer. *The New England Journal of Medicine*, *347*(25), 1999–2009. doi:10.1056/NEJMoa021967

van 't Veer, L. J., Dai, H., van de Vijver, M. J., He, Y. D., Hart, A. A., & Mao, M. (2002). Gene expression profiling predicts clinical outcome of breast cancer. *Nature*, *415*(6871), 530–536. doi:10.1038/415530a

van Vliet, M. H., Klijn, C. N., Wessels, L. F., & Reinders, M. J. (2007). Module-based outcome prediction using breast cancer compendia. *PLoS ONE*, *2*(10), e1047. doi:10.1371/journal.pone.0001047

Vizcaino, J. A., Cote, R., Reisinger, F., Foster, J. M., Mueller, M., & Rameseder, J. (2009). A guide to the Proteomics Identifications Database proteomics data repository. *Proteomics*, *9*(18), 4276–4283. doi:10.1002/pmic.200900402

Wachi, S., Yoneda, K., & Wu, R. (2005). Interactome-transcriptome analysis reveals the high centrality of genes differentially expressed in lung cancer tissues. *Bioinformatics (Oxford, England)*, *21*(23), 4205–4208. doi:10.1093/bioinformatics/bti688

Wang, Y., Klijn, J. G., Zhang, Y., Sieuwerts, A. M., Look, M. P., & Yang, F. (2005). Gene-expression profiles to predict distant metastasis of lymph-node-negative primary breast cancer. *Lancet*, *365*(9460), 671–679.

Wong, S. Y., Haack, H., Kissil, J. L., Barry, M., Bronson, R. T., & Shen, S. S. (2007). Protein 4.1B suppresses prostate cancer progression and metastasis. *Proceedings of the National Academy of Sciences of the United States of America*, *104*(31), 12784–12789. doi:10.1073/pnas.0705499104

Wu, Z., Zhao, X., & Chen, L. (2009). Identifying responsive functional modules from protein-protein interaction network. *Molecules and Cells*, *27*(3), 271–277. doi:10.1007/s10059-009-0035-x

Xu, L., Geman, D., & Winslow, R. L. (2007). Large-scale integration of cancer microarray data identifies a robust common cancer signature. *BMC Bioinformatics*, *8*, 275. doi:10.1186/1471-2105-8-275

Yu, G. Z., Chen, Y., Long, Y. Q., Dong, D., Mu, X. L., & Wang, J. J. (2008). New insight into the key proteins and pathways involved in the metastasis of colorectal carcinoma. *Oncology Reports*, *19*(5), 1191–1204.

Yu, K., Ganesan, K., Tan, L. K., Laban, M., Wu, J., & Zhao, X. D. (2008). A precisely regulated gene expression cassette potently modulates metastasis and survival in multiple solid cancers. *PLOS Genetics*, *4*(7), e1000129. doi:10.1371/journal.pgen.1000129

Zhang, Y., Sieuwerts, A. M., McGreevy, M., Casey, G., Cufer, T., & Paradiso, A. (2009). The 76-gene signature defines high-risk patients that benefit from adjuvant tamoxifen therapy. *Breast Cancer Research and Treatment*, *116*(2), 303–309. doi:10.1007/s10549-008-0183-2

Zhou, Y., Yau, C., Gray, J. W., Chew, K., Dairkee, S. H., & Moore, D. H. (2007). Enhanced NF kappa B and AP-1 transcriptional activity associated with antiestrogen resistant breast cancer. *BMC Cancer*, *7*, 59. doi:10.1186/1471-2407-7-59

ADDITIONAL READING

Alexe, G., Bhanot, G., Venkataraghavan, B., Ramaswamy, R., Lepre, J., & Levine, A. J. (2005). A robust meta-classification strategy for cancer diagnosis from gene expression data. *Proceedings / IEEE Computational Systems Bioinformatics Conference, CSB. IEEE Computational Systems Bioinformatics Conference*, 322–325.

Barrett, A. B., Phan, J. H., & Wang, M. D. (2008). Combining multiple microarray studies using bootstrap meta-analysis. *Conference Proceedings; ... Annual International Conference of the IEEE Engineering in Medicine and Biology Society. IEEE Engineering in Medicine and Biology Society. Conference*, *2008*, 5660–5663.

Bertucci, F., & Birnbaum, D. (2009). Distant metastasis: not out of reach any more. *Journal of Biology*, *8*(3), 28. doi:10.1186/jbiol128

Cahan, P., Rovegno, F., Mooney, D., Newman, J. C., St Laurent, G. III, & McCaffrey, T. A. (2007). Meta-analysis of microarray results: challenges, opportunities, and recommendations for standardization. *Gene*, *401*(1-2), 12–18. doi:10.1016/j.gene.2007.06.016

Choi, H., Shen, R., Chinnaiyan, A. M., & Ghosh, D. (2007). A latent variable approach for meta-analysis of gene expression data from multiple microarray experiments. *BMC Bioinformatics*, *8*, 364. doi:10.1186/1471-2105-8-364

Choi, J. K., Choi, J. Y., Kim, D. G., Choi, D. W., Kim, B. Y., & Lee, K. H. (2004). Integrative analysis of multiple gene expression profiles applied to liver cancer study. *FEBS Letters*, *565*(1-3), 93–100. doi:10.1016/j.febslet.2004.03.081

DeConde, R. P., Hawley, S., Falcon, S., Clegg, N., Knudsen, B., & Etzioni, R. (2006). Combining results of microarray experiments: a rank aggregation approach. *Stat Appl Genet Mol Biol*, *5*, Article15.

Dobbin, K. K., & Simon, R. M. (2007). Sample size planning for developing classifiers using high-dimensional DNA microarray data. *Biostatistics (Oxford, England)*, *8*(1), 101–117. doi:10.1093/biostatistics/kxj036

Dobbin, K. K., Zhao, Y., & Simon, R. M. (2008). How large a training set is needed to develop a classifier for microarray data? *Clinical Cancer Research*, *14*(1), 108–114. doi:10.1158/1078-0432.CCR-07-0443

Ma, S., & Huang, J. (2009). Regularized gene selection in cancer microarray meta-analysis. *BMC Bioinformatics*, *10*, 1. doi:10.1186/1471-2105-10-1

Park, T., Yi, S. G., Shin, Y. K., & Lee, S. (2006). Combining multiple microarrays in the presence of controlling variables. *Bioinformatics (Oxford, England)*, *22*(14), 1682–1689. doi:10.1093/bioinformatics/btl183

Pihur, V., & Datta, S. (2008). Finding common genes in multiple cancer types through meta-analysis of microarray experiments: a rank aggregation approach. *Genomics*, *92*(6), 400–403. doi:10.1016/j.ygeno.2008.05.003

Rhodes, D. R., Barrette, T. R., Rubin, M. A., Ghosh, D., & Chinnaiyan, A. M. (2002). Meta-analysis of microarrays: interstudy validation of gene expression profiles reveals pathway dysregulation in prostate cancer. *Cancer Research*, *62*(15), 4427–4433.

Warnat, P., Oberthuer, A., Fischer, M., Westermann, F., Eils, R., & Brors, B. (2007). Cross-study analysis of gene expression data for intermediate neuroblastoma identifies two biological subtypes. *BMC Cancer*, *7*, 89. doi:10.1186/1471-2407-7-89

KEY TERMS AND DEFINITIONS

Gene Signature: list of genes correlated with a given phenotype.

Interactome: Large scale gene interaction map that can be physical or functional, and inferred experimentally or by in silico analysis.

Meta-Analysis: integrated simultaneous analysis of multiple datasets.

Metastatic Relapse: Relapse of cancer after treatment with spreading to distant organs.

Robustness: Stability, Repeatability.

Chapter 20
Using Systems Biology Approaches to Predict New Players in the Innate Immune System

Bin Li
Merrimack Pharmaceuticals, USA

ABSTRACT

Toll-like receptors (TLRs) are critical players in the innate immune response to pathogens. However, transcriptional regulatory mechanisms in the TLR activation pathways are still relatively poorly characterized. To address this question, the author of this chapter applied a systematic approach to predict transcription factors that temporally regulate differentially expressed genes under diverse TLR stimuli. Time-course microarray data were selected from mouse bone marrow-derived macrophages stimulated by six TLR agonists. Differentially regulated genes were clustered on the basis of their dynamic behavior. The author then developed a computational method to identify positional overlapping transcription factor (TF) binding sites in each cluster, so as to predict possible TFs that may regulate these genes. A second microarray dataset, on wild-type, $Myd88^{-/-}$ and $Trif^{-/-}$ macrophages stimulated by lipopolysaccharide (LPS), was used to provide supporting evidence on this combined approach. Overall, the author was able to identify known TLR TFs, as well as to predict new TFs that may be involved in TLR signaling.

DOI: 10.4018/978-1-60960-491-2.ch020

INTRODUCTION

The innate immune system provides the first line of defense against microbial pathogens (Ye et al., 2002; Beutler, 2004; Oda & Kitano, 2006). Toll-like receptors (TLRs) are important components of the innate immune system, which recognize foreign invaders and activate pathogen-specific immune response through the fine-tuned regulation of transcription factors (TFs). In recent years, much progress has been made in discovering new components and understanding their interactions within the Myd88-dependent and Trif-dependent TLR signaling pathways, with NFkBs and IRFs as downstream transcription factors, respectively (Beutler, 2004; Oda & Kitano, 2006). On the contrary, only a limited number of transcription factors have thus far been identified to be involved in the TLR signaling network, namely the NFkB, AP1, IRFs, and CREB families.

Recent technological developments in high-throughput experiments, such as microarrays (Draghici, Khatri, Eklund, & Szallasi, 2006), massively parallel signature sequencing (Stolovitzky et al., 2005), and Chromatin immunoprecipitation coupled to microarray hybridization (Buck & Lieb, 2004), enable the collection of data across an entire genome, making it possible to gain knowledge at the systems level. In addition, databases such as TRANSFAC (Matys et al., 2006) and JASPAR (Sandelin, Alkema, Engstrom, Wasserman, & Lenhard, 2004) have collected TF binding site information in the form of positional weight matrices, enabling computational scanning of known TF binding sites (TFBSs). It is possible to combine these genome-wide data to systematically predict novel TFs in biological systems. Indeed, we recently adopted such an approach to predict and validate ATF3 as a regulator of Lipopolysaccharide (LPS) induced innate immune responses (Gilchrist et al., 2006).

The success of this systematic approach is dependent on finding shared patterns of TFBSs among co-regulated genes. One strategy for defining shared TFBS patterns requires that the predicted TFBSs form a spatial cluster on the DNA (Frith, Li, & Weng, 2003). Additional constraints may be added so that predicted TFBSs must occur in the same 5'-3' order on the DNA, or that distances among the predicted TFBSs be conserved. Here, we developed a novel computational method for identifying positional overlapping TF binding sites to search TFBSs that are at similar distances from transcription start sites among putative co-regulated genes.

To better identify putative co-regulated genes, we selected a dataset from a series of microarray experiments with six different stimuli (LPS, PAM2, PAM3, Poly I:C, R848, and CpG), each at six time points (0 min, 20 min, 40 min, 60 min, 80 min and 120 min) (Ramsey et al., 2008). We utilized this fine-grained and relatively short time-course in order to circumvent positive and negative feedback loops and thus to define direct transcriptional targets. Using this set of time-course data, we identified differentially regulated genes for each stimulus and then clustered genes based on their dynamic behavior. We then applied a stringent computational method to predict possible regulators for genes in each cluster. As a proof-of-principle, we were able to recapitulate roles for Nfkb and Irf which are well-known in the TLR signaling system. More interestingly, we identified a novel regulatory role for Egr1 in regulating an early transient gene cluster. Overall, the novel TF predictions and the high-quality microarray dataset represent a useful resource for the research community.

METHODS

Microarray Expression Measurements

Femurs from the C57BL/6 (Jackson Laboratories) mice were flushed with complete RPMI (RPMI 1640 supplemented with 10% FBS, 2mM

L-glutamine, 100 IU/mL penicillin and 100 mg/mL streptomycin. All of these reagents were from Cellgro, Mediatech, except that the FBS was from Hyclone). Bone marrow cells were plated on non-tissue culture treated plastic in complete RPMI supplemented with recombinant human M-CSF (rhM-CSF) at 50 ng/mL (gift from Chiron). On day 4, the cells were washed two times with RPMI with no additions and then allowed to grow for 2 more days in complete RPMI supplemented with 50 ng/mL of rhM-CSF. On day 6, the cells were lifted from the non-tissue culture treated plastic, counted and plated at a density of 1×10^5 cells/cm2 (1×10^6 cells per well in a 6-well dish) on tissue culture-treated plastic. On day 7, cells were stimulated with TLR agonists at the proper concentrations (Ramsey et al., 2008), without changing the media. Stimulation of the cells was verified by the presence of TNFa in the culture supernatants detected by ELISA (Duoset ELISA Assay Development System, R&D Systems). Total RNA was isolated using TRIzol (Invitrogen) and analyzed for overall quality using an Agilent 2100 Bioanalyzer. mRNA was labeled using the Affymetrix One-Cycle Target Labeling protocol and reagents for eukaryotic target preparation. The labeled cRNA was hybridized to an Affymetrix GeneChip Mouse Genome 430 2.0 array using standard protocols and reagents from Affymetrix. Probe intensities were measured using the Affymetrix GeneChip Scanner 3000 and processed into CEL files using Affymetrix GeneChip Operating Software.

Affymetrix GeneChip Analysis

The time-course microarray dataset was part of a previous research (Ramsey et al., 2008), which was based on the Affymetrix Genechip Mouse Genome 430 2.0 Array (ArrayExpress, http://www.ebi.ac.uk/microarray-as/ae/, ID: E-TABM-310). A quantile normalization (rma) was performed using Bioconductor (Gentleman et al., 2004). Determination of p-values for time-course microarray data on each stimulus was performed using Gaussian kernel density estimation (Hwang et al., 2005). Significantly regulated genes under each stimulus were identified using a p-value cutoff of 0.01, to only include the most differentially regulated genes for clustering analysis and TFBS prediction.

The second microarray dataset is on wild-type, Myd88$^{-/-}$ and Trif$^{-/-}$ stimulated by LPS at 1 and 4 hours (Matsushita et al., 2009), using Affymetrix Genechip Mouse Genome 430A 2.0 Array (ArrayExpress, http://www.ebi.ac.uk/microarray-as/ae/, ID: E-GEOD-14890) (Matsushita et al., 2009). We first normalized this second microarray dataset using RMA method from Bioconductor (Gentleman et al., 2004), then calculated $\log_2$(Fold change) on stimulated wild-type or knock-out cases against un-stimulated wild-type intensities for each probeset.

Clustering Analysis

Clustering analysis was done using absolute correlation as the dissimilarity measure, and the selection of absolute correlation (instead of correlation) is data driven. The number of clusters is determined based on both pseudo T^2- and F-statistics (SAS/STAT User's Guild, 1989). Ward is used as the linkage method (Hwang, Stephanopoulos, & Chan, 2004).

MotifLocator Scanning

MotifLocator (Version 3.1) was used to scan all genes with all matrices (http://homes.esat.kuleuven.be/~thijs/download.html).

MotifLocator was provided with a 2nd order background model derived from 5K base pair upstream regions, from the first 496 genes on chromosome 17. The background data file (bModel.fa) was created using CreateBackgroundModel (downloaded from the same URL as for MotifLocator).

MotifLocator uses an extension of the classical position-weight matrix scoring scheme, please see details in our previous publication (Gilchrist et al., 2006). A complete description of the scoring scheme can be found in: Thijs G: Probabilistic methods to search for regulatory elements in sets of coregulated genes. *PhD thesis* Katholieke Universiteit Leuven, Faculty of Applied Sciences 2003. [INSERT FIGURE 001]

TFBS Prediction

Our transcription factor predictions have the following steps:

1. We downloaded the 2 kb 5' promoter regions for all the differentially regulated genes in each cluster from the Ensemble database (http://www.ensembl.org/Multi/martview).
2. We applied Motiflocator to scan the TF binding sites for each promoter region. Each identified TF binding site was extended by 8 bp on each side. The constant scores obtained from Motiflocator in the TF binding region were Fourier transferred to generate a bell-shape binding segment for calculating the positional overlap score below. This way, exactly overlapping binding sites will get the highest overlap score, while slightly shifted overlapping binding sites will get a lower score.
3. We aligned the transcription start sites of genes in the cluster and added TFBS scores from individual genes in the cluster at the same upstream positions to get the overlapping score. The score is divided by the number of genes in the cluster to generate the density scores. The density curve along the 2 kb up-stream region is used to identify common regulators for each cluster of genes: If the density curve has a peak above the statistical cutoff, we will make a prediction that the corresponding TF is a potential regulator of this cluster of genes.
4. We generated two control datasets for statistical analysis: (a) we randomly picked 100 differentially regulated genes from different clusters and obtained their 2 kb promoter regions as a real-gene control dataset and (b) we shuffled the promoter sequences of these 100 genes to get a shuffled-sequence control dataset.
5. We selected a list of 79 motifs, to represent the 360 mouse motifs currently in the TRANSFAC database (Matys et al., 2006) (professional version 8.4). For example, we chose the IRF_Q6 matrix among several similar IRF matrices in the TRANSFAC database.
6. For each cluster of genes, we randomly picked the same number of sequences from each of the control datasets, and scanned each of the 79 motifs in our motif list. MotifLocator (http://homes.esat.kuleuven.be/~thijs/download.html) was used to scan all genes with all matrices. This process was repeated 100 times to generate 200,000 samples (2000 bp × 100) on each control dataset. A value corresponding to the 99.9 percentile (p-value 0.001) was picked as the cutoff value for each control dataset and the bigger one of the two cutoffs was used as a final cutoff. This small p-value cutoff was chosen to reduce false-positive predictions, as well as to address potential multiple testing problem.
7. We scanned the target cluster of genes using the same 79 motifs and used the final cutoff to identify statistically overrepresented TFs. Since the final cutoff was 99.9 percentile or more on both control datasets, the identified TFs were statistically significant against both random sequences and randomly picked real sequences, therefore were more likely to be the TFs regulating each cluster of genes.

The software is available upon request from the author.

Generating the list of predicted TFs for the TLR signaling network. We applied an additional intensity filter when generating the predicted TFs list (Table 2). Using RT-PCR and ELISA experiments, we found that Myod1, which has $\log_2$ (gene expression intensity) of 4.7, may not be present in macrophages (data not shown). On the other hand, Creb1, as a known regulator in innate immune system, has a maximum $\log_2$ (gene expression intensity) of 6.88 under LPS at 2 hrs. Therefore, we chose a maximum $\log_2$ (gene expression intensity) value of 6.5 as a cutoff to filter out predicted TFs that had low expression levels.

Cytoscape. We used Cytoscape (http://www.cytoscape.org/) to visualize networks and overlay cDNA expression profiles on a matrix of protein-protein and protein-DNA interaction networks.

RESULTS

To make reliable predictions of new transcription factors in a system, we selected a high-quality microarray dataset with multiple TLR agonists (Ramsey et al., 2008); performed a specially designed two-step clustering analysis driven by features of known TLR signaling pathways; developed a new computational method to identify positionally overlapping transcription factor binding sites; recaptured several individual predictions as well-known TFs in the innate immune system; obtained supporting evidence from a second (knock-out) microarray dataset on TLR signaling; and provided a list of the TF predictions.

Clustering TLR Regulated Genes using Time-Course Microarray Data

Time-course microarray data were generated on mouse bone marrow-derived macrophages stimulated with LPS (a TLR4 agonist), PAM2 (a TLR2/6 agonist), PAM3 (a TLR2/1 agonist), Poly I:C (a TLR3 agonist), R848 (a TLR7 agonist), or CpG (a TLR9 agonist). Measurements were taken at time points of 0, 20, 40, 60, 80, and 120 minutes for each stimulus, with two or three biological replicates for each condition. We identified differentially regulated genes for each stimulus based on the overall behavior among the 6 time points (see methods). There were 393, 205, 512, 150, 288 and 139 genes differentially regulated by LPS, PAM2, PAM3, Poly I:C, R848 and CpG, respectively. Many differentially regulated genes were shared by different stimuli, while each stimulus also regulated some unique genes. In total, there were 655 differentially regulated genes for all stimuli, 70 of which were shared by all 6 stimuli.

Cluster analysis is a commonly applied technique to find co-expressed genes based on all tested conditions. The first step of our analysis was based on this approach. Genes were clustered among all 36 conditions (6 stimuli, 6 time points) based on the absolute correlation of $\log_2$-fold changes relative to the unstimulated cases (the use of absolute correlation to separate clusters is data driven). Thus, each resulting cluster has both up- and down-regulated genes. However, here we chose to only pay attention to up-regulated genes based on the following reasons. First, TLR stimuli normally up-regulate both positive responders (like Tnf and Irf3/7) and negative responders (like A20 and IκB) to guide innate immune responses. Second, upon TLR stimulation, the number of up-regulated genes far exceeds that of down-regulated genes, which results in very small down-regulated gene clusters and less reliable predictions on common transcription factor binding sites. Lastly, a large fold-change for down-regulated genes only represent a small decrease in gene expression level, making it harder to experimentally validate down-regulated genes compared to up-regulated ones.

Four clusters were identified from this first clustering analysis. Figure 1 shows only the mostly differentially expressed genes in each cluster. This clustering analysis groups the genes that are co-expressed in all the tested conditions, while genes in the same cluster may have different behavior under different TLR stimuli. For example, cluster IV appeared to include genes that were uniquely regulated by Poly I:C. By contrast, cluster I genes were quickly regulated by LPS and PAMs, but slowly regulated by Poly I:C (Figure 1A).

We further refined the clustering analysis as follows: If two or more TLRs use a common signaling pathway, gene clusters deduced from datasets acquired by stimulation with the individual TLR ligands should show similar dynamical behavior. Conversely, similar dynamic behavior suggests shared signaling pathways among different TLRs. Therefore, we implemented clustering analysis based solely on the dynamic behavior of the TLR-regulated genes, disregarding the stimulus other than as a row label in conjunction with each gene. All differentially expressed gene-stimulus profiles (e.g. Tnf-LPS, Tnf-PAM2) were clustered using the six time points (Figure 1B). Thus, each resulting cluster contains genes that have similar dynamic behavior, which may be shared by several stimuli.

The first clustering step produces co-expressed groups of genes not necessarily having the same

Figure 1. Clustering of differentially regulated genes following TLR activation. (A) Identifying co-expressed groups of genes based on all tested conditions. (B) Clustering based on dynamic behavior. Any differentially regulated gene under any stimulus can be a row (e.g. TNF for LPS as a row, while TNF for PAM2 was another row), and be clustered over the six time points. The purpose of this two-step clustering analysis is to find groups of genes that are not only co-expressed (from step A), but also likely to be controlled by the shared signaling pathways (from step B).

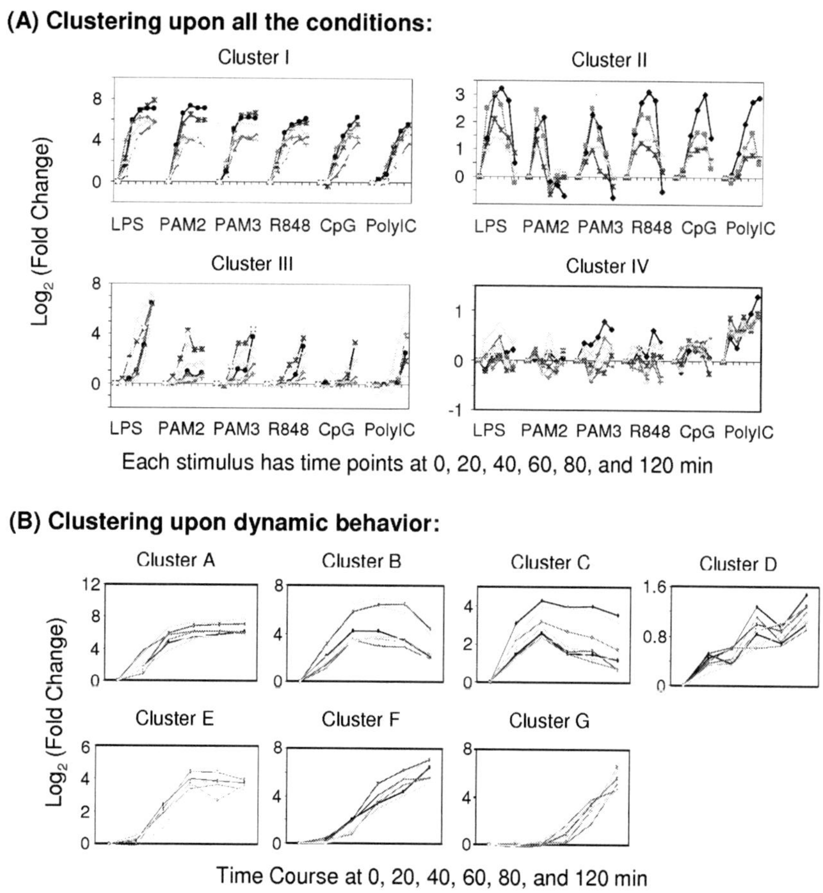

Table 1. Co-regulated genes from two-step clustering results on TLR time-course arrays

	A	B	C	D	E	F	G
I	CpG (7) LPS (27) PAM2 (9) PAM3 (13) R848 (25)	LPS (16) PAM2 (12) PAM3 (15) R848 (12)	PAM2 (12)		CpG (6) LPS (14) PAM3 (18) R848 (16)	CpG (20) Poly I:C (24) R848 (9)	CpG (5) Poly I:C (20)
II		R848 (5)					
III	LPS (11) PAM2 (12) PAM3 (7) R848 (5)			LPS (6) PAM2 (8)	LPS (9) PAM2 (48) PAM3 (22)	CpG (13) LPS (121) PAM2 (47) PAM3 (97) R848 (68)	CpG (56) LPS (83) PAM2 (8) PAM3 (48) Poly I:C (50) R848 (53)
IV				Poly I:C (14)			

Note: Each cell in the table represented an intersection of the two-step clustering results, indicating a putative co-regulated group of genes. The rows in the table correspond to the first clustering step; the columns correspond to the second clustering step. The first clustering step produces co-expressed groups of genes not necessarily having the same dynamic behavior under different stimuli, while the second clustering step groups together the genes that have the same dynamic behavior. Therefore, each intersection of these two clustering results contains genes that are not only co-expressed, but also likely to be controlled by a shared signaling pathway. The numbers in parentheses are the number of differentially regulated genes for each agonist.

dynamic behavior under different stimuli, while the second clustering step groups together genes that have the same dynamic behavior. Therefore, each intersection of these two clustering results contains genes that are not only co-expressed, but also likely to be controlled by a shared signaling pathway.

Comparisons of Gene Regulation Programs among Different TLRs Based on the Clustering Analysis

From the clustering analysis results, we can obtain an overview of the similarities, as well as specificities, of the differentially regulated genes under different TLR signaling pathways. Each cell in Table 1 represents an intersection of the two clustering results described above, indicating a putative co-regulated group of genes. The rows in Table 1 correspond to the first clustering approach; the columns correspond to the second clustering approach. Hereafter, we label the sets of genes represented in Table 1 using the shorthand: stimulus_column_row. For example, LPS_A_I represents the group of LPS regulated genes that is both in cluster A and in cluster I (see Figure 1). Omission of the stimulus term corresponds to shared clusters; thus the term "A_I" represents all the A_I type clusters. The numbers in parentheses are the number of differentially regulated genes for each agonist. The full set of genes in each cluster is enumerated in Table 3 which is available upon request from the author.

There are several clusters shared by most of the TLRs signaling pathways. Genes in the A_I and B_I clusters are regulated quickly, though A_I genes tend to stay induced while the B_I genes revert back toward their unstimulated states (Figure 1 and Table 1). In addition, clusters F_III and G_III are shared by most of the TLRs and contain late-regulated genes. These clusters may represent shared TLR signaling pathways. In addition, the A_I cluster was commonly regulated by several different stimuli (Table 1). The genes belonging to this cluster include common genes such as IκB, Jun, Socs3 (Table 3), which are well-known to be the key output and/or regulator genes for several TLRs.

In addition, the clustering results also revealed facets of specificity in the TLR regulatory network.

The PAM2_C_I cluster contains genes largely overlapping with LPS_B_I and PAM3_B_I genes (Table 3), suggesting that this group of genes may be co-regulated by these three TLRs, although this behavior was more transient with PAM2. The same trend is found in G_III clusters: PAM2 has a very small number of genes compared to any other TLR agonist. In summary, the clustering results suggest that PAM2 regulates shorter-lived responses than other TLR agonists. In addition, PAM2 and PAM3 have more genes in cluster E_III than LPS, suggesting a TLR2-specific program. However, only CpG and Poly I:C regulate cluster G_I (Table 1), possibly reflecting the intracellular localization of TLR9 and TLR3 or their shared anti-viral response (Barton, Kagan, & Medzhitov, 2006; Takeda, Kaisho, & Akira, 2003).

Computational Predictions of TFs Involved in TLR Signaling Pathways

A Positional Overlap Method to Predict Common TFBSs of Co-Regulated Genes

We designed and implemented a computational prediction algorithm, to discover potential regulatory TFs in each cluster. This method identified positionally overlapping TF binding sites among promoter regions of putative co-regulated genes derived from the clustering analysis (Table 1). As illustrated in Figure 2A, RNA polymerase II interacts with regulating TFs and initiates gene expression from the transcription start site in mammalian cells. Therefore, the structure of

Figure 2. Biological insights driving the development of a TFBS prediction method. (A) Illustrated is the RNA polymerase II complex, showing that both protein-DNA and protein-protein interactions are important for co-regulated genes. (B) The structure of protein complex demands positional overlap of TF binding sites for co-regulated genes.

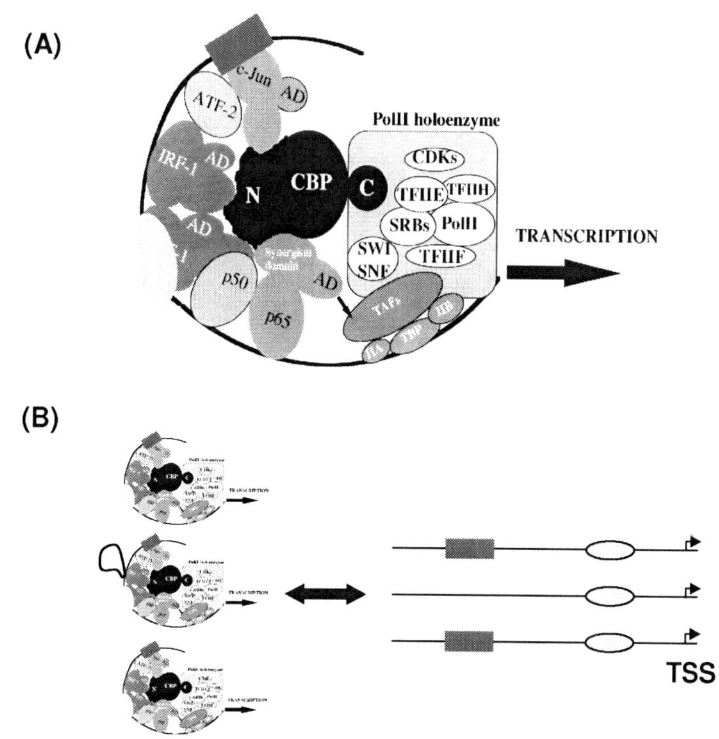

protein complex demands positional overlap of TF binding sites for co-regulated genes (Figure 2B). In other words, we assumed that when putatively co-regulated genes were aligned based on their transcription start sites, TFBSs for their common regulators (TFs) will likely to be positionally overlapped (Figure 2B).

To identify these positionally overlapping TFBSs, the 2 kb up-stream regions of genes in each cluster were aligned based on their transcription start sites, and a density curve for each TRANSFAC matrix of interest was calculated by adding the normalized TFBS scanning scores among these genes. Two random sequence datasets of the same size as target gene cluster were generated to calculate the *p*-value of positional overlap scores. A statistical test was performed to detect the common TFBS binding sites and the potential regulator(s), for the target cluster of genes.

Recapturing Nfkb

Nfκb is a well-known TF involved in the TLR signaling system (Leung, Hoffmann, & Baltimore, 2004; Yoshimura, Ohishi, Aki, & Hanada, 2004), thus represents a good test of this TFBS prediction approach, for the prediction of itself as a TF and its known target genes, such as IκBα, Jun, and Socs3 (Leung et al., 2004; Yoshimura et al., 2004) (Figure 3A).

From our clustering analysis, genes in cluster LPS_A_I are rapidly induced (Figure 3B) and include well-known Nfκb regulated genes (Figure 3A and Table 3 (Leung et al., 2004; Yoshimura et al., 2004). Using the TF prediction method, we found a statistically significant Nfκb peak (using a *p*-value cutoff of 0.001) on its density plot over a 2000 bp promoter region, suggesting common Nfκb binding sites at –75 to –36 bp upstream of the transcription start sites for this cluster of genes (Figure 3C, the dashed line showing *p*-value cutoff of 0.001). Thus, this TF prediction method

Figure 3. Recapturing Nfκb as a TF regulating early TLR-response genes. (A) Known biology of Myd88-dependent signaling pathway and its downstream genes. Details on the genes in this cluster can be found in Table 3 (B) Genes in LPS_A_I cluster were correlatively regulated. (C) Density curve of Nfκb binding sites in the promoter regions, which measures the overlapping Nfκb binding sites among genes in this cluster. The area above the red dashed line represents common Nfκb binding sites.

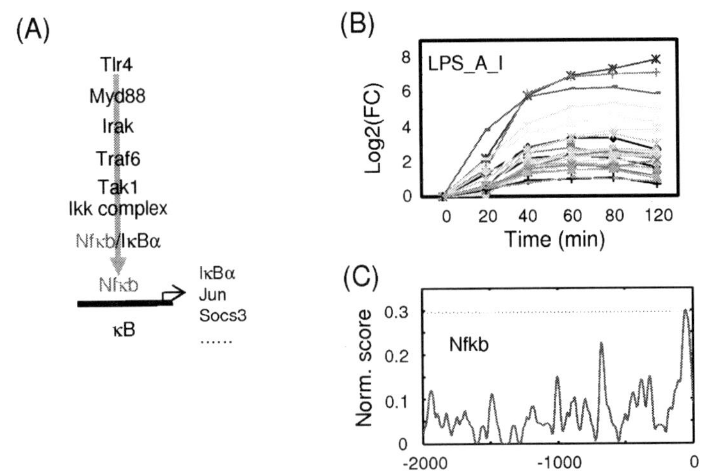

successfully predicts a known TF and links it to its known target genes.

Recapturing Cooperative Regulation by Nfκb/Irf

We further sought to assess whether this TF prediction method is able to identify multiple TFs that may cooperatively regulate a set of genes. In TLR signaling pathways, it is known that two adaptor molecules, Myd88 and Trif mediate the activation of Nfκb- and Irf-dependent transcription pathways, respectively (Akira & Takeda, 2004) (Figure 4A). Whereas TLR2 agonists only activate Myd88 and TLR3 agonists only activate Trif, TLR4 agonists activate both pathways (Akira & Takeda, 2004) (Figure 4A). This dual pathway of TLR4 activity has been shown to crosstalk, and, to a lesser extent, induces superadditive effects (Hoebe et al., 2003). We wished to test the TF prediction method in order to explore this superadditive effect. We selected a group of genes that exhibited greater induction upon TLR4 activation than twice the sum of the fold-changes induced by TLR3 and TLR2 agonists alone. Twelve genes that met this criterion (Figure 4B) were co-expressed, and thus potentially co-regulated. Our computational method predicted Irf and Nfκb as possible regulators of this group of genes (Figure 4C), in good agreement with previous observations (Akira & Takeda, 2004).

Recapturing Atf3

We previously identified Atf3 as a regulator of TLR signaling, albeit based on a longer time-course array dataset with LPS as the only stimulus (Gilchrist et al., 2006). Here, we obtained a similar early transiently regulated cluster upon LPS stimulation (LPS_B_I cluster in Table 1), and recaptured Atf3 as a potential regulating transcription factor

Figure 4. Case study of a group of known Nfκb and Irf co-affected genes. (A) The Myd88- and Trif-dependent signaling pathways and their possible interactions. Nfκb and Irf are known transcription factors downstream of these signaling pathways. (B) Twelve genes were superadditively regulated upon LPS stimulation, which exhibited greater induction upon TLR4 activation than twice the sum of the fold-changes induced by TLR3 and TLR2 agonists. (C) Predicted common TFBSs for these genes. Nfκb and Irf (colored in red) were recaptured to be possible regulators for this group of genes.

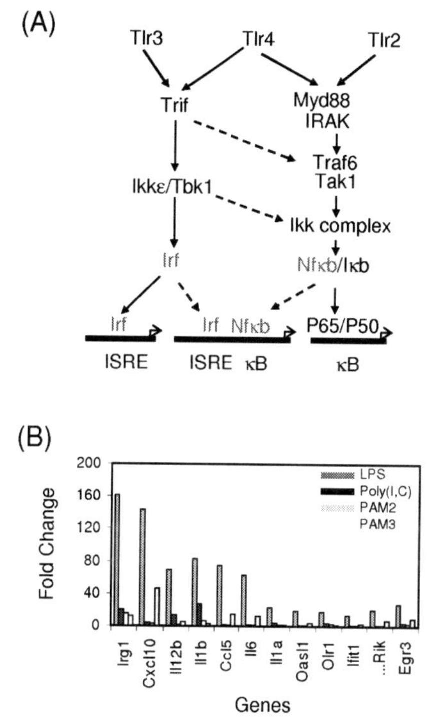

(C)

Predicted common TF sites

matrices	cutoff	start	end	peak
HMGIY_Q6	0.35	-87	-78	0.36
IRF_Q6	0.33	-147	-124	0.34
NFκB_01	0.36	-88	-79	0.37
SRY_01	0.40	-620	-599	0.46
TBP_01	0.33	-38	-14	0.40
YY1_Q6	0.30	1534	1525	0.31

of this cluster of genes. This group of genes has common ATF3 binding sites between -126 bp and -138 bp upstream of the transcription start site. Interestingly, although the individual genes were somewhat different between the previous and current clusters (they did have some common genes), the common Atf3 binding sites were highly conserved, occurring between -121 and -153 bp upstream of the transcription start sites in the previous study (Gilchrist et al., 2006).

A Second Microarray Dataset Supports the Combined Clustering/TFBS Prediction Approaches

In addition to recapturing individual known TFs in the innate immune systems as a way to support the combined clustering and TFBS prediction approaches, we also searched the microarray databases and found a second dataset(s) on wild-type, Myd88$^{-/-}$ and Trif$^{-/-}$ mice cells stimulated by LPS at 1 and 4 hours (Matsushita et al., 2009). This knock-out microarray dataset was generated by a different research team, and provides a good opportunity to check (on a systems level) our clustering and TFBS prediction results.

We first normalized this second microarray dataset using RMA method from Bioconductor (Gentleman et al., 2004), then calculated $\log_2$(Fold change) on stimulated wild-type or knock-out cases against un-stimulated wild-type intensities for each probeset. Since the purpose was to use the second microarray dataset to evaluate the clustering and TFBS prediction results from the first time-course microarray dataset, we simply matched the LPS clustering results from the time-course microarray dataset (Table 3 to the second dataset. More than half of the genes in cluster III from the original paper of the second dataset (Matsushita et al., 2009), which is Myd88-dependent, were found in cluster LPS_A_I here (Figure 7. Also, genes in this cluster are predicted

Figure 5. Prediction and validation of Egr1 as a possible regulator in TLR signaling pathway. (A) An up-and-down regulated cluster of genes under LPS stimulation. (B) Density curve of Egr1 binding sites in the promoter region (2 kb upstream), measuring the overlapping of Egr1 binding sites among genes in this cluster. The area above the red dashed line represents common Egr1 binding sites. (C) Microarray observed fold changes for members of Egr family. (D) Protein-protein interaction network linking Egr1 (in blue) and its target genes (in green) in this cluster.

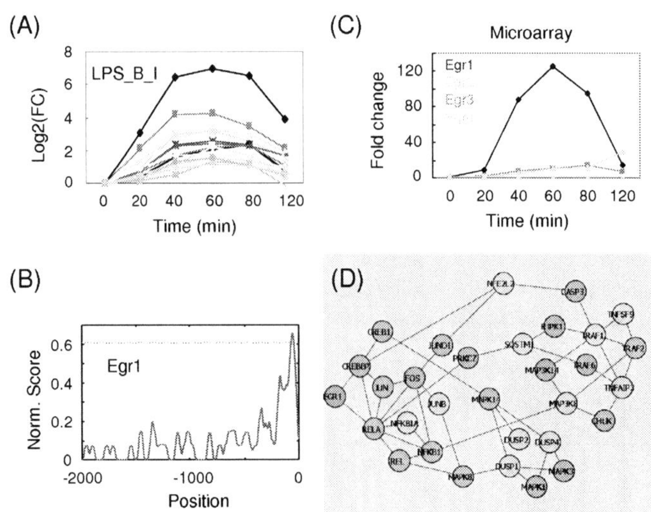

to be regulated by Nfkb, consistent with the fact that they are Myd88-dependent (Figure 3C). Interestingly, without using the knock-out information, the time-course clustering result was able to distinguish genes responding differently to LPS in Trif$^{-/-}$ mice cells (Figure 7A and B, both overlapping with cluster III of the original paper (Matsushita et al., 2009)). In addition, the time-course clustering results were also able to separate Trif-dependent genes into two sub-groups (Figure 7C and D).

Predicting Egr1 as a New Regulator in TLR Signaling Pathways

To test the utility of the TFBS prediction method, genes in the LPS_ B_I cluster (Figure 5A and Table 3 were investigated and the method found an over-abundance of Egr binding sites from -67 to -42 bp upstream of the transcription start sites (Figure 5B). We hypothesized that Egr may be a TF regulating this group of genes. In addition, Among the Egr family members, Egr1 was

Table 2. Summary of predicted TFs that may be involved in TLR signaling network (predicted TFs and their corresponding clusters colored in red were discussed in the text)

TF matrix	Gene name	Found in clusters
Predicted to be involved in TLR signaling network:		
IK1_01	Kcnn4	CpG_F_I, PAM3_F_III
MEF2_02	Mef2a	PAM2/CpG_F_III, PAM3_A_I, Poly I:C_D_IV
NFE2_01	Nfe2l2	LPS_A_III
PU1_Q6	Sfpi1	LPS_G_III, LPS_E_III
HMGIY_Q6	Hmga1	PAM2_D_III, PAM3_A_III, Poly I:C_G_III
SP1_Q6	Sp1	LPS/PAM2/PAM3_B_I, PAM2/R848_A_I, LPS_E_I
YY1_Q6	Yy1	PAM2_B_I, R848_F_III, CpG_G_III
NFY_Q6_01	Nfya	PAM2/PAM3_E_III
NFAT_Q4_01	Nfatc2	CpG_F_I, CpG_F_III, PAM3_E_I, R848_B_I
LYF1_01	Znfn1a1	PAM2_F_III, R848_G_III
SMAD_Q6_01	Madh1	LPS_F_III, R848_G_III
E2F_03	E2f1	LPS/PAM3_F_III, LPS_B_I, CpG_F_I
USF_Q6_01	Usf1	PAM3_B_I
MAZ_Q6	Maz	LPS/PAM3_F_III, PAM2/PAM3/R848_B_I, LPS_E_I, PAM2_D_III, R848_A_I
TBP_01	Tbp	LPS/Poly I:C_G_III, PAM3_E_III
GABP_B	Gabpb1	LPS/PAM2_B_I, PAM3_E_I, Poly I:C_G_I
FOXO1_01	Foxo1/3	PAM2/PAM3_A_III, LPS_D_III
AHR_Q5	Ahr	PAM3_B_I
EGR_Q6	Egr1	LPS/PAM2/PAM3_B_I
Known TFs recaptured in TLR signaling network:		
NFKAPPAB_01	Rela	LPS/PAM2/PAM3/R848_A_I, PAM2_C_I, Poly I:C_F_I
ATF3_Q6	Atf3	LPS/PAM3_B_I
IRF_Q6	Irf3	LPS/CpG_F_III, Poly I:C/PAM3_G_III
AP1_C	Junb	CpG_A_I, PAM2_E_III
CREB_01	Creb1	LPS/PAM3_B_I

the most highly induced under LPS-stimulation (Figure 5C). Therefore, Egr1 is predicted to be a potential regulator in TLR signaling pathways.

Having predicted that Egr1 might be a novel transcription factor for genes in the LPS_B_I cluster, we decided to explore how Egr1 regulates its target genes. We reasoned that knowledge from studies on other cell types and/or conditions may provide hypotheses for our study on macrophage TLR signaling pathways, as well as serve as additional evidence for our predictions. Previously, we applied a protein-protein interaction information search (HPRD, http://www.hprd.org/) to our Atf3 prediction, and hypothesized then validated that Atf3 forms a complex with Hdac and inhibits Il6/Il12b transcription via chromatin remodeling (Gilchrist et al., 2006). Here, by using Egr1 and LPS_B_I genes as "seeds", we identified their first neighbors as a way to construct an Egr1-related protein-protein interaction network (Figure 5D). The resulted network suggests that Egr1 may functionally interact with possible co-factors, Nfκb, Jun, or Creb1, to regulate genes in the LPS_B_I cluster (Figure 5D).

Systematic Prediction of TFs Involved in TLR Signaling Pathways

We applied the same TF prediction method to all the clusters identified in Table 1, to systematically predict TFs that may be involved in TLR signaling pathways (Table 2).

As TFBSs often represent potential binding sites for families of similar transcription factors, it is difficult to predict which TF actually binds to the target genes. We used additional evidence to select the most likely TFs for the predictions. For example, in a previous study, we found common Atf binding sites in a cluster of LPS-regulated genes, and picked Atf3 as the main target since it is the only member in the Atf family that was differentially regulated (Gilchrist et al., 2006). In addition, we also applied a gene expression cutoff to keep only the TFs that were considered to be present in the macrophages, using both Affymetrix microarrays and massively parallel signature sequencing technology (Stolovitzky et al., 2005) data.

Table 2 lists the predicted TFs and the corresponding clusters where they may play a regulatory role. Based on the clusters where each TF was predicted (Table 2) and genes in each cluster (Table 3, hypotheses can be derived for future study on individual TF(s). For example, PAM2 and PAM3 shared common enrichment in cluster E_III (Table 1), and Nfya was predicted to be a regulator for both PAM2_E_III and PAM3_E_III (Table 2). Therefore, a reasonable hypothesis might be that Nfya is a special regulator for PAMs. Moreover, if two or more TFs were predicted to regulate the same cluster of genes, these TFs may work together in regulating the target cluster of genes.

To help generate hypotheses on how predicted TFs may be linked to TLR signaling networks, we built a protein-protein interaction map linking known TLR signaling components and predicted TFs. The Human protein reference database (HPRD, http://www.hprd.org/) was adopted as a reliable source of known protein-protein interactions. We collected a literature-based TLR signaling network based on previous publications. Using the known TLR components and nine representative predicted TFs as seeds, we created a network containing the known TLR signaling components, predicted transcription factors, and their possible "bridges" (Figure 6). As an example, known links from predicted TF Ahr to known TLR signaling components are highlighted in red. Those links may or may not be "true" in TLR signaling, and they only indicate that previous studies have observed these interactions in some cell-types under certain conditions. Nonetheless, we believe that such information may lead to reasonable hypotheses for detailed studies on each of the predicted TFs.

From this interaction network (Figure 6), we found that Ahr may interact, directly or through other proteins, with several known TLR compo-

Figure 6. Protein-Protein interaction network derived from Human Protein Reference Database (HPRD), linking known TLR signaling components (in blue) and predicted TFs (in magenta). The figure is arranged roughly from TLRs on the top, then all the way down to transcription factors (e.g. TLRs -> Myd88 -> Traf6 -> IKK -> IkBs -> TFs). The predicted TFs are placed at the bottom of the network, and different links between predicted TFs to known TFs and upstream TLR components are displayed at different levels. Possible links between Ahr and known TLR components are highlighted in thick red lines.

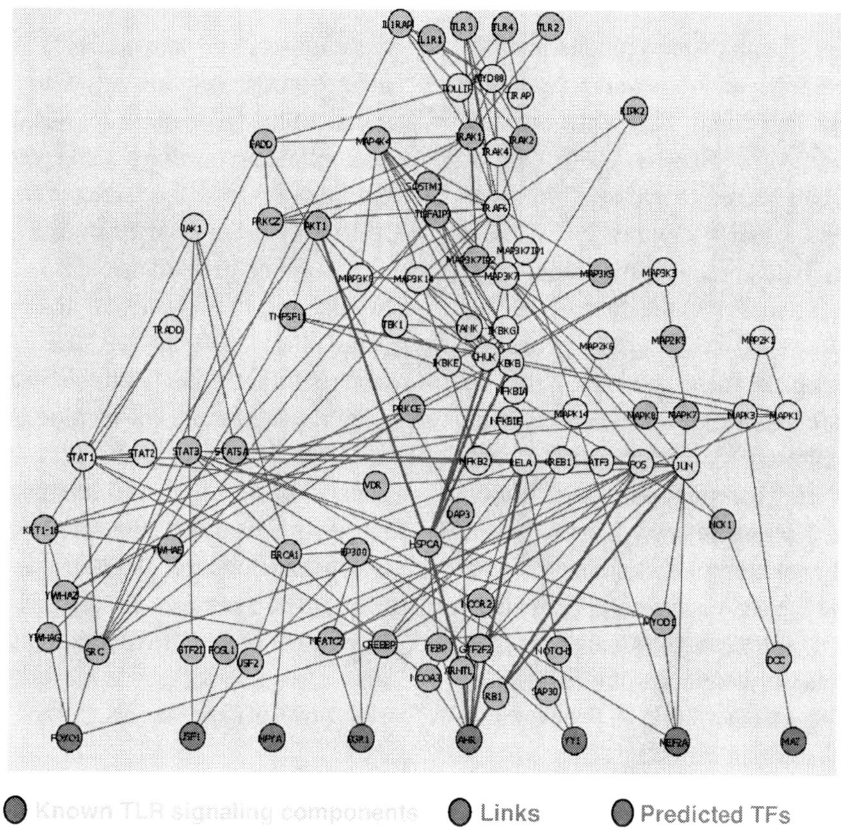

nents (red thick lines in Figure 6). Ahr has direct interaction with Rela; a possible path might be Ahr – Rb1 – Jun/Fos; another path may be Ahr – Arnt – Hspca – Ikkα/Ikkβ (Chuk and Ikbkb in Figure 6). These links suggest that Ahr may work downstream of the Myd88 signaling pathway commonly used by several TLRs. Alternatively, it is possible that Ahr links the TLR signaling pathway through Ahr – Arntl – Hspca – Akt1 (Figure 6). Since Akt1 is well known to be a key player in PI3K signaling pathways, the protein-protein interaction network suggests a testable hypothesis that Ahr may be involved in a signaling pathway such as TLR -> PI3k -> Akt1 -> … -> Ahr. Additional hypotheses on other predicted TFs can be generated in a similar way.

DISCUSSION

Clustering Analysis to Identify Putative Co-Regulated Genes

Clustering analysis is commonly used as a first step to treat systematic data like multiple condition microarray datasets, with selection of the clustering method largely dependent on the features of the system being studied (D'Haeseleer, 2005). Our clustering analysis here has features in common with bi-clustering analysis (Reiss, Baliga, & Bonneau, 2006), while using a supervised two-step approach to capture the special feature of TLR signaling pathways. For example, it is well-known that different TLR stimuli share common signaling pathways (like the Myd88-dependent signaling pathway), while individual TLR may activate along unique pathways (Beutler, 2004; Oda & Kitano, 2006). We therefore pursued the two-step approach and grouped together the dynamic behavior of each TLR stimulus in the second step. For this clustering analysis, the first clustering step produces co-expressed groups of genes not necessarily having the same dynamic behavior under different stimuli, while the second clustering step groups together the genes that have the same dynamic behavior. Therefore, each intersection of these two clustering results contains genes that are not only co-expressed, but also likely to be controlled by the shared signaling pathway.

This clustering approach gives insight into different gene-regulatory patterns among TLR signaling pathways (Figure 1 and Table 1). For example, the clustering result that different TLRs quickly up-regulate a core set of genes (the A_I cluster, Table 1) is in good agreement with the fact that most TLRs utilize the Myd88-dependent signaling pathway for innate immune responses (Akira & Takeda, 2004). However, PAM2 and PAM3 have more genes in cluster E_III than LPS (Table 1), leading to a biological interesting hypothesis that these genes may be specially regulated by TLR2. Overall, the clustering analysis here established a basis for systematic predictions of new regulators in TLR signaling.

Rationale for Developing a Positional Overlap TFBS Prediction Method

A new method to predict TFBS, based on positional overlap on putative co-regulated genes aligned upon their transcription start sites, was developed to incorporate biological information and to make more reliable predictions. In eukaryotic cells, RNA polymerase II interacts with regulating TFs and initiates gene expression from a transcription start site. Therefore, we hypothesized that when putatively co-regulated genes aligned based on their transcription start sites, TFBSs identified in their promoter regions will likely overlap. Computationally, the method applies stringent criteria by implicitly requiring that the predicted TFs have binding sites with the same order and at the same relative distance from the transcription start sites.

Experimental evidence shows that positionally overlapping TFBSs exist among co-regulated genes. Previously, Ihmels and co-workers demonstrated that mitochondrial ribosomal proteins displayed a strongly correlated expression pattern in *Candida albicans*, which was resulted from a conserved cis-regulatory element at about -110 bp upstream of the corresponding transcription start sites (Ihmels et al., 2005). In addition, Shen et al. found a group of interleukin-17 target genes having common C\EBP binding sites about -180 bp upstream of the transcription start sites (Shen, Hu, Goswami, & Gaffen, 2006).

Systematically Predicting Transcription Factors in TLR Signaling Network

Our goal in the current research, as well as in systematic predictions as a whole, is to both make reliable predictions and provide good overall coverage of the predicted TFs in the system. By making reliable predictions, we recaptured both Nfκb and ATF3 as key regulators in the innate immune system. For overall coverage of the target system, we were able to recapture all known TLR related TFs, such as Nfkb, Irf, Ap1, Creb and Atf3 (Table 2). In comparison, a similar attempt to systematically predict TFs, which used a microarray dataset with a single stimulus (LPS) and different computational TFBS prediction methods, did not have any Nfκb or Ap1 motifs in their top 20 predicted motif list (Nilsson et al., 2006).

From the systematic prediction, we identified some unique TFs that may be novel regulators of the TLR signaling network (Table 2). These predicted TFs can be divided into two categories: some have supporting evidence but are not yet widely accepted as key players in the TLR signaling network; others may be totally new to the innate immune system. Egr1 is a good example for the first category. Previous studies have linked Egr1 to individual gene(s) upon LPS stimulation, including increased TNF-alpha production (Shi, Kishore, McMullen, & Nagy, 2002), and regulating both the basal and LPS-induced activity of the SOCS-1 promoter (Mostecki, Showalter, & Rothman, 2005). Similarly, evidence can be found to support possible roles of Yy1 (Gordon, Saleque, & Birshtein, 2003) and Foxo1 (Seoane, Le, Shen, Anderson, & Massague, 2004; Stitt et al., 2004) in the immune system. Maz and Nfya are examples of "totally new" TF predictions. Hypotheses on how these "new" TFs may interact with TLR signaling pathways can be generated using the protein-protein interaction network (Figure 6).

CONCLUSION

In summary, we obtained sets of putative co-regulated genes based on temporal expression patterns. For each cluster of genes, we applied very stringent criteria to computationally predict regulatory TFs. Protein-protein interaction information was used to link the predicted TFs to the known TLR signaling components, so as to generate hypotheses on how these new TFs may be involved in TLR signaling pathways. The systematic predictions represent a valuable resource for the research community to further expand and define the TLR transcriptional network.

FUTURE WORK

As an example of interdisciplinary research in the field of computational and systems biology, this work focused on combining biological insights and computational method developments to systematically predict new transcription factors in the innate immune system.

There are additional studies that can be done to improve our current research on both biological and computational aspects. On the biological side, recent publications found that TLR4 requires internalization to recruit TRIF in the endosome (Mollen et al., 2008; Tanimura, Saitoh, Matsumoto, Akashi-Takamura, & Miyake, 2008). Therefore, there might be a delay in the induced expression of genes related to the TLR4-TRIF sub-pathway. A micorarray time course dataset using LPS to stimulate TRIF$^{-/-}$ mice would be able to capture special features of this signaling program. Moreover, the current dataset focuses on early events after TLR stimulation (up to 2 hrs), it would be very interesting to extend the work to longer time courses. Also, if available, Chromatin immunoprecipitation coupled with microarray hybridization (Buck & Lieb, 2004) datasets can provide direct TF binding information on a systematic scale.

On the computational side, a comparison of our positional overlap method to other TFBS prediction methods, such as PAP (Chang, Fontaine, Stormo, & Nagarajan, 2007), CLOVER (Frith et al., 2004), oPOSSUM (Ho Sui, Fulton, Arenillas, Kwon, & Wasserman, 2007), and PASTAA (Roider, Manke, O'Keeffe, Vingron, & Haas, 2009), would be informative. CpG stratification might be considered to classify the promoters of clustered genes. It may be beneficial to re-run the analysis in the future, since databases (like HPRD and TRANSFAC) are periodically updated. In addition, the current work uses a curated database HPRD to build protein-protein interaction networks, which was mainly driven by the fact that HPRD is free. There are commercial curated databases available, like Ingenuity (http://ingenuity.com/) and GeneGo (http://www.genego.com). Incorporating the rich information from Ingenuity and/or GeneGo may help to further increase the biological insights one may obtain from protein-protein interaction networks.

ACKNOWLEDGMENT

We thank Dr. Alan Aderem, Dr. Daehee Hwang, Dr. Hamid Bolouri, Dr. Mark Gilchrist and Dr. Vesteinn Thorsson for their supports and helpful discussion.

REFERENCES

Akira, S., & Takeda, K. (2004). Toll-like receptor signalling. *Nature Reviews. Immunology*, *4*(7), 499–511. doi:10.1038/nri1391

Barton, G. M., Kagan, J. C., & Medzhitov, R. (2006). Intracellular localization of Toll-like receptor 9 prevents recognition of self DNA but facilitates access to viral DNA. *Nature Immunology*, *7*(1), 49–56. doi:10.1038/ni1280

Beutler, B. (2004). Inferences, questions and possibilities in Toll-like receptor signalling. *Nature*, *430*(6996), 257–263. doi:10.1038/nature02761

Buck, M. J., & Lieb, J. D. (2004). ChIP-chip: Considerations for the design, analysis, and application of genome-wide chromatin immunoprecipitation experiments. *Genomics*, *83*(3), 349–360. doi:10.1016/j.ygeno.2003.11.004

Chang, L.W., Fontaine, B.R., Stormo, G.D. & Nagarajan, R. (2007). PAP: A comprehensive workbench for mammalian transcriptional regulatory sequence analysis. *Nucleic Acids Research*, *35*(Web Server issue), W238-244.

D'Haeseleer, P. (2005). How does gene expression clustering work? *Nature Biotechnology*, *23*(12), 1499–1501. doi:10.1038/nbt1205-1499

Draghici, S., Khatri, P., Eklund, A. C., & Szallasi, Z. (2006). Reliability and reproducibility issues in DNA microarray measurements. *Trends in Genetics*, *22*(2), 101–109. doi:10.1016/j.tig.2005.12.005

Frith, M. C., Fu, Y., Yu, L., Chen, J. F., Hansen, U., & Weng, Z. (2004). Detection of functional DNA motifs via statistical over-representation. *Nucleic Acids Research*, *32*(4), 1372–1381. doi:10.1093/nar/gkh299

Frith, M. C., Li, M. C., & Weng, Z. (2003). Cluster-buster: Finding dense clusters of motifs in DNA sequences. *Nucleic Acids Research*, *31*(13), 3666–3668. doi:10.1093/nar/gkg540

Gentleman, R. C., Carey, V. J., Bates, D. M., Bolstad, B., Dettling, M., & Dudoit, S. (2004). Bioconductor: Open software development for computational biology and bioinformatics. *Genome Biology*, *5*(10), R80. doi:10.1186/gb-2004-5-10-r80

Gilchrist, M., Thorsson, V., Li, B., Rust, A. G., Korb, M., & Kennedy, K. (2006). Systems biology approaches identify ATF3 as a negative regulator of Toll-like receptor 4. *Nature*, *441*(7090), 173–178. doi:10.1038/nature04768

Gordon, S. J., Saleque, S., & Birshtein, B. K. (2003). Yin Yang 1 is a lipopolysaccharide-inducible activator of the murine 3' Igh enhancer, hs3. *Journal of Immunology (Baltimore, MD.: 1950)*, *170*(11), 5549–5557.

Ho Sui, S.J., Fulton, D.L., Arenillas, D.J., Kwon, A.T. & Wasserman, W.W. (2007). oPOSSUM: Integrated tools for analysis of regulatory motif over-representation. *Nucleic Acids Research*, *35*(Web Server issue), W245-252.

Hoebe, K., Du, X., Georgel, P., Janssen, E., Tabeta, K., & Kim, S. O. (2003). Identification of Lps2 as a key transducer of MyD88-independent TIR signalling. *Nature*, *424*(6950), 743–748. doi:10.1038/nature01889

Hwang, D., Smith, J. J., Leslie, D. M., Weston, A. D., Rust, A. G., & Ramsey, S. (2005). A data integration methodology for systems biology: Experimental verification. *Proceedings of the National Academy of Sciences of the United States of America*, *102*(48), 17302–17307. doi:10.1073/pnas.0508649102

Hwang, D., Stephanopoulos, G., & Chan, C. (2004). Inverse modeling using multi-block PLS to determine the environmental conditions that provide optimal cellular function. *Bioinformatics (Oxford, England)*, *20*(4), 487–499. doi:10.1093/bioinformatics/btg433

Ihmels, J., Bergmann, S., Gerami-Nejad, M., Yanai, I., McClellan, M., & Berman, J. (2005). Rewiring of the yeast transcriptional network through the evolution of motif usage. *Science*, *309*(5736), 938–940. doi:10.1126/science.1113833

Leung, T. H., Hoffmann, A., & Baltimore, D. (2004). One nucleotide in a kappaB site can determine cofactor specificity for NF-kappaB dimers. *Cell*, *118*(4), 453–464. doi:10.1016/j.cell.2004.08.007

Matsushita, K., Takeuchi, O., Standley, D. M., Kumagai, Y., Kawagoe, T., & Miyake, T. (2009). Zc3h12a is an RNase essential for controlling immune responses by regulating mRNA decay. *Nature*, *458*(7242), 1185–1190. doi:10.1038/nature07924

Matys, V., Kel-Margoulis, O. V., Fricke, E., Liebich, I., Land, S., & Barre-Dirrie, A. (2006). TRANSFAC and its module TRANSCompel: Transcriptional gene regulation in eukaryotes. *Nucleic Acids Research*, *34*(Database issue), D108–D110. doi:10.1093/nar/gkj143

Mollen, K. P., Gribar, S. C., Anand, R. J., Kaczorowski, D. J., Kohler, J. W., & Branca, M. F. (2008). Increased expression and internalization of the endotoxin coreceptor CD14 in enterocytes occur as an early event in the development of experimental necrotizing enterocolitis. *Journal of Pediatric Surgery*, *43*(6), 1175–1181. doi:10.1016/j.jpedsurg.2008.02.050

Mostecki, J., Showalter, B. M., & Rothman, P. B. (2005). Early growth response-1 regulates lipopolysaccharide-induced suppressor of cytokine signaling-1 transcription. *The Journal of Biological Chemistry*, *280*(4), 2596–2605. doi:10.1074/jbc.M408938200

Nilsson, R., Bajic, V. B., Suzuki, H., di Bernardo, D., Bjorkegren, J., & Katayama, S. (2006). Transcriptional network dynamics in macrophage activation. *Genomics*, *88*(2), 133–142. doi:10.1016/j.ygeno.2006.03.022

Oda, K., & Kitano, H. (2006). A comprehensive map of the toll-like receptor signaling network. *Molecular Systems Biology*, *2*, E1–E16. doi:10.1038/msb4100057

Ramsey, S. A., Klemm, S. L., Zak, D. E., Kennedy, K. A., Thorsson, V., & Li, B. (2008). Uncovering a macrophage transcriptional program by integrating evidence from motif scanning and expression dynamics. *PLoS Computational Biology*, *4*(3), e1000021. doi:10.1371/journal.pcbi.1000021

Reiss, D. J., Baliga, N. S., & Bonneau, R. (2006). Integrated biclustering of heterogeneous genome-wide datasets for the inference of global regulatory networks. *BMC Bioinformatics*, *7*, 280. doi:10.1186/1471-2105-7-280

Roider, H. G., Manke, T., O'Keeffe, S., Vingron, M., & Haas, S. A. (2009). PASTAA: Identifying transcription factors associated with sets of co-regulated genes. *Bioinformatics (Oxford, England)*, *25*(4), 435–442. doi:10.1093/bioinformatics/btn627

Sandelin, A., Alkema, W., Engstrom, P., Wasserman, W. W., & Lenhard, B. (2004). JASPAR: An open-access database for eukaryotic transcription factor binding profiles. *Nucleic Acids Research*, *32*(Database issue), D91–D94. doi:10.1093/nar/gkh012

Seoane, J., Le, H. V., Shen, L., Anderson, S. A., & Massague, J. (2004). Integration of Smad and forkhead pathways in the control of neuroepithelial and glioblastoma cell proliferation. *Cell*, *117*(2), 211–223. doi:10.1016/S0092-8674(04)00298-3

Shen, F., Hu, Z., Goswami, J., & Gaffen, S. L. (2006). Identification of common transcriptional regulatory elements in interleukin-17 target genes. *The Journal of Biological Chemistry*, *281*(34), 24138–24148. doi:10.1074/jbc.M604597200

Shi, L., Kishore, R., McMullen, M. R., & Nagy, L. E. (2002). Chronic ethanol increases lipopolysaccharide-stimulated Egr-1 expression in RAW 264.7 macrophages: Contribution to enhanced tumor necrosis factor alpha production. *The Journal of Biological Chemistry*, *277*(17), 14777–14785.

Stitt, T. N., Drujan, D., Clarke, B. A., Panaro, F., Timofeyva, Y., & Kline, W. O. (2004). The IGF-1/PI3K/Akt pathway prevents expression of muscle atrophy-induced ubiquitin ligases by inhibiting FOXO transcription factors. *Molecular Cell*, *14*(3), 395–403. doi:10.1016/S1097-2765(04)00211-4

Stolovitzky, G. A., Kundaje, A., Held, G. A., Duggar, K. H., Haudenschild, C. D., & Zhou, D. (2005). Statistical analysis of MPSS measurements: Application to the study of LPS-activated macrophage gene expression. *Proceedings of the National Academy of Sciences of the United States of America*, *102*(5), 1402–1407. doi:10.1073/pnas.0406555102

Takeda, K., Kaisho, T., & Akira, S. (2003). Toll-like receptors. *Annual Review of Immunology*, *21*, 335–376. doi:10.1146/annurev.immunol.21.120601.141126

Tanimura, N., Saitoh, S., Matsumoto, F., Akashi-Takamura, S., & Miyake, K. (2008). Roles for LPS-dependent interaction and relocation of TLR4 and TRAM in TRIF-signaling. *Biochemical and Biophysical Research Communications*, *368*(1), 94–99. doi:10.1016/j.bbrc.2008.01.061

Ye, H., Arron, J. R., Lamothe, B., Cirilli, M., Kobayashi, T., & Shevde, N. K. (2002). Distinct molecular mechanism for initiating TRAF6 signalling. *Nature*, *418*(6896), 443–447. doi:10.1038/nature00888

Yoshimura, A., Ohishi, H. M., Aki, D., & Hanada, T. (2004). Regulation of TLR signaling and inflammation by SOCS family proteins. *Journal of Leukocyte Biology*, *75*(3), 422–427. doi:10.1189/jlb.0403194

ADDITIONAL READING

Aerts, S., van Helden, J., Sand, O., & Hassan, B. A. (2007). Fine-tuning enhancer models to predict transcriptional targets across multiple genomes. *PLoS ONE*, *2*(11), e1115. doi:10.1371/journal.pone.0001115

Akira, S., & Takeda, K. (2004). Toll-like receptor signalling. *Nature Reviews. Immunology*, *4*(7), 499–511. doi:10.1038/nri1391

Bais, A. S., Grossmann, S., & Vingron, M. (2007). Incorporating evolution of transcription factor binding sites into annotated alignments. *Journal of Biosciences*, *32*(5), 841–850. doi:10.1007/s12038-007-0084-2

Berezikov, E., Guryev, V., & Cuppen, E. (2007). Exploring conservation of transcription factor binding sites with CONREAL. *Methods in Molecular Biology (Clifton, N.J.)*, *395*, 437–448. doi:10.1007/978-1-59745-514-5_27

Draghici, S., Khatri, P., Eklund, A. C., & Szallasi, Z. (2006). Reliability and reproducibility issues in DNA microarray measurements. *Trends in Genetics*, *22*(2), 101–109. doi:10.1016/j.tig.2005.12.005

Gentleman, R. C., Carey, V. J., Bates, D. M., Bolstad, B., Dettling, M., & Dudoit, S. (2004). Bioconductor: open software development for computational biology and bioinformatics. *Genome Biology*, *5*(10), R80. doi:10.1186/gb-2004-5-10-r80

Gilchrist, M., Thorsson, V., Li, B., Rust, A. G., Korb, M., & Kennedy, K. (2006). Systems biology approaches identify ATF3 as a negative regulator of Toll-like receptor 4. *Nature*, *441*(7090), 173–178. doi:10.1038/nature04768

Hwang, D., Smith, J. J., Leslie, D. M., Weston, A. D., Rust, A. G., & Ramsey, S. (2005). A data integration methodology for systems biology: experimental verification. *Proceedings of the National Academy of Sciences of the United States of America*, *102*(48), 17302–17307. doi:10.1073/pnas.0508649102

Kim, N. K., Tharakaraman, K., Marino-Ramirez, L., & Spouge, J. L. (2008). Finding sequence motifs with Bayesian models incorporating positional information: an application to transcription factor binding sites. *BMC Bioinformatics*, *9*, 262. doi:10.1186/1471-2105-9-262

Laurila, K., Yli-Harja, O., & Lahdesmaki, H. (2009). A protein-protein interaction guided method for competitive transcription factor binding improves target predictions. *Nucleic Acids Research*, *37*(22), e146. doi:10.1093/nar/gkp789

Levitsky, V. G., Ignatieva, E. V., Ananko, E. A., Turnaev, I. I., Merkulova, T. I., & Kolchanov, N. A. (2007). Effective transcription factor binding site prediction using a combination of optimization, a genetic algorithm and discriminant analysis to capture distant interactions. *BMC Bioinformatics*, *8*, 481. doi:10.1186/1471-2105-8-481

Liu, G. E., Weirauch, M. T., Van Tassell, C. P., Li, R. W., Sonstegard, T. S., & Matukumalli, L. K. (2008). Identification of conserved regulatory elements in mammalian promoter regions: a case study using the PCK1 promoter. *Genomics, Proteomics & Bioinformatics*, *6*(3-4), 129–143. doi:10.1016/S1672-0229(09)60001-2

Mahadevan, R., Yan, B., Postier, B., Nevin, K. P., Woodard, T. L., & O'Neil, R. (2008). Characterizing regulation of metabolism in Geobacter sulfurreducens through genome-wide expression data and sequence analysis. *OMICS: A Journal of Integrative Biology*, *12*(1), 33–59. doi:10.1089/omi.2007.0043

Matsushita, K., Takeuchi, O., Standley, D. M., Kumagai, Y., Kawagoe, T., & Miyake, T. (2009). Zc3h12a is an RNase essential for controlling immune responses by regulating mRNA decay. *Nature*, *458*(7242), 1185–1190. doi:10.1038/nature07924

Matys, V., Kel-Margoulis, O. V., Fricke, E., Liebich, I., Land, S., & Barre-Dirrie, A. (2006). TRANSFAC and its module TRANSCompel: transcriptional gene regulation in eukaryotes. *Nucleic Acids Research*, *34*(Database issue), D108–D110. doi:10.1093/nar/gkj143

Mollen, K. P., Gribar, S. C., Anand, R. J., Kaczorowski, D. J., Kohler, J. W., & Branca, M. F. (2008). Increased expression and internalization of the endotoxin coreceptor CD14 in enterocytes occur as an early event in the development of experimental necrotizing enterocolitis. *Journal of Pediatric Surgery*, *43*(6), 1175–1181. doi:10.1016/j.jpedsurg.2008.02.050

Oda, k., & Kitano, H. (2006). "A comprehensive map of the toll-like receptor signaling network". *Mol Syst Biol, 2*, E1-E16.

Oh, Y. M., Kim, J. K., Choi, Y., Choi, S., & Yoo, J. Y. (2009). Prediction and experimental validation of novel STAT3 target genes in human cancer cells. *PLoS ONE*, *4*(9), e6911. doi:10.1371/journal.pone.0006911

Ramsey, S. A., Klemm, S. L., Zak, D. E., Kennedy, K. A., Thorsson, V., & Li, B. (2008). Uncovering a macrophage transcriptional program by integrating evidence from motif scanning and expression dynamics. *PLoS Computational Biology*, *4*(3), e1000021. doi:10.1371/journal.pcbi.1000021

Reddy, T. E., Shakhnovich, B. E., Roberts, D. S., Russek, S. J., & DeLisi, C. (2007). Positional clustering improves computational binding site detection and identifies novel cis-regulatory sites in mammalian GABAA receptor subunit genes. *Nucleic Acids Research*, *35*(3), e20. doi:10.1093/nar/gkl1062

Shen, F., Hu, Z., Goswami, J., & Gaffen, S. L. (2006). Identification of common transcriptional regulatory elements in interleukin-17 target genes. *The Journal of Biological Chemistry*, *281*(34), 24138–24148. doi:10.1074/jbc.M604597200

Steinhoff, C., Paulsen, M., Kielbasa, S., Walter, J., & Vingron, M. (2009). Expression profile and transcription factor binding site exploration of imprinted genes in human and mouse. *BMC Genomics*, *10*, 144. doi:10.1186/1471-2164-10-144

Takeda, K., Kaisho, T., & Akira, S. (2003). Toll-like receptors. *Annual Review of Immunology*, *21*, 335–376. doi:10.1146/annurev.immunol.21.120601.141126

Tanimura, N., Saitoh, S., Matsumoto, F., Akashi-Takamura, S., & Miyake, K. (2008). Roles for LPS-dependent interaction and relocation of TLR4 and TRAM in TRIF-signaling. *Biochemical and Biophysical Research Communications*, *368*(1), 94–99. doi:10.1016/j.bbrc.2008.01.061

Vega, V. B., Lin, C. Y., Lai, K. S., Kong, S. L., Xie, M., & Su, X. (2006). Multiplatform genome-wide identification and modeling of functional human estrogen receptor binding sites. *Genome Biology*, *7*(9), R82. doi:10.1186/gb-2006-7-9-r82

von Rohr, P., Friberg, M. T., & Kadarmideen, H. N. (2007). Prediction of transcription factor binding sites using genetical genomics methods. *Journal of Bioinformatics and Computational Biology*, *5*(3), 773–793. doi:10.1142/S0219720007002680

Wang, T., Furey, T. S., Connelly, J. J., Ji, S., Nelson, S., & Heber, S. (2009). A general integrative genomic feature transcription factor binding site prediction method applied to analysis of USF1 binding in cardiovascular disease. *Human Genomics*, *3*(3), 221–235.

Whitington, T., Perkins, A. C., & Bailey, T. L. (2009). High-throughput chromatin information enables accurate tissue-specific prediction of transcription factor binding sites. *Nucleic Acids Research*, *37*(1), 14–25. doi:10.1093/nar/gkn866

Zadissa, A., McEwan, J. C., & Brown, C. M. (2007). Inference of transcriptional regulation using gene expression data from the bovine and human genomes. *BMC Genomics*, *8*, 265. doi:10.1186/1471-2164-8-265

KEY TERMS AND DEFINITIONS

CpG: It is from bacteria or viruses and being used as a TLR9 agonist.

Lipopolysaccharide (LPS): It is from Gram-negative bacteria and being used as a TLR4 agonist.

PAM2Cys-SKKKK (PAM2): It is from bacteria and being used as a TLR2/6 agonist.

PAM3Cys-SKKK (PAM3): It is from bacteria and being used as a TLR2/1 agonist.

Poly I:C: It is used to represent viruses and being used as a TLR3 agonist.

Resimiquod (R848): It is a small synthetic compound and being used as a TLR7 agonist.

TFBS Prediction: The prediction of potential transcription factor binding sites (TFBS) with reasonable accuracy.

Toll-Like Receptors: A class of proteins that play a key role in the innate immune system. They are single, membrane-spanning, non-catalytic receptors that recognize structurally conserved molecules derived from microbes.

Transcription Factor: A protein that binds to specific DNA sequences, thereby controlling the transfer (or transcription) of genetic information from DNA to mRNA.

APPENDIX

Figure 7. Matching time-course clustering results with knock-out microarray data. (A) to (D) represents different clusters of genes

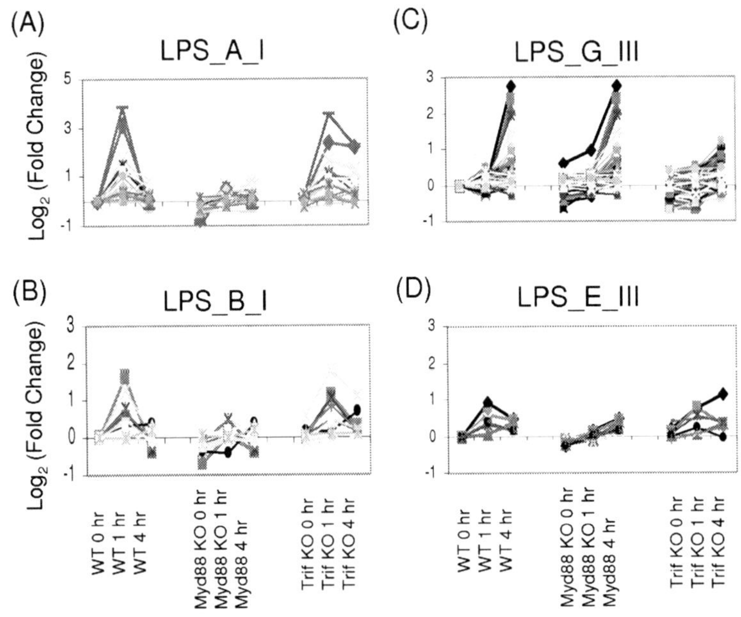

Table 3. Details of clustering results corresponding to Table 1

stimulus	Gene Symbols	LocusID	second_step_clustering	first_step_clustering
CpG	B430217B02Rik	319544	F	I
CpG	Gdf15	23886	F	I
CpG	Pim1	18712	F	I
CpG	Sgk	20393	F	I
CpG	Dusp4	319520	F	I
CpG	Snk	20620	F	I
CpG	Nfkbia	18035	F	I
CpG	Fbxw1b	103583	F	I
CpG	C630016O21Rik	210105	F	I
CpG	Idb2	15902	F	I
CpG	Marcks	17118	F	I
CpG	Ccnl1	56706	F	I
CpG	Myc	17869	F	I
CpG	Nfe2l2	18024	F	I
CpG	Creb5	231991	F	I
CpG	Chk	12660	F	I

continued on following page

Table 3. continued

stimulus	Gene Symbols	LocusID	second_step_clustering	first_step_clustering
CpG	LOC233400	233400	F	I
CpG	Fosl2	14284	F	I
CpG	Taf7	24074	F	I
CpG	Zfp36l1	12192	F	I
CpG	Gadd45b	17873	F	III
CpG	Il10	16153	F	III
CpG	Sh3bgrl	56726	F	III
CpG	Ccl7	20306	F	III
CpG	Ccl2	20296	F	III
CpG	Cxcl1	14825	F	III
CpG	Clecsf9	56619	F	III
CpG	Ccl4	20303	F	III
CpG	---	240672	F	III
CpG	BC036563	230738	F	III
CpG	Arhe	74194	F	III
CpG	Ccl3	20302	F	III
CpG	4930431B09Rik	74645	F	III
CpG	Axud1	215418	G	I
CpG	Adamts1	11504	G	I
CpG	Ets2	23872	G	I
CpG	Copeb	23849	G	I
CpG	Erbb2ip	59079	G	I
CpG	Cd83	12522	G	III
CpG	Jag1	16449	G	III
CpG	Gem	14579	G	III
CpG	Ptgs2	19225	G	III
CpG	Cd69	12515	G	III
CpG	Pde4b	18578	G	III
CpG	Maff	17133	G	III
CpG	BC038313	216850	G	III
CpG	4921515A04Rik	268301	G	III
CpG	Egr3	13655	G	III
CpG	1300002F13Rik	74155	G	III
CpG	Ccrn4l	12457	G	III
CpG	Prdm1	12142	G	III
CpG	Ifrd1	15982	G	III
CpG	Ptger4	19219	G	III
CpG	Skil	20482	G	III
CpG	Rab20	19332	G	III

continued on following page

Table 3. continued

stimulus	Gene Symbols	LocusID	second_step_clustering	first_step_clustering
CpG	Ccl12	20293	G	III
CpG	Cflar	12633	G	III
CpG	Irg1	16365	G	III
CpG	Cxcl10	15945	G	III
CpG	4732452O09Rik	320292	G	III
CpG	Cias1	216799	G	III
CpG	Flrt3	71436	G	III
CpG	Cited2	17684	G	III
CpG	Gpr84	80910	G	III
CpG	Tank	21353	G	III
CpG	BC010311	209212	G	III
CpG	2010306G19Rik	67035	G	III
CpG	H2-Q7	15018	G	III
CpG	Trps1	83925	G	III
CpG	Mlp	17357	G	III
CpG	Odc	18263	G	III
CpG	Mmp13	17386	G	III
CpG	Mmd	67468	G	III
CpG	---	231462	G	III
CpG	Stx11	74732	G	III
CpG	Nfkbie	18037	G	III
CpG	Etv3	27049	G	III
CpG	Rel	19696	G	III
CpG	Tlr2	24088	G	III
CpG	Icam1	15894	G	III
CpG	Arl8	75869	G	III
CpG	Mdm2	17246	G	III
CpG	Il1b	16176	G	III
CpG	Zfp263	74120	G	III
CpG	Mad	17119	G	III
CpG	D7Ertd458e	52118	G	III
CpG	Birc2	11796	G	III
CpG	Tgm2	21817	G	III
CpG	Klf7	93691	G	III
CpG	Gpr85	64450	G	III
CpG	Ripk2	192656	G	III
CpG	Malt1	240354	G	III
CpG	Gspt1	14852	G	III
CpG	Il12b	16160	G	III

continued on following page

Table 3. continued

stimulus	Gene Symbols	LocusID	second_step_clustering	first_step_clustering
CpG	Egr1	13653	A	I
CpG	Cxcl2	20310	A	I
CpG	Dusp1	19252	A	I
CpG	Ier3	15937	A	I
CpG	Sqstm1	18412	A	I
CpG	Map3k8	26410	A	I
CpG	AI467484	98376	A	I
CpG	Btg2	12227	E	I
CpG	Dusp2	13537	E	I
CpG	2410011G03Rik	66414	E	I
CpG	1110035O14Rik	59027	E	I
CpG	Osm	18413	E	I
CpG	Txnrd1	50493	E	I
LPS	Cd83	12522	F	III
LPS	Jag1	16449	F	III
LPS	BC031781	208768	F	III
LPS	Pde4b	18578	F	III
LPS	Ccrn4l	12457	F	III
LPS	Ptgs2	19225	F	III
LPS	Il10	16153	F	III
LPS	1300002F13Rik	74155	F	III
LPS	BC038313	216850	F	III
LPS	Sh3bgrl	56726	F	III
LPS	Prdm1	12142	F	III
LPS	4921515A04Rik	268301	F	III
LPS	Clecsf9	56619	F	III
LPS	---	240672	F	III
LPS	Sele	20339	F	III
LPS	Mmp13	17386	F	III
LPS	Mlp	17357	F	III
LPS	Ptger4	19219	F	III
LPS	Rel	19696	F	III
LPS	Icam1	15894	F	III
LPS	Skil	20482	F	III
LPS	Cd69	12515	F	III
LPS	Cflar	12633	F	III
LPS	Gpr85	64450	F	III
LPS	4732452O09Rik	320292	F	III
LPS	H2-Q7	15018	F	III

continued on following page

Table 3. continued

stimulus	Gene Symbols	LocusID	second_step_clustering	first_step_clustering
LPS	BC036563	230738	F	III
LPS	Csf2	12981	F	III
LPS	Arl8	75869	F	III
LPS	4930534K13Rik	75234	F	III
LPS	Gpr84	80910	F	III
LPS	Trps1	83925	F	III
LPS	Tlr2	24088	F	III
LPS	Arhc	74194	F	III
LPS	Stx11	74732	F	III
LPS	Mmd	67468	F	III
LPS	Cias1	216799	F	III
LPS	Ccl12	20293	F	III
LPS	Rab20	19332	F	III
LPS	Il10ra	16154	F	III
LPS	Relb	19698	F	III
LPS	Itga5	16402	F	III
LPS	Tnfaip2	21928	F	III
LPS	Ppfibp1	67533	F	III
LPS	2010306G19Rik	67035	F	III
LPS	Mdm2	17246	F	III
LPS	Nupr1	56312	F	III
LPS	---	231462	F	III
LPS	8430412F05	242860	F	III
LPS	D430024K22Rik	214855	F	III
LPS	Gspt1	14852	F	III
LPS	2310016C08Rik	69573	F	III
LPS	D7Ertd458e	52118	F	III
LPS	Pcdh7	54216	F	III
LPS	Tgm2	21817	F	III
LPS	Mad	17119	F	III
LPS	Nfil3	18030	F	III
LPS	Odc	18263	F	III
LPS	BC010311	209212	F	III
LPS	Birc2	11796	F	III
LPS	Ccl2	20296	F	III
LPS	1500041J02Rik	67876	F	III
LPS	E130307H12Rik	320404	F	III
LPS	Tiparp	99929	F	III
LPS	Bcl2l11	12125	F	III

continued on following page

Table 3. continued

stimulus	Gene Symbols	LocusID	second_step_clustering	first_step_clustering
LPS	4933428L19Rik	71198	F	III
LPS	Ehd1	13660	F	III
LPS	Bcl2a1a	12044	F	III
LPS	Fyb	23880	F	III
LPS	Bcl6	12053	F	III
LPS	Gja1	14609	F	III
LPS	0610012A05Rik	67434	F	III
LPS	BC035207	240396	F	III
LPS	Vcam1	22329	F	III
LPS	Nfkbib	18036	F	III
LPS	Il1rn	16181	F	III
LPS	Sod2	20656	F	III
LPS	B930060C03	217578	F	III
LPS	Rrs1	59014	F	III
LPS	Gch	14528	F	III
LPS	Xrcc1	22594	F	III
LPS	B630005N14Rik	101148	F	III
LPS	Kctd12	239217	F	III
LPS	Bsf3	56708	F	III
LPS	2210412D01Rik	70178	F	III
LPS	Prg	19073	F	III
LPS	Tgfb1i4	21807	F	III
LPS	E130119J17Rik	212168	F	III
LPS	4732496O19Rik	99470	F	III
LPS	AU021107	229055	F	III
LPS	Sdc4	20971	F	III
LPS	Slfn2	20556	F	III
LPS	---	74588	F	III
LPS	2010109K11Rik	72123	F	III
LPS	Tgfb1i4	21807	F	III
LPS	Samsn1	67742	F	III
LPS	Cebpb	12608	F	III
LPS	D17Ertd808e	52040	F	III
LPS	Dscr1	54720	F	III
LPS	D030023K18	327987	F	III
LPS	A030007L17Rik	68252	F	III
LPS	Zfp216	22682	F	III
LPS	D19Wsu55e	28000	F	III
LPS	Ccng2	12452	F	III

continued on following page

Table 3. continued

stimulus	Gene Symbols	LocusID	second_step_clustering	first_step_clustering
LPS	Cxcl16	66102	F	III
LPS	Ccr3	12771	F	III
LPS	Cdkn1a	12575	F	III
LPS	5830415L20Rik	68152	F	III
LPS	5430427O19Rik	71398	F	III
LPS	1810013L24Rik	69053	F	III
LPS	Casp4	12363	F	III
LPS	Plek	56193	F	III
LPS	2610103J23Rik	67154	F	III
LPS	Pdcd1lg1	60533	F	III
LPS	Siah2	20439	F	III
LPS	Snag1	170625	F	III
LPS	Igsf6	80719	F	III
LPS	Irg1	16365	F	III
LPS	Il1b	16176	F	III
LPS	Cxcl10	15945	F	III
LPS	Egr3	13655	F	III
LPS	Il1a	16175	G	III
LPS	Il6	16193	G	III
LPS	Il12b	16160	G	III
LPS	2510004L01Rik	58185	G	III
LPS	Oasl1	231655	G	III
LPS	Ccl5	20304	G	III
LPS	Olr1	108078	G	III
LPS	Ifit1	15957	G	III
LPS	Nfkbie	18037	G	III
LPS	Malt1	240354	G	III
LPS	Hspa1a	193740	G	III
LPS	Csf1	12977	G	III
LPS	4930488L10Rik	319710	G	III
LPS	1200009I06Rik	74190	G	III
LPS	Ptprj	19271	G	III
LPS	BC022623	224093	G	III
LPS	G1p2	53606	G	III
LPS	BC006779	229003	G	III
LPS	1600029O10Rik	72239	G	III
LPS	Ripk2	192656	G	III
LPS	Slc4a7	218756	G	III
LPS	Ch25h	12642	G	III

continued on following page

Table 3. continued

stimulus	Gene Symbols	LocusID	second_step_clustering	first_step_clustering
LPS	Flrt3	71436	G	III
LPS	Trim13	66597	G	III
LPS	Ifnb	15977	G	III
LPS	Plagl1	22634	G	III
LPS	Peli1	67245	G	III
LPS	Icosl	50723	G	III
LPS	Dusp16	70686	G	III
LPS	Pdzgef1	76089	G	III
LPS	Cdk5r	12569	G	III
LPS	Serpinb2	18788	G	III
LPS	D8Ertd82e	244418	G	III
LPS	Serpine1	18787	G	III
LPS	Slc2a6	227659	G	III
LPS	9530058O11Rik	208449	G	III
LPS	Klf7	93691	G	III
LPS	Slc11a2	18174	G	III
LPS	Tnfrsf5	21939	G	III
LPS	Cdc42ep4	56699	G	III
LPS	Slamf7	75345	G	III
LPS	Ell2	192657	G	III
LPS	Baz1a	14896	G	III
LPS	1700055P21Rik	73379	G	III
LPS	Rabgef1	56715	G	III
LPS	Nfkb2	18034	G	III
LPS	Jundm2	81703	G	III
LPS	AI195350	106878	G	III
LPS	Ifi205	226695	G	III
LPS	2510010F15Rik	67921	G	III
LPS	Tcfec	21426	G	III
LPS	Plscr1	22038	G	III
LPS	Ifit2	15958	G	III
LPS	A730041O15Rik	269717	G	III
LPS	Tnip1	57783	G	III
LPS	Gbp3	55932	G	III
LPS	Ifi203	15950	G	III
LPS	Trex1	22040	G	III
LPS	Pfkp	56421	G	III
LPS	1190002H23Rik	66214	G	III
LPS	Zfp263	74120	G	III

continued on following page

Table 3. continued

stimulus	Gene Symbols	LocusID	second_step_clustering	first_step_clustering
LPS	Kcna3	16491	G	III
LPS	Saa3	20210	G	III
LPS	Gbp2	14469	G	III
LPS	Bcl3	12051	G	III
LPS	Plaur	18793	G	III
LPS	Pumag	80885	G	III
LPS	Plagl2	54711	G	III
LPS	Ddx6	13209	G	III
LPS	Slc7a2	11988	G	III
LPS	Rab12	19328	G	III
LPS	Siah2	20439	G	III
LPS	A630077B13Rik	215900	G	III
LPS	1110019C08Rik	224250	G	III
LPS	Marcks	17118	G	III
LPS	Nck1	17973	G	III
LPS	Bcor	71458	G	III
LPS	Mapkapk2	17164	G	III
LPS	AI597013	100182	G	III
LPS	Nap4	192157	G	III
LPS	P2ry2	18442	G	III
LPS	AW539457	99382	G	III
LPS	Hivep2	15273	G	III
LPS	Egr1	13653	B	I
LPS	Btg2	12227	B	I
LPS	Ier3	15937	B	I
LPS	Ier2	15936	B	I
LPS	2410011G03Rik	66414	B	I
LPS	Osm	18413	B	I
LPS	Gdf15	23886	B	I
LPS	Chk	12660	B	I
LPS	S100a10	20194	B	I
LPS	Spag9	70834	B	I
LPS	2610103N14Rik	72472	B	I
LPS	Ccnl1	56706	B	I
LPS	Mbtd1	103537	B	I
LPS	Txnrd1	50493	B	I
LPS	Taf7	24074	B	I
LPS	Phlda1	21664	B	I
LPS	Egr2	13654	B	I

continued on following page

Table 3. continued

stimulus	Gene Symbols	LocusID	second_step_clustering	first_step_clustering
LPS	Socs3	12702	A	I
LPS	Sqstm1	18412	A	I
LPS	AA408868	80859	A	I
LPS	Cxcl2	20310	A	I
LPS	Tnf	21926	A	I
LPS	Tnfaip3	21929	A	I
LPS	Ccrl2	54199	A	I
LPS	Tnfsf9	21950	A	I
LPS	Junb	16477	A	I
LPS	Traf1	22029	A	I
LPS	Myd116	17872	A	I
LPS	Dusp1	19252	A	I
LPS	B430217B02Rik	319544	A	I
LPS	IkappaBNS	243910	A	I
LPS	Sgk	20393	A	I
LPS	Pim1	18712	A	I
LPS	Dusp2	13537	A	I
LPS	Nfkbia	18035	A	I
LPS	Dusp4	319520	A	I
LPS	Idb3	15903	A	I
LPS	Nfe2l2	18024	A	I
LPS	1110035O14Rik	59027	A	I
LPS	Pabpc1	18458	A	I
LPS	Map3k8	26410	A	I
LPS	Idb2	15902	A	I
LPS	Zfp36l1	12192	A	I
LPS	Gadd45b	17873	A	III
LPS	Ccl4	20303	A	III
LPS	Cxcl1	14825	A	III
LPS	Ifrd1	15982	A	III
LPS	Ccl3	20302	A	III
LPS	Cited2	17684	A	III
LPS	Ccl7	20306	A	III
LPS	Bmp2	12156	A	III
LPS	D10Ertd749e	52696	A	III
LPS	Dtr	15200	A	III
LPS	Rnf2	19821	A	III
LPS	Zfp131	72465	D	III
LPS	Dusp14	56405	D	III

continued on following page

Table 3. continued

stimulus	Gene Symbols	LocusID	second_step_clustering	first_step_clustering
LPS	Pik3ca	18706	D	III
LPS	2010305C02Rik	69891	D	III
LPS	Ifi16	15951	D	III
LPS	Dnajb5	56323	D	III
LPS	Snk	20620	E	I
LPS	LOC233400	233400	E	I
LPS	C630016O21Rik	210105	E	I
LPS	Adamts1	11504	E	I
LPS	Erbb2ip	59079	E	I
LPS	Nfat5	54446	E	I
LPS	Copeb	23849	E	I
LPS	Fbxw1b	103583	E	I
LPS	Myc	17869	E	I
LPS	Hrb	15463	E	I
LPS	Stx6	58244	E	I
LPS	Tgif	21815	E	I
LPS	A530090O15Rik	211770	E	I
LPS	Pmaip1	58801	E	I
LPS	Gem	14579	E	III
LPS	Maff	17133	E	III
LPS	Tank	21353	E	III
LPS	Atf3	11910	E	III
LPS	F3	14066	E	III
LPS	Etv3	27049	E	III
LPS	Aebp2	11569	E	III
LPS	Pnrc1	108767	E	III
LPS	Clcn7	26373	E	III
PAM2	Prdm1	12142	F	III
PAM2	Irg1	16365	F	III
PAM2	Rel	19696	F	III
PAM2	BC036563	230738	F	III
PAM2	Rab20	19332	F	III
PAM2	Icam1	15894	F	III
PAM2	4930534K13Rik	75234	F	III
PAM2	Slc4a7	218756	F	III
PAM2	Malt1	240354	F	III
PAM2	Skil	20482	F	III
PAM2	Trim13	66597	F	III
PAM2	Tnfaip2	21928	F	III

continued on following page

Table 3. continued

stimulus	Gene Symbols	LocusID	second_step_clustering	first_step_clustering
PAM2	Stx11	74732	F	III
PAM2	Dusp16	70686	F	III
PAM2	Icosl	50723	F	III
PAM2	4930488L10Rik	319710	F	III
PAM2	Birc2	11796	F	III
PAM2	BC010311	209212	F	III
PAM2	Pdzgcf1	76089	F	III
PAM2	Tlr2	24088	F	III
PAM2	Ehd1	13660	F	III
PAM2	Bcl2a1a	12044	F	III
PAM2	Tcfec	21426	F	III
PAM2	Arhe	74194	F	III
PAM2	A730041O15Rik	269717	F	III
PAM2	Olr1	108078	F	III
PAM2	Gpr84	80910	F	III
PAM2	Tgm2	21817	F	III
PAM2	D8Ertd82e	244418	F	III
PAM2	Ripk2	192656	F	III
PAM2	Relb	19698	F	III
PAM2	9530058O11Rik	208449	F	III
PAM2	Ptprj	19271	F	III
PAM2	A530088I07Rik	212167	F	III
PAM2	1200009I06Rik	74190	F	III
PAM2	Plscr1	22038	F	III
PAM2	Cebpb	12608	F	III
PAM2	Itga5	16402	F	III
PAM2	Slfn2	20556	F	III
PAM2	Tnfrsf5	21939	F	III
PAM2	Lig3	16882	F	III
PAM2	Tnip1	57783	F	III
PAM2	Hspa1a	193740	F	III
PAM2	Ednrb	13618	F	III
PAM2	Zfp263	74120	F	III
PAM2	Stk38l	232533	F	III
PAM2	Serpine1	18787	F	III
PAM2	Csf1	12977	G	III
PAM2	Sema6d	214968	G	III
PAM2	Herpud1	64209	G	III
PAM2	Bcl3	12051	G	III

continued on following page

Table 3. continued

stimulus	Gene Symbols	LocusID	second_step_clustering	first_step_clustering
PAM2	Jundm2	81703	G	III
PAM2	Igsf6	80719	G	III
PAM2	Slc11a2	18174	G	III
PAM2	Ahr	11622	G	III
PAM2	Cxcl2	20310	C	I
PAM2	Tnf	21926	C	I
PAM2	Btg2	12227	C	I
PAM2	Phlda1	21664	C	I
PAM2	Myd116	17872	C	I
PAM2	Dusp1	19252	C	I
PAM2	Ier3	15937	C	I
PAM2	Dusp2	13537	C	I
PAM2	LOC233400	233400	C	I
PAM2	Ier2	15936	C	I
PAM2	Ccnl1	56706	C	I
PAM2	1110035O14Rik	59027	C	I
PAM2	Egr2	13654	B	I
PAM2	Tnfaip3	21929	B	I
PAM2	Tnfsf9	21950	B	I
PAM2	IkappaBNS	243910	B	I
PAM2	Axud1	215418	B	I
PAM2	Sgk	20393	B	I
PAM2	Dusp4	319520	B	I
PAM2	Snk	20620	B	I
PAM2	Copeb	23849	B	I
PAM2	Chk	12660	B	I
PAM2	2610103N14Rik	72472	B	I
PAM2	Spag9	70834	B	I
PAM2	Socs3	12702	A	I
PAM2	AA408868	80859	A	I
PAM2	Ccrl2	54199	A	I
PAM2	Junb	16477	A	I
PAM2	Traf1	22029	A	I
PAM2	Sqstm1	18412	A	I
PAM2	Pim1	18712	A	I
PAM2	Nfkbia	18035	A	I
PAM2	Map3k8	26410	A	I
PAM2	BC031781	208768	A	III
PAM2	Sh3bgrl	56726	A	III

continued on following page

Table 3. continued

stimulus	Gene Symbols	LocusID	second_step_clustering	first_step_clustering
PAM2	Ptger4	19219	A	III
PAM2	Ccl4	20303	A	III
PAM2	Mad	17119	A	III
PAM2	Pnrc1	108767	A	III
PAM2	Cited2	17684	A	III
PAM2	Ccl3	20302	A	III
PAM2	2010306G19Rik	67035	A	III
PAM2	Aebp2	11569	A	III
PAM2	Etv3	27049	A	III
PAM2	Atf3	11910	A	III
PAM2	Egr3	13655	D	III
PAM2	4933428L19Rik	71198	D	III
PAM2	Rnf2	19821	D	III
PAM2	Syk	20963	D	III
PAM2	Mcl1	17210	D	III
PAM2	D10Ertd749e	52696	D	III
PAM2	Thbs1	21825	D	III
PAM2	2810474O19Rik	67246	D	III
PAM2	Cd83	12522	E	III
PAM2	Jag1	16449	E	III
PAM2	Gadd45b	17873	E	III
PAM2	Pde4b	18578	E	III
PAM2	Gem	14579	E	III
PAM2	Ccrn4l	12457	E	III
PAM2	1300002F13Rik	74155	E	III
PAM2	Il1b	16176	E	III
PAM2	Cd69	12515	E	III
PAM2	4921515A04Rik	268301	E	III
PAM2	BC038313	216850	E	III
PAM2	Mmp13	17386	E	III
PAM2	Trps1	83925	E	III
PAM2	4732452O09Rik	320292	E	III
PAM2	Nfkbie	18037	E	III
PAM2	Ppfibp1	67533	E	III
PAM2	Cflar	12633	E	III
PAM2	Tank	21353	E	III
PAM2	Il10ra	16154	E	III
PAM2	Clecsf9	56619	E	III
PAM2	Gpr85	64450	E	III

continued on following page

Table 3. continued

stimulus	Gene Symbols	LocusID	second_step_clustering	first_step_clustering
PAM2	Bcl2l11	12125	E	III
PAM2	Cias1	216799	E	III
PAM2	D430024K22Rik	214855	E	III
PAM2	Arl8	75869	E	III
PAM2	Ifrd1	15982	E	III
PAM2	Cxcl10	15945	E	III
PAM2	Il1a	16175	E	III
PAM2	Flrt3	71436	E	III
PAM2	Nfil3	18030	E	III
PAM2	Fyb	23880	E	III
PAM2	H2-Q7	15018	E	III
PAM2	Ccng2	12452	E	III
PAM2	E130307H12Rik	320404	E	III
PAM2	Xrcc1	22594	E	III
PAM2	Dscr1	54720	E	III
PAM2	0610012A05Rik	67434	E	III
PAM2	Samsn1	67742	E	III
PAM2	D7Ertd458e	52118	E	III
PAM2	Fabp4	11770	E	III
PAM2	Mdm2	17246	E	III
PAM2	B630005N14Rik	101148	E	III
PAM2	Ch25h	12642	E	III
PAM2	5830415L20Rik	68152	E	III
PAM2	Cdk5r	12569	E	III
PAM2	Nupr1	56312	E	III
PAM2	Baz1a	14896	E	III
PAM2	Mlp	17357	E	III
PAM3	Cd83	12522	F	III
PAM3	Jag1	16449	F	III
PAM3	Gadd45b	17873	F	III
PAM3	Pde4b	18578	F	III
PAM3	Cd69	12515	F	III
PAM3	Prdm1	12142	F	III
PAM3	Il1b	16176	F	III
PAM3	Sh3bgrl	56726	F	III
PAM3	Irg1	16365	F	III
PAM3	Ccrn4l	12457	F	III
PAM3	Egr3	13655	F	III
PAM3	4921515A04Rik	268301	F	III

continued on following page

Table 3. continued

stimulus	Gene Symbols	LocusID	second_step_clustering	first_step_clustering
PAM3	Olr1	108078	F	III
PAM3	Rel	19696	F	III
PAM3	Mmp13	17386	F	III
PAM3	Il1a	16175	F	III
PAM3	Skil	20482	F	III
PAM3	Ptger4	19219	F	III
PAM3	Trps1	83925	F	III
PAM3	H2-Q7	15018	F	III
PAM3	Tnfaip2	21928	F	III
PAM3	Gpr85	64450	F	III
PAM3	Pnrc1	108767	F	III
PAM3	Rab20	19332	F	III
PAM3	Flrt3	71436	F	III
PAM3	BC036563	230738	F	III
PAM3	Clecsf9	56619	F	III
PAM3	Slc4a7	218756	F	III
PAM3	Icam1	15894	F	III
PAM3	Arl8	75869	F	III
PAM3	Csf1	12977	F	III
PAM3	Trim13	66597	F	III
PAM3	Malt1	240354	F	III
PAM3	Arhe	74194	F	III
PAM3	Bcl2l11	12125	F	III
PAM3	Birc2	11796	F	III
PAM3	4930488L10Rik	319710	F	III
PAM3	Stx11	74732	F	III
PAM3	Cxcl10	15945	F	III
PAM3	Vcam1	22329	F	III
PAM3	Dusp16	70686	F	III
PAM3	4930534K13Rik	75234	F	III
PAM3	Cias1	216799	F	III
PAM3	Tlr2	24088	F	III
PAM3	Ptprj	19271	F	III
PAM3	Gpr84	80910	F	III
PAM3	Ripk2	192656	F	III
PAM3	Nfil3	18030	F	III
PAM3	4930431B09Rik	74645	F	III
PAM3	Tcfec	21426	F	III
PAM3	9530058O11Rik	208449	F	III

continued on following page

Table 3. continued

stimulus	Gene Symbols	LocusID	second_step_clustering	first_step_clustering
PAM3	D7Ertd458e	52118	F	III
PAM3	Bcl2a1a	12044	F	III
PAM3	2610103J23Rik	67154	F	III
PAM3	Klf7	93691	F	III
PAM3	Slamf7	75345	F	III
PAM3	BC010311	209212	F	III
PAM3	0610012A05Rik	67434	F	III
PAM3	Tgm2	21817	F	III
PAM3	Sh3bp5	24056	F	III
PAM3	3110043O21Rik	73205	F	III
PAM3	D10Ertd749e	52696	F	III
PAM3	BC035207	240396	F	III
PAM3	Slfn2	20556	F	III
PAM3	A530088I07Rik	212167	F	III
PAM3	B630005N14Rik	101148	F	III
PAM3	Fyb	23880	F	III
PAM3	Mad	17119	F	III
PAM3	2010109K11Rik	72123	F	III
PAM3	Ccr3	12771	F	III
PAM3	Ednrb	13618	F	III
PAM3	Cdk5r	12569	F	III
PAM3	Igsf6	80719	F	III
PAM3	2010305C02Rik	69891	F	III
PAM3	Zfp263	74120	F	III
PAM3	Plscr1	22038	F	III
PAM3	Prg	19073	F	III
PAM3	Nfkbib	18036	F	III
PAM3	Rabgef1	56715	F	III
PAM3	Gch	14528	F	III
PAM3	Ehd1	13660	F	III
PAM3	Nupr1	56312	F	III
PAM3	Tgfb1i4	21807	F	III
PAM3	Mlp	17357	F	III
PAM3	2810474O19Rik	67246	F	III
PAM3	Rnf2	19821	F	III
PAM3	2810036K01Rik	67222	F	III
PAM3	Sod2	20656	F	III
PAM3	Rnf19	30945	F	III
PAM3	Mmd	67468	F	III

continued on following page

Table 3. continued

stimulus	Gene Symbols	LocusID	second_step_clustering	first_step_clustering
PAM3	E130119J17Rik	212168	F	III
PAM3	Odc	18263	F	III
PAM3	Ell2	192657	F	III
PAM3	Snag1	170625	F	III
PAM3	BC006779	229003	F	III
PAM3	Cebpb	12608	F	III
PAM3	Casp4	12363	F	III
PAM3	Il6	16193	G	III
PAM3	Serpine1	18787	G	III
PAM3	Hspa1a	193740	G	III
PAM3	Nfkbie	18037	G	III
PAM3	1200009I06Rik	74190	G	III
PAM3	D8Ertd82e	244418	G	III
PAM3	Pdzgef1	76089	G	III
PAM3	Lig3	16882	G	III
PAM3	Ch25h	12642	G	III
PAM3	Slc7a11	26570	G	III
PAM3	Icosl	50723	G	III
PAM3	Sema6d	214968	G	III
PAM3	Tnfrsf5	21939	G	III
PAM3	Samsn1	67742	G	III
PAM3	A730041O15Rik	269717	G	III
PAM3	Relb	19698	G	III
PAM3	Tnip1	57783	G	III
PAM3	Herpud1	64209	G	III
PAM3	Slc11a2	18174	G	III
PAM3	AI504432	229694	G	III
PAM3	Bcl6	12053	G	III
PAM3	Plagl2	54711	G	III
PAM3	Arg2	11847	G	III
PAM3	C230060M08Rik	232314	G	III
PAM3	Cdc42ep4	56699	G	III
PAM3	4631422O05Rik	78749	G	III
PAM3	Ifi203	15950	G	III
PAM3	Slco4a1	108115	G	III
PAM3	Spata13	219140	G	III
PAM3	2210412D01Rik	70178	G	III
PAM3	Ddx6	13209	G	III
PAM3	Stk38l	232533	G	III

continued on following page

Table 3. continued

stimulus	Gene Symbols	LocusID	second_step_clustering	first_step_clustering
PAM3	Serpinb2	18788	G	III
PAM3	Slc2a6	227659	G	III
PAM3	Nck1	17973	G	III
PAM3	Jundm2	81703	G	III
PAM3	Pumag	80885	G	III
PAM3	Saa3	20210	G	III
PAM3	BC022623	224093	G	III
PAM3	Tnfrsf1b	21938	G	III
PAM3	P2ry2	18442	G	III
PAM3	Irak2	108960	G	III
PAM3	Itga5	16402	G	III
PAM3	Gpr132	56696	G	III
PAM3	1810029B16Rik	66282	G	III
PAM3	Bcor	71458	G	III
PAM3	Il12b	16160	G	III
PAM3	Bcl3	12051	G	III
PAM3	Egr1	13653	B	I
PAM3	Btg2	12227	B	I
PAM3	Phlda1	21664	B	I
PAM3	Myd116	17872	B	I
PAM3	Dusp1	19252	B	I
PAM3	Snk	20620	B	I
PAM3	Ier3	15937	B	I
PAM3	Dusp2	13537	B	I
PAM3	Spag9	70834	B	I
PAM3	Gdf15	23886	B	I
PAM3	C030048B08Rik	269623	B	I
PAM3	Chk	12660	B	I
PAM3	Osm	18413	B	I
PAM3	1110035O14Rik	59027	B	I
PAM3	Ccnl1	56706	B	I
PAM3	Socs3	12702	A	I
PAM3	AA408868	80859	A	I
PAM3	Cxcl1	14825	A	I
PAM3	Tnf	21926	A	I
PAM3	Cxcl2	20310	A	I
PAM3	Traf1	22029	A	I
PAM3	Ccrl2	54199	A	I
PAM3	Junb	16477	A	I

continued on following page

Table 3. continued

stimulus	Gene Symbols	LocusID	second_step_clustering	first_step_clustering
PAM3	Sqstm1	18412	A	I
PAM3	Nfkbia	18035	A	I
PAM3	Map3k8	26410	A	I
PAM3	Nfe2l2	18024	A	I
PAM3	2610103N14Rik	72472	A	I
PAM3	Ccl4	20303	A	III
PAM3	Ccl3	20302	A	III
PAM3	Cited2	17684	A	III
PAM3	2010306G19Rik	67035	A	III
PAM3	Etv3	27049	A	III
PAM3	Ccng2	12452	A	III
PAM3	Sdc4	20971	A	III
PAM3	Tnfaip3	21929	E	I
PAM3	Egr2	13654	E	I
PAM3	Tnfsf9	21950	E	I
PAM3	Axud1	215418	E	I
PAM3	B430217B02Rik	319544	E	I
PAM3	Pim1	18712	E	I
PAM3	IkappaBNS	243910	E	I
PAM3	Sgk	20393	E	I
PAM3	LOC233400	233400	E	I
PAM3	Dusp4	319520	E	I
PAM3	Erbb2ip	59079	E	I
PAM3	Hrb	15463	E	I
PAM3	C630016O21Rik	210105	E	I
PAM3	Nfat5	54446	E	I
PAM3	Copeb	23849	E	I
PAM3	A530090O15Rik	211770	E	I
PAM3	Stx6	58244	E	I
PAM3	Fbxw1b	103583	E	I
PAM3	Ptgs2	19225	E	III
PAM3	Gem	14579	E	III
PAM3	BC038313	216850	E	III
PAM3	1300002F13Rik	74155	E	III
PAM3	Maff	17133	E	III
PAM3	---	240672	E	III
PAM3	4732452O09Rik	320292	E	III
PAM3	Il10	16153	E	III
PAM3	Sele	20339	E	III

continued on following page

Table 3. continued

stimulus	Gene Symbols	LocusID	second_step_clustering	first_step_clustering
PAM3	Tank	21353	E	III
PAM3	Ppfibp1	67533	E	III
PAM3	Ifrd1	15982	E	III
PAM3	Acbp2	11569	E	III
PAM3	Il10ra	16154	E	III
PAM3	E130307H12Rik	320404	E	III
PAM3	D430024K22Rik	214855	E	III
PAM3	Atf3	11910	E	III
PAM3	Mdm2	17246	E	III
PAM3	Dscr1	54720	E	III
PAM3	Fabp4	11770	E	III
PAM3	Syk	20963	E	III
PAM3	AU021107	229055	E	III
PolyIC	Egr1	13653	F	I
PolyIC	AA408868	80859	F	I
PolyIC	Cxcl2	20310	F	I
PolyIC	Tnf	21926	F	I
PolyIC	Cxcl1	14825	F	I
PolyIC	Phlda1	21664	F	I
PolyIC	Btg2	12227	F	I
PolyIC	Dusp1	19252	F	I
PolyIC	Ifrd1	15982	F	I
PolyIC	Ier3	15937	F	I
PolyIC	Junb	16477	F	I
PolyIC	Traf1	22029	F	I
PolyIC	Ier2	15936	F	I
PolyIC	Etv3	27049	F	I
PolyIC	2410011G03Rik	66414	F	I
PolyIC	Nfkbia	18035	F	I
PolyIC	Gdf15	23886	F	I
PolyIC	Sqstm1	18412	F	I
PolyIC	Dusp2	13537	F	I
PolyIC	Pim1	18712	F	I
PolyIC	Myc	17869	F	I
PolyIC	Idb2	15902	F	I
PolyIC	Itga5	16402	F	I
PolyIC	Marcks	17118	F	I
PolyIC	Socs3	12702	G	I
PolyIC	Tnfaip3	21929	G	I

continued on following page

Table 3. continued

stimulus	Gene Symbols	LocusID	second_step_clustering	first_step_clustering
PolyIC	Ccrl2	54199	G	I
PolyIC	Tnfsf9	21950	G	I
PolyIC	Myd116	17872	G	I
PolyIC	Axud1	215418	G	I
PolyIC	Egr2	13654	G	I
PolyIC	1110035O14Rik	59027	G	I
PolyIC	Sgk	20393	G	I
PolyIC	IkappaBNS	243910	G	I
PolyIC	Spag9	70834	G	I
PolyIC	Adamts1	11504	G	I
PolyIC	Snk	20620	G	I
PolyIC	C630016O21Rik	210105	G	I
PolyIC	B430217B02Rik	319544	G	I
PolyIC	Dusp4	319520	G	I
PolyIC	Zfp36l1	12192	G	I
PolyIC	Taf7	24074	G	I
PolyIC	Ets2	23872	G	I
PolyIC	Erbb2ip	59079	G	I
PolyIC	Cd83	12522	G	III
PolyIC	Il6	16193	G	III
PolyIC	Cxcl10	15945	G	III
PolyIC	Ifnb	15977	G	III
PolyIC	Ccl4	20303	G	III
PolyIC	Il10	16153	G	III
PolyIC	Maff	17133	G	III
PolyIC	Cflar	12633	G	III
PolyIC	Ccl7	20306	G	III
PolyIC	Sh3bgrl	56726	G	III
PolyIC	Egr3	13655	G	III
PolyIC	Ptgs2	19225	G	III
PolyIC	BC038313	216850	G	III
PolyIC	Ccl3	20302	G	III
PolyIC	Ccl2	20296	G	III
PolyIC	---	240672	G	III
PolyIC	Malt1	240354	G	III
PolyIC	Ch25h	12642	G	III
PolyIC	Pde4b	18578	G	III
PolyIC	Clecsf9	56619	G	III
PolyIC	Ccrn4l	12457	G	III

continued on following page

Table 3. continued

stimulus	Gene Symbols	LocusID	second_step_clustering	first_step_clustering
PolyIC	Cflar	12633	G	III
PolyIC	Rab20	19332	G	III
PolyIC	Gem	14579	G	III
PolyIC	Mmp13	17386	G	III
PolyIC	Cited2	17684	G	III
PolyIC	Ccl12	20293	G	III
PolyIC	Copeb	23849	G	III
PolyIC	Ptger4	19219	G	III
PolyIC	Mlp	17357	G	III
PolyIC	Skil	20482	G	III
PolyIC	4921515A04Rik	268301	G	III
PolyIC	1300002F13Rik	74155	G	III
PolyIC	2010306G19Rik	67035	G	III
PolyIC	Kctd12	239217	G	III
PolyIC	D7Ertd458e	52118	G	III
PolyIC	Arhe	74194	G	III
PolyIC	D7Ertd413e	52325	G	III
PolyIC	Tiparp	99929	G	III
PolyIC	Ccr3	12771	G	III
PolyIC	BC035207	240396	G	III
PolyIC	Gpr85	64450	G	III
PolyIC	2010109K11Rik	72123	G	III
PolyIC	Mmd	67468	G	III
PolyIC	AA407452	57867	G	III
PolyIC	Gja1	14609	G	III
PolyIC	Cpeb4	67579	G	III
PolyIC	Peli1	67245	G	III
PolyIC	8430412F05	242860	G	III
PolyIC	Ccl5	20304	G	III
PolyIC	2610018G03Rik	70415	D	IV
PolyIC	Taf15	70439	D	IV
PolyIC	Birc4	11798	D	IV
PolyIC	5830484J08Rik	76131	D	IV
PolyIC	4930503L19Rik	269033	D	IV
PolyIC	Abca1	11303	D	IV
PolyIC	1200008O12Rik	74107	D	IV
PolyIC	1110001A05Rik	56376	D	IV
PolyIC	Clecsf10	56620	D	IV
PolyIC	5830433M19Rik	67770	D	IV

continued on following page

Table 3. continued

stimulus	Gene Symbols	LocusID	second_step_clustering	first_step_clustering
PolyIC	Hnrpdl	50926	D	IV
PolyIC	8030446C20Rik	100715	D	IV
R848	Idb3	15903	F	I
R848	Dusp4	319520	F	I
R848	Adamts1	11504	F	I
R848	C630016O21Rik	210105	F	I
R848	A530090O15Rik	211770	F	I
R848	Ets2	23872	F	I
R848	Nfat5	54446	F	I
R848	Eif4e	13684	F	I
R848	Fosl2	14284	F	I
R848	Cd83	12522	F	III
R848	Gadd45b	17873	F	III
R848	BC031781	208768	F	III
R848	Gem	14579	F	III
R848	Egr3	13655	F	III
R848	Ptgs2	19225	F	III
R848	Pde4b	18578	F	III
R848	Maff	17133	F	III
R848	Ccrn4l	12457	F	III
R848	BC038313	216850	F	III
R848	Cd69	12515	F	III
R848	---	240672	F	III
R848	Sh3bgrl	56726	F	III
R848	4921515A04Rik	268301	F	III
R848	Prdm1	12142	F	III
R848	1300002F13Rik	74155	F	III
R848	Ifrd1	15982	F	III
R848	Cflar	12633	F	III
R848	Skil	20482	F	III
R848	4732452O09Rik	320292	F	III
R848	Tank	21353	F	III
R848	Rab20	19332	F	III
R848	BC036563	230738	F	III
R848	Mmp13	17386	F	III
R848	H2-Q7	15018	F	III
R848	Cias1	216799	F	III
R848	Clecsf9	56619	F	III
R848	Arhe	74194	F	III

continued on following page

Table 3. continued

stimulus	Gene Symbols	LocusID	second_step_clustering	first_step_clustering
R848	Pnrc1	108767	F	III
R848	Aebp2	11569	F	III
R848	Gpr84	80910	F	III
R848	---	231462	F	III
R848	BC010311	209212	F	III
R848	Odc	18263	F	III
R848	4930431B09Rik	74645	F	III
R848	Etv3	27049	F	III
R848	Il10ra	16154	F	III
R848	4930534K13Rik	75234	F	III
R848	Mlp	17357	F	III
R848	E130307H12Rik	320404	F	III
R848	Cited2	17684	F	III
R848	2010306G19Rik	67035	F	III
R848	Mdm2	17246	F	III
R848	D430024K22Rik	214855	F	III
R848	Birc2	11796	F	III
R848	Dscr1	54720	F	III
R848	Zfp216	22682	F	III
R848	Gpr85	64450	F	III
R848	Pdcd1lg1	60533	F	III
R848	D10Ertd749e	52696	F	III
R848	0610012A05Rik	67434	F	III
R848	2310016C08Rik	69573	F	III
R848	Mmd	67468	F	III
R848	Sdc4	20971	F	III
R848	Dtr	15200	F	III
R848	Rnf19	30945	F	III
R848	Prg	19073	F	III
R848	Plek	56193	F	III
R848	Nfkbib	18036	F	III
R848	Kctd12	239217	F	III
R848	Bcl6	12053	F	III
R848	Btg1	12226	F	III
R848	Bcl2l11	12125	F	III
R848	D17Ertd808e	52040	F	III
R848	Pumag	80885	F	III
R848	Sod2	20656	F	III
R848	Mcl1	17210	F	III

continued on following page

Table 3. continued

stimulus	Gene Symbols	LocusID	second_step_clustering	first_step_clustering
R848	Fabp4	11770	F	III
R848	Jag1	16449	G	III
R848	Irg1	16365	G	III
R848	Il1b	16176	G	III
R848	Flrt3	71436	G	III
R848	Stx11	74732	G	III
R848	Rel	19696	G	III
R848	Hspa1a	193740	G	III
R848	Olr1	108078	G	III
R848	Nfkbie	18037	G	III
R848	Bcl2a1a	12044	G	III
R848	Icam1	15894	G	III
R848	Trps1	83925	G	III
R848	Nfil3	18030	G	III
R848	Cxcl10	15945	G	III
R848	Malt1	240354	G	III
R848	Tlr2	24088	G	III
R848	Mad	17119	G	III
R848	Tgm2	21817	G	III
R848	4930488L10Rik	319710	G	III
R848	Cpeb4	67579	G	III
R848	Spic	20728	G	III
R848	Ripk2	192656	G	III
R848	Il1rn	16181	G	III
R848	Arl8	75869	G	III
R848	Klf7	93691	G	III
R848	Dusp16	70686	G	III
R848	Trim13	66597	G	III
R848	Plagl1	22634	G	III
R848	AI195350	106878	G	III
R848	Bhlhb2	20893	G	III
R848	Serpinb2	18788	G	III
R848	Tnfaip2	21928	G	III
R848	Zfp263	74120	G	III
R848	Cdk5r	12569	G	III
R848	Ehd1	13660	G	III
R848	Rabgef1	56715	G	III
R848	Cebpb	12608	G	III
R848	Slc4a7	218756	G	III

continued on following page

Table 3. continued

stimulus	Gene Symbols	LocusID	second_step_clustering	first_step_clustering
R848	Slfn2	20556	G	III
R848	1200009I06Rik	74190	G	III
R848	Relb	19698	G	III
R848	Plagl2	54711	G	III
R848	9530058O11Rik	208449	G	III
R848	A530088I07Rik	212167	G	III
R848	Il12b	16160	G	III
R848	Sema6d	214968	G	III
R848	D8Ertd82e	244418	G	III
R848	Slc11a2	18174	G	III
R848	Ptprj	19271	G	III
R848	Il1a	16175	G	III
R848	Ddx6	13209	G	III
R848	Icosl	50723	G	III
R848	Pdzgef1	76089	G	III
R848	Egr1	13653	B	I
R848	Btg2	12227	B	I
R848	Ier2	15936	B	I
R848	Gdf15	23886	B	I
R848	Dusp2	13537	B	I
R848	Osm	18413	B	I
R848	2410011G03Rik	66414	B	I
R848	1110035O14Rik	59027	B	I
R848	S100a10	20194	B	I
R848	C030048B08Rik	269623	B	I
R848	Creb5	231991	B	I
R848	Zfp36l1	12192	B	I
R848	Klf2	16598	B	II
R848	Fos	14281	B	II
R848	Jun	16476	B	II
R848	Rgs1	50778	B	II
R848	Mapk6	50772	B	II
R848	Socs3	12702	A	I
R848	AA408868	80859	A	I
R848	Cxcl1	14825	A	I
R848	Tnf	21926	A	I
R848	Cxcl2	20310	A	I
R848	Phlda1	21664	A	I
R848	Junb	16477	A	I

continued on following page

Table 3. continued

stimulus	Gene Symbols	LocusID	second_step_clustering	first_step_clustering
R848	Ccrl2	54199	A	I
R848	Dusp1	19252	A	I
R848	Egr2	13654	A	I
R848	Axud1	215418	A	I
R848	Traf1	22029	A	I
R848	Ier3	15937	A	I
R848	Sgk	20393	A	I
R848	Sqstm1	18412	A	I
R848	Pim1	18712	A	I
R848	Nfkbia	18035	A	I
R848	Idb2	15902	A	I
R848	Nfe2l2	18024	A	I
R848	Tgif	21815	A	I
R848	Map3k8	26410	A	I
R848	Marcks	17118	A	I
R848	Copeb	23849	A	I
R848	Itga5	16402	A	I
R848	Ndel1	83431	A	I
R848	Ccl4	20303	A	III
R848	Ptger4	19219	A	III
R848	Atf3	11910	A	III
R848	Ccl3	20302	A	III
R848	Ccl7	20306	A	III
R848	Tnfaip3	21929	E	I
R848	Tnfsf9	21950	E	I
R848	Myd116	17872	E	I
R848	B430217B02Rik	319544	E	I
R848	Spag9	70834	E	I
R848	Snk	20620	E	I
R848	IkappaBNS	243910	E	I
R848	Fbxw1b	103583	E	I
R848	Chk	12660	E	I
R848	Erbb2ip	59079	E	I
R848	Ccnl1	56706	E	I
R848	Myc	17869	E	I
R848	LOC233400	233400	E	I
R848	Ier5	15939	E	I
R848	Stx6	58244	E	I
R848	Txnrd1	50493	E	I

Chapter 21
Dynamic Modeling and Parameter Identification for Biological Networks:
Application to the DNA Damage and Repair Processes

Fortunato Bianconi
University of Perugia, Italy

Gabriele Lillacci
University of Perugia, Italy

Paolo Valigi
University of Perugia, Italy

ABSTRACT

DNA damage and repair processes are key cellular phenomena that are being intensely studied because of their implications in the onset and therapy of cancer. This chapter introduces a general dynamic model of gene expression, and proposes a genetic network modeling framework based on the interconnection of a continuous-time model and a hybrid model. This strategy is applied to a network built around the p53 gene and protein, which detects DNA damage and activates the downstream nucleotide excision repair (NER) network, which carries out the actual repair tasks.

Then, two different parameter identification techniques are presented for the proposed models. One is based on a least squares procedure, which treats the signals provided by a high gain observer; the other one is based on a Mixed Extended Kalman Filter. Prior to the estimation phase, identifiability and sensitivity analyses are used to determine which parameters can be and/or should be estimated. The procedures are tested and compared by means of data obtained by in silico experiments.

DOI: 10.4018/978-1-60960-491-2.ch021

INTRODUCTION

Systems biology is emerging as a new interdisciplinary subject, and several tools are being developed within this framework. It is largely recognized that new systems-level knowledge is required in order to achieve a better understanding of biological phenomena (Sontag, 2005), and to allow for further insights and predictions for already well understood problems (Brazma, 2006).

Among the several problems tackled by systems biology, this chapter focuses on the dynamic modeling and parameter analysis and identification. In particular, we study a dynamic model for DNA damage sensing and repair. The sensing phase is carried out by p53 action, while the repair phase is performed by nucleotide excision repair (NER). The model is based on ordinary differential equations and hybrid systems, and allows us to study the time evolution of NER, from DNA damage sensing, through NER activation, to damaged single strand DNA excision.

Based on the proposed models, sensitivity analysis and identifiability analysis are carried out and parameter estimation techniques are introduced and evaluated by means of *in silico* simulations. These problems are generally recognized as key issues in systems biology (Aldridge et al., 2006; Kitano, 2001). Some simulation results are presented and discussed. Finally, the Appendix provides some biological background.

BACKGROUND

When in 1953 James Watson and Francis Crick described the DNA double helix, its structure seemed so solid and stable that research on notions such as DNA damage and repair was initially hindered (Friedberg, 2003). Actually, DNA is continuously exposed to several types of damage. The most relevant kind of DNA damage is helix-distorting lesions throughout the genome. These lesions are generated by different causes, including the formation of DNA adducts after administration of common drugs used in cancer therapy (Fayad et al., 2009), such as cisplatin and other platinum-based compounds. Another type of damage occurs when internal or environmental factors, such as exposition to radiation, cause breakages in both strands of the DNA helix. This event, called double strand breaks (DSB) can be lethal to the whole organism if not properly treated, since they can induce cancer and hereditary diseases (Bolderson et al., 2009).

Cells react to DNA damage in three basic ways. If the damage level is very low it is *tolerated*: the cell has specific structures to operate DNA replication and transcription even in presence of lesions. If the damage is more serious, it is *repaired*: the cellular growth is arrested (cell cycle arrest) and one of the several repair mechanisms available is started. If the damage level is too high to be effectively treated, *apoptosis* is started: this is a programmed death through which the cell eliminates itself from a population that might otherwise suffer the serious pathological consequences of the transmission of disrupted genetic material (Letai et al., 2008).

Sensing DNA damages is a very complex process, which involves a large number of pathways. A genetic network built upon the p53 gene and protein plays a key role in this process. First discovered in 1979, the actual tumor suppressing function of p53 was clarified only twenty years later (Vogelstein et al., 2000). The fact that DSBs induce an increase in p53 levels which, in turn, induce apoptosis has been demonstrated for the first time by Yonish-Rouach et al. (1991), and it is now generally accepted (Meek, 2009).

Among a number of different mechanisms for DNA repair, in this chapter, we consider *nucleotide excision repair* (NER). It is a versatile DNA repair mechanism that enables cells to eliminate helix-distorting lesions throughout the genome (Cleaver et al., 2009).

DNA damage is investigated by means of dynamic mathematical modeling, in the framework of systems biology. Mathematical modeling of biological systems often require the judicious choice of parameters and variables, which remains a challenging problem.

The parameter identification problem for biological systems has been addressed by several authors using a number of different approaches. Identification techniques based on global optimization tools are studied by Moles et al. (2003), where several stochastic and deterministic methods were compared. Several other approaches, such as the *multilevel coordinate search* algorithm, a *differential evolution* method, an *unconstrained evolution strategy* method, the *stochastic ranking evolution strategy*, and an additional evolution strategy approach were also compared. The study reached the conclusion that, at least for the specific system studied there ("a three-step pathway" that describes the variation of the metabolite concentrations with time), the best results were achieved by evolution strategies, although they required a considerable amount of computational efforts.

Model identification is also addressed by Gadkar et al. (2005), where a detailed identifiability analysis is carried out, and subsequently an identification method is proposed and studied for the Varner caspase activation model and the Schoeberl model for the MAPK system. The model identification methodology comprises three major steps: a model complexity analysis, aimed at measurements determination; a state estimation phase, to estimate the state not available for measurements, and finally a true parameter estimation phase, based on a least-squares algorithm.

The identification approach proposed by McKinney et al. (2006) is based on hybrid grammar and unscented Kalman filtering: the grammar-based component, by means of a genetic algorithm, is used to search the space of model architectures, and an unscented Kalman Filter is used for parameter estimation. Kalman filter approaches have also been applied to genetic bistable switches (Dunlop & Murray, 2006), where it is compared to a nonlinear least squares method for parameter estimation on a genetic bistable switch. The study also addressed the issue of optimal input selection for identification purposes. Kalman-filter-based techniques have also been successfully applied to transcriptional regulatory networks, including cascade pathways (Quach et al., 2007), feedback modules (Sun et al., 2008), and general models for gene expression (Wang et al., 2009).

Considerable interest has also been raised by statistical methods, especially the ones based on Bayesian inference (Wilkinson, 2007). They present significant advantages of being able to infer the whole probability density of the unknown parameters, as opposed to just a point estimate. Their application, however, tend to be limited to small-dimensional systems (Toni et al., 2009), due to the computational issues that arise in their implementation.

When dealing with parameter estimation, experimental design can benefit from a detailed analysis of parameter identifiability. In other words, it is important to determine whether the quantities that are measured will yield all the necessary information to compute, at least in an approximate manner, the unknown parameters. This problem is a very challenging one in itself, and a global answer is difficult to obtain for general biological systems. Some results have been derived by Evans et al. (2002), and applied to biological models by Evans et al. (2004). Other approaches, based on the transformation of the original model into a more suitable form, are presented by Farina et al. (2006) and Fey et al. (2008).

Sensitivity analysis is another useful tool that can be used to study the dependence of the system's behavior on changes in the parameter values. The application of sensitivity analysis to chemical and biological networks has a long history (e.g., Rabitz, 1987; Rabitz et al., 1983). Sensitivities have been extensively used in the study of metabolic pathways (e.g., Cornish-Bowden, 1995; Heinrich et al., 2003 and the references therein). Sensitivity analysis can also help in evaluating which parameters should be identified: since in most practical cases, the available data are not enough to estimate all the parameters in a given ODE model, it is common to select for estimation the ones that have the most effect on the model's behavior. This issue is addressed in (Cho et al., 2003) and the references therein.

INTRODUCTION TO MODELING GENE EXPRESSION

Preliminary Remarks

The entire life of an organism is controlled by the way its genes are expressed, that is the way in which the genetic information stored in its DNA is used to direct the synthesis of proteins. The process involves two main steps, transcription and translation. During transcription, an enzyme called RNA polymerase reads a segment of DNA and produces a messenger RNA (mRNA) strand. mRNA is then exported from the cellular nucleus (assuming eukaryotic cells) and transported to ribosomes where it is converted into a protein. This step is called translation because it involves a change of language from the nucleotide sequence of DNA and RNA into the amino acid sequence of the proteins (Bruce et al., 2008).

Dynamic mathematical models of genetic networks are based on several approximations, among which the commonly accepted assumption that all the reactions occur in a homogeneous chemical solution, so that mean concentrations can be used instead of actual number of molecules. This implies that the cell's spatial structure is completely ignored (Sontag, 2004).

A mix of qualitative and quantitative modeling is often adopted in biological phenomena. Large classes of signaling systems may be profitably studied by first decomposing them into several subsystems, each of which is endowed with certain qualitative mathematical properties. These qualitative properties, in conjunction with a relatively small amount of quantitative data, allow the behavior of the entire reconstituted system to be easily deduced from the behavior of its interconnected parts. This paradigm of decomposition and reconnection has been one of the basic principles in systems theory and control engineering (Bianconi, 2010).

In addition, cellular networks exhibit combinations of both discrete and continuous behaviors. The continuous dynamics describe the protein concentrations and activity within a single cell while the activation or deactivation of portions of pathways are triggered by switches encoding protein concentrations reaching given thresholds. From a mathematical point of view, continuous dynamics are usually described by Ordinary Differential Equations (ODEs) while hybrid systems, which combine continuous ODE dynamics with discrete control events, represent an ideal framework to analyze processes with switching elements.

Among the large number of contributions in the field of hybrid models for systems biology, we recommend (Neogi, 2003; Ghosh and Tomlin, 2004; Belta et al., 2004; Lincoln and Tiwari., 2004; De Jong et al., 2003) for interested readers.

A General ODE Model for Transcription-Translation

A specific genetic process can be represented by means of a collection of transcription-translation cycles, together with some feedback loops among different cycles. Such a biological network can be turned into a system of Ordinary Differential Equations modeling the time behavior of the concentrations of mRNAs and proteins. Let m_i denote the i-th type of mRNA that will be translated into protein p_i, and let p_i^* denote the "activated" form of p_i (several proteins do not interact with other components unless activated in some manner, e.g., through phosphorylation). Then, the generic transcription-translation cycle can be described by:

$$\begin{cases} \dot{[m_i]} = \lambda_i - \mu_i[m_i] \pm TF_{i,j} \\ \dot{[p_i]} = \alpha_i[m_i] - \gamma_i[p_i] + \nu_i[p_i^*] - A_i \pm PPI_i \\ \dot{[p_i^*]} = A_i - \nu_i[p_i^*] \pm PPI_i^* \end{cases} \quad (1)$$

Where $[x]$ denotes the concentration of species x, the $TF_{i,j}$ terms model transcription factors (see below for details), the PPI_i and PPI_i^* terms model protein-protein interactions, and A_i terms model protein activation.

The ODE system is based on the assumption that the production of mRNA m_i is governed by a constant rate λ_i, by a spontaneous degradation with constant rate μ_i (modeling also the variety of regulating mechanisms, such as posttranscriptional repression, exerted by small RNAs, see e.g., (Liu, 2008)), and by transcription factors $TF_{i,j} = TF_i(p_j, p_j^*)$ which directly bind to DNA and regulate the transcription of a gene. The latter term depends on the concentration of some protein p_j acting as a transcription factor (in active or inactive form). The most common example for a TF function is based on Hill-type kinetics, for some parameter h_j and integer n:

$$TF_{i,j} = \frac{[p_j^*]^n}{h_j^n + [p_j^*]^n}.$$

Common protein-protein interaction functions include linear combinations of Michaelis-Menten terms (for enzymatic interactions) and mass action kinetic terms (for direct bindings or uncertain mechanisms), such as:

$$PPI_i = \frac{k_{cat,j}[p_j][p_i]}{k_{M,j} + [p_i]} + \frac{k_{cat,\ell}[p_\ell][p_i]}{k_{M,\ell} + [p_i]} \pm k[p_i][p_m]$$

with protein p_j and p_ℓ acting as enzymes for p_i, and protein p_m directly binding to p_i. The PPI_i term concerning active protein interactions is denoted using these terns.

In cells, transcription typically requires 15-30 minutes, while translation is usually faster (about 2-3 minutes for a typical gene). This leads to the consideration that, for modeling purposes, the duration of the whole process can be entirely ascribed to transcription (Monk, 2003). In some biological processes, this requires the introduction of a suitable *transcriptional time delay* into the model (see, e.g., Monk, 2003; Mihalas et al., 2000; and Tiana et al., 2002). The dynamic models for biological phenomena considered in this chapter assume a negligible value for the delay parameter, either because the transcription dynamics is much faster that the other interactions involved in the process, or because no transcriptional regulation is modeled.

In this second case, the mRNA equation can be discarded from (1), thus yielding a lower-dimensional model. Similarly, if the protein does not need an activation to participate in the network's interactions, the active protein's equation can be omitted.

The equation structure (1) can be replicated for each gene/protein pair under investigation, and the interactions can be identified between the various species according to the *PPI*, *PPI** and *TF* terms, thus obtaining a complete ODE model. This procedure also allows us to derive the model of single dynamic modes of a hybrid system, whenever the underlying biological phenomenon has a relevant switching behavior.

The biological process discussed in this chapter comprises the p53 network that monitors DNA damage, and the NER mechanism that carries out the actual repair task. While the p53 network operates continuously and therefore can be described by a system of "classical" ODEs obtained using the general model (1), the NER mechanism, which is made up of clearly defined steps occurring in order, is described by a hybrid system. Therefore, the complete model of DNA damage detection and repair used in this chapter is made up of a continuous-time p53 subsystem connected to a hybrid NER subsystem.

In the following sections, these two key elements will be referred to as the "p53 sub-model" and the "NER sub-model"'. Furthermore, to avoid cumbersome notation, concentrations will be

Box 1.

Ordinary Differential Equations:

The ODE of p53 in Figure 8 are:

$$\dot{m}_{P53} = \lambda_{p53} - \mu_{p53} m_{P53}$$

$$\dot{P53} = \alpha_{P53} m_{P53} - \mu_{P53} P53 + \nu_{P53} P53^* - \frac{k_1 ATM^* P53}{k_{M1} + P53} - \frac{k_{cat} MDM2\ P53}{\alpha K_B + P53}$$

$$\dot{P53}^* = \frac{k_1 ATM^*\ P53}{k_{M1} + P53} - \nu_{P53} P53^* - \frac{k_{cat}^* MDM2 P53^*}{\alpha K_B + P53^*}$$

$$\dot{m}_{mdm2} = \lambda_{mdm2} - \mu_{mdm2} m_{mdm2} + \varphi_{mdm2} \frac{P53^*(t)^{n_1}}{P53_0^{n_1} + P53^*(t)^{n_1}}$$

$$\dot{MDM2} = \alpha_{MDM2} m_{mdm2} - \mu_{MDM2} MDM2 - \frac{k_2 ATM^* MDM2}{k_{M2} + MDM2} - k_4 ARF\ MDM2$$

$$\dot{m}_{E2F1} = \lambda_{e2f1} - \mu_{e2f1} m_{E2F1}$$

$$\dot{E2F1} = \alpha_{e2f1} m_{e2f1} - \mu_{e2f1} E2F1 + \nu_{E2F1} E2F1^* - \frac{k_3 ATM^* E2F1}{k_{M3} + E2F1}$$

$$\dot{E2F1}^* = \frac{k_3 ATM^*\ E2F1}{k_{M3} + E2F1} - \nu_{E2F1} E2F1^* - k_5 ARF E2F1^*$$

$$\dot{m}_{arf} = \lambda_{arf} - \mu_{arf} m_{arf} + \varphi_{arf} \frac{E2F1^*(t)^{n_2}}{E2F1_0^{n_2} + E2F1^*(t)^{n_2}}$$

$$\dot{ARF} = \alpha_{ARF} m_{arf} - \mu_{ARF} ARF - k_4 ARF\ MDM2 - k_5 ARF\ E2F1^*$$

denoted simply by the name of the species (i.e., in all the equations x will stand for $[x]$).

p53 Sub-Model

The p53 genetic network plays a key role in cell's response to DNA damage. While p53 is known to interact with hundreds of other pathways (Meek, 2009), in the present chapter, we will focus our attention on the four major genes involved in the network. These can be thought of as the "DNA damage sensing module" of the p53 network. By applying the general scheme (1) to these four genes (see Appendix A), the model reported in Box 1 can be obtained. The construction of the p53 sub-model is further detailed in (Lillacci et al., 2006a; Lillacci et al., 2006b).

In the equations of Box 1, the variable m_P is the mRNA coding for the protein P, e.g., m_{P53} is the mRNA coding for the P53 protein. The model describes the interaction of mRNA and proteins P53, MDM2, E2F1 and ARF. Parameters with lower case subscripts refer to mRNA molecules (e.g., λ_{p53}), while parameters with upper case subscripts refer to proteins (e.g., α_{p53}).

NER Sub-Model

The NER mechanism is a versatile DNA repair pathway that enables cells to eliminate helix-distorting lesions throughout the genome (see Cleaver et al., 2009 and Appendix B). Among therapeutic drugs, cisplatin is a widely used chemotherapeutic agent which damages DNA by forming helix-distorting lesions called *adducts*. Cisplatin-DNA adducts are removed primarily by the NER pathway which is usually described by a hybrid model (see Figure 1), consisting of a set Q of five discrete modes: $Q = (q_1, q_2, q_3, a_A, q_S)$.

The key equations of the proposed NER hybrid model are reported in Box 2. The continuous state space has six state variables: the concentration of proteins *P53**, *XPC** and *XPA*; *ssDNA*, which is the variable describing the opening of the double helix; the substances *XPGssDNA* and *ndsDNA* are described in the following paragraphs. Hence, $x_{NER}=(P53^*, XPC^*, XPA, ssDNA, XPCGssDNA, ndsDNA)^T$.

The discrete mode q_1 describes the dynamics of repair activation; the mode q_2 describes the dynamics of double helix unwinding around the damage site; the mode q_3 describes the dynam-

Figure 1. Formal representation of the NER hybrid model

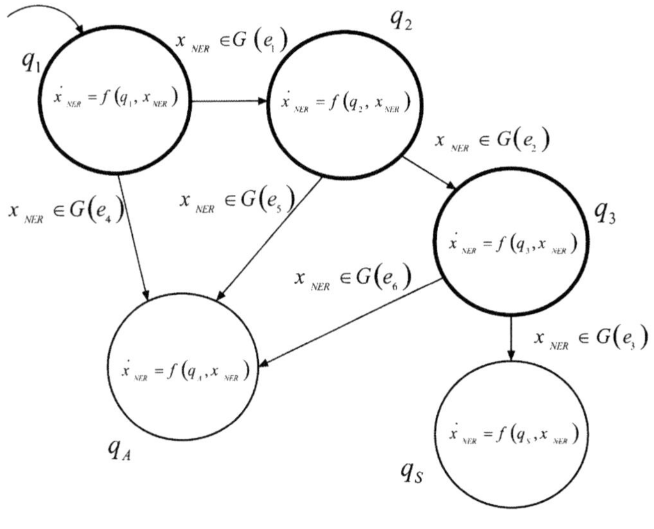

Box 2.

Ordinary Differential Equations:
The ODE of NER in Figure 9 are:

$$\dot{XPC^*} = \frac{k_{A2}P53^*\left(XPC_T - XPC^*\right)}{k_{MA2} + XPC_T - XPC^*} - \frac{k_{D2}XPC^*}{k_{MD2} + XPC^*}$$

$$\dot{ssDNA} = k_3 GB_{XPB^*}\left(ssDNA\right) + k_4 GB_{XPD^*}\left(ssDNA\right) + \left(k_5 + k_7\right)XPC^* - k_6 ssDNA - k_8 ssDNA \cdot XPA$$

$$\dot{XPA} = k_9 XPC^* - k_{10} XPA$$

$$\dot{XPGssDNA} = k_{11} GB_{XPG^*}\left(XPGssDNA\right) + k_{12} ssDNA \cdot XPA - k_{13} XPGssDNA$$

$$\dot{ndsDNA} = k_{14} GB_{ERCC1^*}\left(ndsDNA\right) + k_{15} ssDNA \cdot XPA - k_{16} ndsDNA$$

where

$$GB\left(k_a S, k_d, \frac{k_{Ma}}{R_T}, \frac{k_{Md}}{R_T}\right) = \frac{k_d\left(1 + \frac{k_{Ma}}{R_T}\right) + k_a S \frac{k_{Md}}{R_T}}{2\left(k_d - k_a S\right)} + \frac{\sqrt{\left[\left(1 + \frac{k_{Ma}}{R_T}\right) + k_a S \frac{k_{Md}}{R_T}\right]^2 + 4\frac{k_{Md}}{R_T}\left(k_d - k_a S\right)k_a S}}{2\left(k_d - k_a S\right)}$$

is the Goldbeter-Koshland functions.

ics of excision of single-strand damaged DNA segments. Two additional modes have also been introduced: the mode q_S for DNA synthesis and the mode q_A for apoptosis (although their dynamics are not modeled since the main interest here is on the NER mechanism).

Within each mode q, the dynamics of state variables are described by the vector field $f(q,x_{NER})$. Usually in hybrid systems, the transition from one discrete mode to another is modeled by the occurrence of an event e, triggered by the state satisfying a guard condition $G(e)$: i.e., a transition occurs whenever $x_{NER} \in G(e)$. In view of the characteristics of the NER mechanism, the hybrid model is only described in terms of proteins, while mRNA dynamics are neglected. This is because the transcription and translation processes are slow dynamics and the NER interactions are protein-protein interactions that happen much faster (millisecond timescale vs. second timescale). This means that we invoke the steady state assumption for mRNAs in Box 2 (Szallasi et al., 2006) and mRNA's dynamics are not modeled.

First Mode: Repair Activation

The pathway of repair activation is shown in Figure 2(A). It is assumed that cisplatin enters the cell and it generates DNA-adducts as in Figure 9. Hence, the p53 network is activated, and active protein

Figure 2. Continuous dynamic pathways of the hybrid system. (A) First mode: activation of XPC-HR23B. (B) Second mode: combination of robust and two parallel positive feedbacks modules. (C) Third mode: two parallel positive feedbacks for ERCC1-XPF and XPG.

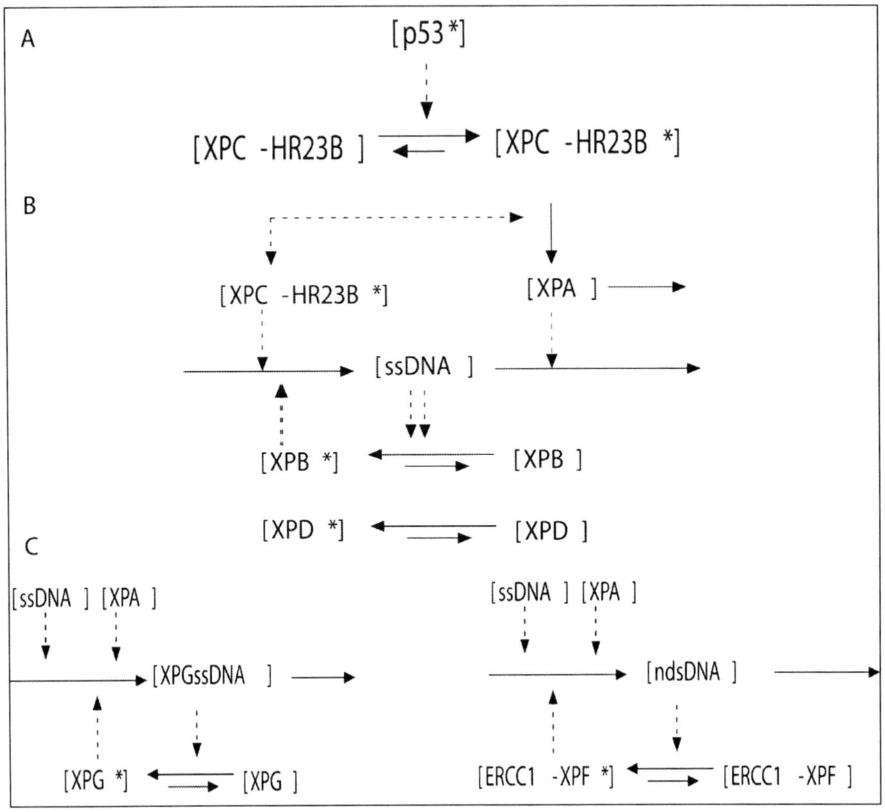

P53* acts as a stimulus for the "sigmoidal module" (Tyson et al., 2003). The first equation in Box 2 describes mode q_1 dynamics where P53* protein is governed by the corresponding sub-model. Using the principle of mass conservation, the total XPC-HR23B concentration in the cell is assumed to be constant, and equal to XPC_T. Parameters k_{A2} and k_{D2} indicate activation and deactivation kinetics constants, respectively, while k_{MA2} and k_{MD2} are Michaelis-Menten constants of activation and deactivation reactions, respectively.

Second Mode: Helix Unwinding

The diagram in Figure 2(B) illustrates the pathways of XPA protein and XPB and XPD helicases in the DNA duplex opening mechanism around the damaged nucleotide. Following (Tyson et al., 2003), the combination of two control mechanisms is assumed: "robust module" and "positive feedback". The opening of the single strand DNA (ssDNA) is modeled in the following way. XPA confirms the location of damage while the two helicases operate in parallel to break hydrogen bonds among the nucleotides. The activation of the enzymes XPB and XPD is described by means of the Goldbeter-Koshland function (Goldbeter and Koshland, 1981).

The mode behavior is described by the second and third equations in Table 3, where $GB_{XPB^*}(ssDNA)$ and $GB_{XPD^*}(ssDNA)$ denote the Goldbeter-Koshland functions, whose general expression is shown in the sixth equation in

Table 3, where k_a and k_d are the activation and deactivation kinetic constants, respectively; k_{Ma} and k_{Md} are Michaelis-Menten constants of activation and deactivation reactions, respectively; R_T is the total protein and S is the stimuli protein. In particular, for mode q_2, $GB_{XPB^*}(ssDNA) = G\left(k_{A3}ssDNA, k_{D3}, \frac{k_{MA3}}{XPB_T}, \frac{k_{MD3}}{XPB_T}\right)$, where the function arguments are the kinetics parameters k_{A3} and k_{D3} as the concentration and total concentration, respectively, and the Michaelis-Menten constants k_{MA3} and k_{MD3}. Additional details on the analytic model can be found in (Bianconi, 2006; Bianconi et al., 2006).

Third Mode: Excision

The role played by the helicases XPG and ERCC1 in the pathways under investigation is shown in Figure 2(C). The positive feedback module is also used in this case. The substances XPGssDNA and ndsDNA arise from XPA and ssDNA. XPGssDNA is the opening of DNA tied with XPG helicase in position 3', while ndsDNA represents the non damaged single DNA, where ERCC1 has been cut in position 5' (the heterodimer XPF-ERCC1 (see Appendix B) is denoted by ERCC1). The mode dynamics are described by fourth and fifth equations in Box 2. As in mode q_2, it is assumed that the activation of the enzymes XPG and ERCC1 can be described by means of Goldbeter-Koshland functions. In particular, note that the total concentrations of XPG and ERCC1, respectively, appear as parameters of the two functions $GB_{XPG^*}\left(XPGssDNA\right)$ and $GB_{ERRC1^*}\left(ndsDNA\right)$.

Mode Transitions of the Hybrid System

As for the discrete dynamics of the switching among modes, i.e, for the guard conditions mentioned in Figure 1, it is important to underline that:

- $G(e_1)$ indicates that transition to mode q_1 (i.e., the transition forcing NER activation) occurs whenever *P53** is activated and *XPC** reaches the threshold representing the recognition of damaged sites;
- $G(e_2)$ indicates the opening of the double helix and XPC protein has confirmed all damages;
- $G(e_3)$ indicates the removing of single strand DNA segments with damaged nucleotides;
- $G(e_4)$, $G(e_5)$ and $G(e_6)$ are the transitions to apoptosis state from the states q_1, q_2 and q_3, respectively .

Setting the Wild-Type Behavior

Choosing reasonable initial values for the many parameters included in the models is one of the key problems in systems biology. In the case of the p53 sub-model, this has been achieved by requiring the model to reproduce experimentally observed behaviors, like the ones listed below.

- It is well established, both by biological experiments and by other models, that the main dynamic property of the p53 network is that it can display a periodic behavior (Ma et al., 2005; Monk 2003; Lev Bar-Or et al.,2000; Tiana et al., 2002; Geva-Zatorsky et al.,2006).
- Since the model refers to a single cell, the P53 and MDM2 concentrations are expected to display a sustained oscillatory behavior (Ma et al., 2005; Lev Bar-Or et al., 2000), that persists for as long as DNA damage is present (and therefore the input signal is active, see Appendix).
- Oscillations' period should be about 400 minutes (Lahav et al., 2004).
- A reference value for steady-state P53 concentration is 0.35 μM (Ma et al., 2005).

- In damaged cells, P53 concentration can rise up to 16 times the basal level (Tiana et al., 2002).

A similar approach has been used to select suitable parameters for the NER hybrid model. Based on (Politi et al., 2006) and the references therein, and by making use of some functions of the *Systems Biology Toolbox* (see Schmidt and Jirstrand, 2006), the set of parameters achieving the best fit between the model solutions and the behaviors from the literature has been selected. Typical behaviors for both the continuous and the hybrid model are reported in Figure 3. These simulations will be used later for the generation of the *in silico* data for parameter estimation.

Model Notation

The dynamics models proposed above will also be used in the subsequent sections of the chapter to describe tools for parameter identification.

To avoid cumbersome notation, the dynamic models in Box 1 for p53 and single mode NER models in Box 2 will be described by means of the following general compact form:

$$\dot{x} = f\left(x(t), \pi\right),\ x(0) = x_0 \tag{2}$$

Figure 3. Results of a typical model simulation. (A) P53 network. (B) NER simulation.

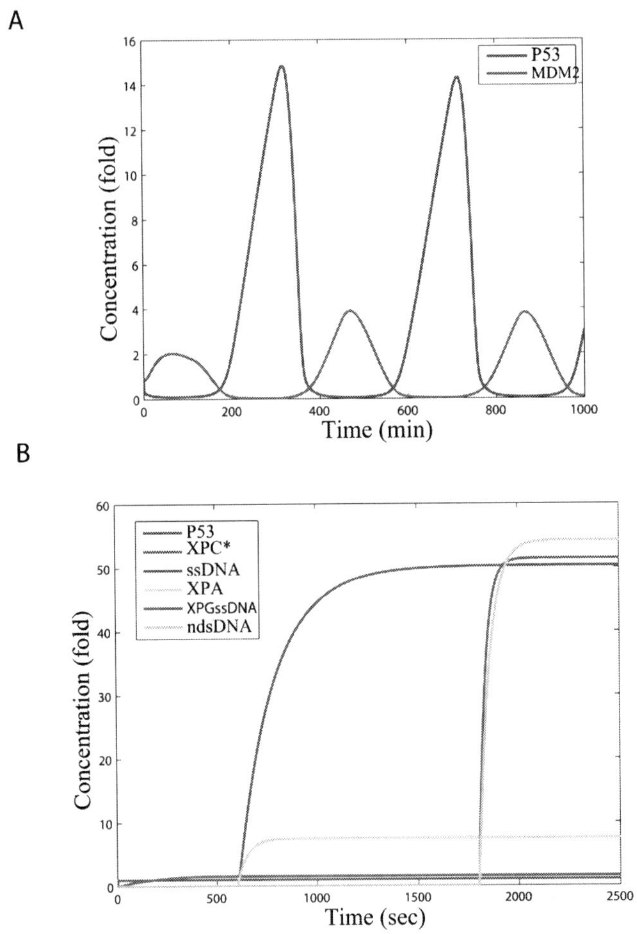

$$y(t,\pi) = H(x(t),\pi), \tag{3}$$

where $x \in R^n$ is the state vector, $y \in R^r$ is the vector of measured output, whose dependence on state variables and model parameters is described by the output map $h(x(t),\pi)$ and $\pi \in \Omega \subset R^{n_p}$ is the vector of relevant model parameters (e.g., the ones unknown, hence to be identified).

As far as the output function is concerned, quite often in biological experiments, measurements of protein concentration cannot distinguish among the different forms of a protein. This implies that one is usually able to only measure the sum of $P+P^*$. Hence, for each measured protein, $P+P^*$, a corresponding output signal y will be available $y = P+P^*$. Since protein concentrations are actually state variables, a linear output map $y = Cx$ will be possible. C is a matrix of suitable dimension. Linear output maps are desirable, as they simplify computation.

It is emphasized that the issue of time behavior measurement for biological dynamic processes is also a key research topic, and several new approaches are being proposed and examined (Chin, 2008; Longo & Hasty, 2006).

While dealing with identification procedures, an extended state space vector $z = (x^T, \pi^T)^T$ will be used. In this case, the dynamics in (2) and (3) can be rewritten as in the more compact form:

$$\dot{z} = f_z(z),\ z(0) = z_0 \tag{4}$$

$$Y_z(z) = h_z(z), \tag{5}$$

assuming the dynamics $\dot{\pi} = 0$ for unknown parameters.

In the following sections, either the form (2) and (3) or the more compact ones in (4) and (5) will be used.

SENSITIVITY AND IDENTIFIABILITY ANALYSIS

Sensitivity analysis can help in evaluating the relevance of parameters on system behaviors, and can be used to select the set of parameters to be identified in a given ODE model.

In addition, in this chapter, identifiability analysis of crucial parameters has been carried out as well. Such an analysis allows a more accurate design of identification schemes and yields a deeper understanding of biological system properties.

Identifiability analysis has been carried out based on the approach developed in (Evans et al., 2003) and in (Evans et al., 2004). Both sensitivity and identifiability analyses have been applied to the dynamic model of the P53 network and to single hybrid modes of the NER submodel.

In this chapter, attention will be given to the first order steady-state normalized sensitivities with respect to small changes in single parameters (Varma et al., 2005). For the i-th state x_i, with respect to a variation $\Delta\pi_j$ to the j-th parameter π_j, the sensitivity is defined as:

$$S_{i,j} = \frac{\left| x_{SS_i}\left(\pi_j + \Delta\pi_j\right) - x_{SS_i}\left(\pi_j\right)\right|}{\Delta\pi_j} \tag{6}$$

where $x_{SS_i}\left(\bar{\pi}_j\right)$ is the steady-state value for the i-th state variable, evaluated at the given value $\bar{\pi}_j$ of the j-th parameter. The normalized form is:

$$S_{n_{i,j}} = \frac{\pi_j}{x_{SS_i}} S_{i,j}. \tag{7}$$

To introduce the notion of identifiability, consider the model in (2) and (3). Given two points π and $\bar{\pi}$ in the parameter set Ω, they are said to be *indistinguishable,* if they give rise to identical system output response (for the same identical conditions), i.e.:

$$y\left(t,\pi\right) = y\left(t,\bar{\pi}\right) \quad \forall t \geq 0.$$

Basically, if two points of the parameter space are indistinguishable, then it follows that the knowledge of the output function over a given time-interval does not allow us to discriminate between them, hence it does not allow us to select the true value. Therefore, the presence of indistinguishable pairs implies that the non identifiability of the whole parameter vector based on the given output function, i.e., based on the given measurement vector.

Indistinguishability is related to the observability of the system (2, 3) which in turn can be checked (Evans et al., 2005; Isidori, 1995) by studying the rank of the Jacobian matrix:

$$J = \left[\frac{\partial h_1}{\partial x}^T \quad \cdots \quad \frac{\partial h_r}{\partial x}^T \quad \frac{\partial \varphi_1}{\partial x}^T \quad \cdots \quad \frac{\partial \varphi_{n-r}}{\partial x}^T \right]^T \tag{8}$$

Where h_i denotes the i-th component of the output map and $\varphi_1,\ldots,\varphi_{n-r}$ are n-r functions suitably chosen among the components of the following Lie derivatives of the output:

$$L_f^i h\left(x,\pi\right) = \frac{\partial L_f^{i-1} h(x,\pi)}{\partial x} f(x,\pi); \quad i = 1,2,\ldots \tag{9}$$

$$L_f^0 h\left(x,\pi\right) = h(x,\pi). \tag{10}$$

Notice that a non-observability situation could be dealt with using additional measurements, provided this is possible from a biological and/or technological point of view. In addition, the observability test can be used to select a proper set of measured signals, which can be actually measured from a biological standpoint.

The identifiability approach used here (Evans et al., 2005) is based on the computation of a function $\eta: R^n \rightarrow R^n$, defined in a suitable neighborhood of the initial condition x_0, for some arbitrary parameter vector $\bar{\pi}$, satisfying:

$$h_i(\eta(x),\pi) = h_i(x,\bar{\pi}), \quad i = 1,\ldots,r \tag{11}$$

$$\varphi_i(\eta(x),\pi) = \varphi_i(x,\bar{\pi}), \quad i = 1,\ldots,n-r. \tag{12}$$

Then, the set of indistinguishable parameters can be defined (Evans, 2005) as the set $L(\pi)$ of all the points $\bar{\pi} \in \Omega$ such that, for some $\tau>0$:

$$\eta(x_0) = x_0, \tag{13}$$

$$f(\eta(x(\bar{\pi})),\pi) = \frac{\partial \eta}{\partial x}\left(x\left(t,\bar{\pi}\right)\right) f\left(x\left(t,\bar{\pi}\right),\bar{\pi}\right), \tag{14}$$

for all $t \in [0,\tau)$. If the set $L(\pi)$ is empty, then the output function allows us to identify any point in the parameter space.

PARAMETER ESTIMATION

In this section, two different parameter identification algorithms will be presented and discussed. The first approach is based on an modified Extended Kalman Filter (EKF), the second one is based on the joint use of asymptotic observers for output time derivative estimation and nonlinear least square method for map inversion.

State and Parameter Estimation using Extended Kalman Filtering

In this section, an Extended Kalman Filter (EKF), is designed for the compact form of the models given by (4) and (5), with the goal of estimating the state x and the greatest possible number of parameters.

The design of an Extended Kalman Filter for a continuous time nonlinear biological model is a challenging problem. On one hand, the strong nonlinearity appearing in the models in Boxes 1 and 2 make it difficult to derive (approximate) discrete time dynamics. On the other hand, a common issue in biology is that measurements are only available at discrete time instants. Such a problem is handled here by designing a Mixed Extended Kalman Filter (MEKF) which estimates continuous-time models with discrete-time measurements.

Mathematically, one has a continuous-time process which is measured at discrete time instants. In terms of the compact notation in (4) and (5), one has:

$$\begin{cases} \dot{z} = f_z(z) \\ y_k = h_z(z(t_k)) \end{cases}$$

where t_k denotes the generic discrete time at which the k-th measurement is taken.

The classical EKF for discrete-time models is made up of two parts, the *predictor* and the *estimator*. Let $\hat{z}_{k|k-1}$ denotes the estimate of state z at time k conditioned on measurements upon time $k-1$, and similarly for $\hat{z}_{k-1|k-1}$. The predictor uses the system's model to predict the current state estimate $\hat{z}_{k|k-1}$ from the previous estimate $\hat{z}_{k-1|k-1}$ and uses the discrete Riccati equation to predict the current error covariance $P_{k|k-1}$ from the previous error covariance $P_{k-1|k-1}$ Then, the estimator computes the gain L_k using the current measurement and adjusts the predicted values with the innovation term, yielding the actual state estimate $\hat{z}_{k|k}$ and the actual error covariance $P_{k|k}$ at the current time step k.

The proposed MEKF is still based on the prediction phase and estimation phase. The estimator part uses the measurement available at the sampling instant k (which occurs at time t_k), just like in the discrete-time EKF:

$$\hat{z}_{k|k} = \hat{z}_{k|k-1} - L_k(h(\hat{z}_{k|k-1}) - y_k)$$

$$P_{k|k} = P_{k|k-1} - L_k H_k P_{k|k-1}$$

$$L_k = P_{k|k-1} H_k^T W_k$$

$$W_k = (H_k P_{k|k-1} H_k^T + R_k)^{-1} \quad H_k = \left.\frac{\partial h}{\partial z}\right|_{z=\hat{z}_{k|k-1}} .$$

The predictor part is different, since it integrates the continuous-time process in the time interval $[t_{k-1},t_k]$ between two sampling instants using the previous state estimate as initial condition, and takes as the predicted state estimate the value of the solution $z_{t|t_{k-1}}$ at time t_k. To obtain a prediction for the current error covariance, the predictor integrates a differential Riccati equation using the previous error covariance as initial condition, and takes the value of the solution $P_{t|t_{k-1}}$ at time t_k as the predicted covariance.

$$t \in \left[t_{k-1}, t_k\right] \quad \dot{\hat{z}}_{t|t_{k-1}} = f(\hat{z}_{t|t_{k-1}})$$

$$\hat{z}_{t_{k-1}|t_{k-1}} = \hat{z}_{k-1|k-1}$$

$$t \in \left[t_{k-1}, t_k\right] \quad \dot{P}_{t|t_{k-1}} = FP_{t|t_{k-1}} + P_{t|t_{k-1}} F^T + Q$$

$$P_{t|t_{k-1}} = P_{k-1|k-1}$$

$$\hat{z}_{k|k-1} = \hat{z}_{t_k|t_{k-1}}$$

$$P_{k|k-1} = P_{t_k|t_{k-1}}$$

$$F = \left.\frac{\partial f}{\partial x}\right|_{x=\hat{x}_{k-1|k-1}}$$

This approach can easily accommodate non-uniform sampling times.

Parameter Estimation Using Derivative Estimator and Non Linear Least Squares

The second identification algorithm (shortly referred to as DENLS) relies on state observers for derivative estimation, and it basically comprises the inversion of a suitable nonlinear map between the Lie derivatives of the output and the parameter vector. Consider systems (2) and (3). By taking Lie derivatives of the output function up to a given order, one can construct a set of functions $\zeta_i(x,\pi)$, depending on parameter vector π:

$$y(t) = h =: \zeta_1(x,\pi), \tag{15}$$

$$\dot{y}(t) = L_f h =: \zeta_2(x,\pi), \tag{16}$$

$$y^{(m-1)}(t) = L_f^{m-1} h =: \zeta_m(x,\pi). \tag{17}$$

If the output function and its time derivatives are known, one could compute the vector π (and the state x) by inversion of the following nonlinear map, provided it is invertible:

$$\begin{pmatrix} y(t) \\ \dot{y}(t) \\ \vdots \\ y^{(m-1)}(t) \end{pmatrix} = \begin{pmatrix} \zeta_1(x,\pi) \\ \zeta_2(x,\pi) \\ \vdots \\ \zeta_m(x,\pi) \end{pmatrix} =: \zeta(x,\pi). \tag{18}$$

The inversion theorem suggests to select derivation order (m-1) large enough to ensure, if possible, the Jacobian matrix of the map ζ to have rank equal to $n+n_p$, i.e., equal to the dimension of the extended state $z = [x^T,\pi^T]^T$.

In order to compute time derivatives of measured output function y, we propose the following practical observer, used in (Nicosia et al., 1991) to solve a different kind of nonlinear inversion problem:

$$\begin{aligned} \dot{\eta}_1 &= \frac{k_1}{\varepsilon}(y-\eta_1)+\eta_2 \\ \dot{\eta}_2 &= \frac{k_2}{\varepsilon^2}(y-\eta_1)+\eta_3 \\ &\vdots \\ \dot{\eta}_m &= \frac{k_m}{\varepsilon^m}(y-\eta_1) \end{aligned} \tag{19}$$

where ε is a small positive gain, and design constants $k_1, k_2,\ldots, k_m$ are such that polynomial $\lambda^m + k_m\lambda^{m-1} + k_2\lambda + k_1$ is Hurwitz. Selection of design parameters is, as usual, a tradeoff between rate of convergence and measurement noise rejection. In particular, ε is a scale factor for the eigenvalues of the estimation error dynamics yielded by observer (19).

Then, based on output derivative estimates, parameter identification can be achieved by inversion of the nonlinear map (18). Here we make use of nonlinear least square approach (as implemented in the *lsqnonlin MATLAB*© command), to cope with measurement noise and estimation error.

RESULTS

To evaluate the effectiveness of the identification schemes proposed in the previous Section, *in silico* experiments will be used (Edelman et al., 2009), which allow us to test algorithms and other mathematical tools in a "protected" environment before running them on real systems. *In silico* models, i.e., mathematical models solved through simulation software packages, allow us to generate realistic signals that can be used, e.g., to drive identification algorithms "as if" they were actually applied to a real biological system. At the same time, *in silico* models made further information available, information which cannot be measured in real systems, such as time behavior of all the state variables as well as true parameter values. Such quantities can be used to accurately assess the performance of, e.g., identification approaches, and to accurately plan and design real biological experiments. It is evident that *in silico* models are approximate and will never cover the complexity of a real biological system. Nevertheless, as is common in all the system theory applications, the use of mathematical models allow us to reduce time and cost in developing algorithms, and also allow us to increase robustness of the design.

Three indexes are used to assess the behavior of the proposed identification algorithms by means of *in silico* experiments. The first one is called *State Estimation Error* (SEE):

$$SEE = \frac{\left\|\hat{x} - x\right\|}{\left\|x\right\|}100, \tag{20}$$

where $\|\cdot\|$ denotes Euclidean norm, $\hat{x}$ and x are the estimated state and the actual state, respectively, when the system is at a steady state. The second index, associated to each single parameter π_i is called *parameter estimation error* (PEE):

$$PEE_{\pi_i} = \frac{|\hat{\pi}_i - \pi_i|}{\pi_i}100, \tag{21}$$

where $\hat{\pi}_i$ is the estimated parameter value and π_i is the true one. Finally, a *mean parameter estimation error* (MPEE) is used:

$$MPEE = \frac{PEE_{\pi_1} + \cdots + PEE_{\pi_{n_p}}}{n_p} \tag{22}$$

Where n_p is the number of estimated parameters.

The proposed methodology has been tested by means of output measurement signals, generated via *in silico* experiments both on the simplified p53 sub-model, and on the first three modes of the NER hybrid model reported in Box 2. The output signals derived from these *in silico* experiments have been corrupted by a Gaussian random process, so as to describe additive measurement errors.

The simplified version of the p53 sub-model only cover the "strict" P53 dynamics, hence the state variables are $x = (m_{p53}, P53, P53^*)^T$. The model does not use any TF terms, and uses a PPI term of Michaelis-Menten type in describing the phosphorylation of P_{53}, i.e.:

$$\begin{cases} \dot{x}_1 = \lambda_{p53} - \mu_{p53}x_1 \\ \dot{x}_2 = \alpha_{p53}x_1 - \gamma_{p53}x_2 + \nu_{p53}x_3 - k_1\dfrac{x_2}{1+x_2} \\ \dot{x}_3 = k_1\dfrac{x_2}{1+x_2} - \nu_{p53}x_3 \end{cases} \tag{23}$$

Identifiability Analysis

The first step in the proposed general approach is the identifiability analysis. As for the simplified p53 sub-model in (23), assume it is possible to measure p53 mRNA and proteins in both active and non active forms, then there is $y=Cx$ with C as a diagonal matrix. Therefore, observability is trivially verified. With the choice $\eta(x)=x$, it follows $L(\pi)=\varnothing$ (all the parameters can be identified).

In typical biological experiments, however, only the total protein concentrations (i.e. the sum of activated and non activated protein) can be measured, and in this case y=[011]x. In such a situation, the only identifiable parameter is λ_{p53} (remarkably, according to the sensitivity analysis below, this turns out to be the most significant parameter for the considered model).

As for mode q_1 of the NER hybrid model, similar computation yields:

$$x^{q_1} = \left[P53^*, XPC^*\right]^T, \quad C^{q_1} = \begin{bmatrix} 1 & 0 \\ 0 & 1 \end{bmatrix}, \pi^{q_1} = \left[k_{A2}, k_{D2}\right]^T, \tag{24}$$

where x^{q_1}, C^{q_1} and π^{q_1} denote the state vector, the output matrix, and the parameter vector, respectively. It is assumed that the total concentrations of *XPC* and *P*53 active proteins can be measured. As for the set of indistinguishable parameters, it is easy to find:

$$\eta^{q_1} = \left[P53^*, XPC^*\right]^T, \; L^{q_1} = \varnothing, \tag{25}$$

hence identifiability is assured.

As for the identifiability of the second mode of the hybrid NER model, to keep the analysis to a tractable level of complexity, the following two approximations on Goldbeter-Koshland functions are introduced: $GB_{XPB^*}(\cdot) = XPB_T$ and $GB_{XPD^*}(\cdot) = XPD_T$, where XPB_T and XPD_T are total protein concentrations. Such an approximation is not too restrictive, since protein mean quantity in the cell during the NER process can be measured. As for kinetic constants, it is assumed that $k_{5T} = (k_5+k_7)$ and $k_3 \approx k_4$, which is reasonable taking into account that the two helicases XPB and XPD have a similar activation reaction. Hence, in the second mode, the state vector, output matrix, parameter vector, and indistinguishable set, respectively, are given by:

$$x^{q_2} = \left[ssDNA, XPA^*\right]^T, \quad C^{q_2} = \begin{bmatrix} 1 & 0 \\ 0 & 1 \end{bmatrix}, \pi^{q_2} = \left[k_3, k_{5T}, k_6, k_8, k_9, k_{10}\right]^T \tag{26}$$

$$\eta^{q_2} = \left[ssDNA, XPA^*\right]^T, L^{q_2} = \varnothing, \tag{27}$$

hence identifiability is assured also in this case.

Finally, concerning the third mode, with similar notation:

$$x^{q_3} = \left[XPGssDNA, ndsDNA\right]^T, C^{q_3} = \begin{bmatrix} 1 & 0 \\ 0 & 1 \end{bmatrix}, \pi^{q_3} = \left[k_{12}, k_{13}, k_{15}, k_{16}\right]^T \tag{28}$$

$$\eta^{q_3} = \left[XPGssDNA, ndsDNA\right]^T, L^{q_3} = \varnothing, \tag{29}$$

hence the four parameters can be identified.

Identifiability analysis for the NER model measuring only one state variable for each mode indicates that in such a situation the model is not identifiable.

Sensitivity Analysis

Sensitivity analysis carried out on the simplified p53 model in (23), with parameter perturbations of 5%, 20% and 50%, yields that the most significant parameters are λ_{p53} α_{p53} and k_1 (see Figure 4).

Sensitivity analysis over the first three modes of the NER hybrid model, with the same relative perturbation as above has also been carried out. For the first mode, it turns out that the most "important" parameters are the kinetic constants k_{A2} and k_{D2}, as reported in Figure 4.

For the second mode, the parameters have been collected into two separate sets. The first one comprises the parameters k_{A3}, k_{MA3}, k_{D3}, k_{MD3}, k_{A4}, k_{MA4}, k_{D4} and k_{MD4}, that is, those not having a clear influence on the *ssDNA* and *XPA* state variables. The second set comprises parameters k_3, k_{5i}, k_6,

k_8, k_9 and k_{10}, having considerable influence on the above state variables (see Figure 4). Kinetic constants k_9 and k_{10} affect both continuous variables *XPA* and *ssDNA*.

Sensitivity analysis for the third mode q_3 allowed us to identify four important parameters, k_{12}, k_{13}, k_{15}, and k_{16} (see Figure 4).

Parameter Estimation

The two approaches for parameter identification have been studied under several operating conditions. The dynamic models proposed above have been used to generate time behaviour for total protein concentration in the p53 model (23) and for the (P53, XPC, XPA, ssDNA and XPG) protein in the NER modes. Also, the *in silico* models have been used in such a way to simulate the use of several sampling times, and several measurement noise power (i.e., noise variance). For each sampling time and for each noise variance, SEE and PEE indicators have been recorded.

The results achieved by means of the MEKF scheme for the case of the p53 model are reported in Figure 5. In particular, for the state variables m_{p53}, *P*53 and *P*53*, Figure 5 (A) illustrates SEE versus sampling time for experiments with noise variance equal to 10% of true output signal, and Figure 5(B) reports SEE versus relative noise variance with a sampling time equal to 10 min-

Figure 4. Sensitivity analysis for the case of kinetic constants with 20% relative perturbation on all model parameters. (A) p53 simplified model (23). (B) NER mode q_1; parameters (kA2,KMA2,kD2,kMD2); states: P53 and XPC. (C) NER mode q_2; (most significant) parameters: (k3,k5i,k6,k8,k9,k10); states: ssDNA and XPC. (D) NER mode q_3; parameters: (k12,k13,k14,k15,k16); states: XPGssDNA and ndsDNA.*

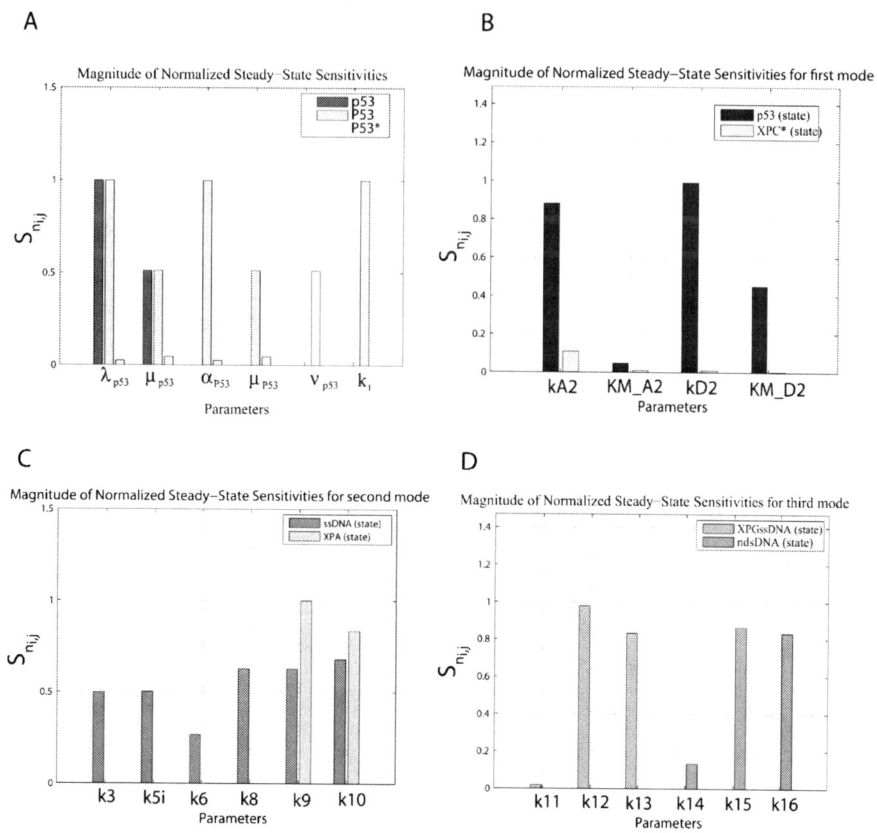

utes. The same type of results for parameter α_{p53} are reported in Figure 5(C) and Figure 5(D), for the PEE index.

The results achieved by the DENLS scheme, based on derivative estimation and nonlinear least square, are illustrated in Figure 6, still for the simplified p53 model. Based on sensitivity analysis results, the most relevant parameters are λ_{p53} α_{p53} and k_1. For these parameters, Figure 6(A) reports *PEE* and MPEE indexes versus sampling time for data experiments with an additive noise having a variance equal to 10% of true measurements. Similarly, Figure 6(B) illustrates indexes *PEE* and MPEE versus percentage noise variance with sampling time equal to 10 minutes. The results for the same indexes and the same noise variance and sampling time, for the case in which parameters have to be identified are shown in Figure 6(C) and Figure 6(D), for the case of *PEE* and MPEE indexes, respectively.

As for the NER hybrid model, *PEE* and *MPEE* indexes have been studied versus sampling time and percentage noise variance for all three modes, according to the results of the sensitivity analysis. The results for mode q_1 are reported in Figure 7(A) and Figure 7(B): the estimated parameters are $p^{q_1} = \left[k_{A2}, k_{D2}\right]$ (Notice that for the NER model, the time scale is in seconds).

As far as the mode q_2 of hybrid NER model is concerned, two cases have been considered: in

Figure 5. MEKF performance in terms of state and parameter estimation for the p53 simplified model. (A) SEE vs. sampling time and data experiments with 10% noise variance. (B) SEE vs. relative noise variance and sampling time at 10 minutes. (C) PEE vs sampling time and data experiments with 10% noise variance. (D) PEE vs. noise variance and sampling time at 10 minutes.

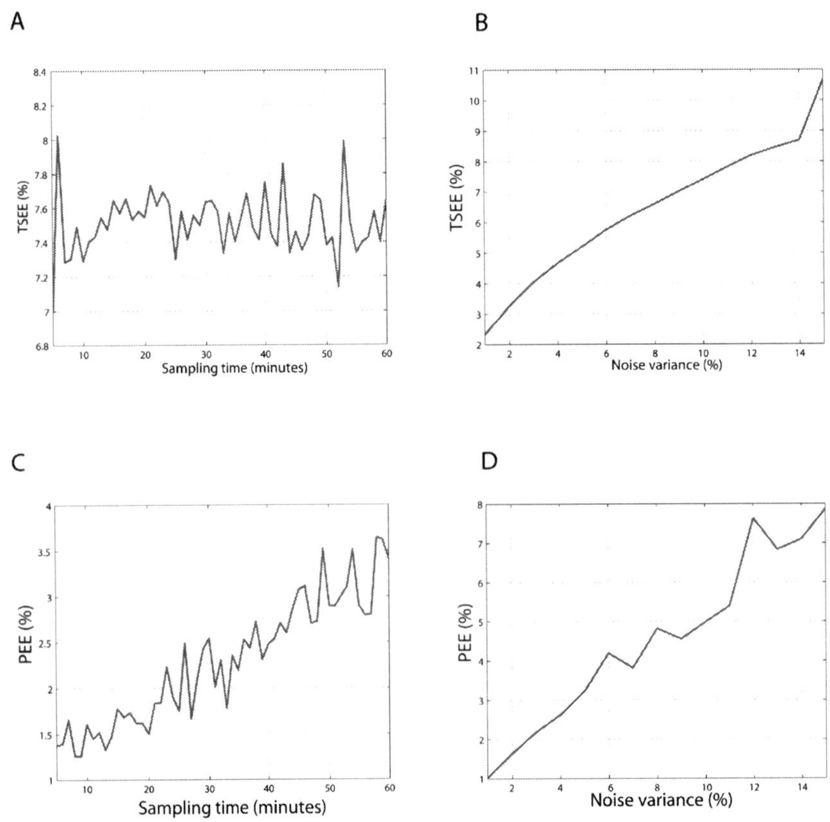

Figure 6. DENLS performance for parameter estimation for p53 simplified model. (A) PEE for three parameters (λ_{p53}, α_{p53} and k_1) and MPEE vs. sampling time and data experiments with 10% relative noise variance. (B) PEE_{π_i} for three parameters (λ_{p53}, α_{p53} and k_1) and MPEE vs. perceptual relative noise variance and sampling time at 10 minutes. (C) PEE for six parameters (λ_{p53}, μ_{p53}, α_{p53}, μ_{P53}, v_{P53} and k_1) and MPEE vs sampling time and data experiments with 5% relative noise variance. (D) PEE for six parameters (λ_{p53}, μ_{p53}, α_{p53}, μ_{P53}, v_{P53} and k_1) and MPEE vs. perceptual relative noise variance and sampling time at 10 minutes.

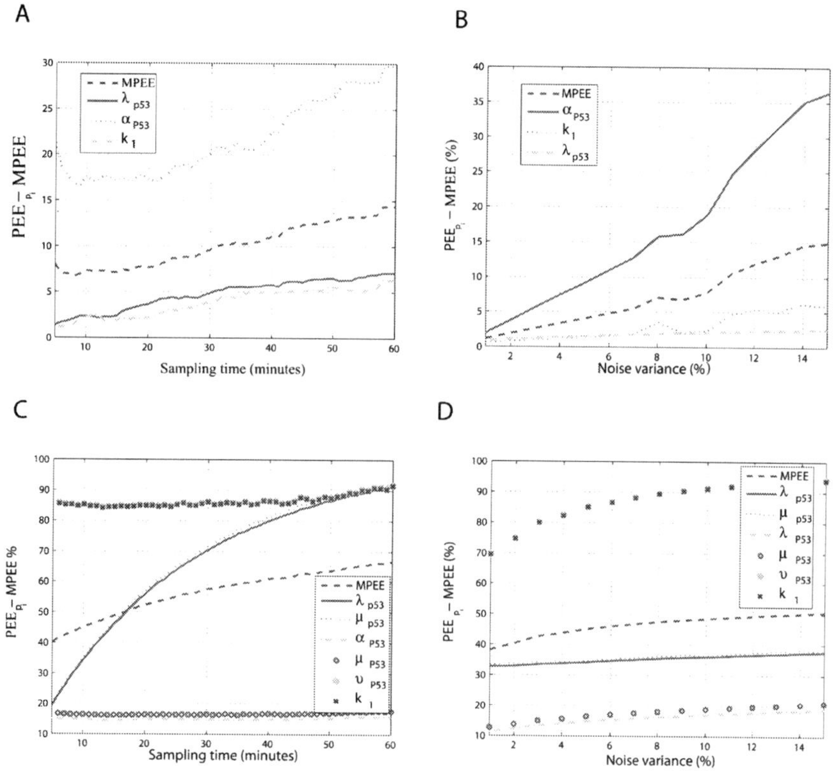

case a), estimation is restricted to the set $p^{q_2} = [k_9, k_{10}]$; in case b), estimation covers the whole set of six parameters in (27). In the first case, results are similar to those obtained in the first mode. In the second case, the achieved performance is similar to the case of six parameter for the p53 simplified model.

The third mode requires estimation of four parameters: $p^{q_3} = \left[k_{12}, k_{13}, k_{15}, k_{16}\right]$. The results, in terms of *PEE* and MPEE indexes, are depicted in Figure 7(C) versus sampling time (noise variance equal 5%) and in Figure 7(D) *PEE* and MPEE versus noise variance (sampling time equal to 15 seconds).

Discussion

In the previous two subsections, we presented two different parameter estimation algorithms, MEKF, based on mixed-mode extended Kalman filter, and DENLS, based on derivative estimation and nonlinear least-squares. The MEKF scheme has a major advantage: it estimates three states variables (*p*53, *P*53 and *P*53*) and one parameter (α_{p53} which has an important role from sensitivity

Figure 7. DENLS performance for parameter estimation for NER hybrid model. (A) PEE for two parameters (k_{A2} and k_{D2}) and MPEE vs. sampling time and data experiments with 5% relative noise variance. (B) PEE for two parameters (k_{A2} and k_{D2}) and MPEE vs. perceptual relative noise variance and sampling time at 20 seconds. (C) PEE for four parameters (k_{12}, k_{13}, k_{15} and k_{16}) and MPEE vs. sampling time and data experiments with 5% relative noise variance. (D) PEE for four parameters (k_{12}, k_{13}, k_{15} and k_{16}) and MPEE vs. perceptual relative noise variance and sampling time at 15 seconds.

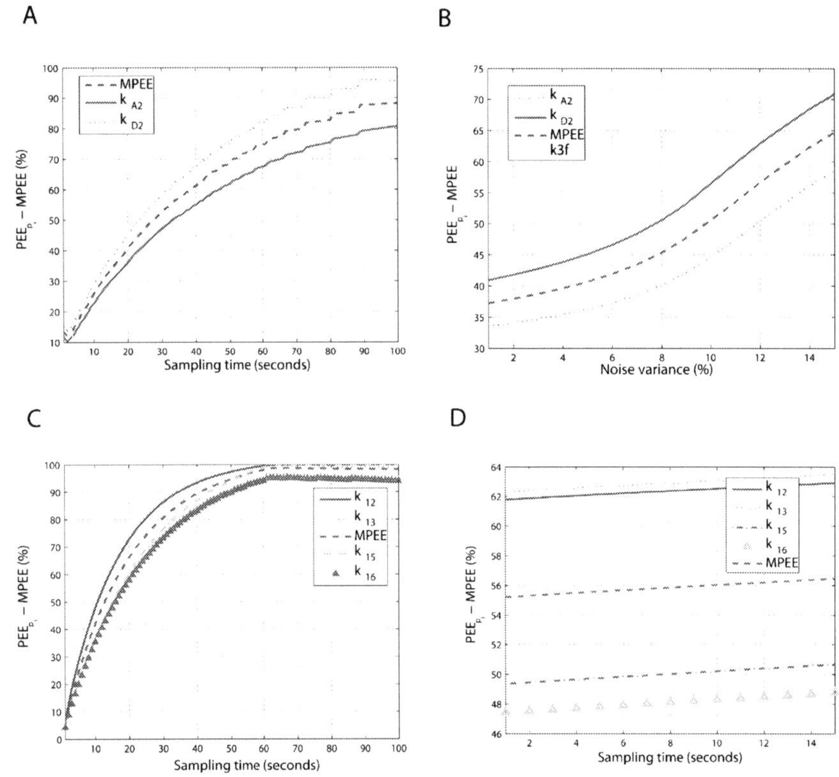

analysis) measuring total protein concentration. This is very important in systems biology since this implies that one can estimate mRNA level only measuring total concentration of P53 protein. SEE versus sampling time appears approximately constant around 7.5%, while SEE versus noise variance increases up to 11% (see Figure 5 (A)) and Figure 5(B)). Figure 5(C) and Figure 6(A) (continuous line) allows us to compare PEE versus sampling time for the two estimation schemes. MEKF seems to achieve a better estimation of α_{p53} than the DENLS scheme. A similar statement holds for PEE versus perceptual noise variance (Figure 5(A) and Figure 6(B), continuous line), where MEKF seems to be less sensible to noise measurement. Comparing the PEE indexes achieved by MEKF (see Figure 5(A) and Figure 5(D)) with the MPEE (Figure 6(A) (dotted line) and Figure 6(B) (continuous line)), we note that their values are quite similar.

Nevertheless, due to a loss of observability in the system, the MEKF can only estimate one parameter beyond the state, and the choice was on α_{p53} because it gives the relationship between mRNA and protein. This is very important in biological experiments, because it gives us an idea about how many times an mRNA molecule can be translated and, consequently, how much protein can be made from it. Sensitivity analysis further underlines the importance of α_{p53}.

The parameters estimation using the DENLS scheme, although needs to measure all the state variables, has the important advantage that it yields estimates for a larger number of parameters. The Figure 6(C) and Figure 6(D) show that MPEE (dotted line) for all p53 simplified model parameters is larger than 70% for a sampling time equal to 60 minutes. In real experiments, an acceptable sampling time is 20 minutes and in this case the MPEE index is 50%.

As anticipated above and shown by the above discussion, the behavior of PEE and MPEE indexes (see Figures in appendix) can have important application in planning of biological experiments. *In silico* experiments can be run, to evaluate the PEE index versus sampling time thus allowing selection of crucial experiment parameters and reducing cost in planning and design of actual biological experiments.

For the specific models discussed in this chapter, *in silico* experiments will be extremely useful because novel inhibitors of both p53 and NER are being discovered and tested. Recently, Barakat et al. (2009a) identified a set of compounds that have been computationally predicted to ultimately activate the p53 pathway in tumor cells retaining the wild-type protein. Also, they studied the inhibitory dynamic pharmacophore for the ERCC1-XPA interaction using a combined molecular dynamics and virtual screening approach (Barakat et al. 2009b). Therefore, a combination of *in silico* and *in vitro* studies should help to predict new inhibitors and test model predictions.

FUTURE RESEARCH DIRECTIONS

The chapter gives an example of a typical approach for a systems biology application in DNA related modeling.

Future research activities will cover extension of the damage model to consider additional features, and additional relevant hybrid modes. The identifiability issue and the identification problem both require additional work, and feasibility analysis for real biological experiments will be investigated. In addition, model validation/invalidation is also a key issue.

Identifiability analysis based on nonlinear observability criterion for a suitably extended dynamical systems is currently under investigation.

CONCLUSION

In this chapter, we discussed mathematical dynamic modeling of genetic networks, and the related issue of parameters identification. In particular, the cases of dynamic modeling of a simplified p53 network and of the NER repair mechanism are considered. The parameter identification problem for these model is addressed to gain a deeper understanding on the roles of model parameters. In this perspective, sensitivity analysis and identifiability analysis were also carried out.

As for the specific algorithms proposed in this chapter, the *in silico* analysis allows to draw some conclusion.

The algorithm based on Extended Kalman filter, MEKF, allows for a better estimate of state variables, and of some parameters. Sensitivity to variance in noise measurements is not very high. At the same time, if only total protein concentration can be measured, the MEKF algorithm allow us to estimate only one parameter. Here we choose the one showing a larger sensitivity, namely the parameter describing protein production in the p53 model.

The algorithm based on output derivative estimation and least square, DENLS, has worst performance as for state estimation, while giving better results from the point of view of parameter estimation. In particular, for sampling time in the order of 20 minutes, which appear reasonable in real biological applications, the mean parameter estimation error is lower than 50% over the whole set of parameters for the simplified p53 model.

The comparative analysis of the two identification algorithms yields similar conclusion for the case of the single mode within the hybrid dynamic model of NER.

ACKNOWLEDGMENT

The authors would like to thank the medical and biological staff of the Oncology Department, Ospedale S. Maria della Misericordia, Perugia, for fruitful discussions and suggestions.

The work has been partly supported by funds from Consorzio per lo Sviluppo del Polo di Terni-Progetto di Sviluppo 2007, through Polo Scientifico Didattico di Terni.

REFERENCES

Adimoolam, S., & Ford, J. M. (2003). p53 and regulation of DNA damage recognition during nucleotide excision repair. *DNA Repair*, *2*(9), 947–954. doi:10.1016/S1568-7864(03)00087-9

Aldridge, B., Burke, J., Lauffenburger, D., & Sorger, P. (2006). Physicochemical modelling of cell signalling pathways. *Nature Cell Biology*, *8*(11), 1195–1203. doi:10.1038/ncb1497

Audoly, S., Bellu, G., D'Angiò, L., Saccomani, M., & Cobelli, C. (2001). Global identifiability of nonlinear models of biological systems. *IEEE Transactions on Bio-Medical Engineering*, *48*(1), 55–65. doi:10.1109/10.900248

Bakkenist, C. J., & Kastan, M. B. (2003). DNA damage activates ATM through intermolecular autophosphorylation and dimer dissociation. *Nature*, *421*(6922), 499–506. doi:10.1038/nature01368

Barakat, K., Huzil, J. T., Dumontet, C., Jordheim, L., & Tuszynski, J. A. (2009). Characterization of an inhibitory pharmacophore for the ERCC1-XPA interaction using a combined molecular dynamics and virtual screening approach. *Journal of Molecular Graphics & Modelling*, *28*(2), 113–130. doi:10.1016/j.jmgm.2009.04.009

Barakat, K., Mane, J., Friesen, D., & Tuszynski, J. A. (2009). Ensemble-based virtual screening reveals dual-inhibitors for the p53-MDM2/MDMX interactions. *Journal of Molecular Graphics & Modelling*, *28*(6), 555–568. doi:10.1016/j.jmgm.2009.12.003

Belta, C., Finin, P., Habets, L. C., Halasz, G. J. M., Imielinski, A. M. M., Kumar, R. V., et al. (2004). *Dynamic partitioning of large discrete event biological systems for hybrid simulation and analysis*. Paper presented at the 7th International Workshop Hybrid Systems Computation and Control, *2993*, 111-125.

Bianconi, F. (2006). *A hybrid model of nucleotide excision repair in neoplastic diseases and in vitro experiments*. Master Degree Thesis, Department of Electronic and Information Engineering, University of Perugia.

Bianconi, F. (2010). *Dynamic modeling, parameter estimation and experiment design in systems biology with applications to oncology*. PhD thesis, Department of Electronic and Information Engineering, University of Perugia.

Bianconi, F., Valigi, P., & Crinò, L. Ludovini, V., Piattoni, S., Orleth, A., et al. (2006). *A hybrid model of nucleotide excision repair in neoplastic diseases and in vitro experiments*. Tech. Rep. RT-003-06, Department of Electronic and Information Engineering, University of Perugia.

Bolderson, E., Richard, D. J., Zhou, B.-B. S., & Khanna, K. K. (2009). Recent advances in cancer therapy targeting proteins involved in DNA double-strand break repair. *Clinical Cancer Research*, *15*(20), 6314–6320. doi:10.1158/1078-0432.CCR-09-0096

Brazma, A., Krestyaninova, M., & Sarkans, U. (2006). Standards for systems biology. *Nature Reviews. Genetics*, *7*, 593–605. doi:10.1038/nrg1922

Bruce, A., Johnson, A., Lewis, J., Raff, M., Roberts, K., & Walter, P. (2008). *Molecular biology of the cell* (5th ed.). Garland Science.

Chin, C. S., Chubukov, V., Jolly, E. R., DeRisi, J., & Li, H. (2008). Dynamics and design principles of a basic regulatory architecture controlling metabolic pathways. *PLoS Biology*, *6*(6), e146. doi:10.1371/journal.pbio.0060146

Cho, K., Shin, S., Kolch, W., & Wolkenhauer, O. (2003). Experimental design in systems biology, based on parameter sensitivity analysis using a Monte Carlo method: A case study for the TNF(alpha)-mediated NF-(kappa) b signal transduction pathway. *Simulation*, *79*(12), 726–739. doi:10.1177/0037549703040943

Cleaver, J. E., Lam, E. T., & Revet, I. (2009). Disorders of nucleotide excision repair: The genetic and molecular basis of heterogeneity. *Nature Reviews. Genetics*, *10*(11), 756–768. doi:10.1038/nrg2663

Cornish-Bowden, A. (1995). *Fundamentals of enzyme kinetics*. Portland Press.

Costa, R. M. A., Chigancas, V., da Silva Galhardo, R., Carvalho, H., & Menck, C. F. M. (2003). The eukaryotic nucleotide excision repair pathway. *Biochimie*, *85*(11), 1083–1099. doi:10.1016/j.biochi.2003.10.017

De Jong, H., Gouz, J. L., Hernandez, C., Page, M., Sari, T., & Geiselmann, J. (2003). *Hybrid modeling and simulation of genetic regulatory networks: A qualitative approach*. Paper presented at the 6th International Workshop Hybrid Systems Computation and Control, 2623, 267-282.

Dunlop, M., & Murray, R. (2006). *Towards biological system identification: Fast and accurate estimates of parameters in genetic regulatory networks*. Paper presented at the 45th IEEE Conference on Decision and Control.

Edelman, L.B., Eddy, J.A. & Price N.D. (2009). In silico models of cancer. *Wiley Interdisciplinary Reviews: Systems Biology and Medicine*, 1939-5094.

Evans, N. D., Chapman, M. J., Chappell, M. J., & Godfrey, K. R. (2002). Identifiability of uncontrolled nonlinear rational systems. *Automatica*, *38*(10), 1799–1805. doi:10.1016/S0005-1098(02)00094-8

Evans, N. D., Errington, R. J., Shelley, M., Feeney, G. P., Chapman, M. J., & Godfrey, K. R. (2004). A mathematical model for the in vitro kinetics of the anti-cancer agent topotecan. *Mathematical Biosciences*, *189*(2), 185–217. doi:10.1016/j.mbs.2004.01.007

Farina, M., Findeisen, R., Bullinger, E., Bittanti, S., Allgower, F., & Wellstead, P. (2006). *Results towards identifiability properties of biochemical reaction networks*. Paper presented at the 45th IEEE Conference on Decision and Control, 2104–2109.

Fayad, W., Brnjic, S., Berglind, D., Blixt, S., Shoshan, M. C., & Berndtsson, M. (2009). Restriction of cisplatin induction of acute apoptosis to a subpopulation of cells in a three-dimensional carcinoma culture model. *International Journal of Cancer*, *125*(10), 2450–2455. doi:10.1002/ijc.24627

Fey, D., Findeisen, R., & Bullinger, E. (2008). *Parameter estimation in kinetic reaction models using nonlinear observers is facilitated by model extensions*. Paper presented at the 17th IFAC World Congress.

Friedberg, E. C. (2003). DNA damage and repair. *Nature*, *421*(6921), 436–440. doi:10.1038/nature01408

Furuta, T., Ueda, T., Aune, G., Sarasin, A., Kraemer, A., & Pommier, Y. (2002). Transcription-coupled nucleotide excision repair as a determinant of cisplatin sensitivity of human cells. *Cancer Research*, *62*(17), 4899–4902.

Gadkar, K., Varner, J., & Doyle, F. III. (2005). Model identification of signal transduction networks from data using a state regulator problem. *Systems Biology*, *2*(1), 17–30. doi:10.1049/sb:20045029

Geva-Zatorsky, N., Rosenfeld, N., Itzkovitz, S., Milo, R., Sigal, A., & Dekel, E. (2006). Oscillations and variability in the p53 network. *Molecular Systems Biology*, *2*, 33. doi:10.1038/msb4100068

Ghosh, R., & Tomlin, C. (2004). Symbolic reachable set computation of piecewise affine hybrid automata and its application to biological modelling: Delta-notch protein signaling. Paper presented at *the IEE Proceedings. Systems Biology*, *1*(1), 170–183. doi:10.1049/sb:20045019

Gillespie, D. T. (1977). Exact stochastic simulation of coupled chemical reactions. *Journal of Physical Chemistry*, *81*(25), 2340–2361. doi:10.1021/j100540a008

Goldbeter, A., & Koshland, D. E. (1981). An amplified sensitivity arising from covalent modification in biological systems. *Proceedings of the National Academy of Sciences of the United States of America*, *78*(11), 6840–6844. doi:10.1073/pnas.78.11.6840

Hanawalt, P. C. (2002). Subpathways of nucleotide excision repair and their regulation. *Oncogene*, *21*(21), 8949–8956. doi:10.1038/sj.onc.1206096

Heinrich, R., & Schuster, S. (2003). *The regulation of cellular systems*. Springer.

Isidori, A. (1995). *Nonlinear control systems*. Springer.

Jones, S. N., Roe, A. E., Donehower, L. A., & Bradley, A. (1995). Rescue of embryonic lethality in MDM2-deficient mice by absence of p53. *Nature*, *378*(6553), 206–208. doi:10.1038/378206a0

Kamijo, T., Zindy, F., Roussel, M. F., Quelle, D. E., Downing, J. R., & Ashmun, R. A. (1997). Tumor suppression at the mouse INK4a locus mediated by the alternative reading frame product p19. *Cell*, *91*(5), 649–659. doi:10.1016/S0092-8674(00)80452-3

Kitano, H. (2001). *Foundations of systems biology*. MIT Press.

Lahav, G., Rosenfeld, N., Sigal, A., Geva-Zatorsky, N., Levine, A. J., & Elowitz, M. B. (2004). Dynamics of the p53-MDM2 feedback loop in individual cells. *Nature Genetics*, *36*(2), 147–150. doi:10.1038/ng1293

Letai, A. G. (2008). Diagnosing and exploiting cancer's addiction to blocks in apoptosis. *Nature Reviews. Cancer*, *8*(2), 121–132. doi:10.1038/nrc2297

Lev Bar-Or, R., Maya, R., Segel, L. A., Alon, U., Levine, A. J., & Oren, M. (2000). Generation of oscillations by the p53-MDM2 feedback loop: A theoretical and experimental study. *Proceedings of the National Academy of Sciences of the United States of America*, *97*(21), 11250–11255. doi:10.1073/pnas.210171597

Lillacci, G., Boccadoro, M., & Valigi, P. (2006). *In silico analysis of p53 response to DNA damage*. Paper presented at the 6th IFAC symposium on Modelling and Control in Biomedical Systems (including Biological Systems), 507-512.

Lillacci, G., Boccadoro, M., & Valigi, P. (2006). *The p53 network and its control via MDM2 inhibitors: Insights from a dynamic model*. Paper presented at the 45th IEEE Conference on Decision and Control, 2110-2115.

Lincoln, P., & Tiwari, A. (2004). *Symbolic systems biology: Hybrid modeling and analysis of biological networks*. Paper presented at the 7th International Workshop Hybrid System Computation and Control, 2993, 660-672. Berlin/Heidelberg: Springer.

Liu, J. (2008). Control of protein synthesis and mRNA degradation by microRNA. *Current Opinion in Cell Biology*, *20*(2), 214–222. doi:10.1016/j.ceb.2008.01.006

Longo, D., & Hasty, J. (2006). Dynamics of single-cell gene expression. *Molecular Systems Biology*, 28.

Ma, L., Wagner, J., Rice, J. J., Hu, W., Levine, A. J., & Stolovitzky, G. A. (2005). A plausible model for the digital response of p53 to DNA damage. *Proceedings of the National Academy of Sciences of the United States of America*, *102*(40), 14266–14271. doi:10.1073/pnas.0501352102

McKinney, B., Crowe, J., Voss, H. Jr, Crooke, P., Barney, N., & Moore, J. (2006). Hybrid grammar-based approach to nonlinear dynamical systems identification from biological time series. *Physical Review E: Statistical, Nonlinear, and Soft Matter Physics*, *73*(021912), 1–7.

Meek, D. W. (2009). Tumour suppression by p53: A role for the DNA damage response? *Nature Reviews. Cancer*, *9*(10), 714–723.

Mihalas, G. I., Simon, Z., Balea, G., & Popa, E. (2000). Possible oscillatory behavior in p53-MDM2 interaction computer simulation. *Journal of Biological System*, *8*(1), 21–29. doi:10.1142/S0218339000000031

Moles, C. G., Mendes, P., & Banga, J. R. (2003). Parameter estimation in biochemical pathways: A comparison of global optimization methods. *Genome Research*, *13*(11), 2467–2474. doi:10.1101/gr.1262503

Monk, N. A. (2003). Oscillatory expression of Hes1, p53, and NF- B driven by transcriptional time delays. *Current Biology*, *13*(16), 1409–1413. doi:10.1016/S0960-9822(03)00494-9

Neogi, N. A. (2004). *Dynamic partitioning of large discrete event biological systems for hybrid simulation and analysis*. Paper presented at the 7th International Workshop Hybrid Systems: Computation and Control, 2993, 463-476.

Nicosia, S., Tornambé, A., & Valigi, P. (1991). A solution to the generalized problem of nonlinear map inversion. *Systems & Control Letters*, *17*(5), 383–394. doi:10.1016/0167-6911(91)90138-5

Politi, A., Monè, M. J., Houtsmuller, A. B., Hoogstraten, D., Vermeulen, W., & Heinrich, V. (2005). Mathematical modeling of nucleotide excision repair reveals efficiency of sequential assembly strategies. *Molecular Cell*, *19*(5), 679–690. doi:10.1016/j.molcel.2005.06.036

Quach, M., Brunel, N., & d'Alche Buc, F. (2007). Estimating parameters and hidden variables in non-linear state-space models based on ODEs for biological networks inference. *Bioinformatics (Oxford, England)*, *23*, 3209–3216. doi:10.1093/bioinformatics/btm510

Rabitz, H. (1987). Chemical dynamics and kinetics phenomena as revealed by sensitivity analysis techniques. *Chemical Reviews*, *87*, 101. doi:10.1021/cr00077a006

Rabitz, H., Kramer, M., & Dacol, D. (1983). Sensitivity analysis in chemical kinetics. *Annual Review of Physical Chemistry*, *34*, 419. doi:10.1146/annurev.pc.34.100183.002223

Raj, A., & van Oudenaarden, A. (2009). Single-molecule approaches to stochastic gene expression. *Annual Review of Biophysics*, *38*(1), 255–270. doi:10.1146/annurev.biophys.37.032807.125928

Schmidt, H., & Jirstrand, M. (2006). Systems biology toolbox for MATLAB: A computational platform for research in systems biology. *Bioinformatics (Oxford, England)*, *22*(4), 514–515. doi:10.1093/bioinformatics/bti799

Sontag, E. D. (2005). Molecular systems biology and control. *European Journal of Control, 11*, 1–40. doi:10.3166/ejc.11.396-435

Sugasawa, K., Ng, J. M. Y., Masutani, C., Iwai, S., van der Spek, P. J., & Eker, A. P. M. (1998). Xeroderma pigmentosum group c protein complex is the initiator of global genome nucleotide excision repair. *Molecular Cell, 2*(2), 223–232. doi:10.1016/S1097-2765(00)80132-X

Sun, X., Jin, L., & Xiong, M. (2008). Extended Kalman filter for estimation of parameters in nonlinear state-space models of biochemical networks. *PLoS ONE, 3*, e3758. doi:10.1371/journal.pone.0003758

Szallasi, Z., Stelling, J., & Periwal, V. (2006). *System modeling in cellular biology from concepts to nuts and bolts*. The MIT Press.

Thattai, M., & van Oudenaarden, A. (2001). Intrinsic noise in gene regulatory networks. *Proceedings of the National Academy of Sciences of the United States of America, 98*(15), 8614–8619. doi:10.1073/pnas.151588598

Tiana, G., Jensen, M., & Sneppen, K. (2002). Time delay as a key to apoptosis induction in the p53 network. *The European Physical Journal B, 29*(1), 135–140. doi:10.1140/epjb/e2002-00271-1

Toni, T., Welch, D., Strelkowa, N., Ipsen, A., & Stumpf, M. P. H. (2009). Approximate Bayesian computation scheme for parameter inference and model selection in dynamical systems. *Journal of the Royal Society, Interface, 6*, 187–202. doi:10.1098/rsif.2008.0172

Tyson, J. J., Chen, K. C., & Novak, B. (2003). Sniffers, buzzers, toggles and blinkers: Dynamics of regulatory and signaling pathways in the cell. *Current Opinion in Cell Biology, 15*(2), 221–231. doi:10.1016/S0955-0674(03)00017-6

Varma, A., Morbidelli, M., & Wu, H. (2005). *Parametric sensitivity in chemical systems*. Cambridge University Press.

Vogelstein, B., Lane, D., & Levine, A. J. (2000). Surfing the p53 network. *Nature, 408*(6810), 307–310. doi:10.1038/35042675

Volker, M., Mone, M. J., Karmakar, P., van Hoffen, A., Schul, W., & Vermeulen, W. (2001). Sequential assembly of the nucleotide excision repair factors in vivo. *Molecular Cell, 8*(1), 213–224. doi:10.1016/S1097-2765(01)00281-7

Wang, D., Hara, R., Singh, G., Sancar, A., & Lippard, S. (2003). Nucleotide excision repair from site-specifically platinum-modified nucleosomes. *Biochemistry, 42*(22), 6747–6753. doi:10.1021/bi034264k

Wang, Z., Liu, X., Liu, Y., Liang, J., & Vinciotti, V. (2009). An extended Kalman filtering approach to modelling nonlinear dynamic gene regulatory networks via short gene expression time series. *IEEE/ACM Transactions on Computational Biology and Bioinformatics, 6*(3), 410–419. doi:10.1109/TCBB.2009.5

Wilkinson, D. J. (2007). Bayesian methods in bioinformatics and computational systems biology. *Briefings in Bioinformatics, 8*(2), 109–116. doi:10.1093/bib/bbm007

Yonish-Rouach, E., Resnftzky, D., Lotem, J., Sachs, L., Kimchi, A., & Oren, M. (1991). Wild-type p53 induces apoptosis of myeloid leukaemic cells that is inhibited by interleukin-6. *Nature, 352*(6333), 345–347. doi:10.1038/352345a0

ADDITIONAL READING

Alon, U. (2006). *An Introduction to Systems Biology: Design Principles of Biological Circuits*. Chapman & Hall/CRC.

Camacho, D., & Collins, J. (2009). Systems Biology Strikes Gold. *Cell*, *137*(1), 24–26. doi:10.1016/j.cell.2009.03.032

Cho, C. R., Labow, M., Reinhardt, M., Oostrum, J. V., & Peitsch, M. C. (2006). The application of systems biology to drug discovery. *Current Opinion in Chemical Biology*, *10*(4), 294–302. doi:10.1016/j.cbpa.2006.06.025

Deamer, D. (2009). On the origin of system: Systems biology, synthetic biology and the origin of life. *EMBO Reports*, *10*, S1–S4. doi:10.1038/embor.2009.117

Endy, D. (2005). Foundations for Engineering Biology. *Nature*, *438*, 449–453. doi:10.1038/nature04342

Hood, L., Heath, J. R., Phelps, M. E., & Lin, B. (2004). Systems Biology and New Technologies Enable Predictive and Preventative Medicine. *Science*, *306*(5696), 640–643. doi:10.1126/science.1104635

Hucka, M. (2003). The systems biology markup language (SBML): a medium for representation and exchange of biochemical network models. *Bioinformatics (Oxford, England)*, *19*(4), 524–553. doi:10.1093/bioinformatics/btg015

Khalil, H. K. (1996). *Nonlinear systems* (2nd ed.). Prentice Hall.

Kirschner, M. (2005). The Meaning of Systems Biology. *Cell*, *121*(4), 503–504. doi:10.1016/j.cell.2005.05.005

Kitano, H. (2002). Systems Biology: A Brief Overview. *Science*, *295*(5560), 1662–1664. doi:10.1126/science.1069492

Ljung, L., & Söderström, T. (1987). *Theory and Practice of Recursive Identification*. MIT Press.

Moraal, P. E., & Grizzle, J. W. (1995). Observer design for nonlinear systems with discrete-time measurements. *IEEE Transactions on Automatic Control*, *40*(3), 395–404. doi:10.1109/9.376051

Otero, J. M., & Nielsen, J. (2009). Industrial systems biology. *Biotechnology and Bioengineering*, *105*(3), 439–460. doi:10.1002/bit.22592

Snyder, M., & Gallaghera, J. E. G. (2009). Systems biology from a yeast omics perspective. *FEBS Letters*, *583*(24), 3895–3899. doi:10.1016/j.febslet.2009.11.011

Tsinias, J. (1989). Observer design for nonlinear systems. *Systems & Control Letters*, *13*(2), 135–142. doi:10.1016/0167-6911(89)90030-3

Vidyasagar, M. (2002). *Non-linear Systems Analysis*. SIAM. doi:10.1137/1.9780898719185

Weston, A. D., & Hood, L. (2004). Systems Biology, Proteomics, and the Future of Health Care: Toward Predictive, Preventative, and Personalized Medicine. *Journal of Proteome Research*, *3*(2), 179–196. doi:10.1021/pr0499693

KEY TERMS AND DEFINITIONS

DNA Repair: Refers to a collection of processes by which a cell identifies and corrects damage to the DNA molecules that encode its genome. DNA repair process is constantly active as it responds to damage in the DNA structure. When normal repair processes fail, and when cellular apoptosis does not take effect, "Irreparable DNA Damage" can occur, including double-strand breaks and DNA cross linkages

Extended Kalman Filter: An efficient recursive filter that estimates the state of a non linear dynamic system from a series of noisy measurements.

Hybrid Model: In control theory, is a dynamical system with interacting continuous-time dynamics (modeled, for example, by differential equations) and discrete-event dynamics (modeled, for example, by automata).

Identifiability Analysis: Concerned with whether the unknown parameters can be uniquely determined from perfect, noise-free, and continuous data.

Nucleotide Excision Repair (NER): A DNA repair mechanism by cell can prevent unwanted mutations by removing the vast majority of UV-induced DNA damage (mostly in the form of thymine dimers and 6-4-photoproducts).

Ordinary Differential Equation (ODE): A relation that contains functions of only one independent variable, and one or more of its derivatives with respect to that variable.

Sensitivity Analysis: The study of how uncertainty in the output of a model (numerical or otherwise) can be apportioned to different sources of uncertainty in the model input.

State Estimation: Concerns the problem of estimating the time behavior of the internal state of a process which is not directly measurable or accessible.

State Observer: In control theory, is a system that models a real system in order to provide an estimate of its internal state, given measurements of the input and output of the real system. An asymptotic observer is an observer that converges to the internal state of the system for time that goes to infinity.

APPENDIX BASIC BIOLOGY

A. Biology of the p53 Network

The presence of DNA damage is sensed by a genetic network built upon the p53 gene and protein. p53 plays a key role in cell's response to DNA damage. First discovered in 1979, this gene was initially identified as an oncogene. Its actual tumor suppressing function has been clarified only twenty years later (Vogelstein et al., 2000). The fact that DSBs induce an increase in p53 levels which, in turn, induces apoptosis has been demonstrated for the first time by E. Yonish-Rouach et al. in 1991 (Yonish-Rouach et al., 1991) and it's now generally accepted (Meek et al., 2009).

Since then the gene, and the protein it codes for, have been intensely investigated to determine their function in apoptosis induction. p53 is now known to be involved in a great number of cellular processes, but in many cases the exact mechanisms still need to be elucidated. A failure in p53 bears critical consequences: approximately 50% of all human cancers displays p53 mutations, and in the vast majority of the other cases the gene's functionality is otherwise impaired.

p53 interacts with many proteins, genes and transcription factors. In our description, we will consider only the main known components and the fundamental interactions among them, a block diagram of which is shown in Figure 8.

Before a damage is detected, the network will be in a steady state. In such conditions, P53 is present in very small amounts. This is necessary to avoid triggering the system in undamaged cells, and is achieved by a strong downregulation of P53 by MDM2. MDM2 can rapidly degrade P53 through a pathway known as *proteolytic degradation*. MDM2 attaches several ubiquitin molecules to P53. These act as labels that target it for destruction. P53, in turn, can act as a transcription factor and enhance mdm2 mRNA transcription. Thus, if P53 levels go up, more MDM2 is produced and excess P53 is destroyed. The two components form a feedback loop.

In eukaryotic cells, DNA is contained only in the cellular nucleus, packed in a structure called *chromatin*. When damage occurs, the whole chromatin shape is thought to be altered. This event is detected by *ATM*, the sensor protein of the network, which undergoes a rapid autophosphorylation at serine 1981. Bakkenist and Kastan (2003) have shown that in response to ionizing radiation (IR) doses as low as 0.5 Gy such an activation involves most of the ATM molecules in the cell. The concentration of phosphorylated ATM (which from our point of view is the active form of ATM) produces a step-like digital signal, with a rapid transition between two discrete states. In our model, active ATM will be the input signal of the network. Defective ATM functionality is associated with the hereditary disease *ataxia telangiectasia*, characterized by a strong predisposition to cancer and extreme cellular sensitivity to radiation.

Active ATM phosphorylates *P53*. It also phosphorylates *MDM2* at serine 395. These modifications alter the P53-MDM2 equilibrium described above. P53 becomes more stable and MDM2 becomes less effective in promoting P53 degradation, so P53 levels are allowed to rise and induce cell cycle arrest and apoptosis. There is more than one mechanism through which MDM2 hinders P53 functionality, but we will consider ubiquitin-mediated proteolysis only, which is considered to be the major one. Impaired p53 and mdm2 mRNA's functionality is associated with serious consequences: Li-Fraumeni syndrome (lack of p53), p53 mRNA mutations and mdm2 mRNA overexpression are associated with the onset cancer (Vogelstein et al., 2003), while mdm2 mRNA elimination in mice results in early embryonic lethality (Jones et al., 1995).

Figure 8. Main components and interactions in the p53 network. Continuous lines mean positive actions, dashed lines negative actions, red circular arrows feedback loops.

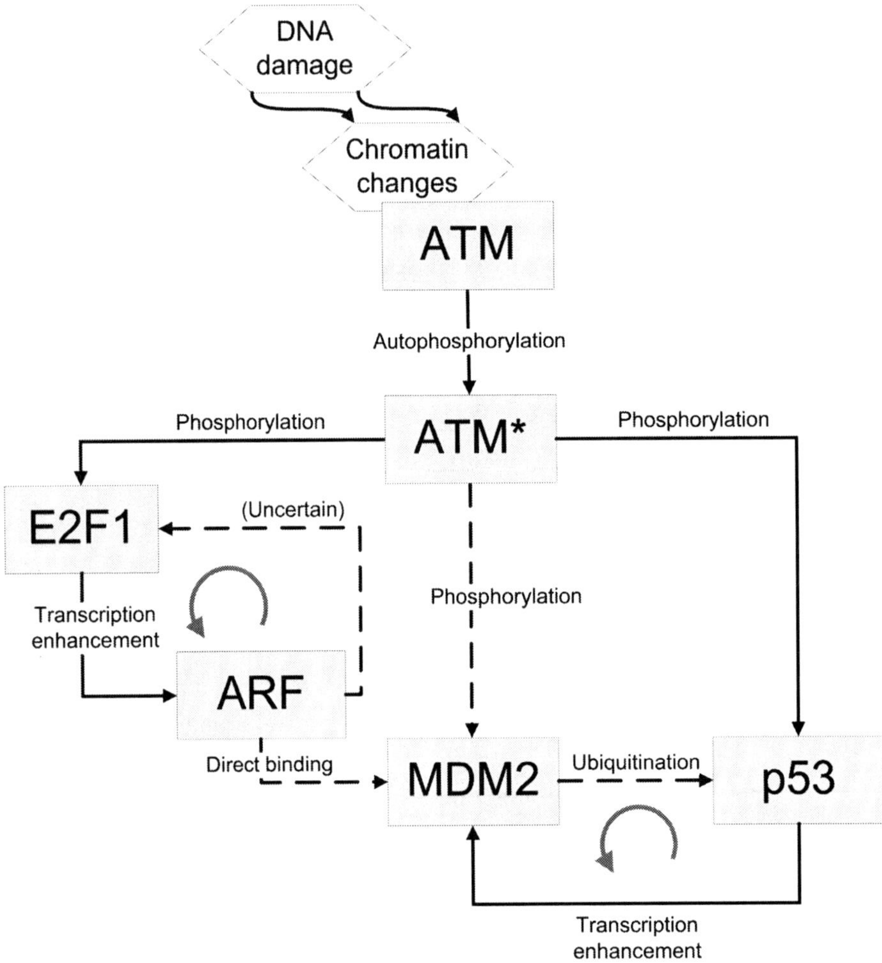

Other components of the p53 network include E2F1 and ARF. *E2F1* is a transcription factor phosphorylated and stabilized by active ATM. It has been suggested that E2F1 can enhance arf transcription. *ARF* is a small protein called $p14^{ARF}$ in man and $p19^{ARF}$ in mice. ARF is a negative regulator of MDM2 and it seems to be rather important in the p53 network: mice without arf display almost the same predisposition to cancer as the ones without p53 (Kamijo et al., 1997). In man, arf deletion is associated with the onset of breast, brain and lung cancer (Vogelstein et al, 2003). ARF directly binds to MDM2, blocking its ability to ubiquitinate P53. ARF is also a negative regulator of E2F1: this two components also form a feedback loop. arf, just like p53, is expressed at very low levels when no stress in present.

Looking again at Figure 8, it is seen that the p53 network can be regarded as the result of the interaction of two feedback loops, whose equilibrium is perturbed by active ATM.

Figure 9. Nucleotide excision repair pathways: the DNA damage repaired through GG-NER (global genomic)

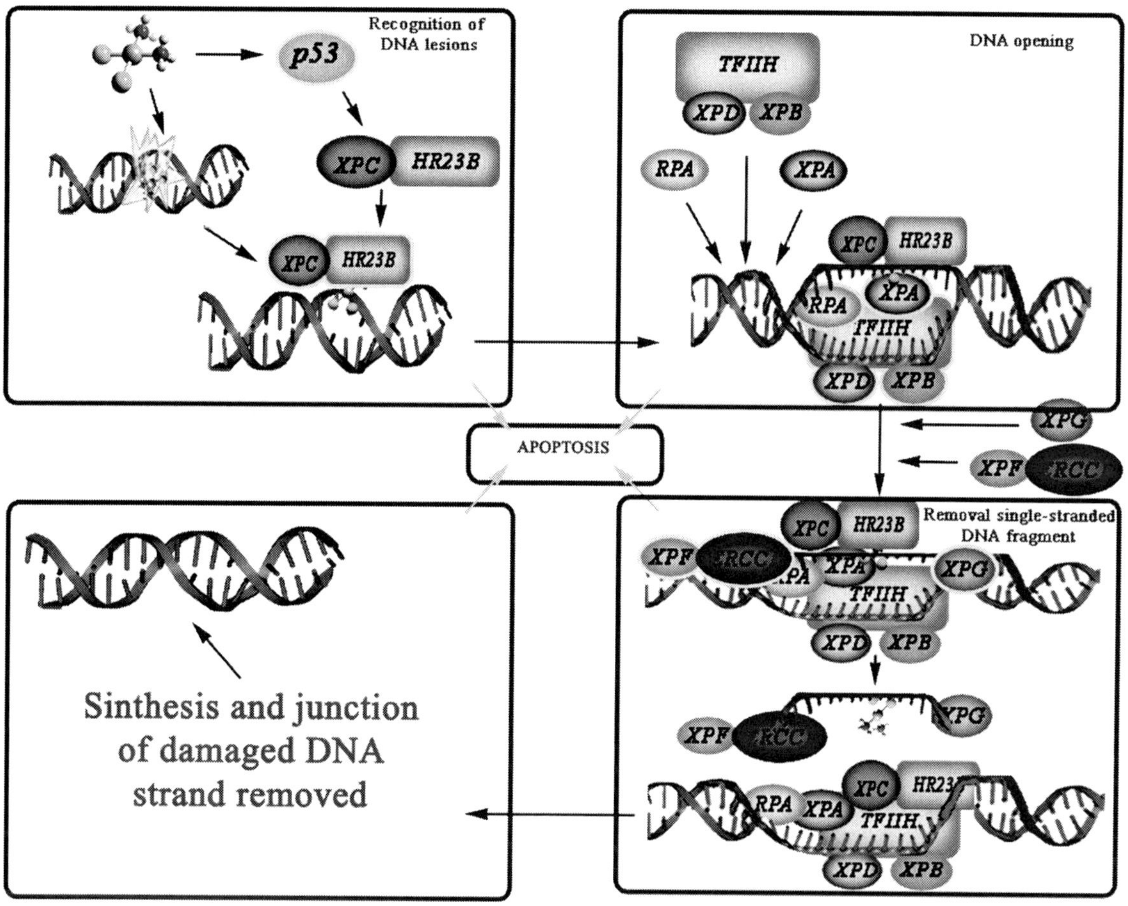

B. Biology of the NER Mechanism

Nucleotide excision repair (NER) is the major DNA repair pathway used in the repair of bulky DNA damage generated by most environmental insults and therapeutic drugs. Among therapeutic drugs, cisplatin is a widely used chemotherapeutic agent which damages DNA by forming helix-distorting lesions called *adducts*. Cisplatin-DNA adducts are removed primarily by the NER pathway *in vitro* and *in vivo* (Hanawalt & Schuster, 2003; Wang et al., 2003, Furuta et al., 2002).

The p53 pathway can partially mediate cisplatin cytotoxicity. p53 interacts with several components of the nucleotide excision repair machinery such as XPC-HR23B (*Xeroderma Pigmentosum, complementation group C*) and CSB (*Cockayne Syndrome B*), both involved in damage recognition (Adimoolam and Ford, 2003).

Nucleotide excision repair requires the concerted action of many different proteins that assemble at sites of damaged DNA in different steps: the recognition of DNA lesions, excision and removal of

≈30-base-pair single-stranded DNA fragments that contain the lesions, and re-synthesis and ligation of the newly synthesized repair patch to the pre-existing strand (Costa et al., 2003).

There are two distinct NER pathways: *transcription-coupled NER* (TC-NER) and *global genome NER* (GG-NER). In this paper we focus on GG-NER since there is a more comprehensive biological understanding concerning the proteins recognizing DNA damage in such pathway. At the same time, TC-NER is currently subject of investigation in the literature, although key factors of its operation remain unclear. As for the proposed model, it appears possible to add new discrete modes and continuous state variables to describe new additional NER's key aspects (e.g. extension to TC-NER).

In GG-NER, during the DNA damage recognition, p53 is activated and XPC-HR23B complex binds to damaged DNA (Sugasawa et al., 1998). After the different modes of initial recognition, the two sub-pathways share common subsequent repair events; XPA (*Xeroderma Pigmentosum, complementation group A*) and RPA (*Replication Protein A*) are recruited and they act as another damage recognition complex that might participate in lesion verification. TFIIH (*Transcription Factor IIH*) is then recruited. This protein complex is involved in both transcription and DNA repair. XPB (*Xeroderma Pigmentosum, complementing group B*) and XPD (*Xeroderma Pigmentosum, complementing group D*), which compose the TFIIH helicase, unwind the portion of DNA duplex that surrounds the DNA adduct before the incision step (Volker et al., 2001). The XPG protein makes a 3' incision, which is followed by a 5' incision made by the XPF/ERCC1, resulting in a single-stranded gap of 27-32 bases (Hanawalt et al., 2002). The DNA polymerases fill the gap and the DNA ligase seals the gap to complete the DNA repair process. After a single repair event (which takes several minutes) the entire complex is disassembled again. These steps are irreversible.

Chapter 22
Granger Causality:
Its Foundation and Applications in Systems Biology

Tian Ge
Fudan University, China

Jianfeng Feng
Fudan University, China & University of Warwick, UK

ABSTRACT

As one of the most successful approaches to uncover complex network structures from experimental data, Granger causality has been widely applied to various reverse engineering problems. This chapter first reviews some current developments of Granger causality and then presents the graphical user interface (GUI) to facilitate the application. To make Granger causality more computationally feasible and satisfy biophysical constraints for dealing with increasingly large dynamical datasets, two attempts are introduced including the combination of Granger causality and Basis Pursuit when faced with non-uniformly sampled data and the unification of Granger causality and the Dynamic Causal Model as a novel Unified Causal Model (UCM) to bring in the notion of stimuli and modifying coupling. Several examples, both from toy models and real experimental data, are included to demonstrate the efficacy and power of the Granger causality approach.

INTRODUCTION

With the rapid progress in the development of experimental techniques, more and more high-throughput datasets measuring temporal behavior of hundreds of or even thousands of proteins or genes are offering rich opportunities for researchers. In order to exploit the full potential of these approaches, we have to be able to convert the resulting data into the most appropriate framework to account for the functioning of the underlying biological system. Over the past two decades, a variety of attempts have been carried out in this field and reverse engineering approaches to uncover network structures in genes, proteins,

DOI: 10.4018/978-1-60960-491-2.ch022

neurons and brain areas are still one of the hottest topics in computational systems biology.

Causality analysis based upon experimental data has become one of the most powerful and valuable tools in discovering connections between different elements in complex biological systems (Cantone et al., 2009; Camacho & Collins, 2009). Comparing approaches including information theory, control theory or Bayesian statistics, here we focus on another successful approach: Granger causality, which is based upon simple ideas and has a concise theory but is even more powerful to capture the nature and dynamics of a biological system. As an example, in one recent comment on a paper in Cell, we have demonstrated that Granger causality outperforms all the other approaches the authors had employed to build causal networks (Zou et al., 2009).

The basic idea of Granger causality can be traced back to Wiener (Wiener, 1956) who put forward the notion that if the prediction of one process can be improved by incorporating the past information of the second process, then the second process causes the first one. Later, Granger followed this point and formalized it in the context of linear regression models (Granger, 1969). Geweke's decomposition of a vector autoregressive process endowed Granger causality with a spectral representation (Geweke, 1982, 1984) and made the interpretation more informative in that interactions in different frequency bands could be clearly figured out instead of only in a single number. Recently, a series of papers based upon its original formalism have been published to make Granger causality suitable to address biological and computational issues in different situations. These useful extensions include partial Granger causality (Guo et al., 2008) which is able to eliminate the influences of exogenous inputs and latent variables; complex Granger causality (Ladroue et al., 2009) which can uncover the interactions among groups of time-series and harmonic Granger causality (Wu et al. 2008) which introduces a model with an oscillating external input and puts special emphasis on environmental effects. These methods can be combined to identify interactions in the time and frequency domains in local and global networks. Furthermore, detailed and intensive comparisons between Granger causality and Bayesian networks have also been carried out (Zou & Feng, 2009). In this chapter, we first apply well established Granger causal analysis approaches to microarray data from *Arabidopsis thaliana* (Arabidopsis) to recover a well-known gene circuit. Our graphical user interface (GUI) is also presented to facilitate the application. These will show the power of Granger causality and its convenient implementation.

In spite of all the successful extensions and applications of Granger causality mentioned above, some limitations still exist which restrict its application on a broader basis. The first issue we encounter, whatever the approach is to be applied to a set of data, is preprocessing. Preprocessing should not be ignored and can sometimes play a critical role in determining final conclusions. A brute-force application of Granger causality could simply result in false or erroneous conclusions. Two general approaches of preprocessing in dealing with temporal data are down-sampling and up-sampling, after filtering out noise or extreme points. We have found that both techniques are useful and are commonly implemented before doing further analysis. The choice of down-sampling or up-sampling depends on the nature of data. In neurophysiology, the original data are usually sampled at a very high frequency, for example, 2 kHz. Even if we fit the data with an autoregressive model with an order of 20, the model only covers a time window of 10 milliseconds, a very short duration. As a result, information in the low frequency band could be lost and the features of slow oscillations such as theta rhythms (4-8 Hz) are difficult to be captured. On the other hand, in gene microarray experiments, since each chip is costly, we have to sample the data as sparse as possible. This generally results in with too few time points or non-uniformly sampled data and here up-sampling is frequently needed, as discussed (Bar-Joseph, 2004). One might argue that there are many existing approaches to tackle the issue: for example, using splines as reported in the literature. However, a more detailed look tells us that spline interpolation, which uses only local information, will fail to preserve the periods and rhythms in general. Unfortunately, periodic activities in a dataset are the key reasons for us to perform temporal recordings and missing them implies that we have lost our core information. To overcome this shortcoming, in this chapter, we first introduce an approach based on Basis Pursuit (Chen et al., 2001) to obtain a continuous representation of stationary time-series and thus deal with non-uniformly sampled data. The method is tested in toy models to illustrate its superiority over the traditional cubic spline interpolation.

There are a number of approaches to uncover the network structures from recorded data. As mentioned above, Granger causality seems to have an upper hand in comparison with all the conventional approaches. Nevertheless, there is still ongoing debate on the choice of Granger causality and the Dynamic Causal Model (Friston, 2009) in the literature especially in neuroimaging data analysis. In the current chapter, we unify the two seemingly different causal models: the Granger Causal Model (GCM) and the model-based Dynamic Causal Model (DCM) (Friston et al., 2003) to create a more general model: the Unified Causal Model (UCM) which includes both GCM and DCM as special cases. This method is first tested in toy models and then applied to local field potential (LFP) data recorded in the sheep inferotemporal cortex (IT) of both left and right hemispheres before and after they learned a visual face discrimination task to exemplify the efficacy.

GENE CIRCUIT OF ARABIDOPSIS LEAF

We carried out a study on experimental data of microarray experiments. The gene data were collected from two cases of Arabidopsis leaf: the mock (normal) case and the infected case with the plant pathogen *Botrytis cinerea*. A total of 31,000 genes were measured with a time interval of two hours, with a total of 24 sampling points (two days) and four replicates. Conditional Granger causality was applied to recover the structure of a well-known circadian circuit which contains 7 genes: PRR7, GI, PRR9, ELF4, LHY, CCA1 and TOC1. A bootstrapping method was used to construct 95% confidence intervals. Two network structures for two different cases are shown in Figure 1. It is clearly evident that the global patterns for the mock case and the infected case are different. Some of the features for the mock case are in good accordance with the results in the literature. For example, it is known that PRR7 and LHY share a feedback loop. In other words, there should be two directed arcs connected from node 1 (PRR7) to node 5 (LHY) and from node 5 to node 1. The network structures derived from Granger causality are in agreement with this known relationship. Also, it is known that ELF4 has some interactions with both LHY and CCA1. There should be some connections between node 4 (ELF4) and node 5 (LHY), and between node 4 (ELF4) and node 6 (CCA1). Granger causality also detected these connections, which are in agreement with the known structure in the literature (Locke et al., 2006; Ueda 2006).

The application of well established Granger causal analysis approaches as well as preprocessing techniques can be implemented using our graphical user interface (GUI). The software can be downloaded from http://www.dcs.warwick.ac.uk/~feng/causality/ along with all the source codes. After loading the data, commonly used preprocessing techniques such as different types of filters, down-sampling and finite order differ-

Figure 1. Granger causality approaches applied on experimental data. The experiment measures the intensity of 7 genes in two cases of Arabidopsis Leaf: mock (normal) and infected with 4 realizations of 24 time points. The time interval is 2 hours. The network structures both for the mock case (a) and the infected case (b) are shown. A bootstrapping method was used to construct 95% confidence intervals.

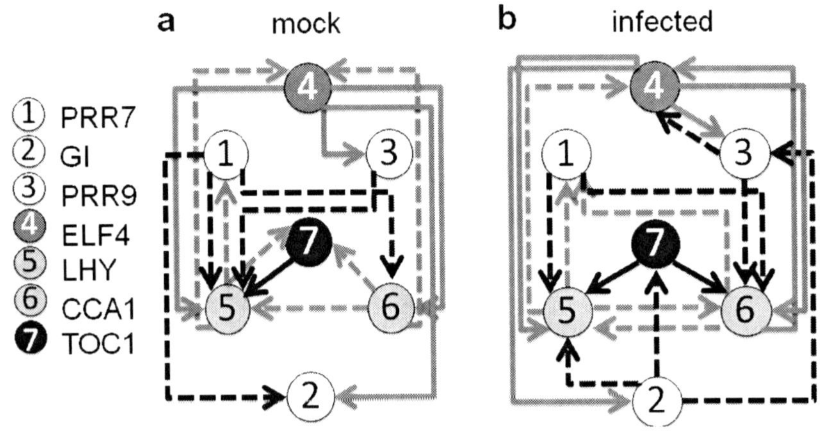

ence (ARIMA model) can be easily carried out. Then, conditional Granger causality or partial Granger causality can be applied to the data to detect causal influences. Bootstrap methods can be employed to construct confidence intervals. For further information about the application of Granger causality and this software, we refer the readers to the above website.

Preprocessing: Basis Pursuit

A majority of the current available data are non-uniformly sampled. As reported in (Zou & Feng, 2009), we have acquired microarray data from the whole genome of *Arabidopsis* during senescence. The data were collected for 22 days (one sample per half day for one day and then left the next day out). For this set of data, all dynamical activities with a time scale of day length are captured, but more detailed dynamics such as those at hours are certainly missed out. On the other hand, we have also performed analysis of the other experiment that spans two days sampled every two hours. So the data we have are typically non-uniformly sampled. In situations like this, a direct application of Granger causality will certainly lead to a false result due to the limited time points and, something more vital, the disorder of the sequence. Up-sampling is therefore indispensable before any further analysis.

Assume that there are N uniformly distributed time points denoted as $1, 2, \cdots, N$. However, the data may be non-uniformly sampled.

Let $s = (s_{t_1}, s_{t_2}, \cdots, s_{t_n}), t_i \in Z, 1 \le t_i \le N$ be a time-series sampled at time $t_1, t_2, \cdots, t_n$. This may also be viewed as a vector of length n in R^n. We want to reconstruct the signal using superposition of elementary waveforms without losing its global properties and estimate the unobserved time points. One traditional method of reconstruction involves the use of n orthogonal bases such as the Fourier bases, various discrete cosine transform bases and orthogonal wavelet bases which viewed in R^n are linearly independent and the representation of the signal as a linear combination of these waveforms is unique. In order to obtain a more precise continuous representation of the time-series, we introduce a method called Basis Pursuit which represents the time-series with an over-complete dictionary.

A dictionary is a collection of parameterized waveforms $D=(\phi_\gamma:\gamma\in\Gamma)$ and the waveforms ϕ_γ are time-series of length n called atoms. Depending on the dictionary, the parameter γ can have the interpretation of indexing frequency, indexing time / scale jointly or of indexing time / frequency jointly. The dictionaries are complete or over-complete in which case they contain exactly n atoms or more than n atoms.

Here we construct a dictionary as follows: we take $\gamma = (\omega,\varphi)$ where $\omega\in[0,\pi]$ is a frequency, $\varphi\in[0,\pi)$ is a phase and consider the atoms:

$$\phi_\gamma = \cos(\omega t+\varphi).$$

Such atoms consist of different frequencies and phases. Discrete dictionaries can be built from lattices $\omega_k = k\Delta\omega$ and $\varphi_l = l\Delta\varphi$ and here we use

$$\omega_k = \frac{2\pi k}{nd}, k = 0,1,2,\cdots,[\frac{nd}{2}]$$

and

$$\varphi_l = \frac{\pi l}{p}, l = 1,2,\cdots,p-1$$

where d and p are both whole numbers larger than 1 which determine the precision of the representation. With sufficient precision, the dictionary is obviously over-complete.

Assuming that the dictionary is over-complete, there are in general many representations $s=\sum_\gamma \alpha_\gamma \phi_\gamma$ and the principle of Basis Pursuit is to find a representation of the time-series whose coefficients have minimal l^1 norm. Formally, we can summarize the problem as:

$$\begin{aligned} &\min \|\alpha\|_1 \\ &s.t.\Phi\alpha = s \end{aligned} \qquad (1)$$

where Φ is an n by q matrix, q is the number of atoms in the dictionary and Φ_{ij} is the value of the j-th atom at time t_i.

Although this problem involves nonlinear optimization, it can be equivalently reformulated as a linear program in the standard form. Since α can be decomposed as $\alpha = u-v$, $u\geq 0$, $v\geq 0$ and if we define $\tilde{\alpha} = [u^T, v^T]^T$, then

$$\|\alpha\|_1 = \|u-v\|_1 \leq \|u\|_1 + \|v\|_1 = \|\tilde{\alpha}\|_1 .$$

Hence, define $f=[1,1,\cdots,1]^T$ and we can rewrite the system (1) as:

$$\min \|\tilde{\alpha}\|_1 = f^T\tilde{\alpha}$$

$$s.t.[\Phi,-\Phi]\tilde{\alpha} = s, \tilde{\alpha} \geq 0$$

After obtaining the coefficients vector $\tilde{\alpha}$ we can set a critical value to pick up those dominant frequency components in the time-series and estimate the unobserved time points by the reconstructed continuous representation.

Numerical Examples

The toy model we used here comes from a traditional vector autoregression (VAR) model which has been extensively applied in tests of Granger causality (Gourvitch et al., 2006). We generated the time series according to the following equations:

$$\begin{aligned} x_1(t) &= 0.95\sqrt{2}x_1(t-1) - 0.9025x_1(t-2) + \varepsilon_1(t) \\ x_2(t) &= -0.5x_1(t-1) + 0.5x_3(t-2) + \varepsilon_2(t) \\ x_3(t) &= -0.5x_2(t-1) + 0.5x_3(t-1) + \varepsilon_3(t) \end{aligned} \qquad (2)$$

where $\varepsilon_i(t), i = 1,2,3$ were zero mean uncorrelated Gaussian noise with variances 0.5. Inspection of the above equations reveals that $x_1(t)$ is a direct source to $x_2(t)$, $x_2(t)$ and $x_3(t)$ share a feedback loop. There is no direct connection between the remaining pairs of the state variables.

Simulation was performed to generate three time-series of 200 data points. Then, we randomly selected 80 points from the three time-series respectively and hence the new time-series can be regarded as non-uniformly sampled.

Figure 2 shows the amplitudes of the coefficients corresponding to different atoms with different frequencies and phases in the dictionary we constructed when we recovered the three time-series respectively. Here we used d=3 and p=4 in our algorithm so that there were in all 1200 different atoms. It is clear that the representations of the generated time-series are sparse and we can easily find out the dominant rhythms of the data.

Figure 3 shows the results when we used our Basis Pursuit method (upper panel) and the traditional cubic spline interpolation (lower panel) to recover the original data x_1 (middle panel). It can be easily found that the time-series is stationary and consisted of some periodic fluctuations which are quite common cases in real data. Sampled time points are marked by small circles on the Time axis. Two intervals where Basis Pursuit significantly outperforms cubic spline interpolation are marked by large ovals. We can see that the recovered time-series using Basis Pursuit preserves the global properties of the original data quite well and could fit the original time-series in general while the time-series recovered by cubic spline is not so ideal especially when there is a large gap between two observed time points. In fact, this can be anticipated since traditional interpolation methods recover the data using only local information but ignore the global properties of the data. Hence, it will be easy to lose effectiveness when a large number of data points are missed locally.

Besides the great performance in recovering the data, Basis Pursuit also preserves the causal relationships of the original time-series quite well which makes it more useful. Figure 4 shows the accuracy of causality prediction using Basis Pursuit and Cubic Spline interpolation. For different sampling numbers, we simulated the above autoregressive model (2) for 100 times and sampled the generated data randomly. Then we recovered the three time-series from the sampled data using Basis Pursuit and Cubic Spline interpolation and

Figure 2. The amplitudes of the coefficients corresponding to different atoms in the dictionary when recovering the three time-series using Basis Pursuit respectively

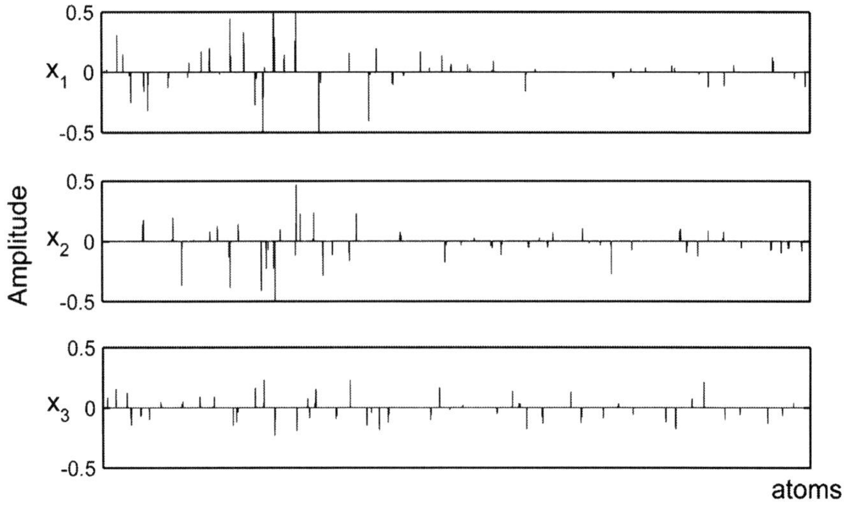

Figure 3. The time-series recovered by Basis Pursuit and Cubic Spline interpolation along with the original data x_1. The solid line in the middle is the generated time-series x_1 with 200 time points while the upper and lower dashed lines are the recovered time-series from 80 randomly sampled points using Basis Pursuit (upper panel) and Cubic Spline (lower panel) respectively. Sampling points are marked by small circles on the Time axis. Two intervals where Basis Pursuit significantly outperforms cubic spline are marked by large ovals. Two dashed lines are shifted upward and downward for visualization purpose.

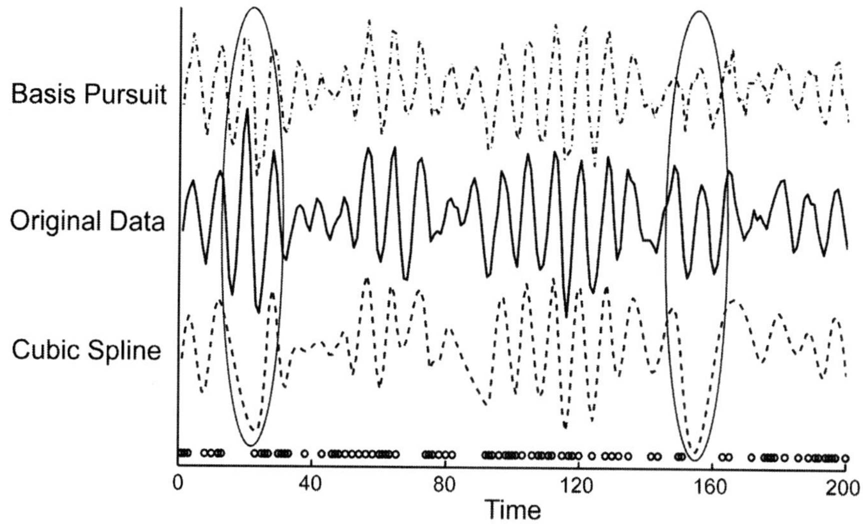

detected the causal influence with Conditional Granger causality. The accuracy of the causality prediction for both methods is defined as the percentage that the correct causal network is recovered. It is obvious that Basis Pursuit outperforms Cubic Spline interpolation. With fewer than 140 sampling points, i.e. less than 70% of the original data, the prediction accuracy of Basis Pursuit is more than 80% while the Cubic Spline interpolation method requires more than 80% of the data to reach the same level. This property makes it valuable to combine Basis Pursuit and Granger causality when we are faced with non-uniformly sampled data which are present in the majority of current databases.

UNIFIED CAUSAL MODEL

The Granger Causal Model (GCM) and Dynamic Causal Model (DCM) are two prominent techniques that have been introduced to address temporal dependencies and indentify directed causal influences from functional magnetic resonance imaging (fMRI) data and various other data sources. In spite of their common use and successful applications, these two models have always been considered to differ radically from each other (Friston, 2009). DCM establishes state variables in the observed data and is believed to be a causal model in a true sense. On the other hand, GCM is a phenomenological model which just tests statistical dependencies among the observations to determine how the data may be caused (Friston et al., 2003; Friston, 2009).

Generally speaking, both GCM and DCM have their own characteristics and advantages. GCM based upon vector autoregressive models naturally introduces time delays into the model which are quite ubiquitous in biological systems, no matter whether considering gene, protein, metabolic or neuronal networks. A single number measuring

Figure 4. The accuracy of the causality prediction using Basis Pursuit and Cubic Spline interpolation with different sampling numbers

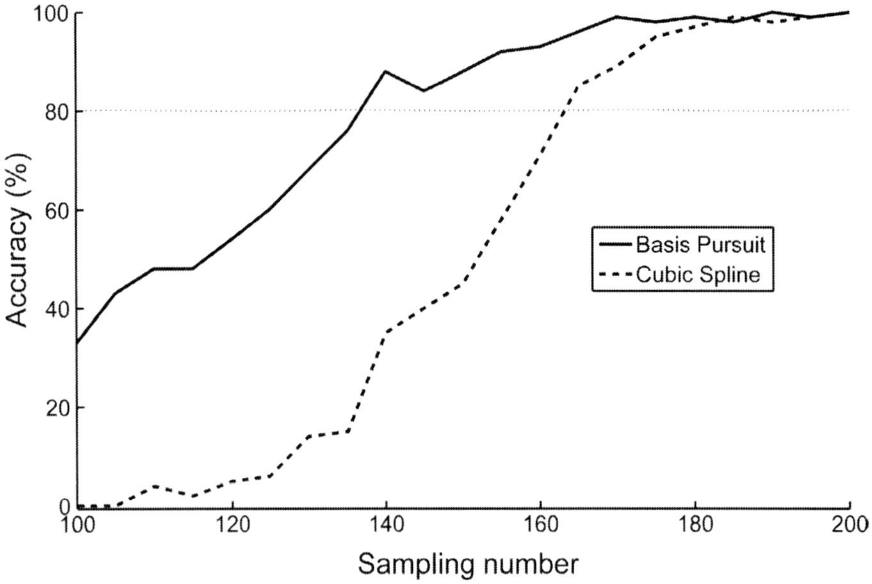

the strength of directed causal influence in the time domain makes the approach quite transparent and easy to understand while a frequency domain decomposition, on the other hand, provides the detailed patterns of interactions between different frequency bands which makes GCM more informative. In contrast, as a biologically informed model based upon Bayesian inferences, DCM includes state variables and observation variables as well as deterministic inputs which makes the model more realistic and biophysically constrained and also endows the parameters with strong interpretability. Hence, the significance of a unified causal model that has the advantages of both approaches is obvious. We would expect that it could represent a powerful new tool in systems and computational biology, particularly in association with increasingly powerful genomic, proteomic and metabolic methodologies allowing time-series measurements of large numbers of putatively interacting molecules.

We consider the Dynamic Causal Model:

$$\begin{cases} d\vec{x}(t) = f(\vec{x}, u, \theta)dt + \sigma dB_t \\ \vec{y}(t) = \vec{g}(\vec{x}(t)) + \vec{\varepsilon}(t) \end{cases}$$

where $\vec{x}(t) = (x_1, \cdots, x_N)^T$ are state variables and $(\cdot)^T$is a transpose of vector, *u*(*t*) is the known deterministic input corresponding to designed experimental effects, θ is the set of parameters to estimate, σ is the diffusion matrix (could depend on time) and B_t is the Brownian motion (or in general, it could be a martingale). The state variables $\vec{x}(t)$ enter a specific model to produce the outputs $\vec{y}(t)$ with the observation noise $\vec{\varepsilon}(t)$. Here we mainly focus on the bilinear approximation of the Dynamic Causal Model which is the most parsimonious but useful form (Friston K., 2009):

$$\begin{cases} d\vec{x}(t) = [A + u(t)B]\vec{x}(t)dt + u(t)\vec{c}dt + \sigma dB_t \\ \vec{y}(t) = \vec{g}(\vec{x}(t)) + \vec{\varepsilon}(t) \end{cases}$$

where $A = \frac{\partial f}{\partial x}, B = \frac{\partial^2 f}{\partial x \partial y}, \vec{c} = \frac{\partial f}{\partial u}$ are parameters that mediate the intrinsic coupling among states, allow the inputs to modulate the coupling, and elicit the influence of extrinsic inputs on the states respectively. For simplicity, we have expanded the state equation around $\vec{x} = 0$ and assumed that $f(0)=0$.

On the other hand, the traditional and widely used Granger Causal Model takes the form:

$$\vec{x}(t) = A_1\vec{x}(t-1) + \cdots A_p\vec{x}(t-p) + \vec{\varepsilon}(t)$$

where $\vec{x}(t) \in R^N, A_i, i = 1, \cdots, p$ are coefficient matrices, $\vec{\varepsilon}(t)$ is the noise, and the model has a vector autoregressive representation with an order up to p.

The bilinear form of DCM uses nonlinear differential equations to describe the dynamics of state variables, while GCM is formulated in discrete time and the dependencies among state variables are approximated by a linear mapping over time-lags which seem to be quite different. However, if we first assume that there are no deterministic inputs $u(t)$ and the observation variables are identical to the state variables in DCM, we can find that it can be regarded as a VAR(1) model which is a special case of the GCM represented by a VAR(p) model. On the other hand, the GCM with autoregressive representation always takes the past information into consideration while the bilinear approximation of DCM has no time-lags included in the differential equations. So, if we alter DCM to the form:

$$\begin{cases} d\vec{x}(t) = dt\int_0^t \left\{[A + u(t-\tau)B]\vec{x}(t-\tau) + u(t-\tau)\vec{c}\right\} k(\tau)d\tau + \sigma dB_t \\ \vec{y}(t) = \tilde{g}(\vec{x}(t)) + \vec{\varepsilon}(t) \end{cases}$$

where $k(\cdot)$is a kernel function. Then DCM also shares the feature of GCM. Hence, we can find that GCM has a strong connection with DCM and no essential contradiction exists to discriminate these two approaches. By taking deterministic inputs as well as observation variables into GCM and following the Volterra expansion to achieve a more accurate and biophysical constrained representation of biological systems, we unify DCM and GCM as the following form:

$$\begin{cases} \vec{x}(t) = [A_1 + u(t-1)B_1]\vec{x}(t-1) + \cdots + [A_p + u(t-p)B_p]\vec{x}(t-p) + v(t-1)\vec{c} + \vec{\varepsilon}(t) \\ \vec{y}(t) = \tilde{g}(\vec{x}(t)) + \vec{\varepsilon}^*(t) \end{cases}$$

where $u(t)$ and $v(t)$ are deterministic inputs, $\vec{y}(t)$ are the observation variables which are the function $\tilde{g}$ of the state variables, B_i, $i = 1,\cdots,p$ and $\vec{c}$ are the coefficients that allow the inputs to modulate the coupling of the state variables $\vec{x}(t)$ $\vec{\varepsilon}(t)$ and $\vec{\varepsilon}^*(t)$ are intrinsic and observation noise and are mutually independent.

Since UCM allows the inclusion of observation variables as well as deterministic inputs, the conventional schemes used in GCM for estimating regression coefficients are abandoned and we introduce an algorithm to estimate the state variables as well as all its parameters which will give us the first inference of the connection of the state variables.

UCM Algorithm

Let

$$\vec{X}(t) = \left(\vec{x}(t-1)^T, \cdots, \vec{x}(t-p+1)^T\right)^T$$
$$\vec{Y}(t) = \left(\vec{y}(t-1)^T, \cdots, \vec{y}(t-p+1)^T\right)^T$$
$$\vec{U}(t) = \left(u(t-1), \cdots, u(t-p+1)\right)^T$$

Box 1.

$$\vec{X}(t+1)=\begin{pmatrix} A_1+u(t)B_1 & A_2+u(t-1)B_2 & \cdots & A_p+u(t-p+1)B_p \\ I & 0 & \cdots & 0 \\ \vdots & \ddots & \ddots & \vdots \\ 0 & \cdots & I & 0 \end{pmatrix}\vec{X}(t)$$

$$+v(t)\begin{pmatrix} \vec{c} \\ 0 \\ \vdots \\ 0 \end{pmatrix}+\vec{w}(t)=A(\vec{\theta},\vec{U}(t))\vec{X}(t)+C(\vec{\theta})v(t)+\vec{w}(t)=\vec{f}(\vec{X}(t),\vec{U}(t),v(t),\vec{\theta})+\vec{w}(t)$$

$$\vec{Y}(t)=\vec{h}(\vec{X}(t))+\vec{s}(t)$$

we can find that the VAR(p) model can be reduced to a VAR(1) model shown in Box 1.

Where $\vec{\theta}$ is the parameter vector to estimate, $\vec{w}(t)$ and $\vec{s}(t)$ are both zero-mean uncorrelated Gaussian noise with covariance matrix $Q(t)$ and $R(t)$ respectively. In order to estimate both the states and parameters of the model from input variables and noise observations, we regard the parameters as special state variables and recursively approximate the nonlinear system by a linear model and use the traditional Kalman filter for the linearized model.

Let $\vec{\xi}=(\vec{X}^T,\vec{\theta}^T)^T$

$$\vec{\xi}(t+1)=\begin{pmatrix} \vec{X}(t+1) \\ \vec{\theta}(t+1) \end{pmatrix}=\begin{pmatrix} \vec{f}(\vec{X}(t),\vec{U}(t),v(t),\vec{\theta}) \\ \vec{\theta}(t) \end{pmatrix}+\begin{pmatrix} \vec{w}(t) \\ \vec{\eta}(t) \end{pmatrix}$$
$$=\begin{pmatrix} A(\vec{\theta},\vec{U}(t))\vec{X}(t)+C(\vec{\theta})v(t) \\ \vec{\theta}(t) \end{pmatrix}+\vec{\zeta}(t)=\vec{g}(\vec{\xi}(t),\vec{U}(t),v(t))+\vec{\zeta}(t)$$

where $\vec{\eta}(t)$ is uncorrelated Gaussian noise with covariance matrix $Z(t)$.

Define

$$\hat{\xi}_{t|t}=E[\xi(t)\mid\vec{Y}(t),\vec{U}(t),v(t)]$$
$$\hat{\xi}_{t+1|t}=E[\xi(t+1)\mid\vec{Y}(t),\vec{U}(t),v(t)]$$
$$\Omega_{t|t}=E[(\xi(t)-\hat{\xi}_{t|t})(\xi(t)-\hat{\xi}_{t|t})^T\mid\vec{Y}(t),\vec{U}(t),v(t)]$$
$$\Omega_{t+1|t}=E[(\xi(t+1)-\hat{\xi}_{t+1|t})(\xi(t+1)-\hat{\xi}_{t+1|t})^T\mid\vec{Y}(t),\vec{U}(t),v(t)]$$

where

$$\hat{\xi}_{t|t}=\begin{pmatrix} \hat{X}_{t|t} \\ \hat{\theta}_{t|t} \end{pmatrix},\hat{\xi}_{t+1|t}=\begin{pmatrix} \hat{X}_{t+1|t} \\ \hat{\theta}_{t+1|t} \end{pmatrix}$$

The algorithm for dual estimation consists of two steps: prediction and updating.

Prediction: predict the state variables and the covariance matrix of prediction error of the system at time t+1 from the estimated state $\hat{\xi}_{t|t}$, the observation $\vec{Y}(t)$ and inputs $\vec{U}(t)$ and $v(t)$. (see Box 2)

Updating: Update the system with the new observations $\vec{Y}(t+1)$ at time t+1.

$$\hat{\xi}_{t+1|t+1}=\hat{\xi}_{t+1|t}+G(t+1)[\vec{Y}(t+1)-\vec{h}(\hat{X}_{t+1|t})]$$
$$\Omega_{t+1|t+1}=[I-G(t+1)H(t+1)]\Omega_{t+1|t}$$

where

$$G(t+1)=\Omega_{t+1|t}H^T(t+1)[H(t+1)\Omega_{t+1|t}H^T(t+1)+R(t+1)]^{-1}$$
$$H(t+1)=\left.\begin{pmatrix} \dfrac{\partial h}{\partial X^T} & 0 \end{pmatrix}\right|_{\hat{X}_{t+1|t}}$$

After recovering the state variables using the UCM algorithm above, we can define the causal-

Box 2.

$$\hat{\xi}_{t+1|t} = E[\xi(t+1) \mid \vec{Y}(t), \vec{U}(t), v(t)] = [\vec{g}(\xi(t), \vec{U}(t), v(t)) + \vec{\zeta}(t) \mid \vec{Y}(t), \vec{U}(t), v(t)]$$
$$\approx E[\vec{g}(\hat{\xi}_{t|t}, \vec{U}(t), v(t)) + \frac{\partial \vec{g}}{\partial \xi^T}(\xi(t) - \hat{\xi}_{t|t}) \mid \vec{Y}(t), \vec{U}(t), v(t)]$$
$$= \vec{g}(\hat{\xi}_{t|t}, \vec{U}(t), v(t)) = \begin{pmatrix} A(\hat{\theta}_{t|t}, \vec{U}(t)) & 0 \\ 0 & I \end{pmatrix} \hat{\xi}_{t|t} + \begin{pmatrix} C(\hat{\theta}_{t|t}) \\ 0 \end{pmatrix} v(t)$$

where

$$\Omega_{t+1|t} = E[(\xi(t+1) - \hat{\xi}_{t+1|t})(\xi(t+1) - \hat{\xi}_{t+1|t})^T \mid \vec{Y}(t), \vec{U}(t), v(t)]$$
$$\approx E[(\frac{\partial \vec{g}}{\partial \xi^T}(\xi(t) - \hat{\xi}_{t|t}) + \vec{\zeta}(t))(\frac{\partial \vec{g}}{\partial \xi^T}(\xi(t) - \hat{\xi}_{t|t}) + \vec{\zeta}(t))^T \mid \vec{Y}(t), \vec{U}(t), v(t)]$$
$$= F_t \Omega_{t|t} F_t^T + \Psi(t)$$

$$\Psi(t) = \begin{pmatrix} Q(t) & 0 \\ 0 & Z(t) \end{pmatrix} \text{ and } F_t = \begin{pmatrix} \frac{\partial f}{\partial X^T} & \frac{\partial f}{\partial \theta^T} \\ 0 & I \end{pmatrix} = \begin{pmatrix} A(\vec{\theta}, \vec{U}(t)) & \frac{\partial}{\partial \theta^T}[A(\vec{\theta}, \vec{U}(t))\vec{X} + C(\vec{\theta})v(t)] \\ 0 & I \end{pmatrix}_{\theta = \hat{\theta}_{t|t}, X = \hat{X}_{t|t}}$$

ity with the idea proposed by Granger as long as we take two deterministic inputs $u(t)$ and $v(t)$ into consideration. Causality could be defined in both time and frequency domains.

Time Domain Formulation

For the simplicity of notation, here we only formulate UCM for two time series X_t and Y_t. Generalization to multi time-series case could be easily made.

Assume that X_t and Y_t have the following representation:

$$X_t = \sum_{j=1}^{p} [a_{1j} + u(t-j)b_{1j}]X_{t-j} + C_{1x}v(t-1) + \varepsilon_{1t}$$
$$Y_t = \sum_{j=1}^{p} [d_{1j} + u(t-j)e_{1j}]Y_{t-j} + C_{1y}v(t-1) + \varepsilon_{2t}$$

A joint autoregressive representation in UCM which includes the past information of both processes takes the form:

$$X_t = \sum_{j=1}^{p} [a_{2j} + u(t-j)b_{2j}]X_{t-j} + \sum_{j=1}^{p} [d_{2j} + u(t-j)e_{2j}]Y_{t-j} + C_{2x}v(t-1) + \varepsilon_{3t}$$
$$Y_t = \sum_{j=1}^{p} [f_{2j} + u(t-j)g_{2j}]X_{t-j} + \sum_{j=1}^{p} [h_{2j} + u(t-j)k_{2j}]Y_{t-j} + C_{2y}v(t-1) + \varepsilon_{4t}$$

where p is the maximum number of lagged observations included in the model. $\varepsilon_{it}, i = 1,2,3,4$ are prediction errors with variances Σ_i and are uncorrelated over time. Then, according to the causality definition of Granger, if the prediction of one process can be improved by incorporating the past information of the second process, then the second process causes the first process. So, in UCM, we define that if the variance of prediction error for the process X_t is reduced by the inclusion of the past information of the process Y_t, then a causal relation from Y_t to X_t exists. This can be quantified as:

$$F_{Y \to X} = \ln \frac{\Sigma_1}{\Sigma_3}$$

If $F_{Y \to X} > 0$, there is causal influence from Y_t to X_t and if $F_{Y \to X} = 0$, there isn't. Similarly, we can define the causal influence from X_t to Y_t as:

$$F_{X \to Y} = \ln \frac{\Sigma_2}{\Sigma_4}$$

Frequency Domain Formulation

Define the lag operator L to be $LX_t = X_{t-1}$ and assume here that the input $u(t)$ is a constant, i.e. $u(t)=u$ to avoid the appearance of nonlinearity. In this case, the joint representation of both processes can be expressed as:

$$X_t = \sum_{j=1}^{p} \tilde{a}_{2j} X_{t-j} + \sum_{j=1}^{p} \tilde{b}_{2j} Y_{t-j} + C_{2x} v(t-1) + \varepsilon_{3t}$$
$$Y_t = \sum_{j=1}^{p} \tilde{c}_{2j} X_{t-j} + \sum_{j=1}^{p} \tilde{d}_{2j} Y_{t-j} + C_{2y} v(t-1) + \varepsilon_{4t}$$

Rewrite it in terms of lag operator, we obtain:

$$\begin{pmatrix} \tilde{a}_2(L) & \tilde{b}_2(L) \\ \tilde{c}_2(L) & \tilde{d}_2(L) \end{pmatrix} \begin{pmatrix} X_t \\ Y_t \end{pmatrix} = \begin{pmatrix} \varepsilon_{3t} \\ \varepsilon_{4t} \end{pmatrix} + v(t-1) \begin{pmatrix} C_{2x} \\ C_{2y} \end{pmatrix}$$

where $\tilde{a}_2(0) = 1, \tilde{b}_2(0) = 0, \tilde{c}_2(0) = 0, \tilde{d}_2(0) = 1$.

Since what we really care about is the causal relationship caused by the intrinsic connection of the state variables rather than the outside driving force, i.e. the input $v(t)$, after fitting the model and getting the covariance matrix of the prediction error, it's reasonable to eliminate $v(t)$ and just focus on the intrinsic causal influence in the frequency domain. After normalizing the equation following the transformation proposed by Geweke (Geweke, 1982, 1984) to eliminate the cross term in the spectra and applying Fourier transform to both sides, we have:

$$\begin{pmatrix} \bar{a}_2(\omega) & \bar{b}_2(\omega) \\ \bar{c}_2(\omega) & \bar{d}_2(\omega) \end{pmatrix} \begin{pmatrix} X(\omega) \\ Y(\omega) \end{pmatrix} = \begin{pmatrix} \bar{E}_x(\omega) \\ \bar{E}_y(\omega) \end{pmatrix}$$

Recasting this equation into the transfer function format, we obtain:

$$\begin{pmatrix} X(\omega) \\ Y(\omega) \end{pmatrix} = \begin{pmatrix} H_{xx}(\omega) & H_{xy}(\omega) \\ H_{xy}(\omega) & H_{yy}(\omega) \end{pmatrix} \begin{pmatrix} \bar{E}_x(\omega) \\ \bar{E}_y(\omega) \end{pmatrix}$$

After proper ensemble averaging we have the spectral matrix:

$$S(\omega) = H(\omega) \Sigma H^*(\omega) = \begin{pmatrix} S_{xx} & S_{xy} \\ S_{xy} & S_{yy} \end{pmatrix}$$

where * denotes the complex conjugate and matrix transpose and $\Sigma = \begin{pmatrix} \Sigma_{xx} & \Sigma_{xy} \\ \Sigma_{xy} & \Sigma_{yy} \end{pmatrix}$ is the covariance matrix of the prediction errors. Hence, we can define the causal influence from Y_t to X_t at frequency ω as:

$$f_{Y \to X}(\omega) = \ln \frac{S_{xx}(\omega)}{H_{xx}(\omega) \Sigma_{xx} H^*_{xx}(\omega)}$$

Note that although here we just provide the definition of pairwise Granger causality for UCM, it's obvious that similar methods can be easily applied to the definition of conditional, partial or complex Granger causality mentioned in the INTRODUCTION part in both time and frequency domains. Since the explicit meaning of the parameters in the UCM (i.e. the intrinsic coupling among state variables, the strength of the inputs to modulate the coupling and the influence of the inputs on the state variables directly), we can also get an idea of the connection of the state variables and how the inputs affect them from the fitted model before we translate it into a single number.

Numerical Examples

Example 1

The first example tests the performance of UCM algorithm to estimate the parameters as well as the state variables. The model comes from a traditional VAR model which is similar to the model used in the part Basis Pursuit and has been extensively applied to the tests of Granger causality where we added deterministic inputs *u*(*t*) and *v*(*t*) as well as observation variables to it. *u*(*t*) was assumed to a constant while *v*(*t*) was assumed to be a harmonic oscillator and had the form of sinusoidal function to model the biological rhythms which are a common phenomenon in various biological systems. Observation variables were designed to be nonlinear functions of the state variables since it's a real challenge to uncover state variables with nonlinear mapping from states to measurements. Specifically, we generated the time-series according to the equations in Box 3.

Where $\varepsilon_{it}, i = 1,2,3$were zero mean uncorrelated Gaussian noise with variances 0.5, 0.8 and 0.6 respectively and uncorrelated with $\varepsilon_{it}, i = 4.5.6$ whose variances were all 0.1. The additional inputs did not change the causal relationships of the state variables, i.e., $x_1(t)$ is a direct source to $x_2(t)$, $x_2(t)$ and $x_3(t)$ share a feedback loop. According to the framework of UCM, we have the equation in Box 4.

Simulations were performed for 2 seconds with sampling rate 1000 Hz. UCM algorithm was then used to estimate all the parameters $\tilde{A}_1, A_2, \vec{c}$ and state variables from deterministic inputs and noise observations.

Figure 5 (upper panel) shows that the parameters converged to their true values with only small fluctuations after several hundred data points. No prior knowledge was included here and all the initial values of the parameters were assigned to zeros. Since the covariance matrix *Z*(*t*) of the noise will affect the convergence of the parameters significantly, it was carefully controlled to ensure the speed and accuracy of the parameters. Specifically, it was first set to decay slowly to achieve faster convergence and then set to decay faster after two hundred time points to ensure a better accuracy.

Causality was calculated in both time and frequency domains and used bootstrap to construct confidence intervals. Specifically, we simulated the fitted model to generate a data set of 1000 realizations of 2000 time points and used 3 sigma as the confidence intervals. In this result, a causal connection was illustrated as part of the network if, and only if, the lower bound of the

Box 3.

$$\begin{cases} x_1(t) = 0.95\sqrt{2}x_1(t-1) - 0.9025x_1(t-2) + 0.1x_1(t-1) + 0.1\cos[\frac{2\pi}{50}(t-1)] + \varepsilon_1(t) \\ x_2(t) = -0.5x_1(t-1) + 0.5x_3(t-2) - 0.8 \times 0.1x_2(t-1) + \varepsilon_2(t) \\ x_3(t) = -0.5x_2(t-1) + 0.5x_3(t-1) + 1.2 \times 0.1x_3(t-1) - 0.1\cos[\frac{2\pi}{50}(t-1)] + \varepsilon_3(t) \end{cases}$$

$$\begin{cases} y_1(t) = x_1(t) + \varepsilon_4(t) \\ y_2(t) = x_2(t) + x_3(t) + \varepsilon_5(t) \\ y_3(t) = x_1(t) \cdot x_3(t) + \varepsilon_6(t) \end{cases}$$

Box 4.

$$A_1=\begin{pmatrix}a_{11}^1 & a_{12}^1 & a_{13}^1\\ a_{21}^1 & a_{22}^1 & a_{23}^1\\ a_{31}^1 & a_{32}^1 & a_{33}^1\end{pmatrix}=\begin{pmatrix}0.95\sqrt{2} & 0 & 0\\ -0.5 & 0 & 0\\ 0 & -0.5 & 0.5\end{pmatrix},$$

$$A_2=\begin{pmatrix}a_{11}^2 & a_{12}^2 & a_{13}^2\\ a_{21}^2 & a_{22}^2 & a_{23}^2\\ a_{31}^2 & a_{32}^2 & a_{33}^2\end{pmatrix}=\begin{pmatrix}-0.9025 & 0 & 0\\ 0 & 0 & 0.5\\ 0 & 0 & 0\end{pmatrix},$$

$$B_1=\begin{pmatrix}b_{11}^1 & b_{12}^1 & b_{13}^1\\ b_{21}^1 & b_{22}^1 & b_{23}^1\\ b_{31}^1 & b_{32}^1 & b_{33}^1\end{pmatrix}=\begin{pmatrix}1 & 0 & 0\\ 0 & -0.8 & 0\\ 0 & 0 & 1.2\end{pmatrix}, B_2=0, \vec{c}=\begin{pmatrix}1\\ 0\\ -1\end{pmatrix},$$

$$\tilde{A}_1=\begin{pmatrix}\tilde{a}_{11}^1 & \tilde{a}_{12}^1 & \tilde{a}_{13}^1\\ \tilde{a}_{21}^1 & \tilde{a}_{22}^1 & \tilde{a}_{23}^1\\ \tilde{a}_{31}^1 & \tilde{a}_{32}^1 & \tilde{a}_{33}^1\end{pmatrix}=A_1+uB_1=\begin{pmatrix}1.4435 & 0 & 0\\ -0.5 & -0.08 & 0\\ 0 & -0.5 & 0.62\end{pmatrix},$$

$$u(t)\equiv 0.1, v(t)=0.1\cos[\frac{2\pi}{50}(t-1)]$$

confidence interval of the causality was greater than zero. The results show that UCM can detect the causal relationship correctly in both time and frequency domains.

Example 2

It is quite common that the state variables of a biological system are affected by some stimulus that we could not include into the model due to the complexity of the stimulus and the limitation of computational efforts. Since traditional GCM is not tailored particularly well for biological experiments where we are often faced with the case of the data being recorded with and without a stimulus present, we could only treat different situations separately and sometimes this will lead to different network structures as the case for the gene network data considered in (Cantone et al., 2009; Camacho & Collins, 2009). However, this is not so convincing since in many cases the time gap between two adjacent stimuli is very short and we would expect the network structure to remain unchanged during the whole experiment and there exists a common and true structure. The second example shows that when we are faced with a system affected by intermittent inputs, it is quite probable to get a misleading structure if we ignore the influence of these inputs and use a traditional VAR model to detect the causality even if we use a high-order model. Furthermore, as a more biophysically constrained model with the inclusion of inputs, UCM could solve the problem to some extent.

We still use the same connection coefficients between the three state variables but add an additional simple constant input function p:

$$\begin{aligned}x_1(t)&=0.95\sqrt{2}x_1(t-1)-0.9025x_1(t-2)+p+\varepsilon_1(t)\\ x_2(t)&=-0.5x_1(t-1)+0.5x_3(t-2)+p+\varepsilon_2(t)\\ x_3(t)&=-0.5x_2(t-1)+0.5x_3(t-1)-p+\varepsilon_3(t)\end{aligned} \quad (3)$$

Figure 5. (Upper panel) Examples of the estimated parameters. All initial values were set to zero. The covariance matrix Z(t) is first set to decay slowly to achieve faster convergence and then set to decay faster after two hundred time points to ensure a better accuracy. (Lower panel) Frequency decomposition of all pairs of state variables.

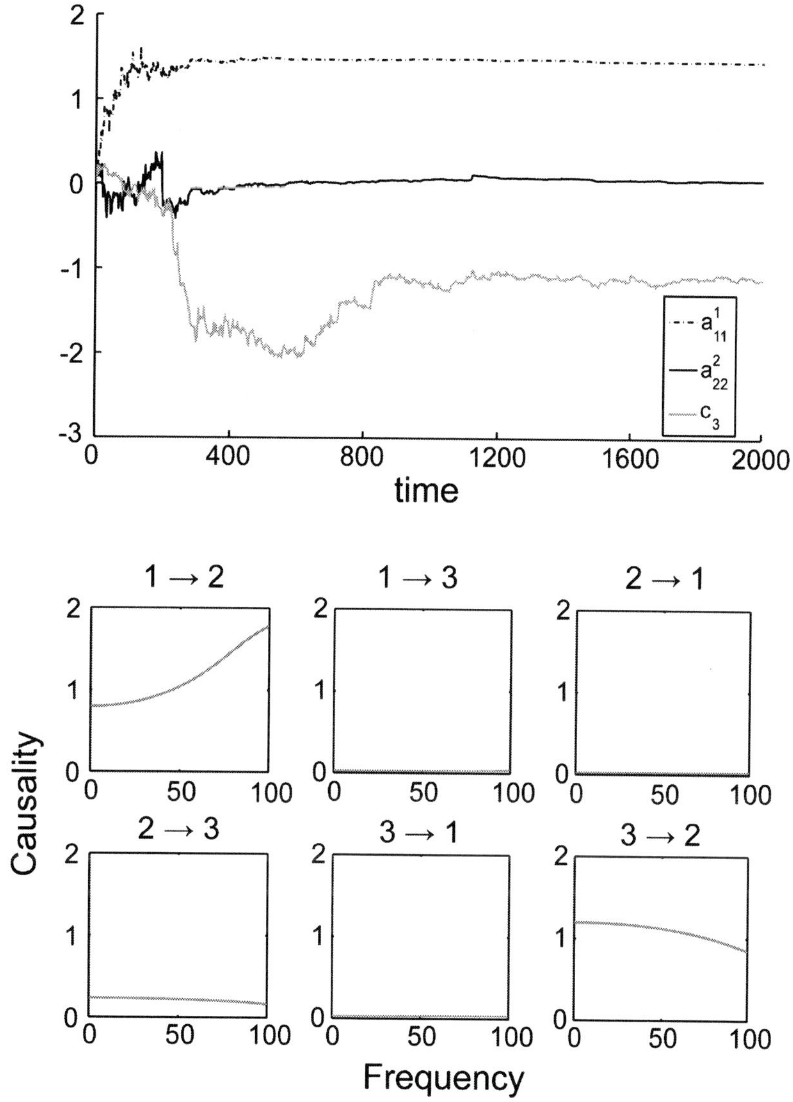

where $\varepsilon_i(t), i = 1,2,3$ were zero mean uncorrelated Gaussian noise with variances 0.5, 0.8 and 0.6 respectively and p=0.5. The observation variables $y_i(t), i = 1,2,3$ were identical to the state variables with observation noise 0.1. It is self-evident that the network structure is the same as the examples mentioned above but a direct application of traditional GCM led to the misleading structure shown in the upper panel of Figure 6 with confidence intervals where two additional causal relationships (i.e. 1→3 and 3→1) are presented.

To exemplify the application of UCM to establishing causal influence in a intermittently perturbed system, we generated a time series of

Figure 6. (Upper panel) Confidence intervals of all links between units. The data were generated with Equation (3) (the AR model with a constant input), but traditional VAR(10) was applied to detect the causal influence. (Lower panel) Confidence intervals of all links between units. Significant causal relationships are highlighted on the X axis as bold text. The data were generated with Equation (4) where p~N(0,1) and c_i ~ N(0,1), i=1,2,3.

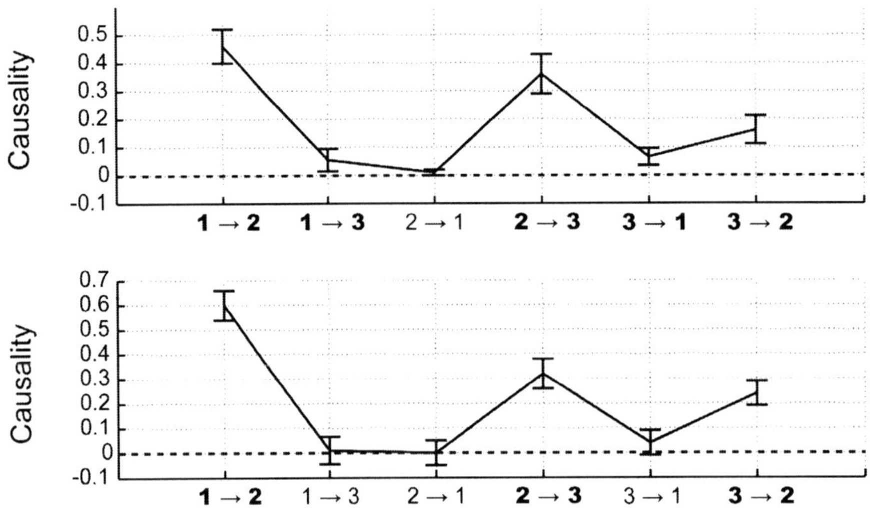

10000 time points which was composed of 10 segments with equal length, i.e., 1—1000, 1001—2000, · · ·, 9001—10000. Each segment took the form:

$$\begin{pmatrix} x_1(t) \\ x_2(t) \\ x_3(t) \end{pmatrix} = A_1 \begin{pmatrix} x_1(t-1) \\ x_2(t-1) \\ x_3(t-1) \end{pmatrix} + A_2 \begin{pmatrix} x_1(t-2) \\ x_2(t-2) \\ x_3(t-2) \end{pmatrix} + p \begin{pmatrix} c_1 \\ c_2 \\ c_3 \end{pmatrix} + \begin{pmatrix} \varepsilon_1(t) \\ \varepsilon_2(t) \\ \varepsilon_3(t) \end{pmatrix} \quad (4)$$

where A_1 and A_2 were the same matrices used in the last example, $\varepsilon_i(t), i = 1,2,3$ were zero mean uncorrelated Gaussian noise with variances 0.5, 0.8 and 0.6 respectively. The five segments 1—1000, 2001—3000, · · ·, 8001—9000 were generated according to the above toy model without input, i.e., p=0, while the remaining five segments were assumed to include input of random intensity which would also affect the state variables randomly. Specifically, within each segment, p was assigned a random value which was generated with the normal distribution p~$N(0,1)$, and the same was the case with c_i: $c_i \sim N(0,1)$, i=1,2,3. Observation variables were still assumed to be identical to the state variables with the variation of observation as 0.1. Hence, the network structure of the three state variables remains unchanged while each state variable is affected by some input that we don't know the intensity of. The lower panel of Figure 6 shows that UCM can recover the network structure correctly in this situation.

To exemplify the application of Granger causality and UCM to establishing causality in a specific biological system, we have applied it to microarray data as well as local field potential (LFP) data.

LFP from Left and Right Brain Hemisphere

To exemplify the direct application of UCM to establish causality in a specific biological system, we have applied it to local field potential (LFP) data recorded in the sheep inferotemporal cortex (IT) of both left and right hemispheres before and after learning a visual face discrimination task. Local field potential data were obtained from 64-channel multielectrode arrays implanted in the right and left inferior temporal cortices of three sheep (one sheep only had electrodes in the right hemisphere) as previously described (Kendrick et al., 2009). Recordings were made while the animals were presented with pairs of faces which they were required to discriminate between using an operant response in order to obtain a food reward. In between face pair presentations the animals were presented with a visual fixation stimulus (a white spot on a black screen). Recordings were made during sessions of 20–40 trials where they were either still learning the discrimination or had successfully achieved the learning criterion (>75% correct choice of rewarded face).

There is considerable interest in establishing functional differences between the ways the left and right brain hemispheres interact and process information (Shinohara et al., 2008). Recent research has revealed the asymmetrical processing of faces in the sheep brain similar to humans (Kendrick, 2006). Learning also alters both local and population based encoding in sheep IT as well as frequency oscillations in both hemispheres (Kendrick et al., 2009). It has been hypothesized that the left hemisphere specializes in controlling routine and tends to focus on local aspects of the stimulus while the right hemisphere specializes in responding to unexpected stimuli and tends to deal with the global environment (Turgeon, 1994). The direct application of our UCM model will be helpful to test this hypothesis.

Two main problems exist to detect the network structure from the electrodes in both hemispheres. One is that due to the large number of the electrodes, at least several thousand parameters need to be fit to recover the global structure of the network. The other is that the duration of visual stimulus is quite short (1—3 seconds) and it can be expected that the connection will not change in such a short time. However, if we treat the time-series before and after the stimulus separately, two different structures will occur. To avoid these two issues, we randomly selected 3 time series from each region respectively for each session and detected the network structure for the six electrodes and repeated this for 100 times for each session to get an overall illustration of the network. At the same time, we used UCM to detect the unified network structure and regarded the stimulus as a constant input. The inclusion of the stimulus signal will certainly make the model more reasonable.

Figure 7 shows the mean connections within and between the left and right IT calculated using UCM and as a function of learning. The results clearly demonstrate an asymmetry between the hemispheres. There was a significant decrease in the number of connections from the left to the right and an increase in connections within the right during the process of learning. Furthermore, a strong negative correlation between the number of left to right connections and the number within the right IT was discovered for both animals which suggests that the left to right IT connections may exert some form of inhibitory control over the number within the right IT, something needs to be weakened for new face discriminations to be learned.

The frequency decomposition of UCM offers us a chance to obtain a more detailed picture of causal interactions. We concentrate on the two main frequency bands: theta band (4—8Hz) and gamma band (30—70Hz) present in the IT recording data and which have been extensively linked to mechanisms of learning and memory (Kendrick et al., 2009). To discover the effect of learning on different frequency bands, we looked at the following two quantities:

mean ratio = mean interaction in the theta band /

mean interaction in the gamma band

Figure 7. Asymmetry between left and right hemisphere in the time domain. The mean connections from left hemisphere to right hemisphere, right hemisphere to left hemisphere and within both regions with the three bars corresponding to the results before learning (left bar), after learning (middle bar), and one month after learning (right bar) in Sheep B. Significant changes after t-test are marked by arrows (right to left, all pairs are not significant, as indicated by "none"; within the right hemisphere, all pairs are significant, marked by "all"). For Sheep C, an additional bar (one week after learning) is added (the third bar). Only significant changes from left to right and within the right hemisphere are indicated by arrows.

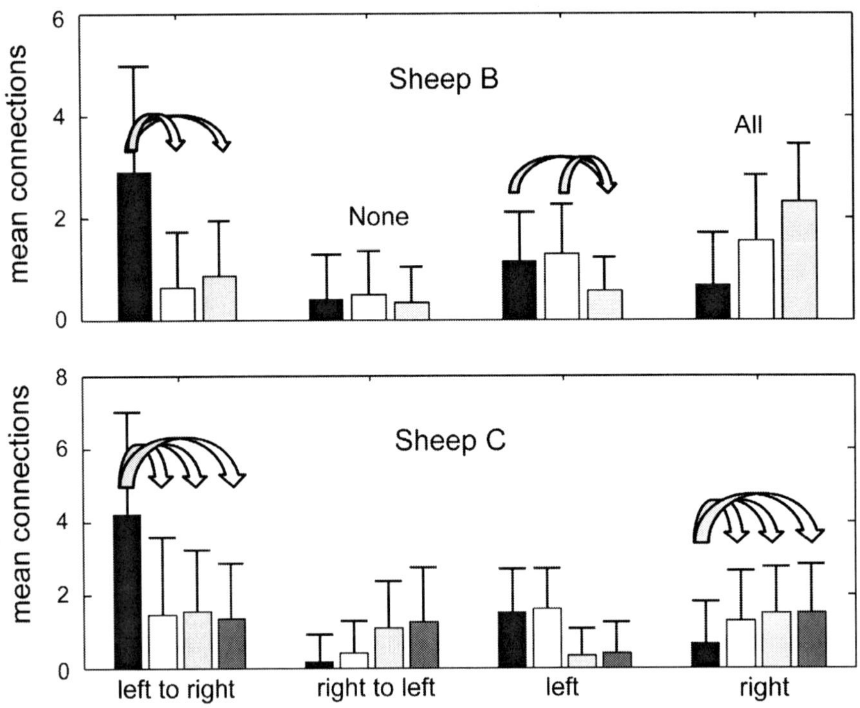

max ratio = max interaction in the theta band / max interaction in the gamma band

The left panel of Figure 8 shows the mean and maximum ratio integrating the data from all the three sheep in the experiment and at different stages of learning. It can be seen that both the mean and maximum ratios in the right hemisphere IT are about double those in the left hemisphere. This clearly indicates that for the right hemisphere, the theta band interaction is more dominant, i.e., the right hemisphere deals more with signals of lower frequency.

In order to provide a deeper insight into this frequency story, we computed the ratios at different stages of learning. The right panel of Figure 8 shows the mean and maximum ratio in two sheep before learning, after learning and a month after learning. An additional set of data, one week after learning, is for sheep C only. The most noticeable change is the reduction in the interactions in the theta band (low frequencies) in the right IT which occurs after learning and is maintained subsequently. This may suggest that during the course of learning new faces the right IT uses a

Figure 8. Asymmetry in the frequency domain interactions. (Left panel) Mean and maximum ratio using all the three sheep before and after learning. (Right panel) Mean and maximum ratio of sheep B before learning, after learning and one month after learning and mean and maximum ratio of sheep C before learning (the first bar), immediately after learning (the second bar), one week after learning and one month after learning (the third and the fourth bar).

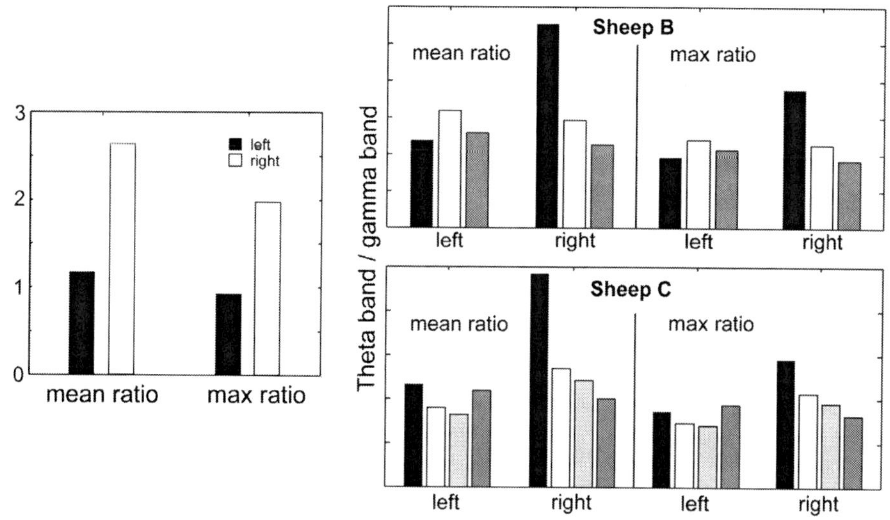

more global mode of encoding to promote more rapid learning and that once learning has successfully occurred the right IT shifts to a more localized encoding strategy for maintaining learning.

All these results are in broad agreement with recent proposals that the left hemisphere is more involved in local encoding and the right in global encoding (MacNeilage et al., 2009). The results of UCM analysis of IT LFP data provide the first evidence for connectivity changes between and within left and right ITs as a result of face recognition learning which shows the power of UCM.

CONCLUSION

In this chapter, the origin and some current developments of Granger causality are reviewed with applications in molecular biology and physiology. A GUI is presented to facilitate the applications of Granger causality.

When dealing with the ever-increasing large amount of data, preprocessing of these datasets including up-sampling and down-sampling is usually needed based on different features of the data from various sources. To make Granger causality more applicable, Basis Pursuit is introduced to obtain continuous representation of stationary time-series from an over-complete dictionary of waveforms. Its great ability to recover missed sample points and preserve causal relationships makes it a good choice to deal with non-uniformly sampled data. We can also expect that it could be further applied to the joint analysis of time-series with different time scales and from various sources.

We also unify two seemingly different causal models: Granger Causal Model (GCM) and Dynamic Causal Model (DCM) to create a more general model: Unified Causal Model (UCM) which includes both GCM and DCM as special cases. UCM shares the advantages of both models which makes it more biophysical constrained and full of interpretability power. It offers a chance to have deeper insight into different kinds of systems especially those affected by some stimulus that could not be formulated explicitly.

These approaches are intensively tested in toy models and applied to different datasets. The gene circuit of Arabidopsis recovered from Granger causality is in broad agreement with the known structure in the literature. The application of UCM to local field potential data provides the first evidence for connectivity changes between and within left and right inferotemporal cortexes as a result of face recognition learning. All these examples show that the development of Granger causality will help us take full advantage of high-throughput data in biology and obtain more reliable results.

REFERENCES

Bar-Joseph, Z. (2004). Analyzing time series gene expression data. *Bioinformatics (Oxford, England)*, *20*(16), 2493. doi:10.1093/bioinformatics/bth283

Camacho, D., & Collins, J. (2009). Systems biology strikes gold. *Cell*, *137*(1), 24–26. doi:10.1016/j.cell.2009.03.032

Cantone, I., Marucci, L., Iorio, F., Ricci, M., Belcastro, V., & Bansal, M. (2009). A yeast synthetic network for in vivo assessment of reverse-engineering and modeling approaches. *Cell*, *137*(1), 172–181. doi:10.1016/j.cell.2009.01.055

Chen, S., Donoho, D., & Saunders, M. (2001). Atomic decomposition by basis pursuit. *SIAM Review*, *43*(1), 129–159. doi:10.1137/S003614450037906X

Friston, K. (2009). Causal modelling and brain connectivity in functional magnetic resonance imaging. *PLoS Biology*, *7*(2). doi:10.1371/journal.pbio.1000033

Friston, K., Harrison, L., & Penny, W. (2003). Dynamic causal modelling. *NeuroImage*, *19*(4), 1273–1302. doi:10.1016/S1053-8119(03)00202-7

Ge, T., Kendrick, K. & Feng, J. (2009). *A novel extended Granger causal model approach demonstrates brain hemispheric differences during face recognition learning*.

Geweke, J. (1982). Measurement of linear dependence and feedback between multiple time series. *Journal of the American Statistical Association*, *77*(378), 304–313. doi:10.2307/2287238

Geweke, J. (1984). Measures of conditional linear dependence and feedback between time series. *Journal of the American Statistical Association*, *79*(388), 907–915. doi:10.2307/2288723

Gourvitch, B., Bouquin-Jeanns, R., & Faucon, G. (2006). Linear and nonlinear causality between signals: Methods, examples and neurophysiological applications. *Biological Cybernetics*, *95*(4), 349–369. doi:10.1007/s00422-006-0098-0

Granger, C. (1969). Investigating causal relations by econometric models and cross-spectral methods. *Econometrica: Journal of the Econometric Society*, *37*(3), 424–438.

Guo, S., Seth, A., Kendrick, K., Zhou, C., & Feng, J. (2008). Partial Granger causality-Eliminating exogenous inputs and latent variables. *Journal of Neuroscience Methods*, *172*(1), 79–93. doi:10.1016/j.jneumeth.2008.04.011

Guo, S., Wu, J., Ding, M., & Feng, J. (2008). Uncovering interactions in the frequency domain. *PLoS Computational Biology*, *4*(5). doi:10.1371/journal.pcbi.1000087

Ladroue, C., Guo, S., Kendrick, K. & Feng, J. (2009). *Beyond element-wise interactions: Identifying complex interactions in biological processes*.

Locke, J., Kozma-Bognar, L., Gould, P., & Feher, B., Kevei, Nagy, F., et al. (2006). Experimental validation of a predicted feedback loop in the multi-oscillator clock of Arabidopsis thaliana. *Molecular Systems Biology*, *2*(1).

Ueda, H. (2006). Systems biology flowering in the plant clock field. *Molecular Systems Biology*, *2*(1).

Wiener, N. (1956). The theory of prediction. *Modern Mathematics for Engineers*, *1*, 125–139.

Wu, J., Liu, X., & Feng, J. (2008). Detecting causality between different frequencies. *Journal of Neuroscience Methods*, *167*(2), 367–375. doi:10.1016/j.jneumeth.2007.08.022

Zou, C., & Feng, J. (2009). Granger causality vs. dynamic Bayesian network inference: A comparative study. *BMC Bioinformatics*, 10.

Zou, C., Kendrick, K.M. & Feng, J. (2009). The fourth way: Granger causality is better than the three other reverse-engineering approaches. *Cell*.

ADDITIONAL READING

Chen, S., Donoho, D., & Saunders, M. (2001). Atomic decomposition by basis pursuit. *SIAM Review*, *43*(1), 129–159. doi:10.1137/S003614450037906X

Friston, K. (2009). Causal modelling and brain connectivity in functional magnetic resonance imaging. *PLoS Biology*, *7*(2). doi:10.1371/journal.pbio.1000033

Friston, K., Harrison, L., & Penny, W. (2003). Dynamic causal modelling. *NeuroImage*, *19*(4), 1273–1302. doi:10.1016/S1053-8119(03)00202-7

Ge, T., Kendrick, K., & Feng, J. (2009). A Novel Extended Granger Causal Model Approach Demonstrates Brain Hemispheric Differences during Face Recognition Learning.

Geweke, J. (1982). Measurement of linear dependence and feedback between multiple time series. *Journal of the American Statistical Association*, *77*(378), 304–313. doi:10.2307/2287238

Geweke, J. (1984). Measures of conditional linear dependence and feedback between time series. *Journal of the American Statistical Association*, *79*(388), 907–915. doi:10.2307/2288723

Granger, C. (1969). Investigating causal relations by econometric models and cross-spectral methods. *Econometrica: Journal of the Econometric Society*, *37*(3), 424–438.

Guo, S., Seth, A., Kendrick, K., Zhou, C., & Feng, J. (2008). Partial Granger causality--Eliminating exogenous inputs and latent variables. *Journal of Neuroscience Methods*, *172*(1), 79–93. doi:10.1016/j.jneumeth.2008.04.011

Guo, S., Wu, J., Ding, M., & Feng, J. (2008). Uncovering interactions in the frequency domain. *PLoS Computational Biology*, *4*(5). doi:10.1371/journal.pcbi.1000087

Ladroue, C., Guo, S., Kendrick, K., & Feng, J. (2009). Beyond element-wise interactions: Identifying complex interactions in biological processes.

Wiener, N. (1956). The theory of prediction. *Modern mathematics for engineers. Series*, *1*, 125–139.

Wu, J., Liu, X., & Feng, J. (2008). Detecting causality between different frequencies. *Journal of Neuroscience Methods*, *167*(2), 367–375. doi:10.1016/j.jneumeth.2007.08.022

Zou, C., & Feng, J. (2009). Granger causality vs. dynamic Bayesian network inference: a comparative study. *BMC Bioinformatics*, 10.

Zou, C., Kendrick, K. M., & Feng, J. (2009). The fourth way: Granger causality is better than the three other reverse-engineering approaches. *Cell*, http://www.cell.com/ comments/S0092-8674(09)00156-1.

KEY TERMS AND DEFINTIONS

Basis Pursuit: A technique to obtain a continuous representation of a signal by decomposing it into a superposition of elementary waveforms with sparse coefficients.

Causal Network: A directed network which illustrates the causal dependencies of all the components in the network.

Complex Granger Causality: An extension of Granger causality for determining the causal relationship between groups of time series.

Conditional Granger Causality: An extension of Granger causality for determining whether the causal relationship from one time series to another is direct or mediated by a third time series.

Dynamic Causal Modeling: The aim of Dynamic Causal Modeling (DCM) is to make inferences and estimate the causal architecture of coupled or distributed dynamical systems. It relies on comparing models of how data are generated, where these Dynamic Causal Models are formulated in terms of stochastic or ordinary differential equations. These equations model the dynamics of hidden states in the nodes of a probabilistic graphical model, where conditional dependencies are parameterized in terms of directed effective connectivity.

Granger Causality: A technique for determining whether one time series is the cause of another one.

Partial Granger Causality: An extension of Granger causality to eliminate the effect of common input from latent variables when detecting the causal relationships among several time series.

Reverse Engineering: The process of discovering the technological principles of a device, object or system through analysis of its structure, function and operation.

Chapter 23
Connecting Microbial Population Genetics with Microbial Pathogenesis:
Engineering Microfluidic Cell Arrays for High-throughput Interrogation of Host-Pathogen Interaction

Palaniappan Sethu
University of Louisville, USA

Kalyani Putty
University of Louisville, USA

Yongsheng Lian
University of Louisville, USA

Awdhesh Kalia
University of Louisville, USA

ABSTRACT

A bacterial species typically includes heterogeneous collections of genetically diverse isolates. How genetic diversity within bacterial populations influences the clinical outcome of infection remains mostly indeterminate. In part, this is due to a lack of technologies that can enable contemporaneous systems-level interrogation of host-pathogen interaction using multiple, genetically diverse bacterial strains. This chapter presents a prototype microfluidic cell array (MCA) that allows simultaneous elucidation of molecular events during infection of human cells in a semi-automated fashion. It shows that infection of human cells with up to sixteen genetically diverse bacterial isolates can be studied simultaneously. The versatility of MCAs is enhanced by incorporation of a gradient generator that allows interrogation of host-pathogen interaction under four different concentrations of any given environmental variable at the same time. Availability of high throughput MCAs should foster studies that can determine how differences in bacterial gene pools and concentration-dependent environmental variables affect the outcome of host-pathogen interaction.

DOI: 10.4018/978-1-60960-491-2.ch023

INTRODUCTION

"...organisms of the most different sorts are constructed from the very same battery of genes. The diversity of life forms results from small changes in the regulatory systems that govern expression of these genes."

—Francois Jacob, Of Flies, Mice and Men

Decades of population-based molecular epidemiological studies have revealed three general aspects of bacterial pathogenesis: (1) there is extensive genetic diversity within most bacterial species; (2) sibling bacterial species, and even strains within species, differ in their virulence potential; and (3) that normally harmless bacteria can be pathogenic in an injured or immuno-compromised host (Dykhuizen & Kalia, 2008). Thus, for several decades, studies aimed at understanding the mechanisms of bacterial pathogenesis have focused on addressing two major questions: (1) what are the molecular and evolutionary mechanisms that generate and maintain genetic diversity among natural bacterial populations? And (2) what are the mechanisms employed (or exploited) by bacterial pathogens to cause human disease? While several strides have been made in our understanding of both aspects, we have only just begun to comprehend the impressive complexity in the molecular nature of host-pathogen interactions, and the resulting clinical outcomes. Even infection with genetically related bacterial isolates can result in multiple clinical outcomes that range from asymptomatic and benign carriage to development of severe and fatal disease in the human host. These complex interactions are perhaps best illustrated by the two bacterial pathogens: *Helicobacter pylori* and *Streptococcus pyogenes*. Both these bacterial species are considered obligate human-specific pathogens and are responsible for a wide-variety of clinical disorders (Atherton & Blaser, 2009; Carapetis JR, Steer AC, Mulholland EK, & M., 1995). For example, *H. pylori* can cause gastric cancer, peptic ulcers, duodenal ulcers, atrophic gastritis, and *S. pyogenes* can cause a spectrum of diseases that range from mild (strep throat) to severe invasive diseases (e.g. TSLS) and autoimmune syndromes such rheumatic fever. Paradoxically, both *H. pylori* and *S. pyogenes* can also establish long-term colonization in their human host without causing any significant disease. Disconcertingly, there are no reliable markers that can predict the likelihood of a particular clinical outcome following infection by either *H. pylori* or *S. pyogenes*. Thus, understanding the molecular basis for such clearly distinct clinical outcomes of infections with genetically related bacterial isolates is a fundamental problem in pathogenesis and, which is necessary to address in order to develop predictive models for infectious disease risks.

Systems approaches to develop predictive network models of disease are based on the notion that disease-perturbed gene and protein regulatory networks differ from their normal counterparts. Host cellular response to an infectious agent likely involves interactions at multiple gene loci and environmental cues in particular, contact with bacterial cell and/or its secreted products. The response might include fluxes in gene regulatory networks that integrate dynamically changing signals and activate batteries of genes mediating physiological responses via proteins that may function alone, in complexes, or in networks that arise from protein interactions (Nicholson, Holmes, Lindon, & Wilson, 2004). Any host response invokes an equivalent "negotiating response" from the bacterium (Figure 1). This dynamical interaction of cellular entities is representative of the heterogeneity that characterizes biology - from the differences in how individual cells respond to bacterial stimulus, to the diversity of cell types and environments within real tissues and their associated microbiomes.

In addition, many infectious diseases, in particular those that result from chronic infections, have a multifactorial etiology (Ewald, 2004);

however, majority of studies aimed at characterizing bacterial-host interaction typically use one standard laboratory strain under a limited set of test conditions thereby providing only a snap-shot of the pathogenetic mechanisms and thus ignoring the genetic heterogeneity that characterizes any bacterial species. Thus, constructing globally predictive network models integrating host responses to infection with their subsequent management by bacteria has remained intractable (Hood, Heath, Phelps, & Lin, 2004). In order to develop biomarkers that reliably predict the clinical outcome of infection (eg., how likely is an individual to develop gastric cancer following *H. pylori* infection?) we must begin to address the following two additional questions: (1) what are the global differences in host response to infection with genetically diverse isolates (of any given bacterial species)? And conversely, (2) how do differences in host responses shape bacterial life-styles (pathogenic or benign)? In order to understand the complex interplay between host and pathogen we must incorporate factors such as (a) bacterial heterogeneity (b) host heterogeneity, as well as (c) identify the signaling pathways that are engaged in the host, by different bacterial strains (virulent vs. benign) and study these interactions contemporaneously. However, studying global host cell responses to multiple genetically characterized bacterial isolates alone can become prohibitively expensive, and are difficult to perform reproducibly under experimental conditions and in a contemporaneous manner. Thus, techniques are needed that are highly parallel, allow for multiple types of measurements, and are ideally miniaturized (to keep costs low) and automated (to enhance reproducibility).

Microfluidics is the study of fluids in the microscale (Melin & Quake, 2007). Using simple soft-lithographic techniques (Whitesides, Ostuni, Takayama, Jiang, & Ingber, 2001) based on polymer molding, fabrication of microscale devices can be accomplished to perform several functions, some of which are not possible using

Figure 1. The dynamic interplay between bacterial and host diversity can result in diverse outcomes of infection that range from asymptomatic long-term colonization to severe disease. Diversity in bacterial populations can arise from mutation, recombination in coding sequences, and from regulatory differences that govern the timing and level of gene expression. We suggest that the trajectory of clinical outcome following infection depends on the 'signals' exchanged between the interacting partners (i.e., bacterial pathogen and the host), and that elucidating such signals to develop predictive models of infectious disease risk depends on incorporating bacterial genetic heterogeneity into the experimental design.

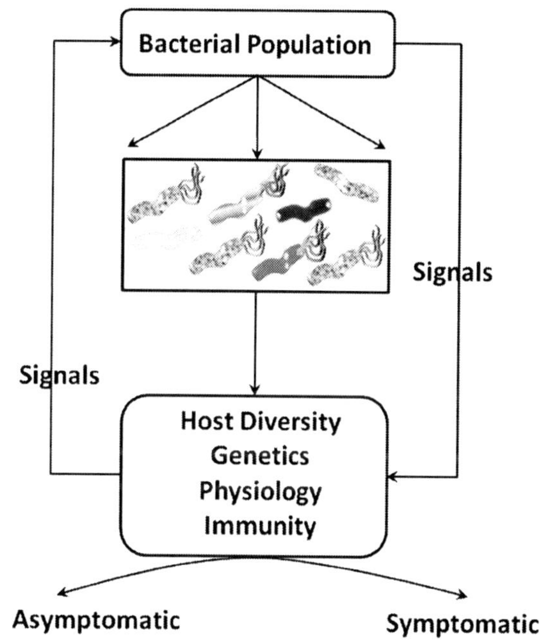

conventional macroscale techniques. By exploiting flow phenomenon and scaling, microfluidic devices can provide enabling technology for several biomedical applications (Einav, et al., 2008; Fan, Blumenfeld, El-Sayed, Chueh, & Quake, 2009; Gerber, Maerkl, & Quake, 2009; Gulati, et al., 2009; Kane, Zinner, Yarmush, & Toner, 2006; Lee, Snyder, & Quake, 2010). Here, we present a prototype high throughput microfluidic cell array (MCA) to characterize host gene regulatory networks following infection with genetically diverse bacterial strains. The MCA incorporates a gradient generator and a flow-encoded switch (FES) (King KR, Wang S, Jayaraman A, Yarmush ML, & M., 2008) thereby enabling simultaneous elucidation of host gene expression fluxes in response to combinations of concentration dependent (e.g. pH, salt concentration and lipid content in the culture medium) and time dependent (e.g. pro and anti- inflammatory cytokines) variables.

What Does Genetic Heterogeneity in Bacterial Populations Mean, and How Might It Influence Host Cellular Responses?

Designing specific interventions to prevent, control, treat and potentially eradicate infectious diseases depends critically on our understanding of pathogen population dynamics and structure in the context of their ability to cause human disease. Similarly, to identify components of host signaling pathways that can be utilized as predictive, diagnostic or therapeutic reagents, we must first understand how differences in a bacterial population structure translate to differences in host cellular responses. The two fundamental forces of evolution that shape bacterial populations structure are random genetic drift and natural selection (Dykhuizen & Kalia, 2008). Since population structure is determined by birth and death processes in finite populations, it is closely associated with random genetic drift (or fluctuations in allele frequencies simply by chance), and the effects of random drift are same genome-wide. Natural selection on the other hand favors certain alleles of a gene over others and thus can change the pattern of variation expected from random drift. To put in the context of pathogenesis, most bacterial proteins ('virulence factors') that are secreted to the bacterial membrane or extracellularly are targeted by natural selection and variants of such proteins are only retained in the bacterial genomes if they confer a survival advantage (eg., escaping host immune response or edge over other competing siblings) (Aspholm-Hurtig, et al., 2004). How such genome-wide variations in bacterial proteins affect host cell responses is less well understood.

Infectious diseases can be categorized as epidemic or endemic and acute or chronic. These categories imply different underlying differences in pathogen population structures (Gupta & Maiden, 2001). An infectious disease epidemic is likely to be caused by a bacterial clone that has rapidly infected many individuals within a locality. Thus the epidemic pathogen population often consists of a group of related clones that have caused epidemics in the host population in the past, and distant clones that exist at low frequencies in the host population. Deciphering, which host signaling pathways are most commonly deregulated by epidemic pathogen clones (by comparing them to low-frequency clones) can help minimize the impact of such epidemic outbreaks. In contrast, the population of an endemic pathogen is likely to consist of multiple distantly related clones, and two people infected at about the same time in the same general location are often colonized with different distantly related clones. This gives more opportunity for recombination and thus for clonal divergence promoted by recombination in endemic than in epidemic species. Infections with such diverse clones should induce dramatically different host cellular responses, which may correlate with the observed severe or benign clinical outcomes. Pathogens that establish chronic or persistent infections are expected to have more within-host diversity, and consequently even more recombination. In extreme cases, if the within-host diversity is high enough, a pattern of linkage equilibrium (lack clonal population structure because of random association of alleles) may be observed. In such cases, each individual isolate is likely to induce and manage host cell responses differently.

All bacterial reproduce clonally by binary fission. Bacteria may occasionally exchange their genomic content (recombination) with related or distant members using transduction, transformation or conjugation (Ochman, Lawrence, & Groisman, 2000). Thus, despite clonal reproduction, bacterial populations differ from each other in terms of their overall genetic diversity (Figure 2).

Bacterial populations that show little or no recombination among strains are largely clonal (*eg. M. tuberculosis, S. typhi*), and genetic differences among isolates arise from sequentially accumulated mutations following descent from recent ancestors (Holt, et al., 2008). These organisms can be epidemic or endemic and can cause acute or chronic infections. In many normally endemic, chronic pathogens an epidemic phase will produce transient clonality due to higher fitness of particular clones in otherwise recombining populations (*eg. S. pyogenes, N. meningitidis*) (Beres, et al., 2010). Alternatively, strains can be panmictic, or quite freely recombining, with very little evidence of clonality or epidemic spread (*eg. H. pylori* and also *N. gonorrhoeae*) (Feil & Spratt, 2001). Thus knowing the dynamics and constraints on the evolution of any given bacterial species is crucial towards developing an understanding of how genetic variation in such populations may differentially affect host cellular responses and thereby shape the observed differential clinical outcomes.

Microfluidic Cell Arrays: Historical Background of Cell Arrays

Studies at the cellular level provide highly specific information with regards to molecular signaling events. Measurement of temporal signaling patterns is currently limited to techniques such as northern blotting, western blotting, reverse transcription polymerase chain reaction (RT-PCR) and enzyme linked immunosorbent assays (ELISA). Recent emergence of high-throughput genomics and proteomics allows for simultaneous analysis

Figure 2. Despite clonal reproduction bacterial species differ in terms of genetic diversity accumulated over time. MLST data-sets for each bacterial species or group shown above were downloaded from www.MLST.net and nucleotide diversity per site was calculated using the Jukes-Cantor correction implemented in DNASP ver. 5.1. Vertical bars indicate standard error.

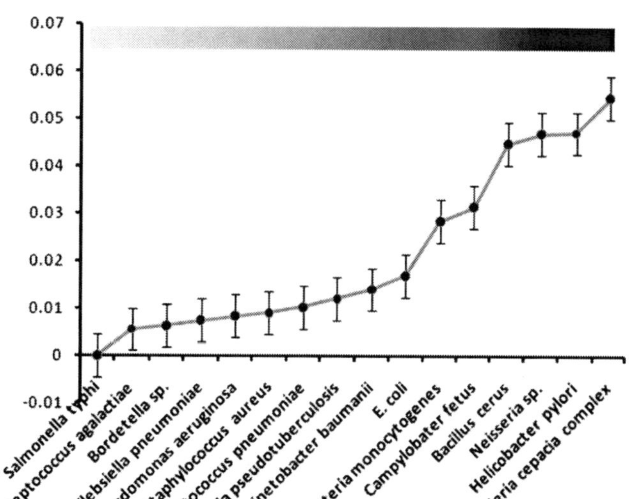

of several genes and proteins but are less suitable for contemporaneous interrogation of time dependent behavior under multiple conditions. Further, scaling the number of inputs or conditions is not possible with these techniques. Therefore there exists a critical need for high throughput technologies capable of creating complex patterns of soluble stimuli that simulate the dynamic cellular microenvironment in a highly parallel fashion to facilitate systematic characterization of cell responses and underlying signaling mechanisms. Application of microfluidics technology to cell biology offers an attractive platform for massively parallel integration of cell culture and multiple and complex stimuli.

Microfluidic Cell Arrays (MCAs) offer the potential to facilitate interrogation of cells in a high-throughput format. A single 8 X 8 array provides 64 unique data sets representing a significant saving in terms of costs and time while ensuring uniformity in experimental conditions. Further, complexity not possible in conventional cell culture systems in terms of integration of components including delivery of inputs (spatial (0.5-10μm), temporal, concentration), organization of cells and extra-cellular matrix (ECM) to study signaling (cell-cell interactions, cell-ECM interactions, soluble factors) and monitoring of outputs (microscopy, fluorescence, electrical signals, electrochemical signals) can be integrated.

Few versions of MCAs have been developed over the past years (Khademhosseini A, et al., 2005). MCA provide a scalable tool with sufficient throughput, generation and delivery of complex stimuli and control over cellular and extra-cellular environment in ways not possible using conventional cell culture techniques. Unlike conventional cell culture systems and multi-well plates, MCAs can accomplish delivery of concentration gradients, time varying signals and spatially defined signals, organization of cells and ECM for culture or co-culture, and measurements using microscopy or secreted factors. A single 8 X 8 MCA seeded with cells at the same time and cultured using identical conditions provides uniformity across the entire array. Delivery of unique signals or stimuli to each well of the array results in 64 unique events representing a significant savings in terms of time and costs in comparison to conventional cell culture systems and multi-well plates.

MCA Applications: Limitations of MCAs for Host-Pathogen Studies

Though technology development has taken place at a fairly fast pace, actual application of these arrays has been limited to proof-of-concept demonstration and to model systems to study inflammatory responses using reporter gene transfected cell lines (Thompson, et al., 2004) and the effect of hepatocyte co-culture with mesenchymal cell types (King KR, et al., 2007). Widespread use for practical or research applications has been limited due to complexities associated with seeding, culture and extraction of information from cells cultured within the arrays.

Despite significant advantages in comparison to conventional cell culture and multi-well plates, MCAs have yet to be used to study host-pathogen interactions. This can be attributed to several restrictions as a consequence of cell culture in the path of fluid flow. Current MCAs are limited to anchorage dependent cells that can withstand a significant amount of fluid flow induced shear stress thereby excluding culture of cells that cannot withstand high shear stress, cells in suspension and cells cultured in 3D as spheroid bodies. Evaluation of physiological response of cells within these arrays is also limited to direct immunofluorescence, reporter gene expression or analysis of secreted factors. No MCA to date has accomplished removal of cells from individual wells for off-chip molecular expression analysis. Further, setup and operation of these arrays requires trained experts adept at manipulating syringe pumps for generation of gradients and controlling experimental conditions. Overcoming these limitations while at the same time maintaining functional complexity is critical in ensuring widespread application for cell-based studies, particularly the evaluation of host-pathogen interactions.

MCAs to Study Host-Pathogen Interactions: Design Requirements

Host-pathogen studies require a system where the interaction of a small number of cells with pathogens can be studied in a highly controlled environment. In addition to controlling the environment, the following factors require careful consideration:

1. Ability to seed cells and infect with different pathogens or strains of pathogens within the same array w/o cross-contamination
2. Cell culture without significant shear stress to prevent shear induced apoptosis
3. Ensure changes in culture conditions to result in immediate local changes in each well
4. Allow easy and straightforward retrieval of the contents of individual wells without cell loss or contamination

To fulfill the aforementioned criteria, functional complexities of existing MCAs need to be conserved while allowing manual seeding, infection and retrieval of cells. A feasible design for such an MCA consists of 3 reversibly sealed layers sandwiched between two polycarbonate plates (Figure 3 A-C). The first layer made of silicone (poly) dimethylsiloxane (PDMS) is 7.5 cm x 3.5 cm and 5 mm thick and precut into 5mm x 5 mm islands at locations below each cell culture well. The second layer is also made of PDMS and is 7.5 cm x 3.5 and 6 mm thick with 3 mm diameter holes for cell culture wells punched through the layer. The third and final layer contains the fluidic channels for delivery of complex signals. Layers 1 and 2 are assembled and cell seeding can be accomplished. The dimensions of the wells ensure culture of ~ 3000 cells/well. Following seeding and cell attachment, infections can be accomplished manually and then the third layer can be assembled on top. The entire setup can be sandwiched between the two poly carbonate plates and clamped to ensure hermetic sealing. Following completion of the study, the setup can be disassembled and the cells on the islands on layer 3 are used for analysis.

Device Fabrication

Fabrication of the MCA platform utilizes simple soft-lithographic techniques (Melin & Quake, 2007). This process typically involves creating a layout of the desired features using layout software and printing onto a mask. The mask is then used to selectively define features by patterning a photo-definable polymer called a photoresist on a silicon master. The silicon master is then used to mold a specific layer of the device. PDMS molding is accomplished by mixing the PDMS pre-polymer with a cross-linking agent in a 100:1 ratio and curing at 80^{0}C for 3 hours. For the MCA platform described earlier, the master for the first layer contained patterns for creation of the islands, the master for the second layer contained patterns

Figure 3. Schematic of a MCA platform to study hostpathogen interactions. A-C. Layer by layer assembly and irreversible sealing ensure functional complexity possible using microfluidics while at the same time allowing for manual seeding, infection and retrieval.

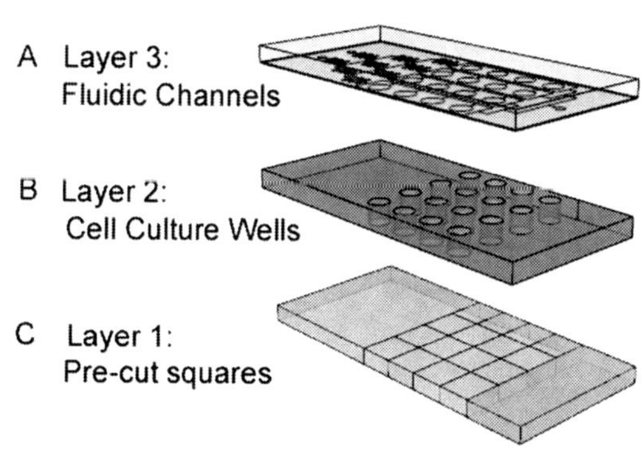

to punch through holes to define culture wells and the master for the 3rd layer contained patterns for the fluidic channels, inlets, outlets, gradient generator and flow encoded switch.

Device Set-up for Cell Seeding and Infection

The first and second layers were assembled together on top of a standard large glass slide. The wells were pre-treated with 50 ng/ml of fibronectin for 12 hrs at 37°C to promote cell adhesion (for most anchorage dependent mammalian cells). Following pre-treatment, the fibronectin was removed and the wells were washed with 1X saline solution. Cells were trypsinized from larger flasks and re-suspended in cell culture medium at a concentration of 1.2 x 10^5 cells/ml. 25 μl of solution from this was metered into each cell culture well resulting in ~ 3000 cells/well. Cells were allowed to attach for 4 hrs and cell culture medium was replaced. Following 24 hrs in culture, the medium was replaced with serum free medium for 12 hrs following which infections were performed. Immediately following infections, the top layer containing the microfluidic gradient generator and FES were assembled carefully to prevent introduction of air bubbles and set in perfusion. Using the two inlets for the concentration gradient generation, any gradient (type of) gradient can be generated using programmable syringe pumps. Similarly, the FES can be programmed to actuate at any predetermined time points. Following culture within the cell array, the top two layers are disassembled and the islands (squares) containing cells from each well can be individually extracted for subsequent analysis.

Gradient Generator and Flow-Encoded Switch

Concentration and time dependency are the most commonly evaluated variables in cell biology. The gradient generator is an extremely useful tool to evaluate concentration dependent signals (Irimia, Geba, & Toner, 2006). Integration of the gradient generator with the MCA platform allows individual columns to experience a unique concentration gradient. Gradient generators have been previously integrated with MCAs to generate inputs of different concentrations. The gradient generator typically consists of two inlets, one for delivery of the desired chemical/biomolecule and the other for delivery of a dilution buffer. The channel structures are designed for step by step binary mixing using a Christmas-tree like structure to generate different concentrations. The mixing channels need to be sufficiently long for a particular width to ensure complete mixing of the required molecule via diffusion. Apart from linear gradients, generation of complex gradients is also possible. The FES has also been previously integrated with MCAs (King KR, et al., 2007). On initiation of fluid flow through the flow-encoded switch results in perfusion of a complete row with the input solution. Increase in flow rate and reduction in the flow rate from the gradient generator side results in additional rows being perfused with the input solution. This switch can be used to perfuse cells cultured within the MCA with a time dependent signal and can be delivered by sequentially increasing the flow rate. An operational gradient generator and FES is shown in Figure 4a and b.

Shear Stress and Mass Transport

Shear stress is an important determinant that affects cell structure and function. Most mammalian cells are shear protected and are not directly exposed to shear stress. Highly sensitive cells survive at flow rates of ~ 0.01 μl/min. This corresponds to a wall shear stress of ~ 40 μN/m^2. This value serves as the threshold for device design for successful culture of mammalian cells. For microfluidic channels with aspect ratios > 2:1 (height: width), interesting flow phenomenon occurs within the wells. Computational fluid dynamics (CFD) modeling

Figure 4. A. Schematic diagram showing integration of a gradient generator and a flow encoded switch with cell culture wells is a MCA. **B**. *Visualization of the gradient generator (i) and operation of the FES showing sequential delivery of stimulus to each row of wells (ii-v).*

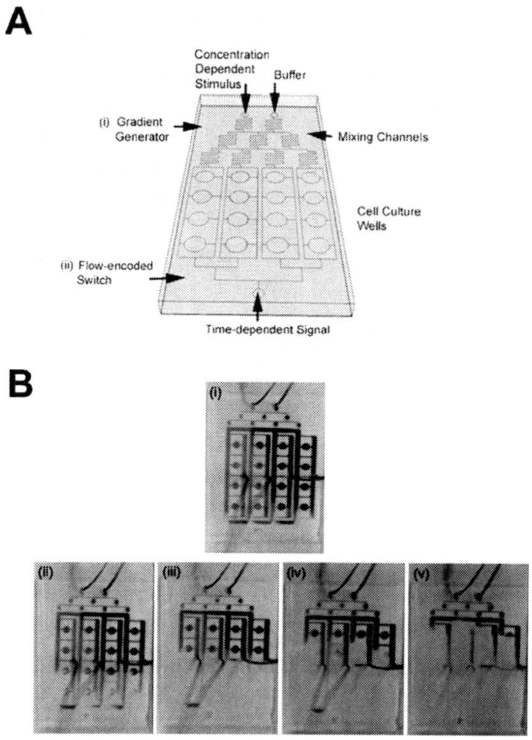

shows that the fluid flow in the channel network barely affects the cells at the bottom of the wells whereas at the bottom of the wells, vortices are generated which result in shear stress ~ 1 to 2 orders of magnitude smaller than in the channel network. CFD results of streamlines of fluid flow and the shear stress generated at different locations within the device are shown in Figure 5.

This has two interesting implications (1) The fluid shear stress at the bottom of the cell culture wells is extremely small and does not affect the culture of anchorage dependent mammalian cells, (2) generation of the vortex and constant recirculation results in rapid mass transport ensuring changes in flow conditions are reflected in immediate changes in the well conditions.

Application of MCA to Contemporaneously Study Host Cell Responses to Infection with Genetically Diverse *H. Pylori* Strains

"Evolution proceeds like a tinkerer who, during millions of years, has slowly modified his products, retouching, cutting, lengthening, using all opportunities to transform and create." –F. Jacob, The Possible and the Actual

Genetic Diversity in *H. Pylori*

Chronic *H. pylori* infection results in diverse clinical outcomes that range from asymptomatic gastritis to peptic ulcers and gastric cancer. Bacterial and host factors that determine the trajectory of possible clinical outcomes of long-term *H. pylori* infection are not fully understood. Thus, there is no diagnostic marker that can reliably predict the outcome of *H. pylori* infection in particular, the risk of developing gastric cancer. Several factors contribute to the lack of reliable predictive marker/s for *H. pylori*-induced gastric pathology: 1) natural *H. pylori* populations show extraordinary genetic diversity and clinical isolates are seldom genetically identical (Achtman, et al., 1999; N. Akopyanz, Bukanov, Westblom, & Berg, 1992; N. Akopyanz, Bukanov, N.O., Westblom, T.U., Kresovich S., and Berg, D.E., 1992); 2) *H. pylori* populations are geographically structured and different *H. pylori* genotypes predominate in different human populations (Achtman, et al., 1999; Covacci, Telford, Giudice, Parsonnet, & Rappuoli, 1999; Dykhuizen & Kalia, 2008; Falush, et al., 2003; Kalia, et al., 2004); 3) clinical outcome of *H. pylori* infection itself varies geographically: e.g., while the risk of gastric cancer is the highest in Japan, duodenal ulcers are more likely in India and South Africa, and

Figure 5. Streamlines and normalized shear stress contours at the lowest flow speed of 3.3 mm/s

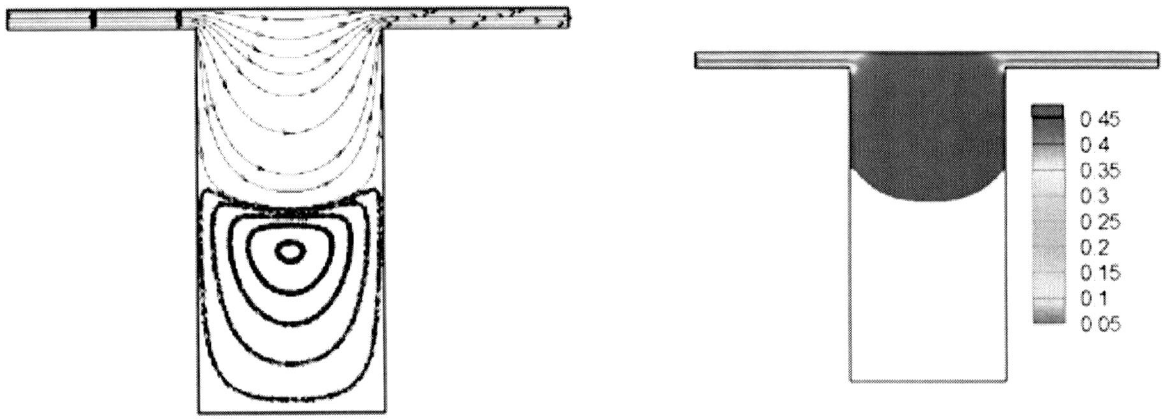

peptic ulcers are more likely in the Amazonians (Brown, 2000; Cover, Berg, Blaser, & Mobley, 2001; Kate, 1998); and 4) *H. pylori* strains adapt rapidly to genetic and physiologic differences in local human populations (Aspholm-Hurtig, et al., 2004; Ogura, et al., 2007)[see below]. Thus, how do human population-specific *H. pylori* adaptations differentially affect host cellular responses and result in geographically variable clinical outcomes of infection? Investigating this aspect of *H. pylori* pathogenesis is greatly impeded further by the current lack of technologies that would enable studying multiple clinical isolates contemporaneously under varying environmental stimuli.

Human Population-Specific Adaptations in *H. Pylori*: The Example of *hepC* Evolution

The molecular and evolutionary mechanisms that contribute to *H. pylori*'s persistence and virulence are fields of active research. Key contributors to *H. pylori*'s persistence likely include mechanisms for avoiding or subverting innate and adaptive immune responses, and for coping with or avoiding inflammatory responses (Atherton, 2006). *H. pylori* produce several putative effectors that help modulate its microenvironment and/or directly damage certain host cells, although host molecular targets are known for only a handful. Compared to *H. pylori* housekeeping genes, alleles of *H. pylori* virulence factors show much greater amino acid sequence variation and geographic partitioning (Figure 6A). For example, a comparison of *hepC* DNA sequences with DNA sequences of seven housekeeping genes from diverse *H. pylori* isolates revealed a dramatically altered evolutionary pattern: the Japanese-type *hepC* alleles were strikingly divergent from their Korean counterparts (Figure 6A, circled cluster) whereas the housekeeping genes from the same populations were mixed, as expected (Figure 6B, shaded cluster) (Ogura, et al., 2007).

Thus, what was driving the unexpected evolution of *hepC* alleles in the Japanese population? Phylogenetic analyses revealed that *hepC* evolution was driven by strong pressure to diversify rapidly presumably as an adaptive response to differences in local populations (Darwinian selection; adaptive evolution) (Ogura, et al., 2007). Specific amino acids under Darwinian selection in the Japanese lineage identified using an empirical Bayesian analyses were present on HepC surface (Ogura, et al., 2007). Since most amino acids undergoing Darwinian evolution were surface exposed, the HepC variants from different populations interacted with their cognate host partner with different affinities (*Kalia A et al,*

Figure 6. Natural selection favors different allelic variants of the hepC gene in geographically distinct H. pylori populations. (A) Comparison of hepC and housekeeping gene phylogenies showing distinct and unexpected clustering patterns among Japanese and Korean hepC alleles. phylogenetic tree reconstructed from 27 full-length hepC sequences from geographically distinct H. pylori isolates (shown in different colors) and (B) phylogenetic tree of concatenated sequences of six housekeeping genes from H. pylori isolates.

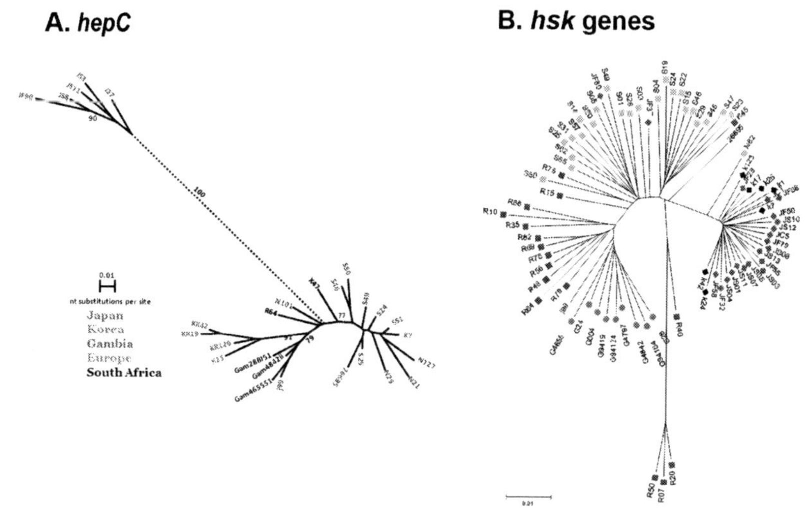

unpublished data; in preparation). Such differences in the strength of interaction change the downstream signaling cascades, and can therefore differentially affect host-pathogen interaction. However, little is known about how geographically distinct *H. pylori* strains differentially affect host cellular responses and signaling cascades.

Infection of AGS Cells with Diverse *H. Pylori* Strains Using a Prototype 4 X 4 Microfluidic Cell Arrays

A prototype 4 X 4 MCA was constructed to characterize seeding and culture conditions for propagation and maintenance of AGS cells. The prototype MCA device is described above. Devices were autoclaved to ensure sterility. The devices were pretreated with 50 ng/ml of fibronectin for 12 hrs at 37°C and then washed with 1X phosphate buffered saline (PBS). AGS were then introduced into the device at a concentration of 2 X 10^3 cells /ml and allowed to grow in an antibiotic free medium containing DMEM supplemented with 10% FBS for 24 hrs in a 5% CO_2 incubator (Figure 7A). After cell culture became confluent, AGS cells were maintained for 12 hrs in fresh antibiotic and serum-free DMEM prior to infection with *H. pylori* strains. Prior to infection each MCA well contained 4 – 5 X 10^3 cells; each well was then infected with *H. pylori* strains J99 (European), 26695 (European), JS7 (Japanese), and G27MA (cell-culture adapted) at a multiplicity of infection (MOI) of 100. *H. pylori* induce rapid cytoskeletal changes in AGS cells, which begin to become apparent following 6 hrs of infection in conventional cell culture analyses. We observed no difference in the dynamics of *H. pylori* infection in MCAs compared to conventional cell culture assays (Figure7B and *Putty K, Sethu P and Kalia, A, in preparation*). Thus, at 6 hrs post-infection, AGS cells were collected from each MCA well

Figure 7. Infection of AGS cells with genetically diverse H. pylori strains using MCA. (A) Schematic design of the prototype 4 x 4 MCA used in infection assays, with 2X, 10X and 20X, respectively, magnified view of the MCA chambers (in the inset). (B) Cytoskeletal dynamics observed following six hrs of H. pylori infection mimic those observed in conventional scale cell culture; formation of cell extensions following infection with geographically distinct H. pylori strains is shown. (C) The assessment of RNA integrity with Agilent 2100 bioanalyzer shows the electropherogram of extracted RNA samples from ~5 X 10³ cells compared (Lanes 1, 2 4, 6, and 8) compared with RNA extracted from 0.5 X 10⁶ cells from conventional cell culture-based infection assay (lanes 3, 5, 7 and 10). (D) RIN visualization using the Agilent expert software for representative samples from 5 X 10³ cells (lane 2) and 0.5 X 10⁶ cells (lane 5). (E) Validation that the extracted RNA was suitable for gene expression analyses using RT-PCR.

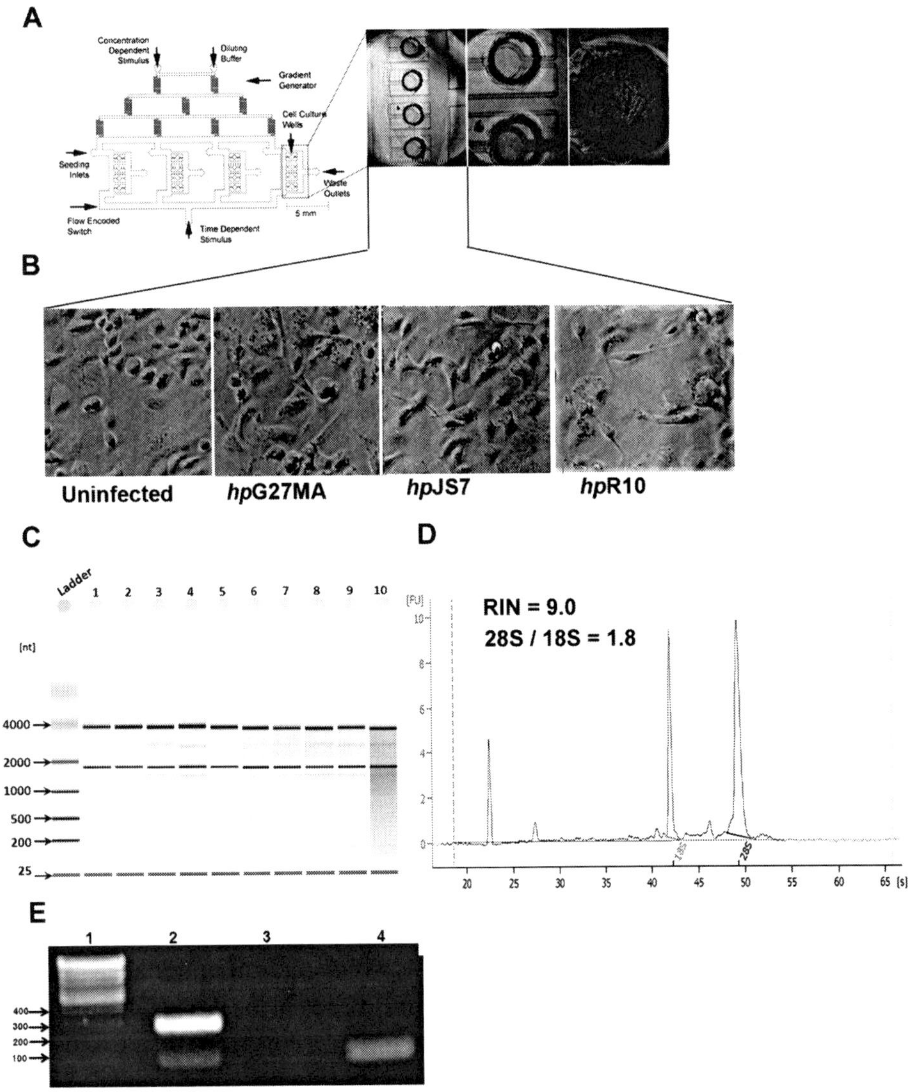

and then processed for downstream host gene expression analyses.

Extraction of Total RNA and Analyses of RNA Integrity and Quality

Given that each MCA well contained only 4 – 5 X 10^3 cells one major concern is regarding the quantity and quality of extracted RNA. Total RNA was extracted and purified from cells using the Qiagen mRNA extraction kit and then analyzed for integrity using the Agilent Bioanalyzer 2100. Determining RNA integrity is a critical step in gene expression analysis. Thus we determined RNA integrity and quality using three different measures, which included, electropherogram analysis, determination of rRNA ratio and the RNA integrity number (RIN) (Schroeder, et al., 2006). RIN is a powerful new tool and is expressed as a scale that ranges from 1 – 10. RIN values closer to 10 indicate grater RNA integrity, and thus potentially high reproducibility for high throughput gene expression analysis (eg., microarrays). These analyses showed that consistent RNA yields of 35 – 50 ng / µl were obtained from ~4 – 5 X 10^3 infected cells and that the extracted RNA had RIN numbers that ranged from 7.5 to 9.2, and equivalent to that extracted from AGS cells infected in conventional culture in terms of its integrity (Figure 7C). cDNA was then synthesized using the purified total RNA using the Qiagen cDNA synthesis kit and used as template for reverse-transcription PCR. RT-PCR results show that the cDNA synthesized from extracted total RNA was of sufficient quality for downstream transcript analyses of host cellular response to *H. pylori* infection (Figure 7D). Thus, we conclude that the MCA platform provides a suitable platform for contemporaneous analyses of AGS cell responses to infection with geographically diverse *H. pylori* strains.

FUTURE DIRECTIONS AND DEVELOPMENTAL PROSPECTS

MCAs offer throughput associated with a multi-well plate with the ability to integrate functional complexity and expose cells cultured within the array to a wide array of concentration and time dependent signals. The deconstructed array used for host-pathogen studies further simplifies aspects of seeding, infection and extraction of cells from wells. The array in its current state represents the basic minimum in terms of functionality. Microfluidic techniques can be used to further enhance the capabilities the system. Cell patterning techniques can be used to create organized co-cultures of different cell populations within a culture well to mimic tissue level organization prior to infection. The environment within a culture well can also be actively monitored or changed through integration of sensors, detectors, electrodes, heating/cooling elements etc. Localized or even sub cellular delivery of signaling agents can be accomplished by integrating controlled release valves or droplet based microfluidics. The gradient generator can be designed and constructed to seamlessly deliver any type of gradient (exponential, logarithmic, parabolic, step etc.) Integration of the entire array on a motorized stage of a microscope or plate reader allows automated analysis via spectrophotometry or microscopy. Here we have shown that a prototype 4 X 4 MCA can be successfully used to study the interaction of up to 16 distinct *H. pylori* strains at the same time under one highly environmental condition or alternatively, up to 4 different *H. pylori* strains in 4 tightly controlled environmental conditions. This microscale interrogation yields high purity RNA that is suitable for downstream analyses such as pathway-specific PCR-arrays or direct RNA-seq. A future application will be to attain seamless integration of such analytical methods with the MCA.

ACKNOWLEDGMENT

This work is supported by a Multidisciplinary Research Grant (MRG-2009) from the University of Louisville (A.K.), and by a National Science Foundation under EPSCoR Grant No. 0814194 (P.S.)

REFERENCES

Achtman, M., Azuma, T., Berg, D. E., Ito, Y., Morelli, G., & Pan, Z.-J. (1999). Recombination and clonal groupings within Helicobacter pylori from different geographical regions. *Molecular Microbiology*, *32*(3), 459–470. doi:10.1046/j.1365-2958.1999.01382.x

Akopyanz, N., Bukanov, N. O., Westblom, T. U., & Berg, D. E. (1992). PCR-based RFLP analysis of DNA sequence diversity in the gastric pathogen Helicobacter pylori. *Nucleic Acids Research*, *20*(23), 6221–6225. doi:10.1093/nar/20.23.6221

Akopyanz, N., Bukanov, N. O., Westblom, T. U., Kresovich, S., & Berg, D. E. (1992). DNA diversity among clinical isolates of Helicobacter pylori detected by PCR-based RAPD fingerprinting. *Nucleic Acids Research*, *20*(19), 5137–5142. doi:10.1093/nar/20.19.5137

Aspholm-Hurtig, M., Dailide, G., Lahmann, M., Kalia, A., Ilver, D., & Roche, N. (2004). Functional adaptation of BabA, the H. pylori ABO blood group antigen binding adhesion. *Science*, *305*(5683), 519–522. doi:10.1126/science.1098801

Atherton, J. C. (2006). The pathogenesis of Helicobacter pylori induced gastro-duodenal diseases. *Annual Review of Pathology: Mechanisms of Disease*, *1*(1), 63–96. doi:10.1146/annurev.pathol.1.110304.100125

Atherton, J. C., & Blaser, M. J. (2009). Coadaptation of Helicobacter pylori and humans: Ancient history, modern implications. *The Journal of Clinical Investigation*, *119*(9), 2475–2487. doi:10.1172/JCI38605

Beres, S. B., Carroll, R. K., Shea, P. R., Sitkiewicz, I., Martinez-Gutierrez, J. C., & Low, D. E. (2010). Molecular complexity of successive bacterial epidemics deconvoluted by comparative pathogenomics. *Proceedings of the National Academy of Sciences of the United States of America*, *107*(9), 4371–4376. doi:10.1073/pnas.0911295107

Brown, L. M. (2000). Helicobacter Pylori: Epidemiology and routes of transmission. *Epidemiologic Reviews*, *22*(2), 283–297.

Carapetis, J.R., Steer, A.C., & Mulholland, E.K. & M.W. (1995). The global burden of group A streptococcal diseases. *The Lancet Infectious Diseases*, *5*(11), 685–694. doi:10.1016/S1473-3099(05)70267-X

Covacci, A., Telford, J. L., Giudice, G. D., Parsonnet, J., & Rappuoli, R. (1999). Helicobacter pylori Virulence and Genetic Geography. *Science*, *284*(5418), 1328–1333. doi:10.1126/science.284.5418.1328

Cover, T. L., Berg, D. E., Blaser, M., & Mobley, H. L. T. (2001). *H. pylori pathogenesis*. New York: Academic Press.

Dykhuizen, D. E., & Kalia, A. (2008). *Population genetics of pathogenic bacteria* (2nd ed.). Oxford University Press.

Einav, S., Gerber, D., Bryson, P. D., Sklan, E. H., Elazar, M., & Maerkl, S. J. (2008). Discovery of a hepatitis C target and its pharmacological inhibitors by microfluidic affinity analysis. *Nature Biotechnology*, *26*(9), 1019–1027. doi:10.1038/nbt.1490

Ewald, P. W. (2004). Evolution of virulence. *Infectious Disease Clinics of North America*, *18*(1), 1–15. doi:10.1016/S0891-5520(03)00099-0

Falush, D., Wirth, T., Linz, B., Pritchard, J. K., Stephens, M., & Kidd, M. (2003). Traces of human migrations in Helicobacter pylori populations. *Science*, *299*(5612), 1582–1585. doi:10.1126/science.1080857

Fan, H.C., Blumenfeld, Y.J., El-Sayed, Y.Y., Chueh, J. & Quake, S.R. (2009). Microfluidic digital PCR enables rapid prenatal diagnosis of fetal aneuploidy. *American Journal of Obstetrics and Gynecology, 200*(5), 543, e541-543, e547.

Feil, E. J., & Spratt, B. G. (2001). Recombination and the population structures of bacterial pathogens. *Annual Review of Microbiology*, *55*(1), 561–590. doi:10.1146/annurev.micro.55.1.561

Gerber, D., Maerkl, S. J., & Quake, S. R. (2009). An in vitro microfluidic approach to generating protein-interaction networks. *Nature Methods*, *6*(1), 71–74. doi:10.1038/nmeth.1289

Gulati, S., Rouilly, V., Niu, X., Chappell, J., Kitney, R. I., & Edel, J. B. (2009). Opportunities for microfluidic technologies in synthetic biology. *Journal of the Royal Society, Interface*, *6*(Suppl 4), S493–S506. doi:10.1098/rsif.2009.0083.focus

Gupta, S., & Maiden, M. C. J. (2001). Exploring the evolution of diversity in pathogen populations. *Trends in Microbiology*, *9*(4), 181–185. doi:10.1016/S0966-842X(01)01986-2

Holt, K. E., Parkhill, J., Mazzoni, C. J., Roumagnac, P., Weill, F.-X., & Goodhead, I. (2008). High-throughput sequencing provides insights into genome variation and evolution in Salmonella Typhi. *Nature Genetics*, *40*(8), 987–993. doi:10.1038/ng.195

Hood, L., Heath, J. R., Phelps, M. E., & Lin, B. (2004). Systems biology and new technologies enable predictive and preventative medicine. *Science*, *306*(5696), 640–643. doi:10.1126/science.1104635

Irimia, D., Geba, D. A., & Toner, M. (2006). Universal microfluidic gradient generator. *Analytical Chemistry*, *78*(10), 3472–3477. doi:10.1021/ac0518710

Kalia, A., Mukhopadhyay, A. K., Dailide, G., Ito, Y., Azuma, T., & Wong, B. C. Y. (2004). Evolutionary dynamics of insertion sequences in Helicobacter pylori. *Journal of Bacteriology*, *186*(22), 7508–7520. doi:10.1128/JB.186.22.7508-7520.2004

Kane, B. J., Zinner, M. J., Yarmush, M. L., & Toner, M. (2006). Liver-specific functional studies in a microfluidic array of primary mammalian hepatocytes. *Analytical Chemistry*, *78*(13), 4291–4298. doi:10.1021/ac051856v

Kate, V., Ananthakrishnan, N., Badrinath, S., & Ratnakar, C. (1998). Prevalence of Helicobacter pylori infection in disorders of the upper gastrointestinal tract in south India. *The National Medical Journal of India*, *11*(1), 5–8.

Khademhosseini, A., Yeh, J., Eng, G., Karp, J., Kaji, H., & Borenstein, J. (2005). Cell docking inside microwells within reversibly sealed microfluidic channels for fabricating multiphenotype cell arrays. *Lab on a Chip*, *5*(12), 1380–1386. doi:10.1039/b508096g

King, K.R., Wang, S., Irimia, D., Jayaraman, A., & Toner, M. & M.L.Y. (2007). A high-throughput microfluidic real-time gene expression living cell array. *Lab on a Chip*, *7*(1), 77–85. doi:10.1039/b612516f

King, K.R., Wang, S., Jayaraman, A., & Yarmush, M.L. & M.T. (2008). Microfluidic flow-encoded switching for parallel control of dynamic cellular microenvironments. *Lab on a Chip*, *8*(1), 107–116. doi:10.1039/b716962k

Lee, C.-C., Snyder, T. M., & Quake, S. R. (2010). A microfluidic oligonucleotide synthesizer. *Nucleic Acids Research*, 92.

Melin, J., & Quake, S. R. (2007). Microfluidic large-scale integration: The evolution of design rules for biological automation. *Annual Review of Biophysics and Biomolecular Structure*, *36*(1), 213–231. doi:10.1146/annurev.biophys.36.040306.132646

Nicholson, J. K., Holmes, E., Lindon, J. C., & Wilson, I. D. (2004). The challenges of modeling mammalian biocomplexity. *Nature Biotechnology*, *22*(10), 1268–1274. doi:10.1038/nbt1015

Ochman, H., Lawrence, J. G., & Groisman, E. A. (2000). Lateral gene transfer and the nature of bacterial innovation. *Nature*, *405*(6784), 299–304. doi:10.1038/35012500

Ogura, M., Perez, J. C., Mittl, P. R. E., Lee, H.-K., Dailide, G., & Tan, S. (2007). Helicobacter pylori evolution: Lineage-specific adaptations in homologs of eukaryotic sel1-like genes. *PLoS Computational Biology*, *3*(8), e151. doi:10.1371/journal.pcbi.0030151

Schroeder, A., Mueller, O., Stocker, S., Salowsky, R., Leiber, M., & Gassmann, M. (2006). The RIN: An RNA integrity number for assigning integrity values to RNA measurements. *BMC Molecular Biology*, *7*(1), 3. doi:10.1186/1471-2199-7-3

Thompson, D. M., King, K. R., Wieder, K. J., Toner, M., Yarmush, M. L., & Jayaraman, A. (2004). Dynamic gene expression profiling using a microfabricated living cell array. *Analytical Chemistry*, *76*(14), 4098–4103. doi:10.1021/ac0354241

Whitesides, G. M., Ostuni, E., Takayama, S., Jiang, X., & Ingber, D. E. (2001). Soft lithography in biology and biochemistry. *Annual Review of Biomedical Engineering*, *3*(1), 335–373. doi:10.1146/annurev.bioeng.3.1.335

KEY TERMS AND DEFINITIONS

Clonal Bacterial Population: Bacterial population characterized by very little or no genetic diversity among isolates prevalent in the host population at any given time eg., Salmonella Typhi.

Epidemic Bacterial Population: Bacterial population characterized by rapid spread (epidemic) of one or few genotypes in the host population, even though other genotypes exist at low levels.

Genetic Drift: Fluctuations in gene or allele frequency purely by chance.

Microfluidics: Study of fluids at microscale.

Panmictic Bacterial Population: Bacterial population characterized by extreme genetic diversity whereby each clinical isolate can be genetically distinct from another.

Virulence: The potential of a pathogenic microorganism to cause human pathology.

Section 4
Structural and Mathematical Modeling

This section contains five chapters, including three chapters on the structural modeling of biological molecules and two chapters on the mathematical modeling of specific biological phenomena. Chapter 24 reviews state-of-the-art methods for non-coding RNA identification based on structural alignment of RNAs and with full consideration of pseudoknots. Chapter 25 presents an original research paper on the computational modeling of RNA folding based on both folding kinetics and energetic considerations. Chapter 26 presents an original research paper demonstrating a novel method for a reduced representation of protein structure in the application of ligand binding site modeling and screening. Chapter 27 presents an original research paper on modeling the rolling of a cell on the surface of the extracellular matrix by simulating the successive attachment and detachment processes. Chapter 28 presents an original research paper describing the modeling of chemotactic axon guidance, an important neurological process, at both microscopic and macroscopic scales.

Chapter 24
Structural Alignment of RNAs with Pseudoknots

Thomas K. F. Wong
The University of Hong Kong, Hong Kong

S. M. Yiu
The University of Hong Kong, Hong Kong

ABSTRACT

Non-coding RNAs (ncRNAs) are found to be critical for many biological processes. However, identifying these molecules is very difficult and challenging due to the lack of strong detectable signals such as opening read frames. Most computational approaches rely on the observation that the secondary structures of ncRNA molecules are conserved within the same family. Aligning a known ncRNA to a target candidate to determine the sequence and structural similarity helps in identifying de novo ncRNA molecules that are in the same family of the known ncRNA. However, the problem becomes more difficult if the secondary structure contains pseudoknots. Only until recently, many of the existing approaches could not handle structures with pseudoknots. This chapter reviews the state-of-the-art algorithms for different types of structures that contain pseudoknots including standard pseudoknot, simple non-standard pseudoknot, recursive standard pseudoknot, and recursive simple non-standard pseudoknot. Although none of the algorithms is designed for general pseudoknots, these algorithms already cover all known ncRNAs in both Rfam and PseudoBase databases. The evaluation of the algorithms also shows that the approach is useful in identifying ncRNA molecules in other species, which are in the same family of a known ncRNA.

INTRODUCTION

A non-coding RNA (ncRNA) is a functional RNA molecule that is not translated into a protein.

DOI: 10.4018/978-1-60960-491-2.ch024

There are many different types of ncRNAs such as tRNAs, rRNAs, snoRNAs, microRNAs, and siRNAs. These RNA molecules have been found to be involved in many biological processes such as gene regulation, chromosome replication and RNA modification (Frank and Pace, 1998; Nguyen *et al.*,

2001; Yang *et al.*, 2001). Some are found to be related to cancers and other diseases as well. Similar to proteins, ncRNAs also appear to form a highly structured network that regulates gene expression and translation in the cell (Esquela-Kerscher and Slack, 2006). The number of ncRNAs within the human genome was underestimated before, but recently some databases reveal over 212,000 ncRNAs (He *et al.*, 2007) and more than 1,300 ncRNA families (Griffiths-Jones *et al.*, 2003). Data accumulated on ncRNAs and their families show that ncRNAs may be as diverse as protein molecules (Eddy, 2001).

Identifying ncRNAs is an important problem in the system biological studies. However, this process is very difficult and challenging. Although it is known that some ncRNAs do have promoters and terminators, it is generally believed that ncRNA genes do not contain signals such as open reading frames and ribosome binding sites, which can be easily detected (Argaman *et al.*, 2001). Many different computational approaches have been proposed to solve this problem. There are few possible approaches to identify ncRNAs along the genome. Since it is known that the secondary structure of an ncRNA molecule usually plays an important role in its biological functions, for example, the hairpin structures for miRNA precursors and cloverleaf structures for tRNAs, some researches attempted to identify ncRNAs by considering the stability of secondary structures formed by the substrings of a given genome (Le *et al.*, 1990). However, this method is not effective because a random sequence with high GC composition also allows for an energetically favorable secondary structure (Rivas and Eddy, 2000).

Another promising method is the comparative approach. The idea is to make use of some known ncRNAs and try to identify ncRNA candidates along the genome. Along this direction, some authors (Lowe and Eddy, 1997; Nawrocki *et al.*, 2009) use a set of ncRNAs from the same family to train a model (e.g. covariance model). Then, they employ this model to scan the genome and identify potential regions that are ncRNA candidates of that family. The information to be captured from the known ncRNAs depends on how the model is defined. However, in some cases, there are not enough known members in a given family to reliably train a model.

Since the primary sequence and the secondary structure of ncRNA are evolutionary conserved, the ncRNAs of the same family share similar sequence and structure. Another approach is to use a known ncRNA and identify the regions along the genome whose sequence and structure are similar to that of the ncRNA. The resulting regions are the potential ncRNAs candidates of the same family. The key of this approach is to compute the structural alignment between the folded ncRNA (*query*) and the unfolded region (*target*). The unfolded sequence will be folded and aligned simultaneously to the folded ncRNA. The alignment score represents their sequence and structural similarity. The methods like PHMMTS-based method (Sakakibara, 2003), RSEARCH (Klein and Eddy, 2003) and FASTR (Zhang *et al.*, 2005) belong to this category.

However, these methods do not support *pseudoknots*. Given two base pairs at positions (i,j) and (i_0,j_0), where $i<j$ and $i_0<j_0$, pseudoknots are base pairs crossing each other, i.e. $(i<i_0<j<j_0)$ or $(i_0<i<j_0<j)$, as shown in Figure 1. A structure without pseudoknot is regarded as *regular*. Secondary structures including pseudoknots are found in some telomerases (Chen & Greider, 2005), and self-splicing introns (Adams *et al.*, 2004). Pseudoknot structures appear to be key players in some catalytic reactions as well (Dam *et al.*, 1992). For example, pseudoknot structure in the human telomerase RNA is conserved in all vertebrates and is essential for telomerase activity (Chen & Greider, 2005). Structural alignment of ncRNAs containing pseudoknots is believed to be NP-complete (which means the problem cannot be solved in polynomial time) and the problem is computationally demanding. Therefore, researchers started to develop approaches for specific types

Figure 1. Example of a pseudoknot. The base pair (1,12) crosses another base pair (9,14).

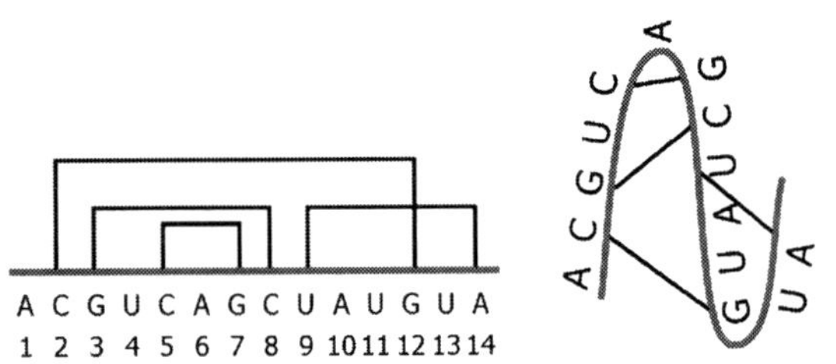

of pseudoknots. Matsui *et al.* (2005) developed a method for computing the alignment according to only the structures between a folded query and an unfolded target. The method is based on tree adjoining alphabets and supports the structure of recursive standard pseudoknot (i.e. pseudoknot/regular structures exist within another standard pseudoknot structure) of degree $k \leq 4$ (the formal definition of recursive standard pseudoknot of degree k will be mentioned later). The time complexity is $O(mn^5)$ where m is the length of the query sequence and n is the length of the target sequence. Since this algorithm only considers the similarity between the structures during the alignment, the resulting alignment score does not reflect the sequence similarity between them. Han *et al.* (2008) developed PAL to compute the structural alignment between the folded query and the unfolded target according to both the sequences and structures. Their algorithm supports structures with standard pseudoknot of degree k and runs in $O(kmn^k)$ with space $O(mn^k)$. To enhance the applicability of the method, Wong *et al.* (2008) proposed a memory-efficient algorithm for solving the same structural alignment problem. For standard pseudoknot of degree k, their method can reduce the space complexity to $O(n^k)$ while maintaining the same time complexity of $O(mn^k)$.

All these previous works focus on (recursive) standard pseudoknots. In 2009 Wong *et al.* (2009) identified and expanded their analysis on two types of complex pseudoknots that were not considered before: *simple non-standard pseudoknot* (as in Figure 2a) and *recursive simple non-standard pseudoknot* (as in Figure 2b). Simple non-standard pseudoknot is a structure which allows for some restricted cases with 3 base pairs mutually crossing each other (i.e. any two of them are crossing). The authors also developed structural alignment algorithms to support these complex structures based on the same scoring model. Their algorithm which is designed for simple non-standard pseudoknot, runs in $O(kmn^k)$ time for degree k with $O(mn^k)$ space. For recursive simple non-standard pseudoknots (i.e. pseudoknot/regular structures exist within another simple non-standard pseudoknot structure), their algorithm runs in $O(kmn^{k+2})$ time with $O(mn^k)$ space. In Rfam 9.1 database (Griffiths-Jones et al., 2003), among 71 pseudoknotted families, 18 of them have complex pseudoknot structure. In the PseudoBase database (van Batenburg et al., 2000), among 304 pseudoknot RNAs, 8 of them have complex pseudoknot structures.

Table 1 summarizes the time and space complexity of all the structural alignment algorithms for different types of pseudoknots.

In this chapter, we will focus on the problem of structural alignment between a folded ncRNA sequence (query) with a pseudoknot structure and an unfolded sequence (target). We will first look into the algorithm designed for regular structure. The knowledge of the algorithm for regular structure is useful for understanding the algorithms for complex pseudoknot structures like a pseudoknot structure with a regular structure inside. Then we will present the algorithms for different types of pseudoknot structure, like standard pseudoknot, simple non-standard pseudoknot and recursive simple non-standard pseudoknot. Although none of these algorithms is designed for generic pseudoknots, these methods already cover all known ncRNAs in both Rfam 9.1 and PseudoBase databases. Finally, by employing experimental data we will demonstrate that the algorithms are useful in identifying ncRNA molecules in other

Figure 2. (a) The secondary structure of RF00140 from Rfam 9.1 database (Griffiths-Jones et al., 2003). Consider three base pairs: one from region 1, one from region 2 and one from region 4, they are mutually crossing each other (i.e. any two of them are crossing). (b) The secondary structure of self-cleaving ribozymes of hepatitis delta virus from (Ferre-D'Amare et al., 1998) (i.e. RF00094 from Rfam 9.1 database).

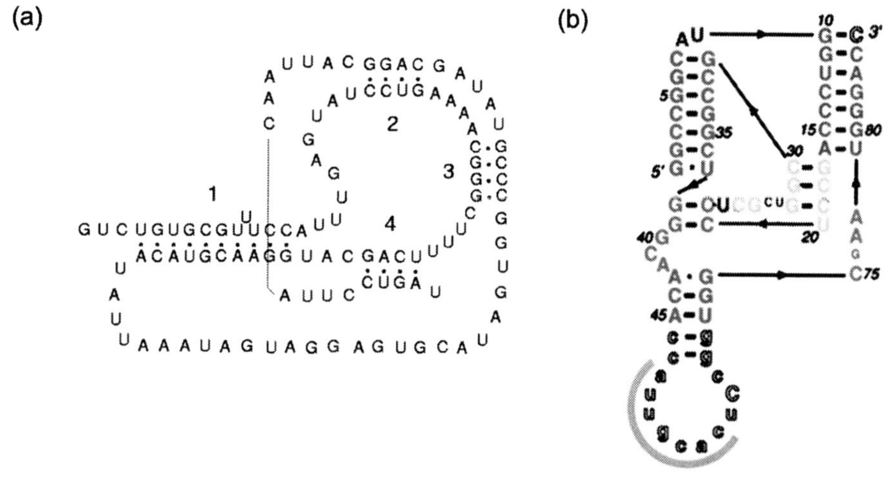

Table 1. Summary of time and space complexity of the structural alignment algorithms for different types of pseudoknots

Algorithm	Psuedoknot type	Complexity	
		Time	Space
Matsui *et al.* (2005)*	Recursive standard pseudoknot of degree $k \leq 4$	$O(mn^5)$	$O(mn^4)$
Han *et al.* (2008)	Standard pseudoknot of degree k	$O(kmn^k)$	$O(mn^k)$
Wong *et al.* (2008)	Standard pseudoknot of degree k	$O(kmn^k)$	$O(n^k)$
Wong *et al.* (2009)	Simple non standard pseudoknot of degree k	$O(kmn^k)$	$O(mn^k)$
Wong *et al.* (2009)	Recursive simple non standard pseudoknot of degree k	$O(kmn^{k+2})$	$O(mn^k)$

* The alignment algorithm provided by Matsui *et al.* (2005) only considers the structure but not sequence, while the other alignment algorithms consider both sequence and structure.

species which are in the same family of a known ncRNA.

PSEUDOKNOT DEFINITIONS

Let $A = a_1a_2... a_m$ be a length-m ncRNA sequence and M be the secondary structure of A. M is represented as a set of base pair positions, i.e. $M=\{(i, j)|1 \leq i < j \leq m, (a_i,a_j)$ *is a base pair*$\}$. Let $M_{\{x,y\}} \subseteq M$ be the set of base pairs within the subsequence $a_xa_{x+1}...a_y$, $1 \leq x < y \leq m$, i.e., $M_{x,y}=\{(i, j)\in M \mid x\leq i<j\leq y\}$ with $M=M_{1,m}$. We assume that there is no two base pairs sharing the same position, i.e., for any $(i_1,j_1),(i_2,j_2)\in M$, $i_1\neq j_2$, $i_2\neq j_1$, and $i_1=i_2$ if and only if $j_1=j_2$.

Definition 1. $M_{x,y}$ is a *regular structure* if there do not exist two base pairs $(i, j),(k, l)\in M_{x,y}$ such that $i<k<j<l$ or $k<i<l< j$. An empty set is also considered as a regular structure.

A regular structure is one without pseudoknots. A structure is a standard pseudoknot of degree k if the RNA sequence can be divided into k consecutive regions (see Figure 3(a)) such that base pairs must

Figure 3. (a) Standard pseudoknot of degree k. (b) Simple non-standard recursive pseudoknot of degree k (Type I). (c) Simple non-standard recursive pseudoknot of degree k (Type II). (d) Recursive non-standard pseudoknot of degree k (the region $[a_1,b_1]$ is a recursive region).

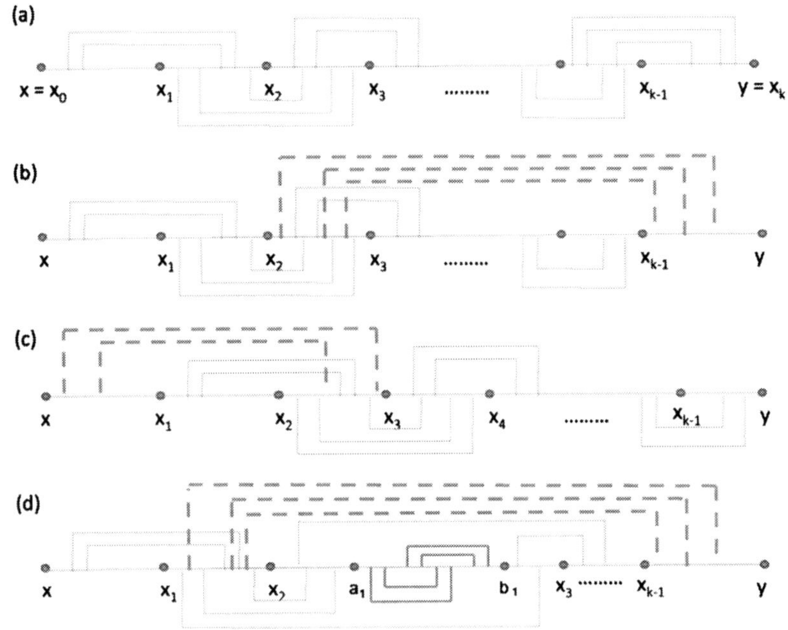

have end points in adjacent regions and base pairs that are in the same adjacent regions do not cross each other. The formal definition is as follows.

Definition 2. $M_{x,y}$ is a standard pseudoknot of degree $k \geq 3$ if there exists a set of pivot points $x_1, x_2, \ldots, x_{k-1}$ ($x = x_0 < x_1 < x_2 < \ldots < x_{k-1} < x_k = y$) that satisfy the following condition. Let $M_w(1 \leq w \leq k-1) = \{(i,j) \in M_{x,y} \mid x_{w-1} \leq i < x_w \leq j < x_{w+1}\}$. Note that we allow $j = x_k$ for M_{k-1} to resolve the boundary case.

- For each $(i,j) \in M_{x,y}$, $(i,j) \in M_w$ for some $1 \leq w \leq k-1$.
- $M_w(1 \leq w \leq k-1)$ is a regular structure.

A standard pseudoknot of degree 3 is usually referred as a *simple pseudoknot*. Now, we define a *simple non-standard pseudoknot* to include some structures with three base pairs crossing each other. For a simple non-standard pseudoknot of degree k, similar to a standard pseudoknot, the RNA sequence can be divided into k regions with the region at one of the ends (say, the right end) designated as the special region. Base pairs with both end points in the first $k-1$ regions have the same requirements as in a standard pseudoknot. And there is an extra group of base pairs that can start in one of the first $k-2$ regions and end at the last special region, and, again, these pairs do not cross each other (see Figure 3(b) and the formal definition below).

Definition 3. $M_{x,y}$ is a simple non-standard pseudoknot of degree $k \geq 4$ (Type I) if there exist $x_1, \ldots, x_{k-1}$ and t where $x = x_0 < x_1 < \ldots < x_{k-1} < x_k = y$ and $1 \leq t \leq k-2$ that satisfy the following. Let $M_w(1 \leq w \leq k-2) = \{(i;j) \in M_{x,y} \mid x_{w-1} \leq i < x_w \leq j < x_{w+1}\}$. Let $X = \{(i;j) \in M_{x,y} \mid x_{t-1} \leq i < x_t, x_{k-1} \leq j \leq y\}$.

- For each $(i,j) \in M_{x,y}$, either $(i,j) \in M_w(1 \leq w \leq k-2)$ or $(i,j) \in X$.
- M_w and X are regular structures.

Type II simple non-standard pseudoknots (see Figure 3(c)) are symmetrical to Type I simple non standard pseudoknots with the special region on the left end. In the rest of the chapter, we will only consider Type I simple non-standard pseudoknots and simply refer to it as simple non-standard pseudoknots.

Lastly, we define recursive standard pseudoknot and recursive simple non-standard pseudoknot (see Figure 3(d)).

Definition 4. $M_{x,y}$ is a recursive standard pseudoknot of degree $k \geq 3$ if $M_{x,y}$ is either regular or standard pseudoknot of degree k, or $\exists a_1, b_1, \ldots, a_s, b_s$ $(x \leq a_1 < b_1 < \ldots < a_s < b_s \leq y)$ that satisfy the following. Each M_{a_i,b_i} is called a recursive region.

- M_{a_i,b_i}, for $1 \leq i \leq s$, is a recursive standard pseudoknot of degree $\leq k$.
- $(M_{x,y} - \bigcup_{1 \leq i \leq s} M_{a_i,b_i})$ is either regular structure or standard pseudoknot of degree $\leq k$.

Definition 5. $M_{x,y}$ is a recursive simple non-standard pseudoknot of degree $k \geq 3$ if $M_{x,y}$ is either regular, standard pseudoknot of degree k or simple non-standard pseudoknot of degree k (if $k \geq 4$), or $\exists a_1, b_1, \ldots, a_s, b_s$ $(x \leq a_1 < b_1 < \ldots < a_s < b_s \leq y)$ that satisfy the following. Each M_{a_i,b_i} is called a recursive region.

- M_{a_i,b_i}, for $1 \leq i \leq s$, is a recursive simple non-standard pseudoknot of degree $\leq k$.
- $(M_{x,y} - \bigcup_{1 \leq i \leq s} M_{a_i,b_i})$ is either regular structure, standard pseudoknot of degree $\leq k$ or simple non-standard pseudoknot of degree $\leq k$.

Note that only the regular structure, recursive simple pseudoknot and recursive non-standard pseudoknot allow branching, which means there may exist base pairs $(i_0,j_0),(i_1,j_1),(i_2,j_2) \in M$ such that $(i_1,j_1),(i_2,j_2)$ are both inside $[i_0 \ldots j_0]$, but (i_1,j_1) is not inside $[i_2 \ldots j_2]$ and (i_2,j_2) is not inside $[i_1 \ldots j_1]$.

DEFINITION OF STRUCTURAL ALIGNMENT

Let $S[1...m]$ be a query sequence with known secondary structure M, and $T[1...n]$ be a target sequence with unknown secondary structure. S and T are both sequences of $\{A,C,G,U\}$. A structural alignment between S and T is a pair of sequences $S'[1...r]$ and $T'[1...r]$ where $r \geq m,n$, S' is obtained from S and T' is obtained from T with spaces inserted to make both of them of the same length. A space cannot appear in the same position of S' and T'. The score of the alignment, which determines the sequence and structural similarity between S' and T', is defined as follows (Han *et al.*, 2008).

$$score = \sum_{i=1}^{r} \gamma(S'[i], T'[i]) + \sum_{\substack{i,j \text{ s.t. } (\eta(i),\eta(j)) \in M \\ S'[i],S'[j],T'[i],T'[j] \neq `\_'}} \delta(S'[i], S'[j], T'[i], T'[j])$$

where $\eta(i)$ is the corresponding position in S according to the position i in S'; $\gamma(t_1,t_2)$ and $\delta(x_1,y_1,x_2,y_2)$ where $t_1,t_2 \in \{A,C,G,U,`\_'\}$ and $x_1,x_2,y_1,y_2 \in \{A,C,G,I\}$ are scores for character similarity and for base pair similarity, respectively.

The calculation of structural alignment score is not restricted to any kind of secondary structure. The problem of structural alignment is to find an alignment to maximize the score. Higher score represents higher similarity between the two sequences according to their sequences and structures. Also, if the score is high, then the alignment can reasonably reveal the secondary structure of the target sequence.

ALGORITHM FOR REGULAR STRUCTURE

Zhang *et al.* (2005) developed dynamic programming for regular structure. The dynamic programming is as follows: let $S[1\ldots m]$ be an ncRNA sequence with regular structure and $T[1\ldots n]$ be a target sequence with unknown structure. Define $A(p,q,e,f)$ to be the score of the optimal structural alignment between $S[p\ldots q]$ and $T[e\ldots f]$. Note that the optimal alignment score between $S[1\ldots m]$ and $T[1\ldots n]$ is $A(1,m,1,n)$.

There are three conditions: (I) (p,q) is a base pair; (II) there exists q' where $p < q' < q$ such that (p,q') is a base pair; (III) p is a single base. We will look into each case one by one. If (p,q) is a base pair (i.e. $(p,q) \in M$), then there are four cases: $\text{MATCH}_{\text{both}}$ – aligning (p,q) with (e,f); $\text{MATCH}_{\text{single}}$ – aligning only one of the bases in (p,q) with the corresponding base in (e,f); INSERT – inserting a space in S; DELETE – deleting (p,q) from S. The lemma 1 summarizes these cases:

Lemma 1. Let $A(p,q,e,f)$ be the score of the optimal structural alignment between $S[p\ldots q]$ and $T[e\ldots f]$. If (p,q) is a base pair, then

$$A(p,q,e,f) = \max\begin{cases} // \text{ MATCH}_{\text{both}} \\ A(p+1,q-1,e+1,f-1)+\gamma(S[p],T[e]) \\ +\gamma(S[q],T[f])+\delta(S[p],S[q],T[e],T[f]), \\ // \text{ MATCH}_{\text{single}} \\ A(p+1,q-1,e+1,f)+\gamma(S[p],T[e]) \\ +\gamma(S[q],`\_'), \\ A(p+1,q-1,e,f-1)+\gamma(S[p],`\_') \\ +\gamma(S[q],T[f]), \\ // \text{ INSERT} \\ A(p,q,e+1,f)+\gamma(`\_',T[e]), \\ A(p,q,e,f-1)+\gamma(`\_',T[f]), \\ // \text{ DELETE} \\ A(p+1,q-1,e,f)+\gamma(S[p],`\_')+\gamma(S[q],`\_') \end{cases}$$

If there exists q' where $p<q'<q$ such that (p,q') is a base pair, then there should be f' where $e<f'<f$ such that the sum of optimal alignment score between $S[p\ldots q']$ and $T[e\ldots f']$, and that between $S[q'+1\ldots q]$ and $T[f'+1\ldots f]$ is maximum, and $A(p,q,e,f)$ should equal to the value of the sum.

Lemma 2. Let $A(p,q,e,f)$ be the score of the optimal structural alignment between $S[p\ldots q]$ and $T[e\ldots f]$. If there exists q' where $p<q'<q$ such that (p,q') is a base pair, then

$$A(p,q,e,f) = \max_{f'}\{A(p,q',e,f') \\ +A(q'+1,q,f'+1,f)\}$$

If p is a single base (i.e. there does not exist q' such that (p,q') is a base pair), then there are also three cases: MATCH – aligning $S[p]$ with $T[e]$; INSERT – inserting a space in S; DELETE – deleting $S[p]$.

Lemma 3. Let $A(p,q,e,f)$ be the score of the optimal structural alignment between $S[p\ldots q]$ and $T[e\ldots f]$. If p is a single base, then

$$A(p,q,e,f) = \max\begin{cases} // \text{ MATCH} \\ A(p+1,q,e+1,f)+\gamma(S[p],T[e]), \\ // \text{ INSERT} \\ A(p,q,e+1,f)+\gamma(`\_',T[e]), \\ // \text{ DELETE} \\ A(p+1,q,e,f)+\gamma(S[p],`\_') \end{cases}$$

For each entry of the dynamic programming table, it needs $O(n)$ time for computation due to the equation in Lemma 2. To fill the dynamic programming table, not all entries for all possible subranges of S need to be filled in. In the following, we define a function $\rho(p,q)$ to determine for which set of subranges in S, we need to fill in the corresponding A entries. (see Box 1)

Box 1.

$$\rho(p,q) = \begin{cases} \{(p-1,q+1)\}, \text{ if } (p,q) \text{ is a base pair} \\ \{(p,q'),(q'+1,q)\}, \text{ if } \exists q' \text{ where } p < q' < q \text{ s.t. } (p,q') \text{ is a base pair} \\ \{(p-1,q)\}, \text{ if } p \text{ is a single base} \end{cases}$$

Figure 4. (a) Consider a query S[x...y] with standard pseudoknot of degree 3, subregion R(S,(i,j,k))=[x...i]∪[j...k], where $x \leq i < x_1 \leq j < x_2 \leq k \leq y$. Note that the structure of the subregion is also a standard pseudoknot of degree 3. (b) Subregion R(S,(x_1-1,x_1,y)) represents the whole pseudoknot region of S[x...y].

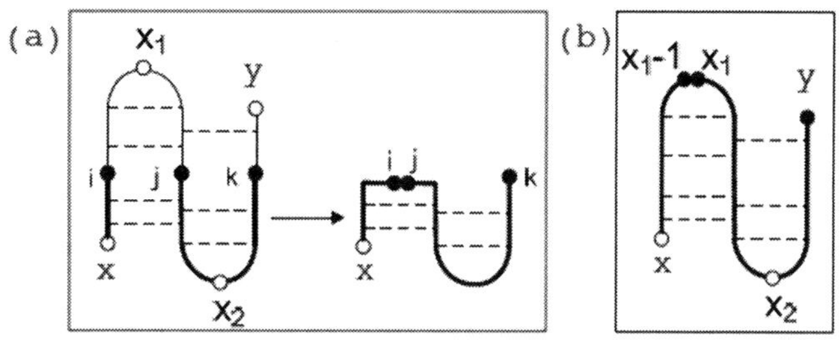

We only need to fill in the entries for A provided $(1,m)$ can be obtained from (p,q) by applying ρ function repeatedly. Intuitively, ρ guides which recursion formula to be used. And there are only $O(m)$ such (p,q) values. The following theorem summarizes the time complexity of this algorithm.

Theorem 1. For any sequence $S[1...m]$ with regular structure and any sequence $T[1...n]$ with unknown structure, the optimal alignment score between $S[1...m]$ and $T[1...n]$ can be computed in $O(mn^3)$.

ALGORITHM FOR STANDARD PSEUDOKNOTS

Han *et al.* (2008) solved the problem using dynamic programming. The key is to define a substructure to enable finding the solution recursively. We use simple standard pseudoknots of degree 3 for illustration. The result can be easily extended to general k. For easily understanding what a substructure is, we draw the standard pseudoknot of degree 3 using another approach (see Figure 4).

First, we formally define a substructure. Let $S[x...y]$ be an ncRNA sequence with a standard pseudoknot of degree 3. Note that there exists $x \leq x_1 < x_2 \leq y$ such that the structure M can be divided into two regular structures M_1 and M_2 as mentioned before. Let $v=(i,j,k)$ be a triple with $x \leq i < x_1 \leq j < x_2 \leq k \leq y$. Define the subregion $R(S,v) = [x...i] \cup [j...k]$, as shown in Figure 4(a). Let $Struct(R)=\{(i,j)\in M \mid i,j \in R\}$ where R is a subregion. We say that a subregion R defines a valid substructure ($Struct(R)$) of M if there does not exist $(i,j) \in M$ such that one endpoint of (i,j) is in R and the other one is outside the region. Obviously, $Struct(R)$ is also a standard pseudoknot of degree 3.

Let $S[1...m]$ be a query sequence with known secondary structure M, and $T[1...n]$ be a target sequence with unknown secondary structure. Note that the pivot points x_1, x_2 for S are known. We can apply the definitions of R to T. For any

Box 2.

$$B\left(R_x, R_y\right) = max \begin{cases} // \text{ MATCH}_{both} \\ B(R(S,(p-1,q+1,r)), R(T,(e-1,f+1,g))) \\ +\gamma(S[p],T[e]) + \gamma(S[q],T[f]) + \delta(S[p],S[q],T[e],T[f]), \\ // \text{ MATCH}_{single} \\ B(R(S,(p-1,q+1,r)), R(T,(e-1,f,g))) + \gamma(S[p],T[e]) + \gamma(S[q],`\_'), \\ B(R(S,(p-1,q+1,r)), R(T,(e,f+1,g))) + \gamma(S[p],`\_') + \gamma(S[q],T[f]), \\ // \text{ INSERT} \\ B(R(S,(p,q,r)), R(T,(e-1,f,g))) + \gamma(`\_',T[e]), \\ B(R(S,(p,q,r)), R(T,(e,f+1,g))) + \gamma(`\_',T[f]), \\ B(R(S,(p,q,r)), R(T,(e,f,g-1))) + \gamma(`\_',T[g]), \\ // \text{ DELETE} \\ B(R(S,(p-1,q+1,r)), R(T,(e,f,g))) + \gamma(S[p],`\_') + \gamma(S[q],`\_') \end{cases}$$

$v'=(e,f,g)$ such that $1 \leq e < f < g \leq n$, we define the subregion $R(T,v')=[1\ldots e]\cup[f\ldots g]$.

Define $B(R_x,R_y)$ to be the score of the optimal alignment between a subregion R_x in S with substructure $Struct(R_x)$ and a subregion R_y in T. The score of the optimal alignment between S and T can be obtained by setting $v^*=(1,x_1-1,x_1,m)$ that includes the whole query sequence S (as shown in Figure 4(b)). The entry $\max_{x_1'}\{B(R(S,v^*), R(T,v'=(1,x_1'-1,x_1',n)))\}$ provides the answer.

The value of $B(R_x,R_y)$ can be computed recursively. Let $R_x=R(S,(p,q,r))$ and $R_y=R(T,(e,f,g))$. If (p,q) is a base pair in $Struct(R_x)$, there are four cases to consider. Case 1: $\text{MATCH}_{\text{both}}$ – aligning the base pair (p,q) of S with (e,f) of T; Case 2: $\text{MATCH}_{\text{single}}$ – aligning only one of the bases in (p,q) with the corresponding base in (e,f); Case 3: INSERT – insert a space on S; Case 4: DELETE – delete the base-pair (p,q) from S. Lemma 4 summarizes these cases. The other condition, i.e. (q,r) is a base pair, is similar.

Lemma 4. Let $v=(p,q,r)$ and $v'=(e,f,g)$. Let $R_x=R(S,v)$ and $R_y=R(T,v')$. If (p,q) is a base pair, then (see Box 2.)

On the other hand, if none of these are base pairs, assume that $p-1 \geq 1$ and $S[p]$ is a single base (i.e. the positions which do not belong to any base pair), then we can compute $B(R_x,R_y)$ recursively according to another three cases. Case 1: Match – aligning $S[p]$ with $T[e]$; Case 2: INSERT – insert a space on S; Case 3: Delete – delete $S[p]$.

Lemma 5. Let $v=(p,q,r)$ and $v'=(e,f,g)$. Let $R_x=R(S,v)$ and $R_y=(S,v')$. If $1 \leq p < x_1$ and $S[p]$ is a single base, then

$$B\left(R_x, R_y\right) = max \begin{cases} // \text{ MATCH} \\ B(R(S,(p-1,q,r)), R(T,(e-1,f,g))) + \gamma(S[p],T[e]), \\ // \text{ INSERT: same as the one defined in Lemma 4} \\ // \text{ DELETE} \\ B(R(S,(p-1,q,r)), R(T,(e,f,g))) + \gamma(S[p],`\_') \end{cases}$$

The other conditions, i.e. $S[q]$ is a single base or $S[r]$ is a single base, are similar. Note that if none of these are base pairs and there exist more than one single base, we only need to follow the recursion on one of the bases. To fill in the dynamic programming table, not all entries for all possible subranges of S need to be filled in. In the

Figure 5. Substructure of a simple non-standard pseudoknot

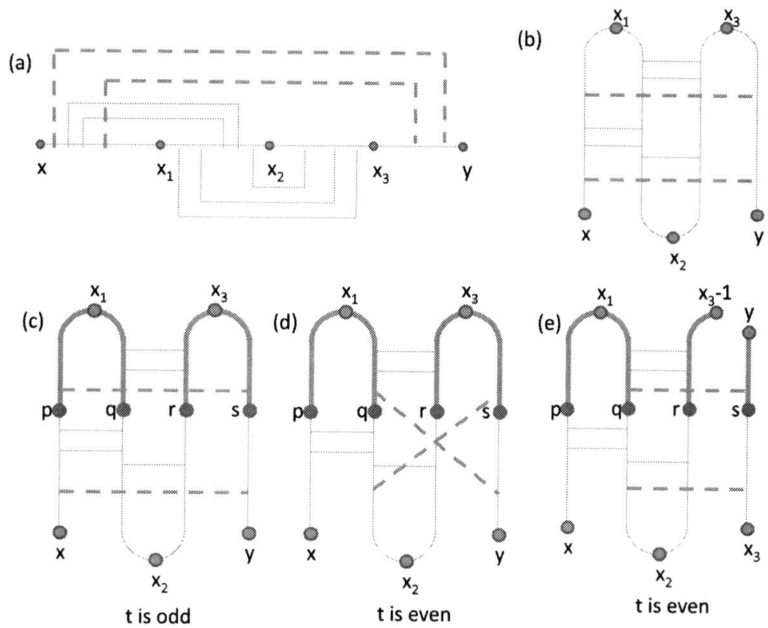

following, we define a function $\zeta(v)$ to determine for which subregions in S we need to fill in the corresponding B entries.

Case 1. If (p,q) or (q,r) is a base pair, then

$$\zeta(v)=\begin{cases}(p-1,q+1,r), \text{ if (p,q) is a base pair}\\(p,q+1,r-1), \text{ if (q,r) is a base pair}\end{cases}$$

Case 2. If none of (p,q) or (q,r) is base pair, then

$$\zeta(v)=\begin{cases}(p-1,q,r), \text{ if } p \text{ is a single base and } 1\le p<x_1\\(p,q+1,r), \text{ else if } q \text{ is a single base and } x_1\le q<x_2\\(p,q,r-1), \text{ else if } r \text{ is a single base and } x_2\le r\le m\end{cases}$$

It is obvious that if v defines a subregion with a valid substructure, $\zeta(v)$ also defines a valid substructure. Let $v^*=(1,x_1-1,x_1,m)$. We only need to fill in the entries for B provided v^* can be obtained from v by applying ζ function repeatedly. Intuitively, ζ guides which recursion formula can be used. And there are only $O(m)$ such v values. The following lemma summarizes the time complexity of this algorithm.

Lemma 6. For any sequence $S[1\ldots m]$ with standard pseudoknot of degree 3 and any sequence $T[1\ldots n]$ with unknown structure, the optimal structural alignment score between $S[1\ldots m]$ and $T[1\ldots n]$ can be computed in $O(mn^3)$.

The algorithm can be easily extended to simple non-standard pseudoknot of degree k.

Theorem 2. For any sequence $S[1\ldots m]$ with standard pseudoknot of degree k and any sequence $T[1\ldots n]$ with unknown structure, the optimal alignment score between $S[1\ldots m]$ and $T[1\ldots n]$ can be computed in $O(kmn^k)$.

ALGORITHM FOR SIMPLE NON-STANDARD PSEUDOKNOTS

Wong *et al.* (2009) developed a dynamic programming algorithm to solve this problem. We use simple non-standard pseudoknot of degree 4 for illustration. The result again can be easily extended to general k. Figure 5(b) shows the same pseudoknot structure as in Figure 5(a). By drawing the pseudoknot structure this way, the base pairs can be drawn without crossing and can be ordered from the top to bottom. According to this ordering, we can define a substructure based on four points on the sequence (see Figure 5(c) in which the substructure is highlighted in bold) such that all base pairs are either with both end points inside or outside the substructure. Note that in Figure 5(c), t=1 (t is odd), if t=2 (t is even), we have to use a slightly different definition for substructures, otherwise base pairs cannot be ordered from top to bottom without crossing each other (see Figure 5(d) and (e). Note that the two base pairs that cross in Figure 5(d) are due to the way we draw the pseudoknot (they do not actually cross each other). These are the only cases we need to consider.

Now, we formally define what a substructure is for simple non-standard pseudoknot. Let $S[x\ldots y]$ be a RNA sequence with known simple non-standard pseudoknot structure M of degree 4. Note that x_1, x_2, x_3 and t are known. Let $v=(p,q,r,s)$ be a quadruple with $x \leq p < x_1 \leq q < x_2 \leq r < x_3 \leq s \leq y$. If t is odd, define the subregion $R_{odd}(S,v)=[p,q]\cup[r,s]$. Otherwise, define the subregion $R_{even}(S,x_3,v)=[p,q]\cup[r,x_3-1]\cup[s,y]$. Note that x_3 is not a parameter, but a fixed value for S. Let $Struct(R_x)=\{(i,j)\in M \mid i,j\in R_x\}$ where R_x is a subregion. We say that a subregion R_x defines a valid substructure ($Struct(R_x)$) of M if these does not exist $(i,j)\in M$ such that one endpoint of (i,j) is in R_x and the other is outside the region. Obviously, $Struct(R_x)$ is also a simple non-standard pseudoknot structure.

Let $S[1\ldots m]$ be the query sequence with known structure M and $T[1\ldots n]$ be the target sequence with unknown structure. Note that the pivot points x_1, x_2, x_3 and t for S is known. We can apply the definitions of R_{odd} and R_{even} to T. If t is odd, for any $v'=(e,f,g,h)$ such that $1 \leq e < f < g < h \leq n$, we define the subregion $R_{odd}(T,v')=[e\ldots f]\cup[g\ldots h]$. If t is even, for any $v'=(e,f,g,h)$ and x_3' such that $1 \leq e < f < g < x_3' \leq h \leq n$ we define the subregion $R_{even}(T,x_3',v')=[e\ldots f]\cup[g\ldots x_3'\text{-}1]\cup[h\ldots n]$. Note that since the structure of T is unknown, x_3' is a parameter.

Define $C(R_x,R_y)$ to be the score of the optimal alignment between a subregion R_x in S with substructure $Struct(R_x)$ and a subregion R_y in T. The score of the optimal alignment between S and T can be obtained as follows. If t is odd, setting $v^*=(1,x_2-1,x_2,m)$ includes the whole query sequence S, the entry $\max_{x_2'}\{C(R_{odd}(S,v^*),R_{odd}(T,v'=(1,x_2'-1,x_2',n)))\}$ provides the answer. On the other hand, if t is even, setting $v^*=(1,x_2-1,x_2,x_3)$, the optimal score can be obtained from the entry $\max_{x_2'}\max_{x_3'>x_2'}\{C(R_{even}(S,x_3,v^*),$ $R_{even}(T,x_3',v'=(1,x_2'-1,x_2',x_3')))\}$.

The value of $C(R_x,R_y)$ can be computed recursively. Assume that t is odd. Let $R_x=R_{odd}(S,(p,q,r,s))$ and $R_y=R_{odd}(T,(e,f,g,h))$. If (p,q) is a base pair in $Struct(R_x)$, there are four cases to consider: Case 1: MATCH$_{both}$ – aligning the base pair (p,q) of S with (e,f) of T (as shown in Figure 6); Case 2: MATCH$_{single}$ – aligning only one of the bases in (p,q) with the corresponding base in (e,f); Case 3: INSERT – insert a space on S; Case 4: DELETE – delete the base-pair (p,q) from S. Lemma 7 summarizes these cases.

The other cases, where (q,r) is a base pair or (p,s) is a base pair, are similar. Note that if more than one such base pair exists (e.g. both (q,r) and (p,s) are base pairs), we only need to follow the recursion on one of the pairs. However, we cannot pick any of them in an arbitrary manner, since otherwise, when we fill the dynamic programming table, we need to fill in all entries for all possible regions of S. We will address this issue in the later part of this section.

Lemma 7. Let $v=(p,q,r,s)$ and $v'=(e,f,g,h)$. Let t be odd. And $R_x = R_{odd}(S,v)$, $R_y=R_{odd}(T,v')$. If (p,q) is a base pair, then see Box 3.

On the other hand, if none of these are base pairs, assume that $p+1<x_1$ and $S[p]$ is a single

Figure 6. Illustration of the computation of $C(R_x,R_y)$ for the case $MATCH_{both}$ – aligning the base pair (p,q) of S with (e,f) of T when (p,q) is a base pair

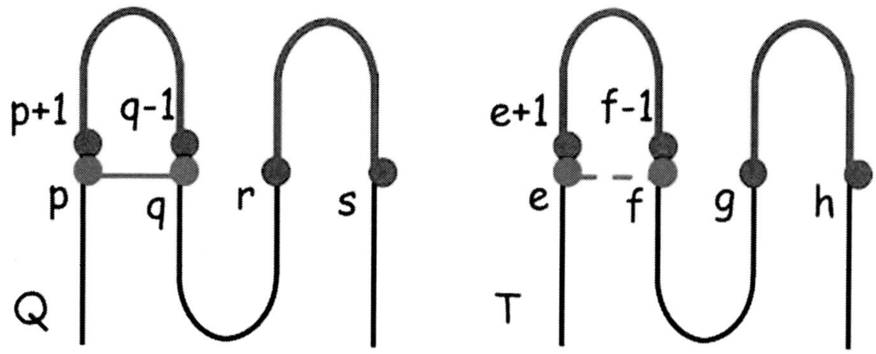

Box 3.

$$C\left(R_x,\ R_y\right) = max\begin{cases} //\ \text{MATCH}_{both} \\ C(R_{odd}(S,(p+1,q-1,r,s)),R_{odd}(T,(e+1,f-1,g,h))) \\ +\gamma(S[p],T[e])+\gamma(S[q],T[f])+\delta(S[p],S[q],T[e],T[f]), \\ //\ \text{MATCH}_{single} \\ C(R_{odd}(S,(p+1,q-1,r,s)),R_{odd}(T,(e+1,f,g,h)))+\gamma(S[p],T[e])+\gamma(S[q],`\_'), \\ C(R_{odd}(S,(p+1,q-1,r,s)),R_{odd}(T,(e,f-1,g,h)))+\gamma(S[p],`\_')+\gamma(S[q],T[f]), \\ //\ \text{INSERT} \\ C(R_{odd}(S,(p,q,r,s)),R_{odd}(T,(e+1,f,g,h)))+\gamma(`\_',T[e]), \\ C(R_{odd}(S,(p,q,r,s)),R_{odd}(T,(e,f-1,g,h)))+\gamma(`\_',T[f]), \\ C(R_{odd}(S,(p,q,r,s)),R_{odd}(T,(e,f,g+1,h)))+\gamma(`\_',T[g]), \\ C(R_{odd}(S,(p,q,r,s)),R_{odd}(T,(e,f,g,h-1)))+\gamma(`\_',T[h]), \\ //\ \text{DELETE} \\ C(R_{odd}(S,(p+1,q-1,r,s)),R_{odd}(T,(e,f,g,h)))+\gamma(S[p],`\_')+\gamma(S[q],`\_') \end{cases}$$

base. Then we can compute $C(R_x\_R_y)$ recursively according to other three cases. Case 1: Match – aligning $S[p]$ with $T[e]$; Case 2: INSERT – insert a space on S; Case 3: Delete – delete $S[p]$.

Lemma 8. Let $v=(p,q,r,s)$ and $v'=(e,f,g,h)$. Let t be odd. And $R_x=R_{odd}(S,v)$, $R_y=R_{odd}(T,v')$. If $p+1<x_1$ and $S[p]$ is a single base, then see Box 4.

If t is even, we consider whether (p,q),(q,r), and (q,s) are base pairs in $Struct(R_x)$ and we have to consider all possible cases for x'_3 since the structure of T is unknown (i.e. the pivot points are unknown).

To fill in the dynamic programming table, not all entries for all possible subranges of S need to be filled in. For any given subregion $v=(p,q,r,s)$ in S, we first define $pair_{min}(v)$ and $single_{min}(v)$ as follows. If there exists a set of base pairs, say $\{(i_1,j_1),\ldots,(i_d,j_d)\}$, such that all $i_k,j_k(1\leq k\leq d)$ equal to p (if $x\lneq p<x_1$), q (if $x_1\leq q<x_2$), r (if $x_2\leq r<x_3$) or s (if $x_3\leq s\leq y$), then $pair_{min}(v)$ is the pair with minimum value of i. Also, if there exists a

Box 4.

$$C\left(R_x, R_y\right) = max\begin{cases} C(R_{odd}(S,(p+1,q,r,s)), R_{odd}(T,(e+1,f,g,h))) + \gamma(S[p],T[e]) \text{ // MATCH} \\ \text{// INSERT: same as the one defined in Lemma 4} \\ C(R_{odd}(S,(p+1,q,r,s)), R_{odd}(T,(e,f,g,h))) + \gamma(S[p],\text{`}\_\text{'}) \text{ // DELETE} \end{cases}$$

set of single bases (i.e. the positions which do not belong to any base pair), say $\{u_1, ..., u_d\}$, such that all $u_k (1 \leq k \leq d)$ equal to p (if $x \leq p \leq x_1$), q (if $x_1 \leq q < x_2$), r (if $x_2 \leq r < x_3$) or s (if $x_3 \leq s \leq y$), then $single_{min}(v)$ is the one with the minimum value.

Now, we define a function $\psi(v)$ to determine for which subregions in S, we need to fill in the corresponding C entries.

Case 1. If $(i,j)=pair_{min}(v)$ exists, then

$$\psi(v) = \begin{cases} (p+1, q-1, r, s), & \text{if } (i,j) = (p,q) \\ (p, q-1, r+1, s), & \text{if } (i,j) = (q,r) \\ (p+1, q, r, s-1), & \text{if } (i,j) = (p,s) \text{ i.e. } t \text{ is odd} \\ (p, q-1, r, s+1), & \text{if } (i,j) = (q,s) \text{ i.e. } t \text{ is even} \end{cases}$$

Case 2. If $pair_{min}(v)$ does not exist, then $u=single_{min}(v)$ should exist and

$$\psi(v) = \begin{cases} (p+1, q, r, s), & \text{if } u = p \\ (p, q-1, r, s), & \text{if } u = q \\ (p, q, r+1, s), & \text{if } u = r \\ (p, q, r, s-1), & \text{if } u = s \text{ and } t \text{ is odd} \\ (p, q, r, s+1), & \text{if } u = s \text{ and } t \text{ is even} \end{cases}$$

It is obvious that if v defines a subregion with a valid substructure, $\psi(v)$ also defines a valid substructure. If t is odd, let $v^*=(1,x_2-1,x_2,m)$. We only need to fill in the entries for C provided v^* can be obtained from v by applying the ψ function repeatedly. If t is even let $v^*=(1,x_2-1,x_2,x_3)$. Intuitively, ψ guides which recursion formula to be employed. And there are only $O(m)$ such v values. The following lemma summarized the time complexity for this algorithm.

Lemma 9. For any sequence $S[1...m]$ with simple non-standard pseudoknot of degree 4 and any sequence $T[1...n]$, with c the max length of $[x'_3...n]$, the optimal alignment score between $S[1...m]$ and $T[1...n]$ can be computed in $O(cmn^4)$.

Note that the factor c is only needed when t is even due to the extra parameter x'_3. Wong *et al.* (2009) examined all the sequences in Rfam 9.1 and PseudoBase, and they found that c is usually between 5 and 7. So, one can assume that c is a constant. The algorithm can be easily extended to simple non-standard pseudoknot of degree k.

Theorem 3. For any sequence $S[1...m]$ with simple non-standard pseudoknot of degree k and any sequence $T[1...n]$, the optimal alignment score between $S[1...m]$ and $T[1...n]$ can be computed in $O(kmn^k)$.

ALGORITHM FOR 2-LEVEL RECURSIVE PSEUDOKNOTS

In this section, we describe the algorithm (Wong *et al.*, 2009) designed to handle a special type of recursive pseudoknots in which each recursive region is a regular structure and where, after excluding all recursive regions, the remaining base pairs form a simple pseudoknot or a simple non-standard pseudoknot. We refer to this recursive pseudoknot as *2-level pseudoknot with regular recursive regions*. We use a recursive pseudoknot of degree 4 with simple non-standard pseudoknot to illustrate the algorithm. The algorithm for simple pseudoknot is simpler and the approach can be easily extended to general k.

Let $S[1..m]$ be the query sequence with recursive pseudoknot structure M. Recall the definition of a recursive pseudoknot. There can be disjoint recursive regions, namely $M_{a1,b1}, \ldots, M_{as,bs}$, in M. By removing all these recursive regions, the remaining structure $M-(M_{a1,b1} \cup \ldots \cup M_{as,bs})$ together with the remaining sequence $S[1..a_1-1]S[b_1+1..a_2-1] \ldots S[b_s+1..m]$ are referred to as level-0. For each removed recursive region $M_{ai,bi}$, we can apply the same procedure to define level-1, level-2,..., level-l structures. In our case, we only have level-1 structures (see Figure 7 for an example).

Let $T[1..n]$ be the target sequence. Define $H[a_i,b_i,x',y']$ to be the score of the optimal alignment between the recursive region $S[a_i..b_i]$ with structure $M_{ai,bi}$ and $T[x'..y']$, where $1 \leq x' < y' \leq n$. We now show how to compute the score of the optimal alignment between S and T recursively for Type-I simple non-standard pseudoknot (i.e. t is odd). The case when t is even is similar. Let $v = (p, q, r, s)$ be a quadruple that defines a substructure of S. Let $S[p..y_p]$ be a recursive region. The following lemma shows how to compute $C(R_x,R_y)$, the score of the optimal alignment between R_x and R_y where $R_x = R_{odd}(S, v)$ and $R_y = R_{odd}(T, v')$.

Lemma 10. Let $v=(p,q,r,s)$ and $v'=(e,f,g,h)$. Assume that t is odd. $R_x=R_{odd}(S,v)$ and $R_y=R_{odd}(T,v')$. If $S[p...y_p]$ is a recursive region, then see Box 5.

Other cases, when $S[x_q...q]$ or $S[r...y_r]$ or $S[x_s...s]$ is a recursive region, can be handled in a similar way. Again, we do not need to compute C for all possible values of (p,q,r,s). We need to determine for which subregions in S we have to fill in the corresponding C entries. So, we enhance ψ function as follows.

Consider a quadruple $v=(p,q,r,s)$ in a region $S[a'...b']$ where the structure is a simple non-standard pseudoknot of degree 4 if all the next-level subregions inside (i.e. $S[a_1'...b_1']$, $S[a_2'...b_2']$,..., $S[a_t'...b_t']$) are excluded. Let us define $subregion_{min}(v)$ as follows: if there exists a set of next-level subregions, say $\{[i_1..j_1],\ldots,[i_d..j_d]\}$ where $x \leq i_k < j_k \leq y$ for all $1 \leq k \leq d$ such that either i_k or j_k equals to p (if $x \leq p < x_1$), q (if $x_1 \leq q < x_2$), r (if $x_2 \leq r < x_3$)) or s (if $x_3 \leq s \leq y$), then let $subregion_{min}(v)$ be the region with minimum value of i. We add the following case to ψ func-

Figure 7. (a) An example of a 2-level pseudoknot with regular recursive regions. (b) Shows the level-0 and level-1 structures. (c) Shows another view for the same example.

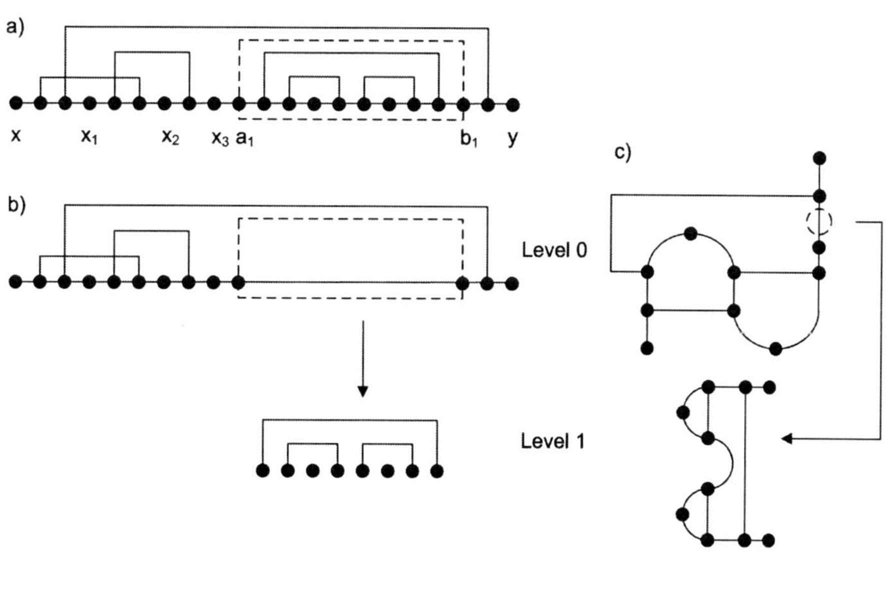

Figure 8. Illustration of the computation of $C(R_x,R_y)$ for the case MATCH – aligning the recursive region (p,y_p) of S with (e,w) of T when (p,y_p) is a recursive region

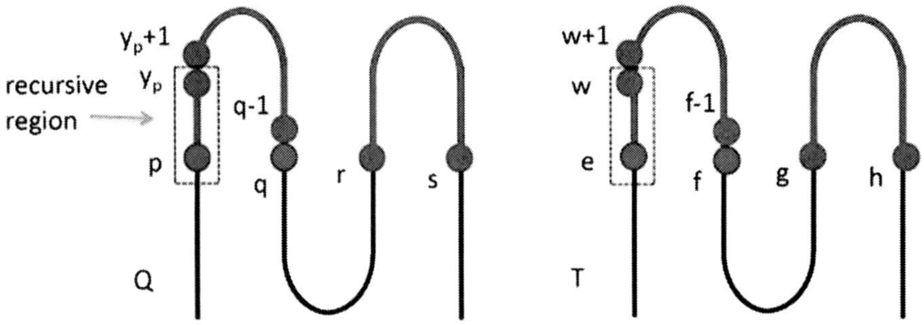

Box 5.

$$C(R_x, R_y) = \max \begin{cases} // \text{ MATCH (as illustrated in Figure 7)} \\ \max_{e \le w \le f} \left\{ C(R_{odd}(S,(y_p+1,q,r,s)), R_{odd}(T,(w+1,f,g,h))) + H(p,y_p,e,w) \right\} \\ // \text{ INSERT} \\ \text{same as INSERT defined in Lemma 7} \\ // \text{ DELETE} \\ \mathrm{C}(\mathrm{R}_{odd}(S,(y_p+1,q,r,s)), R_{odd}(T,(e,f,g,h))) + \sum_{p \le w \le y_p} \gamma(S[w], `\_') \end{cases}$$

tion. Note that the t value refers to the structure of $S[x..y]$ excluding all next-level subregions.

Case 0 of $\psi(v)$: If $[i...j] = subregions_{min}(v)$ exists, then

$$\psi(v) = \begin{cases} (j+1,q,r,s), & \text{if } i = p \\ (p,i-1,r,s), & \text{if } j = q \\ (p,q,j+1,s), & \text{if } i = r \\ (p,q,r,i-1), & \text{if } j = s \text{ // i.e. } t \text{ is odd} \\ (p,q,r,j+1), & \text{if } i = s \text{ // i.e. } t \text{ is even} \end{cases}$$

There are at most $O(m)$ v values we need to consider. So, assuming all $H()$ values that have been computed, it takes $O(mn^5)$ time to fill in all C entries. Since the recursive region is a regular structure, we can make use of the previous algorithm for regular region to compute all $H()$ values for all possible subregions of T in $O(mn^3)$ time (because for each region $S[a_i'...b_i']$, after computation of $A(a_i',b_i',1,n)$, all values of $H(a_i',b_i',e,f)$ for all $1 \le e < f \le n$ become available in the matrix A, and sum of lengths of all regions $S[a_i'...b_i'] \le m$). The following theorem summarizes the result of this section. The algorithm presented in this section can be extended to general recursive pseudoknots with more than 2 levels and with recursive regions having other structures as defined in Definition 5 with an increase of $O(n)$ factor in the time complexity

Theorem 4. For any sequence $S[1...m]$ with 2-level recursive simple non-standard pseudoknot of degree k with regular recursive regions and any sequence $T[1...n]$, the optimal alignment score between $S[1...m]$ and $T[1...n]$ can be computed in $O(kmn^{k+1})$ time.

Table 2. The details of the ncRNA families used in the experiments

Family	Query Sequence ID	Length of Query Sequence	Number of members
RF00094	AB037947/685-775	91	432
RF00140	AM286690/459188-459298	111	164
RF00622	AAGV01475186/596-519	78	47
RF01075	AY787207/2721-2816	96	23
RF01084	AF325738/2039-2167	129	263
RF01085	K01776/83-200	118	8
RF00176	D00719/409-499	91	80

EXPERIMENTAL RESULTS

We now focus on the experimental results based on *Wong et al. (2009)* for simple non-standard pseudoknot and recursive simple non-standard pseudoknot. For the evaluation of other methods, please refer to the relevant publications. They implemented algorithms in C++. By inputting a query ncRNA sequence (*Q*) and its secondary structure, the program can scan a long RNA sequence (*T*) and output the score for every region in *T*. Higher score indicates that the sequence and the structure of the region is more similar to those of *Q*. To evaluate the effectiveness of their algorithm, a set of families in Rfam 9.1 database was selected for which the structures of the ncRNAs in these families are either simple non-standard pseudoknot, recursive standard pseudoknot or recursive simple non-standard pseudoknot. For each family, they selected one of the seed members (in Rfam 9.1 database, for each family, there is a set of reliable members which are regarded as seed members) as the query sequence *Q*. To demonstrate the power of structural alignment, the query sequence selected has the lowest sequence similarity with the other seed members. The details of the families including the sequence selected as the query, the length of this sequence, and the number of members in each family are given in Table 2.

A long random genome sequence of length about 300 times the length of the query sequence was constructed. Then, all the ncRNA sequences (seed members or non-seed members) of the family except the query sequence were embedded into this long random sequence in arbitrary positions. The resulting sequence is *T*. For every region in *T* with length similar to that of the query sequence (i.e. the length of each region equals the length of the query plus 20), the structural alignment score of the region and the query sequence was computed. The scoring matrix RIBOSUM85-60 created by Klein and Eddy (2003) was used. The running time required for the computation of the structural alignment score of the region and the query sequence is summarized in Table 3.

It was assumed that the regions other than the real members of the family are false hits as they are not likely to be members of the family. Figure 9 shows the distribution of the alignment scores of the true hits (real members) and false hits. From Figure 9, it is quite clear that the real members can be easily distinguished from the false hits except for the family RF00176 based on the alignment scores. The authors have investigated why the approach does not work well for RF00176. They have found that the length of the query sequence is much longer than that of the other member sequences. Since the method was de-

Table 3. The time required for aligning between the query sequence and a region in T of similar length

Family	Pseudoknot Type	Length of Query Sequence	Time
RF00094	Degree-4 recursive simple non-standard	91	15 min
RF00140	Degree-4 simple non-standard	111	32 min
RF00622	Degree-4 recursive simple non-standard	78	9 min
RF01075	Degree-3 recursive standard	96	51 sec
RF01084	Degree-3 recursive standard	129	2 min
RF01085	Degree-3 recursive standard	118	1 min 40 sec
RF00176	Degree-3 recursive standard	91	40 sec

signed for global alignment, the big differences in length lead to a big penalty score in the method and thus the resulting structural alignment scores became very low. To verify this observation, they identified a conserved region inside the multiple sequence alignment of the family. Then they only used the corresponding conserved region (the length is 37 while the total length of the sequence is 91) of the selected query sequence as the new query sequence. The result has been improved substantially (see Figure 10). From this observation, it is believed that developing a tool for local structural alignment is desirable.

Since there is no existing software which is available freely for performing structural alignment for complex pseudoknot structures, Wong and co-authors (2009) followed the evaluation method of (Han et al., 2008) to compare the performance of this method with BLAST. That is, they wanted to compare the effectiveness of considering only sequence similarity (based on BLAST) and their method which also considers the structural similarity. They used default parameters for BLAST except that the wordsize was set to 7 to increase its sensitivity. For each family, they used the same query sequence and the random sequence *T* as in the above experiment. Again, for each region in *T*, they computed a BLAST score between the region and the query sequence. To compute the effectiveness of this method and BLAST, a threshold was set as the maximum score of the false hits. It was assumed that the method finds a real hit if the score of the region is larger than this threshold. Thus a real hit will be missed if the computed score is smaller than or equal to this threshold. Different thresholds were examined and the results appeared to be similar.

Table 4 compares the results from the two methods. Note that we have omitted the family RF00176 in this comparison. For most of the families, the Wong's algorithm does not have as many misses as BLAST does. For example, in Family RF01084, the algorithm misses only 22 sequences, while BLAST misses 202 sequences. It is clear that this method is more effective than BLAST demonstrating that considering structural similarity is important for the ncRNA alignment.

Figure 11 shows the detailed scores for Wong's method and BLAST for family RF00094. For the sake of clarity, we only show the scores for the seed members. Among 32 seed members (except the one selected as query sequence), BLAST missed 13 of them. However, all the regions of these 32 members got the highest scores if using Wong's algorithm and, thus, none of them was missed. In scrutinizing the missed cases for BLAST, we found out that the missed sequence is usually not similar to the query sequence based only on the sequence similarity while the corresponding structure is similar to that of the query

Figure 9. The distribution of alignment scores of true hits and false hits

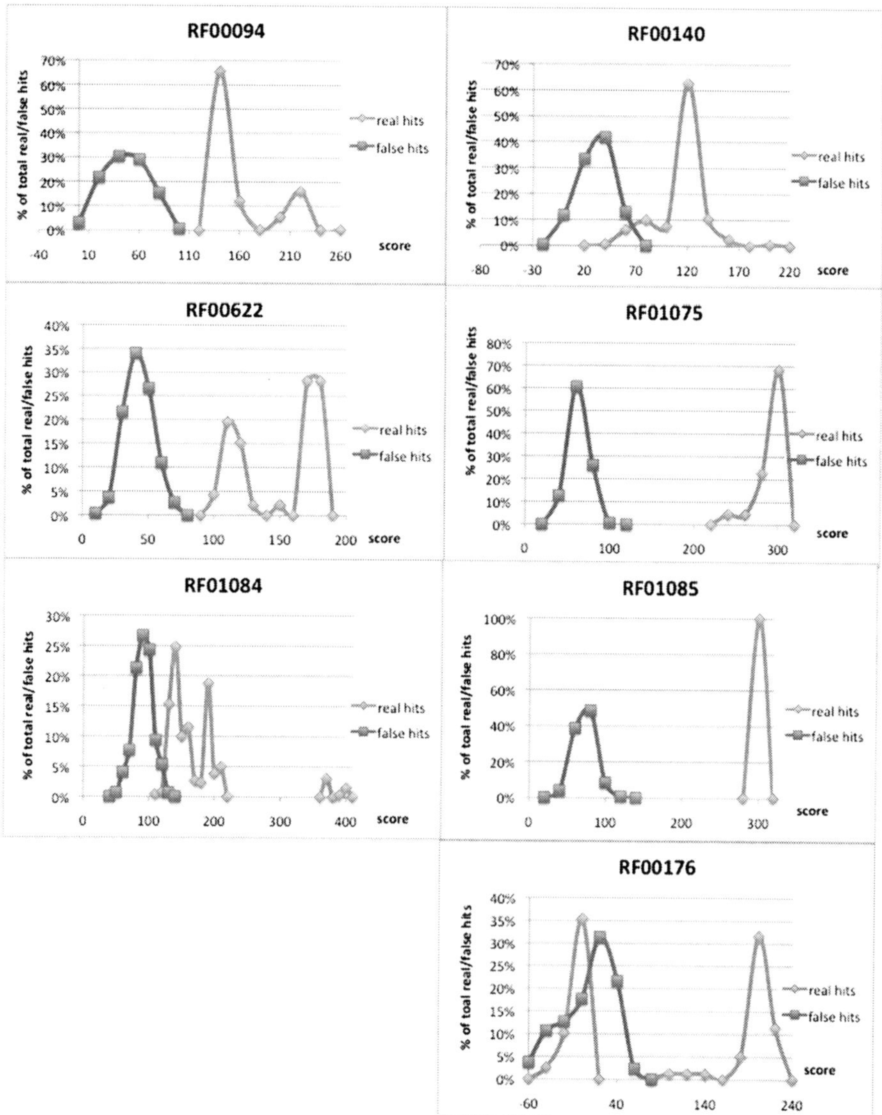

Figure 10. The distribution of alignment scores of true hits and false hits for the family RF00176 after considering the conserved region

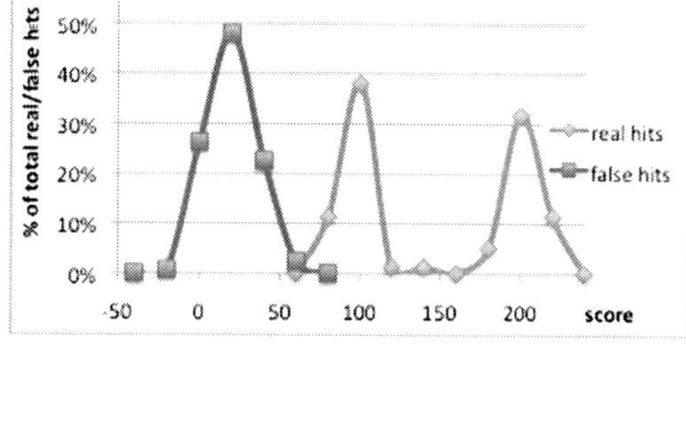

Table 4. Summary of comparison on results between BLAST and Wong's method

Family	# of Real Hits	BLAST misses #	Wong's method misses #
RF00094	431	334	0
RF00622	46	15	0
RF00140	163	8	9
RF01084	262	202	22
RF01075	22	0	0
RF01085	7	0	0

Figure 11. Comparison between resulting scores from the program and BLAST of family RF00094. The squares represent the positions of real hits, the triangles represent the positions of BLAST hits and the line represents the score outputted by the method along different positions. Among 32 real hits, BLAST missed 13 of them. However, Wong's method (the method which considers both sequence and structure similarity) did not miss any of them.

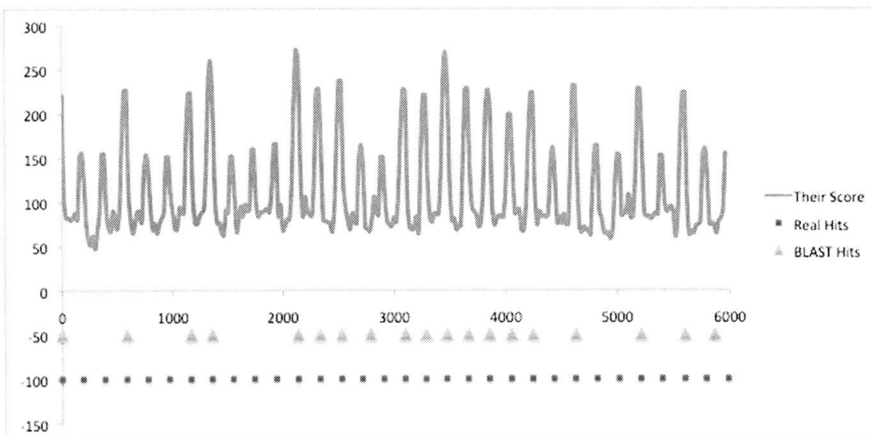

sequence. Figure 11 shows some examples of these cases. The top one, circled by red line, is the query sequence, and the others are those missed by BLAST but still can be identified by the algorithm. We can see that although the similarity between the query sequence and the other sequences are not high enough, all of their secondary structures are highly conserved.

CONCLUSION

In the chapter, we highlighted the difficulties in developing computational approaches for identifying the de novo ncRNAs due to the presence of pseudoknots in the secondary structures of the ncRNAs. We reviewed the set of algorithms that handle structural alignment of RNA with different structures including regular structure, standard pseudoknot, simple non-standard pseudoknot, recursive standard pseudoknot and recursive simple non-standard pseudoknot. The algorithms

Figure 12. A multiple sequence alignment of some members in the family RF01084 with recursive standard pseudoknot of degree 3. By using the selected query sequence (which is circled in red), BLAST cannot locate the other member sequences. However, since both the structure of the query sequence and that of the member sequences are highly conserved, the method which considers both sequence and structure similarity can locate all of them.

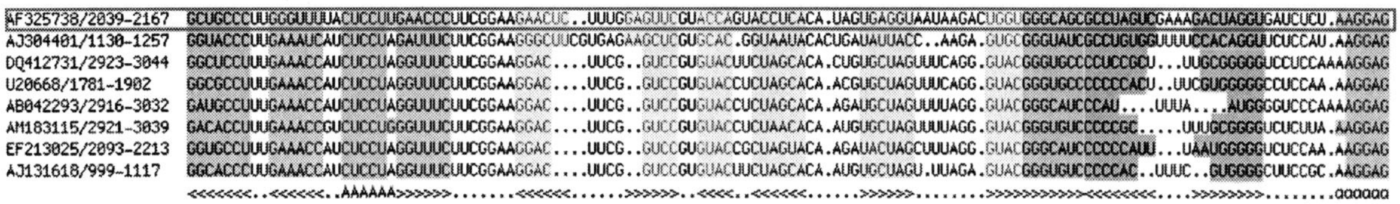

have been shown to be effective when using one known ncRNA to identify the other ncRNAs of the same family along the genome.

Although this structural alignment problem is believed to be NP-hard for general pseudoknots, with the recent development in this area, we believe that a promising direction is to carefully analyze the secondary structures of those ncRNAs found in nature.There may be other types of complex pseudoknot structures which are still tractable computationally so that computational biologists can help developing efficient algorithms to identify ncRNAs with these structures. Besides the algorithms for structural alignment of ncRNAs, there are still a lot of issues to be further investigated. For example, the scoring tables for structural alignment, which form a core component for these computational approaches, require an in-depth study for different types of pseudoknot structures. A related issue is to consider the energy/entropy of formation of the secondary structure of the potential ncRNA candidate which may provide a good measure to further confirm whether it could be an ncRNA molecule. In fact, it is also possible to consider or combine the energy with the scoring tables. Also, how to set a good threshold based on the similarity scores to distinguish real ncRNAs from others is a challenging, but a practical problem. Similar to other computational problems (e.g. motif finding), setting up benchmark evaluation platform may help the biologists to compare and identify appropriate computational tools for their study. And speeding up all structural alignment algorithms and developing local structural alignment algorithms are certainly desirable.

REFERENCES

Adams, P. L., Stahley, M. R., Kosek, A. B., Wang, J., & Strobel, S. A. (2004). Crystal structure of a self-splicing group I intron with both exons. *Nature, 430*(6995), 45–50. doi:10.1038/nature02642

Argaman, L., Hershberg, R., Vogel, J., Bejerano, G., Wagner, E. G., & Margalit, H. (2001). Novel small RNA-encoding genes in the intergenic regions of Escherichia coli. *Current Biology, 11*(12), 941–950. doi:10.1016/S0960-9822(01)00270-6

Chen, J.-L., & Greider, C. W. (2005). Functional analysis of the pseudoknot structure in human telomerase RNA. *Proceedings of the National Academy of Sciences of the United States of America, 102*(23), 8080–8085. doi:10.1073/pnas.0502259102

Dam, E., Pleij, K., & Draper, D. (1992). Structural and functional aspects of RNA pseudoknots. *Biochemistry, 31*(47), 11665–11676. doi:10.1021/bi00162a001

Eddy, S. R. (2001). Non-coding RNA genes and the modern RNA world. *Nature Reviews. Genetics, 2*(12), 919–929. doi:10.1038/35103511

Esquela-Kerscher, A., & Slack, F. J. (2006). Oncomirs-microRNAs with a role in cancer. *Nature Reviews. Cancer*, *6*(4), 259–269. doi:10.1038/nrc1840

Frank, D. N., & Pace, N. R. (1998). Ribonuclease P: Unity and diversity in a tRNA processing ribozyme. *Annual Review of Biochemistry*, *67*, 153–180. doi:10.1146/annurev.biochem.67.1.153

Griffiths-Jones, S., Bateman, A., Marshall, M., Khanna, A., & Eddy, S. R. (2003). Rfam: An RNA family database. *Nucleic Acids Research*, *31*(1), 439–441. doi:10.1093/nar/gkg006

Han, B., Dost, B., Bafna, V., & Zhang, S. (2008). Structural alignment of pseudoknotted RNA. *Journal of Computational Biology*, *15*(5), 489–504. doi:10.1089/cmb.2007.0214

He, S., Liu, C., Skogerbø, G., Zhao, H., Wang, J., & Liu, T. (2008). NONCODE v2.0: Decoding the non-coding. *Nucleic Acids Research*, *36*(Database issue), D170–D172. doi:10.1093/nar/gkm1011

Klein, R. J., & Eddy, S. R. (2003). RSEARCH: Finding homologs of single structured RNA sequences. *BMC Bioinformatics*, *4*, 44. doi:10.1186/1471-2105-4-44

Le, S. Y., Chen, J. H., & Maizel, J. V. (1990). Efficient searches for unusual folding regions in RNA sequences. *Structure and Methods: Human Genome Initiative and DNA Recombination*, *1*, 127–136.

Lowe, T. M., & Eddy, S. R. (1997). tRNAscan-SE: A program for improved detection of transfer RNA genes in genomic sequence. *Nucleic Acids Research*, *25*(5), 955–964. doi:10.1093/nar/25.5.955

Matsui, H., Sato, K., & Sakakibara, Y. (2005). Pair stochastic tree adjoining grammars for aligning and predicting pseudoknot RNA structures. *Bioinformatics (Oxford, England)*, *21*(11), 2611–2617. doi:10.1093/bioinformatics/bti385

Nawrocki, E. P., Kolbe, D. L., & Eddy, S. R. (2009). Infernal 1.0: Inference of RNA alignments. *Bioinformatics (Oxford, England)*, *25*(10), 1335–1337. doi:10.1093/bioinformatics/btp157

Nguyen, V. T., Kiss, T., Michels, A. A., & Bensaude, O. (2001). 7SK small nuclear RNA binds to and inhibits the activity of CDK9/cyclin T complexes. *Nature*, *414*(6861), 322–325. doi:10.1038/35104581

Rivas, E., & Eddy, S. R. (2000). Secondary structure alone is generally not statistically significant for the detection of noncoding RNAs. *Bioinformatics (Oxford, England)*, *16*(7), 583–605. doi:10.1093/bioinformatics/16.7.583

Sakakibara, Y. (2003). Pair hidden Markov models on tree structures. *Bioinformatics (Oxford, England)*, *19*(Suppl 1), i232–i240. doi:10.1093/bioinformatics/btg1032

van Batenburg, F. H., Gultyaev, A. P., Pleij, C. W., Ng, J., & Oliehoek, J. (2000). PseudoBase: A database with RNA pseudoknots. *Nucleic Acids Research*, *28*(1), 201–204. doi:10.1093/nar/28.1.201

Wong, T., Chiu, Y. S., Lam, T. W., & Yiu, S. M. (2008). A memory efficient algorithm for structural alignment of RNAs with embedded simple pseudoknots. *Proceedings of the 6th Asia-Pacific Bioinformatics Conference*, 89-99.

Wong, T., Lam, T.W., Sung, W.K. & Yiu, S.M. (2009). Structural alignment of RNA with complex pseudoknot structure. *Algorithms in Bioinformatics*, 403-414.

Yang, Z., Zhu, Q., Luo, K., & Zhou, Q. (2001). The 7SK small nuclear RNA inhibits the CDK9/cyclin T1 kinase to control transcription. *Nature*, *414*, 317–322. doi:10.1038/35104575

Zhang, S., Haas, B., Eskin, E., & Bafna, V. (2005). Searching genomes for noncoding RNA using FastR. *IEEE/ACM Transactions on Computational Biology and Bioinformatics*, *2*(4), 366–379. doi:10.1109/TCBB.2005.57

ADDITIONAL READING

Han, B., Dost, B., Bafna, V., & Zhang, S. (2008). Structural alignment of pseudoknotted RNA. [structural alignment for standard pseudoknot]. *Journal of Computational Biology*, *15*(5), 489–504. doi:10.1089/cmb.2007.0214

Klein, R. J., & Eddy, S. R. (2003). RSEARCH: Finding homologs of single structured RNA sequences. [alignment using covariance model for regular structure]. *BMC Bioinformatics*, *4*, 44. doi:10.1186/1471-2105-4-44

Matsui, H., Sato, K., & Sakakibara, Y. (2005). Pair stochastic tree adjoining grammars for aligning and predicting pseudoknot RNA structures. Bioinformatics (Oxford, England), 21(11), 2611–2617. (alignment according to structure only for recursive standard pseudoknot of degree $k \leq 4$)

Wong, T., Chiu, Y. S., Lam, T. W., & Yiu, S. M. (2008). A memory efficient algorithm for structural alignment of RNAs with embedded simple pseudoknots. *Proceedings of the 6th Asia-Pacific Bioinformatics Conference*, 89-99. (memory efficient algorithm of structural alignment for standard pseudoknot)

Wong, T., Lam, T. W., Sung, W. K., & Yiu, S. M. (2009). Structural alignment of RNA with complex pseudoknot structure. *Algorithms in Bioinformatics*, 403-414. (Structural alignment for simple non-standard pseudoknot and recursive simple non-standard pseudoknot)

Zhang, S., Haas, B., Eskin, E., & Bafna, V. (2005). Searching genomes for noncoding RNA using FastR. [structural alignment for regular structure]. *IEEE/ACM Transactions on Computational Biology and Bioinformatics*, *2*(4), 366–379. doi:10.1109/TCBB.2005.57

KEY TERMS AND DEFINITIONS

Non-Coding RNA: A non-coding RNA (ncRNA) is an RNA molecule that does not translate into a protein.

Pseudoknot: Given two base pairs at positions (i,j) and (i_0,j_0), where $i<j$ and $i_0<j_0$, pseudoknots are base pairs crossing each other, i.e. $(i<i_0<j<j_0)$ or $(i_0<i<j_0<j)$.

Regular Structure: The structure is regular if there does not exist any pseudoknot.

Structural Alignment: Structural alignment between the folded ncRNA (*query*) and the unfolded region (*target*).The unfolded sequence will be folded and aligned simultaneously to the folded ncRNA. The alignment score represents their sequence and structural similarity.

Chapter 25
Finding Attractors on a Folding Energy Landscape

Wilfred Ndifon
Princeton University, USA & Weizmann Institute of Science, Israel

Jonathan Dushoff
McMaster University, Canada

ABSTRACT

RNA sequences fold into their native conformations by means of an adaptive search of their folding energy landscapes. The energy landscape may contain one or more suboptimal attractor conformations, making it possible for an RNA sequence to become trapped in a suboptimal attractor during the folding process. How the probability that an RNA sequence will find a given attractor before it finds another one depends on the relative positions of those attractors on the energy landscape is not well understood. Similarly, there is an inadequate understanding of the mechanisms that underlie differences in the amount of time an RNA sequence spends in a particular state. Elucidation of those mechanisms would contribute to the understanding of constraints operating on RNA folding. This chapter explores the kinetics of RNA folding using theoretical models and experimental data. Discrepancies between experimental predictions and expectations based on prevailing assumptions about the determinants of RNA folding kinetics are highlighted. An analogy between kinetic accessibility and evolutionary accessibility is also discussed.

INTRODUCTION

Biopolymers such as proteins and RNA play significant functional roles in living organisms; they are crucial for, among other things, locomotion, protection against disease, regulation of gene expression, and catalysis of biochemical reactions (Alberts *et al*., 2002; Yen *et al*., 2004; Serganov *et al*., 2006). This functionality is often mediated by specific native conformations. The attainment of such native conformations involves a search of an astronomically large conformation space, via a series of elementary structural rearrangements. At

DOI: 10.4018/978-1-60960-491-2.ch025

the molecular level, these elementary structural rearrangements include the formation/dissociation of hydrogen bonds between ribonucleotides and the formation/dissociation of hydrogen, van der Waals, and disulfide bonds between amino acid residues. If the search for the native conformation were a simple random process, it would take a biopolymer a biologically unrealistic length of time to find its native conformation following its biosynthesis. For example, assuming that during folding an RNA stays at each of its possible non-native secondary structures no longer than the duration of an atomic oscillation, a simple calculation shows that it would take $\sim 4.5\times 10^{6}$ years for the phenylalanil-tRNA from the yeast *Saccharomyces cereviseae* to complete a random search for its native conformation. This discrepancy between the time scale required for a random search of the conformation space and the time scale required for biological functionality was originally pointed out by Levinthal (1969) and is known as Levinthal's paradox.

There is, however, evidence that the search for the native conformation is not random but it proceeds along an energy gradient induced by differences in the thermal stabilities of the possible biopolymer conformations (Onuchic *et al.*, 2000, Wolynes, 2005). A folding biopolymer "moves" on an "energy landscape", from regions of relatively high free energy to regions of lower free energy. A number of models applying this energy landscape perspective to analyze the RNA folding process have been published. Molecular dynamics models (Pan & Mackerell, 2003; Sorin *et al.*, 2004) are mostly suitable only for the analysis of very short RNA sequences due to the very large number of computations involved. Analytical models (Zhang & Chen, 2002, 2006) are similarly limited in the scope of their application because they require the enumeration of an astronomical number of possible conformations of the RNA sequence under consideration. Monte Carlo models (Flamm *et al.*, 2000; Isambert & Siggia, 2000; Xayaphoumine *et al.*, 2003; Ndifon, 2005) and models based on genetic algorithms (Gultyaev *et al.*, 1995; Shapiro *et al.*, 2001) partially circumvent these computational limitations and they are consequently applicable to RNA sequences of varying lengths. Important drawbacks of the genetic algorithms include the facts that (1) they do not allow the inference of physically relevant RNA folding times, and (2) they employ certain parameters (e.g., mutation and crossover rates) that do not relate to any physical aspects of RNA folding dynamics and whose values are determined by means of trial and error (Higgs, 2000).

There are important open questions concerning the shape of the energy landscapes of natural RNAs and how this shape determines the kinetics of RNA folding. To motivate the particular questions addressed in this chapter, it is useful to recall that some folding paths of certain RNA sequences (e.g., the SV11 RNA sequence) lead to suboptimal attractor conformations that, once realized, take a very long time to unravel. In some cases, the suboptimal attractors are functional within the time window required for their unraveling (e.g., see Gultyaev *et al.*, 1995). However, in many cases the suboptimal attractors simply extend the waiting time until an RNA sequence can attain its optimal conformation and thereby become functional. This chapter investigates the above aspects of the kinetics of RNA folding using both theoretical models and experimental data.

RNA FOLDING MODEL

In this section, the model of RNA folding kinetics used in this work is described in detail. First, a working definition of RNA sequences and their conformations is given. Methods for calculating the free energies of RNA conformations conditioned on the underlying RNA sequence are also described. The model of RNA folding kinetics is then presented.

Figure 1. (a) Sequence and graphical representations of the (b) secondary and (c) tertiary structures of the phenylalanil-tRNA (PDB ID: 1EVV) from yeast. The tertiary structure is displayed at atomic resolution. (d) A simple string representation of the tRNA's secondary structure. In the string representation, each unpaired base x_a is denoted by "." while paired bases are denoted by a pair of matching parentheses. Figure (b) was drawn using mfold (Zuker, 2003), while (c) was downloaded from the Protein Data Bank (Berman et al., 2000).

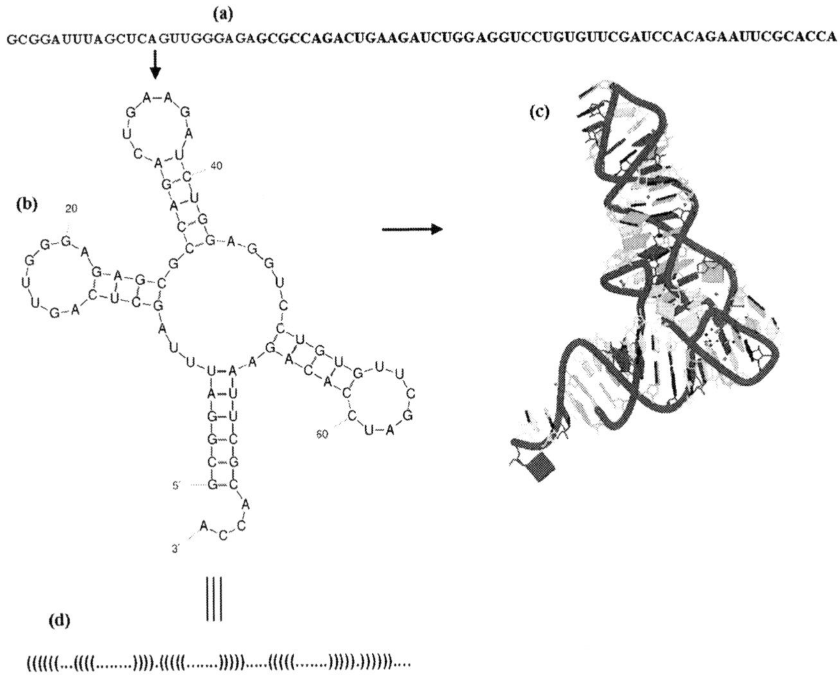

RNA Sequence and Conformation

An RNA sequence X of length n is denoted by a string [INSERT FIGURE 001] defined over the nucleotide alphabet $\{A, C, G, U\}$. The letters A, C, G, and U denote the nucleotides (or simply bases) adenine, cytosine, guanine, and uracil, respectively. Complementary bases in X have a propensity to pair (or form bonds) with each other. A base pair formed by the bases x_a and $x_{a'}$, $1 \leq a < a' \leq n$, is denoted by (a, a'). [This refers to the canonical nucleotide complementarity rules: Uracil (U) is complementary to both adenine (A) and guanine (G), while guanine (G) is complementary to cytosine (C). Non-canonical base pairs are not considered]. Two base pairs (a, a') and (b, b'), with $a<\underline{b}$ are said to be compatible if either $a < b < b' < a'$ or $a' < b$. A secondary structure of X refers to the two-dimensional topology of compatible base pairs from X, while a tertiary structure refers to the three-dimensional topology of both compatible and incompatible base pairs (see Figure 1). Incompatible base pairs can form pseudoknots. The secondary structure serves as the geometric, thermodynamic, and kinetic scaffold for formation of the tertiary structure (Flamm *et al.*, 2000; Bailor *et al.*, 2010): the importance of secondary structure for RNA function is shown by its important influence on RNA evolution (Higgs, 2000). The kinetics of formation of the secondary structure is of primary interest in this work.

Figure 2. Schematic representation of various components of an RNA conformation. Arrows indicate terminal mismatched pairs. An energetic contribution is computed for each base pair (bp) using the Turner 2.3 energy rules. The energetic contributions depend on the structural context of each bp. For example, in the presence of the bp labeled 4, the energetic contribution due to bp 5 is the sum of the stacking energy between bp 4 and bp 5, the stacking energy between bp 5 and the adjacent, terminal mismatched pair (i.e., A-G), and the (entropic) cost of closing the adjacent hairpin loop, which is of size 6. Similarly, the energetic contribution due to bp 4 is the sum of the stacking energy between bp 4 and bp 5, the stacking energy between bp 4 and the adjacent, terminal mismatched pair, and the cost of closing the adjacent internal loop, which is of size 2. These energetic contributions are read directly from pre-computed energy tables (http://www.bioinfo.rpiscrews.us/zukerm/cgi-bin/efiles.cgi?T=37).

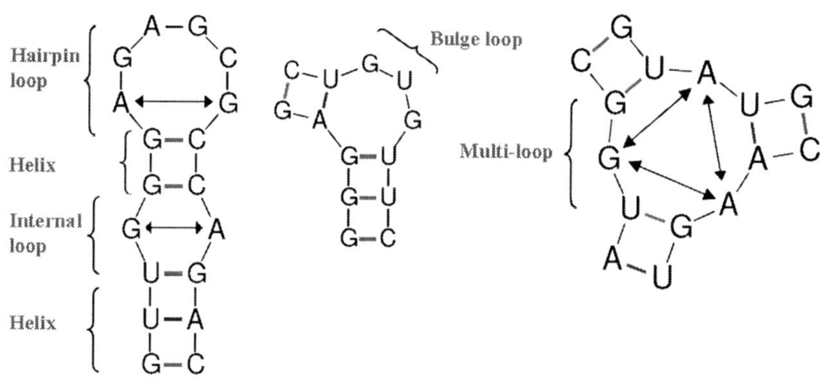

Energetics of RNA Folding

Experimental studies (Freier *et al.*, 1986; Jaeger *et al.*, 1989; Wu *et al.*, 1995; Mathews *et al.*, 1998) of the thermodynamics of RNA folding strongly suggest that RNA conformations are stabilized by interactions between adjacent base pairs; between base pairs and adjacent unpaired pairs of bases (also called "mismatched pairs") that delimit internal and hairpin loops; and between base pairs and adjacent unpaired bases that delimit multi-loops (see Figure 2). (Note that stabilizing stacking interactions result from the overlap of the p-orbitals of adjacent, aromatic bases). On the other hand, RNA conformations are destabilized by the formation of various types of loops. The formation of each base pair involves additive contributions from both stabilizing (i.e., free energy decreasing) and destabilizing (i.e., free energy increasing) processes. For example, the formation of a base pair that nucleates a hairpin loop involves (Jaeger *et al.*, 1989; Tinoco & Bustamante, 1999): (1) an increase in free energy due to a decrease in entropy, and (2) a decrease in free energy due to stacking interactions between the base pair closing the hairpin loop and the adjacent mismatched pair of bases. These energetic contributions associated with base pairs are computed using the free energy parameters of Turner and co-workers (Freier *et al.*, 1986; Jaeger *et al.*, 1989; Wu *et al.*, 1995[REMOVED HYPERLINK FIELD]). More specifically, the computed energetic contributions include those derived from interactions between the focal base pair and adjacent base pairs; between the focal base pair and terminal mismatched pairs of bases occurring in hairpins and interior loops; between the focal base pair and the unpaired base found in an adjacent bulge loop of size one; and the entropic costs associated with formation of hairpin, internal, and bulge loops (Figure 2). Multi-loops are treated in the same way as are internal loops. Entropic costs for loops longer than 30 bases are computed in the usual way (Jaeger *et al.*, 1989): $\Delta G(n>30) = \Delta G(30) + 1.75RT\ln(n/30)$, where n is the length of the loop, R is the gas constant, and T is the absolute temperature.

Kinetics of RNA Folding

The folding process is modeled as a time series of steps taken on the energy landscape, starting from an initial conformation and proceeding until a prescribed target conformation is found. In each time step all base pair formation and dissociation events (denoted structural rearrangements, SRs) that can occur in the current conformation are determined. These feasible SRs are subsequently partitioned into clusters, such that any two feasible SRs belonging to different (resp. the same) clusters are mutually independent (resp. dependent). Two feasible SRs are considered dependent if either they are mutually incompatible (see *RNA Sequence and Conformation*) or their constituent bases are close enough to engage in the types of energetic interactions described in the preceding section (i.e., the SRs are energetically dependent). The rate $\bar{r}$ at which a feasible SR, involving the formation (resp. dissociation) of the base pair $\left(a, a'\right)$ occurs, is defined as $\exp\left(-\Delta G^{aa'} / 2RT\right)$ [resp. $\exp\left(\Delta G^{aa'} / 2RT\right)$], where $\Delta G^{aa'}$ denotes the energetic contributions (measured in kcal/mol) associated with the base pair $\left(a, a'\right)$ (see *Energetics of RNA Folding* for details). Because of their mutual independence, there is no physical constraint on the ability of two feasible SRs from different clusters to occur in a given time step other than their individual probability of occurring. Therefore, in each time step one feasible SR is selected at random from each cluster with probability proportional to $\bar{r}$. All the selected feasible SRs are realized, the current conformation is updated, and the folding time is incremented by an amount equal to the ratio of the number of realized SRs to the sum of the rates of all feasible SRs. The above procedure is then repeated in the next time step. The definition of the probability of feasible SRs used in the above model is due to Kawasaki (1966); use of the (usual) Metropolis equation (1953) for the probability of feasible SRs resulted in computationally less efficient folding simulations.

RESULTS AND DISCUSSION

The odds that a bistable RNA sequence (i.e., an RNA sequence that folds into one of two possible attractor conformations) will find one attractor conformation before it finds the other one—the relative kinetic accessibility of the former attractor—is explored using the theoretical model described in the preceding section. The theoretical predictions are then compared to experimental data. The dependence of the relative kinetic accessibility on the difference between the free energies of the focal attractors is investigated, as is the influence of differences in the mean length of folding paths leading to the two attractors. This analysis makes use of experimental data on the folding kinetics of 14 bistable RNA sequences (Nagel, 2003; Nagel *et al.*, 2006). In the experiments cited above, each of the 14 RNA sequences was allowed to fold for a short period of time and the conformation realized by the sequence was immediately trapped by rapidly cooling the folding medium to 0°C. The relative kinetic accessibility of the two conformations attained by each sequence was defined as the number of folding experiments (resp. simulations) in which the sequence folded into one conformation before it could fold into the other (second) conformation, divided by the number of experiments (resp. simulations) in which it first folded into the other conformation. The results are presented in Table 1.

Sequences, attractor conformations, free energies, and experimental estimates for relative kinetic accessibilities from (Nagel, 2003; Nagel et

Table 1. Relative kinetic accessibilities for RNA sequences

RNA		Free energy, 0°C	Relative kinetic accessibility		
				Theory	
ID	Sequence & attractor conformations		Exp't	1SR	Multiple SRs
JN1C	CUGUUUUUGCAGCAAAAGCUGCAAAAGCAGCUUUUGUUG (((((((((((((....)))))))))))))...........((....))((((((((((((....)))))))))))).	-32.71 -32.25	1.3-1.4	1.2	**1.3**
JN2C	CUGUUUUUGCAGCAGAAGCUGCAGAAGCAGCUUCUGUUG (((((((((((((....)))))))))))))...........((....))((((((((((((....)))))))))))).	-33.09 -35.74	.7-.9	1.9	2.1
JN2D	CUGUUUUUGCAGCGGAAGCUGCAGAAGCAGCUUCCGUUG (((((((((((((....)))))))))))))...........((....))((((((((((((....)))))))))))).	-36.71 -37.42	.6-1.6	**1.0**	**1.0**
JN1H	GGGUGGAACCACGAGGUUCCACGAGGAACCACGAGGUUCCUCCC ..((((((((....))))))))((((((((....))))))))..(((.((((((.((.((((((....)))))).)).)))))).)))	-50.04 -44.50	1.8-2.1	1.6	2.5
JN2H	GUGGAACCACGAGGUUCCACGAGGAACCACGAGGUUCCUC ((((((((....))))))))((((((((....))))))))..((((((.((.((((((....)))))).)).))))))..	-45.95 -37.23	2.2-3.0	1.9	**2.3**
JN3H	GGGUGGAACCACGAGGUUCCGCGAGGAACCACGAGGUUCCUCCC ..((((((((....))))))))((((((((....))))))))..(((.((((((.((.((((((....)))))).)).)))))).)))	-49.26 -47.68	1.7-5.7	1.6	**2.4**
JN4H	GUGGAACCACGAGGUUCCGCGAGGAACCACGAGGUUCCUC ((((((((....))))))))((((((((....))))))))..((((((.((.((((((....)))))).)).))))))..	-45.17 -40.41	3.6-6.1	1.6	2.1
JN3A	CUGUUUUUGCAGUGAAAGCUGCGAAAGCAGCUUUUAUUGU (((((((((((((....)))))))))))))..................((((((((((((((....))))))))))))))	-32.70 -37.12	.7-1.2	**.8**	**.9**
JN3B	GUUGUUUUUGCAGUGAAAGCUGCGAAAGCAGCUUUUAUUG ((((((((((((((....))))))))))))))..........(((....)))(((((((((((....)))))))))))..	-35.48 -34.25	.5-1.2	**.9**	**.9**
JN4A	CUGUUUUUGCAGCAAAAAGCUGCAAAAGCAGCUUUUUGUUG (((((((((((((.....)))))))))))))............((....))(((((((((((((....))))))))))))).	-32.71 -33.85	1.1-1.3	.9	**1.2**
JN4B	CUGUUUCUGCAGCAAGAAGCUGCAGAAGCAGCUUUUUGUUG (((((((((((((.....)))))))))))))............((....))(((((((((((((....))))))))))))).	-36.20 -34.23	.7-.9	**.8**	1.1
JN5A	AAGUGUUUUUGGGCGGGAGCGCGGGAGCGCUUUUGCC ((((((((((((.((....)).))))))))))))..............(((((((((((....)))))))))))	-29.60 -34.40	.8-1.4	.3	.3
JN5B	AAGUGCUUUUGGGCGAGAGCGCGAGAGCGCUUUUGUC ((((((((((((.((....)).))))))))))))..............(((((((((((....)))))))))))	-32.78 -31.06	.6-1.0	**.6**	.5
JN6A	GUCUUAUGCUGCUUUCUGCAGCGUGAGGCUGCAGAAAGU ((((((((((((.....)))))))))))).((.....))((.....)).((((((((((((.....))))))))))))	-34.01 -33.90	1.7-3.4	1.0	1.0

al., 2006). See Figure 1 for the format in which the conformations are expressed. The lower and upper bounds of experimental predictions are given by $(f_1-d_1)/(f_2+d_2)$ and $(f_1+d_1)/(f_2-d_2)$, respectively, where $f_1 \pm d_1$ (resp. $f_2 \pm d_2$) is the fraction of sequences that folded into the first (resp. second) conformation indicated. Each theoretical estimate for the relative kinetic accessibility was obtained by running 1,500 folding simulations at 0°C, using the folding model described in the text. Each folding simulation was initialized with the unfolded conformation of the sequence under consideration and it was run until one of the two possible attractor conformations of the sequence was found. Theoretical predictions that lie within the range of corresponding experimental predictions are highlighted.

A qualitative comparison of the experimental and theoretical relative kinetic accessibilities shows that the theoretical predictions are in agreement with the experimental predictions in 11 out of 14 cases, when either only one SR or multiple SRs are allowed by the folding model (see Table 1). [Note that in the qualitative comparisons, a

Figure 3. Relationship between the relative kinetic accessibility and the relative mean path length. The relative kinetic accessibilities are given in Table 1. Both the relative kinetic accessibilities and the relative mean path lengths were obtained using the folding model described in the text, while allowing only 1SR to occur per time step.

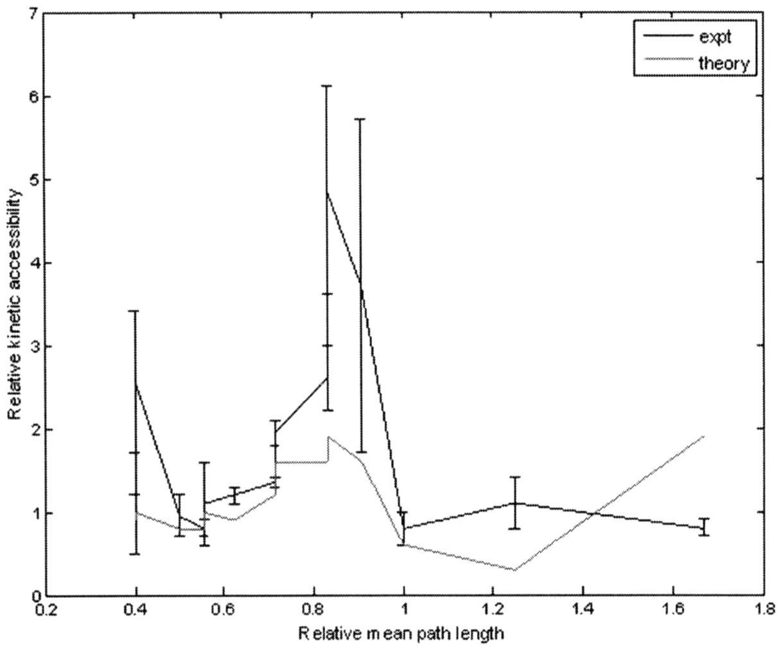

theoretical prediction is considered successful if it is greater (resp. less) than 1 when the experimental prediction (i.e., the mean) is also greater (resp. less) than 1]. This suggests that the fundamental assumptions of the folding model are biologically reasonable. However, only 7 (resp. 5) out of 14 theoretical predictions are in quantitative agreement with the corresponding experimental predictions when multiple (resp. only one) SRs are allowed by the folding model (see Table 1). [Note that in the quantitative comparisons, a theoretical prediction is considered successful if it lies within the range of the values predicted experimentally]. Some of the experimental and theoretical predictions are consistent with expectations based on current assumptions about the sequence and structural features that influence folding kinetics (e.g., Tinoco & Bustamante, 1999; Flamm *et al.*, 2000), but other predictions are not. For example, the helices of the two attractors for the JN1C and JN2C sequences all have identical features (the sequence of the hairpin loop is GCAAAAGC for both attractors of JN1C, and it is GCAGAAGC for both attractors of JN2C) and, as expected, both the experimental and the theoretical relative kinetic accessibilities for those attractors do not differ significantly from 1 (see Table 1). Similarly, the attractors for the JN1H, JN2H, JN3H and JN4H sequences differ only in the number of their helix-nucleation points and, as expected, the attractor with the greater number of nucleation points is generally significantly more accessible as predicted by both experimental data and the folding model (see Table 1).

In contrast, it is currently assumed that the formation of helices nucleated by the GC base pair would be kinetically favored over the formation of helices nucleated by the wobbly GU base pair.

But the two attractors for the JN3A and JN3B sequences, which differ in their helix-nucleating base pair (one has the GC base pair while the other has the GU base pair), do not differ significantly in their kinetic accessibilities as predicted by both experimental data and the folding model (see Table 1). In addition, it was expected that the formation of shorter hairpin loops would be kinetically favored over the formation of longer hairpin loops, all else being equal (Flamm *et al.*, 2000). However, the attractors for the JN4A and JN4B sequences, which contain hairpin loops of different sizes, do not differ significantly in their kinetic accessibilities as predicted by both experimental data and the folding model (see Table 1).

Furthermore, for the 14 RNA sequences analyzed neither the experimental nor the theoretical relative kinetic accessibilities correlate significantly with the ratio of the mean length of paths leading to one of the two possible attractors of each sequence (quantified as the waiting time to that attractor) to the mean length of paths leading to the other attractor (see Figure 3). Similarly, there is only a weak correlation between both the experimental and the theoretical relative kinetic accessibilities and differences in folding free energies between the two attractors of each RNA sequence (see Table 1).

There is a useful analogy between the kinetic accessibility of RNA conformations and the evolutionary accessibility of biological phenotypes. A metaphor that has proved very useful to the study of the evolutionary dynamics of biological populations is that of a population moving on a fitness landscape, which results from mapping the space of possible genotypes of the population to the fitnesses of those genotypes. If the shape of the fitness landscape is fixed under the existing environmental conditions and the product of the population's size and the mutation rate is small, then the trajectory of the population on the fitness landscape can be approximated by the trajectory of the population's modal phenotype. In this case, the evolutionary dynamics of the population is analogous to the folding kinetics of an RNA sequence that is moving on an energy landscape induced by the mapping from the different possible conformations of the sequence to the free energies of those conformations. Differences between conformations correspond to the number of base pair differences (analogous to mutations) between those conformations. Environment-induced changes to the fitness landscape can be modeled as perturbations to the ionic concentration or the temperature of the folding environment. The above model could be used to gain insight(s) and generate hypotheses about how the shape of the fitness landscape influences the ability of an evolving population to find adaptive phenotypes, the probability that the population will become trapped in a suboptimal phenotype, etc.

It is useful to contrast the theoretical approach used here, to predict the rate of finding an attractor conformation, to an alternative approach used in some recent studies (Sonick & Pan, 2004; Nkwanta & Ndifon, 2009). The former approach is direct in the sense that it explicitly simulates the folding process, whereas the latter is indirect -- it assumes that the rate of finding an attractor can be accurately predicted using only information about the topology of the attractor, either without (Sosnick & Pan, 2004) or with (Nkwanta & Ndifon, 2009) additional information about the underlying RNA sequence. This assumption is supported by recent observations (Bailor *et al.*, 2010). The indirect approach has the important advantage that it is computationally very efficient, and it can be applied to very large RNA sequences whose folding kinetics cannot be explicitly simulated. However, in contrast to the direct approach, it does not provide detailed insight into the folding energy landscape, including, when applicable, information about metastable conformations that delay the access to an attractor of interest. Which of the above two broad approaches will be more useful in practice depends on the particular question being addressed.

CONCLUSION

Theoretical predictions of the relative kinetic accessibilities of attractor conformations of bistable RNA sequences were compared to the corresponding experimental predictions. Agreement between the two sets of predictions was very good qualitatively, but only modest quantitatively. These results supported the biological realism of the fundamental assumptions underlying the theoretical analysis of folding kinetics. The analysis of relative kinetic accessibilities in relation to both the lengths of paths, leading to one of two possible attractors, and the free energies of those attractors suggested that a particular attractor may be preferentially reached from a prescribed starting point on the energy landscape even if that attractor is further away than the other attractor, and even if the focal attractor is thermodynamically less stable.

ACKNOWLEDGMENT

The authors were supported by a graduate fellowship from Princeton University (W.N.) and by the U.S. Defense Advanced Research Projects Agency under grants HR0011-05-1-0057 & HR0011-09-1-0055 (W.N. & J.D.).

REFERENCES

Alberts, B., Johnson, A., Lewis, J., Raff, M., Roberts, K., & Walter, P. (2002). *Molecular biology of the cell.* Garland.

Bailor, M. H., Sun, X., & Al-Hashimi, H. M. (2010). Topology links RNA secondary structure with global conformation dynamics and adaptation. *Science, 327*(5962), 202–206. doi:10.1126/science.1181085

Berman, H. M., Westbrook, J., Feng, Z., Gilliland, G., Bhat, T. N., & Weissig, H. (2000). The protein data bank. *Nucleic Acids Research, 28*(1), 235–242. doi:10.1093/nar/28.1.235

Flamm, C., Fontana, W., Hofacker, I. L., & Schuster, P. (2000). RNA folding at elementary step resolution. *RNA (New York, N.Y.), 6*(3), 325–338. doi:10.1017/S1355838200992161

Freier, S. M., Kierzek, R., Jaeger, J. A., Sugimoto, N., Caruthers, M. H., & Neilson, T. (1986). Improved parameters for predictions of RNA duplex stability. *Proceedings of the National Academy of Sciences of the United States of America, 83*(24), 9373–9377. doi:10.1073/pnas.83.24.9373

Gultyaev, A. P., van Batenburg, F. H. D., & Pleij, C. W. A. (1995). The influence of a metastable structure in plasmid primer RNA on antisense RNA binding kinetics. *RNA (New York, N.Y.), 23*(18), 3718–3725.

Higgs, P. G. (2000). RNA secondary structure: Physical and computational aspects. *Quarterly Reviews of Biophysics, 33*(3), 199–253. doi:10.1017/S0033583500003620

Isambert, H., & Siggia, E. D. (2000). Modeling RNA folding paths with pseudoknots: Application to hepatitis delta virus ribozyme. *Proceedings of the National Academy of Sciences of the United States of America, 97*(12), 6515–6520. doi:10.1073/pnas.110533697

Jaeger, J. A., Turner, D. H., & Zuker, M. (1989). Improved predictions of secondary structures for RNA. *Proceedings of the National Academy of Sciences of the United States of America, 86*(20), 7706–7710. doi:10.1073/pnas.86.20.7706

Kawasaki, K. (1966). Diffusion constants near the critical point for time-dependent Ising models. *Physical Review, 145*(1), 224–230. doi:10.1103/PhysRev.145.224

Levinthal, C. (1969) How to fold graciously. In J.T.P. DeBrunner & E. Munck (Eds.), *Mossbauer spectroscopy in biological systems: Proceedings of a meeting held at Allerton House, Monticello, Illinois,* (pp. 22-24). University of Illinois Press.

Mathews, D. H., Sabina, J., Zuker, M., & Turner, D. H. (1998). Expanded sequence dependence of thermodynamic parameters provides robust prediction of RNA secondary structure. *Journal of Molecular Biology*, *288*(5), 911–940. doi:10.1006/jmbi.1999.2700

Metropolis, N., Rosenbluth, A., Rosenbluth, M., Teller, A., & Teller, E. (1953). Equation of state calculations by fast computing machines. *The Journal of Chemical Physics*, *21*(6), 1087–1092. doi:10.1063/1.1699114

Nagel, J. H. A. (2003). *A study of metastable structures in RNA*. Unpublished doctoral dissertation, University of Leiden, The Netherlands.

Nagel, J. H. A., Flamm, C., Hofacker, I. L., Franke, K., de Smit, M. H., & Schuster, P. (2006). Structural parameters affecting the kinetics of RNA-hairpin formation. *Nucleic Acids Research*, *34*(12), 3568–3576. doi:10.1093/nar/gkl445

Ndifon, W. (2005). A complex adaptive systems approach to the kinetic folding of RNA. *Bio Systems*, *82*(3), 257–265. doi:10.1016/j.biosystems.2005.08.004

Nkwanta, A., & Ndifon, W. (2009). A contact-waiting-time metric and RNA folding rates. *FEBS Letters*, *583*(14), 2392–2394. doi:10.1016/j.febslet.2009.06.038

Onuchic, J. N., Nymeyer, H., Garcia, A. E., Chahine, J., & Socci, N. D. (2000). The energy landscape theory of protein folding: Insights into folding mechanisms and scenarios. *Advances in Protein Chemistry*, *53*, 87–152. doi:10.1016/S0065-3233(00)53003-4

Pan, J., Thirumalai, D., & Woodson, S. A. (1997). Folding of RNA involves parallel pathways. *Journal of Molecular Biology*, *273*(1), 7–13. doi:10.1006/jmbi.1997.1311

Serganov, A., Polonskaia, A., Phan, A. T., Breaker, R. R., & Patel, D. J. (2006). Structural basis for gene regulation by a thiamine pyrophosphate-sensing riboswitch. *Nature*, *441*(7097), 1167–1171. doi:10.1038/nature04740

Shapiro, B. A., Wu, J. C., Bengali, D., & Potts, M. J. (2001). The massively parallel genetic algorithm for RNA folding: MIMD implementation and population variation. *Bioinformatics (Oxford, England)*, *17*(2), 137–148. doi:10.1093/bioinformatics/17.2.137

Sorin, E. J., Nakatani, B. J., Rhee, Y. M., Jayachandran, G., Vishal, V., & Pande, V. S. (2004). Does native state topology determine the RNA folding mechanism? *Journal of Molecular Biology*, *337*(4), 789–797. doi:10.1016/j.jmb.2004.02.024

Sosnick, T. R., & Pan, T. (2004). Reduced contact order and RNA folding rates. *Journal of Molecular Biology*, *342*(5), 1359–1365. doi:10.1016/j.jmb.2004.08.002

Tinoco, I. Jr, & Bustamante, C. (1999). How RNA folds. *Journal of Molecular Biology*, *293*(2), 271–281. doi:10.1006/jmbi.1999.3001

Wolynes, P. G. (2005). Recent successes of the energy landscape theory of protein folding and function. *Quarterly Reviews of Biophysics*, *38*(4), 405–410. doi:10.1017/S0033583505004075

Wu, M., McDowell, J. A., & Turner, D. H. (1995). A periodic table of symmetric tandem mismatches in RNA. *Biochemistry*, *34*(10), 3204–3211. doi:10.1021/bi00010a009

Xayaphoummine, A., Bucher, T., Thalmann, F., & Isambert, H. (2003). Prediction and statistics of pseudoknots in RNA structures using exactly clustered stochastic simulations. *Proceedings of the National Academy of Sciences of the United States of America*, *100*(26), 15310–15315. doi:10.1073/pnas.2536430100

Yen, L., Svendsen, J., Lee, J., Gray, J. T., Magnier, M., & Baba, T. (2004). Exogenous control of mammalian gene expression through modulation of RNA self-cleavage. *Nature*, *431*(7007), 471–476. doi:10.1038/nature02844

Zhang, W., & Chen, S. J. (2002). RNA-hairpin-folding kinetics. *Proceedings of the National Academy of Sciences of the United States of America*, *99*(4), 1931–1936. doi:10.1073/pnas.032443099

Zhang, W., & Chen, S. J. (2006). Exploring the complex folding kinetics of RNA-hairpins: I. General folding kinetics analysis. *Biophysical Journal*, *90*(3), 765–777. doi:10.1529/biophysj.105.062935

ADDITIONAL READING

On Amount of Time Spent in an Attractor in Relation to Timescale of Biological Function

Solomatin, S. V., Greenfeld, M., Chu, S., & Herschlag, D. (2010). Multiple native states reveal persistent ruggedness of an RNA folding landscape. *Nature*, *463*(7281), 681–684. doi:10.1038/nature08717

KEY TERMS AND DEFINITIONS

Attractor: A local minimum found on the folding energy landscape.

Conformation/Genotype/Fitness/Folding Energy Space: A space in which the objects of interest are *conformations/Genotypes/Fitnesses/Folding Free Energies:* See *Space.*

Fitness Landscape: A mapping from genotype space onto fitness space.

Folding Energy Landscape: A mapping from conformation space onto folding energy space.

Genetic Algorithm: A heuristic method for finding solutions to an optimization problem that takes advantage of evolutionary principles; different possible solutions to the problem are iteratively subjected to "replication", "mutation" and "selection" processes. In order to illuminate its general principles a simple instance of the method is described below. In the context of RNA folding, the genetic algorithm might start with a randomly generated set of conformations that are compatible with the RNA sequence being folded. Then, in each iteration of the algorithm, multiple copies of each conformation are made (the replication step); more copies are made for conformations with lower free energies. The copying process is not perfect, but it introduces "mutations", which may involve the creation/destruction of base pairs or entire helices. Following replication and mutation, a subset of the resulting conformations is selected based on their free energies and subsequently subjected to the next round of replication, mutation, and selection. Eventually, the obtained conformations would be enriched for those with free energies approaching the lowest possible free energy for the RNA sequence being folded.

Molecular Dynamics Simulation: A simulation of the movement of the atoms of a molecule occurringduring the process under study (e.g. folding). Both the distance and the direction in which each atom moves in the time interval $t+\delta t$ are calculated using Newton's laws of motion and the position of the atom at time t.

Space (of Objects of a Particular Kind): Given a set S of objects and a metric d defined on S, a space (S,d) is obtained by arranging the objects into a structure such that $d(s_1,s_2)$ is the distance between the locations of any two objects s_1 and s_2 found in the obtained structure.

Chapter 26
Visualization of Protein 3D Structures in 'Double-Centroid' Reduced Representation:
Application to Ligand Binding Site Modeling and Screening

Vicente M. Reyes
School of Life Sciences, College of Science, Rochester Institute of Technology, Rochester, USA

Vrunda Sheth
School of Life Sciences, College of Science, Rochester Institute of Technology, Rochester, USA

ABSTRACT

This article is of two parts: (a) the development of a protein reduced representation and its implementation in a Web server; and (b) the use of the reduced protein representation in the modeling of the binding site of a given ligand and the screening for the model in other protein 3D structures. Current methods of reduced protein 3D structure representation such as the Cα trace method not only lack essential molecular detail, but also ignore the chemical properties of the component amino acid side chains. This chapter describes a reduced protein 3D structure representation called "double-centroid reduced representation" and presents a visualization tool called the "DCRR Web Server" that graphically displays a protein 3D structure in DCRR along with non-covalent intra- and intermolecular hydrogen bonding and van der Waals interactions. In the DCRR model, each amino acid residue is represented as two points: the centroid of the backbone atoms and that of the side chain atoms; in the visualization Web server, they and the non-bonded interactions are color-coded for easy identification. The visualization tool in this chapter is implemented in MATLAB and is the first for a reduced protein representation as well as one that simultaneously displays non-covalent interactions in the molecule. The DCRR model reduces the atomicity of the protein structure by ~75% while capturing the essential chemical properties

DOI: 10.4018/978-1-60960-491-2.ch026

of the component amino acids. The second half of this chapter describes the application of this reduced representation to the modeling and screening of ligand binding sites using a data model termed the "tetrahedral motif." This type of ligand binding site modeling and screening presents a novel type of pharmacophore modeling and screening, one that depends on a reduced protein representation.

INTRODUCTION

There exist different methods of protein 3D structure representation and visualization methods, the most popular of which being the all-atom representation (AAR), ribbon or 'spaghetti' representations, and space-filling models (please see refs. Sayle & Milner, 2000; DeLano, 2002; Guex, et al., 1999; Schwede, et al., 2003; Richardson & Richardson, 1992). The AAR model such as the wireframe and ball-and-stick models display every atom of the protein. But, even though all chemical information of the component amino acid residues are accounted for, the display is too crowded and overwhelming. On the other hand, van der Waals (VDW) surface representations such as the space-filling model are a good way to view the surface properties of the protein and locate shape complementarity involved in protein interactions but they fail to clearly show secondary structures, loops, functional sites and non-covalent interactions. Finally, ribbon and spaghetti models and the like provide a good view of the secondary structures and loops but do not show any side chain structural elements. In this paper we describe "double-centroid reduced representation" (DCRR), a reduced protein representation wherein amino acid residues in the protein are represented by two point coordinates: the centroid of the backbone atoms (N, Cα C' and O), and the centroid of the side chain atoms (CB and beyond). This method is similar to, but not identical to and independently conceived from, that proposed by Kolinski (2004) and Liwo, et al. (1997). In these two models, the Cα position is used instead of the centroid of the backbone atoms, and additionally in the Liwo, at al. (1997) method, a 'united peptide group' is inserted between two consecutive Cα atoms, to which the corresponding 'united sidechain group' is attached by a virtual bond.

We further develop a graphical visualization tool implemented in MATLAB that displays the reduced representation of the input protein PDB file, while simultaneously showing the intramolecular H-bonds and VDW interactions, as well as intermolecular ones with any bound ligands and water molecules. Our other aim was to develop a way of modeling ligand binding sites and to screen for these models in other proteins. We thought that this might find applications in pharmacophore modeling and screening that is quite different or even improved relative to current methods (Guner et al., 2004; Guner, 2005; Hopfinger, 2000; Khedkar et al., 2007; Mason, et al., 2001; Sun, 2008).

We thus proceeded to apply the DCRR method to the modeling of ligand binding sites (LBS) in proteins. Our LBS model is composed of the four most dominant amino acid centroids of the protein (in DCRR) which interact with the ligand atoms. These interactions may be in the form of hydrogen bonds or van der Waals interactions. The four centroids form a tetrahedron in 3D space, hence we term the model 'tetrahedral motif' model.

Finally we developed a screening method for the tetrahedral motif in any given protein, in order to predict whether the given protein would bind the ligand whose binding site tetrahedral motif is being sought. The screening procedure is composed of a series of Fortran programs that takes in two inputs, namely, a protein PDB structure file in DCRR, and the dimensions and centroid identities of the tetrahedral motif under query. The programs then either outputs the coordinates of four centroids in the protein that closely matches the tetrahedral motif if it finds one, or outputs null if it does not.

METHODS

This report essentially consists of two parts, namely (a.) the implementation of the protein double-centroid reduced representation and the creation of a web server for it, and (b.) the development of a ligand binding site modeling method and screening for such model in any given protein 3D structure.

Part (a.): We have developed a new model of protein structure representation that provides a balance between too much information and too little information and at the same time captures the chemical information of the side-chain is needed. We have implemented a visualization interface that displays the protein in a reduced representation along with displaying its H-bond and van der Waals interactions. We call the representation the Double Centroid Reduced Representation (DCRR) as each amino acid is represented as 2 data points: the centroid of the backbone and the centroid of the side-chain. DCRR reduces the atomicity of the protein highlighting just enough chemical information embodied in the side-chains.

We also developed a Web Server wherein users can enter a PDB id or upload a model protein and get the co-ordinates as well as the structure of the protein in DCRR.

Converting AAR to DCRR

The protein coordinate file from the PDB (Berman et al., 2002) is converted from its all-atom representation to double-centroid reduced representation by calculating (1) the centroid of the backbone atoms N, CA C' and O, of each amino acids, and (2) the centroid of the side chain atoms CB and beyond of each amino acid. No weights (such as atomic weights) were used in the calculation of the centroids; only the atomic positions (x,y,z) were considered in calculating the centroids.

Calculation of Hydrogen Bonds and Van Der Waals Interactions

Nearest neighbor analysis was used to identify H-bonding and VDW interactions. A sphere typically of radius 5.0 - 6.0 Å was constructed around every atom in the protein as center; all other atoms found in the interior of such a sphere is considered 'neighbors' of the central atom. Hydrogen bonds were taken to be those that are within close neighborhood of 2.80 Å between central atom and neighbor (see below), with the compatible chemical identities (those involving P, O, N and/or S). As for van der Waals interaction, we considered only those of the C-H• • • • H-C type and whose distances between the carbon atoms are within close neighborhood of 3.38 Å (see below).

Optimization of H-Bond and VDW Distances for Display

We next determined the appropriate number of H-bonds and van der Waals interactions to show in the display that is not too many to crowd the display and not too few to miss the important ones.

We designed a window around the ideal H-bond and van der Waal distances and the upper and lower limits in both were varied until an optimal number of interactions for display were obtained. The ideal H-bond length is 2.80 Å (Jeffrey, 1997) and the ideal C-H• • • • H-C van der Waals distance is 3.38 Å (Bondi, 1964; Kuzmin & Katzer, 2005; Nyburg & Faerman, 1985). After performing several trials, we designate a recommended range of 2.73 Å to 3.22 Å for H-bonds and 3.20 Å to 3.85 Å for VDW interactions. In the Web Server user has have a choice of using the recommended limits above, as well as wide limits or narrow limits. The wide and narrow limits for H-bonds are 2.66 Å - 3.36 Å and 2.75 Å - 3.00 Å, respectively. The wide and narrow limits for VDW interactions are 3.10 Å - 3.95 Å and 3.30 Å - 3.75 Å, respectively.

The DCRR Web Server

To make the DCRR method freely available to the scientific community, we have created the "Protein DCRR Web Server" at the Rochester Institute of Technology's Bioinformatics Division, and its URL is http://tortellini.bioinformatics.rit.edu/vns4483/dcrr.php. The web server interfaces a database which contains the DCRR co-ordinates of well over 50,000 protein structures from the PDB, as well as the MATLAB image of the proteins in reduced representation, along with intermolecular and intramolecular H-bonds and VDW interactions. The image also shows any bound ligands and the ordered water molecules in x-ray crystallographic structures. In the future we plan to include NMR structures in our database.

Part (b.): We have developed a method for the modeling of the binding site of any given ligand in a protein based on the 'double-centroid reduced representation' (DCRR) of the protein. We designate the model as 'tetrahedral motif' as it is composed of four points in space. An algorithm has also been developed to search for this motif in any given protein 3D structure.

The Tetrahedral Motif Data Model

Our initial objective was to model the binding site of any given ligand using a reduced protein representation so that a computationally economical and general screening procedure for the said model could be developed. Since the protein DCRR coordinates as well as the H-bonds and van der Waals interactions have already been pre-computed in the our database, modeling a given ligand will simply involve the determination of the four most dominant H-bonds and/or van der Waals interactions between protein and ligand atoms. The result of such determination is the 'tetrahedral motif' model for the ligand in question. Ideally this procedure is performed on several proteins containing the same ligand, in order to arrive at a consensus motif.

Constructing the Tetrahedral LBS Motif

We shall describe the procedure only briefly here, as a more comprehensive manuscript describing the application of the present method to LBS and pharmacophore modeling is in the process of publication elsewhere (V.M. Reyes, in preparation). First a training set consisting of protein 3D structures with bound ligand of interest is selected from the PDB. For each training structure, nearest protein atom neighbors of each ligand atom is determined using a nearest-neighbor Fortran program that finds all protein atoms within a sphere of radius ca. 6.0 Å of each ligand atom. From these nearest neighbors, those which are judged to be H-bonds or -CH...HC- van der Waals interactions are further selected; selection is based on atom identities of the neighbors and their distances from each other, and implemented using a Fortran program. Then the four most dominant H-bond or VDW interactions are selected as the vertices of the tetrahedron. One vertex (usually the most dominant) is arbitrarily designated as the "root", and the other three as "node1", "node2" and "node3" (R, n1, n2 and n3, respectively, for short). A "dominant" interaction is one that either occurs most frequently in the training structures, and/or has the most ideal H-bonding or VDW distance between the protein-ligand atom neighbors involved. The validity of such feature extraction from a set of heterogeneous proteins binding the same ligand has its roots from the work of Kobayashi and Go (1997a, 1997b), who showed that the LBS for ATP have nearly identical or very similar architectures in a set of heterogeneous ATP-binding proteins; they showed a similar phenomenon for a set of heterogeneous GTP-binding proteins.

Screening for the Tetrahedral Motif

Once the tetrahedral motif - or, more preferably, a consensus tetrahedral motif - is determined, the next logical step would be to find out if it occurs as well in other proteins; if it does, then those proteins where it occurs would be potential receptors for the ligand in question. If the proteins are functionally unannotated, this may be considered as a way of assigning function to those proteins, since knowing what ligand(s) a protein binds gives us a clue about its biological function.

We developed a five-step search algorithm for the tetrahedral motif (or consensus motif). They are all implemented in either Fortran 77 or 90, and are illustrated and discussed in the next section.

RESULTS AND DISCUSSION

Part (a.)

Protein DCRR Visualization

The image of an all-α protein, 1RFY, in DCRR with bound ordered water molecules as well as intra- and intermolecular H-bonds and VDW interactions, is shown in Figure 1. The H-bonds are shown as blue dashed lines and the VDW interactions as red dashed lines. Black solid lines connect adjoining backbone centroids, while solid orange lines connect side chain centroids with their respective backbone centroids. Ligands are shown as bright green triangles and bound ordered water molecules as blue squares. Every amino acid is labeled using a single letter code and color coded according to its polarity (hydrophilicity). The image includes a legend located on the right hand side allowing easy identification of the amino acids and their interactions.

For completeness, we also show the images of three more proteins in DCRR: 1RIE, an all-β protein, in Figure 2; 4OVO, an α/β protein, in Figure 3; and 1HMK, an α+β protein, in Figure 4, which has a calcium binding site where the Ca^{+} ion is readily visible in green.

The DCRR Web Server

The image of the DCRR Web Server is shown in Figure 5; it is available to the public at URL http://tortellini.bioinformatics.rit.edu/vns4483/dcrr.php. Users simply enter the PDB ID of the structure they wish to view in DCRR. If the protein is not deposited in the PDB, they may upload its structure coordinates by clicking the "Browse" button; they will then obtain a link containing the DCRR of the protein. Result will also be e-mailed to the users if they provide an e-mail address.

Part (b.)

The Tetrahedral Motif Data Model

The 'tetrahedral motif' model of a ligand binding site is shown in Figure 6. It is composed of four points, namely, a unique root, R, and three different nodes, n_1, n_2 and n_3. Each corresponds to an amino acid backbone or side chain centroid in the protein, and each has a set of (x,y,z) coordinates. These amino acids are in H-bonding and/or VDW interaction with ligand atoms. Also included in the data model are the lengths of the six sides of the tetrahedron, namely, the three branches Rn_1, Rn_2 and Rn_3, and the three node-edges, n_1n_2, n_2n_3 and n_1n_3, all in Angstrom units, Å. Note what we call 'branches' are root-to-node edges, while 'node-edges' are node-to-node edges. Thus the tetrahedral motif may be considered to be a data model that contains 14 parameters, of which eight qualitative and six are quantitative. The eight qualitative parameters are the amino acid identities of the four centroids, in combination with their being backbone or side chain centroids (4 x 2 = 8), while the six quantitative parameters are the lengths of the six edges mentioned above.

In developing the data model, we tried (a.) three points in space, or a 'plane triangular' model', and (b.) five points in space, or a 'pentahedral' model. The plane triangular model has low specificity and produced many false positives.

Figure 1. Image of protein 1RFY, an all-α protein in DCRR with the hydrogen bonds, van der Waals interactions as well as the ligands and the bound ordered water molecules

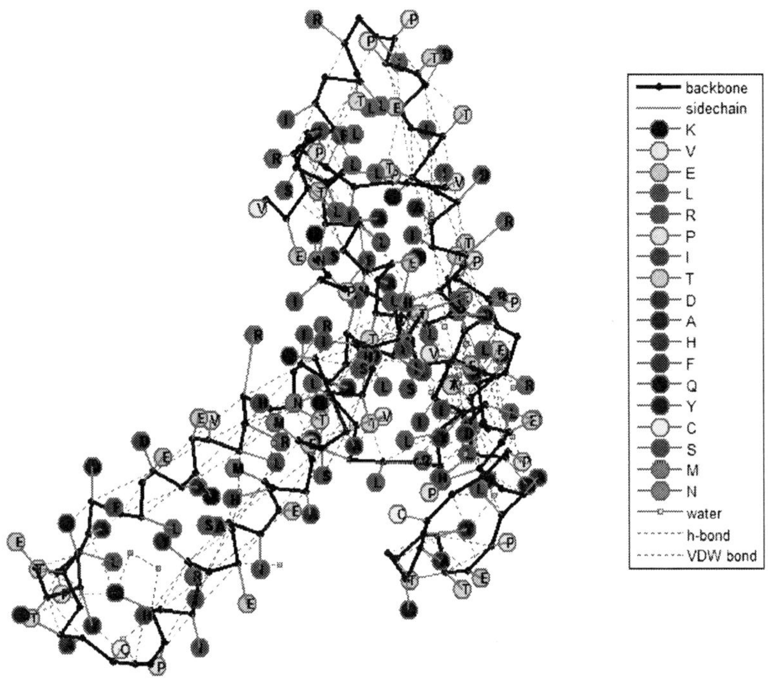

Figure 2. Image of protein 1RIE, an all-β protein in DCRR with the hydrogen bonds, van der Waals interactions as well as the ligands and the bound ordered water molecules

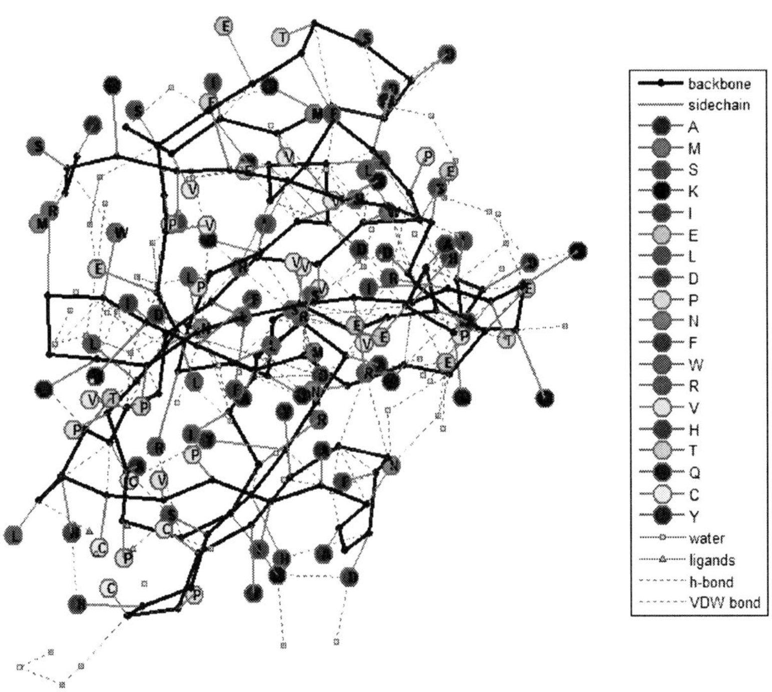

Figure 3. Image of protein 1HMK, an α+β protein in DCRR with the hydrogen bonds, van der Waals interactions as well as the ligands and the bound ordered water molecules

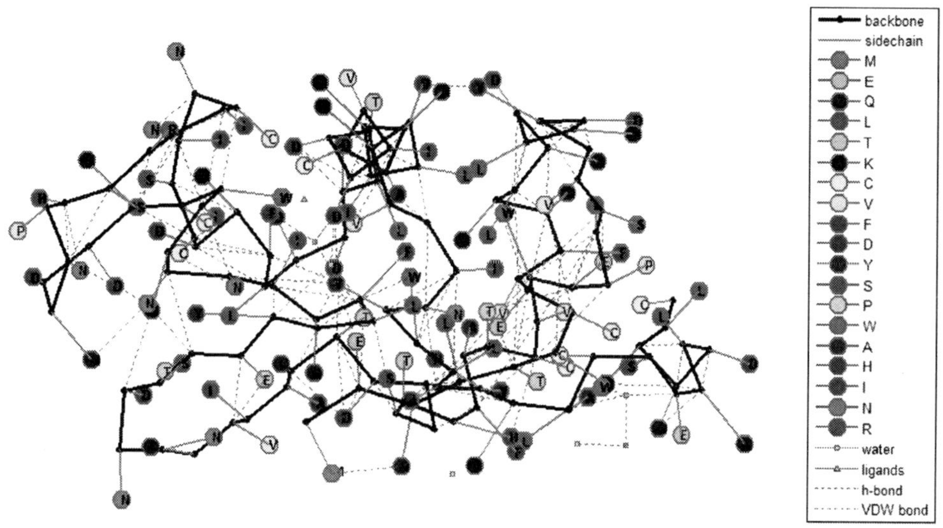

Figure 4. Image of protein 4OVO, an α/β protein in DCRR with the hydrogen bonds, van der Waals interactions as well as the ligands and the bound ordered water molecules

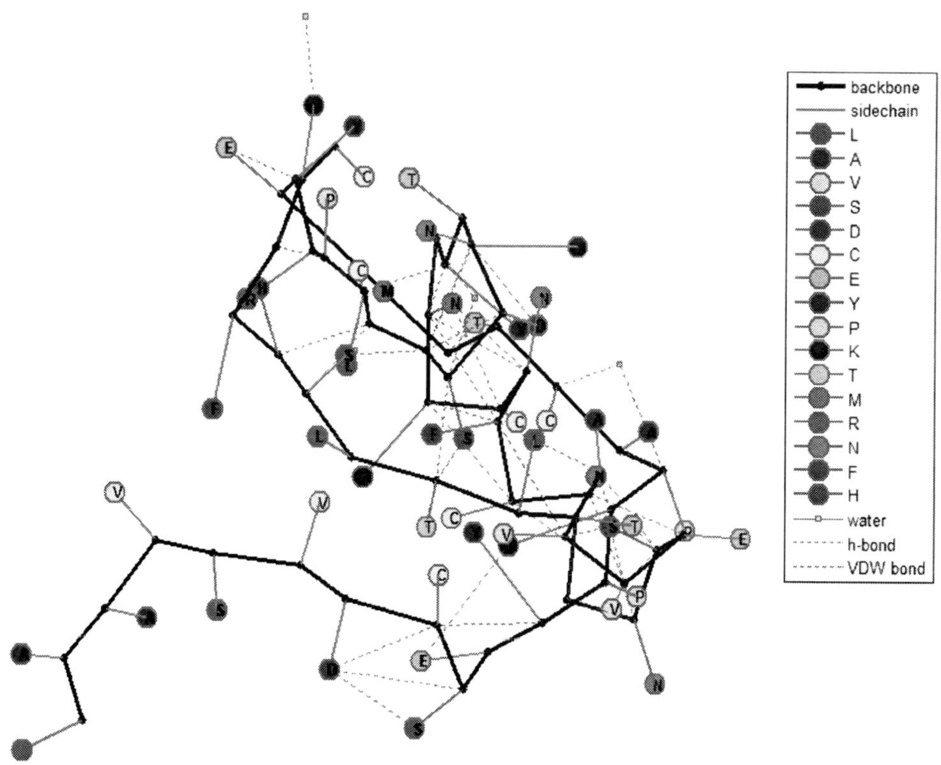

Figure 5. Image of the protein DCRR Web server

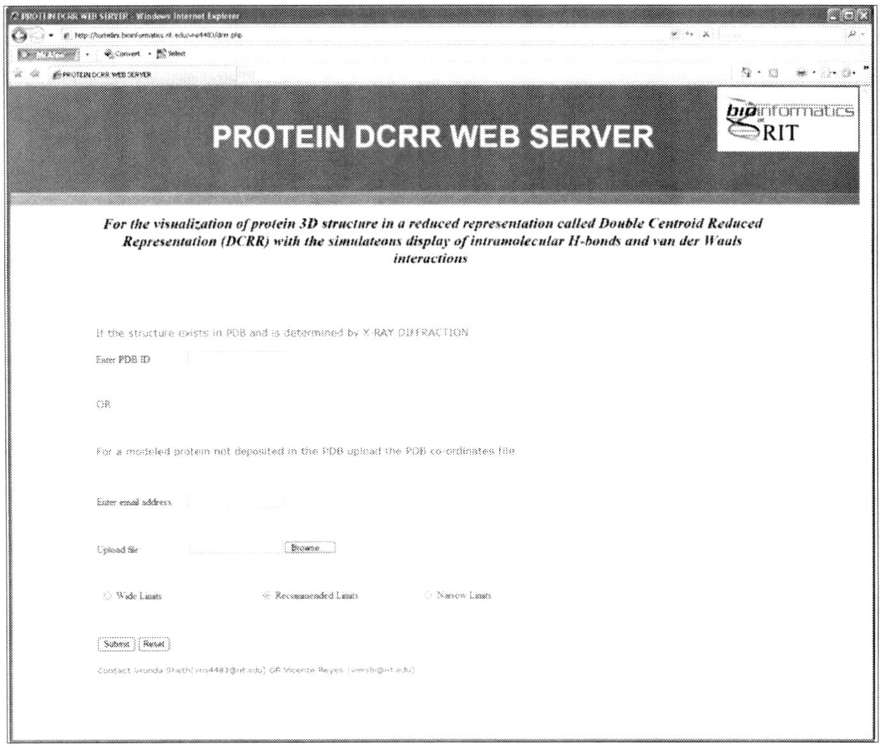

The pentahedral model, on the other hand, was too computationally cumbersome. The tetrahedral motif proved to be the optimal model.

Screening for the Tetrahedral Motif

The procedure for screening for the binding site tetrahedral motif of a given protein in a protein 3D structure is outlined in Table 1 and illustrated step-by-step in Figures 7-12. When used in screening, the tetrahedral motif is oftentimes called the '3D search motif' (3D SM).

The screening procedure is written as a series of Fortran programs that takes in two inputs, namely, a protein PDB structure file in DCRR, and the dimensions and centroid identities of the tetrahedral motif under query (Figure 7). We first sequester the amino acid residues in the query protein that are found in the 3D SM; in the example, the 3D SM contains the vertices Fb (phe backbone centroid), Es (glu side chain centroid), Ds (asp side chain centroid) and As (ala side chain centroid), we would sequester all F, E, D and A residues from the query protein (Figure 8, step #1). Then from the sequestered set of residues above, the appropriate centroids for the backbone or side chain are selected; in the example, the F side chain centroids are discarded, and so are the E, D and A backbone centroids, retaining only Fb, Es, Ds and As centroids, which are precisely what the 3D SM contains (Figure 9, step #2); this reduces the size of the sequestered group in half. Next, the distances between centroids in the sequestered group are calculated, and only those falling within limits of the corresponding branches in the 3D SM are retained (Figure 10, step #3). For example, if the Fb-Es branch is 8.80 Å, then only Fb-Es lengths in the sequestered group falling within 8.80 Å ± ε are retained; ε is the "fuzzy margin" and we usually set it at 1.0 to 1.5 Å. Then,

Figure 6. The tetrahedral motif data model for ligand binding sites

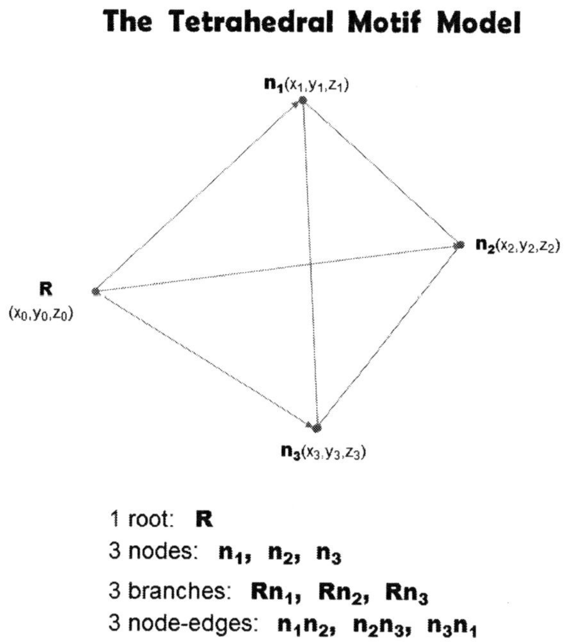

roots that are associated with exactly the three nodes in the 3D SM are chosen, further reducing the size of the sequestered group. We call each such combination of one root and three nodes a 'group', and each is a candidate LBS in eth query protein (Figure 11, step #4). Finally, the lengths of the node-edges in each group are computed, and those falling within limits of the corresponding edges in the 3D SM are chosen (Figure 12, step #5); we use a similar fuzzy margin ε of around 1.0 to 1.50 Å here. What is left at this point are potential LBS(s) in the query protein, as it/they have similar parameters as the 3D SM.

The set of steps above were all coded in Fortran 77 or 90 programs and are currently in the process of being published in a biological program source code repository journal/database.

The above approach of modeling and then screening for the binding site of any given ligand whose 3D structure in complex with its cognate receptor protein using a reduced protein representation is novel. It may be applied not only to pharmacophore modeling and screening but also to

Table 1. Algorithm for screening a protein 3D structure for a tetrahedral 3D search motif

Step #:	
0	Start with protein 3D structure in DCRR and the 3D search motif
1	Sequester amino acid residues in protein which are in 3D search motif
2	Select backbone or side chain centroids according to 3D search motif
3	Calculate distances and select those within limits of sides of 3D search motif
4	Select roots associated with three nodes as specified in 3D search motif
5	Select node-edges with lengths within limits of those in 3D search motif

Figure 7. Screening for a tetrahedral motif model: starting point: Step #0

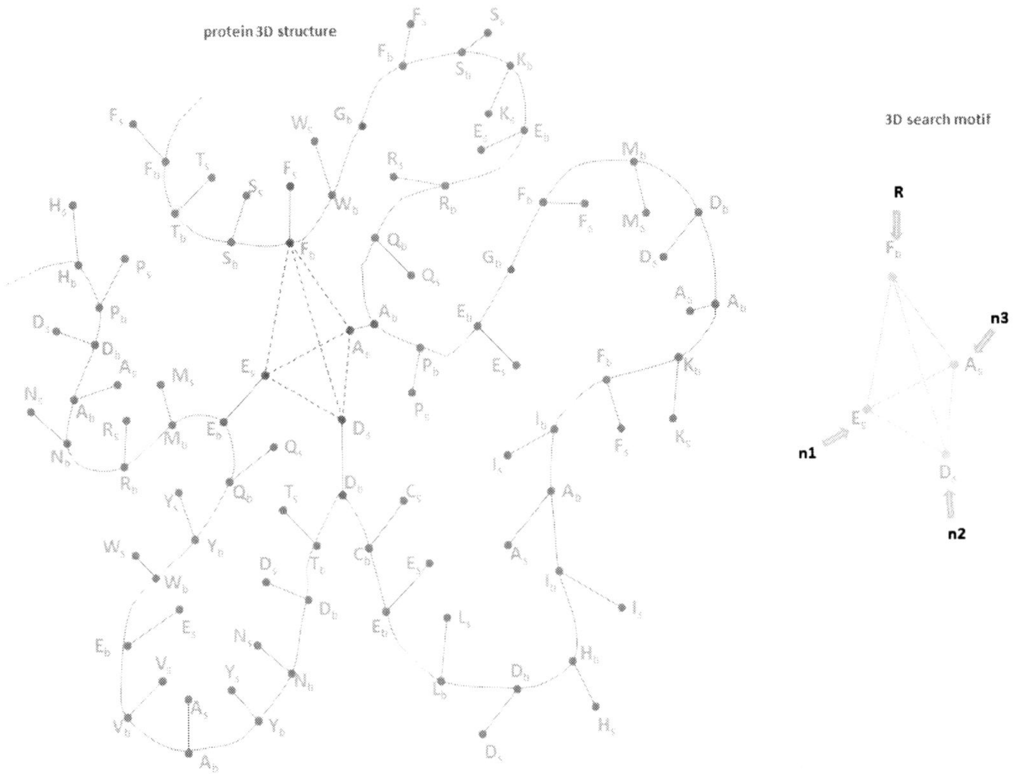

Start with protein 3D Structure in DCRR and 3D Search Motif

Figure 8. Screening for a tetrahedral motif model: Step #1

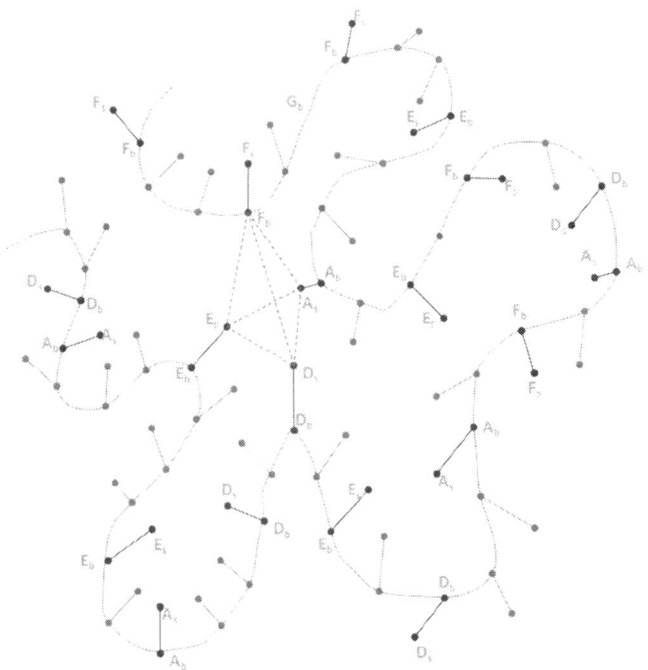

Step #1: Sequester amino acid residues in protein included in 3D Search Motif

Figure 9. Screening for a tetrahedral motif model: Step #2

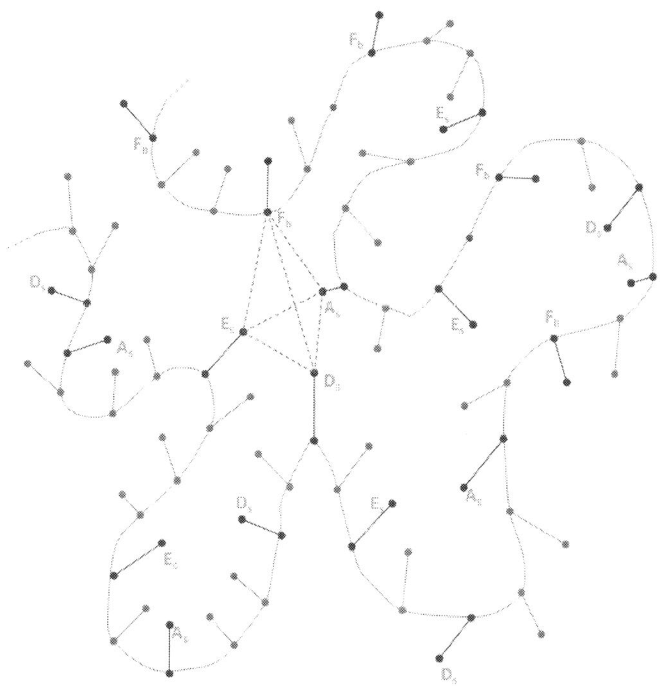

Step #2: Select backbone or side chain centroids according to 3D Search Motif

Figure 10. Screening for a tetrahedral motif model: Step #3

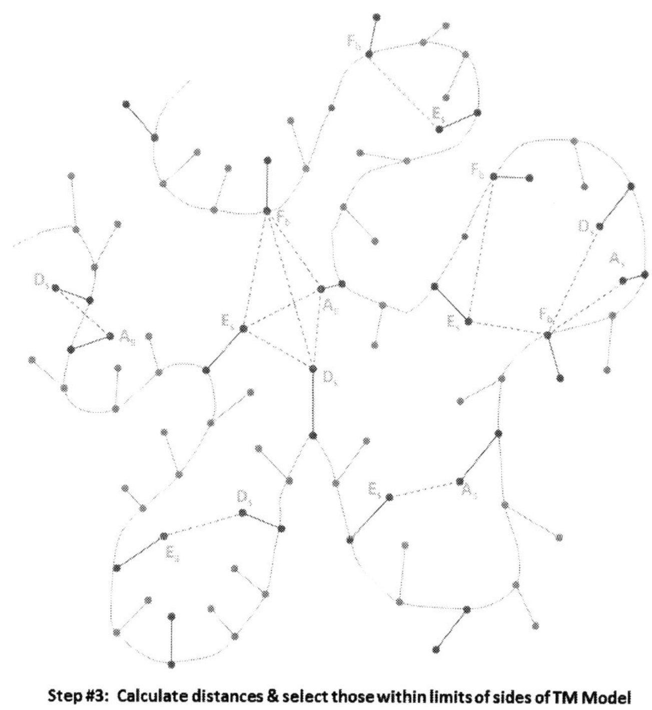

Step #3: Calculate distances & select those within limits of sides of TM Model

Figure 11. Screening for a tetrahedral motif model: Step #4

Step #4: Select roots associated with three nodes as specified in the 3D Search Motif

Figure 12. Screening for a tetrahedral motif model: Step #5

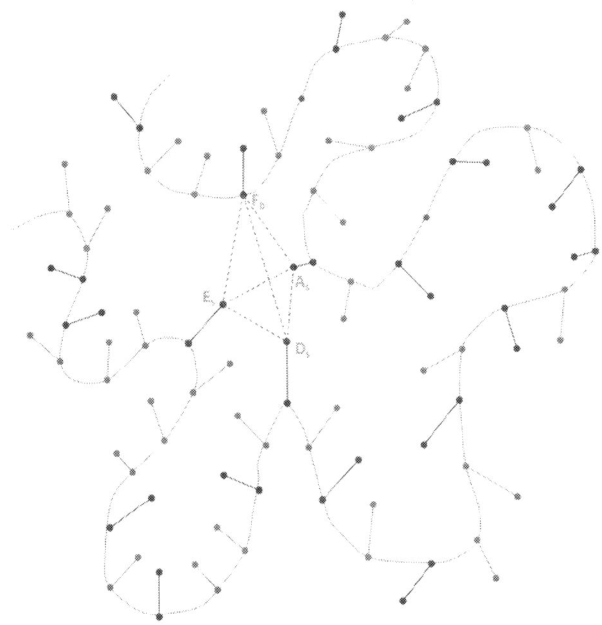

Step #5: Select node-edges with lengths within limits of those in the 3D Search Motif

structure-based protein function assignment, since the growth of functionally unannotated protein structures in the PDB has been significant due to the many structural genomics studies (Levitt, 2004) currently in operation.

We have applied the above procedure to the modeling of the ATP binding site in the ser/thr protein kinase family as well as the GTP binding site in the small, Ras-like G-protein family. We only describe a summary of results here as a more comprehensive manuscript describing the application of the present method to LBS and pharmacophore modeling is in the process of publication elsewhere (V. M. Reyes, in preparation). Briefly, a training set of ATP-binding proteins composed of structures 1B38, 1B39, 1FIN, 1GOL, 1HCK, 1JST, 1PHK, 1QL6, 1QMZ and 2PHK, and a training set of GTP-binding proteins composed of 1E96, 1N6L, 1NVU, 1LOO, 1M7B, 1O3Y and 2RAP were used. Tetrahedral LBS motif models were then built (as described in the Methods section) for each protein family. The performance of each model in screening for ATP and GTP LBSs were validated using a set of 15 'unseen' positive control structures for each family; a set of 30 negative controls were used for both families. The screening algorithm yielded a sensitivity of ~60% and a success rate of ~87% for the ATP-binding family, while it yielded a sensitivity of ~93% and a success rate of ~97% for the GTP-binding family; both have specificities close to 100%. Thus the ATP and GTP 3D SMs built from their respective training sets may be considered robust.

Using the models, ~800 solved protein structures in the PDB but without functional annotation were screened for the ATP- and GTP-binding site models, thus assigning potential functions to these structures. The results of this study will be published in a separate submission (ibid.).

CONCLUSION AND FUTURE DIRECTIONS

The idea for DCRR came from the motivation that we needed a simplified yet chemically meaningful protein structure representation. For example, using the all-atom representation for molecular dynamics and protein-protein interaction studies is too computationally uneconomical due to the exceedingly large memory requirements for manipulating and analyzing the sheer number of data points in a protein. The double-centroid reduced representation of proteins is quite appropriate for these types of work: it makes possible the drastic simplification of the protein structure information as each amino acid is represented by two data points, the centroid of the backbone and the centroid of the side-chain, reducing the overall data points by as much as 75%. Since DCRR contains both the backbone and the side-chain information, the essential biochemical information is captured, which is unlike the Cα-trace method where all side chain information is lost.

The MATLAB visualization tool has been utilized for the visualization of the protein DCRR. MATLAB is an excellent visualization interface allowing users to rotate, zoom in, zoom out and translate the protein for a better 3D view. Our DCRR visualization tool allows simultaneous display of the secondary structures of the protein as well as the H-bonds and the van der Waals interactions. It also shows any ligand(s) bound to the protein as well as bound ordered water molecules. Our DCRR tool provides a good view of the ligand binding site as well as the protein-water and the protein-ligand interactions. The visualization script has been programmed to include a legend for easy identification of amino acids and ligands. Unlike other protein visualization tools, our DCRR tool allows easy identification of the amino acid residues at each site as each side chain centroid is labeled and color-coded according to its polarity and hydrophobicity. We also think it is more user-friendly compared to other visualization tools where the user needs to click on each point to identify the amino acid. Another advantage of using MATLAB for visualization is the fact that it is freely available in most if not all academic and research institutions, and all undergraduate science and engineering students are proficient in it, as it is also used in mathematics, statistics and engineering courses. Moreover, most if not all colleges and universities offer free MATLAB tutorials to all of their entering students precisely for the above purpose.

Our Protein DCRR Web Server contains a database of the pre-computed DCRR coordinates of most of the x-ray crystallographic structures currently deposited in the PDB, save for a few thousand structures containing segment breaks which our wrapper script cannot handle currently; we shall rectify this minor "bug" in the near future. Our DCRR web server also currently does not contain NMR structures, but we plan to include them in future versions of the site.

Finally, using the DCRR method, we have also developed and implemented a method to model the binding site of a given ligand, and then screen any protein 3D structure for the presence of that LBS model. We term the model 'tetrahedral motif' as it composed of four points corresponding to backbone or side chain centroids of the amino acids contacting the ligand at the ligand binding site. This combined modeling and screening approach is novel in the sense that it employs a reduced protein representation, namely the 'double-centroid reduced representation' or DCRR. The entire algorithm is written in Fortran 77/90 code and run on a UNIX platform, and is thus fast and especially amenable to batch, high-throughput implementations. The set of programs are currently in the process of being deposited in a journal/repository for biologically applicable source codes.

Future directions of the present research would include (a.) application of the method (modeling and screening) to biologically important ligands other than ATP and GTP, (b.) use of two tetrahedral motifs for large ligands, and (c.) extension of the method (applied here to protein-ligand interactions) to protein-protein interactions (in progress).

ACKNOWLEDGMENT

We thank Ryan Lewis and Paul Mezzanini from the RIT Department of Research Computing headed by Dr. Gurcharan Khanna, as well as the Kyle Dewey from the RIT Department of Biology for their valuable computational assistance. We also thank Drs. Gary Skuse and Paul Craig from the RIT College of Science for serving as M.S. thesis committee members for V.S. V.M.R. thanks the RIT College of Science for a F.E.A.D. early faculty award during the summer of 2009.

REFERENCES

Berman, H. M., Westbrook, J., Feng, Z., Gilliland, G., Bhat, T. N., & Weissig, H. (2000). The Protein Data Bank. *Nucleic Acids Research*, *28*(1), 235–242. doi:10.1093/nar/28.1.235

Bondi, A. (1964). van der Waals volumes and radii. *Journal of Physical Chemistry*, *68*(3), 441–451. doi:10.1021/j100785a001

Guex, N., Diemand, A., & Peitsch, M. C. (1999). Protein modelling for all. *Trends in Biochemical Sciences*, *24*(9), 364–367. doi:10.1016/S0968-0004(99)01427-9

Güner, O., Clement, O., & Kurogi, Y. (2004). Pharmacophore modeling and three dimensional database searching for drug design using catalyst: Recent advances. *Current Medicinal Chemistry*, *11*(22), 2991–3005.

Guner, O. F. (2005). The impact of pharmacophore modeling in drug design. *IDrugs*, *8*(7), 567–572.

Hopfinger, A. J., & Duca, J. S. (2000). Extraction of pharmacophore information from high-throughput screens. *Current Opinion in Biotechnology*, *11*(1), 97–103. doi:10.1016/S0958-1669(99)00061-0

Jeffrey, G. A. (1997). *An introduction to hydrogen bonding*. Pittsburgh: Oxford Univ. Press.

Khedkar, S. A., Malde, A. K., Coutinho, E. C., & Srivastava, S. (2007). Pharmacophore modeling in drug discovery and development: An overview. *Medicinal Chemistry (Shariqah, United Arab Emirates)*, *3*(2), 187–197. doi:10.2174/157340607780059521

Kobayashi, N., & Go, N. (1997). A method to search for similar protein local structures at ligand binding sites and its application to adenine recognition. *European Biophysics Journal*, *26*(2), 135–144. doi:10.1007/s002490050065

Kobayashi, N., & Go, N. (1997). ATP binding proteins with different folds share a common ATP-binding structural motif. *Nature Structural Biology*, *4*(1), 6–7. doi:10.1038/nsb0197-6

Kolinski, A. (2004). Protein modeling and structure prediction with a reduced representation. *Acta Biochimica Polonica*, *51*(2), 349–371.

Kuzmin, V. S., & Katser, S. B. (2005). Calculations of van der Waals volumes of Organic Molecules. *Russian Chemical Bulletin*, *41*(4), 720–727.

Levitt, M. (2007). Growth of novel protein structural data. *Proceedings of the National Academy of Sciences of the United States of America*, *104*(9), 3183–3188. doi:10.1073/pnas.0611678104

Liwo, A., Oldziej, S., Pincus, M. R., Wawak, R. J., Rackovsky, S., & Scheraga, H. A. (1997). A united-residue force field for off-lattice protein-structure simulations. I. Functional forms and parameters of long-range side-chain interaction potentials from protein crystal data. *Journal of Computational Chemistry*, *18*(7), 849–873. doi:10.1002/(SICI)1096-987X(199705)18:7<849::AID-JCC1>3.0.CO;2-R

Mason, J. S., Good, A. C., & Martin, E. J. (2001). 3-D Pharmacophores in drug discovery. *Current Pharmacology Descriptions*, *7*(7), 567–597. doi:10.2174/1381612013397843

Nyburg, S. C., & Faerman, C. H. (1985). A revision of van der Waals atomic radii for molecular crystals: N, O, F, S, Cl, Se, Br, and I bonded to carbon. *Acta Crystallographica. Section B, Structural Science*, *41*(4), 274–279. doi:10.1107/S0108768185002129

Richardson, D. C., & Richardson, J. S. (1992). The kinemage: A tool for scientific communication. *Protein Science*, *1*(1), 3–9. doi:10.1002/pro.5560010102

Sayle, R. A., & Milner, J. E. (2000). Rasmol: Biomolecular graphics for all. *Trends in Biochemical Sciences*, *20*(9), 374–376. doi:10.1016/S0968-0004(00)89080-5

Schwede, T., Kopp, J., Guex, N., & Petsch, M. C. (2003). SWISS-MODEL: An automated protein homology-modeling server. *Nucleic Acids Research*, *31*(13), 3381–3385. doi:10.1093/nar/gkg520

Sun, H. (2008). Pharmacophore-based virtual screening. *Current Medicinal Chemistry*, *15*(10), 1018–1024. doi:10.2174/092986708784049630

KEY TERMS AND DEFINITIONS

3D Motif: A specific local 3D arrangement of specific protein atoms (from its backbone or side chains) created when they are brought close together in space by protein folding; the residues involved may or may not be contiguous in primary sequence

3D Search Motif: A 3D motif that is encoded in a computer program to be used for screening structures, usu. of proteins, using a search algorithm; in the present context, this is the tetrahedral motif corresponding to a LBS

Centroid: In the sense used this article, the unweighted geometric centroid of a group of neighboring atoms, considering only their x-, y-, and z-coordinates, and without consideration of their atomic masses

Double-Centroid Representation: A protein 3D reduced representation wherein each amino acid is represented by two centroids: that of the backbone atoms (N, Cα, C', O) and that of the sidechain atoms (Cβ and beyond)

Ligand Binding Site: The specific site on a protein, usually a crevice or a pocket of varying depth, where a ligand binds in a specific geometry and orientation, and with high affinity

Pharmacophore: A subset of the 3D structural features of a ligand that are specifically recognized at its binding site in its cognate protein receptor molecule and are essential for its biological action(s)

Pharmacophore Modeling: The extraction of the essential geometric and electrostatic (i.e., chemical) properties of a ligand, preferably in the form of a specific data structure for computational input, that are essential for its biological function; an important step in LBS screening and drug design

Reduced Representation: A method of representing macromolecules, usually with a visual component, where the atomicity (number of coordinates) is significantly reduced compared to the all-atom representation (usually derived from an x-ray crystallographic model)

Tetrahedral Motif: the reduced representation of a particular LBS composed of four points which are centroids of the backbone or sidechain atoms in the protein contacting the ligand via H-bonding or VDW interaction at its LBS; on vertex is denoted as the 'root', and the other three 'nodes' 1, 2 and 3.

Chapter 27
Mechanical Models of Cell Adhesion Incorporating Nonlinear Behavior and Stochastic Rupture of the Bonds:
Concepts and Preliminary Results

Jean-François Ganghoffer
LEMTA – ENSEM, France

ABSTRACT

The rolling of a single biological cell is analysed using modelling of the local kinetics of successive attachment and detachment of bonds occurring at the interface between a single cell and the wall of an ECM (extracellular matrix). Those kinetics correspond to a succession of creations and ruptures of ligand-receptor molecular connections under the combined effects of mechanical, physical (both specific and non-specific), and chemical external interactions. A three-dimensional model of the interfacial molecular rupture and adhesion kinetic events is developed in the present contribution. From a mechanical point of view, this chapter works under the assumption that the cell-wall interface is composed of two elastic shells, namely the wall and the cell membrane, linked by rheological elements representing the molecular bonds. Both the time and space fluctuations of several parameters related to the mutual affinity of ligands and receptors are described by stochastic field theory; especially, the individual rupture limits of the bonds are modelled in Fourier space from the spectral distribution of power. The bonds

DOI: 10.4018/978-1-60960-491-2.ch027

are modelled as macromolecular chains undergoing a nonlinear elastic deformation according to the commonly used freely joined chains model, while the cell membrane facing the ECM wall is modelled as a linear elastic plate. The cell itself is represented by an equivalent constant rigidity. Numerical simulations predict the sequence of broken bonds, as well as the newly established connections on the 'adhesive part' of the interface. The interplay between adhesion and rupture entails a rolling phenomenon. In the last part of this chapter, a model of the deformation induced by the random fluctuation of the protrusion force resulting from the variation of affinity with chemiotactic sources is calculated, using stochastic finite element methods in combination with the theory of Gaussian random variables.

INTRODUCTION

Cell adhesion is an important phenomenon in biology, especially in immune defence, wound healing, and the growth of tissues. The ability of the cell to divide and give birth of daughter cells is certainly a fundamental feature of the living cell. It is well known that this property strongly depends on adhesion phenomena: most cells can indeed proliferate only if they adhere to a convenient surface. The development and functioning of multicellular organisms includes very often migration of cells on surfaces; this motion relies on the coordination of attachment and detachment processes of molecular bonds (Ndri, 2001). Adhesion is further a key element for the development of vectors for the targeted delivery of medicaments – such as liposomes – which are lipidic pockets transporting active elements (Marques, 2001). Adhesion is a multistep process (Bongrand & Benoliel, 1999; Bongrand, 1982, Limberg, 2002), involving the approach of the cell towards the wall, followed by the critical phase of creation of the first bond. This step is followed by a consolidation step, consisting of an adaptation of the membrane shape, the concentration of receptors in the contact zone, and eventually the reinforcement of the cell membrane in the vicinity of the adhesion zone (Bongrand & Benoliel, 1999).

Cell adhesion is a multiscale phenomenon involved in *cell rolling* and cell migration (Bongrand et al., 1982; Bongrand & Benoliel, 1999), due to the induced *cell motility*. Cell adhesion involves complex phenomena that intervene in various biological processes such as the growth of the tissues and the immune response, due to the motion of leukocyte cells (Bongrand et al., 1982). The motion of cells along a wall occurs by two different mechanisms: rolling (e.g., the movement of leukocytes along a blood vein), due to the action of the fluid flow around the cell on the contact interface (Figure 1), and protrusion and retraction, resulting from a modification of the cytoskeleton structure (Bongrand & Benoliel, 1999; Sagvolden et al., 1999). Protrusion is usually associated to the creation of lamellipods and *focal contacts* on the adherent part of the contact interface (Figure 1).

Rolling and active deformation of the cell occur, for instance, during immune defence due to the action of leukocytes, which are transported by plasma flow, captured by the wall by rolling and further move towards the infected zone by active deformations (Jones, 1996).

The creation of new connections is the result of the junction between free and specific molecules, the ligands and the receptors (Bongrand, 1982; Marques, 2001); the failure of existing connections occurs by pulling effects due to the cell motion, which result from the action of various forces. Several models have been developed in the literature in order to describe the adhesion kinetics or the deformation of the cell during the adhesion process: those models can generally be classified into probabilistic approaches (Roberts et al. 1990; Haussy and Ganghoffer 2001) or deterministic modelling strategies (Combs et al. 2004). Other modelling or experimental studies

Figure 1. Top view: Rolling of a cell connected to an ECM by ligand-receptor pairs. Bottom view: distinct zones of bond adhesion and rupture at the cell-ECM wall interface.

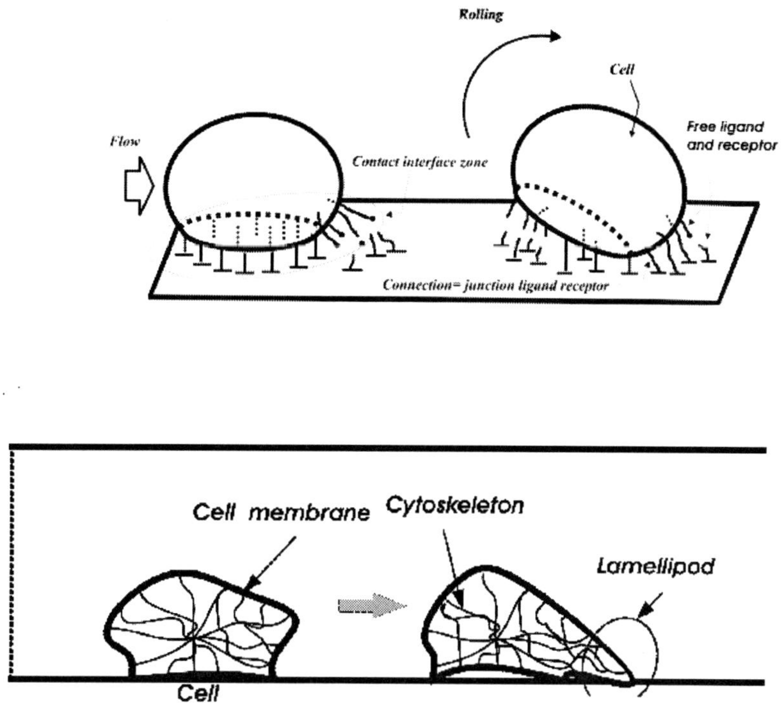

concerning the deformation mechanisms during cell adhesion have been developed by several authors, see (Sagvolden et al, 1999, Lenormand, 2001, Edelman, 1976; Evans & Needham, 1987).

Several works in the literature describe cell adhesion, either from an experimental or a theoretical point of view. One of the most widespread experimental techniques is the micropipette technique, which is used to determine the adhesion force (Skalak and Evans, 1984; Evans and Needham, 1987; Evans, 1992; Evans and Ritchie, 1997; Zhao et al., 2001). As an alternative, the microscopy technique has been used by (Bruinsma and Sackman, 2001; Simon, 2002) to describe the cell display. As far as the theoretical approaches are concerned, one may distinguish the probabilistic and the deterministic methods. In the probabilistic approaches, some or all parameters of the model are the result of a probabilistic calculus (contrary to the deterministic approaches). A more detailed account of the kinetic models in the literature shall be given in the next section.

Despite this early occurrence of probabilistic models of cell adhesion since the eighties, those probabilistic models are rather of a global nature (they provide the number of broken or adhering bonds versus time and temperature), as they do not describe the individual rupture and adhesion events: hence a detailed picture of the kinetics of bond association and dissociation relying on a model of the underlying biological mechanisms is still lacking.

The main objective of this contribution is then to provide models and simulations of the local *kinetics of bond rupture and formation* for a single cell adhering to a rigid substrate, within a *stochastic framework*. The modeling of the behaviour of the set of bonds (defining an interface) linking the cell membrane to the ECM (extra cellular matrix) wall shall essentially aim at describing three mechanisms:

- The motion of the interface under the action of the fluid and other physical forces;
- The time evolution of the rupture of the contact interface and its influence on its motion (rolling);
- The junction of complementary adhesive molecules and the creation of new connections.

As an essential aspect of this work, the description of the kinetics of adherence of bonds relies on a separation of the interface into a zone of bond rupture (the initial configuration is that of initially existing firm bonds) and a distinct zone of bond adhesion (Figure 1).

The outline of the present contribution is the following: a review of the existing deterministic and statistical models of the kinetics of cell adhesion is presented in section 2. The physical interactions between ligands and receptors are described in section 3, including the stochastic modelling of the bond rupture. The adhesion of new connections (association of ligand-receptor pairs) is described in section 4. Simulation results are provided in section 5. Section 6 is devoted to a model of cell protrusion. A summary and a list of perspectives are given in the last section 7.

LITERATURE SURVEY OF DETERMINISTIC AND PROBABILISTIC CELL ADHESION MODELS

Deterministic Approaches

Envisaging first deterministic approaches, the reference model in (Bell et al, 1984) is based on the coupling of the cell membrane (supposed to be elastic) with the kinetic equations of the adhesion molecules. The surface densities of the free and bound molecules, viz the quantities $A_f(s,t)$, $A_b(s,t)$ resp., with s,t the distance and time resp., inform the adhesion kinetics:

$$A_f \xrightarrow{K_f(Y)} A_b$$

$$A_f \xleftarrow[K_r(Y)]{} A_b$$

The parameters K_f, K_r represent the molecular formation and dissociation constants resp., and depend upon the vertical separation distance y between the membrane and the wall. The authors consider that the total number of molecules is given by

$$A_{total} = A_b + A_f$$

with the continuity equation for the density of bound molecules given by

$$\frac{\partial A_b}{\partial s} = K_f(Y)A_{total} - (K_r(Y) + K_f(Y))A_b .$$

They model the bonds as springs having a free energy under traction or compression up to length y given by

$$G_h = G_{h0} + 0.5\kappa(Y-\lambda)^2,$$

G_{h_0} being the free energy of a free spring and κ the spring rigidity. The affinity $K_{eq} = K_f(\lambda)/K_r(\lambda)$ in the initial state (λ is the initial cell / wall separation distance) is expressed as

$$\begin{aligned} K_f(Y) / K_r(Y) &= \exp(-G_h / K_B T_a) \\ &= K_{eq} \exp(-0.5\kappa((Y-\lambda)^2) / K_B T_a) \end{aligned}.$$

The authors further assume that the difference between the transition period and the period characterized by a stable bond is solely due to a variation of rigidity, hence

$$K_f(Y) = K_f(\lambda)\exp(-0.5\kappa_{ts}((Y-\lambda)^2) / K_bT_a)$$

$$K_r(Y) = K_r(\lambda)\exp(0.5(\kappa-\kappa_{ts})(Y-\lambda)^2 / K_BT_a)$$

with κ_{ts} the transient spring rigidity. This model highlights the influence of κ_{ts} on the molecular formation and dissociation constants, as κ_{st} is inversely proportional to K_f.

Relying on some parameters of the Bell model, (Dong and Lei, 2000) studied the rolling of white blood cells under the flow of the plasma; the authors combined an experimental set up (flow chamber) for the determination of the adhesion kinetics with a finite element (FE) model to quantify the cell deformation and rolling as well as the length of the adhesion zone. The adhesion kinetics relies on contributions of (Dembo et al., 1998), whereby the continuity equation is written in the case of rolling without sliding

$$v_c\frac{\partial N_b}{\partial s} + K_f(N_l - N_{b)})(N_r - N_b) - K_rN_b = 0$$

with N_b the density of bonds, N_r the density of receptors on the cell, N_l the density of ligands on the wall and s the curvilinear position. The energy contributions include the flow, the dissipation of the cytoplasm identified to the flow around the cell, and adhesion. FE simulations show that increasing the shear flow increases the rolling velocity, whereas the ligand density is inversely proportional to the rolling velocity.

(Hammer and Lauffenburger, 1987) modeled the adhesion kinetics from a dynamical model of the adhesion of cells using the association and dissociation constant; the equilibrium of the density of bonds and free receptors is described by the two equations

$$\frac{dN_b}{dt} = k_fN_{l0}N_a - k_rN_b;$$

$$\frac{dN_a}{dt} = -k_fN_{l0}N_a + k_rN_b + \Delta(N_{c0} - N_a),$$

with N_a,N_b the density of bonds and free receptors respectively, Δ a coefficient accounting for the accumulation of free receptors in the contact zone, and N_{c0} the initial density of receptors on the cell.

The time variable intervenes in an explicit manner, e.g., through the period of formation of the non-stretched bonds. This model leads to an asymptotic behavior (over long time periods) of the affinity versus the number of receptors, according to the interplay between chemical and mechanical factors at the interface.

Xiao and Truskey (1996) have also studied the influence of the ligand-receptor affinity on the strength of adhesion of an endothelial cell. The ligand receptor complexes are modeled as elastic springs. The critical distance representing the threshold for which half of the bonds have disappeared is determined from (Dembo et al, 1998), and is shown to increase in a non-linear manner with the ligand density, according to

$$\Delta_{crit} = \sqrt{\frac{2k_bT_a}{\kappa}\ln(\frac{K_{d0}}{N_{lr0.5\max}})},$$

with κ, K_{d_0}, $N_{l_{R_{0.5max}}}$ the bonds' rigidity, the dissociation constant for the non active bonds and the decrease of bonds by 50%. As to the purely kinetic approach, the variation of the critical distance is inversely proportional to the initial dissociation constant and is sensitive to the density of bonds. Simulations show that the force supported by a fiber is an increasing nonlinear function of the ligand density.

Other families of deterministic models have been developed, either using wetting theory or thermodynamics. In (Bell et al, 1984), thermodynamics is the framework for the analysis of adhesion between two cells, relying on the following assumption: adhesion can occur only when the distance between both cells is small enough that the receptors on both surfaces enter into contact. The free energy is chosen depending upon the bond and receptor densities, the chemical potential of the free receptors and the repulsive and attractive potential on the area of contact. The repulsion potential generated by the presence of negative charge at the surface and by steric effects has an asymptotic behavior, between the variation of the contact area (linear) and the variation of the number of bonds or receptors. An inverse behavior is obtained for the variation of the separation distance versus the number of receptors.

Using a different approach, Lipowsky and Seifert (1991) studied the adhesion of vesicles, assuming closed surfaces. The considered energy F includes the curvature, the adhesion free energy, and two terms corresponding to external loadings. The tension energy is neglected since the membrane is assumed to exhibit a fluid behavior:

$$F = (\frac{\kappa}{2})\oint (C_1 + C_2 - C_0)^2 dA - WA^* + PV + \sum A,$$

with C_0 the spontaneous curvature, A* the area of vesicle-substrate contact, P the pressure and V the volume of the vesicle. The adhesion energy W is obtained from Youg-Dupré law. This model ignores the kinetic aspect of adhesion, but considers a purely energetic point of view; the time variable does not appear explicitly as in many other models.

Another purely mechanical approach has been developed in (Naili and Yasmineh, 2001), inspired by the theory of curvilinear media, also introducing the proportion of active links, variable β, in the adhesive contact. This variable is spatially distributed according to a differential equation obtained from considerations tied to both the thermodynamics and the mechanics of curvilinear media. This model seems not adequate for cellular adhesion, since it ignores the physics and chemistry of biological phenomena, the plasma flow.

A more numerical analysis has been performed by (Ndri et al, 2001), considering flow as a macroscopic problem, whereas adhesion is modeled at the microscopic scale using the Dembo et al (1998) model. The cell is abstracted to a droplet and the membrane influence is neglected; the force applied to the cell is obtained by solving the Laplace-Young equation with the boundary element method. This model does not incorporate membrane deformation and undulations.

Boundary integral methods have been further used by Agresar et al (1998) to simulate adhesion between two cells, in the context of fluid flow. The contribution of mechanical forces due to the bonds (modeled as elastic springs) and non-specific forces has been considered. This model allows one to follow the evolution versus time of adhesion, but does not consider electrostatic repulsion.

A last class of works have focused on the mechanical behavior of biological cells submitted to various solicitations (flows), discarding the adhesion mechanisms (Eggleton and Popel, 1998; Wache et al., 2001; Bruinsma and Sackmann, 2001; Walter et al., 2001; Ramanuajan and Pozrikidis, 1998).

For cells lacking a nucleus such as red blood cells, their abstraction as droplets surrounded by a thin membrane leads to large deformation under flow. Amongst the widely used constitutive laws in the framework of hyperelasticity, the Mooney-Rivlin and the Skalak et al (1984) models can be mentioned: the density of this last model is given by

$$W^{MR} = \frac{G_s}{4}\left(I_1^2 + 2I_1 - 2I_2 + C\, I_2\right),$$

with I_1,I_2 the two first strain invariants, and C a parameter characterizing the resistance of the surface to surface stretch. Those works have incorporated either constitutive law into numerical methods (FE or boundary integral methods) for the simulation of cells under flow (Ndri et al, 2001; Eggleton and Popel, 1998; Ramanujan and Pozrikidis, 1998) amongst others.

Probabilistic Approaches

In this second class of models, several parameters are considered as having a random origin, allowing the use of tools from probability theory. Pioneering work is done in (Cozens et al, 1990), considering a Markovian system; the model includes deterministic parameters such as the association and dissociation constants of bonds. The density of ligands is supposed to be much higher than the density of receptors, and forces due to flow are neglected. The rupture phenomenon is modeled in a similar manner to adhesion, leading to Fokker-Planck equations, allowing one to study the average number of bonds versus the attachment time. This model eliminates the fluctuations appearing in deterministic models for small numbers of molecules. This work only deals with the adhesion kinetics and does not consider the cell influence on adhesion.

(Evans and Ritchie, 1997) studied the dynamics of molecular adhesion forces, considering rupture as a random phenomenon. In the same period, (Chesla et al, 1998) quantified the dependency of the adhesion probability versus the duration of the contact and the density of receptor-ligands bonds. The reversible chemical reaction involving v_l ligands, v_r receptors and leading to the formation of v_b bonds is given by

$$v_r M_r + v_l M_l \underset{k_r^0}{\overset{k_f^0}{\Leftrightarrow}} v_b M_b$$

Those authors further measured the formation and dissociation constants of the bonds, relative to the mutual adhesion of two cells, for several adhesion molecules: it appears that the adhesion probability increases with the contact time.

In contrast to those global kinetic models, the present description of the cell-ECM wall bond kinetics relies on a view of individual bonds subjected to forces, as described next. Vectors are denoted by boldface symbols in the following discussion.

SKETCH OF THE CELL-WALL INTERACTIONS

Rolling of biological cells corresponds to the slowing down of a moving cell (along the wall of an extracellular matrix or a substrate), followed by the capture of the cell by the ECM wall (e.g., an arterial wall). Rolling is the result of complex molecular kinetic events of simultaneous creation and rupture of molecular connections, namely the ligand-receptor bonds, involving specific proteins. These connections are subject to the action of the fluid flow, Van der Waals attractive interactions, electrostatic repulsion (Bongrand et al., 1982) and affinity forces, in combination with their extensibility. The fluid flow (force) around the cell is transmitted from the cell membrane to the contact interface (Mefti, 2006; Mefti et al., 2006; Bell et al., 1984), as shown in Figure 6.

Since the specific *signalisation* during adhesion leads to a local increase of the cell membrane stiffness (in the interface zone, Figure 1), consecutive to the change of the cytoskeleton structure (Bongrand et al., 1982), we model the cell membrane (of the contact interface zone, Figure 1) and the wall as two elastic plates, endowed with high stiffness. The plates are linked by rheological elements, representing the molecular connections (Zhu, 2000).

The cell-ECM wall interactions are traduced in terms of forces, the equilibrium of which shall be formulated by the principle of virtual work to express the dynamical equilibrium of the interface; the external forces acting on the interface are:

- The fluid force: in the case of blood, the action of plasma is periodic and generates a periodic force characterized by two adjustable parameters, its intensity F_0 and inclination, viz the two angles $\hat{\beta}_{rand}(t), \hat{\theta}_{rand}(t)$;
- The Van der Waals forces (attractive) are obtained by the derivation of the expression of the Van der Waals energy between two elastic plates (Tadmor, 2001), as

$$F_{attr} = \frac{A_0\, l_{a_0}}{6\pi}\left(\frac{1}{(S+D_0)^3} - \frac{1}{S^3}\right), \tag{1}$$

with A_0 the Hamacker constant, l_{a_0} the length of the interaction zone, S the cell-wall distance (measured from the center of gravity of the cell membrane), and D_0 the thickness of the cell membrane;
- The electrostatic force – of a repulsive nature – (Bongrand et al., 1982), resulting from the presence of negative charges on the cell membrane. This force is obtained by the minimization of the phenomenological energy (Bell et al., 1984):

$$F_{rep} = -\frac{\chi}{S} e^{-S/\tau}\left(\frac{1}{S} + \frac{1}{\tau}\right), \tag{2}$$

with χ a compressive parameter (a force) describing the ease with which the connections may be compressed, and τa thickness parameter (the mean length of the ligand). The main *non-specific interactions* considered are Van der Waals forces and electrostatic forces; other forces, such as the solvent influence, the influence of the surrounding ions and hydrogen bonds (Bongrand, 1982), are here discarded.
- The aspiration force (mimicking the micropipette aspiration test), assumed to be expressed vertically proportional to the aspiration length Δl as (the algebraic force is here considered)

$$F_{asp} = K_c \Delta l, \tag{3}$$

with K_c the equivalent cell rigidity, in general itself depending upon the cell membrane stretch (hence upon $\Delta 1$). This rigidity coefficient will be assumed constant in a first approach.
- In addition to elasticity of the bonds, a viscous behavior is modeled by the viscous friction force proportional to the point velocity

$$\mathbf{F}_{visc} = -c\mathbf{V}(M,t), \tag{4}$$

with c the damping coefficient.

Mechanics of the Ligand-Receptor Bonds

The ligands and receptors are macromolecules attached to the cell and the ECM and characterized from a mechanical viewpoint by a nonlinear entropic elasticity, as initially formalized in (Kuhn and Grun, 1942). This is the most common model: a single polymeric chain can be described as a sequence of N rigid segments of equal length l, called the Kuhn length, such that $L=N.l$, called the *contour length*. The chain deformation resulting from applied forces is defined as the end-to-end distance r, or rather as the relative stretch $\lambda=r/L$ (no dimension). For a specific chain, the kinematic variable λ changes when the chain is submitted to varying forces, according to a competition between the natural tendency to increase the chain entropy (if the chain were free, it would adopt a global configuration maximizing the number of possible local configurations) and the tendency to increase its energy due to the external forces. From a statistical point of view, one can distinguish non-correlated chains (freely joined chains) and correlated chains corresponding to

Figure 2. A) Top view: Analogical model of the cell and cell-ECM interfacial bonds. B) Bottom view: Sketch of the kinematic variables.

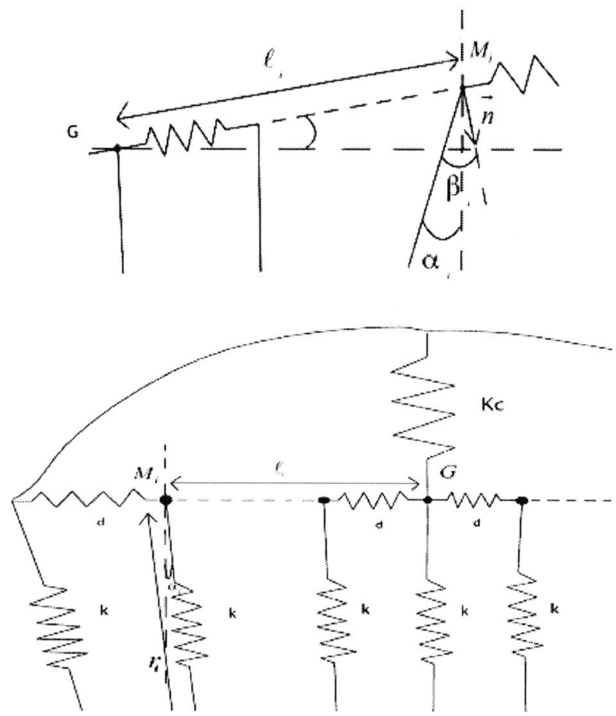

the model of curved chains. We presently adopt the first assumption, thereby considering that the chain may adopt all possible configurations, without correlation between the directions of the successive segments. Adopting this scheme and a Gaussian probability law, Kuhn and Grun (1942) have shown that the retraction force of the freely joined chain is expressed as

$$f=kTNL^{-1}(\lambda), \quad (5)$$

with $L(\lambda)=coth(\lambda)-1/\lambda$ the Langevin function, T the absolute temperature, and k Boltzmann constant; see also (Kuhl et al., 2005).

The shape of the response function shows that the Gaussian model is an approximation of the nonlinear response for small stretches. The slope of the chain response at the origin gives an estimate of the bond rigidity, viz $k=10^{12}N/m$. From a kinematic point of view, the bonds are further endowed with a flexional degree of freedom at their extremity, allowing a rotation of the bond relative to the membrane orientation.

The cell membrane is not explicitly described in the present model, but is instead represented by an equivalent rigidity K_c, supposed to be constant; the part of the cell membrane adhering to the ECM wall is modeled as a linear elastic plate, using linear springs with uniform rigidities d; a sketch of the analogical model of the cell and interfacial bonds is pictured in Figure 2A.

The kinematic variables for the discretized geometry of the cell membrane and interfacial bonds are shown in Figure 2B; a set of Cartesian vectors ($\mathbf{e}_x$,$\mathbf{e}_y$,$\mathbf{e}_z$) is selected, with the second vector $\mathbf{e}_y$ pointing upwards in the plane of the same figure. The size of the contact interface zone is more important than the size of the connections (Ndri *et al.* 2001).

Box 1.

$$m_c \frac{dV_G}{dt} = \frac{A_0 l_{a_0}}{6\pi}\left[\frac{1}{\left(S+D_0\right)^3} - \frac{1}{\left(S\right)^3}\right] + \frac{\chi}{S}\left[\frac{1}{S}\exp\left(-S/\tau\right) + \frac{1}{\tau}\exp\left(-S/\tau\right)\right] + K_c \Delta l$$
$$-kTN_g \frac{1}{\coth\left(S/L_g\right) - L_g/S} - \sum_{i=1}^{N} kTN_i \cos\left(\alpha_i\right) \frac{1}{\coth\left(\frac{S + l_i \sin\theta}{L_i \cos\left(\alpha_i\right)}\right) - \frac{L_i \cos\alpha_i}{S + l_i \sin\theta}} - V_G \sum_{i=1}^{n} c_i \quad (8)$$

A spatial discretization of the cell membrane geometry is performed according to the presence of bonds attached to the cell membrane, with all discretization points on the membrane initially equidistant (distance d), and each discrete point M_i is separated from the gravity center G by a distance l_i. All bonds are supposed to have identical stiffness *k* (although a certain dispersion may be considered). A generic point M_i is linked to the wall by a bond with length r_i. The rolling angle θ represents the angle made by the cell membrane with respect to the horizontal fixed vector $\mathbf{e}_x$ in a Cartesian basis. Two angles related to the bond attached to point M_i are introduced, namely α_i the angle between the vertical and the bond direction, and β_i, the angle between the bond and the exterior normal to the cell membrane (Figure 2B). The three angles α_i,β_i,θ are algebraically related by

$$\alpha_i = \beta_i + \theta. \quad (6)$$

The distance between a generic point M_i and the ECM wall, viz the variable r_i, is given from the geometry of the problem by

$$r_i = \frac{S + l_i \sin\theta}{\cos\alpha_i}. \quad (7)$$

The geometrical relationships (3.7) and (3.8) are inserted into the set of mechanical balance equations to be formulated next.

Governing Equations of Mechanical Equilibrium

The set of field equations required to follow the evolution versus time of the set of kinematic variables (S,θ,l_i,β_i) resulting from the spatially discretized geometry is the following:

1. Conservation of momentum expressed at the gravity center *G*, viz $m_c \frac{d\mathbf{V}_G}{dt} = \sum \mathbf{F}^{ext}$, gives after straightforward calculations. (see Box 1)
   With *N* the initial number of bonds, m_c the cell mass, $N_g \equiv N$ the number of segments of the bond at point G, $N_i \equiv N$ the number of segments of the bond at point M_i, L_g,L_i the initial (free) length of the chain at point G and M_i respectively. The scalar V_G is the algebraic value of the velocity vector $\mathbf{V}_G$.
2. Kinetic theorem at the center of mass G, viz $\frac{d\sigma\left(G,t\right)}{dt} = \mathbf{\Gamma}_{ext}(G,t)$: This theorem states that the time derivative of the angular momentum $\sigma(G,t)$ at the center of mass G is equal to the momentum of all external forces at the same point, quantity $\Gamma_{ext}(G,t)$; accounting for all previous forces and considering the geometry of the problem (Figure 2B), this balance law specializes to Box 2.
3. Momentum balance at a generic membrane point M_i: let point O be defined as the inter-

Box 2.

$$\frac{m_c}{n-1}\sum_{i=2}^{N}\left(\dot{l}_i V_G \cos\theta + l_i \frac{dV_G}{dt}\cos\theta + 2l_i\dot{l}_i\dot{\theta} - l_i V_G\dot{\theta}\sin\theta + \ddot{\theta}l_i^2\right) + \sum_{i=2}^{N} c_i\left(\cos\theta\frac{dV_G}{dt} + l_i\dot{\theta}\right) = \sum_{i=2}^{N}\left(-l_i kTNL^{-1}\left(\frac{S + l_i\sin\theta}{L_i\left(\cos\theta\cos\beta_i - \sin\theta\sin\beta_i\right)}\right)\cos\beta_i\right) + \sum_{i=2}^{N} C_f\beta_i \quad (9)$$

section of the bond emanating from the center of mass G with the ECM wall. One can then write the following relation

$$\mathbf{V}_i = \frac{d}{dt}\mathbf{OM}_\mathrm{i} = \frac{d}{dt}\mathbf{OG} + \frac{d}{dt}\mathbf{GM}_\mathrm{i}$$

Hence, deriving once more previous vectorial decomposition allows one to relate the acceleration of M_i to that of point G, viz

$$\frac{d\mathbf{V}_i}{dt} = \frac{dV_G}{dt}\mathbf{e}_y + \left(\ddot{l}_i\cos\theta - 2\dot{l}_i\dot{\theta}\sin\theta - \ddot{\theta}l_i\sin\theta - \dot{\theta}^2 l_i\cos\theta\right)\mathbf{e}_x + \left(\ddot{l}_i\sin\theta + 2\dot{l}_i\dot{\theta}\cos\theta - \dot{\theta}^2 l_i\sin\theta + \ddot{\theta}l_i\cos\theta\right)\mathbf{e}_y \quad (10)$$

Expressing next the forces in the set of Cartesian basis vectors ($\mathbf{e}_x$,$\mathbf{e}_y$) results in the dynamical equation of motion projected along the direction $\mathbf{e}_x$ (see Box 3) and along the direction $\mathbf{e}_y$. (see Box 4)

Preliminary calculations show that the flexion rigidity C_f has a negligible influence on the global cell behavior.

Failure of the Cell-Ligand Connections

Several mechanisms are responsible of the failure of the molecular connections, which can be either active or passive (Bongrand et al., 1999). Rupture is assumed to occur only by a pulling effect: thus, the ith connection fails if the traction force it sustains exceeds a threshold value, viz

$$F\left(x_i, y_i, t\right) \geq \tilde{F}_{rupt}^{rand}\left(x_i, y_i\right). \quad (13)$$

$F(x_i,y_i,t)$ is the total external force acting on the ith connection having the (discrete) position (x_i,y_i), while $\tilde{F}_{rupt}^{rand}(x_i, y_i)$ is the limit of failure of the same connection.

Box 3.

$$\frac{m_c}{n}\left(\frac{dV_G}{dt} + \ddot{l}_i\sin\theta + 2\dot{l}_i\dot{\theta}\cos\theta - \dot{\theta}^2 l_i\sin\theta + \ddot{\theta}l_i\cos\theta\right) + c_i\left(V_G + \ddot{l}_i\sin\theta + \ddot{\theta}l_i\cos\theta\right) = \frac{A_0 l_{a_0}}{6\pi}\left[\frac{1}{\left(S+D_0\right)^3} - \frac{1}{\left(S\right)^3}\right] + \frac{\chi}{S}\left[\frac{1}{S}\exp\left(-S/\tau\right) + \frac{1}{\tau}\exp\left(-S/\tau\right)\right] + kTN\left(\sin\beta_i\sin\theta - \cos\beta_i\cos\theta\right)L^{-1}\left(\frac{S + l_i\cos\theta}{L\left(\cos\theta\cos\beta_i - \sin\theta\sin\beta_i\right)}\right) + d\sin\theta\left(\Delta l_{i+1} - \Delta l_i\right) \quad (11)$$

Box 4.

$$\frac{m_c}{n}\left(\ddot{l}_i\cos\theta - 2\dot{l}_i\dot{\theta}\sin\theta - \dot{\theta}^2 l_i\cos\theta - \ddot{\theta} l_i\sin\theta\right) + c_i\left(\dot{l}_i\cos\theta - \dot{\theta} l_i\sin\theta\right) =$$
$$kTN\left(\sin\beta_i\cos\theta + \cos\beta_i\sin\theta\right)L^{-1}\left(\frac{S + l_i\sin\theta}{L\left(\cos\theta\cos\beta_i - \sin\theta\sin\beta_i\right)}\right) + d\cos\theta\left(\Delta l_{i+1} - \Delta l_i\right) \quad (12)$$

It is natural to consider that the intensity of the forces of the fluid applied to the connections decreases in the direction of the flow: hence, we assume a linear decrease of the fluid forces. The spatial distribution of forces involves the notion of rows: the support of each force (point of application) is identified by its position in the corresponding transversal row. The junctions between the connections and the cell membrane correspond to the nodes. The following node indexing is used: a given node is identified by two parameters (i,j), which correspond respectively to the position of the transversal and the longitudinal row. Accordingly, the connection forces are indexed as $F_i^{(j)}(t)$.

Using the equilibrium equations and the calculated shape of the rupture force distribution gives

$$F_i^{(1)}(t) = \frac{F_{fluid}(t)}{n^{(1)} + \sum_{j=1}^{NR-1} n^{(j)}\left(\frac{l_{j+1}}{l_j}\right)}. \quad (14)$$

The forces acting on the successive longitudinal rows are given by

$$F_i^{(2)} = \left(\frac{l_2}{l_1}\right)F_i^{(1)};\ F_i^{(3)} = \left(\frac{l_3}{l_1}\right)F_i^{(1)};\ F_i^{(NR-1)} = \left(\frac{l_{NR-1}}{l_1}\right)F_i^{(1)} \quad (15)$$

The set of Equations (14), (15) gives the forces applied to the molecular connection under the effect of the net external force $F(t)$ evaluated as:

$$F(t) = F_{fluid}(t) + F_{rep} - F_{attr} \quad (16)$$

Since the established connections are the result of the junction between adhesion molecules (receptors on the cell and ligands on the ECM wall) occurring during the rolling, one may assume that these junctions are established under non stationary processes; consequently, these variations may lead to spatial fluctuations of the connection properties. In order to simplify the model, we assume that the fluctuation in the limit of failure is characterized by a single scalar quantity (we are not considered multidimensional situations), which is modelled within the *theory of stochastic fields*.

A first simple strategy for the modeling of the fluctuation of the intensity of adhesion will be briefly evoked, although not used in the present contribution: as reported in (Bongrand and Benoliel 1999), the fluctuation of the limit of rupture is described by the following expression (Ayyub et al. 2005):

$$\tilde{F}_i^{rupt}(x_i) = d_i(b - a) + a,$$

where the sequence d_i denote pseudo-random numbers (normal distribution between *0* and *1*), *b* and *a* are parameters identified as the maximum and minimum value of the limit of failure respectively. The pseudo-random numbers d_i (with a uniform distribution in the interval [*0 1*]) are obtained by the use of the congruential algorithms (Newman and Odell, 1971; Matsumoto and Nishimura 1998): however, in the case of the normal distribution, the Lindeberg-Levy theorem can be used. This kind of formulation (pseudo random sequence) does not involve any spatial correlation between the limit of rupture of the connections, contrary to the random process technique that we shall follow in the sequel, which allows considering such a correlation.

The spatial distribution of the limit of failure is given in normalized form as

$$\tilde{F}_{rupt}^{rand}\left(x_i, y_i\right) = F_0\left(1 + \delta_{\mathrm{p}}\frac{f_{rand}\left(x_i, y_i\right)}{f_{\max}}\right) \quad (17)$$

The amplitude F_0 is the limit of failure of a connection in the case of a uniform distribution (it can be measured, for instance in the classical micropipette aspiration test), δ_p the value of the maximal fluctuation (an input data) of the Gaussian stochastic field $f_{rand}(x_i,y_i)$, and f_{max} its maximal value (also an input data). Independent simulations have shown that the proportionality parameter δ_p has the largest influence on the distribution of the rupture forces, as it determines the variation between the average rupture force and its maximum.

The Gaussian stochastic field (spatial fluctuation) is obtained by the spectral approach of (Shinozuka 1971-1972, Shinozuka and Deotadis, 1991; Shinozuka et al., 1999; Shinozuka and Lenoe, 1976; Fenton, 1990), involving a decomposition of a Gaussian process into M_1 (resp. M_2) harmonics in the direction x (resp. y):

$$f_{rand}(x_i, y_i) = \sqrt{2}\,\mathrm{Re}\left\{ \sum_{m=0}^{M_1-1}\sum_{n=0}^{M_2-1}\sqrt{4S_{00}(k_{1m}, k_{2n})} \times \sqrt{\Delta k_1 \Delta k_2}\, e^{i\varphi_{mn}} e^{i(k_{1m}x + k_{2n}y)} \right\} \quad (18)$$

with the discrete spatial positions of the bonds given by

$x_i = p\Delta x$, $p = 1,2,3...M_1$; $y_i = q\Delta y$, $q = 1,2,3...M_2$.

The function $S_{00}(k_{1m},k_{2n})$ in (18) represents the *Spectral Density of Power* (SDP), accounting for the long range correlations between the values of the failure limit; the SDP is the Fourier transform of the self-correlation function (according to the Bochner-Wienwer-Khintchine theorem for stationary processes). The expression (18) is in fact the fast Fourier transform realization of the Gaussian process $f_{rand}(x_i,y_i)$. We presently use the SDP given in (Nour et al., 2003):

$$S_{00}(k_1, k_2) = \sigma^2\,\frac{\mathrm{b}_1 b_2}{4\pi}\,\mathrm{e}^{\left[-\left(\frac{\mathrm{b}_1 k_1}{2}\right)^2 - \left(\frac{b_2 k_2}{2}\right)^2\right]}. \quad (19)$$

The *correlation lengths* (b_1,b_2) determine the extent of the spatial range of interactions in the directions x and y respectively, and σ is the standard deviation (Nour et al., 2003). For values of those lengths approaching zero, one recovers a purely local process.

The spatial discretization steps $(\Delta x,\Delta y)$ depend on the periodicity lengths (L_x,L_y) (Shinozuka et *al.*, 1999):

$$\Delta x = \frac{L_x}{M_1};\ \Delta y = \frac{L_y}{M_2}.$$

In (19), (k_{1m},k_{2n}) are the wave numbers in the directions x,y resp. and $(\Delta k_{1m},\Delta k_{2n})$ their increments, determined from two integers N_1,N_2, as

$$\Delta k_{1m} = \frac{k_{1u}}{N_1};\ \Delta k_{2n} = \frac{k_{2u}}{N_2}$$

Decreasing the discretization parameters N_1,N_2 of the wave number increments lead to a similar decrease of the variation of the rupture threshold between two adjacent bonds. The parameters k_{1u},k_{2u} therein are resp. the upper limits of the wave numbers in the directions x,y, such that

Box 5.

$$\Delta F^{j}_{i+1}(t+\Delta t) = \frac{F^{j}_{i}(t)}{4+(R_D - L^{j,j}_{i,i+1})} \frac{1}{(4R_D - 4L^{j,j}_{i,i+1} - L^{j,j-1}_{i+1,i-1} - L^{j,j+1}_{i+1,i-1}} \frac{}{-L^{j,j-1}_{i+1,i+1} - L^{j,j+1}_{i+1,i+1}}$$

$-k_{1u} \le k_{1m} \le +k_{1u}$; $-k_{2u} \le k_{2n} \le +k_{2u}$.

The upper limit of the wave numbers, quantities k_{1u}, k_{2u}, are determined from the approximate condition

$$\int_0^{k_u} S_{00}(k_{1m}, k_{2n}) dk_1 dk_2 \approx \int_0^{\infty} S_{00}(k_{1m}, k_{2n}) dk_1 dk_2$$

Observe that also the SDP involves a set of parameters, only (b_1, b_2), σ, and the upper wavelengths limits k_{1u}, k_{2u} are true physical parameters, which shall ideally be determined from measurements of the stochastic signal; the quantities $L_x, L_y, M_1, M_2, N_1, N_2$ are parameters that determine the accuracy and quality of the discretization scheme. All input data relative to the SDP used in the modelling of rupture forces shall be given in section 6 devoted to simulation results. It is not the purpose of this contribution to study the sensitivity of the model to the discretization parameters; this has been done at length in (Mefti, 2006) to which the reader is referred. It is nevertheless clear that beyond the pure numerical aspects related to sensitivity and robustness of the model, the identification of the SDP from suitable measurements is a challenging problem.

When the rupture of the connection (i,j) occurs, the applied force $F^{j}_{i}(t)$ is transferred to the other connections located in a circular influence zone of radius R_D, resulting in a force jump ΔF^{l}_{k} at the nodal position (i,k). Connections close enough to the connection (i,j) recover a proportion of the initial force $F^{j}_{i}(t)$ being redistributed. Introducing $L^{k,l}_{j,j}$ as the distance between the connections (i,j) and (k,l), and assuming that the force transfer is linear and decreasing, the equilibrium equations give the force jump in Box 5.

The set of input data is given in Table 1.

The previous governing equations are discretized in both space and time using a finite difference scheme, with the spatial discretization naturally associated to the set of bonds (their attachment points on the cell membrane define the discretization nodes); a fine enough time discretization has to be selected to accurately follow

Table 1. Geometrical and mechanical input data for the simulations

Parameter	Value
Transversal length of the interface	$l_{a_0} = 3000nm$
Thickness of the cell membrane	D_0=7nm
Room temperature	T=293°K
Rigidity of the bonds	k=$10^{-12}n/m$
Compressibility parameter of the bonds	χ=10^{-11}N
Mass of the cell (red human blood cell)	m_c=$10^{-4}nkg$
Hamaker constant	A_0=$10^{-20}J$
Spring constant of the membrane	d=$10^{-12}N$/m
Flexion modulus of the bonds	C_f=1=$^{-12}N/m$
Equivalent rigidity of the cell	K_c=2*$10^{-9}N/m$
Cut off separation cell-ECM wall distance	τ=10nm
Average aspiration (detachment) force	F_{may}=0.07nN
Reference curvilinear length of the chain	L_{ref}=2.125nm
Number of chain segments	N_{ref}=20

Figure 3. Evolution of the membrane-ECM wall distance vs. the aspiration length. Cell equivalent rigidity K_c=8.10⁻⁹N/m.

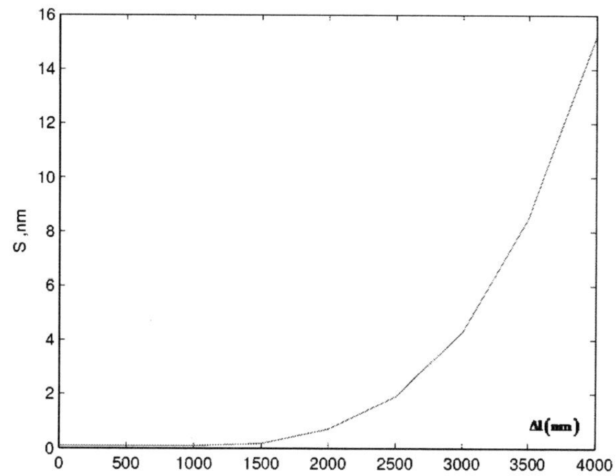

the discrete rupture and adhesion events related to the kinetics of bond rupture.

As a global result, the evolution of the membrane displacement versus the aspiration length (Figure 3) shows a nonlinear behavior with an increasing slope, due to the increased softening of the interface resulting from the successive rupture events.

The displacement of the centre of gravity G versus the aspiration length (simulating thereby the micropipette aspiration test) is pictured in Figure 3: the cell is attracted towards the substrate (the balance of forces is negative) until an aspiration distance of 1500nm (no rupture of bonds occurs), after which the balance of forces favours a repulsive behavior, and a nonlinear increase of the distance of the cell to substrate versus the aspiration length is obtained.

ADHESION OF MOLECULAR CONNECTIONS

The generation of molecular bonds is the result of the close contact of complementary adhesion molecules – ligands and receptors – on a portion of the cell outer surface, (Bongrand and Benoliel, 1999). We assume that the ligand is subjected to the action of the thermal agitation and to the molecular specific interactions (Figure 4), namely Van der Waals forces, electrostatic interactions, the solvent influence and the influence of the surrounding ions and hydrogen bonds (Bongrand et al., 1999).

The Brownian force is a quasi-random phenomenon, according to the physical picture of forces resulting from the shocks between the fluid particles and the ligands. We model this action relatively to each ligand by a periodic force, characterized by both a random orientation and intensity (Mefti et al., 2006). In order to simplify the problem, the selected specific interactions being considered are restricted to Van der Waals forces. Consequently, the intensities of the external forces applied to the ith ligand are:

$$F_{aff_j} = \frac{4}{k_T T}\left(\frac{\mu_a \mu_b}{4\pi\varepsilon_0}\right)\frac{1}{r^7}; \qquad (20)$$
$$F_{brow_j} = F'_{0_j}\sin(\omega_{brow} t)$$

The parameters μ_b,μ_b therein are the dipolar moments of the ligands and receptors respectively, which can be measured (Bamba and

Figure 4. Forces acting on free adhesion molecules with the ith ligand-receptor (Marques, 2001)

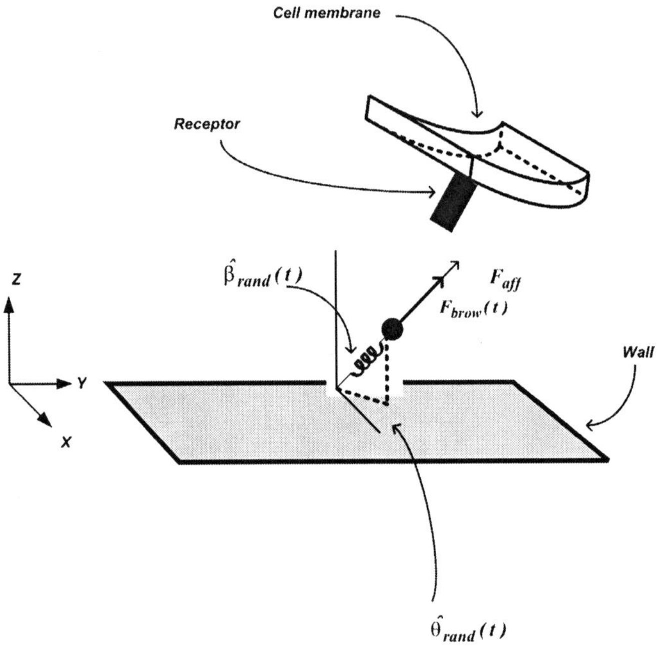

N'guessan, 2003) for the adhesion molecules (LFA1…); k_T is the Boltzmann constant, T the temperature, ε_0 the dielectric constant; F_{0_j} the maximal value of the Brownian force. The considered random parameters are the orientation angles of the ligand $\hat{\beta}_{rand}(t), \hat{\theta}_{rand}(t)$ (Figure 4) and the pulsation of the solicitation ω_{brow}, both represented in normalized form as:

$$\hat{\beta}_{rand}(t) = \hat{\beta}_{\max} \frac{f_{\beta}^{rand}(t)}{f_{\beta}^{\max}}; \quad \hat{\theta}_{rand}(t) = \hat{\theta}_{\max} \frac{f_{\theta}^{rand}(t)}{f_{\theta}^{\max}}; \quad \hat{\omega}_{brow}(t) = \hat{\omega}_{\max} \frac{f_{\omega}^{rand}(t)}{f_{\omega}^{\max}} \tag{21}$$

The quantities f_{β}^{rand}, f_{θ}^{rand}, f_{ω}^{rand} therein are Gaussian stochastic processes (time dependent processes), given by the following series expression obtained by the spectral approach of (Shinozuka and Deotadis, 1991):

$$f^{rand}(t) = \mathrm{Re}\left\{ \sum_{n=0}^{M-1} B_n \exp\left[i \left(n\ \Delta\omega \right) \left(p\ \Delta t \right) \right] \right\} \tag{22}$$

with the amplitudes therein given by

$$B_n = \sqrt{2} A_n e^{i\phi_n}; \quad A_n = (2 S_{f_0 f_0}(n\Delta\omega)\Delta\omega)^{1/2}$$

and the time step such that

$t = p\Delta t$; $p = 0,1,2,\ldots,M\text{-}1$ (M is the number of spectral discretization points).

The function $S_{f_0 f_0}$ in (4.3) is the SDP of this Gaussian process, given by (Figure 5)

Figure 5. Shape of the SDP with b=1s

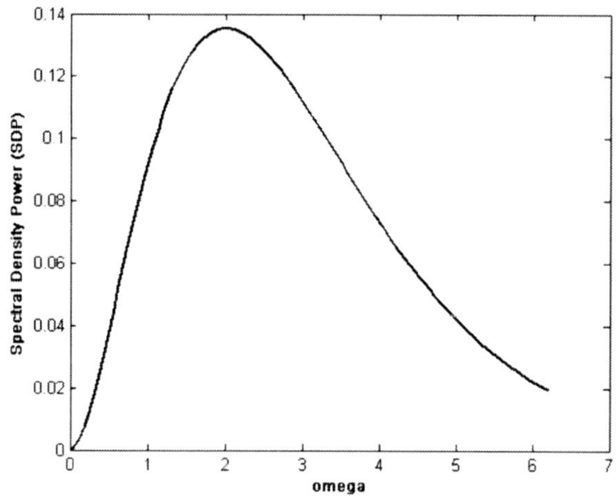

$$S_{f_0 f_0} = \frac{1}{4}\sigma^2 b^3 \omega^2 e^{-b|\omega|}. \tag{23}$$

The parameter σ therein is the standard deviation, b the *correlation time*, and ω the pulsation. The frequential increment Δω is given by: $\Delta\omega = {}^{\omega_u}\!/_{N}$, with the condition $N \leq \frac{1}{2}M$, in order to avoid aliasing. The parameter ω_u corresponds to the upper cut-off frequency. Furthermore, M has to be even, while Δt has to satisfy the Heisenberg-Gabor inequality (Shinozuka, 1999), (Shinozuka and Deotadis, 1991):

$$\Delta t \leq \frac{2\pi}{\omega_u}. \tag{24}$$

The junction between a ligand and a receptor (setting up the bond) is described by a kinematic criterion of minimal proximal distance (Mefti et al., 2006). The motion of the receptors is the result of the rolling, itself being induced by the sequence of rupture / adhesion events of the existing connections. The dynamical equations of equilibrium of the ith ligand are solved with a finite difference scheme. The damping of molecular connections represents the internal friction occurring at a microscopic scale; the Rayleigh method is presently used for the determination of the damping coefficient, in terms of the mass and rigidity, viz

$c_i = \alpha_d + \beta_d m_i$.

CELL PROTRUSION DUE TO CYTOSKELETON POLYMERIZATION

The polymerization of the cytoskeleton (consisting in the disappearance of monomer fibers) in response to the signaling events due to the cell environment – especially in the vicinity of focal points – (mechanotransduction due to the existence of chemiotactic sources) is associated with active deformations of the cell membrane occurring in the adhesion zone (Figure 1). In the opposite zone of the cell membrane, an opposite process of polymerization takes place within the cytoskeleton, leading globally to a motion of the cell directed towards the signaling zone. This change of the cytoskeleton structure is a random process (Peeters, 2004). First results of the influence of the protrusion force on the cell deformation will be next presented.

The spatial fluctuation of the internal pressure is a stochastic process, which can be modeled in the frame of Gaussian random processes. Before the onset of active deformations, one may assume an equilibrium state between the internal and external pressures applied to the cell membrane. The active deformation of the cell resulting from the polymerization leads to a fluctuation of the internal pressure exerted on the inner surface of the cell; this unbalance of force is associated with a protrusion force exerted on the inner surface of the cell (Figure 6).

The cell, regarding its geometry, is modelled as a 3D spherical shell adhering to its substrate in a stable state; the cell membrane is endowed with extension, pure flexion and shear deformation modes, as observed for the human red blood cell (Lenormand, 2001).

The determination of the new equilibrium configuration of the cell resulting form the fluctuation of the inner pressure is obtained as the solution of a mechanical problem involving a stochastic force field, using the stochastic finite elements method (abbreviated SFE method). This method derives from the deterministic FE method, and constitutes an extension to account for a variability of the response for structures exhibiting material, loading and / or geometrical uncertainties (Nour et al, 2004; Fenton, 1990, Ghanem, 2003). The most common SFE methods are based on a representation of the stochastic fields by a series of random variables, such that the weighted Integral method, the Local Average method, polynomial chaos, Neumann expansion, or perturbation methods (Fenton, 1990; Papadrakakis and Papadopoulos, 1996; Ghanem, 2003) can be applied. Notwithstanding those variants, the global procedure consists of assembling the elementary contributions into global vectors and matrices (with suitable boundary conditions), and solving an algebraic dynamical system

$$[K]\,\langle U\rangle=\langle F\rangle, \qquad (25)$$

with $\left[K\right]=\sum_{e}[K^{e}]$ the global rigidity matrix (here a deterministic quantity), $\langle U\rangle$ the vector of nodal displacements (a stochastic output), and $\langle F\rangle$ the vector of internal nodal forces, a mixed deterministic / stochastic quantity (its resultant is known, but it is spatially distributed as a random Gaussian fluctuation).

The dynamical aspect arises here from the time variation of the protrusion force, itself due to either the motility of the chemiotactic source or the variation of its intensity (Stéphanou and Tracqui, 2002). The applied protrusion force on the right hand side of (25) is written in a system of local spherical (angular) coordinates $\phi_i, \overline{\theta}_i$

Figure 6. Sketch of the internal and external pressures before (a) and after the (b) polymerization of the cytoskeleton. Uniform distribution of internal forces (a) Stochastic fluctuation of internal forces (b)

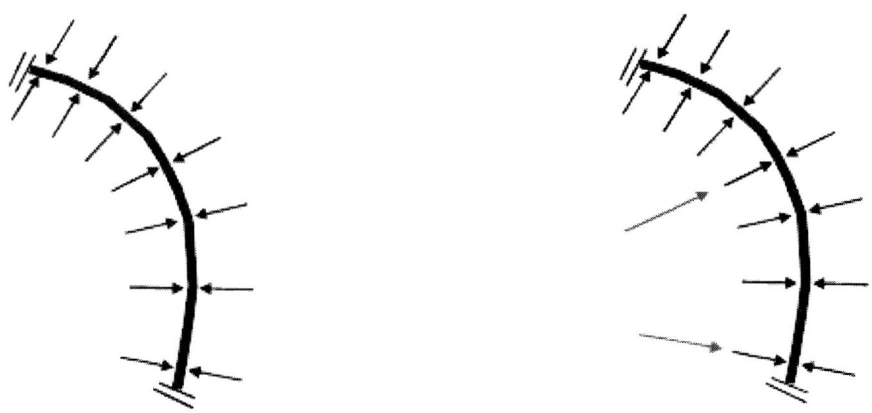

$$\left\langle \tilde{F}_{p_i}^{rand}\left(\phi_i,\bar{\theta}_i\right)\right\rangle = \left\langle F_{0_i}^{p} + \frac{\Delta f^{rand}\left(\phi_i,\bar{\theta}_i\right)}{\Delta f_{\max}}\ \mathrm{F}_0^{\mathrm{p}}\ \mathrm{F}_{\mathrm{prop}}\right\rangle \tag{26}$$

resulting from the following decomposition of the protrusion force around its average:

$$F_p^{rand}\left(\phi,\bar{\theta}\right) = F_0^p + \Delta F_p^{rand}\left(\phi,\bar{\theta}\right).$$

A normalized random process is then generated according to the maximum value of the fluctuation

$$\Delta F_p^{rand}\left(\phi,\bar{\theta}\right) = \Delta F_p^{norm}\left(\phi,\bar{\theta}\right)\ \Delta\mathrm{F}_{\max}^{\det},$$

with $\Delta F_{\max}^{\det}$ the maximal (deterministic quantity) of the fluctuation relative to the average force. This leads to the expression of the increase of the polymerization:

$$F_p^{rand}\left(\phi,\bar{\theta}\right) = F_0^p + \frac{\Delta f^{rand}\left(\phi,\bar{\theta}\right)}{\Delta f_{\max}}\ \mathrm{F}_0^{\mathrm{p}}\ \mathrm{F}_{\mathrm{prop}}. \tag{27}$$

The previous expression is conveniently re-written in discrete form using the Fast Fourier Transform:

$$F_{p_i}^{rand}\left(\phi_i,\bar{\theta}_i\right) = F_{0_i}^p + \frac{\Delta f^{rand}\left(\phi_i,\bar{\theta}_i\right)}{\Delta f_{\max}}\ \mathrm{F}_0^{\mathrm{p}}\ \mathrm{F}_{\mathrm{prop}}\ . \tag{28}$$

The polymerization force is directly related to the force of affinity (of chemical origin) with the chemiotactic molecules (Stéphani and Tracqui, 2002); we consider both F_0^p and F_{prop} as given input data. The random fluctuation $\Delta f^{rand}\left(\phi_i,\bar{\theta}_i\right)$ expressed in normalized form is decomposed in a series of harmonics for a stationary two-dimensional process (Shinuzoka and Lence, 1976)

$$\Delta f^{rand}\left(\phi,\bar{\theta}\right) = \sqrt{2}\sum_{m=1}^{N_1}\sum_{n=1}^{N_2}\sqrt{4S_{f_0f_0}\left(\Delta\omega_\phi\Delta\omega_{\bar{\theta}}\right)}\ \cos\left(\omega_{\phi_m}\phi + \omega_{\bar{\theta}_n}\bar{\theta} + \varphi_{mn}\right) \tag{29}$$

The SDP in (29) is selected as the following isotropic function (Nour et al, 2003)

$$S_{f_0f_0}\left(\omega_{\bar{\theta}},\omega_\phi\right) = \frac{\sigma^2\ \bar{\alpha}^2}{4\pi}\ \exp\left(\left(\frac{-\bar{\alpha}^2}{4}\right)\ \left(\omega_{\bar{\theta}}^2 + \omega_\phi^2\right)\right) \tag{30}$$

with $\sigma^2,\bar{\alpha}$ resp. the mean square deviation and a real constant to tune the shape of $S_{f_0f_0}\left(\omega_{\bar{\theta}},\omega_\phi\right)$. Results of the cell protrusion shall be shown in the next section.

SIMULATION RESULTS

The numerical simulations give the time evolution and the localisation of the rupture / adhesion of molecular bonds, as opposed to global kinetic models described in section 2. Considering leukocytes, we simulate the behaviour of an interface composed of 60 initially existing connections; the rupture of the bonds depends of the spatial distribution of their limit of failure, according to the chosen SDP function. We consider an interface having a circular shape. The statistical parameters are given the following values, relying mostly on the average measured rupture force; the other parameters are tied to the discretization, and are in fact not physical parameters:

$$F_0 = 0.07nN;\ \delta_p = 14\%;\ M_1 = M_2 = 64;$$
$$N_1 = N_2 = 16;\ \sigma = 1nm;\ b_1 = b_2 = 100nm;$$
$$\omega = 4.71rd/s;\ k_{1u} = k_{2u} = 0.07\ \mathrm{rd/nm};$$
$$\beta_{\max} = 0.8rd;\ \theta_{\max} = 6.28rd$$

The correlation lengths are conceived as parameters, and are taken equal for simplicity

Figure 7. Time evolution of the number of broken bonds. Cell rigidity: K_c=8.10^{-9}N/m. Fluid resultant: 0.1nN

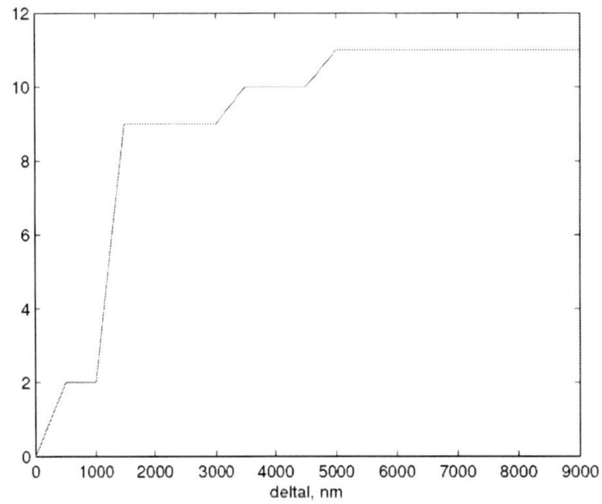

reasons, viz $b_1=b_2$; the measurements of correlation distances is in fact quite challenging, and is out of the scope of the present contribution. The algorithm for the solution procedure of the set of governing equations is the following:

1. Initialize the kinematic variables (S,θ,l_i,β_i).
2. Increase the aspiration length by Δl.
3. Solve the momentum equation for the membrane to wall distance $S(t)$.
4. Solve the kinetic moment equation at point G for the rolling angle $\theta(t)$.
5. Solve the momentum equation projected on y at each point M_i for the distance l_i between M_i and the centre of gravity of the membrane G.
6. Solve the momentum equation projected on x at each point M_i for the angle β_i.
7. Test the rupture of all remaining bonds.
8. Stop if all bonds are broken, otherwise go to step 2.

All the governing equations are discretized with a finite difference scheme in space and time.

The interface is further subjected to a fluid force having a magnitude of 0.1nN. The global simulated rupture force is 0.07nN; this value is in good agreement with the range of measured adhesion forces in the interval [1.7pN, 6.7nN], (Bongrand and Benoliel, 1999). The simulations show that the adhesive zone failure occurs gradually during the cell aspiration (Figure 7).

In contrast to this, an avalanche behavior has been observed when the cell is considered infinitely stiff (Mefti et al., 2006); this difference is due to the cell elasticity, which delays rupture by absorbing part of the external work (aspiration). Several parameters, such as the number of discretization steps M_1,M_2 and the correlation length b_1,b_2, influence the fluctuation frequency. We notice for example that increasing the correlation distances b_1,b_2 lead to a more uniform fluctuation, due to a strongest correlation (of the limit of failure) between the connections (Figure 8).

From the obtained SPD, the total failure of the contact interface still occurs after *1s* of solicitation, even if the shape of the fluctuation is different: this is due to the intensity of the external solicitation applied to the cell (fluid: *0.01nN*).

The adhesion of connections is next modelled, considering 25 potential pairs of free adhesion molecules (ligands-receptors pairs) present on the

Figure 8. Spatial fluctuation of the limit of rupture $b_1=b_2=1000nm$

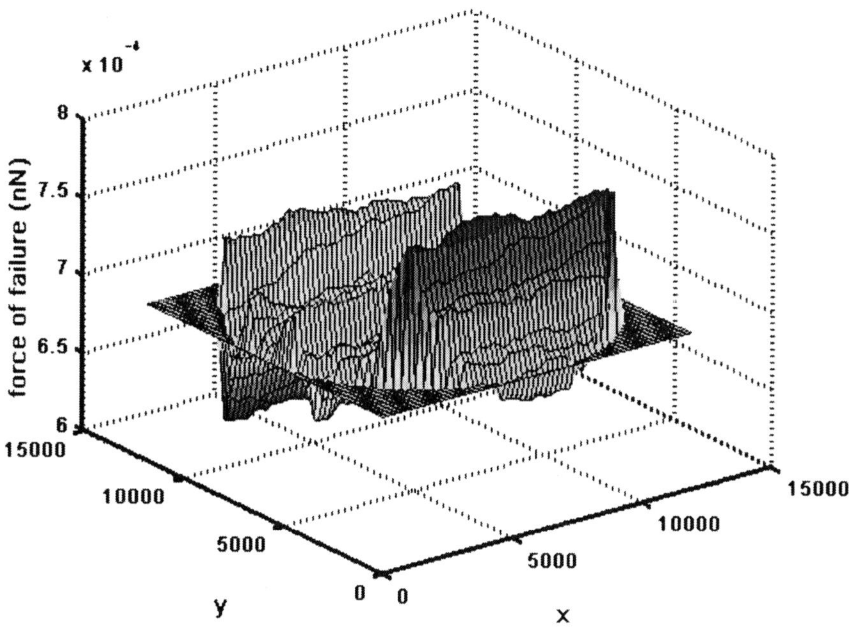

adhesion zone (Figure 1). The time evolution of the stochastic parameters describing the orientation of the first ligand is obtained from previous equations of motion.

The coupling between the thermal agitation and the specific interactions characterized by a Van der Waals force expressed in (3.20) with $\mu_a=\mu_b=1$*debye* leads to the fast junction of *15* couples of molecules (Figure 9A, B), in both cases of damped and undamped ligand-receptor junctions.

The adhesion kinetics includes two steps: the transitive step corresponds to a period without adhesion (*5s* for the damped system, insert A, and

Figure 9. Time evolution of the creation of new bonds with (a) and without damping (b)

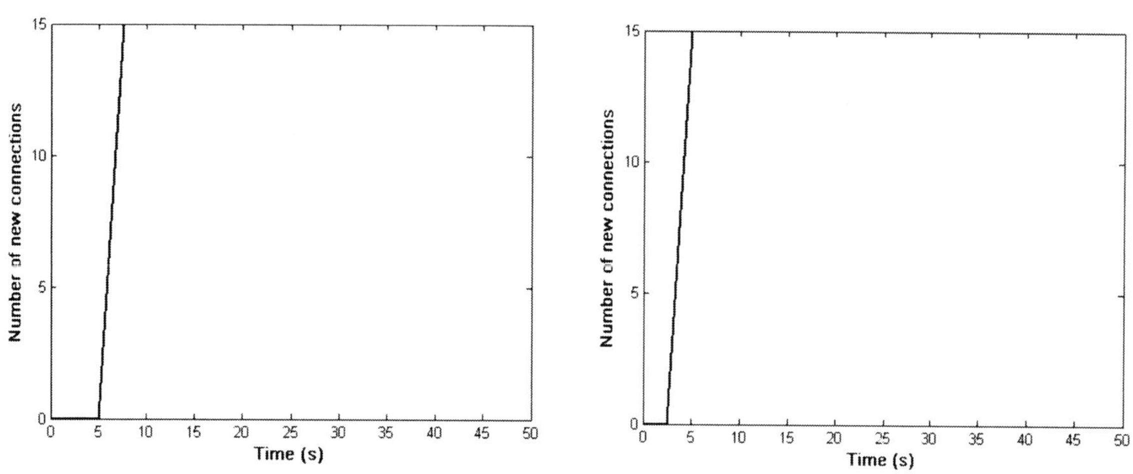

2s for the undamped system, insert B), followed by the junction between the free adhesion molecules (after respectively *5s* and *2s* for the damped and undamped system). The damping delays the onset of adhesion due to lower vibration amplitude of the ligands, but it does not modify the evolution of the kinetics of adhesion thereafter.

Results related to the protrusion of the cell under the action of the internal pressure distributed in a stochastic manner (see previous section) are next highlighted. The stochastic discretization parameters are selected as follows:

$$\begin{vmatrix} N_1 = N_2 = 4; M_1 = M_2 = 11 \\ \sigma = 1 \\ b_1 = b_2 = 1rd \\ k_\phi = k_\theta = 12.6rd^{-1} \end{vmatrix}$$

The cell membrane is modeled as an elastic shell with Young modulus 7.$\mu N/m^2$ and Poisson's ratio *0.3* (Lenormand, 2001); the protrusion force is selected with an average of 0.03nN, and its spatial distribution is selected to evolve in a self similar manner in time for a total duration of 40" (only the initial and final distributions are shown). The cell membrane has been discretized with 25*25 nodes (along the axes defined by the angular coordinates), and the inner surface submitted to protrusion forces represents 15% of the cell surface. The distribution of the inner pressure is drawn schematically on Figure 9. The resulting strain distribution on the polymerization area is shown in Figure 10, with peak values around 5% in zones of highest protrusion forces.

Further simulations (not presented here) show a great sensitivity of the cell deformation to the value of the cell mechanical properties, leading in some cases to large localized deformations. This preliminary result is of importance as it paves the way for dynamical simulations of the cell protrusion due to the modification of the cytoskeleton in response to external signaling. Especially, those protrusions are the mechanisms by which the cell will move on the ECM.

FUTURE RESEARCH DIRECTIONS

The present three-dimensional modelling of the adherence behavior of a single cell to the wall of an ECM describes the rupture and creation of molecular bonds. The cell membrane and the interfacial bonds have been modelled in an analogous manner, employing an equivalent spring for the cell and a set of parallel nonlinear springs for the ligand-receptor bonds; a viscous damping force may be additionally considered, modelled as proportional to the velocity. The freely joined chains model has been considered to model the nonlinear extensional behavior of the bonds. Both cases of undamped and damped motion of the cell have been considered.

The individual adhesion and rupture events responsible of the kinetics of attachment / detachment of the cell are the results of the interplay between the specific interactions considered in the model – Van der Waals attraction, the affinity between ligands and receptors, electrostatic repulsion – and mechanical retraction forces. The rupture of bonds is based on the competition between the external forces applied to the connections and their limit of resistance under simple traction. The fluctuation of the limit of failure is described by a stochastic field approach in a spatial context, whereby the failure limits are distributed according to a Gaussian process by the method of Shinozuka. The creation of new connections is described by a kinematic proximity criterion, similarly using a stochastic process description for the forces of affinity and the force of Brownian motion exerted by the surrounding fluid particles.

A preliminary description of the cell deformation in response to protrusion forces arising from the mechanotransduction due to chemiotactic sources has been made; the average protrusion force is spatially distributed as a 2D Gaussian

Figure 10. Equivalent strain distribution on the polymerization zone. Cell radius: 9μm. Membrane thickness:20 nm.

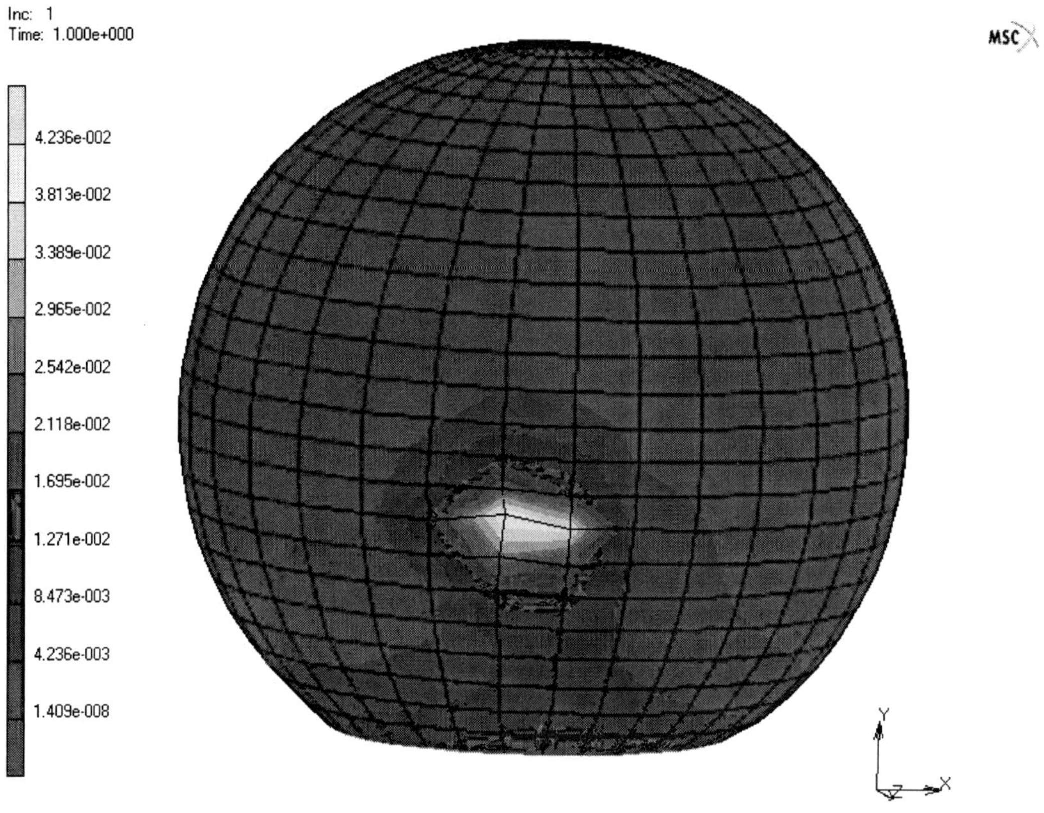

random process, and the stochastic FE method evaluates the cell response, which can be analyzed in terms of the membrane protrusion.

The model has first a conceptual interest, since the effect of parameters is difficult to access from an experimental point of view, and can be presently quantitatively analyzed. The simulation results can be classified according to the considered scale, namely the evolution of the force required to detach the cell from its substrate, as the consequence of the local rupture and adhesion events.

From our point of view, the novel aspect advocated in this contribution relies first on a stochastic description of the local events of bond association and dissociation. Furthermore, a relatively complete picture of the forces governing the physics of the cell to ECM wall interactions has been given. No claim of completeness is made here, since we have reported preliminary results to highlight the interest and potential of the model.

Perspectives of development of the present model of cell adhesion include the following (non-exhaustive) list of points:

- A spatial distribution of the initial length is likely to be the true biological situation, rather than considering, as in the present model, a uniform distribution. Hence, the initial length shall be modeled as a stochastic field.
- Consideration of the evolution of the equivalent cell rigidity from FE calculations of the cell membrane deformation. It is likely that the active deformation of the cell leads to large deformation.

- The description and modelling of the stochastic variability inherent to biological processes involves a model for the spectral distribution of power in the Fourier space. Instead of starting from the Fourier picture, one could rely on a direct physical representation based on the knowledge of measurements (when possible); especially, the estimate of the correlation time or length needs proper measurements techniques specific to the very small scales involved.
- Analysis of instabilities, according to the ratio of the equivalent cell rigidity to the bond rigidity, K_c/K; preliminary results in (Haussy and Ganghoffer, 2005) indeed show that this parameter strongly influences the global stability of the interfacial behavior, also depending upon the parameters of the chosen probability model.
- In the same line of thoughts, displacement control methods – such as the arc length method – shall be employed rather than force control based methods, to closely follow the development of rupture and adhesion induced instabilities (Haussy and Ganghoffer, 2005).
- The coupling between the membrane motion and deformation and the kinetics of bond rupture at the interface is a challenging problem of mechanobiology, requiring a proper understanding of the sequence of events and the pathways of the interactions between chemical signaling and the cell mechanical response.
- As previously mentioned, cell motion may occur by different mechanisms (rolling, active deformation of the cell); the discrimination of the situations in which the cell may adopt one preferential mechanism is a challenging problem in understanding the true mechanisms behind the chosen mode of motion.

Experimental developments are also required in order to determine more realistic values of various parameters of the model. A combined theoretical-experimental approach is necessary in order to both identify the parameters of the model (bond elasticity; specific forces; Brownian forces) and to develop a specific SDP for the biological phenomena, using the FFT approach (Lee et al., 1997).

REFERENCES

Agrasar, G., Linderman, J. J., Tryggvason, G., & Powell, K. G. (1998). An adaptative, Cartesian, front-tracking method for the motion, deformation, and adhesion of circulating cells. *Journal of Computational Physics*, *143*, 346–380. doi:10.1006/jcph.1998.5967

Badal, A., & Sempan, J. (2006). A package of linux scripts for the parallelization of Monte Carlo simulation. *Computer Physics Communications*, *175*, 440–450. doi:10.1016/j.cpc.2006.05.009

Bamba, E., & N'guessan, Y. (2003). Nouvelle approche de détermination du moment dipolaire en solution dans des solvants polaires. *Revue Ivoirienne des Sciences et Technologies.*, *4*, 25–33.

Bell, G. (1978). Models for specific adhesion of cell to cell. *Science*, *200*, 618–627. doi:10.1126/science.347575

Bell, G. I., Dembo, M., & Bongrand, P. (1984). Cell adhesion: Competition between non-specific repulsion and specific bonding. *Biophysical Journal*, *45*, 1051–1064. doi:10.1016/S0006-3495(84)84252-6

Bongrand, P. (1982). Ligand-receptor interactions. *Reports on Progress in Physics*, *62*, 921–968. doi:10.1088/0034-4885/62/6/202

Bongrand, P., & Benoliel, A. M. (1999). Adhésion cellulaire. *RSTD*, *44*, 167–178.

Bongrand, P., Capo, C., & Depied, R. (1982). Physics of cell adhesion. *Progress in Surface Science*, *12*, 217–286. doi:10.1016/0079-6816(82)90007-7

Bruinsma, R., & Sackmann, E. (2001). Bioadhesion and the dewetting transition. *Comptes Rondus de l'Academie Sciences Paris*, *2*(4), 803–815.

Chesla, C., Selvaraj, P., & Zhu, C. (1998). Measuring two-dimensional receptor-ligand binding-kinetic by micropipette. *Biophysical Journal*, *75*, 1553–1572. doi:10.1016/S0006-3495(98)74074-3

Coombs, D., Dembo, M., Wosfy, C., & Goldstein, B. (2004). Equilibrium thermodynamics of cell-cell adhesion mediated by multiple ligand-receptor pairs. *Biophysical Journal*, *86*, 1408–1423. doi:10.1016/S0006-3495(04)74211-3

Cozens, C., Lauffenburger, D., & Quinn, J. A. (1990). Receptor-mediated cell attachment and detachment kinetics-probabilistc model and analysis. *Biophysical Journal*, *58*, 841–856. doi:10.1016/S0006-3495(90)82430-9

Dembo, M., Torney, D. C., Saxman, K., & Hammer, D. (1998). The reaction-limited kinetics of membrane-to-surface adhesion and detachment. *Proceedings. Biological Sciences*, *234*, 55–83. doi:10.1098/rspb.1988.0038

Dong, C., & Lei, X. (2000). Biomechanics of cells rolling: Shear flow, cell-surface adhesion, and cell deformability. *Journal of Biomechanics*, *33*, 35–43. doi:10.1016/S0021-9290(99)00174-8

Edelman, G. (1976). Surface modulation in cell recognition and cell growth. *Science*, *192*, 219–226. doi:10.1126/science.769162

Eggleton, C. D., & Popel, A. S. (1998). Large deformation of red blood cell in a simple shear flow. *Physics of Fluids*, *10*(8). doi:10.1063/1.869703

Evans, E. (1992). Equilibrium wetting of surfaces by membrane-covered vesicles. *Advances in Colloid and Interface Science*, *39*, 103–128. doi:10.1016/0001-8686(92)80057-5

Evans, E., & Needham, D. (1987). Physical properties of surfactant bilayer membranes: Thermal, transition, elasticity, rigidity, cohesion, and colloidal interactions. *Journal of Physical Chemistry*, *91*, 4219–4228. doi:10.1021/j100300a003

Evans, E., & Ritchie, K. (1997). Dynamic strength of molecular adhesion bonds. *Biophysical Journal*, *72*, 1541–1555. doi:10.1016/S0006-3495(97)78802-7

Fenton, G. A. (1990). *Simulation and analysis of random fields*. PhD thesis, Princeton University.

Ghanem, R., & Spanos, P. D. (2003). *Stochastic finite elements: A spectral approach*. Dover Publications Inc.

Hammer, D. A., & Lauffenburger, D. A. (1987). A dynamical model for receptor-mediated cell adhesion to surfaces. *Biophysical Journal*, *52*, 475–487. doi:10.1016/S0006-3495(87)83236-8

Haussy, B. & Ganghoffer, J.F. (2005). Probabilistic mechanisms of adhesive contact formation and interfacial processes. *Archives of Applied Mechanics, 75*, 2006, 338-354.

Jones, A. D., Smith, C. W., & McIntire, L. (1996). Leukocyte adhesion under flow conditions: principles important in tissue engineering. *Biomaterials*, *17*, 337–347. doi:10.1016/0142-9612(96)85572-4

Kuhl, E., Garikipati, K., Arruda, E. M., & Grosh, K. (2005). Remodeling of biological tissues: Mechanically induced reorientation of a transversely isotropic chain network. *Journal of the Mechanics and Physics of Solids*, *53*, 1552–1573. doi:10.1016/j.jmps.2005.03.002

Kuhn, W., & Grün, F. (1942). Beziehungen zwischen elastischen Konstanten und Dehnungsdoppelbrechung hochelastischer Stoffe. *Kolloid-Zeitschrift*, *101*, 248–271. doi:10.1007/BF01793684

Lavalle, P., Stoltz, J. F., Senger, B., Voegel, J. C., & Schaaf, P. (1996). Red blood cell adhesion on a solid/liquid interface. *Proceedings of the National Academy of Sciences of the United States of America*, *93*, 15136–15140. doi:10.1073/pnas.93.26.15136

Lee, H. S., Gheysel, E., & Bell, W. R. (1997). *Seasonal time series and autocorrelation function estimation*. Série scientifique CIANO 97s-35, Montréal, Canada.

Lenormand, G. (2001). *Elasticité du globule rouge humain-une étude par pinces optiques*, Thèse de doctorat de l'Université Paris VI.

Lipowsky, R., & Seifert, U. (1991). Adhesion of membranes: A theoretical perspective. *Langmuir*, *7*, 1867–1873. doi:10.1021/la00057a009

Marques, C. M. (2001). Le récepteur, le ligand et sa chaine polymère, peut-on contrôler l'adhésion cellulaire. *CNRS Infos*, *396*, 21–22.

Matsumoto, M., & Nishimura, T. (1998). *Mersenne twister: A 623-dimensionally equidistributed uniform pseudorandom number generator*. ACM Transactions in Modeling and Computer Simulations.

Mefti, N. (2006). *Mise en oeuvre d'un modèle mécanique de l'adhésion cellulaire: Approche stochastique*. Thèse de Doctorat de l'INPL. Nancy, France.

Mefti, N., Haussy, B., & Ganghoffer, J. F. (2006). Mechanical modelling of the rolling phenomenon at the cell scale. *International Journal of Solids and Structures*, *43*(24), 7378–7392. doi:10.1016/j.ijsolstr.2006.05.006

Mochizuki, A. (2002). Pattern formation of the cone mosaic in the Zebrafish retina: A cell rearrangement model. *Journal of Theoretical Biology*, *215*, 345–361. doi:10.1006/jtbi.2001.2508

Mochizuki, A., Ywasha, Y., & Takeda, Y. (1996). A stochastic model for cell sorting and measuring cell-cell adhesion. *Journal of Theoretical Biology*, *179*, 129–146. doi:10.1006/jtbi.1996.0054

Naili, S., & Yasmineh, S. (2001). Un modèle de l'adhésion pour les milieux curvilignes. *Comptes Renuds de l'Academie Sciences, Paris*, *2*(2), 161–167.

Ndri, N., & Udaykumar, W. S. & Tay, R.T.S. (2001). Computational modeling of cell adhesion and movement using continuum-kinetic approach. *Proceedings of the Bioengineering Conference ASME, 50*, 367-368.

Newman, T.J. & Odell, P.L. (1971). *The generation of random variate. Griffin's statistical monograph and courses.*

Nour, A., Slimani, A., Laouami, N., & Afra, H. (2003). Finite element model for probabilistic seismic response of heterogeneous soil profile. *Soil Dynamics and Earthquake Engineering*, *23*, 331–348. doi:10.1016/S0267-7261(03)00036-8

Oliver, T., Lee, J., & Jacobson, K. (1994). Forces exerted by locomoting cells. *Seminars in Cell Biology*, *5*, 139–147. doi:10.1006/scel.1994.1018

Papadrakakis, M., & Papadopoulos, V. (1996). Robust and efficient methods for stochastic finite element analysis using Monte Carlo simulation. *Computational Methods of Applied Mechanics*, *134*, 325–340. doi:10.1016/0045-7825(95)00978-7

Peeters, E. (2004). *Biomechanics of single cells under compression*. PhD thesis, University of Endhoven.

Ramanujan, S., & Pozrikidis, C. (1998). Deformation of liquid capsules enclosed by elastic membranes in simple shear flow: large deformations and effect of fluid viscosities. *Journal of Fluid Mechanics*, *361*, 117–143. doi:10.1017/S0022112098008714

Richert, L., Engler, A. J., Discher, D. E., & Picart, C. (2004). *Surface measurement of the elasticity of native and cross linked polyelectrolyte multilayer film*. XXIXCongrès de la Société de Biomécanique. France: Créteil.

Roberts, C., Lauffenburger, D. F., & Quinn, J. A. (1990). Receptor mediated cell attachment and detachment kinetics I: Probabilistic model and analysis. *Biophysical Journal*, *58*, 841–856. doi:10.1016/S0006-3495(90)82430-9

Sagvolden, G., Giaver, I., Pettersen, E. O., & Feder, J. (1999). Cell adhesion force microscopy. *Proceedings of the National Academy of Sciences USA*, 471-476.

Shinozuka, M. (1971). Simulation of multivariate and multidimensional random process. *The Journal of the Acoustical Society of America*, *49*(1), 357–367. doi:10.1121/1.1912338

Shinozuka, M. (1972). Monte Carlo solution of structural dynamics. *Computers & Structures*, *2*, 855–874. doi:10.1016/0045-7949(72)90043-0

Shinozuka, M. (1972). Digital simulation of random processes and its applications. *Journal of Sound and Vibration*, *25*(1), 111–128. doi:10.1016/0022-460X(72)90600-1

Shinozuka, M., Deodatis, G., Zhang, R., & Papageoriou, A. R. (1999). Modeling, synthesis and engineering application of strong earthquake wave motion. *Soil Dynamics and Earthquake Engineering*, *18*, 209–228. doi:10.1016/S0267-7261(98)00045-1

Shinozuka, M., & Deotadis, G. (1991). Simulation of stochastic process by spectral representation. *Applied Mechanics Reviews*, *44*(4), 191–203. doi:10.1115/1.3119501

Shinozuka, M., & Lenoe, E. (1976). A probabilistic model for spatial distribution of material properties. *Engineering Fracture Mechanics*, *8*, 217–227. doi:10.1016/0013-7944(76)90087-4

Simon, A. (2002). *Intérêt de la microscopie à force atomique sur la biofonctionnalisation de matériaux: caractérisation du greffage et de l'adhésion cellulaire*. Thèse de doctorat, Université Bordeaux I.

Skalak, R., & Evans, E. A. (1984). *Mechanics and thermodynamics of biomembranes*. Boca Raton, FL: CRC Press Inc.

Stéphanou, A., & Tracqui, P. (2002). Cytomechanics of cell deformation and migration: From models to experiments. *Current Review of Biology*, *325*, 295–308.

Tadmor, R. (2001). The London–van der Waals interactions between objects of various geometries. *Journal of Physics Condensed Matter*, *13*, 195–202. doi:10.1088/0953-8984/13/9/101

Takano, R., Mochizuki, A., & Iwasa, Y. (2003). Possibility of tissue separation caused by cell adhesion. *Journal of Theoretical Biology*, *221*, 459–474. doi:10.1006/jtbi.2003.3193

Turner, S., & Sherrat, J. A. (2002). Intercellular adhesion and cancer invasion: a discrete simulation using the extended potts model. *Journal of Theoretical Biology*, *216*, 85–100. doi:10.1006/jtbi.2001.2522

Wache, P., Giddens, D. P., & Wang, X. (2001). *Couplage fluide-solide. Analyse 3D de l'état de contrainte d'une cellule endothéliale dans un écoulement*. 15ème Congrès Français de Mécanique, Nancy.

Walter, A., Rehage, H., & Leonhard, H. (2001). Shear induced deformation of microcapsules: Shape oscillations and membranes folding. *Colloids and Surfaces*, *183*, 123–132. doi:10.1016/S0927-7757(01)00564-7

Williams, T., & Bjerknes, R. (1972). Stochastic model for abnormal clone spread through epithelial basal layer. *Nature*, *236*, 19–21. doi:10.1038/236019a0

Xiao, Y., & Truskey, G. (1996). An effect of receptor-ligand affinity on the strength of endothelial cell adhesion. *Biophysical Journal*, *71*, 2869–2884. doi:10.1016/S0006-3495(96)79484-5

Zhao, H., Stoltz, J. F., Zhuang, F., & Wang, X. (2001). *Etude dynamique de l'interaction entre molécules d'adhésion à la surface cellulaire*. 15ème Congrès Français de Mécanique, Nancy.

Zhu, C., Bao, G., & Wang, N. (2000). Cell mechanics: Mechanical response, cell adhesion, and molecular deformation. *Annual Review of Biomedical Engineering*, *2*, 189–226. doi:10.1146/annurev.bioeng.2.1.189

KEY TERMS AND DEFINITIONS

Cell Adhesion: Cellular adhesion is the binding of a cell to a surface, extracellular matrix or another cell using cell adhesion molecules such as selectins, integrins, and cadherins.

Ligand-Receptor Connections: The ligand (*Latin* ligare = to bind) is a substance that is able to bind to and form a complex with a biomolecule to serve a biological purpose. In a narrower sense, it is a signal triggering molecule, binding to a site on a target protein. The binding occurs by intermolecular forces, such as ionic bonds, hydrogen bonds and Van der Waals forces. Biochemical receptors are large protein molecules that can be activated by the binding of a ligand (such as a hormone or drug). Receptors can be membrane-bound, occurring on the cell membrane of cells, or intracellular, such as on the nucleus or mitochondrion. Binding occurs as a result of non-covalent interaction between the receptor and its ligand, at a location called the binding site on the receptor. A receptor may contain one or more binding sites for different ligands.

Rolling: Like velcro, ligands on the circulating leukocytes bind to selectin molecules on the inner wall of the vessel. This causes the leukocytes to slow down and begin rolling along the inner surface of the vessel wall. During this rolling motion, transitory bonds are formed and broken between selectins and their ligands.

Rupture: The rupture phenomenon for a physical entity (presently one bond) is the separation in two (or more) pieces under the action of a force or a stress.

Statistical Nonlocality: From a general point of view, nonlocality is a direct influence of one object on another distant object. The spectral distribution of power used to model the spatial distribution of rupture thresholds of the ligand-receptor bonds exhibits some nonlocality, as the rupture limit of one bond may be influenced by that of the adjacent and non-adjacent bonds.

Stochastic Fields: In probability theory, a stochastic process, sometimes called a random process, is the counterpart to a deterministic process. Instead of dealing with only one possible reality of how the process might evolve over time (as is the case, for example, for solutions of an ordinary differential equation), in a stochastic or random process there is some indeterminacy in its future evolution described by probability distributions. This means that even if the initial condition (or starting point) is known, there are many possible paths the process might take, but some paths may be more probable and others less. In the simplest possible case (discrete time), a stochastic process amounts to a sequence of random variables known as a time series (for example, see Markov chain). Another basic type of a stochastic process is a random field, whose domain is a region of space, in other words, a random function whose arguments are drawn from a range of continuously changing values.

Viscoelasticity: This refers to the property of materials that exhibit both viscous and elastic characteristics when undergoing deformation. Viscous materials resist shear flow and strain linearly with time when a stress is applied. Elastic materials strain instantaneously when stretched and just as quickly return to their original state once the stress is removed. Viscoelastic materials have elements of both of these properties and, as such, exhibit time dependent strain.

Chapter 28
A Multiscale Computational Model of Chemotactic Axon Guidance

Giacomo Aletti
University of Milan, Italy

Paola Causin
University of Milan, Italy

Giovanni Naldi
University of Milan, Italy

Matteo Semplice
University of Insubria, Italy

ABSTRACT

In the development of the nervous system, the migration of neurons driven by chemotactic cues has been known since a long time to play a key role. In this mechanism, the axonal projections of neurons detect very small differences in extracellular ligand concentration across the tiny section of their distal part, the growth cone. The internal transduction of the signal performed by the growth cone leads to cytoskeleton rearrangement and biased cell motility. A mathematical model of neuron migration provides hints of the nature of this process, which is only partially known to biologists and is characterized by a complex coupling of microscopic and macroscopic phenomena. This chapter focuses on the tight connection between growth cone directional sensing as the result of the information collected by several transmembrane receptors, a microscopic phenomenon, and its motility, a macroscopic outcome. The biophysical hypothesis investigated is the role played by the biased re-localization of ligand-bound receptors on the membrane, actively convected by growing microtubules. The results of the numerical simulations quantify the positive feedback exerted by the receptor redistribution, assessing its importance in the neural guidance mechanism.

DOI: 10.4018/978-1-60960-491-2.ch028

INTRODUCTION

The ability of cells of responding to chemical signals present in the environment is of utmost importance for life, for example to recognize peers or locating food sources. Chemical cues also serve to mark pathways, which lead cells to a target (attractive cues) as well as repel them from selected regions (repulsive cues). Pathfinding by chemical cues is a key mechanism in the embryo, where sets of cells have to organize and reach specific areas to form the different body tissues. Cells crawl along the concentration gradient, towards (or away from) the direction of increasing diffusible chemical signal, moving from the peripheries to the source. This phenomenon is known as chemotaxis and its discovery dates back to the 18th century, allowed by the invention of the microscopy. An interesting example of chemotaxis is found in the developing nervous system, where axons, long and slender projections of nerve cells, find the targets they will innervate navigating in the extracellular environment through a chemotactic guidance mechanism (see, *e.g.*, Tessier-Lavigne & Goodman, 1996; Mueller, 1999; Song & Poo, 2001). Detection and transduction of navigational cues in chemotactic axon guidance is mediated by the growth cone (GC), a highly dynamic structure located at the axon tip (see, *e.g.*, Guan & Rao, 2003 and refer to Figure 1). From the microscopic point of view, directional sensing is initiated by differential binding with the extracellular ligand (the chemical cue) of the specialized receptors located on the opposite sides of the GC membrane. In order to respond to the very shallow ligand gradients observed in nature, the GC must optimize concentration measurements, overcoming the surrounding noise. Several mathematical models investigate this concept. In the seminal work of Berg & Purcell (1977), each receptor is considered as a "measuring device" which provides an estimation of the local ligand concentration based on its average time of permanence in the binding state during a certain period. In Mortimer et al. (2009b), it is shown that, if -in addition- the number of unbound-to-bound transitions is also signalled by the receptor, a more precise measure of the ligand concentration is yielded. When coming to consider the complete pool of receptors present on the GC, a strategy to weight the whole set of binding measurements should also be envisaged. In Mortimer et al. (2009a), it is shown that the optimal measuring strategy gives to each receptor a weight proportional to its distance from the geometrical center of the GC. In the present work, we propose a modification of this latter concept, introducing a weighting strategy depending on the distance of the single receptor from the center of mass of the receptor pool, a quantity dynamically varying according to the activity level of each single receptor (as will be defined more thoroughly in the following). The biophysical fact which motivates this hypothesis is the recent finding of Bouzigues et al. (2007a, 2007b) that, in presence of an attractive gradient of the diffusible cue GABA, ligand-bound GC receptors undergo two fundamental types of motion on the membrane: the first kind of motion is free diffusion, which is present even under an uniform external field, while the second kind of motion is a biased drift toward the side facing the attractive ligand source. This latter motion is driven by the physical interaction of bound receptors with the GC microtubules, which serve as conveyor belts (Saxton, 1994; Saxton & Jacobson, 1997). The overall effect of this mechanism is the establishment of an autocatalytic loop: bias in receptor localization induces, via internal polarization of molecules, preferential growth of the microtubules toward the leading edge of the GC and this, in turn, enhances convey of receptors on that same side (Bouzigues et al., 2007b). Once a weighting strategy for the receptor measurements is established, one should model the subsequent internal polarization chain leading to motion. Mathematical models in this context most often do not enter into the details of the extremely complex biochemical signalling cascade, but rather adopt phenomenological simplified descriptions that provide a "black box" information of the functional behaviour of the system. A first class of approaches (see, *e.g.*, Buettner et al., 1994; Maskery & Shinbrot, 2005) is based on persistent random walk models. The GC trajectory is typically described by a system of ordinary differential equations accounting for a deterministic velocity field and random "kicks" arising from stochastic terms, macroscopically representing fluctuations in gradient sensing and signal transduction. Evolutions of these models are presented in Hentschel & van Ooyen (2000) and further in Aletti & Causin (2008), where the GC trajectory is described by more sophisticated stochastic partial differential systems of equations, including diffusion and inertia contributions. A second class of models are investigated in Aeschlimann & Tettoni (2001); Goodhill & Urbach (1999); Goodhill et al. (2004); Xu et al. (2005), where there is the attempt of introducing a description of the intracellular chain. Namely, the probability of finding a transmembrane receptor at a certain angular position on the GC is supposed to be linked to some significant intracellular parameter, for example the local concentration of ionic calcium in Aeschlimann & Tettoni (2001). We also refer to the mathematical models presented in a series of works by some of the Authors of this Chapter: in particular, we refer to Aletti et al. (2008a), where a novel modelling of ligand-receptor binding was proposed, introducing Markov chains to describe the state of receptors; to Causin & Facchetti (2009), where a detailed analysis of the internal polarization chain triggered by the receptor redistribution was carried out, performing an analysis of the amplification steps occurring in the chain; to Aletti & Causin (2008b), where the random walk models were used to study axon trajectories and to infer the internal organization of the GC.

Figure 1. Schematic representation of a neural cell, with its long slender projection, the axon. At the axon tip, it is located the growth cone, an ameboid structure which extends and retracts seeking out external signals.

BIOLOGICAL BACKGROUND

The GC and the Chemotactic Assay

The GC moves forward through the extracellular matrix, as a consequence both of being pushed and of being pulled. The pushing effect stems from the synthesis of microtubules and by the arrival at the GC of material transported along the axon from the cell body (see Figure 1), pulling stems by the GC own membrane extensions, the filopodia. The filopodia extend in various directions from the GC membrane and adhere to the surrounding substratum. Within each filopodium, actin microfilaments are synthesised. As the microfilaments contract, the filopodia also contract, pulling the rest of the GC along.

In the absence of instructive cues, GCs progress along a relatively straight path. Observe that the rate of advance of the GC is about 10–40 μm per hour (Tessier-Lavigne & Goodman, 1996), which is significantly less than the typical velocity observed in eukaryotic cell movements (Mortimer et al., 2009b). In the presence of an external guidance cue gradient, the direction of movement is biased by information transduced through directional sensing. This latter is the mechanism by which the asymmetry in the external signal is detected and transformed into an intracellular polarization, leading to a cytoskeletal reorganization and substrate adhesion (the so-called motility, see Mortimer et al., 2008), capable of generating a biased movement. In order to gain insights into the complex phenomenon of axon guidance, a chemotactic assay is performed in *in vitro* experiments. This assay studies the response of GCs exposed to steady graded concentrations of a single attractive/repulsive ligand released by a pipette (Zheng et al., 1994; Ming et al., 1997; Rosoff et al., 2004). Axon turning angles are measured after a certain time interval, usually 1 h, from the onset of the gradient (see Figure 2), repeating on several different axons the same test. The gradient assay models a well defined situation, simple but already sufficient to gain an insight into the axon guidance mechanism. This kind of experiment, as described in the above cited references, is a 2D framework: axons move on a flat substrate and the cue they are attracted by is also established on a 2D plane; the real *in vivo* situation is of course a 3D framework, but dramatically few quantitative data are available in this case, which is, moreover, a too rich and complex environment to be mathematically studied, at the present level of knowledge.

Figure 2. Chemotactic assay in axon guidance: the pipette, located in the right corner on the top of the figures, establishes a graded field of a chemoattractant. The axon moves towards it, in the direction of the increasing gradient. An indicative time is indicated in each panel. Experiments record the final turning angle γ for several axon trajectories, as depicted in the rightmost panel of the bottom row.

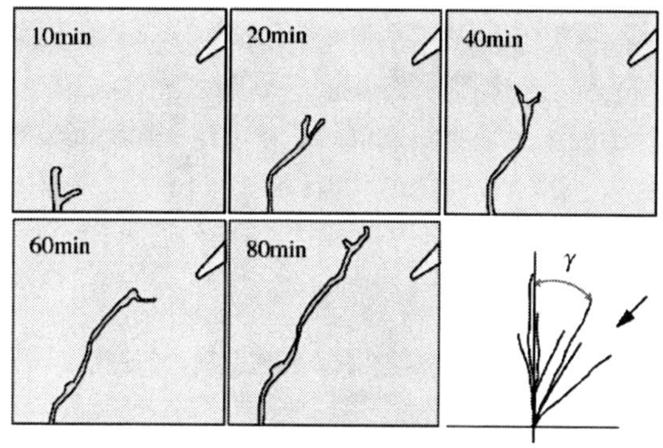

Experimental Techniques

We briefly review here the experimental techniques which produced the biological evidences at the basis of our modeling assumptions. We refer the interested Reader to the cited references for a more detailed description. Bouzigues & Dahan (2007a); Bouzigues et al. (2007b) combined a GC chemotactic assay with single Quantum Dot (QD) imaging techniques (Bouzigues et al., 2007c; Courty et al., 2006) to study the lateral dynamics of individual $GABA_A$ receptors when placed in a GABA gradient. They labeled the γ_2 subunits, known to be present in functional receptors, with biotinylated antibodies and streptavidin-coated QDs such that 10–30 individual QDs (identified by their fluorescence intermittency) could be detected simultaneously in a GC. No internalization of the tagged receptors was observed. After 30 min upon application of the external gradient, cells were rapidly fixed and immunostained with a primary antibody, followed by fluorescent secondary antibodies. The ratio of the fluorescence signal between the proximal and distal region was 1.9±0.1, significantly larger than the value in control conditions. This result unambiguously demonstrated that the redistribution observed with QDs was not an artifact of the experimental approach and reflected a physiological situation. Then, to investigate the mechanisms underlying the asymmetric redistribution of receptors, they first measured the lateral diffusion of neural cell adhesion molecule (N-CAM), an unrelated membrane-bound protein, and found that, in a GABA gradient, N-CAM molecules remained symmetrically localized. This ruled out that receptors redistribution resulted from of a membrane flow and argued in favor of an active and specific transport mechanism. Finally, pharmacological treatments provided additional information on the mechanisms controlling the distribution receptors. They tested the role of microtubules, which are crucial in axon growth and in directional. They found that the asymmetric redistribution of the receptors was suppressed when microtubules were depolymerized, thus assessing the role of this feedback mechanism.

MATHEMATICAL BACKGROUND

In the following, we briefly address two mathematical topics, which knowledge is important for the understanding of the present mathematical model. We refer the interested Reader to the classical textbooks by Norris (1998), Revuz & Yor (1999) and Borodin & Salminen (2002) for a more comprehensive exposition.

Brownian Motion

Brownian motion is the temperature-dependent perpetual, irregular motion of the particles immersed in a fluid, caused by their continuous bombardment by the surrounding molecules of much smaller size. The atoms or molecules that make up the fluid are in constant thermal motion, and their velocity distribution is determined by the temperature of the system. The impact of the fluid tiny molecules makes the particles move. The net effect is an erratic, random motion of the particle through the fluid. The mathematical model of the Brownian motion is a so-called Wiener Process, a time-continuous stochastic process with independent, stationary, Gaussian distributed increments with zero mean and variance equal to the length of the increment's time-interval. Formally, a stochastic process $\{W_t, t \geq 0\}$ is a Wiener process if it satisfies:

- *Property 1:* $W_0=0$;
- *Property 2:* W_t-W_s is independent on W_r-W_q, for any $0 \leq q \leq r \leq s \leq t$;
- *Property 3:* W_t is distributed as W_{t-s};
- *Property 4:* W_t is distributed as a Gaussian variable of mean 0 and variance *t*.
- *Property 5:* Brownian motion can be represented by continuous paths

The time discretization of a Wiener process, needed for computer simulation, is a random walk $\{W_i, i=0,1,2,\ldots\}$. To introduce time discretization, we establish a grid of time steps $0=t_0 \leq t_1 \leq \ldots \leq t_n$, such that $t_i-t_j=\Delta t$, $i,j=0,1,2,\ldots$ and start with $W_0=0$ (Property 1). Then, Property 2 says that the increment $W_{t0}-W_{t1}$ is a normal random variable and Property 4 says that its mean and variance are 0 and t_1, respectively. Therefore, a realization of the random variable $W_{t1}= Wt_0 +Y_1= Y_1$ can be made by choosing for Y a sample from the normal distribution $N(0\,t_1)=\Delta t^{0.5}N(0,1)$, in other words by multiplying a standard normal random number by $\Delta t^{0.5}$, and so on. The sequence of increments Y_1,Y_2, … is a sequence of independent and identically distributed random increments. The random walk at step n is just the sequence $W_n=W_0+ Y_1+Y_2+\ldots+Y_n$ of accumulated steps. In Matlab™, we can write an approximation to the Brownian motion by using the built-in normal random number generator randn. For example, using n time steps of length deltat, we have the following implementation

```
sqdel=sqrt(deltat);
W(1)=0;
for i=1:n
Y=sqdel*randn;
W(i+1)=W(i)+Y;
end
```

Markov Chains

Let us consider a particle that jumps from one of M possible states to another one at each time step $i=1,...,n$. The mathematical process that describes this phenomenon is called Markov chain if, at any time t_i, the probability of moving from the k-th state to the l-th one only depends on the k–th state and not on the whole history that had moved the particle till state k. Roughly speaking, a process is called Markovian if its futures does only depend on its present status and not on the past. When the Markov chain is stationary, the probability of moving from one state (state k) to another (state l) is denoted by p_{kl} and is constant in time. Some of the transitions are impossible, and hence they have probabilities equal to zero. The M by M matrix of all such probabilities is called a transition matrix, and is denoted by T. It has nonnegative entries, and its rows sum up to one.

When a time-continuous model is used, the transition matrix can be characterized by the corresponding intensity matrix Q. In fact, the probability of being in the state l at time $s+t$ conditioned on being in the state k at time s, is given by the entry (k,l) of the matrix $T_t = exp(Qt)$. The rows of Q sum up to zero. All the non-diagonal elements q_{kl} are nonnegative and are proportional to the probability of reaching each of the states l, once we leave the state k. The diagonal negative elements of Q are related to the mean occupancy time τ_k of each state k by the relation $q_{kk}=-1/\tau_k$. In order to perform a computer simulation, we discretize the time interval, as already described above. Starting from the intensity matrix Q, a realization of the Markov process $X_{t+\Delta t}$ conditioned on being in the state k at time t, can be made in Matlab™ by using the built-in functions expm (to get from Q the transition probability matrix T), cumsum (to compute the cumulative function of $X_{t+\Delta t}$ given $X_t=k$) and rand to simulate the new state.

For example, using n time steps of length deltat and starting from the first state (k=1), we have the following implementation

```
T=expm(Q*deltat); % transition matrix
FT=cumsum(T,2);   % compute cumulative
                   functions on each row
X(1)=1;           % starting state
for i=1:n
     Fpk=FT(X(i),:);
X(i+1)=find(rand>[0,Fpk],1,'last');
end
```

METHOD: THE MATHEMATICAL MODEL

Preliminaries

Motivated by the above considerations on the biological problem, in our mathematical model, we assume that the GC is a 2D disk of radius R, $O = O(t)$ being the position of its center in a fixed cartesian system $x = (x_1,x_2)$. We also introduce a second coordinate system $X = (X_1,X_2)$ moving along with the center of the GC (see Figure 3). On the GC surface, we distribute N receptors $\omega_1,\omega_2,...,\omega_N$, whose position is conveniently identified with respect to the local X system. The trajectory on the membrane of each receptor is modelled as a stochastic process $X_t^{(i)}= X_t(\omega_i)$, $i = 1,...,N$, with the constraint with the constraint that the receptor must restrain its trajectory inside the boundaries of the GC, that is $||X_t^{(i)}||^2 \leq R^2 \forall t \geq 0$. We endow each receptor of an equivalent mass $m^{(i)}$, normalized such that $\sum_{i=1}^{N} m^{(i)} = 1$, which represents the weight of the receptor to the GC turning decision, as will be detailed in the following. The ligand concentration is given in each point by the function $C(x)=C_0\exp(\alpha(x_1\sin\varphi + x_2\cos\varphi))$, C_0 being the concentration at the origin, φ the angle of the ligand gradient with respect to the x axis and α the gradient steepness (see Figure 3). In the rest of our discussion, we denote by $I_S(z)$ the indicator function of argument z on the set S, (*i.e.*, the function which is =1 if z is contained in

S and =0 elsewhere), by I_d the identity matrix and by $\langle \cdot,\cdot \rangle$ the standard scalar product in $\mathfrak{R}^2$.

Model of Receptor Dynamics

The probability of a receptor to be bound with a ligand particle is given by the Michaelis-Menten-type law

$$p_b(x) = \frac{C(x)}{C(x) + K_d}, \tag{1}$$

where K_d is the ligand dissociation constant and where we have highlighted the dependence on the spatial position of the concentration. Observe that local coordinates can be used in the above equation, upon the coordinate change $X=X(x)$. Receptors can undergo two types of motion:

- a free Brownian motion, so that the receptor position is governed by the following stochastic diffusion

$$dX_t = D_c\, dW_t, \tag{2}$$

where D_c is the diffusion coefficient and W_t a 2D Wiener process;

- a convected motion, so that the receptor movement is governed by the law

$$dX_t = v_c\, e_X\, dt, \tag{3}$$

where e_X is the unit vector of the direction of motion and v_c the drift velocity modulus connected to microtubule transport. Brownian diffusion, in principle, exists even when motion of type (3) is activated, but in this case its contribution can be considered as negligible (Bouzigues 2007a, 2007b).

Each receptor moves on the membrane following one of the above laws, according to its binding status. We identify three receptor binding states:

Figure 3. Mathematical representation of the growth cone and notation for the coordinate systems and concentration field. The blue arrow indicates the direction of the increasing cue concentration.

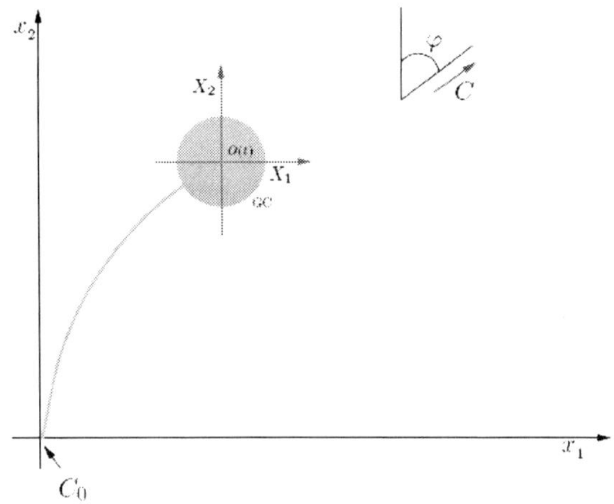

- the receptor is free, unbound both from the ligand and from the cytoskeleton (*f*). Its movement is governed by the law (2);
- the receptor is bound to the ligand but not bound to the cytoskeleton (*l*). Again, its movement is governed by the law (2);
- the receptor is bound to the ligand and convected by the cytoskeleton (*c*). In this case, its movement is governed by the law (3).

The transition process from one state to another is modelled as a Markov chain M_t, whose intensity matrix Q is built relying on the following hypotheses:

- when the receptor is in the state *f*, it can jump only to the bounded state *l* after an exponential random variable T_f of rate k_f, *i.e.* the probability density of the random binding time T_f is $f_{T_f}(t) = k_f \exp(-k_f t)$ and thus the Q_{fl} entry equals k_f; moreover, since the receptor cannot jump directly to the state c from the state f, we set the Q_{fc} entry equal to 0;

- when the receptor is in the state l, it can jump to both the states f and c. We set $Q_{lf} = k_c$ (inverse of the free diffusion mean time) and $Q_{lc} = k_l$;
- when the receptor is in the state c, it can jump only to the state f after an exponential random variable of rate k_u (inverse of the drift motion duration time), and hence the density of the random time T_u the receptor is bound to the ligand and convected by the cytoskeleton is $f_{T_u}(t) = k_u \exp(-k_u t)$. Accordingly, we set $Q_{cl} = 0$ and $Q_{cf} = k_u$.

Summing up, we obtain

$$Q = \begin{pmatrix} -k_f & k_f & 0 \\ k_c & -k_c - k_l & k_l \\ k_u & 0 & -k_u \end{pmatrix}, \tag{4}$$

where the rows refer to the states f, l and c, respectively. We model the state of each receptor as an independent realization M_t of the above Markov process. We observe that once the receptor is bound, the average binding time τ_b does not depend on the concentration field, that is on the spatial position, but only on the intrinsic biophysical characteristics of the receptors (for this point, see also Mortimer et al., 2009b). Conversely, the mean time τ_f a receptor remains free will depend on the concentration field. Moreover, following Berg & Purcell (1977), using in (4) the approximation that the binding probability is the ratio between the binding time over the total time, that is $p_b \cong \frac{\tau_b}{\tau_b + \tau_f}$, we straightforwardly obtain the average unbound main time τ_f between two successive binding events as $\tau_f(x) = \frac{\tau_b(1-p_b)}{p_b} = \tau_b K_d (1 + \frac{K_d}{C(x)})$. It is straightforward to compute $\tau_f = 1/k_f$, while the quantity $\tau_f = \frac{1}{k_l + k_c} + \frac{1}{k_u}\frac{k_l}{k_l + k_c} = \frac{k_l + k_u}{k_u(k_l + k_c)}$ is obtained by noting that, once the state l is reached, the receptor spends an average time of $1/(k_l+k_c)$ in that state and then an additional time of mean $1/k_u$ if it jumps to the state f, event of probability $k_l/(k_l+k_c)$.

The position on the membrane of each receptor is then described by the system

$$\begin{cases} dX_t = (\chi_1(t,X_t)v_c e_X(t) - v_{GC}e(t))dt + \chi_2(t,X_t)D_c I_d dW_t, \\ \chi_1(t,X_t) = I_{[0,\Re^2)}\left(\|X_t\|^2\right) I_c(M_t) - I_{[\Re^2,+\infty)}\left(\|X_t\|^2\right), \\ \chi_2(t,X_t) = \left(I_l(M_t) + I_f(M_t)\right) I_{[0,\Re^2)}\left(\|X_t\|^2\right), \end{cases} \tag{5}$$

obtained by appropriately combining (2) and (3) through the indicator functions, and where $e_X(t)$ is the unit vector of the microtubule velocity, expressed in the X coordinate system, $e(t)$ is the unit vector of the GC growth velocity v_{CG} in the x coordinate system, and the functions $\chi_1(t,X_t)$ and $\chi_2(t,X_t)$ are switches for the state of the receptor at time t. Observe that system (5) and the Markov process (4) are mutually coupled through the dependence of the binding probability on the spatial position.

Model of GC Dynamics

We suppose that the microtubules grow as a finger-shaped structure, symmetrically dispersed with respect to the direction of GC motion. Moreover, we suppose that the microtubules are densely distributed on the GC disk, so that in every point a receptor is located, it may bind to a microtubule. If $e(t)$ represents the macroscopic direction of the GC and we assume it as the symmetry axis of the fan, we model the local direction of the microtubule $e_X(t)$ to depend on position X by (refer to Figure 4)

$$e_X(t) = \frac{e(t) + \frac{k_e}{R}(X - \langle X, e(t)\rangle e(t))}{\left\| e(t) + \frac{k_e}{R}(X - \langle X, e(t)\rangle e(t)) \right\|}, \tag{6}$$

where $k_e \in [0,1]$ is a parameter which modulates the amplitude of the fan, to be calibrated, which modulates the amplitude of the fan. Observe that in the above relation we are supposing that direction e_X is a linear combination of the GC direction $e(t)$ and of the perpendicular component of the position vector X with respect to $e(t)$ itself. Normalization is necessary in order to obtain a unit vector.

We now have to relate the GC trajectory $e(t)$, which is a macroscopic parameter, to the microscopic movement of the receptors described above. We proceed as follows. First, we introduce the weighted position of the mass center of the receptors $b(t)$

$$b(t) = \sum_{i=1}^{N} m^{(i)} X_t^{(i)} \quad \forall t \geq 0. \tag{7}$$

At time $t+dt$, the GC will move along the new unit vector $e(t+dt)$, obtained as the weighted sum

Figure 4. Notation for the geometrical quantities appearing in Equation(6)

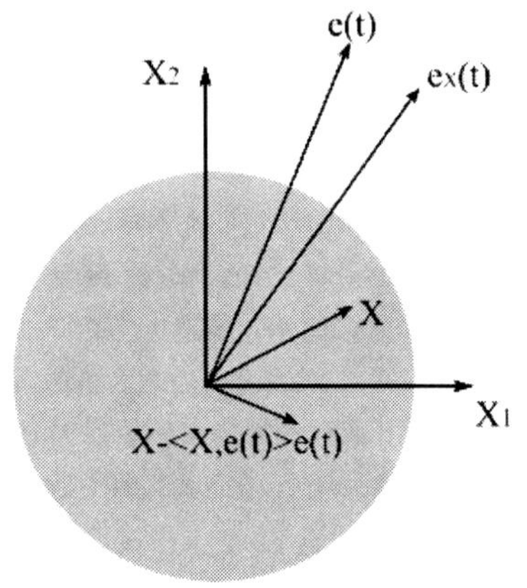

$$e(t + dt) = ((1 - \lambda_1)e(t) + \lambda_1 b^{\perp}(t) / v_{GC})\, dt, \tag{8}$$

where $\lambda_1 \in [0,1]$ is a proportionality constant (modeling GC inertia), and $b^{\perp}(t)$ the component of $b(t)$ perpendicular to $e(t)$ which is assumed to be the "pull" which deviates the GC trajectory (observe that in our model no variations take place along the direction parallel to the trajectory). The mass $m^{(i)}$ of the i-th receptor is chosen to vary according to the activity of each receptor, so that it becomes bigger as the receptor is more "directionally active", that is, as longer as it remains in the c state.

Table 1. Definition and value of physical parameters in the mathematical model

Parameter	Definition	Value	Ref.
D_c	diffusion coefficient	0.22 μm² s⁻¹	Bouzigues et al. (2007a)
k_c	receptor-cytoskeleton binding rate	0.25 s⁻¹	Bouzigues et al. (2007a)
K_d	ligand dissociation constant	0.1 nM	Goodhill (1997)
k_u	receptor-cytoskeleton unbinding rate	2.375 s⁻¹	Bouzigues et al. (2007a)
N	receptor number	100	Goodhill (1997)
R	GC radius	10 μm	Goodhill (1997)
τ_b	ligand-receptor binding time	1 s	Berg and Purcell (1977)
v_c	microtubule velocity	0.29 μm s⁻¹	Bouzigues et al. (2007a)
v_{GC}	GC velocity modulus	20 μm h⁻¹	Zheng et al. (1994)

The activity of the receptor ω_i is mathematically described by the function

$$L^{(i)}(t) = L(X_t(\omega_i)) = I_c(M_t^{(i)}), \tag{9}$$

while the total receptor activity $\underline{L}(t)$ is the weighted sum

$$\underline{L}(t) = \sum_{i=1}^{N} m^{(i)} L^{(i)}(t). \tag{10}$$

Eventually, the mass of each receptor evolves according to the difference between the actual activity level of the receptor and the average (total) activity level

$$dm^{(i)}(t) = k_m\, m^{(i)}(t)\, (L^{(i)}(t) - \underline{L}(t))\, dt, \tag{11}$$

where $k_m \in [0,1]$ is a constant (stiffness) which modulates the "mass transfer rate" and the weights $m^{(i)}$ appear here to ensure mass conservation over the receptor pool.

Figure 5. Time evolution of the distribution of the receptors under a gradient different angles ($\varphi = 45°$ top left, $\varphi = 90°$ top right, $\varphi = 135°$ bottom). Receptors are represented as dots with area proportional to their equivalent mass at the specified time. The blue point indicates the position of the GC center of mass, the triangle the position of the source.

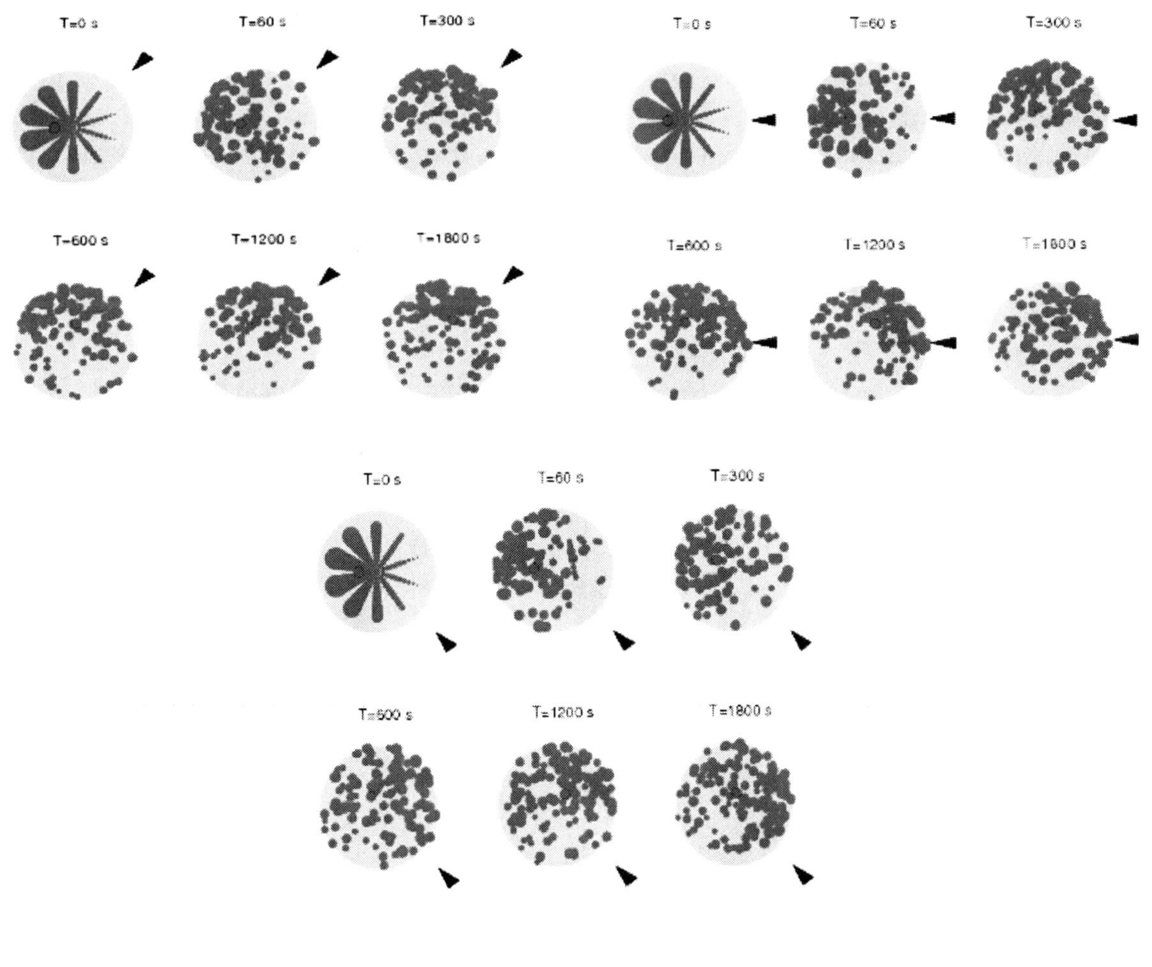

RESULTS

Simulation Parameters and Numerical Algorithm

In all the computer simulations, we have set the initial position of the GC at the origin of the fixed coordinate system, the initial direction of the GC at $e(0)=(0,1)^T$, and we have initialized the Markov chain at a random state. As for the ligand concentration field, we have considered $\alpha=0.1$ (a typical value found *in vivo*), and $C_0=K_d$ and φ as specified in each test. We have set the other physical parameters as in Table 1, except for the parameters λ_1, k_m and k_e, which are internal to the model and which choice is discussed in the following. In order to reduce the computational time of each simulation, we consider the temporal multiscale nature of the phenomenon we are studying. Namely, while the receptors undergo rapid drift/diffusion processes on the membrane, the GC as a global structure has a rather slow temporal dynamics. For this reason, the position $X_t^{(i)}$ of the single receptor is updated every 1/30 s by solving Equation (5), coupled with the evolution of the Markov chain governed by (4), keeping the GC fixed, while the GC trajectory, given by Equation(8), is updated at macro-steps of 10 s, according to the receptor distribution attained at the beginning of each macro-step. The computer code is written in Matlab™; the computational time to simulate on a standard Pentium PC a GC trajectory is about 40 s for T=4 min and 4 min for T=1 h.

Test 1: Receptor Redistribution in a Pausing GC

As a first study, we simulate the redistribution of the receptors. In this simulation, the GC does not alter its trajectory in response to the receptor asymmetry. This situation is obtained by setting $\lambda_1=0$ in Equation (8) and conceptually corresponds to the GC "pausing-state" considered in Bouzigues et al. (2007a). Figure 5 represents the time evolution of the position of the receptors, depicted as dots with area proportional to their equivalent mass. The initial distribution of receptors is asymmetrical, favoring the left (thus opposite) side of the GC. The three panels in differ for the inclination of the ligand gradient, indicated by the black triangle (φ = 45°, 90°, 135°, respectively). The blue point indicates the position of the GC mass center. In any case, after a few minutes of simulated exposure to the cue, the model achieves a receptor distribution, which reflects the actual

Figure 6. Simulation of the GC trajectories obtained over 50 runs for 1h of exposition to an attractive cue

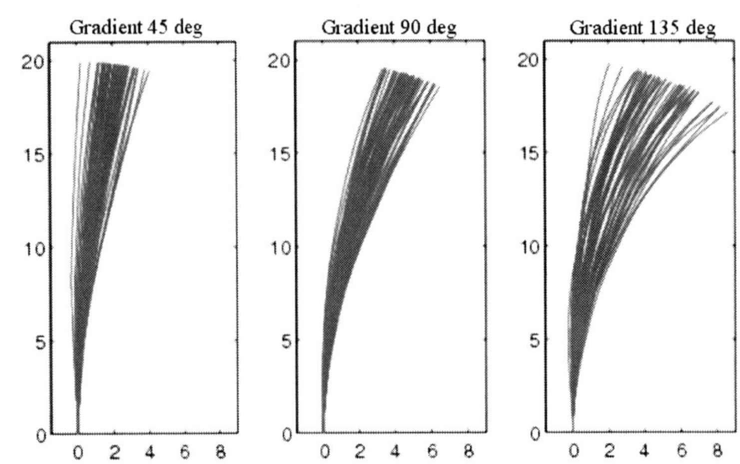

gradient orientation. The parameters k_m and k_e are calibrated in order to fit the experimental results of Bouzigues et al. (2007a), so that after 10 min the ratio between the number of receptors on the GC side facing the source and on the side opposing it is about 5.

Test 2: GC Trajectories

We study here the GC trajectory under the exposition to an attractive cue for 1h. We have run simulations with different values of λ_1: the value that best fits the experimental observations is a very low one, corresponding to a contribution of the receptors of the sole 0.5% in the update of the GC barycenter position (see also the results in Goodhill et al., 2004). Figure 6 represents the geometrical trajectories obtained over 50 runs. These results are to be compared with the experimental assays in Zheng et al. (1994).

In Figure 7, we represent the turning angles of the GCs as a function of time, computed as the angle between the vertical direction and the final position of the GC. The results are obtained from the above-mentioned 50 runs, with $\varphi = 45°$, $90°$, $135°$, respectively. The red central line represents

Figure 7. Turning angles as a function of time for GC exposed to a gradient with $\varphi = 45°, 90°, 135°$ (panels from top left to bottom). The red central line represents the average turning angle, while the top and bottom black lines are the standard deviations.

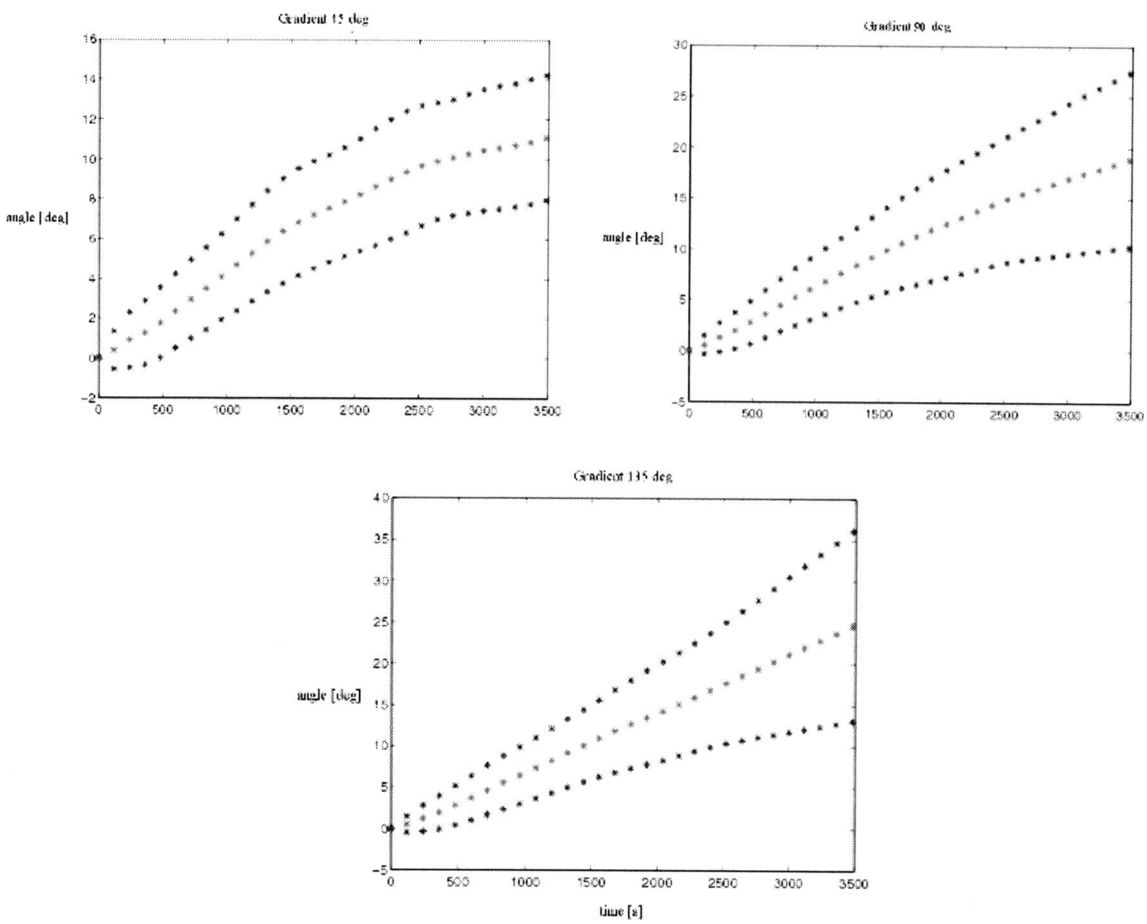

Figure 8. Axon trajectories obtained over 50 run for ligand gradient angle +φ (blue) and -φ (red)

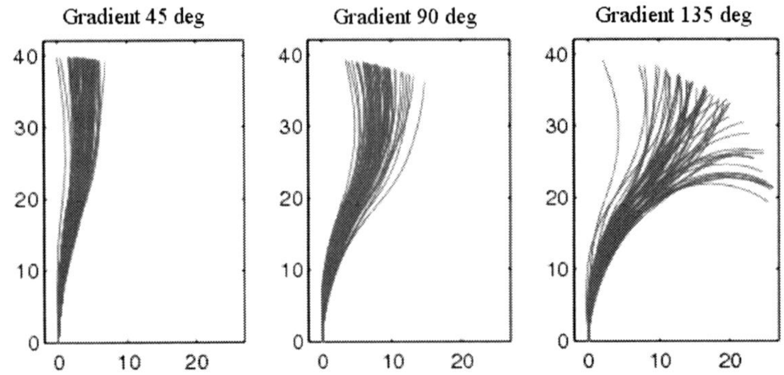

Figure 9. Turning angles as a function of time for GC exposed to a gradient with +φ for 1h and then with -φ for another 1h (panels for left to right refer to φ=45°, 90°, 135°, respectively). The red central line represents the average turning angle, while the top and bottom black lines are the standard deviations.

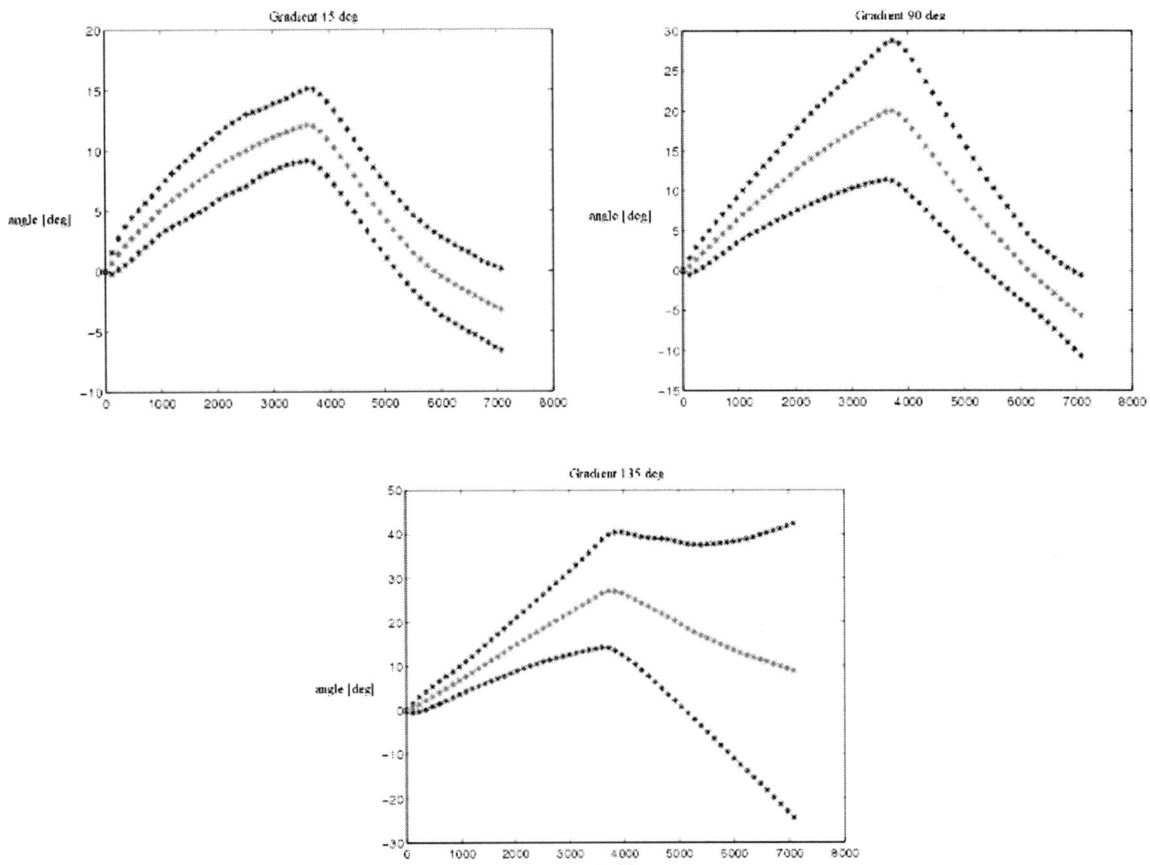

the average turning angle, while the top and bottom black lines are the standard deviations.

Test 3: GC Trajectories under a Time-Dependent Ligand Gradient

An interesting situation arises when an axon is exposed first to an attractive cue and then to a repulsive one. This happens *in vivo*, for example, when axons project to the central midline of the nervous system and once they have reached it, they are repulsed from the line itself (Kidd et al., 1999). We have numerically modeled this situation by running tests where the axon is exposed for 1h to an attractive gradient with angle $+\varphi$ and then for another 1h with angle $-\varphi$.

In Figure 8, we represent the trajectories obtained over 50 runs, for $\varphi = 45°$, $90°$ and $135°$, respectively. The blue part of the trajectories correspond to the angle $+\varphi$, the red part to $-\varphi$. Observe as the mathematical model correctly reproduces the "reversibility" of the response, that is, the flexibility of the GC to reorganize itself to react to a dynamically varying environment.

In Figure 9 we plot for the above runs the average turning angle (red lines) and the standard (black lines).

DISCUSSION AND CONCLUSION

We have proposed a mathematical model to study the chemotactic motion triggered by the exposure of GCs to a graded diffusible cue, taking as a paradigm the *in vitro* chemotactic assay. The key problem we focus on is the strategy adopted by the GC to process signals coming from the entire pool of receptors on its membrane and then its outcome on the macroscopic trajectory. These two phenomena take place at different scales, the former being microscopic and rapid, the latter macroscopic and slow.

The biological hypothesis at the basis of the microscopic model is that receptors undergo cycles of unbinding and binding states. When receptors are in this condition, they further alternate two states, one characterized by random motion, and the other by a biased convected drift. Our hypothesis is that the information about the cue gradient direction is weighted according to the time spent by receptors in the convected motion. This mechanism allows for achieving a step of signal amplification in presence of the very shallow gradients encountered in nature, under which a unbiased receptor distribution would result in a binding probability on the two sides of the GC which is practically equal. Moreover, we assess the fact that spatial bias in receptors at the microscale is a key precursor event for the macroscale chemotactic response, so that signal detection and downstream cytoskeleton dynamics cannot be decoupled. Our main contribution is to quantitatively characterize via the mathematical model the consistency of such a connection and to show how receptor redistribution is a flexible mechanism for responding to the external environment.

REFERENCES

Aeschlimann, M., & Tettoni, L. (2001). Biophysical model of axonal pathfinding. *Neurocomputing, 38-40*, 87–92. doi:10.1016/S0925-2312(01)00539-2

Aletti, G., & Causin, P. (2008b). Mathematical characterization of the transduction chain in growth cone pathfinding. *IET Systems Biology, 2*(3), 150–161. doi:10.1049/iet-syb:20070059

Aletti, G., Causin, P., & Naldi, G. (2008a). A model for axon guidance: Sensing, transduction, and. ovement. In *Collective Dynamics: Topics on Competition and Cooperation in the Biosciences: A Selection of Papers in the Proceedings of the BIOCOMP2007 International Conference*, AIP Conference Proceedings: Vol. 1028

Berg, H. C., & Purcell, E. M. (1977). Physics of chemoreception. *Biophysical Journal*, *20*(2), 193–219. doi:10.1016/S0006-3495(77)85544-6

Borodin, A. N., & Salminen, P. (2002). *Handbook of Brownian motion-facts and formulae* (2nd ed.). Basel: Birkhäuser Verlag.

Bouzigues, C., & Dahan, M. (2007a). Transient directed motions of $GABA_A$ receptors in growth cones detected by a speed correlation index. *Biophysical Journal*, *92*(2), 654–660. doi:10.1529/biophysj.106.094524

Bouzigues, C., Lévi, S., Triller, A., & Dahan, M. (2007c). Single quantum dot tracking of membrane receptors. *Methods in Molecular Biology (Clifton, N.J.)*, *374*, 81–91.

Bouzigues, C., Morel, M., Triller, A., & Dahan, M. (2007b). Asymmetric redistribution of GABA receptors during GABA gradient sensing by nerve growth cones analyzed by single quantum dot imaging. *Proceedings of the National Academy of Sciences of the United States of America*, *104*(11), 251–256.

Buettner, H. M., Pittman, R. N., & Ivins, J. (1994). A model of neurite extension across regions of nonpermissive substrate: Simulations based on experimental measurements of growth cone motility and filopodial dynamics. *Developmental Biology*, *163*(2), 407–422. doi:10.1006/dbio.1994.1158

Causin, P., & Facchetti, G. (2009). Amplification and polarization in chemotaxis: Addressing the specificity of neural cells via mathematical modelling and numerical simulation. *PLoS Computational Biology*, *5*(8), e1000479.

Courty, S., Bouzigues, C., Luccardini, C., Ehrensperger, M. V., Bonneau, S., & Dahan, M. (2006). Tracking individual proteins in living cells using single quantum dot imaging. *Methods in Enzymology*, *414*, 211–228. doi:10.1016/S0076-6879(06)14012-4

Goodhill, G. J. (1997). Diffusion in axon guidance. *European Journal of Neurology*, *9*(7), 1414–1421.

Goodhill, G. J., Gu, M., & Urbach, J. S. (2004). Predicting axonal response to molecular gradients with a computational model of filopodial dynamics. *Neural Computation*, *16*(11), 2221–2243. doi:10.1162/089976604194 1934

Goodhill, G. J., & Urbach, J. S. (1999). Theoretical analysis of gradient detection by growth cones. *Journal of Neurobiology*, *41*(2), 230–241. doi:10.1002/(SICI)1097-4695(19991105)41:2<230::AID-NEU6>3.0.CO;2-9

Guan, K. L., & Rao, Y. (2003). Signalling mechanisms mediating neuronal responses to guidance cues. *Nature Reviews. Neuroscience*, *4*(12), 941–956. doi:10.1038/nrn1254

Hentschel, H. G. E., & van Ooyen, A. (2000). Dynamic mechanisms for bundling and guidance during neural network formation. *Physica A*, *288*(1-4), 369–379. doi:10.1016/S0378-4371(00)00434-9

Kidd, T., Bland, K., & Goodman, C. (1999). Slit is the midline repellent for the Robo receptor in Drosophila. *Cell*, *96*(6), 785–594. doi:10.1016/S0092-8674(00)80589-9

Maskery, S. M., & Shinbrot, T. (2005). Deterministic and stochastic elements of axonal guidance. *Annual Review of Biomedical Engineering*, *7*, 187–221. doi:10.1146/annurev.bioeng.7.060804.100446

Ming, G., Song, H. J., Berninger, B., Holt, C., & Tessier-Lavigne, M. (1997). cAMP-dependent growth cone guidance by netrin-1. *Neuron*, *19*(6), 1225–1235. doi:10.1016/S0896-6273(00)80414-6

Mortimer, D., Dayan, P., Burrage, K., & Goodhill, G. J. (2009b). Optimizing chemotaxis by measuring unbound-bound transitions. *Physica D. Nonlinear Phenomena*, *239*(9), 477–484. doi:10.1016/j.physd.2009.09.009

Mortimer, D., Feldner, J., Vaughan, T., Vetter, I., Pujic, Z., & Rosoff, W. J. (2009a). A Bayesian model predicts the response of axons to molecular gradients. *Proceedings of the National Academy of Sciences of the United States of America, 106*(25), 10296–10301. doi:10.1073/pnas.0900715106

Mortimer, D., Fothergill, T., Pujic, Z., Richard, L. J., & Goodhill, G. J. (2008). Growth cone chemotaxis. *Trends in Neurosciences, 31*(2), 90–98. doi:10.1016/j.tins.2007.11.008

Mueller, B. (1999). Growth cone guidance: First steps towards a deeper understanding. *Annual Review of Neuroscience, 22*, 351–601. doi:10.1146/annurev.neuro.22.1.351

Norris, J. R. (1998). *Markov chains*. Cambridge, UK: Cambridge University Press.

Revuz, D., & Yor, M. (1999). *Continuous martingales and Brownian motion* (3rd ed.). Berlin: Springer-Verlag.

Rosoff, W. J., Urbach, J. S., Esrick, M. A., McAllister, R. G., Richards, L. J., & Goodhill, G. J. (2004). A new chemotaxis assay shows the extreme sensitivity of axons to molecular gradients. *Nature Neuroscience, 7*(6), 678–682. doi:10.1038/nn1259

Saxton, M. J. (1994). Single-particle tracking: Models of directed transport. *Biophysical Journal, 67*(5), 2110–2119. doi:10.1016/S0006-3495(94)80694-0

Saxton, M. J., & Jacobson, K. (1997). Single-particle tracking: Application to membrane dynamics. *Annual Review of Biophysics and Biomolecular Structure, 26*, 373–399. doi:10.1146/annurev.biophys.26.1.373

Song, H. J., & Poo, M. M. (2001). The cell biology of neuronal navigation. *Nature Cell Biology, 3*(3), E81–E88. doi:10.1038/35060164

Tessier-Lavigne, M., & Goodman, C. (1996). The molecular biology of axon guidance. *Science, 274*(5290), 1123–1133. doi:10.1126/science.274.5290.1123

Xu, J., Rosoff, W. J., Urbach, J., & Goodhill, G. J. (2005). Adaptation is not required to explain the long-term response of axons to molecular gradients. *Proceedings of the National Academy of Sciences of the United States of America, 132*, 4545–4562.

Zheng, J. Q., Felder, M., Connor, J. A., & Poo, M. (1994). Turning of nerve growth cone induced by neurotransmitters. *Nature, 368*(6467), 140–144. doi:10.1038/368140a0

ADDITIONAL READING

Chien, C. B., Rosenthal, D. E., Harris, W. A., & Holt, C. E. (1993). Navigational errors made by growth cones without filopodia in the embryonic Xenopus brain. *Neuron, 11*, 237–251. doi:10.1016/0896-6273(93)90181-P

Cojoc, D., Difato, F., Ferrari, E., Shahapure, R. B., Laishram, J., & Righi, M. (2007). Properties of the Force Exerted by Filopodia and Lamellipodia and the Involvement of Cytoskeletal Components. *PLoS ONE, 2*(10), e1072. doi:10.1371/journal.pone.0001072

Goldberg, J. L. (2003). How does an axon grow? *Genes & Development, 17*, 941–958. doi:10.1101/gad.1062303

Goodhill, G. J. (1998). Mathematical guidance for axons. *Trends in Neurosciences, 21*(6), 226–231. doi:10.1016/S0166-2236(97)01203-4

Gordon-Weeks, P. R. (2000). *Neuronal growth cones*. Cambridge University Press. doi:10.1017/CBO9780511529719

Heidemann, S. R., Lamoreux, P., & Buxbaum, R. E. (1990). Growth cone behaviour and production of traction force. *The Journal of Cell Biology*, *111*, 1949–1957. doi:10.1083/jcb.111.5.1949

Huber, A. B., Kolodkin, A. L., Ginty, D. D., & Cloutier, J.-F. (2003). Signaling at the growth cone: ligand-receptor complexes and the control of axon growth and guidance. *Annual Review of Neuroscience*, *155*(3), 509–563. doi:10.1146/annurev.neuro.26.010302.081139

Krottje, J. K., & van Ooyen, A. (2007). A mathematical framework for modelling axon guidance. *Bulletin of Mathematical Biology*, *69*, 3–31. doi:10.1007/s11538-006-9142-4

Lauffenburger, D. A., & Linderman, J. J. (1993, 2nd edition 1996). *Receptors: Models for Binding, Trafficking and Signalling*, Oxford University Press.

Levchenko, A., & Iglesias, P. (2002). Models of eukaryotic gradient sensing: application to chemotaxis of amoebae and neutrophils. *Biophysical Journal*, *82*(1), 50–63. doi:10.1016/S0006-3495(02)75373-3

Li, S., Guan, J., & Chein, S. (2005). Biochemistry and biomechanics of cell motility. *Annual Review of Biomedical Engineering*, *7*, 105–150. doi:10.1146/annurev.bioeng.7.060804.100340

Myers, P. Z., & Bastiani, J. (1993). Growth cone dynamics during the migration of an identified commissural growth cone. *The Journal of Neuroscience*, *13*(1), 127–143.

Ueda, M., Sako, Y., Tanaka, T., Devreotes, P., & Yanagida, T. (2001). Single-molecule analysis of chemotactic signaling in Dictyostelium cells. *Science*, *294*(5543), 864–867. doi:10.1126/science.1063951

KEY TERMS AND DEFINITIONS

Axon Guidance: Axon guidance (also called axon pathfinding) is the process by which neurons send out axons to reach the correct targets to wire up the nervous system. Axon guidance is driven by extracellular signal, called guidance cues, which can be fixed in place or diffusible; they can attract or repel axons.

Brownian Diffusion: Brownian diffusion is the chaotic and irregular movement of a particle immersed in a fluid, caused by its collisions with the surrounding molecules of much smaller size. The mathematical model of this random path is called Wiener process, and consists of a real-valued centred time-homogeneous Gaussian process that starts at zero and has independent increments.

Chemotactic Assay: Experimental tool for evaluation of the chemotactic ability of cells. A wide variety of techniques are known and applied. Some of them qualitative and investigator can determine whether the cells prefer or not the tested chemical, others are quantitative and we can get information about the intensity of the responses in a more detailed way (for example from the angle of deviation of the trajectory).

Drift: Transport mechanism of a substance or of particles by an external field in a particular direction. The field motion in advection is described mathematically as a vector field, and the material transported is typically described as a scalar value.

Growth Cone: A specialized structure at the end of a growing axon that guides the neuron to its destination during the development of the nervous system by means of interaction with signaling molecules in its surroundings and its own motile mechanism.

Markov Chain: Sequence of random objects $X_1, X_2, X_3, \ldots$ taking values in a set of states with the Markov property, namely that, given the present state, the future and past states are independent. Discrete-time (resp. continuous-time) homogeneous Markov chains are characterized by their transition probability matrix (resp. intensity matrix).

Microtubule: One of the components of the cytoskeleton. Microtubules serve as structural components within cells and are involved in many cellular processes. Microtubules also act as conveyor belts inside the cells. They move vesicles, granules, organelles like mitochondria, and chromosomes via special attachment proteins.

Multiscale: Field of solving physical problems that have important features at multiple scales, particularly multiple spatial and(or) temporal scales. Multiscale modeling in physics is aimed to calculation of material properties or system behaviour on one level using information or models from different levels.

Compilation of References

Abagyan, R., & Totrov, M. (2001). High-throughput docking for lead generation. *Current Opinion in Chemical Biology*, *5*(4), 375–382. doi:10.1016/S1367-5931(00)00217-9

Abersold, R., & Mann, M. (2003). Mass spectrometry-based proteomics. *Nature*, *422*(6928), 198–207. doi:10.1038/nature01511

Achtman, M., Azuma, T., Berg, D. E., Ito, Y., Morelli, G., & Pan, Z.-J. (1999). Recombination and clonal groupings within Helicobacter pylori from different geographical regions. *Molecular Microbiology*, *32*(3), 459–470. doi:10.1046/j.1365-2958.1999.01382.x

Ackermann, M., & Strimmer, K. (2009). A general modular framework for gene set enrichment analysis. *BMC Bioinformatics*, *10*, 47. doi:10.1186/1471-2105-10-47

Acquisti, A., & Grossklags, J. (2005). Privacy and Rationality in Individual Decision Making. *IEEE Security & Privacy*, *3*(1), 26–33. doi:10.1109/MSP.2005.22

Adamcsek, B. (2006). CFinder: Locating cliques and overlapping modules in biological networks. *Bioinformatics (Oxford, England)*, *22*(8), 1021–1023. doi:10.1093/bioinformatics/btl039

Adams, P. L., Stahley, M. R., Kosek, A. B., Wang, J., & Strobel, S. A. (2004). Crystal structure of a self-splicing group I intron with both exons. *Nature*, *430*(6995), 45–50. doi:10.1038/nature02642

Addona, T. A., Abbatiello, S. E., & Schilling, B. (2009). Multi-site assessment of the revision and reproducibility of multiple reaction monitoring-based measurements of proteins in plasma. *Nature Biotechnology*, *27*(7), 633–641. doi:10.1038/nbt.1546

Adimoolam, S., & Ford, J. M. (2003). p53 and regulation of DNA damage recognition during nucleotide excision repair. *DNA Repair*, *2*(9), 947–954. doi:10.1016/S1568-7864(03)00087-9

Adjei, A. A., & Hidalgo, M. (2005). Intracellular signal transduction pathway proteins as targets for cancer therapy. *Journal of Clinical Oncology*, *23*(23), 5386–5403. doi:10.1200/JCO.2005.23.648

Aebersold, R., Auffray, C., Baney, E., Barillot, E., Brazma, A., & Brett, C. (2009). Report on EU-USA workshop: how systems biology can advance cancer research (27 October 2008). *Molecular Oncology*, *3*(1), 9–17. doi:10.1016/j.molonc.2008.11.003

Aerts, S. (2006). Gene prioritization through genomic data fusion. *Nature Biotechnology*, *24*(5), 537–544. doi:10.1038/nbt1203

Aeschlimann, M., & Tettoni, L. (2001). Biophysical model of axonal pathfinding. *Neurocomputing*, *38-40*, 87–92. doi:10.1016/S0925-2312(01)00539-2

Agrasar, G., Linderman, J. J., Tryggvason, G., & Powell, K. G. (1998). An adaptative, Cartesian, front-tracking method for the motion, deformation, and adhesion of circulating cells. *Journal of Computational Physics*, *143*, 346–380. doi:10.1006/jcph.1998.5967

Agrawal, R., Imielinski, T., & Swami, A. (1993). *Mining association rules between sets of items in large databases.* SIGMOD.

Agrawal, R., & Srikant, R. (1994). Fast algorithms for mining association rules. In. *Proceedings of VLDB*, *94*, 487–499.

Ahmed, S., Thomas, G., Ghoussaini, M., Healey, C. S., Humphreys, M. K., & Platte, R. (2009). Newly discovered breast cancer susceptibility loci on 3p24 and 17q23.2. *Nature Genetics*, *41*(5), 585–590. doi:10.1038/ng.354

Ahmed, A., & Xing, E. P. (2009). Recovering time-varying networks of dependencies in social and biological studies. *Proceedings of the National Academy of Sciences of the United States of America*, *106*(29), 11878–11883. doi:10.1073/pnas.0901910106

Ajioka, R. S., Phillips, J. D., & Kushner, J. P. (2006). Biosynthesis of heme in mammals. *Biochimica et Biophysica Acta*, *1763*(7), 723–736. doi:10.1016/j.bbamcr.2006.05.005

Akira, S., & Takeda, K. (2004). Toll-like receptor signalling. *Nature Reviews. Immunology*, *4*(7), 499–511. doi:10.1038/nri1391

Akopyanz, N., Bukanov, N. O., Westblom, T. U., & Berg, D. E. (1992). PCR-based RFLP analysis of DNA sequence diversity in the gastric pathogen Helicobacter pylori. *Nucleic Acids Research*, *20*(23), 6221–6225. doi:10.1093/nar/20.23.6221

Akopyanz, N., Bukanov, N. O., Westblom, T. U., Kresovich, S., & Berg, D. E. (1992). DNA diversity among clinical isolates of Helicobacter pylori detected by PCR-based RAPD fingerprinting. *Nucleic Acids Research*, *20*(19), 5137–5142. doi:10.1093/nar/20.19.5137

Ala, U., Piro, R. M., Grassi, E., Damasco, C., Silengo, L., & Oti, M. (2008). Prediction of human disease genes by human-mouse conserved coexpression analysis. *PLoS Computational Biology*, *4*(3), e1000043. doi:10.1371/journal.pcbi.1000043

Alberts, B., Johnson, A., Lewis, J., Raff, M., Roberts, K., & Walter, P. (2002). *Molecular biology of the cell*. Garland.

Aldridge, B., Burke, J., Lauffenburger, D., & Sorger, P. (2006). Physicochemical modelling of cell signalling pathways. *Nature Cell Biology*, *8*(11), 1195–1203. doi:10.1038/ncb1497

Aletti, G., & Causin, P. (2008b). Mathematical characterization of the transduction chain in growth cone pathfinding. *IET Systems Biology*, *2*(3), 150–161. doi:10.1049/iet-syb:20070059

Aletti, G., Causin, P., & Naldi, G. (2008a). A model for axon guidance: Sensing, transduction, and. ovement. In *Collective Dynamics: Topics on Competition and Cooperation in the Biosciences: A Selection of Papers in the Proceedings of the BIOCOMP2007 International Conference*, AIP Conference Proceedings: Vol. 1028

Alles, M., Gardiner-Garden, M., Nott, D., Wang, Y., Foekens, J., & Sutherland, R. (2009). Meta-analysis and gene set enrichment relative to er status reveal elevated activity of MYC and E2F in the basal breast cancer subgroup. *PLoS ONE*, *4*(3). doi:10.1371/journal.pone.0004710

Almer, A., Rudolph, H., Hinnen, A., & Horz, W. (1986). Removal of positioned nucleosomes from the yeast PHO5 promoter upon PHO5 induction releases additional upstream activating DNA elements. *The EMBO Journal*, *5*(10), 2689–2696.

Al-Shahrour, F., Minguez, P., Tárraga, J., Montaner, D., Alloza, E., & Vaquerizas, J. M. M. (2003). BABELOMICS: A systems biology perspective in the functional annotation of genome-scale experiments. *Nucleic Acids Research*, *34*, W472–W476. doi:10.1093/nar/gkl172

Altschul, S. F., Gish, W., Miller, W., Myers, E. W., & Lipman, D. J. (1990). Basic local alignment search tool. *Journal of Molecular Biology*, *215*(3), 403–410.

Altshuler, D., Daly, M. J., & Lander, E. S. (2008). Genetic mapping in human disease. *Science*, *322*(5903), 881–888. doi:10.1126/science.1156409

Altshuler, D., Brooks, L. D., Chakravarti, A., Collins, F. S., Daly, M. J., & Donnelly, P. (2005). A haplotype map of the human genome. *Nature*, *437*, 1299–1320. doi:10.1038/nature04226

Alves, N. (2007). Unveiling community structures in weighted networks. *Physical Review E: Statistical, Nonlinear, and Soft Matter Physics*, *76*(3), 036101. doi:10.1103/PhysRevE.76.036101

America, A. H. P., & Cordewener, J. H. G. (2008). Comparative LC-MS: A landscape of peaks and valleys. *Proteomics*, *8*(4), 731–749. doi:10.1002/pmic.200700694

Anand, S., Prasad, M.V., Yadav, G., Kumar, N., Shehara, J., Ansari, M.Z., et al. (2010). SBSPKS: Structure based sequence analysis of polyketide synthases. *Nucleic Acids Research, 38*(Web server issue), W487-496.

Anderlik, M. R., & Rothstein, M. A. (2001). Privacy and confidentiality of genetic information: what rules for the new science? *Annual Review of Genomics and Human Genetics*, *2*(1), 401–433. doi:10.1146/annurev.genom.2.1.401

Anders, C. K., Acharya, C. R., Hsu, D. S., Broadwater, G., Garman, K., & Foekens, J. A. (2008). Age-specific differences in oncogenic pathway deregulation seen in human breast tumors. *PLoS ONE*, *3*(1), e1373. doi:10.1371/journal.pone.0001373

Anderson, N. L., & Anderson, N. G. (1998). Proteome and proteomics. New technologies, new concepts, and new words. *Electrophoresis*, *19*(11), 1853–1861. doi:10.1002/elps.1150191103

Anderson, N. L., Anderson, N. G., Haines, L. R., Hardie, D. B., Olafson, R. W., & Pearson, T. W. (2004). Mass spectrometric quantitation of peptides and proteins using stable isotope standards and capture by anti-peptide antibodies (SISCAPA). *Journal of Proteome Research*, *3*(2), 235–244. doi:10.1021/pr034086h

Andre, F.E. (1990). Overview of a 5-year clinical experience with a yeast-derived hepatitis B vaccine. *Vaccine, 8 Suppl*, S74-78; discussion S79-80.

Andrews, L. B., Fullarton, J. E., Holtman, N. A., & Motulsky, A. G. (1994). *Assessing genetic risks: implications for health and social policy*. Washington, DC: National Academy Press.

Androulakis, I. P., Yang, E., & Almon, R. R. (2007). Analysis of time-series gene expression data: Methods, challenges, and opportunities. *Annual Review of Biomedical Engineering*, *9*, 205–228. doi:10.1146/annurev.bioeng.9.060906.151904

Angeli, D., & Sontag, E. D. (2003). Monotone control systems. *IEEE Transactions on Automatic Control*, *48*(10), 1684–1698. doi:10.1109/TAC.2003.817920

Angell, M. (2000). Is academic medicine for sale? *The New England Journal of Medicine*, *342*(20), 1516–1518. doi:10.1056/NEJM200005183422009

Annas, G. J., Glantz, L. H., & Roche, P. A. (1995). Drafting the Genetic Privacy Act: Science, Policy, and Practical Considerations. *The Journal of Law, Medicine & Ethics*, *23*, 360. doi:10.1111/j.1748-720X.1995.tb01378.x

Anonymous,. (1966). Microfiche system saves time, cuts storage by 98 per cent. *Modern Hospital*, *107*, 66–67.

Ansari, M. Z., Sharma, J., Gokhale, R. S., & Mohanty, D. (2008). In silico analysis of methyltransferase domains involved in biosynthesis of secondary metabolites. *BMC Bioinformatics*, *9*, 454. doi:10.1186/1471-2105-9-454

Ansari, M.Z., Yadav, G., Gokhale, R.S. & Mohanty, D. (2004). NRPS-PKS: A knowledge-based resource for analysis of NRPS/PKS megasynthases. *Nucleic Acids Research, 32*(Web Server issue), W405-413.

Anstey, K. J., Lipnicki, D. M., & Low, L. F. (2008). Cholesterol as a risk factor for dementia and cognitive decline: A systematic review of prospective studies with meta-analysis. *The American Journal of Geriatric Psychiatry*, *16*(5), 343–354.

Aqvist, J., Medina, C., & Samuelsson, J. E. (1994). A new method for predicting binding affinity in computer-aided drug design. *Protein Engineering*, *7*(3), 385–391. doi:10.1093/protein/7.3.385

Aranda, B., Achuthan, P., Alam-Faruque, Y., Armean, I., Bridge, A., & Derow, C. (2010). The IntAct molecular interaction database in 2010. *Nucleic Acids Research*, *38*(Database issue), D525–D531. doi:10.1093/nar/gkp878

Arbeitman, M. N., Furlong, E. E., Imam, F., Johnson, E., Null, B. H., & Baker, B. S. (2002). Gene expression during the life cycle of Drosophila melanogaster. *Science*, *297*(5590), 2270–2275. doi:10.1126/science.1072152

Argaman, L., Hershberg, R., Vogel, J., Bejerano, G., Wagner, E. G., & Margalit, H. (2001). Novel small RNA-encoding genes in the intergenic regions of Escherichia coli. *Current Biology*, *11*(12), 941–950. doi:10.1016/S0960-9822(01)00270-6

Ashburner, M., Ball, C. A., Blake, J. A., Botstein, D., Butler, H., & Cherry, J. M. (2000). Gene ontology: Tool for the unification of biology. The Gene Ontology Consortium. *Nature Genetics*, *25*(1), 25–29. doi:10.1038/75556

ASHG. (1996). American Society of Human Genetics ASHG report: statement on informed consent for genetic research. *American Journal of Human Genetics*, *59*, 471–474.

Ashmore, J. (2008). Cochlear outer hair cell motility. *Physiological Reviews*, *88*(1), 173–210. doi:10.1152/physrev.00044.2006

Ashworth, L., & Free, C. (2006). Marketing Dataveillance and Digital Privacy: Using Theories of Justice to Understand Consumers Online Privacy Concerns. *Journal of Business Ethics*, *67*(2), 107–123. doi:10.1007/s10551-006-9007-7

Asiago, V. M., Gowda, G. A. N., Zhang, S., Shanaiah, J. C., & Raftery, D. (2008). Use of EDTA to minimize ionic strength dependent frequency shifts in the H NMR spectra of urine. *Metabolomics*, *4*(4), 328–336. doi:10.1007/s11306-008-0121-7

Askland, K., Read, C., & Moore, J. H. (2009). Pathway-based analyses of whole-genome association study data in bipolar disorder reveal genes mediating ion channel activity and synaptic neurotransmission. *Human Genetics*, *125*, 63–79. doi:10.1007/s00439-008-0600-y

Aspholm-Hurtig, M., Dailide, G., Lahmann, M., Kalia, A., Ilver, D., & Roche, N. (2004). Functional adaptation of BabA, the H. pylori ABO blood group antigen binding adhesion. *Science*, *305*(5683), 519–522. doi:10.1126/science.1098801

Atchley, W. R., Wollenberg, K. R., Fitch, W. M., Terhalle, W., & Dress, A. W. (2000). Correlations among amino acid sites in bHLH protein domains: An information theoretic analysis. *Molecular Biology and Evolution*, *17*(1), 164–178.

Aten, J. E., Fuller, T. F., Lusis, A. J., & Horvath, S. (2008). Using genetic markers to orient the edges in quantitative trait networks: the NEO software. *BMC Systems Biology*, *2*, 34. doi:10.1186/1752-0509-2-34

Atherton, J. C. (2006). The pathogenesis of Helicobacter pylori induced gastro-duodenal diseases. *Annual Review of Pathology: Mechanisms of Disease*, *1*(1), 63–96. doi:10.1146/annurev.pathol.1.110304.100125

Atherton, J. C., & Blaser, M. J. (2009). Coadaptation of Helicobacter pylori and humans: Ancient history, modern implications. *The Journal of Clinical Investigation*, *119*(9), 2475–2487. doi:10.1172/JCI38605

Audoly, S., Bellu, G., D'Angiò, L., Saccomani, M., & Cobelli, C. (2001). Global identifiability of nonlinear models of biological systems. *IEEE Transactions on Bio-Medical Engineering*, *48*(1), 55–65. doi:10.1109/10.900248

Austin, M. B., Saito, T., Bowman, M. E., Haydock, S., Kato, A., & Moore, B. S. (2006). Biosynthesis of Dictyostelium discoideum differentiation-inducing factor by a hybrid type I fatty acid-type III polyketide synthase. *Nature Chemical Biology*, *2*(9), 494–502. doi:10.1038/nchembio811

Avruch, J. (2007). MAP kinase pathways: The first twenty years. *Biochimica et Biophysica Acta*, *1773*(8), 1150–1160. doi:10.1016/j.bbamcr.2006.11.006

Axler, R. E., Irvine, R., Lipworth, W., Morrell, B., & Kerridge, I. H. (2008). Why might people donate tissue for cancer research? Insights from organ/tissue/blood donation and clinical research. *Pathobiology*, *75*(6), 323–329. doi:10.1159/000164216

Bachmann, B. O., & Ravel, J. (2009). Chapter 8. Methods for in silico prediction of microbial polyketide and nonribosomal peptide biosynthetic pathways from DNA sequence data. *Methods in Enzymology*, *458*, 181–217. doi:10.1016/S0076-6879(09)04808-3

Badal, A., & Sempan, J. (2006). A package of linux scripts for the parallelization of Monte Carlo simulation. *Computer Physics Communications*, *175*, 440–450. doi:10.1016/j.cpc.2006.05.009

Badano, J. L., & Katsanis, N. (2002). Beyond Mendel: An evolving view of human genetic disease transmission. *Nature Reviews. Genetics*, *3*, 779–789. doi:10.1038/nrg910

Bader, G. D., & Hogue, C. W. (2003). An automated method for finding molecular complexes in large protein interaction networks. *BMC Bioinformatics*, *4*, 2. doi:10.1186/1471-2105-4-2

Bader, S., Kühner, S., & Gavin, A.-C. (2008). Interaction networks for systems biology. *FEBS Letters*, *582*(8), 1220–1224. doi:10.1016/j.febslet.2008.02.015

Baerga-Ortiz, A., Popovic, B., Siskos, A. P., O'Hare, H. M., Spiteller, D., & Williams, M. G. (2006). Directed mutagenesis alters the stereochemistry of catalysis by isolated ketoreductase domains from the erythromycin polyketide synthase. *Chemistry & Biology*, *13*(3), 277–285. doi:10.1016/j.chembiol.2006.01.004

Bailor, M. H., Sun, X., & Al-Hashimi, H. M. (2010). Topology links RNA secondary structure with global conformation dynamics and adaptation. *Science*, *327*(5962), 202–206. doi:10.1126/science.1181085

Bakkenist, C. J., & Kastan, M. B. (2003). DNA damage activates ATM through intermolecular autophosphorylation and dimer dissociation. *Nature*, *421*(6922), 499–506. doi:10.1038/nature01368

Baldi, P., & Long, A. (2001). A Bayesian framework for the analysis of microarray expression data: Regularized t-test and statistical inferences of gene changes. *Bioinformatics (Oxford, England)*, *17*, 509–516. doi:10.1093/bioinformatics/17.6.509

Balding, D. J. (2006). A tutorial on statistical methods for population association studies. *Nature Reviews. Genetics*, *7*(10), 781–791. doi:10.1038/nrg1916

Baltimore, D. (2001). Our genome unveiled. *Nature*, *409*(6822), 814–816. doi:10.1038/35057267

Baltz, R. H. (2006). Molecular engineering approaches to peptide, polyketide and other antibiotics. *Nature Biotechnology*, *24*(12), 1533–1540. doi:10.1038/nbt1265

Bamba, E., & N'guessan, Y. (2003). Nouvelle approche de détermination du moment dipolaire en solution dans des solvants polaires. *Revue Ivoirienne des Sciences et Technologies.*, *4*, 25–33.

Banks, E., Nabieva, E., Peterson, R., & Singh, M. (2008). NetGrep: Fast network schema searches in interactomes. *Genome Biology*, *9*, R138. doi:10.1186/gb-2008-9-9-r138

Banskota, A. H., McAlpine, J. B., Sorensen, D., Aouidate, M., Piraee, M., & Alarco, A. M. (2006). Isolation and identification of three new 5-alkenyl-3,3(2H)-furanones from two streptomyces species using a genomic screening approach. *The Journal of Antibiotics*, *59*(3), 168–176. doi:10.1038/ja.2006.24

Banskota, A. H., McAlpine, J. B., Sorensen, D., Ibrahim, A., Aouidate, M., & Piraee, M. (2006). Genomic analyses lead to novel secondary metabolites. Part 3. ECO-0501, a novel antibacterial of a new class. *The Journal of Antibiotics*, *59*(9), 533–542. doi:10.1038/ja.2006.74

Barabasi, A. L., & Oltvai, Z. N. (2004). Network biology: Understanding the cell's functional organization. *Nature Reviews. Genetics*, *5*(2), 101–113. doi:10.1038/nrg1272

Barakat, K. H., Torin Huzil, J., Luchko, T., Jordheim, L., Dumontet, C., & Tuszynski, J. (2009). Characterization of an inhibitory dynamic pharmacophore for the ERCC1-XPA interaction using a combined molecular dynamics and virtual screening approach. *Journal of Molecular Graphics & Modelling*, *28*(2), 113–130. doi:10.1016/j.jmgm.2009.04.009

Barakat, K., Mane, J., Friesen, D.,& Tuszynski, J. (2009). Ensemble-based virtual screening reveals dual-inhibitors for the p53-MDM2/MDMX interactions. *Journal of Molecular Graphics Models*.

Bar-Joseph, Z. (2004). Analyzing time series gene expression data. *Bioinformatics (Oxford, England)*, *20*(16), 2493. doi:10.1093/bioinformatics/bth283

Barkow, S., Bleuer, S., Prelic, A., Zimmermann, P., & Zitzler, E. (2006). BicAT: A biclustering analysis toolbox. *Bioinformatics (Oxford, England)*, *22*(10), 1282–1283. doi:10.1093/bioinformatics/btl099

Barnidge, D. R., Hall, G. D., Stocker, D. C., & Muddiman, D. C. (2004). Evaluation of a cleavable stable isotope labeled synthetic peptide for absolute protein quantification using LC-MS/MS. *Journal of Proteome Research*, *3*(3), 658–661. doi:10.1021/pr034124x

Barnum, D., Greene, J., Smellie, A., & Sprague, P. (1996). Identification of common functional configurations among molecules. *Journal of Chemical Information and Computer Sciences*, *36*(3), 563–571. doi:10.1021/ci950273r

Barreiro, L. B., Henriques, R., & Mhlanga, M. M. (2009). High-throughput SNP genotyping: combining tag SNPs and molecular beacons. *Methods in Molecular Biology (Clifton, N.J.)*, *578*, 255–276. doi:10.1007/978-1-60327-411-1_17

Barrett, J. C., Hansoul, S., Nicolae, D. L., Cho, J. H., Duerr, R. H., & Rioux, J. D. (2008). Genome-wide association defines more than 30 distinct susceptibility loci for Crohn's disease. *Nature Genetics*, *40*(8), 955–962. doi:10.1038/ng.175

Barrett, T., Troup, D. B., Wilhite, S. E., Ledoux, P., Rudnev, D., & Evangelista, C. (2007). NCBI GEO: Mining tens of millions of expression profiles-database and tools update. *Nucleic Acids Research*, *35*(Database issue), D760–D765. doi:10.1093/nar/gkl887

Barrett, T., Troup, D. B., Wilhite, S. E., Ledoux, P., Rudnev, D., & Evangelista, C. (2009). NCBI GEO: Archive for high-throughput functional genomic data. *Nucleic Acids Research*, *37*(Database issue), D885–D890. doi:10.1093/nar/gkn764

Barrows, R. C., & Clayton, P. D. (1996). Privacy, Confidentiality, and Electronic Medical Records. *Journal of the American Medical Informatics Association*, *3*(2), 139–148.

Barry, W. T. (2005). Significance analysis of functional categories in gene expression studies: A structured permutation approach. *Bioinformatics (Oxford, England)*, *21*(9), 1943–1949. doi:10.1093/bioinformatics/bti260

Barski, A., Cuddapah, S., Cui, K., Roh, T. Y., Schones, D. E., & Wang, Z. (2007). High-resolution profiling of histone methylations in the human genome. *Cell*, *129*(4), 823–837.

Barton, G. M., Kagan, J. C., & Medzhitov, R. (2006). Intracellular localization of Toll-like receptor 9 prevents recognition of self DNA but facilitates access to viral DNA. *Nature Immunology*, *7*(1), 49–56. doi:10.1038/ni1280

Bate, N., Bignell, D. R., & Cundliffe, E. (2006). Regulation of tylosin biosynthesis involving SARP-helper activity. *Molecular Microbiology*, *62*(1), 148–156. doi:10.1111/j.1365-2958.2006.05338.x

Bates, B. R., Lynch, J. A., & Bevan, J. L. (2005). Condit CM: Warranted concerns, warranted outlooks: a focus group study of public understandings of genetic research. *Social Science & Medicine*, *60*, 331–344. doi:10.1016/j.socscimed.2004.05.012

Bauer-Mehren, A., Furlong, L. I., & Sanz, F. (2009). Pathway databases and tools for their exploitation: Benefits, current limitations and challenges. *Molecular Systems Biology*, *5*, 290. doi:10.1038/msb.2009.47

Baxter, C. A., Murray, C. W., Clark, D. E., Westhead, D. R., & Eldridge, M. D. (1998). Flexible docking using Tabu search and an empirical estimate of binding affinity. *Proteins*, *33*(3), 367–382. doi:10.1002/(SICI)1097-0134(19981115)33:3<367::AID-PROT6>3.0.CO;2-W

Beer, D. G., Kardia, S. L., Huang, C. C., Giordano, T. J., Levin, A. M., & Misek, D. E. (2002). Gene-expression profiles predict survival of patients with lung adenocarcinoma. *Nature Medicine*, *8*(8), 816–824.

Behar, M., Dohlman, H. G., & Elston, T. C. (2007). Kinetic insulation as an effective mechanism for achieving pathway specificity in intracellular signaling networks. *Proceedings of the National Academy of Sciences of the United States of America*, *104*(41), 16146–16151. doi:10.1073/pnas.0703894104

Beißbarth, T., & Speed, T. P. (2004). GOstat: Find statistically overrepresented gene ontologies within a group of genes. *Bioinformatics Applications Note*, *20*(9), 1464–1465.

Bell, G. (1978). Models for specific adhesion of cell to cell. *Science*, *200*, 618–627. doi:10.1126/science.347575

Bell, G. I., Dembo, M., & Bongrand, P. (1984). Cell adhesion: Competition between non-specific repulsion and specific bonding. *Biophysical Journal*, *45*, 1051–1064. doi:10.1016/S0006-3495(84)84252-6

Belshaw, P. J., Walsh, C. T., & Stachelhaus, T. (1999). Aminoacyl-CoAs as probes of condensation domain selectivity in nonribosomal peptide synthesis. *Science*, *284*(5413), 486–489. doi:10.1126/science.284.5413.486

Belta, C., Finin, P., Habets, L. C., Halasz, G. J. M., Imielinski, A. M. M., Kumar, R. V., et al. (2004). *Dynamic partitioning of large discrete event biological systems for hybrid simulation and analysis*. Paper presented at the 7th International Workshop Hybrid Systems Computation and Control, *2993*, 111-125.

Bender, A., Jenkins, J. L., Scheiber, J., Sukuru, S. C., Glick, M., & Davies, J. W. (2009). How similar are similarity searching methods? A principal component analysis of molecular descriptor space. *Journal of Chemical Information and Modeling*, *49*(1), 108–119. doi:10.1021/ci800249s

Ben-Dor, A. (2000). Tissue classification with gene expression profiles. *Journal of Computational Biology*, *7*, 559–584. doi:10.1089/106652700750050943

Ben-Dor, A., Chor, B., Karp, R., & Yakhini, Z. (2003). Discovering local structure in gene expression data: The order-preserving submatrix problem. *Journal of Computational Biology*, *10*(3-4), 373–384. doi:10.1089/10665270360688075

Benito, M. (2004). Adjustment of systematic microarray data biases. *Bioinformatics (Oxford, England)*, *20*(1), 105–114. doi:10.1093/bioinformatics/btg385

Bennett, C. J. (1995). *The Political Economy of Privacy: A Review of the Literature*. Hackensack, NJ: Center for Social and Legal Research.

Bennett, S. T., Barnes, C., Cox, A., Davies, L., & Brown, C. (2005). Toward the 1,000 dollars human genome. *Pharmacogenomics*, *6*, 373–382. doi:10.1517/14622416.6.4.373

Benson, S. D., Bamford, J. K., Bamford, D. H., & Burnett, R. M. (2004). Does common architecture reveal a viral lineage spanning all three domains of life? *Molecular Cell*, *16*(5), 673–685. doi:10.1016/j.molcel.2004.11.016

Bentley, D. R. (2006). Whole-genome re-sequencing. *Current Opinion in Genetics & Development*, *16*, 545–552. doi:10.1016/j.gde.2006.10.009

Berardini, T. Z., Li, D., Huala, E., Bridges, S., Burgess, S., & McCarthy, F. (2010). The gene ontology in 2010: Extensions and refinements. *Nucleic Acids Research*, *38*(Database issue), D331–D335. doi:10.1093/nar/gkp1018

Beres, S. B., Carroll, R. K., Shea, P. R., Sitkiewicz, I., Martinez-Gutierrez, J. C., & Low, D. E. (2010). Molecular complexity of successive bacterial epidemics deconvoluted by comparative pathogenomics. *Proceedings of the National Academy of Sciences of the United States of America*, *107*(9), 4371–4376. doi:10.1073/pnas.0911295107

Berg, H. C., & Purcell, E. M. (1977). Physics of chemoreception. *Biophysical Journal*, *20*(2), 193–219. doi:10.1016/S0006-3495(77)85544-6

Berger, S. I., & Iyengar, R. (2009). Network analyses in systems pharmacology. *Bioinformatics (Oxford, England)*, *25*(19), 2466–2472. doi:10.1093/bioinformatics/btp465

Bergmann, S., Ihmels, J., & Barkai, N. (2003). Iterative signature algorithm for the analysis of largescale gene expression data. *Physical Review E: Statistical, Nonlinear, and Soft Matter Physics*, *67*(3 Pt 1), 03190201–03190218.

Berman, S. D., West, J. C., Danielian, P. S., Caron, A. M., Stone, J. R., & Lees, J. A. (2009). Mutation of p107 exacerbates the consequences of Rb loss in embryonic tissues and causes cardiac and blood vessel defects. *Proceedings of the National Academy of Sciences of the United States of America*, *106*(35), 14932–14936. doi:10.1073/pnas.0902408106

Berman, H. M., Westbrook, J., Feng, Z., Gilliland, G., Bhat, T. N., & Weissig, H. (2000). The Protein Data Bank. *Nucleic Acids Research*, *28*(1), 235–242. doi:10.1093/nar/28.1.235

Bernhardt, B. A., Tambor, E. S., Fraser, G., Wissow, L. S., & Geller, G. (2002). Parents' and children's attitudes toward the enrollment of minors in genetic susceptibility research: Implications for informed consent. *American Journal of Medical Genetics*, *116A*(4), 315–323. doi:10.1002/ajmg.a.10040

Bernstein, B. E., Liu, C. L., Humphrey, E. L., Perlstein, E. O., & Schreiber, S. L. (2004). Global nucleosome occupancy in yeast. *Genome Biology*, *5*(9), R62.

Bernstein, B. E., Meissner, A., & Lander, E. S. (2007). The mammalian epigenome. *Cell*, *128*(4), 669–681.

Bernstein, B. E., Mikkelsen, T. S., Xie, X., Kamal, M., Huebert, D. J., & Cuff, J. (2006). A bivalent chromatin structure marks key developmental genes in embryonic stem cells. *Cell*, *125*(2), 315–326.

Berriz, G. F., King, O. D., Bryant, B., Sander, C., & Roth, F. P. (2003). Characterizing gene sets with FuncAssociate. *Bioinformatics Applications Note*, *19*(18), 2502–2504.

Bertucci, F., Finetti, P., Cervera, N., & Birnbaum, D. (2008). Prognostic classification of breast cancer and gene expression profiling. *Medecine Sciences*, *24*(6-7), 599–606.

Beskow, L. M., Burke, W., Merz, J. F., Barr, P. A., & Terry, S., V.B., P., et al. (2001). Informed consent for population-based research involving genetics. *Journal of the American Medical Association*, *286*, 2315–2321. doi:10.1001/jama.286.18.2315

Betzi, S., Restouin, A., Opi, S., Arold, S. T., Parrot, I., & Guerlesquin, F. (2007). Protein protein interaction inhibition (2P2I) combining high throughput and virtual screening: Application to the HIV-1 Nef protein. *Proceedings of the National Academy of Sciences of the United States of America*, *104*(49), 19256–19261. doi:10.1073/pnas.0707130104

Beutler, B. (2004). Inferences, questions and possibilities in Toll-like receptor signalling. *Nature*, *430*(6996), 257–263. doi:10.1038/nature02761

Bhattacharjee, A., Richards, W. G., Staunton, J., Li, C., Monti, S., & Vasa, P. (2001). Classification of human lung carcinomas by mRNA expression profiling reveals distinct adenocarcinoma subclasses. *Proceedings of the National Academy of Sciences of the United States of America*, *98*(24), 13790–13795. doi:10.1073/pnas.191502998

Bianco, A. R. (2004). Targeting c-erbB2 and other receptors of the c-erbB family: Rationale and clinical applications. *Journal of Chemotherapy (Florence, Italy)*, *16*(4), 52–54.

Bianconi, F. (2006). *A hybrid model of nucleotide excision repair in neoplastic diseases and in vitro experiments*. Master Degree Thesis, Department of Electronic and Information Engineering, University of Perugia.

Bianconi, F. (2010). *Dynamic modeling, parameter estimation and experiment design in systems biology with applications to oncology.* PhD thesis, Department of Electronic and Information Engineering, University of Perugia.

Bianconi, F., Valigi, P., & Crinò, L. Ludovini, V., Piattoni, S., Orleth, A., et al. (2006). *A hybrid model of nucleotide excision repair in neoplastic diseases and in vitro experiments.* Tech. Rep. RT-003-06, Department of Electronic and Information Engineering, University of Perugia.

Bicknell, R., & Harris, A. L. (2004). Novel angiogenic signaling pathways and vascular targets. *Annual Review of Pharmacology and Toxicology*, *44*, 219–238. doi:10.1146/annurev.pharmtox.44.101802.121650

Bidaut, G., & Stoeckert, C. J. Jr. (2009). Large scale transcriptome data integration across multiple tissues to decipher stem cell signatures. *Methods in Enzymology*, *467*, 229–245. doi:10.1016/S0076-6879(09)67009-9

Bidaut, G., Suhre, K., Claverie, J. M., & Ochs, M. F. (2006). Determination of strongly overlapping signaling activity from microarray data. *BMC Bioinformatics*, *7*, 99. doi:10.1186/1471-2105-7-99

Bidaut, G., & Stoeckert, C. J., Jr. (2009). Characterization of unknown adult stem cell samples by large scale data integration and artificial neural networks. *Pacific Symposium on Biocomputing*, 356-367.

Bild, A. H., Yao, G., Chang, J. T., Wang, Q., Potti, A., & Chasse, D. (2006). Oncogenic pathway signatures in human cancers as a guide to targeted therapies. *Nature*, *439*(7074), 353–357. doi:10.1038/nature04296

Bird, A. (2002). DNA methylation patterns and epigenetic memory. *Genes & Development*, *16*(1), 6–21.

Bissantz, C., Folkers, G., & Rognan, D. (2000). Protein-based virtual screening of chemical databases. 1. Evaluation of different docking/scoring combinations. *Journal of Medicinal Chemistry*, *43*(25), 4759–4767. doi:10.1021/jm001044l

Bittner, M., Meltzer, P., Chen, Y., Jiang, Y., Seftor, E., & Hendrix, M. (2000). Molecular classification of cutaneous malignant melanoma by gene expression profiling. *Nature*, *406*(6795), 536–540. doi:10.1038/35020115

Blackstock, W. P., & Weir, M. P. (1999). Proteomics: Quantitative and physical mapping of cellular proteins. *Trends in Biotechnology*, *17*(3), 121–127. doi:10.1016/S0167-7799(98)01245-1

Blondel, V. (2008). Fast unfolding of communities in large networks. *Journal of Statistical Mechanics*, (10): P10008. doi:10.1088/1742-5468/2008/10/P10008

Bock, C., & Lengauer, T. (2008). Computational epigenetics. *Bioinformatics (Oxford, England)*, *24*(1), 1–10.

Bock, C., Paulsen, M., Tierling, S., Mikeska, T., Lengauer, T., & Walter, J. (2006). CpG island methylation in human lymphocytes is highly correlated with DNA sequence, repeats, and predicted DNA structure. *PLOS Genetics*, *2*(3), e26.

Bock, C., Walter, J., Paulsen, M., & Lengauer, T. (2008). Inter-individual variation of DNA methylation and its implications for large-scale epigenome mapping. *Nucleic Acids Research*, *36*(10), e55.

Boekhoff-Falk, G. (2005). Hearing in drosophila: Development of Johnston's organ and emerging parallels to vertebrate ear development. *Developmental Dynamics: An Official Publication of the American Association of Anatomists*, *232*(3), 550–558.

Bohm, H. J. (1992). The computer program LUDI: A new method for the de novo design of enzyme inhibitors. *Journal of Computer-Aided Molecular Design*, *6*(1), 61–78. doi:10.1007/BF00124387

Bok, S. (1989). *Secrets: On the Ethics of Concealment and Revelation*. New York: Vintage.

Bolderson, E., Richard, D. J., Zhou, B.-B. S., & Khanna, K. K. (2009). Recent advances in cancer therapy targeting proteins involved in DNA double-strand break repair. *Clinical Cancer Research*, *15*(20), 6314–6320. doi:10.1158/1078-0432.CCR-09-0096

Bolstad, B. M., Irizarry, R. A., Astrand, M., & Speed, T. P. (2003). A comparison of normalization methods for high density oligonucleotide array data based on variance and bias. *Bioinformatics (Oxford, England)*, *19*(2), 185–193. doi:10.1093/bioinformatics/19.2.185

Bondi, A. (1964). van der Waals volumes and radii. *Journal of Physical Chemistry*, *68*(3), 441–451. doi:10.1021/j100785a001

Bonenfant, D., Towbin, H., Coulot, M., Schindler, P., Mueller, D. R., & van Oostrum, J. (2007). Analysis of dynamic changes in post-translational modifications of human histones during cell cycle by mass spectrometry. *Molecular & Cellular Proteomics*, *6*(11), 1917–1932. doi:10.1074/mcp.M700070-MCP200

Bongrand, P. (1982). Ligand-receptor interactions. *Reports on Progress in Physics*, *62*, 921–968. doi:10.1088/0034-4885/62/6/202

Bongrand, P., & Benoliel, A. M. (1999). Adhésion cellulaire. *RSTD*, *44*, 167–178.

Bongrand, P., Capo, C., & Depied, R. (1982). Physics of cell adhesion. *Progress in Surface Science*, *12*, 217–286. doi:10.1016/0079-6816(82)90007-7

Bonifaci, N., Berenguer, A., Díez, J., Reina, O., Medina, I., & Dopazo, J. (2008). Biological processes, properties and molecular wiring diagrams of candidate low-penetrance breast cancer susceptibility genes. *BMC Medical Genomics*, *1*, 62. doi:10.1186/1755-8794-1-62

Bonneau, R. (2008). Learning biological networks: From modules to dynamics. *Nature Chemical Biology*, *4*(11), 658–664. doi:10.1038/nchembio.122

Boone, C., Howard Bussey, H., & Andrews, B. J. (2007). Exploring genetic interactions and networks with yeast. *Nature Reviews. Genetics*, *8*, 437–449. doi:10.1038/nrg2085

Borodin, A. N., & Salminen, P. (2002). *Handbook of Brownian motion-facts and formulae* (2nd ed.). Basel: Birkhäuser Verlag.

Bossi, A., & Lehner, B. (2009). Tissue specificity and the human protein interaction network. *Molecular Systems Biology*, *5*, 260. doi:10.1038/msb.2009.17

Botstein, D., & Risch, N. (2003). Discovering genotypes underlying human phenotypes: Past successes for Mendelian disease, future approaches for complex disease. *Nature Genetics*, *33*(Supplement), 228–237. doi:10.1038/ng1090

Botstein, D., White, R. L., Skolnick, M., & Davis, R. W. (1980). Construction of a genetic linkage map in man using restriction fragment length polymorphisms. *American Journal of Human Genetics*, *32*(3), 314–331.

Bouzigues, C., & Dahan, M. (2007a). Transient directed motions of $GABA_A$ receptors in growth cones detected by a speed correlation index. *Biophysical Journal*, *92*(2), 654–660. doi:10.1529/biophysj.106.094524

Bouzigues, C., Lévi, S., Triller, A., & Dahan, M. (2007c). Single quantum dot tracking of membrane receptors. *Methods in Molecular Biology (Clifton, N.J.)*, *374*, 81–91.

Bouzigues, C., Morel, M., Triller, A., & Dahan, M. (2007b). Asymmetric redistribution of GABA receptors during GABA gradient sensing by nerve growth cones analyzed by single quantum dot imaging. *Proceedings of the National Academy of Sciences of the United States of America*, *104*(11), 251–256.

Boyle, E. I., Shuai, W., Jeremy, G., Heng, J., David, B., & Michael, C. J. (2004). GO: TermFinder-open source software for accessing Gene Ontology information and finding significantly enriched Gene Ontology terms associated with a list of genes. *Bioinformatics (Oxford, England)*, *20*(18), 3710–3715. doi:10.1093/bioinformatics/bth456

Brader, G., Djamei, A., Teige, M., Palva, E. T., & Hirt, H. (2007). The MAP Kinase Kinase MKK2 affects disease resistance in Arabidopsis. *Molecular Plant-Microbe Interactions*, *20*(5), 589–596. doi:10.1094/MPMI-20-5-0589

Brandes, U. (2007). On modularity clustering. *IEEE Transactions on Knowledge and Data Engineering*, *20*(2), 172–188. doi:10.1109/TKDE.2007.190689

Brazma, A., Krestyaninova, M., & Sarkans, U. (2006). Standards for systems biology. *Nature Reviews. Genetics*, *7*, 593–605. doi:10.1038/nrg1922

Breiman, L. (1996). Bagging predictors. *Machine Learning*, *24*, 123–140. doi:10.1007/BF00058655

Breiman, L. (2001). Random forests. *Machine Learning*, *45*, 5–32. doi:10.1023/A:1010933404324

Breiman, L., Friedman, J., Olshen, R., & Stone, C. (1984). *Classification and regression trees*. New York: Chapman & Hall.

Breitkreutz, B. J., Stark, C., Reguly, T., Boucher, L., Breitkreutz, A., & Livstone, M. (2008). The BioGRID interaction database: 2008 update. *Nucleic Acids Research*, *36*(Database issue), D637–D640. doi:10.1093/nar/gkm1001

Breitling, R. (2004). Iterative Group Analysis (iGA): A simple tool to enhance sensitivity and facilitate interpretation of microarray experiments. *BMC Bioinformatics*, *5*, 34. doi:10.1186/1471-2105-5-34

Breitling, R., Armengaud, P., Amtmann, A., & Herzyk, P. (2004). Rank products: A simple, yet powerful, new method to detect differentially regulated genes in replicated microarray experiments. *FEBS Letters*, *573*(1-3), 83–92. doi:10.1016/j.febslet.2004.07.055

Brenk, R., Naerum, L., Gradler, U., Gerber, H. D., Garcia, G. A., & Reuter, K. (2003). Virtual screening for submicromolar leads of tRNA-guanine transglycosylase based on a new unexpected binding mode detected by crystal structure analysis. *Journal of Medicinal Chemistry*, *46*(7), 1133–1143. doi:10.1021/jm0209937

Brenner, S. E. (2001). A tour of structural genomics. *Nature Reviews. Genetics*, *2*(10), 801–809. doi:10.1038/35093574

Brettschneider, J., Collin, F., Bolstad, B. M., & Speed, T. P. (2008). Rejoinder for quality assessment for short oligonucleotide microarray data. *Technometrics*, *50*(3), 279–283. doi:10.1198/004017008000000389

Broadhurst, R. W., Nietlispach, D., Wheatcroft, M. P., Leadlay, P. F., & Weissman, K. J. (2003). The structure of docking domains in modular polyketide synthases. *Chemistry & Biology*, *10*(8), 723–731. doi:10.1016/S1074-5521(03)00156-X

Brohee, S., & van Helden, J. (2006). Evaluation of clustering algorithms for protein-protein interaction networks. *BMC Bioinformatics*, *7*, 488. doi:10.1186/1471-2105-7-488

Bron, C., & Kerbosch, J. (1973). Algorithm 457: Finding all cliques of an undirected graph. *Communications of the ACM*, *16*(9), 575–577. doi:10.1145/362342.362367

Brosch, R., Gordon, S. V., Garnier, T., Eiglmeier, K., Frigui, W., & Valenti, P. (2007). Genome plasticity of BCG and impact on vaccine efficacy. *Proceedings of the National Academy of Sciences of the United States of America*, *104*(13), 5596–5601. doi:10.1073/pnas.0700869104

Broughton, H.B. (2000). A method for including protein flexibility in protein-ligand docking: improving tools for database mining and virtual screening. *Journal of Molecular Graphics Models, 18*(3), 247-257, 302-244.

Brown, S. M. (2002). *Essentials of Medical Genomics*. Hoboken, NJ: John Wiley & Sons, Inc.doi:10.1002/0471483087

Brown, M. P., Grundy, W. N., Lin, D., Cristianini, N., Sugnet, C. W., & Furey, T. S. (2000). Knowledge-based analysis of microarray gene expression data by using support vector machines. *Proceedings of the National Academy of Sciences of the United States of America*, *97*, 262–267. doi:10.1073/pnas.97.1.262

Brown, L. M. (2000). Helicobacter Pylori: Epidemiology and routes of transmission. *Epidemiologic Reviews*, *22*(2), 283–297.

Bruce, A., Johnson, A., Lewis, J., Raff, M., Roberts, K., & Walter, P. (2008). *Molecular biology of the cell* (5th ed.). Garland Science.

Bruggeman, F. J., & Westerhoff, H. V. (2007). The nature of systems biology. *Trends in Microbiology*, *15*(1), 45–50. doi:10.1016/j.tim.2006.11.003

Bruinsma, R., & Sackmann, E. (2001). Bioadhesion and the dewetting transition. *Comptes Rondus de l'Academie Sciences Paris*, *2*(4), 803–815.

Bruse, S., Moreau, M., Azaro, M., Zimmerman, R., & Brzustowicz, L. (2008). Improvements to bead-based oligonucleotide ligation SNP genotyping assays. *BioTechniques*, *45*(5), 559–571. doi:10.2144/000112960

Buchholz, T. J., Geders, T. W., Bartley, F. E. III, Reynolds, K. A., Smith, J. L., & Sherman, D. H. (2009). Structural basis for binding specificity between subclasses of modular polyketide synthase docking domains. *ACS Chemical Biology*, *4*(1), 41–52. doi:10.1021/cb8002607

Buck, M. J., & Lieb, J. D. (2004). ChIP-chip: Considerations for the design, analysis, and application of genome-wide chromatin immunoprecipitation experiments. *Genomics*, *83*(3), 349–360. doi:10.1016/j.ygeno.2003.11.004

Bueno-de-Mesquita, J. M., van Harten, W. H., Retel, V. P., van't Veer, L. J., van Dam, F. S., & Karsenberg, K. (2007). Use of 70-gene signature to predict prognosis of patients with node-negative breast cancer: A prospective community-based feasibility study (RASTER). *The Lancet Oncology*, *8*(12), 1079–1087. doi:10.1016/S1470-2045(07)70346-7

Buettner, H. M., Pittman, R. N., & Ivins, J. (1994). A model of neurite extension across regions of nonpermissive substrate: Simulations based on experimental measurements of growth cone motility and filopodial dynamics. *Developmental Biology*, *163*(2), 407–422. doi:10.1006/dbio.1994.1158

Bugg, T. (2004). *An introduction to enzyme and coenzyme chemistry*. Oxford, Cambridge, MA: Blackwell Science. doi:10.1002/9781444305364

Bui, C. T., Babon, J. J., Lambrinakos, A., & Cotton, R. G. (2003). Detection of mutations in DNA by solid-phase chemical cleavage method. A simplified assay. *Methods in Molecular Biology (Clifton, N.J.)*, *212*, 59–70.

Buness, A., Huber, W., Steiner, K., Sltmann, H., & Poustka, A. (2005). arrayMagic: Two-colour cDNA microarray quality control and preprocessing. *Bioinformatics (Oxford, England)*, *21*, 554–556. doi:10.1093/bioinformatics/bti052

Bunet, R., Mendes, M. V., Rouhier, N., Pang, X., Hotel, L., & Leblond, P. (2008). Regulation of the synthesis of the angucyclinone antibiotic alpomycin in Streptomyces ambofaciens by the autoregulator receptor AlpZ and its specific ligand. *Journal of Bacteriology*, *190*(9), 3293–3305. doi:10.1128/JB.01989-07

Bunger, M. K., Cargile, B. J., Ngunjiri, A., Bundy, J. L., & Stephenson, J. L. Jr. (2008). Automated proteomics of E. coli via top-down electron-transfer dissociation mass spectrometry. *Analytical Chemistry*, *80*(5), 1459–1467. doi:10.1021/ac7018409

Bureau, A., Dupuis, J., Hayward, B., Falls, K., & Van Eerdewegh, P. (2003). Mapping complex traits using random forests. *BMC Genetics*, *4*, S64. doi:10.1186/1471-2156-4-S1-S64

Burger, L., & van Nimwegen, E. (2008). Accurate prediction of protein-protein interactions from sequence alignments using a Bayesian method. *Molecular Systems Biology*, *4*, 165. doi:10.1038/msb4100203

Burton, J., Ijjaali, I., Petitet, F., Michel, A., & Vercauteren, D. P. (2009). Virtual screening for cytochromes: Successes of machine learning filters. *Combinational Chemistry and High Throughput Screening*, *12*(4), 369–382. doi:10.2174/138620709788167935

Bush, W. S., Dudek, S. M., & Ritchie, M. D. (2009). BioFilter: A knowledge-integration system for the multi-locus analysis of genome-wide association studies. *Pacific Symposium on Biocomputing*, 368-379.

Busygin, S., Boyko, N., Pardalos, P. M., Bewernitz, M., & Ghacibeh, G. (2007). Biclustering EEG data from epileptic patients treated with vagus nerve stimulation. *AIP Conference Proceedings*, *953*, 220. doi:10.1063/1.2817345

Bylesjo, M., Rantalainen, M., Cloarec, O., Nicholson, J. K., Holmes, E., & Trygg, J. (2007). OPLC discriminant analysis: Combining the strengths of PLS-DA and SIMCA classification. *Journal of Chemometrics*, *20*, 341–351. doi:10.1002/cem.1006

C.B.R., Subramanian, J. & Sharma, S.D. (2009). Managing protein flexibility in docking and its applications. *Drug Discovery Today*, *14*(7-8), 394–400. doi:10.1016/j.drudis.2009.01.003

Cabusora, L., Sutton, E., Fulmer, A., & Forst, C. (2005). Differential network expression during drug and stress response. *Bioinformatics (Oxford, England)*, *21*(12), 2898–2905. doi:10.1093/bioinformatics/bti440

Caffrey, P. (2003). Conserved amino acid residues correlating with ketoreductase stereospecificity in modular polyketide synthases. *ChemBioChem*, *4*(7), 654–657. doi:10.1002/cbic.200300581

Caldas, J., & Kaski, S. (2008). Bayesian biclustering with the plaid model. *In Proceedings of the IEEE International Workshop on Machine Learning for Signal Processing*, 291-296.

Caldas, J., & Kaski, S. (2010). Generative tree biclustering for information retrieval and microRNA biomarker discovery. *In Proceedings of RECOMB 2010*, April 25-28, Lisbon, Portugal.

Calvano, S., Xiao, W., Richards, D., Felciano, R., Baker, H., & Cho, R. (2005). A network-based analysis of systemic inflammation in humans. *Nature*, *437*(7061), 1032–1037. doi:10.1038/nature03985

Calvo, S., Jain, M., Xie, X., Sheth, S. A., Chang, B., & Goldberger, O. A. (2006). Systematic identification of human mitochondrial disease genes through integrative genomics. *Nature Genetics*, *38*(5), 576–582. doi:10.1038/ng1776

Camacho, D., & Collins, J. (2009). Systems biology strikes gold. *Cell*, *137*(1), 24–26. doi:10.1016/j.cell.2009.03.032

Camp, L. J. (1999). Web security and privacy: An American perspective. [Article]. *The Information Society*, *15*(4), 249–256. doi:10.1080/019722499128411

Campbell, A. V. (2007). The ethical challenges of genetic databases: safeguarding altruism and trust. *King's Law Journal*, *18*, 227–246.

Campone, M., Campion, L., Roche, H., Gouraud, W., Charbonnel, C., & Magrangeas, F. (2008). Prediction of metastatic relapse in node-positive breast cancer: Establishment of a clinicogenomic model after FEC100 adjuvant regimen. *Breast Cancer Research and Treatment*, *109*(3), 491–501. doi:10.1007/s10549-007-9673-x

Cantone, I., Marucci, L., Iorio, F., Ricci, M., Belcastro, V., & Bansal, M. (2009). A yeast synthetic network for in vivo assessment of reverse-engineering and modeling approaches. *Cell*, *137*(1), 172–181. doi:10.1016/j.cell.2009.01.055

Caplan, A. L. (2009). What No One Knows Cannot Hurt You: The Limits of Informed Consent in the Emerging World of Biobanking. In Solbakk, J. H., Holm, S., & Hofmann, B. (Eds.), *The Ethics of Research Biobanking* (pp. 25–32). Springer. doi:10.1007/978-0-387-93872-1_2

Capocci, A. (2005). Detecting communities in large networks. *Physica A. Statistical and Theoretical Physics*, *352*(2-4), 669–676. doi:10.1016/j.physa.2004.12.050

Carapetis, J.R., Steer, A.C., & Mulholland, E.K. & M.W. (1995). The global burden of group A streptococcal diseases. *The Lancet Infectious Diseases*, *5*(11), 685–694. doi:10.1016/S1473-3099(05)70267-X

Carbon, S., Ireland, A., Mungall, C. J., Shu, S., Marshall, B., & Lewis, S. (2009). AmiGO hub, Web presence working group. AmiGO: Online access to ontology and annotation data. *Bioinformatics (Oxford, England)*, *25*(2), 288–289. doi:10.1093/bioinformatics/btn615

Carlson, H. A., Masukawa, K. M., Rubins, K., Bushman, F. D., Jorgensen, W. L., & Lins, R. D. (2000). Developing a dynamic pharmacophore model for HIV-1 integrase. *Journal of Medicinal Chemistry*, *43*(11), 2100–2114. doi:10.1021/jm990322h

Carlson, C. S. (2006). Agnosticism and equity in genome-wide association studies. *Nature Genetics*, *38*(6), 605–606. doi:10.1038/ng0606-605

Carosati, E., Mannhold, R., Wahl, P., Hansen, J. B., Fremming, T., & Zamora, I. (2007). Virtual screening for novel openers of pancreatic K(ATP) channels. *Journal of Medicinal Chemistry*, *50*(9), 2117–2126. doi:10.1021/jm061440p

Carvalho, P. C. (2008). PatternLab for proteomics: A tool for differential shotgun proteomics. *BMC Bioinformatics*, *9*, 316–329. doi:10.1186/1471-2105-9-316

Cassa, C. A. (2008). *Privacy and identifiability in clinical research, personalized medicine, and public health surveillance*: PhD thesis, Massachusetts Institute of Technology.

Castilho, M. S., Postigo, M. P., de Paula, C. B., Montanari, C. A., Oliva, G., & Andricopulo, A. D. (2006). Two- and three-dimensional quantitative structure-activity relationships for a series of purine nucleoside phosphorylase inhibitors. *Bioorganic & Medicinal Chemistry*, *14*(2), 516–527. doi:10.1016/j.bmc.2005.08.055

Caudill, E. M., & Murphy, P. E. (2000). Consumer Online Privacy: Legal and Ethical Issues. *Journal of Public Policy & Marketing*, *19*(1), 7–19. doi:10.1509/jppm.19.1.7.16951

Causin, P., & Facchetti, G. (2009). Amplification and polarization in chemotaxis: Addressing the specificity of neural cells via mathematical modelling and numerical simulation. *PLoS Computational Biology*, *5*(8), e1000479.

Cavallo, F., Calogero, R. A., & Forni, G. (2007). Are oncoantigens suitable targets for anti-tumour therapy? *Nature Reviews. Cancer*, *7*(9), 707–713. doi:10.1038/nrc2208

Cavasotto, C. N., & Phatak, S. S. (2009). Homology modeling in drug discovery: Current trends and applications. *Drug Discovery Today*, *14*(13-14), 676–683. doi:10.1016/j.drudis.2009.04.006

Cavoukian, A. (2009). *Privacy by Design: Take the Challenge*. Ontario, Canada: Information and Privacy Commissioner of Ontario.

Ceol, A., Chatr Aryamontri, A., Licata, L., Peluso, D., Briganti, L., & Perfetto, L. (2009). MINT, the molecular interaction database: 2009 update. *Nucleic Acids Research*, *38*(Database issue), D532–D539. doi:10.1093/nar/gkp983

Cervino, A., Li, G., Edwards, S., Zhu, J., Laurie, C., & Tokiwa, G. (2005). Integrating QTL and high-density SNP analyses in mice to identify Insig2 as a susceptibility gene for plasma cholesterol levels. *Genomics*, *86*(5), 505–517. doi:10.1016/j.ygeno.2005.07.010

Challis, G. L., Ravel, J., & Townsend, C. A. (2000). Predictive, structure-based model of amino acid recognition by nonribosomal peptide synthetase adenylation domains. *Chemistry & Biology*, *7*(3), 211–224. doi:10.1016/S1074-5521(00)00091-0

Chalmers, D. (2007). International co-operation between biobanks: the case for harmonisation of guidelines and governance. In Stranger, M. (Ed.), *Human Biotechnology & Public Trust: Trends, Perceptions and Regulation* (pp. 237–246). Hobart: Centre for Law and Genetics.

Champion, A., Picaud, A., & Henry, Y. (2004). Reassessing the MAP3K and MAP4K relationships. *Trends in Plant Science*, *9*(3), 123–129. doi:10.1016/j.tplants.2004.01.005

Chang, J. T., Carvalho, C., Mori, S., Bild, A. H., Gatza, M. L., & Wang, Q. (2009). A genomic strategy to elucidate modules of oncogenic pathway signaling networks. *Molecular Cell*, *34*(1), 104–114. doi:10.1016/j.molcel.2009.02.030

Chang, H. Y., Sneddon, J. B., Alizadeh, A. A., Sood, R., West, R. B., & Montgomery, K. (2004). Gene expression signature of fibroblast serum response predicts human cancer progression: Similarities between tumors and wounds. *PLoS Biology*, *2*(2), E7. doi:10.1371/journal.pbio.0020007

Chang, L.W., Fontaine, B.R., Stormo, G.D. & Nagarajan, R. (2007). PAP: A comprehensive workbench for mammalian transcriptional regulatory sequence analysis. *Nucleic Acids Research, 35*(Web Server issue), W238-244.

Chanock, S. (2001). Candidate genes and single nucleotide polymorphisms (SNPs) in the study of human disease. *Disease Markers*, *17*, 89–98.

Chanock, S. J., Manolio, T., Boehnke, M., Boerwinkle, E., Hunter, D. J., & Thomas, G. (2007). Replicating genotype-phenotype associations. *Nature*, *447*(7145), 655–660. doi:10.1038/447655a

Chanrion, M., Negre, V., Fontaine, H., Salvetat, N., Bibeau, F., & MacGrogan, G. (2008). A gene expression signature that can predict the recurrence of tamoxifen-treated primary breast cancer. *Clinical Cancer Research*, *14*(6), 1744–1752. doi:10.1158/1078-0432.CCR-07-1833

Charifson, P. S., Corkery, J. J., Murcko, M. A., & Walters, W. P. (1999). Consensus scoring: A method for obtaining improved hit rates from docking databases of three-dimensional structures into proteins. *Journal of Medicinal Chemistry*, *42*(25), 5100–5109. doi:10.1021/jm990352k

Charlesworth, J. C., Peralta, J. M., Drigalenko, E., Göring, H. H., Almasy, L., & Dyer, T. D. (2009). Toward the identification of causal genes in complex diseases: A gene-centric joint test of significance combining genomic and transcriptomic data. *BMC Proceedings*, *3*(Supplement 7), S92. doi:10.1186/1753-6561-3-s7-s92

Charo, R. A. (2007). Body of Research - Ownership and use of Human Tissue. *The New England Journal of Medicine*, *355*, 1517–1519. doi:10.1056/NEJMp068192

Chaudhuri, R., Ahmed, S., Ansari, F. A., Singh, H. V., & Ramachandran, S. (2008). MalVac: Database of malarial vaccine candidates. *Malaria Journal*, *7*, 184. doi:10.1186/1475-2875-7-184

Chaurasia, G., Malhotra, S., Russ, J., Schnoegl, S., Hänig, C., & Wanker, E. E. (2009). UniHI 4: New tools for query, analysis and visualization of the human protein–protein interactome. *Nucleic Acids Research*, *37*, D657–D660. doi:10.1093/nar/gkn841

Chautard, E., Thierry-Mieg, N., & Ricard-Blum, S. (2009). Interaction networks: From protein functions to drug discovery. A review. *Pathologie Biologie*, *57*(4), 324–333. doi:10.1016/j.patbio.2008.10.004

Cheatham, M. A., Zheng, J., Huynh, K. H., Du, G. G., Gao, J., & Zuo, J. (2005). Cochlear function in mice with only one copy of the prestin gene. *The Journal of Physiology*, *569*(Pt 1), 229–241. doi:10.1113/jphysiol.2005.093518

Chen, J. Y., Yan, Z., Shen, C., Fitzpatrick, D. P., & Wang, M. (2007). A systems biology approach to the study of cisplatin drug resistance in ovarian cancers. *Journal of Bioinformatics and Computational Biology*, *5*(2a), 383–405. doi:10.1142/S0219720007002606

Chen, C. Y., Chang, Y. H., Bau, D. T., Huang, H. J., Tsai, F. J., & Tsai, C. H. (2009). Ligand-based dual target drug design for H1N1: Swine flu-a preliminary first study. *Journal of Biomolecular Structure & Dynamics*, *27*(2), 171–178.

Chen, H., Lyne, P. D., Giordanetto, F., Lovell, T., & Li, J. (2006). On evaluating molecular-docking methods for pose prediction and enrichment factors. *Journal of Chemical Information and Modeling*, *46*(1), 401–415. doi:10.1021/ci0503255

Chen, Z., Li, H. L., Zhang, Q. J., Bao, X. G., Yu, K. Q., & Luo, X. M. (2009). Pharmacophore-based virtual screening versus docking-based virtual screening: a benchmark comparison against eight targets. *Acta Pharmacologica Sinica*, *30*(12), 1694–1708. doi:10.1038/aps.2009.159

Chen, Y., Zhu, J., Lum, P. Y., Yang, X., Pinto, S., & MacNeil, D. J. (2008). Variations in DNA elucidate molecular networks that cause disease. *Nature*, *452*(7186), 429–435. doi:10.1038/nature06757

Chen, X., Lie, C. T., Zhang, M., & Zhang, H. (2007). A forest based approach to identifying gene and gene-gene interactions. *Proceedings of the National Academy of Sciences of the United States of America*, *104*, 19199–19203. doi:10.1073/pnas.0709868104

Chen, K. C., Csikasz-Nagy, A., Gyorffy, B., Val, J., Novak, B., & Tyson, J. J. (2000). Kinetic analysis of a molecular model of the budding yeast cell cycle. *Molecular Biology of the Cell*, *11*(1), 369–391.

Chen, R. E., & Thorner, J. (2007). Function and regulation in MAPK signaling pathways: Lessons learned from the yeast Saccharomyces cerevisiae. *Biochimica et Biophysica Acta (BBA)-. Molecular Cell Research*, *1773*(8), 1311–1340.

Chen, S., Donoho, D., & Saunders, M. (2001). Atomic decomposition by basis pursuit. *SIAM Review*, *43*(1), 129–159. doi:10.1137/S003614450037906X

Chen, J.-L., & Greider, C. W. (2005). Functional analysis of the pseudoknot structure in human telomerase RNA. *Proceedings of the National Academy of Sciences of the United States of America*, *102*(23), 8080–8085. doi:10.1073/pnas.0502259102

Cheng, K. O., Law, N. F., Siu, W. C., & Lau, T. H. (2007). Bivisu: Software tool for bicluster detection and visualization. *Bioinformatics (Oxford, England)*, *23*, 2342–2344. doi:10.1093/bioinformatics/btm338

Cheng, Y., & Church, G. M. (2000). Biclustering of expression data. *Proceedings of the International Conference on Intelligent Systems for Molecular Biology*, *8*, 93–103.

Chesla, C., Selvaraj, P., & Zhu, C. (1998). Measuring two-dimensional receptor-ligand binding-kinetic by micropipette. *Biophysical Journal*, *75*, 1553–1572. doi:10.1016/S0006-3495(98)74074-3

Chiang, Y. M., Szewczyk, E., Davidson, A. D., Keller, N., Oakley, B. R., & Wang, C. C. (2009). A gene cluster containing two fungal polyketide synthases encodes the biosynthetic pathway for a polyketide, asperfuranone, in Aspergillus nidulans. *Journal of the American Chemical Society*, *131*(8), 2965–2970. doi:10.1021/ja8088185

Chin, C. S., Chubukov, V., Jolly, E. R., DeRisi, J., & Li, H. (2008). Dynamics and design principles of a basic regulatory architecture controlling metabolic pathways. *PLoS Biology*, *6*(6), e146. doi:10.1371/journal.pbio.0060146

Cho, K., Shin, S., Kolch, W., & Wolkenhauer, O. (2003). Experimental design in systems biology, based on parameter sensitivity analysis using a Monte Carlo method: A case study for the TNF(alpha)-mediated NF-(kappa) b signal transduction pathway. *Simulation*, *79*(12), 726–739. doi:10.1177/0037549703040943

Chodavarapu, R. K., Feng, S., Bernatavichute, Y. V., Chen, P. Y., Stroud, H., & Yu, Y. (2010). Relationship between nucleosome positioning and DNA methylation. *Nature*, *466*(7304), 388–392.

Chopra, T., Banerjee, S., Gupta, S., Yadav, G., Anand, S., & Surolia, A. (2008). Novel intermolecular iterative mechanism for biosynthesis of mycoketide catalyzed by a bimodular polyketide synthase. *PLoS Biology*, *6*(7), e163. doi:10.1371/journal.pbio.0060163

Christinat, Y., Wachmann, B., & Zhang, L. (2008). Gene expression data analysis using a novel approach to biclustering combining discrete and continuous data. *IEEE/ACM Transactions on Computational Biology and Bioinformatics*, *5*(4), 583–593. doi:10.1109/TCBB.2007.70251

Chua, H. N. (2006). Exploiting indirect neighbours and topological weight to predict protein function from protein-protein interactions. *Bioinformatics (Oxford, England)*, *22*(13), 1623–1630. doi:10.1093/bioinformatics/btl145

Chuang, H. Y., Lee, E., Liu, Y. T., Lee, D., & Ideker, T. (2007). Network-based classification of breast cancer metastasis. *Molecular Systems Biology*, *3*, 140. doi:10.1038/msb4100180

Chung, H.J., Park, C.H., Han, M.R., Lee, S., Ohn, J.H., Kim, J., et al. (2005). ArrayXPath II: Mapping and visualizing microarray gene-expression data with biomedical ontologies and integrated biological pathway resources using Scalable Vector Graphics. *Nucleic Acids Research, 33*(Web server issue), W621-6.

Church, G. M. (2005). The personal genome project. *Mol Syst Biol, 1*, 2005 0030.

Churchill, G. A. (2002). Fundamentals of experimental design for cDNA microarrays. *Nature Genetics*, *32*(Supplement), 490–495. doi:10.1038/ng1031

Clark, A. G., Boerwinkle, E., Hixson, J., & Sing, C. F. (2005). Determinants of the success of whole-genome association testing. *Genome Research*, *15*, 1463–1467. doi:10.1101/gr.4244005

Clauset, A. (2004). Finding community structure in very large networks. *Physical Review E: Statistical, Nonlinear, and Soft Matter Physics*, (6): 066111. doi:10.1103/PhysRevE.70.066111

Cleaver, J. E., Lam, E. T., & Revet, I. (2009). Disorders of nucleotide excision repair: The genetic and molecular basis of heterogeneity. *Nature Reviews. Genetics*, *10*(11), 756–768. doi:10.1038/nrg2663

Cline, M. S., Smoot, M., Cerami, E., Kuchinsky, A., Landys, N., & Workman, C. (2007). Integration of biological networks and gene expression data using cytoscape. *Nature Protocols*, *2*(10), 2366–2382. doi:10.1038/nprot.2007.324

Clote, P., & Backhofen, R. (2000). *Computational molecular biology: An introduction*. Hoboken, NJ: Wiley.

Clugston, S. L., Sieber, S. A., Marahiel, M. A., & Walsh, C. T. (2003). Chirality of peptide bond-forming condensation domains in nonribosomal peptide synthetases: The C5 domain of tyrocidine synthetase is a (D)C(L) catalyst. *Biochemistry*, *42*(41), 12095–12104. doi:10.1021/bi035090+

Cokus, S. J., Feng, S., Zhang, X., Chen, Z., Merriman, B., & Haudenschild, C. D. (2008). Shotgun bisulphite sequencing of the Arabidopsis genome reveals DNA methylation patterning. *Nature*, *452*(7184), 215–219.

Colditz, G. A., Berkey, C. S., Mosteller, F., Brewer, T. F., Wilson, M. E., & Burdick, E. (1995). The efficacy of bacillus Calmette-Guerin vaccination of newborns and infants in the prevention of tuberculosis: Meta-analyses of the published literature. *Pediatrics*, *96*(1 Pt 1), 29–35.

Cole, J. C., Murray, C. W., Nissink, J. W., Taylor, R. D., & Taylor, R. (2005). Comparing protein-ligand docking programs is difficult. *Proteins*, *60*(3), 325–332. doi:10.1002/prot.20497

Cole, S. T., Brosch, R., Parkhill, J., Garnier, T., Churcher, C., & Harris, D. (1998). Deciphering the biology of Mycobacterium tuberculosis from the complete genome sequence. *Nature*, *393*(6685), 537–544. doi:10.1038/31159

Collier, T. S., Hawkridge, A. M., Georgianna, D. R., Patne, G. A., & Muddiman, D. C. (2008). Top-down identification and quantification of stable isotope labeled proteins from Aspergillus flavus using online nano-flow reversed-phase liquid chromatography coupled to a LTQ-FTICR mass spectrometer. *Analytical Chemistry*, *80*(13), 4994–5001. doi:10.1021/ac800254z

Collins, F. S., Green, E. D., Guttmacher, A. E., & Guyer, M. S. (2003). A vision for the future of genomics research. *Nature*, *422*(6934), 835–847. doi:10.1038/nature01626

Collins, F. S., Guyer, M. S., & Charkravarti, A. (1997). Variations on a theme: Cataloging human DNA sequence variation. *Science*, *278*(5343), 1580–1581. doi:10.1126/science.278.5343.1580

Conlon, E. M., Song, J. J., & Liu, J. S. (2006). Bayesian models for pooling microarray studies with multiple sources of replications. *BMC Bioinformatics*, *7*, 247. doi:10.1186/1471-2105-7-247

Conrad, D. F., Pinto, D., Redon, R., Feuk, L., Gokcumen, O., & Zhang, Y. (2010). Origins and functional impact of copy number variation in the human genome. *Nature*, *464*(7289), 704–712. doi:10.1038/nature08516

Conti, C., Veenstra, D. L., Armstrong, K., Lesko, J. L., & Grosse, S. D. (2010). (in press). Personalized medicine and genomics: challenges and opportunities in assessing effectiveness, cost effectiveness, and future research priorities. *Medical Decision Making*. doi:10.1177/0272989X09347014

Conti, E., Stachelhaus, T., Marahiel, M. A., & Brick, P. (1997). Structural basis for the activation of phenylalanine in the non-ribosomal biosynthesis of gramicidin S. *The EMBO Journal*, *16*(14), 4174–4183. doi:10.1093/emboj/16.14.4174

Conti, R., Veenstra, D.L., Armstrong, K., Lesko, L.J. & Grosse, S.D. (2010). Personalized medicine and genomics: Challenges and opportunities in assessing effectiveness, cost-effectiveness, and future research priorities. *Medical Decision Making.*

Cookson, W., Liang, L., Abecasis, G., Moffatt, M., & Lathrop, M. (2009). Mapping complex disease traits with global gene expression. *Nature Reviews. Genetics*, *10*(3), 184–194. doi:10.1038/nrg2537

Coombs, D., Dembo, M., Wosfy, C., & Goldstein, B. (2004). Equilibrium thermodynamics of cell-cell adhesion mediated by multiple ligand-receptor pairs. *Biophysical Journal*, *86*, 1408–1423. doi:10.1016/S0006-3495(04)74211-3

Cordell, H. J. (2009). Genome-wide association studies: Detecting gene-gene interactions that underlie human diseases. *Nature Reviews. Genetics*, *10*(6), 392–404. doi:10.1038/nrg2579

Core, L. J., & Lis, J. T. (2008). Transcription regulation through promoter-proximal pausing of RNA polymerase II. *Science*, *319*(5871), 1791–1792.

Cornish-Bowden, A. (1995). *Fundamentals of enzyme kinetics*. Portland Press.

Cosgrove, B. D., King, B. M., Hasan, M. A., Alexopoulos, L. G., Farazi, P. A., & Hendriks, B. S. (2009). Synergistic drug-cytokine induction of hepatocellular death as an in vitro approach for the study of inflammation-associated idiosyncratic drug hepatotoxicity. *Toxicology and Applied Pharmacology*, *237*(3), 317–330. doi:10.1016/j.taap.2009.04.002

Costa, R. M. A., Chigancas, V., da Silva Galhardo, R., Carvalho, H., & Menck, C. F. M. (2003). The eukaryotic nucleotide excision repair pathway. *Biochimie*, *85*(11), 1083–1099. doi:10.1016/j.biochi.2003.10.017

Coulibaly, I., & Page, G. P. (2008). Bioinformatics tools for inferring functional information from plant microarray data II: Analysis beyond single gene. *International Journal of Plant Genomics*, (893941): 13.

Courty, S., Bouzigues, C., Luccardini, C., Ehrensperger, M. V., Bonneau, S., & Dahan, M. (2006). Tracking individual proteins in living cells using single quantum dot imaging. *Methods in Enzymology*, *414*, 211–228. doi:10.1016/S0076-6879(06)14012-4

Couzin, J., & Kaiser, J. (2007). Genome-wide association: Closing the net on common disease genes. *Science*, *316*, 820–822. doi:10.1126/science.316.5826.820

Covacci, A., Telford, J. L., Giudice, G. D., Parsonnet, J., & Rappuoli, R. (1999). Helicobacter pylori Virulence and Genetic Geography. *Science*, *284*(5418), 1328–1333. doi:10.1126/science.284.5418.1328

Cover, T. L., Berg, D. E., Blaser, M., & Mobley, H. L. T. (2001). *H. pylori pathogenesis*. New York: Academic Press.

Cox, R. J. (2007). Polyketides, proteins and genes in fungi: Programmed nano-machines begin to reveal their secrets. *Organic & Biomolecular Chemistry*, *5*(13), 2010–2026. doi:10.1039/b704420h

Cozens, C., Lauffenburger, D., & Quinn, J. A. (1990). Receptor-mediated cell attachment and detachment kinetics-probabilistc model and analysis. *Biophysical Journal*, *58*, 841–856. doi:10.1016/S0006-3495(90)82430-9

Cramer, R. D. III, Patterson, D. E., & Bunce, J. D. (1989). Recent advances in comparative molecular field analysis (CoMFA). *Progress in Clinical and Biological Research*, *291*, 161–165.

Creighton, C., & Hanash, S. (2003). Mining gene expression databases for association rules. *Bioinformatics (Oxford, England)*, *19*(1), 79–86. doi:10.1093/bioinformatics/19.1.79

Crockford, D. J., Holmes, E., Lindon, J. C., Plumb, R. S., Zirah, S., & Bruce, S. J. (2006). Statistical heterospectroscopy, (SHY), an approach to the integrated analysis of NMR and UPLC-MS data sets: Application in metabonomic toxicology studies. *Analytical Chemistry*, *78*(2), 363–371. doi:10.1021/ac051444m

Cruciani, G., Pastor, M., & Guba, W. (2000). VolSurf: A new tool for the pharmacokinetic optimization of lead compounds. *European Journal of Pharmaceutical Sciences*, *11*(Suppl 2), S29–S39. doi:10.1016/S0928-0987(00)00162-7

Csardi, G., Kutalik, Z., & Bergmann, S. (2010). Modular analysis of gene expression data with R. *Bioinformatics Applications Note*, *26*(10), 1376–1377.

Culnan, M. J. (1993). 'How Did They Get My Name'? An Exploratory Investigation of Consumer Attitudes toward Secondary Information Use. *Management Information Systems Quarterly*, *17*(3), 341–364. doi:10.2307/249775

Culnan, M. J. (1995). Consumer Awareness of Name Removal Procedures: Implication for Direct Marketing. *Journal of Interactive Marketing*, *9*, 10–19.

Curtis, R. K., Oresic, M., & Vidal-Puig, A. (2005). Pathways to the analysis of microarray data. *Trends in Biotechnology*, *23*, 429–435. doi:10.1016/j.tibtech.2005.05.011

Daikos, G. K. (2007). History of medicine: our Hippocratic heritage. *International Journal of Antimicrobial Agents*, *29*(6), 617–620. doi:10.1016/j.ijantimicag.2007.01.008

Daily, J. P., Le Roch, K. G., Sarr, O., Ndiaye, D., Lukens, A., & Zhou, Y. (2005). In vivo transcriptome of Plasmodium falciparum reveals overexpression of transcripts that encode surface proteins. *The Journal of Infectious Diseases*, *191*(7), 1196–1203. doi:10.1086/428289

Dam, E., Pleij, K., & Draper, D. (1992). Structural and functional aspects of RNA pseudoknots. *Biochemistry*, *31*(47), 11665–11676. doi:10.1021/bi00162a001

Das, R., Dimitrova, N., Xuan, Z., Rollins, R. A., Haghighi, F., & Edwards, J. R. (2006). Computational prediction of methylation status in human genomic sequences. *Proceedings of the National Academy of Sciences of the United States of America*, *103*(28), 10713–10716.

Dasika, M. S., Burgard, A., & Maranas, C. D. (2006). A computational framework for the topological analysis and targeted disruption of signal transduction networks. *Biophysical Journal*, *91*(1), 382–398. doi:10.1529/biophysj.105.069724

Davies, H. (2002). Mutations of the BRAF gene in human cancer. *Nature*, *417*, 949–954. doi:10.1038/nature00766

Davis, A.P., et al. (2009). Comparative toxicogenomics database: A knowledgebase and discovery tool for chemical-gene-disease networks. *Nucleic Acids Research, 37*(database issue), D786-792.

Davis, R. (1995). Online medical records raise privacy fears. *USA Today March 22.*

Day-Richter, J., Harris, M. A., & Haendel, M. (2007). Gene ontology OBO-edit working group, OBO-edit-an ontology editor for biologists. *Bioinformatics (Oxford, England)*, *23*(16), 2198–2200. doi:10.1093/bioinformatics/btm112

De Crecy-Lagard, V., Marliere, P., & Saurin, W. (1995). Multienzymatic non ribosomal peptide biosynthesis: Identification of the functional domains catalysing peptide elongation and epimerisation. *Comptes Rondus de l'Academie des Sciences III*, *318*(9), 927–936.

De Jong, H., Gouz, J. L., Hernandez, C., Page, M., Sari, T., & Geiselmann, J. (2003). *Hybrid modeling and simulation of genetic regulatory networks: A qualitative approach.* Paper presented at the 6th International Workshop Hybrid Systems Computation and Control, 2623, 267-282.

de la Fuente van Bentem, S., Anrather, D., Dohnal, I., Roitinger, E., Csaszar, E., & Joore, J. (2008). Site-specific phosphorylation profiling of Arabidopsis proteins by mass spectrometry and peptide chip analysis. *Journal of Proteome Research*, *7*(6), 2458–2470. doi:10.1021/pr8000173 de Lavallade, H., Finetti, P., Carbuccia, N., Khorashad, J. S., Charbonnier, A., & Foroni, L. (2010). A gene expression signature of primary resistance to imatinib in chronic myeloid leukemia. *Leukemia Research*, *34*(2), 254–257. doi:10.1016/j.leukres.2009.09.026

Decramer, S. (2006). Predicting the clinical outcome of congenital unilateral ureteropelvic junction obstruction in newborn by urinary proteome analysis. *Nature Medicine*, *12*(4), 398–400. doi:10.1038/nm1384

Deluca, T. F., Wu, I. H., Pu, J., Monaghan, T., Peshkin, L., & Singh, S. (2006). Roundup: A multi-genome repository of orthologs and evolutionary distances. *Bioinformatics (Oxford, England)*, *22*(16), 2044–2046. doi:10.1093/bioinformatics/btl286

Dembo, M., Torney, D. C., Saxman, K., & Hammer, D. (1998). The reaction-limited kinetics of membrane-to-surface adhesion and detachment. *Proceedings. Biological Sciences*, *234*, 55 83. doi:10.1098/rspb.1988.0038

Dempsey, K., Currall, B., Hallworth, R., & Ali, H. (In press). An intelligent data-centric approach toward identification of conserved motifs in protein sequences. In *ACM International Conference on Bioinformatics and Computational Biology2010*.

Deng, M. (2003). Prediction of protein function using protein-protein interaction data. *Journal of Computational Biology*, *10*(6), 947–960. doi:10.1089/106652703322756168

Dennis, G. Jr, Sherman, B. T., Hosack, D. A., Yang, J., Gao, W., & Lane, H. C. (2003). DAVID: Database for Annotation, Visualization, and Integrated Discovery. *Genome Biology*, *4*(5), 3. doi:10.1186/gb-2003-4-5-p3

DeRisi, J., Penland, L., Brown, P. O., Bittner, M. L., Meltzer, P. S., & Ray, M. (1996). Use of a cdna microarray to analyse gene expression patterns in human cancer. *Nature Genetics*, *14*(4), 457–460. doi:10.1038/ng1296-457

DeRisi, J. L., Iyer, V. R., & Brown, P. O. (1997). Exploring the metabolic and genetic control of gene expression on a genomic scale. *Science*, *278*(5338), 680–686. doi:10.1126/science.278.5338.680

DesJarlais, R. L., Seibel, G. L., Kuntz, I. D., Furth, P. S., Alvarez, J. C., & Ortiz de Montellano, P. R. (1990). Structure-based design of nonpeptide inhibitors specific for the human immunodeficiency virus 1 protease. *Proceedings of the National Academy of Sciences of the United States of America*, *87*(17), 6644–6648. doi:10.1073/pnas.87.17.6644

Desmedt, C., Piette, F., Loi, S., Wang, Y., Lallemand, F., & Haibe-Kains, B. (2007). Strong time dependence of the 76-gene prognostic signature for node-negative breast cancer patients in the TRANSBIG multicenter independent validation series. *Clinical Cancer Research*, *13*(11), 3207–3214. doi:10.1158/1078-0432.CCR-06-2765

Devlin, B., & Roeder, K. (1999). Genomic control for association studies. *Biometrics*, *55*(4), 997–1004. doi:10.1111/j.0006-341X.1999.00997.x

Devos, D., & Russell, R. B. (2007). A more complete, complex, and structured interactome. *Current Opinion in Structural Biology*, *17*(3), 370–377. doi:10.1016/j.sbi.2007.05.011

Dezso, Z., Nikolsky, Y., Nikolskaya, T., Miller, J., Cherba, D., & Webb, C. (2009). Identifying disease-specific genes based on their topological significance in protein networks. *BMC Systems Biology*, *3*, 36. doi:10.1186/1752-0509-3-36

D'haeseleer, P. (2005). How does gene expression clustering work? *Nature Biotechnology*, *23*(12), 1499–1501. doi:10.1038/nbt1205-1499

Dhillon, I. S. (2001). Co-clustering documents and words using bipartite spectral graph partitioning. *In Proceedings of the 7th ACM SIGKDD International Conference on Knowledge Discovery and Data Mining*, 269–274.

Díaz-Uriarte, R., & Alvarez de Andrés, S. (2006). Gene selection and classification of microarray data using random forest. *BMC Bioinformatics*, *7*, 3. doi:10.1186/1471-2105-7-3

Diella, F., Niall, H., Chica, C., Budd, A., Michael, S., & Brown, N. P. (2008). Understanding eukaryotic linear motifs and their role in cell signaling and regulation. *Frontiers in Bioscience*, *13*, 6580–6603. doi:10.2741/3175

Dietmann, S., Georgii, E., Antonov, A., Tsuda, K., & Mewes, H. W. (2009). The DICS repository: Module-assisted analysis of disease-related gene lists. *Bioinformatics (Oxford, England)*, *25*(6), 830–831. doi:10.1093/bioinformatics/btp055

Dion, M. F., Altschuler, S. J., Wu, L. F., & Rando, O. J. (2005). Genomic characterization reveals a simple histone H4 acetylation code. *Proceedings of the National Academy of Sciences of the United States of America*, *102*(15), 5501–5506.

Dittrich, M., Klau, G., Rosenwald, A., Dandekar, T., & Müller, T. (2008). Identifying functional modules in protein-protein interaction networks: An integrated exact approach. *Bioinformatics (Oxford, England)*, *24*(13), i223–i231. doi:10.1093/bioinformatics/btn161

Divina, F., & Aguilar-Ruiz, J. S. (2006). Biclustering of expression data with evolutionary computation. *IEEE Transactions on Knowledge and Data Engineering*, *18*, 590–602. doi:10.1109/TKDE.2006.74

Dixon, A. L., Liang, L., Moffatt, M. F., Chen, W., Heath, S., & Wong, K. C. (2007). A genome-wide association study of global gene expression. *Nature Genetics*, *39*(10), 1202–1207. doi:10.1038/ng2109

Dobbin, K., & Simon, R. (2007)... *Biostatistics (Oxford, England)*, *8*, 101–117. doi:10.1093/biostatistics/kxj036

Dobbin, K., Zhao, Y. D., & Simon, R. (2008)... *Clinical Cancer Research, 14*, 108–114. doi:10.1158/1078-0432.CCR-07-0443

Dobbin, K. K., Beer, D. G., Meyerson, M., Yeatman, T. J., Gerald, W. L., & Jacobson, J. W. (2005). Interlaboratory comparability study of cancer gene expression analysis using oligonucleotide microarrays. *Clinical Cancer Research, 11*(2 Pt 1), 565–572.

Dojer, N., Gambin, A., Mizera, A., Wilczyński, B., & Tiuryn, J. (2006). Applying dynamic Bayesian networks to perturbed gene expression data. *BMC Bioinformatics, 7*, 249. doi:10.1186/1471-2105-7-249

Doman, T. N., McGovern, S. L., Witherbee, B. J., Kasten, T. P., Kurumbail, R., & Stallings, W. C. (2002). Molecular docking and high-throughput screening for novel inhibitors of protein tyrosine phosphatase-1B. *Journal of Medicinal Chemistry, 45*(11), 2213–2221. doi:10.1021/jm010548w

Doms, A. & Schroeder, M. (2005). GoPubMed: Exploring PubMed with the gene ontology. *Nucleic Acids Research, 33*(Web Server issue), W783-786.

Donadio, S., & Katz, L. (1992). Organization of the enzymatic domains in the multifunctional polyketide synthase involved in erythromycin formation in Saccharopolyspora erythraea. *Gene, 111*(1), 51–60. doi:10.1016/0378-1119(92)90602-L

Donaldson, T., & Dunfee, W. T. (1999). *Ties that Bind: A Social Contracts Approach to Business Ethics*. Cambridge, MA: Harvard Business School Press.

Donetti, L., & Muñoz, M. (2004). Detecting network communities: A new systematic and efficient algorithm. *Journal of Statistical Mechanics, 2004*, P10012. doi:10.1088/1742-5468/2004/10/P10012

Dong, J., Olano, J. P., McBride, J. W., & Walker, D. H. (2008). Emerging pathogens: Challenges and successes of molecular diagnostics. *The Journal of Molecular Diagnostics, 10*(3), 185–197. doi:10.2353/jmoldx.2008.070063

Dong, C., & Lei, X. (2000). Biomechanics of cells rolling: Shear flow, cell-surface adhesion, and cell deformability. *Journal of Biomechanics, 33*, 35–43. doi:10.1016/S0021-9290(99)00174-8

Donnelly, P. (2008). Progress and challenges in genome-wide association studies in human. *Nature, 456*(7223), 728–731. doi:10.1038/nature07631

Dorwart, M. R., Shcheynikov, N., Yang, D., & Muallem, S. (2008). The solute carrier 26 family of proteins in epithelial ion transport. *Physiology (Bethesda, MD), 23*, 104–114. doi:10.1152/physiol.00037.2007

Dost, B., Shlomi, T., Gupta, N., Ruppin, E., Bafna, V., & Sharan, R. (2008). Qnet: A tool for querying protein interaction networks. *Journal of Computational Biology, 15*, 1–15. doi:10.1089/cmb.2007.0172

Downward, J. (2003). Targeting RAS signalling pathways in cancer therapy. *Nature Reviews. Cancer, 3*(1), 11–22. doi:10.1038/nrc969

Draghici, S., Khatri, P., Bhavsar, P., Shah, A., Krawetz, S. A., & Tainsky, M. A. (2003, July). Onto-Tools, the toolkit of the modern biologist: Onto-Express, Onto-Compare, Onto-Design and Onto-Translate. *Nucleic Acids Research, 31*(13), 3775–3781. doi:10.1093/nar/gkg624

Draghici, S., Khatri, P., Tarca, A. L., Amin, K., Done, A., & Voichita, C. (2007). A systems biology approach for pathway level analysis. *Genome Research, 17*(10), 1537–1545. doi:10.1101/gr.6202607

Draghici, S., Khatri, P., Eklund, A. C., & Szallasi, Z. (2006). Reliability and reproducibility issues in DNA microarray measurements. *Trends in Genetics, 22*(2), 101–109. doi:10.1016/j.tig.2005.12.005

Drews, J. (2000). Drug discovery: A historical perspective. *Science, 287*(5460), 1960–1964. doi:10.1126/science.287.5460.1960

Du, Y., Parks, B. A., Sohn, S., Kwast, K. E., & Kelleher, N. L. (2006). Top-down approaches for measuring expression ratios of intact yeast proteins using Fourier transform mass spectrometry. *Analytical Chemistry, 78*(3), 686–694. doi:10.1021/ac050993p

Du, L., Sanchez, C., Chen, M., Edwards, D. J., & Shen, B. (2000). The biosynthetic gene cluster for the antitumor drug bleomycin from Streptomyces verticillus ATCC15003 supporting functional interactions between nonribosomal peptide synthetases and a polyketide synthase. *Chemistry & Biology, 7*(8), 623–642. doi:10.1016/S1074-5521(00)00011-9

Duarte, N. C. (2007). Global reconstruction of the human metabolic network based on genomic and bibliomic data. *Proceedings of the National Academy of Sciences of the United States of America, 104*(6), 1777–1782. doi:10.1073/pnas.0610772104

Duch, J., & Arenas, A. (2005). Community detection in complex networks using extremal optimization. *Physical Review E: Statistical, Nonlinear, and Soft Matter Physics, 72*(2), 027104. doi:10.1103/PhysRevE.72.027104

Duda, P. (2001). *Pattern classification*. New York: Wiley.

Dudoit, S., Fridlyand, J., & Speed, T. P. (2002). Comparison of discrimination methods for the classification of tumors using gene expression data. *Journal of the American Statistical Association, 97*, 77–87. doi:10.1198/016214502753479248

Dumas, M. E., Canlet, C., Andre, F., Vercauteren, J., & Paris, A. (2002). Metabonomic assessment of physiological disruptions using ^{1}H-^{13}C HMBC-NMR spectroscopy combined with pattern recognition procedures performed on filtered variables. *Analytical Chemistry, 74*(10), 2261–2273. doi:10.1021/ac0156870

Dunlop, M., & Murray, R. (2006). *Towards biological system identification: Fast and accurate estimates of parameters in genetic regulatory networks*. Paper presented at the 45th IEEE Conference on Decision and Control.

Dunn, W. B. (2008). Current trends and future requirements for the mass spectrometric investigation of microbial, mammalian and plant metabolomes. *Physical Biology*, *5*, 1–24. doi:10.1088/1478-3975/5/1/011001

Dunning, M. J., Smith, M. L., Ritchie, M. E., & Tavare, S. (2007). beadarray: R classes and methods for Illumina bead-based data. *Bioinformatics (Oxford, England)*, *23*, 2183–2184. doi:10.1093/bioinformatics/btm311

Dwinell, M. R., et al. (2009). The Rat Genome Database 2009: Variation, ontologies and pathways. *Nucleic Acids Research, 37*(database issue), D744-749.

Dykhuizen, D. E., & Kalia, A. (2008). *Population genetics of pathogenic bacteria* (2nd ed.). Oxford University Press.

Ebert, B., Pretz, J., Bosco, J., Chang, C., Tamayo, P., & Galili, N. (2008). Identification of RPS14 as a 5q- syndrome gene by RNA interference screen. *Nature*, *451*(7176), 335–339. doi:10.1038/nature06494

Eddy, S. R. (2001). Non-coding RNA genes and the modern RNA world. *Nature Reviews. Genetics*, *2*(12), 919–929. doi:10.1038/35103511

Edelman, G. (1976). Surface modulation in cell recognition and cell growth. *Science*, *192*, 219–226. doi:10.1126/science.769162

Edelman, L.B., Eddy, J.A. & Price N.D. (2009). In silico models of cancer. *Wiley Interdisciplinary Reviews: Systems Biology and Medicine,* 1939-5094.

Edgar, R. C. (2004). MUSCLE: A multiple sequence alignment method with reduced time and space complexity. *BMC Bioinformatics*, *5*, 113. doi:10.1186/1471-2105-5-113

Edwards, A. O., Ritter, R. III, Abel, K. J., Manning, A., Panhuysen, C., & Farrer, L. A. (2005). Complement factor H polymorphism and age-related macular degeneration. *Science*, *308*(5720), 421–424. doi:10.1126/science.1110189

Efron, B., & Tibshirani, R. (2007). On testing the significance of sets of genes. *Annual Applied Statistics*, *1*, 107–129. doi:10.1214/07-AOAS101

Eggleton, C. D., & Popel, A. S. (1998). Large deformation of red blood cell in a simple shear flow. *Physics of Fluids*, *10*(8). doi:10.1063/1.869703

Einav, S., Gerber, D., Bryson, P. D., Sklan, E. H., Elazar, M., & Maerkl, S. J. (2008). Discovery of a hepatitis C target and its pharmacological inhibitors by microfluidic affinity analysis. *Nature Biotechnology*, *26*(9), 1019–1027. doi:10.1038/nbt.1490

Ein-Dor, L., Zuk, O., & Domany, E. (2006). Thousands of samples are needed to generate a robust gene list for predicting outcome in cancer. *Proceedings of the National Academy of Sciences of the United States of America*, *103*(15), 5923–5928. doi:10.1073/pnas.0601231103

Eisen, M. B., Spellman, P. T., Brown, P. O., & Botstein, D. (1998). Cluster analysis and display of genome-wide expression patterns. *Proceedings of the National Academy of Sciences of the United States of America*, *95*(25), 14863–14868. doi:10.1073/pnas.95.25.14863

Eldridge, M. D., Murray, C. W., Auton, T. R., Paolini, G. V., & Mee, R. P. (1997). Empirical scoring functions: I. The development of a fast empirical scoring function to estimate the binding affinity of ligands in receptor complexes. *Journal of Computer-Aided Molecular Design*, *11*(5), 425–445. doi:10.1023/A:1007996124545

Elger, B. S., & Caplan, A. L. (2006). Consent and anonymization in research involving biobanks: Differing terms and norms present serious barriers to an international framework. *EMBO Reports*, *7*(7), 661–666. doi:10.1038/sj.embor.7400740

El-Sayed, N. M., Myler, P. J., Bartholomeu, D. C., Nilsson, D., Aggarwal, G., & Tran, A. N. (2005). The genome sequence of Trypanosoma cruzi, etiologic agent of Chagas disease. *Science*, *309*(5733), 409–415. doi:10.1126/science.1112631

Elston, R. C., & Cordell, H. J. (2001). Overview of model-free methods for linkage analysis. *Advances in Genetics*, *42*, 135–150. doi:10.1016/S0065-2660(01)42020-7

Emanuel, E., Wendler, D., & Grady, C. (2000). What makes clinical research ethical? *Journal of the American Medical Association*, *283*(20), 2701–2711. doi:10.1001/jama.283.20.2701

Emilsson, V., Thorleifsson, G., Zhang, B., Leonardson, A. S., Zink, F., & Zhu, J. (2008). Genetics of gene expression and its effect on disease. *Nature*, *452*(7186), 423–428. doi:10.1038/nature06758

Emily, M., Mailund, T., Hain, J., Schauser, L., & Schierup, M. H. (2009). Using biological networks to search for interacting loci in genome-wide association studies. *European Journal of Human Genetics*, *17*(10), 1231–1240. doi:10.1038/ejhg.2009.15

Enright, A. J. (2002). An efficient algorithm for large-scale detection of protein families. *Nucleic Acids Research*, *30*(7), 1575–1584. doi:10.1093/nar/30.7.1575

EOP. (2010). *Grand challenges for the 21st century.* Retrieved February 18, 2010, from http://www.whitehouse.gov/administration/eop/ostp/grand-challenges-request-information.

Erhardt, R. A., Schneider, R., & Blaschke, C. (2006). Status of text-mining techniques applied to biomedical text. *Drug Discovery Today*, *11*(7-8), 315–325. doi:10.1016/j.drudis.2006.02.011

Eriksen, K. (2003). Modularity and extreme edges of the Internet. *Physical Review Letters*, *90*(14). doi:10.1103/PhysRevLett.90.148701

Eriksson, L., Johansson, E., Kettaneh-Wold, N., & Wold, S. (2001). *Multi- and megavariate data analysis*. Umetrics Academy.

Esquela-Kerscher, A., & Slack, F. J. (2006). Oncomirs-microRNAs with a role in cancer. *Nature Reviews. Cancer*, *6*(4), 259–269. doi:10.1038/nrc1840

Esteller, M. (2007). Cancer epigenomics: DNA methylomes and histone-modification maps. *Nature Reviews. Genetics*, *8*(4), 286–298.

Evans, B. J., Flockhart, D. A., & Meslin, E. M. (2004). Creating incentives for genomics research to improve targeting of therapies. *Nature Medicine*, *10*, 1289–1291. doi:10.1038/nm1204-1289

Evans, B. J., & Meslin, E. M. (2006). Encouraging Translational Research Through Harmonization of FDA and Common-Rule Informed Consent Requirements for Research with Banked Specimens. *Journal of Legal Medicine*, *27*, 119–166. doi:10.1080/01947640600716366

Evans, N. D., Chapman, M. J., Chappell, M. J., & Godfrey, K. R. (2002). Identifiability of uncontrolled nonlinear rational systems. *Automatica*, *38*(10), 1799–1805. doi:10.1016/S0005-1098(02)00094-8

Evans, N. D., Errington, R. J., Shelley, M., Feeney, G. P., Chapman, M. J., & Godfrey, K. R. (2004). A mathematical model for the in vitro kinetics of the anti-cancer agent topotecan. *Mathematical Biosciences*, *189*(2), 185–217. doi:10.1016/j.mbs.2004.01.007

Evans, E. (1992). Equilibrium wetting of surfaces by membrane-covered vesicles. *Advances in Colloid and Interface Science*, *39*, 103–128. doi:10.1016/0001-8686(92)80057-5

Evans, E., & Needham, D. (1987). Physical properties of surfactant bilayer membranes: Thermal, transition, elasticity, rigidity, cohesion, and colloidal interactions. *Journal of Physical Chemistry*, *91*, 4219–4228. doi:10.1021/j100300a003

Evans, E., & Ritchie, K. (1997). Dynamic strength of molecular adhesion bonds. *Biophysical Journal*, *72*, 1541–1555. doi:10.1016/S0006-3495(97)78802-7

Everett, M. G., & Borgatti, S. P. (1998). Analyzing clique overlap. *Connections*, *21*, 49–61.

Ewald, P. W. (2004). Evolution of virulence. *Infectious Disease Clinics of North America*, *18*(1), 1–15. doi:10.1016/S0891-5520(03)00099-0

Ewing, R. M., Chu, P., Elisma, F., Li, H., Taylor, P., & Climie, S. (2007). Large-scale mapping of human protein-protein interactions by mass spectrometry. *Molecular Systems Biology*, *3*, 89. doi:10.1038/msb4100134

Falush, D., Wirth, T., Linz, B., Pritchard, J. K., Stephens, M., & Kidd, M. (2003). Traces of human migrations in Helicobacter pylori populations. *Science*, *299*(5612), 1582–1585. doi:10.1126/science.1080857

Fan, S., Zhang, M. Q., & Zhang, X. (2008). Histone methylation marks play important roles in predicting the methylation status of CpG islands. *Biochemical and Biophysical Research Communications*, *374*(3), 559–564.

Fan, C., Oh, D. S., Wessels, L., Weigelt, B., Nuyten, D. S., & Nobel, A. B. (2006). Concordance among gene-expression-based predictors for breast cancer. *The New England Journal of Medicine*, *355*(6), 560–569. doi:10.1056/NEJMoa052933

Fan, H.C., Blumenfeld, Y.J., El-Sayed, Y.Y., Chueh, J. & Quake, S.R. (2009). Microfluidic digital PCR enables rapid prenatal diagnosis of fetal aneuploidy. *American Journal of Obstetrics and Gynecology*, *200*(5), 543, e541-543, e547.

Fanciulli, M., Norsworthy, P. J., Petretto, E., Dong, R., Harper, L., & Kamesh, L. (2007). FCGR3B copy number variation is associated with susceptibility to systemic, but not organ-specific, autoimmunity. *Nature Genetics*, *39*(6), 721–723. doi:10.1038/ng2046

Fang, F., Fan, S., Zhang, X., & Zhang, M. Q. (2006). Predicting methylation status of CpG islands in the human brain. *Bioinformatics (Oxford, England)*, *22*(18), 2204–2209.

Farina, M., Findeisen, R., Bullinger, E., Bittanti, S., Allgower, F., & Wellstead, P. (2006). *Results towards identifiability properties of biochemical reaction networks*. Paper presented at the 45th IEEE Conference on Decision and Control, 2104–2109.

Farley, M. M. (1995). Group B streptococcal infection in older patients. Spectrum of disease and management strategies. *Drugs & Aging*, *6*(4), 293–300. doi:10.2165/00002512-199506040-00004

Fayad, W., Brnjic, S., Berglind, D., Blixt, S., Shoshan, M. C., & Berndtsson, M. (2009). Restriction of cisplatin induction of acute apoptosis to a subpopulation of cells in a three-dimensional carcinoma culture model. *International Journal of Cancer*, *125*(10), 2450–2455. doi:10.1002/ijc.24627

Feher, M. (2006). Consensus scoring for protein-ligand interactions. *Drug Discovery Today*, *11*(9-10), 421–428. doi:10.1016/j.drudis.2006.03.009

Feil, E. J., & Spratt, B. G. (2001). Recombination and the population structures of bacterial pathogens. *Annual Review of Microbiology*, *55*(1), 561–590. doi:10.1146/annurev.micro.55.1.561

Feinberg, A., & Irizarry, R. (2010). Stochastic epigenetic variation as a driving force of development, evolutionary adaptation, and disease. *Proceedings of the National Academy of Sciences, Early Edition.*

Feldman, I., Rzhetsky, A., & Vitkup, D. (2008). Network properties of genes harboring inherited disease mutations. *Proceedings of the National Academy of Sciences of the United States of America*, *105*(11), 4323–4328. doi:10.1073/pnas.0701722105

Feltus, F. A., Lee, E. K., Costello, J. F., Plass, C., & Vertino, P. M. (2006). DNA motifs associated with aberrant CpG island methylation. *Genomics*, *87*(5), 572–579.

Fenn, J. B., Mann, M., Meng, C. K., Wong, S. F., & Whitehouse, C. M. (1989). Electrospray ionization for mass spectrometry of large biomolecules. *Science*, *246*, 64. doi:10.1126/science.2675315

Fenton, G. A. (1990). *Simulation and analysis of random fields*. PhD thesis, Princeton University.

Ferguson, J. T., Wenger, C. D., Metcalf, W. W., & Kelleher, N. L. (2009). Top-down proteomics reveals novel protein forms expressed in Methanosarcina acetivorans. *JASMS*, *20*(9), 1743–1750.

Fey, D., Findeisen, R., & Bullinger, E. (2008). *Parameter estimation in kinetic reaction models using nonlinear observers is facilitated by model extensions*. Paper presented at the 17th IFAC World Congress.

Fiehn, O. (2002). Metabolomics–the link between genotypes and phenotypes. *Plant Molecular Biology*, *48*(1), 155–171. doi:10.1023/A:1013713905833

Field, Y., Fondufe-Mittendorf, Y., Moore, I. K., Mieczkowski, P., Kaplan, N., & Lubling, Y. (2009). Gene expression divergence in yeast is coupled to evolution of DNA-encoded nucleosome organization. *Nature Genetics*, *41*(4), 438–445.

Field, Y., Kaplan, N., Fondufe-Mittendorf, Y., Moore, I. K., Sharon, E., & Lubling, Y. (2008). Distinct modes of regulation by chromatin encoded through nucleosome positioning signals. *PLoS Computational Biology*, *4*(11), e1000216.

Figueroa, M. E., Lugthart, S., Li, Y., Erpelinck-Verschueren, C., Deng, X., & Christos, P. J. (2010). DNA methylation signatures identify biologically distinct subtypes in Acute Myeloid Leukemia. *Cancer Cell*, *17*(1), 13–27. doi:10.1016/j.ccr.2009.11.020

Fine, P. E. (1995). Variation in protection by BCG: Implications of and for heterologous immunity. *Lancet*, *346*(8986), 1339–1345. doi:10.1016/S0140-6736(95)92348-9

Fischbach, M. A., & Walsh, C. T. (2006). Assembly-line enzymology for polyketide and nonribosomal Peptide antibiotics: Logic, machinery, and mechanisms. *Chemical Reviews*, *106*(8), 3468–3496. doi:10.1021/cr0503097

Fischer, E. (1894). Einfluss der Configuration auf die Wirkung der Enzyme. *Ber. Dtsch. Chem. Ges.*, *27*, 2984–2993.

Fishel, I., Kaufman, A., & Ruppin, E. (2007). Meta-analysis of gene expression data: A predictor-based approach. *Bioinformatics (Oxford, England)*, *23*(13), 1599–1606. doi:10.1093/bioinformatics/btm149

Fisher, J., & Henzinger, T. (2007). Executable cell biology. *Nature Biotechnology*, *25*(11), 1239–1249. doi:10.1038/nbt1356

Fisher, R. A. (1925). *Statistical methods for research workers*. London: Edinburg.

Fishman, M. C., & Porter, J. A. (2005). Pharmaceuticals: A new grammar for drug discovery. *Nature*, *437*(7058), 491–493. doi:10.1038/437491a

Fitch, W. M. (1969). Locating gaps in amino acid sequences to optimize the homology between two proteins. *Biochemical Genetics*, *3*(2), 99–108. doi:10.1007/BF00520346

Flamm, C., Fontana, W., Hofacker, I. L., & Schuster, P. (2000). RNA folding at elementary step resolution. *RNA (New York, N.Y.)*, *6*(3), 325–338. doi:10.1017/S1355838200992161

Fleischmann, R. D., Adams, M. D., White, O., Clayton, R. A., Kirkness, E. F., & Kerlavage, A. R. (1995). Whole-genome random sequencing and assembly of Haemophilus influenzae Rd. *Science*, *269*(5223), 496–512. doi:10.1126/science.7542800

Flicek, P., Aken, B. L., Beal, K., Ballester, B., Caccamo, M., & Chen, Y. (2008). Ensembl 2008. *Nucleic Acids Research*, *36*(Database issue), D707–D714. doi:10.1093/nar/gkm988

Flyvbjerg, H., Jobs, E., & Leibler, S. (1996). Kinetics of self-assembling microtubules: An inverse problem in biochemistry. *Proceedings of the National Academy of Sciences of the United States of America*, *93*(12), 5975–5979. doi:10.1073/pnas.93.12.5975

Fortunato, S., & Barthelemy, M. (2007). Resolution limit in community detection. *Proceedings of the National Academy of Sciences of the United States of America*, *104*(1), 36–41. doi:10.1073/pnas.0605965104

Fortunato, S. (2009). *Community detection in graphs*. arXiv:0906.0612.

Foster, L. J. (2006). A mammalian organelle map by protein correlation profiling. *Cell*, *125*(1), 187–199. doi:10.1016/j.cell.2006.03.022

Fournier, M. L., Gilmore, J. M., Martin-Brown, S. A., & Washburn, M. P. (2007). Multidimensional separations-based shotgun proteomics. *Chemical Reviews*, *107*(8), 3654–3686. doi:10.1021/cr068279a

Foxman, E. R., & Kilcoyne, P. (1993). Information Technology, Marketing Practice, and Consumer Privacy: Ethical Issues. *Journal of Public Policy & Marketing*, *12*(1), 106–119.

Francis, N. J., & Kingston, R. E. (2001). Mechanisms of transcriptional memory. *Nature Reviews. Molecular Cell Biology*, *2*(6), 409–421.

Frank, D. N., & Pace, N. R. (1998). Ribonuclease P: Unity and diversity in a tRNA processing ribozyme. *Annual Review of Biochemistry*, *67*, 153–180. doi:10.1146/annurev.biochem.67.1.153

Franke, L., Schwarz, O., Muller-Kuhrt, L., Hoernig, C., Fischer, L., & George, S. (2007). Identification of natural-product-derived inhibitors of 5-lipoxygenase activity by ligand-based virtual screening. *Journal of Medicinal Chemistry*, *50*(11), 2640–2646. doi:10.1021/jm060655w

Franke, L., Bakel, H., Fokkens, L., de Jong, E. D., Egmont-Petersen, M., & Wijmenga, C. (2006). Reconstruction of a functional human gene network, with an application for prioritizing positional candidate genes. *American Journal of Human Genetics*, *78*(6), 1011–1025. doi:10.1086/504300

Franzosa, E., Linghu, B., & Xia, Y. (2009). Computational reconstruction of protein-protein interaction networks: Algorithms and issues. *Methods in Molecular Biology (Clifton, N.J.)*, *541*, 89–100. doi:10.1007/978-1-59745-243-4_5

Free, S. M. (1964). A mathematical contribution to structure–activity studies. *Journal of Medicinal Chemistry*, *7*, 395–399. doi:10.1021/jm00334a001

Freeman, L. (1979). Centrality in social networks conceptual clarification. *Social Networks*, *1*(3), 215–239. doi:10.1016/0378-8733(78)90021-7

Freeman, D. J., Li, A. G., Wei, G., Li, H. H., Kertesz, N., & Lesche, R. (2003). PTEN tumor suppressor regulates p53 protein levels and activity through phosphatase-dependent and -independent mechanisms. *Cancer Cell*, *3*(2), 117–130. doi:10.1016/S1535-6108(03)00021-7

Freier, S. M., Kierzek, R., Jaeger, J. A., Sugimoto, N., Caruthers, M. H., & Neilson, T. (1986). Improved parameters for predictions of RNA duplex stability. *Proceedings of the National Academy of Sciences of the United States of America*, *83*(24), 9373–9377. doi:10.1073/pnas.83.24.9373

Freue, G. V. C., Hollander, Z., Shen, E., Zamar, R. H., Balshaw, R., & Scherer, A. (2007). MDQC: A new quality assessment method for microarrays based on quality control reports. *Bioinformatics (Oxford, England)*, *23*, 3162–3169. doi:10.1093/bioinformatics/btm487

Friedberg, E. C. (2003). DNA damage and repair. *Nature*, *421*(6921), 436–440. doi:10.1038/nature01408

Friedman, N. (2004). Inferring cellular networks using probabilistic graphical models. *Science*, *303*(5659), 799–805. doi:10.1126/science.1094068

Friedman, A., & Perrimon, N. (2006). High-throughput approaches to dissecting MAPK signaling pathways. *Nature*, *40*(3), 262–271.

Friedman, A., & Perrimon, N. (2007). Genetic screening for signal transduction in the era of network biology. *Cell*, *128*(2), 225–231. doi:10.1016/j.cell.2007.01.007

Friedman, N., Linial, M., Nachman, I. & Pe'er, D. (2000). Using Bayesian networks to analyze expression data. *Journal of Computational Biology: A Journal of Computational Molecular Cell Biology*, *7*(3-4), 601-20.

Friesner, R. A., Banks, J. L., Murphy, R. B., Halgren, T. A., Klicic, J. J., & Mainz, D. T. (2004). Glide: A new approach for rapid, accurate docking and scoring. 1. Method and assessment of docking accuracy. *Journal of Medicinal Chemistry*, *47*(7), 1739–1749. doi:10.1021/jm0306430

Friesner, R. A., Murphy, R. B., Repasky, M. P., Frye, L. L., Greenwood, J. R., & Halgren, T. A. (2006). Extra precision glide: Docking and scoring incorporating a model of hydrophobic enclosure for protein-ligand complexes. *Journal of Medicinal Chemistry*, *49*(21), 6177–6196. doi:10.1021/jm051256o

Friston, K. (2009). Causal modelling and brain connectivity in functional magnetic resonance imaging. *PLoS Biology*, *7*(2). doi:10.1371/journal.pbio.1000033

Friston, K., Harrison, L., & Penny, W. (2003). Dynamic causal modelling. *NeuroImage*, *19*(4), 1273–1302. doi:10.1016/S1053-8119(03)00202-7

Frith, M. C., Fu, Y., Yu, L., Chen, J. F., Hansen, U., & Weng, Z. (2004). Detection of functional DNA motifs via statistical over-representation. *Nucleic Acids Research*, *32*(4), 1372–1381. doi:10.1093/nar/gkh299

Frith, M. C., Li, M. C., & Weng, Z. (2003). Cluster-buster: Finding dense clusters of motifs in DNA sequences. *Nucleic Acids Research*, *31*(13), 3666–3668. doi:10.1093/nar/gkg540

Frolkis, A., et al. (2010). SMPDB: The Small Molecule Pathway Database. *Nucleic Acids Research, 38*(database issue), D480-487.

Fuller, T. F., Ghazalpour, A., Aten, J. E., Drake, T. A., Lusis, A. J., & Horvath, S. (2007). Weighted gene coexpression network analysis strategies applied to mouse weight. *Mammalian Genome*, *18*(6-7), 463–472. doi:10.1007/s00335-007-9043-3

Fullerton, S. M., Anderson, N. R., Guzauskas, G., Freeman, D., & Fryer-Edwards, K. (2010). Meeting the Governance Challenges of Next-Generation Biorepository Research. *Science Translational Medicine*, *2*(15), cm3. doi:10.1126/scitranslmed.3000361

Furey, T. S., Cristianini, N., Duffy, N., Bednarski, D. W., Schummer, M., & Haussler, D. (2000). Support vector machine classification and validation of cancer tissue samples using microarray expression data. *Bioinformatics (Oxford, England)*, *16*, 906–914. doi:10.1093/bioinformatics/16.10.906

Furuta, T., Ueda, T., Aune, G., Sarasin, A., Kraemer, A., & Pommier, Y. (2002). Transcription-coupled nucleotide excision repair as a determinant of cisplatin sensitivity of human cells. *Cancer Research*, *62*(17), 4899–4902.

Futreal, P. A., Coin, L., Marshall, M., Down, T., Hubbard, T., & Wooster, R. (2004). A census of human cancer genes. *Nature Reviews. Cancer*, *4*(3), 177–183. doi:10.1038/nrc1299

Gaasterland, T., & Bekiranov, S. (2000). Making the most of microarray data. *Nature Genetics*, *24*, 204–206. doi:10.1038/73392

Gadkar, K., Varner, J., & Doyle, F. III. (2005). Model identification of signal transduction networks from data using a state regulator problem. *Systems Biology*, *2*(1), 17–30. doi:10.1049/sb:20045029

Gan, W., Zhao, G., Xu, H., Wu, W., Du, W., & Huang, J. (2010). Reverse vaccinology approach identify an Echinococcus granulosus tegumental membrane protein enolase as vaccine candidate. *Parasitology Research*, *106*(4), 873–882. doi:10.1007/s00436-010-1729-x

Gan, G. (2007). *Data clustering: Theory, algorithms, and applications*. Society for Industrial and Applied Mathematics. doi:10.1137/1.9780898718348

Ganter, B., Snyder, R. D., Halbert, D. N., & Lee, M. D. (2006). Toxicogenomics in drug discovery and development: Mechanistic analysis of compound/class-dependent effects using the DrugMatrix database. *Pharmacogenomics*, *7*(7), 1025–1044. doi:10.2217/14622416.7.7.1025

Gardiner, W. P., & Gettinby, G. (1998). *Experimental design techniques in statistical practice: A practical software-based approach*. Chichester, W. Sussex, UK: Horwood Pub.

Gardner, M. J., Hall, N., Fung, E., White, O., Berriman, M., & Hyman, R. W. (2002). Genome sequence of the human malaria parasite Plasmodium falciparum. *Nature*, *419*(6906), 498–511. doi:10.1038/nature01097

Gardner, T. S., di Bernardo, D., Lorenz, D., & Collins, J. J. (2003). Inferring genetic networks and identifying compound mode of action via expression profiling. *Science*, *301*(5629), 102–105. doi:10.1126/science.1081900

Gardner, T. S., & Faith, J. J. (2005). Reverse-engineering transcription control networks. *Physics of Life Reviews*, *2*, 65–88. doi:10.1016/j.plrev.2005.01.001

Garey, M. R., & Johnson, D. S. (1979). *Computers and intractability: A guide to the theory of NP-completeness*. W. H. Freeman.

Gargalovic, P. S. (2006). Identification of inflammatory gene modules based on variations of human endothelial cell responses to oxidized lipids. *Proceedings of the National Academy of Sciences of the United States of America*, *103*(34), 12741–12746. doi:10.1073/pnas.0605457103

Garrison, F. H. (1917). *An Introduction to the History of Medicine* (2nd ed.). Philadelphia: W.B. Saunders Company.

Gasch, A. P., Spellman, P. T., Kao, C. M., Carmel-Harel, O., Eisen, M. B., & Storz, G. (2000). Genomic expression programs in the response of yeast cells to environmental changes. *Molecular Biology of the Cell*, *11*(12), 4241–4257.

Gasch, A. P. (2000). Genomic expression programs in the response of yeast cells to environmental changes. *Molecular Biology of the Cell*, *11*(12), 4241–4257.

Gaudet, M., Fara, A. G., Beritognolo, I., & Sabatti, M. (2009). Allele-specific PCR in SNP genotyping. *Methods in Molecular Biology (Clifton, N.J.)*, *578*, 415–424. doi:10.1007/978-1-60327-411-1_26

Gaul, W., & Schader, M. (1996). A new algorithm for two-mode clustering. In Bock, H.-H., & Polasek, W. (Eds.), *Data analysis and Information Systems* (pp. 15–23). Heidelberg: Springer.

Gautier, L., Cope, L., Bolstad, B. M., & Irizarry, R. A. (2004). Affy–analysis of affymetrix genechip data at the probe level. *Bioinformatics (Oxford, England)*, *20*, 307–315. doi:10.1093/bioinformatics/btg405

Gavin, A. C. (2006). Proteome survey reveals modularity of the yeast cell machinery. *Nature*, *440*(7084), 631–636. doi:10.1038/nature04532

Ge, T., Kendrick, K. & Feng, J. (2009). *A novel extended Granger causal model approach demonstrates brain hemispheric differences during face recognition learning.*

Gelsi-Boyer, V., Cervera, N., Bertucci, F., Trouplin, V., Remy, V., & Olschwang, S. (2007). Gene expression profiling separates chronic myelomonocytic leukemia in two molecular subtypes. *Leukemia*, *21*(11), 2359–2362. doi:10.1038/sj.leu.2404805

Gene Ontology. (2010). *Home page*. Retrieved from http://www.geneontology.org

Gentleman, R. C., Carey, V. J., Bates, D. J., Bolstad, B. M., Dettling, M., & Dudoit, S. (2004). Bioconductor: Open software development for computational biology and bioinformatics. *Genome Biology*, *5*, R80. doi:10.1186/gb-2004-5-10-r80

Gentleman, R. C., Garey, V. J., Huber, W., Irizarry, R., & Dudoit, S. (2005). *Bioinformatics and computational biology solutions using R and Bioconductor*. New York: Springer. doi:10.1007/0-387-29362-0

Gentleman, R. C., Carey, V. J., Bates, D. M., Bolstad, B., Dettling, M., & Dudoit, S. (2004). Bioconductor: Open software development for computational biology and bioinformatics. *Genome Biology*, *5*(10), R80. doi:10.1186/gb-2004-5-10-r80

Georgii, E., Dietmann, S., Uno, T., Pagel, P., & Tsuda, K. (2009). Enumeration of condition-dependent dense modules in protein interaction networks. *Bioinformatics (Oxford, England)*, *25*(7), 933–940. doi:10.1093/bioinformatics/btp080

Gerber, S. A., Rush, J., Stemman, O., Kirschner, M. W., & Gygi, S. P. (2003). Absolute quantification of proteins and phosphoproteins from cell lysates by tandem MS. *Proceedings of the National Academy of Sciences of the United States of America*, *100*(12), 6940–6945. doi:10.1073/pnas.0832254100

Gerber, D., Maerkl, S. J., & Quake, S. R. (2009). An in vitro microfluidic approach to generating protein-interaction networks. *Nature Methods*, *6*(1), 71–74. doi:10.1038/nmeth.1289

Getz, G., Levine, E., & Domany, E. (2000). Coupled two-way clustering analysis of gene microarray data. *Proceedings of the National Academy of Sciences of the United States of America*, *97*(22), 12079–12084. doi:10.1073/pnas.210134797

Geva-Zatorsky, N., Rosenfeld, N., Itzkovitz, S., Milo, R., Sigal, A., & Dekel, E. (2006). Oscillations and variability in the p53 network. *Molecular Systems Biology*, *2*, 33. doi:10.1038/msb4100068

Geweke, J. (1982). Measurement of linear dependence and feedback between multiple time series. *Journal of the American Statistical Association*, *77*(378), 304–313. doi:10.2307/2287238

Geweke, J. (1984). Measures of conditional linear dependence and feedback between time series. *Journal of the American Statistical Association*, *79*(388), 907–915. doi:10.2307/2288723

Ghanem, R., & Spanos, P. D. (2003). *Stochastic finite elements: A spectral approach.* Dover Publications Inc.

Ghobrial, I. M., Witzig, T. E., & Adjei, A. A. (2005). Targeting apoptosis pathways in cancer therapy. *CA: a Cancer Journal for Clinicians*, *55*(3), 178–194. doi:10.3322/canjclin.55.3.178

Ghosh, R., & Tomlin, C. (2004). Symbolic reachable set computation of piecewise affine hybrid automata and its application to biological modelling: Delta-notch protein signaling. Paper presented at *the IEE Proceedings. Systems Biology*, *1*(1), 170–183. doi:10.1049/sb:20045019

Giacomini, K. M., Brett, C. M., Altman, R. B., Benowitz, N. L., Dolan, M. E., & Flockhart, D. A. (2007). The pharmacogenetics research network: from SNP discovery to clinical drug response. *Clinical Pharmacology and Therapeutics*, *81*(3), 328–345. doi:10.1038/sj.clpt.6100087

Gibbs, A.J. & McIntyre, G.A. (1970). The diagram, a method for comparing sequences. Its use with amino acid and nucleotide sequences. *European Journal of Biochemistry / FEBS*, *16*(1), 1-11.

Gibson, D. G., Benders, G. A., Andrews-Pfannkoch, C., Denisova, E. A., Baden-Tillson, H., & Zaveri, J. (2008). Complete chemical synthesis, assembly, and cloning of a Mycoplasma genitalium genome. *Science*, *319*(5867), 1215–1220. doi:10.1126/science.1151721

Gibson, D. G., Glass, J. I., Lartigue, C., Noskov, V. N., Chuang, R. Y., & Algire, M. A. (2010). Creation of a bacterial cell controlled by a chemically synthesized genome. *Science*, *329*(5987), 52–56. doi:10.1126/science.1190719

Gibson, T. J. (2009). Cell regulation: Determined to signal discrete cooperation. *Trends in Biochemical Sciences*, *34*(10), 471–482. doi:10.1016/j.tibs.2009.06.007

Gilbert, M. T., Sanchez, J. J., Haselkorn, T., Jewell, L. D., Lucas, S. B., & Van Marck, E. (2007). Multiplex PCR with minisequencing as an effective high-throughput SNP typing method for formalin-fixed tissue. *Electrophoresis*, *28*(14), 2361–2367. doi:10.1002/elps.200600589

Gilbert, D. (2005). Biomolecular interaction network database. *Briefings in Bioinformatics*, *6*(2), 194–198. doi:10.1093/bib/6.2.194

Gilchrist, M., Thorsson, V., Li, B., Rust, A. G., Korb, M., & Kennedy, K. (2006). Systems biology approaches identify ATF3 as a negative regulator of Toll-like receptor 4. *Nature*, *441*(7090), 173–178. doi:10.1038/nature04768

Gillespie, D. (1976). A general method for numerically simulating the stochastic time evolution of coupled chemical reactions. *Journal of Computational Physics*, *22*(4), 403–434. doi:10.1016/0021-9991(76)90041-3

Gillespie, D. T. (1977). Exact stochastic simulation of coupled chemical reactions. *Journal of Physical Chemistry*, *81*(25), 2340–2361. doi:10.1021/j100540a008

Gilman, A. G., Simon, M. I., Bourne, H. R., Harris, B. A., Long, R., & Ross, E. M. (2002). Overview of the Alliance for Cellular Signaling. *Nature*, *420*(6916), 703–706. doi:10.1038/nature01304

Ginolhac, A., Jarrin, C., Robe, P., Perriere, G., Vogel, T. M., & Simonet, P. (2005). Type I polyketide synthases may have evolved through horizontal gene transfer. *Journal of Molecular Evolution*, *60*(6), 716–725. doi:10.1007/s00239-004-0161-1

Girvan, M., & Newman, M. E. (2002). Community structure in social and biological networks. *Proceedings of the National Academy of Sciences of the United States of America*, *99*(12), 7821–7826. doi:10.1073/pnas.122653799

Giuliani, M. M., Adu-Bobie, J., Comanducci, M., Arico, B., Savino, S., & Santini, L. (2006). A universal vaccine for serogroup B meningococcus. *Proceedings of the National Academy of Sciences of the United States of America*, *103*(29), 10834–10839. doi:10.1073/pnas.0603940103

Goeman, J., & Buhlmann, P. (2007). Analyzing gene expression data in terms of gene sets: Methodological issues. *Bioinformatics (Oxford, England)*, *23*(8), 980. doi:10.1093/bioinformatics/btm051

Goethals, B., & Zaki, M. (2003). Advances in frequent itemset mining implementations: Report on FIMI'03. *SIGKDD Explorations*, *6*(1), 109–117. doi:10.1145/1007730.1007744

Goh, K. I., Cusick, M. E., Valle, D., Childs, B., Vidal, M., & Barabasi, A. L. (2007). The human disease network. *Proceedings of the National Academy of Sciences of the United States of America*, *104*(21), 8685–8690. doi:10.1073/pnas.0701361104

Gohlke, H., Hendlich, M., & Klebe, G. (2000). Knowledge-based scoring function to predict protein-ligand interactions. *Journal of Molecular Biology*, *295*(2), 337–356. doi:10.1006/jmbi.1999.3371

Goldbeter, A., & Koshland, D. E. (1981). An amplified sensitivity arising from covalent modification in biological systems. *Proceedings of the National Academy of Sciences of the United States of America*, *78*(11), 6840–6844. doi:10.1073/pnas.78.11.6840

Goldman, B. R., 4, 83. (2005). Pharmacogenomics: Privacy in the Era of Personalized Medicine. *Northwestern Journal of Technology and Intellectual Property*, *4*(1), 140–143.

Goldman, R. E., Kingdon, C., Wasser, J., Clark, M. A., Goldberg, R., Papandonatos, G. D., et al. (2008). Rhode Islanders' attitudes towards the development of a statewide genetic biobank *Personalized Medicine, 5*(4), 339-359.

Goldstein, D. B. (2009). Common genetic variation and human traits. *The New England Journal of Medicine*, *360*(17), 1696–1698. doi:10.1056/NEJMp0806284

Golub, T. R., Slonim, D. K., Tamayo, P., Huard, C., Gaasenbeek, M., & Mesirov, J. P. (1999). Molecular classification of cancer: Class discovery and class prediction by gene expression monitoring. *Science*, *286*(5439), 531–537. doi:10.1126/science.286.5439.531

Golub, G., & Van Loan, C. (1996). *Matrix computations (Johns Hopkins studies in mathematical sciences, 3rd ed.)*. The Johns Hopkins University Press.

Good, A. C., Krystek, S. R., & Mason, J. S. (2000). High-throughput and virtual screening: Core lead discovery technologies move towards integration. *Drug Discovery Today*, *5*(12Suppl 1), 61–69. doi:10.1016/S1359-6446(00)00015-5

Goodarzi, H. (2009). Revealing global regulatory perturbations across human cancers. *Molecular Cell*, *36*(5), 900–911. doi:10.1016/j.molcel.2009.11.016

Goodford, P. J. (1985). A computational procedure for determining energetically favorable binding sites on biologically important macromolecules. *Journal of Medicinal Chemistry*, *28*(7), 849–857. doi:10.1021/jm00145a002

Goodhill, G. J. (1997). Diffusion in axon guidance. *European Journal of Neurology*, *9*(7), 1414–1421.

Goodhill, G. J., Gu, M., & Urbach, J. S. (2004). Predicting axonal response to molecular gradients with a computational model of filopodial dynamics. *Neural Computation*, *16*(11), 2221–2243. doi:10.1162/0899766041941934

Goodhill, G. J., & Urbach, J. S. (1999). Theoretical analysis of gradient detection by growth cones. *Journal of Neurobiology*, *41*(2), 230–241. doi:10.1002/(SICI)1097-4695(19991105)41:2<230::AID-NEU6>3.0.CO;2-9

Goodman, K. W. (Ed.). (1998). *Ethics, Computing and Medicine: Informatics and the Transformation of Health Care*. New York: Cambridge University Press.

Goodman, K. W., & Miller, R. (2006). Ethics and Health Informatics: Users, Standards and Outcomes. In Shortliffe, E. H., Cimino, J., Garber, A. M., Owens, D. K., Singer, S. J., & Enthoven, A. C. (Eds.), *Medical Informatics: Computer Applications in Health Care and Biomedicine* (3rd ed., pp. 379–402). New York: Springer-Verlag.

Goodsaid, F., & Frueh, F. (2006). Process map proposal for the validation of genomic biomarkers. *Pharmacogenomics*, *7*, 773–782. doi:10.2217/14622416.7.5.773

Goodsaid, F., & Frueh, F. (2007). Biomarker qualification pilot process at the US Food and Drug Administration. *The AAPS Journal*, *9*(1), E105–E108. doi:10.1208/aapsj0901010

Goodsell, D. S., & Olson, A. J. (1990). Automated docking of substrates to proteins by simulated annealing. *Proteins*, *8*(3), 195–202. doi:10.1002/prot.340080302

Goodsell, D. S. (1999). The molecular perspective: The ras oncogene. *The Oncologist*, *4*(3), 263–264.

Gordon, S. J., Saleque, S., & Birshtein, B. K. (2003). Yin Yang 1 is a lipopolysaccharide-inducible activator of the murine 3' Igh enhancer, hs3. *Journal of Immunology (Baltimore, MD.: 1950)*, *170*(11), 5549–5557.

GOSt, [http://biit.cs.ut.ee/gprofiler/]

Gourvitch, B., Bouquin-Jeanns, R., & Faucon, G. (2006). Linear and nonlinear causality between signals: Methods, examples and neurophysiological applications. *Biological Cybernetics, 95*(4), 349–369. doi:10.1007/s00422-006-0098-0

Gouveia-Oliveira, R., Roque, F. S., Wernersson, R., Sicheritz-Ponten, T., Sackett, P. W., & Molgaard, A. (2009). InterMap3D: Predicting and visualizing co-evolving protein residues. *Bioinformatics (Oxford, England), 25*(15), 1963–1965. doi:10.1093/bioinformatics/btp335

Gowda, G. N., Zhang, S., Gu, H., Asiago, V., Shanaiah, N., & Raftery, D. (2008). Metabolomics-based methods for early disease diagnostics. *Expert Review of Molecular Diagnostics, 8*, 617–633. doi:10.1586/14737159.8.5.617

Gozal, D. (2009). Two-dimensional differential in-gel electrophoresis proteomic approaches reveal urine candidate biomarkers in pediatric obstructive sleep apnea. *American Journal of Respiratory and Critical Care Medicine, 180*(12), 1253–1261. doi:10.1164/rccm.200905-0765OC

Granger, C. (1969). Investigating causal relations by econometric models and cross-spectral methods. *Econometrica: Journal of the Econometric Society, 37*(3), 424–438.

Graumann, J., Hubner, N. C., Kim, J. B., Ko, K., Moser, M., & Kumar, C. (2008). Stable isotope labeling by amino acids in cell culture (SILAC) and proteome quantitation of mouse embroyonic stem cells to depth of 5,111 proteins. *Molecular & Cellular Proteomics, 7*(4), 672–683. doi:10.1074/mcp.M700460-MCP200

Greene, C. S., Penrod, N. M., Williams, S. M., & Moore, J. H. (2009). Failure to replicate a genetic association may provide important clues about genetic architecture. *Public Library of Science ONE, 4*, e5639.

Greene, C. S., & Moore, J. H. (2008). *Ant colony optimization for genome-wide genetic analysis*. (LNCS 5217), (pp. 27-47).

Greene, C. S., & Moore, J. H. (2009). Solving complex problems in human genetics using nature-inspired algorithms requires strategies which exploit domain-specific knowledge. *Nature Inspired Informatics, 7*, 166-180. Hershey, PA: IGI Global.

Greene, C. S., Gilmore, J. M., Kiralis, J., Andrews, P. C., & Moore, J. H. (2009). *Optimal use of expert knowledge in ant colony optimization for the analysis of epistasis in human disease.* (LNCS 5483), (pp. 92-103).

Griffith, O. L., Pleasance, E. D., Fulton, D. L., Oveisi, M., Ester, M., & Siddiqui, A. S. (2005). Assessment and integration of publicly available sage, cdna microarray, and oligonucleotide microarray expression data for global coexpression analyses. *Genomics, 86*(4), 476–488. doi:10.1016/j.ygeno.2005.06.009

Griffith, O. L., Gao, B., Bilenky, M., Prychyna, Y., Ester, M., & Jones, S. (2009). KiWi: A scalable subspace clustering algorithm for gene expression analysis. In *Proceedings of the 3rd International Conference on Bioinformatics and Biomedical Engineering,* June 11–13, Beijing, China.

Griffiths-Jones, S., Grocock, R. J., van Dongen, S., Bateman, A., & Enright, A. J. (2006). MiRBase: MicroRNA sequences, targets and gene nomenclature. *Nucleic Acids Research, 34*(Database issue), D140–D144. doi:10.1093/nar/gkj112

Griffiths-Jones, S., Bateman, A., Marshall, M., Khanna, A., & Eddy, S. R. (2003). Rfam: An RNA family database. *Nucleic Acids Research, 31*(1), 439–441. doi:10.1093/nar/gkg006

Gronborg, M., Kristiansen, T. Z., & Iwahori, A. (2006). Biomarker discovery from pancreatic cancer secretome using a differential proteomic approach. *Molecular & Cellular Proteomics, 5*(1), 157–171. doi:10.1074/mcp.M500178-MCP200

Gross, H., Stockwell, V. O., Henkels, M. D., Nowak-Thompson, B., Loper, J. E., & Gerwick, W. H. (2007). The genomisotopic approach: A systematic method to isolate products of orphan biosynthetic gene clusters. *Chemistry & Biology, 14*(1), 53–63. doi:10.1016/j.chembiol.2006.11.007

Gu, J., & Liu, J. S. (2008). Bayesian biclustering of gene expression data. *BMC Genomics, 9*(Suppl 1), S4. doi:10.1186/1471-2164-9-S1-S4

Guan, K. L., & Rao, Y. (2003). Signalling mechanisms mediating neuronal responses to guidance cues. *Nature Reviews. Neuroscience, 4*(12), 941–956. doi:10.1038/nrn1254

Guenther, M. G., Levine, S. S., Boyer, L. A., Jaenisch, R., & Young, R. A. (2007). A chromatin landmark and transcription initiation at most promoters in human cells. *Cell, 130*(1), 77–88.

Guex, N., Diemand, A., & Peitsch, M. C. (1999). Protein modelling for all. *Trends in Biochemical Sciences, 24*(9), 364–367. doi:10.1016/S0968-0004(99)01427-9

Guillemin, K., Salama, N. R., Tompkins, L. S., & Falkow, S. (2002). Cag pathogenicity island-specific responses of gastric epithelial cells to Helicobacter pylori infection. *Proceedings of the National Academy of Sciences of the United States of America, 99*(23), 15136–15141. doi:10.1073/pnas.182558799

Guimerà, R. (2004). Modularity from fluctuations in random graphs and complex networks. *Physical Review E: Statistical, Nonlinear, and Soft Matter Physics, 70*(2), 025101. doi:10.1103/PhysRevE.70.025101

Guimerà, R., & Amaral, L. A. N. (2005). Functional cartography of complex metabolic networks. *Nature, 433*(7028), 895–900. doi:10.1038/nature03288

Gulati, S., Rouilly, V., Niu, X., Chappell, J., Kitney, R. I., & Edel, J. B. (2009). Opportunities for microfluidic technologies in synthetic biology. *Journal of the Royal Society, Interface*, *6*(Suppl 4), S493–S506. doi:10.1098/rsif.2009.0083.focus

Gultyaev, A. P., van Batenburg, F. H. D., & Pleij, C. W. A. (1995). The influence of a metastable structure in plasmid primer RNA on antisense RNA binding kinetics. *RNA (New York, N.Y.)*, *23*(18), 3718–3725.

Guner, O. F. (2005). The impact of pharmacophore modeling in drug design. *IDrugs*, *8*(7), 567–572.

Güner, O., Clement, O., & Kurogi, Y. (2004). Pharmacophore modeling and three dimensional database searching for drug design using catalyst: Recent advances. *Current Medicinal Chemistry*, *11*(22), 2991–3005.

Guo, S., Seth, A., Kendrick, K., Zhou, C., & Feng, J. (2008). Partial Granger causality-Eliminating exogenous inputs and latent variables. *Journal of Neuroscience Methods*, *172*(1), 79–93. doi:10.1016/j.jneumeth.2008.04.011

Guo, S., Wu, J., Ding, M., & Feng, J. (2008). Uncovering interactions in the frequency domain. *PLoS Computational Biology*, *4*(5). doi:10.1371/journal.pcbi.1000087

Gupta, S., & Maiden, M. C. J. (2001). Exploring the evolution of diversity in pathogen populations. *Trends in Microbiology*, *9*(4), 181–185. doi:10.1016/S0966-842X(01)01986-2

Gupta, N., & Aggarwal, S. (2008). SISA: Seeded Iterative Signature Algorithm for biclustering gene expression data. *IADIS, European Conference on Data Mining*.

Guttman, M., Amit, I., Garber, M., French, C., Lin, M. F., & Feldser, D. (2009). Chromatin signature reveals over a thousand highly conserved large non-coding RNAs in mammals. *Nature*, *458*(7235), 223–227.

Guvench, O., & MacKerell, A. D. Jr. (2008). Comparison of protein force fields for molecular dynamics simulations. *Methods in Molecular Biology (Clifton, N.J.)*, *443*, 63–88. doi:10.1007/978-1-59745-177-2_4

Guyon, I., Weston, J., Barnhill, S., & Vapnik, V. (2002). Gene selection for cancer classification using support vector machines. *Machine Learning*, *46*, 389–422. doi:10.1023/A:1012487302797

Gygi, S. P., Rist, B., Gerber, S. A., Turecek, F., Gelb, M. H., & Aebersold, R. (1999). Quantitative analysis of complex protein mixtures using isotope-coded affinity tags. *Nature Biotechnology*, *17*(10), 994–999. doi:10.1038/13690

Haas, D., Renbarger, J., Meslin, E. M., Drabiak, K., & Flockhart, D. (2008). Patient attitudes toward genotyping in an urban women's health clinic. *Obstetrics and Gynecology*, *112*, 1023–1028. doi:10.1097/AOG.0b013e318187e77f

Habel, L. A., Shak, S., Jacobs, M. K., Capra, A., Alexander, C., & Pho, M. (2006). A population-based study of tumor gene expression and risk of breast cancer death among lymph node-negative patients. *Breast Cancer Research*, *8*(3), R25. doi:10.1186/bcr1412

Hahn, M., & Stachelhaus, T. (2004). Selective interaction between nonribosomal peptide synthetases is facilitated by short communication-mediating domains. *Proceedings of the National Academy of Sciences of the United States of America*, *101*(44), 15585–15590. doi:10.1073/pnas.0404932101

Hahn, M., & Stachelhaus, T. (2006). Harnessing the potential of communication-mediating domains for the biocombinatorial synthesis of nonribosomal peptides. *Proceedings of the National Academy of Sciences of the United States of America*, *103*(2), 275–280. doi:10.1073/pnas.0508409103

Halkidi, M., Batistakis, Y., & Vazirgiannis, M. (2001). On clustering validation techniques. *Intelligent Information Systems Journal*, *17*(2-3), 107–145. doi:10.1023/A:1012801612483

Hall, N., Karras, M., Raine, J. D., Carlton, J. M., Kooij, T. W., & Berriman, M. (2005). A comprehensive survey of the Plasmodium life cycle by genomic, transcriptomic, and proteomic analyses. *Science*, *307*(5706), 82–86. doi:10.1126/science.1103717

Hall, M. (1999). Correlation-based feature selection for machine learning. Unpublished doctoral thesis, Department of Computer Science, Waikato University, New Zealand.

Halperin, I., Ma, B., Wolfson, H., & Nussinov, R. (2002). Principles of docking: An overview of search algorithms and a guide to scoring functions. *Proteins*, *47*(4), 409–443. doi:10.1002/prot.10115

Halperin, Y., Linhart, C., Ulitsky, I., & Ron Shamir, R. (2009). Allegro: Analyzing expression and sequence in concert to discover regulatory programs. *Nucleic Acids Research*, *37*(5), 1566–1579. doi:10.1093/nar/gkn1064

Hammer, D. A., & Lauffenburger, D. A. (1987). A dynamical model for receptor-mediated cell adhesion to surfaces. *Biophysical Journal*, *52*, 475–487. doi:10.1016/S0006-3495(87)83236-8

Hamosh, A., Scott, A. F., Amberger, J. S., Bocchini, C. A., & McKusick, V. A. (2005). Online Mendelian inheritance in man (OMIM), a knowledgebase of human genes and genetic disorders. *Nucleic Acids Research*, *33*(Database issue), D514–D517. doi:10.1093/nar/gki033

Han, J. D. (2008). Understanding biological functions through molecular networks. *Cell Research*, *18*(2), 224–237. doi:10.1038/cr.2008.16

Han, B., Dost, B., Bafna, V., & Zhang, S. (2008). Structural alignment of pseudoknotted RNA. *Journal of Computational Biology*, *15*(5), 489–504. doi:10.1089/cmb.2007.0214

Hanahan, D., & Weinberg, R. A. (2000). The hallmarks of cancer. *Cell, 100*(1), 57–70. doi:10.1016/S0092-8674(00)81683-9

Hanawalt, P. C. (2002). Subpathways of nucleotide excision repair and their regulation. *Oncogene, 21*(21), 8949–8956. doi:10.1038/sj.onc.1206096

Hanisch, D., Zien, A., Zimmer, R., & Lengauer, T. (2002). Co-clustering of biological networks and gene expression data. *Bioinformatics (Oxford, England), 18*(Suppl 1), S145–S154.

Hansch, C. (1964). p–s–p Analysis. A method for the correlation of biological activity and chemical structure. *Journal of the American Chemical Society, 86*, 1616–1626. doi:10.1021/ja01062a035

Harary, F., & Ross, I. (1957). A procedure for clique detection using the group matrix. *Sociometry, 20*, 205–215. doi:10.2307/2785673

Harbison, C. T. (2004). Transcriptional regulatory code of a eukaryotic genome. *Nature, 431*(7004), 99–104. doi:10.1038/nature02800

Hardman, M., & Makarov, A. A. (2003). Interfacing the orbitrap mass analyzer to an electrospray ion source. *Analytical Chemistry, 75*(7), 1699–1075. doi:10.1021/ac0258047

Hardy, J., & Singleton, A. (2009). Genomewide association studies and human disease. *The New England Journal of Medicine, 360*, 1759–1768. doi:10.1056/NEJMra0808700

Harman, L. B. (2001). *Ethical challenges in the management of health information* (1 ed.): Aspen Pub.

Harper, J. A. (2003). Notch signaling in development and disease. *Clinical Genetics, 64*(6), 461–472. doi:10.1046/j.1399-0004.2003.00194.x

Harris, C. C. (1996). p53 tumor suppressor gene: From the basic research laboratory to the clinic—an abridged historical perspective. *Carcinogenesis, 17*, 1187–1198. doi:10.1093/carcin/17.6.1187

Harsha, H. C., Molina, H., & Pandey, A. (2008). Quantitative proteomics using stable isotope labeling with amino acids in cell culture. *Nature Protocols, 3*(3), 505–516. doi:10.1038/nprot.2008.2

Hart, T. N., & Read, R. J. (1992). A multiple-start Monte Carlo docking method. *Proteins, 13*(3), 206–222. doi:10.1002/prot.340130304

Hart, G. T., Ramani, A. K., & Marcotte, E. M. (2006). How complete are current yeast and human protein-interaction networks? *Genome Biology, 7*(11), 120. doi:10.1186/gb-2006-7-11-120

Hartemink, A., Gifford, D., Jaakkola, T., & Young, R. (2001). Using graphical models and genomic expression data to statistically validate models of genetic regulatory networks. *Pacific Symposium on Biocomputing Pacific Symposium on Biocomputing*, 422-33.

Hartigan, J. A. (1972). Direct clustering of a data matrix. [JASA]. *Journal of the American Statistical Association, 67*(337), 123–129. doi:10.2307/2284710

Hartigan, J. A. (1975). *Clustering algorithms (Probability & mathematical statistics)*. John Wiley & Sons Inc.

Hartwell, L. H. (1999). From molecular to modular cell biology. *Nature, 402*(6761Suppl), C47–C52. doi:10.1038/35011540

Hartwell, L. H., Hopfield, J. J., Leibler, S., & Murray, A. W. (1999). From molecular to modular cell biology. *Nature, 402*(6761Suppl), C47–C52. doi:10.1038/35011540

Hastie, T., Tibshirani, R., & Buja, A. (1994). Flexible discriminant analysis. *Journal of the American Statistical Association, 89*, 1255–1270. doi:10.2307/2290989

Hastie, T., Tibshirani, R., & Friedman, J. (2009). *The elements of statistical learning: Data mining, inference, and prediction*. New York: Springer-Verlag.

Hauenschild, A., Ringrose, L., Altmutter, C., Paro, R., & Rehmsmeier, M. (2008). Evolutionary plasticity of polycomb/trithorax response elements in Drosophila species. *PLoS Biology, 6*(10), e261.

Haussy, B. & Ganghoffer, J.F. (2005). Probabilistic mechanisms of adhesive contact formation and interfacial processes. *Archives of Applied Mechanics, 75*, 2006, 338-354.

Hawkins, R. D., Hon, G. C., & Ren, B. (2010). Next-generation genomics: an integrative approach. *Nature Reviews. Genetics, 11*(7), 476–486.

He, S., Liu, C., Skogerbø, G., Zhao, H., Wang, J., & Liu, T. (2008). NONCODE v2.0: Decoding the non-coding. *Nucleic Acids Research, 36*(Database issue), D170–D172. doi:10.1093/nar/gkm1011

Hedges, S. B., Blair, J. E., Venturi, M. L., & Shoe, J. L. (2004). A molecular timescale of eukaryote evolution and the rise of complex multicellular life. *BMC Evolutionary Biology, 4*, 2. doi:10.1186/1471-2148-4-2

Heinrich, R., & Schuster, S. (2003). *The regulation of cellular systems*. Springer.

Heintzman, N. D., Hon, G. C., Hawkins, R. D., Kheradpour, P., Stark, A., & Harp, L. F. (2009). Histone modifications at human enhancers reflect global cell-type-specific gene expression. *Nature, 459*(7243), 108–112.

Heintzman, N. D., Stuart, R. K., Hon, G., Fu, Y., Ching, C. W., & Hawkins, R. D. (2007). Distinct and predictive chromatin signatures of transcriptional promoters and enhancers in the human genome. *Nature Genetics*, *39*(3), 311–318.

Heiser, L. M., Wang, N. J., Talcott, C. L., Laderoute, K. R., Knapp, M., & Guan, Y. (2009). Integrated analysis of breast cancer cell lines reveals unique signaling pathways. *Genome Biology*, *10*(3), R31. doi:10.1186/gb-2009-10-3-r31

Helft, P. R., Champion, V. L., Eckles, R., Johnson, C. S., & Meslin, E. M. (2007). Cancer patients' attitudes toward future research uses of stored human biological materials. *Journal of Empirical Research on Human Research Ethics; JERHRE*, *2*(3), 15–22. doi:10.1525/jer.2007.2.3.15

Helgadottir, A., Gretarsdottir, S., & St Clair, D. (2005). Association between the gene encoding 5-lipoxygenase-activating protein and stroke replicated in a Scottish population. *American Journal of Human Genetics*, *76*, 505–509. doi:10.1086/428066

Helgesson, G., & Swartling, U. (2008). Views on data use, confidentiality and consent in a predictive screening involving children. *Journal of Medical Ethics*, *34*, 206–209. doi:10.1136/jme.2006.020016

Helleday, T., Petermann, E., Lundin, C., Hodgson, B., & Sharma, R. A. (2008). DNA repair pathways as targets for cancer therapy. *Nature Reviews. Cancer*, *8*(3), 193–204. doi:10.1038/nrc2342

Hellerstein, M. K. (2008). Exploiting complexity and the robustness of network architecture for drug discovery. *The Journal of Pharmacology and Experimental Therapeutics*, *325*(1), 1–9. doi:10.1124/jpet.107.131276

Hendriks, B. S., Cook, J., Burke, J. M., Beusmans, J. M., Lauffenburger, D. A., & de Graaf, D. (2006). Computational modelling of ErbB family phosphorylation dynamics in response to transforming growth factor alpha and heregulin indicates spatial compartmentation of phosphatase activity. *Systems Biology*, *153*(1), 22–33. doi:10.1049/ip-syb:20050057

Hentschel, H. G. E., & van Ooyen, A. (2000). Dynamic mechanisms for bundling and guidance during neural network formation. *Physica A*, *288*(1-4), 369–379. doi:10.1016/S0378-4371(00)00434-9

Hernandez, S., Gomez, A., Cedano, J., & Querol, E. (2009). Bioinformatics annotation of the hypothetical proteins found by omics techniques can help to disclose additional virulence factors. *Current Microbiology*, *59*(4), 451–456. doi:10.1007/s00284-009-9459-y

Hert, J., Willett, P., Wilton, D. J., Acklin, P., Azzaoui, K., & Jacoby, E. (2004). Comparison of topological descriptors for similarity-based virtual screening using multiple bioactive reference structures. *Organic & Biomolecular Chemistry*, *2*(22), 3256–3266. doi:10.1039/b409865j

Hertzog, P. J., O'Neill, L. A., & Hamilton, J. A. (2003). The interferon in TLR signaling: More than just antiviral. *Trends in Immunology*, *24*(10), 534–539. doi:10.1016/j.it.2003.08.006

Hewett, M. (2002). PharmGKB: The Pharmacogenetics Knowledge Base. *Nucleic Acids Research*, *30*(1), 163–165. doi:10.1093/nar/30.1.163

Hickman, G., & Hodgman, T. (2009). Inference of gene regulatory networks using boolean-network inference methods. *Journal of Bioinformatics and Computational Biology*, *7*(6), 1013–1029. doi:10.1142/S0219720009004448

Higgs, P. G. (2000). RNA secondary structure: Physical and computational aspects. *Quarterly Reviews of Biophysics*, *33*(3), 199–253. doi:10.1017/S0033583500003620

Hill, A. M. (2006). The biosynthesis, molecular genetics, and enzymology of the polyketide-derived metabolites. *Natural Product Reports*, *23*(2), 256–320. doi:10.1039/b301028g

Hindorff, L. A., Sethupathy, P., Junkins, H. A., Ramos, E. M., Mehta, J. P., & Collins, F. S. (2009). Potential etiologic and functional implications of genome-wide association loci for human diseases and traits. *Proceedings of the National Academy of Sciences of the United States of America*, *106*(23), 9362–9367. doi:10.1073/pnas.0903103106

Hirschhorn, J. N. (2009). Genomewide association studies – illustrating biologic pathways. *The New England Journal of Medicine*, *360*, 1699–1701. doi:10.1056/NEJMp0808934

Hirschhorn, J. N., Lohmueller, K., Byrne, E., & Hirschhorn, K. (2002). A comprehensive review of genetic association studies. *Genetics in Medicine*, *4*(2), 45–61. doi:10.1097/00125817-200203000-00002

Hirschhorn, J. N., & Daly, M. J. (2005). Genome-wide association studies for common diseases and complex traits. *Nature Reviews. Genetics*, *6*, 95–108. doi:10.1038/nrg1521

Hiscock, S. J., & Allen, A. M. (2008). Diverse cell signalling pathways regulate pollen-stigma interactions: The search for consensus. *The New Phytologist*, *179*(2), 286–317. doi:10.1111/j.1469-8137.2008.02457.x

Hishigaki, H. (2001). Assessment of prediction accuracy of protein function from protein-protein interaction data. *Yeast (Chichester, England)*, *18*(6), 523–531. doi:10.1002/yea.706

Ho Sui, S.J., Fulton, D.L., Arenillas, D.J., Kwon, A.T. & Wasserman, W.W. (2007). oPOSSUM: Integrated tools for analysis of regulatory motif over-representation. *Nucleic Acids Research, 35*(Web Server issue), W245-252.

Hobbs, M. S., & McCall, M. G. (1970). Health statistics and record linkage in Australia. *Journal of Chronic Diseases*, *23*(5), 375–381. doi:10.1016/0021-9681(70)90020-2

Hochreiter, S., Bodenhofer, U., Heusel, M., & Mayr, A. (2010). *FABIA: Factor Analysis for Bicluster Acquisition.* Bioinformatics Advance Access.

Hodge, J. G., Gostin, L. O., & Jacobson, P. D. (1999). Legal Issues Concerning Electronic Health Information: Privacy, Quality, and Liability. *Journal of the American Medical Association*, *282*(15), 1466–1471. doi:10.1001/jama.282.15.1466

Hoebe, K., Du, X., Georgel, P., Janssen, E., Tabeta, K., & Kim, S. O. (2003). Identification of Lps2 as a key transducer of MyD88-independent TIR signalling. *Nature*, *424*(6950), 743–748. doi:10.1038/nature01889

Hoffmann, R., Krallinger, M., Andres, E., Tamames, J., Blaschke, C., & Valencia, A. (2005). Text mining for metabolic pathways, signaling cascades, and protein networks. *Science's STKE*, (283): e21.

Hoffren, A. M., Murray, C. M., & Hoffmann, R. D. (2001). Structure-based focusing using pharmacophores derived from the active site of 17beta-hydroxysteroid dehydrogenase. *Current Pharmaceutical Design*, *7*(7), 547–566. doi:10.2174/1381612013397870

Holman, C. D. J., Bass, A. J., Rouse, I. L., & Hobbs, M. S. T. (1999). Population-based linkage of health records in Western Australia: development of a health services research linked database. *Australian and New Zealand Journal of Public Health*, *23*, 453–459. doi:10.1111/j.1467-842X.1999.tb01297.x

Holt, K. E., Parkhill, J., Mazzoni, C. J., Roumagnac, P., Weill, F.-X., & Goodhead, I. (2008). High-throughput sequencing provides insights into genome variation and evolution in Salmonella Typhi. *Nature Genetics*, *40*(8), 987–993. doi:10.1038/ng.195

Holtzman, N. A., & Watson, M. S. (Eds.). (1998). *Promoting safe and effective genetic testing in the United States: final report of the task force on genetic testing*. Baltimore: Johns Hopkins University Press.

Homer, N., Szelinger, S., Redman, M., Duggan, D., Tembe, W., & Muehling, J. (2008). Resolving individuals contributing trace amounts of DNA to highly complex mixtures using high-density SNP genotyping microarrays. *PLOS Genetics*, *4*(8), e1000167. doi:10.1371/journal.pgen.1000167

Hong, F., & Breitling, R. (2008). A comparison of meta-analysis methods for detecting differentially expressed genes in microarray experiments. *Bioinformatics (Oxford, England)*, *24*(3), 374–382. doi:10.1093/bioinformatics/btm620

Hong, F., Breitling, R., McEntee, C. W., Wittner, B. S., Nemhauser, J. L., & Chory, J. (2006). RankProd: A bioconductor package for detecting differentially expressed genes in meta-analysis. *Bioinformatics (Oxford, England)*, *22*(22), 2825–2827. doi:10.1093/bioinformatics/btl476

Hood, L., Heath, J. R., Phelps, M. E., & Lin, B. (2004). Systems biology and new technologies enable predictive and preventative medicine. *Science*, *306*(5696), 640–643. doi:10.1126/science.1104635

Hooijmans, C. R., & Kiliaan, A. J. (2008). Fatty acids, lipid metabolism and Alzheimer pathology. *European Journal of Pharmacology*, *585*(1), 176–196. doi:10.1016/j.ejphar.2007.11.081

Hopfinger, A. J., & Duca, J. S. (2000). Extraction of pharmacophore information from high-throughput screens. *Current Opinion in Biotechnology*, *11*(1), 97–103. doi:10.1016/S0958-1669(99)00061-0

Hopkins, A. L. (2007). Network pharmacology. *Nature Biotechnology*, *25*(10), 1110–1111. doi:10.1038/nbt1007-1110

Hopkins, A. L. (2008). Network pharmacology: The next paradigm in drug discovery. *Nature Chemical Biology*, *4*(11), 682–690. doi:10.1038/nchembio.118

Hopkins, A. L., & Groom, C. R. (2002). The druggable genome. *Nature Reviews. Drug Discovery*, *1*(9), 727–730. doi:10.1038/nrd892

Horvath, D. (1997). A virtual screening approach applied to the search for trypanothione reductase inhibitors. *Journal of Medicinal Chemistry*, *40*(15), 2412–2423. doi:10.1021/jm9603781

Hosack, D. A., Dennis, G., Sherman, B. T., Lane, H. C., & Lempicki, R. A. (2003). Identifying biological themes within lists of genes with EASE. *Genome Biology*, *4*(10), R70. doi:10.1186/gb-2003-4-10-r70

Hsueh, R. C., Natarajan, M., Fraser, I., Pond, B., Liu, J., & Mumby, S. (2009). Deciphering signaling outcomes from a system of complex networks. *Science Signaling*, *2*(71), ra22. doi:10.1126/scisignal.2000054

Hu, J., Zou, F., & Wright, F. A. (2005). Practical FDR-based sample size calculations in microarray experiments. *Bioinformatics (Oxford, England)*, *21*(15), 3264–3272. doi:10.1093/bioinformatics/bti519

Hu, G., Chong, R. A., Yang, Q., Wei, Y., Blanco, M. A., & Li, F. (2009). MTDH activation by 8q22 genomic gain promotes chemoresistance and metastasis of poor-prognosis breast cancer. *Cancer Cell*, *15*(1), 9–20. doi:10.1016/j.ccr.2008.11.013

Hu, Z., Mellor, J., Wu, J., Kanehisa, M., Stuart, J. M., & DeLisi, C. (2007). Towards zoomable multidimensional maps of the cell. *Nature Biotechnology*, *25*(5), 547–554. doi:10.1038/nbt1304

Hu, Z., Snitkin, E. S., & DeLisi, C. (2008). Visant: An integrative framework for networks in systems biology. *Briefings in Bioinformatics*, *9*(4), 317–325. doi:10.1093/bib/bbn020

Hu, Z., Hung, J.H., Wang, Y., Chang, Y.C., Huang, C.L., Huyck, M., et al. (2009). Visant 3.5: Multi-scale network visualization, analysis and inference based on the gene ontology. *Nucleic Acids Research, 37*(Web Server issue), W115-121.

Huan, T., Wu, X., & Chen, J. Y. (2010). (in press). Systems Biology Visualization Tools for Drug Target Discovery. *Expert Opinion on Drug Discovery*. doi:10.1517/17460441003725102

Huang, D. W., Sherman, B. T., & Lempicki, R. A. (2009). Systematic and integrative analysis of large gene lists using DAVID Bioinformatics Resources. *Nature Protocols*, *4*(1), 44–57. doi:10.1038/nprot.2008.211

Huang, C. Y., & Ferrell, J. E. Jr. (1996). Ultrasensitivity in the mitogen-activated protein kinase cascade. *Proceedings of the National Academy of Sciences of the United States of America*, *93*(19), 10078–10083. doi:10.1073/pnas.93.19.10078

Hubbard, T. J. (2009). Ensembl 2009. *Nucleic Acids Research*, *37*(Database issue), D690–D697. doi:10.1093/nar/gkn828

Hubble, J., Demeter, J., Jin, H., Mao, M., Nitzberg, M., & Reddy, T. B. (2009). Implementation of GenePattern within the Stanford Microarray Database. *Nucleic Acids Research*, *37*(Database issue), D898–D901. doi:10.1093/nar/gkn786

Huettenhain, R., Malmstroem, J., Picotti, P., & Aebersold, R. (2009). Perspectives of targeted mass spectrometry for protein biomarker verification. *Current Opinion in Chemical Biology*, *13*(5-6), 518–525. doi:10.1016/j.cbpa.2009.09.014

Hugot, J. P., Chamaillard, M., Zouali, H., Lesage, S., Cézard, J. P., & Belaiche, J. (2001). Association of NOD2 leucine-rich repeat variants with susceptibility to Crohn's disease. *Nature*, *411*(6837), 599–603. doi:10.1038/35079107

Humgen. (2010). http://www.humgen.org/int/GB2_p.cfm?mod=1.

Hung, J.-H., Whitfield, T., Yang, T.-H., Hu, Z., Weng, Z., & Delisi, C. (2010). Identification of functional modules that correlate with phenotypic difference: The influence of network topology. *Genome Biology*, *11*(2), R23. doi:10.1186/gb-2010-11-2-r23

Hunter, D. J., & Kraft, P. (2007). Drinking from the fire hose–statistical issues in genome-wide association studies. *The New England Journal of Medicine*, *357*(5), 436–439. doi:10.1056/NEJMp078120

Hunter, P. (2000). Signaling–2000 and beyond. *Cell*, *100*(1), 113–127. doi:10.1016/S0092-8674(00)81688-8

Hussain, A., & Abdullah, A. (2006). A new biclustering technique based on crossing minimization. *Neurocomputing Journal*, *69*, 1882–1896. doi:10.1016/j.neucom.2006.02.018

Huttenhower, C., Haley, E. M., Hibbs, M. A., Dumeaux, V., Barrett, D. R., & Coller, H. A. (2009). Exploring the human genome with functional maps. *Genome Research*, *19*(6), 1093–1106. doi:10.1101/gr.082214.108

Hwang, W. C., Zhang, A., & Ramanathan, M. (2008). Identification of information flow-modulating drug targets: A novel bridging paradigm for drug discovery. *Clinical Pharmacology and Therapeutics*, *84*(5), 563–572. doi:10.1038/clpt.2008.129

Hwang, D., Smith, J. J., Leslie, D. M., Weston, A. D., Rust, A. G., & Ramsey, S. (2005). A data integration methodology for systems biology: Experimental verification. *Proceedings of the National Academy of Sciences of the United States of America*, *102*(48), 17302–17307. doi:10.1073/pnas.0508649102

Hwang, D., Stephanopoulos, G., & Chan, C. (2004). Inverse modeling using multi-block PLS to determine the environmental conditions that provide optimal cellular function. *Bioinformatics (Oxford, England)*, *20*(4), 487–499. doi:10.1093/bioinformatics/btg433

Ibrahim, M., Noman, N., & Iba, H. (2009). *Genome Informatics*, December 14-16, Yokohama Pacifico, Japan.

Ichimura, K., Kazuo, S., Guillaume, T., Jen, S., Champion, H. Y., & Martin, A. K. (2002). Mitogen-activated protein kinase cascades in plants: A new nomenclature. *Trends in Plant Science*, *7*(7), 301–308. doi:10.1016/S1360-1385(02)02302-6

Ichimura, K., Mizoguchi, T., Irie, K., Morris, P., Giraudat, J., & Matsumoto, K. (1998). Isolation of ATMEKK1 (a MAP Kinase Kinase Kinase)-interacting proteins and analysis of a MAP Kinase cascade in Arabidopsis. *Biochemical and Biophysical Research Communications*, *253*(2), 532–543. doi:10.1006/bbrc.1998.9796

Ichimura, K., Mizoguchi, T., Yoshida, R., Yuasa, T., & Shinozaki, K. (2000). Various abiotic stresses rapidly activate Arabidopsis MAP kinases ATMPK4 and ATMPK6. *The Plant Journal*, *24*(5), 655–665. doi:10.1046/j.1365-313x.2000.00913.x

Ideker, T. (2002). Discovering regulatory and signaling circuits in molecular interaction networks. *Bioinformatics (Oxford, England)*, *18*(Suppl. 1.), S233–S240.

Ideker, T., & Sharan, R. (2008). Protein networks in disease. *Genome Research*, *18*(4), 644–652. doi:10.1101/gr.071852.107

Ideker, T., Ozier, O., Schwikowski, B., & Siegel, A. (2002). Discovering regulatory and signalling circuits in molecular interaction networks. *Bioinformatics (Oxford, England)*, *18*(Suppl 1), S233–S240.

Ihmels, J., Friedlander, G., Bergmann, S., Sarig, O., Ziv, Y., & Barkai, N. (2002). Revealing modular organization in the yeast transcriptional network. *Nature Genetics*, *31*, 370–377.

Ihmels, J., Bergmann, S., Gerami-Nejad, M., Yanai, I., McClellan, M., & Berman, J. (2005). Rewiring of the yeast transcriptional network through the evolution of motif usage. *Science*, *309*(5736), 938–940. doi:10.1126/science.1113833

International HapMap Consortium. (2005). A haplotype map of the human genome. *Nature*, *437*(7063), 1299–1320. doi:10.1038/nature04226

International Human Genome Sequencing Consortium. (2004). Finishing the euchromatic sequence of the human genome. *Nature*, *431*(7011), 931–945. doi:10.1038/nature03001

IOM. (2001). *Preserving public trust: accreditation and human research participant protection programs*. Washington, DC: Institute of Medicine.

Ioshikhes, I., Bolshoy, A., Derenshteyn, K., Borodovsky, M., & Trifonov, E. N. (1996). Nucleosome DNA sequence pattern revealed by multiple alignment of experimentally mapped sequences. *Journal of Molecular Biology*, *262*(2), 129–139.

Ioshikhes, I. P., Albert, I., Zanton, S. J., & Pugh, B. F. (2006). Nucleosome positions predicted through comparative genomics. *Nature Genetics*, *38*(10), 1210–1215.

Irimia, D., Geba, D. A., & Toner, M. (2006). Universal microfluidic gradient generator. *Analytical Chemistry*, *78*(10), 3472–3477. doi:10.1021/ac0518710

Irizarry, R. A., Warren, D., Spencer, F., Kim, I. F., Biswal, S., & Frank, B. C. (2005). Multiple-laboratory comparison of microarray platforms. *Nature Methods*, *2*(5), 345–350. doi:10.1038/nmeth756

Isambert, H., & Siggia, E. D. (2000). Modeling RNA folding paths with pseudoknots: Application to hepatitis delta virus ribozyme. *Proceedings of the National Academy of Sciences of the United States of America*, *97*(12), 6515–6520. doi:10.1073/pnas.110533697

Isidori, A. (1995). *Nonlinear control systems*. Springer.

Issaq, H. J., Van, Q. N., Waybright, T. J., Muschik, G. M., & Veenstra, T. D. (2009). Analytical and statistical approaches to metabonomics research. *Journal of Separation Science*, *32*, 2183–2199. doi:10.1002/jssc.200900152

Ito, T. (2001). A comprehensive two-hybrid analysis to explore the yeast protein interactome. *Proceedings of the National Academy of Sciences of the United States of America*, *98*(8), 4569–4574. doi:10.1073/pnas.061034498

Ivshina, A. V., George, J., Senko, O., Mow, B., Putti, T. C., & Smeds, J. (2006). Genetic reclassification of histologic grade delineates new clinical subtypes of breast cancer. *Cancer Research*, *66*(21), 10292–10301. doi:10.1158/0008-5472.CAN-05-4414

Iwabe, N., Kuma, K., & Miyata, T. (1996). Evolution of gene families and relationship with organismal evolution: Rapid divergence of tissue-specific genes in the early evolution of chordates. *Molecular Biology and Evolution*, *13*(3), 483–493.

Jacobs, M. R. (2004). Streptococcus pneumoniae: Epidemiology and patterns of resistance. *The American Journal of Medicine*, *117*(Suppl 3A), 3S–15S.

Jaeger, J. A., Turner, D. H., & Zuker, M. (1989). Improved predictions of secondary structures for RNA. *Proceedings of the National Academy of Sciences of the United States of America*, *86*(20), 7706–7710. doi:10.1073/pnas.86.20.7706

Jaenisch, R., & Bird, A. (2003). Epigenetic regulation of gene expression: how the genome integrates intrinsic and environmental signals. *Nature Genetics*, *33*(Supplement), 245–254. doi:10.1038/ng1089

Jain, A. K. (1999). Data clustering: A review. *ACM Computing Surveys*, *31*(3), 264–323. doi:10.1145/331499.331504

Janes, K. A., Albeck, J. G., Gaudet, S., Sorger, P. K., Lauffenburger, D. A., & Yaffe, M. B. (2005). A systems model of signaling identifies a molecular basis set for cytokine-induced apoptosis. *Science*, *310*(5754), 1646–1653. doi:10.1126/science.1116598

Janes, K. A., Kelly, J. R., Gaudet, S., Albeck, J. G., Sorger, P. K., & Lauffenburger, D. A. (2004). Cue-signal-response analysis of TNF-induced apoptosis by partial least squares regression of dynamic multivariate data. *Journal of Computational Biology*, *11*(4), 544–561. doi:10.1089/cmb.2004.11.544

Janes, K. A., & Lauffenburger, D. A. (2006). A biological approach to computational models of proteomic networks. *Current Opinion in Chemical Biology*, *10*(1), 73–80. doi:10.1016/j.cbpa.2005.12.016

Janga, S.C. & Tzakos, A. (2009). Structure and organization of drug-target networks: Insights from genomic approaches for drug discovery. *Molecular Biosystems*.

Jansen, R., Greenbaum, D., & Gerstein, M. (2002). Relating whole-genome expression data with protein-protein interactions. *Genome Research*, *12*(1), 37–46. doi:10.1101/gr.205602

Jeffrey, G. A. (1997). *An introduction to hydrogen bonding*. Pittsburgh: Oxford Univ. Press.

Jenke-Kodama, H., & Dittmann, E. (2009). Bioinformatic perspectives on NRPS/PKS megasynthases: Advances and challenges. *Natural Product Reports*, *26*(7), 874–883. doi:10.1039/b810283j

Jenke-Kodama, H., Sandmann, A., Muller, R., & Dittmann, E. (2005). Evolutionary implications of bacterial polyketide synthases. *Molecular Biology and Evolution*, *22*(10), 2027–2039. doi:10.1093/molbev/msi193

Jensen, L. J., Kuhn, M., Stark, M., Chaffron, S., Creevey, C., & Muller, J. (2009). STRING 8-a global view on proteins and their functional interactions in 630 organisms. *Nucleic Acids Research*, *37*, D412–D416. doi:10.1093/nar/gkn760

Jensen, L. J., Saric, J., & Bork, P. (2006). Literature mining for the biologist: From information retrieval to biological discovery. *Nature Reviews. Genetics*, *7*(2), 119–129. doi:10.1038/nrg1768

Jensen, L.J., Lagarde, J., von Mering, C. & Bork, P. (2004). Arrayprospector: A Web resource of functional associations inferred from microarray expression data. *Nucleic Acids Research, 32*(Web server issue), W445-448.

Jenssen, T. K., Laegreid, A., Komorowski, J., & Hovig, E. (2001). A literature network of human genes for high-throughput analysis of gene expression. *Nature Genetics*, *28*(1), 21–28. doi:10.1038/88213

Jenuwein, T., & Allis, C. D. (2001). Translating the histone code. *Science*, *293*(5532), 1074–1080.

Jeong, H. (2000). The large-scale organization of metabolic networks. *Nature*, *407*(6804), 651–654. doi:10.1038/35036627

Jeong, H., Mason, S., Barabási, A., & Oltvai, Z. (2001). Lethality and centrality in protein networks. *Nature*, *411*(6833), 41–42. doi:10.1038/35075138

Jezequel, P., Campone, M., Roche, H., Gouraud, W., Charbonnel, C., & Ricolleau, G. (2009). A 38-gene expression signature to predict metastasis risk in node-positive breast cancer after systemic adjuvant chemotherapy: A genomic substudy of PACS01 clinical trial. *Breast Cancer Research and Treatment*, *116*(3), 509–520. doi:10.1007/s10549-008-0250-8

Jiang, C., & Pugh, B. F. (2009). Nucleosome positioning and gene regulation: Advances through genomics. *Nature Reviews. Genetics*, *10*(3), 161–172.

Jiang, W., Li, X., Rao, S., Wang, L., Du, L., & Li, C. (2008). Constructing disease-specific gene networks using pair-wise relevance metric: Application to colon cancer identifies interleukin 8, desmin and enolase 1 as the central elements. *BMC Systems Biology*, *2*(1), 72. doi:10.1186/1752-0509-2-72

Jirapech-Umpai, T., & Aitken, S. (2005). Feature selection and classification for microarray data analysis: Evolutionary methods for identifying predictive genes. *BMC Bioinformatics*, *6*, 148. doi:10.1186/1471-2105-6-148

Jirtle, R. L., & Skinner, M. K. (2007). Environmental epigenomics and disease susceptibility. *Nature Reviews. Genetics*, *8*(4), 253–262.

Johnson, W. E., & Li, C. (2007). Adjusting batch effects in microarray expression data using empirical Bayes methods. *Biostatistics (Oxford, England)*, *8*(1), 118–127. doi:10.1093/biostatistics/kxj037

Johnson, S. M., Tan, F. J., McCullough, H. L., Riordan, D. P., & Fire, A. Z. (2006). Flexibility and constraint in the nucleosome core landscape of Caenorhabditis elegans chromatin. *Genome Research*, *16*(12), 1505–1516.

Johri, A. K., Paoletti, L. C., Glaser, P., Dua, M., Sharma, P. K., & Grandi, G. (2006). Group B Streptococcus: Global incidence and vaccine development. *Nature Reviews Microbiology*, *4*(12), 932–942. doi:10.1038/nrmicro1552

Jones, G., Willett, P., & Glen, R. C. (1995a). A genetic algorithm for flexible molecular overlay and pharmacophore elucidation. *Journal of Computer-Aided Molecular Design*, *9*(6), 532–549. doi:10.1007/BF00124324

Jones, G., Willett, P., & Glen, R. C. (1995b). Molecular recognition of receptor sites using a genetic algorithm with a description of desolvation. *Journal of Molecular Biology*, *245*(1), 43–53. doi:10.1016/S0022-2836(95)80037-9

Jones, G., Willett, P., Glen, R. C., Leach, A. R., & Taylor, R. (1997). Development and validation of a genetic algorithm for flexible docking. *Journal of Molecular Biology*, *267*(3), 727–748. doi:10.1006/jmbi.1996.0897

Jones, P. A., & Baylin, S. B. (2002). The fundamental role of epigenetic events in cancer. *Nature Reviews. Genetics*, *3*(6), 415–428.

Jones, D. (2008). Pathways to cancer therapy. *Nature Reviews. Drug Discovery*, *7*(11), 875–876. doi:10.1038/nrd2748

Jones, S., Zhang, X., Parsons, D. W., Lin, J. C., Leary, R. J., & Angenendt, P. (2008). Core signaling pathways in human pancreatic cancers revealed by global genomic analyses. *Science*, *321*(5897), 1801–1806. doi:10.1126/science.1164368

Jones, S. N., Roe, A. E., Donehower, L. A., & Bradley, A. (1995). Rescue of embryonic lethality in MDM2-deficient mice by absence of p53. *Nature*, *378*(6553), 206–208. doi:10.1038/378206a0

Jones, A. D., Smith, C. W., & McIntire, L. (1996). Leukocyte adhesion under flow conditions: principles important in tissue engineering. *Biomaterials*, *17*, 337–347. doi:10.1016/0142-9612(96)85572-4

Jonsson, P., & Bates, P. (2006). Global topological features of cancer proteins in the human interactome. *Bioinformatics (Oxford, England)*, *22*(18), 2291–2297. doi:10.1093/bioinformatics/btl390

Jorgensen, W. L. (2004). The many roles of computation in drug discovery. *Science*, *303*(5665), 1813–1818. doi:10.1126/science.1096361

Joshi-Tope, G. (2005). Reactome: A knowledgebase of biological pathways. *Nucleic Acids Research*, *33*(Database issue), D428–D423. doi:10.1093/nar/gki072

Jung, S. H. (2005). Sample size calculation for multiple testing in microarray data analysis. *Biostatistics (Oxford, England)*, *6*(1), 157–169. doi:10.1093/biostatistics/kxh026

Kabashima, K., Saji, T., Murata, T., Nagamachi, M., Matsuoka, T., & Segi, E. (2002). The prostaglandin receptor EP4 suppresses colitis, mucosal damage and CD4 cell activation in the gut. *The Journal of Clinical Investigation*, *109*(7), 883–893.

Kaestner, K. H., Lee, C. S., Scearce, L. M., Brestelli, J. E., Arsenlis, A., & Le, P. P. (2003). Transcriptional program of the endocrine pancreas in mice and humans. *Diabetes*, *52*(7), 1604–1610. doi:10.2337/diabetes.52.7.1604

Kaiser, S., & Leisch, F. (2008). Biclust-a toolbox for bicluster analysis in R. *In Proceedings of Computational Statistics*.

Kalia, A., Mukhopadhyay, A. K., Dailide, G., Ito, Y., Azuma, T., & Wong, B. C. Y. (2004). Evolutionary dynamics of insertion sequences in Helicobacter pylori. *Journal of Bacteriology*, *186*(22), 7508–7520. doi:10.1128/JB.186.22.7508-7520.2004

Kalichman, M. (2007). Responding to challenges in educating for responsible conduct of research. *Academic Medicine*, *82*, 870–875. doi:10.1097/ACM.0b013e31812f77fe

Kalir, S., & Alon, U. (2004). Using a quantitative blueprint to reprogram the dynamics of the flagella gene network. *Cell*, *117*(6), 713–720. doi:10.1016/j.cell.2004.05.010

Kamiie, J., Ohtsuki, S., & Iwase, R. (2008). Quantitative atlas of membrane transporter proteins: development and application of a highly sensitive simultaneous LC/MS/MS method combined with novel in-silico peptide selection criteria. *Pharmaceutical Research*, *25*(6), 1469–1483. doi:10.1007/s11095-008-9532-4

Kamijo, T., Zindy, F., Roussel, M. F., Quelle, D. E., Downing, J. R., & Ashmun, R. A. (1997). Tumor suppression at the mouse INK4a locus mediated by the alternative reading frame product p19. *Cell*, *91*(5), 649–659. doi:10.1016/S0092-8674(00)80452-3

Kamra, P., Gokhale, R.S. & Mohanty, D. (2005). SEARCHGTr: A program for analysis of glycosyltransferases involved in glycosylation of secondary metabolites. *Nucleic Acids Research, 33*(Web Server issue), W220-225.

Kandpal, R., Saviola, B., & Felton, J. (2009). The era of 'omics unlimited. *BioTechniques*, *46*(5), 351–352, 354–355. doi:10.2144/000113137

Kane, B. J., Zinner, M. J., Yarmush, M. L., & Toner, M. (2006). Liver-specific functional studies in a microfluidic array of primary mammalian hepatocytes. *Analytical Chemistry*, *78*(13), 4291–4298. doi:10.1021/ac051856v

Kanehisa, M., Goto, S., Kawashima, S., Okuno, Y., & Hattori, M. (2004). The kegg resource for deciphering the genome. *Nucleic Acids Research*, *32*(Database issue), D277–D280. doi:10.1093/nar/gkh063

Kanehisa, M., Goto, S., Furumichi, M., Tanabe, M., & Hirakawa, M. (2010). KEGG for representation and analysis of molecular networks involving diseases and drugs. *Nucleic Acids Research*, *38*(Database issue), D355–D360. doi:10.1093/nar/gkp896

Kaplan, N., Hughes, T. R., Lieb, J. D., Widom, J., & Segal, E. (2010, Nov 30). Contribution of histone sequence preferences to nucleosome organization: proposed definitions and methodology. *Genome Biology*, *11*(11), 140.

Kaplan, N., Moore, I. K., Fondufe-Mittendorf, Y., Gossett, A. J., Tillo, D., & Field, Y. (2009). The DNA-encoded nucleosome organization of a eukaryotic genome. *Nature*, *458*(7236), 362–366.

Kaplow, I., Singh, R., Friedman, A., Bakal, C., Perrimon, N., & Berger, B. (2009). RNAiCut: Automated detection of significant genes from functional genomic screens. *Nature Methods*, *6*(7), 476–477. doi:10.1038/nmeth0709-476

Karaoz, U. (2004). Whole-genome annotation by using evidence integration in functional-linkage networks. *Proceedings of the National Academy of Sciences of the United States of America*, *101*(9), 2888–2893. doi:10.1073/pnas.0307326101

Kate, V., Ananthakrishnan, N., Badrinath, S., & Ratnakar, C. (1998). Prevalence of Helicobacter pylori infection in disorders of the upper gastrointestinal tract in south India. *The National Medical Journal of India*, *11*(1), 5–8.

Kathiresan, S., Melander, O., Guiducci, C., Surti, A., Burtt, N. P., & Rieder, M. J. (2008). Six new loci associated with blood low-density lipoprotein cholesterol, high-density lipoprotein cholesterol or triglycerides in humans. *Nature Genetics*, *40*(2), 189–197. doi:10.1038/ng.75

Kathiresan, S., Willer, C. J., Peloso, G. M., Demissie, S., Musunuru, K., & Schadt, E. E. (2009). Common variants at 30 loci contribute to polygenic dyslipidemia. *Nature Genetics*, *41*(1), 56–65. doi:10.1038/ng.291

Katsanis, N. (2009). From association to causality: The new frontier for complex traits. *Genome Medicine*, *1*(2), 23. doi:10.1186/gm23

Katz, S., Irizarry, R. A., Lin, X., Tripputi, M., & Porter, M. W. (2006). A summarization approach for Affymetrix GeneChip data using a reference training set from a large, biologically diverse database. *BMC Bioinformatics*, *7*, 464. doi:10.1186/1471-2105-7-464

Kauffman, S. (1969). Metabolic stability and epigenesis in randomly constructed genetic nets. *Journal of Theoretical Biology*, *22*(3), 437–467. doi:10.1016/0022-5193(69)90015-0

Kauffmann, A., Gentleman, R., & Huber, W. (2009). arrayQualityMetrics-a bioconductor package for quality assessment of microarray data. *Bioinformatics (Oxford, England)*, *25*(3), 415–416. doi:10.1093/bioinformatics/btn647

Kaufman, A., Keinan, A., Meilijson, I., Kupiec, M., & Ruppin, E. (2005). Quantitative analysis of genetic and neuronal multi-perturbation experiments. *PLoS Computational Biology*, *1*(6), e64. doi:10.1371/journal.pcbi.0010064

Kawasaki, K. (1966). Diffusion constants near the critical point for time-dependent Ising models. *Physical Review*, *145*(1), 224–230. doi:10.1103/PhysRev.145.224

Kay, R. G., Gregory, B., Grace, P. B., & Pleasance, S. (2007). The application of ultra-performance liquid chromatography/tandem mass spectrometry to the detection and quantitation of apolipoproteins in human serum. *Rapid Communications in Mass Spectrometry*, *21*(16), 2585–2593. doi:10.1002/rcm.3130

Kaye, J., & Stranger, M. (Eds.). (2009). *Principles and Practice in Biobank Governance*. Surrey, UK: Ashgate.

Keatinge-Clay, A. T. (2007). A tylosin ketoreductase reveals how chirality is determined in polyketides. *Chemistry & Biology*, *14*(8), 898–908. doi:10.1016/j.chembiol.2007.07.009

Keiser, M. J., Setola, V., Irwin, J. J., Laggner, C., Abbas, A. I., & Hufeisen, S. J. (2009). Predicting new molecular targets for known drugs. *Nature*, *462*(7270), 175–181. doi:10.1038/nature08506

Keller, A., Nesvizhskii, A. I., Kolker, E., & Aebersold, R. (2002). Empirical statistical model to estimate the accuracy of peptide identifications made by MS/MS and database search. *Analytical Chemistry*, *74*(20), 5383–5392. doi:10.1021/ac025747h

Keller, A., Backes, C., Gerasch, A., Kaufmann, M., Kohlbacher, O., & Meese, E. (2009). A novel algorithm for detecting differentially regulated paths based on gene set enrichment analysis. *Bioinformatics (Oxford, England)*, *25*(21), 2787–2794. doi:10.1093/bioinformatics/btp510

Kelley, B., Sharan, R., Karp, R., Sittler, T., Root, D., & Stockwell, B. (2003). Conserved pathways within bacteria and yeast as revealed by global protein network alignment. *Proceedings of the National Academy of Sciences of the United States of America*, *100*(20), 11394–11399. doi:10.1073/pnas.1534710100

Kelman, C. W., Bass, A. J., & Holman, C. D. (2002). Research use of linked health data--a best practice protocol. *Australian and New Zealand Journal of Public Health*, *26*, 251–255. doi:10.1111/j.1467-842X.2002.tb00682.x

Kernighan, B. W., & Lin, S. (1970). An efficient heuristic procedure for partitioning graphs. *The Bell System Technical Journal*, *49*, 291–307.

Kerr, M. K., & Churchill, G. A. (2001). Experimental design for gene expression microarrays. *Biostatistics (Oxford, England)*, *2*, 183–201. doi:10.1093/biostatistics/2.2.183

Kerr, G. (2008). Techniques for clustering gene expression data. *Computers in Biology and Medicine*, *38*(3), 283–293. doi:10.1016/j.compbiomed.2007.11.001

Kerrien, S., Alam-Faruque, Y., Aranda, B., Bancarz, I., Bridge, A., & Derow, C. (2007). IntAct-open source resource for molecular interaction data. *Nucleic Acids Research*, *35*(Database issue), D561–D565. doi:10.1093/nar/gkl958

Keshava Prasad, T. S., Goel, R., Kandasamy, K., Keerthikumar, S., Kumar, S., & Mathivanan, S. (2009). Human Protein Reference Database-2009 update. *Nucleic Acids Research*, *37*(Database issue), D767–D772. doi:10.1093/nar/gkn892

Keshet, I., Schlesinger, Y., Farkash, S., Rand, E., Hecht, M., & Segal, E. (2006). Evidence for an instructive mechanism of de novo methylation in cancer cells. *Nature Genetics*, *38*(2), 149–153.

Khademhosseini, A., Yeh, J., Eng, G., Karp, J., Kaji, H., & Borenstein, J. (2005). Cell docking inside microwells within reversibly sealed microfluidic channels for fabricating multiphenotype cell arrays. *Lab on a Chip*, *5*(12), 1380–1386. doi:10.1039/b508096g

Khan, J., Wei, J. S., Ringner, M., Saal, L. H., Ladanyi, M., & Westermann, F. (2001). Classification and diagnostic prediction of cancers using gene expression profiling and artificial neural networks. *Nature Medicine*, *7*, 673–679. doi:10.1038/89044

Khatri, P., & Draghici, S. (2005). Ontological analysis of gene expression data: Current tools, limitations, and open problems. *Bioinformatics (Oxford, England)*, *21*, 3587–3595. doi:10.1093/bioinformatics/bti565

Khedkar, S. A., Malde, A. K., Coutinho, E. C., & Srivastava, S. (2007). Pharmacophore modeling in drug discovery and development: An overview. *Medicinal Chemistry (Shariqah, United Arab Emirates)*, *3*(2), 187–197. doi:10.2174/157340607780059521

Kholodenko, B. N. (2007). Untangling the signaling wires. *Nature Cell Biology*, *9*(3), 247–249. doi:10.1038/ncb0307-247

Kholodenko, B. N. (2009). Spatially distributed cell signaling. *FEBS Letters*, *583*(24), 4006–4012. doi:10.1016/j.febslet.2009.09.045

Kholodenko, B. N. (2000). Negative feedback and ultrasensitivity can bring about oscillations in the mitogen-activated protein kinase cascades. *European Journal of Biochemistry*, *267*(6), 1583–1588. doi:10.1046/j.1432-1327.2000.01197.x

Kholodenko, B. N. (2006). Cell-signalling dynamics in time and space. *Nature Reviews. Molecular Cell Biology*, *7*(3), 165–176. doi:10.1038/nrm1838

Kidd, T., Bland, K., & Goodman, C. (1999). Slit is the midline repellent for the Robo receptor in Drosophila. *Cell*, *96*(6), 785–594. doi:10.1016/S0092-8674(00)80589-9

Kightley, D. A., Chandra, N., & Elliston, K. (2004). Inferring gene regulatory networks from raw data-a molecular epistemics approach. *Pacific Symposium of Biocomputing*, 510-520.

Kim, S. Y., Lee, Y. S., Kang, T., Kim, S., & Lee, J. (2006). Pharmacophore-based virtual screening: The discovery of novel methionyl-tRNA synthetase inhibitors. *Bioorganic & Medicinal Chemistry Letters*, *16*(18), 4898–4907. doi:10.1016/j.bmcl.2006.06.057

Kim, S. Y., & Volsky, D. J. (2005). PAGE: Parametric analysis of gene set enrichment. *BMC Bioinformatics*, *6*, 144. doi:10.1186/1471-2105-6-144

Kim, S. Y., Imoto, S., & Miyano, S. (2003). Inferring gene networks from time series microarray data using dynamic Bayesian networks. *Briefings in Bioinformatics*, *4*(3), 228–235. doi:10.1093/bib/4.3.228

King, A. D. (2004). Protein complex prediction via cost-based clustering. *Bioinformatics (Oxford, England)*, *20*(17), 3013–3020. doi:10.1093/bioinformatics/bth351

King, K.R., Wang, S., Irimia, D., Jayaraman, A., & Toner, M. & M.L.Y. (2007). A high-throughput microfluidic real-time gene expression living cell array. *Lab on a Chip*, *7*(1), 77–85. doi:10.1039/b612516f

King, K.R., Wang, S., Jayaraman, A., & Yarmush, M.L. & M.T. (2008). Microfluidic flow-encoded switching for parallel control of dynamic cellular microenvironments. *Lab on a Chip*, *8*(1), 107–116. doi:10.1039/b716962k

Kingsmore, S. F., Lindquist, I. E., Mudge, J., & Beavis, W. D. (2007). Genome-wide association studies: Progress in identifying genetic biomarkers in common, complex diseases. *Biomarker Insights*, *2*, 283–292.

Kislinger, T. (2006). Global survey of organ and organelle protein expression in mouse: combined proteomic and transcriptomic profiling. *Cell*, *125*(1), 173–186. doi:10.1016/j.cell.2006.01.044

Kitano, H. (2002). Systems biology: A brief overview. *Science*, *295*(5560), 1662–1664. doi:10.1126/science.1069492

Kitano, H. (2001). *Foundations of systems biology*. MIT Press.

Kitchen, D. B., Decornez, H., Furr, J. R., & Bajorath, J. (2004). Docking and scoring in virtual screening for drug discovery: Methods and applications. *Nature Reviews. Drug Discovery*, *3*(11), 935–949. doi:10.1038/nrd1549

Kitteringham, N. R., Jenkins, R. E., Lane, C. S., Elliott, V. L., & Park, B. K. (2009). Multiple reaction monitoring for quantitative biomarker analysis in proteomics and metabolomics. *Journal of Chromatography. B, Analytical Technologies in the Biomedical and Life Sciences*, *877*(13), 1229–1239. doi:10.1016/j.jchromb.2008.11.013

Kittler, J. (1978). Feature set search algorithms. *Pattern recognition and signal processing,* (pp. 41–60).

Klein, T. E., Altman, R. B., Eriksson, N., Gage, B. F., Kimmel, S. E., & Lee, M. T. (2009). Estimation of the warfarin dose with clinical and pharmacogenetic data. *The New England Journal of Medicine*, *360*(8), 753–764. doi:10.1056/NEJMoa0809329

Klein, R. J., Zeiss, C., Chew, E. Y., Tsai, J. Y., Sackler, R. S., & Haynes, C. (2005). Complement factor H polymorphism in age-related macular degeneration. *Science*, *308*(5720), 385–389. doi:10.1126/science.1109557

Klein, R. J., & Eddy, S. R. (2003). RSEARCH: Finding homologs of single structured RNA sequences. *BMC Bioinformatics*, *4*, 44. doi:10.1186/1471-2105-4-44

Kleinstein, S. H. (2008). Getting started in computational immunology. *PLoS Computational Biology*, *4*(8), e1000128.

Klopfer, P. H., & Rubenstein, D. L. (1977). The concept privacy and its biological basis. *The Journal of Social Issues*, *33*, 52–65. doi:10.1111/j.1540-4560.1977.tb01882.x

Kloster, M., Tang, C., & Wingreen, N. S. (2005). Finding regulatory modules through large-scale gene-expression data analysis. *Bioinformatics (Oxford, England)*, *21*, 1172–1179. doi:10.1093/bioinformatics/bti096

Kluger, Y., Barsi, R., Cheng, J. T., & Gerstein, M. (2003). Spectral biclustering of microarray data: Coclustering genes and conditions. *Genome Research*, *13*(4), 703–716. doi:10.1101/gr.648603

Knegtel, R. M., Kuntz, I. D., & Oshiro, C. M. (1997). Molecular docking to ensembles of protein structures. *Journal of Molecular Biology*, *266*(2), 424–440. doi:10.1006/jmbi.1996.0776

Knoppers, B. M. (2005). Consent revisited: points to consider. *Health Law Review*, *13*(2/3), 33–38.

Knoppers, B. M., & Saginur, M. (2005). The Babel of genetic data terminology. *Nature Biotechnology*, *23*(8), 925–927. doi:10.1038/nbt0805-925

Knowles, M. R., Cervino, S., & Skynner, H. A. (2003). Multiplex proteomic analysis by two-dimensional differential in-gel electrophoresis. *Proteomics*, *3*(7), 1162–1171. doi:10.1002/pmic.200300437

Kobayashi, N., & Go, N. (1997). A method to search for similar protein local structures at ligand binding sites and its application to adenine recognition. *European Biophysics Journal*, *26*(2), 135–144. doi:10.1007/s002490050065

Kobayashi, N., & Go, N. (1997). ATP binding proteins with different folds share a common ATP-binding structural motif. *Nature Structural Biology*, *4*(1), 6–7. doi:10.1038/nsb0197-6

Kohler, S., Bauer, S., Horn, D., & Robinson, P. N. (2008). Walking the interactome for prioritization of candidate disease genes. *American Journal of Human Genetics*, *82*(4), 949–958. doi:10.1016/j.ajhg.2008.02.013

Kola, I., & Landis, J. (2004). Can the pharmaceutical industry reduce attrition rates? *Nature Reviews. Drug Discovery*, *3*(8), 711–715. doi:10.1038/nrd1470

Kolch, W., Calder, M., & David, G. (2005). When kinases meet mathematics: The systems biology of MAPK signalling. *FEBS Letters-Systems Biology*, *579*(8), 1891–1895. doi:10.1016/j.febslet.2005.02.002

Kolinski, A. (2004). Protein modeling and structure prediction with a reduced representation. *Acta Biochimica Polonica*, *51*(2), 349–371.

Kondo, T. (2008). Cancer proteomics for biomarker development. *Journal of Proteomics and Bioinformatics*, *1*(9), 477–484. doi:10.4172/jpb.1000055

Kong, X., Mas, V., & Archer, K. (2008). A non-parametric meta-analysis approach for combining independent microarray datasets: Application using two microarray datasets pertaining to chronic allograft nephropathy. *BMC Bioinformatics*, 9.

Kooperberg, C., & Ruczinski, I. (2005). Indentifying interaction SNPs using Monte Carlo logic regression. *Genetic Epidemiology*, *28*, 157–170. doi:10.1002/gepi.20042

Korbel, J. O., Urban, A. E., Affourtit, J. P., Godwin, B., Grubert, F., & Simons, J. F. (2007). Paired-end mapping reveals extensive structural variation in the human genome. *Science*, *318*(5849), 420–426. doi:10.1126/science.1149504

Korkko, J., Milunsky, J., Prockop, D. J., & Ala-Kokko, L. (1998). Use of conformation sensitive gel electrophoresis to detect single-base changes in the gene for COL10A1. *Human Mutation*, (Suppl 1), S201–S203.

Kornberg, R. D. (1974). Chromatin structure: A repeating unit of histones and DNA. *Science*, *184*(139), 868–871.

Kornberg, R. D., & Lorch, Y. (1999). Twenty-five years of the nucleosome, fundamental particle of the eukaryote chromosome. *Cell*, *98*(3), 285–294.

Kornberg, R. D., & Stryer, L. (1988). Statistical distributions of nucleosomes: Nonrandom locations by a stochastic mechanism. *Nucleic Acids Research*, *16*(14A), 6677–6690.

Kouzarides, T. (2007). Chromatin modifications and their function. *Cell*, *128*(4), 693–705.

Kraft, P., & Hunter, D. J. (2009). Genetic risk prediction-are we there yet? *The New England Journal of Medicine*, *360*(17), 1701–1703. doi:10.1056/NEJMp0810107

Kramer, A., Horn, H. W., & Rice, J. E. (2003). Fast 3D molecular superposition and similarity search in databases of flexible molecules. *Journal of Computer-Aided Molecular Design*, *17*(1), 13–38. doi:10.1023/A:1024503712135

Kramer, B., Rarey, M., & Lengauer, T. (1999). Evaluation of the FLEXX incremental construction algorithm for protein-ligand docking. *Proteins*, *37*(2), 228–241. doi:10.1002/(SICI)1097-0134(19991101)37:2<228::AID-PROT8>3.0.CO;2-8

Kreike, B., Halfwerk, H., Kristel, P., Glas, A., Peterse, H., & Bartelink, H. (2006). Gene expression profiles of primary breast carcinomas from patients at high risk for local recurrence after breast-conserving therapy. *Clinical Cancer Research*, *12*(19), 5705–5712. doi:10.1158/1078-0432.CCR-06-0805

Kricka, L. J., Master, S. R., Joos, T. O., & Fortina, P. (2006). Current perspectives in protein array technology. *Annals of Clinical Biochemistry*, *43*(Pt 6), 457–467. doi:10.1258/000456306778904731

Krithika, R., Marathe, U., Saxena, P., Ansari, M. Z., Mohanty, D., & Gokhale, R. S. (2006). A genetic locus required for iron acquisition in Mycobacterium tuberculosis. *Proceedings of the National Academy of Sciences of the United States of America*, *103*(7), 2069–2074. doi:10.1073/pnas.0507924103

Krogan, N. J. (2006). Global landscape of protein complexes in the yeast Saccharomyces cerevisiae. *Nature*, *440*(7084), 637–643. doi:10.1038/nature04670

Krovat, E. M., Fruhwirth, K. H., & Langer, T. (2005). Pharmacophore identification, in silico screening, and virtual library design for inhibitors of the human factor Xa. *Journal of Chemical Information and Modeling*, *45*(1), 146–159. doi:10.1021/ci049778k

Krovat, E. M., & Langer, T. (2004). Impact of scoring functions on enrichment in docking-based virtual screening: An application study on renin inhibitors. *Journal of Chemical Information and Computer Sciences*, *44*(3), 1123–1129. doi:10.1021/ci0342728

Krueger, M., Kratchmarova, I., Blagoev, B., Tseng, Y. H., Kahn, C. R., & Mann, M. (2008). Dissection of the insulin signaling pathway via quantitative phosphoproteomics. *Proceedings of the National Academy of Sciences of the United States of America*, *105*(7), 2451–2456. doi:10.1073/pnas.0711713105

Krueger, M., Moser, M., Ussar, S., Thievessen, I., & Luber, C. A. (2008). SILAC mouse for quantitative proteomics uncovers kindling-3 as an essential factor for red blood cell function. *Cell*, *134*(2), 353–364. doi:10.1016/j.cell.2008.05.033

Krysan, P. J., Jester, P. J., Gottwald, J. R., & Sussman, M. R. (2002). An Arabidopsis mitogen-activated protein kinase gene family encodes essential positive regulators of cytokinesis. *The Plant Cell*, *14*(5), 1109–1120. doi:10.1105/tpc.001164

Ku, M., Koche, R. P., Rheinbay, E., Mendenhall, E. M., Endoh, M., & Mikkelsen, T. S. (2008). Genomewide analysis of PRC1 and PRC2 occupancy identifies two classes of bivalent domains. *PLOS Genetics*, *4*(10), e1000242.

Kuhl, E., Garikipati, K., Arruda, E. M., & Grosh, K. (2005). Remodeling of biological tissues: Mechanically induced reorientation of a transversely isotropic chain network. *Journal of the Mechanics and Physics of Solids*, *53*, 1552–1573. doi:10.1016/j.jmps.2005.03.002

Kuhn, W., & Grün, F. (1942). Beziehungen zwischen elastischen Konstanten und Dehnungsdoppelbrechung hochelastischer Stoffe. *Kolloid-Zeitschrift*, *101*, 248–271. doi:10.1007/BF01793684

Kulp, D. C., & Jagalur, M. (2006). Causal inference of regulator-target pairs by gene mapping of expression phenotypes. *BMC Genomics*, *7*, 125. doi:10.1186/1471-2164-7-125

Kumar, S., Filipski, A., Swarna, V., Walker, A., & Hedges, S. B. (2005). Placing confidence limits on the molecular age of the human-chimpanzee divergence. *Proceedings of the National Academy of Sciences of the United States of America*, *102*(52), 18842–18847. doi:10.1073/pnas.0509585102

Kuntz, I. D. (1992). Structure-based strategies for drug design and discovery. *Science*, *257*(5073), 1078–1082. doi:10.1126/science.257.5073.1078

Kuntz, I. D., Blaney, J. M., Oatley, S. J., Langridge, R., & Ferrin, T. E. (1982). A geometric approach to macromolecule-ligand interactions. *Journal of Molecular Biology*, *161*(2), 269–288. doi:10.1016/0022-2836(82)90153-X

Kurdistani, S. K., & Grunstein, M. (2003). Histone acetylation and deacetylation in yeast. *Nature Reviews. Molecular Cell Biology*, *4*(4), 276–284.

Kuzmin, V. S., & Katser, S. B. (2005). Calculations of van der Waals volumes of Organic Molecules. *Russian Chemical Bulletin*, *41*(4), 720–727.

Kuzyk, M., Smith, D., & Yang, J. (2009). Multiple reaction monitoring-based, multiplexed, absolute quantitation of 45 proteins in human plasma. *Molecular & Cellular Proteomics*, *8*(8), 1860–1877. doi:10.1074/mcp.M800540-MCP200

Kwok, P. Y., & Duan, S. (2003). SNP discovery by direct DNA sequencing. *Methods in Molecular Biology (Clifton, N.J.)*, *212*, 71–84.

Ladroue, C., Guo, S., Kendrick, K. & Feng, J. (2009). *Beyond element-wise interactions: Identifying complex interactions in biological processes*.

LaFramboise, T. (2009). Single nucleotide polymorphism arrays: A decade of biological, computational and technological advances. *Nucleic Acids Research*, *37*(13), 4181–4193. doi:10.1093/nar/gkp552

Lage, K., Karlberg, O. E., Størling, Z. M., Ólason, P. Í., Pedersen, A. G., & Rigina, O. (2007). A human phenome-interactome network of protein complexes implicated in genetic disorders. *Nature*, *25*, 309–316. doi:10.1038/nbt1295

Lage, K., Hansen, N. T., Karlberg, E. O., Eklund, A. C., Roque, F. S., & Donahoe, P. K. (2008). A large-scale analysis of tissue-specific pathology and gene expression of human disease genes and complexes. *Proceedings of the National Academy of Sciences of the United States of America*, *105*(52), 20870–20875. doi:10.1073/pnas.0810772105

Lahana, R. (1999). How many leads from HTS? *Drug Discovery Today*, *4*(10), 447–448. doi:10.1016/S1359-6446(99)01393-8

Lahav, G., Rosenfeld, N., Sigal, A., Geva-Zatorsky, N., Levine, A. J., & Elowitz, M. B. (2004). Dynamics of the p53-MDM2 feedback loop in individual cells. *Nature Genetics*, *36*(2), 147–150. doi:10.1038/ng1293

Lai, P. C., Bahl, G., Gremigni, M., Matarazzo, V., Clot-Faybesse, O., & Ronin, C. (2008). An olfactory receptor pseudogene whose function emerged in humans: A case study in the evolution of structure-function in GPCRs. *Journal of Structural and Functional Genomics*, *9*(1-4), 29–40. doi:10.1007/s10969-008-9043-x

Lamb, J. (2007). The connectivity map: A new tool for biomedical research. *Nature Reviews. Cancer*, *7*(1), 54–60. doi:10.1038/nrc2044

Lamb, J., Crawford, E. D., Peck, D., Modell, J. W., Blat, I. C., & Wrobel, M. J. (2006). The connectivity map: Using gene-expression signatures to connect small molecules, genes, and disease. *Science*, *313*(5795), 1929–1935. doi:10.1126/science.1132939

Lan, F., Nottke, A. C., & Shi, Y. (2008). Mechanisms involved in the regulation of histone lysine demethylases. *Current Opinion in Cell Biology*, *20*(3), 316–325.

Landgraf, R., Fischer, D., & Eisenberg, D. (1999). Analysis of heregulin symmetry by weighted evolutionary tracing. *Protein Engineering*, *12*(11), 943–951. doi:10.1093/protein/12.11.943

Lantermann, A. B., Straub, T., Strålfors, A., Yuan, G. C., Ekwall, K., & Korber, P. (2010). Schizosaccharomyces pombe genome-wide nucleosome mapping reveals positioning mechanisms distinct from those of Saccharomyces cerevisiae. *Nature Structural & Molecular Biology*, *17*(2), 251–257.

Larkin, M. A., Blackshields, G., Brown, N. P., Chenna, R., McGettigan, P. A., & McWilliam, H. (2007). Clustal W and clustal X version 2.0. *Bioinformatics (Oxford, England)*, *23*(21), 2947–2948. doi:10.1093/bioinformatics/btm404

Lasko, T. A., & Vinterbo, S. A. (2010). Spectral Anonymization of Data. *IEEE Transactions on Knowledge and Data Engineering*, *22*(3), 437–446. doi:10.1109/TKDE.2009.88

Laufer, R. S., & Wolfe, M. (1977). Privacy as a Concept and a Social Issue - Multidimensional Developmental Theory. *The Journal of Social Issues*, *33*(3), 22–42. doi:10.1111/j.1540-4560.1977.tb01880.x

Lautru, S., Deeth, R. J., Bailey, L. M., & Challis, G. L. (2005). Discovery of a new peptide natural product by Streptomyces coelicolor genome mining. *Nature Chemical Biology*, *1*(5), 265–269. doi:10.1038/nchembio731

Lavalle, P., Stoltz, J. F., Senger, B., Voegel, J. C., & Schaaf, P. (1996). Red blood cell adhesion on a solid/liquid interface. *Proceedings of the National Academy of Sciences of the United States of America*, *93*, 15136–15140. doi:10.1073/pnas.93.26.15136

Lawrence, C. E., Altschul, S. F., Boguski, M. S., Liu, J. S., Neuwald, A. F., & Wootton, J. C. (1993). Detecting subtle sequence signals: A Gibbs sampling strategy for multiple alignment. *Science*, *262*(5131), 208–214. doi:10.1126/science.8211139

Lazzeroni, L., & Owen, A. (2000). Plaid models for gene expression data. *Statistica Sinica*, *12*, 61–86.

Le, S. Y., Chen, J. H., & Maizel, J. V. (1990). Efficient searches for unusual folding regions in RNA sequences. *Structure and Methods: Human Genome Initiative and DNA Recombination*, *1*, 127–136.

Le Novère, N., Bornstein, B., Broicher, A., Courtot, M., Donizelli, M., & Dharuri, H. (2006). BioModels Database: A free, centralized database of curated, published, quantitative kinetic models of biochemical and cellular systems. *Nucleic Acids Research*, *34*(Database issue), D689–D691. doi:10.1093/nar/gkj092

Leach, A.R., Gillet, V.J., Lewis, R.A. & Taylor, R. (2009). Three-dimensional pharmacophore methods in drug discovery. *Journal of Medical Chemistry*.

Lee, Y., & Lee, C. (2003). Classification of multiple cancer types by multicategory support vector machines using gene expression data. *Bioinformatics (Oxford, England)*, *19*(9), 1132–1139. doi:10.1093/bioinformatics/btg102

Lee, T. I., Rinaldi, N. J., Robert, F., Odom, D. T., Bar-Joseph, Z., & Gerber, G. K. (2002). Transcriptional regulatory networks in Saccharomyces cerevisiae. *Science*, *298*, 799–804. doi:10.1126/science.1075090

Lee, J. T. (2009). Lessons from X-chromosome inactivation: Long ncRNA as guides and tethers to the epigenome. *Genes & Development*, *23*(16), 1831–1842.

Lee, W., Tillo, D., Bray, N., Morse, R. H., Davis, R. W., & Hughes, T. R. (2007). A high-resolution atlas of nucleosome occupancy in yeast. *Nature Genetics*, *39*(10), 1235–1244.

Lee, D. S., Park, J., Kay, K. A., Christakis, N. A., Oltvai, Z. N., & Barabasi, A. L. (2008a). The implications of human metabolic network topology for disease comorbidity. *Proceedings of the National Academy of Sciences of the United States of America*, *105*(29), 9880–9885. doi:10.1073/pnas.0802208105

Lee, H. K., Hsu, A. K., Sajdak, J., Qin, J., & Pavlidis, P. (2004). Coexpression analysis of human genes across many microarray data sets. *Genome Research*, *14*(6), 1085–1094. doi:10.1101/gr.1910904

Lee, I., Lehner, B., Crombie, C., Wong, W., Fraser, A. G., & Marcotte, E. M. (2008b). A single gene network accurately predicts phenotypic effects of gene perturbation in caenorhabditis elegans. *Nature Genetics*, *40*(2), 181–188. doi:10.1038/ng.2007.70

Lee, E., Salic, A., Krüger, R., Heinrich, R., & Kirschner, M. (2003). The roles of APC and Axin derived from experimental and theoretical analysis of the Wnt pathway. *PLoS Biology*, *1*(1), E10. doi:10.1371/journal.pbio.0000010

Lee, B. T., Liew, L., Lim, J., Tan, J. K., Lee, T. C., & Veladandi, P. S. (2008). Candidate List of yoUr Biomarker (CLUB): A Web-based platform to aid cancer biomarker research. *Biomarker Insights*, *3*, 65–71.

Lee, H. K., Braynen, W., Keshav, K., & Pavlidis, P. (2005). ErmineJ: Tool for functional analysis of gene expression data sets. *BMC Bioinformatics*, *6*, 269. doi:10.1186/1471-2105-6-269

Lee, C.-C., Snyder, T. M., & Quake, S. R. (2010). A microfluidic oligonucleotide synthesizer. *Nucleic Acids Research*, 92.

Lee, H. S., Gheysel, E., & Bell, W. R. (1997). *Seasonal time series and autocorrelation function estimation*. Série scientifique CIANO 97s-35, Montréal, Canada.

Lehmann, S. (2008). Biclique communities. *Physical Review E: Statistical, Nonlinear, and Soft Matter Physics*, *78*(1), 016108. doi:10.1103/PhysRevE.78.016108

Lehne, B., & Schlitt, T. (2009). Protein-protein interaction databases: Keeping up with growing interactomes. *Human Genomics*, *3*(3), 291–297.

Leighton, J., Brown, P., & Ellis, A. (2006). Workgroup report: Review of genomics data based on experience with mock submissions-view of the CDER Pharmacology Toxicology Nonclinical Pharmacogenomics Subcommittee. *Environmental Health Perspectives*, *114*(4), 573–578. doi:10.1289/ehp.8318

Leiman, D. A., Lorenzi, N. M., Wyatt, J. C., Doney, A. S. F., & Rosenbloom, S. T. (2008). US and Scottish Health Professionals' Attitudes toward DNA Biobanking. *Journal of the American Medical Informatics Association*, *15*(3), 357–362. doi:10.1197/jamia.M2571

Lemmen, C., & Lengauer, T. (2000). Computational methods for the structural alignment of molecules. *Journal of Computer-Aided Molecular Design*, *14*(3), 215–232. doi:10.1023/A:1008194019144

Lemmen, C., Lengauer, T., & Klebe, G. (1998). FLEXS: A method for fast flexible ligand superposition. *Journal of Medicinal Chemistry*, *41*(23), 4502–4520. doi:10.1021/jm981037l

Lenormand, G. (2001). *Elasticité du globule rouge humain-une étude par pinces optiques*, Thèse de doctorat de l'Université Paris VI.

Lenz, E. M., Bright, J., Wilson, I. D., Morgan, S. R., & Nash, A. F. P. (2003). A ^{1}H NMR-based metabonomic study of urine and plasma samples obtained from healthy human subjects. *Journal of Pharmaceutical and Biomedical Analysis*, *33*(5), 1103–1115. doi:10.1016/S0731-7085(03)00410-2

Lesko, L. J., Salerno, R. A., Spear, B. B., Anderson, D. C., Anderson, T., & Brazell, C. (2003). Pharmacogenetics and pharmacogenomics in drug development and regulatory decision making: Report of the first FDA-PWG-PhRMA-DruSafe Workshop. *Journal of Clinical Pharmacology*, *43*, 342–358. doi:10.1177/0091270003252244

Letai, A. G. (2008). Diagnosing and exploiting cancer's addiction to blocks in apoptosis. *Nature Reviews. Cancer*, *8*(2), 121–132. doi:10.1038/nrc2297

Letovsky, S., & Kasif, S. (2003). Predicting protein function from protein/protein interaction data: A probabilistic approach. *Bioinformatics (Oxford, England)*, *19*(1), i197–i204. doi:10.1093/bioinformatics/btg1026

Lettre, G., & Rioux, J. D. (2008). Autoimmune diseases: Insights from genome-wide association studies. *Human Molecular Genetics*, *17*(R2), R116–R121. doi:10.1093/hmg/ddn246

Leung, T. H., Hoffmann, A., & Baltimore, D. (2004). One nucleotide in a kappaB site can determine cofactor specificity for NF-kappaB dimers. *Cell*, *118*(4), 453–464. doi:10.1016/j.cell.2004.08.007

Lev Bar-Or, R., Maya, R., Segel, L. A., Alon, U., Levine, A. J., & Oren, M. (2000). Generation of oscillations by the p53-MDM2 feedback loop: A theoretical and experimental study. *Proceedings of the National Academy of Sciences of the United States of America*, *97*(21), 11250–11255. doi:10.1073/pnas.210171597

Levchenko, A., Bruck, J., & Sternberg, P. W. (2000). Scaffold proteins may biphasically affect the levels of mitogen-activated protein kinase signaling and reduce its threshold properties. *Proceedings of the National Academy of Sciences of the United States of America*, *97*(11), 5818–5823. doi:10.1073/pnas.97.11.5818

Levinthal, C. (1969) How to fold graciously. In J.T.P. DeBrunner & E. Munck (Eds.), *Mossbauer spectroscopy in biological systems: Proceedings of a meeting held at Allerton House, Monticello, Illinois*, (pp. 22-24). University of Illinois Press.

Levitt, M. (2007). Growth of novel protein structural data. *Proceedings of the National Academy of Sciences of the United States of America*, *104*(9), 3183–3188. doi:10.1073/pnas.0611678104

Lewis, E. B. (1978). A gene complex controlling segmentation in Drosophila. *Nature*, *276*(5688), 565–570.

Li, H. F., Lu, T., Zhu, T., Jiang, Y. J., Rao, S. S., & Hu, L. Y. (2009). Virtual screening for Raf-1 kinase inhibitors based on pharmacophore model of substituted ureas. *European Journal of Medicinal Chemistry*, *44*(3), 1240–1249. doi:10.1016/j.ejmech.2008.09.016

Li, C., & Wong, W. H. (2001). Model-based analysis of oligonucleotide arrays: Model validation, design issues and standard error applications. *Genome Biology*, *2*(8), 1–11.

Li, Q., Fraley, C., Bumgarner, R. E., Yeung, K. Y., & Raftery, A. E. (2005). Donuts, scratches and blanks: Robust model-based segmentation of microarray images. *Bioinformatics (Oxford, England)*, *21*, 2875–2882. doi:10.1093/bioinformatics/bti447

Li, J. (2009). Network-assisted protein identification and data interpretation in shotgun proteomics. *Molecular Systems Biology*, *5*, 303. doi:10.1038/msb.2009.54

Li, M. H., Ung, P. M., Zajkowski, J., Garneau-Tsodikova, S., & Sherman, D. H. (2009). Automated genome mining for natural products. *BMC Bioinformatics*, *10*, 185. doi:10.1186/1471-2105-10-185

Li, L., Weinberg, C. R., Darden, T. A., & Pedersen, L. G. (2001). Gene selection for sample classification based on gene expression data: Study of sensitivity to choice of parameters of the GA/KNN method. *Bioinformatics (Oxford, England)*, *17*(12), 1131–1142. doi:10.1093/bioinformatics/17.12.1131

Liang, S., Zheng, D., Zhang, C., & Zacharias, M. (2009). Prediction of antigenic epitopes on protein surfaces by consensus scoring. *BMC Bioinformatics*, *10*, 302. doi:10.1186/1471-2105-10-302

Liao, J. G., & Chin, K. V. (2007). Logistic regression for disease classification using microarray data: model selection in a large p and small n case. *Bioinformatics (Oxford, England)*, *23*(15), 1945–1951. doi:10.1093/bioinformatics/btm287

Liao, B. Y., & Zhang, J. (2007). Mouse duplicate genes are as essential as singletons. *Trends in Genetics*, *23*(8), 378–381. doi:10.1016/j.tig.2007.05.006

Liaw, A., & Wiener, M. (2003). Classification and regression by randomForest. *R News*, *2/3*, 18–22.

Libioulle, C., Louis, E., Hansoul, S., Sandor, C., Farnir, F., & Franchimont, D. (2007). Novel Crohn disease locus identified by genome-wide association maps to a gene desert on 5p13.1 and modulates expression of PTGER4. *PLOS Genetics*, *3*(4), e58. doi:10.1371/journal.pgen.0030058

Lievens, S., Lemmens, I., & Tavernier, J. (2009). Mammalian two-hybrids come of age. *Trends in Biochemical Sciences*, *34*(11), 579–588. doi:10.1016/j.tibs.2009.06.009

Lillacci, G., Boccadoro, M., & Valigi, P. (2006). *In silico analysis of p53 response to DNA damage*. Paper presented at the 6th IFAC symposium on Modelling and Control in Biomedical Systems (including Biological Systems), 507-512.

Lillacci, G., Boccadoro, M., & Valigi, P. (2006). *The p53 network and its control via MDM2 inhibitors: Insights from a dynamic model*. Paper presented at the 45th IEEE Conference on Decision and Control, 2110-2115.

Lim, J., Hao, T., Shaw, C., Patel, A. J., Szabo, G., & Rual, J. F. (2006). A protein-protein interaction network for human inherited ataxias and disorders of purkinje cell degeneration. *Cell*, *125*(4), 801–814. doi:10.1016/j.cell.2006.03.032

Lin, Z., Owen, A. B., & Altman, R. B. (2004). GENETICS: Genomic Research and Human Subject Privacy. *Science*, *305*(5681), 183. doi:10.1126/science.1095019

Lin, Y., & Moret, B. M. (2008). Estimating true evolutionary distances under the DCJ model. *Bioinformatics (Oxford, England)*, *24*(13), i114–i122. doi:10.1093/bioinformatics/btn148

Lincoln, P., & Tiwari, A. (2004). *Symbolic systems biology: Hybrid modeling and analysis of biological networks*. Paper presented at the 7th International Workshop Hybrid System Computation and Control, 2993, 660-672. Berlin/Heidelberg: Springer.

Linding, R., Jensen, L. J., Pasculescu, A., Olhovsky, M., Colwill, K., & Bork, P. (2008). NetworKIN: A resource for exploring cellular phosphorylation networks. *Nucleic Acids Research*, *36*(Database issue), D695–D699. doi:10.1093/nar/gkm902

Linding, R., Jensen, L. J., Ostheimer, G. J., van Vugt, M. A., Jorgensen, C., & Miron, I. M. (2007). Systematic discovery of in vivo phosphorylation networks. *Cell*, *129*(7), 1415–1426. doi:10.1016/j.cell.2007.05.052

Lindon, J. C., Holmes, E., & Nicholson, J. K. (2006). Metabonomics techniques and applications to pharmaceutical research & development. *Pharmaceutical Research*, *23*, 1075–1088. doi:10.1007/s11095-006-0025-z

Lindon, J. C., Nicholson, J. K., Holmes, E., Antti, H., Bollard, M. E., & Keun, H. (2003). The role of metabonomics in toxicology and its evaluation by the COMET project. *Toxicology and Applied Pharmacology*, *187*, 137–146. doi:10.1016/S0041-008X(02)00079-0

Linghu, B., & Delisi, C. (2010). Phenotypic connections in surprising places. *Genome Biology*, *11*(4), 116. doi:10.1186/gb-2010-11-4-116

Linghu, B., Snitkin, E. S., Holloway, D. T., Gustafson, A. M., Xia, Y., & DeLisi, C. (2008). High-precision high-coverage functional inference from integrated data sources. *BMC Bioinformatics*, *9*, 119. doi:10.1186/1471-2105-9-119

Linghu, B., Snitkin, E. S., Hu, Z., Xia, Y., & Delisi, C. (2009). Genome-wide prioritization of disease genes and identification of disease-disease associations from an integrated human functional linkage network. *Genome Biology*, *10*(9), R91. doi:10.1186/gb-2009-10-9-r91

Linne, U., Stein, D. B., Mootz, H. D., & Marahiel, M. A. (2003). Systematic and quantitative analysis of protein-protein recognition between nonribosomal peptide synthetases investigated in the tyrocidine biosynthetic template. *Biochemistry*, *42*(17), 5114–5124. doi:10.1021/bi034223o

Lipinski, C. A., Lombardo, F., Dominy, B. W., & Feeney, P. J. (2001). Experimental and computational approaches to estimate solubility and permeability in drug discovery and development settings. *Advanced Drug Delivery Reviews*, *46*(1-3), 3–26. doi:10.1016/S0169-409X(00)00129-0

Lipowsky, R., & Seifert, U. (1991). Adhesion of membranes: A theoretical perspective. *Langmuir*, *7*, 1867–1873. doi:10.1021/la00057a009

Lister, R., Pelizzola, M., Dowen, R. H., Hawkins, R. D., Hon, G., & Tonti-Filippini, J. (2009). Human DNA methylomes at base resolution show widespread epigenomic differences. *Nature*, *462*(7271), 315–322.

Listgarten, J., & Emili, A. (2005). Statistical and computational methods for comparative proteomic profiling using liquid chromatography-tandem mass spectrometry. *Molecular & Cellular Proteomics*, *4*(4), 419–434. doi:10.1074/mcp.R500005-MCP200

Liu, M., & Wang, S. (1999). MCDOCK: A Monte Carlo simulation approach to the molecular docking problem. *Journal of Computer-Aided Molecular Design*, *13*(5), 435–451. doi:10.1023/A:1008005918983

Liu, H., Sadygov, R. G., & Yates, J. R. (2004). A model for random sampling and estimation of relative protein abundance in shotgun proteomics. *Analytical Chemistry*, *76*(14), 4193–4201. doi:10.1021/ac0498563

Liu, X., & Wang, L. (2007). Computing the maximum similarity bi-clusters of gene expression data. *Bioinformatics (Oxford, England)*, *23*(1), 50–56. doi:10.1093/bioinformatics/btl560

Liu, Y., Shao, Z., & Yuan, G. C. (2010). Prediction of polycomb target genes in mouse embryonic stem cells. *Genomics*, *96*(1), 17–26.

Liu, M., Liberzon, A., Kong, S., Lai, W., Park, P., & Kohane, I. (2007). Network-based analysis of affected biological processes in type 2 diabetes models. *PLOS Genetics*, *3*(6), e96. doi:10.1371/journal.pgen.0030096

Liu, B. A. (2006). The human and mouse complement of SH2 domain proteins–establishing the boundaries of phosphotyrosine signaling. *Molecular Cell*, *22*(6), 851–868. doi:10.1016/j.molcel.2006.06.001

Liu, M., Matsumura, N., Mandai, M., Li, K., Yagi, H., & Baba, T. (2009). Classification using hierarchical clustering of tumor-infiltrating immune cells identifies poor prognostic ovarian cancers with high levels of COX expression. *Modern Pathology*, *22*(3), 373–384. doi:10.1038/modpathol.2008.187

Liu, Y., & Ringnér, M. (2007). Revealing signaling pathway deregulation by using gene expression signatures and regulatory motif analysis. *Genome Biology*, *8*(5), R77. doi:10.1186/gb-2007-8-5-r77

Liu, J. (2008). Control of protein synthesis and mRNA degradation by microRNA. *Current Opinion in Cell Biology*, *20*(2), 214–222. doi:10.1016/j.ceb.2008.01.006

Liu, J., Yang, J., & Wang, W. (2004). Biclustering in gene expression data by tendency. *IEEE Computational Systems Bioinformatics Conference Proceedings, 182*(193), 16-19.

Liwo, A., Oldziej, S., Pincus, M. R., Wawak, R. J., Rackovsky, S., & Scheraga, H. A. (1997). A united-residue force field for off-lattice protein-structure simulations. I. Functional forms and parameters of long-range side-chain interaction potentials from protein crystal data. *Journal of Computational Chemistry*, *18*(7), 849–873. doi:10.1002/(SICI)1096-987X(199705)18:7<849::AID-JCC1>3.0.CO;2-R

Ljung, L. (1999). *System identification: Theory for the user* (2nd ed.). Paramus, NJ: Prentice Hall.

Lliuk, A., Galan, J., & Tao, W. A. (2009). Playing tag with quantative proteomics. *Analytical and Bioanalytical Chemistry*, *393*(2), 503–513. doi:10.1007/s00216-008-2386-0

Locasale, J. W., Shaw, A. S., & Chakraborty, A. K. (2007). Scaffold proteins confer diverse regulatory properties to protein kinase cascades. *Proceedings of the National Academy of Sciences of the United States of America*, *104*(33), 13307–13312. doi:10.1073/pnas.0706311104

Locke, J., Kozma-Bognar, L., Gould, P., & Feher, B., Kevei, Nagy, F., et al. (2006). Experimental validation of a predicted feedback loop in the multi-oscillator clock of Arabidopsis thaliana. *Molecular Systems Biology*, *2*(1).

Lohmussaar, E., Gschwendtner, A., & Mueller, J. C. (2005). ALOX5AP gene and the PDE4D gene in a central European population of stroke patients. *Stroke*, *36*, 731–736. doi:10.1161/01.STR.0000157587.59821.87

Loi, S., Haibe-Kains, B., Desmedt, C., Wirapati, P., Lallemand, F., & Tutt, A. M. (2008). Predicting prognosis using molecular profiling in estrogen receptor-positive breast cancer treated with tamoxifen. *BMC Genomics*, *9*, 239. doi:10.1186/1471-2164-9-239

Longo, D., & Hasty, J. (2006). Dynamics of single-cell gene expression. *Molecular Systems Biology*, 28.

Lopanik, N. B., Shields, J. A., Buchholz, T. J., Rath, C. M., Hothersall, J., & Haygood, M. G. (2008). In vivo and in vitro trans-acylation by BryP, the putative bryostatin pathway acyltransferase derived from an uncultured marine symbiont. *Chemistry & Biology*, *15*(11), 1175–1186. doi:10.1016/j.chembiol.2008.09.013

Lowe, T. M., & Eddy, S. R. (1997). tRNAscan-SE: A program for improved detection of transfer RNA genes in genomic sequence. *Nucleic Acids Research*, *25*(5), 955–964. doi:10.1093/nar/25.5.955

Lubovac, Z. (2006). Combining functional and topological properties to identify core modules in protein interaction networks. *Proteins*, *64*(4), 948–959. doi:10.1002/prot.21071

Lubovac, Z., Olsson, B., Jonsson, P., Laurio, K., & Anderson, M. L. (2001). Biological and statistical evaluation of clusterings of gene expression profiles. In C.E. D'Attellis, V.V. Kluev & N.E. Mastorakis, (Eds.), *Proc. Mathematics and Computers in Biology and Chemistry (MCBC '01),* (pp. 149–155). Skiathos Island, Greece, September.

Luce, R., & Perry, A. (1949). A method of matrix analysis of group structure. *Psychometrika*, *14*(2), 95–116. doi:10.1007/BF02289146

Luger, K., Mader, A. W., Richmond, R. K., Sargent, D. F., & Richmond, T. J. (1997). Crystal structure of the nucleosome core particle at 2.8. A resolution. *Nature*, *389*(6648), 251–260.

Lunshof, J. E., Chadwick, R., Vorhaus, D. B., & Church, G. M. (2008). From genetic privacy to open consent. *Nature Reviews. Genetics*, *9*(5), 406–411. doi:10.1038/nrg2360

Luo, B., Cheung, H. W., Subramanian, A., Sharifnia, T., Okamoto, M., & Yang, X. (2008). Highly parallel identification of essential genes in cancer cells. *Proceedings of the National Academy of Sciences of the United States of America*, *105*(51), 20380–20385. doi:10.1073/pnas.0810485105

Lyne, P. D. (2002). Structure-based virtual screening: An overview. *Drug Discovery Today*, *7*(20), 1047–1055. doi:10.1016/S1359-6446(02)02483-2

Ma, W. W., & Adjei, A. A. (2009). Novel agents on the horizon for cancer therapy. *CA: a Cancer Journal for Clinicians*, *59*(2), 111–137. doi:10.3322/caac.20003

Ma, L., Wagner, J., Rice, J. J., Hu, W., Levine, A. J., & Stolovitzky, G. A. (2005). A plausible model for the digital response of p53 to DNA damage. *Proceedings of the National Academy of Sciences of the United States of America*, *102*(40), 14266–14271. doi:10.1073/pnas.0501352102

Macek, B., Waanders, L. F., Olsen, J. V., & Mann, M. (2006). Top-down protein sequencing and MS3 on a hybrid linear quadrupole ion trap-orbitrap mass spectrometer. *Molecular & Cellular Proteomics*, *5*(5), 949–958. doi:10.1074/mcp.T500042-MCP200

Machanavajjhala, A., Kifer, D., Gehrke, J., & Venkitasubramaniam, M. (2007). L-diversity: Privacy beyond k-anonymity. *ACM Trans. Knowl. Discov. Data*, *1*(1), 3. doi:10.1145/1217299.1217302

Mackay, T. F. C. (2004). The genetic architecture of quantitative traits: Lessons from Drosophila. *Current Opinion in Genetics & Development*, *14*(3), 253–257. doi:10.1016/j.gde.2004.04.003

MacQueen, J. (1967). *Some methods for classification and analysis of multivariate observations.* Paper presented at the Proceedings of the Fifth Berkeley Symposium on Mathematical Statistics and Probability. Volume I: Statistics.

Madeira, S. C., & Oliveira, A. L. (2004). Biclustering algorithms for biological data analysis: A survey. *IEEE/ACM Transactions on Computational Biology and Bioinformatics*, *1*(1), 24–45. doi:10.1109/TCBB.2004.2

Madej, T., Panchenko, A. R., Chen, J., & Bryant, S. H. (2007). Protein homologous cores and loops: Important clues to evolutionary relationships between structurally similar proteins. *BMC Structural Biology*, *7*, 23. doi:10.1186/1472-6807-7-23

Maere, S., Heymans, K., & Kuiper, M. (2005). BiNGO: A cytoscape plugin to assess overrepresentation of gene ontology categories in biological networks. *Bioinformatics (Oxford, England)*, *21*, 3448–3449. doi:10.1093/bioinformatics/bti551

Maher, B. (2008). Personal genomes: The case of the missing heritability. *Nature*, *456*(7218), 18–21. doi:10.1038/456018a

Mahfouz, M.A. & Ismail, M.A. (2009). BIDENS: Iterative density based biclustering algorithm with application to gene expression analysis. *Proceedings of World Academy of Science, Engineering and Technology*, *37*(2070-3740), 342–348.

Mailman, R.B. & Murthy, V. (2009). Third generation antipsychotic drugs: Partial agonism or receptor functional selectivity? *Current Pharmaceutical Design.*

Maione, D., Margarit, I., Rinaudo, C. D., Masignani, V., Mora, M., & Scarselli, M. (2005). Identification of a universal Group B streptococcus vaccine by multiple genome screen. *Science*, *309*(5731), 148–150. doi:10.1126/science.1109869

Malin, B. A., & Sweeney, L. (2004). How (not) to protect genomic data privacy in a distributed network: using trail re-identification to evaluate and design anonymity protection systems. *Journal of Biomedical Informatics*, *37*(3), 179–192. doi:10.1016/j.jbi.2004.04.005

Mandal, S., Moudgil, M., & Mandal, S. K. (2009). Rational drug design. *European Journal of Pharmacology*, *625*(1-3), 90–100. doi:10.1016/j.ejphar.2009.06.065

Mani, K. M., Lefebvr, C., Wang, K., Lim, W. K., Basso, K., & Dalla-Favera, R. (2008). A systems biology approach to prediction of oncogenes and molecular perturbation targets in B-cell lymphomas. *Molecular Systems Biology*, *4*, 169. doi:10.1038/msb.2008.2

Manolio, T. A., Collins, F. S., Cox, N. J., Goldstein, D. B., Hindorff, L. A., & Hunter, D. J. (2009). Finding the missing heritability of complex diseases. *Nature*, *461*(7265), 747–753. doi:10.1038/nature08494

Marchler-Bauer, A., Anderson, J. B., Cherukuri, P. F., DeWeese-Scott, C., Geer, L. Y., & Gwadz, M. (2005). CDD: A Conserved Domain Database for protein classification. *Nucleic Acids Research*, *33*(Database issue), D192–D196. doi:10.1093/nar/gki069

Mardis, E. R. (2008). The impact of next-generation sequencing technology on genetics. *Trends in Genetics*, *24*(3), 133–141.

Margulies, M. (2005). Genome sequencing in microfabricated high-density picolitre reactors. *Nature*, *437*, 376–380.

Markel, S., & Leon, D. (2003). *Sequence analysis in a nutshell.* Sebastopol, CA: O'Reilly.

Marks, A. D., & Steinberg, K. K. (2002). The Ethics of Access to Online Genetic Databases: Private or Public? *American Journal of Pharmacogenomics*, *2*, 207–212. doi:10.2165/00129785-200202030-00006

Marks, F., Klingmüler, U., & Müller-Decker, K. (Eds.). (2009). *Cellular signal processing: An introduction to the molecular mechanisms of signal transduction.* New York: Garland Science.

Marques, C. M. (2001). Le récepteur, le ligand et sa chaine polymère, peut-on contrôler l'adhésion cellulaire. *CNRS Infos*, *396*, 21–22.

Martens, J. H., O'Sullivan, R. J., Braunschweig, U., Opravil, S., Radolf, M., & Steinlein, P. (2005). The profile of repeat-associated histone lysine methylation states in the mouse epigenome. *The EMBO Journal*, *24*(4), 800–812.

Martin, Y. C. (2005). A bioavailability score. *Journal of Medicinal Chemistry*, *48*, 3164–3170. doi:10.1021/jm0492002

Martin, Y. C., Bures, M. G., Danaher, E. A., DeLazzer, J., Lico, I., & Pavlik, P. A. (1993). A fast new approach to pharmacophore mapping and its application to dopaminergic and benzodiazepine agonists. *Journal of Computer-Aided Molecular Design*, *7*(1), 83–102. doi:10.1007/BF00141577

Martin, D., Brun, C., Remy, E., Mouren, P., Thieffry, D., & Bernard Jacq, B. (2004). GOToolBox: Functional analysis of gene datasets based on gene ontology. *Genome Biology*, *5*(12). doi:10.1186/gb-2004-5-12-r101

Marx, G. T. (1999). What's in a name? Some reflections on the sociology of anonymity. [Article]. *The Information Society*, *15*(2), 99–112. doi:10.1080/019722499128565

Maskery, S. M., & Shinbrot, T. (2005). Deterministic and stochastic elements of axonal guidance. *Annual Review of Biomedical Engineering*, *7*, 187–221. doi:10.1146/annurev.bioeng.7.060804.100446

Mason, J. S., Good, A. C., & Martin, E. J. (2001). 3-D Pharmacophores in drug discovery. *Current Pharmacology Descriptions*, *7*(7), 567–597. doi:10.2174/1381612013397843

Mathews, D. H., Sabina, J., Zuker, M., & Turner, D. H. (1998). Expanded sequence dependence of thermodynamic parameters provides robust prediction of RNA secondary structure. *Journal of Molecular Biology*, *288*(5), 911–940. doi:10.1006/jmbi.1999.2700

Mathivanan, S., Periaswamy, B., Gandi, T., Kandasamy, K., Suresh, S., Mohmood, R., et al. (2006). An evaluation of human protein-protein interaction data in the public domain. *BioMed Central Bioinformatics, 7*, Suppl 5S19.

Matsui, H., Sato, K., & Sakakibara, Y. (2005). Pair stochastic tree adjoining grammars for aligning and predicting pseudoknot RNA structures. *Bioinformatics (Oxford, England)*, *21*(11), 2611–2617. doi:10.1093/bioinformatics/bti385

Matsumoto, M., & Nishimura, T. (1998). *Mersenne twister: A 623-dimensionally equidistributed uniform pseudorandom number generator*. ACM Transactions in Modeling and Computer Simulations.

Matsushita, K., Takeuchi, O., Standley, D. M., Kumagai, Y., Kawagoe, T., & Miyake, T. (2009). Zc3h12a is an RNase essential for controlling immune responses by regulating mRNA decay. *Nature*, *458*(7242), 1185–1190. doi:10.1038/nature07924

Matthews, L., Gopinath, G., Gillespie, M., Caudy, M., Croft, D., & de Bono, B. (2009). Reactome knowledgebase of human biological pathways and processes. *Nucleic Acids Research*, *37*(Database issue), D619–D622. doi:10.1093/nar/gkn863

Matys, V., Kel-Margoulis, O. V., Fricke, E., Liebich, I., Land, S., & Barre-Dirrie, A. (2006). TRANSFAC and its module TRANSCompel: Transcriptional gene regulation in eukaryotes. *Nucleic Acids Research*, *34*(Database issue), D108–D110. doi:10.1093/nar/gkj143

Mavrich, T. N., Ioshikhes, I. P., Venters, B. J., Jiang, C., Tomsho, L. P., & Qi, J. (2008). A barrier nucleosome model for statistical positioning of nucleosomes throughout the yeast genome. *Genome Research*, *18*(7), 1073–1083.

McAlpine, J. B., Bachmann, B. O., Piraee, M., Tremblay, S., Alarco, A. M., & Zazopoulos, E. (2005). Microbial genomics as a guide to drug discovery and structural elucidation: ECO-02301, a novel antifungal agent, as an example. *Journal of Natural Products*, *68*(4), 493–496. doi:10.1021/np0401664

McCarroll, S. A., Kuruvilla, F. G., Korn, J. M., Cawley, S., Nemesh, J., & Wysoker, A. (2008). Integrated detection and population-genetic analysis of SNPs and copy number variation. *Nature Genetics*, *40*(10), 1166–1174. doi:10.1038/ng.238

McCarthy, M. I., Abecasis, G. R., Cardon, L. R., Goldstein, D. B., Little, J., & Ioannidis, J. P. (2008). Genome-wide association studies for complex traits: Consensus, uncertainty and challenges. *Nature Reviews. Genetics*, *9*(5), 356–369. doi:10.1038/nrg2344

McGann, M. R., Almond, H. R., Nicholls, A., Grant, J. A., & Brown, F. K. (2003). Gaussian docking functions. *Biopolymers*, *68*(1), 76–90. doi:10.1002/bip.10207

McGary, K. L., Lee, I., & Marcotte, E. M. (2007). Broad network-based predictability of saccharomyces cerevisiae gene loss-of-function phenotypes. *Genome Biology*, *8*(12), R258. doi:10.1186/gb-2007-8-12-r258

McKelvie, J. R., Yuk, J., Xu, Y., Simpson, A. J., & Simpson, M. J. (2009). ^{1}H NMR and GC-MS metabonomics of earthworm responses to sub-lethal DDT and endosulfan exposure. *Metabolomics*, *5*(1), 84–94. doi:10.1007/s11306-008-0122-6

McKinney, B. A., Reif, D. M., White, B. C., Crowe, J. E. Jr, & Moore, J. H. (2007). Evaporative cooling feature selection for genotypic data involving interactions. *Bioinformatics (Oxford, England)*, *23*, 2113–2120. doi:10.1093/bioinformatics/btm317

McKinney, B., Crowe, J., Voss, H. Jr, Crooke, P., Barney, N., & Moore, J. (2006). Hybrid grammar-based approach to nonlinear dynamical systems identification from biological time series. *Physical Review E: Statistical, Nonlinear, and Soft Matter Physics*, *73*(021912), 1–7.

McKusick, V. A. (2007). Mendelian inheritance in man and its online version, OMIM. *American Journal of Human Genetics*, *80*(4), 588–604. doi:10.1086/514346

McLafferty, F. W., Breuker, K., & Jin, M. (2007). Top-down MS, a power complement to the high capabilities of proteolysis proteomics. *The FEBS Journal*, *274*(24), 6256–6268.

McLendon, (2008). Comprehensive genomic characterization defines human glioblastoma genes and core pathways. *Nature*, *455*(7216), 1061–1068. doi:10.1038/nature07385

McQuarrie, D. A. (1967). Stochastic approach to chemical kinetics. *Journal of Applied Probability*, *4*(3), 413–478. doi:10.2307/3212214

Medema, R. H., & Bos, J. L. (1993). The role of p21-ras in receptor tyrosine kinase signaling. *Critical Reviews in Oncogenesis*, *4*, 615–661.

Meek, D. W. (2009). Tumour suppression by p53: A role for the DNA damage response? *Nature Reviews. Cancer*, *9*(10), 714–723.

Mefti, N., Haussy, B., & Ganghoffer, J. F. (2006). Mechanical modelling of the rolling phenomenon at the cell scale. *International Journal of Solids and Structures*, *43*(24), 7378–7392. doi:10.1016/j.ijsolstr.2006.05.006

Mefti, N. (2006). *Mise en oeuvre d'un modèle mécanique de l'adhésion cellulaire: Approche stochastique*. Thèse de Doctorat de l'INPL. Nancy, France.

Meissner, A., Mikkelsen, T.S., Gu, H., Wernig, M., Hanna, J., Sivachenko, A., et al. (2008). Genome-scale DNA methylation maps of pluripotent and differentiated cells. *Nature*.

Melin, J., & Quake, S. R. (2007). Microfluidic large-scale integration: The evolution of design rules for biological automation. *Annual Review of Biophysics and Biomolecular Structure*, *36*(1), 213–231. doi:10.1146/annurev.biophys.36.040306.132646

Mendoza, L., Thieffry, D., & Alvarez-Buylla, E. (1999). Genetic control of flower morphogenesis in Arabidopsis thaliana: A logical analysis. *Bioinformatics (Oxford, England)*, *15*(7-8), 593–606. doi:10.1093/bioinformatics/15.7.593

Menges, M., Dóczi, R., Ökrész, L., Morandini, P., Mizzi, P., & Soloviev, M. (2008). Comprehensive gene expression atlas for the Arabidopsis MAP kinase signalling pathways. *The New Phytologist*, *179*(3), 643–662. doi:10.1111/j.1469-8137.2008.02552.x

Meslin, E. M., & Quaid, K. A. (2004). Ethical Issues in the Storage, Collection and Research Use of Human Biological Materials. *The Journal of Laboratory and Clinical Medicine*, *144*, 229–234. doi:10.1016/j.lab.2004.08.003

Meslin, E. M. (2010). The Value of Using Top-Down and Bottom-Up Approaches for Building Trust and Transparency in Biobanking. *Public Health Genomics, 13*(4).

Meslin, E. M., & Goodman, K. W. (2010). An Ethics and Policy Agenda for Biobanks and Electronic Health. *Science Progress, February 25th*, http://www.scienceprogress.org/2010/2002/bank-on-it/.

Mestres, J., Rohrer, D.C. & Maggiora, G.M. (1997). A molecular field-based similarity approach to pharmacophoric pattern recognition. *Journal of Molecular Graphics and Modelling, 15*(2), 114-121, 103-116.

Meszaros, T., Helfer, A., Hatzimasoura, E., Magyar, Z., Serazetdinova, L., & Rios, G. (2006). The Arabidopsis MAP kinase kinase MKK1 participates in defence responses to the bacterial elicitor flagellin. *The Plant Journal*, *48*(4), 485–498. doi:10.1111/j.1365-313X.2006.02888.x

Metropolis, N., Rosenbluth, A., Rosenbluth, M., Teller, A., & Teller, E. (1953). Equation of state calculations by fast computing machines. *The Journal of Chemical Physics*, *21*(6), 1087–1092. doi:10.1063/1.1699114

Michell, A. W., Mosedale, D., Grainger, D. J., & Barker, R. A. (2008). Metabolomic analysis of urine and serum in Parkinson's disease. *Metabolomics*, *4*(3), 191–201. doi:10.1007/s11306-008-0111-9

Michiels, S., Koscielny, S., & Hill, C. (2005). Prediction of cancer outcome with microarrays: A multiple random validation strategy. *Lancet*, *365*(9458), 488–492. doi:10.1016/S0140-6736(05)17866-0

Mickels, P. A., & Rigden, D. J. (2006). Evolutionary analysis of fructose 2,6-bisphohate metabolism. *International Union of Biochemistry and Molecular Biology (IUBMB). Life (Chicago, Ill.)*, *58*(3), 133–141.

Miele, V., Vaillant, C., d'Aubenton-Carafa, Y., Thermes, C., & Grange, T. (2008). DNA physical properties determine nucleosome occupancy from yeast to fly. *Nucleic Acids Research*, *36*(11), 3746–3756.

Miele, L. (2006). Notch signaling. *Clinical Cancer Research*, *12*(4), 1074–1078. doi:10.1158/1078-0432.CCR-05-2570

Mihalas, G. I., Simon, Z., Balea, G., & Popa, E. (2000). Possible oscillatory behavior in p53-MDM2 interaction computer simulation. *Journal of Biological System*, *8*(1), 21–29. doi:10.1142/S0218339000000031

Mikkelsen, T. S., Ku, M., Jaffe, D. B., Issac, B., Lieberman, E., & Giannoukos, G. (2007). Genome-wide maps of chromatin state in pluripotent and lineage-committed cells. *Nature*, *448*(7153), 553–560.

Miller, M. D., Kearsley, S. K., Underwood, D. J., & Sheridan, R. P. (1994). FLOG: A system to select 'quasi-flexible' ligands complementary to a receptor of known three-dimensional structure. *Journal of Computer-Aided Molecular Design*, *8*(2), 153–174. doi:10.1007/BF00119865

Miller, L. D., Smeds, J., George, J., Vega, V. B., Vergara, L., & Ploner, A. (2005). An expression signature for p53 status in human breast cancer predicts mutation status, transcriptional effects, and patient survival. *Proceedings of the National Academy of Sciences of the United States of America*, *102*(38), 13550–13555. doi:10.1073/pnas.0506230102

Millstein, J., Zhang, B., Zhu, J., & Schadt, E. E. (2009). Disentangling molecular relationships with a causal inference test. *BMC Genetics*, *10*, 23. doi:10.1186/1471-2156-10-23

Millstein, J., Conti, D. V., Gilliland, F. D., & Gauderman, J. W. (2006). A testing framework for identifying susceptibility genes in the presence of epistasis. *American Journal of Human Genetics*, *78*, 15–27. doi:10.1086/498850

Ming, G., Song, H. J., Berninger, B., Holt, C., & Tessier-Lavigne, M. (1997). cAMP-dependent growth cone guidance by netrin-1. *Neuron*, *19*(6), 1225–1235. doi:10.1016/S0896-6273(00)80414-6

Minowa, Y., Araki, M., & Kanehisa, M. (2007). Comprehensive analysis of distinctive polyketide and nonribosomal peptide structural motifs encoded in microbial genomes. *Journal of Molecular Biology*, *368*(5), 1500–1517. doi:10.1016/j.jmb.2007.02.099

Mishra, N., Basu, A., Jayaprakash, V., Sharon, A., Basu, M., & Patnaik, K. K. (2009). Structure based virtual screening of GSK-3beta: importance of protein flexibility and induced fit. *Bioorganic & Medicinal Chemistry Letters*, *19*(19), 5582–5585. doi:10.1016/j.bmcl.2009.08.042

Mishra, N.S., Tuteja, Renu, Tuteja & Narendra. (2009). Signaling through MAP kinase networks in plants. *Archives of Biochemistry and Biophysics*, *452*(1).

Mitchell, T. (2009). *Machine learning*. New York: McGraw-Hill.

Mitra, S., & Banka, H. (2006). Multi-objective evolutionary biclustering of gene expression data. *Pattern Recognition*, *39*(12), 2464–2477. doi:10.1016/j.patcog.2006.03.003

Mizoguchi, T., Hirayama, I. K., Hayashida, T., Yamaguchi-Shinozaki, N., Matsumoto, K., & Shinozaki, K. (1996). A gene encoding a mitogen-activated protein kinase kinase kinase is induced simultaneously with genes for a mitogen-activated protein kinase and an S6 ribosomal protein kinase by touch, cold, and water stress in Arabidopsis thaliana. *Proceedings of the National Academy of Sciences of the United States of America*, *93*(2), 765–769. doi:10.1073/pnas.93.2.765

Mizoguchi, T., Ichimura, K., Irie, K., Morris, P., Giraudat, J., & Matsumoto, K. (1998). Identification of a possible MAP kinase cascade in Arabidopsis thaliana based on pairwise yeast two-hybrid analysis and functional complementation tests of yeast mutants. *FEBS Letters*, *437*(1-2), 56–60. doi:10.1016/S0014-5793(98)01197-1

Mizutani, M. Y., Tomioka, N., & Itai, A. (1994). Rational automatic search method for stable docking models of protein and ligand. *Journal of Molecular Biology*, *243*(2), 310–326. doi:10.1006/jmbi.1994.1656

MMWR. (1999). Impact of vaccines universally recommended for children--United States, 1990-1998. *Morbidity and Mortality Weekly Report*, *48*(12), 243–248.

Moazed, D. (2009). Small RNAs in transcriptional gene silencing and genome defence. *Nature*, *457*(7228), 413–420.

Mobini, R. (2009). A module-based analytical strategy to identify novel disease-associated genes shows an inhibitory role for interleukin 7 receptor in allergic inflammation. *BMC Systems Biology*, *3*, 19. doi:10.1186/1752-0509-3-19

Mochizuki, A. (2002). Pattern formation of the cone mosaic in the Zebrafish retina: A cell rearrangement model. *Journal of Theoretical Biology*, *215*, 345–361. doi:10.1006/jtbi.2001.2508

Mochizuki, A., Ywasha, Y., & Takeda, Y. (1996). A stochastic model for cell sorting and measuring cell-cell adhesion. *Journal of Theoretical Biology*, *179*, 129–146. doi:10.1006/jtbi.1996.0054

Moffitt, M. C., & Neilan, B. A. (2003). Evolutionary affiliations within the superfamily of ketosynthases reflect complex pathway associations. *Journal of Molecular Evolution*, *56*(4), 446–457. doi:10.1007/s00239-002-2415-0

Mohn, F., Weber, M., Rebhan, M., Roloff, T. C., Richter, J., & Stadler, M. B. (2008). Lineage-specific polycomb targets and de novo DNA methylation define restriction and potential of neuronal progenitors. *Molecular Cell*, *30*(6), 755–766.

Moldenhauer, J., Gotz, D. C., Albert, C. R., Bischof, S. K., Schneider, K., & Sussmuth, R. D. (2010). The final steps of bacillaene biosynthesis in Bacillus amyloliquefaciens FZB42: Direct evidence for beta, gamma dehydration by a trans-acyltransferase polyketide synthase. *Angewandte Chemie International Edition*, *49*(8), 1465–1467.

Moles, C. G., Mendes, P., & Banga, J. R. (2003). Parameter estimation in biochemical pathways: A comparison of global optimization methods. *Genome Research*, *13*(11), 2467–2474. doi:10.1101/gr.1262503

Mollen, K. P., Gribar, S. C., Anand, R. J., Kaczorowski, D. J., Kohler, J. W., & Branca, M. F. (2008). Increased expression and internalization of the endotoxin coreceptor CD14 in enterocytes occur as an early event in the development of experimental necrotizing enterocolitis. *Journal of Pediatric Surgery*, *43*(6), 1175–1181. doi:10.1016/j.jpedsurg.2008.02.050

Monk, N. A. (2003). Oscillatory expression of Hes1, p53, and NF- B driven by transcriptional time delays. *Current Biology*, *13*(16), 1409–1413. doi:10.1016/S0960-9822(03)00494-9

Mook, S., Knauer, M., Bueno-de-Mesquita, J. M., Retel, V. P., Wesseling, J., & Linn, S. C. (2010). Metastatic potential of T1 breast cancer can be predicted by the 70-gene MammaPrint signature. *Annals of Surgical Oncology*, *17*(5), 1406–1413. doi:10.1245/s10434-009-0902-x

Moore, J. H. (2003). The ubiquitous nature of epistasis in determining susceptibility to common human diseases. *Human Heredity*, *56*, 73–82. doi:10.1159/000073735

Moore, J. H. (2009). From genotype to genometype: Putting the genome back in genome-wide association studies. *European Journal of Human Genetics*, *17*(10), 1231–1240. doi:10.1038/ejhg.2009.39

Moore, J. H., Asselbergs, F. W., & Williams, S. M. (2010). Bioinformatics challenges for genome-wide association studies. *Bioinformatics (Oxford, England)*, *26*(4), 445–455. doi:10.1093/bioinformatics/btp713

Moore, J. H., & White, B. C. (2007). Tuning ReliefF for genome-wide genetic analysis. *Lecture Notes in Computer Science*, *4447*, 166–175. doi:10.1007/978-3-540-71783-6_16

Moore, J. H., & Williams, S. M. (2002). New strategies for identifying gene-gene interactions in hypertension. *Annals of Medicine*, *34*, 88–95. doi:10.1080/07853890252953473

Moore, J. H., & Williams, S. M. (2005). Traversing the conceptual divide between biological and statistical epistasis: Systems biology and a more modern synthesis. *BioEssays*, *27*, 637–646. doi:10.1002/bies.20236

Moore, J. H., & Williams, S. M. (2009). Epistasis and its implications for personal genetics. *American Journal of Human Genetics*, *85*(3), 309–320. doi:10.1016/j.ajhg.2009.08.006

Moore, A. D. (2008). Arrangements in the modular evolution of proteins. *Trends in Biochemical Sciences*, *33*(9), 444–451. doi:10.1016/j.tibs.2008.05.008

Mootha, V., Lindgren, C., Eriksson, K.-F., Subramanian, A., Sihag, S., & Lehar, J. (2003). PGC-1alpha-responsive genes involved in oxidative phosphorylation are coordinately downregulated in human diabetes. *Nature Genetics*, *34*(3), 267–273. doi:10.1038/ng1180

Mootz, H. D., Schwarzer, D., & Marahiel, M. A. (2002). Ways of assembling complex natural products on modular nonribosomal peptide synthetases. *ChemBioChem*, *3*(6), 490–504. doi:10.1002/1439-7633(20020603)3:6<490::AID-CBIC490>3.0.CO;2-N

Mora, M., Veggi, D., Santini, L., Pizza, M., & Rappuoli, R. (2003). Reverse vaccinology. *Drug Discovery Today*, *8*(10), 459–464. doi:10.1016/S1359-6446(03)02689-8

Morita, H., Kondo, S., Oguro, S., Noguchi, H., Sugio, S., & Abe, I. (2007). Structural insight into chain-length control and product specificity of pentaketide chromone synthase from Aloe arborescens. *Chemistry & Biology*, *14*(4), 359–369. doi:10.1016/j.chembiol.2007.02.003

Morley, M., Molony, C. M., Weber, T. M., Devlin, J. L., Ewens, K. G., & Spielman, R. S. (2004). Genetic analysis of genome-wide variation in human gene expression. *Nature*, *430*(7001), 743–747. doi:10.1038/nature02797

Morozov, A. V., Fortney, K., Gaykalova, D. A., Studitsky, V. M., Widom, J., & Siggia, E. D. (2009). Using DNA mechanics to predict in vitro nucleosome positions and formation energies. *Nucleic Acids Research*, *37*(14), 4707–4722.

Morozova, O., & Marra, M. A. (2008). Applications of next-generation sequencing technologies in functional genomics. *Genomics*, *92*(5), 255–264. doi:10.1016/j.ygeno.2008.07.001

Morrison, D. K., & Davis, R. J. (2003). Regulation of MAP kinase signaling modules by scaffold proteins in mammals. *Annual Review of Cell and Developmental Biology*, *19*(1), 91–118. doi:10.1146/annurev.cellbio.19.111401.091942

Morsy, M., Gouthu, S., Orchard, S., Thorneycroft, D., Harper, J. F., & Mittler, R. (2008). Charting plant interactomes: Possibilities and challenges. *Trends in Plant Science*, *13*(4), 183–191. doi:10.1016/j.tplants.2008.01.006

Mortimer, D., Dayan, P., Burrage, K., & Goodhill, G. J. (2009b). Optimizing chemotaxis by measuring unbound-bound transitions. *Physica D. Nonlinear Phenomena*, *239*(9), 477–484. doi:10.1016/j.physd.2009.09.009

Mortimer, D., Feldner, J., Vaughan, T., Vetter, I., Pujic, Z., & Rosoff, W. J. (2009a). A Bayesian model predicts the response of axons to molecular gradients. *Proceedings of the National Academy of Sciences of the United States of America*, *106*(25), 10296–10301. doi:10.1073/pnas.0900715106

Mortimer, D., Fothergill, T., Pujic, Z., Richard, L. J., & Goodhill, G. J. (2008). Growth cone chemotaxis. *Trends in Neurosciences*, *31*(2), 90–98. doi:10.1016/j.tins.2007.11.008

Moss, S. J., Martin, C. J., & Wilkinson, B. (2004). Loss of co-linearity by modular polyketide synthases: A mechanism for the evolution of chemical diversity. *Natural Product Reports*, *21*(5), 575–593. doi:10.1039/b315020h

Mostecki, J., Showalter, B. M., & Rothman, P. B. (2005). Early growth response-1 regulates lipopolysaccharide-induced suppressor of cytokine signaling-1 transcription. *The Journal of Biological Chemistry*, *280*(4), 2596–2605. doi:10.1074/jbc.M408938200

Motsinger, A. A., Ritchie, M. D., & Dobrin, S. E. (2006). Clinical applications of whole-genome association studies: Future applications at the bedside. *Expert Review of Molecular Diagnostics*, *6*(4), 551–565. doi:10.1586/14737159.6.4.551

Motsinger-Reif, A. A., Dudek, S. M., Hahn, L. W., & Ritchie, M. D. (2008). Comparison of approaches for machine-learning optimization of neural networks for detecting gene-gene interactions in genetic epidemiology. *Genetic Epidemiology*, *32*, 325–340. doi:10.1002/gepi.20307

Mount, D. B., & Romero, M. F. (2004). The SLC26 gene family of multifunctional anion exchangers. *Pflugers Archive: European Journal of Physiology*, *447*(5), 710–721. doi:10.1007/s00424-003-1090-3

Moxley, R. A., & Duhamel, G. E. (1999). Comparative pathology of bacterial enteric diseases of swine. *Advances in Experimental Medicine and Biology*, *473*, 83–101.

Mu, J., Awadalla, P., Duan, J., McGee, K. M., Keebler, J., & Seydel, K. (2007). Genome-wide variation and identification of vaccine targets in the Plasmodium falciparum genome. *Nature Genetics*, *39*(1), 126–130. doi:10.1038/ng1924

Muegge, I. (2006). PMF scoring revisited. *Journal of Medicinal Chemistry*, *49*(20), 5895–5902. doi:10.1021/jm050038s

Mueller, A. K., Labaied, M., Kappe, S. H., & Matuschewski, K. (2005). Genetically modified Plasmodium parasites as a protective experimental malaria vaccine. *Nature*, *433*(7022), 164–167. doi:10.1038/nature03188

Mueller, B. (1999). Growth cone guidance: First steps towards a deeper understanding. *Annual Review of Neuroscience*, *22*, 351–601. doi:10.1146/annurev.neuro.22.1.351

Mufano, M. R., & Flint, J. (2005). Meta-analysis of genetic association studies. *Trends in Genetics*, *21*(5), 268–269.

Mulder, N. J., Apweiler, R., Attwood, T. K., Bairoch, A., Barrell, D., & Bateman, A. (2003). The InterPro Database, 2003 brings increased coverage and new features. *Nucleic Acids Research*, *31*(1), 315–318. doi:10.1093/nar/gkg046

Muller, H. M., Kenny, E. E., & Sternberg, P. W. (2004). Textpresso: An ontology-based information retrieval and extraction system for biological literature. *PLoS Biology*, *2*(11), e309. doi:10.1371/journal.pbio.0020309

Müller, F.-J., Laurent, L., Kostka, D., Ulitsky, I., Williams, R., & Lu, C. (2008). Regulatory networks define phenotypic classes of human stem cell lines. *Nature*, *455*(7211), 401–405. doi:10.1038/nature07213

Munch, R. (2003). PRODORIC: Prokaryotic database of gene regulation. *Nucleic Acids Research*, *31*, 266–269. doi:10.1093/nar/gkg037

Munro, K. M., & Perreau, V. M. (2009). Current and future applications of transcriptomics for discovery in CNS disease and injury. *Neuro-Signals*, *17*(4), 311–327. doi:10.1159/000231897

Murali, T. M., & Kasif, S. (2003). Extracting conserved gene expression motifs from gene expression data. *In Proceedings of the 8th Pacific Symposium on Biocomputing, 8*, 77-88.

Murphy, J., Scott, J., Kaufman, D., Geller, G., LeRoy, L., & Hudson, K. (2009). Public Perspectives on Informed Consent for Biobanking. *American Journal of Public Health*, *99*(12), 2128–2134. doi:10.2105/AJPH.2008.157099

Musser, J. M., & Shelburne, S. A. III. (2009). A decade of molecular pathogenomic analysis of group A Streptococcus. *The Journal of Clinical Investigation*, *119*(9), 2455–2463. doi:10.1172/JCI38095

Myers, A. J., Kaleem, M., Marlowe, L., Pittman, A. M., Lees, A. J., & Fung, H. C. (2005). The H1c haplotype at the MAPT locus is associated with Alzheimer's disease. *Human Molecular Genetics*, *14*(16), 2399–2404. doi:10.1093/hmg/ddi241

Nabieva, E. (2005). Whole-proteome prediction of protein function via graph-theoretic analysis of interaction maps. *Bioinformatics (Oxford, England)*, *21*(1), i302–i310. doi:10.1093/bioinformatics/bti1054

Nachman, I., Regev, A., & Friedman, N. (2004). Inferring quantitative models of regulatory networks from expression data. *Bioinformatics (Oxford, England)*, *20*(Suppl 1), i248–i256. doi:10.1093/bioinformatics/bth941

Nacu, S., Critchley-Thorne, R., Lee, P., & Holmes, S. (2007). Gene expression network analysis and applications to immunology. *Bioinformatics (Oxford, England)*, *23*(7), 850–858. doi:10.1093/bioinformatics/btm019

Nagel, J. H. A., Flamm, C., Hofacker, I. L., Franke, K., de Smit, M. H., & Schuster, P. (2006). Structural parameters affecting the kinetics of RNA-hairpin formation. *Nucleic Acids Research*, *34*(12), 3568–3576. doi:10.1093/nar/gkl445

Nagel, J. H. A. (2003). *A study of metastable structures in RNA*. Unpublished doctoral dissertation, University of Leiden, The Netherlands.

Naili, S., & Yasmineh, S. (2001). Un modèle de l'adhésion pour les milieux curvilignes. *Comptes Renuds de l'Academie Sciences, Paris*, *2*(2), 161–167.

Nakagami, H., Kiegerl, S., & Hirt, H. (2004). OMTK1, a novel MAPKKK, channels oxidative stress signaling through direct MAPK interaction. *The Journal of Biological Chemistry*, *279*(26), 26959–26966. doi:10.1074/jbc.M312662200

Nandyal, R. R. (2008). Update on group B streptococcal infections: Perinatal and neonatal periods. *The Journal of Perinatal & Neonatal Nursing*, *22*(3), 230–237.

Nariai, N., Kim, S., Imoto, S., & Miyano, S. (2004). Using protein-protein interactions for refining gene networks estimated from microarray data by Bayesian networks. *Pacific Symposium on Biocomputing*, 336-347.

Natarajan, M., Lin, K. M., Hsueh, R. C., Sternweis, P. C., & Ranganathan, R. (2006). A global analysis of cross-talk in a mammalian cellular signalling network. *Nature Cell Biology*, *8*(6), 571–580. doi:10.1038/ncb1418

National Institute on Aging. (2009). Alzheimer's disease genetics facts sheet. Retrieved from http://www.nia.nih.gov/Alzheimers/Publications/geneticsfs.htm

Nawrocki, E. P., Kolbe, D. L., & Eddy, S. R. (2009). Infernal 1.0: Inference of RNA alignments. *Bioinformatics (Oxford, England)*, *25*(10), 1335–1337. doi:10.1093/bioinformatics/btp157

Naylor, S., & Chen, J. Y. (2010). Unraveling Human Complexity and Disease with Systems Biology and Personalized Medicine. *Personalized Medicine*, *7*(3), 275–289. doi:10.2217/pme.10.16

Naylor, E., Arredouani, A., Vasudevan, S. R., Lewis, A. M., Parkesh, R., & Mizote, A. (2009). Identification of a chemical probe for NAADP by virtual screening. *Nature Chemical Biology*, *5*(4), 220–226. doi:10.1038/nchembio.150

NBAC. (2001). *Ethical and Policy Issues in Research Involving Human Participants*. Bethesda, MD: National Bioethics Advisory Commission.

NBAC. (1999). *Research involving human biological materials: ethical issues and policy guidance, vol I: report and recommendations*. Bethesda, MD: National Bioethics Advisory Commission.

Ndifon, W. (2005). A complex adaptive systems approach to the kinetic folding of RNA. *Bio Systems*, *82*(3), 257–265. doi:10.1016/j.biosystems.2005.08.004

Ndri, N., & Udaykumar, W. S. & Tay, R.T.S. (2001). Computational modeling of cell adhesion and movement using continuum-kinetic approach. *Proceedings of the Bioengineering Conference ASME, 50*, 367-368.

Nejentsev, S., Walker, N., Riches, D., Egholm, M., & Todd, J. A. (2009). Rare variants of IFIH1, a gene implicated in antiviral responses, protect against type 1 diabetes. *Science*, *324*(5925), 387–389. doi:10.1126/science.1167728

Nelson, M. R., Kardia, S. L., Ferrell, R. E., & Sing, C. F. (2001). A combinatorial partitioning method to identify multilocus genotypic partitions that predict quantitative trait variation. *Genome Research*, *11*, 458–470. doi:10.1101/gr.172901

Neogi, N. A. (2004). *Dynamic partitioning of large discrete event biological systems for hybrid simulation and analysis*. Paper presented at the 7th International Workshop Hybrid Systems: Computation and Control, 2993, 463-476.

Nesvizhskii, A. I., Keller, A., Kolker, E., & Aebersold, R. (2003). A statistical model for identifying proteins by tandem mass spectrometry. *Analytical Chemistry*, *75*(17), 4646–4658. doi:10.1021/ac0341261

Nesvizhskii, A. I., & Aebersold, R. (2005). Interpretation of shotgun proteomic data: The protein inference problem. *Molecular & Cellular Proteomics*, *4*(10), 1419–1440. doi:10.1074/mcp.R500012-MCP200

Nettles, J. H., Jenkins, J. L., Williams, C., Clark, A. M., Bender, A., & Deng, Z. (2007). Flexible 3D pharmacophores as descriptors of dynamic biological space. *Journal of Molecular Graphics & Modelling*, *26*(3), 622–633. doi:10.1016/j.jmgm.2007.02.005

Neves, M. A., Dinis, T. C., Colombo, G., & Sa e Melo, M. L. (2009). Fast three dimensional pharmacophore virtual screening of new potent non-steroid aromatase inhibitors. *Journal of Medicinal Chemistry*, *52*(1), 143–150. doi:10.1021/jm800945c

Newman, M. E. J. (2004). Fast algorithm for detecting community structure in networks. *Physical Review E: Statistical, Nonlinear, and Soft Matter Physics*, *69*(6), 066133. doi:10.1103/PhysRevE.69.066133

Newman, M. E. J. (2005). A measure of betweenness centrality based on random walks. *Social Networks*, *27*(1), 39–54. doi:10.1016/j.socnet.2004.11.009

Newman, M. E. J. (2006a). Finding community structure in networks using the eigenvectors of matrices. *Physical Review E: Statistical, Nonlinear, and Soft Matter Physics*, *74*(3), 036104. doi:10.1103/PhysRevE.74.036104

Newman, M. E. J. (2006b). Modularity and community structure in networks. *Proceedings of the National Academy of Sciences of the United States of America*, *103*(23), 8577–8582. doi:10.1073/pnas.0601602103

Newman, M. E. J., & Girvan, M. (2004). Finding and evaluating community structure in networks. *Physical Review E: Statistical, Nonlinear, and Soft Matter Physics*, *69*(2), 026113. doi:10.1103/PhysRevE.69.026113

Newman, T.J. & Odell, P.L. (1971). *The generation of random variate. Griffin's statistical monograph and courses.*

Newton-Cheh, C., & Hirschhorn, J. N. (2005). Genetic association studies of complex traits: Design and analysis issues. *Mutation Research*, *573*(1-2), 54–69. doi:10.1016/j.mrfmmm.2005.01.006

Ng, A., et al. (2001). *On spectral clustering: Analysis and an algorithm.* Paper presented at the Advances in Neural Information Processing Systems 14.

Nguyen, T., Ishida, K., Jenke-Kodama, H., Dittmann, E., Gurgui, C., & Hochmuth, T. (2008). Exploiting the mosaic structure of trans-acyltransferase polyketide synthases for natural product discovery and pathway dissection. *Nature Biotechnology*, *26*(2), 225–233. doi:10.1038/nbt1379

Nguyen, V. T., Kiss, T., Michels, A. A., & Bensaude, O. (2001). 7SK small nuclear RNA binds to and inhibits the activity of CDK9/cyclin T complexes. *Nature*, *414*(6861), 322–325. doi:10.1038/35104581

Nicholson, J. K., Lindon, J. C., & Holmes, E. (1999). Metabonomics: Understanding the metabolic responses of living systems to pathophysiological stimuli via multivariate statistical analysis of biological NMR spectroscopic data. *Xenobiotica*, *29*, 1181–1189. doi:10.1080/004982599238047

Nicholson, R. I., Gee, J. M., & Harper, M. E. (2001). EGFR and cancer prognosis. *European Journal of Cancer*, *37*(Suppl 4), S9–S15. doi:10.1016/S0959-8049(01)00231-3

Nicholson, J. K., Holmes, E., Lindon, J. C., & Wilson, I. D. (2004). The challenges of modeling mammalian biocomplexity. *Nature Biotechnology*, *22*(10), 1268–1274. doi:10.1038/nbt1015

Nicol, D., & Critchley, C. (2009). What Benefit Sharing Arrangements Do People Want From Biobanks? A Survey of Public Opinion in Australia. In Kaye, J., & Stranger, M. (Eds.), *Principles and Practice in Biobank Governance* (pp. 17–31). Surrey, UK: Ashgate.

Nicosia, S., Tornambé, A., & Valigi, P. (1991). A solution to the generalized problem of nonlinear map inversion. *Systems & Control Letters*, *17*(5), 383–394. doi:10.1016/0167-6911(91)90138-5

Niehrs, C., & Pollet, N. (1999). Synexpression groups in eukaryotes. *Nature*, *402*(6761), 483–487. doi:10.1038/990025

NIH/CEPH Collaborative Mapping Group. (1992). A comprehensive genetic linkage map of the human genome. *Science*, *258*, 67–86. doi:10.1126/science.1439770

Nikolsky, Y., Nikolskaya, T., & Bugrim, A. (2005). Biological networks and analysis of experimental data in drug discovery. *Drug Discovery Today*, *10*(9), 653–662. doi:10.1016/S1359-6446(05)03420-3

Nilsson, R., Bajic, V. B., Suzuki, H., di Bernardo, D., Bjorkegren, J., & Katayama, S. (2006). Transcriptional network dynamics in macrophage activation. *Genomics*, *88*(2), 133–142. doi:10.1016/j.ygeno.2006.03.022

Nkwanta, A., & Ndifon, W. (2009). A contact-waiting-time metric and RNA folding rates. *FEBS Letters*, *583*(14), 2392–2394. doi:10.1016/j.febslet.2009.06.038

Norinder, U. (2005). In silico modelling of ADMET—a mini review of work from 2000 to 2004. *SAR and QSAR in Environmental Research*, *16*, 1–11. doi:10.1080/10629360412331319835

Norris, J. R. (1998). *Markov chains*. Cambridge, UK: Cambridge University Press.

Norton, P. G., Bain, J., Birtwhistle, R., Davis, D., & Dunn, E. CP, H., et al. (1994). Guidelines for the Dissemination of New Information Discovered by Researchers. In M. A. Stewart, P. G. Norton, M. Bass, E. Dunn & F. Tudiver (Eds.), *Disseminating Research/Changing Practice. Research Methods for Primary Care* (Vol. 6, pp. 87-94). Thousand Oaks, CA: Sage Publications.

Nour, A., Slimani, A., Laouami, N., & Afra, H. (2003). Finite element model for probabilistic seismic response of heterogeneous soil profile. *Soil Dynamics and Earthquake Engineering*, *23*, 331–348. doi:10.1016/S0267-7261(03)00036-8

Nyburg, S. C., & Faerman, C. H. (1985). A revision of van der Waals atomic radii for molecular crystals: N, O, F, S, Cl, Se, Br, and I bonded to carbon. *Acta Crystallographica. Section B, Structural Science*, *41*(4), 274–279. doi:10.1107/S0108768185002129

Ochman, H., Lawrence, J. G., & Groisman, E. A. (2000). Lateral gene transfer and the nature of bacterial innovation. *Nature*, *405*(6784), 299–304. doi:10.1038/35012500

Oda, A., Tsuchida, K., Takakura, T., Yamaotsu, N., & Hirono, S. (2006). Comparison of consensus scoring strategies for evaluating computational models of protein-ligand complexes. *Journal of Chemical Information and Modeling*, *46*(1), 380–391. doi:10.1021/ci050283k

Oda, Y., Huang, K., Cross, F. R., Cowburn, D., & Chait, B. T. (1999). Accurate quantitation of protein expression and site specific phosphorylation. *Proceedings of the National Academy of Sciences of the United States of America*, *96*(12), 6591–6596. doi:10.1073/pnas.96.12.6591

Oda, K., & Kitano, H. (2006). A comprehensive map of the toll-like receptor signaling network. *Molecular Systems Biology*, *2*, E1–E16. doi:10.1038/msb4100057

Ogata, H. (1999). KEGG: Kyoto Encyclopedia of Genes and Genomes. *Nucleic Acids Research*, *27*(1), 29–34. doi:10.1093/nar/27.1.29

Ogura, M., Perez, J. C., Mittl, P. R. E., Lee, H.-K., Dailide, G., & Tan, S. (2007). Helicobacter pylori evolution: Lineage-specific adaptations in homologs of eukaryotic sel1-like genes. *PLoS Computational Biology*, *3*(8), e151. doi:10.1371/journal.pcbi.0030151

Oh, D. C., Gontang, E. A., Kauffman, C. A., Jensen, P. R., & Fenical, W. (2008). Salinipyrones and pacificanones, mixed-precursor polyketides from the marine actinomycete Salinispora pacifica. *Journal of Natural Products*, *71*(4), 570–575. doi:10.1021/np0705155

Okada, Y., Fujibuchi, W., & Horton, P. (2007). Module discovery in gene expression data using closed itemset mining algorithm. *IPSG Transactions in Bioinformatics*, *48*, 39–48.

Okita, K., Ichisaka, T., & Yamanaka, S. (2007). Generation of germline-competent induced pluripotent stem cells. *Nature*, *448*(7151), 313–317.

Okoniewski, M., & Miller, C. (2008). Comprehensive analysis of affymetrix exon arrays using BioConductor. *PLoS Computational Biology*, *4*(2). doi:10.1371/journal.pcbi.0040006

Oldham, M. C. (2006). Conservation and evolution of gene coexpression networks in human and chimpanzee brains. *Proceedings of the National Academy of Sciences of the United States of America*, *103*(47), 17973–17978. doi:10.1073/pnas.0605938103

Oliver, T., Lee, J., & Jacobson, K. (1994). Forces exerted by locomoting cells. *Seminars in Cell Biology*, *5*, 139–147. doi:10.1006/scel.1994.1018

Olsen, R. J., Shelburne, S. A., & Musser, J. M. (2009). Molecular mechanisms underlying group A streptococcal pathogenesis. *Cellular Microbiology*, *11*(1), 1–12. doi:10.1111/j.1462-5822.2008.01225.x

Ong, S.-E., Blagoev, B., Kratchmarova, I., Kristensen, D. B., Steen, H., & Pandey, A. (2002). Stable isotope labelling by amino acids in cell culture, SILAC, as a simple and accurate approach to expression proteomics. *Molecular & Cellular Proteomics*, *1*(5), 376–386. doi:10.1074/mcp.M200025-MCP200

Onuchic, J. N., Nymeyer, H., Garcia, A. E., Chahine, J., & Socci, N. D. (2000). The energy landscape theory of protein folding: Insights into folding mechanisms and scenarios. *Advances in Protein Chemistry*, *53*, 87–152. doi:10.1016/S0065-3233(00)53003-4

Osterberg, F., Morris, G. M., Sanner, M. F., Olson, A. J., & Goodsell, D. S. (2002). Automated docking to multiple target structures: incorporation of protein mobility and structural water heterogeneity in AutoDock. *Proteins*, *46*(1), 34–40. doi:10.1002/prot.10028

Oti, M., & Brunner, H. G. (2007). The modular nature of genetic diseases. *Clinical Genetics*, *71*(1), 1–11. doi:10.1111/j.1399-0004.2006.00708.x

Oti, M., Huynen, M. A., & Brunner, H. G. (2008). Phenome connections. *Trends in Genetics*, *24*(3), 103–106. doi:10.1016/j.tig.2007.12.005

Oti, M., Snel, B., Huynen, M. A., & Brunner, H. G. (2006). Predicting disease genes using protein-protein interactions. *Journal of Medical Genetics*, *43*(8), 691–698. doi:10.1136/jmg.2006.041376

Oura, T., Matsui, S., & Kawakami, K. (2009). Sample size calculations for controlling the distribution of false discovery proportion in microarray experiments. *Biostatistics (Oxford, England)*, *10*(4), 694–705. doi:10.1093/biostatistics/kxp024

Ozsolak, F., Song, J. S., Liu, X. S., & Fisher, D. E. (2007). High-throughput mapping of the chromatin structure of human promoters. *Nature Biotechnology*, *25*(2), 244–248.

Pagel, P., Kovac, S., Oesterheld, M., Brauner, B., Dunger-Kaltenbach, I., & Frishman, G. (2005). The MIPS mammalian protein-protein interaction database. *Bioinformatics (Oxford, England)*, *21*(6), 832–834. doi:10.1093/bioinformatics/bti115

Pages, F., Galon, J., Dieu-Nosjean, M. C., Tartour, E., Sautes-Fridman, C., & Fridman, W. H. (2009). Immune infiltration in human tumors: A prognostic factor that should not be ignored. *Oncogene*, *29*(8), 1093–1102. doi:10.1038/onc.2009.416

Paik, S., Tang, G., Shak, S., Kim, C., Baker, J., & Kim, W. (2006). Gene expression and benefit of chemotherapy in women with node-negative, estrogen receptor-positive breast cancer. *Journal of Clinical Oncology*, *24*(23), 3726–3734. doi:10.1200/JCO.2005.04.7985

Palla, G., Derenyi, I., Farkas, I., & Vicsek, T. (2005). Uncovering the overlapping community structure of complex networks in nature and society. *Nature*, *435*(7043), 814–818. doi:10.1038/nature03607

Palla, G. (2005). Uncovering the overlapping community structure of complex networks in nature and society. *Nature*, *435*(7043), 814–818. doi:10.1038/nature03607

Palsson, B. (2002). In silico biology through "omics". *Nature Biotechnology*, *20*(7), 649–650. doi:10.1038/nbt0702-649

Palsson, B. (2006). *Systems biology: properties of reconstructed networks*. Cambridge, New York: Cambridge University Press. doi:10.1017/CBO9780511790515

Palsson, B. O., Price, N. D., & Papin, J. A. (2003). Development of network-based pathway definitions: The need to analyze real metabolic networks. *Trends in Biotechnology*, *21*(5), 195–198. doi:10.1016/S0167-7799(03)00080-5

Pan, J., Thirumalai, D., & Woodson, S. A. (1997). Folding of RNA involves parallel pathways. *Journal of Molecular Biology*, *273*(1), 7–13. doi:10.1006/jmbi.1997.1311

Pan, W. (2008). Network-based model weighting to detect multiple loci influencing complex diseases. *Human Genetics*.

Papadrakakis, M., & Papadopoulos, V. (1996). Robust and efficient methods for stochastic finite element analysis using Monte Carlo simulation. *Computational Methods of Applied Mechanics*, *134*, 325–340. doi:10.1016/0045-7825(95)00978-7

Pardalos, P., & Xue, J. (1994). The maximum clique problem. *Journal of Global Optimization*, *4*(3), 301–328. doi:10.1007/BF01098364

Park, M. Y., & Hastie, T. (2008). Penalized logistic regression for detecting gene interactions. *Biostatistics (Oxford, England)*, *9*(1), 30–50. doi:10.1093/biostatistics/kxm010

Park, J., Lee, D. S., Christakis, N. A., & Barabasi, A. L. (2009). The impact of cellular networks on disease comorbidity. *Molecular Systems Biology*, *5*, 262. doi:10.1038/msb.2009.16

Parker, J. S., Mullins, M., Cheang, M. C., Leung, S., Voduc, D., & Vickery, T. (2009). Supervised risk predictor of breast cancer based on intrinsic subtypes. *Journal of Clinical Oncology*, *27*(8), 1160–1167. doi:10.1200/JCO.2008.18.1370

Parkinson, H., Kapushesky, M., Kolesnikov, N., Rustici, G., Shojatalab, M., & Abeygunawardena, N. (2009). ArrayExpress update-from an archive of functional genomics experiments to the atlas of gene expression. *Nucleic Acids Research*, *37*(Database issue), D868–D872. doi:10.1093/nar/gkn889

Parman, C. & Halling, C. (2005). affyQCReport: QC report generation for affyBatch objects. *R package,* version 1.17.0.

Parsons, H. M., Ludwig, C., Gunther, U. L., & Viant, M. R. (2007). Improved classification accuracy in 1- and 2-dimensional NMR metabolomics data using the variance stabilizing generalized logarithm transformation. *BMC Bioinformatics*, *8*, 234. doi:10.1186/1471-2105-8-234

Parsons, D. W., Jones, S., Zhang, X., Lin, J. C., Leary, R. J., & Angenendt, P. (2008). An integrated genomic analysis of human glioblastoma multiforme. *Science*, *321*(5897), 1807–1812. doi:10.1126/science.1164382

Patil, A., & Nakamura, H. (2005). Filtering high-throughput protein-protein interaction data using a combination of genomic features. *BMC Bioinformatics*, *6*, 100. doi:10.1186/1471-2105-6-100

Patterson, D. E., Cramer, R. D., Ferguson, A. M., Clark, R. D., & Weinberger, L. E. (1996). Neighborhood behavior: a useful concept for validation of "molecular diversity" descriptors. *Journal of Medicinal Chemistry*, *39*(16), 3049–3059. doi:10.1021/jm960290n

Pattin, K. A., & Moore, J. H. (2009). Role for protein-protein interaction databases in human genetics. *Expert Review of Proteomics*, *6*, 647–659. doi:10.1586/epr.09.86

Pavlidis, P. (2004). Using the gene ontology for microarray data mining: A comparison of methods and application to age effects in human prefrontal cortex. *Neurochemical Research*, *29*(6), 1213–1222. doi:10.1023/B:NERE.0000023608.29741.45

Pavlopoulos, G. A. G., Wegener, A. L. A., & Schneider, R. R. (2008). A survey of visualization tools for biological network analysis. *BioData Mining*, *1*(1), 12. doi:10.1186/1756-0381-1-12

Pawitan, Y., Bjohle, J., Amler, L., Borg, A. L., Egyhazi, S., & Hall, P. (2005). Gene expression profiling spares early breast cancer patients from adjuvant therapy: Derived and validated in two population-based cohorts. *Breast Cancer Research*, *7*(6), R953–R964. doi:10.1186/bcr1325

Pawson, T., & Nash, P. (2003). Assembly of cell regulatory systems through protein interaction domains. *Science*, *300*, 445–452. doi:10.1126/science.1083653

Pearl, J. (2000). *Causality: Models, reasoning, and inference*. Cambridge, UK: Cambridge University Press.

Pearson, T. A., & Manolio, T. A. (2008). How to interpret a genome-wide association study. *Journal of the American Medical Association*, *299*(11), 1335–1344. doi:10.1001/jama.299.11.1335

Peckham, H. E., Thurman, R. E., Fu, Y., Stamatoyannopoulos, J. A., Noble, W. S., & Struhl, K. (2007). Nucleosome positioning signals in genomic DNA. *Genome Research*, *17*(8), 1170–1177.

Pedley, K. F., & Martin, G. B. (2005). Role of mitogen-activated protein kinases in plant immunity. *Current Opinion in Plant Biology*, *8*(5), 541–547. doi:10.1016/j.pbi.2005.07.006

Peeters, E. (2004). *Biomechanics of single cells under compression*. PhD thesis, University of Endhoven.

Pei, J., & Grishin, N. V. (2001). AL2CO: Calculation of positional conservation in a protein sequence alignment. *Bioinformatics (Oxford, England)*, *17*(8), 700–712. doi:10.1093/bioinformatics/17.8.700

Pellegrino, E. D., & Thomasma, D. C. (1984). *For the Patient's Good: The Restoration of Beneficence in Health Care*. New York: Oxford.

Peng, G., Luo, L., Siu, H., Zhu, Y., Hu, P., & Hong, S. (2010). Gene and pathway-based second-wave analysis of genome-wide association studies. *European Journal of Human Genetics*, *18*(1), 111–117. doi:10.1038/ejhg.2009.115

Peri, S., Navarro, J. D., Amanchy, R., Kristiansen, T. Z., Jonnalagadda, C. K., & Surendranath, V. (2003). Development of human protein reference database as an initial platform for approaching systems biology in humans. *Genome Research*, *13*(10), 2363–2371. doi:10.1101/gr.1680803

Perlman, Z. E., Slack, M. D., Feng, Y., Mitchison, T. J., Wu, L. F., & Altschuler, S. J. (2004). Multidimensional drug profiling by automated microscopy. *Science*, *306*(5699), 1194–1198. doi:10.1126/science.1100709

Perou, C. M., Sorlie, T., Eisen, M. B., van de Rijn, M., Jeffrey, S. S., & Rees, C. A. (2000). Molecular portraits of human breast tumours. *Nature*, *406*(6797), 747–752. doi:10.1038/35021093

Perrin, B. E., Ralaivola, L., Mazurie, A., Bottani, S., Mallet, J., & d'Alche-Buc, F. (2003). Gene networks inference using dynamic Bayesian networks. *Bioinformatics (Oxford, England)*, *19*(Suppl 2), ii138–ii148. doi:10.1093/bioinformatics/btg1071

Peterson, E. L., Kondev, J., Theriot, J. A., & Phillips, R. (2009). Reduced amino acid alphabets exhibit an improved sensitivity and selectivity in fold assignment. *Bioinformatics (Oxford, England)*, *25*(11), 1356–1362. doi:10.1093/bioinformatics/btp164

Petri, A., Fleckner, J., & Matthiessen, M. W. (2004). Array-a-lizer: A serial DNA microarray quality analyzer. *BMC Bioinformatics*, *5*, 12. doi:10.1186/1471-2105-5-12

Petyuk, V. A., Qian, W.-J., Smith, R. D., & Smith, D. J. (2010). Mapping protein abundance patterns in the brain using voxelation combined with liquid chromatography and mass spectrometry. *Methods (San Diego, Calif.)*, *50*(2), 77–84. doi:10.1016/j.ymeth.2009.07.009

Philippopoulos, M., & Lim, C. (1999). Exploring the dynamic information content of a protein NMR structure: Comparison of a molecular dynamics simulation with the NMR and X-ray structures of Escherichia coli ribonuclease HI. *Proteins*, *36*(1), 87–110. doi:10.1002/(SICI)1097-0134(19990701)36:1<87::AID-PROT8>3.0.CO;2-R

Pietrogrande, M. C., Marchetti, N., Dondi, F., & Righetti, P. G. (2006). Decoding 2D-PAGE complex maps: Relevance to proteomics. *Journal of Chromatography B – Analytical. Technological, and Biomedical Life Sciences*, *833*(1), 51–62. doi:10.1016/j.jchromb.2005.12.051

Piette, J., Neel, H., & Marechal, V. (1997). Mdm2: Keeping p53 under control. *Oncogene*, *15*(9), 1001–1010. doi:10.1038/sj.onc.1201432

Pisitkun, T., Johnstone, R., & Knepper, M. A. (2006). Discovery of unrinary biomarkers. *Molecular & Cellular Proteomics*, *5*(10), 1760–1771. doi:10.1074/mcp.R600004-MCP200

Pitluk, Z., & Khalil, I. (2007). Achieving confidence in mechanism for drug discovery and development. *Drug Discovery Today*, *12*(21-22), 924–930. doi:10.1016/j.drudis.2007.10.001

Pitman, M. C., Huber, W. K., Horn, H., Kramer, A., Rice, J. E., & Swope, W. C. (2001). FLASHFLOOD: A 3D field-based similarity search and alignment method for flexible molecules. *Journal of Computer-Aided Molecular Design*, *15*(7), 587–612. doi:10.1023/A:1011921423829

Pizza, M., Scarlato, V., Masignani, V., Giuliani, M. M., Arico, B., & Comanducci, M. (2000). Identification of vaccine candidates against serogroup B meningococcus by whole-genome sequencing. *Science*, *287*(5459), 1816–1820. doi:10.1126/science.287.5459.1816

Plant, N. (2007). The human cytochrome P450 sub-family: transcriptional regulation, inter-individual variation and interaction networks. *Biochimica et Biophysica Acta*, *1770*(3), 478–488.

Poirot, O., Suhre, K., Abergel, C., O'Toole, E., & Notredame, C. (2004). 3DCoffee@igs: A Web server for combining sequences and structures into a multiple sequence alignment. *Nucleic Acids Research*, *32*, W37-40. doi:10.1093/nar/gkh382

Pokholok, D. K., Harbison, C. T., Levine, S., Cole, M., Hannett, N. M., & Lee, T. I. (2005). Genome-wide map of nucleosome acetylation and methylation in yeast. *Cell*, *122*(4), 517–527.

Poland, G. A., Ovsyannikova, I. G., & Jacobson, R. M. (2008). Personalized vaccines: The emerging field of vaccinomics. *Expert Opinion on Biological Therapy*, *8*(11), 1659–1667. doi:10.1517/14712598.8.11.1659

Politi, A., Monè, M. J., Houtsmuller, A. B., Hoogstraten, D., Vermeulen, W., & Heinrich, V. (2005). Mathematical modeling of nucleotide excision repair reveals efficiency of sequential assembly strategies. *Molecular Cell*, *19*(5), 679–690. doi:10.1016/j.molcel.2005.06.036

Pollard, J., Butte, A., Hoberman, S., Joshi, M., Levy, J., & Pappo, J. (2005). A computational model to define the molecular causes of type 2 diabetes mellitus. *Diabetes Technology & Therapeutics*, *7*(2), 323–336. doi:10.1089/dia.2005.7.323

Pons, P., & Latapy, M. (2005). *Computing communities in large networks using random walks*. In (LNCS 3733). (pp. 284-293).

Popescu, S. C., Popescu, G. V., Bachan, S., Zhang, Z., Gerstein, M., & Snyder, M. (2009). MAPK target networks in Arabidopsis thaliana revealed using functional protein microarrays. *Genes & Development*, *23*(1), 80–92. doi:10.1101/gad.1740009

Popescu, S. C., Popescu, G. V., Bachan, S., Zhang, Z., Seay, M., & Gerstein, M. (2007). Differential binding of calmodulin-related proteins to their targets revealed through high-density Arabidopsis protein microarrays. *Proceedings of the National Academy of Sciences of the United States of America*, *104*(11), 4730–4735. doi:10.1073/pnas.0611615104

Popescu, S. C., Popescu, G. V., Snyder, M., & Dinesh-Kumar, S. P. (2009). Integrated analysis of co-expressed MAP kinase substrates in Arabidopsis thaliana. *Plant Signaling & Behavior*, *4*(6), 524–527. doi:10.4161/psb.4.6.8576

Pradines, J., Farutin, V., Rowley, S. & Dancík, V. (2005). Analyzing protein lists with large networks: Edge-count probabilities in random graphs with given expected degrees. *Journal of Computational Biology: A Journal of Computational Molecular Cell Biology*, *12*(2), 113-28.

Prasad, T. S. K., Goel, R., Kandasamy, K., Keerthikumar, S., Kumar, S., & Mathivanan, S. (2009). Human protein reference database-2009 update. *Nucleic Acids Research*, *37*, D411–D414.

Prelic, A., Bleuler, S., Zimmermann, P., Wille, A., Buhlmann, P., & Gruissem, W. (2006). A systematic comparison and evaluation of biclustering methods for gene expression data. *Bioinformatics (Oxford, England)*, *22*, 1122–1129. doi:10.1093/bioinformatics/btl060

Presson, A. P. (2008). Integrated weighted gene co-expression network analysis with an application to chronic fatigue syndrome. *BMC Systems Biology*, *2*, 95. doi:10.1186/1752-0509-2-95

Price, A. L., Patterson, N. J., Plenge, R. M., Weinblatt, M. E., Shadick, N. A., & Reich, D. (2006). Principal components analysis corrects for stratification in genome-wide association studies. *Nature Genetics*, *38*, 904–909. doi:10.1038/ng1847

Pritchard, J. K., & Rosenberg, N. A. (1999). Use of unlinked genetic markers to detect population stratification in association studies. *American Journal of Human Genetics*, *65*(1), 220–228. doi:10.1086/302449

Project, H. G. (2010). *Human Genome Project*: www.ornl.gov/hgmis/home.shtml.

Pujana, M. A., Han, J. D., Starita, L. M., Stevens, K. N., Tewari, M., & Ahn, J. S. (2007). Network modeling links breast cancer susceptibility and centrosome dysfunction. *Nature Genetics*, *39*(11), 1338–1349. doi:10.1038/ng.2007.2

Puntervoll, P., Linding, R., Gemund, C., Chabanis-Davidson, S., Mattingsdal, M., & Cameron, S. (2003). ELM server: A new resource for investigating short functional sites in modular eukaryotic proteins. *Nucleic Acids Research*, *31*(13), 3625–3630. doi:10.1093/nar/gkg545

Pusch, W., Wurmbach, J. H., Thiele, H., & Kostrzewa, M. (2002). MALDI-TOF mass spectrometry-based SNP genotyping. *Pharmacogenomics*, *3*(4), 537–548. doi:10.1517/14622416.3.4.537

Qiu, W., & Lee, M. T. (2006). SPCalc: A Web-based calculator for sample size and power calculations in micro-array studies. *Bioinformation*, *1*(7), 251–252.

Quach, M., Brunel, N., & d'Alche Buc, F. (2007). Estimating parameters and hidden variables in non-linear state-space models based on ODEs for biological networks inference. *Bioinformatics (Oxford, England)*, *23*, 3209–3216. doi:10.1093/bioinformatics/btm510

Quackenbush, J. (2002). Microarray data normalization and transformation. *Nature Genetics*, *32*, 496–501. doi:10.1038/ng1032

Querec, T. D., Akondy, R. S., Lee, E. K., Cao, W., Nakaya, H. I., & Teuwen, D. (2009). Systems biology approach predicts immunogenicity of the yellow fever vaccine in humans. *Nature Immunology*, *10*(1), 116–125. doi:10.1038/ni.1688

Quest, D., Dempsey, K., Shafiullah, M., Bastola, D., & Ali, H. (2008). MTAP: The motif tool assessment platform. *BMC Bioinformatics*, *9*(9), S6. doi:10.1186/1471-2105-9-S9-S6

Rabitz, H. (1987). Chemical dynamics and kinetics phenomena as revealed by sensitivity analysis techniques. *Chemical Reviews*, *87*, 101. doi:10.1021/cr00077a006

Rabitz, H., Kramer, M., & Dacol, D. (1983). Sensitivity analysis in chemical kinetics. *Annual Review of Physical Chemistry*, *34*, 419. doi:10.1146/annurev.pc.34.100183.002223

Rahnenführer, J., Domingues, F., Maydt, J., & Lengauer, T. (2004). Calculating the statistical significance of changes in pathway activity from gene expression data. *Statistical Applications in Genetics and Molecular Biology*, 3.

Raj, A., & van Oudenaarden, A. (2009). Single-molecule approaches to stochastic gene expression. *Annual Review of Biophysics*, *38*(1), 255–270. doi:10.1146/annurev.biophys.37.032807.125928

Rajcevic, U., Petersen, K., & Knol, J. C. (2009). iTRAQ-based protoemics profiling reveals increased metabolic activity and cellular cross-talk in angiogenic compared with invasive glioblastoma phenotype. *Molecular & Cellular Proteomics*, *8*(11), 2595–2612. doi:10.1074/mcp.M900124-MCP200

Ramakrishnan, S. R. (2009). Mining gene functional networks to improve mass-spectrometry-based protein identification. *Bioinformatics (Oxford, England)*, *25*(22), 2955–2961. doi:10.1093/bioinformatics/btp461

Ramani, A. K., Bunescu, R. C., Mooney, R. J., & Marcotte, E. M. (2005). Consolidating the set of known human protein-protein interactions in preparation for large-scale mapping of the human interactome. *Genome Biology*, *6*(5), R40. doi:10.1186/gb-2005-6-5-r40

Ramanujan, S., & Pozrikidis, C. (1998). Deformation of liquid capsules enclosed by elastic membranes in simple shear flow: large deformations and effect of fluid viscosities. *Journal of Fluid Mechanics*, *361*, 117–143. doi:10.1017/S0022112098008714

Ramaswamy, S., & Golub, T. R. (2002). DNA microarrays in clinical oncology. *Journal of Clinical Oncology*, *20*(7), 1932–1941.

Ramsey, P. (1960). *For the Patient's Good*. Princeton University Press.

Ramsey, S. A., Klemm, S. L., Zak, D. E., Kennedy, K. A., Thorsson, V., & Li, B. (2008). Uncovering a macrophage transcriptional program by integrating evidence from motif scanning and expression dynamics. *PLoS Computational Biology*, *4*(3), e1000021. doi:10.1371/journal.pcbi.1000021

Rando, O. J., & Chang, H. Y. (2009). Genome-wide views of chromatin structure. *Annual Review of Biochemistry*, *78*, 245–271.

Rangan, V. S., & Smith, S. (1997). Alteration of the substrate specificity of the malonyl-CoA/acetyl-CoA: Acyl carrier protein S-acyltransferase domain of the multifunctional fatty acid synthase by mutation of a single arginine residue. *The Journal of Biological Chemistry*, *272*(18), 11975–11978. doi:10.1074/jbc.272.18.11975

Rao, K. V. G., Chand, P. P., & Murthy, M. V. R. (2007). A neural network approach in medical decision systems. *Journal of Theoretical and Applied Information Technology*, *3*(4).

Rarey, M., Kramer, B., Lengauer, T., & Klebe, G. (1996). A fast flexible docking method using an incremental construction algorithm. *Journal of Molecular Biology*, *261*(3), 470–489. doi:10.1006/jmbi.1996.0477

Rausch, C., Hoof, I., Weber, T., Wohlleben, W., & Huson, D. H. (2007). Phylogenetic analysis of condensation domains in NRPS sheds light on their functional evolution. *BMC Evolutionary Biology*, *7*, 78. doi:10.1186/1471-2148-7-78

Rausch, C., Weber, T., Kohlbacher, O., Wohlleben, W., & Huson, D. H. (2005). Specificity prediction of adenylation domains in nonribosomal peptide synthetases (NRPS) using transductive support vector machines (TSVMs). *Nucleic Acids Research*, *33*(18), 5799–5808. doi:10.1093/nar/gki885

Ravasz, E. (2002). Hierarchical organization of modularity in metabolic networks. *Science*, *297*(5586), 1551–1555. doi:10.1126/science.1073374

Ravasz, E., Somera, A. L., Mongru, D. A., Oltvai, Z. N., & Barabasi, A. L. (2002). Hierarchical organization of modularity in metabolic networks. *Science*, *297*(5586), 1551–1555. doi:10.1126/science.1073374

Raychaudhuri, S., Plenge, R. M., Rossin, E. J., Ng, A. C., Purcell, S. M., & Sklar, P. (2009). Identifying relationships among genomic disease regions: Predicting genes at pathogenic SNP associations and rare deletions. International Schizophrenia Consortium. *PLOS Genetics*, *5*(6), e1000534. doi:10.1371/journal.pgen.1000534

Rea, T. J., Brown, C. M., & Sing, C. F. (2006). Complex adaptive system models and the genetic analysis of plasma HDL-cholesterol concentration. *Perspectives in Biology and Medicine*, *49*(4), 490–503. doi:10.1353/pbm.2006.0063

Reddy, A. S., Pati, S. P., Kumar, P. P., Pradeep, H. N., & Sastry, G. N. (2007). Virtual screening in drug discovery-a computational perspective. *Current Protein & Peptide Science*, *8*(4), 329–351. doi:10.2174/138920307781369427

Reddy, P. S., Legault, H. M., Sypek, J. P., Collins, M. J., Goad, E., & Goldman, S. J. (2008). Mapping similarities in mTOR pathway perturbations in mouse lupus nephritis models and human lupus nephritis. *Arthritis Research & Therapy*, *10*(6), R127. doi:10.1186/ar2541

Redon, R., Ishikawa, S., Fitch, K. R., Feuk, L., Perry, G. H., & Andrews, T. D. (2006). Global variation in copy number in the human genome. *Nature*, *444*(7118), 444–454. doi:10.1038/nature05329

Reeves, C. D., Murli, S., Ashley, G. W., Piagentini, M., Hutchinson, C. R., & McDaniel, R. (2001). Alteration of the substrate specificity of a modular polyketide synthase acyltransferase domain through site-specific mutations. *Biochemistry*, *40*(51), 15464–15470. doi:10.1021/bi015864r

Reichardt, J., & Bornholdt, S. (2004). Detecting fuzzy community structures in complex networks with a Potts model. *Physical Review Letters*, *93*(21), 218701. doi:10.1103/PhysRevLett.93.218701

Reichardt, J., & Bornholdt, S. (2006). When are networks truly modular? *Physica D. Nonlinear Phenomena*, *224*(1-2), 20–26. doi:10.1016/j.physd.2006.09.009

Reif, D. M., White, B. C., & Moore, J. H. (2004). Integrated analysis of genetic, genomic and proteomic data. *Expert Review of Proteomics*, *1*, 1095–1104. doi:10.1586/14789450.1.1.67

Reiss, D. J., Baliga, N. S., & Bonneau, R. (2006). Integrated biclustering of heterogeneous genome-wide datasets for the inference of global regulatory networks. *BMC Bioinformatics*, *7*, 280. doi:10.1186/1471-2105-7-280

Ren, B., Robert, F., Wyrick, J. J., Aparicio, O., Jennings, E. G., & Simon, I. (2000). Genome-wide location and function of DNA binding proteins. *Science*, *290*(5500), 2306–2309.

Revuz, D., & Yor, M. (1999). *Continuous martingales and Brownian motion* (3rd ed.). Berlin: Springer-Verlag.

Reyal, F., Stransky, N., Bernard-Pierrot, I., Vincent-Salomon, A., de Rycke, Y., & Elvin, P. (2005). Visualizing chromosomes as transcriptome correlation maps: Evidence of chromosomal domains containing co-expressed genes-a study of 130 invasive ductal breast carcinomas. *Cancer Research*, *65*(4), 1376–1383. doi:10.1158/0008-5472.CAN-04-2706

Reynolds, K. J., Yao, X., & Fenselau, C. (2002). Proteolytic 18O labeling for comparative proteomics: Evaluation of endoprotease glu-C as the catalytic agent. *Journal of Proteome Research*, *1*(1), 27–33. doi:10.1021/pr0100016

Rhodes, D. R., Yu, J., Shanker, K., Deshpande, N., Varambally, R., & Ghosh, D. (2004). ONCOMINE: A cancer microarray database and integrated data-mining platform. *Neoplasia (New York, N.Y.)*, *6*(1), 1–6.

Richardson, D. C., & Richardson, J. S. (1992). The kinemage: A tool for scientific communication. *Protein Science*, *1*(1), 3–9. doi:10.1002/pro.5560010102

Richert, L., Engler, A. J., Discher, D. E., & Picart, C. (2004). *Surface measurement of the elasticity of native and cross-linked polyelectrolyte multilayer film*. XXIXCongrès de la Société de Biomécanique. France: Créteil.

Richmond, T. J., & Davey, C. A. (2003). The structure of DNA in the nucleosome core. *Nature*, *423*(6936), 145–150.

Richter, C. D., Nietlispach, D., Broadhurst, R. W., & Weissman, K. J. (2008). Multienzyme docking in hybrid megasynthetases. *Nature Chemical Biology*, *4*(1), 75–81. doi:10.1038/nchembio.2007.61

Rider, M. H. (2004). 6-Phosphofructo-2-kinase/fructose-2,6-bisphosphatase: Head-to-head with a bifunctional enzyme that controls glycolysis. *The Biochemical Journal*, *381*, 561–579. doi:10.1042/BJ20040752

Rifai, N., Gillette, M. A., & Carr, S. A. (2006). Protein biomarker discovery and validation: The long and uncertain path to clinical utility. *Nature Biotechnology*, *24*(8), 971–983. doi:10.1038/nbt1235

Ringnér, M., & Peterson, C. (2003). Microarray-based cancer diagnosis with artificial neural networks. *BioTechniques*, *34*, S30–S35.

Ringrose, L., Rehmsmeier, M., Dura, J. M., & Paro, R. (2003). Genome-wide prediction of Polycomb/Trithorax response elements in Drosophila melanogaster. *Developmental Cell*, *5*(5), 759–771.

Rinn, J. L., Kertesz, M., Wang, J. K., Squazzo, S. L., Xu, X., & Brugmann, S. A. (2007). Functional demarcation of active and silent chromatin domains in human HOX loci by noncoding RNAs. *Cell*, *129*(7), 1311–1323.

Ripley, B. (1996). *Pattern recognition and neural networks*. Cambridge, UK: Cambridge University Press.

Risch, N., & Teng, J. (1998). The relative power of family-based and case-control designs for linkage disequilibrium studies of complex human diseases I. DNA pooling. *Genome Research*, *8*(12), 1273–1288.

Risch, N. J., & Merikangas, K. R. (1996). The future of genetic studies of complex human disease. *Science*, *273*, 1516–1517. doi:10.1126/science.273.5281.1516

Ritchie, M. D. (2009). Using prior knowledge to and genome-wide association to identify pathways involved in multiple sclerosis. *Genome Medicine*, *1*(6), 65. doi:10.1186/gm65

Ritchie, M. D., Hahn, L. W., & Moore, J. H. (2003). Power of multifactor dimensionality reduction for detecting gene-gene interactions in the presence of genotyping error, missing data, phenocopy, and genetic heterogeneity. *Genetic Epidemiology*, *24*, 150–157. doi:10.1002/gepi.10218

Rivals, I., Personnaz, L., Taing, L., & Potier, M. C. (2007). Enrichment or depletion of a GO category within a class of genes: Which test? *Bioinformatics (Oxford, England)*, *23*, 401–407. doi:10.1093/bioinformatics/btl633

Rivas, E., & Eddy, S. R. (2000). Secondary structure alone is generally not statistically significant for the detection of noncoding RNAs. *Bioinformatics (Oxford, England)*, *16*(7), 583–605. doi:10.1093/bioinformatics/16.7.583

Rix, U., Fischer, C., Remsing, L. L., & Rohr, J. (2002). Modification of post-PKS tailoring steps through combinatorial biosynthesis. *Natural Product Reports*, *19*(5), 542–580. doi:10.1039/b103920m

Robert, F., Pokholok, D. K., Hannett, N. M., Rinaldi, N. J., Chandy, M., & Rolfe, A. (2004). Global position and recruitment of HATs and HDACs in the yeast genome. *Molecular Cell*, *16*(2), 199–209.

Roberts, C., Lauffenburger, D. F., & Quinn, J. A. (1990). Receptor mediated cell attachment and detachment kinetics I: Probabilistic model and analysis. *Biophysical Journal*, *58*, 841–856. doi:10.1016/S0006-3495(90)82430-9

Robertson, J. A. (2001). Consent and privacy in pharmacogenetic testing. *Nature Genetics*, *28*(3), 207–209. doi:10.1038/90032

Rodriguez-Pinilla, S. M., Jones, R. L., Lambros, M. B., Arriola, E., Savage, K., & James, M. (2007). MYC amplification in breast cancer: A chromogenic in situ hybridisation study. *Journal of Clinical Pathology*, *60*(9), 1017–1023. doi:10.1136/jcp.2006.043869

Rognan, D., Lauemoller, S. L., Holm, A., Buus, S., & Tschinke, V. (1999). Predicting binding affinities of protein ligands from three-dimensional models: application to peptide binding to class I major histocompatibility proteins. *Journal of Medicinal Chemistry*, *42*(22), 4650–4658. doi:10.1021/jm9910775

Roh, T. Y., Cuddapah, S., & Zhao, K. (2005). Active chromatin domains are defined by acetylation islands revealed by genome-wide mapping. *Genes & Development*, *19*(5), 542–552.

Roider, H. G., Manke, T., O'Keeffe, S., Vingron, M., & Haas, S. A. (2009). PASTAA: Identifying transcription factors associated with sets of co-regulated genes. *Bioinformatics (Oxford, England)*, *25*(4), 435–442. doi:10.1093/bioinformatics/btn627

Rollinger, J. M., Haupt, S., Stuppner, H., & Langer, T. (2004). Combining ethnopharmacology and virtual screening for lead structure discovery: COX-inhibitors as application example. *Journal of Chemical Information and Computer Sciences*, *44*(2), 480–488. doi:10.1021/ci030031o

Rollinger, J. M., Hornick, A., Langer, T., Stuppner, H., & Prast, H. (2004). Acetylcholinesterase inhibitory activity of scopolin and scopoletin discovered by virtual screening of natural products. *Journal of Medicinal Chemistry*, *47*(25), 6248–6254. doi:10.1021/jm049655r

Ronaghi, M. (2003). Pyrosequencing for SNP genotyping. *Methods in Molecular Biology (Clifton, N.J.)*, *212*, 189–195.

Ronen, M., Rosenberg, R., Shraiman, B. I., & Alon, U. (2002). Assigning numbers to the arrows: Parameterizing a gene regulation network by using accurate expression kinetics. *Proceedings of the National Academy of Sciences of the United States of America*, *99*(16), 10555–10560. doi:10.1073/pnas.152046799

Rosoff, W. J., Urbach, J. S., Esrick, M. A., McAllister, R. G., Richards, L. J., & Goodhill, G. J. (2004). A new chemotaxis assay shows the extreme sensitivity of axons to molecular gradients. *Nature Neuroscience*, *7*(6), 678–682. doi:10.1038/nn1259

Ross, P. L., Huang, Y. N., Marchese, J. N., Williamson, B., & Parker, K. (2004). Multiplexed protein quantitation in Saccharomyces cerevisiae using amine-reactive isobaric tagging reagents. *Molecular & Cellular Proteomics*, *3*(12), 1154–1169. doi:10.1074/mcp.M400129-MCP200

Rothstein, M. A., & Epps, P. G. (2001). Ethical and legal implications of pharmacogenomics. *Nature Reviews. Genetics*, *2*(3), 228–231. doi:10.1038/35056075

Rual, J. F., Venkatesan, K., Hao, T., Hirozane-Kishikawa, T., Dricot, A., & Li, N. (2005). Towards a proteome-scale map of the human protein-protein interaction network. *Nature*, *437*(7062), 1173–1178. doi:10.1038/nature04209

Ruppert, D., Nettleton, D., & Hwang, J. T. G. (2007). Exploring the information in p-values for the analysis and planning of multiple-test experiments. *Biometrics*, *63*(2), 483–495. doi:10.1111/j.1541-0420.2006.00704.x

Ruths, D. A., Nakhleh, L., Iyengar, M. S., Reddy, S. A., & Ram, P. T. (2006). Hypothesis generation in signaling networks. *Journal of Computational Biology*, *13*(9), 1546–1557. doi:10.1089/cmb.2006.13.1546

Sabatti, C., Service, S. K., Hartikainen, A. L., Pouta, A., Ripatti, S., & Brodsky, J. (2009). Genome-wide association analysis of metabolic traits in a birth cohort from a founder population. *Nature Genetics*, *41*(1), 35–46. doi:10.1038/ng.271

Saccani, S., & Natoli, G. (2002). Dynamic changes in histone H3 Lys 9 methylation occurring at tightly regulated inducible inflammatory genes. *Genes & Development*, *16*(17), 2219–2224.

Saccone, S. F., Saccone, N. L., Swan, G. E., Madden, P. A., Goate, A. M., & Rice, J. P. (2008). Systematic biological prioritization after a genome-wide association study: An application to nicotine dependence. *Bioinformatics (Oxford, England)*, *24*, 1805–1811. doi:10.1093/bioinformatics/btn315

Sachidanandam, R. (2001). A map of human genome sequence variation containing 1.42 million single nucleotide polymorphisms. *Nature*, *409*, 928–933. doi:10.1038/35057149

Sachs, K., Perez, O., Pe'er, D., Lauffenburger, D., & Nolan, G. (2005). Causal protein-signaling networks derived from multiparameter single-cell data. *Science*, *308*(5721), 523–529. doi:10.1126/science.1105809

Saez-Rodriguez, J., Alexopoulos, L., Epperlein, J., Samaga, R., Lauffenburger, D., & Klamt, S. (2009). Discrete logic modelling as a means to link protein signalling networks with functional analysis of mammalian signal transduction. *Molecular Systems Biology*, *5*, 331. doi:10.1038/msb.2009.87

Saez-Rodriguez, J., Simeoni, L., Lindquist, J., Hemenway, R., Bommhardt, U., & Arndt, B. (2007). A logical model provides insights into T cell receptor signaling. *PLoS Computational Biology*, *3*(8), e163. doi:10.1371/journal.pcbi.0030163

Sagvolden, G., Giaver, I., Pettersen, E. O., & Feder, J. (1999). Cell adhesion force microscopy. *Proceedings of the National Academy of Sciences USA*, 471-476.

Saha, S., Harrison, S. H., & Chen, J. Y. (2009). Dissecting the human plasma proteome and inflammatory response biomarkers. *Proteomics*, *9*(2), 470–484. doi:10.1002/pmic.200800507

Saha, S., Harrison, S. H., Shen, C., Tang, H., Radivojac, P., & Arnold, R. J. (2008). HIP2: An online database of human plasma proteins from healthy individuals. *BMC Medical Genomics*, *1*, 12. doi:10.1186/1755-8794-1-12

Sakakibara, Y. (2003). Pair hidden Markov models on tree structures. *Bioinformatics (Oxford, England)*, *19*(Suppl 1), i232–i240. doi:10.1093/bioinformatics/btg1032

Salazar, E. J., Veléz, A. C., Parra, C. M., & Ortega, O. (2002). A cluster validity index for comparing non-hierarchical clustering methods. In *Memorias Encuentro de Investigaci'on sobre Tecnologias de Informacion Aplicadas a la Soluci'on de Problemas* (EITI2002), Medell'ın, Colombia, 2002.

Salerno, R. A., & Lesko, L. J. (2004). Pharmacogenomic data: FDA voluntary and required submission guidance. *Pharmacogenomics*, *5*, 503. doi:10.1517/14622416.5.5.503

Salgado, H. (2006). RegulonDB (version 5.0): Escherichia coli K-12 transcriptional regulatory network, operon organization, and growth conditions. *Nucleic Acids Research*, *34*, D394–D397. doi:10.1093/nar/gkj156

Salwinski, L., Miller, C. S., Smith, A. J., Pettit, F. K., Bowie, J. U., & Eisenberg, D. (2004). The database of interacting proteins: 2004 update. *Nucleic Acids Research*, *32*(Database issue), D449–D451. doi:10.1093/nar/gkh086

Sandelin, A., Alkema Engström, W. P., Wasserman, W. W., & Lenhard, B. (2004). JASPAR: An open-access database for eukaryotic transcription factor binding profiles. *Nucleic Acids Research*, *32*, D91–D94. doi:10.1093/nar/gkh012

Sandusky, P., & Raftery, D. (2005). Use of semiselective TOCSY and the Pearson correlation for the metabonomic analysis of biofluid mixtures: Application to urine. *Analytical Chemistry*, *77*, 7717–7723. doi:10.1021/ac0510890

Sankaranarayanan, R. (2006). A type III PKS makes the difference. *Nature Chemical Biology*, *2*(9), 451–452. doi:10.1038/nchembio0906-451

Santamaría, R., Therón, R., & Quintales, L. (2008a). A visual analytics approach for understanding biclustering results from microarray data. *BMC Bioinformatics*, *9*, 247. doi:10.1186/1471-2105-9-247

Santamaría, R., Therón, R., & Quintales, L. (2008b). A tool for bicluster visualization. *Bioinformatics (Oxford, England)*, *24*, 1212–1213. doi:10.1093/bioinformatics/btn076

Santos, S. D., Verveer, P. J., & Bastiaens, P. I. (2007). Growth factor-induced MAPK network topology shapes Erk response determining PC-12 cell fate. *Nature Cell Biology*, *9*(3), 324–330. doi:10.1038/ncb1543

Satchwell, S. C., Drew, H. R., & Travers, A. A. (1986). Sequence periodicities in chicken nucleosome core DNA. *Journal of Molecular Biology*, *191*(4), 659–675.

Saude, E. J., & Sykes, B. D. (2007). Urine stability for metabolomic studies: Effects of preparation and storage. *Metabolomics*, *3*(1), 19–27. doi:10.1007/s11306-006-0042-2

Saunders, A. M., Strittmatter, W. J., Schmechel, D., George-Hyslop, P. H., Pericak-Vance, M. A., & Joo, S. H. (1993). Association of apolipoprotein E allele epsilon 4 with late-onset familial and sporadic Alzheimer's disease. *Neurology*, *43*(8), 1467–1472.

Saxton, M. J. (1994). Single-particle tracking: Models of directed transport. *Biophysical Journal*, *67*(5), 2110–2119. doi:10.1016/S0006-3495(94)80694-0

Saxton, M. J., & Jacobson, K. (1997). Single-particle tracking: Application to membrane dynamics. *Annual Review of Biophysics and Biomolecular Structure*, *26*, 373–399. doi:10.1146/annurev.biophys.26.1.373

Sayers, E. W., Barrett, T., Benson, D. A., Bolton, E., Bryant, S. H., & Canese, K. (2010). Database resources of the National Center for Biotechnology Information. *Nucleic Acids Research*, *38*(Database issue), D5–D16. doi:10.1093/nar/gkp967

Sayle, R. A., & Milner, J. E. (2000). Rasmol: Biomolecular graphics for all. *Trends in Biochemical Sciences*, *20*(9), 374–376. doi:10.1016/S0968-0004(00)89080-5

Scalbert, A., Brennan, L., Fiehn, O., Hankemeier, T., Kristal, B. S., & Ommen, B. V. (2009). Mass-spectrometry-based metabolomics: limitations and recommendations for future progress with particular focus on nutrition research. *Metabolomics*, *5*, 435–458. doi:10.1007/s11306-009-0168-0

Scarselli, M., Giuliani, M. M., Adu-Bobie, J., Pizza, M., & Rappuoli, R. (2005). The impact of genomics on vaccine design. *Trends in Biotechnology*, *23*(2), 84–91. doi:10.1016/j.tibtech.2004.12.008

Schadt, E., Li, C., Eliss, B., & Wong, W. H. (2002). Feature extraction and normalization algorithms for high-density oligonucleotide gene expression array data. *Journal of Cellular Biochemistry*, *84*(S37), 120–125. doi:10.1002/jcb.10073

Schadt, E., Li, C., Su, C., & Wong, W. H. (2001). Analyzing highdensity oligonucleotide gene expression array data. *Journal of Cellular Biochemistry*, *80*, 192–202. doi:10.1002/1097-4644(20010201)80:2<192::AID-JCB50>3.0.CO;2-W

Schadt, E. E., Lamb, J., Yang, X., Zhu, J., Edwards, S., & Guhathakurta, D. (2005). An integrative genomics approach to infer causal associations between gene expression and disease. *Nature Genetics*, *37*(7), 710–717. doi:10.1038/ng1589

Schadt, E. E., Monks, S. A., Drake, T. A., Lusis, A. J., Che, N., & Colinayo, V. (2003). Genetics of gene expression surveyed in maize, mouse and man. *Nature*, *422*(6929), 297–302. doi:10.1038/nature01434

Schadt, E. E. (2009). Molecular networks as sensors and drivers of common human diseases. *Nature*, *461*(7261), 218–223. doi:10.1038/nature08454

Schaechinger, T. J., & Oliver, D. (2007). Nonmammalian orthologs of prestin (SLC26A5) are electrogenic divalent/chloride anion exchangers. *Proceedings of the National Academy of Sciences of the United States of America*, *104*(18), 7693–7698. doi:10.1073/pnas.0608583104

Schaefer, C., Anthony, K., Krupa, S., Buchoff, J., Day, M., & Hannay, T. (2009). PID: The pathway interaction database. *Nucleic Acids Research*, *37*(Database issue), D674–D679. doi:10.1093/nar/gkn653

Schapire, R., Freund, Y., Bartlett, P., & Lee, W. S. (1998). Boosting the margin: A new explanation for the effectiveness of voting methods. *Annals of Statistics*, *26*, 1651–1686. doi:10.1214/aos/1024691352

Schena, M., Shalon, D., Davis, R. W., & Brown, P. O. (1995). Quantitative monitoring of gene expression patterns with a complementary DNA microarray. *Science*, *270*(5235), 467–470. doi:10.1126/science.270.5235.467

Schevchenko, A., Chernushevich, I., & Ens, W. (1997). Rapid de novo peptide sequencing by a combination of nanoelectrospray, isotopic labeling and a quadrupole/time-of-flight mass spectrometer. *Rapid Communications in Mass Spectrometry*, *11*(9), 1015–1024. doi:10.1002/(SICI)1097-0231(19970615)11:9<1015::AID-RCM958>3.0.CO;2-H

Schlesinger, Y., Straussman, R., Keshet, I., Farkash, S., Hecht, M., & Zimmerman, J. (2007). Polycomb-mediated methylation on Lys27 of histone H3 pre-marks genes for de novo methylation in cancer. *Nature Genetics*, *39*(2), 232–236.

Schlessinger, J. (2000). Cell signaling by receptor tyrosine kinases. *Cell*, *103*(2), 277–280. doi:10.1016/S0092-8674(00)00114-8

Schmidt, M., Bohm, D., von Torne, C., Steiner, E., Puhl, A., & Pilch, H. (2008). The humoral immune system has a key prognostic impact in node-negative breast cancer. *Cancer Research*, *68*(13), 5405–5413. doi:10.1158/0008-5472.CAN-07-5206

Schmidt, H., & Jirstrand, M. (2006). Systems biology toolbox for MATLAB: A computational platform for research in systems biology. *Bioinformatics (Oxford, England)*, *22*(4), 514–515. doi:10.1093/bioinformatics/bti799

Schneider, G., & Bohm, H. J. (2002). Virtual screening and fast automated docking methods. *Drug Discovery Today*, *7*(1), 64–70. doi:10.1016/S1359-6446(01)02091-8

Schoeman, F. D. (1984). *Philosophical Dimensions of Privacy: an Anthology*. Cambridge, New York: Cambridge University Press. doi:10.1017/CBO9780511625138

Schones, D. E., Cui, K., Cuddapah, S., Roh, T. Y., Barski, A., & Wang, Z. (2008). Dynamic regulation of nucleosome positioning in the human genome. *Cell*, *132*(5), 887–898.

Schroeder, A., Mueller, O., Stocker, S., Salowsky, R., Leiber, M., & Gassmann, M. (2006). The RIN: An RNA integrity number for assigning integrity values to RNA measurements. *BMC Molecular Biology*, *7*(1), 3. doi:10.1186/1471-2199-7-3

Schuettengruber, B., Chourrout, D., Vervoort, M., Leblanc, B., & Cavalli, G. (2007). Genome regulation by polycomb and trithorax proteins. *Cell*, *128*(4), 735–745.

Schwartz, P. H., & Meslin, E. M. (2008). The Ethics of Information: Absolute Risk Reduction and Patient Understanding of Screening. *Journal of General Internal Medicine*.

Schwarzer, D., Finking, R., & Marahiel, M. A. (2003). Nonribosomal peptides: From genes to products. *Natural Product Reports*, *20*(3), 275–287. doi:10.1039/b111145k

Schwecke, T., Aparicio, J. F., Molnar, I., Konig, A., Khaw, L. E., & Haydock, S. F. (1995). The biosynthetic gene cluster for the polyketide immunosuppressant rapamycin. *Proceedings of the National Academy of Sciences of the United States of America*, *92*(17), 7839–7843. doi:10.1073/pnas.92.17.7839

Schwede, T., Kopp, J., Guex, N., & Petsch, M. C. (2003). SWISS-MODEL: An automated protein homology-modeling server. *Nucleic Acids Research*, *31*(13), 3381–3385. doi:10.1093/nar/gkg520

Schwikowski, B. (2000). A network of protein-protein interactions in yeast. *Nature Biotechnology*, *18*(12), 1257–1261. doi:10.1038/82360

Sciabola, S., Morao, I., & de Groot, M. J. (2007). Pharmacophoric fingerprint method (TOPP) for 3D-QSAR modeling: Application to CYP2D6 metabolic stability. *Journal of Chemical Information and Modeling*, *47*(1), 76–84. doi:10.1021/ci060143q

Sedel, F., Turpin, J. C., & Baumann, N. (2007). Neurological presentations of lysosomal diseases in adult patients. *Revista de Neurologia*, *163*(10), 919–929. doi:10.1016/S0035-3787(07)92635-1

Segal, E., Fondufe-Mittendorf, Y., Chen, L., Thastrom, A., Field, Y., & Moore, I. K. (2006). A genomic code for nucleosome positioning. *Nature*, *442*(7104), 772–778.

Segal, E., & Widom, J. (2009). What controls nucleosome positions? *Trends in Genetics*, *25*(8), 335–343.

Segal, E., Friedman, N., Koller, D., & Regev, A. (2004). A module map showing conditional activity of expression modules in cancer. *Nature Genetics*, *36*(10), 1090–1098. doi:10.1038/ng1434

Seidman, S. (1980). Clique-like structures in directed networks. *Journal of Social and Biological Structures*, *3*, 43–54. doi:10.1016/0140-1750(80)90019-6

Seidman, S. (1983a). Internal cohesion of LS sets in graphs. *Social Networks*, *5*(2), 97–107. doi:10.1016/0378-8733(83)90020-5

Seidman, S. (1983b). Network structure and minimum degree. *Social Networks*, *5*, 269–287. doi:10.1016/0378-8733(83)90028-X

Seidman, S., & Foster, B. (1978). A graph-theoretic generalization of the clique concept. *The Journal of Mathematical Sociology*, *6*, 139–154. doi:10.1080/0022250X.1978.9989883

Sekinger, E. A., Moqtaderi, Z., & Struhl, K. (2005). Intrinsic histone-DNA interactions and low nucleosome density are important for preferential accessibility of promoter regions in yeast. *Molecular Cell*, *18*(6), 735–748.

Seligson, D. B., Horvath, S., Shi, T., Yu, H., Tze, S., & Grunstein, M. (2005). Global histone modification patterns predict risk of prostate cancer recurrence. *Nature*, *435*(7046), 1262–1266.

Seoane, J., Le, H. V., Shen, L., Anderson, S. A., & Massague, J. (2004). Integration of Smad and forkhead pathways in the control of neuroepithelial and glioblastoma cell proliferation. *Cell*, *117*(2), 211–223. doi:10.1016/S0092-8674(04)00298-3

Serganov, A., Polonskaia, A., Phan, A. T., Breaker, R. R., & Patel, D. J. (2006). Structural basis for gene regulation by a thiamine pyrophosphate-sensing riboswitch. *Nature*, *441*(7097), 1167–1171. doi:10.1038/nature04740

Serruto, D., Adu-Bobie, J., Capecchi, B., Rappuoli, R., Pizza, M., & Masignani, V. (2004). Biotechnology and vaccines: Application of functional genomics to Neisseria meningitidis and other bacterial pathogens. *Journal of Biotechnology*, *113*(1-3), 15–32. doi:10.1016/j.jbiotec.2004.03.024

Shah, N. H., & Fedoroff, N. V. (2004). CLENCH: A program for calculating Cluster ENriCHment using the gene ontology. *Bioinformatics (Oxford, England)*, *20*(7), 1196–1197. doi:10.1093/bioinformatics/bth056

Shannon, P., Markiel, A., Ozier, O., Baliga, N., Wang, J., & Ramage, D. (2003). Cytoscape: A software environment for integrated models of biomolecular interaction networks. *Genome Research*, *13*(11), 2498–2504. doi:10.1101/gr.1239303

Shapero, M. H., Leuther, K. K., Nguyen, A., Scott, M., & Jones, K. W. (2001). SNP genotyping by multiplexed solid-phase amplification and fluorescent minisequencing. *Genome Research*, *11*(11), 1926–1934.

Shapiro, B. A., Wu, J. C., Bengali, D., & Potts, M. J. (2001). The massively parallel genetic algorithm for RNA folding: MIMD implementation and population variation. *Bioinformatics (Oxford, England)*, *17*(2), 137–148. doi:10.1093/bioinformatics/17.2.137

Sharan, R., Suthram, S., Kelley, R. M., Kuhn, T., McCuine, S., & Uetz, P. (2005). Conserved patterns of protein interaction in multiple species. *Proceedigns of the National Academy of Sciences USA*, *102*(6), 1974–1979. doi:10.1073/pnas.0409522102

Sharan, R. (2005). Identification of protein complexes by comparative analysis of yeast and bacterial protein interaction data. *Journal of Computational Biology*, *12*(6), 835–846. doi:10.1089/cmb.2005.12.835

Sharan, R. (2007). Network-based prediction of protein function. *Molecular Systems Biology*, *3*, 88. doi:10.1038/msb4100129

Sharff, A., & Jhoti, H. (2003). High-throughput crystallography to enhance drug discovery. *Current Opinion in Chemical Biology*, *7*(3), 340–345. doi:10.1016/S1367-5931(03)00062-0

Sharma, S., Kelly, T. K., & Jones, P. A. (2010). Epigenetics in cancer. *Carcinogenesis*, *31*(1), 27–36.

Shaw, A. S., & Filbert, E. L. (2009). Scaffold proteins and immune-cell signalling. *Nature Reviews. Immunology*, *9*(1), 47–56. doi:10.1038/nri2473

Shedden, K., Taylor, J. M., Enkemann, S. A., Tsao, M. S., Yeatman, T. J., & Gerald, W. L. (2008). Gene expression-based survival prediction in lung adenocarcinoma: A multi-site, blinded validation study. *Nature Medicine*, *14*(8), 822–827. doi:10.1038/nm.1790

Shen, B. (2003). Polyketide biosynthesis beyond the type I, II and III polyketide synthase paradigms. *Current Opinion in Chemical Biology*, *7*(2), 285–295. doi:10.1016/S1367-5931(03)00020-6

Shen, F., Hu, Z., Goswami, J., & Gaffen, S. L. (2006). Identification of common transcriptional regulatory elements in interleukin-17 target genes. *The Journal of Biological Chemistry*, *281*(34), 24138–24148. doi:10.1074/jbc.M604597200

Shendure, J. (2005). Accurate multiplex polony sequencing of an evolved bacterial genome. *Science*, *309*, 1728–1732. doi:10.1126/science.1117389

Sheridan, D. L., Kong, Y., Parker, S. A., Dalby, K. N., & Turk, B. E. (2008). Substrate discrimination among mitogen-activated protein kinases through distinct docking sequence motifs. *The Journal of Biological Chemistry*, *283*(28), 19511–19520. doi:10.1074/jbc.M801074200

Sherry, S. T., Ward, M. H., Kholodov, M., Baker, J., Phan, L., & Smigielski, E. M. (2001). dbSNP: the NCBI database of genetic variation. *Nucleic Acids Research*, *29*(1), 308–311. doi:10.1093/nar/29.1.308

Sheta, E. A., Appel, S. H., & Goldknopf, I. L. (2006). 2D gel blood serum biomarkers reveal differential clinical proteomics of the neurodegenerative diseases. *Expert Review of Proteomics*, *3*(1), 45–62. doi:10.1586/14789450.3.1.45

Shi, J., & Malik, J. (2000). Normalized cuts and image segmentation. *IEEE Transactions on Pattern Analysis and Machine Intelligence*, *22*(8), 888–905. doi:10.1109/34.868688

Shi, Z. (2010). Co-expression module analysis reveals biological processes, genomic gain, and regulatory mechanisms associated with breast cancer progression. *BMC Systems Biology*, *4*, 74. doi:10.1186/1752-0509-4-74

Shi, L., Reid, L. H., Jones, W. D., Shippy, R., Warrington, J. A., & Baker, S. C. (2006). The MicroArray Quality Control (MAQC) project shows inter- and intra-platform reproducibility of gene expression measurements. *Nature Biotechnology*, *24*(9), 1151–1161. doi:10.1038/nbt1239

Shi, L., Kishore, R., McMullen, M. R., & Nagy, L. E. (2002). Chronic ethanol increases lipopolysaccharide-stimulated Egr-1 expression in RAW 264.7 macrophages: Contribution to enhanced tumor necrosis factor alpha production. *The Journal of Biological Chemistry*, *277*(17), 14777–14785.

Shinozuka, M. (1971). Simulation of multivariate and multidimensional random process. *The Journal of the Acoustical Society of America*, *49*(1), 357–367. doi:10.1121/1.1912338

Shinozuka, M. (1972). Monte Carlo solution of structural dynamics. *Computers & Structures*, *2*, 855–874. doi:10.1016/0045-7949(72)90043-0

Shinozuka, M. (1972). Digital simulation of random processes and its applications. *Journal of Sound and Vibration*, *25*(1), 111–128. doi:10.1016/0022-460X(72)90600-1

Shinozuka, M., Deodatis, G., Zhang, R., & Papageoriou, A. R. (1999). Modeling, synthesis and engineering application of strong earthquake wave motion. *Soil Dynamics and Earthquake Engineering*, *18*, 209–228. doi:10.1016/S0267-7261(98)00045-1

Shinozuka, M., & Deotadis, G. (1991). Simulation of stochastic process by spectral representation. *Applied Mechanics Reviews*, *44*(4), 191–203. doi:10.1115/1.3119501

Shinozuka, M., & Lenoe, E. (1976). A probabilistic model for spatial distribution of material properties. *Engineering Fracture Mechanics*, *8*, 217–227. doi:10.1016/0013-7944(76)90087-4

Shmulevich, I., Dougherty, E., Kim, S., & Zhang, W. (2002). Probabilistic Boolean networks: A rule-based uncertainty model for gene regulatory networks. *Bioinformatics (Oxford, England)*, *18*(2), 261–274. doi:10.1093/bioinformatics/18.2.261

Shoichet, B. K., Stroud, R. M., Santi, D. V., Kuntz, I. D., & Perry, K. M. (1993). Structure-based discovery of inhibitors of thymidylate synthase. *Science*, *259*(5100), 1445–1450. doi:10.1126/science.8451640

Shrabanek, L., Saini, H. K., Bader, G. D., & Enright, A. J. (2007). Computational prediction of protein-protein interactions. *Molecular Biotechnology*, *38*, 1–17. doi:10.1007/s12033-007-0069-2

Shujiro, O., Takuji, Y., Masami, H., Masumi, I., Toshiaki, K., & Peer, B. (2008). KEGG atlas mapping for global analysis of metabolic pathways. *Nucleic Acids Research*, *36*(2), W423.

Sidorov, I. A., Hosack, D. A., Gee, D., Yang, J., Cam, M. C., & Lempicki, R. A. (2002). Oligonucleotide microarray data distribution and normalization. *Information Sciences*, *146*, 65–71. doi:10.1016/S0020-0255(02)00215-3

Silva, J. C., Denny, R., & Dorschel, C. A. (2005). Quantitative proteomic analysis by accurate mass retention time pairs. *Analytical Chemistry*, *77*(7), 2187–2200. doi:10.1021/ac048455k

Silva, J. M., Marran, K., Parker, J. S., Silva, J., Golding, M., & Schlabach, M. R. (2008). Profiling essential genes in human mammary cells by multiplex RNAi screening. *Science*, *319*(5863), 617–620. doi:10.1126/science.1149185

Simeone, R., Constant, P., Guilhot, C., Daffe, M., & Chalut, C. (2007). Identification of the missing trans-acting enoyl reductase required for phthiocerol dimycocerosate and phenolglycolipid biosynthesis in Mycobacterium tuberculosis. *Journal of Bacteriology*, *189*(13), 4597–4602. doi:10.1128/JB.00169-07

Simon, R. M., McShane, L. M., Korn, E. L., & Radmacher, M. D. (2003). *Design and analysis of DNA microarray investigations*. New York: Springer.

Simon, J. A., & Kingston, R. E. (2009). Mechanisms of polycomb gene silencing: Knowns and unknowns. *Nature Reviews. Molecular Cell Biology*, *10*(10), 697–708.

Simon, A. (2002). *Intérêt de la microscopie à force atomique sur la biofonctionnalisation de matériaux: caractérisation du greffage et de l'adhésion cellulaire*. Thèse de doctorat, Université Bordeaux I.

Simonsen, I. (2004). Diffusion on complex networks: A way to probe their large-scale topological structures. *Physica A. Statistical and Theoretical Physics*, *336*(1-2), 163–173. doi:10.1016/j.physa.2004.01.021

Sing, C. F., Standard, J. H., & Kardia, S. L. (2003). Genes, environment, and cardiovascular disease. *Arteriosclerosis, Thrombosis, and Vascular Biology*, *23*, 1190–1196. doi:10.1161/01.ATV.0000075081.51227.86

Sing, A., Pannell, D., Karaiskakis, A., Sturgeon, K., Djabali, M., & Ellis, J. (2009). A vertebrate Polycomb response element governs segmentation of the posterior hindbrain. *Cell*, *138*(5), 885–897.

Siracusano, A., Teggi, A., & Ortona, E. (2009). Human cystic echinococcosis: Old problems and new perspectives. *Interdisciplinary Perspectives on Infectious Diseases*, *2009*, 474368. doi:10.1155/2009/474368

Sirois, S., Wei, D. Q., Du, Q., & Chou, K. C. (2004). Virtual screening for SARS-CoV protease based on KZ7088 pharmacophore points. *Journal of Chemical Information and Computer Sciences*, *44*(3), 1111–1122. doi:10.1021/ci034270n

Siuti, N., & Kelleher, N. L. (2007). Decoding protein modifications using top-down mass spectrometry. *Nature Methods*, *4*(10), 817–821. doi:10.1038/nmeth1097

Skalak, R., & Evans, E. A. (1984). *Mechanics and thermodynamics of biomembranes*. Boca Raton, FL: CRC Press Inc.

Sladek, R., Rocheleau, G., Rung, J., Dina, C., Shen, L., & Serre, D. (2007). A genome-wide association study identifies novel risk loci for type 2 diabetes. *Nature*, *445*(7130), 881–885. doi:10.1038/nature05616

Slanina, F., & Zhang, C. (2005). Referee networks and their spectral properties. *Acta Physica Polonica B*, *36*(9), 2797.

Smith, J. G., & Newton-Cheh, C. (2009). Genome-wide association study in humans. *Methods in Molecular Biology (Clifton, N.J.)*, *573*, 231–258. doi:10.1007/978-1-60761-247-6_14

Smock, R. G., & Gierasch, L. M. (2009). Sending signals dynamically. *Science*, *324*, 198–203. doi:10.1126/science.1169377

Sohler, F., & Zimmer, R. (2005). Identifying active transcription factors and kinases from expression data using pathway queries. *Bioinformatics (Oxford, England)*, *21*(Suppl 2), ii115–ii122. doi:10.1093/bioinformatics/bti1120

Song, Y., La, T., Phillips, N. D., Bellgard, M. I., & Hampson, D. J. (2009). A reverse vaccinology approach to swine dysentery vaccine development. *Veterinary Microbiology*, *137*(1-2), 111–119. doi:10.1016/j.vetmic.2008.12.018

Song, H. J., & Poo, M. M. (2001). The cell biology of neuronal navigation. *Nature Cell Biology*, *3*(3), E81–E88. doi:10.1038/35060164

Sontag, E., Kiyatkin, A., & Kholodenko, B. N. (2004). Inferring dynamic architecture of cellular networks using time series of gene expression, protein and metabolite data. *Bioinformatics (Oxford, England)*, *20*(12), 1877–1886. doi:10.1093/bioinformatics/bth173

Sontag, E. D. (2007). Monotone and near-monotone biochemical networks. *Systems and Synthetic Biology*, *1*(2), 59–87. doi:10.1007/s11693-007-9005-9

Sontag, E. D. (2005). Molecular systems biology and control. *European Journal of Control*, *11*, 1–40. doi:10.3166/ejc.11.396-435

Soong, T.-t., Wrzeszczynski, K. O., & Rost, B. (2008). Physical protein-protein interactions predicted from microarrays. *Bioinformatics (Oxford, England)*, *24*(22), 2608–2614. doi:10.1093/bioinformatics/btn498

Sorin, E. J., Nakatani, B. J., Rhee, Y. M., Jayachandran, G., Vishal, V., & Pande, V. S. (2004). Does native state topology determine the RNA folding mechanism? *Journal of Molecular Biology*, *337*(4), 789–797. doi:10.1016/j.jmb.2004.02.024

Sorlie, T., Perou, C. M., Tibshirani, R., Aas, T., Geisler, S., & Johnsen, H. (2001). Gene expression patterns of breast carcinomas distinguish tumor subclasses with clinical implications. *Proceedings of the National Academy of Sciences of the United States of America*, *98*(19), 10869–10874. doi:10.1073/pnas.191367098

Sosnick, T. R., & Pan, T. (2004). Reduced contact order and RNA folding rates. *Journal of Molecular Biology*, *342*(5), 1359–1365. doi:10.1016/j.jmb.2004.08.002

Sotiriou, C., Wirapati, P., Loi, S., Harris, A., Fox, S., & Smeds, J. (2006). Gene expression profiling in breast cancer: Understanding the molecular basis of histologic grade to improve prognosis. *Journal of the National Cancer Institute*, *98*(4), 262–272. doi:10.1093/jnci/djj052

Sparmann, A., & van Lohuizen, M. (2006). Polycomb silencers control cell fate, development and cancer. *Nature Reviews. Cancer*, *6*(11), 846–856.

Spellman, P. T., Sherlock, G., Zhang, M. Q., Iyer, V. R., Anders, K., & Eisenm, M. B. (1998). Comprehensive identification of cell cycle-regulated genes of the yeast Saccharomyces cerevisiae by microarray hybridization. *Molecular Biology of the Cell*, *9*(12), 3273–3297.

Spencer, C. C., Su, Z., Donnelly, P., & Marchini, J. (2009). Designing genome-wide association studies: Sample size, power, imputation, and the choice of genotyping chip. *Public Library of Science Genetics*, *5*(5), e1000477.

Spirin, V., & Mirny, L. A. (2003). Protein complexes and functional modules in molecular networks. *Proceedings of the National Academy of Sciences of the United States of America*, *100*(21), 12123–12128. doi:10.1073/pnas.2032324100

Spirin, V., & Mirny, L. A. (2003). Protein complexes and functional modules in molecular networks. *Proceedings of the National Academy of Sciences of the United States of America*, *100*(21), 12123–12128. doi:10.1073/pnas.2032324100

Stachelhaus, T., Mootz, H. D., & Marahiel, M. A. (1999). The specificity-conferring code of adenylation domains in nonribosomal peptide synthetases. *Chemistry & Biology*, *6*(8), 493–505. doi:10.1016/S1074-5521(99)80082-9

Stahura, F. L., & Bajorath, J. (2004). Virtual screening methods that complement HTS. *Combinatorial Chemistry & High Throughput Screening*, *7*(4), 259–269.

Stanley, F. J., Croft, M. L., Gibbins, J., & Read, A. W. (2008). A population database for maternal and child health research in Western Australia using record linkage. *Paediatric and Perinatal Epidemiology*, *8*(4), 433–447. doi:10.1111/j.1365-3016.1994.tb00482.x

Stanley, F. J., & Meslin, E., M. (2007). Australia Needs a Better System for Health Care Evaluation. *The Medical Journal of Australia*, *186*, 220–221.

Starcevic, A., Zucko, J., Simunkovic, J., Long, P. F., Cullum, J., & Hranueli, D. (2008). ClustScan: An integrated program package for the semi-automatic annotation of modular biosynthetic gene clusters and in silico prediction of novel chemical structures. *Nucleic Acids Research*, *36*(21), 6882–6892. doi:10.1093/nar/gkn685

States, D.J., Omenn, G.S., Blackwell, T.W., & Fermin, D., Eng., J., Speicher, D.W., et al. (2006). Challenges in deriving high-confidence protein identifications from data gathered by a HUPO plasma proteome collaborative study. *Nature Biotechnology*, *24*(3), 333–338. doi:10.1038/nbt1183

Stefansson, H., Helgason, A., Thorleifsson, G., Steinthorsdottir, V., Masson, G., & Barnard, J. (2005). A common inversion under selection in Europeans. *Nature Genetics*, *37*(2), 129–137. doi:10.1038/ng1508

Steinthorsdottir, V. (2007). A variant in CDKAL1 influences insulin response and risk of type 2 diabetes. *Nature Genetics*, *39*, 770–775. doi:10.1038/ng2043

Stelzl, U., Worm, U., Lalowski, M., Haenig, C., Brembeck, F. H., & Goehler, H. (2005). A human protein-protein interaction network: A resource for annotating the proteome. *Cell*, *122*(6), 957–968. doi:10.1016/j.cell.2005.08.029

Stéphanou, A., & Tracqui, P. (2002). Cytomechanics of cell deformation and migration: From models to experiments. *Current Review of Biology*, *325*, 295–308.

Stewart, J. J., White, J. T., Yan, X., Collins, S., Drescher, C. W., & Urban, N. D. (2006). Proteins associated with cisplatin resistance in ovarian cancer cells identified by quantitative proteomic technology and integrated with mRNA expression levels. *Molecular & Cellular Proteomics*, *5*(3), 433–443. doi:10.1074/mcp.M500140-MCP200

Stitt, T. N., Drujan, D., Clarke, B. A., Panaro, F., Timofeyva, Y., & Kline, W. O. (2004). The IGF-1/PI3K/Akt pathway prevents expression of muscle atrophy-induced ubiquitin ligases by inhibiting FOXO transcription factors. *Molecular Cell*, *14*(3), 395–403. doi:10.1016/S1097-2765(04)00211-4

Stolovitzky, G., Prill, R., & Califano, A. (2009). Lessons from the DREAM2 challenges. *Annals of the New York Academy of Sciences*, *1158*, 159–195. doi:10.1111/j.1749-6632.2009.04497.x

Stolovitzky, G. A., Kundaje, A., Held, G. A., Duggar, K. H., Haudenschild, C. D., & Zhou, D. (2005). Statistical analysis of MPSS measurements: Application to the study of LPS-activated macrophage gene expression. *Proceedings of the National Academy of Sciences of the United States of America*, *102*(5), 1402–1407. doi:10.1073/pnas.0406555102

Strahl, B. D., & Allis, C. D. (2000). The language of covalent histone modifications. *Nature*, *403*(6765), 41–45.

Stranger, B. E., Forrest, M. S., Dunning, M., Ingle, C. E., Beazley, C., & Thorne, N. (2007). Relative impact of nucleotide and copy number variation on gene expression phenotypes. *Science*, *315*(5813), 848–853. doi:10.1126/science.1136678

Stuart, J. M. (2003). A gene-coexpression network for global discovery of conserved genetic modules. *Science*, *302*(5643), 249–255. doi:10.1126/science.1087447

Sturtevant, A. H. (1913). The linear arrangement of six sex-linked factors in Drosophila, as shown by their mode of association. *The Journal of Experimental Zoology*, *14*, 43–59. doi:10.1002/jez.1400140104

Suarez-Rodriguez, M. C., Adams-Phillips, L., Liu, Y., Wang, H., Su, S.-H., & Jester, P. J. (2007). MEKK1 is required for flg22-Induced MPK4 activation in Arabidopsis plants. *Plant Physiology*, *143*(2), 661–669. doi:10.1104/pp.106.091389

Subramanian, A., Tamayo, P., Mootha, V. K., Mukherjee, S., Ebert, B. L., & Gillette, M. A. (2005). Gene set enrichment analysis: A knowledge-based approach for interpreting genome-wide expression profiles. *Proceedings of the National Academy of Sciences of the United States of America*, *102*(43), 15545–15550. doi:10.1073/pnas.0506580102

Sugar, C. A., & James, G. M. (2003). Finding the number of clusters in a data set: An information theoretic approach. *Journal of the American Statistical Association*, *98*, 750–763. doi:10.1198/016214503000000666

Sugasawa, K., Ng, J. M. Y., Masutani, C., Iwai, S., van der Spek, P. J., & Eker, A. P. M. (1998). Xeroderma pigmentosum group c protein complex is the initiator of global genome nucleotide excision repair. *Molecular Cell*, *2*(2), 223–232. doi:10.1016/S1097-2765(00)80132-X

Suh, Y., & Vijg, J. (2005). SNP discovery in associating genetic variation with human disease phenotypes. *Mutation Research*, *573*(1-2), 41–53. doi:10.1016/j.mrfmmm.2005.01.005

Sun, H. (2008). Pharmacophore-based virtual screening. *Current Medicinal Chemistry*, *15*(10), 1018–1024. doi:10.2174/092986708784049630

Sun, X., Jin, L., & Xiong, M. (2008). Extended Kalman filter for estimation of parameters in nonlinear state-space models of biochemical networks. *PLoS ONE*, *3*, e3758. doi:10.1371/journal.pone.0003758

Suntharalingam, G., Perry, M. R., Ward, S., Brett, S. J., Castello-Cortes, A., & Brunner, M. D. (2006). Cytokine storm in a phase 1 trial of the anti-CD28 monoclonal antibody TGN1412. *The New England Journal of Medicine*, *355*(10), 1018–1028. doi:10.1056/NEJMoa063842

Supper, J., Strauch, M., Wanke, D., Harter, K., & Zell, A. (2007). EDISA: Extracting biclusters from multiple time-series of gene expression profiles. *BMC Bioinformatics*, *8*, 334. doi:10.1186/1471-2105-8-334

Suthram, S., Dudley, J., Chiang, A., Chen, R., Hastie, T., & Butte, A. (2010). Network-based elucidation of human disease similarities reveals common functional modules enriched for pluripotent drug targets. *PLoS Computational Biology*, *6*(2). doi:10.1371/journal.pcbi.1000662

Svaren, J., & Horz, W. (1997). Transcription factors vs nucleosomes: Regulation of the PHO5 promoter in yeast. *Trends in Biochemical Sciences*, *22*(3), 93–97.

Swanson, D. R. (1986). Fish oil, Raynaud's syndrome, and undiscovered public knowledge. *Perspectives in Biology and Medicine*, *30*(1), 7–18.

Swarbreck, D., Wilks, C., & Lamesch, P. (2008). The Arabidopsis Information Resource (TAIR): Gene structure and function annotation. *Nucleic Acids Research*, *36*, D1009–D1014. doi:10.1093/nar/gkm965

Sweeney, L., & Sweeney, L. (2002). Achieving K-Anonymity Privacy Protection Using Generalization and Suppression. *International Journal on Uncertainty, Fuzziness and Knowledge-based Systems*, *10*, 2002.

Syka, J. E., Coon, J. J., Schroeder, M. J., Shabanowitz, J., & Hunt, D. F. (2004). Peptide and protein sequence analysis by electron transfer dissociation mass spectrometry. *Proceedings of the National Academy of Sciences of the United States of America*, *101*(26), 9528–9533. doi:10.1073/pnas.0402700101

Symons, A., Beinke, S., & Ley, S. C. (2006). MAP kinase kinase kinases and innate immunity. *Trends in Immunology*, *27*(1), 40–48. doi:10.1016/j.it.2005.11.007

Szallasi, Z., Stelling, J., & Periwal, V. (2006). *System modeling in cellular biology from concepts to nuts and bolts*. The MIT Press.

Szymkowski, D. E. (2005). Creating the next generation of protein therapeutics through rational drug design. *Current Opinions in Drug Discovery and Development*, *8*(5), 590–600.

Tabor, H. K., Risch, N. J., & Myers, R. M. (2002). Opinion: Candidate-gene approaches for studying complex genetic traits: Practical considerations. *Nature Reviews. Genetics*, *3*, 391–397. doi:10.1038/nrg796

Tadmor, R. (2001). The London–van der Waals interactions between objects of various geometries. *Journal of Physics Condensed Matter*, *13*, 195–202. doi:10.1088/0953-8984/13/9/101

Tae, H., Kong, E. B., & Park, K. (2007). ASMPKS: An analysis system for modular polyketide synthases. *BMC Bioinformatics*, *8*, 327. doi:10.1186/1471-2105-8-327

Tahira, T., Suzuki, A., Kukita, Y., & Hayashi, K. (2003). SNP detection and allele frequency determination by SSCP. *Methods in Molecular Biology (Clifton, N.J.)*, *212*, 37–46.

Takahashi, K., & Yamanaka, S. (2006). Induction of pluripotent stem cells from mouse embryonic and adult fibroblast cultures by defined factors. *Cell*, *126*(4), 663–676.

Takano, R., Mochizuki, A., & Iwasa, Y. (2003). Possibility of tissue separation caused by cell adhesion. *Journal of Theoretical Biology*, *221*, 459–474. doi:10.1006/jtbi.2003.3193

Takeda, K., Kaisho, T., & Akira, S. (2003). Toll-like receptors. *Annual Review of Immunology*, *21*, 335–376. doi:10.1146/annurev.immunol.21.120601.141126

Tam, O. H., Aravin, A. A., Stein, P., Girard, A., Murchison, E. P., & Cheloufi, S. (2008). Pseudogene-derived small interfering RNAs regulate gene expression in mouse oocytes. *Nature*, *453*(7194), 534–538. doi:10.1038/nature06904

Tanay, A., Sharan, R., & Shamir, R. (2002). Discovering statistically significant biclusters in gene expression data. *Bioinformatics (Oxford, England)*, *18*, S136–S144.

Tanay, A., Sharan, R., & Shamir, R. (2006). Biclustering algorithms: A survey. In Aluru, S. (Ed.), *Handbook of computational molecular biology* (pp. 26-1–26-17). Chapman and Hall/CRC Press.

Tang, Z., Hu, Y. J., & Smith, M. D. (2008). Gaining Trust Through Online Privacy Protection: Self-Regulation, Mandatory Standards, or Caveat Emptor. *Journal of Management Information Systems*, *24*(4), 153–173. doi:10.2753/MIS0742-1222240406

Tang, Y., Tsai, S. C., & Khosla, C. (2003). Polyketide chain length control by chain length factor. *Journal of the American Chemical Society*, *125*(42), 12708–12709. doi:10.1021/ja0378759

Taniguchi, C. M., Emanuelli, B., & Kahn, C. R. (2006). Critical nodes in signaling pathways: Insights into insulin action. *Nature Reviews. Molecular Cell Biology*, *7*(2), 85–96. doi:10.1038/nrm1837

Tanimura, N., Saitoh, S., Matsumoto, F., Akashi-Takamura, S., & Miyake, K. (2008). Roles for LPS-dependent interaction and relocation of TLR4 and TRAM in TRIF-signaling. *Biochemical and Biophysical Research Communications*, *368*(1), 94–99. doi:10.1016/j.bbrc.2008.01.061

Tao, P., & Lai, L. (2001). Protein ligand docking based on empirical method for binding affinity estimation. *Journal of Computer-Aided Molecular Design*, *15*(5), 429–446. doi:10.1023/A:1011188704521

Tatsuno, S., Arakawa, K., & Kinashi, H. (2007). Analysis of modular-iterative mixed biosynthesis of lankacidin by heterologous expression and gene fusion. *The Journal of Antibiotics*, *60*(11), 700–708. doi:10.1038/ja.2007.90

Taylor, J. S., & Burnett, R. M. (2000). DARWIN: A program for docking flexible molecules. *Proteins*, *41*(2), 173–191. doi:10.1002/1097-0134(20001101)41:2<173::AID-PROT30>3.0.CO;2-3

TCGA. (2008). Comprehensive genomic characterization defines human glioblastoma genes and core pathways. *Nature*, *455*(7216), 1061–1068.

Tchabo, N. E., Liel, M. S., & Kohn, E. C. (2005). Applying proteomics in clinical trials: assessing the potential and practical limitations in ovarian cancer. *American Journal of Pharmacogenomics*, *5*(3), 141–148. doi:10.2165/00129785-200505030-00001

Tchagang, A. B., Bui, K. V., McGinnis, T., & Benos, P. V. (2009). Extracting biologically significant patterns from short time series gene expression data. *BMC Bioinformatics*, *10*, 255. doi:10.1186/1471-2105-10-255

Tchagang, A. B., Gawronski, A., Bérubé, H., Phan, S., Famili, F., & Pan, Y. (2010). GOAL: A software tool for assessing biological significance of genes group. *BMC Bioinformatics*, *11*, 229. doi:10.1186/1471-2105-11-229

Tchagang, A. B., & Tewfik, A. H. (2006). DNA microarray data analysis: A novel biclustering algorithm approach. *EURASIP Journal on Applied Signal Processing*, *59809*, 12.

Tchagang, A. B., Tewfik, A. H., Skubitz, K. M., DeRycke, M. S., & Skubitz, A. P. N. (2008). Early detection of ovarian cancer using group biomarkers. *Molecular Cancer Therapeutics*, *7*(1), 27–37. doi:10.1158/1535-7163.MCT-07-0565

Tchagang, A. B., Tewfik, A. H., & Benos, P. V. (2008). Biological evaluation of biclustering algorithms using gene ontology and ChIP-chip data. *In Proceedings of IEEE, International Conference on Acoustics, Speech and Signal Processing*, Las Vegas, Nevada.

Tefft, S. K. (1980). *Secrecy, a Cross-Cultural Perspective*. New York, N.Y.: Human Sciences Press.

Tegner, J., Yeung, M. K., Hasty, J., & Collins, J. J. (2003). Reverse engineering gene networks: Integrating genetic perturbations with dynamical modeling. *Proceedings of the National Academy of Sciences of the United States of America*, *100*(10), 5944–5949. doi:10.1073/pnas.0933416100

Teige, M., Scheikl, E., Eulgem, T., Doczi, R., Ichimura, K., & Shinozaki, K. (2004). The MKK2 pathway mediates cold and salt stress signaling in Arabidopsis. *Molecular Cell*, *15*(1), 141–152. doi:10.1016/j.molcel.2004.06.023

Teixeira, M. C. (2006). The YEASTRACT database: A tool for the analysis of transcription regulatory associations in Saccharomyces cerevisiae. *Nucleic Acids Research*, *34*, D446–D451. doi:10.1093/nar/gkj013

Templeton, A. R. (2000). Epistasis and complex traits. In Wolf, J., Wade, M., & Brodie, B. III, (Eds.), *Epistasis and evolutionary process*. New York: Oxford University Press.

Teng, J., & Risch, N. (1999). The relative power of family-based and case-control designs for linkage disequilibrium studies of complex human diseases. II. Individual genotyping. *Genome Research*, *9*(3), 234–241.

Teng, L. & Chan, L. (2007). *Order preserving clustering by finding frequent orders in gene expression data*. (LNCS 4774).

Terp, G. E., Johansen, B. N., Christensen, I. T., & Jorgensen, F. S. (2001). A new concept for multidimensional selection of ligand conformations (MultiSelect) and multidimensional scoring (MultiScore) of protein-ligand binding affinities. *Journal of Medicinal Chemistry*, *44*(14), 2333–2343. doi:10.1021/jm0010901

Tessier-Lavigne, M., & Goodman, C. (1996). The molecular biology of axon guidance. *Science*, *274*(5290), 1123–1133. doi:10.1126/science.274.5290.1123

Tettelin, H. (2009). The bacterial pan-genome and reverse vaccinology. *Genome Dynamics*, *6*, 35–47. doi:10.1159/000235761

Tettelin, H., Masignani, V., Cieslewicz, M. J., Eisen, J. A., Peterson, S., & Wessels, M. R. (2002). Complete genome sequence and comparative genomic analysis of an emerging human pathogen, serotype V Streptococcus agalactiae. *Proceedings of the National Academy of Sciences of the United States of America*, *99*(19), 12391–12396. doi:10.1073/pnas.182380799

Tewfik, A. H., Tchagang, A. B., & Vertatschitsch, L. (2006). Parallel identification of gene biclusters with coherent evolutions. *IEEE Transactions on Signal Processing*, *54*, 2408–2417. doi:10.1109/TSP.2006.873720

Thattai, M., Burak, Y., & Shraiman, B. I. (2007). The origins of specificity in polyketide synthase protein interactions. *PLoS Computational Biology*, *3*(9), 1827–1835. doi:10.1371/journal.pcbi.0030186

Thattai, M., & van Oudenaarden, A. (2001). Intrinsic noise in gene regulatory networks. *Proceedings of the National Academy of Sciences of the United States of America*, *98*(15), 8614–8619. doi:10.1073/pnas.151588598

The Gene Ontology Consortium. (2000). Gene ontology: Tool for the unification of biology. *Nature Genetics*, *25*(1), 25–29. doi:10.1038/75556

The Gene Ontology Consortium. (2009). The gene ontology's reference genome project: A unified framework for functional annotation across species. *PLoS Computational Biology*, *5*(7), e1000431. doi:10.1371/journal.pcbi.1000431

Thomas, R. K. (2006). Sensitive mutation detection in heterogeneous cancer specimens by massively parallel picoliter reactor sequencing. *Nature Medicine*, *12*, 852–855. doi:10.1038/nm1437

Thomas, P., Campbell, M., Kejariwal, A., Mi, H., Karlak, B., & Daverman, R. (2003). PANTHER: A library of protein families and subfamilies indexed by function. *Genome Research*, *13*(9), 2129–2141. doi:10.1101/gr.772403

Thompson, D. C., Humblet, C., & Joseph-McCarthy, D. (2008). Investigation of MM-PBSA rescoring of docking poses. *Journal of Chemical Information and Modeling*, *48*(5), 1081–1091. doi:10.1021/ci700470c

Thompson, D. M., King, K. R., Wieder, K. J., Toner, M., Yarmush, M. L., & Jayaraman, A. (2004). Dynamic gene expression profiling using a microfabricated living cell array. *Analytical Chemistry*, *76*(14), 4098–4103. doi:10.1021/ac0354241

Thorn, C. F., Klein, T. E., & Altman, R. B. (2010). Pharmacogenomics and bioinformatics: PharmGKB. *Pharmacogenomics*, *11*(4), 501–505. doi:10.2217/pgs.10.15

Thornton-Wells, T. A., Moore, J. H., & Haines, J. L. (2004). Genetics, statistics and human disease: Analytical retooling for complexity. *Trends in Genetics*, *20*, 640–647. doi:10.1016/j.tig.2004.09.007

Tian, Y., Tan, A., Sun, X., & Olson, M. T. (2009). Quantitative proteomic analysis of ovarian cancer cells identified mitochondrial proteins associated with paclitaxel resistance. *Proteomics: Clinical Applications*, *3*(11), 1288–1295. doi:10.1002/prca.200900005

Tian, L., Greenberg, S., Kong, S., Altschuler, J., Kohane, I., & Park, P. (2005). Discovering statistically significant pathways in expression profiling studies. *Proceedings of the National Academy of Sciences of the United States of America*, *102*(38), 13544–13549. doi:10.1073/pnas.0506577102

Tiana, G., Jensen, M., & Sneppen, K. (2002). Time delay as a key to apoptosis induction in the p53 network. *The European Physical Journal B*, *29*(1), 135–140. doi:10.1140/epjb/e2002-00271-1

Tibshirani, R., Hastie, T., Narasimhan, B., & Chu, G. (2002). Diagnosis of multiple cancer types by shrunken centroids of gene expression. *Proceedings of the National Academy of Sciences of the United States of America*, *99*, 6567–6572. doi:10.1073/pnas.082099299

Tiffin, N., Adie, E., Turner, F., Brunner, H. G., van Driel, M. A., & Oti, M. (2006). Computational disease gene identification: A concert of methods prioritizes type 2 diabetes and obesity candidate genes. *Nucleic Acids Research*, *34*, 3067–3081. doi:10.1093/nar/gkl381

Tinoco, I. Jr, & Bustamante, C. (1999). How RNA folds. *Journal of Molecular Biology*, *293*(2), 271–281. doi:10.1006/jmbi.1999.3001

Tintori, C., Magnani, M., Schenone, S., & Botta, M. (2009). Docking, 3D-QSAR studies and in silico ADME prediction on c-Src tyrosine kinase inhibitors. *European Journal of Medicinal Chemistry*, *44*(3), 990–1000. doi:10.1016/j.ejmech.2008.07.002

Tirosh, I., Berman, J., & Barkai, N. (2007). The pattern and evolution of yeast promoter bendability. *Trends in Genetics*, *23*(7), 318–321.

Todd, J. A. (2006). Statistical false positive or true disease pathway? *Nature Genetics*, *38*(7), 731–733. doi:10.1038/ng0706-731

Tohyama, S., Kakinuma, K., & Eguchi, T. (2006). The complete biosynthetic gene cluster of the 28-membered polyketide macrolactones, halstoctacosanolides, from Streptomyces halstedii HC34. *The Journal of Antibiotics*, *59*(1), 44–52. doi:10.1038/ja.2006.7

Tompa, M., Li, N., Bailey, T. L., Church, G. M., De Moor, B., & Eskin, E. (2005). Assessing computational tools for the discovery of transcription factor binding sites. *Nature Biotechnology*, *23*(1), 137–144. doi:10.1038/nbt1053

Toni, T., Welch, D., Strelkowa, N., Ipsen, A., & Stumpf, M. P. H. (2009). Approximate Bayesian computation scheme for parameter inference and model selection in dynamical systems. *Journal of the Royal Society, Interface*, *6*, 187–202. doi:10.1098/rsif.2008.0172

Tonon, G., Wong, K. K., Maulik, G., Brennan, C., Feng, B., & Zhang, Y. (2005). High-resolution genomic profiles of human lung cancer. *Proceedings of the National Academy of Sciences of the United States of America*, *102*(27), 9625–9630. doi:10.1073/pnas.0504126102

Totrov, M., & Abagyan, R. (1997). Flexible protein-ligand docking by global energy optimization in internal coordinates. *Proteins*, (Supplement 1), 215–220. doi:10.1002/(SICI)1097-0134(1997)1+<215::AID-PROT29>3.0.CO;2-Q

Toufighi, K., Siobhan, M., Brady, R. A., Ly, E., & Provart, N. J. (2005). The botany array resource: e-Northerns, expression angling, and promoter analyses. *The Plant Journal*, *43*(1), 153–163. doi:10.1111/j.1365-313X.2005.02437.x

Trivedi, O. A., Arora, P., Vats, A., Ansari, M. Z., Tickoo, R., & Sridharan, V. (2005). Dissecting the mechanism and assembly of a complex virulence mycobacterial lipid. *Molecular Cell*, *17*(5), 631–643. doi:10.1016/j.molcel.2005.02.009

Trutwein, B., Holman, C. D., & Rosman, D. L. (2006). Health data linkage conserves privacy in a research-rich environment. *Annals of Epidemiology*, *16*(4), 279–280. doi:10.1016/j.annepidem.2005.05.003

Tsatsanis, C., & Spandidos, D. A. (2000). The role of oncogenic kinases in human cancer [Review]. *International Journal of Molecular Medicine*, *5*(6), 583–590.

Tsukiyama, S. (1977). A new algorithm for generating all the maximal independent sets. *SIAM Journal on Computing*, *6*(3), 505–517. doi:10.1137/0206036

Tu, Z. D., Wang, L., Arbeitman, M., Chen, T., & Sun, F. Z. (2006). An integrative approach for causal gene identification and gene regulatory pathway inference. *Bioinformatics (Oxford, England)*, *22*, e489–e496. doi:10.1093/bioinformatics/btl234

Turner, C. E., & Dasgupta, S. (2003). Privacy on the Web: An Examination of User Concerns, Technology, and Implications for Business Organizations and Individuals. *Information Systems Management*, (Winter): 8–18. doi:10.1201/1078/43203.20.1.20031201/40079.2

Turner, H. L., Bailey, T. C., Krzanowski, W. J., & Hemingway, C. A. (2005). Biclustering models for structured microarray data. *IEEE/ACM Transactions on Computational Biology and Bioinformatics*, *2*(4), 316–329. doi:10.1109/TCBB.2005.49

Turner, S., & Sherrat, J. A. (2002). Intercellular adhesion and cancer invasion: a discrete simulation using the extended potts model. *Journal of Theoretical Biology*, *216*, 85–100. doi:10.1006/jtbi.2001.2522

Twyman, R. M. (2004). SNP discovery and typing technologies for pharmacogenomics. *Current Topics in Medicinal Chemistry*, *4*(13), 1423–1431. doi:10.2174/1568026043387656

Tyson, J. J., Chen, K. C., & Novak, B. (2003). Sniffers, buzzers, toggles and blinkers: Dynamics of regulatory and signaling pathways in the cell. *Current Opinion in Cell Biology*, *15*(2), 221–231. doi:10.1016/S0955-0674(03)00017-6

Ueda, H. (2006). Systems biology flowering in the plant clock field. *Molecular Systems Biology*, *2*(1).

Uetz, P. (2000). A comprehensive analysis of protein-protein interactions in Saccharomyces cerevisiae. *Nature*, *403*(6770), 623–627. doi:10.1038/35001009

Uhlik, M. T., Abell, A. N., Cuevas, B. D., Nakamura, K., & Johnson, G. L. (2004). Wiring diagrams of MAPK regulation by MEKK1, 2, and 3. *Biochemistry and Cell Biology*, *82*, 658–663. doi:10.1139/o04-114

Ulitsky, I., & Shamir, R. (2007). Identification of functional modules using network topology and high-throughput data. *BMC Systems Biology*, *1*, 8. doi:10.1186/1752-0509-1-8

Ulitsky, I., & Shamir, R. (2009). Identifying functional modules using expression profiles and confidence-scored protein interactions. *Bioinformatics (Oxford, England)*, *25*(9), 1158–1164. doi:10.1093/bioinformatics/btp118

US Food and Drug Administration. (FDA). (2005). *Drug-diagnostic co-development concept paper.* Retrieved January 3, 2010, from http://www.fda.gov/downloads/Drugs/ScienceResearch/ResearchAreas/Pharmacogenetics/UCM116689.pdf

US Food and Drug Administration. (FDA). (2006). *Guidance for industry—pharmacogenomic data submissions*. Retrieved January 3, 2010, from http://www.fda.gov/downloads/RegulatoryInformation/Guidances/ucm126957.pdf

Valenzano, C. R., Lawson, R. J., Chen, A. Y., Khosla, C., & Cane, D. E. (2009). The biochemical basis for stereochemical control in polyketide biosynthesis. *Journal of the American Chemical Society*, *131*(51), 18501–18511. doi:10.1021/ja908296m

van Batenburg, F. H., Gultyaev, A. P., Pleij, C. W., Ng, J., & Oliehoek, J. (2000). PseudoBase: A database with RNA pseudoknots. *Nucleic Acids Research*, *28*(1), 201–204. doi:10.1093/nar/28.1.201

van de Vijver, M. J., He, Y. D., van't Veer, L. J., Dai, H., Hart, A. A., & Voskuil, D. W. (2002). A gene-expression signature as a predictor of survival in breast cancer. *The New England Journal of Medicine*, *347*(25), 1999–2009. doi:10.1056/NEJMoa021967

van Driel, M. A., Bruggeman, J., Vriend, G., Brunner, H. G., & Leunissen, J. A. (2006). A text-mining analysis of the human phenome. *European Journal of Human Genetics*, *14*(5), 535–542. doi:10.1038/sj.ejhg.5201585

Van Heyningen, V., & Yeyati, P. L. (2004). Mechanisms of non-Mendelian inheritance in genetic disease. *Human Molecular Genetics*, *13*(2), R225–R233. doi:10.1093/hmg/ddh254

van Iterson, M., 't Hoen, P. A., Pedotti, P., Hooiveld, G. J., den Dunnen, J. T., & van Ommen, G. J. (2009). Relative power and sample size analysis on gene expression profiling data. Find Similar. *BMC Genomics*, *10*(1), 439. doi:10.1186/1471-2164-10-439

Van Laere, S. J., Van den Eynden, G. G., Van der Auwera, I., Vandenberghe, M., van Dam, P., & Van Marck, E. A. (2006). Identification of cell-of-origin breast tumor subtypes in inflammatory breast cancer by gene expression profiling. *Breast Cancer Research and Treatment*, *95*(3), 243–255. doi:10.1007/s10549-005-9015-9

Van Lanen, S. G., & Shen, B. (2006). Microbial genomics for the improvement of natural product discovery. *Current Opinion in Microbiology*, *9*(3), 252–260. doi:10.1016/j.mib.2006.04.002

van Noort, V. (2004). The yeast coexpression network has a small-world, scale-free architecture and can be explained by a simple model. *EMBO Reports*, *5*(3), 280–284. doi:10.1038/sj.embor.7400090

van Reeuwijk, J., Brunner, H. G., & van Bokhoven, H. (2005). Glyc-o-genetics of Walker-Warburg syndrome. *Clinical Genetics*, *67*(4), 281–289. doi:10.1111/j.1399-0004.2004.00368.x

van 't Veer, L. J., Dai, H., van de Vijver, M. J., He, Y. D., Hart, A. A., & Mao, M. (2002). Gene expression profiling predicts clinical outcome of breast cancer. *Nature*, *415*(6871), 530–536. doi:10.1038/415530a

van Vliet, M. H., Klijn, C. N., Wessels, L. F., & Reinders, M. J. (2007). Module-based outcome prediction using breast cancer compendia. *PLoS ONE*, *2*(10), e1047. doi:10.1371/journal.pone.0001047

Vangrevelinghe, E., Zimmermann, K., Schoepfer, J., Portmann, R., Fabbro, D., & Furet, P. (2003). Discovery of a potent and selective protein kinase CK2 inhibitor by high-throughput docking. *Journal of Medicinal Chemistry*, *46*(13), 2656–2662. doi:10.1021/jm030827e

Varady, J., Wu, X., Fang, X., Min, J., Hu, Z., & Levant, B. (2003). Molecular modeling of the three-dimensional structure of dopamine 3 (D3) subtype receptor: Discovery of novel and potent D3 ligands through a hybrid pharmacophore- and structure-based database searching approach. *Journal of Medicinal Chemistry*, *46*(21), 4377–4392. doi:10.1021/jm030085p

Varma, A., Morbidelli, M., & Wu, H. (2005). *Parametric sensitivity in chemical systems*. Cambridge University Press.

Vasgird, D. (2007). Prevention Over Cure: The Administrative Rationale for Education in the Responsible Conduct of Research. *Academic Medicine*, *82*, 835–837. doi:10.1097/ACM.0b013e31812f7e0b

Vaszar, L. T., Cho, M. K., & Raffin, T. A. (2003). Privacy issues in personalized medicine. *Pharmacogenomics*, *4*(2), 107–112. doi:10.1517/phgs.4.2.107.22625

Vaughn, C. P., Crockett, D. K., Lim, M. S., & Elenitoba-Johnson, K. S. J. (2006). Analytical characteristics of cleavable isotope-coded affinity tag-LC-tandem mass spectrometry for quantitative proteomic studies. *The Journal of Molecular Diagnostics*, *8*(4), 513–520. doi:10.2353/jmoldx.2006.060036

Vazquez, A. (2003). Global protein function prediction from protein-protein interaction networks. *Nature Biotechnology*, *21*(6), 697–700. doi:10.1038/nbt825

Veatch, R. M. (1981). *A Theory of Medical Ethics*. New York: Basic Books.

Velec, H. F., Gohlke, H., & Klebe, G. (2005). DrugScore(CSD)-knowledge-based scoring function derived from small molecule crystal data with superior recognition rate of near-native ligand poses and better affinity prediction. *Journal of Medicinal Chemistry*, *48*(20), 6296–6303. doi:10.1021/jm050436v

Verdonk, M. L., Cole, J. C., Hartshorn, M. J., Murray, C. W., & Taylor, R. D. (2003). Improved protein-ligand docking using GOLD. *Proteins*, *52*(4), 609–623. doi:10.1002/prot.10465

Veyrieras, J. B., Kudaravalli, S., Kim, S. Y., Dermitzakis, E. T., Gilad, Y., & Stephens, M. (2008). High-resolution mapping of expression-QTLs yields insight into human gene regulation. *PLOS Genetics*, *4*(10), e1000214. doi:10.1371/journal.pgen.1000214

Vinga, S., & Almeida, J. (2003). Alignment-free sequence comparison-a review. *Bioinformatics (Oxford, England)*, *19*(4), 513–523. doi:10.1093/bioinformatics/btg005

Visscher, P. M., & Hill, W. G. (2009). The limits of individual identification from sample allele frequencies: theory and statistical analysis. *PLOS Genetics*, *5*(10), e1000628. doi:10.1371/journal.pgen.1000628

Visscher, P. M. (2008). Sizing up human height variation. *Nature Genetics*, *40*(5), 489–490. doi:10.1038/ng0508-489

Visscher, P. M., Hill, W. G., & Wray, N. R. (2008). Heritability in the genomics era-concepts and misconceptions. *Nature Reviews. Genetics*, *9*(4), 255–266. doi:10.1038/nrg2322

Visscher, P. M., Medland, S. E., Ferreira, M. A., Morley, K. I., Zhu, G., & Cornes, B. K. (2006). Assumption-free estimation of heritability from genome-wide identity-by-descent sharing between full siblings. *PLOS Genetics*, *2*(3), e41. doi:10.1371/journal.pgen.0020041

Vissers, L.E., Veltman, J.A., van Kessel, A.G. & Brunner, H.G. (2005). Identification of disease genes by whole genome CGH arrays. *Human Molecular Genetics, 14*(Spec No. 2), R215-23.

Viswanathan, G. A., Seto, J., Patil, S., Nudelman, G., & Sealfon, S. C. (2008). Getting started in biological pathway construction and analysis. *PLoS Computational Biology*, *4*(2), e16. doi:10.1371/journal.pcbi.0040016

Vivona, S., Bernante, F., & Filippini, F. (2006). NERVE: New enhanced reverse vaccinology environment. *BMC Biotechnology*, *6*, 35. doi:10.1186/1472-6750-6-35

Vivona, S., Gardy, J. L., Ramachandran, S., Brinkman, F. S., Raghava, G. P., & Flower, D. R. (2008). Computer-aided biotechnology: from immuno-informatics to reverse vaccinology. *Trends in Biotechnology*, *26*(4), 190–200. doi:10.1016/j.tibtech.2007.12.006

Vizcaino, J. A., Cote, R., Reisinger, F., Foster, J. M., Mueller, M., & Rameseder, J. (2009). A guide to the Proteomics Identifications Database proteomics data repository. *Proteomics*, *9*(18), 4276–4283. doi:10.1002/pmic.200900402

Voet, D., Voet, J. C., & Pratt, C. W. (2008). *Fundamentals of biochemistry: Life at the molecular level*. Hoboken, NJ: Wiley.

Vogelstein, B., Lane, D., & Levine, A. J. (2000). Surfing the p53 network. *Nature*, *408*(6810), 307–310. doi:10.1038/35042675

Volker, M., Mone, M. J., Karmakar, P., van Hoffen, A., Schul, W., & Vermeulen, W. (2001). Sequential assembly of the nucleotide excision repair factors in vivo. *Molecular Cell*, *8*(1), 213–224. doi:10.1016/S1097-2765(01)00281-7

von Mering, C., Jensen, L., Kuhn, M., Chaffron, S., Doerks, T., & Krüger, B. (2007). STRING 7-recent developments in the integration and prediction of protein interactions. *Nucleic Acids Research*, *35*(Database issue), D358–D362. doi:10.1093/nar/gkl825

von Minckwitz, G., Harder, S., Hovelmann, S., Jager, E., Al-Batran, S. E., & Loibl, S. (2005). Phase I clinical study of the recombinant antibody toxin scFv(FRP5)-ETA specific for the ErbB2/HER2 receptor in patients with advanced solid malignomas. *Breast Cancer Research*, *7*(5), R617–R626. doi:10.1186/bcr1264

Waanders, L. F., Hanke, S., & Mann, M. (2007). Top-down quantitation and characterization of SILAC-labeled proteins. *Journal of the American Society for Mass Spectrometry*, *18*(11), 2058–2064. doi:10.1016/j.jasms.2007.09.001

Wache, P., Giddens, D. P., & Wang, X. (2001). *Couplage fluide-solide. Analyse 3D de l'état de contrainte d'une cellule endothéliale dans un écoulement*. 15ème Congrès Français de Mécanique, Nancy.

Wachi, S., Yoneda, K., & Wu, R. (2005). Interactome-transcriptome analysis reveals the high centrality of genes differentially expressed in lung cancer tissues. *Bioinformatics (Oxford, England)*, *21*(23), 4205–4208. doi:10.1093/bioinformatics/bti688

Waddington, C. (1942). The epigenotype. *Endeavour*, *1*, 18–20.

Wagner, H., Morgenstern, B., & Dress, A. (2008). Stability of multiple alignments and phylogenetic trees: An analysis of ABC-transporter proteins family. *Algorithms for Molecular Biology; AMB*, *3*, 15. doi:10.1186/1748-7188-3-15

Walter, A., Rehage, H., & Leonhard, H. (2001). Shear induced deformation of microcapsules: Shape oscillations and membranes folding. *Colloids and Surfaces*, *183*, 123–132. doi:10.1016/S0927-7757(01)00564-7

Wang, J., Wolf, R. M., Caldwell, J. W., Kollman, P. A., & Case, D. A. (2004). Development and testing of a general amber force field. *Journal of Computational Chemistry*, *25*(9), 1157–1174. doi:10.1002/jcc.20035

Wang, R., Lai, L., & Wang, S. (2002). Further development and validation of empirical scoring functions for structure-based binding affinity prediction. *Journal of Computer-Aided Molecular Design*, *16*(1), 11–26. doi:10.1023/A:1016357811882

Wang, J., Reijmers, T., Chen, L., Heijden, R. V. D., Wang, M., & Peng, S. (2009). System toxicology study of doxorubicin on rats using ultra performance liquid chromatography coupled with mass spectrometry based metabolomics. *Metabolomics*, *5*, 407–418. doi:10.1007/s11306-009-0165-3

Wang, W. Y., Barratt, B. J., Clayton, D. G., & Todd, J. A. (2005). Genome-wide association studies: Theoretical and practical concerns. *Nature Reviews. Genetics*, *6*, 109–118. doi:10.1038/nrg1522

Wang, A., Kurdistani, S. K., & Grunstein, M. (2002). Requirement of Hos2 histone deacetylase for gene activity in yeast. *Science*, *298*(5597), 1412–1414.

Wang, L., Brown, J. L., Cao, R., Zhang, Y., Kassis, J. A., & Jones, R. S. (2004). Hierarchical recruitment of polycomb group silencing complexes. *Molecular Cell*, *14*(5), 637–646.

Wang, L. (2008). An integrated approach for the analysis of biological pathways using mixed models. *PLOS Genetics*, *4*(7), e1000115. doi:10.1371/journal.pgen.1000115

Wang, L. (2009). A unified mixed effects model for gene set analysis of time course microarray experiments. *Statistical Applications in Genetics and Molecular Biology*, *8*(1), 47. doi:10.2202/1544-6115.1484

Wang, X. (2008). Gene module level analysis: Identification to networks and dynamics. *Current Opinion in Biotechnology*, *19*(5), 482–491. doi:10.1016/j.copbio.2008.07.011

Wang, Y., Klijn, J. G., Zhang, Y., Sieuwerts, A. M., Look, M. P., & Yang, F. (2005). Gene-expression profiles to predict distant metastasis of lymph-node-negative primary breast cancer. *Lancet*, *365*(9460), 671–679.

Wang, D., Hara, R., Singh, G., Sancar, A., & Lippard, S. (2003). Nucleotide excision repair from site-specifically platinum-modified nucleosomes. *Biochemistry*, *42*(22), 6747–6753. doi:10.1021/bi034264k

Wang, Z., Liu, X., Liu, Y., Liang, J., & Vinciotti, V. (2009). An extended Kalman filtering approach to modelling nonlinear dynamic gene regulatory networks via short gene expression time series. *IEEE/ACM Transactions on Computational Biology and Bioinformatics*, *6*(3), 410–419. doi:10.1109/TCBB.2009.5

Wang, H., Wang, W., Yang, J., & Yu, P. S. (2002). Clustering by pattern similarity in large data sets. *Proceedings of 2002 ACMSIGMOD International Conference on the Management of Data*, (pp. 394-405).

Warren, S. D., & Brandeis, D. L. (1890). The Right to Privacy. *Harvard Law Review*, *4*(5), 193–220. doi:10.2307/1321160

Warren, G. L., Andrews, C. W., Capelli, A. M., Clarke, B., LaLonde, J., & Lambert, M. H. (2006). A critical assessment of docking programs and scoring functions. *Journal of Medicinal Chemistry*, *49*(20), 5912–5931. doi:10.1021/jm050362n

Washington, N. L., Haendel, M. A., Mungall, C. J., Ashburner, M., Westerfield, M., & Lewis, S. E. (2009). Linking human diseases to animal models using ontology-based phenotype annotation. *PLoS Biology*, *7*(11), e1000247. doi:10.1371/journal.pbio.1000247

Wasserman, S., & Faust, K. (1994). *Social network analysis: Methods and applications*. Cambridge University Press.

Watts, D. J., & Strogatz, S. H. (1998). Collective dynamics of small-world networks. *Nature*, *393*(6684), 440–442. doi:10.1038/30918

Wei, G., Wei, L., Zhu, J., Zang, C., Hu-Li, J., & Yao, Z. (2009). Global mapping of H3K4me3 and H3K27me3 reveals specificity and plasticity in lineage fate determination of differentiating CD4+ T cells. *Immunity*, *30*(1), 155–167.

Weinberg, R. A. (2007). *The bology of cancer*. Garland Science.

Weir, R. F., & Horton, J. R. (1995). DNA banking and informed consent -- part 2. *IRB*, *17*(5-6), 1–8.

Weir, B. A., Woo, M. S., Getz, G., Perner, S., Ding, L., & Beroukhim, R. (2007). Characterizing the cancer genome in lung adenocarcinoma. *Nature*, *450*(7171), 893–898. doi:10.1038/nature06358

Weissman, K. J. (2006a). Single amino acid substitutions alter the efficiency of docking in modular polyketide biosynthesis. *ChemBioChem*, *7*(9), 1334–1342. doi:10.1002/cbic.200600185

Weissman, K. J. (2006b). The structural basis for docking in modular polyketide biosynthesis. *ChemBioChem*, *7*(3), 485–494. doi:10.1002/cbic.200500435

Weissman, K. J., & Muller, R. (2008). Protein-protein interactions in multienzyme megasynthetases. *ChemBioChem*, *9*(6), 826–848. doi:10.1002/cbic.200700751

Welch, W., Ruppert, J., & Jain, A. N. (1996). Hammerhead: Fast, fully automated docking of flexible ligands to protein binding sites. *Chemistry & Biology*, *3*(6), 449–462. doi:10.1016/S1074-5521(96)90093-9

Weljie, A. M., Newton, J., Mercier, P., Carlson, E., & Slupsky, C. M. (2006). Targeted profiling: Quantitative analysis of ^{1}H NMR metabolomics data. *Analytical Chemistry*, *78*(13), 4430–4442. doi:10.1021/ac060209g

Wellcome Trust Case Control Consortium. (2007). Genome-wide association study of 14,000 cases of seven common diseases and 3,000 shared controls. *Nature*, *447*, 661–682. doi:10.1038/nature05911

Wenzel, S. C., Kunze, B., Hofle, G., Silakowski, B., Scharfe, M., & Blocker, H. (2005). Structure and biosynthesis of myxochromides S1-3 in Stigmatella aurantiaca: Evidence for an iterative bacterial type I polyketide synthase and for module skipping in nonribosomal peptide biosynthesis. *ChemBioChem*, *6*(2), 375–385. doi:10.1002/cbic.200400282

Wenzel, S. C., Meiser, P., Binz, T. M., Mahmud, T., & Muller, R. (2006). Nonribosomal peptide biosynthesis: Point mutations and module skipping lead to chemical diversity. *Angewandte Chemie International Edition*, *45*(14), 2296–2301. doi:10.1002/anie.200503737

Wernig, M., Meissner, A., Foreman, R., Brambrink, T., Ku, M., & Hochedlinger, K. (2007). In vitro reprogramming of fibroblasts into a pluripotent ES-cell-like state. *Nature*, *448*(7151), 318–324.

Westerfors, M., Tedebark, U., Andersson, H. O., Ohrman, S., Choudhury, D., & Ersoy, O. (2003). Structure-based discovery of a new affinity ligand to pancreatic alpha-amylase. *Journal of Molecular Recognition*, *16*(6), 396–405. doi:10.1002/jmr.626

Westin, A. F. (1967). *Privacy and Freedom*. New York: Atheneum.

Whisstock, J. C., & Lesk, A. M. (2003). Prediction of protein function from protein sequence and structure. *Quarterly Reviews of Biophysics*, *36*(3), 307–340. doi:10.1017/S0033583503003901

Whiteaker, J. R. (2007). Integrated pipeline for mass spectrometry-based discovery and confirmation of biomarkers demonstrated in a mouse model of breast cancer. *Journal of Proteome Research*, *6*(10), 3962–3975. doi:10.1021/pr070202v

Whitesides, G. M., Ostuni, E., Takayama, S., Jiang, X., & Ingber, D. E. (2001). Soft lithography in biology and biochemistry. *Annual Review of Biomedical Engineering*, *3*(1), 335–373. doi:10.1146/annurev.bioeng.3.1.335

WHO. (2007). Report of the World Health Organization technical consultation on prevention and control of iron deficiency in infants and young children in malaria-endemic areas, Lyon, France, 12-14 June 2006. *Food and Nutrition Bulletin*, *28*(4Suppl), S489–S631.

Widom, J. (2001). Role of DNA sequence in nucleosome stability and dynamics. *Quarterly Reviews of Biophysics*, *34*(3), 269–324.

Wiener, N. (1956). The theory of prediction. *Modern Mathematics for Engineers*, *1*, 125–139.

Wilke, R. A., Mareedu, R. K., & Moore, J. H. (2008). The pathway less traveled: Moving from candidate genes to candidate pathways in the analysis of genome-wide data from large scale pharmacogenetic association studies. *Current Pharmacogenomics and Personalized Medicine*, *6*, 150–159.

Wilkinson, D. J. (2007). Bayesian methods in bioinformatics and computational systems biology. *Briefings in Bioinformatics*, *8*(2), 109–116. doi:10.1093/bib/bbm007

Willer, C. J., Sanna, S., Jackson, A. U., Scuteri, A., Bonnycastle, L. L., & Clarke, R. (2008). Newly identified loci that influence lipid concentrations and risk of coronary artery disease. *Nature Genetics*, *40*(2), 161–169. doi:10.1038/ng.76

Willett, P. (2006). Similarity-based virtual screening using 2D fingerprints. *Drug Discovery Today*, *11*(23-24), 1046–1053. doi:10.1016/j.drudis.2006.10.005

Williams, S. M., Canter, J. A., Crawford, D. C., Moore, J. H., Ritchie, M. D., & Haines, J. L. (2007). Problems with genome-wide association studies. *Science*, *316*, 1840–1842. doi:10.1126/science.316.5833.1840c

Williams, T., & Bjerknes, R. (1972). Stochastic model for abnormal clone spread through epithelial basal layer. *Nature*, *236*, 19–21. doi:10.1038/236019a0

Wilson, I. D., Nicholson, J. K., Castro-Perez, J., Granger, J. H., Johnson, K. A., & Smith, B. W. (2005). High resolution ultra performance liquid chromatography coupled to as-TOF mass spectrometry as a tool for differential metabolic pathway profiling in functional genomic studies. *Journal of Proteome Research*, *4*, 591–598. doi:10.1021/pr049769r

Wizemann, T. M., Heinrichs, J. H., Adamou, J. E., Erwin, A. L., Kunsch, C., & Choi, G. H. (2001). Use of a whole genome approach to identify vaccine molecules affording protection against Streptococcus pneumoniae infection. *Infection and Immunity*, *69*(3), 1593–1598. doi:10.1128/IAI.69.3.1593-1598.2001

Wolber, G., & Langer, T. (2005). LigandScout: 3-D pharmacophores derived from protein-bound ligands and their use as virtual screening filters. *Journal of Chemical Information and Modeling*, *45*(1), 160–169. doi:10.1021/ci049885e

Wolf, V. A. d., Sieber, J. E., Steel, P. M., & Zarate, A. O. (2005). Part I: What Is the Requirement for Data Sharing? *IRB: Ethics and Human Research*, *27*(6), 12–16. doi:10.2307/3563537

Wolford, J. K., Blunt, D., Ballecer, C., & Prochazka, M. (2000). High-throughput SNP detection by using DNA pooling and denaturing high performance liquid chromatography (DHPLC). *Human Genetics*, *107*(5), 483–487. doi:10.1007/s004390000396

Wolynes, P. G. (2005). Recent successes of the energy landscape theory of protein folding and function. *Quarterly Reviews of Biophysics*, *38*(4), 405–410. doi:10.1017/S0033583505004075

Wong, D. J. (2008). Revealing targeted therapy for human cancer by gene module maps. *Cancer Research*, *68*(2), 369–378. doi:10.1158/0008-5472.CAN-07-0382

Wong, S. Y., Haack, H., Kissil, J. L., Barry, M., Bronson, R. T., & Shen, S. S. (2007). Protein 4.1B suppresses prostate cancer progression and metastasis. *Proceedings of the National Academy of Sciences of the United States of America*, *104*(31), 12784–12789. doi:10.1073/pnas.0705499104

Wong, T., Chiu, Y. S., Lam, T. W., & Yiu, S. M. (2008). A memory efficient algorithm for structural alignment of RNAs with embedded simple pseudoknots. *Proceedings of the 6th Asia-Pacific Bioinformatics Conference*, 89-99.

Wong, T., Lam, T.W., Sung, W.K. & Yiu, S.M. (2009). Structural alignment of RNA with complex pseudoknot structure. *Algorithms in Bioinformatics*, 403-414.

Woo, C. J., Kharchenko, P. V., Daheron, L., Park, P. J., & Kingston, R. E. (2010). A region of the human HOXD cluster that confers polycomb-group responsiveness. *Cell*, *140*(1), 99–110.

Wu, F. Y. (1982). The Potts model. *Reviews of Modern Physics*, *54*(1), 235. doi:10.1103/RevModPhys.54.235

Wu, X., Jiang, R., Zhang, M. Q., & Li, S. (2008). Network-based global inference of human disease genes. *Molecular Systems Biology*, *4*, 189. doi:10.1038/msb.2008.27

Wu, J., Smith, L. T., Plass, C., & Huang, T. H. (2006). ChIP-chip comes of age for genome-wide functional analysis. *Cancer Research*, *66*(14), 6899–6902. doi:10.1158/0008-5472.CAN-06-0276

Wu, N., Cane, D. E., & Khosla, C. (2002). Quantitative analysis of the relative contributions of donor acyl carrier proteins, acceptor ketosynthases, and linker regions to intermodular transfer of intermediates in hybrid polyketide synthases. *Biochemistry*, *41*(15), 5056–5066. doi:10.1021/bi012086u

Wu, Z., Zhao, X., & Chen, L. (2009). Identifying responsive functional modules from protein-protein interaction network. *Molecules and Cells*, *27*(3), 271–277. doi:10.1007/s10059-009-0035-x

Wu, J., Liu, X., & Feng, J. (2008). Detecting causality between different frequencies. *Journal of Neuroscience Methods*, *167*(2), 367–375. doi:10.1016/j.jneumeth.2007.08.022

Wu, M., McDowell, J. A., & Turner, D. H. (1995). A periodic table of symmetric tandem mismatches in RNA. *Biochemistry*, *34*(10), 3204–3211. doi:10.1021/bi00010a009

Xayaphoummine, A., Bucher, T., Thalmann, F., & Isambert, H. (2003). Prediction and statistics of pseudoknots in RNA structures using exactly clustered stochastic simulations. *Proceedings of the National Academy of Sciences of the United States of America*, *100*(26), 15310–15315. doi:10.1073/pnas.2536430100

Xiang, W., Yang, C., Yang, Q., Xue, H., Tang, N. L., & Yu, W. (2009). MegaSNPHunter: A learning approach to detect disease predisposition SNPs and high level interactions in genome wide association studies. *BioMed Central Bioinformatics*, *10*, 13.

Xiao, Y., & Truskey, G. (1996). An effect of receptor-ligand affinity on the strength of endothelial cell adhesion. *Biophysical Journal*, *71*, 2869–2884. doi:10.1016/S0006-3495(96)79484-5

Xie, L., Li, J., Xie, L., & Bourne, P. (2009). Drug discovery using chemical systems biology: Identification of the protein-ligand binding network to explain the side effects of CETP inhibitors. *PLoS Computational Biology*, *5*(5). doi:10.1371/journal.pcbi.1000387

Xing, E. P., Jordan, M. I., & Karp, R. M. (2001). Feature selection for high-dimensional genomic microarray data. *In Proceedings of the Eighteenth International Conference on Machine Learning*, 601–608.

Xiong, H., Callaghan, D., Jones, A., Walker, D. G., Lue, L. F., & Beach, T. G. (2008). Cholesterol retention in Alzheimer's brain is responsible for high beta- and gamma-secretase activities and abeta production. *Neurobiological Discoveries*, *29*(3), 422–437. doi:10.1016/j.nbd.2007.10.005

Xu, H., Teo, H. H., Tan, B. C. Y., & Agarwal, R. (2010). The Role of Push-Pull Technology in Privacy Calculus: The Case of Location-Based Services. *Journal of Management Information Systems*, *26*(3), 137–176.

Xu, P., Widmer, G., Wang, Y., Ozaki, L. S., Alves, J. M., & Serrano, M. G. (2004). The genome of Cryptosporidium hominis. *Nature*, *431*(7012), 1107–1112. doi:10.1038/nature02977

Xu, R., & Wunsch, D. (2005). Survey of clustering algorithms. *IEEE Transactions on Neural Networks*, *16*(3), 645–678. doi:10.1109/TNN.2005.845141

Xu, L., Geman, D., & Winslow, R. L. (2007). Large-scale integration of cancer microarray data identifies a robust common cancer signature. *BMC Bioinformatics*, *8*, 275. doi:10.1186/1471-2105-8-275

Xu, J., Rosoff, W. J., Urbach, J., & Goodhill, G. J. (2005). Adaptation is not required to explain the long-term response of axons to molecular gradients. *Proceedings of the National Academy of Sciences of the United States of America*, *132*, 4545–4562.

Yadav, G., Gokhale, R. S., & Mohanty, D. (2003a). Computational approach for prediction of domain organization and substrate specificity of modular polyketide synthases. *Journal of Molecular Biology*, *328*(2), 335–363. doi:10.1016/S0022-2836(03)00232-8

Yadav, G., Gokhale, R. S., & Mohanty, D. (2003b). SEARCHPKS: A program for detection and analysis of polyketide synthase domains. *Nucleic Acids Research*, *31*(13), 3654–3658. doi:10.1093/nar/gkg607

Yadav, G., Gokhale, R. S., & Mohanty, D. (2009). Towards prediction of metabolic products of polyketide synthases: An in silico analysis. *PLoS Computational Biology*, *5*(4), e1000351. doi:10.1371/journal.pcbi.1000351

Yamada, R., & Ueda, H. (2009). Problems in analysis of large-scale data: Gene expression microarray analysis. *Tanpakushitsu Kakusan Koso*, *54*(10), 1307–1315.

Yang, X., Yang, H., Zhou, G., & Zhao, G. P. (2008). Infectious disease in the genomic era. *Annual Review of Genomics and Human Genetics*, *9*, 21–48. doi:10.1146/annurev.genom.9.081307.164428

Yang, Y. H., & Speed, T. P. (2003). *Design and analysis of comparative microarray experiments. Statistical analysis of gene expression microarray data.* Chapman & Hall.

Yang, Y. H., & Speed, T. (2002). Design issues for cDNA microarray experiments. *Nature Reviews. Genetics*, *3*, 579–588.

Yang, X. (2004). DBParser: Web-based software for shotgun proteomic data analyses. *Journal of Proteome Research*, *3*(5), 1002–1008. doi:10.1021/pr049920x

Yang, Z., Zhu, Q., Luo, K., & Zhou, Q. (2001). The 7SK small nuclear RNA inhibits the CDK9/cyclin T1 kinase to control transcription. *Nature*, *414*, 317–322. doi:10.1038/35104575

Yang, J., Wang, W., Wang, H., & Yu, P. S. (2002). δ-clusters: Capturing subspace correlation in a large data set. *In ICDE*, 517-528.

Yang, J., Wang, W., Wang, H., & Yu, P. S. (2003). Enhanced biclustering on expression data. *Proceedings of the Third IEEE Conference on Bioinformatics and Bioengineering*, 321-327.

Yasuda, K., Miyake, K., Horikawa, Y., Hara, K., Osawa, H., & Furuta, H. (2008). Variants in KCNQ1 are associated with susceptibility to type 2 diabetes mellitus. *Nature Genetics*, *40*(9), 1092–1097. doi:10.1038/ng.207

Yates, N. A., Deyanova, E. G., Geissler, W., & Wiener, M. C. (2007). Identification of peptidase substrates in human plasma by FTMS based differential mass spectrometry. *International Journal of Mass Spectrometry*, *259*(1-3), 174–183. doi:10.1016/j.ijms.2006.09.020

Yates, J. R., Sepp, T., Matharu, B. K., Khan, J. C., Thurlby, D. A., & Shahid, H. (2007). Complement C3 variant and the risk of age-related macular degeneration. *The New England Journal of Medicine*, *357*(6), 553–561. doi:10.1056/NEJMoa072618

Ye, T., Mo, H., Shanaiah, N., Gowda, G. A. N., Zhang, S., & Raftery, D. (2009). Chemoselective ^{15}N Tag for sensitive and high-resolution nuclear magnetic resonance profiling of the carboxy-containing metabolome. *Analytical Chemistry*, *81*(12), 4882–4888. doi:10.1021/ac900539y

Ye, H., Arron, J. R., Lamothe, B., Cirilli, M., Kobayashi, T., & Shevde, N. K. (2002). Distinct molecular mechanism for initiating TRAF6 signalling. *Nature*, *418*(6896), 443–447. doi:10.1038/nature00888

Yeger-Lotem, E., Riva, L., Su, L. J., Gitler, A. D., Cashikar, A. G., & King, O. D. (2009). Bridging high-throughput genetic and transcriptional data reveals cellular responses to alpha-synuclein toxicity. *Nature Genetics*, *41*(3), 316–323. doi:10.1038/ng.337

Yen, L., Svendsen, J., Lee, J., Gray, J. T., Magnier, M., & Baba, T. (2004). Exogenous control of mammalian gene expression through modulation of RNA self-cleavage. *Nature*, *431*(7007), 471–476. doi:10.1038/nature02844

Yeung, K. Y., & Ruzzo, L. W. (2001). Principal component analysis for clustering gene expression data. *Bioinformatics (Oxford, England)*, *17*(9), 763–774. doi:10.1093/bioinformatics/17.9.763

Yeung, M. K., Tegner, J., & Collins, J. J. (2002). Reverse engineering gene networks using singular value decomposition and robust regression. *Proceedings of the National Academy of Sciences of the United States of America*, *99*(9), 6163–6168. doi:10.1073/pnas.092576199

Yildirim, M. A., Goh, K. I., Cusick, M. E., Barabasi, A. L., & Vidal, M. (2007). Drug-target network. *Nature Biotechnology*, *25*(10), 1119–1126. doi:10.1038/nbt1338

Yonish-Rouach, E., Resnftzky, D., Lotem, J., Sachs, L., Kimchi, A., & Oren, M. (1991). Wild-type p53 induces apoptosis of myeloid leukaemic cells that is inhibited by interleukin-6. *Nature*, *352*(6333), 345–347. doi:10.1038/352345a0

Yoo, C. B., & Jones, P. A. (2006). Epigenetic therapy of cancer: Past, present and future. *Nature Reviews. Drug Discovery*, *5*(1), 37–50.

Yoshimura, A., Ohishi, H. M., Aki, D., & Hanada, T. (2004). Regulation of TLR signaling and inflammation by SOCS family proteins. *Journal of Leukocyte Biology*, *75*(3), 422–427. doi:10.1189/jlb.0403194

Yoshioka, K. (2004). Scaffold proteins in mammalian MAP Kinase cascades. *Journal of Biochemistry*, *135*(6), 657–661. doi:10.1093/jb/mvh079

Young, S. S., & Ge, N. (2005). Recursive partitioning analysis of complex disease pharmicogenetic studies I. motivation and overview. *Pharmicogenetics*, *6*, 65–75. doi:10.1517/14622416.6.1.65

Yu, J., Vodyanik, M. A., Smuga-Otto, K., Antosiewicz-Bourget, J., Frane, J. L., & Tian, S. (2007). Induced pluripotent stem cell lines derived from human somatic cells. *Science*, *318*(5858), 1917–1920.

Yu, G. Z., Chen, Y., Long, Y. Q., Dong, D., Mu, X. L., & Wang, J. J. (2008). New insight into the key proteins and pathways involved in the metastasis of colorectal carcinoma. *Oncology Reports*, *19*(5), 1191–1204.

Yu, K., Ganesan, K., Tan, L. K., Laban, M., Wu, J., & Zhao, X. D. (2008). A precisely regulated gene expression cassette potently modulates metastasis and survival in multiple solid cancers. *PLOS Genetics*, *4*(7), e1000129. doi:10.1371/journal.pgen.1000129

Yu, L., & Liu, H. (2004). Redundancy based feature selection for microarray data. *Proceedings of the Tenth ACM SIGKDD international conference on Knowledge discovery and data mining*, Seattle, WA, USA.

Yuan, G. C. (2009). Targeted recruitment of histone modifications in humans predicted by genomic sequences. *Journal of Computational Biology*, *16*(2), 341–355.

Yuan, G. C., & Liu, J. S. (2008). Genomic sequence is highly predictive of local nucleosome depletion. *PLoS Computational Biology*, *4*(1), e13.

Yuan, G. C., Liu, Y. J., Dion, M. F., Slack, M. D., Wu, L. F., & Altschuler, S. J. (2005). Genome-scale identification of nucleosome positions in S. cerevisiae. *Science*, *309*(5734), 626–630.

Yvert, G., Brem, R. B., Whittle, J., Akey, J. M., Foss, E., & Smith, E. N. (2003). Trans-acting regulatory variation in Saccharomyces cerevisiae and the role of transcription factors. *Nature Genetics*, *35*(1), 57–64. doi:10.1038/ng1222

Zazopoulos, E., Huang, K., Staffa, A., Liu, W., Bachmann, B. O., & Nonaka, K. (2003). A genomics-guided approach for discovering and expressing cryptic metabolic pathways. *Nature Biotechnology*, *21*(2), 187–190. doi:10.1038/nbt784

Zeggini, E., Scott, L. J., Saxena, R., Voight, B. F., Marchini, J. L., & Hu, T. (2008). Meta-analysis of genome-wide association data and large-scale replication identifies additional susceptibility loci for type 2 diabetes. *Nature Genetics*, *40*(5), 638–645. doi:10.1038/ng.120

Zhang, J., Sui, J., Ching, C. B., & Chen, W. N. (2008). Protein profile in neuroblastoma cells incubated with S- and R-enantiomers of ibuprofen by iTRAQ-coupled 2-D LC-MS/MS analysis: Possible action of induced proteins on Alzheimer's disease. *Proteomics*, *8*(8), 1595–1607. doi:10.1002/pmic.200700556

Zhang, X. G., Lu, X., Xu, X. Q., Leung, H. E., Wong, W. H., & Liu, J. S. (2006). RSVM: A SVM based strategy for recursive feature selection and sample classification with proteomics mass-spectrometry data. *BMC Bioinformatics*, *7*, 197. doi:10.1186/1471-2105-7-197

Zhang, Y., & Liu, J. S. (2007). Bayesian inference of epistatic interactions in case-control studies. *Nature Genetics*, *39*(9), 1167–1173. doi:10.1038/ng2110

Zhang, Y., Moqtaderi, Z., Rattner, B. P., Euskirchen, G., Snyder, M., & Kadonaga, J. T. (2009). Intrinsic histone-DNA interactions are not the major determinant of nucleosome positions in vivo. *Nature Structural & Molecular Biology*, *16*(8), 847–852.

Zhang, S., & Cao, J. (2009). A close examination of double filtering with fold change and t test in microarray analysis. *BMC Bioinformatics*, *10*, 402. doi:10.1186/1471-2105-10-402

Zhang, B. (2007). Proteomic parsimony through bipartite graph analysis improves accuracy and transparency. *Journal of Proteome Research*, *6*(9), 3549–3557. doi:10.1021/pr070230d

Zhang, B. (2008). From pull-down data to protein interaction networks and complexes with biological relevance. *Bioinformatics (Oxford, England)*, *24*(7), 979–986. doi:10.1093/bioinformatics/btn036

Zhang, R., & Lin, Y. (2009). Deg 5.0, a database of essential genes in both prokaryotes and eukaryotes. *Nucleic Acids Research*, *37*(Database issue), D455–D458. doi:10.1093/nar/gkn858

Zhang, T., Liu, Y., Yang, T., Zhang, L., Xu, S., & Xue, L. (2006). Diverse signals converge at MAPK cascades in plant. *Plant Physiology and Biochemistry*, *44*(5-6), 274–283. doi:10.1016/j.plaphy.2006.06.004

Zhang, M. Q., & Wilkinson, B. (2007). Drug discovery beyond the rule-of-five. *Current Opinion in Biotechnology*, *18*, 1–11. doi:10.1016/j.copbio.2007.10.005

Zhang, Y., Sieuwerts, A. M., McGreevy, M., Casey, G., Cufer, T., & Paradiso, A. (2009). The 76-gene signature defines high-risk patients that benefit from adjuvant tamoxifen therapy. *Breast Cancer Research and Treatment*, *116*(2), 303–309. doi:10.1007/s10549-008-0183-2

Zhang, S., Haas, B., Eskin, E., & Bafna, V. (2005). Searching genomes for noncoding RNA using FastR. *IEEE/ACM Transactions on Computational Biology and Bioinformatics*, *2*(4), 366–379. doi:10.1109/TCBB.2005.57

Zhang, W., & Chen, S. J. (2002). RNA-hairpin-folding kinetics. *Proceedings of the National Academy of Sciences of the United States of America*, *99*(4), 1931–1936. doi:10.1073/pnas.032443099

Zhang, W., & Chen, S. J. (2006). Exploring the complex folding kinetics of RNA-hairpins: I. General folding kinetics analysis. *Biophysical Journal*, *90*(3), 765–777. doi:10.1529/biophysj.105.062935

Zhang, J., Wang, J. J., & Yan, H. (2008). A neural-network approach for biclustering of gene expression data based on the plaid model. *International Conference on Machine Learning and Cybernetics, 2*(2008), 1082-1087.

Zhao, J., Sun, B. K., Erwin, J. A., Song, J. J., & Lee, J. T. (2008). Polycomb proteins targeted by a short repeat RNA to the mouse X chromosome. *Science*, *322*(5902), 750–756.

Zhao, X., Weir, B. A., LaFramboise, T., Lin, M., Beroukhim, R., & Garraway, L. (2005). Homozygous deletions and chromosome amplifications in human lung carcinomas revealed by single nucleotide polymorphism array analysis. *Cancer Research*, *65*(13), 5561–5570. doi:10.1158/0008-5472.CAN-04-4603

Zhao, H., Stoltz, J. F., Zhuang, F., & Wang, X. (2001). *Etude dynamique de l'interaction entre molécules d'adhésion à la surface cellulaire.* 15ème Congrès Français de Mécanique, Nancy.

Zheng, W., Long, J., Gao, Y. T., Li, C., Zheng, Y., & Xiang, Y. B. (2009). Genome-wide association study identifies a new breast cancer susceptibility locus at 6q25.1. *Nature Genetics*, *41*(3), 324–328. doi:10.1038/ng.318

Zheng, T., Wang, H., & Lo, S. H. (2006). Backward genotype-trait association (BGTA) - based dissection of complex traits in case-control design. *Human Heredity*, *62*, 196–212. doi:10.1159/000096995

Zheng, J., Shen, W., He, D. Z., Long, K. B., Madison, L. D., & Dallos, P. (2000). Prestin is the motor protein of cochlear outer hair cells. *Nature*, *405*(6783), 149–155. doi:10.1038/35012009

Zheng, J. Q., Felder, M., Connor, J. A., & Poo, M. (1994). Turning of nerve growth cone induced by neurotransmitters. *Nature*, *368*(6467), 140–144. doi:10.1038/368140a0

Zhong, S., Storch, F., Lipan, O., Kao, M. J., Weitz, C., & Wong, W. H. (2004). GoSurfer: A graphical interactive tool for comparative analysis of large gene sets in gene ontology space. *Applied Bioinformatics*, *3*(4), 1–5.

Zhong, Q., Simonis, N., Li, Q. R., Charloteaux, B., Heuze, F., & Klitgord, N. (2009). Edgetic perturbation models of human inherited disorders. *Molecular Systems Biology*, *5*, 321. doi:10.1038/msb.2009.80

Zhong, W., & Sternberg, P. W. (2006). Genome-wide prediction of c. Elegans genetic interactions. *Science*, *311*(5766), 1481–1484. doi:10.1126/science.1123287

Zhou, Z., Felts, A. K., Friesner, R. A., & Levy, R. M. (2007). Comparative performance of several flexible docking programs and scoring functions: enrichment studies for a diverse set of pharmaceutically relevant targets. *Journal of Chemical Information and Modeling*, *47*(4), 1599–1608. doi:10.1021/ci7000346

Zhou, F., Galan, J., Geahlen, R. L., & Tao, W. A. (2007). A novel quantitative proteomics strategy to study phosphorylation-dependent peptide-protein interactions. *Journal of Proteome Research*, *6*(1), 133–140. doi:10.1021/pr0602904

Zhou, M., Conrads, T. P., & Veenstra, T. D. (2005). Proteomics approaches to biomarker detection. *Briefings in Functional Genomics & Proteomics*, *4*(1), 69–75. doi:10.1093/bfgp/4.1.69

Zhou, V. W., Goren, A., & Bernstein, B. E. (2011). Charting histone modifications and the functional organization of mammalian genomes. *Nature Reviews. Genetics*, *12*(1), 7–18.

Zhou, X., Kao, M.-C., Huang, H., Wong, A., Nunez-Iglesias, J., & Primig, M. (2005). Functional annotation and network reconstruction through cross-platform integration of microarray data. *Nature Biotechnology*, *23*(2). doi:10.1038/nbt1058

Zhou, Y., Yau, C., Gray, J. W., Chew, K., Dairkee, S. H., & Moore, D. H. (2007). Enhanced NF kappa B and AP-1 transcriptional activity associated with antiestrogen resistant breast cancer. *BMC Cancer*, *7*, 59. doi:10.1186/1471-2407-7-59

Zhou, H., & Lipowsky, R. (2004). *Network Brownian motion: A new method to measure vertex-vertex proximity and to identify communities and subcommunities*. In (LNCS 3038). (pp. 1062-1069).

Zhu, J., & Zhang, M. Q. (1999). SCPD: A promoter database of the yeast Saccharomyces cerevisiae. *Bioinformatics (Oxford, England)*, 607–611. doi:10.1093/bioinformatics/15.7.607

Zhu, X., Gerstein, M., & Snyder, M. (2007). Getting connected: Analysis and principles of biological networks. *Genes & Development, 21*(9), 1010–1024. doi:10.1101/gad.1528707

Zhu, J., Zhang, B., & Schadt, E. (2008). A systems biology approach to drug discovery. *Advances in Genetics*, *60*, 603–635. doi:10.1016/S0065-2660(07)00421-X

Zhu, X., Yu, F., Li, X. C., & Du, L. (2007). Production of dihydroisocoumarins in Fusarium verticillioides by swapping ketosynthase domain of the fungal iterative polyketide synthase Fum1p with that of lovastatin diketide synthase. *Journal of the American Chemical Society*, *129*(1), 36–37. doi:10.1021/ja0672122

Zhu, C., Bao, G., & Wang, N. (2000). Cell mechanics: Mechanical response, cell adhesion, and molecular deformation. *Annual Review of Biomedical Engineering*, *2*, 189–226. doi:10.1146/annurev.bioeng.2.1.189

Zien, A., Kuffner, R., Zimmer, R., & Lengauer, T. (2000). Analysis of gene expression data with pathway scores. *Proceedings of the International Conference on Intelligent Systems for Molecular Biology, ISMB International Conference on Intelligent Systems for Molecular Biology*.

Zimmer, J. S., Monroe, M. E., Qian, W. J., & Smith, R. D. (2006). Advances in proteomics data analysis and display using an accurate mass and time tag approach. *Mass Spectrometry Reviews*, *25*(3), 450–482. doi:10.1002/mas.20071

Zirkle, R., Black, T. A., Gorlach, J., Ligon, J. M., & Molnar, I. (2004). Analysis of a 108-kb region of the Saccharopolyspora spinosa genome covering the obscurin polyketide synthase locus. *DNA Sequencing*, *15*(2), 123–134.

Zoete, V., Grosdidier, A., & Michielin, O. (2009). Docking, virtual high throughput screening and in silico fragment-based drug design. *Journal of Cellular and Molecular Medicine*, *13*(2), 238–248. doi:10.1111/j.1582-4934.2008.00665.x

Zou, M., Baitei, E. Y., Alzahrani, A. S., Al-Mohanna, F., Farid, N. R., & Meyer, B. (2009). Oncogenic activation of MAP kinase by BRAF pseudogene in thyroid tumors. *Neoplasia (New York, N.Y.)*, *11*(1), 57–65.

Zou, C., & Feng, J. (2009). Granger causality vs. dynamic Bayesian network inference: A comparative study. *BMC Bioinformatics*, 10.

Zou, C., Kendrick, K.M. & Feng, J. (2009). The fourth way: Granger causality is better than the three other reverse-engineering approaches. *Cell*.

Zubarev, R. A., Kelleher, N. L., & McLafferty, F. W. (1998). Electron capture dissociation of multiply charged protein cations. A nonergodic process. *Journal of the American Chemical Society*, *120*(13), 3265–3266. doi:10.1021/ja973478k

Zwick, D., & Dholakia, N. (2004). Whose Identity Is It Anyway? Consumer Representation in the Age of Database Marketing. *Journal of Macromarketing*, *24*(1), 31–43. doi:10.1177/0276146704263920

About the Contributors

Limin Angela Liu, PhD, obtained her BSc degree from Tsinghua University, Beijing and her PhD degree from Carnegie Mellon University, USA. After postdoctoral research at Johns Hopkins University, USA, she became Associate Professor at Shanghai Jiao Tong University. Her recent work includes the establishment of an *ab initio* method for the prediction of transcription factor binding sites and a novel "tethered-hopping model" for describing the effects of protein-protein interactions on the formation and stability of ternary protein-DNA complexes.

Dongqing Wei, PhD, is the acting head of the Department of Bioinformatics and Biostatistics, College of Life Science and Biotechnology, Shanghai Jiao Tong University, Shanghai, China, the editor-in-Chief of the journal "Interdisciplinary Sciences - Computational Life Sciences," and the chairman of the International Association of Scientists in the Interdisciplinary Areas (IASIA). Prof. Wei's research is in the general area of structural bioinformatics. He is best known for his ground-breaking work on theory of complicated liquids. He, along with Prof. Gren Patey, has found that strongly interacting dipolar spheres can form a ferroelectric nematic phase. This was the first demonstration that dipolar forces alone can create an orientationally ordered liquid state. It is also the first time that the existence of a ferroelectric nematic phase has been established for a model liquid. This discovery solved a long standing problem in theoretical physics, and created a new direction in search for new liquid crystal materials (Phys. Rev. Lett. 68, 2043, 1992, cited about 180 times). In recent years, Prof. Wei has developed tools of molecular simulation and applied them to study biological systems with relevance to computer-aided drug design and structural biology. With more than 150 journal papers and greater than 2000 citations (Science Citation Index), he is becoming a leading figure in the area of structural bioinformatics.

Yixue Li, PhD, was born in Xinjiang, China. Currently, he is the director in Shanghai Center for Bioinformation Technology, vice director and a full research professor of Key Laboratory of Systems Biology at Shanghai Institutes for Biological Sciences, Chinese Academy of Sciences. Dr. Li received his BSc. and Msc. degrees in theoretical physics from Xinjiang University, China, in 1982 and 1987, respectively, and his PhD degree in theoretical physics from the University of Heidelberg, Germany, in 1996. After Dr. Li got his PhD degree he worked as a bioinformatics research staff in European Molecular Biology Laboratory (EMBL) from 1997-2000, and came back to Shanghai, China in the middle of 2000. Dr. Li's research interests include bioinformatics, systems biology and computational biology. Dr. Li has published more than 100 journal papers in various international scientific journals, such as Science, Nature Genetics, Nature Biotechnology, PNAS, Bioinformatics, NAR, Plos Computational Biology, Plos One, Molecular Systems Biology, Molecular Cellular Proteomics, Oncogene, BMC Bioinformatics, Genome Biology, et cetera, and his research results have been cited by more than 1500 researchers worldwide in books, theses, journal and conference papers. Dr. Li has served as an editorial board member for 5 scientific journals.

Huimin Lei, MD, obtained her degree from Inner Mongolia University of Science and Technology, China in 2004. She then became a lecturer and academic advisor for medical students at Baotou Health School,

China. Since 2008, she became an Assistant Editor for the journal "Interdisciplinary Sciences – Computational Life Sciences" and an office administrator of Prof. Dongqing Wei's lab at Shanghai Jiao Tong University. She has served on the organizing committees of several international conferences, including "Theory and Applications of Computational Chemistry – 2008" (TACC2008) and the annual "International Conference on Computational and Systems Biology" (ICCSB) meeting series.

***

Giacomo Aletti, PhD, is a Mathematician. In 2001 he gained a PhD in Probability Theory and Mathematical Statistics working on set-indexed stochastic processes. His current research is devoted to both theoretical aspects and applications. The former ones concern metrics and topologies in different probability spaces, random reinforced urn models, survival analysis in set-valued stochastic processes and the general theory of stochastic geometric processes, while the latter are focused on modelling of social behaviour and biological phenomena, collaboration with medical research (applied/methodological statistics, e.g. statistical planning and modelling) and with numerical research for interdisciplinary approaches. Currently, he is Assistant Professor at Università degli Studi di Milano, Italy.

Hesham H. Ali, PhD, is professor of computer science and the Lee and Wilma Seaman Distinguished Dean of the College of Information Science and Technology at the University of Nebraska at Omaha. He is also the deputy director for computational sciences of the Nebraska Informatics for Life Center and a member of Nebraska Center for Bio-security. He received his PhD from the University of Nebraska-Lincoln in 1988, and his BS and MS in Computer Science from the University of Alexandria, in 1982 and 1985. He has published numerous articles in various IT areas including scheduling, distributed systems, wireless networks, and Bioinformatics. He has also published two books in scheduling and graph algorithms, and several book chapters in Bioinformatics. He is currently serving as the PI or Co-PI of several projects funded by NSF, NIH and Nebraska Research Initiative in the areas of wireless networks and Bioinformatics. He leads a Bioinformatics Research Group at UNO that focuses on developing innovative computational approaches to identify and classify biological organisms.

Swadha Anand was born in New Delhi, India in 1983. She received her BSc in Bio-chemistry from Sri Venkateswara College, University of Delhi and completed her Master's degree in Biotechnology from the Indian Institute of Technology, Mumbai in 2004. She is presently pursuing her PhD in the area of bioinformatics and computational biology at National Institute of Immunology, New Delhi. Her research work involves in silico analysis of protein interaction & regulatory networks in secondary metabolite biosynthetic pathways. She is using a variety of structure and sequence based bioinformatics approaches to understand how complex networking of individual catalytic domains brings about the large diversity in chemical structures of natural products.

Khaled H. Barakat received his BEng with distinction in Electrical Engineering from Cairo University (Egypt) in 2001. He received his M.SC degree in Engineering Physics from Cairo University in 2006. Mr. Barakat is currently a PhD candidate at the department of Physics, University of Alberta (Canada). As a member of Prof. Jack Tuszynski's computational group, his current focus is on developing accurate virtual screening (VS) protocols that can be used in the early stages of the rational drug design process.

Panayiotis (Takis) Benos, PhD, is an Associate Professor at the Department of Computational and Systems Biology, University of Pittsburgh while he holds joint appointments at the University of Pittsburgh Cancer Institute (UPCI) and the Department of Biomedical Informatics. Dr. Benos' background is in Mathematics

(BSc), and he earned a PhD degree in molecular biology and evolution. His post- graduate work includes genome analysis of Drosophila melanogaster with Prof. Michael Ashburner at EMBL-EBI, Cambridge, U.K. and the development of probabilistic algorithms for modeling protein-DNA interactions with Prof. Gary Stormo at Washington University in St. Louis. He joined University of Pittsburgh in 2002 as Assistant Professor and became Associate Professor in 2007. He is interested in the computational modeling of gene regulatory networks and the study of their evolution. More recently, he became interested in the evolution of the RNA viruses. His work has been published in many peer-reviewed journals such as Nature, Science, Genome Research, Genome Biology, and PLoS Computational Biology.

François Bertucci, MD, PhD, is a Professor in Oncology at Institut Paoli Calmettes – Université de la Méditerranée. He is responsible for the Genomics platform in the Department of Molecular Oncology at the CRCM. His research activity is now focusing on improvement of systemic treatments of cancer, mainly breast cancer and sarcoma, through both translational (identification of prognostic and predictive markers by use of genomics) and clinical research projects.

Fortunato Bianconi, PhD, was born in 1981. He received his Ph.D. degree in Information Engineering from University of Perugia, Italy in 2010, where he also received the MSc (Laurea) in Information and Communication Technology Engineering in 2006. He worked at University of California San Francisco as Junior Specialist at El-Samad Systems Biology Lab (2008-2009). His research interests are mainly related to systems biology, with a focus on the application of theoretical and mathematical tools from control engineering to the study of genetic networks. His research focuses on the systems biology of cancer.

Ghislain Bidaut, PhD, holds a doctorate in bioinformatics from the Université de la Méditerranée, with a focus on gene expression analysis and pattern recognition for his research work done with Michael Ochs's group (Fox Chase Cancer Center). Later on, he was a postdoctoral fellow at the University of Pennsylvania (Chris Stoeckert's group) working on large scale data integration in stem cell research. He did a second postdoctoral internship at the Institut Pasteur (Benno Schwikowski's group) before joining the CRCM in January 2008 to run the Integrative Bioinformatics group. He is now focusing on large scale network analysis and heterogeneous data integration, databases and LIMS, and multiparametric flow cytometry analysis to discover novel prognostic markers in cancer.

Daniel Birnbaum, MD, PhD, is head of the molecular biology lab at the Centre de Recherche en Cancérologie de Marseille. His research aims at characterizing gene alterations in breast and colon cancers, and in malignant hematopoietic diseases.

Christoph Brockel, PhD, leads the Translational and Bioinformatics function within Research Business Technologies at Pfizer Inc. since 2008. He is responsible for computational solutions that support target identification, interpretation of data to form mechanistic hypotheses and translational research within Pfizer. Prior to his current role, he was the head of US bioinformatics at Sanofi-Aventis and responsible for the development and application of gene expression and pathway analysis platforms. He has a Ph.D. in biophysics from the Universite Louis Pasteur in Strasbourg, France.

Paola Causin, PhD, is an Aerospace Engineer. In 2003 she earned a PhD in Numerical Analysis, during which, she worked in the field of the numerical simulation of fluid-dynamics with applications to continuum mechanics problems. During her post-doctoral work, she was involved in an European Research Project working on fluid-structure interaction problems applied to physiological flows. Currently, she is Assistant Professor of Numerical Analysis at Università degli Studi di Milano, Italy. Her present scientific interests are devoted, on the one hand, to the theoretical study of innovative numerical methodologies and, on the other hand, to more application-oriented topics, connected with the mathematical modelling of biological phenomena and, namely, to the numerical simulation of cartilage formation process in perfusion bioreactors and of axon chemotaxis in neuron development.

Jake Y. Chen, PhD, is an associate professor of informatics and computer science at the Indianapolis joint campus of Indiana University and Purdue University, where he teaches database systems, bioinformatics, and computational systems biology. He is the founding director of the Indiana Center for Systems Biology and Personalized Medicine, an associate editor of BMC Systems Biology, an ACM senior member, an IEEE senior member, and the central Indiana section chair of the IEEE Engineering in Biology and Medicine Society. He has more than 80 scientific publications that span broadly over biological data management, biological data mining, bioinformatics, systems biology, and personalized medicine. He has given more than 100 invited talks nationwide in bioinformatics. He also has six years of biotech R&D research and management experience, and has been active in high-tech entrepreneurship in both Silicon Valley and Indiana.

Adam W. Culbertson graduated from Indiana University with a Bachelor's of Science in Biology with a concentration in Chemistry. Additionally, he completed a certificate in the Managing in the Life Sciences Program from the Indiana University Professional Development Program. Mr. Culbertson is the author/co-author of multiple publications on the subject of personalized medicine. He has held numerous positions in the healthcare and biopharmaceutical industries which range in size from large fortune 500 companies to a small startup. Currently, he is a graduate student at Indiana University of Bloomington in Informatics in Human-Computer Interaction Design and is also a Student Associate for the Kelly School of Business Center for the Business of the Life Sciences at Indiana University of Bloomington.

Benjamin B. Currall is nearing completion of his graduate studies at Creighton University's School of Medicine. He has studied under the tutelage of both Drs. Richard Hallworth and David Z. He is researching the structure-function relationship of prestin, the mammalian motor protein. These studies have included research in bioinformatics (examining sequence analysis), function (using electrophysiology), and structure (protein-protein interactions using molecular biology and advanced microscopy) of this unique motor protein. Before attending Creighton, Mr. Currall operated a therapeutic drug monitoring laboratory designing drug analysis methods using mass spectrometry at an HIV research clinic in Los Angeles. Mr. Currall received his BS in Biology and BA in Philosophy degrees at Santa Clara University.

Kathryn M. Dempsey is a PhD student at the University of Nebraska Medical Center in the Bioinformatics Specialty track. She graduated in December 2007 with a BS in Bioinformatics from the University of Nebraska at Omaha (UNO), having completed just over two years of research under the supervision of the Nebraska INBRE program and UNO Bioinformatics Research Group. Kate has been honored with multiple Student Travel awards and most recently, a Best Poster award in October 2009 for original research investigating inner ear protein function with in silico analysis. She has coauthored several papers related to motif finding tools in Bioinformatics. She is currently working on a chapter related to advanced sequence analysis techniques. Her current research focuses on the use of correlation networks to discover new relationships among various biological elements, particularly in the domain of aging research.

Wei Ding received his PhD from the State University of New York at Stony Brook in 1998. After working as a Fogarty Fellow in the National Center for Biotechnology Information (NCBI), he joined the Bioinformatics group at the Schering-Plough Research Institute (now Merck Research Laboratories) in 1999. He is also an adjunct professor in the Department of Biological Sciences at Kean University. His research interests include biomarker discovery, development and validation, -omics data mining and analysis, and systems biology. He is also responsible for the development of statistical methods and applications for gene expression, proteomics, metabolomics, pharmcogenomics data analysis. Dr. Ding has authored dozens of research publications and holds several U.S. patents.

Jonathan Dushoff is Associate Professor of Biology at McMaster University in Hamilton, Ontario. He is a theoretical biologist with broad interests, and a particular focus on the evolution and spread of infectious diseases of humans. He is from Philadelphia, Pennsylvania. In addition to USA and Canada, he has also lived in Swaziland and Taiwan.

Fazel Famili is a Group Leader for the Knowledge Discovery group, working at the Institute for Information Technology (IIT) of the National Research Council of Canada, where he has been working for the past 24 years. Fazel has been actively involved in the fields of artificial intelligence, data mining, and bioinformatics, and successful applications of these technologies. He has a strong data mining and bioinformatics team within IIT that is currently engaged in unique research and development in data mining for genomics, proteomics, and health care. His research has been on data mining, machine learning, and bioinformatics, and their applications to real world problems in various data rich environments, such as life sciences.

Jianfeng Feng, PhD, received all his academic degrees from the Department of Probability and Statistics, Peking University. Since 2005, he has been a Professor at Warwick University, UK and since 2008, he has been the Director at the Centre for Computational Systems Biology, Fudan University, PR China. His research interests include computational neuroscience and computational cellular biology. He now works closely with biologists to improve the healthcare of human beings. He has published over 150 papers in top tier journals in biology, mathematics and physics. His modeling work on a 'trust' hormone has attracted wide media interests and was reported in BBC News, Washington Post and Reuters etc.

Wenqing Feng obtained her PhD degree from Rutgers University in 1997, in the area of NMR structure determination of proteins. Prior to joining the Department of Drug Metabolism and Pharmacokinetics at Schering-Plough in 1999, Wenqing was a postdoctoral fellow in the Department of Structure Chemistry at Schering-Plough Research Institute. Her expertise resides in using NMR methods to solve problems of the pharmaceutical industry, including structure identification of organic molecules, quantitation, and NMR-based metabonomics. She chaired the North Jersey NMR group of American Chemical Society from 2006-2007, and a session in Eastern Analytical Symposium, 2008. Wenqing is currently a Principal Consultant at Accela Sciences, LLC.

Pascal Finetti earned a Master's degree in Biochemistry at the Université de la Méditerranée (Marseille) in 1998. Next, he joined the group of Dr. Daniel Birnbaum as lab technician in the Molecular oncology laboratory at Institut Paoli-Calmettes (Marseille) where he has worked under Pr. Bertucci's supervision until now. He manages an oligonucleotide-based microarray platform he set up in 2004 with the aim to profile tumors for the discovery of their taxonomy and markers. Furthermore, he is involved in research programs of the department to analyse large-scale genomic data at the RNA and/or DNA level.

Jean-François Ganghoffer, PhD, is a full Professor in Applied Mathematics and Mechanics of Materials. He received his PhD from Ecole des Mines in Nancy, France in 1992, and worked afterwards as a research fellow at CNRS. He was appointed to full Professor in 2000 at the Institut National Polytechnique de Lorraine, in Nancy. His present interests include biomechanics, covering growth of biological tissues and mechanobiology of the cell, symmetries in continuum mechanics, and the mechanics of fibrous materials. He has also been active in the fields of mechanics of interfaces, nonlocal mechanics, higher order gradient damage and plasticity. He is the author or coauthor of about 80 scientific publications and as many Conference papers.

Maxime Garcia holds a master's degree in bioinformatics. During 2008, he followed an internship at Technological Advances for Genomics and Clinics (TAGC), working on the TranscriptomeBrowser. He joined the Integrative Bioinformatics team at the Centre de Recherche en Cancérologie de Marseille (CRCM) in March 2009 for his final internship. In November 2009, he began his PhD training on discovering biomarkers in breast cancer by interactome-transcriptome integration (ITI project). He is responsible for welcoming CRCM's new students. He is also the webmaster of a student association (Hippo'Thèse) involved within his doctoral school (Ecole Doctorale des Sciences de la Vie et de la Santé).

Tian Ge is a PhD student in the School of Mathematical Sciences and Centre for Computational Systems Biology at Fudan University, Shanghai, People's Republic of China. He received his bachelor's degree in Mathematics from Fudan University in 2009. He will be a joint PhD student in the Centre of Scientific Computing at the University of Warwick, United Kingdom from 2010 to 2012 under a scholarship from the State Scholarship Fund. His research interests include computational neuroscience, systems biology and dynamical systems.

Richard Hallworth, PhD, was born in the United Kingdom, raised in Australia, and educated at the University of Melbourne, where he obtained bachelors and masters degrees in mechanical engineering. After a period working in the semiconductor industry, he moved to the United States, where he obtained the PhD degree in neuroscience from Baylor College of Medicine in Houston, Texas. After post-doctoral research in Houston and Chicago, he was appointed as Assistant Professor in the Department of Otolaryngology-Head and Neck Surgery of the University of Texas Health Science Center at San Antonio, Texas. He is now professor in the Department of Biomedical Sciences, Creighton University, in Omaha, NE.

Rui-Ru Ji received a PhD in Molecular Biology and an MS in Computer Science, both from Purdue University in West Lafayette, Indiana. She joined Celera Genomics in 2000 and is one of the co-authors of the human and mouse genome publications in the journal of Science. In 2002, Rui-Ru moved to New Jersey and joined Purdue Pharma L.P. where she built the bioinformatic infrastructure for the Discovery Research site. In 2005, Rui-Ru joined Bristol-Myers Squibb. She has developed a number of algorithms for data analysis, including a novel methodology for dose response transcriptional profiling, a MANOVA-based approach for gene set enrichment analysis, and methods for analyzing co-expression network transcriptional modules. Her current interests include next-generation sequencing analysis, GWAS, copy number analysis, and miRNA. She is now working closely with scientists in the Oncology and Immunology areas to identify and validate new targets for therapeutic interventions.

Awdhesh Kalia, PhD, is an Associate Professor of Microbiology at the University of Louisville. His work addresses the following two questions: (1) what are the molecular and evolutionary forces that shape and maintain genetic diversity in bacterial species? And (2) how does genetic diversity in bacterial species shape the outcome of host-pathogen interaction? Dr. Kalia graduated from the All India Institute of Medical Sciences, New Delhi and underwent postdoctoral training at Yale University and Washington University School of Medicine. He is a recipient of the Ralph Powe junior faculty enhancement award from ORAU, and a Young Investigator award from the International Chemotherapy Congress. Dr. Kalia has authored or co-authored over 35 research articles and book chapters.

Bin Li, PhD, worked as an experimental biologist, holds three patents, and published eight international papers during his PhD training in China. In 1999, he came to the United States and switched to computational work during his postdoctoral training at the University of Washington in Seattle, publishing six papers on molecular dynamics simulations of biomolecules. In 2003, Dr. Li joined Institute for Systems Biology as a senior scientist to work on the systematic study of large biological networks, focusing on statistical analysis and associated software development on high-throughput data such as microarray and ChIP-chip. Dr. Bin Li became a senior scientist at Merrimack Pharmaceuticals in 2007, working on statistic and mechanistic models to help drug discovery.

Yongsheng Lian, PhD, is currently an assistant professor in the Mechanical Engineering Department at the University of Louisville. He works on the simulation of aerodynamics, bio-fluids, and sustainable energy. Dr. Lian obtained his PhD degree in aerospace engineering from the University of Florida.

Gabriele Lillacci earned his M. Sc. (Laurea) degree in Electronic Engineering from the University of Perugia, Italy in 2005. He is currently a PhD candidate in the Department of Mechanical Engineering at the University of California, Santa Barbara. His research interests include mathematical modeling of gene regulatory networks in several biologically relevant contexts, such as DNA damage and repair processes in mammalians. His current work focuses on parameter estimation and model selection methods for computational biology, combining techniques from engineering and statistics.

Bolan Linghu, PhD, is currently a research scientist in the Biomarker Development group at Novartis Institutes for BioMedical Research. Her research projects at Novartis focus on developing and applying computational tools for the analysis of high-throughput data from Next Generation Sequencing. Before joining Novartis, Dr. Linghu worked as a Senior Scientist in the Electronic Biology group at Boehringer-Ingelheim Pharmaceuticals with her research focusing on identifying novel drug targets for inflammation diseases by mining diverse types of biological data. Dr. Linghu received her PhD in Bioinformatics from Boston University in 2008, where she worked in Dr. Charles DeLisi's lab on developing computational methods to identify novel disease genes and predict functions for unknown genes via integration of diverse functional genomics data.

Guohui Liu, PhD, received his BS degree in mathematics and MS degree in Biostatistics in China. He received his PhD degree in statistics from University of Maryland, Baltimore County in 2006, where his research focused on the optimal experimental designs for early phase clinical trials. Dr. Liu is currently a principal biostatistician in Millennium pharmaceuticals, where he is providing statistical support to multiple clinical oncology trials.

Wei Liu, PhD, graduated from Peking University in Applied Chemistry in 1987. He then obtained his M.Sc. in Polymer Chemistry from the Institute of Chemistry of the Chinese Academy of Sciences in 1990. He studied protein structure and dynamics in solution using fluorescence spectroscopy at Louisiana State University starting in 1991, and obtained his PhD in Biophysics in 1996. He subsequently performed his postdoctoral training at the University of California, Berkeley, in the laboratory of Dr. Stu Linn, studying the DNA replication and repair process. He joined Wyeth Bioinformatics in 2000, and has been focusing on the integrative, cross-platform data-mining and text-mining analytics to help move forward the drug discovery programs at Wyeth. He moved to Wyeth Systems Biology in 2008, supporting multivariate phenotypic profiling of autophagy-inducing compounds, and the genome-wide RNAi knockdown studies to look for new opportunities in drug combinatorial therapy. He joined Agios Pharmaceuticals in 2010, and is now leading an integrated Informatics team to support the drug discovery and development programs in Cancer Metabolism.

Yan-Hui Liu, PhD, is a Senior Principle Scientist at Merck. She received her PhD from University of Michigan in 1996. After one and a half years of post-doctoral work at Schering-Plough Research Institute (SPRI), she joined Mass Spectrometry/Structural Chemistry group in 1997. Dr. Liu is currently working at Merck Research Laboratories on protein mass spectrometry to characterize recombinant proteins and antibodies for drug targets / therapeutic purposes. She is also working on applying proteomic methods for drug toxicity and disease biomarker identification. She is the author of over 30 research publications in the area of mass spectrometry, including several book chapters.

Yingchun Liu, PhD, is a Bioinformatics Scientist at the Department of Medical Oncology in Dana-Farber Cancer Institute / Harvard Medical School, USA. She has worked extensively in research involving identifying unknown subtypes of cancers, identifying biological pathways underlying cancers, and analyzing high-throughput genomic data. She developed a powerful method to identify biological pathways that are dysregulated in different types of cancer and a statistical software application for DIGE data analysis. She has also made significant contributions to the identification of novel molecules that regulate epigenetic modifications during embryonic stem cell development. She earned her PhD in Computational Biology in 2007 from Lund University, Sweden, and her MS in Bioinformatics in 2002 from Chalmers University of Technology, Sweden. She did her postdoctoral research at the Department of Biostatistics in Dana-Farber Cancer Institute / Harvard School of Public Health, USA.

Jonathan Y. Mane, PhD, is a postdoctoral researcher in Dr. J. Tuszynski's research group at the University of Alberta, Edmonton, Canada. He was born in Laguna, Philippines in 1974. He received both his MSc and PhD degrees in chemistry from the University of Alberta. During his graduate studies, he developed a pseudopotential basis set for quantum molecular simulations. He also developed computational tools for large molecular systems integrating quantum mechanics, molecular mechanics and molecular dynamics methods. Currently, he is developing and applying different computational techniques for accurate calculations of protein-ligand interactions. He also has interest in high-performance computing and scientific software and computer platforms.

Patricio Manque, PhD, is Professor and Director of the Center of Genomics at Universidad Mayor, Chile. He earned his PhD in microbiology and immunology at Universidade Federal de São Paulo (UNIFESP), Brazil. During his doctoral training, he studied the mechanisms of invasion of the protozoan parasite Trypanosoma cruzi. He successfully completed postdoctoral training in molecular parasitology under the supervision of Dr. Jose Franco da Silveira in UNIFESP, Brazil and in genomics and functional genomics of pathogens in Dr. Gregory Buck's lab at Virginia Commonwealth University, USA. His research interests include genomics, vaccine development and the study of molecular mechanisms associated with pathogenicity of parasitic protozoans.

Eric Meslin, PhD, is the Founding Director of the Indiana University Center for Bioethics, Associate Dean for Bioethics and Professor of Medicine, Medical and Molecular Genetics, Public Health and Philosophy. On May 9, 2007, he was appointed a Knight of the National Order of Merit by the President of France. Prior to joining Indiana University in 2001, he had been Executive Director of the National Bioethics Advisory Commission (NBAC) appointed by President Bill Clinton, and a Program Director in the Ethical, Legal and Social Implications (ELSI) program at the National Human Genome Research Institute. He has been a consultant to the World Health Organization, the US Observer Mission to UNESCO, the Canadian Institutes of Health Research and sits on several boards and committees. Dr. Meslin received his BA in Philosophy from York University in Toronto, and both his M.A. and PhD from the Bioethics Program in Philosophy at the Kennedy Institute of Ethics at Georgetown University. He has held many academic positions, including at the University of Toronto (1988-96) and at Oxford University (1994-95). He has more than 100 publications on topics ranging from international health research to science policy.

Debasisa Mohanty, PhD, has a Master's degree in Physics from Indian Institute of Technology, Kanpur and PhD in computational biophysics from Indian Institute of Science, Bangalore. After completing his PhD in 1995, Dr. Mohanty joined Hebrew University of Jerusalem, Israel for postdoctoral training. In 1997, Dr.

Mohanty moved to Scripps Research Institute, La Jolla, USA as a research associate. His postdoctoral work involved development of computational methods for ab initio folding and de novo simulation of folding thermodynamics. Since 1998, Dr. Mohanty is leading a research group in Bioinformatics and Computational Biology at National Institute of Immunology, New Delhi, India. His research at NII is focused on the development of knowledge based computational methods for identification of novel biosynthetic pathways and protein interaction networks. Dr. Mohanty was elected as a Fellow of The National Academy of Sciences, India in 2008 and was given National Bioscience Award by Department of Biotechnology, Government of India in 2009.

Jason Moore, PhD, is a Frank Lane Research Scholar in Computational Genetics, a professor of Genetics and Community and Family Medicine at Dartmouth Medical School, and the associate director of Bioinformatics of the Norris Cotton Cancer Center at Dartmouth Hitchcock Medical Center in Lebanon, NH. His research focuses on understanding the role of genetic information in predicting susceptibility to common human diseases. His research program aims to develop, evaluate, distribute, and apply powerful computer algorithms and software for identifying combinations of genetic and environmental factors that are associated with complex clinical endpoints.

Stuart Murray, PhD, completed his PhD research at the University of Newcastle-upon-Tyne, UK, studying the regulation of hormone receptors. His Post-doctoral research was carried out at the Albert Einstein College of Medicine, Bronx, NY. During his post-doctoral work, he identified and characterized basal transcription factors and studied the role transcription factors play in cellular differentiation. He then joined Wyeth Research's Information Management group where he pioneered literature informatics by introducing text-mining technologies to Wyeth Research. Following a transition to the Systems Biology Group, he worked to fully combine literature analytics with bioinformatics analytics to create an integrated analytics platform. More recently, he has had the opportunity to join a dynamic biotechnology company to develop integrated analytics in cancer metabolism research.

Giovanni Naldi, PhD, is a Mathematician. He earned a PhD in Applied Mathematics in 1993. He has been a visiting Professor in institutions in Germany, Japan and USA. Since 2001, he is full professor in Numerical Analysis at University of Milano, Italy. He is Scientific coordinator of national and international research projects and Director of the ADAMSS (ADvanced Applied Mathematical and Statistical Sciences) Center of University of Milano; he serves on the editorial board of several international Journals. His research interests include numerical and theoretical analysis of mathematical models in physiology and neurophysiology, statistical models in epidemiology, wavelet bases for image processing and partial differential equations, numerical methods for kinetic equations, and mathematical models of cell chemotaxis.

Madhusudan (Madhu) Natarajan, PhD, is a Principal Scientist at Pfizer in the Quantitative Biotherapeutics Modeling group in Cambridge, MA. He uses systems biology approaches to develop mechanistic understanding of disease indications and leverages that provide insights to the design of biotherapeutics. Madhu's initial training was in Electronics and Communication Engineering, and he went on to graduate studies in Biomedical Engineering and Neurobiology. He received his Ph.D. from Northwestern University, IL, where he investigated sources of sympathetic rhythm generation, which forms the basis of mammalian cardiovascular control. As a member of the research faculty in the Department of Pharmacology at the University of Texas Southwestern Medical Center (UTSWMC), Madhu was part of the Alliance for Cellular Signaling (AfCS) - a multi-investigator multi-university research collaboration whose goal was to comprehensively address how cells interpret signals in a context-dependent manner. His subsequent work at UTSWMC with Dr. Rama Ranganathan applied analysis of information transduction within proteins to engineer protein chimeras with novel function.

Wilfred Ndifon, PhD, is a Postdoctoral Fellow in Immunology at the Weizmann Institute of Science in Rehovot, Israel. His primary interest is in the development of immunologically grounded approaches to controlling the spread of disease.

Youlian Pan, PhD, is a Research Officer and Project Leader in the Knowledge Discovery group, Institute for Information Technology, National Research Council (NRC), Canada. Prior to joining NRC, Youlian was a Lecturer in Biology at Saint Mary's University, Halifax, Canada; Postdoctoral Research Associate in Marine Biomedicine and Environmental Sciences at the Medical University of South Carolina, Charleston, USA; and Research Associate at the Institute of Oceanology, Chinese Academy of Science, Qingdao, China. He received his M.Sc. in computer science and Ph.D. in Biology from Dalhousie University, Halifax, Canada in 2002 and 1994, respectively. His research interests include bioinformatics, functional genomics, transcription regulation, systems biology, data mining, and machine learning. Currently, Youlian serves in numerous editorial boards of international journals, such as Current Bioinformatics, The Open Medical informatics, The Open Applied Informatics Journal, and The Open Bioinformatics Journal. Youlian also serves on numerous national and international research grant review panels.

Kristine Pattin, PhD, received her BS degree from Boston College in biology with a minor in environmental studies. In 2010, she received her PhD in genetics at Dartmouth College where she investigated approaches to ease the computational burden of detecting epistasis, or gene-gene interactions, in genome-wide studies. Specifically, she explored approaches that integrate expert knowledge from protein-protein interaction (PPI) databases into the analysis process. Other research experience has brought Kristine to work with IDEXX Laboratories in Westbrook, ME, in the area of laser immunodiagnostics and at Enanta Pharmaceuticals in Watertown, MA, in discovery biology. She is currently a research associate at Dartmouth College participating in the nation-wise eagle-I consortium for discovering and making research resources visible across the country.

Victoria Petri, PhD, is a Research Scientist at the Rat Genome Database (RGD), Bioinformatics Program, Human and Molecular Genetics Center, Medical College of Wisconsin. Before joining RGD, she was a post-doctoral fellow in the Chemistry Department at Northwestern University. She holds a diploma from a European Conservatory of Music, a Master in Library and Information Science from Columbia University and a Ph.D. in Biochemistry from Albert Einstein College of Medicine; her interests span many areas of research. At RGD, she has initiated and developed the pathway project; as such, she is interested in understanding how the structure-function correlations of biological macromolecules mold their reactions, recognitions, and interactions, as well as how these events entwine into complex molecular networks, how these networks integrate to shape the behavior of biological systems, and how malfunctioning in parts of the system can lead to the diseased phenotype.

George V. Popescu, PhD, received a PhD degree from Rutgers University in 2001. He is currently a senior researcher at the University Politehnica of Bucharest, Romania. He was with IBM TJ Watson Research Center between 2001 and 2004, performing research in the System Modeling and Optimization group. Between 2004 and 2006, he was a postdoctoral researcher at the Center for Excellence in Genomics Sciences at Yale University, New Haven. His main research interest is analyzing the complexity and dynamics of cell signaling and transcription networks. He is conducting research on stochastic modeling for epigenetics, chromosomal variation analysis and cellular differentiation. He is currently a member of the International Society for Computational Biology, Association for Computing Machinery and Society for Industrial and Applied Mathematics and has formerly been a member of IEEE and INFORMS.

Sorina Popescu, PhD, received her MS/BS degrees in Biology from University of Bucharest, Romania, in 1993, and her PhD degree in Plant Molecular Biology from Rutgers University in 2003. She completed her postdoctoral studies in the Molecular, Cellular and Developmental Biology Department at Yale University between 2003 and 2008. She is currently an Assistant Scientist at Boyce Thompson Institute for Plant Research and Adjunct Professor in the Department of Plant Biology at Cornell University. During her postdoctoral work, Dr. Sorina Popescu spearheaded the development of a large scale methodology for functional characterization of plant proteins-the Arabidopsis Functional Protein Microarray. Her current research interest focuses on the identification and analysis of plant signal transduction pathways activated during interactions between plants and environmental factors. Dr. Popescu is a member of American Society of Plant Biologists, American Chemical Society and the New York Academy of Sciences.

Kalyani Putty is a PhD candidate in the Department of Biology at University of Louisville. Her research work focuses on understanding the pathogenesis of gastric pathogen Helicobacter pylori, and the role of bacterial genetic diversity in geographical differences seen in the clinical outcome of gastric disease. Ms. Putty holds a Bachelors of Veterinary Science degree from Acharya N.G. Ranga Agricultural University, Hyderabad, India. She is a recipient of a travel award from the Center for Genetics and Molecular Medicine at University of Louisville.

Ping Qiu, PhD, is currently a senior principal scientist at Merck Research Laboratories (previous Schering-Plough). Dr. Qiu joined Schering-Plough in 1999 as a senior scientist in Bioinformatics group. He has worked on many different projects in the areas of sequence analysis and annotation, biomarker discovery, comparative genomics, system biology and pharmacogenomics. He is an associate editor for BMC Bioinformatics. Prior to joining Schering-Plough, Dr. Qiu was a bioinformatics scientist in Cadus Pharmaceutical. His main responsibility was to design and implement corporate research database and drug target mining for the research pipeline. Dr. Qiu received his PhD in molecular biology from Nanjing University in 1995. He did his postdoctoral research with Dr. Shubha Govin on studying the function of Drosophila IkB protein in hematopoiesis and cell mediated immune response, and he established the role of Toll/Cactus pathway in Drosophila hematopoiesis. He also holds a MS degree with high honor in computer science from CUNY.

Padmalatha Reddy, PhD, received her PhD in Molecular Biology from C.C.M.B., India. She worked on mapping various substrate and inhibitor binding sites on E. coli RNA polymerase using fluorescence spectroscopy. Her postdoctoral work was done at DIBIT, Milan, Italy and at Boston University School of Medicine, Boston, MA. Her research focused on Quality Control Mechanisms in the secretion of Immunoglobulins, in particular thiol-mediated retention. She joined Wyeth Bioinformatics and has worked in the area of functional, structural and evolutionary genomics supporting many target and biomarker discovery programs for Inflammation. In recent years, she has taken an integrative approach to the analysis and mining of 'omics data. She is presently at Pfizer and continues to support target and biomarker discovery programs for inflammation and immunology.

Vicente M. Reyes, PhD, holds BS degrees in chemistry (magna cum lade) and mathematics (magna cum lade) from the University of the Philippines in Diliman, Quezon City, the Philippines, and a Ph.D. degree in chemistry, with concentration in molecular biology and biochemistry, from the California Institute of Technology in Pasadena, CA. He did postdoctoral research at the National Cancer Institute/NIH, University of California-San Diego, and The Scripps Research Institute, La Jolla, CA, in the fields of HIV molecular biology, protein x-ray crystallography, rational drug design, and bioinformatics, before joining the Rochester Institute of Technology's Department of Biological and Medical Sciences as an assistant professor.

Matteo Semplice, PhD, is a Mathematician. He has gained a PhD in Mathematics in 2002 working in the field of mathematical physics. Currently, he is a post-doc at the Università dell'Insubria (Como, Italy). His research is devoted to the study of novel numerical algorithms and to their application to simulations with mathematical models in many areas of science. Among the theoretical aspects, he has been working on relaxation approximation of diffusion equations and on adaptive algorithms for conservation laws. The applications range from the diffusion of pollutants in the environment to chemotaxis-driven phenomena at the macroscale (embryo vascularization, axon guidance), and within a single cell (cell polarization), to quantifying the degradation of marble monuments by atmospheric pollutants.

Palaniappan Sethu, PhD, is an Assistant Professor in the Department of Bioengineering at the University of Louisville. His research work focuses on the application of microfluidics based technologies to isolate culture and extract functional information from functionally viable cells for various applications in biology and medicine. Dr. Sethu has a PhD in Biomedical Engineering from the University of Michigan and trained as a postdoctoral associate within the Center for Engineering in Medicine at Harvard University, Massachusetts General Hospital and Shriners Burns Hospital. He received the Wallace H. Coulter Foundation Early Career Award for Translational Research and a Young Investigator Award from the Center for Environmental Genomics and Integrative Biology. He has authored or co-authored over 25 journal and book publications.

Vrunda Sheth, MS, obtained her bachelor's degree in bioinformatics from Vellore Institute of Technology, Vellore, India. She came to the U.S. in 2007 to pursue a master's degree in bioinformatics from the Department of Biological and Medical Sciences at the Rochester Institute of Technology, which she earned in October 2009 with her M.S. thesis, "Visualization of protein 3D structure in a reduced representation using Double Centroid Reduced Representation," working with research adviser, V. M. Reyes. She currently works as Scientist-2 at Life Technologies in Beverly, MA where she works on analyzing data from the next-generation sequencing platform, SOLiD.

Pan Shi is a PhD candidate in the College of Information Sciences and Technology at the Pennsylvania State University. She received her Bachelor's degree in Electronic Engineering from Tsinghua University, China in 2005, and her Master's degree in Computer Science from Chinese Academy of Sciences in 2008. Her current research projects focus on usability and design of privacy and security technologies.

Zhiao Shi received a PhD degree in computer science from the University of Tennessee at Knoxville in 2006. He is currently a research assistant professor in the Department of Electrical Engineering and Computer Science at Vanderbilt University. He is also an Education and Outreach Liaison in the Advanced Computing Center for Research and Education (ACCRE) at Vanderbilt. His main research interests include parallel and distributed computing, computational biology and high performance biological network analysis algorithms.

Olivier Stahl has been a bioinformatics engineer since September 2006. He holds an MS from the Faculty of Sciences of Luminy after his work on scientific literature mining in Bernard Jacq's group (IBDML). After becoming an expert in the development of HMI (Human-Machine Interfaces), he joined the INRA to work on a multiple genome comparison browser (NARCISSE project). In October 2008, he joined the CRCM's bioinformatics group to develop analysis and data storage tools. He is also the bioinformatics system administrator and the main developer of the CMS-based DJEEN project.

Alain Tchagang, PhD, is a Researcher at the Knowledge Discovery Group, Institute for Information Technology, National Research Council Canada (NRC). He is also an Adjunct Professor at the School of Information Technology and Engineering, and a member of the Faculty of Graduate and Postdoctoral Studies at the University of Ottawa. Prior to joining NRC, Alain was a Postdoctoral Associate in Computational Biology at the Department of Computational and Systems Biology, School of Medicine at the University of Pittsburgh, and received a PhD degree in Biomedical Engineering and an MS degree in Electrical Engineering from the University of Minnesota, in 2007 and 2004, respectively. Alain's research interests include computational and systems biology, biomedical signal processing, control theory, and robustness in biological systems. Alain is a member of the Institute of Electrical and Electronics Engineers (IEEE), the International Society of Computational Biology, and the IEEE Engineering in Medicine and Biology Society.

Ahmed Tewfik, PhD, received his BS degree from Cairo University, Cairo, Egypt, in 1982 and his M.S., E.E., and S.D. degrees from the Massachusetts Institute of Technology, Cambridge, MA, in 1984, 1985, and 1987, respectively. Dr. Tewfik, the E. F. Johnson Professor of Electronic Communications at the University of Minnesota has been named the new chair of the Cockrell School of Engineering Electrical and Computer Engineering Department at The University of Texas at Austin, effective October 1st, 2010. He has served as a consultant and worked with many companies such as Texas Instruments. He is a Fellow of the IEEE. He was a distinguished Lecturer of the IEEE Signal Processing Society in 1997–1999. He received the IEEE Third Millennium Award in 2000. Dr. Tewfik's current active projects focus on wearable sensors for cardiac monitoring, body area networks, non-invasive and invasive sensing of neural activity, bioinformatics, cognitive radio networks, and wireless networks.

Jack A. Tuszynski, PhD, is the Allard Chair and a professor in the Department of Oncology. The major thrust of his computational biophysics group is in silico drug design for cancer chemotherapy applications and in vitro testing. His research interests are strongly linked to the protein tubulin and the microtubules assembled from it. Due to its prominent role played in eukaryotic cell division, tubulin is an important target for anti-cancer cytotoxic treatments. His on-going research aim is to identify variants of known compounds showing greater tubulin isotype-specifc effects, which could potentially lead to more efficacious chemotherapy treatments with lower side effects. Other studies in his group have examined microtubule electrical, structural, and mechanical properties; proteins that bind to microtubules (MAPs); and the motor proteins in cells that travel along microtubules and actin filaments. The group is also developing physiologically-based models and simulations for pharmacokinetic and pharmacodynamic applications.

Paolo Valigi, PhD, was born in 1961. He received the Laurea degree in 1986 from University of Rome La Sapienza and the PhD degree from University of Rome Tor Vergata in 1991. He was with Fondazione Ugo Bordoni from 1990 to 1994. From 1994 to 1998, he was a research assistant at University of Rome Tor Vergata. From 1998 to 2004, he has been associate professor at University of Perugia, where since 2004 he is full professor of System Theory, at the Department of Electronics and Informatics Engineering. He is the coordinator of the Engineering management program. His research interests are in the field of systems biology, robotics, and distributed control and optimization. He has authored or co-authored more than 100 hundred journal and conference papers and book chapters.

Ute Woehlbier, PhD, obtained her PhD at the University of Heidelberg in Germany, working in the development of a subunit vaccine for malaria in the group of Dr. Hermann Bujard. During a two-year postdoctoral training in the lab of Dr. Gregory Buck at Virginia Commonwealth University, USA, she studied host-pathogen interactions during cryptosporidiosis. Currently she is receiving further postdoctoral training focused on understanding mechanisms of protein misfolding leading to neurodegenerative diseases at the Institute of Biomedical Sciences in Dr. Claudio Hetz's lab at the University of Chile.

Thomas K.F. Wong is a PhD candidate of the Department of Computer Science in the University of Hong Kong. His research interest is bioinformatics. His recent focus is on areas related to non-coding RNA, structural alignment and structural prediction for pseudoknot structure.

Yu (Brandon) Xia, PhD, received his BS in Chemistry (major) and Computer Science (minor) from Peking University, and his PhD in Chemistry from Stanford University. While at Stanford, he worked on computational structural biology as a Howard Hughes Medical Institute Predoctoral Fellow. Following that, he worked on protein bioinformatics as a Jane Coffin Childs Postdoctoral Fellow at Yale University. He is currently an Assistant Professor in the Bioinformatics Program and the Department of Chemistry at Boston University, with a secondary appointment in the Department of Biomedical Engineering. He has published over 40 research articles, scientific reviews, and book chapters. His research interests include the prediction and analysis of protein structures and networks.

Heng Xu, PhD, holder of the endowed PNC Technologies Career Development Professorship, is an assistant professor in the College of Information Sciences and Technology at the Pennsylvania State University. She leads the Privacy Assurance Lab (PAL), an inter-disciplinary research group working on a diverse set of projects related to understanding and assuring information privacy. She received her Ph.D. degree in information systems from the National University of Singapore in 2005. Her current research focus is on the interplay between social and technological issues associated with privacy assurance. Her research in some of these areas has been funded by grants from the National Science Foundation and National Security Agency. She has published journal articles and conference papers on information privacy and security, human-computer interaction, and technology innovation adoption.

S.M. Yiu received his PhD degree in computer science from the University of Hong Kong and is currently an Assistant Professor at the same university. His research interests include bioinformatics and computational biology.

Guo-Cheng Yuan, PhD, is an Assistant Professor at Department of Biostatistics and Computational Biology at Dana-Farber Cancer Institute and at Harvard School of Public Health. Dr. Yuan obtained B.S. and M.A. in Applied Mathematics from Peking University and Ph.D. in Mathematics from University of Maryland at College Park, USA. He did postdoctoral research at Brown University and later at Harvard University. Dr. Yuan's main research interest is in computational epigenomics, with the long term goal to understand the systems-level regulatory mechanisms underlying various biological processes.

Bing Zhang, PhD, received BS and MS degrees in biology from Nanjing University, China in 1993 and 1996, respectively. In 1999, he received a PhD degree in Molecular Genetics from the Shanghai Institute of Plant Physiology, Chinese Academy of Sciences. From 1999 to 2005, he worked as a Postdoctoral Research Fellow in the University of Tennessee at Knoxville and Oak Ridge National Laboratory, where he spent three years doing wet-lab functional genomics research, followed by three years of dry-lab bioinformatics research. Since 2006, he has been an Assistant Professor in the Department of Biomedical Informatics at the Vanderbilt University School of Medicine. His current work focuses on the development and application of systems biology approaches to the study of complex diseases. His research interests include modeling and analysis of biological networks, biological data exploration and integration, and translational bioinformatics.

Daniel Ziemek, PhD, is the "Biological Systems Domain Lead" of the Computational Sciences Center of Emphasis at Pfizer Inc. since 2008. His primary research interest is the development of innovative analysis methods for gene or protein expression data leveraging prior knowledge in the form of biological networks. He received his diploma (MSc) in computer science from the University of Bonn, Germany in 2000 and his PhD from the Ludwig-Maximilians-Universität München (LMU), Germany in 2004 (with summa cum laude). In February 2004, he joined the pharmaceutical company Sanofi-Aventis to work in the field of pathway informatics. He was the scientific lead of an internal enterprise-wide pathway solution platform and contributed to many projects including ODE-based simulation of cardiac arrythmias and statistical evaluation of high-throughput screening (HTS) results. In October 2008, he joined the Computational Sciences Center of Emphasis at Pfizer Inc. and worked on diverse projects ranging from target discovery and toxicity prediction to patient stratification.

Index

D

E

F

G

H

I

K

N

O

P

Q

R

S

T

U

V

W

X

Breinigsville, PA USA
03 April 2011
258719BV00003BA/1/P